The History of Composites

Explanation of pictures on front cover

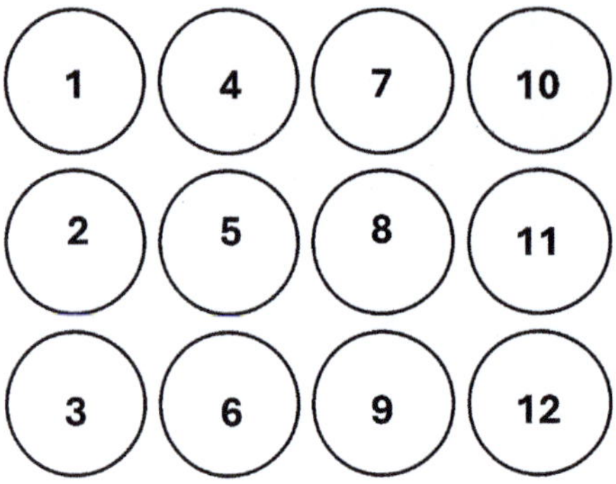

The front cover is inspired by the transverse section of a unidirectional composite. It shows a selection of important persons and key developments in the four areas of the subtitle of this book, from left to right: People, science, technology, and applications in society.

1. Leo Baekeland (1863–1944), Belgium-born inventor of Velox photographic paper and of phenolic resins (Bakelite). He mentioned the concept of fibrous reinforcements in his 1907 USA patent on phenolic resins *(Copyright via Wikimedia Commons Paralaloa, CC BY-SA 4.0)*
2. Shindo Akio (1926–2016), inventor of the method for producing carbon fibers by oxidation and carbonization of PAN fibers, in 1959 at GIRIO laboratories in Osaka (Japan) *(Copyright obtained from AIST, Japan)*
3. Steve Tsai (b. 1929) introduced thousands of scientists and engineers into the micro- and meso-mechanics of composites (laminate plate theory, LTP), developed numerical tools to analyze and simulate their mechanical behavior and tirelessly promoted their use worldwide *(Copyright granted by Steve Tsai)*
4. Transmission electron microscopy (TEM) image of a multiwalled carbon nanotube (CNT), *reproduced from Carbon 49: 4650 (2011), with permission from Elsevier.*
5. Unidirectional carbon fibre composite, X-ray computed tomography image by Rui Guo at Composite Materials Group, KU Leuven (Belgium), *reproduced from Composites Science and Technology 244 110278 (2023), with permission from Elsevier.*
6. Woven glass fibre reinforcement for composites, X-ray CT image by I. Straumit at Composite Materials Group, KU Leuven (Belgium), *reproduced from Comp Struct 179: 568 (2017), with permission from Elsevier.*
7. Largest autoclave in the world, by ASC Process Systems *(Copyright via Wikimedia Commons)*
8. Robot-assisted press for thermoplastic composites at Sirris Leuven-Gent Composites Applications Lab in Leuven (Belgium) *(Copyright granted by Sirris-SLC).*
9. Automated fibre placement (ATP) at Netherlands Aerospace Centre (NLR) *(Copyright granted by NLR)*
10. Aerospace has been an important driving force in the development of composite materials and structures *(Copyright Shutterstock).*
11. Lotus 108 track bike, a monocoque design with an inflatable bladder (1992) *(Copyright via Wikimedia Commons, CC BY 2.0)*
12. Composite wind turbine blades *(Copyright via Wikimedia Commons)*

The History of Composites

People, science, technology and society

Edited by

Ignaas Verpoest
Composite Materials Group, Department of Materials Engineering, Katholieke Universiteit Leuven (KU Leuven), Leuven, Belgium

Co-edited by

Stepan V. Lomov
Composite Materials Group, Department of Materials Engineering, Katholieke Universiteit Leuven (KU Leuven), Leuven, Belgium

Yentl Swolfs
Composite Materials Group, Department of Materials Engineering, Katholieke Universiteit Leuven (KU Leuven), Leuven, Belgium

Mahoor Mehdikhani
Composite Materials Group, Department of Materials Engineering, Katholieke Universiteit Leuven (KU Leuven), Leuven, Belgium

ELSEVIER

Elsevier
Radarweg 29, PO Box 211, 1000 AE Amsterdam, Netherlands
125 London Wall, London EC2Y 5AS, United Kingdom
50 Hampshire Street, 5th Floor, Cambridge, MA 02139, United States

ISBN: 978-0-443-34119-9

For information on all Elsevier publications visit our website at https://www.elsevier.com/books-and-journals

Publisher: Matthew Deans
Acquisitions Editor: Gwen Jones
Editorial Project Manager: Sara Greco
Production Project Manager: Vijayaraj Purushothaman
Cover Designer: Ignaas Verpoest and Mark Rogers

Typeset by TNQ Tech

Contents

Contributors

Suresh G. Advani Department of Mechanical Engineering, University of Delaware, Neward, DE, United States

Albertino Arteiro INEGI, Faculdade de Engenharia, Universidade do Porto, Rua Dr. Roberto Frias, Porto, Portugal

Joaquim Barros ISISE/ARISE, Univ. of Minho, Campus de Azurém, Guimarães, Portugal

Brahim Benmokrane Department of Civil & Building Engineering, University of Sherbrooke, Sherbrooke, QC, Canada

Christophe Binetruy Nantes Université, Ecole Centrale de Nantes, CNRS, GeM, UMR 6183, Nantes, France

Alain Bourmaud Université Bretagne Sud, UMR CNRS 6027, IRDL, Lorient, France

Pedro P. Camanho INEGI, Faculdade de Engenharia, Universidade do Porto, Rua Dr. Roberto Frias, Porto, Portugal

Thanasis Chatziathanasiou Department of Materials Engineering, KU Leuven, Leuven, Belgium

João R. Correia CERIS, Instituto Superior Técnico, Universidade de Lisboa, Lisboa, Portugal

Klaus Drechsler Technical University of Munich, Munich, Germany

Patrick Enjuto The Boeing Co., Arlington, VA, United States

Marc Fette Composite Technology Centre/CTC GmbH - An Airbus Company, Stade, Germany

Christ Glorieux Department of Physics and Astronomy, KU Leuven, Leuven, Belgium; Department of Materials Engineering, KU Leuven, Leuven, Belgium

John Griffith The Boeing Co., Arlington, VA, United States

Gwen M. Gross The Boeing Co., Arlington, VA, United States

Axel Herrmann Airbus Operation GmbH, Stade, Germany

Paula Hohoff Polymer Engineering Center, Department of Mechanical Engineering, University of Wisconsin–Madison, Madison, WI, United States

Seyed Arman Hosseini Department of Civil & Building Engineering, University of Sherbrooke, Sherbrooke, QC, Canada

Jan Ivens Department of Materials Engineering, KU Leuven, Sint-Katelijne-Waver, Belgium

Jae-Moon Jeong Department of Aeronautics and Astronautics, Massachusetts Institute of Technology, Cambridge, MA, United States

Nobuya Kawamura Toyota Motor Corporation, Toyota, Aichi, Japan; Scalable Advanced Composite Technology, Nisshin, Aichi, Japan; Joby Aviation, Santa Cruz, CA, United States

Thomas Keller CCLab, École Polytechnique Fédérale de Lausanne, EPFL ENAC IIC CCLAB, Lausanne, Switzerland

John Keune The Boeing Co., Arlington, VA, United States

Michelle Korwin-Edson Owens Corning Science & Technology, Granville, OH, United States

David Leach Composite Material Solutions, Walnut Creek, CA, United States

Pieter Jan Lemstra School of Chemical and Materials Engineering, Jiangnan University, Wuxi, China; PlemPolco B.V., Veldhoven, the Netherlands

Hans Lilholt Wind Energy Materials and Components Division, Department of Wind and Energy Systems, Technical University of Denmark, Risø Campus, Roskilde, Denmark

Stepan V. Lomov Composite Materials Group, Department of Materials Engineering, Katholieke Universiteit Leuven (KU Leuven), Leuven, Belgium

Bo Madsen Wind Energy Materials and Components Division, Department of Wind and Energy Systems, Technical University of Denmark, Risø Campus, Roskilde, Denmark

Mahoor Mehdikhani Composite Materials Group, Department of Materials Engineering, Katholieke Universiteit Leuven (KU Leuven), Leuven, Belgium

Véronique Michaud Laboratory for Processing of Advanced Composites (LPAC), Institute of Materials (IMX), Ecole Polytechnique Fédérale de Lausanne (EPFL), Lausanne, Switzerland

Eric Moussiaux Exel Composites Free University of Brussels (VUB); European Composites Industry Association (EuCIA); European Pultrusion Technology Association (EPTA)

Jörg Müssig The Biological Materials Group, Biomimetic, HSB – City University of Applied Sciences, Bremen, Germany

Arnt Offringa GKN Aerospace, Hoogeveen, the Netherlands

Tim A. Osswald Polymer Engineering Center, Department of Mechanical Engineering, University of Wisconsin–Madison, Madison, WI, United States

Dimitrios G. Papageorgiou School of Engineering and Materials Science, Queen Mary University of London, London, United Kingdom

Ton Peijs Centre for Polymer and Composites (CPC), WMG, University of Warwick, Coventry, United Kingdom

Helge Pfeiffer Department of Materials Engineering, KU Leuven, Leuven, Belgium

Anthony Pickett IFB (Institute of Aircraft Design) University of Stuttgart, Stuttgart, Germany

R. Byron Pipes Purdue University, West Lafayette, IN, United States

Kevin Potter Bristol Composites Institute, University of Bristol, Bristol, United Kingdom

Marino Quaresimin Department of Management and Engineering, University of Padova, Vicenza, Italy

Chris Skinner Owens Corning Headquarters, Toledo, OH, United States

Jeroen Soete Department of Materials Engineering, KU Leuven, Leuven, Belgium

Yentl Swolfs Composite Materials Group, Department of Materials Engineering, Katholieke Universiteit Leuven (KU Leuven), Leuven, Belgium

Ramesh Talreja Texas A&M University, College Station, TX, United States

Fumihiko Tanaka Composite Materials Research Laboratories, Toray Industries, Inc., Masaki-cho, Ehime, Japan

Helena Teixidó Laboratory for Processing of Advanced Composites (LPAC), Institute of Materials (IMX), Ecole Polytechnique Fédérale de Lausanne (EPFL), Lausanne, Switzerland

Holly Thomas The Boeing Co., Arlington, VA, United States

Tia Benson Tolle The Boeing Co., Arlington, VA, United States

Stephen W. Tsai Department of Aeronautics & Astronautics, Stanford University, Stanford, CA, United States

Panayiotis Tsokanas Composite Materials Group, Department of Materials Engineering, Katholieke Universiteit Leuven (KU Leuven), Leuven, Belgium

Ignaas Verpoest Composite Materials Group, Department of Materials Engineering, Katholieke Universiteit Leuven (KU Leuven), Leuven, Belgium

Brian L. Wardle Department of Aeronautics and Astronautics, Massachusetts Institute of Technology, Cambridge, MA, United States; Department of Mechanical Engineering, Massachusetts Institute of Technology, Cambridge, MA, United States

Martine Wevers Department of Materials Engineering, KU Leuven, Leuven, Belgium

Robert J. Young National Graphene Institute and Department of Materials, University of Manchester, Manchester, United Kingdom

Biography

Prof. Ignaas Verpoest received his Master's degree (1972) and PhD (1982) in Materials Engineering from the Katholieke Universiteit Leuven (KU Leuven), Belgium. As a Full Professor (from 1990 until 2013), he guided a group that consisted of eight postdoctoral researchers and 25 PhD students (average 2008–2013), carrying out research on the meso-mechanics of (textile-based) composites, nano-engineered composites, natural fibre-reinforced (bio)polymers, and advanced production methods for composites. Since September 2013, he has been an Emeritus Professor at KU Leuven. He continues to be involved in research projects on carbon fibre and natural fibre-reinforced composites, and he also has a special interest in "the history of composites" and the relationship between composites and design(ers). Over the past few years, he has organized several exhibitions on "composites and design(ers)" at the Design Museum in Ghent and has collaborated with JEC World (Paris) for several exhibitions on the history of composites.

Prof. Verpoest was the holder of the Toray Chair in Composite Materials at KU Leuven and was Chairman of the European Scientific Committee of the Alliance of European Flax-Linen and Hemp (formerly CELC). He was president of the European Society for Composite Materials (ESCM) and the International Committee on Composite Materials (ICCM).

Prof. Ignaas Verpoest won several awards, amongst others the Descartes Prize for Science Communication of the European Commission (2004) and the "Medal of Excellence in Composite Materials" of the University of Delaware, USA (2014). He is also Cofounder of the company Econcore, producer of innovative honeycomb cores, and was instrumental in creating the startups Rein4ced, producing hybrid carbon-steel fibre-reinforced thermoplastic composite bicycle frames, and BioFibix, producing innovative flax preforms for composites.

Prof. Stepan V. Lomov graduated with an MSc in Physics-Mechanics from Leningrad Polytechnic Institute (1978) and then went on to complete his PhD in 1985. He was awarded his Dr. Hab. on textile materials science in 1995 from St. Petersburg State University of Technology and Design. Since 1999, Professor Lomov has been working at KU Leuven, Belgium, in the Department of Materials Engineering. He was coordinator of the Composite Materials Group from 2013 to 2020. He was the holder of the Toray Chair in Composite Materials at KU Leuven between 2013 and 2022 and has been an Emeritus Professor since 2020. His research areas include composites and textiles science and engineering: internal structure, manufacturing, in-service mechanical behaviour, nanocomposites, experimental damage mechanics, and micro- and meso-level geometrical and mechanical models.

Prof. Yentl Swolfs obtained his Master's degree in Materials Engineering from KU Leuven in 2010. In January 2015, he obtained his PhD in Materials Engineering. The title of his thesis was: *Hybridisation of Self-reinforced Composites: Modelling and Verifying a Novel Hybrid Concept.* This PhD was performed under the framework of the EU-FP7 project HIVOCOMP. After completing his PhD, he spent a year at Imperial College London on a Marie Skłodowska-Curie Fellowship working with Prof. Silvestre Pinho. He then went on to hold a 3-year FWO postdoctoral fellowship at KU Leuven working on fibre-hybrid composites with high-performance polymer fibres. He was appointed Research Professor in the Department of Materials Engineering in 2019, where he leads the research line on "Micro- and Meso-structural design of composites." He is the Coordinator of the Composite Materials Group at KU Leuven and Director of the XCT Core Facility of KU Leuven. In 2022, he won the ESCM award for most promising researcher under 35 years of age. In 2025, he was awarded the Kelly Prize of ICCM, for composite scientists in the early stage of their professional careers. He currently holds an ERC Consolidator grant on off-axis cracking in composites.

Dr. Mahoor Mehdikhani earned a BSc in Materials Science and Engineering from Sharif University of Technology, Iran (2012), followed by an MSc (2014) and PhD (2018) in Materials Engineering from KU Leuven. His PhD focused on the effect of voids in fibre-reinforced composites. He is currently a Senior FWO Postdoctoral Fellow at KU Leuven, specializing in *in situ* multiscale damage monitoring, XCT image processing in composites, and computational micromechanics. He has completed research stays at the Pontificia Universidad Católica de Chile in Santiago and at the University of Manchester in the UK. He supervises multiple PhD, master's, and bachelor's students and has secured research funding from FWO, KU Leuven, and the China Scholarship Council. In July 2024, he was elected to the Council of the European Society for Composite Materials (ESCM), and in November 2024, he was named runner-up for the Composites Science and Technology Young Researcher Award.

Preface

Whenever you embark on a big adventure like editing a book, you should always ask yourself, "Why should I do this?". Is a book on *The History of Composites* really necessary? Can it make a relevant contribution to either the history of science and technology or to the scientific literature on composites? Maybe other composite researchers or science historians have already addressed this topic?

The answer to the last question is quite straightforward: thus far, no comprehensive, scientific description and analysis of the history of composites has been published. So, there is a 'market opportunity', also because composite science and technology has reached a level of maturity in the past decade that justifies a reflection on how we got there. Moreover, embarking upon a history of composites gave some researchers and engineers who made pioneering contributions during the 'turbulent decades' of 1960–1990 the opportunity to share their experience, either through an interview or by writing a testimonial.

Did the idea of writing and publishing this book come all of a sudden? Certainly not. It is the culmination of a career-long interest in science communication about composites as an important vector of my research career. In 1981, after completing a PhD thesis in the Department of Metals Engineering (MTM) at KU Leuven on 'the fatigue of steel wires' used in radial tires, I was asked to evaluate the fatigue behavior of tennis rackets. Belgian companies Donnay and Snauwaert were leading manufacturers of wooden tennis rackets at the time and did not want to miss the disruptive innovation of using carbon fibre composites. Such materials were rather unknown at KU Leuven, but nevertheless I was invited to start a postdoctoral project on composites. As a result, from the outset of my career, I worked in an environment not familiar with composites. The use of these materials had to be explained, even defended against some skepticism, which led to a continuous reflection on the essence of the composites concept and its fundamental differences (advantages?) compared to metals and ceramics, let alone plastics.

While the attitude of fellow materials scientists and engineers was rather defensive, or at least wait-and-see, there was curiosity and fascination in the world of designers and artists. Already in the 1960s, sculptors like Niki de Saint Phalle were discovering glass fibre reinforced composites (GFRP) to replace bronze or stone with 'polyester', as they called it, and furniture designers like Verner Panton had been experimenting with 'fiberglass', yet another misnomer of GFRP. Along with their increasing use in sporting goods, the general public thus came in contact with composites but was often not aware of the true nature of this new material class. Breakthrough innovations in aerospace were not visible to them, and hence the idea arose to start up some activities to raise public awareness, in line with the European Commission's funding programs aiming at opening up the world of science and technology to a broader public.

Combining art and design with explaining the composites concept was original but, in fact, an evident choice. The challenge was to use a language that was scientifically correct, yet understandable to interested laypeople, and appealing to scientists and engineers from other disciplines. During the preparation of exhibitions on composites in design in 2002, 2015 and 2018 in the Design Museum Gent (Belgium), which I curated together with Lut Pil, a colleague from the LUCA School of Arts at KU Leuven, it became clear that we should also explain the origins and early developments of composites, and not only the current status—hence the title of the first exhibition and catalogue, *From Bakelite to Composite.*

This somewhat outside-the-mainstream initiative was picked up by JEC, which organizes JEC World, an annual global trade show in Paris on composite materials and their applications. During several JEC World events, previews of the 'composites in design' exhibitions were shown, and when the JEC organization wanted to celebrate their 60th anniversary, we were invited by Nelly Baron, then Vice-President of JEC, to create an exhibition on *The History of Composites* inside the JEC World halls in 2023 and 2024 and to write some accompanying texts, resulting in two brochures, *Milestones in the History of Composites* and *The Last 50 Years and the Future*.' [1]

The reactions to these brochures were very positive, and many interested people suggested and even encouraged us to extend them into a broader scientific book on the history of composites. Hence the idea of writing this book was born.

As mentioned earlier, a comprehensive and scientifically substantiated book on the history of composites does not exist yet. One can find numerous, often very short and superficial historical 'sketches' on webpages of composite companies, focusing on very specific topics like the history of carbon or glass fibres or the use of composites in aerospace. Composite organisations such as SAMPE Japan and SAMPE International have published booklets on the history of composites in Japan and in honour of the centenary of the first flight of the Wright brothers, respectively, both focusing on the evolution of composites in aerospace. On the more scientific side, in the unparalleled six-book series *Comprehensive Composite Materials*, edited by Anthony Kelly and Carl Zweben in 2000, some but not all of the 182 chapters in total start by placing their subject in a historical context. This impressive collection of six books was, however, clear evidence that composite science and technology had reached a level of maturity, presenting a detailed state of the art at that moment in time.

Rare are the 'testimonial' type of papers, where composite scientists and engineers provide firsthand information on how they contributed to and experienced the developments in their particular field. Almost all such articles that can be found in open literature are referenced in Part I of this book.

[1] Available at About Composites - JEC (https://www.jeccomposites.com/discover-composites/about-composites)

One particular case is the book '*Engineered Stability. The History of Composite Materials in the 19th and 20th Centuries*' by Andreas T. Haka.[2] Contrary to what the title might suggest, it covers almost exclusively the history of composites in Germany, but links with the international developments are provided where necessary or relevant, albeit in highly varying detail.

Hence, a long chapter on the 'general history of composites' would be an asset. My first idea was to extend the two JEC brochures mentioned above towards a more complete, scientifically substantiated story, from the very beginning of composites in prehistoric times up to the most recent developments. The aim was to write a text that could be read in a few hours and would provide a global overview that is also understandable for non-specialists in composites, focusing on breakthrough and disruptive developments. This became Part I of this book.

During discussions with colleagues at KU Leuven, I suggested the idea of adding a collection of shorter chapters, covering specific topics and written by international experts in each particular field. Soon, an 'editorial team' was formed, consisting of Prof. Stepan V. Lomov, Prof. Yentl Swolfs, and Dr. Mahoor Mehdikhani, and a list of about 30 subjects was drawn up. The challenge, however, was to convince the contributing authors not to write yet another 'state of the art' review paper but to present a fact-based history of the main developments or breakthroughs in their specific subject, also with a personal touch. We asked the contributing authors to analyze the dynamics of innovations, illustrating how fundamental discoveries have led to the creation of new materials, processes, and products or have drastically improved existing ones.

This has resulted in a rich variety of 'writing styles', from detailed, meticulously documented reconstructions of scientific breakthroughs and technological milestones to very personal stories from experts who actively contributed themselves. We believe that we have convinced all invited international experts to write a real 'historical' chapter. For this, many of them had to leave their comfort zone of materials or processing research and, in a sense, become historians. The result is 28 stories, grouped in four divisions: materials, processes, applications, and the theoretical and experimental foundations of composites.

This book presents a 'collection of facts' and will therefore be considered in the history of science and technology community as a first level of historical studies. In a large number of chapters, attempts have been made to find connections between the collected facts and to elucidate causal or coincidental relations, which brings these contributions to a deeper analytical level. The third level, however, was beyond the scope, namely to investigate whether historical innovations in composites follow similar patterns as developments in other scientific fields; this could be a future task for 'real' historians of science and technology.

[2] Haka, A. T. (2023). Engineered Stability – The History of Composite Materials in the 19th and 20th Centuries, Springer, https://doi.org/10.1007/978-3-658-41408-5.

In concluding this preface, I would first like to thank my co-editors Stepan V. Lomov, Yentl Swolfs, and Mahoor Mehdikhani. They have creatively suggested ideas for the development of this book, contacted their share of contributing authors, and were instrumental in reviewing not only these contributing chapters but also the longer 'general history of composites' in Part I. All of them also contributed at least one chapter in Part II of the book and co-authored the Part I chapter on 'Challenges and opportunities for composites in the next decades', the latter together with Panayiotos Tsokanas, another postdoctoral researcher at the Composite Materials Group of KU Leuven. It was inspiring and enriching to work with all of you.

Second, I am deeply indebted to all the 55 authors who wrote the 28 chapters in Part II of the book. Their efforts to dig into the history of their specific composites field have resulted in amazing stories and well-documented, detailed analyses of the gradual and sometimes groundbreaking innovations in composites science and technology. For all of them, it was necessary to go a step further than their daily scientific practice, to go back in time, to evaluate the relevant inventions, and, where appropriate, to relate them to developments in other scientific and technological fields and in the society as a whole. All contributing authors enabled the making of this book, the first of its kind, mapping the history of composites in a scientifically sound manner, and hopefully becoming an important reference for anyone interested in this subject.

This book would not have been possible without the highly professional guidance of Sara Greco and Gwen Jones at Elsevier Publishing Company. Their feedback in the conceptual phase of the book, their professionalism during the contract negotiation, and their responsiveness once the manuscripts were coming in and hundreds of questions from contributing authors (and from us) had to be answered were highly appreciated.

Nelly Baron, from JEC, played a crucial role in the beginning of this adventure. Her invitation to create an exhibition on the history of composites for the JEC World trade show was the necessary impetus to start studying the history of composites in a more systematic manner. Without her warm and compassionate collaboration, this book would never have come about.

The feedback from colleagues, researchers, and students on the first JEC brochures, as well as the reactions during the many lectures I have been invited to present since then, motivated me to take up this project. I received many suggestions for additional historical information, sources unknown to me, or people I should contact to gain more insight into certain developments.

Finally, I would like to thank wholeheartedly my beloved family, my wife, children, and grandchildren, and my dear friends for the countless hours I was physically or mentally absent, hidden behind my computer screen to search for new documents or websites, and thus reveal yet another aspect of the history of composites. Your support and your interest in my 'passion' will never be forgotten.

Ignaas Verpoest
Leuven, May 2025

Part I

An introduction to the book

General history of composites, a "bird's eye view" 1.1

Ignaas Verpoest
Composite Materials Group, Department of Materials Engineering, Katholieke Universiteit Leuven (KU Leuven), Leuven, Belgium

1. Introduction

The history of composite materials has, up to now, only been written in a very fragmented way. Mostly, the focus was on specific materials, processes, or application developments, or writings only covered a limited period of time. Some papers or books were written from a rather narrow national perspective, like the excellent book by Haka, which presents maybe the most in-depth historical study on composites, be it mainly from a German point of view, with limited but appropriate references to developments in other parts of the world (Haka, 2023).

This introductory chapter to the book "The history of Composites. People, science, technology and society" tries to depict a more global picture. Not only scientific material and process-related inventions will be presented, but also the steps toward industrialisation and the development of innovative applications will be discussed. In this text, the focus will be on disruptive innovations, not really on incremental ones. The role of people (inventors, designers, entrepreneurs) will be highlighted, and the historical, economic, and societal environment, within which these breakthrough innovations have taken place, will as well be critically assessed.

Section 2 describes how history started some 8000 years ago when the basic concept of reinforcing one material with another (already at that time mostly a fibrous one) was invented and applied. After limiting the scope of this chapter (and of the whole book) to fibre-reinforced polymers (Section 3), the early history of composites is depicted in Section 4, covering the work of the "founding fathers" *(no mothers yet)* in the period 1650 to 1930. The 17th century, the era of the first scientific revolution, saw the birth of glass fibres, the initial inventions of carbon fibres go back to the 19th century, and in the very early 20th century, Baekeland invented and commercialised, as an entrepreneur, the first synthetic polymer "bakelite."

The 1930s saw the first real breakthroughs of the 20th century: the invention of most of the synthetic polymers (or plastics) which are still predominantly used today (polyester, epoxy, polypropylene, polyamide ...), and almost simultaneously the start of industrial production of glass fibres (Section 5). This all led to the first two decades (1940–60) of hesitating enthusiasm *(engineers)* and (un)limited imagination *(designers)* developing a plethora of applications, not only in aerospace but also in automotive, marine, sports, and even consumer goods (furniture), as described in Section 6.

The History of Composites. https://doi.org/10.1016/B978-0-443-34119-9.00001-2

The second breakthrough came in the late 1950s, and early 1960s, with the invention of carbon fibres. Because this was of crucial importance for the further development and industrialisation of composites in general, the invention of carbon fibres is discussed in detail, mostly from a scientific but also from an industrial and applications point of view. An attempt is made to provide a well-documented answer to the question of who played the key roles in Japan, the United States, the United Kingdom, and also the USSR, and to the even more intriguing question of what was contributed by whom in the development of the PAN precursor-based carbon fibre manufacturing, as we know it today (Section 7). At the end of this section, a side-step is made to pitch-based carbon fibres, boron, silicon carbide, and oxide fibres and high-performance polymeric fibres.

All these inventions resulted in an impressive number (and diversity) of new composite applications. They are described in five different parts of Section 8, namely aerospace, transportation (including marine and land transport), consumer goods (including sports and leisure), construction and architecture, and finally energy, ending with an intriguing story on how a Danish adult education school was building one of the first composite wind turbines, performing better than the ones developed by NASA.

Section 9 describes the period up to the year 2000: the era of further innovations in fibres, matrices and preforms and in manufacturing processes, specific for composites. The steady growth of applications, both in number and complexity, is discussed in the last part of this section. The developments during the first two decades of the 21st century conclude this "bird's eye view" on the history of composites (Section 10).

This is a "bird's eye view," so not all important, breakthrough innovations have been treated into the same level of detail. Often, it reflected the personal judgement of what the author found most relevant and/or most interesting. Writing "*a history of*..." is never absolutely neutral nor objective, but an honest attempt was made to be as objective and complete as possible. Facts and data are not simply presented as such, but are contextualised, and causal links between them are discussed.

In many places in this introductory part of the book, reference is made to one of the 28 chapters in Part II of this book, written by more than 40 experts from all over the world. They cover 28 different aspects of materials science, processes, and application developments, as well as theoretical and experimental foundations of composites science and engineering. These chapters should be considered as necessary complements to this part I text.

2. The prehistory, distinguishing myths and facts

Almost every website, article or book that, even superficially, attempts to put composites in a historical perspective, starts with references to Egyptians, Mongolians, or Mesopotamians. It is striking that, in such texts, very loose definitions of composites are used, ranging from just a combination of two materials via a fibre- or rod-reinforced construction material up to a laminated structure. But what we now call "the composite concept" has been emerging in several early civilisations around the world.

It would need more in-depth historical study to find out in which civilisation the very first practical realisation of the composite concept, namely mud bricks reinforced with natural fibres (straw, grass, fine branches of trees, horse hair, ...) has been developed. Has it been invented in one place, and then spreading out over the globe or, more probably, independently invented in different civilisations?

Fibre-reinforced mud bricks indeed closely approach one of the actual concepts of composites, namely adding a fibrous material to act as a crack stopper in intrinsically brittle materials. Many studies suggest that the origin of fibre-reinforced mud bricks might be found in the Middle East. Aernoudt mentions in his book "Materials make history" (Aernoudt, 2003) that already in 9000 BCE in the city of Catal Hüyük (Turkey) bricks reinforced with straw have been produced. Rosenberg et al. (2020) investigated the mud-brick structures of Tel Tsaf in the Jordan Valley (Southern Levant), dating back to the Pre-Pottery Neolitic period (5200–4700 BCE, about 7200 years ago). In these well-preserved sites, they found sun-dried mudbricks formed from "*a mixture of sand, clay, water and frequently a temper (e.g., chopped straw and chaff branches).*" The use of the latter word "temper" in this paper, coming from the realm of historians, not material scientists, is interesting because it describes in an almost poetic way the effect of adding fibrous materials to the intrinsically brittle dried mud: the fibres "temper" the devastating effect of cracks. "Tempering" in the metallurgical realm is used for a heat-treatment process to make alloys less crack-sensitive.

Dried mud is often also referred to as "adobe," a Spanish adaptation of the Arabic word for brick, "toob" (Emery, 2011). In Emery's paper, different compositions of Egyptian adobe are discussed, an optimum one containing one part (in weight) of chopped straw and five parts of "earth," a mixture of sand, clay, and silt; often, also dung is added. The bricks are formed mostly in single moulds (see Fig. 1.1.1) and simply dried in the sun. Alternatively, the mud mixture can be cast in a formwork *(similar to cast concrete)*, typically in layers of about 40 cm thick, building up in this way walls which can rise up to 15 m. Dumarçay (2005) describes in detail how this technique (*also called* "pisé") has been used from the third millennium BC in the

Figure 1.1.1 Mud bricks near a house under construction in Kyrgizistan in 2007.

Indus region but has spread out over the whole of South and South-East Asia. It was also present in China (since 5000 BCE), not only for houses and buildings but also in the Great Wall (Rammed Earth, n.d.). Still today, adobe architecture is alive, like the famous Djenné Mosque in Mali (Fig. 1.1.2), the largest mud building in the world (Morris and Blier, 2004).

Since dried mud structures are susceptible to degradation by rain and erosion by wind, dried mud bricks were gradually complemented by the more resistant fired bricks. However, in a recent study, it was shown that both the cooling capacity of dried mud bricks in tropical areas is superior and their environmental impact is considerably lower compared with other building materials used in affordable housing (like fired bricks and hollow cement blocks) (Udawattha and Halwatura, 2017).

Another often-mentioned early application is the use of several types of reeds (like papyrus) for boats, as extensively described by Gaudet (2014). From 5000 BCE, and well before wooden boats, papyrus boats have been used on the river Nile. They were used for all kinds of transport in the shallowest parts of the river and in swamps because they were "floating" on the water. This effect was realised, according to Gaudet, by "*tying the bundles in several places in order to trap air inside the reeds*." Thor Heyerdahl collected in 1969 papyrus reed from Lake Tana in Ethiopia to build the Ra I, with which he unsuccessfully tried to cross to ocean from Marocco to Barbados; he was, however, successful during a second attempt with the Ra II in 1970.

Although many internet sites mention that, not only in the case of Heyerdahl but also in Egyptian times, the papyrus fibrous bundles would have been impregnated with a kind of bitumen or asphalt to make them water resistant, a scientifically reliable reference was hard to find. The availability of bitumen around the Dead Sea is well documented in Oron et al. (2015). In this paper, even reference is made to the Bible, where Moses's ark (Exodus 2.3) is described as "*a clay-coated gomeh structure, waterproofed with zefet*" (Hebrew for "bitumen" or "pitch"). Confirming this, Schwartz and Hollander (2004) found bitumen artifacts, from the Uruk period (3800 BCE), in Hacinebi (Mesopotamia). On these artifacts, imprints of the reed bundles and the binding ropes can be observed, proving that the bitumen was used to protect the reed against water; however it appears that the bitumen did not really impregnate the reed bundles, hence it acted only as a superficial protection, and not as a matrix in a fibre reinforced composite. Till the 19th century, this technique

Figure 1.1.2 The Djenné Mosque in Mali. Djenné by Ruud Zwart, CC BY-SA 2.5 NL, via Wikimedia Commons.

was used to build even larger boats, as witnessed around 1850 by British explorer Chesney: "*Chesney witnessed the construction of a 20-ton reed boat by Arabs in southern Mesopotamia near the bitumen source of Flit, which was one of the major, if not the major source of bitumen in Mesopotamia in antiquity and the recent past.*" (Schwartz and Hollander, 2004), also mentioning that this boat was waterproofed both inside and outside.

The use of papyrus, linen (flax) fibres, or paper layers impregnated with natural resins or plaster to make ancient Egyptian funerary masks or coffins has been reported as well, from the First Intermediate Period (c. 2181–2055 BCE) to the Roman era (332 BCE–395 AD). This technique looks similar to "papier mâché" (literally: chewed paper) but differs from it because there is no pulping of the fibrous substrate.

Papier maché seems to have been invented during the Han dynasty in China. It is another example of how the composites concept was, already 2000 years ago, intrinsically understood and experimentally realised (Bawden, 1990). Mainly two methods were used: either gluing many layers of paper (also a Chinese invention) to each other or dissolving (by cooking) paper in a glue-containing liquid (water), so that individual cellulose fibres would, upon drying and compressing into a shape, reinforce the adhesive glue. Mankind was not the first to make papier maché! Wasps create the material in their mouth by chewing a mixture of wood fibres and their saliva and then build layer-by-layer their often amazing nest structures (Hansell, 1989) (Fig. 1.1.3). In fact, wasps collect the wood fibres mainly from fallen and disintegrating trees, hence playing a significant role in maintaining the ecological equilibrium.

Figure 1.1.3 A wasp nest made from a "natural" composite: wasp papier maché. Wasp Nest by Kathy Jones, CC BY-SA 2.5, via Wikimedia Commons.

One of the technically most outspoken composite innovations might be the composite bow, made from a combination of wood, bamboo, bone, cattle tendons, horn, and silk bonded with natural pine resin. It was invented by Mongolians around 1200 BCE, or by Parthians and Tartars, according to Gordon in "Structures" (Gordon, 1978), and is well described by Hardy in "Longbow" (Hardy, 1976). Aernoudt (2003) explains how the "composite bow," already in 1700 BCE, was composed of a wooden core (tough wood like maple was often used), to which tendons (or sinews) at the tension side and horn bone at the compression side had been adhesively bonded using bitumen, found in wells in the desert. From the detailed description (Fig. 1.1.4a) of the sophisticated Assyrian bow (Hickman, 1959) and Mongolian bow (Reisinger, 2010), it is evident that Assyrians and Mongolians very well understood the different

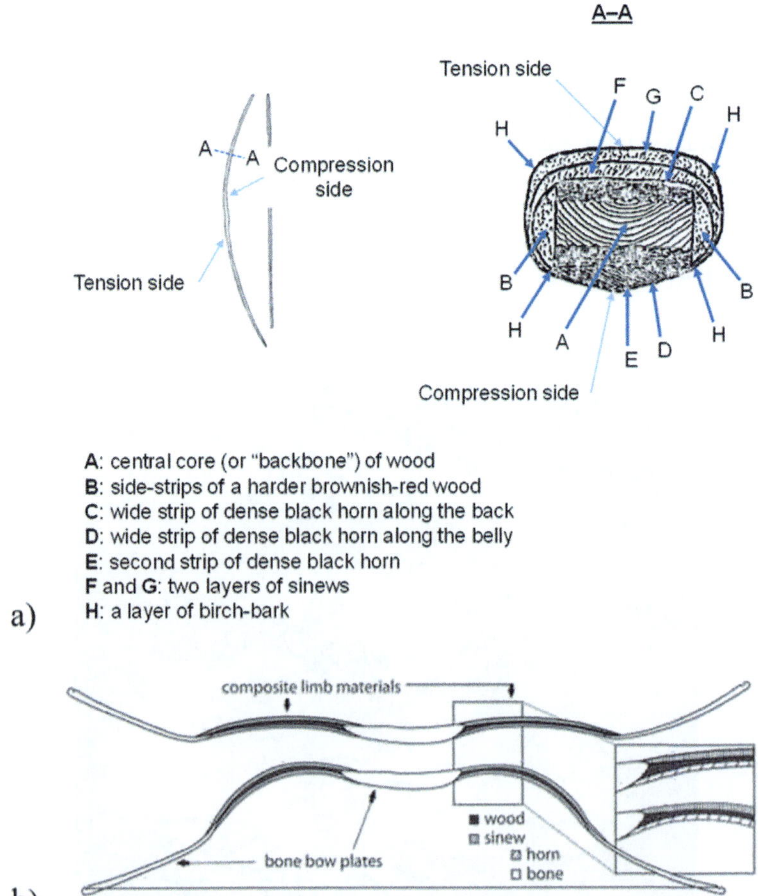

Figure 1.1.4 Longbow construction: (a) Assyrian composite bow, 1275 to 1208 BCE; (b) the intricate construction of a Mongolian bow (*top*: nontensed, *bottom*: tensed).
(a) Redrawn from Hickman (1959); (b) Reisinger (2010).

responses of materials to tensile and compressive loads: tendons, positioned at the tensile side of the tensed bow can reach tensile strength values up to 250 MPa (Burgio et al., 2022), whereas the compressive strength of horn bone, at the opposite side, exceeds that of wood by a factor 3.5 (Reisinger, 2010).

Another aspect of modern composites is their laminated (meso)structure: layers with different fibre orientations are being stacked into a "laminate," to respond as effectively as possible to the imposed loads. This concept is similar to plywood, where thin layers of wood with different orientations (often at 90-degree angles to each other) are glued to form one solid panel. Again, Egyptians are quoted as the first to have documented the concept of cross-grained plywood, as is depicted on a carving in Egyptian Thebes, dating from 3400 BCE (Wilk, 2017). It is interesting to notice that the first theories on the mechanical behaviour of laminates have been developed by Lekhnitskii for plywood laminates (Lekhnitskii, 1947, 1968).

3. A definition for composites?

From this short review on the prehistory of composites, it is obvious that the concept of "composites" is broader than what it is most commonly used for at the beginning of the 21st century, namely fibre-reinforced polymers (or plastics). The "papier mâché" and the composite bow come closest to this 21st century definition, but mud or baked clay reinforced with straw come close to 21st century's fibre-reinforced concrete or ceramic matrix composites. Also, since the mid-1950s, metals are being reinforced with particles or fibres, so the question can be raised which definition of "composites" will be used in this book.

Looking at the worldwide turnover of different types of composite materials, it is obvious that "fibre-reinforced polymers" (FRPs) are the predominant family. It includes both thermoplastic and thermoset matrices (and even elastomers), and they can be reinforced with continuous or short fibres, particle-reinforced plastics being on the borderline, as their behaviour is very close to that of "unreinforced" plastics. Metal and ceramic matrix composites are briefly mentioned further in the text, where the scientific developments of oxide and carbide fibres are discussed. However, their industrial production and application are largely inferior to that of FRPs, both in terms of tonnage and of value, $ or €'s.

3.1 Two special cases

Two special cases, however, should be mentioned separately.

Firstly, tyres for cars, but also for trucks and aeroplanes, have been for decades reinforced with different types of (continuous) fibres, be it polymeric fibres and/or metallic wires or cables; a comprehensive introduction to tyres can be found in Mukai and Morii (2000). Tonnage- and value-wise, they are certainly the most important composites, as on an annual basis more than 1 billion (10^9) radial tyres are produced. Already soon after the invention of the inflatable (or "pneumatic") tyres by J.B.

Dunlop in 1887, later perfected to become "detachable" by A&E Michelin in 1891, these tyres have been reinforced with fibres, initially cotton or other natural fibres.

Up to World War II, these reinforcements were laid in a "bias" or diagonal fashion (in about 45 degrees to the radial direction of the tyre), but this led to too much heating of the tyre because of internal friction of the two (or more) crossing fibre layers. The first radial tyre designs were patented in 1914 by Hamilton and Sloper (1914) and in 1916 by Arthur W. Savage, a tyre manufacturer (1915–19), firearm designer and inventor in San Diego (Savage, 1916). No actual products were created.

In 1946, the laboratories of Michelin in France developed the industrial production of the first radial tyre, in which steel cords (cables of fine steel wires) were laid in another direction, from one side of the rim to the other (see Fig.1.1.5). The radial structure reinforces and stabilises the tyre, whereas the parallel orientation avoids heating by friction during intensive use.

In this way, radial tyres are very close to polymer composites: the wear resistant and flexible rubber is reinforced with fine, stiff steel wires to realise a wear-resistant tyre with tuneable and hence optimised stiffness/flexibility characteristics. And as in polymeric composites, the adhesion between the reinforcements and the rubber, as well as the fatigue behaviour of the fibres, in this case steel wires (Verpoest et al., 1985), will ultimately control the durability of the tyre.

A second special case is concrete. To cope with tensile loads, (brittle) concrete has been reinforced already in a very early stage with (more ductile) steel rebars, presumably since the mid-19th century. As such, it is a perfect example of a "macro-composite," as the reinforcing rebars are not of micron but of millimetres and even centimetres size. By adding ductility to the brittle concrete the steel rebars create a new material with improved properties. If we however concentrate on reinforcements that are closer to the ones used in fibre-reinforced plastics (FRPs), even then there have been interesting developments in the last 50 years.

In steel fibre reinforced concrete, thin flattened steel wires are intimately mixed with concrete, mortar, or grout. The steel fibres act mainly as crack stoppers because they are ductile and hence can withstand tensile forces, in contrast to the brittle concrete, mortar, or grout. Recently, fine steel wires from recycled tyres have been used as well (Liew and Akbar, 2020).

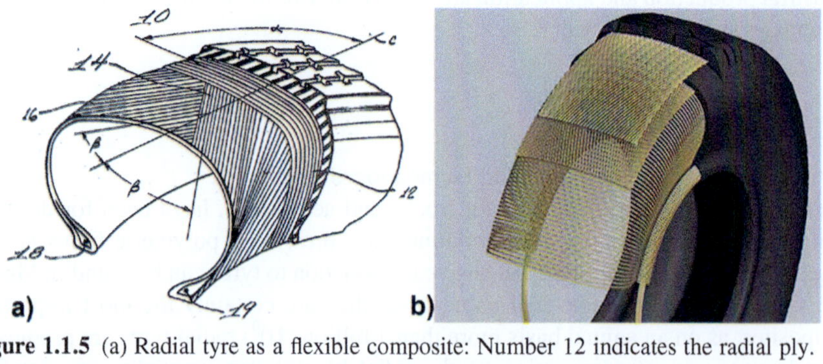

Figure 1.1.5 (a) Radial tyre as a flexible composite: Number 12 indicates the radial ply. Numbers 14 and 16 are bias plies. (b) Today's radial tyre construction.
(a) Radial tyre by Duduk, public domain, via Wikimedia Commons. (b) Courtesy Bekaert Co.

Fibre-reinforced polymer composites are, in certain applications, gradually replacing steel rebars, mainly for three reasons. First, steel rebars can corrode, certainly if they are positioned too close to the concrete surface, and even more in aggressive environments (like marine, industrial, etc.). Second, FRP rebars can slightly reduce the total weight of the construction, but more importantly, they are also easier to handle during the construction phase. Third, when holes must be cut or drilled in the solidified concrete, the presence of glass (or even more carbon) fibre-reinforced rebars is less a hindrance than the more ductile steel rebars. Many different types of mainly glass-fibre reinforced rebars have been introduced to the market, almost all of them produced by a pultrusion process (Exel, n.d.). Adhesion between the rebar and the concrete is an issue, and several inventive solutions have been realised (for instance, adding a roughness on the pultruded rod, as is present on steel rebars, see a more detailed discussion in section 9.3.4, and in chapter 2.3 of part II of this book).

Recently, in the search for always lighter concrete structures, glass or carbon fibre reinforcements, in the form of 2D or 3D weaves, have been directly introduced into solid and sandwich-type concrete panels (Alein and Bhuvaneshwari, 2022; Sudhakar and Muthusubramanian, 2022). The latter developments come closer to the predominant concept of "composites," used in this chapter.

3.2 A delimitating definition of "composites"

At this stage in the text, a delimitating definition of the term "composites," as we will use it further in this chapter *(and in the book as a whole)*, might be useful. We will limit "composites" to materials in which a polymer matrix is reinforced with a fibrous material. Each word in this definition is important.

- **Polymer** (or plastic) **matrices** are, as mentioned earlier, most important in volume and value terms, and hence will be the main focus, although metal and ceramic (including concrete and geopolymer) matrix composites will be briefly mentioned as well.
- "**Fibrous**" reinforcements have a considerable length-over-diameter ratio (or "aspect ratio"); in this way, spherical or ellipsoidal "inclusions" will be excluded. In fact, when these are added to polymers, the resulting materials are called "particle-reinforced" materials. Continuous fibres have an infinite aspect ratio, while discontinuous or short fibres have finite aspect ratios (ranging from 10 to many 1000s); higher aspect ratios result in higher stiffness and strength of the composite.
- "**Reinforcement**" indicates that this phase has mechanical properties superior to those of the matrix, in terms of stiffness and/or strength and/or toughness, or can positively influence other mechanical properties like fatigue or impact behaviour. Physical characteristics (electrical conductivity, thermal expansion coefficient, etc.) might be affected as well and will be addressed where appropriate.

4. Early history: the founding fathers: 1650–1930

4.1 Early glass fibres

In many glass-fibre-related internet sites, the invention of glass fibres has been attributed to Phoenicians or Egyptians. Even in one of the first well-documented papers on the history of composites, Seymour and Deanin (1987) mention that "glass fibres were used by Phoenicians to produce bottles," without further explanation. The process is explained briefly by Cheilakou et al. (2012): "the core-forming technique is so-called because the vessel was made by trailing viscous molten glass around a core of fire-proof material i.e., clay, which had been roughly shaped around a metallic rod in the desired form." In fact, this technique is close to the filament winding process used today in composites processing (Fig. 1.1.6). The earliest known examples of core-formed vessels date from the 16th to 15th century BC in Mesopotamia and Egypt, and were generally small and used as containers for cosmetics, perfumes, and medicines. Such glass filaments could be coloured in different ways, adding to the aesthetics of the vessel, but they are, in fact, too coarse to really call them "fibres." J. Gordon Cook describes how these filaments are made (Gordon Cook, 1959): "The ancient Egyptians were skilled in the art of drawing coarse filament from rods of glass made by rolling glass with a metal bar. The rods, about as thick as a pencil, were reheated at one end, filaments being drawn away from the molten material."

According to a brief historical overview on composites by McMullen (1984) (one of the few available ones, mainly focussing on the United Kingdom and on aircraft applications), the first historical mention of producing glass fibres was by the French Academy of Science (18th century) who claimed that they were able to make thin glass fibres by dipping an arrow in molten glass and firing it! A more complete overview of the early history of glass fibres is presented by Gordon Cook (1984). First, it is revealed that Robert Hooke mentioned already in his Micrographia (1665) the drawing of fine fibres from a heated glass rod. A few decades later (1713), the French scientist René-Antoine Ferchault de Réaumur seems to have produced glass fibres: a glass plate is heated over a fire, and filaments are plucked from the surface and rapidly brought to a fast-rotating spinning wheel. Réaumur, whom we know from one of the temperature scales next to Celsius and Fahrenheit, also proposed to weave them, even if the glass fibres he produced were rather short. In 1836 in Paris, Ignace Dubus-Bonnel and his collaborator Bajeu, filed a patent on "a weave of glass made malleable by steam, pure or mixed with silk, wool, cotton or linen."

Although it is generally accepted that the first industrial process for making glass fibres has been developed in the 1930s in the United States, already in the 19th century initial processes seem to have been demonstrated: Louis Schwabe, a prominent Manchester silk weaver, in 1842 developed a machine for making glass fibres and demonstrated it in Manchester (see Gordon Cook (1984) and Seymour and Deanin (1987)). Schwabe seems to have been the first to extrude molten glass through small orifices

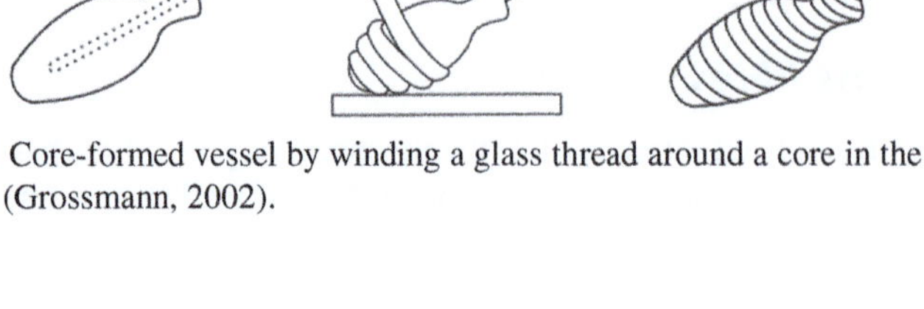

Figure 1.1.6 Core-formed vessel by winding a glass thread around a core in the desired shape of the vessel (Grossmann, 2002).

("spinnerets"), very much like the glass fibre spinning methods of today. According to Verre Online (2004), "*he practices drawing molten glass from a tiny ceramic opening.*" Other sources (Mahapatra and Raichurkar, 2015) mention British inventors as well, like Alexander and David Gordon who invented in 1822 a method to produce incombustible lamp wicks. In Germany, glass fibres were drawn from a single rod composed of two types of glass (lead crystal and plate glass); their difference in thermal expansion results in curly fibres, still today known as "Angel's Hair" (Gordon Cook, 1984) (Fig. 1.1.7).

Skrabec mentions in his book "Edward Drummond Libbey, American glassmaker" (Skrabec, 2011) that in 1893 Libbey presented, together with Mike Owens (*see further on Owens Corning glass fibres*), at the 1893 World's Fair in Chicago the first "spun glass," and several textile products made with it, some with mixed glass–silk yarns. No information could yet be found on the manufacturing process used by Libbey and Owens at that time, except that it took over 37 h to pull 1900 miles of glass filament to create the garment (Nayyar, 2022). Probably the process might have been similar to what Gordon Cook (1984) describes (referring to the 1897 issue of the German magazine Seide): rods of glass are drawn from melting ovens at 1000–1200°C, and wound on a cylinder of 4-metre diameter, revolving at 400 rpm, hence probably similar to Réaumur's method. Elisabeth Teinsky (Austria) filed a patent in 1902, describing a method to produce "weavable glass wool" (Teinsky, 1902). Later on, during World War I, also German inventors would have developed a method to produce glass fibres (McMullen, 1984).

4.2 Carbon fibres

Early explorations also happened for carbon fibres. First, in 1841, Frederick de Moleyns (UK) filed a patent for an incandescent lamp, using powdered charcoal

a)

Figure 1.1.7 Glass fibres for garments: (a) weaving Libbey's glass filaments, (b) glass fibre dress for Infanta (Princess) Eulalia of Spain, presented at the 1893 World's Fair in Chicago (Byrne, 2023).
Both pictures were provided by the Toledo Museum of Art (USA) (Nayyar, 2022).

heated between two platinum wires. Then, in 1860, Joseph Swan (Newcastle, UK) developed a primitive electric light bulb that used a filament of carbonised paper in an evacuated glass bulb (Timmonet, n.d.). However, the lack of good vacuum and an adequate electric source resulted in a short lifetime for the bulb and an inefficient light (Fig. 1.1.8).

Thomas Edison experimented with carbonised natural fibres (a.o. bamboo) in 1879 (Edison, 1879) and later on developed a method to deposit pyrolytic carbon on a carbonized natural fibre by "flashing" it in a hydrocarbon atmosphere, resulting in superior filaments for electric lamps. A few years later, in 1883, Joseph Swan patented the pyrolysis of rayon (Seymour and Deanin, 1987), also to create filaments for light bulbs. Swan and Edison later joined forces to found the Edison-Swan Electric Company, where Lewis Latimer, son of an African slave, developed a reliable carbon filament by encapsulating it in cardboard (Wikipedia, n.d.). Latimer is considered one of the 10 most important black inventors of all time, having also collaborated with Graham Bell in drafting his telephone patent, and having filed many more patents on a wide range of inventions, from safety elevators to a room cooling and disinfecting apparatus.

By the beginning of the 20th century, carbonised filaments were gradually becoming obsolete due to the development of the more durable tungsten filaments, developed by William D. Coolidge at General Electric's laboratories (Ramirez, 2020).

4.3 *From biobased to manmade plastics*

Bensaude-Vincent and Stengers (1993) describe how in the second half of the 19th century, first steps were taken in developing biobased resins/plastics, like celluloid by Parkes in 1856 and cellulose acetate by Paul Schützenberger in 1865. However, it took till the first decades of the 20th century, before the scientific and technical base for polymer matrices for composites had been laid. Leo Baekeland (Fig. 1.1.9), born in Belgium but having worked mainly in the United States, had

Figure 1.1.8 Three early inventors of carbonised filaments: (a) Joseph Swan, (b) Thomas Edison, (c) Lewis Latimer.
(a) Joseph Swan by Public domain, via Wikimedia Commons; (b) Louis Bachrach, Bachrach Studios, restored by Michel Vuijlsteke, Public domain, via Wikimedia Commons; (c) Lewis Latimer by Public domain, via Wikimedia Commons

Figure 1.1.9 (a) Leo Baekeland, inventor of the phenolic resin, named "Bakelite" after him; (b) the lab notebook of Baekeland, used in1907 at the time of the invention.
(a) Paralaloa, CC BY-SA 4.0, via Wikimedia Commons; (b) Brigade Piron, CC BY 3.0, via Wikimedia Commons.

become a wealthy inventor-entrepreneur when he sold his patents on the Velox photographic paper and his company Nepera Chemical to Eastman–Kodak in 1899 (Haka, 2023, pp. 121–157). He invested his fortune in setting up a private laboratory, focussing from 1903 onwards on phenol-formaldehyde reactions, hoping to develop an alternative for the existing celluloid and parkesine materials. Together with four collaborators, he invented the process to make phenolic resins as described in detail by Mercelis (2020). He filed his first patent on February 18, 1907 (US Patent 949,671, 1909), and a second one on July 13, 1907 (US Patent 942,699, 1909).

It is remarkable that in the first patent, Baekeland explains "*According to my invention there is produced synthetically within a fibrous body, and around the fibres or bundles of fibres constituting the same, a hard condensation product of phenols and formaldehyde*," hence focussing on the fact that he describes "*a method of indurating fibrous and cellular materials such as wood, wood-pulp, cotton, asbestos or the like* …". This is a kind of inversion of what we consider nowadays as a polymer composite, namely a polymer reinforced by fibres, because Baekeland "indurates *(=hardens)*" wood, cotton etc. by creating inside them a hard condensation product!

In his second, broader patent (patent US 942,699), Baekeland describes the "heat and pressure" process in more detail, but also mentions here that he mixed the rather brittle phenolic resin with different kinds of fillers "*wood fibre, other fibrous or cellular materials, rubber, casein, lamp black, mica, mineral powders …., pigments, dyes, nitrocellulose, abrasive materials, lime, sulfate of calcium, graphite, cement, powdered horn or bone, pumice stone, talcum, starch,* …" in order "*to yield various valuable products.*" It seems that Baekeland only gradually learnt that these fillers increased the "stability" of the phenolic resin (probably reducing the thermal shrinkage during cooling) but also reduced the cost of the product. Added (biobased) fibres like cotton or flax, moreover could act as crack stoppers. It is unclear at what moment Baekeland and his coworkers understood and optimised this composite reinforcement effect.

Only in February 1909, more than a year and a half after his initial invention, Baekeland felt sufficiently secure to announce his invention publicly and called it from then on Bakelite (Mercelis, 2012).

The idea of adding fibres as crack stoppers to the phenolic resin was further developed by two of his earlier collaborators, Daniel O'Connor and Herbert Faber, who moved away from Baekeland's group to Westinghouse Electric Co (van Dulken, 2002). They impregnated a linen "canvas" textile with phenolic resin to make a laminate that primarily had to be electrically insulating, but also had to have excellent mechanical properties. They called it "Formica," indeed intending to replace mica, a heat-resistant, flake-like mineral. Because Westinghouse had no real interest in commercialising O'Connor's and Faber's product, they started their own "Formica Corp." company. The concept of preimpregnation was described as well in the patent (filed February 1, 1913): "*The process of making a laminated composition that consists in coating one side of each of a plurality of sheets of fibrous material with a phenolic condensation product, superposing the coated sheets, applying heat and pressure to the superposed sheets, then applying a greater degree of heat and a lower pressure.*" Hence this collaborative work of Baekeland, O'Connor and Faber can be considered one of the first realisations of a man-made polymer matrix composite material, as we know it today.

However, already in 1910, George Westinghouse had developed a similar material "micarta," impregnating paper and cotton under high pressure with a phenol-formaldehyde mixture to produce laminates, and one may wonder why he did follow through on this. Haka (2023, pp. 121–157) reveals the role of Baekeland in this development. As mentioned earlier, Baekeland had sold his Nepera Chemical company to Eastman–Kodak in 1899. When Westinghouse was approached by Eastman–Kodak to develop a strong but light material for their camera housings, Baekeland was (like always) eager to promote his products, even when they had been sold or licenced already to other companies. So, they started an intense collaboration, and by 1910 Westinghouse was able to start marketing the micarta material, not only for camera housings but also for many other applications. Interesting to notice is the remark in a 1916 article about the application of micarta in gears (Bakelite-Micarta, 1916): "*… it is as strong as cast iron, is unaffected by atmospheric changes, is vermin proof and can be stored indefinitely without shrinkage or other deterioration.*" Moreover "*… it is quiet in operation … will absorb shocks …*," pointing to an additional advantage compared with metallic gears. Nowadays, the Micarta brand name persists, produced by Norplex-Micarta worldwide; it is popular for knife handles.

It is remarkable that only a few years after the Formica patent and Westinghouse's Micarta development, another Westinghouse employee, Robert Kemp, filed two patents on applying phenolic composites in aeroplanes: first, in March 1916, he described the manufacturing of a composite propeller (US Patent 1,433,728, 1916). Two months later, he filed a very detailed patent on a complete range of structural elements for an aeroplane (US Patent 1,435,244, 1916). Reading them 100 years later, the claims are real eye-openers: "*A vehicle body consisting essentially of layers of fibrous material impregnated with phenolic condensation products and hardened into a homogeneous mass*" and "*In aircraft construction, a hollow fuselage formed at its rear end with integral, hollow vertical and horizontal stabilizers* ..." (Fig. 1.1.10a). Moreover, four different types of composite laminates are described (Fig. 1.1.10b).

Because phenolic resins and fibrous reinforcements are explicitly mentioned, and because Kemp was an employee of Westinghouse at the moment of filing his patents, it was for a long time unclear which would be the link between Leo Baekeland and Robert Kemp. After comparing the information of Kemp's patents with the descriptions in Mercelis' and Haka's books, the puzzle pieces fall into place: both the Formica and the Micarta developments at Westinghouse must have been strongly supported by Baekeland in person!

The fact that the very first composite aeroplane patent by Kemp came out of the laboratories of an electrical company might be surprising, but, as Haka mentions "*Westinghouse Electric Company was one of the first major industrial companies in the United States to establish a scientific research laboratory.*" It was independently operating, and had flexible research structures, both in terms of staffing and research topics. Another example of how "blue sky research," even in an industrial environment, can lead to scientific and technical breakthroughs ...

At this moment, it is however unknown, at least to me, whether Kemp's concepts have ever been turned into real aeroplane prototypes, or even just lab-scale structural elements. Nevertheless, the time seemed ripe for exploring the use of fibre-reinforced, light materials in these early days of aeroplane design and building, both in Europe

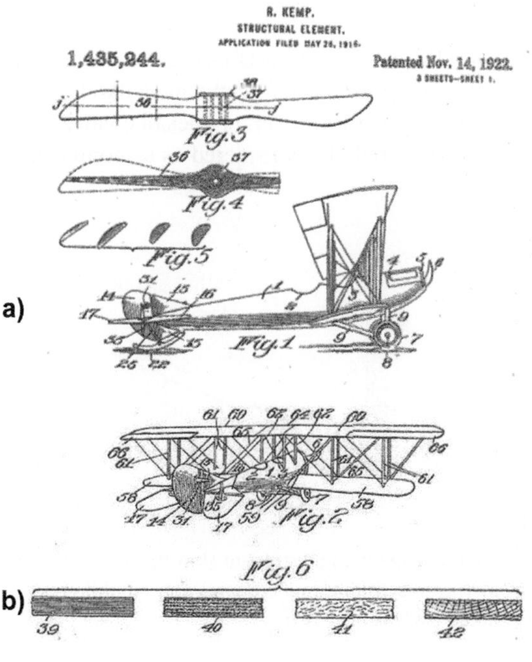

Figure 1.1.10 (a) Robert Kemp's patent, filed May 1916, for a composite aeroplane (US patent 1,435,244). (b) Details on the material types. No. 39: plate material with layers of paper impregnated with phenol and formaldehyde; No. 40 plate material with fabric tapes; No. 41: material with wood-fibre reinforcement and number 42 shows impregnated wood (US Patent 1,435,244 as quoted by Haka (2023)).

and in the United States. Haka (2023, pp. 121–157) describes in detail how, from 1917 on, John Dudley North started designing the P.10 aeroplane at Boulton & Paul (UK). The rear fuselage consisted of a monocoque, a stressed-skin construction with load-bearing "Dilecto" panels on a rolled steel frame. These "Dilecto" panels were produced in the United States by Continental–Diamond, and were yet another commercial product issuing from Baekeland's innovations, namely a phenol-formaldehyde resin reinforced with embedded layers of paper *(similar to the aforementioned Formica and Micarta).* A prototype of the P.10 was shown on the Parisian "Salon de l'Aéronautique" in 1919, and the "Dilecto" was praised for its advantages compared with wood (not hygroscopic, insusceptible to insects, high fire resistance) and metals (no corrosion).

Two years later (1921), in the United States, Albin Kasper Longren designed and built his "New Longren Aeroplane." The fuselage is again a stressed skin, consisting of two complete semimonocoques (left and right halves) made from "vulcanised fibre," as first invented by Thomas Taylor in the United Kingdom already in 1855. In this "vulcanizing" process, rag cotton or wood pulp cellulose paper is saturated with zinc chloride, which acts as a catalyst when bonding the cellulose layers under pressure. The ZnCl is then leached out by water, a lengthy and complex process, and the laminate is dried (Brown, 1999). Longren further elaborated this process in 1922 by pressing the vulcanised fibre layers in between veneer layers, forming in one step the fuselage semimonocoques (Fig. 1.1.11). He tried to get interest from the military, because of the excellent bullet resistance, but failed in commercialising his prototype.

4.4 Laminates

The abovementioned examples by North (UK) and Longren (USA) indicate that from the early 1920s on engineers tried to use existing fibre-reinforced panels like Micarta or Dilecto, coming from Baekeland's realm, or even 19th-century ones like vulcanised fibre—it can be debated whether the latter one could be considered as a "composite"

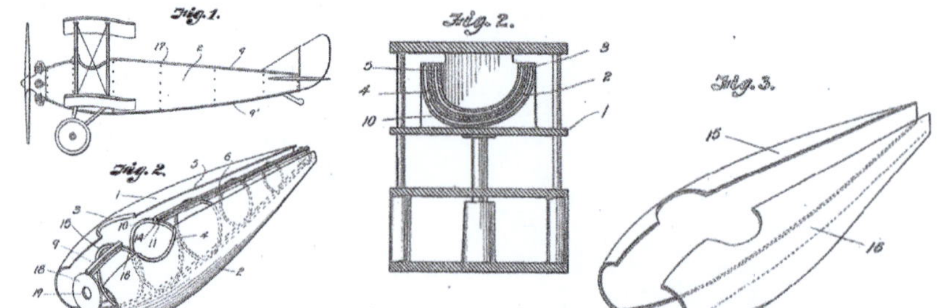

Figure 1.1.11 (a) Longren's concept for an aeroplane with a "stressed skin fuselage"; (b) the semimonocoque fuselage skins, made from vulcanised fibre in a compression moulding process (US Patent 1,541,976, 1922, as quoted by Haka (2023)).

because no matrix is added separately, and impregnated fibres are bonded directly to each other.

Haka describes in detail (Haka, 2023, pp. 121–157) how the German industry has been developing already since 1921 techniques to produce what is nowadays known as "prepregs." The Römmler AG company started replacing "fillers" (particulates, short fibres, etc.) in a compression moldable mass by fibrous reinforcements with well-considered fibre orientations, which can be used to make a multilayered laminate. They filed a patent (1925) describing the production of "*continuous material tapes pre-impregnated with slow-curing resins like cresol*" (later also phenolics and urea). By heating at 120°C, the layers reached an initial hardening state (compared with the nowadays B-staging for epoxies), and could then be laid up and moulded into a final laminate (flat or curved). One of the remarkable applications was interior panels for the passenger lounge and dining room in the luxury Graf Zeppelin LZ 127 airship.

Once Baekeland's original patent expired (1931), many companies started producing fibre-reinforced phenolic matrix composites, more than 100 in Germany alone, according to Haka. Most probably similar developments must have been happening in other European countries and in the United States, but reliable data could not be found.

5. The era of enabling inventions: making material revolutions industrial and creating processes "out of the blue" (1930–1950)

5.1 Materials

The 1930s were a revolutionising era in many respects. They saw the invention of most of the "plastics" we still use today, but also of the first industrial production of the most popular composite reinforcement, namely glass fibres.

5.1.1 Polymers

Many references and schematic overviews can be found on the historical developments of man-made polymers or plastics, a.o. by Palucka and Bensaude-Vincent (2002), Bensaude-Vincent (2011) and Feldman (2008). Some present a simplified timeline from the early developments (Bellis, 2020), while Geyer (2020) concentrates on the developments since 1950, but also highlights societal and environmental aspects of the increasing use of plastics. Presenting a detailed overview of all these developments is out of the scope of this chapter, hence only some crucial innovations will be mentioned briefly.

Carleton Ellis (USA) invented unsaturated polyester as the first "low pressure" thermoset resin, as an alternative for the, at that time, popular phenolic resins, like Bakelite, which required high pressure during curing. In 1936, Ellis filed his first patent in 1940 (US Patent 2,195,362, 1940) describing how glycols reacting with maleic anhydride could be cured by adding a peroxide catalyst (Updegraff, 1982). One year later, he filed a second patent in 1941, which would further improve the applicability of these resins (US Patent 2,255,313, 1941): by adding monomers like vinyl acetate or styrene, the viscosity was reduced, making the application easier, and the curing went faster and more complete.

Epoxy resins came only a few years later: Pierre Castan (Switzerland) filed his first patent on epoxy in 1938 and then licenced it to the Swiss company Ciba (Negoita et al., 2016). Other important developments have been realised by the German scientist Schlack in 1933 and by Greenlee (USA) in 1939, leading to the bisphenol A epoxy, which is the most used epoxy today. Several books on epoxy resins contain a chapter on its history (May, 1988), explaining which developments in organic chemistry in the 1930s led to the invention of epoxy resins, and how they have been further developed during the decades after (Lee and Neville, 1967).

Several thermoplastics have been invented and further elaborated in the same period (polyamide/nylon, polystyrene, polypropylene, ...), but their use as a matrix in composites came much later than that for thermosets and will be addressed later (Section 9.1.4).

5.1.2 Fibres

It may seem like a strange historical coincidence that almost simultaneously with the development of the "low pressure" thermoset resins (polyester and epoxies), the first "upscalable" production process for melt drawing glass fibres has been invented; it was a necessary innovation to replace the small-scale, quasi manual processes in the 18th and 19th century.

Games Slayter (USA), working at Corning–Illinois filed in 1933 a patent to produce glass wool (US Patent 2,133,235, 1933) consisting of discontinuous, slightly oriented glass fibres. The process is rather simple "*The stream of glass ordinarily flows from the supply body through a bushing into the air, the blast of steam being applied at a point beyond the bushing,*" resulting in fine, discontinuous glass fibres, which are then collected on a conveyor belt, making a loose mat (Fig. 1.1.12a).

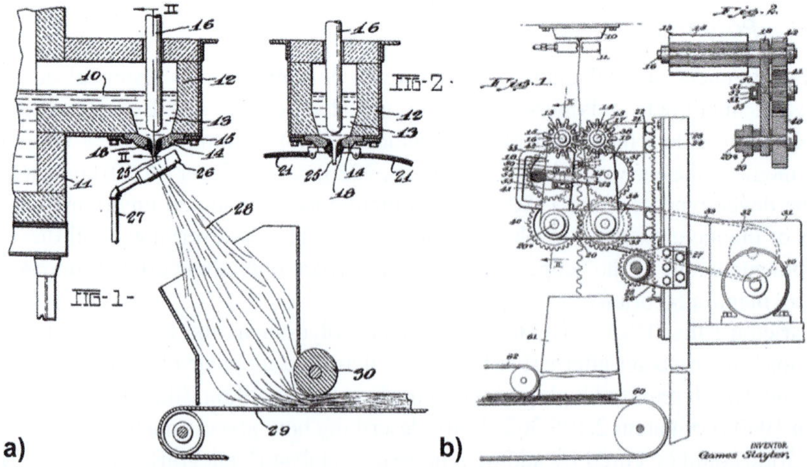

Figure 1.1.12 Process for making glass fibres, according to Slayter's (a) 1933 patent (US 2,133,235) for glass wool and (b) for crimped glass fibres, 1938 patent (US 2,133,235).

Five years later, Slayter filed a second patent, showing how continuous glass fibres could be produced by pulling them down from an orifice (US Patent 2,230,272, 1938). Surprisingly, these fibres were "crimped," according to this patent, because Slayter used several consecutive sets of twin gears, positioned opposite to each other, to pull the solidifying glass flow (Fig. 1.1.12b). Nine out of the 10 claims in this patent relate to such "crimped" glass fibres, and even in the 10th claim, straight fibres are not yet mentioned (only fibres with a "haphazard shape"). It has taken a number of incremental technical innovations to come to straight, continuous glass fibres wound on a bobbin that we know today.

The industrial production started soon. Because the company Owens was also able to produce glass fibres (apparently already in 1932), they joined forces with Corning and created the company Owens-Corning Fibres (from 1935 on). They licenced the technology to Balzaretti Modigliani (Italy), who transferred it in 1939 to Saint-Gobain (France). A more extensive history of the invention and developments of glass fibres is presented in chapter 1.1 of Part II of this book.

5.2 Processes

The first elaborated concepts to make "composites" were not yet using the newly created "low pressure" thermoset resins, but tried to realise what Robert Kemp already had described in his 1916 patents, inspired by Baekeland's patents and by the existence of Micarta and Formica "composites": to preimpregnate a fibrous reinforcement with a phenolic resin, and to "cure" it under high pressure (see Section 4.3).

Norman de Bruyne (born in a Dutch family who later emigrated to Latin America) was not only an engineering don at Cambridge University but also an aerospace enthusiast. He had already built his own experimental aircraft The Snark, using stressed, curved plywood with well-chosen grain orientations for the different veneer layers building up the thickness of the laminate. On January 28, 1937, he presented a paper to the Royal Aeronautical Society in London, stating that "*phenolic resin with suitable reinforcement could have the necessary mechanical properties for aircraft construction, and also the potential to produce lighter-weight components*" (Kinloch, 2000). In this interesting biographical paper by Tony Kinloch, it is further explained that de Bruyne had chosen flattened flax rovings (which he got from an undergraduate student, Mr. Gordon, whose family was in the linen business) to obviously create a unidirectional and hence very efficient reinforcement. Experimental glass "silk" was already available in the USA, but the producer refused to send it to de Bruyne for applications "where it is possibly going to be a failure"!

Although Kemp had already described that curved panels could be produced (in his patents from 1916), the first panels, produced in 1938 by Roger de Bruyne and Malcolm Gordon at the Royal Aircraft Establishment (RAE), were flat panels; they developed an interesting "stop-and-go" semicontinuous process, producing up to 10 m long phenolic composite panels. To achieve this, de Bruyne bought a Siempelkamp press in Germany, not evident in this tense prewar time (while Chamberlain was conferring with Hitler at Bad Godesberg, de Bruyne flew to Dusseldorf to ensure the safe arrival of his press before war broke out). This link with Germany might suggest that de

Bruyne was aware of the technology developed at the Römmler AG company (see above Section 4.4), but no hard proof of this assumption could be found.

The panels were commercialised by their company Aero Research Ltd. as "Gordon Aerolite," a flax-phenolic, later also Kraft paper-phenolic laminate. Norman de Bruyne was also quoted as the inventor of honeycomb (1938) and Redux structural adhesive, a vinyl-modified phenolic, in 1942 (Kinloch, 2000).

Gordon's and de Bruyne's concept was however based on composite structural elements that directly replace metallic ones. It was first tried in 1938, according to McMullen (1984) on the "Blenheim bomber," and then on a "Spitfire" fuselage (Fig. 1.1.13a and b). However, it failed because of the multiple connections typical for a metal-like structure, making the parts too heavy. Moreover, problems were encountered with the hygroscopic nature of the natural fibre reinforcements, and with the high pressures needed during production. Later on, in 1938, Gordon Aerolite was also applied on the Miles Magister (Fig. 1.1.13c), which had a wood structure covered with plywood. As an experiment, 30 flax-phenolic tailplanes were successfully constructed, but they were larger than the original wooden tailplanes to increase the stability of the aircraft.

Almost simultaneously, German aircraft engineers started exploring the potential of phenolic paper and textile-reinforced laminates, initially developed by the Römmler AG company, later on continued in collaboration with Dynamit AG. A remarkable development has been achieved by the Horten brothers, when they combined the innovative idea of a "flying wing" aeroplane concept with the use of Römmler's and Dynamit's laminates. Both brothers, Walter and Reimar, joined the

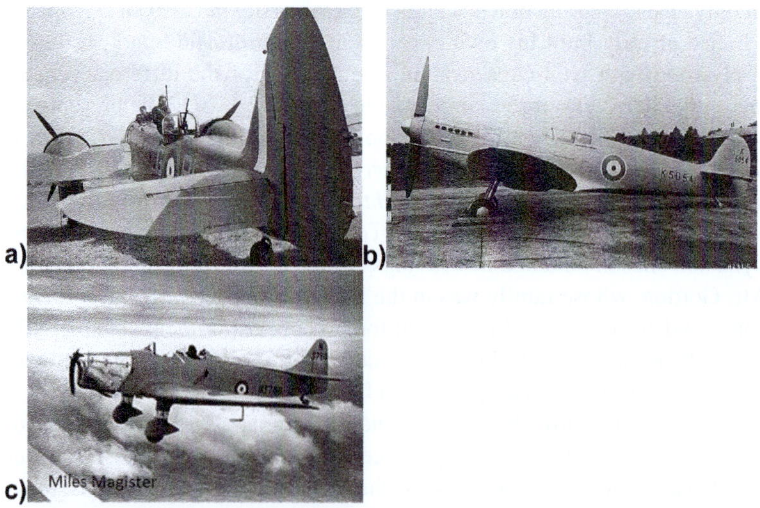

Figure 1.1.13 UK aeroplanes on which the Gordon Aerolite laminates have been used in an experimental phase: (a) the Blenheim Bomber; (b) the Spitfire (UK—Open Government Licence version 1.0); (c) the Miles Magister.
(a) Blenheim Bomber by Stanley Arthur Devon, Public domain, via Wikimedia Commons; (c) Royal Air Force official photographer, Public domain, via Wikimedia Commons.

NDSAP from the very beginning in 1933, three days after Hitler seizing power (Haka, 2023, pp. 231–241). They had good relations with the highest ranks in the nazi regime and its Luftwaffe. As Haka explains in minute detail, Walter Horten was transferred in 1936 from the Luftwaffe (where he served as a pilot) to the Dynamit AG company to "*modify his own design of a tail-less aeroplane*" so that Dynamit's laminates (commercial name Trolitax) could be used. Prototypes of the Horton V-a flying wing, used a metal frame, covered with the Trolitax panels. The concept was presented in 1937 to the Lilienthal Society for Aeronautical Research, where all important German aircraft companies of that time very favourably commented on this innovation. Unfortunately, the prototype crashed during its first test flight, and a second flying wing model (Horten IX) would use multilayered veneer as main structural material.

In parallel, Wolfgang Hütter, brother of Ulrich who would pioneer the use of composites in wind turbine blades in the 1950s, optimised the lay-up of wood veneer layers, impregnated and bonded by a phenolic resin, by using not only 0° and 90 degrees oriented layers, but also ±45 degrees or even 20 degrees layers (Haka, 2023, pp. 197–205). In the patent (Hütter, 1943), not only wood veneer, but also "synthetic panels" are mentioned. Hence, this could be considered as the first deliberate use of unidirectional fibrous layers in a multilayered composite, as shown in Fig. 1.1.14. Hütter designed and built the wings for an all-wood reconnaissance aircraft, the Hü211, at the behest of the Reich Aviation Ministry, but it never surpassed the prototype phase (wings were built, though), as the nazi regime came to its end in 1945. The Horten brothers are on one hand considered as innovative engineers, but their strong links with the nazi regime (even using forced labourers and concentration camp prisoners in their factories) have rightfully drawn much criticism.

Processes for phenolic wood veneer laminates have also been developed in Britain by the de Havilland Aircraft company for their DH91 Albatross and DH 98 Mosquito. In the United States, a similar process "Duramold" was developed in collaboration with the Bakelite Corporation, resulting producing shells for the Hughes H-4 Hercules aeroplane.

The invention, in the mid-1930s, of low-pressure resins and continuous glass fibres would gradually lead to more performant fibre reinforced composites. J.E. Gordon and other colleagues from the RAE (UK) moved to the low-pressure polyester resins, which just had become available. It enabled them to make larger "one piece" parts, or even rather complex shapes, like the composite seat for the Spitfire (McMullen, 1984). Glass fibre reinforcements were studied at RAE from 1946 on, but the advantages of

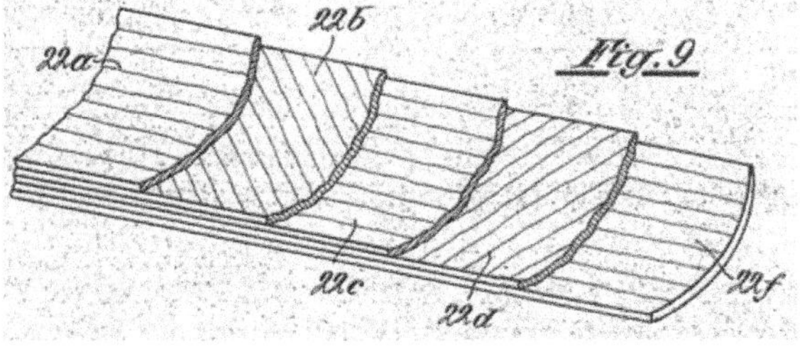

Figure 1.1.14 Plywood orientations in the upper and lower shells of the wing of Hü 211 (Haka 2023, Fig. 4.40).

this material were not sufficiently convincing to intensify research for aeronautical applications, as will be explained in the next section.

Also in Germany, attempts were made to use glass fibre reinforcements. Already in 1935, Schuhman at AEG filed a patent on glass fibre reinforced laminates, intended for insulation purposes (Haka, 2023, pp. 179–183), probably still using phenolic resins. From 1941 on, the DVL (German Aircraft Laboratories) collaborated with P.A. Koch at the Technical University of Dresden to develop "*synthetic resin laminates with glass-fibre fabric inserts*" (Haka, 2011). They obtained their glass fabrics from two German manufacturers (a.o. Thüringische Glaswollindustrie), and from 1942 on, Römmler AG produced such laminates under the brand name Harex.

It is clear that by the early 1940s, paper and flax fabric reinforced phenolic laminates have been gradually replaced by glass fibre reinforced polyester composites.

This logical step toward using low-pressure resins as matrix for composites initiated the development of a range of novel processing methods, dedicated to composites with continuous fibre reinforcements, using bio-based fibres as well as glass fibres: the wet lay-up (or "contact moulding") technique was used from the very beginning, as well as the technique of premixing matrix (phenolic, epoxy or polyester) and fibres (glass or bio-based) into a "moulding compound," which is then compressed at high temperature into the desired shape of the product; the use of such SMC (sheet moulding compound) or BMC (bulk moulding compound) semiproducts was rapidly expanding in the 1950s.

Later on, other new processing methods were invented, like filament winding and pultrusion. Filament winding was initially developed for rocket motors and glass fibre composite pipes (patent J.F. Stephens, 1944–49). Later on, Lubin&Greenberg started their production at the company Basons Industries by (1945). Pultrusion came somewhat later as described in a patent by Meyer and Howell in 1951, and was later perfected by Brandt-Goldsworthy (Composites World, 2023). The historical developments for each of these processes will be more extensively described in part II of this book (Chapter 2.1 by Kevin Potter).

6. The first 2 decades (1940–60) of developing applications with hesitating enthusiasm *(engineers)* and (un)limited imagination *(designers)*

The aforementioned "enabling inventions" came about as a result of both a "technology push" as well as a "market pull." The technology push came from chemists looking for exiting new chemistry which would allow them to create any material they could dream of, whereas materials engineers started to understand that fine, almost defect-free fibres do have an enormous strength potential. As early as 1916, the research of Robert Kemp at the Westinghouse laboratories, and the resulting patents for making a full composite aeroplane, are another strong example of "blue sky" or curiosity-driven research.

But for composites, the driving force was clearly a market pull, namely the need for lighter, yet stronger and stiffer construction materials, which would provide an alternative for aluminium and stressed plywood in the aerospace industry. World War II was a strong catalyst in these developments, and between 1938 and 1945 many experiments were carried out. Developments in the United Kingdom have been briefly mentioned in the previous chapter, and are extensively described in the McMullen paper (McMullen, 1984), where also developments using glass, asbestos, and other fibrous reinforcements are analysed. Simultaneous developments in Germany are discussed in (Haka, 2023).

In the United States, the first plane (Vultee BT15, Fig. 1.1.15) with a glass fibre composite fuselage took off from Wright–Patterson Air Force Base (1943), as described more in detail in chapter 3.2 of Part II of this book. Furthermore, wood veneer-polymer composite wing spars were used on the Hughes' H4 "Spruce Goose" (1945), GFRP wings were installed on USA training aircraft AT-6 and BT-15, etc (Epstein and Heimerdinger, 2003). This publication, at the occasion of the 100th anniversary of the first flight of the Wright Brothers, highlights (in a very detailed way, certainly for the United States) the history of composites and structural adhesives related to air and space flight, for the period 1909–2002.

It is remarkable, but not surprising, that those initially successful innovations did not result in commercial successes. The main reason was the fact that glass fibre composites had a too low stiffness, and hence the weight reductions compared with aluminium were marginal to nonexisting. Moreover, the manufacturing processes were in an early stage of development, and even the material characterisation was in its infancy. For these reasons, the developments of composite (military) aeroplanes stopped in the United Kingdom in 1946 (McMullen, 1984) and only really resumed, both in Europe and in the United States, when the much stiffer carbon fibres became available.

However, civil aircraft found an interesting application of glass fibre composites, exploiting their transparency for frequencies at which their radar operates. The first GFRP radome was installed in 1950 on a Lockheed Constellation aircraft. Grumman Aircraft extended this to their rotating radome on the Hawkeye E-2A. Ten years later, Boeing (in the B727) and Douglas Aircraft (in DC8) reported the renewed use of GFRP in (semi-) structural parts, a.o. spar and vertical tail sections, with Boeing

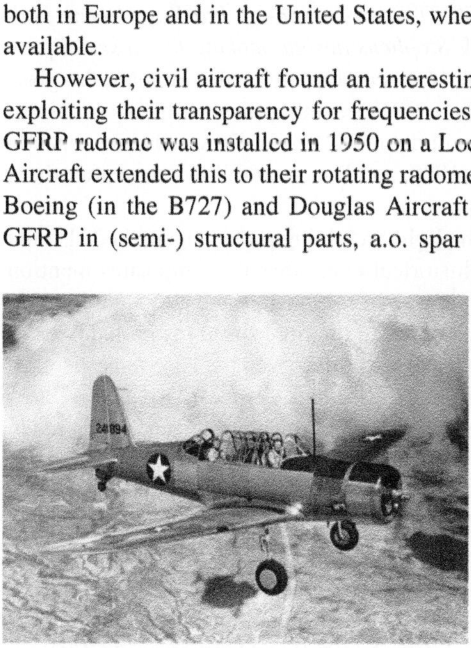

Figure 1.1.15 Vultee BT15, flying over_Laredo_Army Airfield, 1943. U.S. Air Force photo B-39,318 A.C., Public domain, via Wikimedia Commons.

claiming a (surprisingly high) weight reduction in 33% (Epstein and Heimerdinger, 2003). More information on the early developments at Boeing can be found in Chapter 3.2 of Part II of this book.

A special case is gliders. Their development might be related to the political and industrial restrictions imposed to Germany and Japan after World War II by the Allied Forces (Allied Control Council Directive No. 25, 1946). As a consequence, both countries seem to have been frontrunners in the development of "aeroplanes without engines," namely composite gliders. The Phoenix GFRP composite glider was designed and built by Eppler, Naegele, and Lindner in 1957 (Haka, 2023) They were followed by many others, initially using mainly glass fibres, but later on moving toward carbon fibre composites. Helicopter blades are another success story for glass fibres: only 8 years after the opening of the Saint-Gobain glass fibre factory in Chambéry (France, 1950), the first GFRP helicopter blades were produced for the Alouette II (1958) (Palucka and Bensaude-Vincent, 2002).

The corrosion-resistant properties of GFRP immediately got the attention of the marine industry in the United States. Already in 1942, the first "sailing dinghy" was built by R. Greene (Owens-Illinois Glass, after first trials in 1937). In 1946, the first glass fibre surfboards have been manufactured by Pete Peterson, in collaboration with Brandt Goldsworthy (Coldwell, 2006; Gault-Williams, 2020), followed by Bob Simmons in 1949, who built the first sandwich-type surfboard (glass fibre composite skin on a polystyrene foam core) (Marsh, 2006). It is even mentioned that the allied forces used glass fibre composite ships when landing in Normandy (Pil and Verpoest, 2018)! In his interesting overview article, Marsh writes: "*The first fibreglass yacht was probably the Arion ... in 1951. That was a one-off, but Ray Greene had, by 1957, built 175 of a 25 ft Sparkman & Stephens design, making him a series production pioneer.*" It was the start of a most successful application domain for glass fibre composites, continuing to date. Also in Europe, similar developments occurred: Tod's dingy (1951), the Flying Twenty sailboat (1952) ... In France, Jeanneau starts the production of GFRP sailing yachts in 1960, and Beneteau follows soon after them (1963).

Corrosion resistance, but even more the light-weighting potential, sparked also the interest of the automotive industry. All historical overviews on composites mention Henry Ford's "soybean car" (1941) as a pioneer in composite cars, although there remains some confusion whether "soybean" refers to the fibres used (as probably erroneously mentioned in The New York Times at that time) or to the base material used to produce the phenolic matrix of the composite (Pil and Verpoest, 2018).

The concept of a steel frame covered with composite body panels was already applied in a prototype by William Stout and Games Slayter (VP Research at Owens-Corning) in their "Stout Y-46" (Fig. 1.1.16) (Tate, 2015). Whereas Ford's "soybean car" was a unque prototype, the Stout Y-46 can be considered as the first commercially available car. It was later followed by Chevrolet's Corvette, which also has a glass fibre—polyester composite body mounted on a metallic chassis (from 1954 on). At about the same time, East-Germany developed their famous Trabant, using a cotton fibre reinforced phenolic resin (1957) and Citroën used GFRP for the roof of the iconic Citroën DS (1957) (Verpoest et al., 2002).

Figure 1.1.16 The inventor William Stout in his Stout Scarab experimental, also called Y-46.
Rights acquired from Detroit Public Library.

Step by step, it was understood that "everything that moves" could be made lighter, and hence more energy efficient, by using composites. A striking example is the Spacelander bicycle (Fig. 1.1.17), initially conceived in aluminium (1946) by the American designer Benjamin G. Bowden. As in the 50s the knowledge of manufacturing with glass fibre composites was spreading out, Bowden was able to produce a GFRP version of his *(electric!)* bicycle in 1960, but it did not become a commercial success, because it was too heavy (Pil and Verpoest, 2018). It is however interesting to observe that Bowden already exploited, in an intelligent way, one of the unique processing characteristics of composites, namely that the starting material is a highly drapable fibrous cloth, compared with stiff metallic sheets used in the previous design. This enabled him to create a very rounded-off, "organic" design. Moreover, the whole frame consisted of only two parts, mirrored over the longitudinal midplane of the bicycle. They are produced in an open mould, and then glued together; this is the main design and manufacturing concept still used today in carbon fibre composite bicycles.

Figure 1.1.17 The Spacelander bicycle by the American designer Benjamin G. Bowden. Spacelander at Brooklyn Museum, CC BY 3.0, via Wikimedia Commons.

The notion of "drapability," combined with strength and light weight, inspired designers to explore the use of composites in completely different application areas. Historically, the first example is the DAR chair by Charles and Ray Eames in 1948, soon followed by their equally iconic "La Chaise" (Fig. 1.1.18) (Daenens and Bucquoye, 2002). Through history, innovative designers have always had a strong interest in material innovations. Living in Los Angeles in the 1940s, Charles and Ray Eames learnt about composites from friends working as engineers in the Californian aerospace industry, at the time when the first experiments with GFRP were carried out. A similar interaction happened at the American East Coast, where Eero Saarinen and Florence Knoll could convince composite boat builders to produce the shell of their "Womb Chair" in GFRP (1946–48). Saarinen later on also designed the "Tulip Chair" (1956).

However, in all these chairs, only the seating shell was composite, because the legs were technically too difficult to make, and hence remained metallic tubes or rods (DAR, La Chaise, Womb) or aluminium casts (Tulip). The first "full composite" chair was designed by Verner Panton: the "Panton chair" (1959–60) fully exploits on one hand the easy processability and drapability of composites, and on the contrary their strength and stiffness (Fig. 1.18b). In the following decade, an avalanche of composite furniture flooded into the design world, and many more designers were seduced by the "freedom of shape" of composites, leading to playful and colourful designs.

It seems to be a small step from designers to **artists**, but it was only in the late 1960s, more than a decade later, that the first composite sculptures have been made (or "polyester," as they are most often named in art exhibition catalogues). Pioneers were the French artists Niki de Saint-Phalle (Fig. 1.1.19) and Jean Dubuffet and the Spanish artist Joan Miro. But even Henry Moore occasionally used GFRP, mainly in very large sculptures, when bronze or stone would make them simply too heavy. In the 1970s, composites became really popular in the artist's community, as it is extensively illustrated in the book "Les Composites dans l'Art" (CPC, 1992).

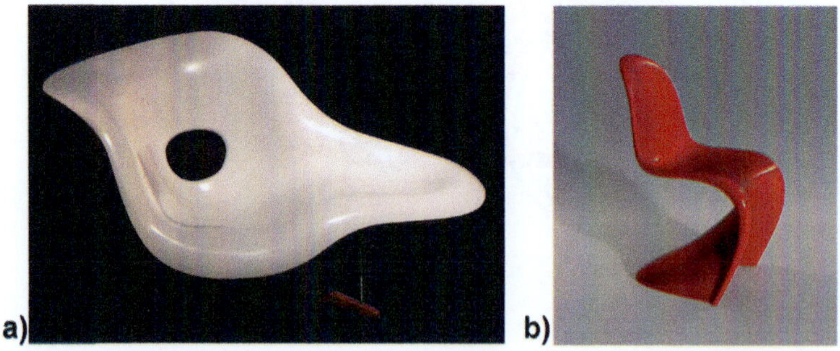

Figure 1.1.18 (a) Charles and Ray Eames "La Chaise," 1948. (b) Verner Panton Chair (1959–60).

(a) Eames Chaise Longue by Sailko, CC BY 3.0, via Wikimedia Commons; (b) Holger Ellgaard Panton, CC BY-SA 3.0, via Wikimedia Commons.

Figure 1.1.19 "Nanas" by Niki de Saint-Phalle.
Nanas by ChristianSchd, CC BY-SA 4.0, via Wikimedia Commons.

7. The second materials breakthrough: carbon fibres (1959–69)

The invention of carbon fibres is to my understanding, and for many others, the pivotal turning point in the history of composite materials, for the simple reason that a material suddenly became available with properties never before realised in other materials: a unique combination of stiffness, strength, and lightness.

Because of its historical importance, the "birth of carbon fibres" has been described in many books, articles, and websites, but often in a sketchy and superficial way. The full, and well documented history still has to be written, but some overviews can be found in "The 25-Year History of SAMPE Japan Chapter" (SAMPE Japan, 2012), in the introductory chapter to JEC's strategic study "Carbon Fibres: history, players and forecast to 2020" (Glowacz, 2015, p. 19–22), and in Akio Shindo's chapter on carbon fibres in Comprehensive Composite Materials, vol. 1 (Shindo, 2000). An impressive, comprehensive series of 25 papers, has been published by Junichi Matsui (Toray), but is up to now only available in Japanese (Matsui, 1997–2006). However, a kind of summary of these papers can be found in Matsui's presentation at the Composites Durability Workshop in Kanazawa, 2010; Matsui, 2010).

In many publications, the history of carbon fibres is studied from a national point of view. A paper by the late Sir Anthony Kelly (Kelly, 2005), in combination with McMullen's earlier mentioned paper, nicely presents the UK-view. The paper of the ACS (American Chemical Society) (ACS, n.d., 2003) highlights the United States contributions. An interesting article by the United Nation's agency WIPO (World Intellectual Property Organisation) (WIPO, n.d.) explains how Shindo's lab handled the IP, patenting and licencing. One of the most complete historical descriptions of what happened between 1955 and 1975 has been written, be it from a mainly Japanese point of view, by Toru Takamatsu from the Osaka School of Economics, in a paper in Technology and Civilisation (Takamatsu, 2000). Even more detailed information on the crucial role of the Japanese industry, and more specifically of Toray, can be found in an article by Katsuo Toma in the book "Industrial innovation in Japan" (Toma,

2008). More recently, Tanaka (2024) has published a well-documented overview of the initial inventions and the gradual performance improvements of carbon fibres. He also authored Chapter 1.2 in Part II of this book on carbon fibres.

When looking further for additional information, some early books on carbon fibres have been very inspirational, as they have been written by researchers who were actively involved in the development of carbon fibres. "*Carbon fibres in composite materials*" (Gill, 1972) was written as early as 1972 by R. M. Gill, who was formerly technical and production director at Morganite Modmor Ltd., one of the UK developers of carbon fibres. More than 10 years later (1985), one of the key persons in the carbon fibre research in the United Kingdom, William Watt, was co-editor (together with B.V. Perov, USSR) of the "*Strong Fibres*" volume of the series "Handbook of Composites" edited by Antony Kelly and Y.N. Rabotnov (Watt and Perov, 1985). The part on carbon fibres (p. 241–606) contains papers by a.o. Konkin, Watt, Johnson, Moreton, and Donnet. The latter, a key carbon specialist from France, wrote at approximately the same time (and together with Roop Chand Bansal, India) the book "*Carbon Fibres*" (Donnet and Bansal, 1984). It gives a clear "state of the art" of carbon fibre manufacturing in the early 1980s. First-hand information is also provided by Roger Bacon, a key inventor in the United States, in his 1986 paper "Carbon fibres: from light bulbs to outer space" (Bacon and Moses, 1986), and by Leslie Phillips, who played a major role in the UK developments at RAE, in his remarkable paper "Carbon fibre reinforced plastics. A personal view" (Phillips, 1968). This paper, published in 1968, in the midst of the competition between Japan, the United States, and the United Kingdom, tries to answer "*... certain queries which are asked again and again ...*" by visitors at the RAE facilities in Farnborough "*... scientists, engineers and designers, -with the occasional military man and business executive.*" This paper very nicely reflects, from an insiders point of view, the hopeful perspectives but also the hesitations and doubts about the future of carbon fibres.

On the development of the use of carbon fibres in composites, and on the gradual improvements of the structural integrity of carbon fibre composites, a comprehensive book has been edited (2016) by Beaumont and Soutis: "*The structural integrity of carbon fibre composites. Fifty years of progress and achievement of the science, development, and applications*" (Beaumont and Soutis, 2016). In the first four chapters, respectively by Adams, Potter, Herakovich, and Jones, some historical information is provided on the development of carbon fibres as such, and on the initial characterisation, testing, and manufacturing of carbon fibre composites.

It is not my aim to write here a "definitive" story on the invention of carbon fibres, as this would need a really in-depth historical study. I only want to mention some important milestones and present a "preliminary" analysis of the role of the main players in this fascinating story. An additional analysis is presented in Chapter 1.2 of Part II of this book, by Fumihiko Tanaka.

In the late 50s and early 60s of last century, the main breakthroughs in carbon fibre research happened in three countries: Japan, the United Kingdom, and the United

States. All the researchers contributed in one way or another; they had moreover mutual contacts and even inspired and motivated each other.

7.1 Preparatory research

The initial example of Edison, carbonising a cellulose-based fibre (bamboo) into its "*all-carbon replica,*" as stated by Bacon in his 1986 paper, has been followed by a number of American researchers in the 1950s, and even earlier. Bacon states that "*already during World War II Union Carbide Corporation ... briefly investigated carbonization of rayon (and PAN) yarns and fabrics ... for control grids in vacuum tube power amplifiers.*" Tanaka (2024) added that this research was performed on rayon and acrylic fibres by dr. Winter in 1942. Soon after World War II, the US National Carbon Company (a division of Union Carbide Company) performed research on using rayon and natural fibres to replace tungsten in vacuum tube power amplifiers.

William F. Abbott, working at the Carbon Wool Corp., seems to have started already in the mid-1950s research on carbonising viscose rayon fibres in an inert atmosphere up to 300–500°F (149–260°C), but he filed his patent only on November 12, 1959 (US Patent 3,053,775, 1959) (Fig. 1.1.20). This patent was based on an earlier (1956) but abandoned US Patent 569,391 (Qiaoting, 2020). A tensile strength up to 11,200 psi (77 MPa) was obtained, a quite low value, obviously because the carbonisation temperature is rather low compared with what would be done later. However, these fibres "*are sufficiently strong to retain their fibrous form upon being subjected to mechanical loads,*" in other words these early carbon fibres have not been developed as "load carrying" in the composites sense, but for "*thermal insulation, filtration processes and the like ....*" Tanaka (2024) quotes even 800°C, for producing "activated carbon fibres," as mentioned in one of the claims of Abbott's patent.

In 1957, the Barnebey-Cheney company, a licensee of Abbot's Carbon Wool corporation, manufactured mats and tows by carbonising cotton and rayon fibres for thermal insulation and filter applications, hence without focus on mechanical properties. A year later, Union Carbide developed a carbonised rayon cloth for the U.S. Air Force as a replacement for glass fibres in rocket nozzle exit cones and re-entry heat shields.

A first real breakthrough happened in Union Carbide Co.'s Parma Technical Centre, when Roger Bacon, in 1958, succeeded in making carbon whiskers, by condensation after sublimation, see Fig. 1.1.21 (US Patent 2,957,756). In fact, he was looking for "the triple point of graphite," where the solid, liquid, and vapour phase are in equilibrium. So, it could be stated that the very first fibre-like and strong carbon was the result of a very fundamental, curiosity-driven research effort. Maybe Bacon produced, without being aware of it, the first (be it rather thick, probably multi-walled) carbon nanotubes, as he mentions "*... the hypothesis that the filaments or "whiskers" made in accord with the invention, consist of large graphite sheets rolled up into a tight tube.*" These were, however, discontinuous fibres, produced in laboratory experiments, and Bacon himself estimated the cost at 20 million US dollar per kilogramme! Bacon published his results only in 1960, in the Journal of Applied Physics (Bacon, 1960). The fact that Bacon even was able to measure their mechanical properties was an important breakthrough: it has now been proven experimentally

3,053,775
METHOD FOR CARBONIZING FIBERS
William F. Abbott, Homer, Alaska, assignor to Carbon Wool Corporation, Ojai, Calif., a corporation of California
Filed Nov. 12, 1959, Ser. No. 852,580
8 Claims. (Cl. 252—421)

The present invention provides for the first time fibers substantially of carbon which are sufficiently strong to retain their fibrous form upon being subjected to mechanical forces.

a)

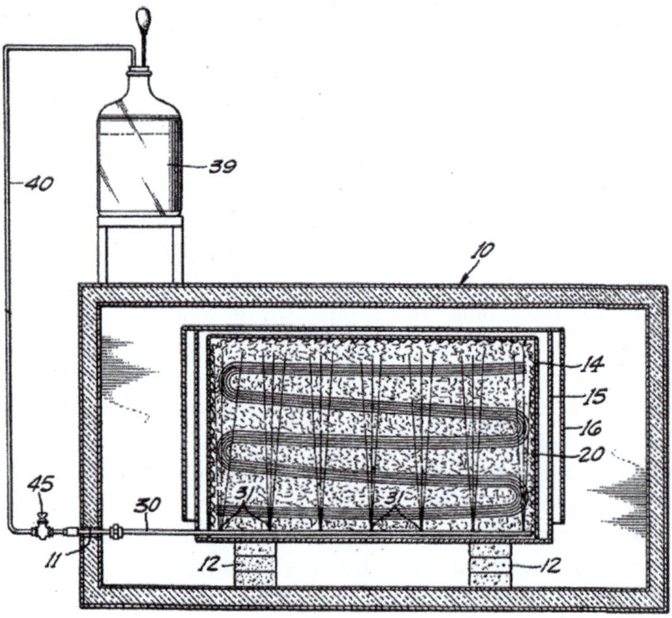

b)

Figure 1.1.20 (a) Abbott's patent for rayon-based carbon fibres, (b) drawing from the patent, showing the furnace to create inert atmosphere at 149–200°C (both from US Patent 3,053,775, accepted and published 1962).

that "carbon" could reach a Young's modulus of 700 GPa and a strength of 20.000 MPa! Hence, this might be considered to be the start of a great interest in using carbon fibres as reinforcement for polymers like phenolics, epoxies, and polyesters.

In 1959, only one year after Bacon's discovery, Ford and Mitchell carbonised, in the same Parma Technical Centre, rayon fibres at temperatures up to 3000°C (ACS, 2003; Ford and Mitchel, 1963), achieving much higher strength properties (up to 900 MPa then those of a commercial lamp filament (200 MPa)). Such high strength was related not only to the higher graphitisation temperature, but mainly to the much smaller fibre/filament diameter (5–7 μm, compared to about 200 μm). In their patent, Ford and Mitchel also described the graphitic structure as having "*a high*

2,957,756

FILAMENTARY GRAPHITE AND METHOD FOR PRODUCING THE SAME

Roger Bacon, Parma, Ohio, assignor to Union Carbide Corporation, a corporation of New York

Filed Mar. 18, 1958, Ser. No. 722,281

2 Claims. (Cl. 23—209.2)

This invention relates to novel articles of manufacture consisting essentially of graphite filaments of unusually high tensile strength and conductivity. The invention is also concerned with methods for producing such articles.

Figure 1.1.21 Patent by Roger Bacon on carbon whiskers (US Patent 2,957,756).

degree of preferred orientation ... throughout the entire filament ...," as they could derive from the measured X-ray diffraction pattern, "*which in part accounts for the superior strength of the monofilament.*"

It was indeed an important further step in the development of carbon fibres from cellulose-based (rayon) precursors, but it took another 5 years until Bacon and Schalamon, in 1964, were able to produce high modulus carbon fibres from rayon by stretching them <u>during</u> the heating process, as will be explained later.

Rayon-based manufacturing of carbon fibres is rather complex.

Already in 1948, Tamaru in Japan (Gill, 1972) discovered that heat-treating cellulose in an inert atmosphere leads to a pyrolytic decomposition involving depolymerisation of the cellulose, but it took until in 1964 that Tang and Bacon published a systematic study describing the four stages of the rayon-to-carbon fibre transformation: physical desorption of water (up to 150°C), dehydration from the cellulose unit (150–240°C), thermal cleavage of C–O and C–C bonds (240–400°C), and finally aromatisation and formation of the graphitic layers (carbonisation up to 1500°C, graphitisation up to 2500°C and higher) (Tang and Bacon, 1964; Bacon and Tang, 1964). These processes involve important weight losses, so that the net yield is only 15%–30%, and considerable shrinkage of the fibres can be observed. This low yield will later on induce a considerable disadvantage compared with PAN-based carbon fibres.

Anyway, 1959 is a crucial year in the carbon fibre history, because it is also the year of Shindo's invention of PAN-based carbon fibres.

7.2 *Shindo's pioneering work on PAN fibres*

In the early 1950s, the Government Industrial Research Institute in Osaka (GIRIO, Japan, now integrated in AIST) had a particular interest in textile research, but also in carbon materials (a.o. for nuclear reactors). When Akio Shindo (Fig. 1.1.22a) joined GIRIO in 1952, he was assigned to the carbon R&D division. Shindo later described the research as follows: "*The degradation of carbon is not necessarily a bad theme depending on the method of research, but it seems that it is unlikely that*

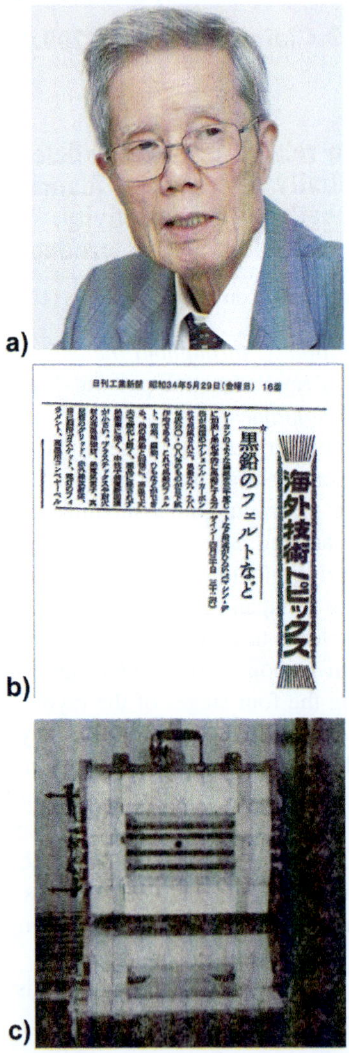

Figure 1.1.22 (a) Shindo Akio. (b) Article in the Japanese business and technology newspaper "Nikkan Kogyo," May 29, 1959. (c) Shindos 1000°C furnace and quartz tubes in GIRIO (Shindo, 1986) (Yoshioka, 2019).
(a) Copyright AIST, Japan, permission for reproduction; (b) approved for reprint by the Nikkan Kogyo Shimbun.

a fruitful result can be expected from the viewpoint of the development of industrial materials." (Takamatsu, 2000). Shindo was indeed, from the very beginning of his research, looking for "*research that would lead to industrialization.*"

As far as I could reconstruct, Akio Shindo was not aware of Bacon's developments, but his attention was drawn to some early developments in the United States on carbon felt and fabrics when reading, on May 29, 1959, an article in the Japanese business and technology newspaper "Nikkan Kogyo", where a reference was made to an article in the US magazine "Machine Design" of April 30, 1959.[1] In this article, a brief description of "*the method to thermochemically process fibres such as rayon to graphite by heating at 3000°C*" was presented, hence most probably pointing at the research on carbonising rayon fibres at Union Carbide Co.'s Parma Technical Centre, probably by Ford and Mitchell, although this is not explicitly mentioned in the journal article. Also, the WIPO-paper (WIPO, n.d.) claims that there was reference made to the developments in the National Carbon Company, which was part of the Union Carbide Company.

This article inspired Shindo to also focus his research on the use of polymeric precursor fibres. Only 3 months later, on September 7, 1959, he filed his patent on using polymeric fibres as precursor for carbon fibres, PAN only, according to the information presented in the Japanese SAMPE overview. In the WIPO paper, it is mentioned that "*by September 1959 dr. Shindo had successfully developed the process to create PAN carbon fibre.*"

Although Shindo himself stated in an interview in 1997 (Takamatsu, 2000, p. 3) that he indeed had been reading the article in Nikkan Kogyo, mentioning the 3000°C heat treatment of rayon to convert it into graphite, it should be noted that the same source mentions that the carbon lab at GIRIO, where Shindo performed his research, was already aware of the American studies on carbonising polymeric fibres in the mid-1950s (Takamatsu, 2000, footnote 6). The question remains from where such information could have come from; it cannot be Ford an Mitchel's work (only started in 1959), and also Abbott's patent on carbonising rayon fibres at Carbon Wool Corp. (see above) had not yet been filed (1959) nor published (1962).

The story of Shindo's invention is interesting, because it can be considered as a textbook example that careful observation during experimental work can lead to important scientific breakthroughs. According to the information, presented in Takamatsu's paper, Shindo bought several types of synthetic fibres in a department store,

[1] The article reads as follows (translation kindly provided by Ichiro Taketa, Toray): "*Overseas technology topics: Graphite felted fabric, etc. National Carbon Co. established the method to thermochemically process fibres such as Rayon to graphite by heating at 3000°C. The one consisting of 99.98% of graphite and 0.08% of ash is under prototyping. This technology enables felted fabric, cloth, knit, string made of graphite. In the same manner as other graphite, it is heat resistant, oxidisation resistant, chemical resistant, strong at thermal shock and small cross-section for neutron capture. The application is expected widely: heat resistant reinforcement for plastics and fireproof material, thermoelectric element, grid of vacuum tube, infrared radiator, self-lubricating gasket, filament for electric light, heat resistant conveyer belt, and so on.*"

but then Shindo started using Dupont's Orlon (PAN) fibres. Tanaka (2024) mentions that "*... at that time, Mr. Shindo's biological younger brother, Eiichi Muramami, who was working for Teijin in the U.S., provided him with commercially available DuPont's "Orlon" acrylic fibres.*"

He first put one single PAN fibre in a tube, flushed it with nitrogen, and heated it up to 1000°C (Fig. 1.1.22b), resulting in a "nonbrittle" carbon fibre *("did not fracture when wound around your finger")*. However, when the same was done with a bundle of fibres, it resulted in a very brittle behaviour. This was related, in Shindo's thinking, to the amount of oxygen remaining in the tube, which was sufficient to create a (partial) oxidation of the single PAN fibre and hence stabilise the polymeric structure so that it can withstand the further temperature increase, and insufficient for the bundle of PAN fibres. With this critical observation, Akio Shindo has laid the basis for one of the most important material (r)evolutions of the 20th century.

Shindo's patent, filed in September 1959, was published in 1962 (patent 37-4405) (Shindo et al., 1959a,b). It stated clearly *"that polyacrylonitrile be heated in an oxidising atmosphere up to 350°C and then to 800°C or higher.*" (Tanaka, 2024). Shindo had already presented some results at a domestic conference in October 1959 (GIRIO, 1959), and published in 1961 the famous GIRIO report n° 317 (Shindo, 1961) where according to McMullen a fibre stiffness of 110 GPa was reported, without mentioning the heat treatment temperature. However, R.M. Gill mentions in his book (p. 41–42) a modulus of 165 GPa, and a strength of 730 MPa, if a PAN homopolymer was used, and even 970 MPa for a co-polymer (and when slowly heating to 1000°C). Johnson refers in Watt's book (p. 405) to the same Shindo report, but mentions that the stiffness was increased from 110 to 160 GPa, when the heat treatment temperature was increased from 1000 to 2000°C. From these conflicting statements, it is clear that translating Shindo's original Japanese patent is urgently needed.

According to Shindo's own text in Comprehensive Composite Materials (Shindo, 2000), attention was drawn to the importance of air oxidation (to stabilise the PAN) and of orientation of the graphitic "molecular units." Shindo's PhD thesis *(doctor of science)* "Studies on graphite fibres," was translated in English, which was rather exceptional at that time. In 1961 and 1962 he published several papers in Japanese journals (a.o. in the Journal of the Ceramic Industry Association), but he also reported on his research during the sixth American Carbon Conference in the United States in 1963. Surprisingly, Shindo waited until 1963 to file an international patent, but it was declined because the deadline for claiming priority was already passed—according to Takamatsu (2000).

Fig. 1.1.23 shows a comparison of stiffness and strength of different fibres, and the development path from Shindo's invention to Toray's fibres. This figure, published in Takamatsu's paper (Takamatsu, 2000), was translated and redrawn; details on the sources of the data can be found in Takamatsu's paper.

The three phases in the PAN-to-carbon fibre process are different from the rayon-based carbon fibre process.

Because PAN fibres as such would melt upon heating to high temperatures, a low-temperature heat treatment under oxidative conditions is applied first. Subsequently,

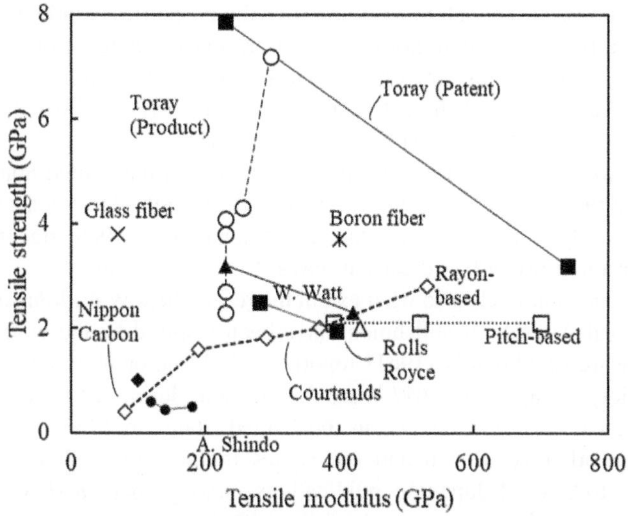

Figure 1.1.23 Carbon fibre performance enhancement path. Comparison of carbon fibre properties with rayon-based carbon fibres, glass, and boron fibres. (Note) Rayon and pitch values are from UCC (redrawn from Takamatsu, 2000).

carbonisation and graphitisation take place under a protective atmosphere. This process has two advantages compared with the rayon-to-carbon fibre process: the yield is considerably higher (40%–50%) and the orientation of the microstructural units can happen at low temperature, namely during the oxidation phase, as will be explained below.

7.3 Who's done it? The invention/discovery of tensioning/restraining the fibres to create orientation of the graphite ...

7.3.1 The question

The "orientation of the graphite molecular units" therefore seems to have become a crucial point in further improving the stiffness (and strength) of carbon fibres. The intriguing and still not really solved question is: who came with this idea first?

Did it happen in Japan, where Shindo already (and possibly unknowingly) applied a "strain" during the oxidation treatment of the PAN fibres (in 1959)? As mentioned earlier, Shindo emphasised the importance of "molecular orientation" in his paper in Comprehensive Composite Materials (2000), and in fact he applied a certain amount of strain by winding and fixing the PAN-precursor fibres on a frame, hence hindered the contraction during the pretreatment stage. Takamatsu (2000) mentions that "*the reason that Shindo gave tension in the pretreatment stage was to make the sample straight,*" because it had to be straight when he wanted to measure afterwards the mechanical properties. The intention to improve the properties by adding some strain during the heat treatment might not have been present in Shindo's mind.

Was it in the United Kingdom, on the contrary, where Watt, Johnson, and Phillips at the Royal Aircraft Establishment (Farnborough, UK) realised a breakthrough in their work on (a different kind of) PAN-fibres (1964)?

Or have Kotlensky et al. (1962) been the very first, in the United States, to study systematically the effect of straining, be it on pyrolytically deposited graphite, followed by the study of Bacon et al. (Bacon and Schalamon, 1969; Bacon and Tang, 1964) in restraining rayon-based carbon fibres ?

It is *for the moment* unclear to what extent the researchers Watt, Johnson, and Phillips at the Royal Aircraft Establishment have been inspired by the work of Shindo. Were they aware of Shindo's (1961) report no. 317 and/or his publication in The Ceramic Society of Japan of 1961? Were they present during his presentation at the sixth American Carbon Conference in the United States in 1963? Did they know the work of Kotlensky, Bacon and colleagues in the United States? Takamatsu (2000) refers to several Japanese publications, stating that "*RAE's research on PAN-based carbon fibres began in the fall of 1963, triggered by Shindo's reports in the United States.*"

7.3.2 Contributions in the United Kingdom

Interestingly, Sir Antony Kelly mentions in two papers (Kelly, 2000, 2005) that, during a meeting in the Royal Society in London (June 1963), he discussed the lack of stiff fibres, although he knew about the carbon whiskers developed by Bacon in the United States in 1958. William Watt, also present at the meeting, told him "*that he could make stiff graphite fibres*." It seems that soon after this (and according to Kelly), Watt, Johnson and Phillips at the Royal Aircraft Establishment (RAE) were able, in a batch process, to produce stiff carbon fibres (modulus of up to 375 GPa) by applying tension during the oxidative stabilisation process. In fact, they initially have been winding the fibres on a frame, very similar to Shindo's practice, making the shrinkage of the fibres impossible and hence effectively "straining" the fibres during the oxidation process, as has been well documented by Gunston in his early paper on carbon fibres (Gunston, 1969). Gunston further explains that Phillips then went to the Atomic Energy Research Establishment (AERE) at Harwell, to ask for help in upscaling the process, because they had furnaces for carbonisation and graphitisation. By 1968, AERE was able to deliver several hundred kilogrammes of high-quality carbon fibre, be it that they still were not continuous (sheets of fibres, 30 cm wide × 120 cm long). So it was a batch process, and continuous processes with in-line stretching would be developed only a few years later—further discussion on these developments can be found in part 2 of this book, Chapter 1.2 by Tanaka).

Watt et al.'s work on studying the effect of straining on the properties of PAN-based carbon fibres was, according to Kelly, "*a step which had been eluded by the Japanese*". William Watt himself presents more detailed data in Chapter IX of his book (Watt, 1985), referring to the research that formed the basis for their patent

(British Patent 1110 791, 1968), filed in April 1965: the effect of "*restraining the PAN-fibres from shrinking during oxidation*" on both stiffness and strength is convincingly shown as a function of the carbonisation/graphitisation temperature. Whereas Shindo's carbon fibres attain a maximum stiffness of 165 GPa, Watt's fibres reach up to 400 GPa. McMullen adds to this that subsequently "*heating to 1000°C and above without restraint develops a carbon structure in which the layer planes are lying closely parallel to the longitudinal axis of the fibre*," hence emphasising that the effect happens by straining during the low-temperature oxidation phase. Moreover, in a later study, Johnson (1985) showed that even just preventing shrinkage during oxidation has a strong effect, whereas additional stretching (and hence effective elongation of the fibre) has almost no effect. Frank Jones explains nicely, in his chapter (Jones, 2016) of Beaumont and Soutis' book, that the shrinkage during the oxidation phase is related to the atactic nature of the PAN molecular chains.

Then, Moreton, Watt, and Johnson at RAE also investigated systematically the effect of higher temperatures during the carbonisation and graphitisation process (Moreton et al., 1967). They found that higher temperatures monotonically lead to higher Young's moduli, but that the strength reaches a maximum at a treatment temperature of about 1500°C. Fig. 1.1.24 has since then appeared in many textbooks on carbon fibres. Interestingly, the authors compared their results with those of Shindo from his famous GIRIO-report 317 from 1961. The fact that Shindo's data are lower, and show a maximum for the stiffness at 2000°C is however not explained, and the reason for the further decrease at higher temperatures "*is obscure*," according to Moreton et al.

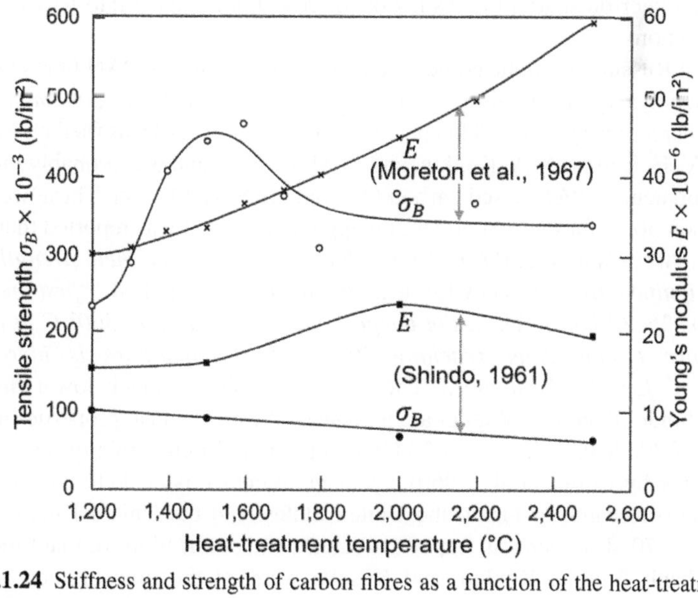

Figure 1.1.24 Stiffness and strength of carbon fibres as a function of the heat-treatment temperature.
Redrawn from Moreton et al. (1967).

Leslie Phillips showed that RAE's fibres were compatible with epoxy and polyester resins and suggested the use of Courtauld's PAN fibres as precursor. The process, patented in 1965, was licenced to three UK companies: Courtaulds, Morganite, and Rolls-Royce.

7.3.3 Developments in the United States

The importance of orientation was, earlier on in 1962, also proven by Kotlensky et al. (1962) in their work on carbon fibres composed of carbon substrates with a pyrolytically deposited graphite layer. They could, by stretching up to 16% at 2750°C, increase the modulus from 16 to 560 GPa. In probably the same year as Watt et al. started their work on straining PAN-precursors during the stabilisation phase, Bacon and Smith (1965) had produced in 1964 rayon-based carbon fibres by stretching them at a temperature of 2800°C, leading to only a stiffness increase of 19%. In a later publication, Bacon and Schalamon (1969) improved the process further by stretching the rayon fibres, not at the maximum temperature, but during heating up to that temperature, reaching a modulus of 110 GPa (at 10% stretch) to 450 GPa (at 50% stretch). For being effective, the stretching during heating of rayon-based carbon fibres happens at higher temperatures than in PAN-based carbon fibres.

7.3.4 Developments in the USSR

One can wonder whether the development of carbon fibres was limited to three countries only: Japan, the United States, and the United Kingdom. KU Leuven colleague Stepan Lomov (2019) drew my attention to the early developments in the USSR/Russia, and later on shared the results of the research he made into Russian patents and publications.

The first Russian patent he could discover was on the use of PAN fibres when carbonising a polyacrylonitrile fibre or weave, to create "*a flame-retardant material with good mechanical properties.*" This patent (Kotina et al., 1960) was filed on March 28, 1960, by V. E. Kotina, A. A. Konkin, and R. M. Kosova, and was probably the start of the development of PAN-based carbon fibres in the Soviet Union. These were, however, carbon fibres for thermal protection applications, as it was reported that the carbon fibres "*were kept at 2500° C for 100 hours, during which their strength did not change significantly.*" A stepwise heat treatment was applied: "*first heating at 200° C for 25–30 hours. Then, the temperature is increased to 300° C at a rate of 1.5–2.0 per 1 min. After reaching 300° C, the temperature is increased to 400–475° C for 15–25 minutes.*" These temperatures are much lower than those used by Shindo, Bacon, or Watt et al., and hence the mechanical properties are lower (strength of 700 MPa, stiffness of 20 GPa, but a high failure strain of 4%).

In a paper by Kotina et al. (1960) extensive reference is made to the carbon (and carbon fibre) research in Japan, the United States, and the United Kingdom in the period 1959–70, discussing all important patents and publications of that time period. Most probably Kotina, Konkin, and Kosova were, at the time of their invention (1960), not aware of Shindo's patent filed in September 1959, which was not publicly available until he published his GIRIO report no. 317 in 1961. Yet, Shindo had presented his invention in a local scientific conference in October 1959, so it is not to be excluded that some of that information quickly reached the Russian researchers.

Almost no information could be found in open sources on the developments in the USSR from the initial low-temperature-treated PAN fibres of Kotina, Konkin, and Kosova in 1960 to the higher modulus and higher strength carbon fibres, which would become available in the USSR from the early 1970s on (see Section 7.4). In a publication only available on a Russian website (Yarovan 2018), it is mentioned that "*the first carbon fibre reinforced plastics were obtained in 1967*," but no information could be found on the production of the carbon fibres themselves. It is clear that the main personalities in that period in the USSR were A. A. Konkin (professor leading the research) and V. E. Kotina, both working in All-Union Scientific-Research Institute for Man-Made Fibres (VNIIV), in Moscow. Interestingly, the 1985 book on "Strong Fibres" of Watt is co-authored by B. V. Perov of the Academy of Science of the USSR, and two chapters in the section on carbon fibres are written by A.A. Konkin.

7.3.5 Conclusion?

So, is there a conclusive answer to the question who first came up with the idea that stretching the fibres was crucial to obtain carbon fibres with high stiffness? As stretching of polymeric fibres during fibre spinning was, at that time, a well-known technology to increase molecular orientation, it seems a rather small and logical step to apply it to these new fibres.

Shindo might have done it unknowingly (in 1959) when he fixed the PAN fibres on a frame to make sure that the resulting carbon fibres would be straight. But Watt et al. in the United Kingdom and Bacon et al. in the United States, both around 1964–65, did it on purpose and studied the effect of different degrees of stretching at different temperatures. Johnson adds to this, in his Chapter X in the book of Watt and Perov (1985) about Shindo's report of 1961: "*It was stated that care was taken to prepare straight fibres for mechanical tests, but it did not say whether tension was used to hold the fibres straight, or more importantly, to eliminate fibre shrinkage at 200° C*". On the contrary, by fixing the fibres, Shindo created a tensile force because he was effectively preventing shrinkage. Still, Shindo was working in a "high T°-thermal insulation" paradigm, not aiming at strong and stiff fibres to be used as reinforcement for composites, as was the case with Watt and Bacon.

Takamatsu (2000) presents a nice story about the moment (in 1965) that Shindo became aware of the importance of stiffness of his carbon fibres: Shindo knew that a stiff carbon fibre was needed when he heard the remarks of Colonel Postelnek of the United States Air Force Materials Research Institute who came to investigate Shindo's research in 1965. "*When we took out the fired sample (I handed the sample to Postelnek), he said "Strong, Strong" while taking each filament between his fingers and pulling with both hands. At that time, I still thought carbon fibre should be more supple. However, he said it should be stiff. At that time, I learned that the high modulus of elasticity is an advantage of carbon fibre.*"

It is obvious that, up to that time, Shindo's team was focused on "*carbon fibre only as a product that added flexibility to existing carbon products*", in contradiction to carbon panels or blocks, as mentioned by Takamatsu. On Shindo's subsequent research, he added: "*With his chemical interests, Shindo's research is not going to enhance mechanical properties, but rather to research on carbon fibre using polyvinyl alcohol and rayon as raw materials*." The latter is confirmed in the books of Gill and Watt, who indeed describe the several attempts of Shindo to find alternative polymer fibres for the rather difficult and expensive PAN fibre.

In contrast, Watt, Phillips, and Johnson seem to have been concentrated on stiff, strong, and lightweight reinforcements for plastics or metals from the very beginning, and they had been engaged by the Royal Aircraft Establishment for that purpose. As mentioned by McMullen (1984, p. 228), this was part of RAE's general focus on making aircraft lighter and hence more performant.

It is clear that there has been, in the 1960s, a lot of interesting parallel developments in Japan, the United States, and the United Kingdom, and little is known about the USSR. An attempt to analyse these innovations, and their interactions, was presented in a paper by Nakamura et al. (2009), published in Synthesiology, a Japanese journal on R&D management. Fig. 1.1.25 shows schematically how these interactions have happened, focussing on the relation between Shindo's scientific innovations, the Japanese industrial developments and the "international" events which were happening in parallel with the Japanese ones, concluding that "*both the top-down*

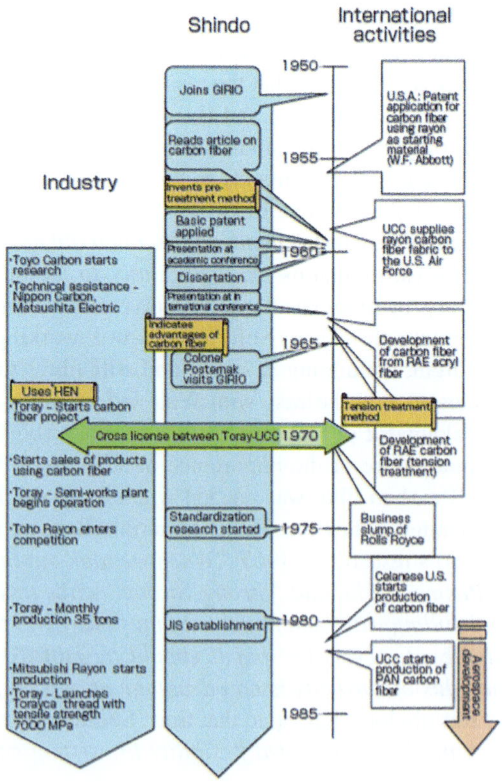

Figure 1.1.25 The flow of carbon fibre development (Nakamura et al., 2009).

management and the personal motivation should be in phase and synergetic with each other" to realise a successful innovation.

7.4 First attempts for commercialisation in Japan early 60s ...

According to the WIPO paper (WIPO, n.d.), Shindo's patent was licenced (*nonexclusive "practicable rights"* were transferred) in 1963 to two Japanese companies active in carbon, namely Nippon Carbon and Tokai Denkyoku (currently Tokai Carbon). Takamatsu (2000) specifies that Nippon Carbon signed a research contract with Shindo, and launched already in 1961 the first commercial carbon fibres "Carbolon" at a cost of 30–50.000 Yen per kilogramme, and having a tensile strength of 1000 MPa and a modulus of 100 GPa. Two years later, Tokai Denkyoku (= "Tokai electrode") launched the "Thermolon" fibre. Both fibres were mainly aiming at applications that have to operate at high temperatures, like insulations and filters, where the flexibility of the fibres would be an advantage compared with solid carbon blocks or panels. However, both companies lacked experience in polymer precursor fibres, and fibre know-how in general, and did not produce themselves PAN fibres.

7.5 Second wave of commercialisation: 1965–75

Two important hurdles, next to many other, smaller ones, still had to be overcome to enter a real industrial production: can the duration of the oxidation process be reduced, and can a continuous process replace the up to that time batch type processes? In the following, a brief overview will be presented of the developments toward industrial production in Japan, USA, UK, USSR, and Europe.

7.5.1 Japan

In 1966, Toray in Japan had started the production of Nylon 66, and Morita had performed research on using aniline in an innovative process to make Nylon 66, known as the Morita-Baylis-Hillman reaction, involved reacting acetaldehyde with acrylonitrile to obtain hydroxyethyl acrylonitrile (HEN) (Takamatsu, 2000, p. 10) (more fsinformation in Tanaka (2024)). This led, in 1968, to patenting a new copolymerised PAN-compound, which would reduce the oxidation time of these new PAN precursor fibres by a factor of 10!

Toray straight away produced PAN-fibres with this new comonomer and delivered it to Nippon Carbon and Tokai Denkyoku who explored the manufacturing advantages and performance of these new carbon fibres. The results were superior, so Morita encouraged in-house commercialisation of the invention and succeeded in starting up a research project on the industrialisation of carbon fibre production. In this endeavour, he was supported by Yoshikazu Ito (later on president of Toray), who was able to set up, in 1969, a project team of 200 people (Tanaka, 2024). Initially, they used a batch process in their studies, but because Toray had no experience in industrial high temperature treatments, the company started a collaboration with Union Carbide C° in the United States (see Fig. 1.1.25, green horizontal bar). However, J.

Matsui, who was joining the carbon fibre team at Toray and had built up experience in Toray's tyre cord research, evaluated the results of the collaboration with UCC and was not at all impressed by the level of technology at UCC.

It was then that Toray decided to acquire the "practicable rights" on Shindo's patent from Nippon Carbon and Tokai Denkyoku in 1970 and the rights on their developments, and to start its own industrial production of PAN-based carbon fibres in Japan (Bacon and Moses, 1986). According to both Takamatsu (2000) and WIPO (n.d.), Toray had already in 1961 opened a carbon fibre research lab and started in 1969, as mentioned earlier, a research project for their own PAN fibre production. In June 1970, a decision was taken to invest in a production plant at Shiga Prefecture, with a capacity of 1 ton per month. In 1971, Toray increased the capacity to 10 ton per year, and effectively produced 300 kg/month of Torayca T200 and T300 (continuous fibre), at a price of 95,000 Yen/kg *(about 800 €/kg in 1971)* compared with Morganite's discontinuous fibres (1 m long) which costed about 2500 €/kg.

At the time of the investment decision, the actual demand for carbon fibres was small, and hence the decision was a daring one. Only a small amount of carbon fibres could be sold, namely to fishing rod manufacturers. So, Toray used the sales capacity of UCC in the United States to commercialise their carbon fibres for use in golf club shafts. Fortunately, as early as in 1973, Gay Bowers did win the Pacific Ocean Masters with a carbon fibre golf club, probably produced by Shakespeare Sporting Goods (USA) in collaboration with Union Carbide. This suddenly increased dramatically the demand for carbon fibres, and hence justified the investment in a production line in Ehime Prefecture, to upgrade the capacity to 60 tons per year (more detailed information in Tanaka's Chapter 1.2 of Part II of this book).

7.5.2 USA

According to the ACS-paper (ACS, 2003), the Union Carbide Company started in 1965 the Thornel 25 production, a rayon-based carbon fibre with a stiffness of 25 million psi (or 175 GPa). This rayon-based Thornel 25 fibre should not be confused with the Thornel P-25 fibre, which is a pitch-based fibre, and is still in production as of today. But the United States had no domestic PAN-based carbon fibre technology. Takamatsu (2000, p. 9) mentions that, in the same year 1965, Boeing has approached Nippon Carbon to produce for them carbon fibres, which resulted (in 1969) in the Carbolon Z-1 fibre. About simultaneously, Tokai Carbon collaborated with the US Air Force, resulting in the production of the Thermolon S fibre.

In the early 1970s, several companies started producing carbon fibres in the United States: Hercules (USA) starts in 1972, Toho Rayon produces Besfight PAN-based carbon fibres in 1973 (transferred/licenced to Celanese, USA, in 1978), Mitsubishi Rayon used Courtaulds' know-how to start a production in 1976 (commercial with home-made acryl precursor in 1983), and finally Toray transferred/licenced its PAN-CF technology to Union Carbide in 1978 (following the somewhat failed cross-licence of 1970). In the 1980s, Toray started a carbon fibre production in France (Soficar) and Toho Rayon in Germany (ENKA).

Looking at these developments, it is unclear at what stage the batch type process, both for PAN- and for rayon-based carbon fibres, has been replaced by a continuous process. With Toray in Japan and Courtaulds in the United Kingdom, two fibre processing companies entered the carbon fibre race, but it is still unclear when exactly the switch from batch to continuous processes occurred.

7.5.3 UK

Gill (1972) refers to a publication by Ezekiel and Spain (1967), describing how stretching was realised by a differential unwinding and rewinding of the (rayon-based) fibres. So they might have been the first to apply the well-known technique of continuous (polymeric) fibre stretching to a carbon fibre manufacturing line. However, in his early paper, one of the key developers in the United Kingdom, Leslie Phillips (1968) gives the impression that in the Royal Aircraft Establishment's laboratory, still a batch process is being used, resulting in "*material of both 14 inch and 39 inch lengths,*" but also mentions that "*the continuous process gives material similar in length to the original textile tow.*" It is however not clear whether, at that moment in history (1968) such a continuous process had already been realised. Also in the paper by Gunston (1969), reference is only made to discontinuous fibres, produced by Rolls-Royce (again, more detailed information can be found in Tanaka's Chapter 1.2, in Part 2 of this book).

As mentioned earlier, Toray produced PAN-based carbon fibres in a continuous process from 1971. But also Rolls-Royce was very active, and "*quite independent from the RAE lab,*" as R. M. Gill mentions (Gill, 1972). The fact that the development of a continuous process was crucial for them is highlighted by the fact that Rolls-Royce filed three consecutive patents, all of them abroad: a Belgian patent in March 1966 on using Courtaulds' Courtelle fibres, a French patent in September 1969 on oxidation under controlled tension, and finally a German patent in November 1969 on a continuous process with "carbon fibre sheets" (sheet of carbon fibres parallel to each other). Thorne (1985) mentions that Rolls-Royce was able to increase their production from 1 ton per year with a batch process in 1967 to 25 ton per year in a continuous process in 1969.

As mentioned earlier, similar production rate increases were realised by Toray in Japan: in November 1972, one line at the Ehime plant in Matsuyama had a capacity of 5 ton per month (or 60 ton per year), and by adding a second line, and further improvements of the line, a capacity of 35 ton per month (or 420 ton per year) had been reached in 1974.

7.5.4 USSR

Following (Yarovan, 2018), work on carbon fibre reinforced polymer composites started in the USSR in mid-1960s in VIAM (Federal Institute for Aviation Materials) with the first carbon fibre composite demonstrated in a defence development exhibition Setun in 1968. The USSR research work was boosted by the 1973 Le Bourget demonstration of the United States using composites in the F-14 and F-15 aeroplanes, but practical realisations in the USSR were strongly hindered by the shortage of carbon fibres and by noneffective centralised resources management. Still, according to (Yarovan, 2018), the first domestic carbon fibre unidirectional (UD) composites were certified (KMU-1) in 1971, with a stiffness of 160 GPa. This reference presents the application area for UD, bi-axial and multiaxial prepregs as "*for structural products (skins, panels, reinforcing elements of the aircraft airframe structure, guides and working blades of gas turbine engines and turbofan engines) operating at temperatures up to 200°C for 1000 h.*"

Carbon fibre composites have been introduced to several Russian aircraft like Antonov's An-22, An-24, and later also to Tupolev's Tu-160 and Tu-204, but "*the industry at this stage was not ready to produce carbon fibre ... in the required quantities.*" Yet, by end 1980s carbon fibres were produced in eight plants in the USSR, including "*the Moscow and Chelyabinsk electrode plants, and the PO Khimvolokno (chemical fibres) plants in Balakovo and Svetlogorsk.*"

7.5.5 Europe

Several books on carbon fibres (Donnet and Bansal, 1984; Gill, 1972; Watt and Perov, 1985) also refer to fundamental studies on carbon and carbon fibres in France, and Gill lists nicely some French and German patents for both rayon- and PAN-based carbon fibres. Finally, the overview of available carbon fibres (in the early 1980s) in Donnet and Bansal (1984) book includes, next to Japanese, British, and American fibres also carbon fibres produced in France (Rigilor by Le Carbone Lorraine) and Germany (Sigrifil by Sigri) (Donnet and Bansal, 1984, p. 3–5).

7.5.6 Conclusion

As a conclusion, it can be stated that three innovations have been crucial for the industrial development of the carbon fibre production: first, the initial patent of Shindo on PAN-based carbon fibres; second, the invention of the PAN comonomer by Morita, which drastically reduced the duration of the oxidation step; and third, the addition of stretching during the oxidation (mainly) and carbonisation, an innovation which could be attributed to Shindo in Japan, Watt et al. in the United Kingdom as well as to Bacon et al. in the United States.

7.6 Two more reinforcing fibres

7.6.1 Pitch-based carbon fibres

A second member of the carbon fibre family are the pitch-based carbon fibres, invented in 1963 by Sugio Ōtani, and further developed by the group of Ōtani at Gunma University (Japan), and by Singer at Union Carbide from 1973 on, as mentioned by Diefendorfer in his chapter in Comprehensive Composite Materials (Diefendorfer, 2000).

Already in the 19th century, there seems to have been research on fibres from carbon black filled pitch, but Diefendorfer cites, as the historical first, a paper by Ōtani in Carbon (Ōtani, 1965) on carbon fibres from unfilled pitch; no properties are given, it is only mentioned that the fibre had "*modest properties.*" In a later paper, Ōtani et al. (1966) report on tests on carbon fibres from different types of pitch (coal tar, petroleum asphalt, PVC-pitch), starting with a heat treatment of the spun fibres at 270–340°C and followed by a carbonisation at 1000°C. They had a strength of only about 1 MPa. B. Rand in his chapter in Watt's book (Rand, 1985) states that these rather poor properties would be related to the isotropic and glassy nature of this pitch.

According to both Rand and Diefendorfer, Ōtani further developed the process by hot stretching at 2500°C, eventually leading to fibres with a modulus of 440 GPa. However, it was discovered that the hot stretching is not needed anymore, probably simultaneously by Ōtani and Singer in the period 1970–75: adding a (low temperature) heat treatment step leads to a mesophase pitch with strong orientation of the liquid crystalline phase. Hence, the complexity of the process was substantially reduced. Singer, at that time working at UCC (later transferred to BP-Amoco and Cytec), showed indeed that the presence of a liquid crystalline mesophase was essential, and produced pitch-based fibres with moduli ranging from 160 to 930 GPa. The work of Ōtani at Gunma University inspired several Japanese companies (Mitsubishi Kasei, Nippon Graphite) to commercialise fibres with similar properties.

The actual process of making pitch-based carbon fibres—a good summary can be found in (Rand 1985)—starts with a pretreatment of the pitch to create a liquid crystalline mesophase. It is then spun into fibres, which are thermoplastic and have already a preferred orientation of the macromolecules along the fibre axis. They then go through an oxidative thermosetting treatment, where the orientation is largely locked-in and retained, followed by a graphitisation process, which occurs at high temperatures, depending on the required stiffness and electrical conductivity properties. The "truly graphitic" nature of these pitch fibres not only enables to reach very high moduli, but also results in a very high electrical conductivity. These two advantageous properties, together with the still higher cost of pitch fibres compared with PAN-based fibres, are the reason why pitch fibres are, as of today, mainly used in applications where ultrahigh modulus and/or high electrical conductivity are required.

7.6.2 Boron, silicon carbide, silicon oxide, and other ceramic fibres

In the same creative 1960s, a range of other high-performance fibres has been developed: boron and SiC-fibres, SiO- and other ceramic fibres. A good overview of these fibres can be found in Chapters 3 to 5 of the previously cited book "Comprehensive Composite Materials" (Kelly and Zweben, 2000). The information provided in this section is hence based on those chapters. As these fibres are used to a rather limited extent in polymeric composites, they will be discussed only briefly.

Boron fibres are produced by chemical vapour deposition of metallic boron onto a tungsten filament or a carbon fibre. The process has been invented by Talley in 1959, when working at Texaco Experiments Inc. (USA), using several types of boron-containing reactive gases.

A similar process was then again used by Talley in 1963 to deposit silicon carbide from a silane-containing reactive gas onto a carbon fibre (tungsten reacts with silicon and carbon, and hence cannot be used as substrate fibre).

Boron and silicon carbide fibres have rather similar properties: boron fibres have a stiffness of 400 GPa, strength of 3400–4200 MPa, at a density of 2.54 g/cm^3; SiC-fibres are slightly heavier (3.0–3.4 g/cm^3), at a stiffness ranging from 395 to 420 GPa, and a strength of 3750–4200 MPa. Because of the CVD-process, both the diameter of the boron and SiC fibres are rather high (100–140 μm), reducing the flexibilty of the filaments. The story goes that this was the reason why boron fibres have been banned from their initial use in tennis rackets: a broken racket with protruding sharp boron fibres would have badly wounded a tennis player!

There is however a different process available to produce silicon carbide fibres, allowing to realise much smaller diameters. The Nicalon fibres, invented by Yajima in 1975 and commercialised by Nippon Carbon, start from a preceramic polymer, which can be spun into a fibre. This is then undergoing a stabilisation by either electron beam radiation or oxidation, before being pyrolysed into a pure Si–C or a Si–C–O–fibre. The moduli range from 270 to 420 GPa, at a strength of about 2700 MPa. Later, several other types of SiC-fibres have been developed (like the Tyranno fibres developed at Ube Industries, Japan), mainly to cope with the harsh processing and in-service conditions when they are used as reinforcements for metal and ceramic matrix composites.

Finally, several types of oxide fibres have emerged in the same period. In Chapter 5 (by Berger and Bunsell) of Volume I of Comprehensive Composite Materials, aluminium oxide (or "alumina") based fibres are described. DuPont de Nemours in the United States was first (1978) to produce a wholly α-alumina (polycrystalline) fibre, by pyrolysing a fibre which was dry-spun out of a α-alumina-containing slurry. This process has been further optimised by companies like Mitsui Mining and 3M (Almax and Nextel fibres). Further improvements, again for applications in metal and ceramic matrix composites, included the development of zirconia reinforced alumina fibres (PRD) and alumina silica fibres (Saffil, Altex, Nextel ...). The mechanical properties at room temperature of all these types of fibres vary greatly with their exact composition and heat treatments, but range typically between 200 and 400 GPa for stiffness and 1000–2500 MPa for strength. However, as these fibres are mostly being used as reinforcements for metal and ceramic matrix composites, their properties at high temperatures are more relevant, but this is out of the scope of this chapter.

7.6.3 High-performance polymeric fibres

The excellent mechanical properties of carbon fibres have been (mainly) related to the orientation of the graphitic layers parallel to the fibre axis. A similar phenomenon was already well known for polymeric fibres, but carbon fibres have the advantage that the graphite layers are two-dimensional, whereas most polymeric fibres have a linear polymer chain structure. Hence, to achieve stiff polymeric fibres, one should combine an almost perfect orientation with a high intrinsic stiffness of the polymer chains and a strong interaction in-between them.

Probably motivated by the success of the development of carbon fibres in the 1960s, researchers in polymer chemistry and physics, and in fibre processing, have looked into the synthesis of new polymers (aramids) and/or have improved the fibre drawing, such as for ultrahigh molecular weight polyethylene, UHMW-PE. This resulted in the 1970s in two new fibres which could become candidates for reinforcement of composites. For aramids, the key contribution came from Stephanie Kwolek at DuPont (Kevlar fibre, patent filed in 1973), although important further innovations have been performed by researchers at the labs of AKZO-Nobel in The Netherlands (Twaron fibre) and of Teijin in Japan (Technora fibre). For the UHMW-PE fibres, the work of A.J. Pennings at DSM (later University of Groningen, The Netherlands) from 1965 on has been crucial.

The story of the aramid fibres is well described by H. H. Yang in Chapter 7 of Volume 1 of Comprehensive Composite Materials (Yang, 2000), whereas I.M. Ward comments on the invention of UHMW-PE fibres in Chapter XVIII of Watt's book on Strong Fibres (Ward, 1985). A schematic overview of the developments until 2000 are discussed in section 9.1.3 of this chapter, whereas a more comprehensive overview of the development of high-performance polymer fibres is presented by Peijs and Lemstra in Chapter 1.3 in Part II of this book.

Being polymers, aramid and polyethylene fibres had the additional advantage of their even lower density then carbon fibres (1.45 resp. 0.95 compared to 1.85 g/cm^3 for carbon fibres). Due to this low density, their specific stiffness can compete with that of (high strength) carbon fibres, but each of them has an additional advantage and (unfortunately also) a disadvantage.

The common advantage over carbon fibres is the higher failure strain, which makes them an excellent choice for ballistic applications, where high amounts of fracture energy have to be absorbed. Unfortunately, aramid molecules are quite polar, and hence easily absorb moisture in humid environments, reducing their mechanical properties and those of the aramid fibre composites. UHMW-PE fibres have, on the contrary, a low softening and melting point, making them unsuited for structural applications where moderate temperatures (>80°C) are to be expected. Finally, both polymeric fibres suffer from a low compressive strength.

As a consequence of all this, the use of aramid and UHMW-PE fibres in composites is currently limited to low temperature and/or low humidity applications, or to applications where energy absorption at fracture predominates. However, large volumes of UHMW-PE fibre ropes are used in marine applications, because of their buoyancy.

7.7 The enabling theoreticians and the evolvement of conferences and "organized communication" about composites (1960–80)

Scientists around the world did not wait until the advent of carbon fibres to try to understand the behaviour of heterogeneous materials (meaning: composed of two or more distinct materials or phases). In his interesting review paper, Herakovich (2012) describes in quite some detail the main steps in the development of the mechanics of composites, from micro-over meso- to macro- or structural mechanics.

After his pioneering work on anisotropic bodies, published in 1935, S.G. Lekhnitshkii (Russia) also developed theories for anisotropic plates (1947) and anisotropic elastic bodies (1950), which all together formed the basis for the later theories on the intrinsically anisotropic composite materials.

For micromechanics of heterogeneous materials, we must go back in history even further. Voigt in 1889 and Reuss in 1929 proposed simple but powerful formulae to predict upper- and lower bounds for the mechanical behaviour of multiphase materials. In 1957, Eshelby provided the basis for inclusion models for stress and strain field calculations. The work by Mori and Tanaka was an important trigger for using the Eshelby approach in composites, because they were able to take into account the interactions between inclusions, which is important because in composites fibres are very close to each other (Mori and Tanaka, 1973). The inclusion models are now being further optimised and extended toward other aspects of the mechanical behaviour of composites (damage development, fatigue …). Cox (1952) elegantly described the mechanics of paper, in fact a randomly oriented short fibre composite; his theory was later on used to describe the strength and stiffness of composites.

Simultaneously with the carbon fibre research in the late 1950s and early 1960s, an avalanche of theoretical work on micro- and meso-mechanics of composites was published: by Hill (in 1963) on the elastic properties of reinforced solids, by Hashin and Shtrikman (also in 1963) on a variational approach for the elastic behaviour of elastic solids, by Antony Kelly and co-workers (Kelly and Tyson, 1965) on strength of unidirectional (initially metal matrix) composites, *just to name a few ….*

Damage mechanics in composites was inspired by the general works of Kachanov in 1958 and Rabotnov in 1968, initially on the application of a continuous damage variable to the creep behaviour of metals. Specifically, for composites, this approach was further developed by Ladevèze in 1983 and 1986 and is treated more in detail by R. Talreja in Chapter 4.4 in Part II of this book.

Already early during the (experimental) development of composites, it was realised that an intelligent choice of fibre orientations was crucial for optimising the mechanical properties of composites for specific loading conditions (McMullen, 1984; Kelly, 2005). Lekhnitshkii's work on anisotropy was a sound basis, but specifically for laminated composites, the "laminate plate theory" or "classical lamination theory" was developed. According to Herakovich, the original analysis on the elastic behaviour of laminates has been provided by Pister and Dong in 1959, Reissner and Stavsky in 1961, and Dong et al. in 1962. For predicting strength, failure criteria had to be formulated: Gol'denblat and Kopnov in 1965, Ashkenazi in 1965, Tsai and Wu in 1971, *just to name a few ….*

As composites gradually entered the real world of manufacturing aeroplanes, cars and other consumer goods, reliability, repeatability and performance enhancement of the manufacturing processes became more and more important. Process modelling was then crucial, first to optimise the autoclave process used in the most "critical" applications, namely aerospace. When newer processes were under development, like

RTM, VARI, and tape laying, process modelling became even an indispensable tool during the development and optimisation phase of such processes (Advani and Sozer, 2010). More information on the history of process modelling is presented in Chapter 2.7 of Part II of this book.

Once the interest in high-performance composites had been sparked by the advent of carbon fibres, not only in academia but also in the industry, the need for better communication in the scientific community, and between academia and industry, was growing. In 1967, Steve Tsai started the Journal of Composite Materials, and in 1968 Tsai, Halpin and Pagano initiated a series of composite workshops for academia and industry. In the same period, the first handbooks on composites have been written, and (inter)national conferences on composites were launched: ICCM in 1975 (in fact in two cities in different continents, namely in Geneva and Boston), ECCM (initially as EACM, in 1985 in France), the ASC (American Society for Composites in 1986), whereas the Japan Society for Reinforced Plastics was already formed (by Hayashi) in 1954. In 1965, the first conference on "glass fibre textiles and reinforced plastics" was held in Paris, later called "Journées Européennes des Composites." This would, in 1979, evolve into the yearly composites trade show JEC World. The history of composites-related organisations and societies is described more in detail by M. Quaresimin in Chapter 4.7 of Part II of this book.

8. Application developments in the 1970s

The start of the industrial production of carbon fibres at the end of the 1960s and beginning of the 1970s was a major stimulus for developing new and more diverse applications of composite materials. The 1970s can be considered as the age where composites really started "conquering the world."

A great number of breakthrough innovations will be described in this section. They will be grouped roughly following the scheme used by JEC World in their annual publication "JEC Observer—Current trends in the global composites industry" (JEC Observer, n.d.), namely aerospace, transportation (incl. marine), consumer goods (incl. sports and leisure, design …), construction and energy. A similar scheme will be followed in the sections 9 and 10, respectively "*till 2000*" and "*from 2000 till now.*" In this way, the economic importance of the innovations described in this text can be quantitatively assessed using the historical overviews and future trend data presented graphically in the JEC Observer publications.

Information on the breakthrough application developments between 1970 and today is abundant on one hand, but very scattered on the other. Some interesting overviews can be found in the following four books: "The challenging world of Composites" (JEC Group, 2013), "Celebrating the 100th anniversary of the Wright's brothers first flight" (Epstein and Heimerdinger, 2003), "From Bakelite to composite" (Verpoest et al., 2002) and "Fibre-Fixed. Composites in Design" (Pil and Verpoest, 2018). However, a significant amount of new information has been added, for which appropriate references will be provided.

8.1 Aerospace

Carbon fibres became available in the late 1960s in experimental quantities in the United States, the United Kingdom, and Japan, but from the early 1970s on also in larger quantities, mainly because of the investments of Toray in their continuous fibre production lines in Japan and the production under licence in the United States (see Section 7.5). In the beginning at a quite high cost, they penetrated different markets at a rather different speed.

Carbon fibres found quickly their way into military aircraft in the United States and Europe, because their aircraft research institutes were directly involved in carbon fibre research, like the Royal Aircraft Establishment (RAE) in the United Kingdom. As early as in 1970, in the United States a CFRP wing tip was installed on the Northrop F5 and in 1972 a CFRP speed brake on the McDonnell F15 (Fig. 1.1.26). General Dynamics followed in 1976 with vertical and horizontal stabiliser skins on the F16 in 1978; the MD F-18 had already 13% of its structural weight in composites. Also in France, in the mid-70s, a CFRP rudder was developed for the Mirage III.

Maybe surprisingly, civil aircraft was not much slower: in 1973 CFRP spoilers appeared on the Boeing 737, which were continuously monitored and evaluated by NASA (Hoffman and Stoecklin, 1980). In 1975, Lockheed designed and built aramid/epoxy composite fairing panels for the L-1001 aircraft while Douglas Aircraft's DC-10 had composite rudders. More information on the early developments in civil aircraft in the United States can be found in Part II, Chapter 3.2 of this book. Likewise, more information on European developments can be found in Chapter 3.1.

In Europe, Airbus introduced its first civil aircraft, the wide-body Airbus 300, in 1972. It was also used as a testbed for CFRP components for nonstructural parts. Many European aerospace-related research institutes (ONERA in France, DLR in Germany and RAE in UK) started huge programs on using carbon fibre composites for structural parts like stabilisers, wings, and even fuselage sections.

In the USSR, the aeronautic applications of CFRP in 1970s and 1980s were primarily in helicopters, in rotor blades and structural elements. Composites in the Ka-50 "Black Shark" amounted 35% in weight of the helicopter. In space, the launcher Energia used 4.500 kg of CFRP, the Buran shuttle had cargo doors with an area of 144 m^2, made of CFRP (Yarovan, 2018).

Figure 1.1.26 McDonnell F-15. F-15 Eagle by arminw, CC0, via Wikimedia commons.

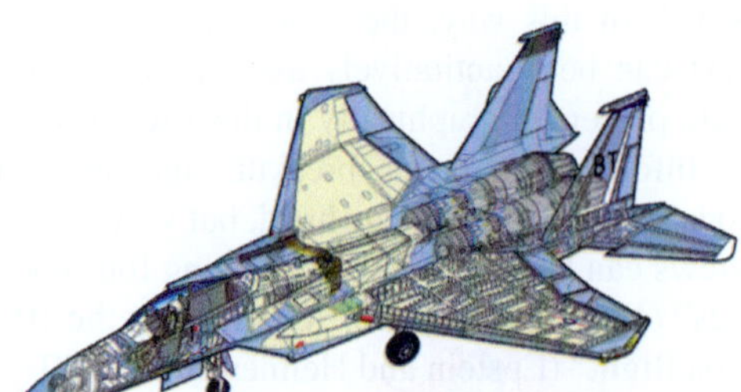

Some graphs and data suggest that the investments in military aeroplanes, and hence in the use of composites in military aeroplanes, would have drastically decreased at the end of the cold war and the opening of the "iron curtain" in Europe, end of the 1980s. Requirements for weight reduction, increased damage tolerance and fatigue resistance in military aircraft have indeed been some of the main driving forces for CFRP applications, and the end of the cold war may have had a negative impact on the whole carbon fibre composite business.

Franck Glowacz indicates in his book on carbon fibres (Glowacz, 2015, p. 24) that in the same period the carbon fibre industry has started to restructure, with several mergers and acquisitions taking place, but the question remains whether this is solely related to the end of the cold war, and/or to the CF-industry getting more mature. Moreover, that period also marks the "real" start of carbon fibre composites in civil aircraft: was this because of the decreased demand for military aircraft? Or because the technology was now sufficiently developed to be used in civil aircraft?

8.2 Transportation

Already in the 1950s the first commercially available car with composite body panels was produced by Chevrolet, namely the Corvette. It had a glass fibre—polyester composite body mounted on a metallic chassis (from 1953 till 1962). At about the same time, East-Germany developed the famous Trabant, using a cotton fibre reinforced phenolic resin (1957) and Citroën used GFRP for the roof of the iconic Citroën DS (1957) (Verpoest et al., 2002). But overall, the use of composite body panels on a metal chassis remained limited to small series sports cars with attractive names like for instance Victress, Warrior, Wildfire, Maverick … in the United States (Victress, n.d.).

A notable exception was the Lotus Type 14 "Elite" (Fig. 1.1.27), produced between 1957 and 1963 (Lotus Elite, n.d.). It has a highly innovative glass fibre composite monocoque construction, with only a steel support to carry the engine and front suspension. In the following model, the Lotus Type 75 Elite, produced between 1974 and 1982, the design returned to the "glass fibre composite body panels on a steel chassis" type, as was done at that time by many other manufacturers of limited series high-performance (or even race) cars, like the Autobianchi Stellina (1963–65), Citroën

Figure 1.1.27 Lotus Type 14 Elite with an innovative glass fibre composite monocoque construction.
Lotus by Sebastian Koppehel, CC BY 4.0, via Wikimedia Commons.

Bijou (1959–64), Daimler SP250 "Dart" (1959–64), Ferrari 308 GTB "Vetroresina" (1975–77), BMW M1 (1978–81), Ford RS200 (1985–86), Mazda Autozam AZ-1 (1992–94), etc… (GFRP Car Bodies, 2021).

In the 1970s, more and more composites were used in "normal" cars, but because of the (still) high cost of carbon fibres, only glass fibres were affordable at that time. An important innovation was the change from hand lay-up open mould manufacturing techniques to compression moulding techniques, which enable higher production speeds and improved quality assurance. In this way, Renault pioneered the use of composites in mass production cars in 1972, when it equipped the popular Renault 5 with glass fibre composite SMC (sheet moulding compound) front and rear ends. Chevrolet introduced, in 1972, their first glass fibre composite SMC's in Corvette body panels, replacing the hand-laid up panels from the 1950s.

8.3 Consumer goods

8.3.1 Sporting goods

Sporting goods happen to be "quick wins" and visibility creators for carbon fibre composites (more detailed information in Part II, Chapter 3.4 of this book). A famous example is the application in golf clubs. In 1969, Frank Thomas made the first "graphite" shaft (Takamatsu, 2000) in the Shakespeare Sporting Goods Co. (USA). Since carbon fibres were still in their infancy at that time, and only very small quantities of (noncontinuous) carbon fibres were produced, this can be considered as a groundbreaking application of these new fibres.

Around the same time, aircraft engineer James Flood developed his version of a carbon fibre composite golf club, which was the start of the Aldila company (1972, San Diego, USA). One year later, in 1973, Gay Bowers did win the Pacific Ocean Masters by using a carbon fibre composite shaft, but it is not clear which brand of golf club he used. Anyway, this remarkable event suddenly increased dramatically the demand for carbon fibres, and a posteriori justified Toray's audacious decision to build, already in 1971, their first industrial plant, producing 300 kg per month of continuous Torayca T200 and T300 fibres.

In a recent, extensive historical-technical study, Taraborrelli et al. (2019) mentioned that initially composites were used as an external reinforcing ply on wooden tennis rackets (Slazenger in 1961; Wilson in 1966). The first full composite tennis rackets would have been the Phantom by Slazenger in 1974 and the TR10 by Rossignol in 1977, first using glass fibre, later on also incorporating carbon fibres. From then on, most tennis racket brands changed to carbon fibre composites, like the Prince Classic in 1976 and the Snauwaert Hi-Ten in 1986, see Fig. 1.1.28. Some of the best known ones are the rackets of Donnay (Björn Borg), Dunlop (John McEnroe, Steffi Graf), Prince (André Agassi, Monica Seles), Wilson (Pete Sampras, Roger Federer), and Head (Novak Djokovic).

Also in badminton, carbon fibres started to be used: in 1978, the first carbon composite rackets were produced by Yonex in Japan: the Carbonex 8 B-8500B (Yonex, 2021), weighing less than 100 g. Since then, most competition badminton rackets

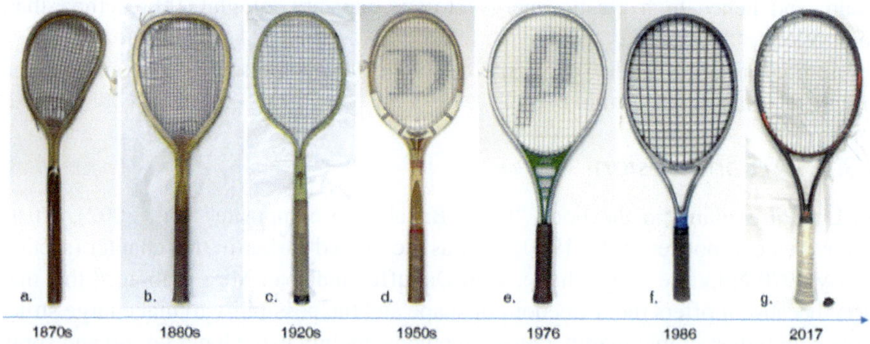

Figure 1.1.28 Seven tennis rackets showing developments over time: (a) Early racket by F. H. Ayres (1870s); (b) Wooden racket by F. H. Ayres (1886); (c) steel racket by Dayton Steel Racket Corporation (1922); (d) Dunlop "Maxply" (1955); (e) Prince "Classic" (1976); (f) Snauwaert Hi-Ten 50 (1986); (g) Head Graphene XT Prestige with the sensor (2017) (Taraborrelli et al., 2019).

use carbon fibre composites, whereas (cheaper) rackets for leisure still stick to glass fibre composites, aluminium or steel.

As happens all too often when a new material becomes available, the frame of the first carbon composite bicycles was a "black copy" of the existing "tube and lug" metallic frames. It is not surprising that the first carbon fibre composite bicycles have been developed in Europe, being the centre of gravity of cycling competitions. The Dutch website "Vintagefiets" (Vintagefiets (n.d.) presents a nice overview of these early developments, and gives the award for the first CFRP bike to Carlton Cycles (UK, later part of Raleigh), who already in 1969 started designing and producing their Carlton Flyer (Road.cc, 2010), using carbon fibre composite tubes glued into aluminium lugs. The two halves of such lugs were screwed together to solve the problem of applying the glue evenly. Similar concepts were later used by Assos (Switzerland) in their Track Bike (1976), using for the first time tubes with oval cross-section, and by Allan (Italy, in collaboration with Toray, 1976). The latter is important, because all these innovations started only a few years after the first industrial production of carbon fibres by Toray in Japan (1970). For instance, UK's Carlton Flyer bike used carbon fibres provided by the RAE, Royal Aircraft Establishment. Vitus (France) produced its Carbon 3 in the early 1980s, and Look followed in 1986 with its KG86, which was used by Greg Lemond when he won the 1986 Tour de France.

Whereas GFRP had become the favourite material for sailing **yachts**, also carbon fibre composites started to be used: from the Olympic games in Münich on (1972), CFRP was used, sometimes in combination with other fibres, in competition boats of all kind (sailing, rowing, ...), and in 1980 the first CFRP trimaran sailing yacht was built (Colt Cars GB) by SP Systems/Gurit on the Isle of Wight (UK).

Just one sport sticked to glass fibre composites. American athlete George Davies broke the pole vault record in 1961 with a "fibreglass pole" (Risepolevault, 2023), and ever since GFRP has been used because of the high strength and high failure strain, and hence high elastic energy storage capacity, of glass fibres (and their composites).

More information on the (early) use of composites in sporting goods can be found in Part II, Chapter 3.4 of this book.

8.3.2 Art and design

As Lut Pil explains in the book "From Bakelite to composite" (Pil, 2002), artists discovered composites in the 1960s, as was mentioned earlier in this chapter (section 6). By 1970 Nikki de Saint-Phalle, Jean Dubuffet and Joan Miro embraced this material because it offers them freedom of shape and the possibility to make large structures with rather simple manufacturing methods, mainly using hand lay-up and spray up techniques (CPC, 1992).

But also furniture designers and architects had discovered the freedom of shape of composites. As has been extensively explained earlier (Section 6), the use of glass fibre composites by furniture designers fully started in the 1960s and boomed in the 70s, with famous designs by Verner Panton, Joe Colombo, Luigi Colani, Eero Arnio, Pierre Paulin and many others (Daenens and Bucquoye, 2002).

8.4 Construction

Composites have been present in buildings and construction since thousands of years, the most well-known example being the dried mud bricks reinforced with straw (see section 2 of this chapter). In the 20th century composites have gradually gained a strong position in the multitude of materials used in buildings and construction. Ten years into the 21st century (2010), already 2.300.000 tons (2.3 Mt) of composites were used in buildings and construction or 27% of all composites used worldwide (JEC Observer, 2022, p. 37). The growth curve in the 20th century is difficult to reconstruct, and it is hard to find out who designed the first composite house, constructed the first composite bridge, who invented composite rebar reinforced concrete, etc. ... But it is clear that in the 1970s glass fibre reinforced composites had a widespread use in buildings, construction and infrastructure. A comprehensive historical overview could not be found, but partial information is presented in papers by Hollaway and Head (2000), Hollaway (2009), Humphreys (2003), and Meier (2012, 2016).

In the 1960s, enthusiast architects started experimenting with glass fibre composites, using the rather easy to handle "wet hand lay-up" technique. The best known example might be the 1968 Futuro house by Matti Suuronen, Fig. 1.1.29 (Futuro House, n.d.), but the "Monsanto House of the Future" (1957) can be considered as the first "plastic" house conceived for mass production (Engelsmann et al., 2010). Widespan, self-supporting GFRP roof structures, for instance by Swiss engineer Hossdorf ("Les échanges" pavilion, Lausanne, 1964), are another example of how architects and engineers discovered the potential of composites.

After this pioneering work, a multitude of projects have been realised in the 1970s, like the Covent Garden flower market in London (1974), a skeletal steel structure, which was infilled with 1000 GFRP modules.

Figure 1.1.29 A Futuro House.
Futuro by Grigur, CC BY-SA 4.0, via Wikimedia Commons.

8.5 Energy

The very first wind turbines for energy generation have been built already in 1888 by C. F. Brush in Cleveland, Ohio (USA) and in 1889 by Poul Le Cour in Askov (Denmark) (Mishnaevsky et al., 2017). The Danish "Gedser wind turbine" (1956–57), designed by Johannes Juul, is mentioned to have three "composite blades," but these seem to be macro composites, a combination of steel spars, aluminium shells and wooden ribs.

The use of polymer matrix composites for making the wind turbine blades was started in the 1950s by the German inventor Ulrich Hütter. who worked already in the 1930s on wind turbines for the Nazi's (they wanted to build 28.500 wind turbines for *"autonomous electricity generation"*) (Hütter, n.d. b). But in 1943, his company was closed because Hitler needed aeroplanes, so Hütter started designing and building aeroplanes, and after World War II, he also built sailplanes … out of glass fibre composites. Finally, he combined his knowledge of composites with his background in sailplanes to design the first glass fibre composite wind turbine in 1958, the StGW-34. In fact, a wind turbine blade is not that different from a sailplane wing! Hütter's composite wind turbine blade was built by the Allgaier Werke in Uhingen (Stuttgart), where Hütter was production manager between 1946 and 1959.

Almost 20 years later, glass fibre composite wind turbines were reinvented, simultaneously in Denmark and in the United States, but in a completely different environment. In 1975, in the wake of the 1972 oil crisis, NASA started a research project in which glass fibre composites, steel, wood and even concrete would be compared as blade materials. The outcome was that glass fibre composites had the best performance. NASA approached the Gougeon Brothers, Inc. to apply their composite boat technology to wind turbines. The 2-blade turbine MOD-0 (100 kW, 38 m diameter) was installed near to the Lewis Research Centre in Sandusky. In 1979, the MOD-1 of General Electric already reached a 20 times higher power (2000 kW, 61 m diameter).

On the other side of the ocean, in the small town of Ulfborg in Denmark, the teachers and students of the vocational school cooperative Tvind initiated research on wind energy, collaborated with the RISOE research centre and DTU (Danish Technical University) for the blade aerodynamics and built in 1978 the "Tvind Kraft,"

Fig. 1.1.30 (Tvindkraft, n.d.). It has been running since then, and by 2013, a new 500 kW frequency converter was installed to fully utilise the capacity of the windmill (González et al., 2013). With a wing diameter of 54 m, the three glass fibre composite turbine blades created a power similar to the American MOD-1. It was the start of the successful Danish wind turbine energy industry (more information on the Danish composite wind turbine history in Part II, Chapter 3.3 of this book).

From the end of the 1970s on, almost all of the wind turbines have been made out of polymer matrix composites, like Vestas' HVK15 (1978) with 10 m rotor and a capacity of 55 kW (Mishnaevsky et al., 2017). Initially, these polymer composite wind turbine blades were manufactured as follows: first two shells are made in open moulds, using a wet hand lay-up technology, and then adhesively bonded.

9. 1970 till 2000: the era of steady growth

New applications, emerging after the carbon fibre boom, inspired research and development in new composite materials and processes. It was an exciting period for researchers in universities, government and private laboratories, because there was a great belief that composite materials would conquer the world in the coming two decades, certainly before the end of the millennium!

9.1 Materials

In the early seventies, most fibrous materials we have available today as reinforcements for polymer matrix composites had been invented. Some industrial production was well established (glass fibres) but others just started (carbon fibres).

9.1.1 Glass fibres

9.1.1.1 1970s

As has been extensively explained earlier in this text (section 5.1.2), glass fibres *as we produce them now* have been invented in 1933, and industrial production started straight away by Owens-Corning in America, followed by Europe (Saint-Gobain in France, Pilkington in the United Kingdom), Japan (Asahi Glass) and others. In the

Figure 1.1.30 (a) Volunteers carrying the Tvind wind turbine blade from the manufacturing site to the installation site; (b) the Tvind wind turbine (permissions received for Chapter 3.3 by Lilholt and Madsen).

beginning of the 1970s, glass fibre production plants were active worldwide, and new fibre compositions were invented for specific applications.

The standard type of glass fibre (E-glass) was supplemented by some new fibre types in the 1970s: alkali-resistant glass was developed at several laboratories and patents were filed in 1972 by Kimura and Yoshimura (Japan) (US Patent 3,861,927, 1972), in 1973 by W. W. Wolf (USA) (US Patent 3,861,925, 1973), and in 1982 by D. R. Cockram and K. M. Fyles (UK) (US Patent 4,330,628, 1979). S2-glass (AGY, 2025) emerged on industrial scale in the same period of time, as an improved version of E-glass. It had a significantly higher modulus (+20%) and strength (+40%). S-glass, with similar properties, had been invented already in the 1960s by Owens-Corning, who licenced its technology in 2014 to AGY (Owens Corning and AGY, 2014).

Two further innovations have revolutionised the traditional glass fibre spinning in the early 70s (Thomason, 2015): the "direct roving" technology allowed for making much heavier tows in one step, whereas chopping glass fibres inline (directly after spinning and applying the sizing) drastically increased the productivity of making chopped fibres.

9.1.1.2 1980s and 1990s

As the use of glass fibres gradually evolved into higher performance composite applications, the adhesion between glass fibres and polymer matrices, both thermoset and thermoplastic, became an important issue (Thomason, 2012). The 1980s were the heyday of interface research, with important developments both in academia and industry (Owens-Corning, Saint-Gobain, …). Scientific conferences dedicated to "interfaces in composites" were started, by Ishida and Koenig (1987) and by Frank R. Jones in Sheffield (UK) (Jones, 1989). J. L. Thomason has written several excellent review papers on glass fibre sizings (Thomason, 2019), but the scope of this text does not allow further detailing on this important subject, which was crucial for the further development and industrialisation of glass fibre composites.

In the 1980s, glass fibres have been continuously improved to meet the demands of the composites industry. An important milestone was realised by Owens-Corning, when they filed their patent on "Boron-free glass fibres" in 1996 (US Patent 5,789,229, 1998), and commercialised it under the trade name "Advantex." Later on, this was followed by patents from PPG Industries (US Patent 6,686,304, 2000) and Johns-Manville (US Patent 7,022,634, 2003).

At its commercial launch in 1997, Advantex claimed to "*combine the electrical insulating and mechanical properties of E-glass, the current industry standard, with the higher heat resistance and acid-corrosion resistance of E-CR glass*" (Owens Corning, 1997).

In the 1990s, a "new" glass fibre started to be used as composite reinforcement: basalt fibres. Whereas glass fibres are produced by melting a mixture of different types of raw materials (silica sand, clay, limestone, dolomite, …), basalt fibres are basically made by just melting basalt rock at temperatures above 1400°C. Specific quarries have to be selected in view of achieving the right chemical composition, as this not only defines the later fibre quality but also the spinnability into thin fibres. Hence, sometimes small amounts of other raw materials are added, whereas in newer technologies the melt can also be purified to reduce the amount of impurities, resulting in a bronze instead of brownish colour.

The first known patent on basalt fibres is by Paul De Ghé, a French engineer living in Paris (1922) (De Ghé, 1922), but this patent only describes a crucible for melting basalt rock. No information could be found on how the technology for making basalt fibres developed, but it can be assumed that the glass fibre spinning technology of the 1930s must have been inspiring. After the Second World War, basalt fibres seem to have been produced experimentally, mainly in Western laboratories connected to the military, and in Russia and Eastern-Europe. Only after the opening of the iron curtain in 1989 and declassification of the basalt fibre technology in the United States in 1995, basalt fibres became widely available (Mahltig, 2018).

Compared with "standard" E-glass fibres, basalt fibres have a somewhat higher modulus (83–87 GPa) at a comparable strength, but they have a higher temperature resistance (continuous use at 700°C is possible). Basalt fibres are sometimes claimed to be "green fibres," because they are just made by melting basalt rock, found in nature.This statement is somewhat misleading, because also glass fibres are "just" made by melting (a mixture of) raw materials found in quarries or on beaches (sand!). Therefore, glass fibres could also claim the "green" label …

9.1.2 Carbon fibres

The story on carbon fibres, after its initial inventions in the 1950s and 1960s, really takes off from 1970 on. Well documented overviews of the further developments of carbon fibres can be found in papers by Matsui, 2010; Tanaka (2020), Takamatsu (2000), Tanaka (2021) and Glowacz (2015. p. 24), and in the time line presented by SAMPE-Japan (SAMPE, 2012). Very recently, an overview of more recent carbon fibre composite developments has been published (Zhang et al., 2023), which contains valuable information on carbon fibre developments as such, but also on the material and processing challenges for past and future application domains. These references will not be repeated in the following texts on carbon fibres and carbon nano-reinforcements, and only additional references will be mentioned.

9.1.2.1 1970s

As mentioned earlier in section 7.5 of this chapter, in 1972 at Toray's Ehime plant in Matsuyama, one line had a capacity of 5 ton per month (or 60 ton per year!), and by adding a second line, and further improvements of the first line, a capacity of 15 ton per month had been reached in 1974. By the end of the decade (1979), a production of 35 ton per month (or 420 ton per year) had been achieved.

Toho Rayon (now Toho Tenax) started their high-volume production in 1973, Mitsubishi Rayon follows in 1976.

In 1966, Courtaulds in the United Kingdom had filed a patent on the continuous production of PAN-based carbon fibres. In the 1970s they continued to bring their Courtelle fibre on the market, despite the setback in 1971, when Rolls-Royce's project to apply carbon fibres in aircraft engine fan blades dramatically failed. Courtaulds ceased carbon fibre production in 1991.

In the United States in the early 1970s, several companies started producing carbon fibres: Hercules (USA) in 1972, Toho Rayon producing Besfight PAN-based carbon fibres in 1973, Mitsubishi Rayon started production in 1976. UCC (Union Carbide Corp.) stopped its rayon precursor based carbon fibre production (1977), and initiates mutual technology exchanges with Toray, leading in 1983 to the start of a PAN-based carbon fibre production, using Toray's PAN precursor fibres.

All these fibres were initially "high strength/low modulus" (type II) carbon fibres, today known as the "T300" type of fibre. Gradually, carbon fibres with improved properties saw light: Toray's T400 (strength of 4000 MPa) and M40 (stiffness of 400 GPa), see Fig. 1.1.31, as well as Toho Rayon's intermediate modulus IM-600 fibres (strength up to 6000 MPa).

9.1.2.2 1980s

The search for even more performant carbon fibres continues (Fig. 1.1.32): Toray starts production of the intermediate modulus T800H fibre in 1984, followed by the T1000 fibre in 1986 (higher strength, intermediate modulus) and the high modulus, medium strength M60 fibre in 1989.

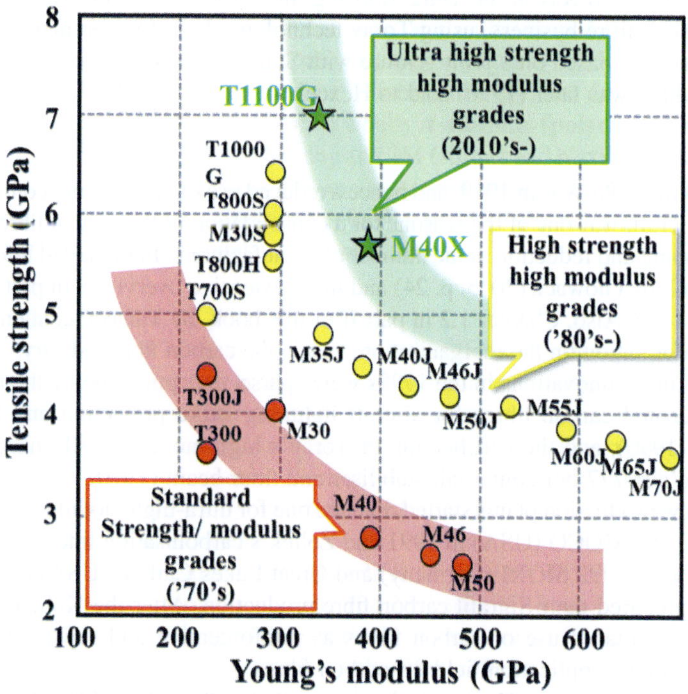

Figure 1.1.31 Overview of the historical development of carbon fibres (as produced by Toray). Redrawn by F. Tanaka from Tanaka (2023).

extensive story on the invention of these high-performance polymeric fibres can be found in (Lemstra, 2022) and in Chapter 1.3 of Part II of this book.

The development of high-performance polymeric fibres continued during the 2 decades after the invention of aramid and UHMW-PE fibres, as has been nicely illustrated by The Society of Fibre Science and Technology Japan in their book "High-Performance and Specialty Fibres," see Fig. 1.1.33 (Murase and Yabuki, 2016).

To the aromatic polyamide Kevlar and Twaron fibres, Teijin added in 1987 the aramid-copolymer fibre Technora, and DSM introduced the Dyneema polyethylene fibre in 1990, based on the earlier mentioned gel spinning technology developed at Technical University of Eindhoven by Smith and Lemstra. Using a similar gel spinning technology, Kuraray was able to produce in 1998 the PVA-fibre Kuralon K-II.

Both these polyethylene fibres are using a flexible chain polymer, in contrast to the rigid chain aromatic polymers, for which a liquid crystalline (solvent) spinning technology is being used. Using a melt spinning technology, Kuraray developed the polyallylate fibre (an aromatic polyester) Vectran (1990). Finally, by the end of the millennium, the Zylon fibre of Toyobo broke the record tensile strength, reaching 6000 MPa. The PBO-fibre Zylon had been initially developed at the USAF and Stanford Research Institute labs by Wolfe, but Dow Chemical obtained the patent and started the commercialisation in collaboration with Toyobo in 1998.

9.1.3.4 Natural/biobased fibres

Fifty years ago, natural or biobased fibres were a nonissue in composites. At the very start of composites they were instrumental in proving the "composite concept": Baekeland in 1907 used cotton and flax fibres to combat the brittleness of his phenolic "bakelite" resin, whereas de Bruyne and Gordon produced in the late 1930s their first flax fibre laminates. Since then, glass and carbon fibres have taken over, because of

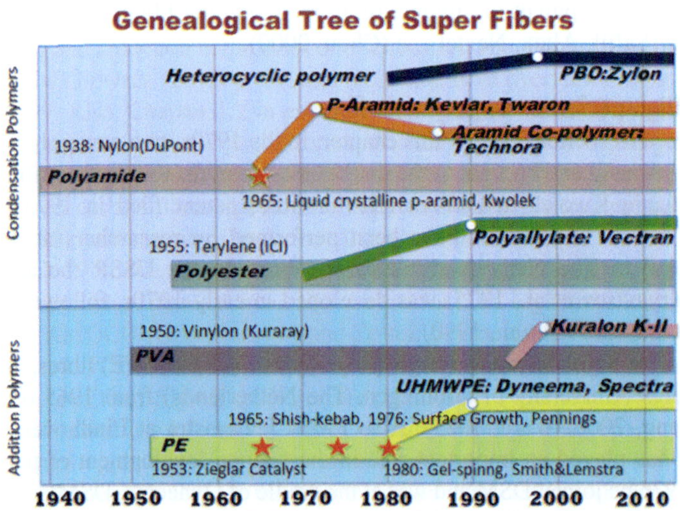

Figure 1.1.33 Historical development of three "families" of high-performance polymeric fibres. White dots: start of (semi-)commercial production (Murase and Yabuki, 2016).

their higher strength and stiffness, and better controlled mechanical properties. Natural fibres were considered to be "something of the past"! It would take another 20 years, till about 1990, before natural fibres would gradually reclaim their position in the composite realm.

In the 1980s and 1990s, natural or biobased fibres were mainly of academic interest. In developing countries though, they were sometimes considered as low cost and/or locally grown alternatives for glass and carbon fibres. It was only in Europe that interest in flax fibres grew, as these fibres had an excellent reputation for being stiff and strong. A big role has been played by CELC, The European Flax and Hemp Confederation, now acting under the name "Alliance for European Flax-Linen and Hemp" (Alliance, n.d.). The progress was well documented in a first reference publication by CELC, jointly with JEC (CELC and JEC, 2012). By the end of the millennium, synergistic collaborations have been created between the world of farmers and natural fibre extractors and processors (mainly flax) on one side, and the industrial world of composites at the other side.

Fig. 1.1.34 presents a general overview of the different families of biobased fibres used as composite reinforcements. Whereas flax and hemp are mainly European fibres, many others like jute, kenaf, ramie and sisal are mainly cultivated in developing countries. They all (except sisal) are stem fibres, which is logical because the very well oriented cellulose crystals in the elementary fibres in the stem have to provide sufficient stiffness and strength to keep up the tiny stem. When the plant is on the field, the stem is experiencing heavy winds and rain, so it must protect the seeds from falling on the ground. As a consequence, all similar plants, like jute, kenaf and ramie, have excellent mechanical properties, only somewhat lower than flax and hemp. The main

Figure 1.1.34 Different families of biobased fibres (Osorio, 2022).

difference is that with flax the extraction and refining processes are much better developed than with hemp. Most other biobased fibres are only available as random mat preforms, or as short fibre reinforcements. On the latter fibres, research will intensify in the next decades, and gradually they will become available as woven fabrics as well.

Bamboo fibres are a special case, because they are extracted from the outer (woody) part of the bamboo culm. Extraction is either mechanical (shaving) or thermal (steam explosion), but Lina Osorio and Eduardo Trujillo developed an elegant (mechanical) method providing clean, undamaged bamboo fibres with excellent mechanical properties, close to those of flax and hemp (Osorio et al., 2011).

A more extensive overview of biobased fibres can be found in Chapter 1.4 of Part II of this book.

9.1.4 Matrices

9.1.4.1 Thermosets

Since both epoxies and unsaturated polyesters (UP) were invented in the 1930s, these thermoset matrices have become the industry standard in the 1970s. Gradual improvements in mechanical properties and processing characteristics helped to make them the predominant resin systems, both for low- and high-pressure processing techniques. Vinylesters also gained some importance, thanks to their superior mechanical properties, despite their higher cost. As of today, unsatured polyesters comprise about 80% of all thermosets, the remainder being epoxies, phenolics, imides, and urethanes (Kandelbauer et al., 2014).

Still, several material innovations tried to address some issues with thermoset resins. Limited use temperatures were addressed by polyimide resins (PMR-15) (Matsui, 2010), developed at NASA in the 1970s. The rubber particle toughening technique was extended to epoxy resin formulations by Sultan and McGarry already in the mid-1960s (Seymour, 1989), and was further improved and commercialised in the 1970s to fight the rather brittle nature of thermosets.

Modification of epoxy resins continues in the 1980–90s, searching for a balance between increased toughness, high stiffness and glass transition temperature, while the resins still should have "easy to use" processing characteristics. For instance, Toray succeeded in keeping the compressive strength after impact sufficiently high by adding thermoplastic particles to Epikote MY720, resulting in the Toray 3900-2 resin (Matsui, 2010), and opening the path to its use by aircraft manufacturers in (semi)structural applications.

Most unsaturated polyesters used styrene as reactive solvent (about 40%). During composite manufacturing, or when the curing reaction was incomplete, styrene evaporates. Because evidence was building up that it is carcinogenic, research was carried out to develop low-styrene or styrene-free polyesters. BASF seems to be one of the pioneers (patent in 1992), followed later by AOC (2004) and Arkema and Cray Valley (Patents Polyester, n.d.).

9.1.4.2 Thermoplastics

Most thermoplastics we know today were invented before or just after the Second World War: low and high density polyethylene (LDPE and HDPE), polyvinylchloride

(PVC), polyurethanes (PU), nylon or polyamide (PA), polyethylenetherephtalate (PET), polypropylene (PP), acrylics (PMMA), …

Up to the 1970s, the use of thermoplastics in composites was limited to short fibre reinforced composites, mainly produced by injection moulding or extrusion, mostly using PP and PA as matrix materials.

The first composites with an acrylic matrix were developed for a rather surprising application domain: dentistry. From the beginning of the 1970s, several types of acrylic resins were mixed with different kinds of ceramic fillers and cured in situ. Extensive research led to gradual improvements in hardness, wear resistance as well as aesthetic characteristics. It would however take several decades before continuous fibre reinforced acrylic matrix composites entered industrial applications, taking advantage of the improved understanding of reactive processing.

The 1980s and 90s are the era of the invention and market penetration of high-performance thermoplastics, which would be used as a composite matrix in highly loaded and/or high temperature applications, and in chemically aggressive environments.

PEEK (polyetheretherketone), a semicrystalline polymer, was developed in the early 1980s at ICI (UK), and was commercialised under the trade name Victrex. It has good abrasion resistance, low flammability and low emission of smoke and toxic gases.

PEI is an amorphous thermoplastic with a high temperature resistance, strength, and stiffness. It was developed by Joseph G. Wirth in the early 1980s at General Electric Plastics Division, and commercialised under the trade name Ultem. SABIC acquired GE Plastics in 2007, and continues to produce it, for a large variety of high temperature applications.

PES (polyethersulfone) most of the times contains aromatic groups, and should hence be named "PAES" (polyarylethersulfone). Invented in the mid-1960s almost simultaneously by 3M Corp. and Union Carbide Corp. (USA) and ICI (UK) (Patents PES, n.d.), they entered the composite market in the 1980s. PES has a high stiffness and strength, combined with high temperature resistance, namely a use temperature up to 180°C and a glass transition temperature up to 230°C.

Initially, the use of these high-performance thermoplastics as matrix for composites was limited to short fibre composites. It would not be until the 1990s that continuous, long fibre composites utilising these high-performance thermoplastics would reach the market, the main problem being the high melt viscosity, hindering an easy impregnation of the fibrous preforms. Several impregnation methods have been developed: film stacking, melt impregnation, powder impregnation, hybrid yarns or co-woven fabrics.

9.2 Processes

9.2.1 1970s

The state of the art of thermoset composite processing in the 1970s has two faces. On the one hand, spray-up and hand lay-up are manual techniques, and are being used for low cost, often small series production, primarily using polyester resins. On the contrary, carbon fibres opened up applications in high(er) performance domains like aerospace, requesting production techniques that would enable making products within narrow performance and quality specifications. For the latter, the autoclave technique was developed, and gradually more preimpregnated fibrous preforms are being used, mostly unidirectional but also specially designed, low crimp woven fabrics.

For higher volume applications, like automotive, sheet moulding compounds (SMC's) emerged. They were developed already in the 1940s, entered the market in the 1950s and were fully operational in the 1970s. A layer of short fibres (mostly glass fibres, later on also carbon fibres) is spread on a belt, is impregnated with a thermoset resin (mostly UP, later on also epoxy), compressed and partly cured. These preforms are then compression moulded into their final shape and cured, allowing for a rather short production cycle. A similar technique uses "bulk moulding compounds (BMCs) allowing easier material flow and hence more complex shapes with larger thickness variations.

Although already developed in the 1940s, two more mechanised (and hence more efficient) production techniques were in full use (JEC Group, 2013): filament winding and pultrusion (see Section 4 of this chapter). Both processes are applied on an industrial scale in the 1970s.

In an excellent review, Lukaszewicz, Ward and Potter (Lukaszewicz et al., 2012) reveal that tape winding and tape placement were already developed in the early 1970s for both thermoset and thermoplastic composites, referring to patents by Chitwood and Howeth in 1971 and by Goldsworthy in 1974. Initial applications were taking off in the 1970s, but real (industrial) breakthrough would only happen during the two following decades.

On the side of thermoplastic composites, the dominant technique is injection moulding of short fibre reinforced thermoplastics. The first patent on injection moulding *as such* was filed in 1872 by Hyatt brothers (US Patent 133,229, 1872), when producing billiard balls from cellulose acetate. The first extrusion screw type machines were built by James Hendry in 1946 (Plastics Hall of Fame, n.d.). The limited stiffening and reinforcing effect of the short fibres is compensated by the high production speed and the "liberty of shapes," typical for injection moulding. Gradually, all fibre types used in continuous fibre composites were also being used in such short fibre injection moulded composites: glass fibres from 1935, carbon fibres from 1959, aramid fibres from 1971 (Ageyeva et al., 2019a) and later on also biobased fibres. Major developments in the 1970s (and 1980s) concentrated on optimising fibre length, fibre-matrix adhesion, and the addition of property and process enhancing additives. This led to a large variety of "short fibre TP compounds," as is more extensively explained in Part II, Chapter 1.9 of this book, on discontinuous fibre composites.

Mat reinforced thermoplastics, initially mainly using short or continuous glass fibres (GMT's), were developed by PPG Industries and Union Carbide in 1967, and introduced under the trade name "Azdel," in a way as an alternative to metal sheets. Upon mild heating (close to or just above the matrix glass transition temperature T_g) the sheet softens and can indeed be thermoformed, similar to (mostly cold) sheet forming in metals, allowing the high production speeds, typically requested in automotive applications. First applications in automotive indeed appeared in the 1970s (front end of Chevrolet Monza, 1975; Ageyeva et al., 2019a).

9.2.2 1980–90s

After the initial development of composite specific manufacturing processes in the decades after World War II, from 1980 on processing innovations for existing techniques like autoclaving, filament winding, pultrusion, compression moulding for SMC and BMC … were driven now by the need for reduced processing time and for increasing reliability and repeatability. In 1983 the body work of the Renault Espace was one of the first high volume applications, proving the potential of combining high performance with large production volumes (Giocosa, 2006).

Such "incremental" innovations are crucial for the further industrialisation of composite manufacturing methods, and for enhanced market penetration for composite products, but as this text concentrates on "breakthrough innovations," they will not be further discussed here.

One technique however really took off in the 1980s: resin transfer moulding (RTM). Pollard and Reese of Bristol Aircraft (UK) filed their patents on RTM between 1952 and 1956, and can be considered as the inventors of the RTM technique, as stated by Potter in a short historical paper (Potter, 1999). After infusing the low-viscosity, uncured unsaturated polyester or epoxy resin into a mould cavity, filled with the dry reinforcement, curing takes place and a dimensionally stable product with low porosity and well controlled fibre volume fraction can be obtained.

At the time of these developments, Bristol Aircraft Ltd. was part of the same group as the Bristol Car Company and the developers of the process were clearly seeing little distinction between the two potential markets. Still, the introduction of the RTM technique in automotive would be faster than in aerospace. Potter explains why the technique did not succeed in being more widely used in aerospace applications right after its invention in the 1950s: despite numerous practical improvements and patents describing them, the fundamental understanding of the resin flow process (under pressure) and its control during the production itself, was insufficient to reach the high-quality requirements for aerospace. Once this knowledge became available, RTM was "rediscovered" in the 1980s, among others by Potter. Only then, the technique started penetrating the aerospace market; automation of the equipment, optimisation of the injection strategies by flow modelling and in-situ observation of the flow front further helped securing high-quality products.

Light RTM or VARI (vacuum assisted resin infusion), are in full use from the early 1990s on. In this infusion technique, one solid mould half is replaced by a semirigid (light-RTM) or a flexible mould half (VARTM), reducing in this way the mould investment cost (Harper, 2009). In the 1990s, light RTM was increasingly used when only one well controlled part surface was needed, for instance in body parts for all kinds of vehicles. The most successful example, from then till now, is the production of wind turbine blades.

Based on the early inventions in the 1970s, the fundamentally new Automated Tape Laying (or ATL) technique gradually evolved into an industrial process. Initially, UD-tapes could be laid only on a flat bed (FTML: Flat Tape Laminating Machine), but in 1984 Cincinnati Milacron introduced the Contour Tape Laminating Machines (CTLM), using a rotating robot-controlled head, allowing laying-up on curved surfaces. From then on, multiple improvements in the tape head design, and in force and movement control have been achieved, turning ATL into a versatile and efficient technique for laying up even large and complex surfaces (Lukaszewicz et al., 2012). Adding local heating and compaction force control opened up, from 1991 on, its application to thermoplastic tape laid composites.

A variant of ATL, namely Automated Fibre Placement (AFP), using narrower slit tapes or even just impregnated tows, allowed for realising more complex, curved fibre paths and fibre layup geometries, hence adding design freedom and improvements in mechanical performance of the composite part.

For thermoplastic composites, several new processing methods saw light. Rubber stamping, hydroforming and vacuum forming were partially borrowed from (nonreinforced) thermoplastic sheet forming, whereas deepdrawing was inspired by well-known metal forming methods. Diaphragm forming was specifically developed for composites, both thermoset and thermoplastic.

More detailed information on innovations in composites processing techniques is provided in Potter's Chapter 2.1 in Part II of this book.

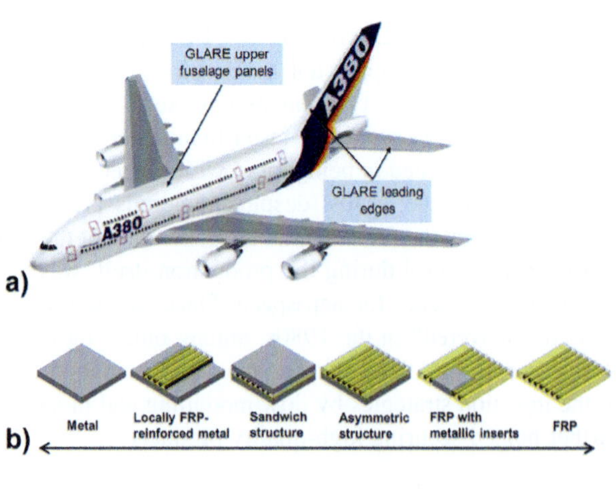

Figure 1.1.35 (a) Application of GLARE in Airbus 380. (b) Various combinations of metal and FRP in metal-FRP hybrid (Min et al., 2022).

For aerospace engineers having some nostalgia for metals, GLARE offered an elegant solution. GLARE "GLAss-REinforced" Fibre Metal Laminate (FML) is a multilayered material, composed of several very thin layers of aluminium interspersed with layers of glass-fibre prepreg. Initially, also aramid fibre prepregs were used (ARALL). The concept of Fibre Metal Laminates was invented by Boud Vogelesang and colleagues at Technical University of Delft (The Netherlands) in collaboration with Fokker, and the GLARE version was patented by AKZO Nobel in 1987 (Vlot, 2001). In 2001 it was applied in the fuselage and in the leading edge of the tail surfaces in the Airbus 380 (Min et al., 2022). Its advantages are: better damage tolerance, impact and fatigue resistance, better corrosion and fire resistance and lower weight, compared with aluminium (Fig. 1.1.35).

9.3 Applications in the 1980s and 1990s

9.3.1 Aerospace

In 1980, the first composite primary structures were created and installed by Boeing: elevators and horizontal stabilisers on the Boeing 727 and Boeing 737 aircraft, and most of the control surfaces (spoilers, ailerons, elevator and rudder) in the Boeing 767 (1981) (Quinlivan and Lovell, 1982). For the latter, a weight saving of 625 kg was quoted. Lockheed followed by designing and certifying a multispar vertical stabiliser for the DC-10 aircraft (see Fig. 1.1.36).

Also in Europe, composites entered civil aircraft: in the mid-1970s, the Airbus 300 had carbon composite fairings, but by the early 1980s the A310 had added spoilers, rudders and airbrakes, resulting in 8% composites out of the total structural weight (Fig. 1.1.37).

An important breakthrough in the use of composites in (civil) aircraft was realised thanks to the development of the "interleaving technology." Toray was the first (1990)

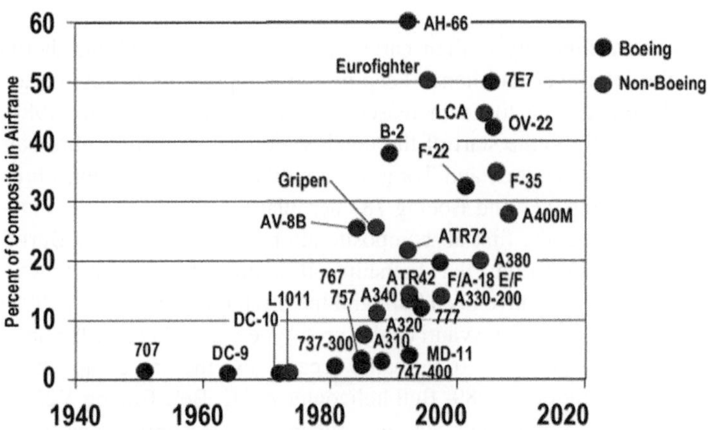

Figure 1.1.36 The use of composites in Boeing and "non-Boeing" aircraft (Roeseler et al., 2007).

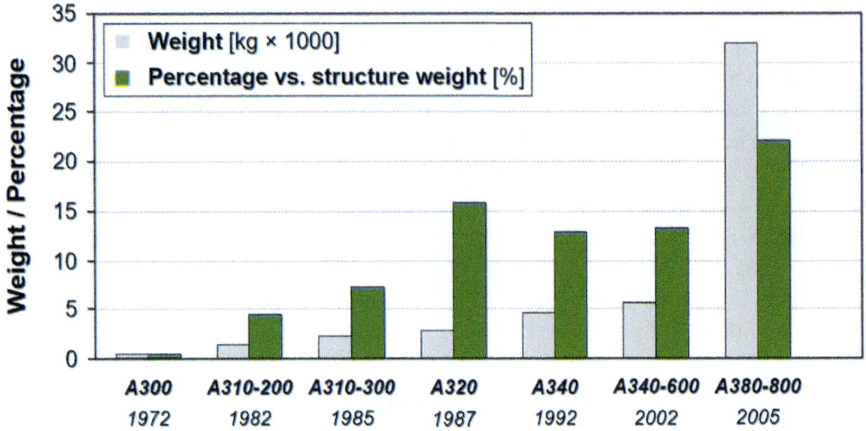

Figure 1.1.37 The increase in the use of composites in Airbus aircraft. Redrawn from Kumpfert (2004).

Figure 1.1.38 Beech Starship, 1986. Beech by Ken Mist/, CC BY 2.0 via Wikimedia Commons.

to introduce this technology in their carbon fibre prepregs, by adding thermoplastic particles at the interlayer and hence keeping the compressive strength after impact at a higher level, leading to an increased toughness (Odagiri et al., 1988, 1991). This is considered to be the start of the much wider use of carbon fibre composites in primary aircraft structures, and Toray became the preferred carbon fibre prepreg partner for the Boeing 777 and Boeing 787 aeroplanes.

In business aircraft, the first all-composite aircraft was realised by Burt Rutan at Beech Aircraft Corporation: the Starship all composite aircraft flew in 1986 (Fig. 1.1.38), but only 50 have been built and the production stopped in 1995 (Smithsonian, n.d.). Piaggio's P.180 Avanti (Italy) prototype was ready in 1986 and full certification was obtained in 1990. In the same decade, also the first composite intensive helicopters were ready (a.o. 1989: Bell helicopter A-6E; Bell-Boeing V22 with 41% composites). Finally, composites also entered into space applications, in satellites, rocket motors, re-entry vehicles etc. ...

9.3.2 Transportation

Chevrolet is, in the United States, probably the first to use GFRP leaf springs, namely a transverse rear spring in the 1981 Corvette, further extended to the front suspension in the 1984 Corvette. This was in continuous production on all Corvette models until 2019, when the C8 version was introduced, which reverts to coil suspension (Brosius, n.d.). Also in 1984, the Mercury Tracer used SMC for the first time in the bumper beam, and General Motors launched its Fiero model, with doors and fenders made using reinforced resin injection moulding (RRIM), while its hood, roof and decklid were made using SMC. By 1995, it was estimated that in the United States, already more than 300 SMC components were used on more than 110 vehicles by 28 manufacturers (CompositesWorld, 2017). Many more GFRP applications in American, Japanese and European cars would follow.

1984 was indeed a year of major innovations, because of Renault's first use of GFRP body panels in a "mass production" car, namely in the Espace (500 cars/day!, Fig. 1.1.39), entering the market in 1984. In fact, Renault had gained experience with composites when building the Alpine A110 (already in 1962), which had a one piece glass fibre polyester chassis on a central (steel) beam type chassis; it only weighed 715 kg!

9.3.3 Consumer goods

After composites had invaded the golf club and tennis racket market, it took another 10 years, till 1983, to see the first CFRP hockey shaft by the Belgian company Snauwaert.

Whereas the 1970s can be considered as the "invasion" of (carbon fibre) composites into golf sticks and tennis racquets, the same can be said for bicycles in the 1980s. As mentioned earlier, in 1986, Greg Lemond was the first Tour de France winner on a carbon fibre composite bike, the Look KG86. The technology used was not yet exploiting the full potential of composites, because straight CFRP tubes were glued into steel (later CFRP) lugs, as was done since 100 years with steel and aluminium bikes.

In 1986, Colnago seems to have been the first to develop a "full carbon" bicycle frame, where not only the tubes but also the lugs and all other connections were CFRP made! Almost simultaneously in the United States (Nelson, 2001), the companies Huffy and Kestrel filed patents (1989) describing a technology where lugs and tubes would be co-cured into a monocoque frame, using inflatable bladders.

Figure 1.1.39 Renault Espace (1984). Renault Espace by Rudolf Stricker, Attribution, via Wikimedia Commons.

Figure 1.1.40 The Kestrel 4000 monocoque CFRP bicycle, from 1989 on (Nelson, 2001).

Figure 1.1.41 Trek 5500 bicycle, with details on CFRP lugs, used by Lance Armstrong during his first Tour de France win in 1999 (Nelson, 2001).

Kestrel produced more than 1000 frames per year for their Kestrel 4000 bike (Nelson, 2001, 2003) (Fig. 1.1.40). Trek further developed this technology, together with Ron Nelson and Dimitrije Milovich at Radius Eng., and filed a patent in 1989 (US Patent 5,264,519, 1989), resulting in the Trek 5500. Their production efficiency being higher, Trek outperformed Kestrel and Lance Armstrong won the 1999 Tour de France, with the improved Trek 5500 OCLV *(Optimum Compaction, Low Void)* (Nelson, 2001, slide 36) (Fig. 1.1.41)—a victory that was, later on, taken away from him because of doping.

Figure 1.1.42 Lotus 108 track bike, a monocoque design with inflatable bladder (1992).
Lotus 108 by Paul Hudson from United Kingdom, CC BY 2.0, via Wikimedia Commons.

The first CFRP monocoque mountain bike was designed by Chris Hornzee-Jones (UK, moved to California) in 1992 (Darlow, 2021). This drastically different frame concept deviated fundamentally from the classical "diamond" (parallelogram with one diagonal) frame and was later sold to Lotus, producing 400 frames in 1993–94. They were "real" monocoques, where the whole frame was produced as a hollow structure (using inflatable bladders) in one step.

Almost simultaneously, Mike Burrows designed his "airfoil" frame, which in 1992 evolved into the famous Lotus 108 track bike (Lotus, n.d.) (Fig. 1.1.42). Its extreme aerodynamics helped Chris Boardman win the gold medal in individual pursuit during the Olympics in Barcelona in 1992. Yet another "monocoque concept" was used, in that first two halves of the frame were laid up and cured, and then adhesively bonded to each other.

Other sports start using carbon fibre composites as well. In 1981, the McLaren MP4/1 (Fig. 1.1.43), driven by James Watson, becomes the first Formula 1 race car to use a carbon fibre composite monocoque structure, and winning the Formula I race in Silverstone. Very quickly all competitors will follow their example. But already in 1965, the first victory with a composite chassis was realised by Jim Hall and Hap Sharp in their Chaparral 2A, in the World Sportscar Championship at the Sebring 12 Hours (Zogg, 2025).

As is well documented in the book "From Bakelite to composite" (Verpoest et al., 2002), the use of glass fibre composites by furniture designers boomed in the 1970s, but by the end of the 1980s carbon fibre composites were finally discovered by the designers community. In 1987, Alberto Meda designed the first CFRP chair, and in

Figure 1.1.43 The McLaren MP4/1 Formula 1 race car during the Silverstone race in 1981.
Mc Laren by David Merrett, CC BY 2, via Wikimedia Commons.

Figure 1.1.44 The C06 chair by designer Pol Quadens, the world's lightest chair (950 grs), 1995.
Courtesy Pol Quadens.

1995, Belgian Pol Quadens designs the C06 chair (Fig. 1.1.44), the lightest chair in the world (950 g). It would be the start of many experiments with carbon composites in all kind of design and consumer good applications.

9.3.4 Construction

After having been discovered by architects in the 1970s, in the following two decades engineers started experimenting with composites as structural elements, either standing by themselves or as a reinforcement for concrete.

The Advanced Composite Construction System (ACSC), invented and marketed by Maunsell Structural Plastics in the early 1980s, consists of self-supporting GFRP structural elements, often interlocking pultruded planks. They were not only used for housing or office buildings, but also for bridges, as for example for the bridge deck of the first composite footbridge across the river Tay in Scotland (1993). However, the very first composite bridges dated from 10 years earlier: the 10 m span bridge constructed at Ginzi, Bulgaria (1981) and the Miyun bridge in Beijing, China, completed in 1982; a good overview of composite bridges is given by Siwowski et al. (Siwowski et al., 2018). Initially, either solid laminates or sandwich plates were used, but pultruded shapes gained importance in the 1990s (Fig. 1.1.45).

Composites in cable stay bridges are another innovation on the 1990s. In 1996, Urs Meier realised the Stork bridge in Winterthur (Switzerland). It has, next to mainly steel cables, also two carbon fibre cables, for which special anchorage heads had to be designed (JEC Group, 2019, p. 57). In the same year, the Tsukuba FRP bridge was erected in Ibaraki (Japan). It is a 20 m long bridge with GFRP pylons, the deck is a GFRP profile strengthened by CFRP lamellas and all the 24 stay cables

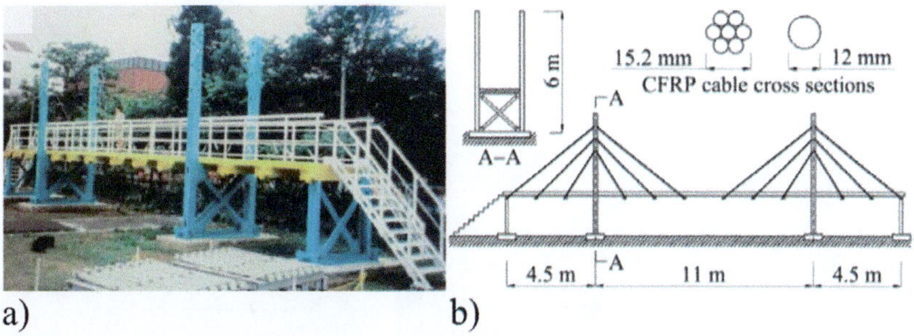

Figure 1.1.45 Tsukuba FRP Bridge: (a) photo, and (b) sketch.
(a) Photo credit: Iwao Sasaki; (b) Liu (2015).

are made of CFRP. Soon, many other bridges using carbon fibre composite cables would be build, using different kinds of cable structures and anchorage systems, as extensively described in the paper by Liu et al. (2015). The construction of full carbon composite bridges would however only start around 2010 (Fig. 1.1.46).

The reinforcement potential of composites is further on also exploited in different kinds of strengthening methods for concrete, either internally or externally. Rebars are internal reinforcements, and have been invented in the 1980s in Japan, initially only using carbon and aramid fibres. However, for cost reasons glass fibre composites are predominantly used, mostly with alkali-resistant glass fibres or sometimes with basalt fibres (Węglowski, 2019). Flexible or rigid UD reinforcements (mainly CFRP) are adhesively bonded as external reinforcement to concrete beams or slabs, or even brickwork, or as retrofitting for damaged structures (for instance after earthquakes). Different types of such external reinforcements for beams, columns or walls, either axially loaded, or loaded in bending, shear or torsion, are systematically reviewed by Naser et al. (2019).

Finally, a large number of smaller construction and infrastructure related applications are gradually evolving: access platforms for industrial, chemical and offshore structures, electricity transmission towers, power poles, light poles, marine structures such as seawalls and fenders, etc. More detailed information can be found in Chapter 3.6 of Part II of this book.

9.3.5 Energy

One of the fastest growing applications over the last 2 decades of the past millennium has been in wind turbine blades. The combination of lightweight, high strength, reasonable stiffness and environmental/corrosion resistance is the perfect mix for the use of glass fibre composites. Gradually, wet hand lay-up techniques are being replaced by vacuum infusion methods. This was a big challenge, because by the end of the millennium the length of the commercially produced blades gradually increased from about 20 m (0.5 MW power) to more than 40 m (2 MW). Special

Figure 1.1.46 Composite rebars (a) glass fibre composite rebars, (b) rebars with different surface modifications for improving adhesion to concrete, (c) installing rebars on a bridge deck, illustrating the lightness of GFRP versus steel rebars (all courtesy of Owens-Corning).

infusion techniques, glass reinforcements and thermoset resins had to be developed to meet this challenge. Eventually, the technology would further evolve into hand lay up of prepregs.

10. The new millennium: 2000–22: continued innovations, huge expansion in applications

10.1 Materials

10.1.1 Glass fibres

At the start of the millennium, due to the increasing importance of renewable energy, there was a strong need for higher performance glass fibres for wind turbine blades. As early as 2005, Owens-Corning filed a patent on high-performance glass fibres, later

Figure 1.1.47 SEM of circular and flat cross-section glass fibres (Thomason, 2023).

followed by 3B Fibreglass (2011) and PPG (2012) (Thomason, 2015). These fibres combine a significantly higher modulus (up to 90 GPa) with a higher strength (+25%), strain energy to failure (+45%) and an improved fatigue resistance (Hiper-Tex, n.d.), all properties being highly relevant for wind turbine applications.

Already in the 1990s, Nittobo (Japan) filed some patents, describing how to produce flat glass fibres (Fig. 1.1.47), later followed by patents from Nippon Electric Glass (Japan) and CPIC (China) (Thomason, n.d.). One of the main advantages of such fibres, achieving a cross-sectional width to thickness ratio of up to 4 (20 by 5 µm), lies in potentially higher fibre volume fractions, and in the reduced warpage and increased dimensional stability for injection moulded short fibre thermoplastic composites (Nittobo, n.d.). Thomason (2023) investigated the effect of fibre flatness on the stress transfer at the interface. He showed that, because of the higher surface-to-cross section ratio, the critical fibre length is higher for flat fibres (as measured in for instance fragmentation tests), and hence the strength of injection moulded discontinuous glass fibre composites increases by 10%–15% when using flat instead of circular cross section glass fibres at a comparable weight volume fraction (typically between 30% and 40%).

10.1.2 Carbon fibres

While carbon fibre manufacturing companies continued to improve the efficiency of their production, trying to reduce costs and increase yield, a breakthrough innovation shocked the carbon community: in 2004, Geim and Novoselov published their paper(s) on the manufacturing, characterisation and properties of single layer graphene (Novoselov et al., 2004), for which they later received the Nobel Prize in Physics.

Several attempts have been made to introduce CNT's, and more recently graphene platelets as well, into the matrix of fibre reinforced composites, to improve some matrix and/or interface dominated composite properties (see an overview in the paper by Gorbatikh and Lomov (2016)). Just to mention one example: Brian L. Wardle in MIT developed in 2000 a "nano-stitch," an interleave of aligned or buckled CNTs in a composite laminate. This (almost) zero-thickness interleave of aligned nanofibres enhances the fracture toughness of the laminate and its load-carrying ability for in-plane loading. The nano-stitch technology was commercialised by N12 Technologies, a spin-off of MIT, and is used in the aeronautic, energy and smart textiles sector (Garcia et al., 2008). More information on carbon nanotubes and graphene can be found in Chapters 1.5 (by Wardle) resp. 1.6 by Young in Part II of this book.

Material innovations in carbon fibres were focussing on gradual improvements, but new perspectives were created when Boeing started the development program for the carbon fibre intensive Boeing 787, which made its first flight in 2009 and entered in service in 2011. Airbus followed almost immediately, with the A350 program starting in 2006, first flight in 2013 and entering in service in 2015. The structure of both aeroplanes contained about 50% of composites by weight. As a consequence, some reorganisations and regroupings occurred in the carbon fibre producing industry. For instance, in 2001, Cytec bought Amoco's carbon fibre division, and was afterwards taken over by Solvay (now Syensqo) in 2015. More examples are being presented in Chapter 1.2 on carbon fibres in Part II of this book.

The increasing demand for carbon fibres in large volume and cost critical applications like automotive and wind turbine blades inspired all companies to develop high-performance large tow (and hence less costly) carbon fibres. In 2011, Mitsubishi Rayon launched its commercial production in the Otake Production Centre (Japan). Zoltek, which had always been focussing on "industrial grade" carbon fibres, became part of the Toray group in 2014 and concentrated even more on the production of large tow carbon fibres, resulting in new investments in additional production capacity (plus 6000 ton in 2021 and 2023). Many other companies expanded their large tow carbon fibre capacities in view of increasing applications in wind energy and automotive. Also China started carbon fibre production, as well as Russia, the latter mainly for internal production (Umatex, n.d.).

The search for even better carbon fibres did not stop. For instance, in 2013, Toray introduced the T1100G fibre with 7000 MPa strength, while maintaining a high stiffness. In 2018, this was followed by the M40X fibre, which has an increased modulus of 380 GPa while maintaining a high strength. Such combinations of strength and stiffness created a new family of high-performance carbon fibres (green area in Fig. 1.1.31).

10.1.3 Polymeric fibres

By the turn of the millennium, a vast range of high-performance polymeric fibres was available on the market. They combined the low density of polymers with the high degree of orientation of the polymer structure, obtained during different types of fibre spinning. This resulted in high specific stiffness and strength values relevant for composites. A comprehensive overview is presented in a review paper by Avci et al. (2019).

The advent of carbon nanotubes, and later of graphene, inspired fibre researchers to attempt including these high stiffness nano-reinforcements into spinnable polymers. High expectations regarding increased mechanical properties have however not been met, as explained by Goutianos and Peijs (2021). So, the advantage of these "composite fibres" might be limited to their multifunctionality, combining reasonable mechanical properties with increased electrical and thermal conductivity.

Electrospinning of polymeric fibres has started in the early 2000s, as explained in the review by Jiang et al. (2018). Submicron fibres can be electrospun using many different polymers (PA, PAN, PI, PLA, ...), and in some of them even CNT's, graphene or other nanoparticles can be included. Mechanical properties can be high, depending on the polymer and reinforcing nano-inclusions used, but they do not reach those of gel, liquid crystalline or melt spun polymer fibres. Moreover, production is not yet upscaled to the tons-range, so their applications are actually limited to those where the ultrafine diameter can be exploited, like in (medical) tissues.

10.1.4 Natural/biobased fibres

As the importance of sustainability steadily increased, composite materials and processes with lower environmental impact became a key issue. Building upon the collaboration between farmers, fibre extractors and traditional textile companies, a new generation of flax preforms for composites was developed, mainly in the three leading flax countries France, Belgium and The Netherlands (Fig. 1.1.48).

First, low twist and low crimp textiles addressed the need for preforms enabling composites with improved stiffness and strength, whereas using random mat preforms created composites with direction-independent in-plane properties. More recently, several research centres and companies jointly developed different types of unidirectional preforms, either as dry layers, as (epoxy resin) prepregs or as thermoplastic UD prepregs. Finally, also hybrid flax-thermoplastic yarns have been developed, which can be used in woven textiles, noncrimp fabrics or knits. All these preforms, their advantages and the applications in which they are being used, are comprehensively

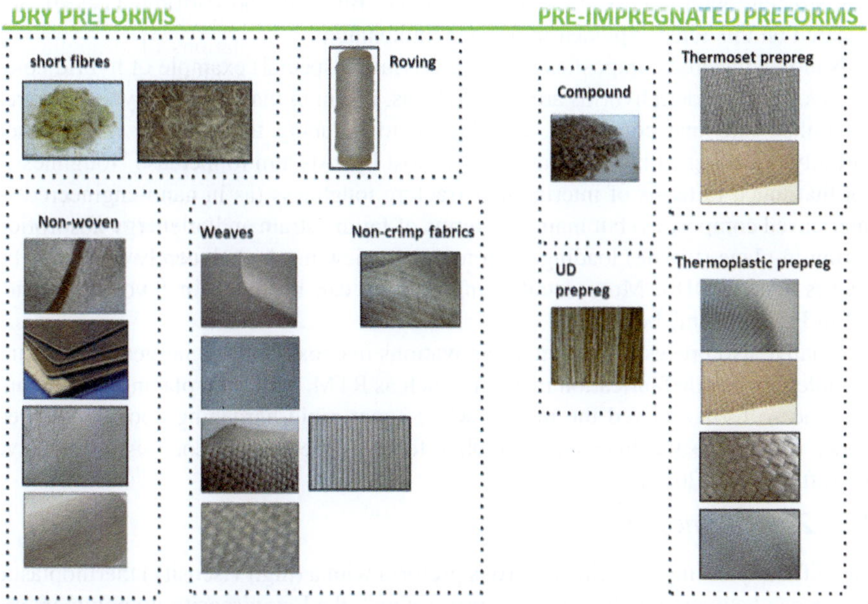

Figure 1.1.48 Overview of different flax fibre preforms.

described in a publication by CELC (now called "The Alliance for Flax, Linnen and Hemp"), jointly with JEC-World (CELC and JEC, 2018).

Biobased polymers, developed in the polymer synthesis and compounding industry, can be used as matrix in composites, but also to produce high-performance reinforcing fibres of different kinds: bacterial or regenerated cellulose fibres, cellulose nano-crystals and polylactic acid (PLA) fibres. They are step by step reaching the market, an excellent overview is presented by Li et al. (2022).

10.2 Matrices

10.2.1 Thermosets

At the start of the millennium, different types of nanoparticles are used to improve the properties of epoxy matrices in composites. Carbon nanotubes, later followed by graphene platelets, were added to improve stiffness and strength, but the big challenge was to achieve a well-controlled distribution inside the tiny matrix areas in between the fibres of a composite (Gorbatikh and Lomov, 2016). Several papers by L. Gorbatikh and co-workers in the Composite Materials Group at KU Leuven, are presenting multiscale models, highlighting the importance of positioning CNT's on and in between the fibres (among others: Romanov et al., 2015). However, no major experimental breakthroughs in improving the mechanical performance of such "hierarchical" composites have been reported yet.

Apart from these carbon-based nanofillers, many different filler materials like metal oxides and clay nanoparticles are being experimented, both for improving mechanical properties (stiffness, toughness) as well as for adding multifunctionality (increased electrical conductivity, sensing properties …) (Frigione and Lettieri, 2020). Furthermore, phase-separation during curing was already invented in the 1980s, to create a nanoscale dual structure with improved toughness (see a.o. Bucknall and Partridge (1983)), and research on this topic exploded in the new millennium.

Nano-engineered composites are just one (rather special) example of hybrid composites. Most modern hybrids are fibre hybrids, meaning that different types of fibres are mixed into one composite. In addition to reducing the cost, by for instance partially replacing carbon by glass fibres, most hybrids aim to increase "toughness," not that much in terms of interlaminar fracture toughness (as in nano-engineered or interleaved composites) but mainly in terms of failure strain and/or energy absorption during in-plane (tensile) loading. A complete review has been recently presented by Swolfs et al. (2018). More detailed information can be found in Swolf's Chapter 1.7 in Part II of this book.

Finally, also processing oriented innovations in epoxy resins emerged: new resins suitable to versatile fabrication methods, such as RTM, RFI, yarn placing, hot stamping, and so forth, entered the market while super rapid hardening epoxies allowed Toray to become the long term supplier for the Boeing 787 composite intensive aircraft (Matsui, 2010).

10.2.2 Thermoplastics

The difficulty of impregnating a fibrous preform with a (high viscosity) thermoplastic matrix can be overcome by "reactive processing": the low-viscosity prepolymers are easily infused into the fibrous preform, after which the polymerisation takes place "in situ." Special attention must be paid to avoid the formation of reactive gases, which can create voids and reduce the final composite properties.

In the quest for making large composite structures with thermoplastic matrix, such "RIM"-type techniques have been developed for different types of thermoplastic polymers. In 2012 Arkema developed Elium, a new thermoplastic acrylic resin (Elium, n.d.). Contrary to other thermoplastic matrices, this resin has the main advantage to be infusible at room temperature, in a similar way to thermoset composite parts. While being a thermoplastic it has the advantages of weldability, recyclability, postforming potential and better toughness compared with (most) thermoset matrices. Moreover, the polymerisation can take place at relatively low temperatures. An extensive overview of this technology can be found in Obande et al. (2021).

Other thermoplastic polymers are also eligible for such types of processes, and since the turn of the millennium, the anionic polyamide 6 (A-PA6) resin and the CBT (butylene terephthalate) resin from Cyclics have been introduced to the market (Arhant and Davies, 2019).

At the start of the millennium, following the trend of lowering the environmental impact of the materials used in composites, a growing interest arose in the development of biobased plastics. Although the first "plastics" in the second half of the 19th century were biobased (such as celluloid, shellac, ...), the advent of cheap and constant quality petroleum as a raw material has caused, in the 1930s, a massive shift away from biobased "renewables." This again changed in the beginning of the new millennium, and new biobased plastics also found their way into composites, both as thermoset and thermoplastic matrix.

It is important to mention that biobased plastics should not be confused with biodegradable plastics (Fig. 1.1.49). For structural applications, like for composites, biodegradability is most often not an advantage, as it is the case for packaging and other

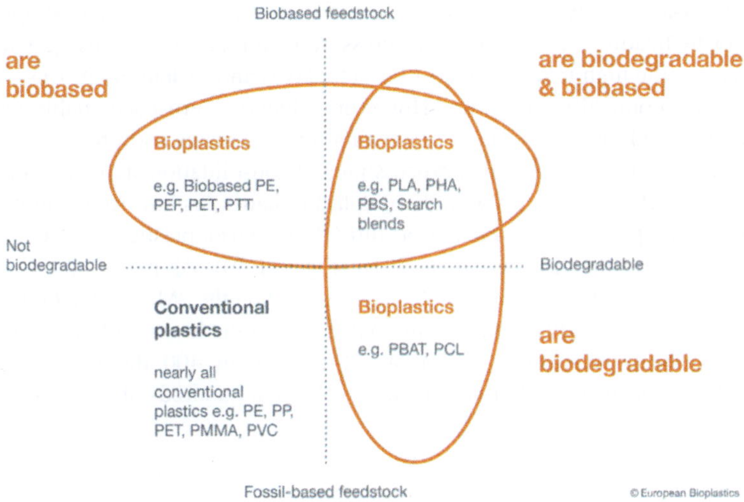

Figure 1.1.49 Biobased/biodegradable plastics.
Source European Bioplastics (n.d.).

"short life" or "single use" applications. Biodegradability can even be a disadvantage, as it often goes hand-in-hand with limited resistance under hot-wet conditions. Hence, for composite matrix applications, only biobased plastics, coming from a renewable resource, should be considered.

The range of biobased plastics has grown spectacularly over the past two decades. Giving an overview of all innovations is out of the scope of this document, only the biogenic raw materials and the most used biobased matrix materials will be briefly mentioned. The biogenic raw materials can be split up in different platforms (Harlin, 2011), related to their chemistry: in the sugar platform biomass carbohydrates are the main renewable resource, providing D-glucose enzymatically produced from starch, from sucrose or even from cellulose in plants, grasses and wood. In the lipid platform fats and oils from vegetable and animal sources are used, whereas in the biogenic olefin platform ethylene can be produced from biogenic ethanol.

A wide range of polymers can be synthesised from the chemicals produced in each of these platforms. A comprehensive but not too complex overview is presented on the website of Bioplastics Europe (Bioplastics Europe, n.d.). Because currently, most biobased plastics are produced from carbohydrate-rich food plants such as corn or sugarcane, or oily plants, R&D efforts are now directed toward the use of second generation "nonedible" feedstock which are not competing with the food production, such as straw, corn stover, bagasse, or even organic waste. The Bioplastics Europe website lists a number of completely (like PLA) or partially (PA, PE, PET, PHA, PBS ...) biobased plastics, and mentions the feedstock (most often corn starch, sugar cane, beet or biomass ...).

10.3 Processes

10.3.1 Out-of-autoclave process

Because also in aircraft applications cost reduction is an important issue, the costly autoclave process (long duration, expensive equipment) is gradually being replaced by the out-of-autoclave process (OOA) (Ekuase et al., 2022). Heat is supplied by an oven or by heating blankets, whereas pressure is limited to 1 bar, using a vacuum bag. Avoiding the high investment cost for autoclaves and its lengthy process, makes OOA parts economically interesting. However, achieving aerospace quality and performance required the modification of existing (epoxy) resins and prepregs.

OOA can also be used as a final processing and consolidation step after automatic tape laying (ATP) of UD tapes with thermoplastic matrix, as has been demonstrated by Stelia Aerospace when making a full-OOA thermoplastic welded primary airframe, receiving the JEC Award in 2018 (Fig. 1.1.50). As part of the European Commission's Clean Sky 2 program, Airbus was leading the "Multi-Functional Fuselage Demonstrator" (MFFD) consortium. In this project the lower half of a fuselage, 8 by 4 m, has been fabricated. It consists of more than 400 thermoplastic fibre-reinforced parts, as well as thousands of spot welds and hundreds of metres of continuous welds (JEC Composites, 2024) (Fig. 1.1.50).

Figure 1.1.50 Full-OoA TP welded primary airframe by Stelia Aerospace (JEC Innovation Award, 2018).

10.3.2 Automated Tape Laying/Fibre Placement

The principles of Automated Tape Laying/Fibre Placement have been explained in a previous section. Nowadays, ATL/AFP uses up-to-date robots like the Coriolis systems (Coriolis, n.d.) and different kinds of tape/fibre laying heads, adapted to the type of material and the geometry of the part. In this way, the lay-up and production time can be reduced, and often also the quality control time and final part cost. More detailed information on recent developments in ATL/AFP can be found in Chapters 2.1 and 2.5 of Part II of this book.

10.3.3 Resin transfer moulding

Resin transfer moulding (RTM) has matured over the past decade. New developments responded to the need for a reduction in processing time (mainly for automotive applications), for simpler and hence less expensive moulds (for medium to low series applications) and for improved quality consistency (mainly for aerospace). All this resulted in different variations of the RTM process, like vacuum assisted RTM (VARTM) and vacuum assisted resin infusion (VARI), which have now been fully accepted as efficient processing methods by the composites industry. More detailed information on these novel infusion processes can be found in Chapter 2.1 of Part II of this book.

The RTM process was also adapted to thermoplastic composites (Ageyeva et al., 2019b). Some thermoplastics can be injected into the mould as prepolymers, and then react "in situ." In the 1980s, innovations in the chemistry of anionic ring opening polymerisation for polymers like PA 6 triggered developments for thermoplastic RTM, but it took until 2010 before good collaboration between chemists and composite processing specialists resulted in commercial success of thermoplastic RTM.

10.3.4 Thin layered laminates

Instead of the standard 125 μm, thicknesses down to 20 μm can be achieved in thin prepregs. They have the potential to increase the failure strain/strength, and even the initiation of damage during static and cyclic loading. They furthermore offer a higher flexibility in design and the possibility of creating pseudo-ductile composite laminates, as is well proven in review articles by Galos (2020). Arteiro et al. (2020) focused on open-hole and notch sensitivity, as well as interlaminar fracture toughness, and showed that thin plies open opportunities to enhance these properties as well (Fig. 1.1.51).

Using thin plies, the labour time for creating a laminate of a certain laminate thickness increases substantially, because of the thinner layers and hence the higher number of plies that are required. However, this been counteracted by the advent of automated tape lying techniques.

Several companies have introduced a range of thin ply preforms and prepregs, like North Thin Ply Technologies (Switzerland) (North Thin Ply, n.d.) and Chomarat (France). Textreme (n.d.) is leading in woven preforms using wide but very thin UD spread tows, hence reducing crimp and enhancing the in-plane stiffness and strength. The thin ply thermoplastic prepregs of Mitsuya (n.d.) are based on the groundbreaking developments of the Industrial Technology Centre of the Fukui Prefecture (Japan).

10.3.5 Continuous processes

Continuous lamination is another highly automated process. The technique is well known for making thermoplastic composite sheets consisting of different layers of thermoplastic prepregs. Such sheets can then be used in sheet forming or stamping processes, as for instance in making the shells for the Samsonite suitcases, using self-reinforced polypropylene sheets (CURV) or more recently their home developed Roxkin sheets, composed of multiple layers of biaxially oriented PP (Samsonite, n.d.).

Continuous lamination is also used for producing composite profiles with V- or hat shaped cross-sections, or even more complex geometries. It can be considered as an alternative for pultrusion. If unimpregnated reinforcements are used, they are first impregnated with resin, but also preimpregnated preforms can be used. Then, they are guided through a set of forming rolls to control the shape, the thickness and the resin content, the latter when unimpregnated reinforcements are used. In case of thermoset resins, the material is passed through a heating zone for curing; in case of

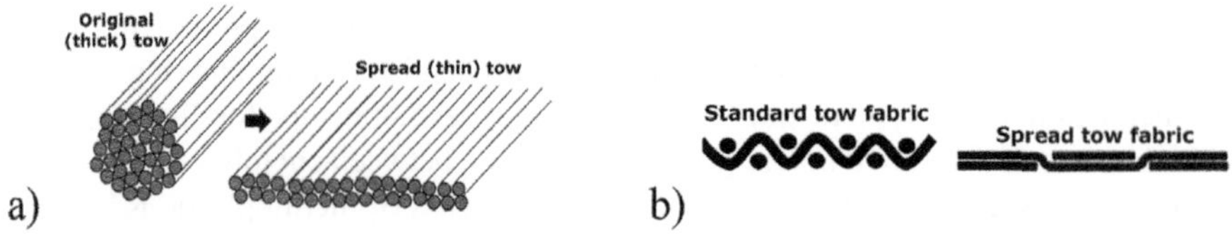

Figure 1.1.51 (a) Tow spreading, (b) spread tows reduce crimp in woven fabrics (Galos, 2020).

thermoplastic composites, the laminates are heated before forming, and afterwards quickly cooled down. Applications similar to pultruded profiles are aimed at, but larger widths can be reached, like truck trailer and RV sidewalls, building panels and other construction industry applications.

Thermoplastics also entered the domain of pultrusion, by using either low-viscosity reactive prepolymers, so that the thermoset pultrusion equipment only needs minor modifications, or specially developed thermoplastic preforms (Minchenkov et al., 2021). For the latter, special thermoplastic prepregs and melt, solution or powder impregnated fibre tows have been developed.

Because of all these process innovations, several new or adapted material preforms had to be developed: special prepregs for out-of-autoclave curing, tapes or rovings for ATL/AFP, but also thermoplastic pellets with short fibre reinforcements, extruded "doughs" and LFT's (long fibre thermoplastics) for compression moulding, "organo-sheets" for sheet forming, etc …

10.3.6 Additive manufacturing (AM) and overmoulding

During the last decade, different types of 3D-printing or additive manufacturing (AM) have been experimented as well, using not only short fibre reinforced filaments but also filaments containing continuous fibres, hence guaranteeing improved mechanical properties (Zelinski, n.d.; AMFG, 2020).They can be considered as a variant for ATL/ATP, but in these technologies, mostly an additional consolidation step is needed (for instance out-of-autoclave, as mentioned earlier), to ensure a high quality, low void content product.

Some 3D-printing methods however deliver a ready-to-use product, without need for postprocessing. In FDM ("fused deposition modelling") an impregnated filament or tape is heated and pressed upon the base plate or mould, or on the previously deposited layers, ensuring a well-bonded and low void product. In FFF (fused filament fabrication) the continuous fibres are impregnated inside the deposition head of the additive manufacturing apparatus.

A comprehensive overview of these techniques is presented by Wagmare et al. (2024). Even for natural fibre composites, 3D-printing offers interesting perspectives (Le Duigou et al., 2020). In recent years, a large variety of 3D-printing or additive manufacturing (AM) machines for continuous fibre composites have been introduced to the market, and multiple prototype productions have been presented.

A further step in process integration is overmoulding or over-injection, well-known in plastics processing where different materials (metals, wood, elastomers …) are being over-moulded or over-injected. In fact, not only traditional injection moulding but also compression moulding of discontinuous long fibre thermoplastics can be used. The aim of this technique is to combine the high mechanical properties of continuous fibre reinforced composites with the shape versatility and high-speed processing of injection or compression moulding. Controlling the process parameters (temperature and pressure profiles) is a major challenge, aiming at optimising the interface adhesion between the continuous fibre composite sheet and the overmolded three-dimensional part, often a flange or rib structure, see Harte and Mc Namara (2007) and Valverde et al. (2020) (Fig. 1.1.52).

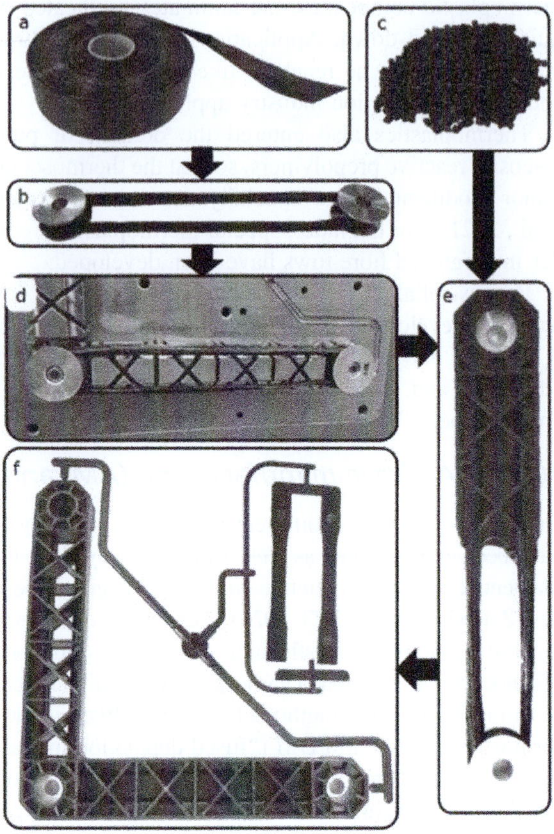

Figure 1.1.52 Continuous fibre tow reinforced hybrid-moulded composite production process. Thermoplastic UD tape (a) is heated, collected, consolidated, and wrapped into a 3D tow reinforcement (b) which is placed into tow allocations within the injection mould (d). Thermoplastic resin (c) is overmolded onto the 3D-tow (e). The resulting part (f) features two legs which are reinforced with tow and have overmolded ribs and a shear web between the 3D-tow reinforcements.
Miller and Manson (2023).

10.3.7 Complex textile preforms

Finally, more complex textile preforms, specifically developed for composites, steadily gained importance: multiaxial noncrimp fabrics, three-dimensional weaves with through-the-thickness reinforcements, 2D- and 3D braids, local stitching of additional reinforcements, etc .... The intense interaction between the textile and the composites world generated some surprising innovations, nonexistent in the "traditional" textile industry. More detailed information and figures showing the intricate structures of such complex textile preforms can be found in S.V. Lomov's Chapter 2.4 of Part II of this book.

In summary, it can be stated that, in the first decades of the 21st century, the composites production technologies have definitely made a big step forward in becoming more efficient, more robust, more flexible.

10.4 Applications (2000 till now ...)

The past 2 decades have been characterised by a stronger increase in the application of composite materials than ever before, both in the number and diversity of application areas as well as in the total amount of composites produced, expressed either in value (Euros, dollars ...) or in tonnage.

This development finally happened because composites have found their way to the market in an increasingly rapid pace. This happened both in "consumer products" purchased by individual consumers (cars, bicycles, sports equipment, furniture) and in public transport, energy generation, infrastructure ... In the latter, consumers acknowledge the advantages of the use of composites, and hence motivate the public and private sector to purchase composite based products.

In the book "Fibre-Fixed. Composites in Design" (Pil and Verpoest, 2018), a larger number of applications is shown, but in this text only the major breakthroughs will be shown, and each time only one or two products will be discussed as an illustration.

10.4.1 Aerospace

From the beginning of the new millennium, the use of (mainly carbon fibre) composites is increasing strongly in civil aircraft, as illustrated in Fig. 1.1.36 (for Boeing) and 37 (for Airbus). In 2006, the Boeing 777 has 1180 kg of its structural weight in composites, whereas the Airbus A380 (2005) has 22% (=32.000 kg!) composites. This development follows the trend set in the military aircraft (see Fig. 1.1.53), where for instance the military A400 (2013) has even 35% composites. The big breakthrough happened when both Boeing 787 (first flight 2009, first commercial flight in 2011 *by ANA*) and Airbus

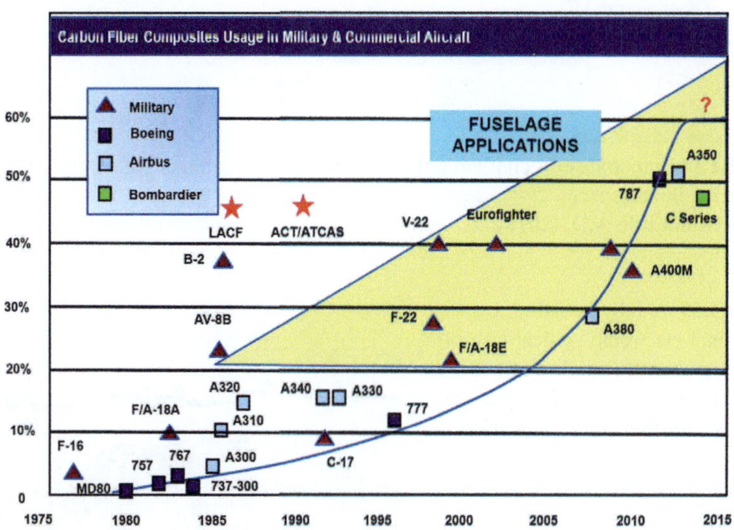

Figure 1.1.53 A global view of the use of composites in military and civil aircraft (Hiken, 2018).

350XWB (2013 resp. 2015 *by Qatar*) have in their structure more than 50% composites (by weight). Due to its 25% fuel reduction, the *Ultra Long Range-version* of the A350 XWB covers nonstop the 18.000 km distance between Singapore and New York (Singapore Airlines). More information on the developments at Boeing and Airbus can be found in Chapters 3.1 and 3.2 of Part II of this book.

The trends of the increasing use of composites in aircraft is illustrated in Fig. 1.1.54. In 2018, commercial aircraft represented about 2/3 of the 29.000 t of composites used in aerospace. The two "composite-intensive" aeroplanes, namely the Airbus A350 and the Boeing 787 alone consume 45% or 13.000 t of composites.

An interesting phenomenon is the emerging trend toward "individual air transportation": flying cars (Pal-V, 2018), air taxis and similar drone-like helicopters (Volocopter, 2016) and small electric aircraft (Lilium, 2018; Fig. 1.1.55). By 2024, several tens of such small aircraft are being designed and produced in prototype versions. Some of them are getting close to the market: Lilium's final assembly is due to start in 2024, whereas Joby (USA) started the final assembly of its eVTOL aircraft in February 2023 (Nehls, 2023). All these electric aircraft have two things in common: a very high percentage of their structural weight is composites, but also a continuous

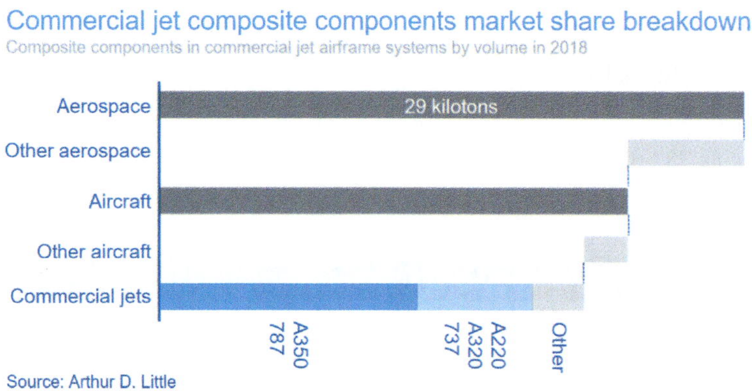

Figure 1.1.54 Importance of civil aircraft ("commercial jets") in the total composites consumption in aerospace.
Redrawn from Little A.D. (2018).

Figure 1.1.55 A Joby Aviation, Inc. experimental electronic vertical take-off and landing aircraft is parked at taxi way following a ground test at Edwards Air Force Base, California, 20 September, 2023.
Wikipedia, https://en.wikipedia.org/wiki/Joby_Aviation.

a)
b)

Figure 1.1.56 (a) Solar Impulse 2, overflying Golden Gate bridge in San Francisco. (b) Solar Impulse 2 in Phoenix (Arizona).
(a and b) Courtesy of Jean Revillard and the Solar Impulse Foundation. Copyright Jean Revillard, the Solar Impulse Foundation and Rezo.ch.

search for additional investors to support the development of this new type of air transport (Lilium, 2020, 2024)! In 2025, several of these start-ups either went bankrupt or were taken over by larger companies.

The Solar Impulse (Fig. 1.1.56) was the first example of an extremely light aeroplane, powered by solar energy, thanks to its low weight (Solar Impulse, n.d.). The Solar Impulse–1 was built in Switzerland by a team led by Andre Borschberg, and coached by Bertrand Piccard, who was also the first to circumnavigate the globe in a balloon. Solar Impulse–1 made its maiden flight in 2009 (Solar Impluse Foundation, n.d.).

The experiences gained during various, even intercontinental flights (to America in 2013), were used as the basis for the design and construction of the Solar Impulse–2. This enormous aeroplane has a wingspan of 71.9 m, slightly under the currently largest operational airliner, the Airbus 380, but weighs only 2300 kg (including 633 kg of battery), which is only 1/200th of the Airbus 380 (500,000 kg). The extremely low weight could be achieved by using mainly carbon fibre composites in the more than 6000 parts realised by the company Solvay (now Syensqo). The light weight, the solar cells on the wings and the lithium-ion batteries for night flights, allowed the Solar Impulse–2 to fly around the earth in 23 days in 2015. The two team leaders, Piccard and Borschberg, divided the 17 stages evenly between them (Solar Impluse Foundation, n.d.).

Experiments like the Solar Impulse have created a momentum to search for alternative propulsion systems for aeroplanes, in this case solar energy: it inspired Airbus to build an extremely light, unmanned aeroplane that could fly at great altitudes for very extended periods of time. In July 2018 the Airbus Zephyr S (Fig. 1.1.56b) remained in uninterrupted flight for 14 days at an altitude of 21,000 m, thus above the clouds, the jet stream, and regular air traffic. This way, the aeroplane approaches the functionality of a satellite, certainly when dealing with observation and communication assignments. The Airbus Zephyr S has a wingspan of 25 m and owing to the use of carbon fibre composites only weighs 75 kg; the wings are entirely covered with solar cells that charge a lithium battery for night-time use, and drive two electric motors (Zephyr, 2018).

Even the Zeppelin has made its come-back for cargo transportation: the (inflated) hull of the Airlander 10 (Fig. 1.1.57), being developed in the United Kingdom and scheduled to fly in 2025, comprises a skin made of a triple-layered combination of composite materials (using Vectran fibres as reinforcement), but also the rigid structural parts are made out of composites (Nehls, 2021).

10.4.2 Transportation

A study by Jun Takahashi (University of Tokyo) (Takahashi, 2006) has shown that, when considering the entire lifecycle, the energy consumption in the use phase dominates all forms of transportation over land. In cars this accounts for 80%, in trucks it ranges from about 90% for a light van up to 95% for a large heavy goods vehicle. Lastly, in buses the percentage climbs even higher as the combined energy consumption for production, maintenance and "end-of-life" amounts to an almost negligible 3%. Because energy consumption in the use phase is highly influenced by the weight of the vehicle, it is only logical that designers also review the use of composite materials for buses and trucks, and even for trains and trams.

The first full composite (electric) buses are on the road, like Evopro's Modulo in Hungary in 2018 (Modulo, n.d.) and Proterra's Catalyst in the United States in 2014, followed in 2019 by the ZX5, actually produced by Phoenix Motorcars (Proterra, n. d.). The Dutch Ebusco 3.0 (2019) has a carbon fibre composite spaceframe, is 33% lighter and has a range of 500 km instead of 350 km compared with its predecessor (Ventre, 2021).

10.4.2.1 Trains

Trains are increasingly using composites. Of course, composites are being used in railway carriages already for many nonstructural components to reduce weight: interior sidewalls, toilet modules, driver's desks, luggage racks, outer protection of the

Figure 1.1.57 The Airlander 10, designed and built by Hybrid Air Vehicle Company (2016).
Airlander by Mikes Photos, CC0, via Wikimedia Commons.

Figure 1.1.58 The composite front end of the French Eurostar and the German ICE high speed trains.
Eurostar en ICE by Rob Dammers, CC BY 2.0, via Wikimedia Commons.

undercarriage, aerodynamic roof hoods, air ducts ... Also the semistructural sandwich floor panels have gained general acceptance, and the eye-catching composite nose adds an important aerodynamic function to high speed trains, like France's Eurostar and TGV (Fig. 1.1.58), Germany's ICE, Japan's Shinkansen and since 2008 also the Chinese HSR (High Speed Rail). Different material combinations are being used: a carbon fibre composite skin over a PMI foam in the Shinkansen, whereas the French TGV and the German ICE use glass fibre composites.

According to an excellent review by Mistry et al. (2021), composites are also gradually being used in structural components of trains. Because the bogey and the car body structure account for 41 resp. 24% of the whole wagon weight, it is logical that these two have been experimented extensively. One of the first carbon composite train bogeys is the CAFIBO (UK, 2019), which uses an interesting combination of recycled and virgin carbon fibres (at a 50/50 rate) (Crosbee et al., 2020), Fig. 1.1.59. In 2022, the Spanish train builder Talgo claims that a weight reduction of 50% has been realised in their carbon fibre composite train bogey, compared with a standard steel structure (Talgo, n.d). In 2014, Kawasaki started the in-use evaluation by different Japanese railway companies of carbon fibre leaf springs for their efWING bogeys; they started the commercialisation in 2016 (Kawasaki, 2016).

The next significant step is the use of carbon fibre composites in the train wagon body structure. Mistry et al. (2021) describe the stepwise innovations, from Schindler's three wagon prototypes (Switzerland, 1995) via the Korean Tilting Train

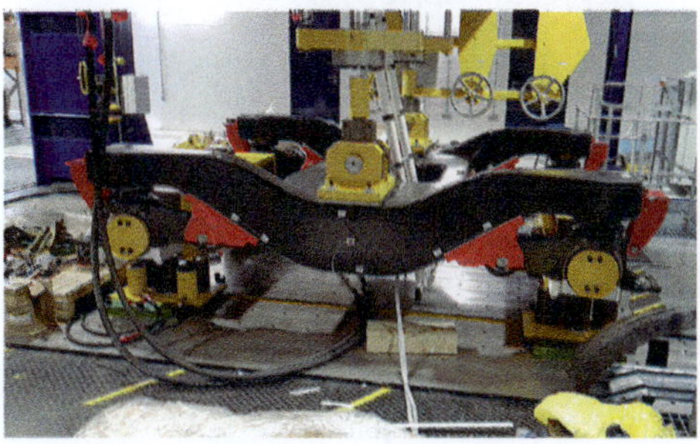

Figure 1.1.59 CAFIBO carbon composite train bogey, on test bed at Huddersfield University (UK) (Crosbee et al., 2020).

eXpress (TTX, 2001) up to the most recent CG Rail wagon, a Chinese-German development (2018). The Korean TTX used autoclave produced carbon fibre/aluminium honeycomb panels, resulting in a 38% weight reduction, whereas CG Rail used large multichamber carbon fibre composite pultrusions, realising a 30% weight reduction. Also in light rail, monorail and metropolitan train systems, carbon and glass composites are increasingly used as structural materials.

10.4.2.2 Trucks and trailers

Already in 1996, Jan Verhaeghe, a Belgian engineer and entrepreneur, built a prototype and tested a full composite trailer, and brought it to the market in 2001. This "Composittrailer" (Fig. 1.1.60a), is built up of adhesively bonded pultruded GFRP profiles. Floors and sidewalls are made from sandwich panels, produced in a self-developed continuous process by adding glass fibre layers to a foam core, and stitching them through the foam. This piece is then drawn through a mould, where it is impregnated with a polyester or vinyl ester resin and then left to harden (Verpoest, 2018).

Fifteen years later, Wabash (2015, USA) has a trailer on the road which still uses mainly glass fibre composites, while several projects in Europe and the United States are developing even lighter trailers using carbon fibre composites. The changing consumer habits (home delivery of goods) sparks a strongly increasing demand for small vans (<3.5 tons) for local transport (2 million vans extra per year, only for Europe). The German company Carbon TT developed in 2018 a part of the chassis, connecting the cab with the rear wheels, weighing only 30 kg, compared to 180 kg for the steel version, at approximately the same purchasing cost (Carbon TT, n.d.).

A next step are the full electric trucks, like the Volta Zero (Fig. 1.1.60b), a large commercial vehicle designed for city-centre deliveries. Production started in 2023, using BComp's flax composites for stiff and lightweight body panels (BComp, 2020a, b). By 2024, 15 trucks are on the European roads for testing. After a filing for bankruptcy in October 2023 (because its battery supplier Proterra (USA) also went bankrupt), the company is now re-emerging and restarted production in 2024, in collaboration with Steyr Trucks in Austria (Zsofia, 2024).

a)  b) 

Figure 1.1.60 (a) The world's first full composite trailer, designed by Jan Verhaeghe, produced at Acrosoma in 2001, (b) the Volta Zero electric truck.
(b) Volta by Alexander-93, CC BY-SA 4.0, via Wikimedia Commons.

10.4.2.3 Cars

Carbon fibre composites are (finally) realising a breakthrough in cars. First, there is the rather massive use of composite components. Examples are the CFRP drive shafts (more than one million produced by Toray since 2006), the GFRP leaf springs in many SUV's and small vans, the natural fibre door trims in many cars, the first CFRP wheels on Ford Mustang GT350R (2015) etc. …

The first full composite cars are, not surprisingly, found in the higher price range: the Mercedes McLaren SLR (2004), the Toyota Lexus LFA (2005–2010), the Lamborghini Aventador (2011) and many others. An increasing number of car manufacturers is using hybrid structures, combining the advantages of steel (or aluminium) and (mostly carbon) composites. Striking examples are the different components in the BMW 7 series (since 2015): B-pilars, roof beam, central channel. Further lightweighting, combined with reducing the carbon footprint, has been pioneered by BComp (BComp, 2023), as they co-developed since 2018, together with several car manufacturers (Volvo, Polestar, BMW, Porsche, KIA, …) not only car interior parts but also semistructural parts like hoods, bumpers and seats using several types of flax fibre reinforced composites (BComp, 2020a,b).

The electric, battery driven BMW i3 (2013) had a full CFRP "life module" on an aluminium base that also contains the batteries. Up to the end of 2021, 220.000 cars have been produced, and expectations were that this would be the real breakthrough for full CFRP cars. But on the short term, the multimaterial concept (steel + aluminium + CFRP) seems to be the preferred solution for most car constructors.

However, many of the experimental electric vehicles, like the Uniti (Sweden, 2018) are full composite cars, to compensate for the rather heavy batteries and to achieve a longer drive range. Unfortunately, Uniti was heavily hit by the COVID pandemic, and had to file for insolvency in April 2022. The futuristic Aptera 3-wheeler (Aptera, n.d.) is a partially solar energy driven, extremely aerodynamic EV car under development in California (Fig. 1.1.61a). Its sleek design is inspired by

a) b)

Figure 1.1.61 Solar cars: (a) the Aptera solar electric car (USA, copyright Aptera), (b) the Infinite of the Innoptus Solar Team of KU Leuven (Belgium) winning the Bridgestone World Solar Challenge in Australia in 2023 (Innoptus Solar Team, n.d.).

the full composite solar cars, built by university teams and worldwide competing in the solar car races. The bi-annual world championship in Australia, the Bridgestone World Solar Challenge, 3021 km from Darwin to Adelaide, was won in 2023 by the team of KU Leuven (Belgium) with their Infinite car (Fig. 1.1.61b) (Innoptus Solar Team, n.d.).

Prototype electric cars continue to propose designs using carbon fibre composites in monocoque type load carrying structures, like the ItoP (Iron to Polymer) electric concept car of Toray Carbon Magic (n.d.). Most other electric vehicle manufacturers are using composites for specific components to compensate for the weight of the battery like the "dragon fly" on the central part of the Polestar One to improve the torsional stiffness (Francis, 2018).

Battery casings are another breakthrough application of composites, as they combine lightweight with corrosion and fire resistance. Recently, Audi developed, together with Polytec Group (The Netherlands) a thermoplastic sandwich panel design for the underride guard, protecting the battery, in the Audi Q8 e-tron (JEC Composites, 2023)˙ A step further is the use of "structural batteries." The structural frame and/or the body panels are carbon composites which at the same time act as batteries (using special matrix and fibre materials), but these technologies are still in an experimental phase (Pandey et al., 2022).

Whereas it is not entirely sure whether hydrogen will be widely used as fuel in personal cars, this technology has a solid future in trucks and busses. Carbon composite hydrogen pressure vessels are hence considered to be one of the future big consumers of carbon fibres. According to Stratview (2023) the market for hydrogen pressure vessels could grow at a CAGR of 50%, reaching US$ 19.8 billion in 2030.

10.4.2.4 Motorbikes

Finally, also motorbikes turned to composites: in 2009, Alan Jenkins designed the Ducati Desmosedici GP9 for the race circuit, followed to road bikes Ducati 1299 Superleggera and BMW HP4 in 2016. The first full carbon fibre composite electric motorbike Saroléa was built in 2014 by the Belgian brothers Robbens (Fig. 1.1.62) (Sarolea, n.d.), for participating at the Isle of Man races. Natural fibre composites are being used in the monocoque structure of the Dutch Be.e electric scooter (Dezeen, 2013).

Figure 1.1.62 Saroléa electric bike (2014) with full carbon fibre composites monocoque frame and swing arms. Courtesy Saroléa.

10.4.3 Consumer goods

10.4.3.1 Sports

In sports, the use of composites has gradually increased, with a remarkable innovation coming from natural fibres: their unique combination of stiffness, lightness ànd vibration damping capacity has inspired many companies to use them in bicycles (Museeuw, 2009), tennis rackets (Artengo, 2010), and a growing number of skis, surfboards, skateboards and so on. The vibration damping characteristics of flax fibre composites has also been used by Tim Duerinck (Duerinck et al., 2019), when designing and manufacturing his composite violins and cellos. In an interesting multidisciplinary study, Duerinck et al. (2020) stated that "*composite materials allow the creation of violins with a large diversity in sounds and therefore offer possibilities to change the sound to the criteria of the player.*"

All road racing bicycles are nowadays made of composites, as has been explained in previous sections of this chapter, and more extensively in section 2.4 on the "composites in sports" Chapter 4.4 of Part II of this book. However, for the same reason as in cars, a trend can be observed to use composites also for electric bicycles. By using carbon fibre composites, KiBiSi designers were able to reduce the weight of the Biomega electric bike to only 18 kg (Denmark, 2015; Brake, 2015), including the battery pack. Even in folding (electric) bikes like the Humingbird by UK designer Petre Craciun in 2016; Malnati (2018), composites enabled the realisation of the lightest bike in its category … at that time.

Since then, further improvements in battery and electric motor technologies, hand-in-hand with optimisations of the carbon fibre composites frame, have enabled designers to drive the weight down even to only 8.5 kg for the HPS Domestique bike (2021) (Hughes, 2021), where the motor and electronics are hidden in the frame, and the battery has the shape of a drinking bottle, positioned in one of the two clamps where one normally finds the bottle!

The hand lay-up and autoclaving manufacturing technique, used in bicycle manufacturing, is labour intensive, as extensively discussed in (Velo, 2023) and hence most composite bicycle frames are produced in Asian countries like Taiwan, China and Vietnam. The Belgian company Rein4ced (Rein4ced, n.d.) realised a triple innovation in 2019: use of thermoplastic matrix and hybrid carbon-steel fibre composites combined with a fully automated, robotised production (ATL, compression moulding). This allows them to reshore bicycle production to Europe, hence they started a plant in Belgium (Leuven) with an annual production capacity of 20.000 frames.

In all water related sports, lightweight and durability in watery environments are crucial. Gradually, carbon fibres have been replacing glass fibres in most boats for sailing, rowing and canoeing competitions, as will be explained in Chapter 4.4 of part II of this book. However, also in more leisure oriented boats, composite innovations continued to happen in the first decades of the new millennium. Canoeing addicts will be delighted with the ONAK foldable canoe (designer Otto Van de Steene, 2016) (Onak, n.d.), who used a full polypropylene sandwich panel, with a folded honeycomb sandwich core (Thermhex by Econcore) and skins made of the same self-reinforced PP composite as has been used by Samsonite in their successful composite suitcases.

10.4.3.2 Travelling

Not being a materials but a consumer goods company, Samsonite can be considered as one of the premium innovators in composites for consumer goods. The "self-reinforced" composite Curv, a polypropylene matrix (PP) reinforced with polypropylene fibres, was invented by Ian Ward and his research team at the University of Leeds (Ward and Hine, 1997). First, a PP film is stretched and cut into tapes, which gives them a higher stiffness (more than 15 GPa) and a higher toughness (elongation at break of more than 20%). The tapes are subsequently woven and pressed at a precise temperature, melting only the thin outer layer of the PP tapes and forming the PP matrix. That way this "self-reinforced" composite (SRPP) combines a high stiffness with a high toughness, properties which are essential in suitcase design.

Researchers at the Katholieke Universiteit Leuven (Belgium) collaborated with Samsonite to further develop this material, to optimise the layered structure, trying to find the best combination of stiffness and impact resistance, and to create different visual effects with colour and texture. A specific production method was invented for this new material: the Curv panels must be heated at a precise temperature and clamped in a controlled way before they are pressed into the shape of a suitcase (KU Leuven Stories, 2025). The Cosmolite suitcases (Fig. 1.1.63) were launched into the market in 2009, and are still among the lightest suitcases available. Together with follow-up models like C-Lite, more than four million have been produced since 2008.

Innovation did not stop there. In 2020, Samsonite developed, again in collaboration with KU Leuven, the Roxkin material (Samsonite, n.d.) (Kandinskaia, 2024) a multi-layered package of BOPP films (biaxially oriented polypropylene), achieving similar and even better properties compared with SRPP, resulting in the Proxis suitcases.

Figure 1.1.63 Composite suitcases: (a) Samsonite Cosmolite suitcase, (b) Samsonite Proxis suitcase.

10.4.3.3 Design

In the previous paragraphs, the word "designer" has popped-up quite often. This is not surprising. Designers are constantly looking for material innovations, to make their designs fit better to the needs of the consumer, but also to the growing ecological expectations of society. It is observed that, in the last decade, there is a renewed interest of designers in the unique potential of composites, as was already illustrated in the many examples above. But also in the field of furniture design, designers have been recently very innovative in experimenting with new composite materials and processes. This has been extensively described by Lut Pil in the catalogue for the exhibition "Fibre-Fixed. Composites in Design" (Pil, 2018a, p. 164–195).

The use of carbon fibre composites in chairs results in sometimes surprising designs. The innovative use of the recent production technologies like automated tape laying (ATL), automated fibre placement (AFP), overmoulding and mouldless curing of inflated structures, is mostly leading to small series productions.

Whereas such CFRP chairs, like the earlier ones of Alberto Meda (1987) and Pol Quadens (1995), are expensive, the availability of short fibre reinforced thermoplastics (SFRT) for injection moulding inspired designers like Philippe Starck and Jasper Morisson to produce, from 2000 on, "affordable" SFRT composite chairs. For his Air-Chair, Jasper Morisson (2000) uses a glass fibre thermoplastic composite in a gas assisted injection moulding process to further reduce the weight of the chair (Magis, n.d.), whereas Philippe Starck's Broom chair (Starck, 2012) is built from 90% recycled industrial waste, namely 75% waste polypropylene, 15% reclaimed wood fibre and 10% glass fibre. Together with Eugeni Quitllet, Starck then designed the Zartan chair (Starck, 2015), where the seat is a compression moulded hemp or jute fibres-recycled PP composite, and the legs are injection moulded using wood powder-recycled PP.

A step further was set by IKEA, when they started selling the Odger chair (2016) (Ikea, n.d.) designed by "From us with Love" and produced with recycled PP reinforced with recycled wood fibres (Fig. 1.1.64).

The emerging application of natural fibre composites in design is another trend of the last decade: the Flax Chair by Christien Meindertsma (2015), the Hemp Chair by Aislinger (2012) and the cute, flax composite baby cod for BBDor by Virginie Breton (2015), are all described in more detail by Pil (2018b) and by Pil et al. (2016). A

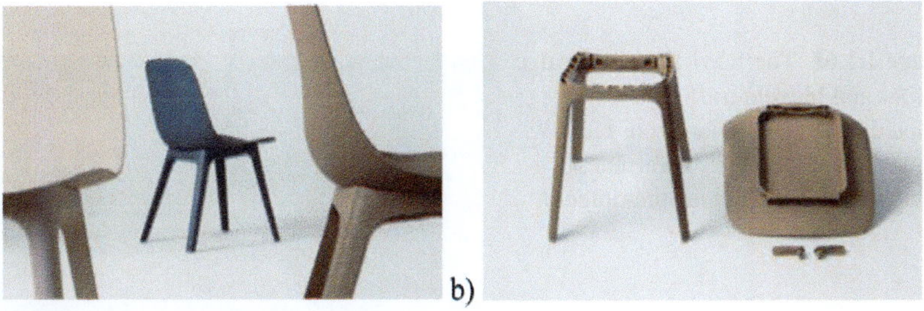
a) b)

Figure 1.1.64 Form Us With Love & Ikea, Odger Chair (2016); (a) the Odger Chair, (b) only three components!.
Picture: Jonas Lindström (http://jonaslindstrom.se/).

disruptive innovation, inspired by designers from the United States and Europe, is the use of (natural fibre reinforced) mycelium, maybe the most convincing example being the Mycelium Chair by Eric Klarenbeek (2013).

10.4.4 Construction

Since the turn of the millennium, composites have been gradually invading the world of buildings and construction. The 2017 JEC book "The future of building: The growing use of composites in construction and architecture" (Maefeld, 2017) illustrates perfectly which are the advantages of using composites in construction and architecture, whereas the 2019 publication "The future of infrastructure" (JEC Group, 2019) highlights recent developments (and future trends) in infrastructural applications of composites.

10.4.4.1 Bridges

For bridges, composites offer durability (less maintenance) and lightness (faster, less costly installation). After the initial experiments with pedestrian bridges in the '90s, the composite bridge industry has fully developed in the last two decades. Not only pedestrian bridges are being built by the hundreds per year, but also heavy-duty road bridges. To guarantee their durability, special GFRP constructions have been invented. Fibrecore (The Netherlands) has developed a remarkable technique that greatly reduces the risk of delamination and thus bridge instability following a possible impact (Fibrecore, n.d.). Foam cores are enveloped using unidirectional or bidirectional glass fibre mats, placed side by side in such a manner that the glass fibre mats overlap and thus create "natural" delamination stoppers. An example is the road bridge in Oosterwolde (The Netherlands, 2010), the first movable composite bridge in Europe (JEC Group, 2019, p. 60). By 2025, more than 1500 composite bridges have been manufactured and installed by Fibrecore.

The "freedom of form" composites offer to designers has also led to a number of striking bridge constructions. The Ooypoort bridge in Nijmegen (The Netherlands, 2014) spans 54 m and is extraordinarily tall to allow houseboats to pass under it. A very unique design was given to the Rhyl Harbour bridge for foot and bicycle traffic in Wales (Mc Callion R., 2016); it consists of two glass fibre composite spans of 30 m each, that can be drawn up from a central pillar. In its raised condition, the bridge resembles a dragon, hence the nickname (Fig. 1.1.65).

Figure 1.1.65 The Rhyl Harbour bridge for foot and bicycle traffic in Wales. Adopted with permission from https://www.themanufacturer.com/articles/composite-materials-enter-the-dragon/.

The first carbon fibre composite "stress ribbon" bridge, offering extreme weight reduction, has been installed in Cuenca (Spain) in 2011 (JEC Group, 2019, p. 48). Even flax fibres are used as reinforcement: after the first experimental (pedestrian) bridge in Eindhoven, end 2019, a flax fibre bridge for bicycles has been installed in Leeuwarden, completing the new ring road (Infra Composites, The Netherlands) (BComp, 2020a, b).

10.4.4.2 Infrastructure

In "The future of infrastructure" publication, a number of composite applications in rail infrastructure are presented (pages 68–75) like tunnel linings, railway platforms, cable troughs, canopies ... Also airports embraced composites: glass fibre composite airport fences have the advantage of being radar transparent. The corrosive environment of waterways and seas asks for corrosion resistant composite structures, like lighthouses, pontoon deckings, lock gates, bridge fenders etc ... (JEC Group, 2019, pp. 82–88). Similarly, the infrastructure for potable and waste water is using more and more composites (JEC Group, 2019, pp. 99–114).

Lightness and insensitivity to corrosive environments are two important reasons for the increasing use of (mainly pultruded) glass fibre composites in "protection" infrastructure like noise barriers, soil retaining walls, or blast walls (JEC Group, 2019, pp. 130–137).

Temporary structures, for instance for podium installations in the concert touring sector, benefit from the lightweight of composites. In 2017, PRG Projects developed the Infinity Spaceframe system for the Beyoncé and Jay-Z "On the Run II" Tour: LED panels are mounted on a carbon fibre based composite frame with an integrated carbon fibre composite wind bracing structure, replacing a heavier steel and aluminium structure (JEC Group, 2019, p. 142; PRG, 2018).

10.4.4.3 Architecture

All the advantages that have stimulated the use of composites in infrastructure, are equally valid for architectural applications. The use of composites for internal (rebars) and external (repair slabs) reinforcements of concrete started already in the '90s, and continues to expand, as is highlighted in the JEC book "The future of building: The growing use of composites in construction and architecture" (Maefeld, 2017). A next step has been realised in Germany, where in 2021 the University of Dresden has realised the first "carbon concrete" building, The Cube (Frearson, 2021). Here, the walls are much thinner than in conventional concrete, where the steel rebars have to be protected from corrosion by a concrete layer of minimum thickness to prevent diffusion of water toward the steel rebars. In "carbon concrete," meshes of (noncorroding) impregnated carbon rovings are directly embedded in concrete, resulting in significantly lower wall thickness and hence a reduced use of concrete.

In chapter "Architecture and the city: designing with composite materials" (Pil, 2018c), Lut Pil has extensively described the use of composites in architecture during the first 2 decades of the new millennium. Two groups of architectural applications can be distinguished—*the examples below are more extensively described in Lut Pil's chapter and/or in the JEC-book*—.

Figure 1.1.66 The Museum of the Future in Dubai.
Lyonerov at Wikipedia.

First, composites are used as façade elements, because of their lightness, freedom of shape, resistance against rain and wind, and the potential to integrate different functions (insulation, fire resistance, …). A complete list of advantages can be found in the above mentioned JEC publication, p. 20-21. The Museum of the Future in Dubai (Bains, 2019) (opened in 2020, Fig. 1.1.66) is a striking example of the versatility of composites, as each of the 1024 fire retardant composite panels has a different shape! Also the extension of the San Francisco Museum of Modern Art (2016, 700 different glass fibre composite profiles), the huge one-piece canopy of the Stedelijk Museum in Amsterdam (2012, carbon and aramid composites) and the 10 petal like volumes of the Singapore Arts Science Museum (2011) are stunning examples. The freedom of shape can also be realised by elastic deformation of (36 m long) pultruded glass composite profiles, as is illustrated in the "Gebouw T" in Gent (2019).

Second, composites are being used as well for the structural elements in buildings, either as a combination of beams or as fully load carrying walls. The roof of the Steve Jobs Theater on the Apple Campus in Cupertino (2017) consists of 44 identical carbon composite panels, forming a round disk of 42 m diameter. Because of its extreme lightness, it is supported by glass walls only.

At the University of Stuttgart, profs. Knippers and Menges have developed a coreless, robotised filament winding technique to create large, self-supporting structures. First realisations were the Elytra Filament Pavillion for the Victoria&Albert Museum in London (2012) and later the ICD/ITKE Research Pavillion for the Bundesgartenschau in Heilbronn (2019) (Menges and Knippers, 2019) (Fig. 1.1.67).

10.4.5 Energy

To increase the power output, the length of off-shore windmill blades has now increased to more than 80 m (Windpower, 2018), for instance in the MHI Vestas V164 reaching a power of 9.5 MW (Vestas, n.d.), and the Siemens Gamesa SG 8.0-167 DD, reaching 8 MW. Even longer blades are being built since 2021, like the Haliade-X off-shore windmill blade built by General Electric, 107 long and realising a power output of 12–13.6 MW (GE Vernova, n.d.) (see Fig. 1.1.68). The Siemens-Gamesa SG-14 236 DD with115 m long blades is delivering an output of

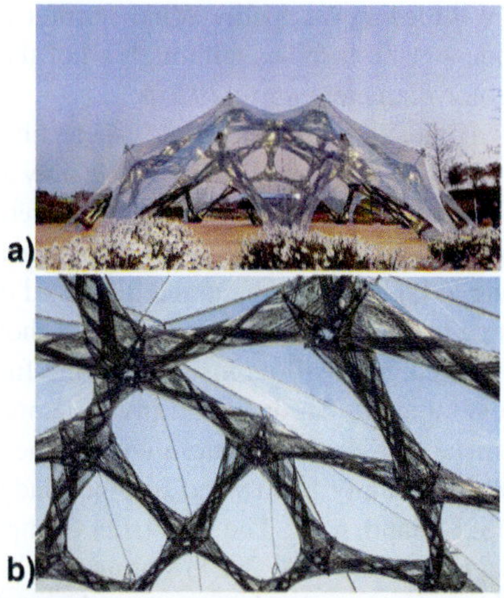

Figure 1.1.67 The ICD/ITKE Research Pavillion for the Bundesgartenschau (BUGA) in Heilbronn; (a) global view, (b) detail of the robotised filament wound structure. Reprinted with permission of ICD/ITKE University of Stuttgart.

Figure 1.1.68 GE's Haliade -X offshore wind turbine.
Haliade by Kees Torn, CC BY-SA 2.0 via Wikimedia Commons.

14 MW (Paleja, 2022)! Wikipedia (Wikipedia Wind Turbines, n.d.) presents a list of even bigger, mainly Chinese wind turbines, although all of them are to date in a prototype phase, reaching rotor diameters up to 292 m.

These extreme lengths require the additional use of carbon fibre composites (often pultruded) at certain critical places, to reduce blade bending, mainly at the tip area. An overall weight reduction helps reduce the centrifugal forces. Moreover, automated tape laying (ATL) or automated fibre placement (AFP) is being used increasingly to automate the production process, aiming at higher throughput and improved quality assurance.

An additional challenge is the end-of-life of the tens of thousands of wind turbine blades that have been installed, and will be installed in the future (SGS, 2024). Two approaches are under development. On one hand investments are underway to build recycling plants, as for instance Siemens-Gamesa's investment (Finland) for preprocessing postconsumer waste (mainly from wind turbine blades) toward recycling in cement kilns (Renewables Finland, n.d.). The European Composites Industry Association EUCIA published a report (EUCIA, 2025) stating that "*each ton of End-of-Life composite waste treated by co-processing in a cement plant delivers an average combined GHG emission reduction of about 0.83 ton of CO2 in comparison with waste incineration with energy recovery (the current common treatment for the majority of End-of-life composite waste)*."

Also new recycling technologies are being developed, as for instance the chemical recycling technology by Vestas in the framework of the CETEC project (Vestas, 2022). On the contrary, thermoplastic wind turbine blades, which are intrinsically easier to recycle, are under development as well: the ZEBRA project, guided by LM Wind Power (part of GE) has already produced (2022) a 62 m blade, using Arkema's infused thermoplastic prepolymer technology (Elium, 2025).

Once electricity is produced, it has to be transported over often large distances, using high voltage transmission. The distance between the cable supporting towers is limited by the sagging of the cables under their own weight. To reduce this effect, aluminium is being used as conducting material (instead of copper), but thermal effects still result in too large sagging. In 2008, Belgian inventor Clement Hiel filed a patent for "Aluminium Conductor Composite Core" (ACCC) (US Patent, 2015): In the centre of the cable, a pultruded carbon fibre composite rod is inserted, increasing the specific stiffness and hence reducing the sagging and the thermal effects on it.

10.5 Conclusion

As a conclusion, it can be stated that composites are, now more than ever, present in society, and that they are playing a positive role in the quest for a more sustainable world. Composites are providing lighter, hence more energy-efficient, and more durable products. They are indispensable to increase the use of renewable energy, as proven by the phenomenal increase in the number of electricity generating wind turbines. In an era where plastics are under pressure, because of their littering and potential pollution problems, this sheds a more positive light on the role of (fibre reinforced) plastics in society.

Acknowledgement

I would like to acknowledge first Nelly Baron *(VP at JEC World for Content, Marketing & Communication)* for the inspiring and constructive collaboration during the concept and writing phase of the preliminary, much shorter version of this text, published by JEC Group, and available at JEC's website (JEC Composites, n.d.). My colleagues Stepan Lomov and Yentl Swolfs at KU Leuven and former collaborator Bart Vangrimde (now at Huntsman, also VP of SAMPE Europe) have reviewed the preliminary JEC version of this text and provided essential feedback.

The extended version in this book had been made possible by the help of Panayiotis Tsokanas, taking care of correct referencing, detailed text corrections and preparing the permissions for figures, next to many other practicalities. The text has then been thoroughly reviewed by Stepan Lomov, who also provided new information on different subjects, not the least on carbon fibre developments in the USSR, the latter together with Sergei B. Sapozhnikov.

For specific subjects, I acknowledge the input of prof. Jim Thomason (Univ. Strathclyde, Scotland) on glass fibres, Fumihiko Tanaka and Ichiro Taketa (Toray) on carbon fibres, Hans Lilholt (retired from Risoe-Denmark) and Frans Van Hulle (retired from ECN-Petten, The Netherlands) on wind turbines and prof. Karel Van Acker (KU Leuven) on sustainability, and many others …

References

ACS, 2003. High Performance Carbon Fibres. American Chemical Society.

ACS, n.d. American Chemical Society National Historic Chemical Landmarks: High Performance Carbon Fibres. http://www.acs.org/content/acs/en/education/whatischemistry/landmarks/carbonfibers.html. Accessed 21 January 2025.

Advani, S.G., Sozer, E.M., 2010. Process Modeling in Composites Manufacturing, second ed. CRC Press. https://doi.org/10.1201/b13142.

Aernoudt, E., 2003. "Materialen maken geschiedenis" (Materials make history), WKW, Antwerpen-Köln, 64-68 (only available in Dutch) See also for a general overview of the history of composites: www.mar-bal.com/language/en/applications/history-of-composites. (Accessed 30 July 2024).

Ageyeva, Barani, Karger-Kocsis, 2019a. Composites. In: Karger-Kocsis, Barani (Eds.), Chapter in "Polypropylene Handbook". Springer, pp. 481–578.

Ageyeva, T., Sibikin, I., Kovács, J.G., 2019b. A review of thermoplastic resin transfer moulding: process modeling and simulation. Polymers 11 (10), 1555. https://doi.org/10.3390/polym11101555.

AGY, 2025. S2 Glass fibre, the strongest glass fibre in the world. https://www.agy.com/s2-glass/. (Accessed 28 January 2025).

Alein, J., Bhuvaneshwari, M., 2022. Textile-reinforced concrete sandwich panels: a review. ACI Struct. J. 119 (5), 207–216. https://doi.org/10.14359/51734899.

Alliance, n.d. Alliance for European Flax-Linen & Hemp. https://allianceflaxlinenhemp.eu/en. Accessed 21 January 2025.

AMFG, 2020. Composite 3D printing: an emerging technology with a bright future, https://amfg.ai/2020/02/25/composite-3d-printing-an-emerging-technology-with-a-bright-future/. Accessed 21 January 2025.

Ando, Y., 2010. Carbon nanotube: the inside story. J. Nanosci. Nanotechnol. 10 (6), 3726–3738.

Aptera, n.d. The Aptera Solar Vehicle. Available at: https://aptera.us/vehicle/. Accessed 21 January 2025.

Arhant, M., Davies, P., 2019. 2 – Thermoplastic matrix composites for marine applications. In: Marine Composites. Woodhead Publishing Series in Composites Science and Engineering, pp. 31–53. https://doi.org/10.1016/B978-0-08-102264-1.00002-9.

Arteiro, A., Furtado, C., Catalanotti, G., Linde, P., Camanho, P.P., 2020. Thin-ply polymer composite materials: a review. Compos. Appl. Sci. Manuf. 132, 105777. https://doi.org/10.1016/j.compositesa.2020.105777.

Avci, H., Hassanin, A., Hamouda, T., Kılıç, A., 2019. High performance fibres: a review on current state of art and future challenges. Eskişehir Osmangazi Üniversitesi Mühendislik Ve Mimarlık Fakültesi Dergisi 27 (2), 130–155. https://doi.org/10.31796/ogummf.537704.

Avrorova, L.V., Volokhina, A.V., Glazunov, V.B., Kudryavtsev, G.I., Makarova, R.A., Oprits, Z. G., Tokarev, A.V., Semenova, A.S., 1990. Third-generation man-made fibres produced in the USSR. Fibre Chem. 21 (4), 298–305. https://doi.org/10.1007/BF00556106.

Bacon and Schalamon, 1969. Quoted in Chapter VIII by Konkin in the Book of Watt, p. 315.

Bacon and Smith, 1965. Quoted in R.M. Gill (1972, p. 39).

Bacon, R., 1960. Growth, structure, and properties of graphite whiskers. J. Appl. Phys. 31 (2), 283–290. https://doi.org/10.1063/1.1735559.

Bacon, R., Moses, C.T., 1986. Carbon fibres, from light bulbs to outer space. In: Seymour, R.B., Kirshenbaum, G.S. (Eds.), High Performance Polymers: Their Origin and Development. Springer, Dordrecht. https://doi.org/10.1007/978-94-011-7073-4_33.

Bacon, R., Tang, M.M., 1964. Carbonization of cellulose fibres—II. Physical property study. Carbon 2 (3), 221–225. https://doi.org/10.1016/0008-6223(64)90036-3. Quoted in R.M. Gill's book, p. 38.

Bains, E., 2019. Museum of the Future: the building designed by an algorithm, www.bbc.com/future/article/20191028-museum-of-the-future-the-building-designed-by-an-algorithm; see also Gardiner (2020): Building the museum of the future, https://www.compositesworld.com/articles/building-the-museum-of-the-future. Accessed 4 February 2025.

Bakelite-Micarta (1916). Bakelite-Micarta gears, electrical age: Natl J. Electr. Pract.; New York, 49 (6) 56.

Bawden, J., 1990. The Art & Craft of Papier-Mâché. Grove Press, London. See also: Wikipedia page. https://en.wikipedia.org/wiki/Papier-mâché.

Bcomp, 2020a. Ritsumasyl Record-Breaking Bio-Composite Bridge. Available at: https://www.bcomp.com/news/ritsumasyl-record-breaking-bio-composite-bridge/. (Accessed 4 February 2025).

BComp, 2020b. Volta Zero made with Bcomp's ampliTex™ and powerRibs™, available at: https://www.bcomp.ch/news/volta-zero-made-with-bcomps-amplitex-and-powerribs/. Accessed 21 January, 2025.

BComp, 2023. Sabelt – Driving Sustainability in Car Seats. https://www.bcomp.com/news/sabelt-driving-sustainability-in-car-seats/. (Accessed 21 January 2025).

Beaumont, P.W.R., Soutis, C. (Eds.), 2016. The Structural Integrity of Carbon Fibre Composites – Fifty Years of Progress and Achievement of the Science, Development, and Applications. Springer, London, p. 953.

Bellis, M. 2020. A Brief History of the Invention of Plastics. Available at: https://www.thoughtco.com/history-of-plastics-1992322, last accessed 5 August 2024.

Bensaude-Vincent, B., 2011. The concept of materials in historical perspective. NTM Zeitschrift für Geschichte der Wissenschaften, Technik und Medizin 19, 107–123. https://doi.org/10.1007/s00048-010-0043-0. Also in Palucka and Bernadette Bensaude-Vincent (2002).

Bensaude-Vincent, B., Stengers, I., 1993. Histoire de la Chimie, Paris: La Découverte. Translated in English: History of Chemistry, vol 1997. Harvard University Press.

Bioplastics Europe, n.d. Bio-plastics. Available at: https://bioplasticseurope.eu/about. Accessed 21 January 2025.

Brake, A. G. 2015. 'KiBiSi introduces lightweight OKO electric bicycle for Biomega' in dezeen, 30 October, 2015, https://www.dezeen.com/2015/10/30/kibisi-oko-lightweight-electric-bicycle-biomega/.

British Patent 1110 791, 1968. Quoted in R.M. Gill (1972, p. 57).

Brosius, D. n.d. Personal communication: "This was in continuous production on all Corvette models until 2019, with the introduction of the C8 version for 2020, which reverts to coil suspension."

Brown, W.F., 1999. Vulcanized fibre-an old material with a new relevancy. Proceedings: Electrical Insulation Conference and Electrical Manufacturing and Coil Winding Conference (Cat. No. 99CH37035), Cincinnati, OH, USA, pp. 309–312. https://doi.org/10.1109/EEIC.1999.826229.

Bucknall, C.B., Partridge, I.K., 1983. Phase separation in epoxy resins containing polyethersulphone. Polymer 24 (5), 639–644. https://doi.org/10.1016/0032-3861(83)90120-9.

Burgio, V., Civera, M., Rodriguez Reinoso, M., Pizzolante, E., Prezioso, S., Bertuglia, A., Surace, C., 2022. Mechanical properties of animal tendons: a review and comparative study for the identification of the most suitable human tendon surrogates. Processes 10, 485. https://doi.org/10.3390/pr10030485.

Carbon TT, n.d. CarbonTT substitutes weight with utility – passionately, available at https://www.carbontt.com/en/. Accessed 4 February 2025.

CELC and JEC, 2012. Flax and Hemp Fibres: A Natural Solution for the Composites Industry, published jointly by CELC and JEC (Paris), 98 pp.

CELC and JEC, 2018. Flax and Hemp Fibre Composites: a Market Reality published by CELC and JEC, Paris, updated in https://proceedings.jeccomposites.com/proceedings/jec-world-2022/JW22_Proceedings_CELC_03-05-2022.pdf. See also: https://eucia.eu/jec-and-eucia-launch-2024-composites-sustainability-report/. Accessed 21 January, 2025.

Byrne, L., 2023. The world's only spun glass dress. https://lynnbyrne.substack.com/p/the-worlds-only-spun-glass-dress. (Accessed 9 February 2025).

Cheilakou, E., Liarokapi, N., Koui, M., 2012. Non destructive characterization by FOM and ESEM-EDX of ancient glass objects from the Aegean with an approach of the manufacturing technique. Mater. Struct. 45, 235–250. https://doi.org/10.1617/s11527-011-9762-5.

T.-W. Chou (ed.), 2000. Chapters 3 to 5 in Comprehensive Composite Materials, vol. 1. Pergamon Press, 107–174.

Coldwell, A., 2006. The History of the Surfboard. Available at: https://clubofthewaves.com/feature/history-of-the-surfboard/, Accessed 1 February 2024.

Composites World, 2023. Brandt Goldsworthy: Composites Visionary. Available at: https://www.compositesworld.com/articles/brandt-goldsworthy-composites-visionary. Accessed 7 February 2025.

Composites World, 2017. Class A composites: A history. Available at: https://www.compositesworld.com/articles/class-a-composites-a-history. Accessed 21 January 2024.

Coriolis, n.d. Available at: Coriolis Composites: The reference in Automated Fibre Placement, https://www.coriolis-composites.com/. Accessed 21 January 2025.

Cox, H.L., 1952. The elasticity and strength of paper and other fibrous materials. Br. J. Appl. Phys. 3 (3), 72–79. https://doi.org/10.1088/0508-3443/3/3/302.

CPC, 1992. Les Composites Dans l'Art. Centre de Promotion des Composites, Paris, p. 208.

Crosbee, D., Rothwell, E., Iwnicki, S., 2020. Developing a carbon fibre railway bogie for passenger trains. Available at: https://www.globalrailwayreview.com/article/102360/carbon-fibre-bogie-passenger-trains-irr/. Accessed January 21 2025.

Daenens, L., Bucquoye, M., 2002. Composites and furniture design, from Leo Baekeland to Maarten Van Severen. In: Verpoest, I., Daenens, L., Bucquoye, M.E. (Eds.), From Bakelite to Composite. Stichting Kunstboek, pp. 133–187.

Darlow, J., 2021. Mountain bike history: world's first carbon monocoque mountain bike. Available at: https://www.mbr.co.uk/news/bike_news/lotus-sport-336454. Accessed 1 January 2025.

De Ghé, P.1922. Patent US 1462446A.

Dezeen, 2013. Be.e electric scooter by Waarmakers. Available at: https://www.dezeen.com/2013/07/13/be-e-electric-scooter-plants-waarmakers/. Accessed 21 January 2025.

Diefendorf, R.J., 2000. Pitch-based carbon fibres. In: Chou, T.-W. (Ed.), Comprehensive Composite Materials, vol. 1. Pergamon Press, pp. 35–83. Chapter 2.

Diefendorfer, R.J., 2000. 1.02 – Pitch precursor carbon fibres. In: Kelly, A., Zweben, C. (Eds.), Comprehensive Composite Materials. Pergamon, pp. 35–83. https://doi.org/10.1016/B0-08-042993-9/00041-3.

Donnet, J.B., Bansal, R.C., 1984. "Carbon Fibres", Volume 3 of the "International Fibre Science and Technology Series". Marcel Dekker Inc., New York and Basel, p. 291.

Duerinck, T., et al., 2019. Music instruments from carbon, glass or flax fibres. JEC Compos. Mag. 129, 63–65.

Duerinck, T., Verberkmoes, G., Fritz, C., Leman, M., Nijs, L., Kersemans, M., Van Paepegem, W., 2020. Listener evaluations of violins made from composites. J. Acoust. Soc. Am. 147 (4), 2647–2655. https://doi.org/10.1121/10.0001159.

Dumarçay, J., 2005. Mud Bricks, Chapter 4 in "Construction Techniques in South and South-East Asia. Brill, Leiden-Boston, pp. 47–58. ISBN 90-04-14126-X.

Edison, T. A.,1879. Electric lamps, US Patent 223 898. Quoted in R.M. Gill's book Carbon Fibres in Composite Materials, 1972, p. 13.

Ekuase, O.A., Anjum, N., Eze, V.O., Okoli, O.I., 2022. A review on the out-of-autoclave process for composite manufacturing. J. Compos. Sci. 6 (6), 172. https://doi.org/10.3390/jcs6060172.

Elium, 2025. 100% Recyclable: World's Largest Thermoplastic Wind Turbine Blade Made in Spain. Available at: https://www.oedigital.com/news/495107-100-recyclable-world-s-largest-thermoplastic-wind-turbine-blade-made-in-spain. Accessed 21 January 2025.

Elium, n.d. Elium – The recyclable liquid thermoplastic resin for all composites. Available at: https://www.arkema.com/files/live/sites/shared_arkema/files/downloads/products-documentations/Brochure_ELIUM_2022-BD.pdf. Accessed 21 January 2025.

Emery, V.L., 2011. Mud-brick. In: W. Wendrich (ed.), UCLA Encyclopedia of Egyptology, Los Angeles, Accessed 26 July 2024, https://escholarship.org/uc/item/7v84d6rh.

Engelsmann, S., Spalding, V., Peters, S., 2010. Plastics in Architecture and Construction. Birkhäuser, Basel.

Epstein and Heimerdinger, 2003. Celebrating the 100th anniversary of the Wright brothers' first flight, SAMPE, USA (highlights in the history of composites and structural adhesives related to air and space flight, 1909-2002, very detailed, certainly for USA).

EUCIA, 2025. European Composites Circularity Solutions. Available at: https://eucia.eu/european-composites-circularity-solutions-january-2025-edition/. Accessed 4 February 2025.

European Bioplastics, n.d. Bioplastics – European Bioplastics e.V. Available at: https://www.european-bioplastics.org/bioplastics/. Accessed 21 January 2025.

Exel. Our Manufacturing Processes. In: https://exelcomposites.com/guide-to-composites/our-manufacturing-processes/. (Accessed 21 January 2025).

Ezekiel, H.M., Spain, R.G., 1967. Preparation of graphite fibres from polymeric fibres. J. Polym. Sci. Polym. Sympo. 19, 249–265. https://doi.org/10.1002/polc.5070190120. Cited by R.M. Gill (1972), p. 39.

Feldman, D., 2008. Polymer history. Des. Monomers Polym. 11 (1), 1–15. https://doi.org/10.1163/156855508X292383.

Fibrecore, n.d. Composite bridges and lock gates. Available at: https://www.fibercore-europe.com/en/. Accessed 4 February 2025.

Ford and Mitchel, 1963. US Patent 3107152A "Fibrous Graphite", Inventors Ford and Mitchell (Union Carbide). Available at: https://patents.google.com/patent/US3107152A/en. Accessed 27 June 2022.

Francis, S., 2018. Carbon fibre components featured in Polestar 1. Available at: https://www.compositesworld.com/news/carbon-fiber-components-featured-in-volvos-polestar-1. Accessed 21 January 2025.

Frearson, A., 2021. The Cube will be "world's first building made of carbon concrete." Available at: https://www.dezeen.com/2021/07/30/henn-tu-dresden-carbon-fibre-concrete-building/. Accessed 21 January 2025.

Frigione, M., Lettieri, M., 2020. Recent advances and trends of nanofilled/nanostructured epoxies. Materials 13 (15), 3415. https://doi.org/10.3390/ma13153415.

Futuro House, n.d., The Futuro House. Available at: https://thefuturohouse.com/welcome.html#google_vignette. Accessed 21 January 2025.

Galos, J., 2020. Thin-ply composite laminates: a review. Compos. Struct. 236, 111920. https://doi.org/10.1016/j.compstruct.2020.111920.

Garcia, E.J., Wardle, B.L., Hart, A.J., 2008. Joining prepreg composite interfaces with aligned carbon nanotubes. Compos. Appl. Sci. Manuf. 39 (6), 1065–1070. https://doi.org/10.1016/j.compositesa.2008.03.011.

Gaudet, J., 2014. Papyrus: The Plant that Changed the World – from Ancient Egypt to Today's Water Wars. Pegasus Books.

Gault-Williams, M. 2020. From wood to foam. Available at: https://legendary-surfers.blogspot.com/2020/07/from-wood-to-foam.html. Accessed 1 February 2024 (for more detailed information on Peterson's collaboration with Brandt Goldsworthy).

GE Vernova, n.d. Haliade-X offshore wind turbine. Available at: https://www.gevernova.com/wind-power/offshore-wind/haliade-x-offshore-turbine. Accessed 21 January 2025.

Geyer, R., 2020. A brief history of plastics. In: Streit-Bianchi, M., Cimadevila, M., Trettnak, W. (Eds.), Mare Plasticum – the Plastic Sea. Springer, Cham. https://doi.org/10.1007/978-3-030-38945-1_2.

GFRP Car Bodies, 2021. Cars With Fibreglass Bodies: All You Need To Know. Available at: https://piketransit.com/cars-with--fiberglass-bodies/. (Accessed 21 January 2025).

Gill, R.M., 1972. Carbon Fibres in Composite Materials. The Plastics Institute, London, p. 38. Contains several references to the work of Tamaru.

Giocosa, A., 2006. Composite parts in Renault cars: the past, the present and the future. Presented during 8th International Conference of Flow Processing (FPCM-8), Douai (France), 11-13 July 2006 (found in. https://ecm-academics.plymouth.ac.uk/jsummerscales/FPCM/FPCM08/FPCM8/papers/FPCM8_giocosa.pdf.

GIRIO, 1959. Atarashi tanso zairyo – kokuen sen'i no seizo" (New carbon material – Manufacture of graphite fibre). Daikoshi News (GIRIO News) 3 (11).

Glowacz, F., 2015. Carbon Fibres: History, Players and Forecast to 2020. JEC Group (Introductory chapter).

González, P.C., Pang, Y., Reigosa, P.D., Dimopoulos, E., Barreras, J.V., Schaltz, E., 2013. Tvindkraft: Implementing a 500 kW 21-IGBT-based frequency converter for a 1.7 MW wind power conversion system. In: International Conference on Renewable Energy Research and Applications (ICRERA), Madrid, Spain, pp. 482–487. https://doi.org/10.1109/ICRERA.2013.6749803.

Gorbatikh, L., Lomov, S.V., 2016. Nano-engineered carbon fibre-reinforced composites: challenges and opportunities. The structural integrity of carbon fibre composites. In: P. Beaumont, C. Soutis (Eds.), Fifty Years of Progress and Achievement of the Science, Development, and Applications. Springer, London, pp. 117–135. https://doi.org/10.1007/978-3-319-46120-5_6.

Gordon, J.E., 1978. Structures or Why Things Do Not Fall Down, vols. 80–84, Ingram Publisher Services. ISBN 9780306812835. "Unabridged replication" of the original edition of 1978, published 2003 by Penguin Books.

Gordon Cook, J., 1959. Handbook of Textile Fibres. Merrow Publ. Co., Watford.

Gordon Cook, J., 1984. Handbook of Textile Fibres, Vol. II – Man-Made Fibres. Woodhead Publ. Co.

Goutianos, S., Peijs, T., 2021. On the low reinforcing efficiency of carbon nanotubes in high-performance polymer fibres. Nanocomposites 7 (1), 53–69. https://doi.org/10.1080/20550324.2021.1917815.

Grossmann, R.A., 2002. Ancient Glass, A Guide to the Yale Collection. Yale University Art Gallery, New Haven (USA). ISBN 0894679554.

Gunston, W.T., 1969. Carbon fibres. Sci. J. (Lond.) 5 (2), 39–49.

Haka, A., 2011. Flügel aus „Schwarzem Gold. N.T.M. 19, 69–105. https://doi.org/10.1007/s00048-011-0047-4.

Haka, A.T., 2023. Engineered stability – the history of composite materials in the 19th and 20th Centuries. Springer. https://doi.org/10.1007/978-3-658-41408-5.

Hamilton, G.H., Sloper, T., 1914. UK patent No. 467.

Hansell, M. H., 1989. Wasp papier-maché. Nat. Hist. 8: 52. ISBN 9780855337681. See also www.discoverwildlife.com/how-to/identify-wildlife/wasp-nests-guide, Accessed January 21, 2025.

Hardy, R., 1976. Longbow, P. Stephens Publ., Cambridge, ISBN 10:0850592186. See also Beukers, A. and Z. van Hinte (1998). Lightness, 010 Publishers, Rotterdam, and: McMullen, P. (1984). Fibre/resin composites for aircraft primary structures: a short history, 1936–1984. Composites 15(3): 222–230 https://doi.org/10.1016/0010-4361(84)90279-9.

Harlin, A., 2011. Biogenic precursors for polyphenol, polyester and polyurethane resins. In: S. Pilla (ed.), Ch. 18 in the "Handbook of Bioplastics and Biocomposites. Engineering Applications." Wiley, 511–553.

Harper, A., 2009. RTM – past, present and future. Reinforc Plast 53 (8), 30–33. https://doi.org/10.1016/S0034-3617(09)70314-5.

Harte, A.M., Mc Namara, J.F., 2007. Overinjection of thermoplastic composites: I. Processing and testing of components. J. Mater. Process. Technol. 182 (1–3), 12–20. https://doi.org/10.1016/j.jmatprotec.2006.06.016.

Herakovich, C.T., 2012. Mechanics of composites: a historical review. Mech. Res. Commun. 41, 1–20. https://doi.org/10.1016/j.mechrescom.2012.01.006.

Hickman, 1959. Journal of the Society of Archer-Antiquaries, vol. 2 *(no pages).*

Hiken, A., 2018. The evolution of the composite fuselage: a manufacturing perspective. In: Aerospace Engineering, Intech. https://doi.org/10.5772/intechopen.82353.

Hiper-Tex, n.d. Technical data sheet for HiPer-tex® fibres from 3B Glass Fibres. Available at: https://www.3b-fibreglass.com/HiPer-tex. Accessed 30 January 2025.

Hoffman and Stoecklin, 1980. 737 Graphite Composite Flight Spoiler Flight Service Evaluation, NASA report NASA-CR-159094.

Hollaway, L.C., 2009. Polymer composites in construction: a brief history. Eng. Comput. Mech. 162 (3), 107–118. https://doi.org/10.1680/eacm.2009.162.3.107.

Hollaway, L.C., Head, P.R., 2000. 6.25 - composite materials and structures in civil engineering. In: Kelly, A., Zweben, C. (Eds.), Comprehensive Composite Materials. Pergamon, pp. 489–527. https://doi.org/10.1016/B0-08-042993-9/00118-2. https://www.compositesworld.com/news/dutch-bridge-claims-to-be-largest-single-span-composite-bridge-in-the-world.

Hughes, A.M., 2021. The world's lightest e-bike? Available at: https://road.cc/content/tech-news/hps-launch-lightest-e-bike-world-280881. Accessed 21 January 2025.

Humphreys, M., 2003. The use of polymer composites in construction. In: Yang, J. (Ed.), Proceedings of the CIB 2003 International Conference on Smart and Sustainable Built Environment. Queensland University of Technology, Australia, Queensland, Brisbane, pp. 1–9.

Hütter, W. (1943), German Patent A 20,029, Haupttragende Flügelschalenkörper insbesondere für Flugzeuge aus verleimten Holzfurnieren und Kunststoffplatten.

Hütter, U., n.d.b. Available at: Ulrich W. Hütter ist der Vater der Windenergie. Ihre Wiege steht in Stutgart (orsted.de). Accessed 21 January 2025.

Ikea, n.d. Odger Chair. Accessible at https://www.ikea.com/be/en/p/odger-chair-white-beige, and https://www.formuswithlove.se/work/odger/. Accessed 21 January 2025.

Innoptus Solar Team, n.d. Bridgestone World Solar Challenge. Available at: https://www.solarteam.be/en/races/bridgestone-world-solar-challenge, and video on https://www.youtube.com/watch?v=9cXu_uYNq8o.

Ishida, H., Koenig, J.L., 1987. Composite Interfaces: International Conference Proceedings. Elsevier Science Ltd, p. 404.

JEC Composites, 2023. BEV battery protection plate in composite design. Available at: https://www.jeccomposites.com/news/by-jec/best-of-2023-the-jec-composites-innovation-awards-winners-revealed. Accessed 4 February 2025.

JEC Composites, 2024. The largest thermoplastic aerostructures successfully manufactured as part of the Multi-Functional Fuselage Demonstrator project. Available at: https://www.jeccomposites.com/news/spotted-by-jec/the-largest-thermoplastic-aerostructures-successfully-manufactured-as-part-of-the-multi-functional-fuselage-demonstrator-project/. Accessed 2 February 2025.

JEC Composites, n.d. About composites. Available at: https://www.jeccomposites.com/discover-composites/about-composites/. Accessed 14 February 2025.

JEC Group, 2013. The challenging world of composites. Paris, ISBN 979190148529. Provides a good overview of different early manufacturing processes.

JEC Group, 2019. The Future of Infrastructure, vol 2019. JEC Group, Paris.

JEC Innovation Award, 2018. In: JEC Composites Magazine 119, Supplement Innovation Report.

JEC Observer, 2022. Current Trends in the Global Composites Industry 2012-2026. JEC World, Paris.

JEC Observer, n.d. Current Trends in the Global Composites Industry. Published yearly by JEC World, Paris.

Jiang, S., Chen, Y., Duan, G., Mei, C., Greiner, A., Agarwal, S., 2018. Electrospun nanofibre reinforced composites: a review. Polym. Chem. 9, 2685–2720. https://doi.org/10.1039/C8PY00378E.

Johnson, W., 1985. Chapter X. In: Watt, W., Perov, B.V. (Eds.), Strong Fibres. Elsevier Science Publ., p. 407

Jones, F.R., 1989. Interfacial phenomena in composite materials. In: Conference Proceedings of the Conference in Sheffield. Butterworth-Heinemann Ltd, p. 320.

Jones, F., 2016. Serendipity in carbon fibres: interfaces and interphases in composites. In: Beaumont, P., Soutis, C. (Eds.), The Structural Integrity of Carbon Fibre Composites, pp. 71–100.

Kandelbauer, et al., 2014. Unsaturated polyesters and vinylesters. In: Handbook of Thermoset Plastics. Elsevier, pp. 111–172.

Kawasaki, 2016. efWING—New-Generation Railway Bogie, Kawasaki Technical Review No. 177 32 September 2016.

Kandinskaia, A., et al., 2024. Time-dependency phenomena in biaxially oriented polypropylene laminates. In: Proceedings of ECCM-21, 1. Nantes, p. 174.

Kelly, A., 2000. Fibre composites: the weave of history. Interdiscip. Sci. Rev. 25 (1), 34–44. https://doi.org/10.1179/isr.2000.25.1.34.

Kelly, A., 2005. Very stiff fibres woven into history – very personal recollections of some of the British scene. Compos. Sci. Technol. 65 (15–16), 2285–2294. https://doi.org/10.1016/j.compscitech.2005.05.010.

Kelly, A., Tyson, W., 1965. Tensile properties of fibre-reinforced metals: copper/tungsten and copper/molybdenum. J. Mech. Phys. Solid. 13 (6), 329–338.

Kelly, A., Zweben, C., 2000. Comprehensive Composite Materials, vol. 6. Pergamon Press.

Kinloch, T., 2000. Norman de Bruyne. In: Biographical Memoirs Fellows of the Royal Society, vol 46, pp. 125–143.

Kobayashi, H., 1999. Recent trends in carbon fibre reinforced composites. J. Jpn Soc. Compos. Mater. 25 (1), 3–8 (in Japanese). Figure translated and redrawn (personal communication Tanaka Fumihiko, Toray, 2024).

Kotina, V.E., Konkin, A.A., Kosova, R.M., 1960. Patent 138324, filed 28.03.1960, published 27.11.1966, https://patentdb.ru/patent/138324.

Kotlensky, W.V., Titus, K.H., Martens, H., 1962. Young's modulus of hot-worked pyrolytic graphite. Nature 193, 1066–1067. Quoted in R.M. Gill (1972, p. 39).

KU Leuven Stories, 2025. On a journey with materials science. Available at: https://stories.kuleuven.be/en/stories/on-a-journey-with-materials-science. Accessed 4 February 2025.

Kudryavtsev, G.I., Tokarev, A.V., Avrorova, L.V., Konstantinov, V.A., 1975. The super-high tenacity high-modulus synthetic fibre SVM. Fibre Chem. 6 (6), 673–674. https://doi.org/10.1007/BF00547742.

Kumpfert, J., 2004. Competing Structure Technologies for NextGeneration AIRBUS Aircrafts. Presentation at KU Leuven.

Le Duigou, A., Correa, D., Ueda, M., Matsuzaki, R., Castro, M., 2020. A review of 3D and 4D printing of natural fibre biocomposites. Mater. Des. 194, 108911. https://doi.org/10.1016/j.matdes.2020.108911.

Lee, H., Neville, K., 1967. Handbook of Epoxy Resins. McGraw-Hill. Chapters I-2 to I-7.

Lekhnitskii, S.G., 1947. Anisotropic Plates (in Russian). OGIZ (ОГИЗ), Moscow – Leningrad.

Lekhnitskii, S.G., 1968. Anisotropic Plates, English translation by S. W. Tsai and T. Cheron, Gordon & Breach.

Lemstra, P.J., 2022. High-performance polyethylene fibres. Adv. Ind. Eng. Poly. Res. 5, 49–59. First of 7 chapters on PE and other high performance fibres in the same issue of this journal.

Lemstra, P. J., n.d. Patents by Inventor Pieter Jan Lemstra. Available at: https://patents.justia.com/inventor/pieter-jan-lemstra. Accessed 21 January 2025.

Li, J., Lemstra, P.J., Ma, P., 2022. Chapter 7: can high-performance fibres be(come) bio-based and also biocompostable? Adv. Ind. Eng. Poly. Res. 5 (2), 117–132. With an interesting epilogue by Lemstra (pp. 133–135).

Liew, K.M., Akbar, A., 2020. The recent progress of recycled steel fibre reinforced concrete. Constr. Build. Mater. 232, 117232. https://doi.org/10.1016/j.conbuildmat.2019.117232.

Lilium, 2020. Carbon composites and the Lilium Jet. Available at: https://lilium.com/newsroom-detail/carbon-composites-and-the-lilium-jet. Accessed 4 February 2025.

Lilium, 2024. Lilium achieves breakthrough in investor search, Available at: https://lilium.com/newsroom-detail/lilium-achieves-breakthrough-in-investor-search. Accessed 4 February 2025.

Little, A.D., 2018. What's Next for Aerospace Composites, report by Arthur D. Little Luxembourg S.A. Available at: https://www.adlittle.com/sites/default/files/viewpoints/adl_aerospace_composites-min.pdf. Accessed 21 January 2025.

Liu, Y., Zwingmann, B., Schlaich, M., 2015. Carbon fibre reinforced polymer for cable structures—a review. Polymers 7 (10), 2078–2099. https://doi.org/10.3390/polym7101501 (Open Access).

Lomov, S., 2019. Personal Communication. KU Leuven, Belgium.

Lotus Elite, n.d. Lotus Elite. Available at: https://en.wikipedia.org/wiki/Lotus_Elite. Accessed 21 January 2025. Also, Lotus Elite - Lotus Cars Media Site.

Lotus, n.d. Lotus 108. Available at: https://en.wikipedia.org/wiki/Lotus_108. Accessed 21 January 2025.

Lukaszewicz, D.H.-J.A., Ward, C., Potter, K.D., 2012. The engineering aspects of automated prepreg layup: history, present and future. Compos. B Eng. 43 (3), 997–1009. https://doi.org/10.1016/j.compositesb.2011.12.003.

Maefeld, A., 2017. The Future of Building. JEC Group, Paris.

Magis, n.d. Air Chair, Jasper Morisson. Available at: https://www.magisdesign.com/product/air-chair/?. Accessed 21 January 2025.

Mahapatra, N.N., Raichurkar, P., 2015. Processing of glass fibres in textile industries. Int J Text. Eng. Processes 1 (2), 35–38.

Mahltig, B., 2018. Basalt fibres. Chapter 9. In: Inorganic and Composite Fibres. Woodhead Publ., pp. 195–217

Malnati, P., 2018. Composites-intensive folding bike. Simplifying multi-modal transportation. Compos World. Available at: https://www.compositesworld.com/articles/composites-intensive-folding-bike-simplifying-multi-modal-transportation. Accessed 4 February 2025.

Marsh, G., 2006. 50 years of reinforced plastic boats. Reinforc Plast 50 (9), 16–19. https://doi.org/10.1016/S0034-3617(06)71125-0.

Matsui, J., 1997–2006. 25 Papers by Junichi Matsui in Kyoka Plastics, vol. 43, no. 5 to vol. 52, no. 12.

Matsui, J., 2010. Composite materials – past and future. In: Proceedings of the Composites Durability Workshop (CDW-15), Kanazawa. Available on. https://wwwr.kanazawa-it.ac.jp/MSRL/Proceedings_cdw/Session%2011/11-5%20Junichi%20Matsui.pdf. (Accessed 27 June 2022).

Mc Callion, R., 2016. Composite materials: enter the dragon. Available at: https://www.themanufacturer.com/articles/composite-materials-enter-the-dragon/. Accessed 4 February 2025.

May, C., 1988. Introduction to epoxy resins. In: Epoxy Resins, Chemistry and Technology, second. Toutledge, doi:10.1201/9780203756713.

McMullen, P., 1984. Fibre/resin composites for aircraft primary structures: a short history, 1936–1984. Composites 15 (3), 222–230. https://doi.org/10.1016/0010-4361(84)90279-9.

Meier, U., 2012. Carbon fibre reinforced polymer cables: why? Why not? What if? Arabian J. Sci. Eng. 37, 399–411. https://doi.org/10.1007/s13369-012-0185-6.

Meier, U., 2016. The life times of polymer composites in construction. AIP Conf. Proc. 1736 (1), 020040. https://doi.org/10.1063/1.4949615.

Menges, A., Knippers, J., 2019. BUGA Fibre Pavilion 2019. Available at: https://www.icd.uni-stuttgart.de/projects/buga-fiber-pavilion/. Accessed 21 January 2025.

Mercelis, J., 2012. Leo Baekeland's Transatlantic Struggle for bakelite: patenting inside and outside of America. Technol. Cult. 53 (2), 366–400.

Mercelis, J., 2020. Beyond Bakelite, Leo Baekeland and the Business of Science and Invention. The MIT Press, Cambridge and London. Chapter 7 Bakelite: Transatlantic Patenting, 137-160. Many more references to this invention, a.o. in Bensaude-Vincent&Sprengers, see above; nice short story in. https://www.k-online.com/en/News/Topic_of_the_Month/Topic_of_the_Month_-_Archive/July_August_2018:_Bakelite_111_years_old_.

Miller, J.D., Manson, J.-A., 2023. Mechanical performance of highly loaded tow reinforced hybrid-molded composite structures. Compos. Struct. 322, 117311. https://doi.org/10.1016/j.compstruct.2023.117311.

Min, J., et al., 2022. Fabrication processes of metal-fibre reinforced polymer hybrid components: a review. Adv. Compos. Hybrid Mater. 5, 651–678. https://doi.org/10.1007/s42114-021-00393-z.

Minchenkov, K., Vedernikov, A., Safonov, A., Akhatov, I., 2021. Thermoplastic pultrusion: a review. Polymers 13 (2), 180. https://doi.org/10.3390/polym13020180.

Mishnaevsky, L., Branner, K., Petersen, H.N., Beauson, J., McGugan, M., Sørensen, B.F., 2017. Materials for wind turbine blades: an overview. Materials 10 (11), 1285. https://doi.org/10.3390/ma10111285.

Mistry, P., Johnson, M., Galappaththi, U., 2021. Selection and ranking of rail vehicle components for optimal lightweighting using composite materials. Proc. Inst. Mech. Eng. F J. Rail Rapid Transit 235 (3), 390–402. https://doi.org/10.1177/0954409720925685.

Mitsuya, n.d. Low cost thin-ply carbon fibre reinforced thermoplastic prepreg. Available at: https://seeds-eng.cfrphwy.jp/Seed/Details/bb78a88e-9d59-4fa4-98a3-5e585089ab18. Accessed 21 January 2025.

Modulo, n.d. Modulo transportation design. Available at: https://www.behance.net/gallery/22944575/Modulo-Bus. Accessed 21 January 2025.

Moreton, R., Watt, W., Johnson, W., 1967. Carbon fibres of high strength and high breaking strain. Nature 213, 690–691. https://doi.org/10.1038/213690A0.

Mori, T., Tanaka, K., 1973. Average stress in matrix and average elastic energy of materials with misfitting inclusions. Acta Metall. 21 (5), 571–574. https://doi.org/10.1016/0001-6160(73)90064-3.

Morris, J., Blier, S.P., 2004. Butabu: Adobe Architecture of West Africa. Princeton Architectural Press.

Mukai, U., Morii, T., 2000. 6.32 – flexible composites: tire and belt. In: Kelly, A., Zweben, C. (Eds.), Comprehensive Composite Materials. Pergamon, pp. 625–644. https://doi.org/10.1016/B0-08-042993-9/00121-2. Quite complete information on the history and recent developments of different types of tires can be found in en.wikipedia.org/wiki/Tire. A well-illustrated historical overview from the viewpoint of Michelin can be found in. www.michelin.com/en/group/heritage.

Murase, H., Yabuki, K., 2016. High-performance and specialty fibres, ed. The Society of Fibre Science and Technology Japan, Springer, chapter "History of Super Fibres: Adventures in Quest of the Strongest Fibre", p. 83–94. Several chapters in this book cover the history and properties of most high performance fibres.

Nakamura, O., Ohana, T., Tazawa, M., Yokota, S., Shinoda, W., Itoh, J., 2009. Study on the PAN carbon fibre innovation for modelling a successful R&D management. Synthesiology (English version) 2(2): 154–164. Supplement to this paper published in Synthesiology (English edition), vol. 4, n°2, pp.119–123 (oct 2011).

Naser, M.Z., Hawileh, R.A., Abdalla, J.A., 2019. Fibre-reinforced polymer composites in strengthening reinforced concrete structures: a critical review. Eng. Struct. 198, 109542. https://doi.org/10.1016/j.engstruct.2019.109542.

Nayyar, R., 2022. The Spun-Glass Dress That Made a Splash at the World's Fair. Article on website Hyperallergic, available at: https://hyperallergic.com/786155/the-spun-glass-dress-that-made-a-splash-at-the-worlds-fair/. (Accessed 28 December 2024).

Negoita, C., Cristache, N., Bodor, M., 2016. The epoxy resin – history and perspectives. Mater. Plast. 53 (3). Also in Palucka and Bernadette Bensaude-Vincent (2002).

Nehls, 2021. Airlander 10: "The future of zero-carbon aviation". Available at: https://www.compositesworld.com/articles/airlander-10-the-future-of-zero-carbon-aviation. Accessed 4 February 2025.

Nehls, G., 2023. Joby begins final assembly of company conforming aircraft. Accessible at: https://www.compositesworld.com/news/joby-begins-final-assembly-of-company-conforming-aircraft. (Accessed 21 January 2025).

Nelson, R., 2001. "All-Composite Bicycle Frames - Past, Present, and Future" Ron Nelson, ClosedMold Composites (Formerly President of Radius Eng.) March 22nd, 2001 |SAMPE Utah, can be found via All-Composite Bicycle Frames -Past, Present, and Future | PPT (slideshare.net).

Nelson, R., 2003. Bike frame races carbon consumer goods forward. Reinforc Plast 47 (7), 36–40. https://doi.org/10.1016/S0034-3617(03)00728-8.

Nippon Steel, 1993. Structure and properties of pitch-based carbon fibres. Nippon Steel Technical report No. 59, 1993.

Nittobo, n.d. Nittobo Flat Fibres Center. Available at: https://www.specialchem.com/polymer-additives/center/nittobo-flat-glass-fibers (accessed 21 January 2025). Or: Owens-Corning: https://dcpd6wotaa0mb.cloudfront.net/mdms/dms/CSB/10023976/10023976-Higher-Performance-Composites-Sell-Sheet.pdf (accessed 21 January 2025).

North Thin Ply, n.d. Our story - North Thin Ply Technology. Available at: https://ntpt.tech/our-story/. Accessed 21 January 2025.

Novoselov, K.S., Geim, A.K., Morozov, S.V., Jiang, D., Zhang, Y., Dubonos, S.V., Grigorieva, I.V., Firsov, A.A., 2004. Electric field effect in atomically thin carbon films. Science 306 (5696), 666–669. https://doi.org/10.1126/science.1102896.

Obande, W., Ó Brádaigh, C.M., Ray, D., 2021. Continuous fibre-reinforced thermoplastic acrylic-matrix composites prepared by liquid resin infusion – a review. Compos. B Eng. 215, 108771. https://doi.org/10.1016/j.compositesb.2021.108771.

Odagiri, N., Muraki, T., Tobukuro, K., 1988. Proc. 33th International SAMPE Symposium, vol 1988, p. 272.

Odagiri, N., Kishi, H., Nakae, T., 1991. Proc. Of the 6th Technical Conference of the American Society for Composites, p. 43.

Onak, n.d. ONAK Cannoe and kayak. Available at: https://www.onakcanoes.com/. Accessed 21 January 2025.

Oron, A., et al., 2015. Early marine activity on the Dead Sea: bitumen harvesting and the possible use of reed watercraft. J Mari Arch 10, 65–88.

Osorio, L., 2022. Microstructural Analysis and Mechanical Behaviour of Bamboo Fibres and Bamboo Fibre Composites. KU Leuven, p. 212. PhD Thesis.

Osorio, L., Trujillo, E., Van Vuure, A.W., Verpoest, I., 2011. Morphological aspects and mechanical properties of single bamboo fibres and flexural characterization of bamboo/epoxy composites. J. Reinforc. Plast. Compos. 30 (5), 396–408. https://doi.org/10.1177/0731684410397683.

Ōtani, Sugio, 1965. On the carbon fibre from the molten pyrolysis products. Carbon 3, 31–34.

Ōtani, S., Yamada, K., Koitabashi, T., Yokoyama, A., 1966. On the raw materials of MP carbon fibre. Carbon 4 (3), 425–432. https://doi.org/10.1016/0008-6223(66)90055-8.

Owens Corning, 1997. Owens Corning Touts Advantex Glass Fibre. Available at: https://www.plasticsnews.com/article/19970210/NEWS/302109969/owens-corning-touts-advantex-glass-fiber. (Accessed 21 January 2025).

Owens Corning and AGY, 2014. Owens Corning, AGY Establish Technology and Service Relationship to Provide High-Performance S-Glass Reinforcements. Available at: https://newsroom.owenscorning.com/all-news-releases/news-details/2014/Owens-Corning-AGY-Establish-Technology-and-Service-Relationship-to-Provide-High-Performance-S-Glass-Reinforcements/default.aspx. (Accessed 21 January 2025).

Paleja, A., 2022. Siemens Gamesa to build the world's longest wind turbine blades. Available at: https://interestingengineering.com/innovation/siemens-gamesa-wind-turbine-blade-377-feet Siemens Gamesa to build the world's longest wind turbine blades. 377 feet long? (interestingengineering.com). Accessed 21 January 2025.

Palucka and Bernadette Bensaude-Vincent, 2002. Materials Research Activities, History of composites – overview. Available at https://authors.library.caltech.edu/5456/1/hrst.mit.edu/hrs/materials/public/composites/Composites Overview.htm, last accessed 2 August 2024.

Pandey, D., Kumar, K.S., Henderson, L.N., Suarez, G., Vega, P., Salvador, H.R., Roberson, L., Thomas, J., 2022. Energized composites for electric vehicles: a dual function energy-storing supercapacitor-based carbon fibre composite for the body panels. Small 18, 2107053. https://doi.org/10.1002/smll.202107053.

Patents PES, n.d. See, for instance, patents GB Patent 1060546, GB Patent 1078234 (Union Carbide) and GB Patent 1153035 (ICI).

Patents Polyester, n.d. By BASF in 1992 (EP0572851 A1), followed in 2004 by AOC (US 2004/0010061 A1) and in 2009 by Arkema (EP 07380309) and Cray Valley (WO 2009 059706 A1).

Phillips, L.N., 1968. Carbon-fibre reinforced plastics—a personal view. Fibre Sci. Technol. 1 (1), 3–10. https://doi.org/10.1016/0015-0568(68)90015-8.

Pil, L., 2002. From mobile home to clone, composites in contemporary art. In: Verpoest, I., Daenens, L., Bucquoye, M.E. (Eds.), From Bakelite to Composite. Stichting Kunstboek, pp. 117–131.

Pil, L., 2018a. Light Light: carbon fibre composite for a lightweight chair. In: Pil, L., Verpoest, I. (Eds.), Fibre-Fixed. Composites in Design. Stichting Kunstboek, pp. 164–195.

Pil, L., 2018b. Biofibres, recycling, unusual shapes. Experimenting with composites. In: Pil, L., Verpoest, I. (Eds.), Fibre-Fixed. Composites in Design. Stichting Kunstboek, pp. 132–163.

Pil, L., 2018c. Architecture and the city: designing with composite materials. In: Pil, L., Verpoest, I. (Eds.), Fibre-Fixed. Composites in Design. Stichting Kunstboek, pp. 100–132.

Pil, L., Verpoest, I., 2018. Fibre-Fixed. Composites in Design, Stichting Kunstboek, Belgium. Critical survey also discussing historical developments in mobility (aircraft, automotive, bicycles, etc…), architecture, interior design, …

Pil, L., Bensadoun, F., Pariset, J., Verpoest, I., 2016. Why are designers fascinated by flax and hemp fibre composites. Composites Part A 83, 193–205.

Plastics Hall of Fame, n.d. Hendry, James R. W. Available at: https://plasticshof.org/members/james-r-w-hendry/. Accessed 21 January 2025.

Potter, K.D., 1999. The early history of the resin transfer moulding process for aerospace applications. Compos. Appl. Sci. Manuf. 30 (5), 619–621. https://doi.org/10.1016/S1359-835X(98)00179-1.

PRG, 2018. PRG announces new innovations for the concert touring sector On the Run II TourProduction. Available at: https://www.prg.com/en/news/on-the-run-ii-tourproduction. Accessed 21 January 2025.

Proterra, n.d. Proterra ZX5. Available at: https://en.wikipedia.org/wiki/Proterra_ZX5. Accessed 21 January, 2025.

Qiaoting, W., 2020. A brief history of carbon fibre: Let's start with the history of the development of carbon fibre technology in the United States today!. Available at: https://okingperformance.com/article/a-brief-history-of-carbon-fibre-lets-start-with-the-history-of-the-development-of-carbon-fibre-technology-in-the-united-states-today-i00004i1.html. Accessed 25 July 2024.

Quinlivan and Lovell, 1982. Material and Process Developments in the Boeing 767, pp. 255–261. ICAS-paper 82-2.2.1.

Ramirez, A., 2020. Tungsten's Brilliant, hidden history. Am. Sci. 108 (2), 88. https://doi.org/10.1511/2020.108.2.88.

Rammed Earth, n.d. Rammed earth. Available at: https://en.wikipedia.org/wiki/Rammed_earth. Accessed 29 July 2024.

Rand, B., 1985. Carbon fibres from mesophase pitch. In: Chapter XIII in W. Watt and B. V. Perov, (eds.), Strong Fibres. Elsevier Science Publ., 497–572.

Rein4ced, n.d. Rein4ced develops more impact resistant composite than carbon. Available at: https://rein4ced.com/; see also https://www.compositesworld.com/articles/cycling-forward-with-bike-frame-materials-and-processes. Accessed 21 January 2025.

Reisinger, M., 2010. New evidence about composite bows and their arrows Inner Asia. The Silk Road 8, 42–62.

Renewables Finland, n.d. The First Wind Turbine Blades were Successfully Recycled in Finland. Available at: https://suomenuusiutuvat.fi/en/the-first-wind-turbine-blades-were-successfully-recycled-in-finland-a-new-domestic-solution-was-created-in-a-joint-project-of-several-actors/.

Risepolevault, 2023. The first fibreglass poles. Available at: https://www.tiktok.com/@risepolevault/video/7250487252344802606. (Accessed 21 January 2025).

Road.cc, 2010. Carlton Flyer – The First Carbon Bike. Youtube. Available at: https://www.youtube.com/watch?v=ugCR53fksjI. (Accessed 21 January 2025).

Roeseler et al., 2007. Composite structures: The first 100 years, presentation at ICCM-7, Kyoto, 2007.

Romano, V., Lomov, S.V., Verpoest, I., Gorbatikh, L., 2015. Stress magnification due to carbon nanotube agglomeration in composites. Compos. Struct. 133, 246–256. https://doi.org/10.1016/j.compstruct.2015.07.069.

Rosenberg, D., Love, S., Hubbard, E., Klimscha, F., 2020. 7,200 years old constructions and mudbrick technology: the evidence from Tel Tsaf, Jordan Valley, Israel. PLoS One 15 (1), e0227288. https://doi.org/10.1371/journal.pone.0227288.

SAMPE, 2012. The 25-Year History of SAMPE Japan Chapter, 1984-2009 (mainly in Japanese), June 2012, p. 183–195.

SAMPE Japan, 2012. The 25-Year History of SAMPE Japan Chapter, pp. 183–195.

Samsonite, n.d. ROXKIN – Samsonite's exclusive new material is the future of luggage. YouTube film: https://www.youtube.com/watch?v=Y-5pBHBAFM8. Accessed 21 January 2025.

Sarolea, n.d. MANX 7. Available at: https://www.sarolea.com/manx7. Accessed 21 January 2025.

Savage, A.W., 1916. Vehicle tire, US Patent 1,203,910.

Schwartz, M., Hollander, D., 2004. Boats, bitumen and Bbrtering: the use of a utilitarian good to track movement and transport in ancient exchange systems. In: Robertson, E.C., Seibert, J.D., Fernandez, D.C., Zender, M.U. (Eds.), Space and Spatial Analysis in Archaeology. Joint Publication by the University of New Mexico Press and University of Calgary Press, Albuquerque, pp. 323–330. ISBN 10: 1-55238-168-4.

Seymour, R.B., 1989. Origin and early development of rubber-toughened plastics. In: Rubber Toughened Plastics, Edited by C. Keith Riew, Advances in Chemistry Series 222, published by American Chemical Society, Washington, DC.

Seymour, R.B., Deanin, R.D., 1987. History of Polymeric Composites, VNU Science Press (Utrecht). Proceedings of the Symposium Held during the 192nd ACS Symposium in Anaheim (Ca, USA), 1986, with a short chapter on "History of fibrous reinforcements".

SGS, 2024. LCA of Composite Waste Processing. Available at: https://eucia.eu/wp-content/uploads/2024/10/LCA-of-Composite-Waste-through-Cement-Co-Processing.pdf. Accessed 21 January 2025.

Shindo, A 1961. Studies on graphite fibre, Osaka Kogyo Giijutsu Shikenjo Hokoku (GIRIO Report 317).

Shindo, A., 1986. Carbon fibre. Ceramics 21 (10), 841.

Shindo, A., Fujii, R., Sengoku, T., 1959a. Tokkyo (Patent application publication) Sho 37-4405 "Akuriru kei gosei kobunshi butsu yori tanso seihin o seizo suru hoho") (Method for manufacturing carbon product from acrylonitrile synthetic macromolecular substance), in Japanese.

Shindo, A., Fujii, R., Takahashi, A., Sengoku, T., 1959b. Kokuen sen'I no kenkyu (dai 1po) netsu sori tmonau kesshoshi no seicho (Study of graphite fibre (1st report) – Growth of crystallite in heat treatment), Kagaku Kankeigaku Kyokai Rengo Shuki Happyokai (Annual Meeting of the Union of Chemistry-related Societies, in Japanese.

Shindo, A., 2000. 1.01 – polyacrylonitrile (PAN)-based carbon fibres. In: A. Kelly, C. Zweben (eds.), Comprehensive Composite Materials. Pergamon, 1–33, https://doi.org/10.1016/B0-08-042993-9/00040-1.

Skrabec, Q.R., 2011. Edward Drummond Libbey, American Glassmaker, McFarland & C°, Jefferson (USA), 85–86.

Smithsonian, n.d. Beached Starship. Available at: https://www.smithsonianmag.com/air-space-magazine/beached-starship-5429731/. Accessed 21 January 2025.

Solar Impulse, n.d. Solar Impulse. Available at: https://en.wikipedia.org/wiki/Solar_Impulse. Accessed 21 January 2025.

Starck, 2012. The Broom chair is built from 90% recycled industrial waste. Available at: https://designwanted.com/broom-chair-philippe-starck-industrial-waste/. Accessed 21 January 2025.

Starck, 2015. Zartan. Available at: https://www.magisdesign.com/wp-content/uploads/2015/02/zartan_raw_eco.pdf. Accessed 4 February 2025.

Stratview, 2023. Carbon Composite Hydrogen Tanks Market. Available at: https://www.stratviewresearch.com/1415/carbon-composite-hydrogen-tanks-market.html. Accessed 21 January 2025.

Siwowski, T., Kaleta, D., Rajchel, M., 2018. Structural behaviour of an all-composite road bridge. Composite Structures 192, 555–567.

Solar Impluse Foundation. 1st round-the-world with a solar aeroplane. https://solarimpulse.com/foundation. (Accessed 21 January 2025).

Sudhakar, A.J., Muthusubramanian, B., 2022. Textile-reinforced concrete sandwich panels: a review. ACI Struct. J. 119 (5), 207–216. https://doi.org/10.14359/51734899.

Swolfs, Y., Verpoest, I., Gorbatikh, L., 2018. Recent advances in fibre-hybrid composites: materials selection, opportunities and applications. Int. Mater. Rev. 64 (4), 181–215. https://doi.org/10.1080/09506608.2018.1467365.

Takahashi, J., 2006. A global perspective for the future development of sustainable land transport (cars, trucks, buses). Presentation at KU Leuven, citing J. Kasai. Int. J. Life Cycle Assess. 5 (5), 316.

Takamatsu, T., 2000. Development of PAN-based carbon fibre. Paper in Japanese in Technology and Civilization 12(1).

Talgo, n.d. Talgo Developing Carbon Fibre Train Bogie for Lighter Trains and Accommodate More Passengers. Available at: https://carbonfibergear.com/blogs/carbonfiber/carbon-fiber-train-bogie. Accessed 21 January 2025.

Tanaka, F., 2020. Development of carbon fibres, past and future. 金属 (Sanzoku, Metal) 90 (6) (in Japanese).

Tanaka, F., 2021. Current status and future prospects for improving mechanical properties of carbon fibres. Chem. Chem. Ind. 74–6, 415–417 (in Japanese).

Tanaka, F., 2024. Pioneering the carbon fibre frontier: a half-century of industry leadership and the road ahead. Compos. B Eng. 281, 111515. https://doi.org/10.1016/j.compositesb.2024.111515.

Tang, M.M., Bacon, R., 1964. Carbonization of cellulose fibres—I. Low temperature pyrolysis. Carbon 2 (3), 211–220. https://doi.org/10.1016/0008-6223(64)90035-1.

Taraborrelli, L., Grant, R., Sullivan, M., Choppin, S., Spurr, J., Haake, S., Allen, T., 2019. Materials have driven the historical development of the tennis racket. Appl. Sci. 9 (20), 4352. https://doi.org/10.3390/app9204352.

Tate, R., 2015. Bill Stout: Engineer to Remember. Available at: https://www.motorcities.org/story-of-the-week/2015/bill-stout-engineer-to-remember. Accessed 21 January 2025.

Teinsky, E., 1902. Verfahren zur Herstellung von webbarer Glaswolle. Kaiserliches-Königliches Patentamt, Österreichisches Patentschrift nr. 6508, filed 30/08/1997, granted 05/01/1902.

Textreme, n.d. Textreme – Technology Thin Ply Available at: textreme.com/technology/thin-ply. Accessed 21 January 21 2025.

Thomason, J.L., 2012. Glass Fibre Sizings: A Review of the Scientific Literature. Blurb Inc.. ISBN 9780957381414.

Thomason, J.L., 2015. Glass Fibre Sizing: A Review of Size Formulation Patents. Blurb Inc., pp. 12–15. ISBN 9780957381438.

Thomason, J.L., 2019. Glass fibre sizing: a review. Compos. Appl. Sci. Manuf. 127, 105619. https://doi.org/10.1016/j.compositesa.2019.105619.

Thomason, J.L., 2023. Flat glass fibres: the influence of fibre cross section shape on composite micromechanics and composite strength. Compos. Appl. Sci. Manuf. 169, 107503. https://doi.org/10.1016/j.compositesa.2023.107503.

Thomason, J. L., n.d. Personal communication, 01/02/2023.

Thorne, D.J., 1985. "Manufacture of carbon fibre from PAN", Chapter XII. In: Watt, W., Perov, B.V. (Eds.), Strong Fibres. Elsevier Science Publ., pp. 476–493

Timmonet, n.d. Joseph Wilson Swan, Electric light and actinic light... a pioneer. Available at: www.timmonet.co.uk/html/body_joseph_swan.htm. Accessed 2 August 2024; a good description of Swan's scientific career can be found in https://en.wikipedia.org/wiki/Joseph_Swan.

Toma, K., 2008. Development of the carbon fibre business in Japan. In: Hara, Kambayashi, Matsushima (Eds.), Industrial Innovation in Japan. Routledge, pp. 82–100.

Toray Carbon Magic, n.d. Concept EV ItoP. Available at: https://www.carbonmagic.com/en/challenge/cha_01.html. Accessed 21 January 2025.

Tvindkraft, n.d. Tvindkraft – how it began. Available at: https://www.tvindkraft.dk/how-it-began-the-history-of-wind-power/. Accessed 21 January 2025.

Udawattha, C., Halwatura, R., 2017. Life cycle cost of different Walling material used for affordable housing in tropics. Case Stud. Constr. Mater. 7, 15–29. https://doi.org/10.1016/j.cscm.2017.04.005.

Umatex, n.d. Alabuga Fibre LLC. Available at: https://rosatom-composites.ru/company/alabugavolokno/. Accessed 21 January 2025.

Updegraff, I.H., 1982. Unsaturated polyester resins. In: Lubin, G. (Ed.), Handbook of Composites. Springer, Boston, MA. https://doi.org/10.1007/978-1-4615-7139-1_2.

US Patent 9,093,191, 2015. Fibre Reinforced Composite Core for an Aluminium Conductor Cable.

US Patent 133,229, 1872. Improvement in Process and Apparatus for Manufacturing Pyroxyline.

US Patent 1,433,728, 1916. Composite Propeller, Inventor R. Kemp.

US Patent 1,435,244, 1916. Structural Elements, inventor R. Kemp.

US Patent 2,195,362, 1940. Glycol-Maleic Acid Resin and Process of Making Same, Inventor C. Ellis.

US Patent 2,255,313, 1941. Ethylenic-Alpha-Beta Synthetic Resins and Process of Making Same, Inventor C. Ellis.

US Patent 3,053,775, 1959. Method for Carbonizing Fibres, by W. F. Abbott, filed 7/11/1959, granted and published 11/09/1962.

US Patent 3,861,925, 1973. Alkali-Resistant, Zirconia-Free Fibrilizable Glass Compositions. Inventors: W.W. Wolf, Owens-Corning Fibreglas Corporation, USA.

US Patent 3,861,927, 1972. Alkali Resistant Glass Fibres", Japan. Inventors: Kimura and Yoshimura.

US Patent 4,330,628, 1979. Alkali-Resistant Glass Fibres. Inventors: D.R. Cockram, K.M. Fyles, Pilkington Brothers Limited.

US Patent 5,264,519, 1989. Inventors R. Nelson, D. Milovich, W. Wilcox and R. Read.

US Patent 5,789,229, 1998. Boron Free Glass Fibres filed 6/6/1996, published 4/8/1998. Inventors Walter L. Eastes, Douglas A. Hofmann; John W. Wingert, all at Owens Corning.
US Patent 942,699, 1909. Filed 13 July 1907, patented 7 December 1909, accessed via Google Patents, latest 02/08/2024.
US Patent 949,671, 1909. Filed 18 February 1908, patented 15 February 1909, accessed via Google Patents, latest 02/08/2024.
US Patent 2,133,235, 1933. Method & Apparatus for Making Glass Wool. Inventor G. Slayter.
US Patent 2,230,272, 1938. Method of Producing Glass Fibres. Inventor G. Slayter.
Valverde, M.A., Kupfer, R., Wollmann, T., Kawashita, L.F., Gude, M., Hallett, S.R., 2020. Influence of component design on features and properties in thermoplastic overmoulded composites. Compos. Appl. Sci. Manuf. 132, 105823. https://doi.org/10.1016/j.compositesa.2020.105823.
van Dulken, S., 2002. Formica. US Patent 1.284.432. In: Inventing the 20th Century. 100 Inventions that Shaped the World. New York University Press, pp. 42–43.
Velo, 2023. Black Magic: How Carbon Fibre Bicycle Frames are Made. Updated 2023, available at: https://velo.outsideonline.com/2018/01/how-carbon-fibre-bicycle-frames-are-made/. (Accessed 21 January 2025).
Ventre, T., 2021. Ebusco 3.0: Redefining Public Transport with Game-Changing Composite Materials. Available in. https://www.azom.com/article.aspx?ArticleID=20459. (Accessed 21 January 2025).
Verpoest, I., 2018. Sustainable mobility. In: Pil, L., Verpoest, I. (Eds.), Fibre-Fixed. Composites in Design. Stichting Kunstboek, p. 44.
Verpoest, I., Aernoudt, E., Deruyttere, A., De Bondt, M., 1985. The fatigue threshold, surface condition and fatigue limit of steel wire. Int. J. Fatig. 7 (4), 199–214.
Verpoest, I., Daenens, L., Bucquoye, M.E., 2002. From Bakelite to Composite. Stichting Kunstboek. Published at the Occasion of the Exhibition in the Design Museum Gent (B).
Verre Online, 2004. Fibre de verre et composites, on the website of the French Institut du Verre. Available at: www.verreonline.fr, Accessed 28 December 2024.
Vestas, 2022. New coalition of industry and academia .... Available at: https://www.vestas.com/en/media/company-news/2021/new-coalition-of-industry-and-academia-to-commercialise-c3347473. Accessed 21 January 2025.
Vestas, n.d. From 1971-1986. Energy Experiments and Brink of Disaster. Available at: https://www.vestas.com/en/about/this-is-vestas/history/from-1971-1986. Accessed 21 January 2025.
Victress, n.d. Category Archives: Victress. Available at: https://www.undiscoveredclassics.com/category/forgotten-fibreglass/victress/. Accessed 21 January 2025. Also, Forgotten Fibreglass Timeline: Table Form | Undiscovered Classics (Accessed 21 January 2025).
Vintagefiets, n.d. The first carbon bike ever. Available at: https://www.vintagefiets.be/en/de-eerste-carbon-fiets-ooit/. Accessed 21 January 2025.
Vlot, A., 2001. Glare: History of the Development of a New Aircraft Material. Springer.
Wagmare, R., Harshe, R., Pednekar, J., et al., 2024. Additive manufacturing of continuous fibre-reinforced polymer composites: current trend and future directions. Progr. Addit. Manuf. https://doi.org/10.1007/s40964-024-00777-9.
Ward, I.M., 1985. Preparation, properties and structure of high-modulus polyolefines and polyoxymethylene. In: Watt, W., Perov, B.V. (Eds.), Strong Fibres. Elsevier Science Publ.
Ward, I.M., Hine, P.J., 1997. Novel composites by hot compaction of fibres. Polym. Eng. Sci. 37 (11), 1809–1814.

Watt, W., 1985. Chemistry and physics of the conversion of PAN fibres into high-modulus carbon fibres. In: Watt, W., Perov, B.V. (Eds.), Strong Fibres. Elsevier Science Publ., pp. 327–386

Watt, W., Perov, B.V., 1985. In: Kelly, A., Rabotnov, Y.N. (Eds.), "Strong Fibres" Volume 1 of the Series "Handbook of Composites. Elevier Science Publ., p. 752

Węglowski, M., 2019. Composite Rebars, 4. Biuletyn Instytutu Spawalnictwa, pp. 7–14.

Wikipedia, n.d. Lewis Howard Latimer, available at https://en.wikipedia.org/wiki/Lewis_Howard_Latimer, last accessed 5 January 2026. See also www.lewislatimerhouse.org/about.

Wikipedia Wind Turbines, n.d. List of most powerful wind turbines. Available at: https://en.wikipedia.org/wiki/List_of_most_powerful_wind_turbines. Accessed 4 February 2025.

Wilk, C., 2017. Plywood: A Material Story. Thames and Hudson/V&A, London, p. 240.

WIPO, n.d. A Patent that Changed an Industry. Available on: https://www.wipo.int/ipadvantage/en/articles/article_0152.html. Accessed 21 January 2025.

Windpower, 2018. Ten of the Biggest Turbines. https://www.windpowermonthly.com/10-biggest-turbines. (Accessed 2 January 2020).

Yang, Hung Man, 2000. Aramid fibres. In: Kelly, A., Zweben, C (Eds.), Composite Materials. Pergamon, vol. 1. Pergamon Press, pp. 199–229.

Yarovan, I.I., 2018. Carbon Fibres. A Glimpse Back. Available at: https://compositesinfo.ru/uglerodnoe-volokno-vchera-i-segodnya/. Accessed 21 January 2025.

Yonex, 2021. Yonex Carbonex 8 B-8500 B worth. Available at: https://www.badmintoncentral.com/forums/index.php?threads/yonex-carbonex-8-b-8500-b-worth.188415/. Accessed 21 January 2025.

Zelinski, P. n.d. How 3D printing will change manufacturing. Available at: https://www.additivemanufacturing.media/articles/how-3d-printing-will-change-composites-manufacturing. Accessed 21 January 2025.

Zephyr, 2018. Airbus press release. Available at: https://www.airbus.com/en/newsroom/press-releases/2018-07-zephyr-s-set-to-break-aircraft-world-endurance-record. Accessed 21 January 2025.

Yoshioka, K., 2019. Development of PAN-based carbon fibre. Past, Present and Future, Slides Presented during JISSE-16 Conference. University of Tokyo.

Zhang, J., Lin, G., Vaidya, U., Wang, H., 2023. Past, present and future prospective of global carbon fibre composite developments and applications. Compos. B Eng. 250, 110463. https://doi.org/10.1016/j.compositesb.2022.110463.

Zogg, M. 2025, Private communication.

Zsofia, P., 2024. Volta Trucks signs manufacturing agreement with Steyr Automotive. Available at: https://trans.info/en/volta-trucks-signs-manufacturing-agreement-with-steyr-automotive-380868. Accessed 4 February 2025.

Challenges and opportunities for composites in the next decades

1.2

Ignaas Verpoest, Stepan V. Lomov, Yentl Swolfs, Mahoor Mehdikhani and Panayiotis Tsokanas
Composite Materials Group, Department of Materials Engineering, Katholieke Universiteit Leuven (KU Leuven), Leuven, Belgium

1. Introduction

The previous chapter provided a "bird's eye view" on the long history of composites. This history is characterized by a large number of disruptive innovations and gradual expansion into diverse application domains. The future is never easy to predict, and that is certainly the case for disruptive innovations that most often come as a surprise. Hence, this chapter could only be grounded in the authors' understanding of the current composite scenery, a critical assessment of major scientific and technological trends, and an almost intuitive analysis of the most important societal needs for the coming decades.

The chapter is divided into four sections. First, some expected and desired innovations in the materials science of composites are presented, not only for polymer matrices and several types of fibres but also for nano-reinforcements, hybrids, bio-based composites, and smart composites. Necessary further developments in processing are then discussed in the second part, covering automation and robotisation and additive manufacturing. The importance of multiscale modelling of materials and processes has gained increasing relevance, and it is believed that this trend will continue in the upcoming decades. In addition, the potential of artificial intelligence is critically assessed. Next, the growth of applications in aerospace, transportation, consumer goods, construction, and energy is sketched in the third section. This chapter concludes with a discussion on the environmental impact of composites and their role in a greener and circular economy. The discussion on expected innovations and enlarging application fields is mainly qualitative; an excellent and quantitative assessment of the dynamics of the future innovations in composite materials can be found in JEC Observer, 2025: "Overview of the global composites market 2024–29" (JEC Observer, 2025).

The History of Composites. https://doi.org/10.1016/B978-0-443-34119-9.00002-4

In industry, several types of **macro-scale hybrids** have recently received a lot of attention. On the one hand, metal–composite hybrids mainly aim at local and lightweight reinforcements for metallic structures. On the contrary, overmoulding of composite parts (mainly flat or slightly curved or formed) with stiffening ribs or connection elements is already realising a breakthrough. Much more can be expected from such "macro-hybrid" structures!

2.5 Biobased and bioinspired composites: saving the planet?

Two decades before the recent rise in ecological awareness (linked to the proven global warming threat), **biobased fibres** had already started ingressing into the composites world. However, some important issues still need to be addressed. Moisture sensitivity seems to be under control for not too harsh environments (Alliance Flax, 2023). The cost of biobased fibre preforms varies largely, depending on the type of fibres used. For the best-performing fibres, namely flax fibres, new preforms are continuously being developed, aiming at either reducing cost or improving mechanical properties or processing characteristics (CELC and JEC, 2018).

The use of **biobased matrices** will depend on the developments toward biobased polymers. Partially biobased thermosets are already on the market, but real breakthroughs are needed to make epoxies and polyester fully biobased. In contrast, thermoplastics could profit from a strong societal push toward "zero-impact" plastics in consumer goods. This might work for polypropylene, polyethylene terephthalate (PET), and polyamide (PA), but specific innovations will be needed to develop biobased, high-performance thermoplastic matrices like polyether ether ketone (PEEK), polyethersulfone (PES), and polyethylenimine (PEI).

Looking at how nature constructs materials stimulates the imagination of materials scientists, and composites offer great possibilities to mimic the best of nature. However, how will scientists transfer their sometimes genius inventions into the reality of the composites industry? Interesting experiments in **biomimicry**, such as morphing materials that react to changes in temperature or humidity, can only be realised with composites. Still, more research is needed to realise the transition from the laboratory to the industrial reality.

2.6 Smart materials and intelligent structures: composites can do both!

A gradual transition from the laboratory and prototype stage to the real-world application of **smart materials** and **intelligent structures** is indeed underway, but how can this process be accelerated and spread out to a wider range of application domains?

Much is expected from **multifunctional composites**. Due to their fibrous and often laminated nature, composites can integrate sensors, actuators, and even batteries inside the structural material instead of having to bond them at the outside, and hence not being part of the load-carrying structure, as is the case with metals. This simple fact has recently led to interesting innovations, but more is needed to fully use their potential in a wide range of application domains, including load-carrying structures in aircraft, electrical, shared, and autonomous cars, continuous monitoring of bridges and other infrastructure, and next-generation sporting goods.

2.7 Ultra-long-term durability

The optimistic statement that carbon fibres do not fatigue may be close to reality for an "operational" fatigue life of one million cycles. The ultra-long-term durability of hundreds of millions and even billions of cycles, which is crucially needed for applications like wind turbines, remains largely a terra incognita. The related phenomenon of creep needs further attention. The problem is intertwined with the effect of defects, which, at void contents below 1% and fibre misalignments of the order of 1 degree, is related to the statistics of low probability events and risks and the physics of very slow processes in polymers. Both are far beyond the abilities of current molecular dynamics modelling and are close to the limits of experimental capabilities.

The micro-mechanisms of fatigue in fibre-reinforced composites are described phenomenologically. It is not well known how fatigue of interfaces develops or whether the micro- or sub-micro-scale features in polymer morphology may lead to the development of fatigue damage. The investigation of these fundamental problems will open a way to ensure a decades-long life for "gigantic" composite structures like windmill blades and safety-driven designed airframes.

3. Challenges and opportunities in composite processing

3.1 Automation, robotisation, and mass production: the real breakthrough of "fast composites"—it has to happen now!

Both material and process innovations have paved the way toward the mass production of thermoset and thermoplastic composite applications. Cost reduction is the driving force in all senses: material and manufacturing cost (together resulting in the product cost), cost of ownership, and environmental cost. More cost-efficient materials and **automated, even robotised production methods** are under full development, and much more is to be expected.

Automated tape laying (**ATL**) and automated fibre placement (**AFP**) have gained a level of maturity that allows them to be used both in relatively small volume applications, in terms of the number of products, such as aircraft wings, fuselage, and other structural parts, as well as in large volume applications like bicycle frames. Faster robots, increased positioning accuracy, and model-based online monitoring will further help the growth of these promising techniques, both for thermoset and thermoplastic composites.

Thermoplastic composites intrinsically have the potential to be processed fast and in an automated way. The massive growth of the number and variety of available thermoplastic-based preforms will stimulate the expansion of thermoplastic composite applications. Carbon, glass, or biobased fibre-reinforced unidirectional tapes, woven, or random mat prepregs are now produced by an ever larger number of companies, raising important hopes for further cost reduction.

Overmoulding has completed its experimental stage and should soon become a standard high-speed production method, combining the advantages of injection moulding, such as speed and complexity of shapes, with those of continuous-fibre laminates, such as mechanical performance.

3.2 Additive manufacturing/3D printing: the new "holy grail" or only for niche markets?

Composite processing could often be considered as a kind of **additive manufacturing** because, in many composite processing techniques, layers of reinforcement are deposited one on top of the other. Additive manufacturing is used for polymers, metals, ceramics, concrete, and even living tissues. Seldom has a new technology emerged so quickly in so many different materials sectors. The challenge of applying this technology to composites is to determine a printing path that efficiently creates the desired shape in the fastest and most reliable way (a key requirement of 3D printing) while also maximising the strength and stiffness of fibre-reinforced composites. The optimal 3D printing path is not necessarily the best one, from a mechanical performance perspective, requiring extensive modelling and optimisation to find the best balance between both factors.

ATL and **AFP** of composites can be considered as the upscale variant of additive manufacturing, so all three should benefit from each other's technological developments.

3.3 Out-of-the-box thinking in process innovation

The advent of new materials often necessitates or provokes the development of new processing methods. This was the case when, in the 1930s, the invention of glass fibres and polyester or epoxy resins gave birth to the development of resin infusion techniques like resin transfer moulding (RTM) and composite-specific production methods like autoclaving, pultrusion, and filament winding. Since then, we have experienced mainly incremental improvements until, by the end of the millennium, robots inspired the invention of ATL, AFP, and other additive manufacturing techniques. It is time to **think out of the box** and invent drastically new composite concepts and processing technologies!

One of the main hurdles in using composites in experimental or rather low-volume products is the high cost of the moulds. Hence, *mouldless* manufacturing techniques would be highly beneficial.

A small personal note (of the first author)
When I was guiding undergraduate students in their bachelor project in the 1990s, we were looking for composite applications of **knitted** glass fabrics. Such fabrics have the unique characteristic that they are extendable and, hence, easily drapable around complex shapes, much better than woven fabrics. We used my children's inflatable swimming pool toy (a duck!) as mold, draped the knitted glass fabric around it, and inflated the duck so that the knit was nicely stretched. Then, we hand-impregnated it with a room-temperature hardening epoxy resin, resulting in a one-off composite product: hollow, light, and complex-shaped! The technique was further elaborated (for instance, by adding a second duck as outer mold and applying resin infusion), but there was no real interest in the composites community...

In the new millennium, such **mouldless manufacturing techniques** have been introduced into the composites community by designers and architects. Dutch designer Marcel Wanders created his *Knotted Chair* (Beukers, 2002) by knotting carbon–aramid hybrid braids, stretching them in a frame, and hand-impregnating them with epoxy. Many more designers followed this example, as described by Pil (2018a). A more sophisticated technique was invented by Profs. Knippers and Menges at the University of Stuttgart (Pil, 2018b), using a robot to wind wet-impregnated carbon or glass rovings in between points of attachment positioned on a simple, open frame. After hardening the epoxy (or polyester/vinylester) matrix, the frame can be easily removed, and the structure becomes a standalone composite product. Such mouldless composite products can be metres long and wide, and they have already been used in several one-off or small series structures. They even received their own name: with a Google search for "**fibrous tectonics**," one will find exciting examples and scientific substantiation of this new family of composite processing methods (Costanza, 2016).

The composites community is waiting for similar out-of-the-box innovations in processing methods! But also, the composite materials as such could benefit from an equally creative approach, a nice example being the *double-double laminates* proposed by the nonagenarian Prof. Steve Tsai (2022)!

3.4 Multiscale modelling at material, process, and product levels: catching up with metals and plastics

Compared with metallic and plastic products, a fully integrated material–process–product simulation chain is still missing. Product simulation has gone a long way, and big progress has been made in the simulation of many processes (like RTM and similar liquid moulding technologies, ATL/AFP, and out-of-autoclave processes). In the material properties prediction, steps are set toward an integrated "nano–micro–meso-approach," but much work is still needed.

Because the microstructure, processing, and product performance are strongly interdependent in composites, an integrated material–process–product simulation chain is absolutely needed to accelerate the market penetration of composites. A particular challenge is to incorporate the effects of defects like voids and wrinkles. While this has already been studied for more than 30 years, it remains a challenge

4.3 Consumer goods

It is expected that the **sports and recreation** market will remain the largest consumer of composites in the consumer goods sector, which in itself represents 8% in volume of the global market for composites, according to JEC Observer (2025, p. 25). Ski, snowboards, and golf shafts are well established, but strong growth is expected in athletics and field sports (e.g., basketball and football) because inserting carbon-fibre composite strips in **shoe soles** enables control over their flexural and torsional rigidity.

Professional and recreational bikers adore their "carbon" bicycle, and this will persist. In the exploding market of battery **electric bicycles**, there is a need for lighter frames to compensate for the weight of the batteries, whereas **cargo bikes** could benefit even more from a composite structure.

Injection moulding of short-fibre thermoplastic composites is expected to further grow for all kinds of consumer goods. These products might not be recognised as "composites" by consumers because they look like ordinary plastics, but the strength and stiffness added by the short fibres are indispensable in many applications. Without people being aware of it, every household must have tens of composite tools, from kitchenware to gardening tools to do-it-yourself equipment to indoor and outdoor chairs and furniture.

4.4 Construction

Ageing infrastructure, both in Europe and the USA, represents a huge opportunity for applying composite repairs to bridges, tunnels, and potable water and sewage pipelines. Furthermore, new structures will increasingly use composites. Especially in areas that are difficult to access, light **composite bridges** can be installed in a shorter time, and they anyway need simpler bridgeheads and supporting structures.

For reinforcing concrete, composite rebars are now well established, but **"carbon concrete"** will allow the use of much less concrete and hence could realise a reduction in the environmental impact of concrete buildings and infrastructure. A further step is the use of **biobased fibres** to create building blocks and insulation that are (almost) CO_2-neutral.

4.5 Renewable energy

With the global warming threat in mind, CO_2-neutral renewable-energy-generating infrastructure will be booming in the next decades. It is expected that between 2023 and 2028, the number of new wind turbine offshore installations (worldwide) will grow with an average annual growth rate of 28%, onshore by 7% (JEC Observer, 2025, p. 53). The composites industry responds to this societal challenge not only by developing ever larger wind turbine blades but also by providing materials for other technologies like tidal or sea-wave energy converting systems. Composite structures for generating tidal and sea-wave energy benefit from the corrosion resistance offered by the glass- or carbon-fibre composites used.

5. Challenges and opportunities on the environmental impact and circular economy

5.1 Materials and processes with reduced environmental impact

Biobased materials are an obvious choice for the future. The trend of using biobased fibres should be further strengthened by developing improved fibre extraction methods for biobased fibres in developing countries and creating more composite-oriented natural fibre preforms.

Whether more biobased matrices will be available in the future will depend mainly on the inventiveness of the chemistry community. Biogenic resources should be either biowaste or nonfood crops in order not to compete with food-oriented agriculture.

Finally, there will be room for more energy-efficient processes. Reduction in the energy content, and hence also of the environmental impact and cost of carbon fibres, remains a challenge. All carbon-fibre manufacturing companies are heavily investing in R&D projects to make breakthrough innovations in that field.

5.2 Composite applications for a greener economy

Composites make products "greener": lighter cars and aeroplanes consume less energy and hence contribute less to global warming. As the pressure to reduce CO_2 emissions persists, the need for composites to replace heavier materials and create lighter structures will remain highly relevant.

Composites are also essential for CO_2-neutral and renewable energy solutions. As mentioned above, wind turbines, tidal or sea-wave energy converting systems, and who knows which other alternatives will take the use of composites to new heights.

5.3 End-of-life: recycling and reuse, end-of-life/repair going for the lowest LCA

In a societal context, **recycling** is evidently the most "visible" challenge for composites. In this regard, important technological progress has been realised, and many new technologies are under development. Pyrolysis is emerging as the standard for carbon-fibre composite recycling, as it enables the reuse of valuable carbon fibres. Solvolysis also has potential, as it allows for matrix recovery, but it still struggles to scale up and recover the right types of monomers.

For glass-fibre composites, several alternatives have been developed or are in the prototype phase, as explained in chapter 1.1 of Part I of this book (section 10.4.5). Additional challenges are now in the logistics and upscaling to ensure these processes become economically viable, as highlighted in recent French (ADEME, 2022) and German (IKK and AKV, 2024) documents.

Maintenance and repair have been addressed extensively in aerospace, but it is questionable whether these solutions can also be applied to other application domains like automotive and marine. An essential part of the maintenance (and potential repair) of composites is the application of damage detection using several kinds of nondestructive testing methods. The challenge is now to make these methods faster and cheaper.

References

Alliance Flax, 2023. Moisture Sensitivity, the Advantages of Flax and Hemp Fibres in Composites, Alliance for Flax Linnen and Hemp. https://allianceflaxlinenhemp.eu/en/european-flax-linen-hemp-media-library/10/moisture-sensitivity-advantages-of-flax-and-hemp-fibres-in-composites.

ADEME, 2022. Guide du recyclage et de l'ecoconception des composites. Technical report by ADEME, France (in French).

Beukers, A., 2002. The unbearable lightness of composite materials. In: Verpoest, I., Daenens, L., Bucquoye, M.E. (Eds.), From Bakelite to Composite. Stichting Kunstboek, pp. 57–67.

Clean Aviation, 2025. Clean Aviation. Who We Are. https://www.clean-aviation.eu/about-us/who-we-are.

Costanza, D.N., 2016. Fibrous tectonics: a rethinking of composite production through innovation and exploration of molding techniques and methodologies. Procedia Eng. 155, 452–463. https://doi.org/10.1016/j.proeng.2016.08.048.

CELC and JEC, 2018. Flax and Hemp Fibre Composites: A Market Reality. CELC and JEC, Paris. In: https://proceedings.jeccomposites.com/proceedings/jec-world-2022/JW22_Proceedings_CELC_03-05-2022.pdf. https://eucia.eu/jec-and-eucia-launch-2024-composites-sustainability-report/.

Czel, G., Jalalvand, M., Wisnom, M.R., Czigany, T., 2017. Design and characterisation of high performance, pseudo-ductile all-carbon/epoxy unidirectional hybrid composites. Compos. B Eng. 111, 348–356. https://doi.org/10.1016/j.compositesb.2016.11.049.

European Federation for Transport & Environment, 2018. Roadmap to Decarbonising European Cars. A Study by the European Federation for Transport & Environment.

Gorbatikh, L., Lomov, S.V., 2017. Nano-engineered carbon fibre-reinforced composites: challenges and opportunities. In: Beaumont, P., Soutis, C., Hodzic, A. (Eds.), The Structural Integrity of Carbon Fiber Composites. Springer, London, pp. 117–135. https://doi.org/10.1007/978-3-319-46120-5_6.

Gray, N., O'Shea, R., Wall, D., Smyth, B., Lens, P.N.L., Murphy, J.D., 2022. Batteries, fuel cells, or engines? A probabilistic economic and environmental assessment of electricity and electrofuels for heavy goods vehicles. Adv. Appl. Energy 8, 100110. https://doi.org/10.1016/j.adapen.2022.100110.

Guo, R., Stubbe, J., Zhang, Y., Schlepütz, C.M., Gomez, C.R., Mehdikhani, M., Breite, C., Swolfs, Y., Villanueva-Perez, P., 2023. Deep-learning image enhancement and fibre segmentation from time-resolved computed tomography of fibre-reinforced composites. Compos. Sci. Technol. 244, 110278. https://doi.org/10.1016/j.compscitech.2023.110278.

Hayashi, T., Koyama, K., Yamazaki, A., Kihira, M., 1972. Development of new material properties by hybrid composition. Fukugo Zairyo (Composite Materials) 1 (1), 18–20.

IATA, 2025. Global Air Passenger Demand Reaches Record High in 2024. https://www.iata.org/en/pressroom/2025-releases/2025-01-30-01.

IKK and AKV, 2024. Composites-Recycling-Study Now Also Available in English. https://composites-united.com/en/composites-recycling-study-now-also-available-in-english/.

JEC Observer, 2025. In: Baron, N. (Ed.), Overview of the Global Composites Market 2024–2029. JEC World, Paris (Special Issue of JEC Magazine).

Patel, P., Furtado, C.F., Cooper, M., Acauan, L., Lomov, S., et al., 2021. Nanoengineered glass fiber reinforced composite laminates with integrated multifunctionality. Proceedings of the American Society for Composites—Thirty-Sixth Technical Conference on Composite Materials. https://doi.org/10.12783/asc36/35897.

Pil, L., 2018a. Light Light: carbon fibre composite for a lightweight chair. In: Pil, L., Verpoest, I. (Eds.), Fibre-Fixed. Composites in Design. Stichting Kunstboek and Design Museum Gent, pp. 166–195.

Pil, L., 2018b. Architecture and the city: designing with composite materials. In: Pil, L., Verpoest, I. (Eds.), Fibre-Fixed. Composites in Design. Stichting Kunstboek and Design Museum Gent, pp. 100–132.

Sapozhnikov, S.B., Swolfs, Y., Lomov, S.V., 2020. Pseudo-ductile unidirectional high modulus/high strength carbon fibre hybrids using conventional ply thickness prepregs. Compos. B Eng. 198, 108213. https://doi.org/10.1016/j.compositesb.2020.108213.

Swolfs, Y., Verpoest, I., Gorbatikh, L., 2019. Recent advances in fibre-hybrid composites: materials selection, opportunities and applications. Int. Mater. Rev. 64 (4), 181–215. https://doi.org/10.1080/09506608.2018.1467365.

Tsai, S.W., 2022. Double-Double, a new perspective in the manufacture and design of composites. In: JEC World and Composites Design Group. Stanford University, p. 267.

Volkswagen, 2025. Electromobility. https://www.volkswagen-group.com/en/tags/electromobility-1052.

Yoshioka, K., 2019. Development of PAN-based carbon fibre. Past, present and future. In: Slides Presented during JISSE-16 Conference. University of Tokyo.

Section 1

Materials

The history of glass fibres as a reinforcement for composites

1.1

Chris Skinner[1] *and Michelle Korwin-Edson*[2]
[1]Owens Corning Headquarters, Toledo, OH, United States; [2]Owens Corning Science & Technology, Granville, OH, United States

1. Introduction

The history of glass fibres as a reinforcement material in composites is a story of innovation, good luck and material science advancements that have helped to shape many modern industries. Glass fibres are widely used as reinforcing materials in composite materials due to their mechanical properties, versatility and, most importantly, cost-effectiveness (Varma and Gupta, 2000).

In turn, glass fibre-reinforced polymer (GFRP) composites have been transformative across many industries due to their unique combination of properties: high strength-to-weight ratio, corrosion resistance, durability and flexibility in manufacturing.

A new system was developed to produce chopped glass fibre reinforcements in 1969, as an in-line process implemented in the Owens Corning factory in Anderson, S.C. This delivered a decrease in the cost to manufacture this product by eliminating the offline chopping process and applying new binder chemistries, allowing the glass fibre reinforcement of polypropylene, nylon and many other thermoplastic engineering resins. Since then, short fibre-reinforced polymer (SFRP) composites have found extensive applications in automobiles, business machines, durable consumer items, sporting goods, electrical industries etc., owing to their low cost, easy processing and their superior mechanical properties over the matrix polymers. Extrusion compounding and injection moulding processes are frequently employed to make SFRP composites. The use of these conventional fabrication techniques to produce large-scale SFRP composite parts makes the manufacturing of these composites efficient and inexpensive compared to the manufacturing of continuous-fibre reinforced polymer composites. Today, the use of short fibre glass reinforcement is the single largest consumer of glass reinforcements in the global market (Mentgesa et al., 2023; Fu et al., 2019).

From the 1970s, the use of glass fibre composites expanded rapidly, driven by their ability to meet the diverse needs of industries ranging from aerospace and automotive to wind energy, construction, infrastructure and consumer goods. This expansion was not only technological but also economic, as the improved production methods and lower costs allowed these advanced materials to penetrate a wider range of applications.

The History of Composites. https://doi.org/10.1016/B978-0-443-34119-9.00003-6

One of the most important and growing applications of glass fibre composites in recent decades has been in the wind energy industry. As the world moves towards renewable energy sources, wind energy has seen rapid expansion, with glass fibre composites playing a central role in the construction of wind turbine blades that are fundamental components in the efficiency of the systems (Kaldellis and Zafirakis, 2011). Evolution in the processing of glass reinforcements coupled with the production of the required composite systems has enabled the continued production of longer and efficient blades (Cognet et al., 2020). As the wind energy industry continues to expand globally, particularly in regions such as Europe, North America and China, the demand for high-performance glass fibre composites is expected to grow alongside the increasing capacity of wind energy installations (Adolphs and Skinner, 2011).

In the construction and infrastructure sectors, glass fibre composites have emerged as an innovative solution for a variety of challenges. Traditional materials, such as steel and concrete, while strong, suffer from limitations such as corrosion, weight and inflexibility. Glass fibre composites offer a way to address these issues while also enabling new architectural designs and engineering possibilities. Notable examples highlighting this utility were the 1974 Trans-Alaska pipeline, shown in Fig. 1.1.1, where glass-reinforced composites provided a special insulation system over the 389-mile pipeline (Alyeska Pipeline Service Company, 2021). In 1978, Owens Corning delivered a 105-acre (425,000 m^2) glass-reinforced roof system for the Hajj Terminal at King Abdulaziz airport in Saudi Arabia, shown in Fig. 1.1.2, based on its experience with other structures such as Pontiac Silverdome in the U.S. (Skidmore, 2024).

Figure 1.1.1 Photograph of the Trans-Alaskan pipeline.
Photo Credit: Luca Galuzzi, Wikimedia Commons.

Figure 1.1.2 Photograph of the Hajj Terminal.
Photo Credit: Wikimedia Commons.

Perhaps the most significant application in the construction space was the conversion of the residential roofing shingle to a glass-reinforced system based on a continuous glass fibre nonwoven mat. In 1977, Owens Corning purchased the Lloyd A. Fry Roofing Company and the Trumbull Asphalt Company and then proceeded to convert all the plants to produce glass fibre-reinforced shingles (Malarky, 2001). Given the performance advantages, the whole industry rapidly converted to this technology. Glass-reinforced composites were also developed in architectural panels, cladding and facades, taking advantage of the combination of durability and the ability to mould into intricate shapes and designs (Bedon, 2016).

The consumer goods industry also embraced glass fibre-reinforced composites with materials found in a wide range of products, from sporting goods to electronics, where their lightweight and strong properties are highly desirable. For example glass fibre composites are used in the construction of tennis rackets, golf clubs, skis and snowboards. The lightweight nature of glass fibres allows for better control and performance, while their durability ensures that these products can withstand the stresses of frequent use. In addition, glass fibre composites offer design flexibility, allowing manufacturers to tailor the stiffness and flexibility of products to meet the specific needs of different sports and athletes (Fan et al., 2024). In electronics, glass fibre-reinforced composite printed circuit boards (PCBs) ensure electronic devices are both reliable and lightweight, supporting the rapid growth of modern consumer electronics, such as smartphones, tablets and laptops (Longobardo, 2010).

The expansion of applications for glass fibre composites has been a remarkable story of modern-day material innovation, driven by the unique properties of glass-reinforced composites and their ability to meet the diverse needs of various industries (Girl et al., 2006). As manufacturing techniques continue to improve and the demand for lightweight, high-performance materials grows, the use of glass fibre composites is likely to expand even further, supporting the future developments in construction, transportation and consumer products in the years to come.

This chapter tells the story of how glass reinforcement fibres were invented, developed and innovated. The first 'products' were curiosities – neckties and parasols—but curiosity and ingenuity led to the young and exciting industry of GFRP and millions of applications.

2. Early developments

Nature has been making glassy fibres since the beginning of time in the form of 'Pele's hair', or attenuated molten lava, named after Pelehonuamea, the Hawaiian Goddess of Fire and Volcanoes. Fine fibres form from various forceful occurrences around molten lava, such as around erupting vents, when ejected lava becomes airborne, or when bubbles in the lava burst. The molten and viscous lava stretches rapidly, and due to surface tension, forms a mix of elongated fibres and shot or beads as shown in Fig. 1.1.3. Pele's hair is commonly glassy or amorphous because the fine-diameter strands cool so quickly. However, the larger diameter strands and rounded beads or shot, known as Pele's tears, may crystallise because they cool more slowly due to their mass (Duffield et al., 1977). Other naturally occurring glasses are

Figure 1.1.3 A photograph of Pele's hair.
Courtesy of National Park Service, Hawai'i Volcanoes National Park; HAVO 20120608 Image Collection (US National Park Service Hawai'i Volcanoes National Park).

fulgurites (formed by lightning strikes), obsidians (volcanic glass) or rhyolites (silica-rich volcanic rock/glass). Their chemical compositions can be found in Morey's book (Morey, 1938, 1960).

Human-made glass fibres are a relatively recent development. Archaeological studies have found that ancient civilisations (~2000–500 BCE) in Egypt, Phoenicia, Mesopotamia, Italy (Venice) and Greece made glass articles with decorations involving coarse fibres. A wonderful table of historic glass compositions can be found in Morey's book (Morey, 1960). Presumably, centimetre-sized glass pieces or rod-like shapes were heated in a fire and stretched manually to create the coarse fibre. Having finessed their craft, Medieval (500–1500 AD) Venetian artefacts show finer fibre being used to adorn glassware.

Shifts in the glass composition over time, with trial and error, made fibre creation a bit easier as glasses with viscosities less sensitive to temperature changes were stumbled upon. Syrian and Italian glass blowers perfected the art over time and continued to use pulled cane to decorate their drinkware pieces.

2.1 *Early patents and industrialisation of glass fibre*

To understand how continuous glass fibre came about, we must learn of the history of the industries and companies that fed into it and how socio-economic events, including Prohibition and the Great Depression, impacted the glass industry.

During the Industrial Revolution, the industrialisation of steel production generated slag byproducts. Slag generally comprises various metal oxides (SiO_2, CaO, MgO, Fe_2O_3) with high levels of impurities (iron, carbon, sulphur). This manmade byproduct (or the natural versions of this material, e.g. basalt) can be melted and attenuated into fibres (and shot). In 1870, a U.S. patent was assigned to John Player (Fig. 1.1.4, describes a method for making vitreous fibre from the slag of blast furnaces (Player, 1870). This was done by using a 'blast of air – by preference, a hot blast' aimed at the stream of molten slag. Sound familiar, Pele? This process led to the development of mineral wool, also called rock wool or slag wool, insulation and with it, an opportunity for improvement in many areas that glass insulation eventually offered, such as little to no shot, a lightweight product and better R-value.

UNITED STATES PATENT OFFICE.

AUGUSTA AMELIA PLAYER AND HENRY McALLISTER, JR., OF PHILADELPHIA, PENNSYLVANIA, ADMINISTRATORS OF JOHN PLAYER, DECEASED.

IMPROVEMENT IN TREATING SLAG FROM IRON AND GLASS FURNACES FOR PRODUCING A USEFUL MATERIAL.

Specification forming part of Letters Patent No. **103,650**, dated May 31, 1870.

Figure 1.1.4 Title page of John Player's 1870 patent.

Meanwhile, others were thinking of glass as a fibre for woven cloth. In France, in 1836, Ignace Dubus-Bonnel patented a method of making glass fibres and weaving cloth with them (Dubus-Bonnel, 1836). In 1880, in a U.S. patent by Hermann Hammesfahr for improvements to glass cloth, he states that 'I am aware that it is not new to spin glass into fibres' and that 'I am also aware that fabrics have been made in which glass fibres have been used to form ornamental portions'. He formed his fibre from ½ in. glass rods heated with a flame and attenuated onto a wheel, and the cloth's only or main component was glass, and not traditional silk, cotton, wool etc., fibres. Fig. 1.1.5 shows his illustration of a glass cloth. The patent suggests the glass fibre cloth be used as fancy clothing, filter media or for insulating telegraph wires (US Patent No. 232122, 1880). Libbey Glass in Toledo, Ohio, acquired Hermann's patent with an interest in the cloth for lampshades, among other things. In 1893, at the World's Fair in Chicago, Edward Libbey hired actress Georgia Cayvan to model a dress and parasol made of the glass cloth. While quite astonishing, it was also heavy and stiff. Up to this

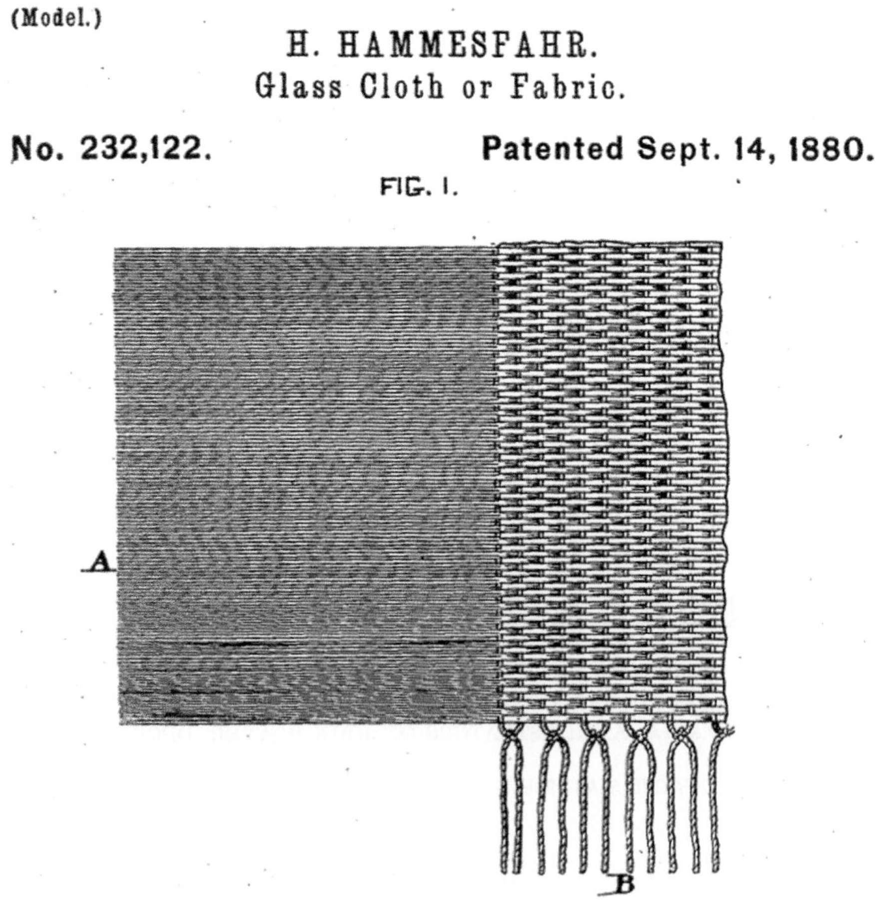

Figure 1.1.5 Sketch of a cloth made of glass as shown in Fig. 1 of Hammesfahr's patent.

point, glass fibre was made by hand or by pseudo-continuous means, one fibre at a time.

2.2 Prohibition throttles the bottle industry, launches glass fibre industry

The first fully automatic bottle machine was patented by Michael J. Owens in 1903, who at the time was an employee at the Libbey Glass Company. In May of 1906, he achieved the first commercial production of bottles in Newark, Ohio, at the American Bottle Company. Through a series of friendly acquisitions and mergers, the American Bottle Company became Owens-Illinois, under the leadership of William Levis, the CEO of the new Owens-Illinois enterprise.

Prohibition (1920–1933) and the Great Depression (1929–1939) all but stopped bottle demand. Companies such as Owens-Illinois needed to find other outlets for their bottle glass, a soda lime silicate composition. In 1930, the Newark, Ohio, plant abruptly shut down due to excess manufacturing capacity, with only a few people remaining on site. At the same time, consultants at Arthur D. Little suggested that Owens-Illinois use the excess glass capacity to manufacture glass mineral wool insulation. In 1931, experimentalists attempted to make the wool in Alton, Illinois, but the result was dense and full of shot. Later that year, Russel Games Slayter, who was hired as a consultant to work on architectural glass block (another use for excess capacity), visited Alton and noticed stray coarse fibres hanging on roof joists and believed they might make a good filtration medium. He also surmised that finer fibre is the key to making good insulation wool. Slayter hired John Thomas as his research assistant in 1931, and, in turn, Thomas hired Dale Kleist in 1932 as his research assistant.

In August of 1932, Kleist was asked to experiment with using a metal layer gun (metalising gun or arc spray gun) fed with milk bottle glass instead of metal to try to weld glass block halves together and create a vacuum-tight seal. Fig. 1.1.6 shows excerpts from a 1927 patent describing the type of metal layer gun likely used in the experiment. The apparatus did not work as expected, and instead of making a stream of glass, it formed fine glass fibres! The men accidentally discovered that applying a jet of compressed air smoothly to a fine stream of molten glass overcomes the problems in coarse glass fibre and mineral wool forming. Fine fibres with low shot content were achieved (Fallen Foran, 1944)

Presumably, the milk glass composition was key in this discovery, as it was different from bottle glass compositions and provided the right viscosity for the temperature at which the metal layer gun operated. In the years that followed, improvements were made to the process, including scale-up, switching to steam jets and new inventions for supporting and auxiliary technologies. An image from one of Slayter's early patents showing a siphoning air jet apparatus to fiberise glass is shown in Fig. 1.1.7.

In 1933, Thomas conducted experiments with Proctor & Schwartz (Philadelphia, Pennsylvania) using glass wool instead of natural fibres in textile processing machinery. Glass wool also started to be sold as electrical wire insulation in place of silk and

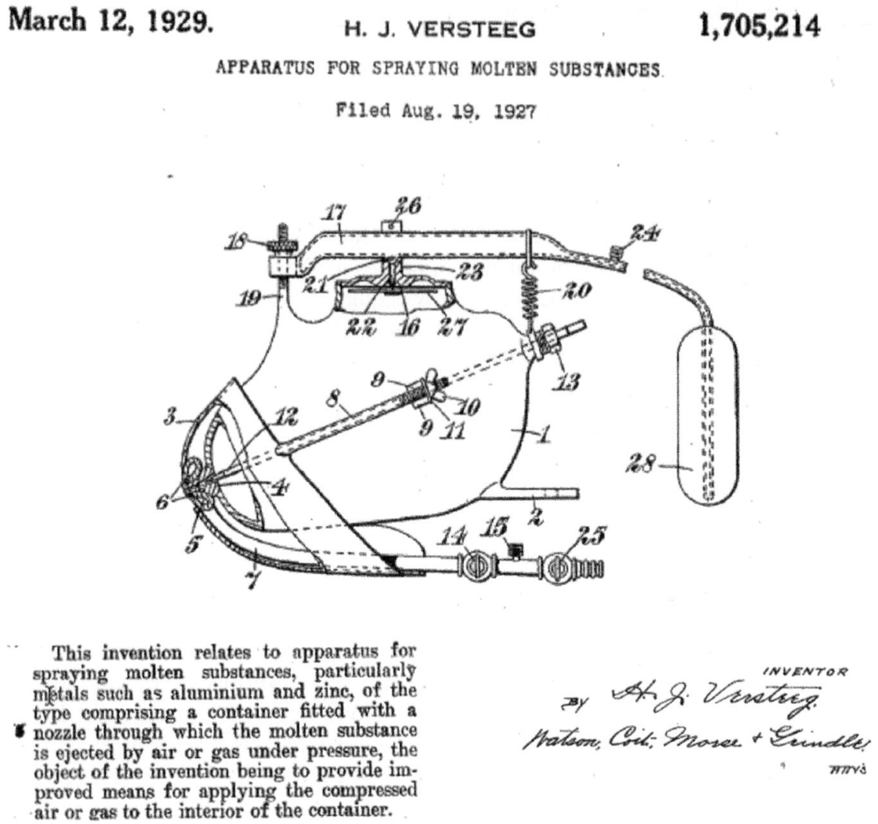

Figure 1.1.6 Excerpts from a 1928 patent describing the kind of metal layer gun used in the fine fibre discovery.

asbestos. Staple glass fibres were formed experimentally into yarn at the same time. Much attention was paid to working the fine glass like cotton to make yarn and textiles, and fine glass wool started being used in applications other than thermal insulation.

In 1935, consultants at A.D. Little and Al Simison combined glass fibre wool and Bakelite polymer, creating the first glass-reinforced plastic material. This work was captured in a 1938 patent entitled ‘Storage Battery-Box Composition’ (Simpson, 1938). Meanwhile, Dr Urban Bowes, Owens-Illinois’ director of research, in a meeting with the Navy, proposed using borosilicate glass formulas for glass fibre insulation. The borosilicate improved the chemical durability or weathering of the glass fibre wool. This was the first time a calcium borosilicate glass was made. These two advances set the stage for future glass formulations and materials combinations for GFRP composites.

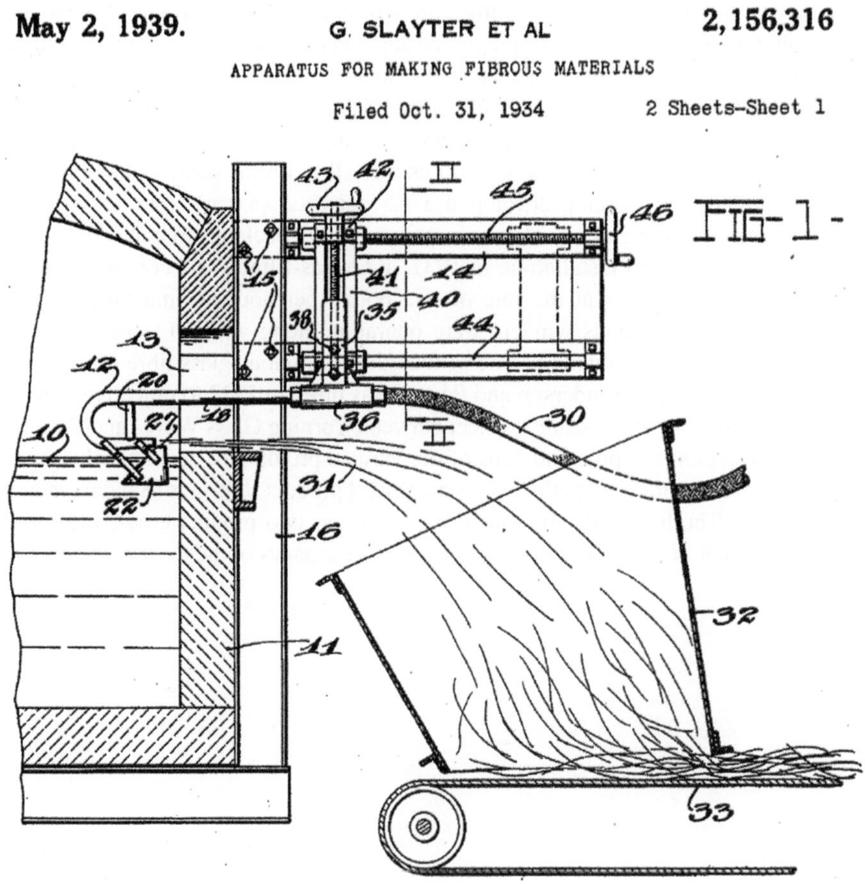

Figure 1.1.7 Siphoning air jet apparatus invented by Slayter from a 1938 patent.

2.3 Partnerships and joint ventures

George Gregory, a consultant to Corning Glass Works, advised Corning of the potential growth of the glass fibre building insulation market, and Owens-Illinois management recognised the potential advantages of tapping Corning's unique knowledge of glass formulation, especially borosilicate glasses. In October 1935, Owens-Illinois and Corning pooled their knowledge and shared the costs of glass fibre development. That same year, Slayter and Thomas conceived of the idea to mechanically pull glass fibres for textile applications; this fibre type would be known as continuous fibre or glass silk.

The first experiments used glass drawn from a thimble-sized bushing over a flannel pad, saturated with starch sizing and wound on a phenolic tube. In November of 1935, Slayter and Thomas visited E.I. duPont de Nemours & Co. in Wilmington, Del., and wrote this in his trip report, 'They are very much interested in our textile fibre and one thing they would like to work with immediately is a thin mat of textile wool or woven cloth from textile material to be impregnated and coated with a resin which they have developed to give a solid sheet of material' (Slayter, 1935).

In 1936, Owens-Illinois helped Corning establish its own experimental glass wool process and other equipment in Corning, NY. Corning mainly focused on developing the staple yarn process, while Owens-Illinois was all-in on the continuous fibre process. Later that year, Corning developed a 12-foot-wide (3.6 m) Norwood stitcher sewing machine to sew glass wool batts to paper or cloth backings to make insulation suitable for vertical wall installation. In 1937, at Owens-Illinois, six continuous fibre winders produced E-glass at the rate of one pound per hour per machine. A few months later, 40 continuous winders were operating along with 10 staple winders. E-glass marbles were produced offline to feed the continuous glass fibre operations.

An engineering firm, Sanderson and Porter, was hired in 1937 to objectively evaluate the joint glass textile research project between Corning Glass Works and Owens-Illinois to determine the pros and cons of the various prototypes, advances that were made by each company, and pathways to successful product offerings (Sanderson and Porter, 1937). The highly detailed analysis suggested two product areas: industrial textiles and decorative, fireproof textiles. The first few pages of their report are shown in Fig. 1.1.8.

SANDERSON & PORTER
Engineers
52 WILLIAM STREET NEW YORK

May 21, 1937

Corning Glass Works,
Corning, N. Y.

and

Owens-Illinois Glass Company,
Toledo, Ohio

Gentlemen:

At your request we have made a study of the developments to date in your glass textile fibre project, for the purpose of assisting you (a) in formulating suitable business and marketing policies and methods for its commercial exploitation; (b) in deciding whether to retain control of the manufacture and sale of products by others from glass yarn produced and sold to them by you; and (c), in view of the licensing and other contracts existing between you, providing for equal participation in the profits from the exploitation of certain glass products including glass textiles, in deciding whether you should together form a single corporation for the manufacture of glass textiles.

We inspected numerous samples of glass fibre in its raw and variously fabricated states and considered the uses to which it may be adapted. We inspected the glass fibre manufacturing processes and practices in your plants, and in the various outside plants where development of fabrication is being undertaken. We examined the policies, personnel, and equipment of your respective research departments; and we interviewed the more important members of those departments, the more important persons in charge of development of fabrication in the outside plants, and various other persons informed about the markets in which glass fibre may eventually compete. We examined, but not critically, the estimated manufacturing costs furnished to us by the respective companies. We made no studies of the financial aspects of the present or prospective glass fibre business.

Figure 1.1.8 First 2 pages of the Sanderson & Porter findings.

3. Mid-20th century advancements in glass reinforcement: A new industry finds its footing

The mid-20th century was a pivotal period in materials science, given the development of a range of new materials. This period was particularly important in the development and use of glass reinforcements. The materials were featured in the 1939 World's Fair in New York, and this, following the incorporation of Owens Corning in the early 1930s, can be seen as the start of the market development of the glass fibre reinforcement and composite industry (Boeschenstein, 1995). Driven by both military demands during World War II and later by several commercial opportunities, glass fibre reinforced composites emerged as a transformative technology in engineering and manufacturing. Key advancements during this period can be traced through military innovations, improvements in glass fibre production techniques, and the commercialisation of glass-reinforced composites in the decades following the war.

3.1 World War II Era: Developments driven by military needs

The outbreak of World War II created an urgent demand for new materials that were lightweight, durable and resistant to environmental stresses, such as temperature, humidity and vibration. Simison's Bakelite-glass fibre composite in 1938 led to a family tree of developing composites. Just prior to the war, the U.S. Navy adopted a glass fibre insulation combined with a phenol formaldehyde impregnated woven glass reinforcement as the insulation for all horizontal and vertical spaces in its warships. By 1944 this product was produced at the scale of 60M ft^2 (5.6M m^2) per annum and was a key contributor to a 1500% increase in sales of glass reinforcements during the prior 5 years (Boeschenstein, 1995).

Radar, an emerging technology during World War II, was critical for detecting enemy aircraft and ships. Early radars were hampered by interference from the very materials used to house them, such as metal coverings, which scattered radar signals and reduced their efficiency. This problem led to the development of radomes—covers for radar systems that allowed electromagnetic waves to pass through without interference. GFRPs proved to be an ideal material for radomes due to their mechanical strength, dielectric properties and corrosion resistance. Glass reinforcements developed for this application allowed Owens Corning Fibreglass to open a textile mill in Ashton, Rhode Island, for the manufacture of textile yarn fibres to produce reinforced materials for body armour, radomes and antennas.

By 1942, the U.S. Air Force had issued a contract to Marco Chemical Co. of Linden, N.J., for the continued development of polyester resins. At this time, the U. S. Air Force was looking for alternative materials for magnesium, plywood and aluminium, and heavily supported the development of low-pressure reinforced laminates. By the end of 1942, low-pressure reinforced plastics with glass reinforcements were being moulded by Uniroyal, Goodyear, Formica, Boeing, Douglas, Gruman, Westinghouse, General Motors, General Electric and Swedlow. The production of these GFRP materials scaled at an unprecedented rate (Milewski, 1981; Babbitt et al., 2012). During this period, the manufacturing technology evolved, and in 1942, the Continental Corp. (Muskegon, Michigan) rolled out a continuous low-pressure polyester that was approved to make high-strength polyester glass laminates for aircraft and missile components.

Glass fibre reinforced structures were first conceived, developed and designed for light airframe structures by the Air Force–Wright Field Structures and Materials Laboratory (Dayton, Ohio). After analysing test results, theoretical calculations indicated that an efficient structure could be designed and fabricated using high-strength glass fibre–polyester laminate faces with a low-density, core material. A survey of available military aircraft to select a structural component to test the feasibility of redesigning into a sandwich structure led to the selection of the aft fuselage of the two-lace Vultee BT-15 basic trainer, shown in Fig. 1.1.9. A balsa wood core was used.

Static tests performed in 1943 on the first fabricated fuselage demonstrated the predicted high structural efficiency. From a strength-to-weight basis, the GFRP sandwich structure was approximately 50% stronger than either the metal or wood type construction. In addition to its high structural efficiency, the sandwich section showed a remarkable absence of skin buckling under high torsional load. This BT-15 with the sandwich fuselage was flown on March 24, 1944, and is considered the first successful major structural component of an airplane using glass reinforcement to be developed and flown (Milewski, 1981).

R&D continued on all types of reinforced materials due to their ability to (1) provide relief to strategic materials and their supply chains, (2) reduce weight in aircraft structures, (3) provide good electrical insulation and signal transparency for radomes. Other, sometimes surprising applications emerged.

As the war continued, composites began to be used as alternatives to traditional materials where shortages were common. For example composites were developed as life jacket fillers, floatation media, various aircraft parts, aeroplane fuel tanks and wire coatings. The expanding portfolio of applications was all made possible

Figure 1.1.9 A photograph of a Vultee BT-15 aircraft from the Stan V. Ballard Special Collection, available at Wikimedia Commons.

by a vibrant research environment focused on the development of these materials that continued throughout the war. In 1944, boat hulls based on GFRP were introduced by Winner Boats Inc. (Trenton, New Jersey) and then produced in volume before the end of the war (Babbitt et al., 2012). Many of these wartime applications also found peacetime uses after the war ended.

The introduction of glass-reinforced plastics as a potential material for automotive applications started in 1941 when Henry Ford swung an ax on an automobile GFRP body to demonstrate the feasibility of these materials, given their strength and durability. Limited commercialisation happened in the automotive sector until 1945 when Owens Corning Fiberglas sponsored the production of an experimental car body with William Stout to develop the Stout-Scarab automobile – the first automobile with a GFRP body and chassis. It was never mass-produced, but it demonstrated feasibility and launched the future of GFRP car bodies that included the Kaiser-Darrin roadster, Brooks Boxer, the Glasspar G2 roadsters and the iconic Chevrolet Corvette (Lamm, 2007).

The glass reinforcement industry faced a major challenge at the end of World War II. The large infrastructure of glass production, transformation and parts production had been constructed nearly entirely for military applications. Within 1 year after the end of hostilities, demand for these materials dropped by more than 50%. The industry, which had grown significantly over a period of 10 years, needed to find new applications and new markets in which to leverage the performance of glass fibre reinforced composites. This called for a period of reinvention. Fortunately, advances in formulation and manufacturing positioned the industry well to pivot toward peacetime applications.

3.2 World War II Era: Advances in glass fibre production techniques

Prior to the late 1930s, all manufactured glass fibre was discontinuous and used for insulation applications. From 1934, with the emergence of applications that required continuous glass fibres (as discussed in the previous section), it became necessary to advance the manufacturing process for glass reinforcements. The combined technical resources of Corning Glass Works and Owens-Illinois focused on solutions for cost-effective production of continuous reinforcement fibres (before Owens Corning formed in 1938) (Boeschenstein, 1995). Prior to the continuous filament process, glass fibres were made using batch processes that limited production capacity and consistency of the resulting fibres. These earlier methods involved manually drawing molten glass into fibres, which were then cut into staple fibres and woven into usable materials.

Research had shown that the production of continuous fibres using refractory orifices was problematic, as the corrosion of the liquid glass required the orifices to be replaced every 8 hours. In 1933, technology was developed to line these orifices with a precious metal insert, or 'bushing', to prevent wear and ensure constant glass stream diameters. This was coupled with innovations around bushing transformer and busbar design, allowing precise regulation of glass flow and viscosity. Bushings were developed into separate assemblies attached beneath refractory orifices, allowing the production of stable and consistent fibres (Grant, 1934–1938).

During this period, key developments in raw material selection, selection of refractories and control systems allowed the rapid development of the industry, as consistent quality and performance of the resultant glass is key for material substitution in different applications.

By the early 1950s, manufacturing was transitioning from the traditional marble melting to a direct melt process. Direct melt is when the glass is melted from the batch directly and fibre is formed in one heating cycle, as opposed to marble melt, where the batch is made in marbles, cooled and then marbles are reheated and fed through another melter to form fibre. The Owens Corning Factory A at Anderson, S.C., was the first of these plants and started production in 1951. This furnace was 5 × 6.5 feet (1.5 × 2 m) in size with a glass depth of 6 inches (15.2 cm) and had a campaign life of 6 months. As production techniques improved, the cost of producing glass fibres dropped, making them more accessible for a wider range of applications.

3.3 Commercialisation: Emergence of new commercial applications in the 1950s and 1960s

The commercialisation of glass-reinforced composites in the 1950s and 1960s marked a significant shift from military to civilian applications. The properties that made glass fibres valuable for military applications – lightweight, high strength, corrosion resistance – were equally beneficial in a range of commercial products. The design flexibility of the material was demonstrated with examples such as the Charles Eames FRP Shell Chair by Herman Miller (like the one shown in Fig. 1.1.10), which won the Museum of Modern Art low-cost furniture competition (Boeschenstein, 1995).

Figure 1.1.10 Eames GFRP chair. Eames Office, LLC. All rights reserved.

An example of a wartime-to-peacetime application is seen in the boat-building industry, which was transformed by the introduction of glass fibre-reinforced materials. Traditional wooden boat hulls were heavy, labour-intensive to construct, and required regular maintenance. Throughout the early 1950s, metal shortages slowed boat production as the National Production Authority limited the amount of steel, aluminium and copper available for consumer goods (Writer, 2014). Glass fibres offered a lightweight, low-maintenance alternative that was easy to shape and mould into complex hull designs, resulting in the rapid growth of GFRP composite adoption. In 1953, 22 boats exhibited at the New York Boat Show contained GFRP. By 1958, 129 of 450 boats in the N.Y. Boat Show were constructed with GFRP. By the 1960s, glass fibre-reinforced composites had become the material of choice for small boat manufacturers, and it remains so today. Coincidentally, during the same period, 90% of all fishing rods sold in the U.S. were manufactured as glass-reinforced composite.

One of the earliest and most successful post-war commercial applications of glass-reinforced composites was in the automotive industry. The introduction of glass fibre-reinforced polymer body panels in automobiles started in 1951 with the Kaiser-Darrin sports car. A 1954 model is shown in Fig. 1.1.11. The glass-reinforced composite material was produced by a hand lay-up process. Only 70 vehicles, which sold for $3,668, were constructed.

The most famous composite automobile was introduced in 1953 (Museum, 2024). The Chevrolet Corvette (a 1954 model shown in Fig. 1.1.12) demonstrated the potential of glass-reinforced composites for mass-market consumer products. These GFRP parts offered a lightweight alternative to traditional metal body panels, which not only improved fuel efficiency but also allowed for more creative and aerodynamic designs

Figure 1.1.11 A 1954 Kaiser-Darrin sports car. '1954 Kaiser-Darrin front concours 6.1.19 der' by Kevauto is licensed under CC BY-SA 4.0.

Figure 1.1.12 A 1954 Chevrolet Corvette with GFRP body. Chevrolet Corvette C1 front-right 2016 'Shanghai Auto Museum' by Morio is licensed under CC BY-SA 2.0.

(Babbitt et al., 2012). Later in 1954, Ford introduced GFRP as a station wagon trim, enabling the production of artificial wood that turned white with age. In the same year, the Ford Thunderbird featured a removable GFRP hard top. In 1961, Corvette introduced the Stingray, and the Studebaker Avanti introduced an all-GFRP body. Glass-reinforced materials found utility in other automotive applications, with materials such as instrument panels and air deflectors introduced by Chrysler.

Not all automotive applications proved successful. In 1967, Goodyear launched their Polyglas tire, which was reinforced with glass cords. The tire used two plies of polyester cords plus two belt plies of glass cord (Stevens and Martin, 2018). The resulting tire delivered improved traction, skid control, low squirm and double the mileage of the previous best-selling Goodyear tires. However, eventual poor glass adhesion to rubber led to issues with fibre-to-rubber separation during flexing and potential tire failure (Contec, 2022).

Domestic applications notwithstanding, the 1950s also saw new developments in military applications using glass reinforcements. In 1957 glass-reinforced composites were used to create the superstructure on the world's first atomic submarine, Nautilus and key aircraft applications included the B-58 Hustler and B-52 Strato-Fortress. At the end of the 1950s, S-glass, a high tensile strength glass, was developed under a contract with the U.S. Navy. This glass composition, with its high tensile strength and high operating temperature, was used for filament-wound rocket motor cases, where a high strength-to-weight ratio was critical. This new glass formulation required new fiberizing technology development due to the elevated processing temperature. This led to the development of the paramelt process, which is still in use today (Machlaa et al., 1960). A paramelt process is one in which a batch is turned into glass fibres in a vertical space of less than 2 feet or 61 cm and typically allows for higher forming temperatures than traditional large-scale melting processes.

By the end of the 1950s, glass-reinforced composites were firmly adopted in commercial aircraft designs. In 1960, the Boeing 727 jet aircraft contained 2000 kg of glass fibre composites with lower cost parts than existing metal parts, and reduced weight by 33%. At the same time the Douglas DC-8 jetliners each contained 1000 kg of composites, including unique glass fibre-reinforced plastic structural parts of spar and vertical tail section.

The construction industry was another early adopter of glass fibre-reinforced materials. In the late 1940s, centrifugal casting of glass-reinforced composites was the first machine-made method to produce pipes suitable for chemical and commercial applications and oil field gathering lines (Fig. 1.1.13). Later, the filament winding process was developed to manufacture pipes with tensioned glass fibres oriented to bear the combination of hoop and axial forces, enabling reinforced piping. It became a viable alternative to protected steel, stainless steel and other costly and high-performance materials by 1950 (Curran, 2013).

The first centrifugally cast glass fibre pipe was produced in Switzerland in the 1950s for use in the crude oil production industry to solve corrosion problems, eventually leading to the formation of the Hobas company in 1957. In the early 1960s, the improved winding technology enabled the development of glass-reinforced storage tanks, where the combination of durability and resistance to corrosion resulted in significant adoption of the solution (Kass et al., 2012). By the mid-60s, glass fibre

Figure 1.1.13 Oil industry GFRP pipe being laid.
Photo Credit: Owens Corning.

composites were accepted for the storage and handling of chemical and underground flammable and combustible liquids in industrial, municipal water, sewage and pulp and paper processing applications. Later, during the 1960s, manufacturers began to develop nationally recognised standards and test methods for glass fibre composite municipal piping systems. Today, there are several nationally recognised standards and specifications for glass fibre composite piping.

While sewer and drainage piping continue to be dominated by concrete, there are many areas where glass fibre-reinforced composite is critical. For example, concrete pipe deteriorates rapidly in sewage from hydrogen sulphide attack. GFRP is unaffected by hydrogen sulphide or purging with caustic or hypochlorite (to suppress sulphide odour). As a result, the industry lines large diameter (48–60 inches, 1.2–1.5 m) concrete pipe with GFRP liners.

The sanitary industry developed the first one-piece shower stalls in 1961, which have become ubiquitous in new housing construction. Other applications, such as bathtubs, shower bases and shower trays, developed quickly.

Perhaps the most iconic development during this period was the introduction of beta yarn in 1963. This fine fibre was used to produce fireproof NASA spacesuits (Orndoff, 2016).

4. Technological innovations

4.1 History of the evolution of glass fibre formulations and the test methods that pushed the envelope

4.1.1 The 1930s and 1940s focused on early technological uses and some decorative techniques

As discussed earlier, the experiments of the 1930s that led to the discovery of technological glass fibres involved glass compositions that came from the bottle/container industry. These compositions – lime silicates and soda lime silicates –were generally not well documented or understood at the time. Only wet chemical analyses techniques were available to ascertain glass chemistries experimentally, and they were slow and required specialised skills to conduct reliably. Once a glass is made in a fine fibre form, due to the high specific surface area involved, the durability (or lack of) quickly becomes apparent. In the mid-late 1930s, Owens-Illinois and Corning started exploring different glass compositions. Corning was working with their #124-HD soda lime bulb glass for most of the staple fibre experiments. These, too, experienced terrible weathering in simple storage conditions.

The wool glass had poor durability, and there had been issues observed when using wool glass in contact with aluminium. In a railcar, wool glass insulation had been used; it settled to the bottom due to the vibration and, in combination with some asbestos-covered brass wire, had corroded holes in the aluminium. It was thought that the leaching of alkali from the glass attacked the aluminium metal. This was highly concerning because this wool material was being supplied to the Navy for battleships. Extensive testing at the Aluminium Co. and both at Newark and Alton labs was carried out. They determined that while in the presence of moisture, wool glass can hydrolyse, forming alkaline solutions and attack aluminium. The attack was exacerbated by the presence of the copper (brass) wire; the effect was much less pronounced with glass alone.

The solution was to improve this product using a new glass formulation. To achieve this and improve the glass manufacturability (advancements in this area were happening rapidly as well as the time), the milk bottle composition was adjusted by increasing CaO and adding B_2O_3 to reduce viscosity, increasing Al_2O_3 to improve durability and to reduce liquidus temperature Na_2O was reduced. This higher durability borosilicate glass, called 23-H, was used in the Navy wool glass products; it is also called 'T-glass' (T for thermal) and was patented by Bowes (1943). C-glass (C for chemical) was also developed around this time and was a refined version of T-glass with lower Na_2O still and higher SiO_2. This was also used for the bulk of the Navy wool glass later. The general naming convention for glass family designation uses a letter indicating its most beneficial property which makes selecting the glass for applications requiring this property straightforward. Besides T- and C- glass, E–, R- and E-CR glasses would soon be developed.

In March of 1936, in a report summarising a call with 'Wire and Cable people', Simpson states, 'If we can do anything by batch composition or method of manufacture to increase the toughness of the fibres and by chemical composition to increase the dielectric strength, then we would have two of the biggest objections licked' (Simpson, 1936). From the fall of 1936 onward, attention was paid to alkali-free compositions which were judged to be clear of the weathering problem. The alkali-free glass also showed superior electrical (low conductivity) properties and better mechanical properties.

At Owens-Illinois, this composition was the first E-glass (E for electrical) formulation, patented by Schoenlaub, with 4.5 wt% MgO, 10% B_2O_3, 14% Al_2O_3, 17.5% CaO and 54% SiO_2 (Schoenlaub, 1943). At Corning the formula was #719-MQ and coined as their standard glass for textile use and worked well to produce staple fibres. From an overall production standpoint, the E-glass eventually proved to be better because of its lower devitrification (crystallisation) temperature, lower bushing operating temperature, lower cost and better formability into marbles. Fig. 1.1.14 is a photograph of a General Research Laboratory report to Schoenlaub, where the E-glass composition was measured after melting in two different environments: a platinum crucible and a larger refractory-lined furnace called a day tank.

Attempts to create glass fibres for colourful, decorative textiles proved unfeasible. Much of the work was either not compatible with the platinum bushings or caused too much damage to the fibre which compromised its strength. However, most of the nomenclature used in the reinforcement glass industry came from the textile industry, hearkening back to early work when Owens-Illinois, Corning and Owens Corning worked closely with various weavers and textile machine manufacturers.

Serendipitously, colouring treatments done by Hyde led to some discoveries of electrical and mechanical performance enhancements by exchanging ions in the glass with those in various salt solutions (Hyde, 1941, 1943). Thus the focus shifted to

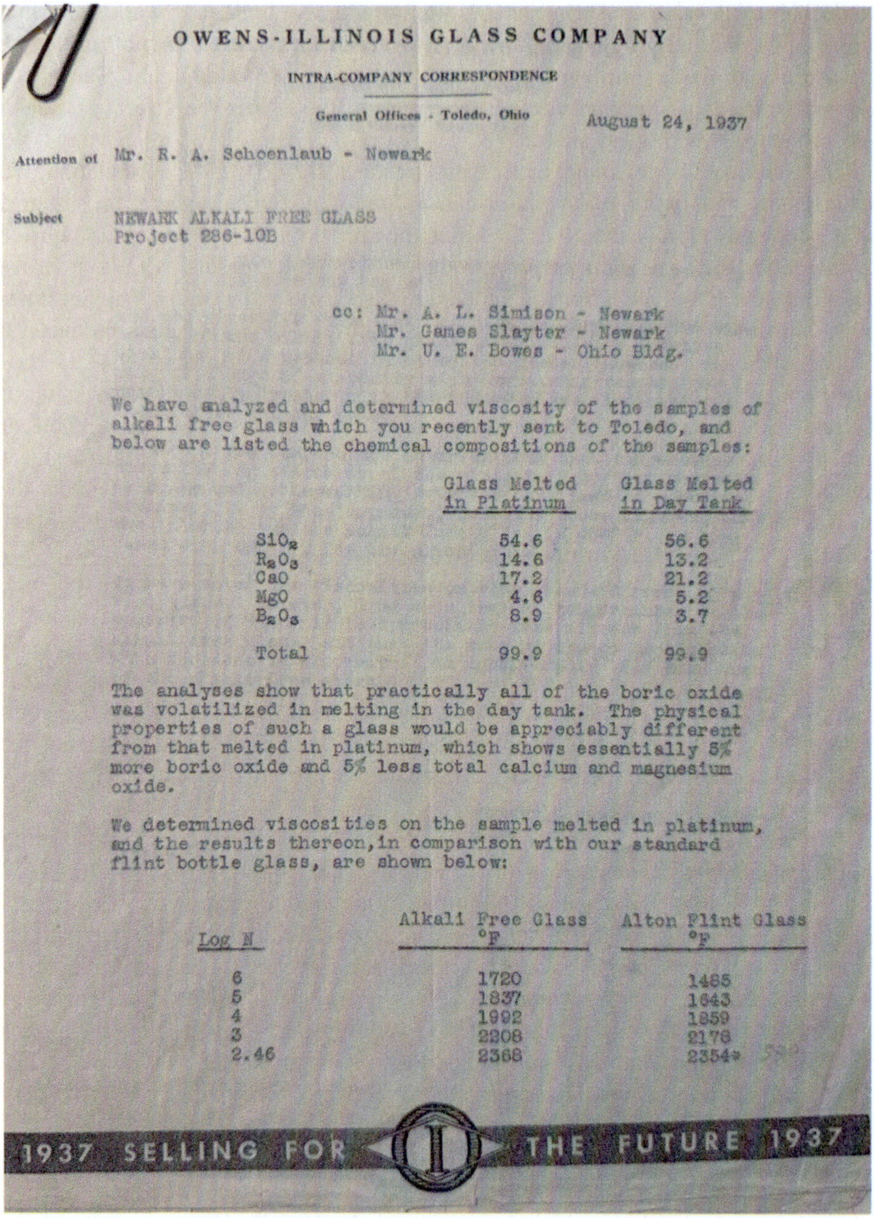

OWENS-ILLINOIS GLASS COMPANY

INTRA-COMPANY CORRESPONDENCE

General Offices - Toledo, Ohio

August 24, 1937

Attention of Mr. R. A. Schoenlaub - Newark

Subject NEWARK ALKALI FREE GLASS
Project 286-10B

cc: Mr. A. L. Simison - Newark
Mr. Games Slayter - Newark
Mr. U. E. Bowes - Ohio Bldg.

We have analyzed and determined viscosity of the samples of alkali free glass which you recently sent to Toledo, and below are listed the chemical compositions of the samples:

	Glass Melted in Platinum	Glass Melted in Day Tank
SiO_2	54.6	56.6
R_2O_3	14.6	13.2
CaO	17.2	21.2
MgO	4.6	5.2
B_2O_3	8.9	3.7
Total	99.9	99.9

The analyses show that practically all of the boric oxide was volatilized in melting in the day tank. The physical properties of such a glass would be appreciably different from that melted in platinum, which shows essentially 5% more boric oxide and 5% less total calcium and magnesium oxide.

We determined viscosities on the sample melted in platinum, and the results thereon, in comparison with our standard flint bottle glass, are shown below:

Log N	Alkali Free Glass °F	Alton Flint Glass °F
6	1720	1485
5	1837	1643
4	1992	1859
3	2208	2178
2.46	2368	2354*

1937 SELLING FOR THE FUTURE 1937

Figure 1.1.14 O-I (Owens-Illinois) General Research Laboratory report regarding E-glass composition and viscosity measurements.

industrial textiles such as filter cloths, acid bottle closures, gas filtration bags, anode bags, electroplating baths etc.

Durability against acids became important in the late 1930s. Owens-Illinois developed a glass called #F-16 containing lithium oxide, high silica and strontium oxide (replacing the calcium fluoride flux). This glass showed superior acid corrosion resistance over the previous glasses. Later in the 1940s, Tiede and Tooley patented another alkali-free glass, also free of MgO, and claimed it had better devitrification tendencies. This glass had only four oxides: 54 wt% SiO_2, 22% CaO, 14% Al_2O_3 and 10% B_2O_3.

In 1948, low dielectric glasses were investigated for use in radomes between Owens Corning and the US Air Force. The designation D-glass (D for dielectric) came about. Companies such as Advanced Glass Yarns (AGY) supply low dielectric glass fibres. AGY was formed in 1998 between Owens Corning and Groupe Porcher Industries of France. Owens Corning sold its half of AGY in 2000.

4.1.2 The 1950s and 1960s focused on applications that became mainstream and mastery of glass chemistry

Between 1960 and 1963, Beta yarn was developed and was used in astronauts' spacesuits thanks to a new, revolutionary bushing design. New glasses were considered but the new bushing design made it possible to use E-glass attenuated to less than 4 μm in diameter. Beta fabric was a dual innovation between Owens Corning and DuPont. Owens Corning contributed the glass, and DuPont supplied the Teflon coating that together made the fabric 'fireproof' and safe for the astronauts to wear. Today, Birdair, originally from Buffalo, N.Y., manufactures architectural fabrics for impressive buildings all over the world (Contribution, 2009), showing the broad versatility of E-glass fibre fabrics.

The standard E-glass composition for decades at Owens Corning, called 200E, was descended from Schoenlaub's original composition, but it contained less B_2O_3. Glasses now encompassing the term 'E-glass' include a wider range of compositions than in Schoenlaub's patent. E-glass was designed for use in electrical applications (E for electrical use) and needed to be low in alkali content. 200E was made into marbles using the Akro-Agate marble machine design and these marbles were used in melters all over Owens Corning. Marbles could easily be inspected for stones and this improved the glass quality in fiberising. Some later research on glass compositions resulted in 621E, which was based on the Tiede/Tooley patent but with lower B_2O_3 and had a lower liquidus temperature than 200E. This was critical to being able to produce this glass using the direct melt process.

The Aerocor superfine insulation material for the aerospace industry was made from 200E and then 621E glass using direct melt. These compositions typically contained B_2O_3 and some amount of fluorine. 200E, 621E and a few other variations were the main glass reinforcement compositions produced at Owens Corning throughout the 1950s, 1960s, 1970s and 80s.

A variation of 621E is still made today. Reducing boron oxide due to its high cost and other compositional changes were forced on the industry due to raw material supply issues over the years. As finer raw materials were developed and as refractories, furnace design and bushing design improved the need for boron to aid in melting and reducing viscosity was no longer required.

As GFRP began to compete with steel in the 1950s, more interest and attention were paid to increasing the glass elastic modulus, the tensile strength and the service temperature. Advancements in measurement capabilities also made detailed research projects more fruitful. Use of tensile testing machines with very sensitive load cells allowed for single glass fibres to be tested and their maximum force and strain to be measured. Many research projects were active at this time in Owens Corning and in Europe, namely at Saint-Gobain Vetrotex, to develop glasses with higher mechanical performance. Several projects were associated with government contracts as well to create more advanced materials for military and aerospace use.

Among the glasses studied, X994 at Owens Corning was of high interest to the U.S. government, and this glass became known as S-glass (S for strength). It was patented in 1968 by Harris and Machlan with a priority date of 1962 (Harris and Machlan, 1968).. S-glass had only three oxides, 65 wt% SiO_2, 25% Al_2O_3, 10% MgO and boasted the highest tensile strength and service temperatures known at that time. While the high-strength glass research went into production, the high modulus research did not take off at that time. Other companies working in this area were Pilkington, Imperial Glass and United Aircraft Research Labs.

The Société du Verre Textile, which became Saint-Gobain Vetrotex, patented an R-glass in 1966 with high Al_2O_3 and MgO contents (Groene, 1966). In French, résistance is strength; in English, so the designation of R-glass for Resistance in Europe took hold. According to the American Society for Testing and Materials (ASTM) D578, R-glasses contain primarily SiO_2, Al_2O_3, MgO and CaO, where S-glasses contain primarily SiO_2, Al_2O_3 and MgO. Both compositions may contain some impurities or minor ingredients.

With the growing popularity of filament winding in military applications, the U.S. Naval Ordnance Laboratory conducted a large research programme on glass fibres in the mid-1960s. One outcome of the research was the development of a sonic single-fibre elastic modulus test instrument. This test technique offers the critical advantage that the specimen form, a fibre, is measured directly, and properties are not inferred from bulk glass forms, which may not have the same density or structure as the same glass in fibre form (Fisher, 1965).

4.1.3 The 1970s, 1980s and 1990s focused on improvements in manufacturing and application refinement

In the 1970s, manufacturers became aware of the need to reduce emissions from manufacturing operations. Owens Corning started to investigate the removal of fluorine and boron from the glass melt. Both constituents are volatile in the melt and tend to escape out of the stack to some extent. It was a happy coincidence that boron was also one of the most expensive batch ingredients, so reducing or removing it altogether was also a cost savings. In 1974, Wolf patented a fluorine-free composition, and that same year also patented the AR-glass composition (AR for alkaline resistance) with high alkaline-condition durability for reinforcing concrete, commonly known as Cem-FIL (Wolf, 1974a, 1974b).

That same year and the following, Erickson and Tiede of Owens Corning developed a few glass compositions without boron oxide and some combination of MgO, TiO_2 and ZnO added (Erickson and Wolf, 1974a, 1974b, 1975). These compositions were the first ECR glass fibres produced and known as Owens Corning glass 3709. ECR or E-CR stands for Electrical with Corrosion Resistance, namely acid corrosion resistance. In 1976, Erickson and Wolf also researched and patented some of the first refractive index-matching compositions for making fibres invisible in various resins (Erickson and Wolf, 1976).

The first specifications for glass fibre compositions came from the ASTM, with ASTM D578 originally appearing in 1940. Fig. 1.1.15 shows the E-glass compositional ranges and the C-glass and S-glass definitions as specified in the ASTM D578 in 1989 (ASTM, 1989). Also shown in the figure is a note about additional glass family definitions being considered at that time. In 1996, Section 4.2.2 was added to the standard as shown in Fig. 1.1.16. The standard then included the ECR definition but continued to require 5%–10% B_2O_3 in E-glass. In other words, an ECR glass could not be an E-glass.

Two years later, the specification was changed to reflect the definitions shown in Fig. 1.1.17. In 1998, the ASTM glass fibre subcommittee adopted a boron- and fluorine-free E-glass, and in so doing, allowed ECR glasses to also be considered E-glasses. Advantex® was the first boron- and fluorine-free E-glass and E-CR glass. Advantex® was a pseudonym combining the words Advantage and Textile. It was invented by Eastes, Hofmann and Wingert that same year (Eastes et al., 1996). The U.S. Department of Defense used ASTM standards as the basis for developing the military-specific standard, MIL-R-60346.

The below three figures are reprinted, with permission, from ASTM D578-XX Standard Specification for Glass Fibre Strands, copyright ASTM International, www.astm.org.

4.1.4 The 2000s until today focused on detailed improvements in key applications

High modulus glass research began again in the 2000s to meet the demand of the wind turbine industry. Turbine blades have been using E-glass and Advantex®, but the drive to make longer blades and generate more electricity requires higher elastic modulus glass reinforcement.

In 2004, Creux and Lecomte of Vetrotex patented H-glass (Creux, 2004). The H-glass terminology is mainly an internal designation to Owens Corning; H stands for High, both high strength and high modulus. Externally, H-glass falls in the R-glass definition per ASTM D578-18. Because most glasses in the R-glass family are used for their high elastic modulus, the R may also be considered to also stand for Rigidity, a synonym for stiff or having a high elastic modulus. In 2005, Hoffmann and McGinnis published Owens Corning's H-glass patent application, and in 2006 Lecomte and Berthereau patented an H-glass containing Li_2O (Hoffmann and McGinnis, 2008,

4. Classification of Glass Fiber

4.1 *"C" Glass*—A family of calcia-boro-alumina-silica glasses with a certified chemical composition which conforms to an applicable material specification and which produces good acid resistance (excluding HF).

4.2 *"E" Glass*—A family of calcia-alumina-silicate glasses which has the following certified chemical composition and which is used for general purposes and most electrical applications.

Chemical	% by Weight
B_2O_3	5 to 10
CaO	16 to 25
Al_2O_3	12 to 16
SiO_2	52 to 56
MgO	0 to 5
Na_2O and K_2O	0 to 2
TiO_2	0 to 1.5
Fe_2O_3	0 to 0.8
F_2	0 to 1.0

4.3 *"S" Glass*—A family of magnesium-alumina-silicate glasses with a certified chemical composition which conforms to an applicable material specification and which produces high mechanical strength.

Note 1—ASTM Subcommittee C14.01 on Nomenclature and Definitions under the jurisdiction of Committee C-14 on Glass and Glass Products is establishing definitions for various families of glasses used in the glass fiber industry. These families include A, AR, C, D, E, R, and S glass families. The description "E-CR-Glass, E-Glass compositions modified for improved resistance to acid and corrosion (excluding HF)" is being considered by Subcommittee C14.01. Once approved, Subcommittee D13.18 on Glass Fiber and Its Products will move to adopt these definitions in Specification D 578.

Figure 1.1.15 Excerpts from ASTM D578-89a, published in 1989, showing the definitions of reinforcement glass fibres at that time.

4.2.2 The nomenclature "E-CR-Glass" is used for boron-free modified E-Glass compositions for improved resistance to corrosion by most acids.

Figure 1.1.16 The added comment about E-CR glass in ASTM D578 in 1996.

2010; Lecomte and Berthereau, 2006). The H-glass competition did not last long as Saint-Gobain Vetrotex and Owens Corning merged in 2007. After the acquisition, Owens Corning manufactured the Li_2O-containing composition because it had patent protection at the time.

Since the creation and rapid advancement of electric vehicles, the price of Li_2O has skyrocketed, and its availability is becoming scarcer. This necessitated the development of lithium-free high modulus glasses. In 2018, Korwin-Edson developed and patented such a glass (Korwin-Edson and McGinnis, 2022). Owens Corning produces this glass today, and several competitors produce their versions of this high modulus glass, some with and some without lithium.

Table 1.1.1 summarises the various glass families and their associated properties. The race to create glass fibre with ever-increasing modulus and specific modulus continues today on a global scale. Presumably, new innovations for GFRP applications will drive more development and extend the 'alphabet' of glass compositions for glass reinforcement.

4.2 Glass fibre surface treatments

One of the first products for glass fibre was air filtration media, and scientists quickly discovered that it needed a starch applied to increase the adhesion of dust to the filter. For glass fibres to be incorporated into a resin matrix, an interface must be applied to the fibre to allow for good adhesion (Loewenstein, 1993). Because glass is an inorganic material, and resins are organic, a sizing agent needs to be applied that has functional groups that are compatible with each material and can create a bond. The development of sizes/sizings for glass fibres happened alongside product development, glass composition, furnace, bushing, process and winding/chopping development. Some very good reviews of size chemistry development can be found in Thomason's papers, so just a brief account of the evolution is shared here (Thomason, 2001, 2015). The term size comes from the textile industry in that a coating applied to a natural fibre was meant to control the dimensional size or diameter of that fibre and keep it from expanding. Glass does not expand in this way, but the terminology was set long ago.

In the glass fibre industry, sizes go to work immediately after they are applied to the glass by lubricating and protecting the fibre as it is handled by forming equipment. Sizes are applied within microseconds after the fibre is formed. After the size is cured, and depending on the product form (chopped, wet/dry, single strand, multi-filament), it holds fibres together, allows them to be separated, assists in resin wet-out and eventually forms the bonding with the resin matrix (just to name a few of its functions). Size formulations are highly complex and dynamic, and each manufacturer has their

4. Classification of Glass Fiber

4.1 *"C" Glass*—A family of glasses composed primarily of the oxides of sodium, calcium, boron, aluminum, and silicon with a certified chemical composition which conforms to an applicable material specification and which produces good acid resistance (excluding HF).

4.2 *"E" Glass*—A family of glasses composed primarily of the oxides of calcium, aluminum, and silicon, which has the following certified chemical compositions.

4.2.1 The following certified chemical composition applies to glass fiber yarn products for printed circuit boards and aerospace.

Chemical	% by Weight
B_2O_3	5 to 10
CaO	16 to 25
Al_2O_3	12 to 16
SiO_2	52 to 56
MgO	0 to 5
Na_2O and K_2O	0 to 2
TiO_2	0 to 0.8
Fe_2O_3	0.05 to 0.4
Fluoride	0 to 1.0

4.2.2 The following certified chemical composition applies to glass fiber products used in general applications.

Chemical	% by Weight
B_2O_3	0 to 10
CaO	16 to 25
Al_2O_3	12 to 16
SiO_2	52 to 62
MgO	0 to 5
Na_2O and K_2O	0 to 2
TiO_2	0 to 1.5
Fe_2O_3	0.05 to 0.8
Fluoride	0 to 1.0

4.2.3 Electrical applications include a wide variety of uses. The composition in 4.2.1 is identical to IPC-EG-140 for printed circuit boards and to MIL-G-55636. Additionally, such fiber glass products often are specified for aerospace applications. Products covered by the composition range in 4.2.2 are used in general applications, such as power company equipment, high voltage devices, residential electric boxes, third rail covers, high voltage standoff rods, electrical pultrusion products, light poles, electrical tool covers, and electrical tape. Other applications include roofing, flooring, filtration, panel rovings, gun rovings, smc rovings, chopped strand reinforcements, paper yarns, and industrial yarns.

4.2.4 The nomenclature "E-CR-Glass" is used for boron-free modified E-Glass compositions for improved resistance to corrosion by most acids.

4.3 *"S" Glass*—A family of glasses composed primarily of the oxides of magnesium, aluminum, and silicon with a certified chemical composition which conforms to an applicable material specification and which produces high mechanical strength.

Figure 1.1.17 ASTM D578 excerpts showing the adoption of an E-CR glass under the E-glass family in 1998.

Table 1.1.1 Types of reinforcement glass fibres and their general properties.

Properties, test conditions (units)	AR-glass	C-glass	D-glass	E-glass	E-CR glass	P-glass	S-glass
Bulk density, unannealed (g/cm^3)	2.73	2.5	2.14	2.62	2.65	2.56	2.44
Refractive index, fibre	1.566	1.509	1.452	1.562	1.567	1.545	1.521
Vickers hardness	617	514	–	620	626	667	651
Pristine fibre tensile strength, 22°C (MPa)	3525	3390	2900	3820	4050	5000	5250
Elongation at break (%)	4.4	6	4.5	5.6	5.9	5.3	6.8
Fibre elastic modulus, 22°C (GPa)	77	68	58	78	82	87	88
Coefficient of thermal expansion, 50–300°C (×10.6 cm/cm°C)	9.9	10.1	–	6.2	6.6	5.3	4.5
Specific heat capacity @ 30°C (J/g°C)	–	0.78	0.73	0.81	0.79	0.75	0.81
Strain point, Log 14.5 poise (°C)	598	487	477	632	680	726	681
Annealing point, Log 13 poise (°C)	642	528	520	670	722	772	727
Thermal conductivity @ 25°C	–	–	–	1.2	1.1	–	1.3
Dielectric Constant @ 1GHz	–	7.06	3.8	6.9	7.16	6.28	5.76
Dissipation factor @ 1GHz	–	0.0097	0.0005	0.0055	0.0061	0.0064	0.0098

own 'secret sauce'. The very early glass yarns had a starch and/or oil size applied for protection and runnability in the loom.

In the 1930s, polyester resins reinforced with glass fibres had less than expected mechanical properties, so a size was used to improve hydrolytic resistance. Research in the 1940s and 1950s led to the discovery of the silane coupling agents to improve glass-to-resin bond (Hyde, 1945). These critical molecules creating the bond are generally organosilanes. At the same time, improvements were needed in runnability such as fuzz reduction and so cationic lubricants entered the formulation. The last critical piece of development for sizings was in the realm of film formers to provide protection to the glass and compatibility with matrix resins. For a complete review of the evolution of sizes for glass fibres please refer to the comprehensive text by Thomason (2015).

5. Challenges and solutions for the use of glass reinforcements in composites

Glass fibre reinforcements are widely used in composite materials due to their excellent mechanical properties. Despite being a relatively mature technology, the production of glass fibres still faces significant technical hurdles, given the nature of the process.

High Energy Consumption: Producing glass fibres requires significant amounts of energy, as it involves melting raw materials such as silica at high temperatures (above 1400°C). The process is energy-intensive, and managing the associated costs, energy efficiency and resulting emissions can be a challenge for manufacturers.

Raw Material Quality: The quality of raw materials used in glass fibre production is critical to the performance of the final product. For example impurities in the silica or other components can lead to defects in the fibre, affecting its mechanical properties such as tensile strength and flexibility.

Consistency in Fibre Diameter: A primary technical challenge in glass fibre production is maintaining consistent fibre diameter. Variations in diameter can lead to performance inconsistencies, making the composite less reliable in demanding application sectors such as aerospace or automotive.

Fibre Breakage and Degradation: Glass fibres are relatively brittle, and mishandling during production can lead to breakage. This compromises the mechanical integrity of the composite material. In addition, glass fibres are prone to surface degradation when exposed to moisture and chemicals, which can diminish their bonding capacity with the polymer matrix in composite materials.

Dust and Safety Concerns: During cutting and machining operations, glass fibres produce fine dust that can be hazardous if inhaled. This poses occupational safety challenges, requiring the use of protective equipment and effective dust collection systems to mitigate health risks.

Recyclability: One of the biggest challenges related to the use of glass fibre reinforcements is the current recyclability. Glass fibre reinforced composites are often thermoset based, meaning they cannot be melted down and remoulded like thermoplastics, posing a significant challenge when it comes to end-of-life management of glass fibre reinforced composites.

6. Current trends and future directions for glass reinforcements

The field of glass fibre reinforcement is experiencing significant innovation, driven by the demand for lightweight, high-performance materials across industries such as automotive, aerospace, construction and renewable energy. Over the next 5–10 years, several key development trends are expected to shape the industry.

Higher-Performance Glass Fibres: One of the primary trends is the development of high-performance glass fibres with enhanced mechanical properties, such as improved tensile strength and stiffness. Researchers are working on novel glass chemistries that offer superior properties compared to traditional E-glass formulations. These new formulations are designed to meet the increasing demands for strong, durable materials in applications where weight reduction and structural integrity are critical, such as wind energy.

Sustainability and Recycling: Sustainability is becoming a critical focus in many areas of the world. Many researchers focus on reducing the impacts of glass fibre production through energy-efficient processes, including the use of hydrogen as an alternative to gas-fired melting. New recycling technologies are being developed to recover and reuse glass fibres from composites at the end-of-life, which will become increasingly important as global regulations and circular economy initiatives evolve.

Improved Resin Compatibility: Another trend is the enhancement of resin compatibility with glass fibres. Surface treatments and coatings are being developed to improve the adhesion between glass fibres and polymer matrices, resulting in strong, durable composites. This research includes the incorporation of nanotechnologies such as graphene. This is particularly relevant in thermoplastic composites, which are becoming more popular for their recyclability and manufacturing advantages.

7. Summary

The history of composites has been enabled by the rapid development of glass reinforcements, enabling GFRP materials with superior performance, durability and cost efficiency compared to traditional materials. Given the requirement for materials to support the energy transition and technologies enabling performance with reducing the negative impact through their life cycle, the role of glass fibre composites – and the glass fibre technology on which they are constructed – will continue to serve as a cornerstone of modern materials engineering. The next decade will see innovations in glass fibre reinforcements that are driven by performance demands, sustainability and the evolution of new composite materials and applications.

References

Adolphs, G., Skinner, C., 2011. Non-crimp fabric composites in wind turbines. In: LOMOV (Ed.), Non-Crimp Fabric Composites – Manufacturing, Properties and Applications. Woodhead Publishing.

Alyeska Pipeline Service Company, 2021. Alyeska Pipeline Service Company. Available: https://www.alyeska-pipe.com/.

ASTM, 1989. Standard specification for glass fibre strands. In: Annual Book of ASTM Standards. ASTM, Philadelph+ia, PA.

Babbitt, B., Berdan, C.I.I., Carlier, E., Chacko, J., Hager, T., Vermilion, D., Wingert, J., Wolf, D., 2012. Compendium of Owens Corning's Technical Achievements - Updated to 2011.

Bedon, C., 2016. Review on the use of FRP composites for façades and building skins. Am. J. Eng. Appl. Sci.

Boeschenstein, W.W., 1995. They didn't know what they couldn't do. In: Glass Industry, July ed.

Bowes, U.E., 1943. Sodium calcium borosilicate glass. In: US patent application.

Cognet, Courrech du Pont, Thiria, 2020. Material optimization of flexible blades for wind turbines. Renewable Energy 160, 1373–1384.

CONTEC, 2022. Rubber and Beyond: What Are Tires Made of? [Online]. Contec. Available: https://contec.tech/rubber-and-beyond-what-are-tires-made-of/.

Contribution, O.T.N., 2009. Tensile Fabrics Enhance Architecture Around The World. NASA Spinoff.

Creux, E.L.S., 2004. Glass Composition. FR patent application FR1435073.

Curran, S.D., 2013. Municipal Fiberglass Pipe: Performance and Design Standards.

Dubus-Bonnel, I., 1836. FR Patent Application 1BA5318.

Duffield, W.A., Gibson, E.K., Heiken, G., 1977. Some characteristics of Pele's hair. J. Res. U. S. Geol. Surv. 5, 93–101.

Eastes, W.L., Hofmann, D.A., Wingert, J.W., 1996. Boron-free glass fibres. US patent application.

Erickson, T., Wolf, W., 1974a. Glass compositions, fibres and methods of making same. US patent application US3847626A.

Erickson, T., Wolf, W., 1974b. Glass compositions, fibres and methods of making same. US patent application US3847627A.

Erickson, T., Wolf, W., 1975. Glass compositions, fibres and methods of making same. US patent application US3876481A.

Erickson, T., Wolf, W., 1976. Glass compositions and their fibres. US patent application US3945838.

Fallen Foran, E., 1944. History of the Research and Development of Fibrelgas April 1931 - December 1939. Owens Corning Owens Corning. Internal Report, Newark.

Fan, Q., Duan, H., Xing, X., 2024. A review of composite materials for enhancing support, flexibility and strength in exercise. Alex. Eng. J. 94, 90–103.

Fisher, Joy, 1965. Glass Fiber Drawing and Measurement Facilities at the US Naval Ordnance Laboratory. Chemistry Research Department, US Naval Ordnance Laboratory.

Fu, S., Lauke, B., Mai, Y., 2019. Science and Engineering of Short fibre-reinforced Polymer Composites. Elsevier.

Girl, V.R., Neelakandan, R., Nadarj, R.N., 2006. Glass fibres - an overview. Synthetic Fibres 35, 16–22.

Grant, J.A., 1934-1938. Early History of Fiberglas in Textile Processes Development. Granville, Ohio. Owens Corning Internal Report.

Groene, E., 1966. Glass Composition. FR patent application FR1435073A.

Harris, R.S., Machlan, G.R., 1968. Glass Composition. US patent application US3402055A.

Hoffmann, D.A., Mcginnis, P.B., 2008. Composition for High Performance Glass, High Performance Glass Fibres and Articles Therefrom. US patent application US20080009403A1.

Hoffmann, D.A., Mcginnis, P.B., 2010. Composition for High Performance Glass, High Performance Glass Fibres and Articles Therefrom. US Patent Application US7799713B2.

Hyde, J.F., 1941. Method of Coloring Glass Fibres. US Patent Application US2245783.
Hyde, J.F., 1943. Treating Glass Fibres. US patent application US2315259.
Hyde, J.F., 1945. Manufacture of Glass Yarn. US patent application US2390370.
Kaldellis, J.K., Zafirakis, D., 2011. The wind energy (r)evolution: a short review of a long history. Renew. Energy 36, 1887–1901.
Kass, M.D., Theiss, T.J., Janke, C.J., Pawel, S.J., 2012. Analysis of Underground Storage Tank System Materials to Increased Leak Potential Associated with E15 Fuel. Oak Ridge National Laboratory.
Korwin-Edson, M., Mcginnis, P.B., 2022. High Performance Fiberglass Composition. US patent application US11214512.
Lamm, Michael, 2007. The Fiberglass Story. Invention and Technology Magazine.
Lecomte, E., Berthereau, A., 2006. Glass Yarns for Reinforcing Organic and/or Inorganic Materials. WO patent application WO2006064164A1.
Loewenstein, K.L., 1993. The Manufacturing Technology of Continuous Glass Fibres
Longobardo, A.V., 2010. Glass fibres for printed circuit boards. In: WALLENBERGER, F.T., BINGHAM, P.A. (Eds.), Fiberglass and Glass Technology: Energy-Friendly Compositions and Applications. Springer US, Boston, MA.
Machlaa, G.R., Stalego, C.J., Tiede, R.L., Isham, A.B., Caramante, D.E., 1960. High modulus, high temperature glass fibres for reinforced plastics. In: CENTRAL, N.M.L.M. (Ed.), Wright Air Development Division.
Malarky, G., 2001. The History of Asphalt Shingles. Accessed 15th April.
Mentgesa, N., Çelika, H., Hopmanna, C., FagerströMB, M., Mirkhalafc, S.M., 2023. Micro mechanical modelling of short fibre composites considering fibre length distributions. Composites, Part B 264, 1–12.
Milewski, J.V., 1981. History of reinforced plastics. J. Macromol. Sci. 15, 1303–1343.
Morey, G., 1938. Properties of Glass. Reinhold Publishing Corporation, New York.
Morey, G., 1960. Properties of Glass. Reinhold Publishing Company, New York.
Museum, N.C., 2024. About Corvette. National Corvette Museum. Available: https://www.corvettemuseum.org/library-archives/. (Accessed 4 October 2024).
Orndoff, E., 2016. Textiles in space. Spec. Fabr. Rev. Available: https://specialtyfabricsreview.com/2016/02/01/textiles-in-space/ Accessed 2024.
Player, J., 1870. Improvement in Treating Slag Iron and Glass Furnaces for Producing a Useful Material. US patent application US103650.
Sanderson & Porter, 1937. Glass Textile Project of Corning Glass Workd and Owens-Illinois Glass Company.
Schoenlaub, R.A., 1943. Glass Composition. US patent application US2334961.
Simpson, D.C., 1936. Calls on Wire and Cable People: Experiments Book No. 3. Owens Corning.
Simpson, A.L., 1938. Storage battery-box composition. US patent application US2148510.
Skidmore, O.M., 2024. King Abdulaziz International Airport – Hajj Terminal. Skidmore, Owings & Merrill. Available: www.som.com/projects/king-abdulaziz-international-airport-hajj-terminal/#:~:text=The%20naturally%20ventilated%20building%20is,light%20over%20the%20support%20area. (Accessed 5 October 2024).
Slayter, G., 1935. Textile Experimental Book No. 1 Memo for Record. Owens Corning. Owens Corning Internal Report.
Stevens, C., Martin, J., 2018. Non-plastic fibre cord reinforcement for tires. Rubber Plast. News 15–16.
Thomason, J., 2001. Sizing up the interphase: an insider's guide to the science of sizing. Composites 32, 313–321.

Thomason, J., 2015. Glass Fibre Sizings: A Review of the Scientific Literature. Indepdent Publishing Platform, Strathmore.

US National Parks. https://www.nps.gov/havo/learn/nature/peles-hair.htm.

Varma, I.K., Gupta, V.B., 2000. A review on composites: selection and its applications. In: Comprehensive Composite Materials.

Wolf, W., 1974a. Fibreizable fluorine-free glass compositions. US patent application US3817764.

Wolf, W., 1974b. Glass Composition. US patent application US3840379.

Writer, S., 2014. 85 Years of Boating History [Online]. boatingindustry.com: Boatingindustry. Available: https://boatingindustry.com/top-stories/2014/06/12/85-years-of-boating-history/. (Accessed 1 October 2024).

History of carbon fibres: Origins and evolution

1.2

Fumihiko Tanaka
Composite Materials Research Laboratories, Toray Industries, Inc., Masaki-cho, Ehime, Japan

1. Origins of carbon fibre

History has its reasons, with the development of carbon fibres being no exception. Created out of the necessity for lightweight materials, carbon fibre industrialisation grew significantly as the understanding of its applications, the fibres themselves, matrix resins and molding methods all evolved. This chapter describes the early research on carbon fibres, which were created and nurtured by innovators. Carbon fibre performance is influenced by the precursor fibres, causing the search for the optimal precursor to last into the 1980s. The research began with rayon fibres, while the later discovery of acrylic fibres' thermal stabilisation via oxidation enabled significant performance improvements. Meanwhile, pitch fibres promised high carbonisation yield, a measure of how efficiently raw material is transformed to final product, leading to increased research on pitch-based carbon fibres. Optimising each precursor fibre type took time, and it was not until the 1990s that the best precursor and the criteria for their use were determined. This chapter details the early research history for each precursor fibre type.

1.1 Rayon-based carbon fibres

Rayon-based carbon fibres, where rayon serves as the precursor fibre, were introduced in 1956 (application Serial No. 569,391, filed March 5, 1956) (Abbott, 1962). Before discussing rayon-based carbon fibres, it is important to discuss rayon fibres themselves. Rayon fibres, known as artificial silk, were not the result of a long-awaited, deliberate search for a silk substitute but were discovered by specialised chemists studying cellulose chemistry. The period from the 1830s to the 1850s saw significant advancements in the understanding of cellulose within organic chemistry. In 1883, Joseph Swan, who competed with Thomas A. Edison over inventing the light bulb, invented a method to obtain continuous, uniform fibres by dissolving nitrocellulose in acetic acid and extruding the resulting colloidal solution through a nozzle using compressed air (Swan, 1883). The British chemical company Courtaulds industrialised this technology and made substantial profits.

Following global trends, Japan also began producing artificial silk. The first production in Japan was by Teijin in 1916. Before then, Japan relied on imports from

The History of Composites. https://doi.org/10.1016/B978-0-443-34119-9.00004-8

Courtaulds, but the economic boom during World War I and the anticipated tariff increases led to the decision to produce rayon domestically. Toyo Rayon (now Toray) was founded in 1926. Thus, the range of clothing materials expanded from expensive natural fibres to chemical fibres.

Attempts to carbonise cellulosic fibres to make filaments for electric light bulbs began in the 1880s and lasted for many years (Edison, 1880). The resulting carbon fibres were porous, brittle and easily degraded by oxidation, although research efforts were able to improve this process. For example, brittle carbon fibres were heated in a chamber containing hydrocarbon gas which coated the fibre surface with new carbonaceous material (Edison, 1881). Graphitising the fibres by heating to over 2300°C was also investigated (Whitney, 1909). Other strategies included washing the carbon fibres with acid as well as heating the fibres in chlorine gas to remove ash. Despite these efforts, research on carbon fibre stagnated once Coolidge successfully made light bulbs using tungsten filament in 1910.

In 1956, Abbott and colleagues at Carbon Wool Corporation invented carbon fibres by heat-treating viscose rayon fibres up to 800°C (Abbott, 1962). This patent was not published until September 1962. In 1958, Curry Ford and Charles Mitchell at Union Carbide Corporation (UCC) invented carbon fibres by heating cellulose fibres up to 2900°C in an inert atmosphere (Ford and Mitchell, 1963). The key point was to slowly raise the temperature between 100 and 400°C. These fibres were intended for use as thermal insulators, acoustic insulators and vibration insulators at high temperatures. Prior to this, Dr. Bacon at Union Carbide had obtained graphite whiskers, with a strength of 20 GPa and a modulus of 700 GPa through arc discharge of graphite under high temperature and pressure (Bacon, 1960). However, due to cost issues, industrialisation efforts focused on rayon-based carbon fibres. Bacon and his team eventually produced two products: 'Thornel' 25 (modulus 175 GPa) in 1964 and 'Thornel' 100 (modulus 700 GPa) (Bacon, 1973; Tang and Bacon, 1964). Advanced research was also conducted to evaluate the carbon fibres' elastic constants, necessary for their use in composite materials (Ruland, 1969; Smith, 1972).

L. S. Singer, who developed pitch-based carbon fibres, and R. Bacon were part of the Pyrolysis group at UCC's Parma Research Laboratory, under the supervision of L. L. Winter, the Assistant Director of the Research Laboratory. They later collaborated with Toray in joint development projects. According to explanations received by Toray from UCC, Union Carbide had been working with acrylic fibres since 1942, during their rayon-based carbon fibre studies and had continued research on a small scale since 1950. For their rayon-based fibres, they developed a graphitisation method under stress by 1964 and a continuous carbonisation process in 1965, whereas only one researcher was studying polyacrylonitrile (PAN)-based carbon fibres (Matsui, 1996). Although they searched for suitable raw materials for carbon fibres among acrylic fibres, no promising candidates were found. Although they researched the pyrolysis of precursor fibres and explored a wide range of raw materials, they did not synthesise any such fibres themselves. The precursor supply for UCC's rayon-based carbon fibres was provided by American Enka.

1.2 PAN-based carbon fibres

1.2.1 Japan

Research on acrylic fibres began in the 1930s by Rein and colleagues at the German company I.G. Farben. In 1948, DuPont commercialised these fibres under the name Orlon. Initially, Orlon was not suited for industrial use due to its poor processability, nor was it practical for clothing because of its poor dyeability. However, in 1954, an improved version of Orlon with better dyeability was introduced to the market, achieving success in knitwear and later expanding into the carpet sector. Companies such as Monsanto, then Chemstrand and A.C.C. exported this technology worldwide.

In Japan, acrylic fibre research began in 1941 when researchers from Nittobo were dispatched to Professor Shu Kanbara at the Tokyo Institute of Technology (Wikipedia: Shu Kanbara; Katayama, 1960). Although the research was halted during the war, it resumed post-war and was driven by the need to replace expensive wool. It took a number of years to commercialise knitwear technology, leading to economic difficulties during the early years. As a result, companies that began the initial investment in industrialising acrylic fibres, including Nittobo, were weeded out of the industry before the knitwear boom began. Nevertheless, acrylic fibre production expanded with the improvement of knitting technology.

The rise of synthetic fibres and the resulting oversupply led to a decline in rayon fibre production, which had previously sold well simply by being produced. By 1960, rayon production was winding down. During this economic downturn, companies diversified their businesses through marketing, establishing new research institutes and other initiatives.

In May 1959, Akio Shindo, a 33-year-old young researcher at the Government Industrial Research Institute, Osaka, came across information about the development of rayon-based carbon fibres (Nakamura et al., 2009). He immediately began heating various synthetic fibres to 1000°C under an inert gas atmosphere. A carbonisation furnace, with electric heaters placed around a quartz tube, was an essential piece of equipment for carbon material researchers. However, the fibres he evaluated either decomposed and disappeared or melted and could not maintain their fibrous shape (Tanaka, 2024).

At that time, his younger brother obtained DuPont's acrylic fibre, Orlon, and sent the fibres to Shindo. Using Orlon, Shindo found that the carbonised fibres retained their shape and exhibited tensile strength incomparable to carbonised fibres from other precursors. Later, he increased the number of fibres being carbonised but, unexpectedly, only obtained brittle materials. This prompted careful examination of the process, which revealed that the nitrogen gas's low flow rate yielded moderately oxidised fibres during the temperature increase process. The oxidation afforded by the nitrogen flow rate is the discovery that enabled the invention (Nakamura et al., 2009; Tanaka, 2024; Shindo, 1993).

Typically, the presence of oxygen causes oxidative decomposition; therefore, air is replaced with nitrogen prior to carbonisation. However, in this case, the attempt to save nitrogen gas successfully stabilised the fibres. The patent filed in September

1959 specified heating polyacrylonitrile in an oxidative atmosphere up to 350°C before heating it above 800°C, covering essential and broad aspects (Shindo et al., 1962). This invention was presented at a conference in Japan in October 1959, garnering significant attention from that time onward (Shindo, 1961).

Shindo and his colleagues not only examined various products made from acrylic fibres but also expanded their choice of raw materials to include polyvinyl alcohol fibres and cellulose fibres, advancing their research on thermal stabilisation (Shindo et al., 1963). They collaborated with carbon material manufacturers, primarily Nippon Carbon and Tokai Electrode, to commercialise their products. However, Shindo and his team were not confident in the value of their own technology (Nakamura et al., 2009), and thus did not file for international patents. Consequently, Tokai Electrode's application in the UK in August 1961 was the first international patent application (Tsunoda, 1962).

The initial patent applications included 'filters for alkaline, acidic, or non-oxidizing high-temperature gases, activated adsorbents for such purposes, packing materials, reinforcement for plastics and ceramics, heat-generating fillers, low-temperature heating elements for heating, and high-temperature heating elements for electric furnaces' (Nakamura et al., 2009; Shindo et al., 1962). Until Shindo heard from the U.S. Air Force Research Laboratory (AFRL) in 1965 that the material was suitable for reinforcing composites, he had little recognition of its mechanical applications and was even attempting to lower the modulus to make the fibres more flexible (Shindo, 1993).

Nippon Carbon, which adopted Shindo's technology, built a plant in 1962 with a monthly production capacity of 0.5 tons of PAN-based carbon fibres and began selling them under the brand name 'Carbolon' (Tanaka, 2024). By 1969, the companies collaborating with Shindo included Matsushita Electric (now Panasonic), Nittobo, Japan Exlan, Kuraray and Sekisui Chemical, primarily focusing on acrylic fibre research. Matsushita and Nittobo obtained licenses, while the others received only technical guidance from Shindo and did not commercialise the technology. These companies were conducting carbon fibre research as part of their efforts to diversify their business operations (Nakamura et al., 2009). Toray also received a patent license in 1970. In Japan, there was no large-scale demand for carbon fibres, and the exploration of applications was still in its initial stages, resulting in limited production growth.

1.2.2 UK

The Royal Aircraft Establishment (RAE) conducted research on military and civilian aircraft, rockets and carbon materials for high-temperature applications and high-temperature gas reactors (nuclear reactors) (Watt, 1972). At the sixth Biennial Carbon Conference held in the United States, June 17–21, 1963 (Nakamura et al., 2009), Shindo presented the results of his carbon fibre research, attracting significant attention from researchers in Europe and the United States. One such researcher, W. Watt of the RAE, began his studies in mid-1963. Watt first confirmed the heat treatment conditions for acrylic fibres that provide thermal stabilisation and investigated the effects of oxidation by tuning oxygen and tension (Watt, 1972). Around 1965, the

Atomic Energy Research Establishment (AERE) joined the carbon fibre research (Langley, 1971). At AERE, precursor fibres were batch-processed for oxidation, carbonisation and graphitisation, resulting in carbon fibre bundles approximately 1 m in length (Fig. 1.2.1) (Gunston, 1969).

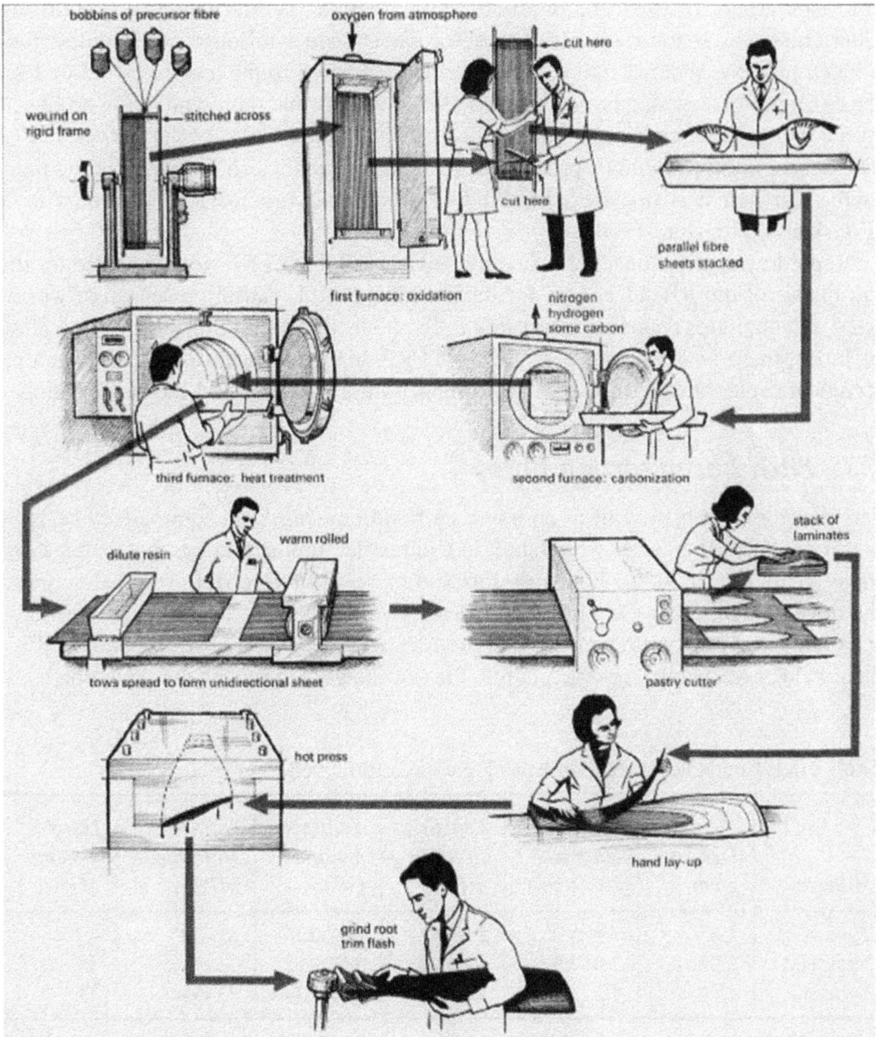

Figure 1.2.1 Batch process for making carbon fibres in the 1960s, including through to the finished composite part (the fan blade for the Rolls-Royce RB 211 engine). The basic technique was devised at the RAE, scaled up at the AERE and turned into a routine industrial process by Rolls-Royce.
Adapted from Gunston (1969).

The National Research Development Corporation (NRDC) promoted the industrialisation of this new material, with Courtaulds and Morganite Research and Development participating (Geoffrey, 2011; Badami et al., 1967; Watt et al., 1966; Johnson et al., 1968; Morris, 1968; Morgan, 2005). P. Morgan, who was with Courtaulds at the time and then with RK Carbon Fibres until 1993 and supported carbon fibre development in the UK from its infancy, has written an excellent book (Morgan, 2005). According to the book, Rolls-Royce conducted its own research and development (Morgan, 2005). In 1967, the production of 'Modmor' by Morganite and 'Hyfil' by Courtaulds was announced. All of these products were 1-m-long fibre bundles from a batch process. Whereas the batch process was limited to the lengths of about 1 m, the existence of a catalog issued in April 1968 confirms that prototypes were produced by a continuous process with total lengths of 270–600 m (Table 1.2.1) (Matsui, 1997). The concept of a continuous process to the final product is also described in the literature, although it is unclear what kind of continuous process it was at this point (Fig. 1.2.2) (Gunston, 1969).

Regarding applications, it was announced in 1967 that CFRP would be used for the fan blades of the RB211 engine for the Lockheed 1011, causing a sensation worldwide, although this endeavor ultimately failed (Gunston, 1969; Matsui, 1997). Unlike in Japan, there was a strong demand for CFRP applications in the UK, where high technical capabilities led to rapid performance improvements and industrialisation.

1.3 Pitch-based carbon fibres

The development history of pitch-based carbon fibres has been summarised by Nippon Graphite Fiber Co., which has led industrial production of these fibres for many years (Arai, 2020). Based on this summary and referencing original sources, the development history is presented below.

Pitch-based carbon fibres were first industrialised in 1970 by Kureha Chemical Industry Co., Ltd., using isotropic pitch as the raw material, based on the invention by S.

Table 1.2.1 Courtaulds carbon fibre catalog issued April 1968.[a]

Material	*Density g/cm^{-3}*	*Diameter 10^{-3} mm*	*Tensile strength MPa*	*Young's modulus GPa*	*ILSS Untreated MPa*	*ILSS Treated MPa*
Type A	1.85	7.5–8.0	1.9–2.3	196–225	61–69	–
Type HM	1.95	7.0–7.5	1.8–2.1	343–412	23–33	55–70
Type HT	1.90	7.5	2.3–2.6	235–274	43–54	72

[a]**Long Staple:** Supplied in 1.2-m lengths of tow, each containing 10,000 parallel filaments of 0.007-mm diameter. One pound of fibre consists of approximately 400 tows. **Pre-impregnated Unidirectional Warp Sheet:** Supplied from all types of fibre in sheets 1.05 × 0.3 m or 1.05 × 0.4 m and of thickness 0.025, 0,05, 0.13 or 0.25 mm. **Continuous 10,000 Filament Roving:** Development quantities available in lengths of 270–600 m in all types of fibre. **Random Fibre Mat:** Available as either 350-mm wide strips in lengths up to 12 m or as pieces 0.7 × 1.2 m. The mat can be supplied as a prepreg or with a light resin binder. **Short Staple and Flock:** Fibre can be supplied cut to any length between 2.54 and 300 mm to suit customer requirements. **Molding Materials:** A range of filled thermoplastic and thermosetting resins are available with various fibre loadings.

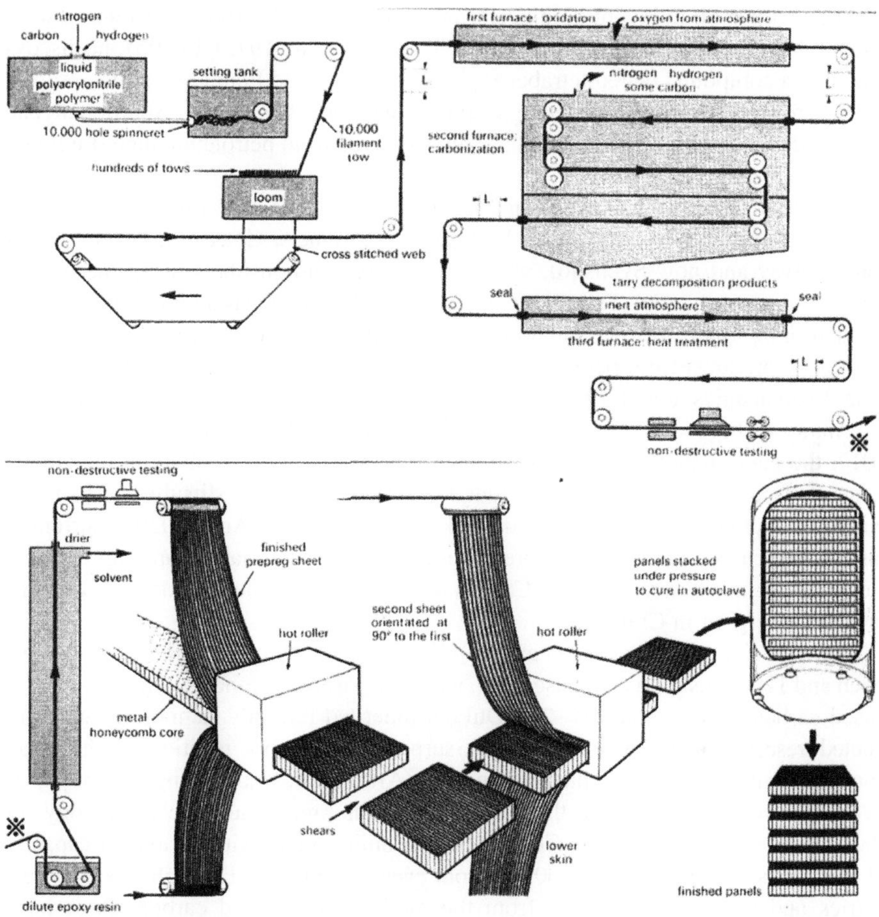

Figure 1.2.2 A concept of continuous process for making carbon fibres proposed in the 1960s, including through to the finished composite panels.
Adapted from Gunston (1969).

Otani at Gunma University in Japan (Otani, 1965, 1966). In their joint research, Otani and Kureha Chemical Industry used PVC pitch obtained by heating polyvinyl chloride to 400°C. Notably, they used a special pitch produced by photopolymerisation at −78°C to improve heat resistance, which was key to achieving both spinnability and thermal stability (Arai, 2020; Otani, 1965).

In tandem, Kureha was also investigating a process to obtain acetylene and ethylene by pyrolyzing crude oil with steam at 2000°C and exploring the effective use of tar pitch, which was produced at 30 wt% (Gomi and Okuda, 1972). Leveraging their knowledge of special pitch, they established a process for short fibres through centrifugal spinning. In 1970, a plant with an annual production capacity of 100 tons was established. Carbon fibres made from this isotropic pitch have a modulus of 40–50 GPa and a tensile strength of 500–1000 MPa (Arai, 2020).

The first person to invent high-performance pitch-based carbon fibres was also S. Otani (Otani et al., 1974; Fujimaki et al., 1975; Otani et al., 1972). In 1967, he discovered that a compound called tetrabenzophenazine exhibited liquid crystalline properties when converted to pitch, making it spinnable (Otani et al., 1974). Additionally, he produced carbon fibres from mesophase pitch derived from petroleum pitch (Fujimaki et al., 1975; Otani et al., 1972).

Industrialisation was achieved in 1975 by L. S. Singer and colleagues at Union Carbide Corporation (later Amoco Performance Products, Cytec Engineered Materials, Solvay and now Syensqo), who industrialised carbon fibres using mesophase pitch as the raw material (Singer, 1977). In the 1980s, there was a boom in the development of high-performance pitch-based carbon fibres aimed at the advanced utilisation of by-products from petroleum refining and coal carbonisation. Although more than 20 companies, mainly from the petroleum and steel industries, entered the market, many withdrew due do the high barrier to industrialisation (Arai, 2020). As of 2017, the main manufacturers of pitch-based carbon fibres are Solvay in the United States, which succeeded UCC, and four Japanese companies: Mitsubishi Chemical, Nippon Graphite Fiber, Kureha and Osaka Gas Chemical (Arai, 2020). Currently, pitch-based carbon fibre manufacturers Dongbang New Material Technology, Hongte Coal Chemical Industry, Novcarb Carbon Materials and Anshan Sonocarb Carbon fibres are emerging in China.

The four companies in Japan have a production capacity of 2000 tons for isotropic pitch and 1200 tons for mesophase pitch, which is extremely minor compared to PAN-based carbon fibres, about 1%–2%. DuPont inherited Exxon's technology and conducted research and development that surpassed Amoco's position at the time, announcing its entry into the business in 1988 (Arai, 2020; Nikkei Sangyo newspaper, 1988). However, DuPont lost the competition with Amoco, and the technology for short fibres was transferred to Conoco, which built a plant with an annual capacity of 3620 tons but withdrew in 2003 without ever operating it. Through repeated new entries and withdrawals to and from the market, pitch-based carbon fibres have continued achieving technical advancements to the present day.

2. Early adopters

From the 1970s, the industrialisation of technologies developed by early innovators began. Numerous companies entered the market, and through mergers and acquisitions (M&A), both corporate consolidation and the convergence of manufacturing technologies progressed (Fig. 1.2.3). As precursor fibres were the most critical technology, acrylic fibre manufacturers became the focal point of M&A activities. In this section, we will organise the main companies that entered the market up to the 1980s, focusing on those that continue to operate today.

2.1 Toray

Following Shindo's report in 1959, Toray began investigating copolymer components for acrylic precursor fibres in 1961 and identified components that promote the

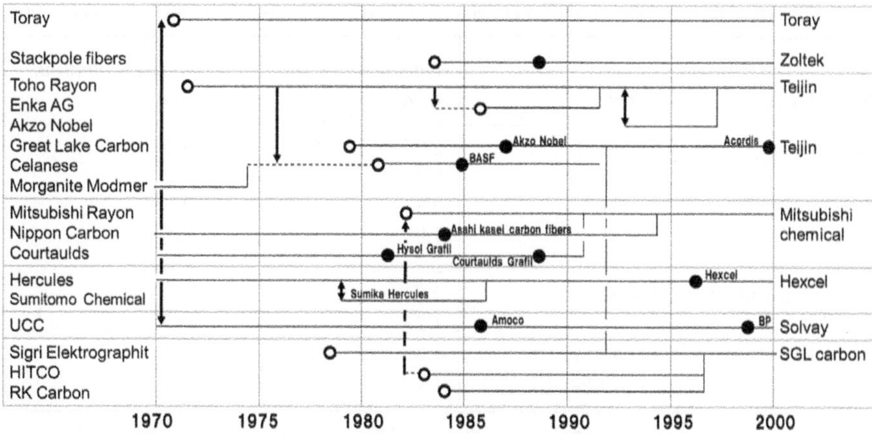

Figure 1.2.3 Carbon fibre manufacturers in production startups and M&A.

stabilisation reaction of polyacrylonitrile (Tanaka, 2024; Matsui, 1997, 1998a). Although Toray initially focused on the production of acrylic fibres and temporarily halted its research on carbon fibres, Kenichi Morita from Toray's Basic Research Laboratory, who was exploring new synthetic routes towards acrylonitrile dimers as intermediates for polyamides, discovered a novel reaction in 1966 (Fig. 1.2.4) (Morita et al., 1968; Wikipedia: Baylis-Hillman reaction). This reaction, later known as the Morita–Baylis–Hillman reaction, involved reacting acetaldehyde with acrylonitrile to obtain hydroxyethyl acrylonitrile (HEN). The copolymerised acrylic fibres produced through this reaction resulted in superior carbon fibres.

Seeking alternative business due to the decline of rayon fibres, Toray intensified its development efforts in acrylic fibres by deploying a team of 200 engineers in 1969 and entered into a technical partnership with UCC in April 1970 (Matsui, 1998a). The acrylic fibres provided by Toray exhibited optimal performance at UCC (Table 1.2.2), leading to a collaboration where Toray supplied acrylic fibre technology and UCC provided carbonisation technology. In 1970, despite having a continuous production process, UCC's production of rayon-based carbon fibres was limited to six strands, resulting in a very small production volume (Matsui, 1998a). However, production capacity at Toray rapidly expanded from 10 tons per year in 1971 at the pilot plant to 60 tons per year in 1973 and 180 tons per year in 1974, establishing a supply

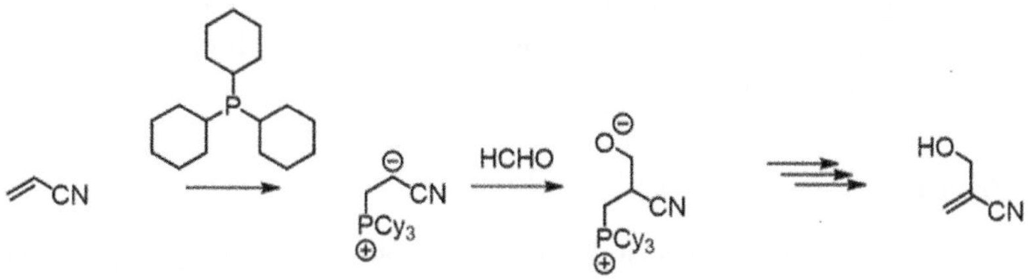

Figure 1.2.4 Synthetic route of 2-(1-hydroxyalkyl) acrylonitrile (HEN).

Table 1.2.2 Results of evaluation of mechanical properties of carbon fibre at Toray as of 1969.

Manufacturer	*Type*	*Tensile strength GPa*	*Young's modulus GPa*
Morganite	Type I	2.0	407
	Type II	3.7	292
Courtaulds	HM	2.3	394
NRDC	HM	1.9	385
RAE	–	2.3	410
Tokai electrode	HM	1.6	298
HITCO	–	1.2	243
UCC	Thornel 50	1.5	304
	Thornel 25	1.2	159

Toray's carbon fibre obtained with HEN had a tensile strength of 3.9 GPa and a Young's modulus of 314 GPa.

system during a period of uncertain demand (Matsui, 1998a). Fortunately, the demand for consumer applications such as golf shafts, tennis rackets and fishing rods significantly increased, balancing supply and demand.

To expand its presence in the European market, Toray established Soficar in Abidos, France, in collaboration with Elf Aquitaine (petrochemicals) and Pechiney (metal smelting). Soficar began producing and selling carbon fibres based on Toray's technology. In 1988, Toray acquired a major share, gaining control over production and sales in Europe. Additionally, in 1991, Toray Composite America was established to supply prepregs for the Boeing 777. Through these efforts, Toray enhanced its supply capabilities in the European and American markets, increasing its global presence.

Separately, former UCC employees established Stackpole Fibers in 1971, producing carbon products out of the former UCC factory in Lowell, Massachusetts. In 1988, Zoltek acquired Stackpole Fibers and began manufacturing low-cost large tow fibres, products with lower strength and modulus (Rumy, 2018). At Stackpole Fibers, carbon fibres for rocket nozzles were produced using a batch process. However, after losing contracts, Zoltek significantly shifted its strategy to pursue low-cost large-tow production of carbon fibres for an industrial market (Rumy, 2018).

Zoltek gradually expanded its scale by acquiring the carbonisation equipment from Courtaulds' Coventry plant, establishing continuous carbonisation technology, introducing precursor fibre technology from DuPont, and constructing plants in Hungary and Mexico, where modular equipment enabled their rapid construction (Rumy, 2018). These efforts enabled Zoltek to secure large projects for wind turbine blades through stable pricing and supply. In 2014, Zoltek was acquired by Toray, making it a company within the Toray Group that possesses both high-performance carbon fibres and low-cost large tow.

2.2 Teijin

In 1964, Toho Rayon began research on carbon fibres, constructing a pilot plant in 1972 and officially entering the carbon fibre market in 1975 (Glowacz, 2015). In

1983, Toho Rayon licensed PAN-based carbon fibre carbonisation technology to Enka AG (Germany), which had ties with Akzo Nobel (The Netherlands) and began manufacturing carbon fibres (Matsui, 1998b). However, due to Akzo Nobel's corporate restructuring, the carbon fibre business was downsized, and in 1992, the major share was sold to Toho Rayon. In 1993, Toho Rayon established a joint venture with Akzo Nobel to sell carbon fibres in North America; in 1997, it acquired the major share (Matsui, 1998b). In 2004, Toho Rayon acquired Acordis' carbon fibre business in Rockwood, Tennessee, expanding its North American business. Great Lakes Carbon began carbon fibre research in 1967 and started production in 1984 in Rockwood, USA. Akzo Nobel acquired its Rockwood plant in 1987, and Acordis later absorbed it in 2000 (Matsui, 1998b). In 1986, Toho Rayon established a carbon fibre production line in Oberbruch, Germany. In 2000, Toho Rayon received capital participation from Teijin, becoming a wholly owned subsidiary in 2007 (Glowacz, 2015).

In 1976, Toho Rayon provided technical support to Celanese, advancing the construction of large-scale manufacturing facilities for precursor fibres via PAN dry-spinning technology. However, before completion, Celanese sold its carbon fibre business to BASF in 1985. BASF initially targeted the automotive industry but found this application was still premature, so BASF redirected its' focus toward aircraft and obtained material certification from Airbus. However, BASF lost interest in continuing the business and its carbon fibre operations ended in 1992.

Celanese's background in the industry also includes Morganite Modmor which was spun off from the R&D department at Morgan Crucible to focus on the carbon fibre business, a company which sold carbon products more broadly. In the U.S., a joint venture between Morganite Modmor and Narmco's parent company, Whittaker, was formed under the name Morganite Modmor. In 1974, Celanese bought the Narmco Materials division of Whittaker, a producer of structural adhesives and graphite composite materials used in aircraft tail assemblies, floorboards, space vehicles, tennis rackets and golf club shafts, among other applications. Thus, although Toho Rayon provided technology to Celanese, the business could not continue based on Toho Rayon's technology.

2.3 Mitsubishi Chemical

In 1965, Mitsubishi Rayon, a Japanese textile manufacturer, began research and development of carbon fibres and started selling its own PAN precursor fibres in 1975 (Glowacz, 2015). In 1976, it introduced technology from Courtaulds and began selling prepregs. In 1982, it partnered with HITCO (formerly H. I. Thompson Glass and Paint Co.) for carbonisation technology, completing the carbonisation facilities (Glowacz, 2015). In 1991, it acquired Courtaulds Grafil in the United States and established Grafil. Courtaulds had established Hysol Grafil in 1982, which became Courtaulds Grafil in 1988 (Matsui, 1998b). In 2010, Mitsubishi Chemical acquired Mitsubishi Rayon.

Nippon Carbon, which was the first to start production of PAN-based carbon fibres in the world, is known for its high-heat-resistant SiC fibres. In the 1970s, its main businesses were artificial graphite electrodes, brushes and carbon black. Nippon

Carbon received PAN precursor fibres from Asahi Kasei and, through a joint venture with Asahi Kasei, the business was succeeded by Asahi Kasei Carbon Fiber. However, in 1994, the business was closed, and the carbon fibre manufacturing facilities were sold to Mitsubishi Rayon (Matsui, 1998b).

2.4 Hexcel

Hercules is a chemical company that started with the manufacture of gunpowder in the United States and has been involved in the production of rocket motor components made of glass fibre composites since 1966 (Wikipedia: Hercules Inc.). The research and development of carbon fibres were conducted under a license from Courtaulds, in collaboration with the Allegany Ballistics Laboratory (ABL), designing and developing carbonisation equipment, and constructing a pilot facility in Bowling Green, Maryland (Wikipedia: Allegany Ballistics Laboratory). With the supply of PAN precursor fibres, the types of carbon fibres at the start of sales in April 1970 were the same as those of Courtaulds: HM-S, HT-S and A (Table 1.2.3). As of 1971, CFRP was supplied to Northrop for the wing tips and flaps of the F5 fighter, to Boeing for spoiler flaps, to IRAD for pressure vessels and other boron fibre composite materials (Matsui, 1998b). After the first pilot facility, a full-scale factory was constructed in Salt Lake City, Utah.

Although precursor fibres were supplied by Courtaulds, the performance of the PAN precursor fibres from Japan Exlan (a joint venture between Toyobo and Sumitomo Chemical) was superior, leading to the establishment of Sumika Hercules (a joint venture between Sumitomo Chemical and Hercules) in 1979, and the construction of a PAN precursor fibre production plant in Decatur, Alabama (Newcomb, 2016). In 1986, Hercules acquired sole ownership from Sumitomo Chemical (Newcomb, 2016). However, during the Reagan administration's recession, the aerospace industry fell into a slump, and the composite materials division was separated and acquired by Hexcel, a prepreg manufacturer (History and timeline of Hexcel).

Hexcel, which derived 70% of its profits from the aerospace industry, faced difficulties around 1992. In 1993, it decided to enter Chapter 11, and after selling off factories and other measures, it emerged from Chapter 11 in 1995. After emerging from bankruptcy protection, it received investments from banks and stocks, and in 1996, it

Table 1.2.3 Hercules carbon fibre catalog issued April 1970 and April 1971.

	Price as of 1970.4 \$/kg	***Price as of 1971.4 \$/kg***	***Tensile Strength GPa***	***Young's modulus GPa***	***ILSS MPa***
Type A	–	340	1.9–2.3	190–220	–
Type HM	500	560	1.8–2.1	340–410	55–70
Type HT	500	515	2.3–2.6	230–270	72

acquired the composite divisions of Hercules and Ciba–Geigy (History and timeline of Hexcel). This acquisition increased its presence in the European market.

2.5 *Solvay/Syensqo*

In the 1970s, UCC imported acrylic fibres from Toray, carbonised them and sold carbon fibres in the United States (Matsui, 1998a). In addition to carbon fibres, UCC also sold carbon products such as artificial graphite electrodes, brushes, carbon black and materials for nuclear reactors. Unlike Hercules, which was actively developing prepregs, UCC focused on the sale of carbon fibres and adopted a different strategy.

After the market for rayon-based carbon fibres shrank, UCC concentrated its development efforts on pitch-based carbon fibres, which it considered the most promising. However, it was unable to gain market share from PAN-based carbon fibres. In 1981, UCC introduced acrylic fibre technology from Toray Industries, Inc. and established a plant in the U.S. to process PAN precursor fibres into carbon fibres (Matsui, 1998b). However, to cover the costs of the accident that occurred in Bhopal, India, in 1985, UCC sold its carbon fibre plant along with its functional resins and composite materials division to Amoco Performance Products, which was later sold to Cytec (later absorbed by Solvay, now Syensqo).

2.6 *SGL Carbon*

SGL Carbon AG originated in 1992 from a merger between SIGRI GmbH (Germany) and Great Lakes Carbon (USA) (Glowacz, 2015). In 1985, Siemens Plania Chemisches Werk Griesheim was renamed SIGRI (Wikipedia: SGL Carbon). In 1997, after acquiring RK Carbon, SGL entered the large-tow business (About SGL carbon at SGL carbon website). Also, in 1997, SGL Carbon acquired HITCO, which was Tier II supplier to the aerospace industry, enabling it to penetrate the aerospace market.

3. Evolution and applications

Carbon fibre innovation in the global market is not just about innovative ideas (seeds), but about matching these ideas with market needs (Fig. 1.2.5). Innovation often seems irrational and unpredictable. In its initial stages, it is hard to show clear economic benefits due to high uncertainty. It is challenging to meet the demands for efficient, economic value creation within social constraints such as efficiency, transparency, accountability, compliance and governance. To overcome these challenges and achieve innovation, it is important to tell a compelling story that goes beyond short-term economic benefits or one that attracts people's interest.

In developing carbon fibre at Toray, the company not only had the world's best technology but also a strong, technology-focused mindset. This mindset was crucial

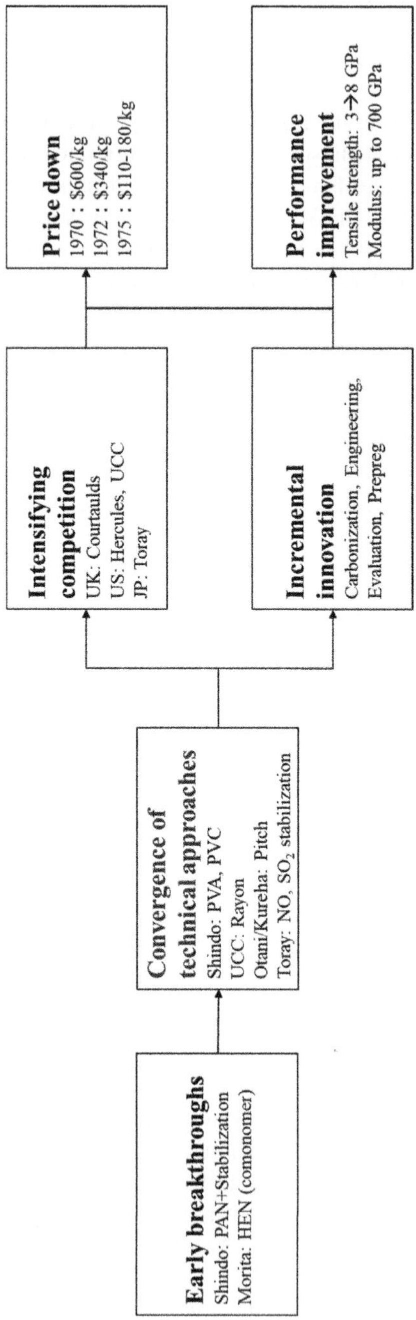

Figure 1.2.5 Innovation history of PAN-based carbon fibre. Adapted from Tanaka (2024), with permission from Elsevier.

for continuing development despite uncertainties about profitability. Additionally, the company advanced due to the support of influential people. Yoshikazu Ito, who became Toray's president in 1981 and was the project leader for carbon fibre development in 1969, had successfully industrialised the photo-nitrosation process for ε-caprolactam in the 1960s (Tanaka, 2024), the raw material for nylon 6. Drawing on this past success, he believed that a new compound for carbon fibres would be a key differentiator (Tanaka, 2024). Toray decided that this new route would enhance its originality and competitiveness as of 1969. Although this alone might not have created competitiveness, Toray developed its own prepreg and woven fabrics with the start of carbon fibre production, opening markets not only for aircraft applications but also for golf shafts and fishing rods, laying the foundation for future growth (Shibata et al., 1976). For example, in the prepreg used for the Boeing 777, in the late 1980s, the company developed a technology to add thermoplastic particles between the prepreg layers to increase compression after impact performance, an important mechanical property. This paved the way for using CFRP in primary structural materials for aircraft (Odagiri et al., 1996).

Doing business with carbon fibre involves more than just producing the fibre itself. As discussed above, it includes making intermediate products using carbon fibre, developing its materials and molding methods, designing laminate structures and creating evaluation and non-destructive testing methods. It also involves collaborating with customers to develop part shapes (Fig. 1.2.6). Additionally, the performance of carbon fibres is being improved by designing their crystalline and amorphous microstructures (Tane et al., 2019; Ishikawa et al., 2023).

These market expansions have led to increased production and lower prices. For example, the price of carbon fibre dropped from \$500/kg in 1970 to \$110/kg in 1975, and to \$40/kg in 1982. This price drops further expanded the market, creating a positive cycle (Sibal, 1984). At the same time, the performance of carbon fibre improved, with tensile strength increasing from 3 GPa in 1970 to 8 GPa in 2023, which has also contributed to market growth (Tanaka, 2024).

4. Sustainability and new prospects

Notably, 70 years have passed since the invention of PAN-based carbon fibre, and it has become widely used in the market. The Japanese Carbon Fiber Manufacturers Association (JCMA) estimates that producing carbon fibre requires 350 MJ/kg of energy, and CO_2 emissions are approximately 20 kg per kg of carbon fibre (Table 1.2.4) (Zhang et al., 2011; Japan Carbon Fibre Manufacturer Association, 2012, 2022). In a total life cycle assessment of carbon fibre use in aircraft, automobiles and windmills, it is estimated that the CO_2 emission reduction effect in aircraft is 700 times that of carbon fibre production, and in windmills, 12,000 times that of carbon fibre production (Japan Carbon Fibre Manufacturer Association 'Lifecycle assessment of aircraft, automobile & windmill').

In addition to carbon fibres' emission savings during use in lightweighting applications, further reduction of CO_2 emissions is being promoted in its

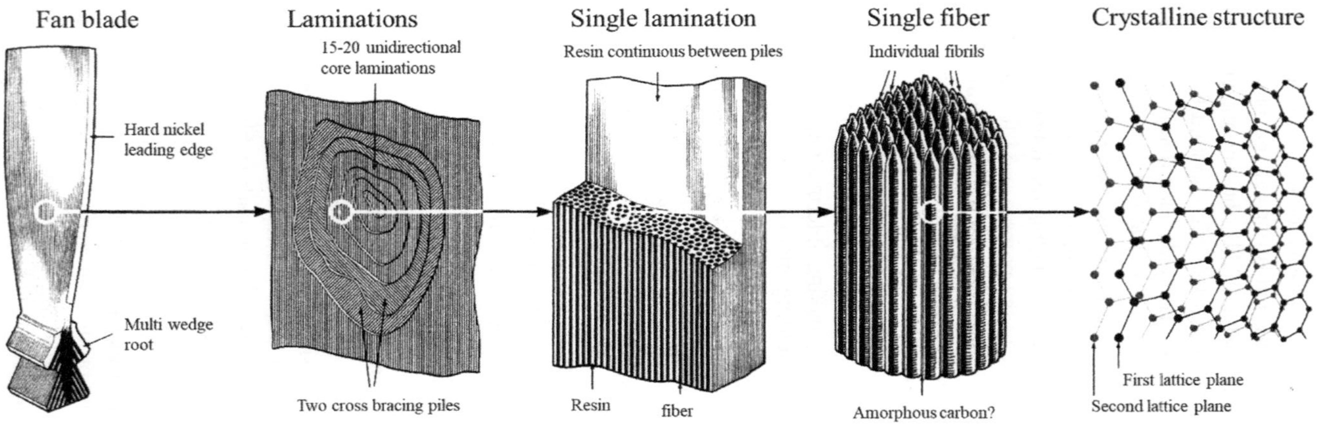

Figure 1.2.6 Structure of carbon fibre products at various hierarchies. Adapted from Gunston (1969), with permission.

Table 1.2.4 Life cycle inventories of carbon fibre published by JCMA.

	Energy (MJ/ kgCF)	*CO_2 (kg/ kgCF)*	*SO_x (kg/ kgCF)*	*NO_x (kg/ kgCF)*
First data from 1999	478.5	29.7	0.068	2.009
Recalculated in 2004	285.9	20.5	0.02	0.146
Recalculated in 2009	286	22.4	0.019	0.121
October 2022 update	350	19.8	0.016	0.035

manufacturing process, with the most advanced efforts focusing on reducing the energy used. Since toxic combustible gases such as HCN are generated during carbon fibre production, pyrolysis using natural gas as a combustion aid is employed to detoxify these gases (Nunna et al., 2019). The combustion of HCN generates heat, and the waste heat is reused to heat various processes. As production volumes increase, the consumption of natural gas is no longer necessary, because HCN alone will provide sufficient combustion temperatures. Additionally, efforts are being made to reduce energy consumption through microwave-based production methods (Kim et al., 2015; Jin et al., 2017).

Efforts to switch to bio-based acrylonitrile are now in the practical application stage (Goyal, 2020). The production of precursors requires a significant amount of energy, and ammonia is used in the production of acrylonitrile, the precursor raw material; ammonia itself consumes enormous amounts of energy. Although there are various synthetic routes for bio-based products, they all require ammonia, which does not solve the problem. Therefore, research on precursors that do not use acrylonitrile has been active, with examples including lignin (Lu et al., 2017) and polyethylene (Barton et al., 2017).

Recycling is also a crucial factor in determining the total manufacturing energy. Carbon fibres perform best when they are in a continuous fibre state, but during recycling, they are cut into discontinuous fibres and used in injection molding and mat press molding. In such cases, recycling methods that minimise performance reduction have been studied. For example, high-performance materials generated during the manufacturing process of the Boeing 787 are being recycled to make laptop computer casings (Fig. 1.2.7) (JEC Composites News, 2023).

As we have seen, carbon fibre is steadily expanding its contribution to society, and this expansion is not stopping with the response to the circular economy. The history of carbon fibre is not only about the discovery that polyacrylonitrile makes an excellent precursor but also about the commercialisation that has accompanied the industrialisation process and the development of applications, which have played a significant role.

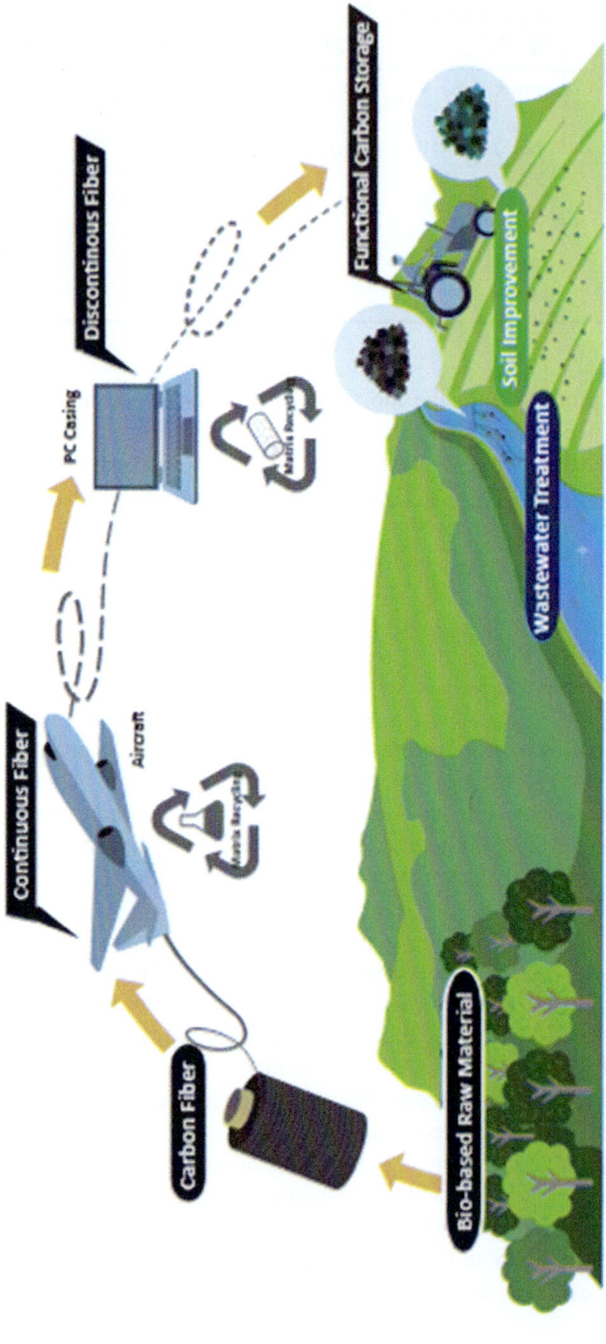

Figure 1.2.7 Carbon fibre circular system.

References

Abbott, W.F., 1962. Method for carbonizing fibers. Patent US3053775A.

About SGL carbon at SGL carbon website, https://www.sglcarbon.com/en/company/about-us/ (Accessed August 25th, 2024).

Arai, Y., 2020. Development and history of Pitch-based carbon fibers. In: Carbon Materials R&D Trends 2020. CPC Research Group Research Report, pp. 16–28.

Bacon, R., 1960. Growth, structure, and properties of graphite whiskers. J. Appl. Phys. 31 (2), 283–290.

Bacon, R., 1973. Carbon fibers from rayon precursors. In: Walker, P.L., Thrower, P.A. (Eds.), Chemistry and Physics of Carbon 9. Marcel Dekker, New York, pp. 1–102.

Badami, D.V., Joiner, J.C., Jones, G.A., 1967. Microstructure of high strength, high modulus carbon fibres. Nature 215 (5099), 386–387.

Barton, B.E., Behr, M.J., Patton, J.T., Hukkanen, E.J., Landes, B.G., Wang, W., Derstine, C., 2017. High-modulus low-cost carbon fibers from polyethylene enabled by boron catalyzed graphitization. Small 13 (36), 1701926.

Edison, T.A., 1880. Electric light. Patent US223898A.

Edison, T.A., 1881. Manufacture of carbons for electric lamps. Patent US248416A.

Ford, C.E., Mitchell, C.V., 1963. Fibrous graphite. Patent US3107152A.

Fujimaki, H, Kodama, T., Sakaguchi, Y., Okuda, K., Otani, S., 1975. Carbon fiber from pitch containing a liquid crystal. Tanso 80, 3–10. https://doi.org/10.7209/tanso.1975.3.

Geoffrey, O., 2011. Corporate strategy and national institutions: the case of the man-made fibres industry. Capital. Soc. De Gruyter 6 (1), 1–59.

Glowacz, F., 2015. Carbon Fibres: History, Players and Forecast to 2020. JEC.

Gomi, S., Okuda, K., 1972. Advances in reinforcing materials for composites. Chem. Eng. Japan 36, 99–104. https://doi.org/10.1252/kakoronbunshu1953.36.1179.

Goyal, A., 2020. Renewable acrylonitrile for carbon fiber production (Accessed August 25th, 2024). https://www.ieabioenergy.com/wp-content/uploads/2020/10/IEA_Bioenergy_eWorkshop_2021_3-4_AmitGoyal.pdf.

Gunston, W.T., 1969. Carbon fibres. Sci. J. (Lond.) 1, 69.

History and timeline of Hexcel, https://www.hexcel.com/About/History-and-Timeline (Accessed August 25th, 2024).

Ishikawa, T., Tanaka, F., Kurushima, K., Yasuhara, A., Sagawa, R., Fujita, T., Yonesaki, R., Iseki, K., Nakamuro, T., Harano, K., Nakamura, E., 2023. Wavy graphene-like network forming during pyrolysis of polyacrylonitrile into carbon fiber. J. Am. Chem. Soc. 145 (22), 12244–12254.

Japan Carbon Fiber Manufacturer Association, 2012. Life Cycle Assessment of Aircraft, Automobile & Windmill: The Japan Carbon Fiber Manufacturer Association's Assessment Model.

Japan Carbon Fiber Manufacturers Association, 2022. Overview of LCI data for carbon fiber (Accessed August 25th, 2024). https://www.carbonfiber.gr.jp/english/tech/pdf/lci2022_en.pdf.

Japan Carbon Fiber Manufacturers Association. Lifecycle assessment of aircraft, automobile & windmill. https://www.carbonfiber.gr.jp/english/tech/lca.html (Accessed August 25th, 2024).

JEC composites news, 2023. Toray carbon fibre recycled from Boeing 787 wing production process applied in Lenovo ThinkPad X1 carbon Gen (Accessed August 25th 2024). https://

www.jeccomposites.com/news/spotted-by-jec/toray-carbon-fibre-recycled-from-boeing-787-wing-production-process-applied-in-lenovo-thinkpad-x1-carbon-gen/.

Jin, S., Guo, C., Lu, Y., Zhang, R., Wang, Z., Jin, M., 2017. Comparison of microwave and conventional heating methods in carbonization of polyacrylonitrile-based stabilized fibers at different temperature measured by an in-situ process temperature control ring. Polym. Degrad. Stabil. 140, 32–41.

Johnson, W., Phillips, L.N., Watt, W., 1968. The production of carbon fibres, GB1110791A.

Katayama, M., 1960. The review of the acrylic synthetic fibers. Yukigoseikyokaishi 18 (3), 135–145.

Kim, S.Y., Kim, S.Y., Choi, J., Lee, S., Jo, S.M., Joo, J., Lee, H.S., 2015. Two step microwave plasma carbonization including low plasma power pre-carbonization for polyacrylonitrile based carbon fiber. Polymer 69, 123–128.

Langley, M., 1971. Carbon fibres- the first five years (carbon fiber reinforced plastics, considering aerospace and medical applications). Flight Int. 100, 406–408.

Lu, C., Blackwell, C., Ren, Q., Ford, E., 2017. Effect of the coagulation bath on the structure and mechanical properties of gel-spun lignin/poly (vinyl alcohol) fibers. ACS Sustain. Chem. Eng. 5 (4), 2949–2959.

Matsui, J., 1996. Tansoseni-no-hanashi. Reinforced plastics, The Japan Reinforced Plastics Society 42 (6), 214–217.

Matsui, J., 1997. Tansoseni-no-hanashi. Reinforced plastics, The Japan Reinforced Plastics Society 43 (8), 296–300.

Matsui, J., 1998a. Tansoseni-no-hanashi. Reinforced plastics, The Japan Reinforced Plastics Society 44 (1), 29–34.

Matsui, J., 1998b. Tansoseni-no-hanashi. Reinforced plastics, The Japan Reinforced Plastics Society 44 (4), 166–172.

Morgan, P., 2005. Carbon Fibers and Their Composites. CRC Press. https://doi.org/10.1201/9781420028744.

Morita, K., Suzuki, Z., Hirose, H., 1968. A tertiary phosphine-catalyzed reaction of acrylic compounds with aldehydes. Bull. Chem. Soc. Jpn. 41 (11), 2815. https://doi.org/10.1246/bcsj.41.2815.

Morris, J.B., 1968. Carbon fibres. Atom 144, 269–278.

Nakamura, O., Ohana, T., Tazawa, M., Yokota, S., Shinoda, W., Itoh, J., 2009. Study on the PAN carbon-fiber-innovation for modeling a successful R&D management—An excited-oscillation management model. Synthesiol. Engl. ed. 2 (2), 154–164.

Newcomb, B.A., 2016. Processing, structure, and properties of carbon fibers. Compos. Appl. Sci. Manuf. 91, 262–282.

Nikkei Sangyo newspaper October 27, 1988, Japan.

Nunna, S., Blanchard, P., Buckmaster, D., Davis, S., Naebe, M., 2019. Development of a cost model for the production of carbon fibres. Heliyon 5 (10), e02698.

Odagiri, N., Kishi, H., Yamashita, M., 1996. Development of TORAYCA prepreg P2302 carbon fiber reinforced plastic for aircraft primary structural materials. Adv. Compos. Mater. 5 (3), 249–254.

Otani, S., 1965. On the carbon fiber from the molten pyrolysis products. Carbon 3 (1), 31–38.

Otani, S., 1966. "Method for producing carbon fiber from molten pyrolysis materials" Japan patent announcement Shouwa 41-15728.

Otani, S., Watanabe, S., Ogino, H., Iijima, K., Koitabashi, T., 1972. High modulus carbon fibers from pitch materials. Bull. Chem. Soc. Jpn. 45 (12), 3710–3714. https://doi.org/10.1246/bcsj.45.3710.

Otani, S., Watanabe, S., Araki, Y., 1974. "Method for manufacturing carbonaceous moldings with large anisotropic properties" Japan patent announcement Shouwa 49-8634.

Ruland, W., 1969. The relationship between preferred orientation and Young's modulus of carbon fibers. Appl. Polym. Symp. 9, 293–301.

Rumy, Z., 2018. The Will to Do, the Soul to Dare. Dorrance Pub Co.

Smith, R.E., 1972. Ultrasonic elastic constants of carbon fibres and their composites. J. Appl. Phys. 43 (6), 2555–2561.

Shibata, N., Nishimura, A., Norita, T., 1976. Graphite fibre's fabric design and composite properties. SAMPE Quart 7 (4), 25–33.

Shindo, A., 1961. Physical properties of graphite fibers. J. Ceram. Soc. Jpn. 69 (786), c195–c199. https://doi.org/10.2109/jcersj1950.69.786_C195.

Shindo, A., Fujii, R., Sengoku, T., 1962. Method for manufacturing carbon products from acrylic synthetic polymers. Japan Patent Announcement Showa 37–4405.

Shindo, A., Fujii, R., Sengoku, T., 1963. Manufacturing method for carbon fiber products. Japan Patent Announcement Shouwa 38–12375.

Shindo, A., 1993. Carbon fibers. Kobunshi 42 (1), 41–42. https://doi.org/10.1295/kobunshi.42.22.

Sibal, A.K., 1984. Carbon fiber development and fabric markets. J. Coat. Fabr. 13 (4), 206–215.

Singer, L.S., 1977. High modulus, high strength carbon fibers produced from mesophase pitch. Patent US4005183A.

Swan, J.W., 1883. Incandescent Lamp: Filament/Threads of Squirted nitro-cellulose. Patent GB5978.

Tanaka, F., 2024. Pioneering the carbon fiber frontier: a half-century of industry leadership and the road ahead. Compos. B Eng. 281, 111515.

Tane, M., Okuda, H., Tanaka, F., 2019. Nanocomposite microstructures dominating anisotropic elastic modulus in carbon fibers. Acta Mater. 166, 75–84.

Tang, M.M., Bacon, R., 1964. Carbonization of cellulose fibers—I. Low temperature pyrolysis. Carbon 2 (3), 211–220.

Tsunoda, Y., 1962. Improvements in or relating to the manufacture of heat resistant and corrosion resistant polyacrylonitrile fibres. GB911542A.

Watt, W.L.N., Phillips, Johnson, W., 1966. Carbon fibres with average tensile strengths. Therm. Eng. 27.

Watt, W., 1972. Carbon work at the royal aircraft establishment. Carbon 10 (2), 121–143.

Whitney, W.R., 1909. Carbon, filament, and method of making the same. US916905A.

Wikipedia: Allegany Ballistics Laboratory, https://en.wikipedia.org/wiki/Allegany_Ballistics_Laboratory (Accessed August 25th, 2024).

Wikipedia: Baylis-Hillman reaction. https://en.wikipedia.org/wiki/Baylis_Hillman_reaction. (Accessed 15 November 2025).

Wikipedia: Hercules Inc., https://en.wikipedia.org/wiki/Hercules_Inc. (Accessed August 25, 2024).

Wikipedia: Shu Kanbara, https://ja.wikipedia.org/wiki/%E7%A5%9E%E5%8E%9F%E5%91%A8 (Accessed August 25th 2024).

Wikipedia: SGL carbon, https://en.wikipedia.org/wiki/SGL_Carbon (Accessed August 25th, 2024).

Zhang, X., Yamouchi, M., Takahashi, J., 2011. Life Cycle Assessment of CFRP in the application of automobile. In: 18th International Conference on Composite Materials. Jeju Island, Korea.

A brief history of high-performance polymer fibres

1.3

Pieter Jan Lemstra [1,2] *and Ton Peijs* [3]
[1]School of Chemical and Materials Engineering, Jiangnan University, Wuxi, China; [2]PlemPolco B.V., Veldhoven, the Netherlands; [3]Centre for Polymer and Composites (CPC), WMG, University of Warwick, Coventry, United Kingdom

1. Introduction

Synthetic polymer fibres revolutionised textiles, manufacturing, and countless industries, including composites, in the 20th century. Their history began in the early 1900s, spurred by the desire to create durable, versatile materials that could mimic or surpass natural fibres like cotton, wool, and silk. The journey started with the development of rayon in the late 19th century, often considered to be the first artificial fibre. Regenerated cellulose fibres derived from natural cellulose in wood pulp or cotton linter, were first commercially developed in the late 19th century as a sustainable alternative to silk. This journey began with the invention of viscose rayon in the 1890s, credited to British scientists Cross and Bevan. Their process transformed cellulose into a viscous solution that could be spun into filaments. While not truly synthetic, rayon paved the way for further experimentation.

The first wholly synthetic fibre, polyamide with the trade name Nylon, was introduced by DuPont in 1935. Developed by Wallace H. Carothers and his team at DuPont, nylon was the first completely synthetic fibre derived from petrochemicals and marked a significant breakthrough in polymer science. Polyester followed in the 1940s, developed by British chemists Whinfield and Dickson. Its durability, wrinkle resistance, and affordability made it a favourite for clothing and industrial applications. Over the decades, other synthetic textile fibres like acrylic, polyolefins, and spandex emerged, each tailored to specific needs for applications in clothing, home furnishings, and general-purpose fabrics. These textile fibres prioritise attributes like comfort, aesthetics, and cost-effectiveness over mechanical properties, which are generally characterised by tensile strengths below 1 GPa and elastic moduli under 15 GPa.

In contrast, high-performance synthetic fibres, such as carbon, aramid, and ultra-high molecular weight polyethylene fibres, are engineered for exceptional mechanical properties. These fibres demonstrate tensile strengths exceeding 3 GPa and moduli surpassing 100 GPa, making them ideal for specialised applications in aerospace, sports equipment, and protective gear. Two primary types of high-performance synthetic fibres can be identified:

The History of Composites. https://doi.org/10.1016/B978-0-443-34119-9.00005-X

- Polymer-based fibres: These include fibres with flexible polymer molecules, such as polyethylene, and those with rigid-rod polymer molecules, such as polyaramids and polybenzoxazoles (PBO).
- Carbon fibres: These include fibres produced through the complete carbonisation of pitch or polymer precursor fibres, primarily polyacrylonitrile (PAN).

Polyethylene, polyaramid, and PBO are so-called 1D (dimensional) fibres viz. polymer chains arc aligned in parallel with the fibre axis, enabling the exploitation of the strong covalent bonds in the chain direction. These fibres combine high strength and stiffness with high toughness and are used in ballistic applications such as bulletproof vests, helmets, etc., as well as cut-resistant gloves and ropes. However, due to their highly anisotropic nature—characterised by strong interactions along the chain direction but relatively weak van der Waals and hydrogen bond interactions between the chains—these fibres lack compressive and shear strength, limiting their use in structural composites.

Carbon fibres are derived from 2D graphite sheets composed of carbon atoms arranged in a hexagonal honeycomb lattice with strong covalent bonds. These sheets are often folded or crumpled, forming an interlocked structure that imparts carbon fibres with exceptional tensile strength and good resistance to shear and compression. This unique combination of properties makes carbon fibres ideal for reinforcing lightweight structural composites. However, their high cost restricts their use to applications where weight reduction is crucial, such as aeronautics and sports equipment. For example, the Boeing Dreamliner incorporates approximately 50% carbon fibre-reinforced plastics, including critical structural elements like the wings and fuselage. In this review, the authors will focus on the development of the 1D polymer fibres. For the history and future of carbon fibres, the reader is referred to a rather recent review article by the authors (Peijs et al., 2022).

2. Fibres based on flexible-chain polymer molecules

2.1 *Nylons and polyesters*

In the 1920s, Hermann Staudinger introduced the concept of a macromolecular polymer chain, but his idea was initially met with scepticism (Morris, 2005). In the 1930s, Wallace H. Carothers at DuPont Central Research validated Staudinger's theory by successfully synthesising the first polyesters and polyamides (Fig. 1.3.1). Within Carothers' group, Julian Hill devoted years to attempting the synthesis of higher molecular weight aliphatic polyesters but eventually abandoned the effort due to persistent polymerisation challenges. These rigorous attempts to produce high molecular weight polyesters are well documented (Hermes, 1996).

Carothers took the next significant step in polymer chemistry by focussing on polyamides, where amide bonds serve as the linking units—similar to the amide bonds found in natural proteins. His team experimented with various diamines and dicarboxylic acids, ultimately identifying pentamethylene diamine and sebacic acid as

Figure 1.3.1 Wallace H. Carothers with another of his inventions, Neoprene rubber (left) and Julian Hill (right) pulling a "fibre" from molten polymer.
Image courtesy of Wikimedia and DuPont.

producing the best polymer for spinnability. However, for economic reasons, R&D manager Elmer Bolton pivoted toward using hexamethylene diamine and adipic acid, leading to the development of nylon 66 in 1935 (Bolton, 1942).

By the end of 1939, a full-scale production facility was operational, and nylon hosiery became available nationwide by mid-1940, marking an impressively short development timeline. Nylon fibres quickly gained immense popularity, particularly as imports of silk from Japan ceased due to political tensions. Nylon 66 fibres offered a timely alternative. However, this public availability was short-lived. Following the United States' entry into World War II in late 1941, nylon production was redirected toward military applications, including parachutes, balloon cloth, glider tow ropes, tyre cords, and military apparel.

Since the pioneering work of Carothers in discovering the first scientific methods for synthesising high-molar-mass polymers, polymer scientists have sought to enhance mechanical properties by orienting chain molecules. As early as 1932, Carothers and Hill published the concept of a perfectly oriented fibre, describing it as "*a single crystal in which the long molecules are in an ordered array parallel to the fibre axis*" (Fig. 1.3.2). This marked the foundational idea behind high-performance fibres (Carothers and Hill, 1932). In the same year, they also reported the process of strengthening polyester fibres through cold-drawing.

Interestingly, Carothers and Hill worked on aromatic polyesters derived from phthalic acid and glycol, which produce a kinked chain molecule, but they did not explore terephthalic acid (Fig. 1.3.3). Terephthalic acid, when combined with glycol, forms polyethylene terephthalate (PET), a polymer that has since become widely

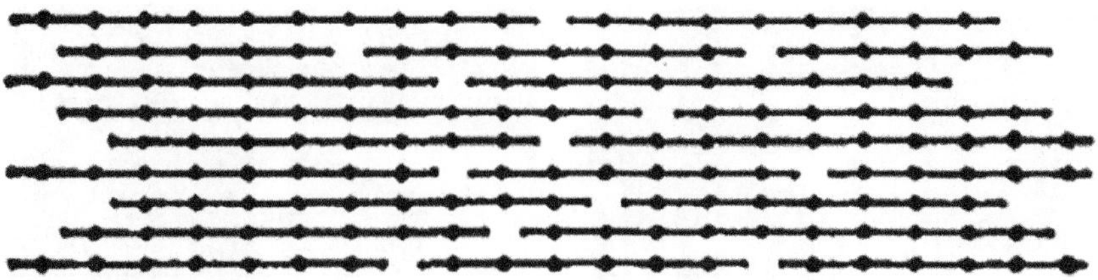

Figure 1.3.2 Ideal polymer chains model for a perfectly oriented fibre. An ideal polymer fibre needs high molecular orientation and chain extension for a high elastic modulus, and few chain-end defects (high molecular weight) for a high tensile strength. This concept was understood and taken to the extreme in the 1951 British satirical science fiction movie "The Man in the White Suit" which is set in a fictional British textile factory and stars Sir Alec Guinness as a brilliant but eccentric scientist who invents a revolutionary, incredibly strong fibre based on aligned polymer molecules of infinite length.
Adapted from Carothers and Hill (1932) with permission from American Chemical Society.

renowned. The development of PET was achieved in 1941 in the UK by John Rex Whinfield and James Tennant. Tragically, Carothers committed suicide by drinking potassium cyanide in 1937, so he did not live to witness the development and success of PET.

In the prewar years, Carothers was allowed to publish his work, including his conclusion that caprolactam was not an ideal monomer for producing nylons. However, at IG Farben AG in Nazi Germany, Paul Schleck took up the challenge and successfully polymerised caprolactam into nylon 6 in 1937, with fibres being produced the following year (Morgan, 1981). After World War II, the Allies dismantled the large IG Farben conglomerate, and as part of the spoils of war, the Dutch companies AKU/Enka (later Akzo) received the technology for nylon 6 fibres, while DSM, also based in the Netherlands, was granted the caprolactam technology. Nowadays, nylon 6 and nylon 66 are the two most significant polyamides, with a combined annual production of nearly 10 million tons. They share roughly equal market shares and are widely recognised as both textile and technical fibres. However, despite their extensive applications, neither nylons nor PET has attained the status of high-performance fibres, as will be discussed later.

In 1960, Treloar wrote a seminal paper in which he showed by a simple calculation that the stiffness of an extended polymer chain like polyethylene or polyamide is about 180 GPa, close to steel (Treloar, 1960). In the 1970s more

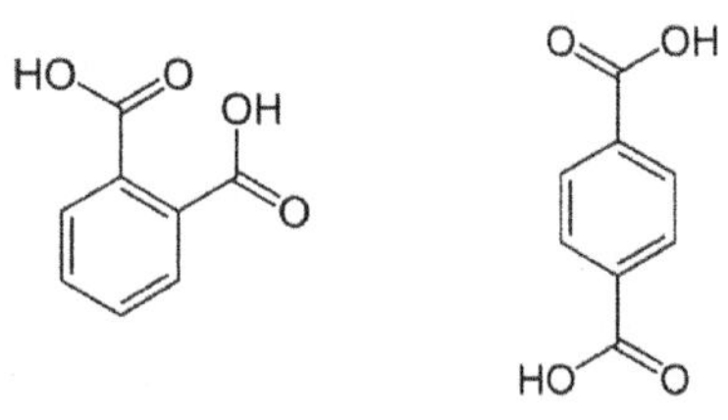

Figure 1.3.3 Phthalic acid (left) and terephthalic acid (right).

sophisticated calculations showed even more impressive data like a tensile strength up to 19 GPa (Zhurkov and Korsukov, 1974; Boudreaux, 1973). These publications triggered many to pursue chain-extension/orientation in polymer fibres.

In a so-called "Gedanken-Experiment," the basic concept for achieving such properties is quite straightforward. Polyethylene, polyamides, and many other polymers are composed of flexible polymer chains. Due to the rotation around the C–C bonds in the main chain, these polymers adopt a "random-coil" conformation in the molten state or in solution, driven by the tendency to maximise entropy. If we grip one end of a polymer chain and pull on the other, deformation or drawing occurs easily. However, when we pull the chain straight (chain extension), we start to effectively load the strong covalent bonds along the main chain. This leads to a sudden increase in stress, as shown in the schematic below (Fig. 1.3.4).

In practice, we are not dealing with single polymer chains but a multitude of highly entangled polymer chains—like cooked spaghetti—in the polymer melt. Chain orientation/extension can be obtained by extruding the polymer melts at high speeds through a capillary followed by high-speed drawing. However, extrusion and melt-drawing do not render high mechanical property values. The reason is that oriented chains that leave the spinneret relax, i.e., want to regain their entropy and hence form random coils again. The relaxation times can be increased by lowering the temperatures, notably in the solid state. But in the solid state (regular) flexible polymer chains like polyethylene and polyamide (nylon) solidify into so-called folded-chain crystals. In the case of polyamide crystals, the strong intermolecular hydrogen bonds

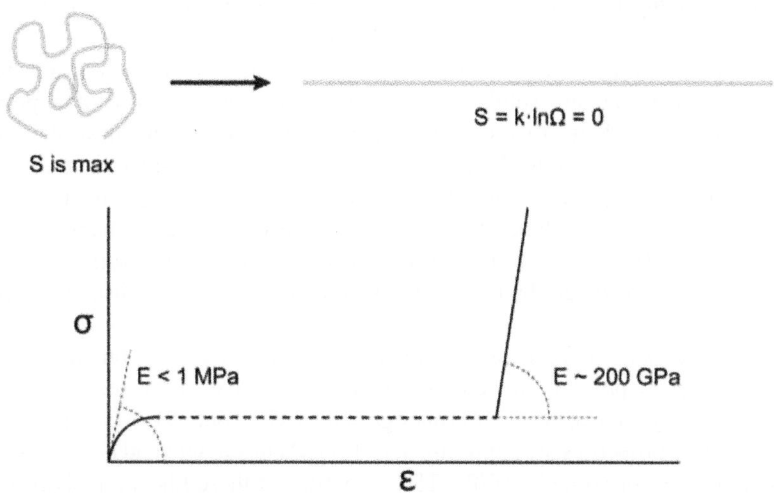

Figure 1.3.4 Deformation of a single isolated polymer chain from its random-coil conformation into an extended chain.
Image courtesy P.J. Lemstra.

prevent unfolding during drawing and act as barriers for chain-orientation, see section on polar molecules below.

The exceptions are apolar polymers like polyethylene (PE) and polypropylene (PP), in which case the crystals are "weak"; and upon drawing, the lamellar crystals unfold in a temperature domain close to but below the melting temperature. However, here chain entanglements can oppose drawing and chain extension, and it took several decades before this was realised, as will be discussed below.

2.2 The route to high-performance PE fibres

Van der Vegt and coworkers melt-extruded linear polyethylenes but encountered the problem that high-density polyethylene (HDPE) melts possess a very low melt strength in contrast with branched low-density polyethylenes (Van der Vegt and Smith, 1967). Once the molten filaments with (partly) oriented chains leave the spinneret, the molecules want to regain their entropy, and due to the low melt strength, this process occurs very rapidly via chain slippage and relaxation. By decreasing the temperature during extrusion, the relaxation process slows down, but ultimately, crystallisation and solidification occur in the capillaries.

A "breakthrough" was achieved by Professor Ian Ward at the University of Leeds, who took a different approach by first solidifying the extruded filaments and then drawing them in a temperature range just below the melting point. This process, known as (semi) solid-state drawing, allows for much shorter relaxation times, making chain orientation permanent. Ward dedicated nearly his entire career to the orientation of PE and other polymers, exploring various methods, including not only melt-spinning but also die-drawing, forcing solid PE through dies (Capaccio and Ward, 1973, 1974; Ciferri and Ward, 1979).

Ward and coworkers knew they were "fighting" against the physical network during melt-extrusion and solid-state drawing. The melt viscosity of a polymer scales with $M_w^{3,4}$ and the rheological behaviour is related to a physical network formed by entangled chains. Moreover, in the solid state, entanglements act as crosslinks, preventing deformation/draw. Based on numerous experiments Ward c. s. realised that they were limited by the melt-viscosity during extrusion, typically a M_w up to 150.000 g/mol, while subsequent drawability in the solid state could be enhanced to some extent by using a hot shroud to invoke crystallisation at higher temperatures.

Ward's process was commercialised in the 1990s by SNIA (Italy) but this company went bankrupt. At present, two companies are still active in the melt-spinning of HDPE fibres; Toyobo (Japan) with Tsunooga and Huvis (South Korea) with Duradron fibres. The fibre properties are approximately 1.5 GPa in strength and a corresponding Young's modulus of around 50 GPa. The properties of these fibres fell short of predictions by Treloar and others, but in retrospect, all calculations regarding the ultimate strength and stiffness of oriented polymer fibres tend to overestimate the achievable values (see intermezzo below).

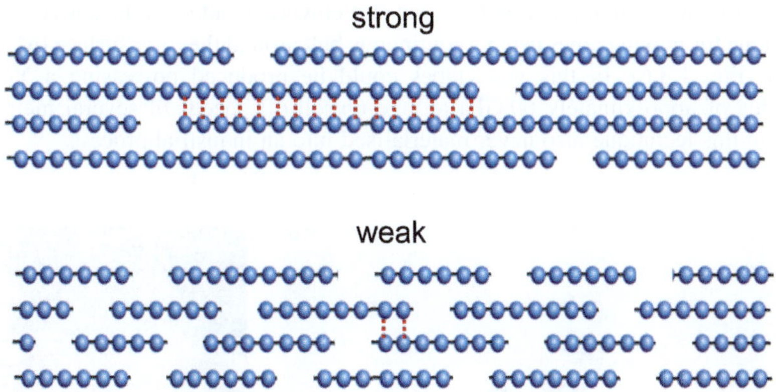

Treloar calculated the strength/stiffness of one single infinite chain. In practice, we are dealing in fibres with a multitude of chains, and the molecules are *finite and not infinite.* In a fibre, supposing full chain alignment, the length of an extended PE molecule of 100.000 g/mol is 5 μm max. whereas in testing at least 10 cm of fibre is required. Consequently, the strength of a fibre is not determined by the strength of the covalent bonds in a single chain but by the small Van der Waals interactions between the many chains, which should prevent chain slippage upon loading. If sufficient Van der Waals interactions are present between overlapping chains (large overlap or stress transfer length) then it can be expected that the covalent bonds in the main chains are effectively addressed upon loading. For this reason, ultra-high-molecular-weight PE (UHMWPE) is needed, but unfortunately, due to its high viscosity, this is not melt-processable. This strengthening concept is related to micromechanics of aligned discontinuous fibres, where the matrix and interface play key roles in stress transfer between fibres. Similar to discontinuous fibre composites, the chain length is more critical for achieving high strength, while high modulus is primarily influenced by chain orientation. The schematic is illustrative only; UHMWPE chains can comprise hundreds of thousands of carbon atoms.

2.3 Solution(gel)-spinning of UHMWPE

In the mid-1970s, Albert Jan Pennings (University of Groningen) and his student Arie Zwijnenburg (Fig. 1.3.5) made a remarkable discovery (Zwijnenburg, 1978). In a Couette type apparatus, a dilute <1% supercooled solution of ultrahigh molecular weight polyethylene (UHMWPE) in xylene was stirred slowly and from the surface of the inner rotor a fibrous, tape-like, PE structure could be drawn by immersing a seed fibre to the surface of the inner rotor and subsequently slow pulling (Fig. 1.3.6). These withdrawn fibrous PE structures possessed a Young's modulus >100 GPa and a tenacity of 3 GPa (Zwijnenburg and Pennings, 1976b). This so-called surface growth technique has many disadvantages with respect to the very slow take-up speed and homogeneity of the PE "fibre", but these experiments proved that oriented UHMWPE structures could be made experimentally possessing superior mechanical properties to melt-spun HDPE fibres.

Attempts were made to scale-up the surface-growth technique by constructing larger rollers and multiple take-up points from the surface but none of these attempts materialised into a real process. Mackley and coworkers adopted the surface-growth technique for making tapes using a continuous belt around the rotor surface (Mackley, 1985) (Fig. 1.3.6). In this way, tapes could be produced possessing a Young's modulus of approximately 60 GPa at take-up/roll-off speeds of several metres per min but this technique also never materialised into an industrial process.

Figure 1.3.5 Prof A.J. Pennings (left) and Dr Arie Zwijnenburg† (right).
Image courtesy of Dyneema.

Arie Zwijnenburg was a Ph.D. student at the University of Groningen. He was a slight man, small enough to fit inside the fume cupboard where he spent countless hours—often entire days and sometimes nights—meticulously pulling "fibres" from the Couette apparatus. His dedication was so intense that he had to be fed from outside, as he couldn't release the fibre from his hands. Tragically, Arie passed away at a much too young age. One might wonder if prolonged exposure to toxic xylene fumes during his Ph.D. years played a role in his untimely death. Despite this, his work laid the experimental foundation for the development of ultra-strong polyethylene (PE) fibres. His contributions were invaluable and should not be forgotten.

At the end of the 1970s, Paul Smith and Pieter Jan Lemstra performed solution-spinning experiments at DSM Central laboratories (Fig. 1.3.7). Using a simple piston-type spinning apparatus they would pump a dilute solution of UHMWPE, typically 1% w/w in decalin, through an orifice of approximately 1 mm. The extruded filament was quenched into a water bath and a gel-like filament was obtained of as little as 1% of UHMWPE polymer in 99% of solvent. Stretching these gel-like filaments at elevated temperatures, but below the dissolution temperature, resulted in UHMWPE fibres with properties like those found by Zwijnenburg and Pennings with typical Young's moduli >100 GPa and a tensile strength of 3 GPa or more (Smith and Lemstra, 1980a). It was recognised immediately that the main advantage over the surface-growth technique was the potential of a continuous solution-spinning process. The high stretchability of the gel-like filaments, with draw ratios typically exceeding 50, was not completely unexpected. Previous studies had demonstrated that polymer

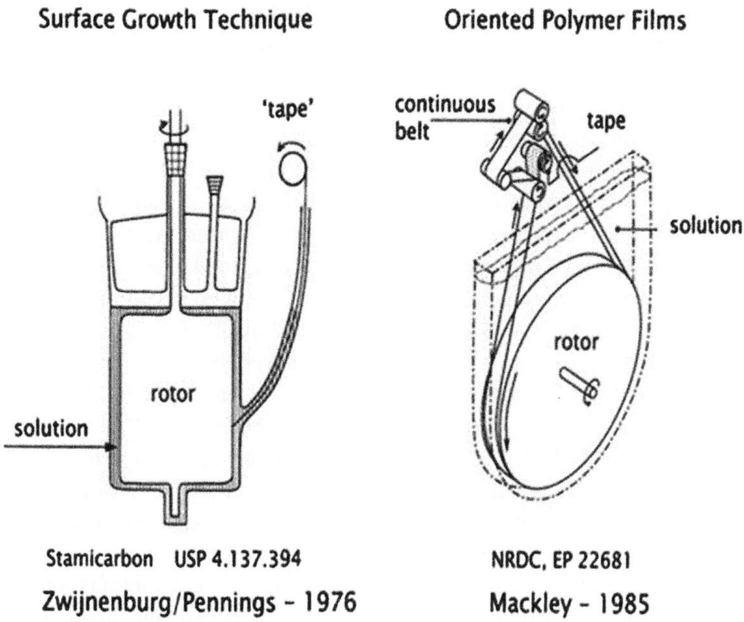

Figure 1.3.6 Surface growth techniques.
Reproduced from Zwijnenburg and Pennings (1976a) with permission from Springer, and Mackley (1985).

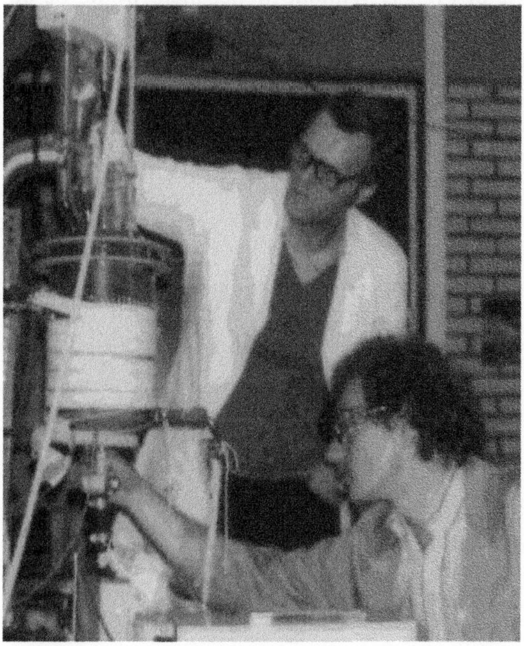

Figure 1.3.7 Paul Smith (right) and Piet Lemstra (left) dissolving UHMWPE in decalin (image courtesy P.J. Lemstra).

gels serve as excellent precursors for achieving highly oriented structures, although UHMWPE had not been specifically investigated in this context (Lemstra et al., 1978). The substantial amount of solvent present was believed to act as a plasticiser during the drawing process.

However, the most remarkable observation occurred when certain as-spun UHMWPE gel fibres that were spun on a Friday and were left undisturbed over the weekend were still stretchable on Monday. Even after the complete removal of solvent from these gel-filaments, the fully dried fibres remained ultradrawable. This suggested that the solvent itself did not play a significant role in the ultradrawing process. Instead, the solvent served to generate a precursor fibre with exceptional drawability. It appeared that the constraints limiting the drawability of melt-crystallised UHMWPE in the solid state—such as those observed by Ward and coworkers in melt-spun HDPE—could be overcome by spinning or casting from dilute solutions (Smith and Lemstra, 1980b). Thus, while the solvent was crucial for processing the highly intractable UHMWPE (as melt processing is impractical due to its extremely high melt viscosity), it also induces a favourable disentangled structure that supports ultradrawing. Importantly, the presence of the solvent was not required during the drawing phase as the material remained in its disentangled state.

A significant challenge in the industrialisation of the gel-spinning process was scaling up from a 1% polymer concentration to an acceptable level. Dissolving UHMWPE in a tank by stirring is impossible due to Weissenberg effects (climb of viscoelastic liquids up a rotating rod). The breakthrough, achieved by Han Meijer (Fig. 1.3.8) during his tenure at DSM, involved the use of a corotating twin-screw extruder (Fig. 1.3.9). This method enabled the creation of homogeneous UHMWPE solutions with concentrations of 30% or higher. Without Meijer there would be no commercial viable process for the solution-spinning UHMWPE. The drawability, λ_{max}, of the as-spun fibres proved to be related to the polymer concentration in solution scaling with the square root of the inverse of the initial polymer concentration: $\lambda_{max} \sim \Phi^{-0.5}$ (Smith et al., 1981). At present the polymer concentration used in commercial processes is approximately 10%.

Figure 1.3.8 Prof Han Meijer (image courtesy TU Eindhoven).

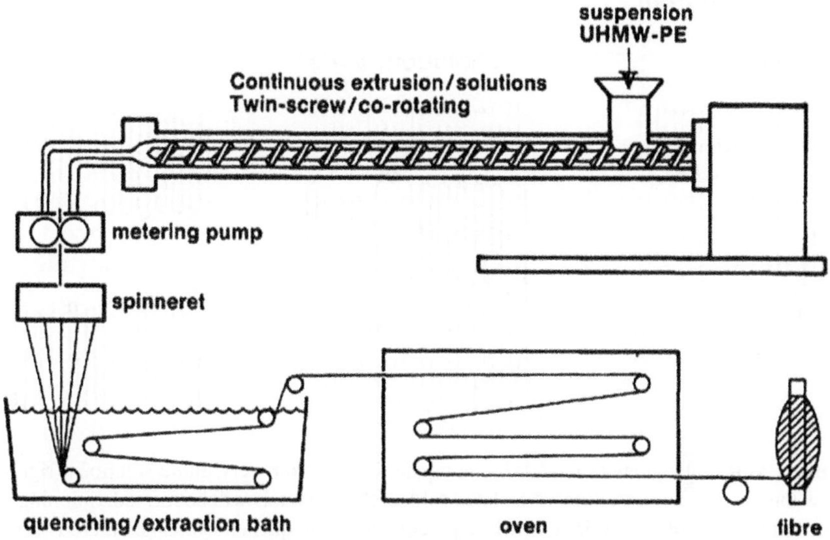

Figure 1.3.9 Schematic representation of the solution (gel) spinning of ultrahigh molecular weight polyethylene (UHMWPE).
Reproduced from Lemstra et al. (1987) with permission from Elsevier.

From these experimental observations it was concluded that the constraints which limit the drawability of UHMWPE in the solid state are the trapped entanglements which act as semipermanent crosslinks upon drawing. Upon dissolution, the number of entanglements is reduced proportional to the degree of dilution. Accordingly, the so-called molar mass between entanglements, Me, scales with $\langle Me \rangle_{sol} \sim \langle Me \rangle / \varphi$.

The homogeneous solutions of UHMWPE in solvents like xylene, decalin, or paraffin oil/wax, as created in a corotating twin-screw extruder, are fed, via a metring pump, to a spinneret containing several hundreds of orifices typically. The next step in the process is the gelation/crystallisation of the UHMWPE, which is done by quenching the as-spun filaments in air, water, or an extraction medium. Crystallisation is an essential step in the process since, at this stage, the low entanglement density structure is "trapped" into the fibre, resulting in a super-drawable precursor fibre. Figure 1.3.10 schematically depicts various possibilities in chain topology. Drawing and removal of remaining solvent are the last steps in the process and they are performed in a temperature range close to but below the melting temperature and above the α-relaxation temperature, which is around 80°C for PE.

During drawing, orientation and chain extension occur, and folded-chain clusters gradually unfold in extended chain clusters. Figure 1.3.11 presents original mechanical property data based on isothermal single-stage drawing at 120°C. Both modulus and tensile strength increase with draw ratio as a result of improved orientation and chain extension. Through optimisation of the polymer grade and drawing procedure, such as multistage in-line drawing, tenacities of 4 GPa and moduli of 150 GPa are now possible on a commercial scale.

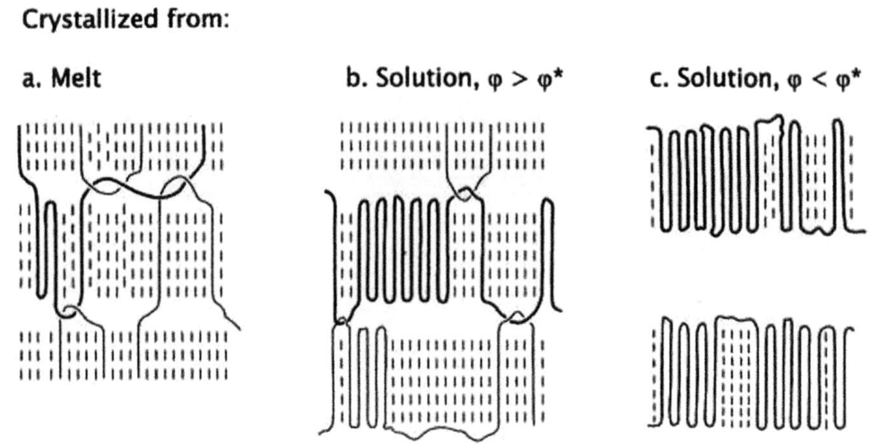

Figure 1.3.10 Entanglement model. Upon crystallisation from very dilute solutions (right) below the overlap concentration φ^* isolated folded-chain crystals will be formed. Spinning of these solutions would result in a powdery precipitate in the quench bath since there is no coherence between the individual crystals and no spinning and drawing is possible. Upon crystallisation from the melt (left) folded-chain crystallisation will occur but ultradrawing is hampered by trapped entanglements. From semidilute solutions (middle) the number of entanglements is significantly reduced and dependent on φ, effectively removing the constraints which limit drawability in melt-crystallised PE. Once this favourable structure with a low entanglement density is formed and fixated by crystallisation the solvent can be removed without losing its super-drawing character.
Reproduced from Lemstra et al. (1987) with permission from Elsevier.

DSM patented the gel-spinning process in the late 1970s. However, the company had no expertise in spinning or fibre technology, and DSM's management initially showed little interest in the project. This negativity only increased after they consulted

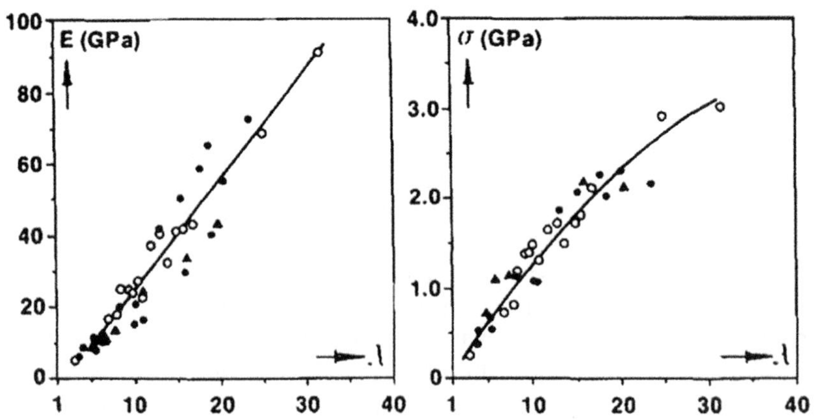

Figure 1.3.11 Young's modulus (left) and tensile strength (right) of gel-spun filaments drawn in a single-stage at 120°C as a function of draw ratio.
Reproduced from Lemstra et al. (1987) with permission from Elsevier.

Akzo, then a world leading fibre producer, also headquartered in the Netherlands. Akzo dismissed the potential of this "candlewax" fibre, citing its low melting point and excessive creep as significant drawbacks. Oriented PE fibres are inherently prone to creep as there is no lock-in mechanism that possesses a limit to the deformation. Creep, and drawing are essentially the same processes governed by the weak Van de Waals interactions between chains, albeit at different temperature and timescales. Following this assessment, DSM initially terminated the project in 1982. Meanwhile, Smith had moved to DuPont and the project was left with Lemstra who together with a small team continued working on the project afterhours. Around that time, Allied Signal (now Honeywell) in the USA obtained a patent strikingly similar to DSM's. After thorough investigations and experiments, the DSM team demonstrated that Allied's fibre was, in fact, identical to the one described in DSM's original patent. Acknowledging this, Allied recognised the need for a licence to use DSM's patent. The resulting licence fees and royalties persuaded the DSM management to invest in the project. Subsequently, DSM formed a joint venture with Toyobo, a leading Japanese synthetic fibre producer, while Allied Signal began manufacturing the material under their tradename Spectra. In Toyobo, DSM found the partner they needed to go commercial themselves. DSM wanted to brand the product Dynema—with just one "e" as it combined the Greek words for strength and fibre. But the Japanese disliked the "ne" as it sounds like two in Japanese (=ni), and they wanted to be number one. So, it was decided to add another "e", and Dyneema was born. Initially, the fibres were produced in Japan, but the Dyneema product portfolio (see Table 1.3.1) is currently produced in Heerlen (Netherlands) and in Greenville, North Carolina (USA). Until 2016, Toyobo had been selling its fibre under the brand name Dyneema in Japan, but to clarify the differences between its own products and products produced by DSM, Toyobo decided to discontinue its Dyneema brand name and replace it with Izanas. DSM sold its Protective Materials business, which includes the Dyneema brand, to Avient Corporation (USA) in April 2022 for nearly $1.5 billion.

Next to these three main suppliers, more than 40 manufacturers of Dyneema-like fibres are active in the field in China, colloquially referred to as "*Chineema*" producers. The Godfather of "*Chineema*" technology is Professor Yimin Wang (Donghua University, Shanghai) (Fig. 1.3.12), who developed in China a process rather similar

Table 1.3.1 Mechanical properties of commercial gel-spun ultrahigh molecular weight polyethylene (UHMWPE) fibres.

	Dyneema					**Spectra**		
Type	SK60	SK65	SK75	SK90	SK99	900	1000	2000
Strength (GPa)	2.7	3.0	3.4	3.8	>4.0	2.1	2.9	3.0
Modulus (GPa)	90	95	110	140	150	80	100	115
Elongation (%)	3.5	3.6	3.8	4.0	4.0	3.6	3.3	3.0

Figure 1.3.12 Prof Yimin Wang. Image courtesy PJ Lemstra.

to the Dyneema process, but he used paraffin oil as a solvent which is much safer to handle than the inflammable decalin. The as-spun fibres are collected in boxes with a perforated bottom and stored for 1 or 2 days, whereupon "syneresis" occurs, the solvent is "sweating out", and the PE concentration rises from typically 10% to appr. 25% w/w (Fig. 1.3.13). Subsequently, these fibres are extracted with heptane or dichloromethane and ultradrawn in hot-air ovens.

Figure 1.3.13 Typical Chinese UHMWPE gel-spinning process, the as-spun fibres are collected in boxes with a perforated bottom and stored for 1–2 days. Due to syneresis the polymer concentration increases from 10% to approximately 25%. Subsequently, the wet fibres are extracted and drawn in sequential ovens.
Reproduced from Lemstra (2022) with permission from Elsevier.

2.4 Solvent-free processing of UHMWPE

UHMWPE is made by using the standard heterogeneous Ziegler/Natta catalysts but can also be made by using homogeneous metallocene-catalysts. In both cases, the polymer chains grow away from the catalyst, and at lower temperatures, the chains will feel a "cold" environment and crystallise. Smith c.s. proposed that at low temperatures, the chain will grow and crystallise into extended chains (Rotzinger et al., 1989) and coined that the UHMWPE obtained was ultradrawable. Whether extended chains are formed is still a matter of debate, but using a metallocene catalyst, the standard folded-chain crystals will form at low temperatures.

Depending on the concentration of the metallocene-catalyst in solution and the relative rate of polymerisation versus nucleation/crystallisation, the chains will not or only slightly entangle at the chain heads, and disentangled UHMWPE reactor powder is obtained (Fig. 1.3.14).

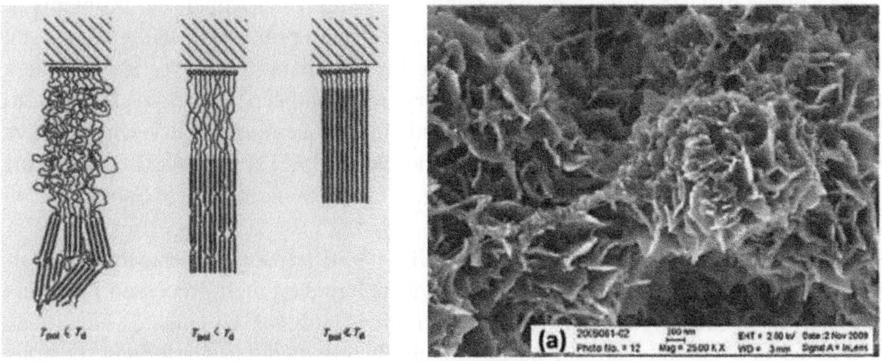

Figure 1.3.14 Model for low-T polymerisation (left). Morphology of nascent reactor powder (right).
Reproduced from Rotzinger et al. (1989) with permission from Elsevier. Image courtesy S. Rastogi.

Around 2010 Teijin Aramid in the Netherlands commercialised the production of UHMWPE tapes and fibres using this solvent-free disentangling technology under the trade name Endumax. UHMWPE reactor powder was consolidated into a film using a continuous belt-press. This film was rolled in a calendar and subsequently drawn to high draw ratios between rollers close to but below the melting temperature (Peters, 2009). When a single-site catalyst such as the FI-catalyst (Mitsui, Japan) was used, films could be obtained with very high stiffness, close to 200 GPa and the slitted tapes possessed a strength of close to 4 GPa. Expectations were running high because for ballistics the packing density of flat tapes in unidirectional (UD) composite laminates is superior to that of fibre (Fig. 1.3.15).

However, the FI-catalyst is a living catalyst producing one UHMWPE chain per molecule of catalyst, and that is expensive. Consequently, Teijin used a commercial

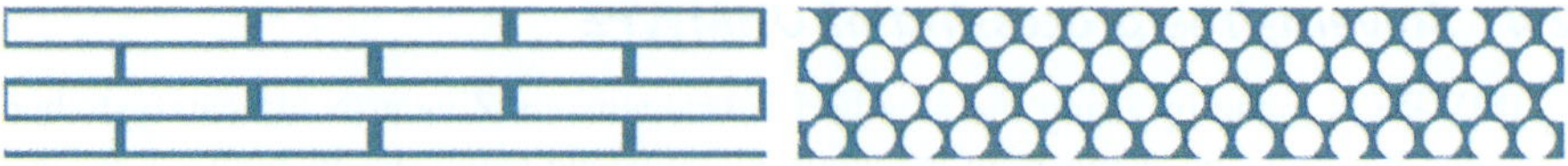

Figure 1.3.15 Stacking of tapes versus fibres in UD-laminates (green is matrix). Reproduced from Bott et al. (2022) with permission from Elsevier.

UHMWPE grade, which showed some disentangled characteristics, and Young's moduli obtained were still very high, but tensile strengths were limited to approximately 2.3 GPa. In ballistics applications, the main issue is the specific energy absorption. The specific energy absorption capability is related to the specific fibre strength and strain at break, $E_{sp} = 0.5 \times \sigma_f$ x ε_b/ρ, effectively the area under the stress-strain curve and the Endumax tapes failed to compete with commercial high-strength polyethylene fibres. The Endumax assets were sold to Barrday (USA) in 2023. DSM also developed UHMWPE tapes and produced for some time the BT10 tapes with a tenacity of 1.7 GPa in Greenville (USA) but also stopped production.

The future of UHMWPE films/tapes is unclear at present, albeit the problems in production and economics have been solved. The main problem of using single-site catalysts is reactor fouling, but recently, Nic Friederichs at Sabic was able to use a supported catalyst and could produce very high molecular weight PE, up to 50×10^6 g/mol (Troisi et al., 2021). In solid-state processing, melt-viscosity plays no role, so very high molar weights are no problem. At ETH Zurich, Theo Tervoort opened new avenues in solid-state processing, in close cooperation with Sabic, viz. 3D products like tubings (Christakopoulos et al., 2024).

While the future of tape technology for UHMWPE remains unclear, this technology is currently the trend in commercial products based on melt-processed polymers like HDPE. In recent years, several self-reinforced polymer composites have been developed and commercialised based on highly oriented tapes, ribbons, or foils (Alcock and Peijs, 2013). The higher volume fraction of reinforcement achievable due to the superior packing of rectangular cross-section geometries, together with the low crimping seen in the resulting composites, makes this geometry particularly attractive for composite applications. In 1993, Hine, Ward, and colleagues at the University of Leeds introduced a simple yet elegant process for creating self-reinforced polymer composites, known as "hot compaction". Hine et al. (1993) described the first application of this technology using melt-spun HDPE fibres but later switched to polypropylene (PP). This pioneering work was commercialised first by BP and later Propex Fabrics GmbH (Germany) under the trade name Curv.

Parallel to the emergence of the Curv technology, a co-extrusion-based processing route for self-reinforced polypropylene (SRPP) was developed by Peijs c.s. at Eindhoven University of Technology (Netherlands), and later at Queen Mary University of London (UK) in collaboration with Lankhorst (Netherlands) (Peijs, 2003). This technology was licenced to Lankhorst and later Milliken (USA). In the PURE process, a typical A-B-A trilayer tape construction is created. The outer PP copolymer layers melt at a lower temperature than the PP homopolymer core, providing a wider processing window (~20°C). Milliken markets these composites under the trademark

Tegris, while PURE is currently owned by DIT Weaving (Netherlands). PURE is available as tape, fabric and preconsolidated sheet material. SRPP are also available from Don & Low under their trade name Torodon (formerly Armordon). SRPP is widely utilised in various products, particularly lightweight suitcases known for their exceptional toughness. Samsonite has manufactured millions of C-Lite suitcases using Curv material, while TUMI (now a part of Samsonite) employed PURE. More recently, Samsonite developed their own proprietary Roxkin material for their Proxis line. Unlike woven tapes, this innovation uses multilayer laminates based on biaxially oriented polypropylene (BOPP) films.

SRPP materials have also been used in sports equipment, hockey skates, protective gear, and armour. NASCAR racing cars use it due to its superior impact resistance and soft failure mode. Unlike carbon fibre composites, it does not splinter when it breaks, reducing the risk of sharp debris, and thereby enhancing driver safety.

Although the first commercial self-reinforced polymer composites focused on polypropylene with relatively modest mechanical properties (strength <600 MPa, modulus <18 GPa), recent developments have focused on other polymers such as PET (Zhang and Peijs, 2010), PLA (Mai et al., 2015), and other biopolymers (Colwell et al., 2024), as well as polymers for much higher performance. The recent work by Lefèvre et al. (2012) and Schaller et al. (2016) on the creation of LCP foils produced by extrusion through filament-fusing dies is a clear example of this. Melt-spun HDPE tapes with a modulus of 60 GPa and tensile strength of 1.2 GPa were initially developed by Novameer in the Netherlands and later by Teijin Monofilament and Compomeer. Self-reinforced PET and PLA composites are currently commercially available from COMFIL, using their commingled yarn technology.

2.5 Intermezzo about entanglements

Thc entanglement model, as shown in Fig. 1.3.10, could explain the dependence of the solid-state drawability on the initial solution concentration of gel-spun fibres and can also explain that in the melt-spun experiments of Ward c.s. slow cooling promotes solid-state drawability by reeling in chains during crystallisation. Entanglements, as such, connect polymer chains in a physical network, and with increasing molecular weight, i.e., more entanglements per molecule, the melt-viscosity rises steeply. As mentioned before, the zero-shear viscosity η_0 scales with $M_w^{3.4}$. De Gennes (1971) proposed his "reptation theory", where chains move like snakes in virtual tubes formed by entanglements with neighbouring chains, and self-diffusion occurs via chain ends. The reptation or terminal time for a chain to move out its original tube and into a new tube scales with M^3.

The question remains, can we benefit from this disentangled state in melt-processing UHMWPE? According to De Gennes' reptation theory, it takes quite some time, in the order of hours, before an equilibrium network of interwoven entangled chains is restored when disentangled UHMWPE is heated above its melting temperature.

Sanjay Rastogi, now at KAUST (Saudi Arabia), has published extensively on the so-called "memory effect" (Rastogi et al., 2005). Upon heating nascent disentangled

UHMWPE reactor powder, it takes several hours before a fully entangled equilibrium melt is restored, as evidenced by the build-up of the G' modulus during rheological experiments and the so-called build-up time scales with M^3, perfect in line with the reptation theory of De Gennes.

However, recent experiments employing advanced rheological techniques, such as the use of a shear rheometer equipped with a partition plate, have demonstrated that the concept put forward by Rastogi c.s. is incorrect and based on experimental artifacts. There are no memory effects, which is surprising to some extent in view of the well-accepted reptation theories. Interestingly, rheological properties like G' are recovered instantaneously upon the heating and melting of disentangled UHMWPE (Christakopoulos et al., 2021). This behaviour is attributed to the "chain explosion" of folded-chain crystals, where the chains regain entropy by transitioning into random coils upon melting. A similar phenomenon was observed much earlier by Bastiaansen et al. (1990), where the advantageous drawing characteristics, as illustrated in Fig. 1.3.16, disappeared immediately when the sample was melted and subsequently recrystallised, as shown by curve S^*. While a detailed discussion of polymer chain dynamics lies beyond the scope of this manuscript, readers interested in exploring this topic further are encouraged to consult recent publications in this field.

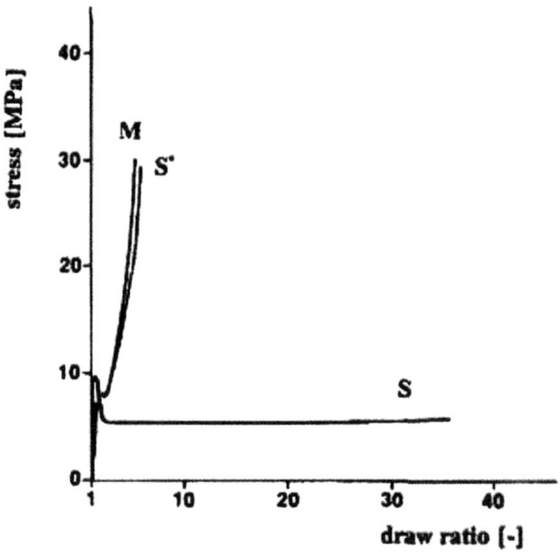

Figure 1.3.16 Stress versus strain curves of drawing of UHMWPE. M = melt-crystallised fibre, showing limited drawability as a result of a dense entangled network; S = solution-crystallised showing ultradrawability as a result of its disentangled state. Reproduced from Bastiaansen et al. (1990) with permission from Elsevier.

3. Fibres based on rigid-rod polymer molecules

3.1 Aromatic polyamides

In the 1960s, nylon and later polyester gradually displaced much of the technical fibre market of rayon, and tensile strengths in the range of 600–700 MPa were the practicable upper limit of fibre strength for many years. Then, in the mid-1960s, research started at DuPont by Kwolek, Blades, and Morgan on extended-chain aromatic polyamides (Morgan, 1977; Kwolek et al., 1977; Bair et al., 1977), while at Monsanto, aromatic polyamide hydrazides were studied by Preston, Black, and others (1973). The DuPont research resulted in the discovery of liquid crystalline solutions of aromatic polyamides which could be spun into a fibre. This fibre, later named Kevlar, exhibited remarkable properties such as a high tensile strength and heat resistance, which immediately attracted attention for potential industrial and military applications.

The DuPont research team addressed the challenge of aligning polymer chains in fibres by sidestepping the difficulties associated with drawing entangled networks of flexible chains. Instead, they focused on synthesising polymer molecules with inherent rigidity. To visualise the difference, think of rigid polymer molecules as uncooked spaghetti, compared to the tangled nature of cooked spaghetti, representing flexible polymer molecules. By linking benzene rings—an established rigid structural unit in chemistry—through amide bonds, they successfully created aromatic polyamides, commonly referred to as (poly)aramids, which would not require unfolding by ultradrawing. An obvious choice initially was PBA (poly-1,4-benzamide), based on the monomer para-amino-benzoic acid, but the final choice was made for PPTA, poly(p-phenylene terephthalamide), the base polymer for the Kevlar fibre (Fig. 1.3.17).

Aromatic polyamides possess very high melting temperatures and cannot be melt-spun because of premature degradation. Consequently, solution-spinning has to be adopted and the only viable solvent proved to be sulphuric acid. These rather rigid polymers possess a unique solution behaviour, as discovered by Ms. Stephanie Kwolek, who worked in a team with Paul Morgan and Herbert Blades at Central Research at DuPont in the 1960s. Kwolek had previously identified the appropriate solvents and polymerisation conditions needed to synthesise poly-m-phenylene isophthalamide. This compound, commercialised by DuPont in 1961 under the trade name Nomex, became known as a flame-resistant fibre. In 1964, Kwolek's group began searching for a strong and lightweight fibre to replace steel cords in tyres. She expanded her research to include PBA and PPTA. These materials displayed highly ordered, rod-like molecular structures in solution, which she meticulously observed and documented.

In contrast with flexible polymer molecules in solution, the more rigid polymer chains are expanded in solution and touch upon each at lower polymer concentrations, giving rise to a steep increase in viscosity. Kwolek discovered that as the polymer concentration increased further, the solution turned opaque and exhibited a significantly reduced viscosity (Fig. 1.3.18). She attributed this phenomenon to the liquid-crystalline behaviour characteristic of PPTA and similar "rigid-rod" polymer

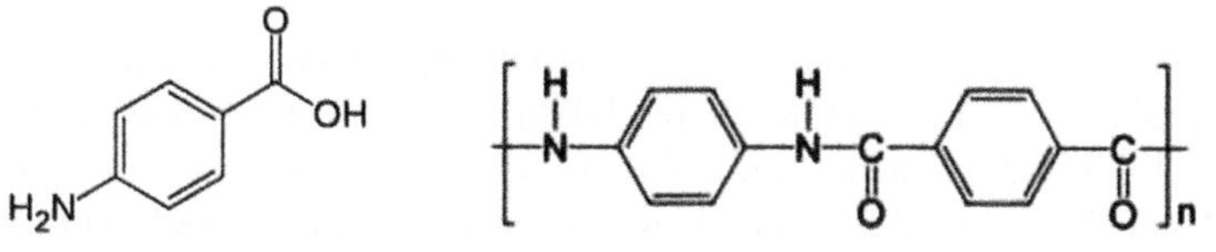

Figure 1.3.17 PBA (poly-1,4-benzamide) and PPTA (poly-p-phenylene terephthalamide).

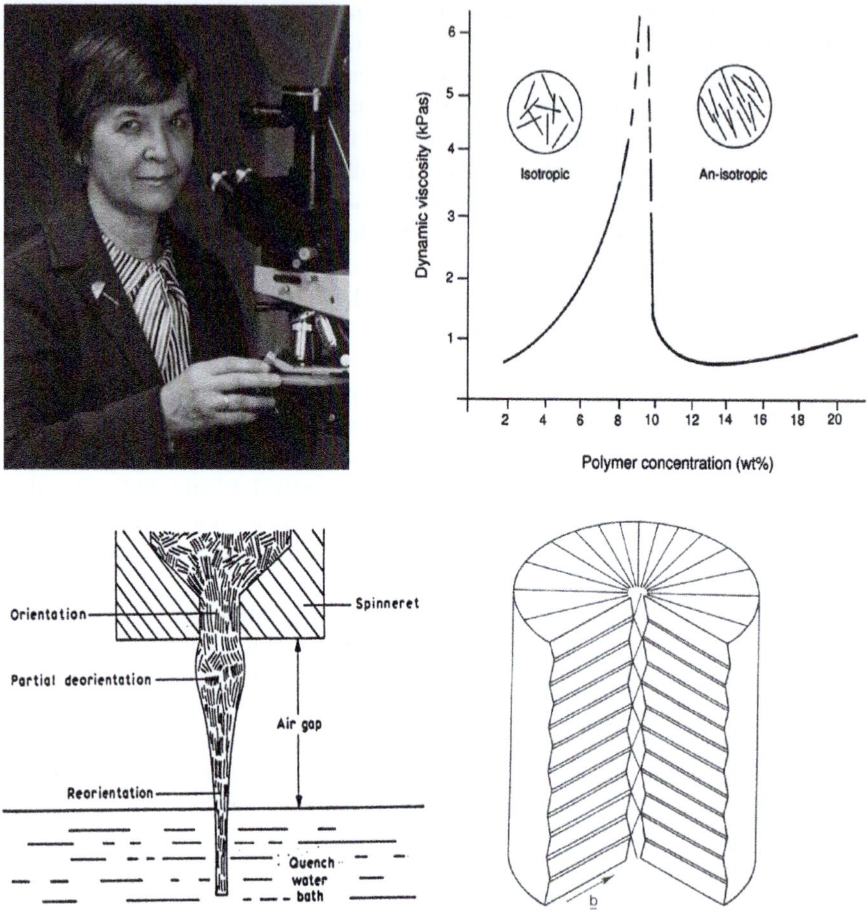

Figure 1.3.18 Ms Stephanie Kwolek. Phase diagram PPTA/H2SO4, showing the reduced viscosity in the nematic phase around a concentration of 20%, spinning from nematic solutions, showing flow alignment of rigid rod PPTA molecules, and radial arrangement of pleated sheets in a PPTA fibre.
Image courtesy of Science History Institute. Adapted from Sikkema (2022) with permission from Elsevier.

molecules. In a 1993 speech, Kwolek recounted her groundbreaking discovery: "The solution was unusually low in viscosity, turbid, stir-opalescent, and had a buttermilk-like appearance. Conventional polymer solutions are typically clear or translucent, with a viscosity comparable to molasses. However, the solution I prepared resembled a dispersion, yet it could be completely filtered through a fine-pore filter. It turned out to be a liquid crystalline solution, though I didn't realise it at the time". Already back in 1956, P.J. Flory published a seminal paper about phase diagrams of rigid rod particles in solution (Flory, 1956), and he was surprised to learn at DuPont that his concepts had now materialised.

The types of cloudy solutions produced by Kwolek were typically discarded, and she was denied access to the spinneret - a device used to produce fibres - because the solution was thought to risk clogging the machine. Undeterred, Kwolek persuaded technician Charles Smullen to test it. To her astonishment, the resulting fibre demonstrated extraordinary properties well beyond those of the engineering fibres of that time. Kwolek's team learnt that the fibres could be made even stronger by heat-treating them (Yang, 1993). This breakthrough caught the immediate attention of DuPont, and a plant with a capacity of 15 million lbs/year started up as early as 1971 and was later expanded to 45 million lbs/year by 1982. The fibres were initially named PRD-29 and 49 and, in the early 1970s, commercialised as Kevlar 29 and 49, later to be followed by higher-tenacity fibres such as Kevlar 129 and 149.

Both Kevlar 29 and Kevlar 49 exhibit tensile strengths of approximately 3 GPa. However, their moduli differ significantly: Kevlar 29 has a modulus of 70 GPa, while Kevlar 49 features a higher modulus of 120 GPa. Postdrawing at elevated temperature can be performed to achieve moduli up to 140 GPa with flattening of the pleated sheets (see Fig. 1.3.18). Due to the presence of hydrogen bonds and much stronger interchain interactions, PPTA molecules of 40.000–50.000 g/mol are sufficient to generate a high strength fibre in contrast with UHMWPE fibre where a molar mass in the millions is needed (Termonia et al., 1985). Spinning of PPTA is performed from so-called nematic solutions in which chains are aligned in domains. Upon entering the water bath, sulfuric acid dissolves, and the PPTA molecules collapse onto each other.

In 1971, Professor Herman Mark, the "Polymer Pope" of his era, advised Akzo Nobel in the Netherlands to position itself as an alternative supplier of PPTA fibres—a move customers were likely to embrace. Akzo decided to enter the para-aramid market and aimed to create a fibre comparable to or better than DuPont's Kevlar, which was launched in 1973. This decision sparked a significant patent dispute, which was ultimately resolved in the mid-1980s. An important aspect of the "war on patents" was the controversy surrounding the polymerisation solvent originally used by DuPont. This solvent, HMPT (hexa-methyl-phosphoric triamide), was found to be carcinogenic. Despite their ongoing disputes, DuPont issued a warning to Akzo about HMPT's dangers. To DuPont's surprise, Akzo responded that they had already developed and patented their own unique solvent: a mixture of NMP (N-methyl-pyrrolidone) and $CaCl_2$, developed by Leo Vollbracht and secured with a patent well before these events unfolded. Twaron was officially introduced in 1986, shortly after Akzo completed its extensive development work. By 2000, Twaron became part of Teijin Aramid.

A notable drawback of PPTA production is that the polymer must be synthesised in a solution that cannot serve as a spinning dope. This polymer is created through the polycondensation of terephthaloyl chloride (TDC) and diaminobenzene, a reaction characterised by its extremely rapid, diffusion-limited kinetics (Fig. 1.3.19).

The high speed of the reaction necessitates precise and immediate mixing of the reactants in a strict 1:1 M ratio, posing significant chemical challenges. Akzo addressed this complexity by employing a Drais reactor; however, achieving a molecular weight suitable for fibre spinning required several years of development and optimisation.

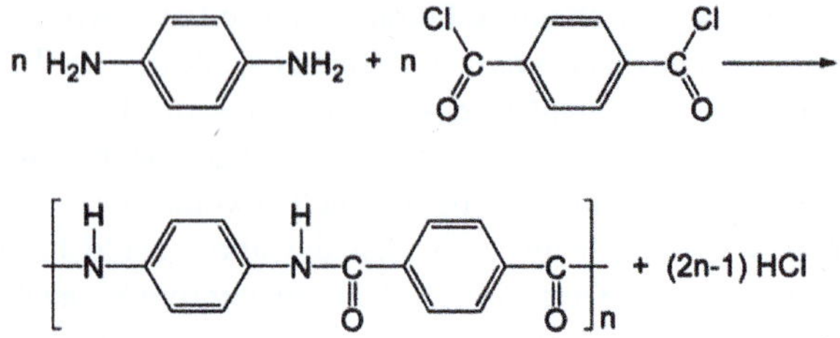

Figure 1.3.19 Polymerisation into PPTA.

3.2 Comparison between para-aramid and PE fibres

Dyneema is advertised as the world's strongest fibre. When measured at room temperature at standard testing speeds, Dyneema is indeed superior in terms of specific strength/stiffness, merely due to its low density of 0.97 g/cm^3 compared with aramid 1.44 g/cm^3 (Fig. 1.3.20).

3.2.1 Creep

Depending on the loading conditions, Dyneema or Spectra fibres are not strong at all. In a creep experiment, the UHMWPE fibre cannot sustain high loads and will "creep" to failure by plastic flow. As mentioned before, UHMWPE fibres are made by ultra-drawing, and there is no lock-in mechanism after drawing, so a creep experiment is the same as drawing, albeit at lower temperatures. Due to its weak intermolecular interactions, a UHMWPE fibre can, in a strict sense, be regarded as a one-dimensional (1D) liquid. Its mechanical response varies with the rate of deformation: at high testing speeds, it behaves like an elastic solid, while at very low testing speeds, it

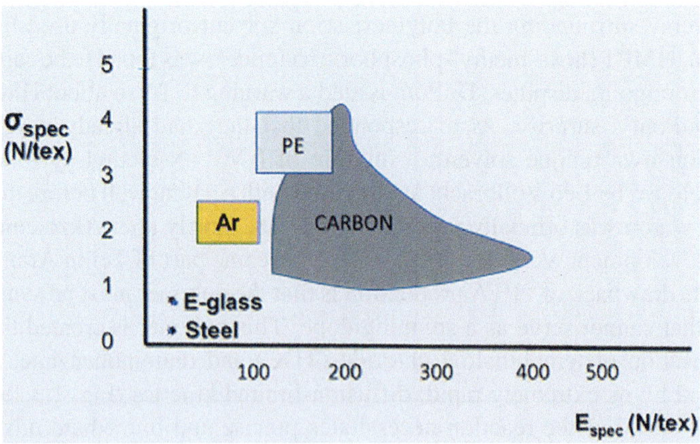

Figure 1.3.20 Specific strength versus specific modulus of high-performance fibres (Ar = aramid fibres) measured at room temperature and standard testing speeds. The unit is N/tex = GPa/ρ where ρ is the density in gram/cm^3.
Reproduced from Lemstra (2022) with permission from Elsevier.

exhibits liquid-like behaviour, i.e., plastic flow (Govaert et al., 1993). This phenomenon is well-documented in polymer rheology and can be understood through the concept of the Deborah number (Reiner, 1964).

Due to their stronger intermolecular interactions (hydrogen bonding compared to Van der Waals forces), aramid fibres do not experience irreversible plastic deformation when subjected to load. These fibres can withstand stresses in the range of 1–1.5 GPa (~50% of their breaking stress) for decades without failure. Additionally, aramid fibres have an exceptionally high melting temperature, estimated to be over 500°C compared to 150 °C for PE fibres, meaning UHMWPE cannot be used in applications like tyre reinforcement.

Research at DSM into the improvement of creep rate, such as the introduction of methyl side-groups, led to the Dyneema Max Technology and the development of the low creep DM20 fibre grade, specially designed for ropes for deep water permanent mooring. At room temperature, DM20 fibre loaded at 10% of the fibre strength showed a creep rate of 0.03% per year, compared to 10% for a standard SK75 fibre grade.

3.2.2 Ballistics

Products for ballistic protection can be grouped as flexible products like bomb blankets and bullet-resistant vests and rigid composite products like protective panels and military and police helmets. Initially, Kevlar became the ultimate fibre for ballistic applications in bullet-resistant vests, helmets and armour and is still synonymous with "life protection" The ballistic protection of a material depends on its ability to absorb energy locally and on the efficiency and speed of transferring the absorbed energy which is realised in so-called unidirectional (UD) laminates either as "soft" inlays or compressed into hard sheets.

The ballistic potential of various high-performance fibres is compared in Fig. 1.3.21 where the sonic velocity ($=\sqrt{E/\rho}$) is plotted against the specific energy absorption capability ($=0.5\sigma^*\varepsilon/\rho$) for several polymer fibres. UHMWPE fibres clearly exhibit a good balance of both these properties. The main advantage of UHMWPE fibres is their 50% lower density and slightly higher strength/stiffness. In principle, UHMWPE tapes like Endumax should perform even better in ballistics due to their improved packing density; however, their tensile strength of around 2 GPa is rather limited up to now.

Allied Signal paved the way for PE fibre in ballistics in the early 1990s (Pervorsek et al., 1991), but currently, Dyneema is leading in this area with their SK99 grade possessing a strength exceeding 4 GPa. In the 1980s, Allied Signal, before its merger in 1999 with Honeywell, developed advanced materials for defence and industrial applications. They developed their Spectra Shield technology—a unidirectional (UD) composite tape where their UHMWPE Spectra fibres are impregnated with a flexible matrix system to create lightweight armour systems. Laminates consisting of cross-plied UD fibres in an elastomeric matrix like Kraton for both hard and soft body armour were later also developed by DSM High Performance Fibres (Dyneema UD-HB), creating composites with higher ballistic efficiency than possible with fabric-reinforced composites (Fig. 1.3.22). Allied Signal also developed related Gold Shield technology based on aramid fibres.

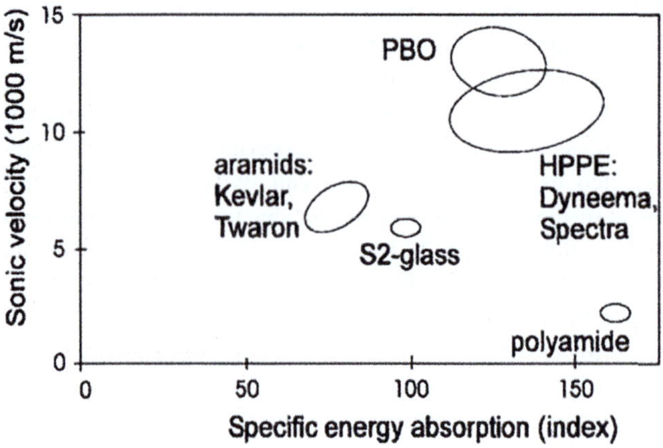

Figure 1.3.21 Basic ballistic performance indicators for various high-strength fibres. Reproduced from Jacobs and Van Dingenen (2001) with permission from Springer.

3.2.3 Structural properties

Both UHMWPE and aramid fibres exhibit highly anisotropic structures. While these fibres demonstrate excellent stiffness and strength along the fibre (chain) direction, their off-axis (transverse and shear) properties are significantly weaker (Peijs et al., 1994, 2018). This anisotropy is a characteristic of all polymer fibres. In aramid fibres, the PPTA molecules are aligned along the fibre axis and held together by covalent bonds, with additional stabilisation provided by hydrogen bonds between the polymer chains. Typical transverse moduli values for aramid fibres are in the range of 5 GPa compared to a longitudinal modulus of 120 GPa. UHMWPE fibres rely on even weaker Van der Waals forces and transverse moduli of only half that of aramid. On

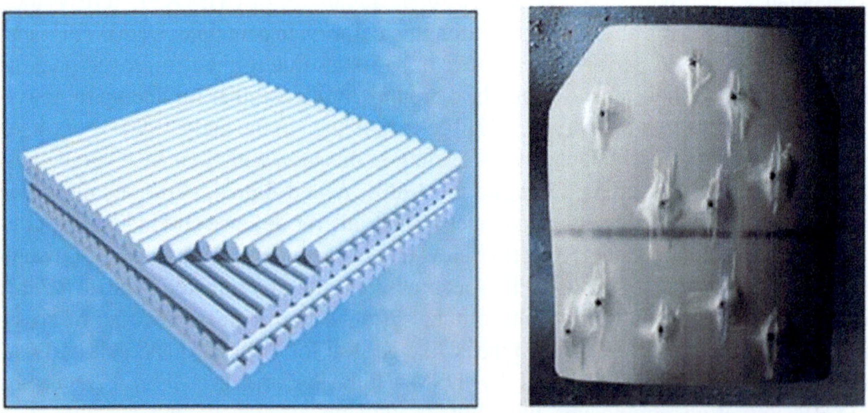

Figure 1.3.22 Dyneema UD is a cross-ply laminate based on a unidirectional polyethylene (PE) composite material. It is available as UD-SB (soft ballistics) and UD-HB (hard ballistics).
Image courtesy DSM Dyneema.

the other hand, glass fibres are typically isotropic materials, exhibiting a uniform modulus of 70 GPa in all directions. The highly anisotropic nature of aramid fibres, and especially polyethylene fibres, presents a significant challenge in structural composite applications. For instance, it has been demonstrated that the maximum interlaminar shear strength of unidirectional UHMWPE composites (~30 MPa) is inherently limited by the low intrinsic shear strength of the fibres (Peijs et al., 1994), rather than by the interface or matrix as in the case of glass or carbon fibres.

In contrast to the high values of tensile strength for UHMWPE and aramid fibres, the compressive strength of these fibres is relatively poor. Oriented polymer fibres fail by the formation of kink bands (Fig. 1.3.23), a process again related to the poor shear properties of the fibre. Clearly, the low compressive strength limits the applicability of these fibres in structural composites. Due to their more anisotropic character, the compressive strength of UHMWPE fibres is very low (50–60 MPa) and only 2% of the tensile strength, compared to 20% in the case of PPTA (600 MPa). Kink band formation occurs already at strain levels <0.1% in UHMWPW, but these kink bands disappear almost completely after subsequent tensile loading and hardly affect the tensile strength, while in the case of PPTA fibres, kink bands lead to a 10% drop in tensile strength. Kink band formation occurs due to buckling and slippage of the polymer chains, processes which are strongly governed by the shear properties of the fibres.

It has been shown that hybridisation can lead to composites where structural properties like compressive, shear and flexural strength are carried by the more structural carbon or glass fibres, while other mechanical functionalities, such as vibration damping, impact resistance, energy absorption, and integrity after impact are provided by the UHMWPE or aramid fibres (Swolfs et al., 2014). For example, carbon/UHMWPE hybrids have shown great potential for applications in impact-resistant structures (Peijs et al., 1990a,b). Here, the carbon fibres provide the necessary compressive and/or flexural strength, while PE fibres improve energy absorption and prevent scattering upon impact. The potential for hybridising carbon fibres with UHMWPE fibres was later also recognised by DSM, which introduced Dyneema Carbon as a commercial product in 2016 (Fig. 1.3.24) (Bouwmeester et al., 2008; Marissen, 2011).

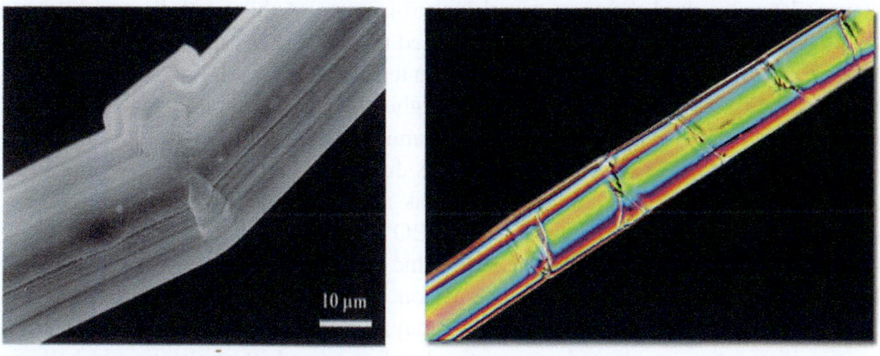

Figure 1.3.23 Scanning electron microscopy (SEM) micrograph (left) and polarised optical micrograph of kink bands in UHMWPE filament (right).
Reproduced from Peijs et al. (1994) with permission from Elsevier.

Figure 1.3.24 Dyneema carbon intralaminar hybrid fabric (left) and application in bicycle frame (right).
Image courtesy DSM Dyneema.

3.3 Other rigid-rod molecules

In textbooks, PPTA is often depicted as a rigid, rod-like molecule resembling a pencil. In its polymer chain, the benzene units are connected by amide bonds, which favour a trans conformation. This structure facilitates some conjugation along the chain and restricts rotation around the main-chain bonds. However, any slender body with a high aspect ratio becomes flexible, viz., a rigid glass rod versus a flexible glass fibre. In dilute solutions, PPTA molecules are semiflexible with a persistence length (length over which a polymer chain appears straight before it starts bending due to thermal energy) of approximately 1/5 of the contour length (chain length at maximum extension). However, upon increasing the concentration, the molecules align parallel in domains in the nematic phase.

3.3.1 PBO

Efforts to develop more rigid chain molecules were notably undertaken in the USA during the 1980s at Wright–Patterson Air Force Base, Stanford Research Institute (SRI) and Dow Chemical, with a focus on PBZT (polybenzobisthiazoles). Toyobo (Japan) later acquired the most promising project out of this portfolio and licenced PBO (Fig. 1.3.25), which they commercialised under the trade name Zylon. PBO exhibits impressive tensile properties, achieving tensile strengths of up to 6 GPa (see Table 1.3.2), although its density is slightly higher than that of aramids (Krause et al., 1988; Adams et al., 1989). However, like aramids, PBO fibres demonstrated unsatisfactory performance in structural composites due to their low compressive and shear strength, stemming from their relatively weak intermolecular interactions.

An even more significant challenge for PBO applications is its dramatic instability to visible light. This issue proved catastrophic in 2003 when a police officer in the USA was fatally shot while wearing a Zylon bulletproof vest. The incident led to Toyobo being fined $66 million USD. Today, Toyobo continues to produce PBO for use in sporting goods and aeronautics but has ceased its application in ballistic protection.

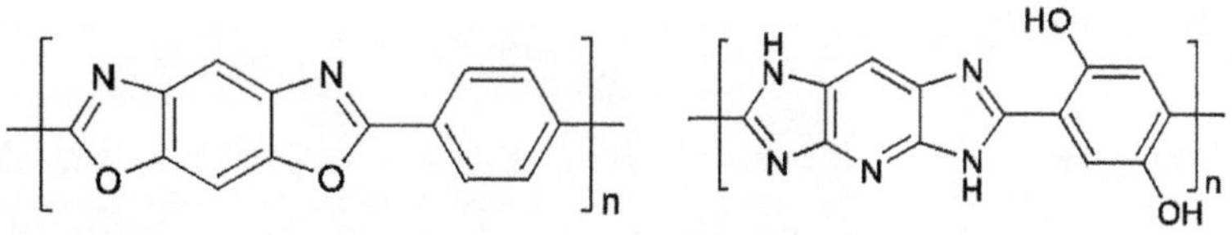

Figure 1.3.25 Chemical structure of PBO (left) and M5 (right).

3.3.2 M5

Doetze Sikkema at Akzo Nobel succeeded in the 1990s in making a polymer from 2,3,5,6, tetra-aminopyridine and 2,5-dihydroxyterephthalic acid, PIPD, abbreviated as monomer M5 (Sikkema, 1998; Afshari et al., 2008). Conventional air gap wet spinning of the as-polymerised nematic solutions in polyphosphoric acid at about 180 °C produced filaments with a diameter of about 8–10 μm and annealing afterwards at high temperatures (>400 °C) gave the final product. Sikkema's aim was to create polymers as rigid rod-like as PBO, but with stronger intermolecular hydrogen bonds, leading to improved lateral strengths of the fibre (Fig. 1.3.26). After much experimentation, the team created a fibre with a crystal structure that features hydrogen bonds in both the x and the y direction (z being the polymer main chain direction), leading to improved shear properties which stabilise the chain in compression as opposed to the planar hydrogen bonding in PPTA.

In 2001 Akzo Nobel divested all its fibre activities. Its crown jewel Twaron was sold to Teijin. M5 went via Magellan company to DuPont but was never brought to fruition. The main reason being that well-established carbon fibre producers already dominated the high-performance fibre market for composites, while DuPont already had its own Kevlar fibre.

3.3.3 Aromatic polyesters

When DuPont came on the market with aramid fibres, a significant effort was undertaken to produce aromatic polyesters, which showed anisotropic liquid-crystalline melts and melt-spinning rather than the cumbersome solution-spinning was envisaged. A pioneer in this area was James Economy, who started with the obvious choice: parahydroxybenzoic acid as Kwolec did with para-amino-benzoic acid (Economy

Table 1.3.2 Mechanical properties of high-performance fibres.

Fibre	Compressive strength (GPa)	Tensile strength (GPa)	Young's modulus (GPa)	Density (g/cm^3)
UHMWPE	<0.06	3–4	130–150	0,96
PPTA	0.4–0.8	2–3	60–140	1.44
PBO	0.2–0.4	5–6	180–280	1.56
M5 (PIPD)	1,7–2	3–6	200–330	1.7
Carbon	1–3	2–7	200–600	1.8

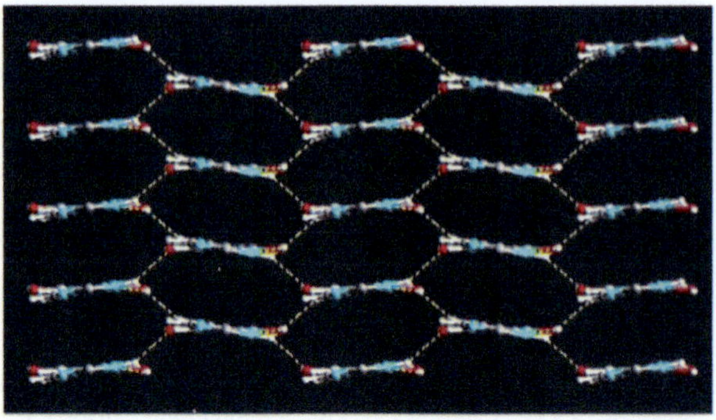

Figure 1.3.26 Dr. Doetze Sikkema and M5 molecules in crystal lattice. The chains are pointing to the reader. Hydrogen bonds are shown as dashed lines and resemble a honey-comb structure and lateral chain-chain interaction. This leads to a high shear modulus and shear strength and, hence, good compressive properties.
Reproduced from Sikkema (2022) with permission from Elsevier.

et al., 1976, 1989). The polymer PHBA is better known under its trade name Ekonol. Still, the melting temperature was prohibitively high for melt-spinning, and companies moved to copolyesters like the one below (Fig. 1.3.27).

Fibres could be made with a strength close to 2 GPa with a corresponding Young's modulus of 120 GPa, and expectations were high that there would be a melt-spun competitor for aramid fibres. Despite major efforts made by the industry, no commercially successful aromatic polyesters have been made. The only commercially successful (co)polyester is Vectran, now manufactured by Kurary. Its melting point is around 350°C; its strength is 2–2.5 GPa, and its corresponding Young's modulus is 50–70 GPa.

3.4 Other flexible-chain molecules

Polyethylene fibres possess a low melting temperature, pronounced creep and lack shear- and compressive strength due to their weak intermolecular Van der Waals forces. So clearly, there is an opportunity for more polar polymers with stronger intermolecular interactions.

3.4.1 Polyamides

An obvious choice is aliphatic polyamides (nylons) with a much higher melting temperature than PE and stronger hydrogen bonding preventing creep. Moreover, very

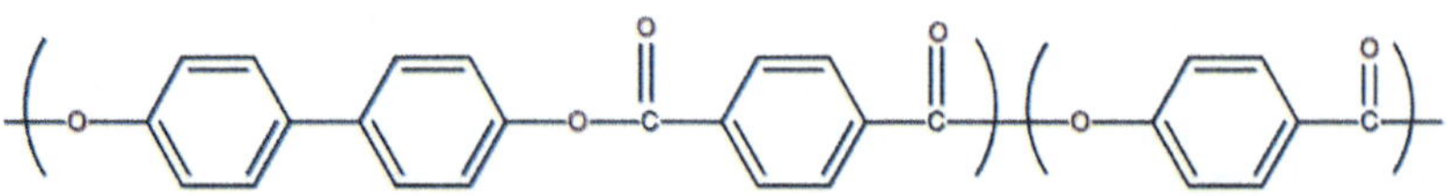

Figure 1.3.27 Copolymer based on PHBA and BPT (biphenol terephthalate).

high molar mass PA6 can be made via anionic polymerisation, so what is the problem? Numerous studies have been devoted to creating high-performance nylon fibres, but the main problem is unfolding the lamellar crystals during solid-state drawing. Polyamide has a "natural" draw ratio limit of approximately 4–5, which impedes optimised ultradrawing and the necessary chain extension for ultrahigh properties. The fundamental reason is the intrinsic "hydrogen-barrier", in the crystalline regions of the as-spun fibres, which contain specific hydrogen bonds between the chains (stems) in the crystal (Postema et al., 1990). These secondary interactions prevent macromolecular translational motion through the crystal lattice, required for chain extension and alignment, during drawing, unlike in the case of PE in which such an effect readily occurs.

A way out is shielding the amide bonds and weakening the hydrogen bonding. In that respect, impressive results have been obtained. In the group of Professor Paul Smith at ETH ZÜrich, Camphor Sulphonic Acid (CSA) was used to form complexes with the amide bonds and "stiffen up" the polymer chains to make them look more like aromatic polyamides (Weenink, 1999). PA66/CSA complexes could be drawn into highly oriented structures, but CSA had to be removed at the end (Fig. 1.3.28). That gives a "naked" oriented polyamide chain, and for entropic reasons, many lower molecular weight chains will recoil back into their random coil conformation and use the still extended chains as nuclei to form the well-known shish kebab structures.

Other attempts have been made to shield the amide bonds by using salts, and an interesting approach was taken by Jules Harings, now at Maastricht University. He makes use of the fact that PA46, a nylon with a melting temperature of around 320°C, dissolves in superheated water and can be mixed with Li–I or Li–Br to shield the amide bonds (Harings et al., 2012). These shielded nylon 46 molecules could be drawn to near-perfect orientation as evidenced by X-ray studies. However, the problem comes when the salts must be removed. The extended nylon 46 chains will have the choice to form extended chain crystals after the removal of the ions by water or to recoil. At this point in time, no high-performance nylon 46 fibres have yet been produced.

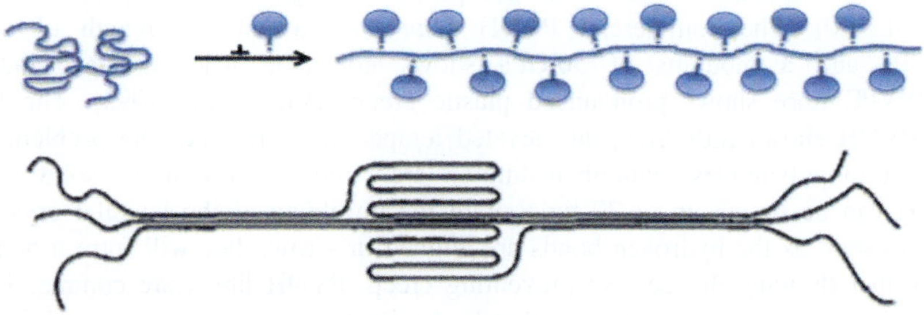

Figure 1.3.28 Artist view of PA66/CSA complexes (top) and shish-kebab structure after removal of CSA (bottom).
Adapted from Weenink (1999).

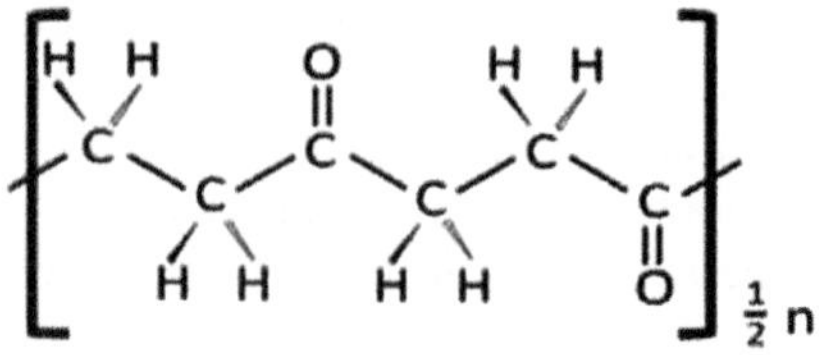

Figure 1.3.29 Structure of polyketone.

3.4.2 Polyketone

In the 1990s, an interesting polymer was discovered by Dr. Eite Drenth at Shell: a strictly alternating copolymer of ethylene and carbon monoxide, polyketone (PK) (Fig. 1.3.29). The big dream of Shell was to enter the market with a novel engineering plastic, which they called Carilon, and Akzo Nobel took an interest in making a fibre out of PK for tyre cord applications, and the leading scientist was Dr. Bert Jan Lommerts (Lommerts et al., 1993). Impressive results have been obtained, such as tensile strengths up to 4 GPa, but the corresponding Young's moduli were surprisingly low at only 50–60 GPa (Lommerts, 2022). Shell sold his patent portfolio to SRI International for 1 US$ (in order to get a huge tax rebate from the Dutch government), and Akzo Nobel divested all his fibre activities around 2001, and both Carilon and the PK fibre are currently shelved.

3.4.3 Polyvinyl alcohol

Another interesting candidate seems to be polyvinyl alcohol (PVOH), which is an atactic polymer, but the small –OH side groups do not prevent crystallisation and the crystal packing density and structure are the same as for PE (Bunn, 1948). The longitudinal crystal modulus and theoretical modulus are similar to PE, but the transverse crystal modulus is much higher: 9 GPa for PVOH versus 3 GPa for PE (Nakamae et al., 1992) due to hydrogen bonding. This increase in lateral interactions between chains results in a higher compressive strength of 180 MPa as compared to 60 MPa for UHMWPE (Peijs et al., 1995).

In comparison with polyamides, the hydrogen bonds are less directed due to the atactic placement of the –OH groups but strong enough to prevent creep (Fig. 1.3.30). The commercial PVOH fibre with a tensile strength of only 1.5 GPa and a modulus of 50 GPa shows no creep whereas the standard UHMWPE fibre shows pronounced plastic creep (Peijs et al., 1995). The fact that PVOH shows little creep at elevated temperatures indicates the problem. In contrast to polyamides, with their limited draw ratio of 4-5 in the solid state, PVOH can be drawn up to 20 times with the unfolding of the lamellar crystals. Still, as soon as the hydrogen bonds are aligned in a row, they will form a barrier for further drawing, hence also preventing creep. PVOH fibres are commercially available from Kuraray in Japan (Kuralon) with tensile strengths up to 1.5 GPa and moduli up to 40 GPa.

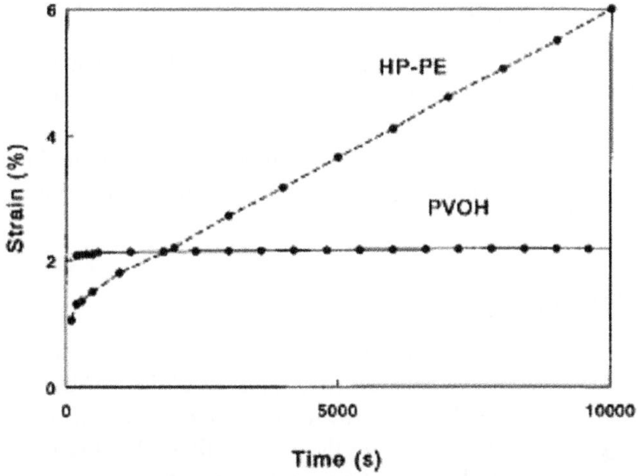

Figure 1.3.30 Creep data at 70°C at a load of 0.5 GPa. The PVOH fibre is a commercial fibre from Kuraray with a strength of 1.5 GPa and a modulus of around 40 GPa. The HP-PE fibre is a Spectra 1000 fibre from Honeywell with a strength of around 3 GPa and a Young's modulus of 100 GPa.
Reproduced from Peijs et al. (1995) with permission from Elsevier.

4. Epilogue and outlook

Currently, both UHMWPE and aramid fibres are well established in the market, with production volumes of approximately 100,000 tons per year for PE fibres and slightly over 100,000 tons per year for aramid fibres. While these figures are modest compared to the global fibre production of 124 million tons, dominated by polyester, which accounts for more than half, they are comparable to the production volume of carbon fibre, which stands at 160,000 tons annually.

The main producers of aramid fibres include DuPont (USA), Teijin (Japan, but with its main production sites in the Netherlands), Yantai (China), Hyosong (South Korea), Toray (Japan), Kolon (South Korea), Huvis (South Korea), Bluestar (China), and X-FIPER (China).

Dyneema, now with Avient (USA), is still the main player in UHMWPE fibres. The original licensee Allied Signal (now Honeywell) is producing their Spectra grade and Toyobo, now separated from DSM, their Izanas grade. But Chinese producers like Xingy are catching up. The advantage of Dyneema is that they produce their own feedstock UHMWPE, so they can modify the molecular structure as evidenced by their low-creep fibre DM20, in which a minute amount of comonomer in the main chain prevents slippage through the crystals. For ballistic applications, achieving ultrahigh strength is the primary goal. Dyneema SK99 grade surpassed the 4 GPa threshold, making it the standout choice. While some Chinese competitors claim to offer similar performance, they have historically faced challenges in maintaining consistent quality.

4.1 Ultimate strength

PE fibres have made some quantum leaps in properties and are still improving in performance (Fig. 1.3.31). It is now generally accepted that failure during tensile loading occurs via chain slippage (O'Connor et al., 2016) and an upper limit in strength of 7 GPa seems realistic. There has been no significant improvement in aramid fibre properties over the last decade related to the nonperfect chain alignment upon entering the water phase.

4.2 Nanotechnology

Driven by the exceptionally high mechanical properties of carbon nanotubes (CNTs) and graphene, with theoretical tensile strength values of 50–100 GPa and moduli of 1000 GPa, extensive research efforts have been devoted to the reinforcement of high-performance polymer fibres using these nanofillers. Because of their 1D character, CNTs seem particularly interesting either as bundles or ropes or as reinforcements for oriented (1D) polymer fibres (Peijs et al., 2022). The introduction of nanofibres like CNTs into oriented polymer fibres could, in principle, mimic the structure of natural fibrous materials, creating "designed" composite fibres with additional levels of hierarchy (Ciselli et al., 2007). Although the prospect of creating the ultimate nanocomposite fibre is intriguing, some basic challenges regarding dispersion, adhesion, and orientation must be addressed. Moreover, since their dimensions are comparable to the crystal cell of PE, their presence may affect the packing density of the polymer chains as well as the mechanical properties. The reinforcing potential of nanofillers in high-modulus polymer matrices might, therefore, be questioned (Van der Werff and Marissen, 2006). To date, despite numerous efforts, improvements in the strength of these fibres have been rather modest even for relatively high nanofiller contents. In a recent article by Goutianos and Peijs (2021) it was shown that the low reinforcing efficiency of such nanofillers, even when perfectly dispersed and bonded,

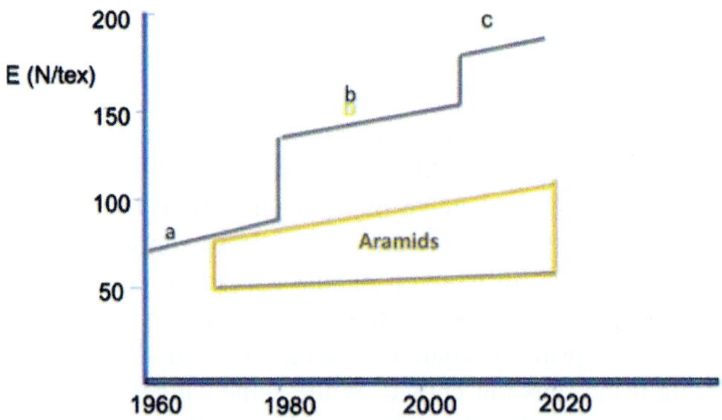

Figure 1.3.31 Development of Young's modulus over the last decades. Aramid versus PE fibres. a) Melt-spinning of HDPE, b) solution(gel)-spinning of UHMWPE, and c) solid-state drawing.

is (again) limited by the intrinsic low shear strength of highly anisotropic polymer fibres, preventing effective stress transfer from the fibre to the nanofiller.

4.3 Biobased

Many companies today strive to cultivate a "green image," and Dyneema is now offered as a biobased fibre to align with this trend. According to Dyneema, the introduction of biobased feedstock generates 5 metric tons less CO_2eq per metric ton of fibre than the same volume of fossil-based Dyneema. In the Netherlands, Teijin Aramid states on its website that it is committed to achieving full circularity with its aramid fibres by recovering and reprocessing used materials. However, a significant portion of used ropes ends up being milled into short fibres, primarily for use in brake pads. Despite this limitation, Teijin Aramid continues to promote its environmentally friendly initiatives and is actively developing a biobased aramid fibre.

The question is whether fully biobased high-performance fibres can be made. Biobased polymers such as polylactide acid (PLA) and polyhydroxyalkanoates have often a very low crystal modulus (Colwell et al., 2024). For example, the theoretical and thus maximum attainable moduli of PLLA and PDLA are 14 GPa, whereas the value for stereocomplex PLA is limited to 20 GPa at room temperature (Lee et al., 2018). A more obvious candidate would be cellulose. Cellulose molecules grow in nature in extended-chain crystals, and the crystal structure of cellulose-I crystals is very high, approximately 130 GPa. To make continuous fibres, cellulose pulp is treated with chemicals in the so-called viscose process to make it soluble and spinnable. But there are nowadays also processes for direct spinning from cellulose dissolved in solvents like NMNO (Lenzing) and phosphoric acid. However, once dissolved and recrystallised, cellulose molecules crystallise into the cellulose-II crystal structure possessing a much lower crystal modulus, approximately 70 GPa and high-performance cellulose fibres never matured.

One of the more promising developments in this field is the regenerated cellulose fibre spun from anisotropic solutions in phosphoric acid, developed by Akzo Nobel. Named "Bocell" after its inventor, Dr. Hanneke Boerstoel, this fibre features highly oriented cellulose crystallites (Northolt et al., 2001). With a tensile strength of 1.3 GPa and an initial modulus of 45 GPa, it outperforms other commercial regenerated cellulose fibres, though its properties remain significantly below those of traditional high-performance fibres.

References

Adams, W.W., Eby, R.K., McLemore, D.E., 1989. The materials science of rigid-rod polymers. In: Materials Research Society Symposium Proceedings, vol 134.

Afshari, M., Sikkema, D.J., Lee, K., Bogle, M., 2008. High performance fibers based on rigid and flexible polymers. Polym. Rev. 48 (2), 230–274.

Alcock, B., Peijs, T., 2013. Technology and development of self-reinforced polymer composites. Polymer composites–polyolefin fractionation–polymeric peptidomimetics–collagens, pp. 1–76.

Bair, T.I., Morgan, P.W., Killian, F.L., 1977. Poly (1, 4-phenyleneterephthalamides). Polymerization and novel liquid-crystalline solutions. Macromolecules 10 (6), 1396–1400.

Bastiaansen, C.W., Meyer, H.E., Lemstra, P.J., 1990. Memory effects in polyethylenes: influence of processing and crystallization history. Polymer 31 (8), 1435–1440.

Black, W.B., Preston, J. (Eds.), 1973. High-Modulus Wholly Aromatic Fibers. Dekker, New York.

Bolton, E.K., 1942. Chemical industry medal. Development of nylon. Ind. Eng. Chem. 34 (1), 53–58.

Boudreaux, D.S., 1973. Calculations of the strength of the polyethylene molecule. J. Polym. Sci. Polym. Phys. Ed 11 (7), 1285–1292.

Bouwmeester, J.G.H., Marissen, R., Bergsma, O.K., 2008. Carbon/dyneema intralaminar hybrids: new strategy to increase impact resistance or decrease mass of carbon fiber composites. In: ICAS2008 Conference Anchorage, Alaska.

Bott, W., Fifield, B., Lemstra, P.J., 2022. High-performance fibers and tapes based on PP and PE. Adv. Ind. Eng. Polym. Res. 5 (2), 60–69.

Bunn, C.W., 1948. Crystal structure of polyvinyl alcohol. Nature 161 (4102), 929–930.

Capaccio, G., Ward, I.M., 1974. Preparation of ultra-high modulus linear polyethylenes; effect of molecular weight and molecular weight distribution on drawing behaviour and mechanical properties. Polymer 15 (4), 233–238.

Capaccio, G., Ward, I.M., 1973. Properties of ultra-high modulus linear polyethylenes. Nat. Phys. Sci. (Lond.) 243 (130), 143.

Carothers, W.H., Hill, J.W., 1932. Studies of polymerization and ring formation. XV. Artificial fibers from synthetic linear condensation superpolymers. J. Am. Chem. Soc. 54 (4), 1579–1587.

Christakopoulos, F., Busato, S.P., Kong, X., Troisi, E., Friederichs, N., Tervoort, T.A., 2024. Solid-state extrusion of nascent disentangled ultra-high molecular weight polyethylene. Polym. Eng. Sci.

Christakopoulos, F., Troisi, E., Friederichs, N., Vermant, J., Tervoort, T.A., 2021. "Tying the knot": enhanced recycling through ultrafast entangling across ultrahigh molecular weight polyethylene interfaces. Macromolecules 54 (20), 9452–9460.

Ciferri, A., Ward, I.M., 1979. Ultra-high modulus polymers. Appl. Sci. Publ.

Ciselli, P., Wang, Z., Peijs, T., 2007. Reinforcing potential of carbon nanotubes in oriented polymer fibres. Mater. Technol. 22 (1), 10–21.

Colwell, J., Halley, P., Varley, R., Heidarian, P., McNally, T., Peijs, T., Vandi, L., 2024. Self-reinforced biodegradable thermoplastic composites. Adv. Compos. Hybrid Mater. 7 (4), 128.

De Gennes, P.G., 1971. Reptation of a polymer chain in the presence of fixed obstacles. J. Chem. Phys. 55 (2), 572–579.

Economy, J., Storm, R.S., Matkovich, V.I., Cottis, S.G., Nowak, B.E., 1976. Synthesis and structure of the p-hydroxybenzoic acid polymer. J. Polym. Sci. Polym. Chem. Ed. 14 (9), 2207–2224.

Economy, J., 1989. Aromatic polyesters of p-hydroxybenzoic acid. Mol. Cryst. Liq. Cryst. 169 (1), 1–22.

Flory, P.J., 1956. Phase equilibria in solutions of rod-like particles. Proc. R. Soc. Lond. A Math. Phys. Sci. 234 (1196), 73–89.

Goutianos, S., Peijs, T., 2021. On the low reinforcing efficiency of carbon nanotubes in high-performance polymer fibres. Nanocomposites 7 (1), 53–69.

Govaert, L.E., Bastiaansen, C.W., Leblans, P.J., 1993. Stress-strain analysis of oriented polyethylene. Polymer 34 (3), 534–540.

Harings, J.A., Deshmukh, Y.S., Hansen, M.R., Graf, R., Rastogi, S., 2012. Processing of polyamides in the presence of water via hydrophobic hydration and ionic interactions. Macromolecules 45 (14), 5789–5797.

Hermes, M.E., 1996. Enough for One Lifetime: Wallace Carothers, Inventor of Nylon. Chemical Heritage Foundation.

Hine, P.J., Ward, I.M., Olley, R.H., Bassett, D.C., 1993. The hot compaction of high modulus melt-spun polyethylene fibres. J. Mater. Sci. 28, 316–324.

Jacobs, M.J.N., Van Dingenen, J.L.J., 2001. Ballistic protection mechanisms in personal armour. J. Mater. Sci. 36, 3137–3142. https://doi.org/10.1023/A:1017922000090.

Krause, S.J., Haddock, T.B., Vezie, D.L., Lenhert, P.G., Hwang, W.F., Price, G.E., Helminiak, T.E., O'brien, J.F., Adams, W.W., 1988. Morphology and properties of rigid-rod poly (p-phenylene benzobisoxazole)(PBO) and stiff-chain poly (2,5(6)-benzoxazole) (ABPBO) fibres. Polymer 29 (8), 1354–1364.

Kwolek, S.L., Morgan, P.W., Schaefgen, J.R., Gulrich, L.W., 1977. Synthesis, anisotropic solutions, and fibers of poly (1,4-benzamide). Macromolecules 10 (6), 1390–1396.

Lee, S., Kimoto, M., Tanaka, M., Tsuji, H., Nishino, T., 2018. Crystal modulus of poly (lactic acid) s, and their stereocomplex. Polymer 138, 124–131.

Lefèvre, J., Feldman, K., Giesbrecht, J., Smith, P., Tervoort, T.A., Meijer, H.E., 2012. "Foil spintrusion" of high-performance polymer films. J. Polym. Sci. B Polym. Phys. 50 (24), 1713–1727.

Lemstra, P.J., 2022. High-performance polyethylene fibers. Adv. Ind. Eng. Polym. Res. 5 (2), 49–59.

Lemstra, P.J., Keller, A., Cudby, M., 1978. Gelation crystallization of poly (vinyl chloride). J. Polym. Sci. Polym. Phys. Ed 16 (8), 1507–1514.

Lemstra, P.J., Kirschbaum, R., Ohta, T., Yasuda, H., 1987. In: Ward, I.M. (Ed.), Developments in Oriented Polymers. Elsevier Applied Science, London, pp. 39–79.

Lommerts, B.J., Klop, E.A., Aerts, J., 1993. Structure and melting of perfectly alternating ethylene-carbon monoxide copolymers. J. Polym. Sci. B Polym. Phys. 31 (10), 1319–1330.

Lommerts, B.J., 2022. High-performance fibers based on flexible polar polymer molecules. Adv. Ind. Eng. Polym. Res. 5 (2), 70–79.

Mackley, M., 1985. Eur Patent 22687.

Mai, F., Tu, W., Bilotti, E., Peijs, T., 2015. Preparation and properties of self-reinforced poly (lactic acid) composites based on oriented tapes. Compos. Appl. Sci. Manuf. 76, 145–153.

Marissen, R., 2011. Design with ultra strong polyethylene fibers. Mater. Sci. Appl. 2 (05), 319.

Morgan, P.W., 1981. Brief history of fibers from synthetic polymers. J. Macromol. Sci.-Chem. 15 (6), 1113–1131.

Morgan, P.W., 1977. Synthesis and properties of aromatic and extended chain polyamides. Macromolecules 10 (6), 1381–1390.

Morris, P.J., 2005. Polymer Pioneers: A Popular History of the Science and Technology of Large Molecules. Chemical Heritage Foundation.

Nakamae, K., Nishino, T., Ohkubo, H., Matsuzawa, S., Yamaura, K., 1992. Studies on the temperature dependence of the elastic modulus of crystalline regions of polymers: 14. Poly (vinyl alcohol) with different tacticities. Polymer 33 (12), 2581–2586.

O'Connor, T.C., Robbins, M.O., 2016. Chain ends and the ultimate strength of polyethylene fibers. ACS Macro Lett. 5 (3), 263–267.

Peijs, A.A., Catsman, P., Govaert, L.E., Lemstra, P.J., 1990. Hybrid composites based on polyethylene and carbon fibres Part 2: influence of composition and adhesion level of polyethylene fibres on mechanical properties. Composites 21 (6), 513–521.

Peijs, A.A., Venderbosch, R.W., Lemstra, P.J., 1990. Hybrid composites based on polyethylene and carbon fibres Part 3: impact resistant structural composites through damage management. Composites 21 (6), 522–530.

Peijs, T., 2018. Comprehensive Composite Materials II, Editors-in-Chief: PWR Beaumont and CH Zweben. In: 1.5 High Performance Polyethylene Fibers, 1, pp. 86–126.

Peijs, T., Kirschbaum, R., Lemstra, P.J., 2022. A critical review of carbon fiber and related products from an industrial perspective. Adv. Ind. Eng. Polym. Res. 5 (2), 90–106.

Peijs, T., Rijsdijk, H.A., De Kok, J.M., Lemstra, P.J., 1994. The role of interface and fibre anisotropy in controlling the performance of polyethylene-fibre-reinforced composites. Compos. Sci. Technol. 52 (3), 449–466.

Peijs, T., Van Vught, R.J., Govaert, L.E., 1995. Mechanical properties of poly (vinyl alcohol) fibres and composites. Composites 26 (2), 83–90.

Peijs, T., 2003. Composites for recyclability. Mater. Today 6 (4), 30–35.

Pervorsek, D.C., Chin, H.B., Kwon, Y.D., Field, J.E., 1991. Strain rate effects in ultrastrong polyethylene fibres and composites. J. Appl. Polym. Sci. 47, 45–66.

Peters, M.W.M.G., 2009. Process for manufacturing polyolefin films WO2009/153318.

Postema, A.R., Smith, P., English, A.D., 1990. Ultra-drawing of polyamides: the hydrogen bond barrier. Polym. Commun. 31 (12), 444–447.

Rastogi, S., Lippits, D.R., Peters, G.W., Graf, R., Yao, Y., Spiess, H.W., 2005. Heterogeneity in polymer melts from melting of polymer crystals. Nat. Mater. 4 (8), 635–641.

Reiner, M., 1964. The deborah number. Phys. Today 17 (1), 62.

Rotzinger, B.P., Chanzy, H.D., Smith, P., 1989. High strength/high modulus polyethylene: synthesis and processing of ultra-high molecular weight virgin powders. Polymer 30 (10), 1814–1819.

Schaller, R., Peijs, T., Tervoort, T.A., 2016. High-performance liquid-crystalline polymer films for monolithic “composites”. Compos. Appl. Sci. Manuf. 81, 296–304.

Sikkema, D.J., 1998. Design, synthesis and properties of a novel rigid rod polymer, PIPD orM5’: high modulus and tenacity fibres with substantial compressive strength. Polymer 39 (24), 5981–5986.

Sikkema, D.J., 2022. Rigid-chain polymers: aromatic polyamides, heterocyclic rigid rod polymers, and polyesters. Adv. Ind. Eng. Polym. Res. 5 (2), 80–89.

Smith, P., Lemstra, P.J., Booij, H.C., 1981. Ultradrawing of high-molecular-weight polyethylene cast from solution. II. Influence of initial polymer concentration. J. Polym. Sci. Polym. Phys. Ed 19 (5), 877–888.

Smith, P., Lemstra, P.J., 1980. Ultra-drawing of high molecular weight polyethylene cast from solution. Colloid Polym. Sci. 258 (7), 891–894.

Smith, P., Lemstra, P.J., 1980. Ultra-high-strength polyethylene filaments by solution spinning/drawing. J. Mater. Sci. 15, 505–514.

Swolfs, Y., Gorbatikh, L., Verpoest, I., 2014. Fibre hybridisation in polymer composites: a review. Compos. Appl. Sci. Manuf. 67, 181–200.

Termonia, Y., Meakin, P., Smith, P., 1985. Theoretical study of the influence of the molecular weight on the maximum tensile strength of polymer fibers. Macromolecules 18 (11), 2246–2252.

Treloar, L.G., 1960. Calculations of elastic moduli of polymer crystals: I. Polyethylene and nylon 66. Polymer 1, 95–103.

Troisi, E., Friederichs, N., Tervoort, T., Chistakopoulos, F., 2021. WO2022/049016A1.

Van der Vegt, A.K., Smith, P.P., 1967. Addresses in polymer science and technology. Soc. Chem. Ind. Monogr. (26), 313.

Van der Werff, H., Marissen, R., 2006. Carbon nanotubes in high strength Dyneema fibres. In: Proceedings of Polymer Fibres 2006, the International Polymer Fibre Conference, Manchester (On CDROM).

Weenink, W.D., 1999. Nylon-sulfonic acid systems phase- and deformation behavior. Doctoral thesis ETH Zurich.

Yang, H.H., 1993. Kevlar Aramid Fiber. John Wiley & Sons.

Zhang, J.M., Peijs, T., 2010. Self-reinforced poly (ethylene terephthalate) composites by hot consolidation of Bi-component PET yarns. Compos. Appl. Sci. Manuf. 41 (8), 964–972.

Zhurkov, S.N., Korsukov, V.E., 1974. Atomic mechanism of fracture of solid polymers. J. Polym. Sci. Polym. Phys. Ed 12 (2), 385–398.

Zwijnenburg, A., 1978. Ph.D. Thesis, University of Groningen.

Zwijnenburg, A., Pennings, A.J., 1976. Longitudinal growth of polymer crystals from flowing solutions III. Polyethylene crystals in Couette flow. Colloid Polym. Sci. 254, 868–881.

Zwijnenburg, A., Pennings, A.J., 1976. Longitudinal growth of polymer crystals from flowing solutions. IV. The mechanical properties of fibrillar polyethylene crystals. J. Polym. Sci. Polym. Lett. Ed. 14 (6), 339–346.

Further reading

Northolt, M.G., Boerstoel, H., Maatman, H., Huisman, R., Veurink, J., Elzerman, H., 2001. The structure and properties of cellulose fibres spun from an anisotropic phosphoric acid solution. Polymer 42 (19), 8249–8264.

The evolution of natural fibre composites through history

1.4

Jörg Müssig[1] *and Alain Bourmaud*[2]
[1]The Biological Materials Group, Biomimetic, HSB – City University of Applied Sciences, Bremen, Germany; [2]Université Bretagne Sud, UMR CNRS 6027, IRDL, Lorient, France

1. Introduction

Natural fibres have a rich history spanning thousands of years, tracing back to ancient civilisations that discovered and took advantage of their versatility and intrinsic properties. This chapter explores the evolution of plant fibres, particularly flax, across various applications from ancient Egypt to today, highlighting the potential of these fibres in modern contexts. Ancient societies developed techniques with flax that have remarkably endured, showcasing natural materials' timeless utility and adaptability. Although similar methods persist, advancements in processing, resource availability, and cultural attitudes have influenced how we process and apply plant fibres today.

Plant fibres were essential for creating textiles, goods, and early composite materials for centuries. In ancient Egypt, artisans skilfully transformed flax into ropes, sails, nets, and mummy wrappings, all crucial to society (see Fig. 1.4.1). Flax sails propelled Egyptian boats along the Nile, representing the ingenuity of early engineering and the connection between trade, transportation, and culture. Beyond practicality, flax also held ceremonial importance; worn sails were repurposed as mummy wrappings, reflecting an early awareness of resource limitations that resonates with today's focus on sustainability. Recycling in ancient times arose from necessity, driven by the high cost and scarcity of materials.

One of the most remarkable aspects of the history of composites is the continuity of plant fibres in marine applications, from sails, nets, and ropes in ancient Egypt, sails, ropes, or sailors' clothing in preindustrial Europe, to flax fibre composite hulls in modern boat building. Throughout different civilisations, fibres such as flax have remained central to boat and sail building, adapting to the advances of each era. This tradition continues today as we look for bio-based, sustainable, and environmentally friendly materials to combat the scarcity of resources. Today, biodegradable, renewable plant fibres are valued components of composite materials for various applications, from automotive parts to high-performance composites in marine construction.

The evolution of plant fibre use also reflects changing perspectives on processing and innovation. While ancient techniques were labor-intensive, their essence has persisted and adapted to modern machinery and scientific advancements. Today, natural fibres are blended with resins and polymers, creating strong and lightweight materials.

The History of Composites. https://doi.org/10.1016/B978-0-443-34119-9.00006-1

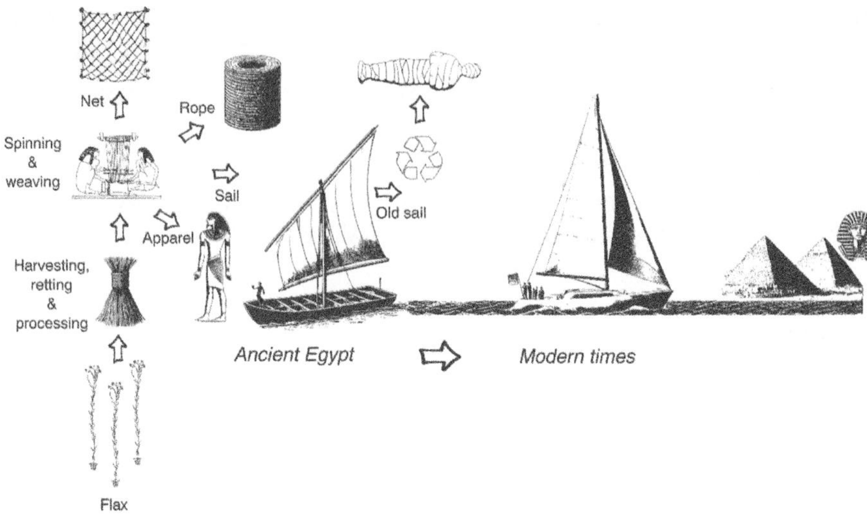

Figure 1.4.1 Schematic illustration tracing the evolution of flax-based materials from ancient Egypt to modern times, highlighting applications in boats, sails, ropes, and textiles. The theme of recycling is represented by reused sails transformed into mummy wrappings, symbolising early sustainable practices.

Though methods have evolved, the core principle of using plant-based materials for durable, versatile products remains.

In summary, the story of natural fibres is one of continuity and transformation. Examining the historical use of flax in naval construction, textiles, and early composites provides insight into the enduring relevance of these materials. The ancient tradition of recycling plant fibres, born of necessity, finds new meaning today as we pursue sustainability. This chapter traces the journey of flax as a case study, exploring the fascinating history of composites from their ancient origins to modern applications. After a brief description of the intrinsic properties of plant fibres, according to their function in the plant, a focus is placed on the evolution of fibre extraction and processing from ancient Egypt to the present day. The third section of the chapter proposes a journey into the uses of plant fibres, especially flax and hemp, in boat manufacturing, demonstrating the constant interest of people in the remarkable properties of plant fibres. Today, the development of new materials is closely linked to their recyclability, and the last section is dedicated to end-of-life and recycling; we demonstrated here that recycling is an old notion, already widely used by old civilisations such as Ancient Egypt.

2. Fibre properties, different fibres, different properties, function in the plant

In the following section, readers will learn how plant fibres' unique properties and functions—such as mechanical properties, weather resistance, and thermal conductivity—are closely linked to their structure, composition, and location within the plant. The section also looks at how local resources influence the availability of fibres such as hemp, flax, jute, coir, and cotton, shaping their historical and modern applications in maritime contexts and beyond. Through detailed comparisons of mechanical properties, lignin content, and fibre length, this section explains why certain fibres are selected for certain applications and examines the relationship between availability, durability, and performance. Natural fibres, both organic and inorganic, are found in nature. This chapter focuses on a selected set of plant fibres with a view to the historical development of natural fibres and natural fibre-reinforced composites—cotton, flax, hemp, jute, and coir—highlighted in Fig. 1.4.2.

There is an enormous variety of plant fibres (see Fig. 1.4.2) whose properties and morphology depend on the location of the fibres in the plant and their function. As shown in Fig. 1.4.3, plant fibres can have very different properties, such as strength and modulus, depending on their function in the plant. The bast fibres of the hemp

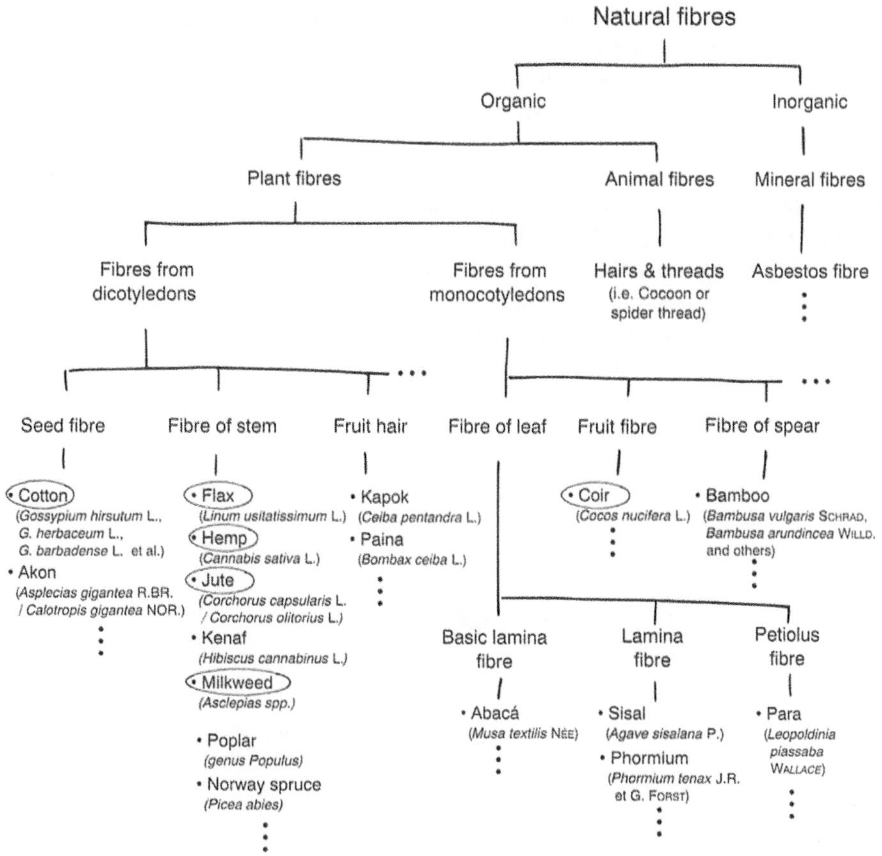

Figure 1.4.2 Overview of natural fibres: the fibres highlighted are discussed in more detail in the chapter.

or flax plant need to be particularly strong and stiff to give the stem a high bending resistance. In particular, the design of a flax plant is remarkable, with unique slenderness (L/d˜300), ensuring the resitance of the plant thanks to the high longitudinal stiffness, length of fibres, and fibre volume fraction (≈95%) at the periphery of the stem. The coconut fibres of *Cocos nucifera* L. are called mesocarp fibres and are used to protect the kernel of the coconut when it falls from the palm and is later transported by sea from one island to another.

However, the most industrially important plant fibre is cotton. Botanically speaking, the cotton fibre is a seed hair, not a fibre, but a trichome according to the botanical classification (Eder, 2010). Its primary function is to disseminate seeds; its mechanical performance is less strong and stiff than flax but has a significantly higher elongation at break. Cotton had a huge industrial development, driven by the Industrial Revolution and the increasing need for textiles, ease in processing, and low production cost.

The structure of plant fibre cell walls resembles a fibre composite, with properties primarily determined by the microfibril arrangement, particularly in the secondary wall's S2 layer, which dominates the cell wall. The microfibril angle (MFA) in this layer significantly influences the tensile strength of the fibre, as illustrated by the impact of increasing MFA from 6 degrees to 25 degrees (Fig. 1.4.4). Bast fibres, like hemp, generally have low MFAs (less than 10 degrees) (Thygesen and Eder, 2007), allowing efficient transfer of cellulose stiffness to the fibre and promoting high longitudinal mechanical properties. The primary wall and middle lamella contribute to the overall structure, with the middle lamella binding fibre bundles together (Desch and Dinwoodie, 1996).

Each layer has its own biochemical composition, ultrastructure, and, consequently, mechanical properties, but S2 remains the most important, representing around 90% of the fibre volume in bast fibres (Goudenhooft et al., 2019). Nevertheless, the middle lamellae has a specific importance and impact on the textile or composite quality.

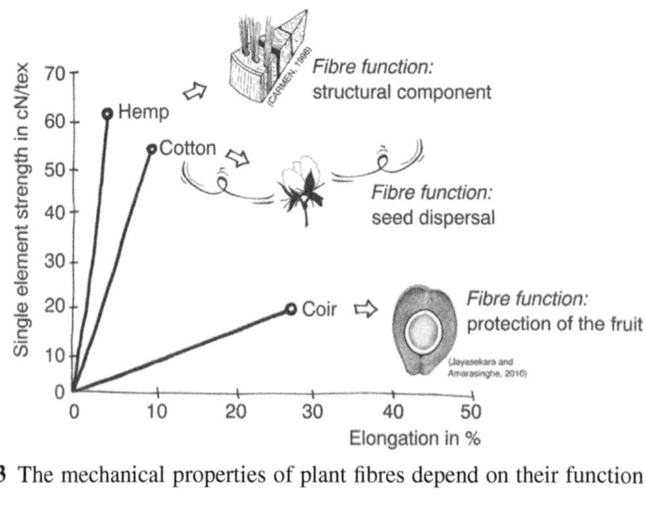

Figure 1.4.3 The mechanical properties of plant fibres depend on their function in the plant.

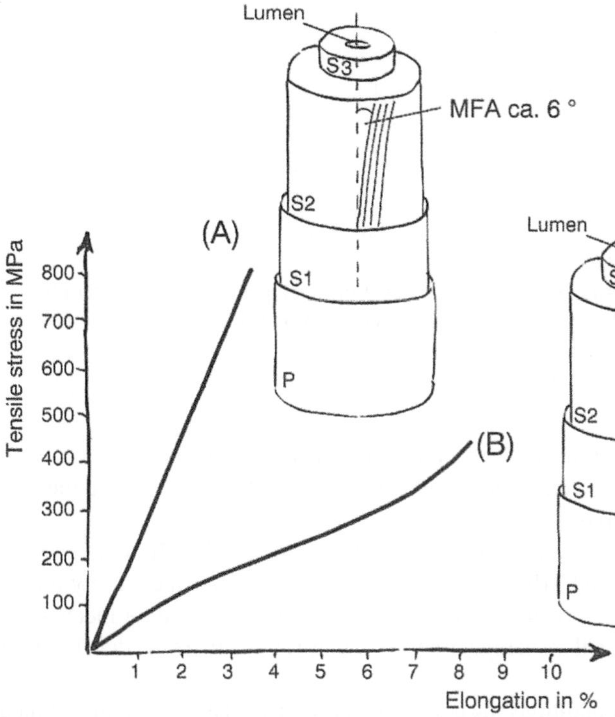

Figure 1.4.4 The cellulose microfibril angle (MFA) influences plant fibres' mechanical properties. (A) A fibre like hemp. (B) A fibre like cotton (Müssig and Hughes, 2012).

Being the glue between the fibres, it drastically impacts their individualisation and fineness. More or less damaged during the retting stage, its residues drive the future quality of the fibre. This point was rapidly assimilated by ancient civilisation; for example, in Egypt, flax was, and is always, water retted in basins along the Nile, providing a unique quality of fibre bundles with high fineness, enabling the design of very fine and famous textiles, sometimes better than what we are able to make today with modern industrial tools (André, 1964).

The properties of natural fibres depend not only on the arrangement of the cellulose fibrils as the main reinforcing element in the cell wall but also on the chemical composition of the cell wall. Natural fibres can vary widely in this respect. Cotton, for example, has a cellulose content of around 90%, while coir has an average cellulose content of 40%. The lignin content of cotton fibres is less than 1%, while that of coir is more than 40% (Bourmaud et al., 2018; Müssig, 2010). These differences in properties are mainly because the coconut has to survive falling from the tree (Graupner et al., 2017), and the sea transport route from one palm island to another would be impossible without the high proportion of lignin in the cell wall.

Generally, fibre mechanical performance is negatively correlated with lignin content and positively with cellulose fraction (Bourmaud et al., 2023), but, as

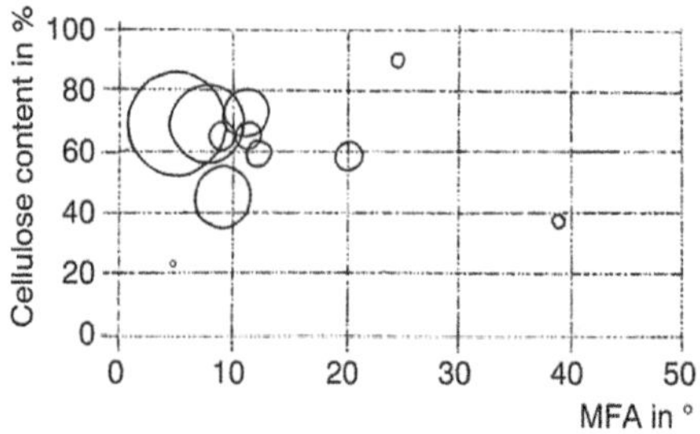

Figure 1.4.5 Young's modulus over the lignin and cellulose content of some natural fibres (cotton, hemp, flax, jute, and coir). The size of the circles is proportional to the single fibre's Young's modulus.
Data from (Bourmaud et al., 2018).

demonstrated in Fig. 1.4.6, biochemical composition is not the only parameter driving the mechanical properties; as explained previously, the MFA is also a key point that must be carefully considered.

Figure 1.4.5 demonstrates that high fibre stiffness is obtained with high cellulose content but also with reduced MFA; these two intrinsic properties are closely linked to the function of the fibres in-planta, the best being the fibres designed for mechanical support of the plant, such as flax, hemp, or nettle. Table 1.4.1 gathers the data used for this graphical illustration.

Cellulose content also conditions the composite properties. This was demonstrated in injection-moulded PP-short plant fibre composite materials using the same fibre volume fraction and the same PP matrix (Bourmaud et al., 2023). Results highlight a remarkable positive correlation between the tensile strength of the injection-moulded composite materials and the glucose content of the plant fibres used, glucose being the main constitutive sugar of cellulose.

Table 1.4.1 MFA, cellulose content, and Young's modulus of a panel of plant fibres.

	MFA in °	**Cellulose content in %**	**Young's modulus in GPa**
Abaca	11	64	18
Bamboo	9	45	36
Coir	39	36	10
Cotton	25	90	9
Flax	5	72	55
Hemp	11	70	30
Jute	9	65	21
Kapok	5	24	3
Kenaf	12	56	19
Ramie	7	68	45
Sisal	20	58	17

Today, world trade enables to obtain fibres from all the world's regions, but of course, due to environmental impacts associated with transportation, local products should be preferred. For ancient civilisations, and over time and depending on the region, the type of plant fibres available was not the same; however, experience and use also contributed to the choice of fibres. Whether for flax, nettle, or hemp, their strong mechanical properties and high cellulose content are in favour of good conservation and justified their selection for textile and rope manufacturing in ancient times. Biochemical composition, especially lignin content, is also a key factor when considering durability. These intrinsic fibre properties were assimilated very early by ancient civilisations, which will be detailed in section 4.

3. Evolution of plant fibre cultivation and associated extraction/separation methods over time

In this section, readers will explore the historical development of plant fibre cultivation in different areas of the world and the associated development of fibre decortication and separation methods, including traditional techniques such as retting and their influence on fibre quality. Advances in textile processing, such as fibre orientation and spinning, will highlight the importance of long flax and hemp for applications such as textiles and ropes. It will also provide insights into the role of manual extraction methods in ancient times, particularly for hemp and flax, and the impact of processing on fibre structure, including kink bands and fibre length.

Flax has been domesticated for a long time, with the first evidence of its use between 9000 and 8000 BCE in the Near East and Fertile Crescent zone (Hopf, 2008; Van Zeist and Bakker-Heeres, 1975), even if (Kvavadze et al., 2009) identified 30,000-year-old artifacts of wild flax fibres from several Upper Palaeolithic layers in the Dzudzuana cave in Georgia. The authors concluded that prehistoric hunter-gatherers made flax cords to fix stone tools, weave baskets, or sew clothing. Nevertheless, identifying these fibres is debated; scientists are unsure if they come from flax or hemp. It seems that independent periods of domestication occurred in different areas, from Asia to the Mediterranean regions.

Flax was used in ancient Egypt to manufacture linen throughout the Pharaonic era (3100 BCE to 332 BCE) but also during the Roman period (until 300 CE). Flax agriculture developed along the Nile Valley between the fifth and fourth millennia (Bats, 2023). Figure 1.4.6 provides an insight into the cultivation practice in ancient Egypt. Today, flax is still cultivated in Egypt; in 2022, Egypt was the fourth country in terms of the cultivated area of flax, with 25,000 ha, behind France, Russia, and Belarus.

In Europe, the first proof of flax use dates from 5500 to 4500 BCE, during the Neolithic period, mainly in lakeside towns in Switzerland (Leuzinger and Rast-Eicher, 2011) but also in Germany, the United Kingdom, Greece, and Italy (Valamoti, 2011). It seems the early uses of flax in Northern Europe were mainly for oil production (Ejstrud et al., 2011), the first evidence of linen fabric in this area dated to 800

Figure 1.4.6 The east wall of Sennedjem's vaulted crypt is decorated with a vignette that illustrates spell number 110 in the book of the dead. Sennedjem and his wife, Iineferti, are shown harvesting grain, sowing seeds, and pulling flax in the abundant fields of the next world. Original decorations are reproduced on a textile in the Museum of Civilisations, Cairo, Egypt (Author's personal photo).

BCE in Denmark or 200–300 CE in Sweden (Andresen and Karg, 2011; Viklund, 2011).

Of course, flax was not the only plant fibre used, but many artifacts are still conserved in museums. Its strong development in Egypt, associated with excellent ageing conditions due to dry climate, is a factor privileging its conservation, even after several millennia. In La Braga, Spain, very ancient nettle fibres (5807-4862 BCE) used for rope application (Herrero-Otal et al., 2023) were found, and also hemp fibres for fine textile manufacturing, especially in Northern Europe in Viking Age Scandinavia (Lukešová et al., 2017).

Until the industrial development of synthetic and cotton fibres in the 20th century, flax strongly developed in Western Europe from the Middle Ages. With notably important trading and commercial activity along the North Sea (Nicholas, 1992), during this time, flax textiles were produced for clothing and also equipment for sailing or fishing activities (e.g., ropes, sails, and bags), but also canvases for painting in the art domain, especially in Italy during the Renaissance (Melelli et al., 2021) or by Flemish and Impressionist painters. Important production centres of flax emerged in Northern Europe, Italy, Poland, Bohemia, Ireland, Belgium, and France. In the French Brittany region, more than 300,000 ha of flax were cultivated during the 17th century, around twice the area currently cultivated in Western Europe. Until the beginning of the 20th century, this market was especially boosted by the luxury market of high-quality fibres developed in Belgium and the Netherlands (Ollerenshaw, 1999).

Fibre separation and associated tools continuously evolved from the early to the industrial period. As suggested by Fig. 1.4.7, Egyptians already used pulling-out

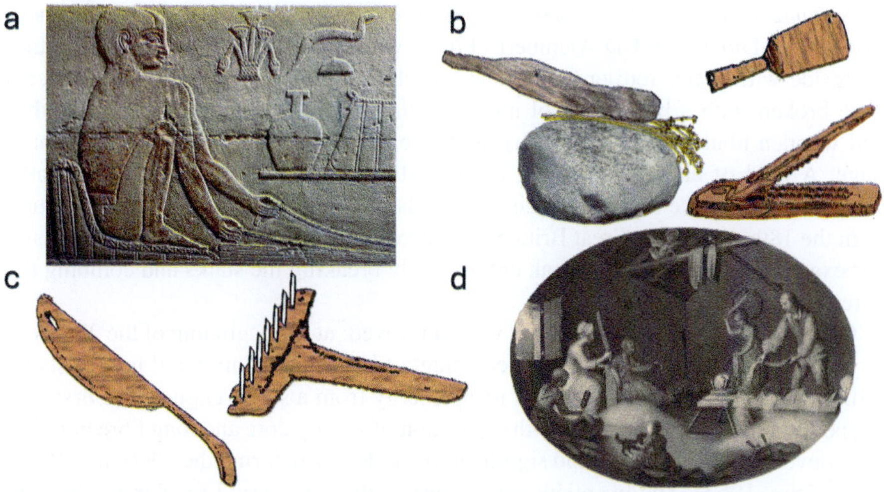

Figure 1.4.7 Illustrations of tools and processes used for fibre extraction and textile processing along time: Details of wall painting from Tombs of Ti, fifth dynasty, Saqqara, Museum of civilisation, Cairo, where a worker is spinning flax fibres (a); reconstruction of wooden tools from Sweden and Norway, used for beating and breaking flax stems (b); schema of a wooden knife found in Gloucestershire (middle bronze age) and reconstruction of a heckle dated between 850 and 1000 AD found in Shetland, Scotland (c) and illustration of a family breaking plants and combing flax fibres (d). From the authors' personal photo (Item X. HSA 318 © National Museums Scotland) and (© British Library Board).

techniques for flax plants, and just after, plants were water retted in basins built very close to the Nile, probably from the Middle Kingdom (Schenkel, 1994). Mesopotamians also used water retting and sun or plant treatment to bleach fibres; *Salsola kali and S. soda* were used to produce soda ash (Musselman, 2012; Soriga, 2017). After retting and possible treatment, details of Egyptian tombs demonstrate that wooden sticks or fans were used to extract fibres from the plant before threading. In Egypt, heckling was optional, depending on the use of fibres for spinning or splicing (Fig. 1.4.7a) (Andersson Strand, 2012).

During the first cultivations of flax in Europe during the Late Neolithic, flax plants were also pulled out, probably water retted (Maier and Schlichtherle, 2011); the finding of tools such as combs and spindles (Tricomi, 2014) assesses the use of specific techniques for textile applications. Findings from Rast-Eicher and Thijsse on Neolithic artifacts give rise to discussion. The textile findings and research series showed that the more gentle 'finger method' was used in addition to the breaking method (Rast-Eicher and Thijsse, 2001). While this method was probably used in the Neolithic period, medieval textiles show characteristics of being worked with the breaker. In Northern Europe, extraction tools such as clubs for stem beating or breakers mounted on a stand (Fig. 1.4.7b) were found, assessing the continuous development and improvement of specific tools (Ejstrud et al., 2011).

A range of details about flax cultivation and extraction were given in l'Encyclopédie of Diderot and D'Alembert (1765), with a description of first soil retting alongside with water retting before sun bleaching, each step lasting 8 days. Stems were broken with a breaking tool mounted on a stand, and a wooden knife with a dull wooden blade (Fig. 1.4.7c) was used to eliminate residues before the combing stage. A kind of alternative retting using boiling seawater, lime, and ash was also described to produce flax fibres using methods similar to cotton. Illustrations dating from the 18th century in Great Britain also represent the value of the flax value chain. For example, Fig. 1.4.7d shows an entire family breaking the stalks and combing the fibres.

Over the centuries, the techniques were improved; at the beginning of the 20th century in France, flax fibre was extracted using a Flemish pedal mill and progressively evolved toward a similar system but using energy from a steam engine. The first prototypes of a scutching machine, with separation of woody core and long fibre bundles, were developed in the 1930s and significantly modernised during the 1940s and 1950s (Malézieux, 1963). The twentieth century was a difficult period for flax production, especially due to rapid scientific and industrial progress, which enabled much greater production capacity than before, leading to major changes in consumer habits and also due to the rise of synthetic fibres such as Nylon (1935) and polyester (1941), as underlined by Searle, describing the issues of the flax industry (Searle, 1946).

During the second half of the 20th century, the globalisation of textile manufacturing and China's entry into the World Trade Organisation were huge bottlenecks for the flax industry's development. Despite this, and due to remarkable European know-how and favourable climate, this region remains the main one for flax fibre production, especially for high-quality markets and technical applications, such as composites in the automotive and aerospace industries. In 2021, 57% of the flax market was devoted to fashion, 28% to decoration, and 15% to technical applications (CELC, 2021).

In response to the growing demand to introduce plant fibres into the composites industry, over the last 20 years, fibre and reinforcement suppliers have developed a comprehensive range of bio-based reinforcements based on flax or hemp fibres that can be used in the manufacture of long-fibre composites. These reinforcements can be supplied in the form of dry preforms, containing only fibres, or preimpregnated preforms in which the fibres are already combined with a matrix. In this section, we will focus solely on flax textile developments, that is, dry preforms.

These preforms and by-products can be classified into seven main families:

i) Nonwovens—in the form of fleeces and needle felts—(Fig. 1.4.8a): these preforms are economical and easy to produce, but their mechanical properties are limited by the random orientation of the fibres. They do, however, offer good drapability.
ii) Unidirectional preforms (UD) (Fig. 1.4.8b, f): all the fibres are oriented in a single direction, making it possible to obtain composites with precise, high mechanical properties (Young's modulus of up to several tens of GPa).
iii) Rovings and yarns (Fig. 1.4.8c): the short flax/hemp fibre bundles are spun to form continuous yarns, necessary for processes such as weaving. A low level of twist improves the mechanical performance of composites.

Figure 1.4.8 Panel of contemporary flax semiproducts and preforms designed for composite industry: (a) Needle felt (© Ecotechnilin—Fibrimat), (b) dry unidirectional preform (© Ecotechnilin—Flaxtape), (c) 400 tex flax roving (© Safilin), (d) roving-based twill 2/2 weaves (© Flipts & Dobbels—171209/FAB twill 2/2), (e) hybrid flax/aramid fibre weave (© Groupe Depestele—Lincore HF), (f) quasi-UD weave (© BComp—ampliTex quasi-UD), (g) Biaxial ±45 degrees noncrimp fabric weave (© BComp—ampliTex qSP5008), (h) dry grid structure on a quasi-UD weave (© BComp—powerRibs).
Figure inspired from (Hendricks, 2018)

iv) Noncrimped fabrics (NCFs) (Fig. 1.4.8g): these fabrics allow varied yarn (fibre) orientations, minimising crimp and maximising mechanical properties.

v) Woven fabrics (Fig. 1.4.8d): Woven fabrics with twisted yarns are also developed, but compared with NCF products, they reduce mechanical performances; twist must be minimised for composite applications. Patterns such as twill provide additional gains in stiffness and strength.

vi) Hybrid reinforcements (Fig. 1.4.8e): plant fibres such as flax and hemp can be combined with industrially produced fibres (carbon, glass, aramid) to create hybrid composites. The resulting composites offer unique properties, such as high stiffness (up to 75 GPa for carbon/flax reinforcements) and good vibration damping.

vii) 0/90 grid structures (Fig. 1.4.8h) that add structural rigidity while being lightweight and economical.

Flax and hemp-based reinforcements continue to gain in complexity and performance thanks to innovations such as reduced torsion, hybrid textiles, and specific architectures. These advances make it possible to meet demanding structural applications, while exploiting the environmental and economic benefits of plant fibres.

As discussed and illustrated here, flax has a long and well-documented history, but hemp fibres have also been used for a long time, as described by Clarke and Merlin. Hemp has been used throughout Eurasia for thousands of years to make rope, cloth, and paper (Clarke and Merlin, 2013). The archaeological and historical record is rich with evidence of the ancient importance of hemp fibre to humans and its spread to different regions. Clarke and Merlin document the widespread introduction of hemp fibre in Central, East, South, and Southwest Asia, Egypt, the Mediterranean, Northern Europe, and finally, the New World. Their research impressively documents the spread, the possible cultivation and processing concepts, and the use of hemp in various products, such as cordage and sailcloth in the Mediterranean region, dating back to the fourth century BCE (Clarke and Merlin, 2013).

4. Case study: marine applications, from ancient Egypt to contemporary sailing races

Throughout history, plant fibres have played a vital role in marine applications, evolving from ancient Egyptian sails and Viking ship ropes to modern bio-based composites used in contemporary sailing, reflecting a continuous adaptation to changing needs and technological advancements.

Ancient writing, particularly from Plinius the Elder in Naturalis Historia (first century, BCE), testifies to the huge interest in the flax plant. A part of the book dedicated to this plant highlights the huge interest in fibre production in ancient times. Their use for sail manufacturing, mainly for trading activities and trips around the Mediterranean Sea, was strongly highlighted (André, 1964). Traces of the use of flax for manufacturing sails were also found in Italy from the pre-Roman period; small quantities of flax were produced in Italy, and the main volumes used were imported from Roman colonies such as Egypt or Gaul (Dimova et al., 2021). A considerable quantity of such sail artifacts have been discovered in Roman and Greek ports (Blue, 2002; Dimova et al., 2021; Whitewright, 2007).

In the naval sector (Fig. 1.4.9), flax fibres were hugely used for ropes and sails manufacturing, but not only, Greek and Egyptian ancient civilisations also developed the use of flax fibres for fishing nets or boat waterproofing; these practices required large quantities of fibres and were widespread between the sixth BCE and second CE centuries. The Greek Gyptis fishing ship is a well-known example, using more than 5 km of flax thread (Pomey and Poveda, 2018). Testimonies from the New Empire period also demonstrate the use of waterproofing in Egypt (Meeks, 1996) through royal correspondence dating from the Amarna period and from the reign of Merenptah, which bears witness to an exchange of naval technology between Mesopotamia and Egypt. This technique was used to secure the sailing and shipping expeditions, which was particularly important to ensure the stability of the commercial exchanges during the Cheops period.

In Egypt, in the archaeological site of Ouadi el-Jarf harbour (Tallet et al., 2012), an impressive diversity and panel of artifacts were found, demonstrating an important

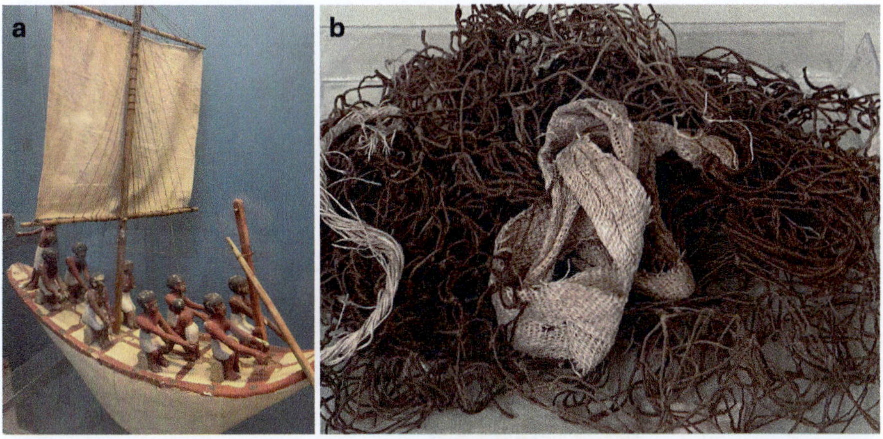

Figure 1.4.9 A painted wooden model of a boat with a flax sail, 12th dynasty, about 1985-1795 BCE (a) and a fishing net made of flax fibres (b) Authors' personal photos, British Museum, London, United Kingdom, and Egyptian Museum, Cairo, Egypt.

range of uses and functionalities for naval activities during the Pharaonic period. Flax artifacts from ropes, sails or clothing linen were found between 2011 and 2021, and archeologist teams excavated more than 260 kg of textiles and ropes (Tallet et al., 2012); most of them being coated with vegetal resin to improve water resistance; this coating protects the fibres from moisture but also increases the strength of the textiles, so these parts can be seen as one of the first bio-based composites, a source of inspiration for future developments. The main part of these textiles is dated from the Cheops period, demonstrating the strong importance of flax in naval and military activities. Flax textiles have also been used for the manufacturing of the Royal solar ship of Cheops (Jenkins, 1983; Lipke, 1984; Nour et al., 1950). This boat was made of Lebanese cedar, the elements of which were held together by linen rods; the sails were certainly also made of flax.

In other parts of the world, other fibres were used for marine applications, it was the case on the Pacific coast of South America, where ancient coastal civilisations developed intense marine activity, especially fishing, and used plant fibres for manufacturing specific tools during the Late Preceramic period in the third millennium BCE (Beresford-Jones et al., 2018; Moseley, 1975). A panel of plants was used for producing ropes and fishing nets. Among plant fibres, those from *Asclepias sp.* (milkweed) stems were widely used due to their suitable properties. Indeed, these fibres exhibit a very interesting fineness and length, well adapted for splicing or spinning (Beresford-Jones et al., 2018; Endress et al., 2014). Moreover, milkweed fibres have suitable mechanical and biochemical properties, such as high lignin content, a key factor for water ageing and naval applications (Whiting, 1943).

Other examples of the use of plant fibres exist in a different time period. With his experimental archaeology and the Kon-Tiki expedition in 1947, Thor Heyerdahl wanted to show that it was technically possible for the pre-Columbian indigenous

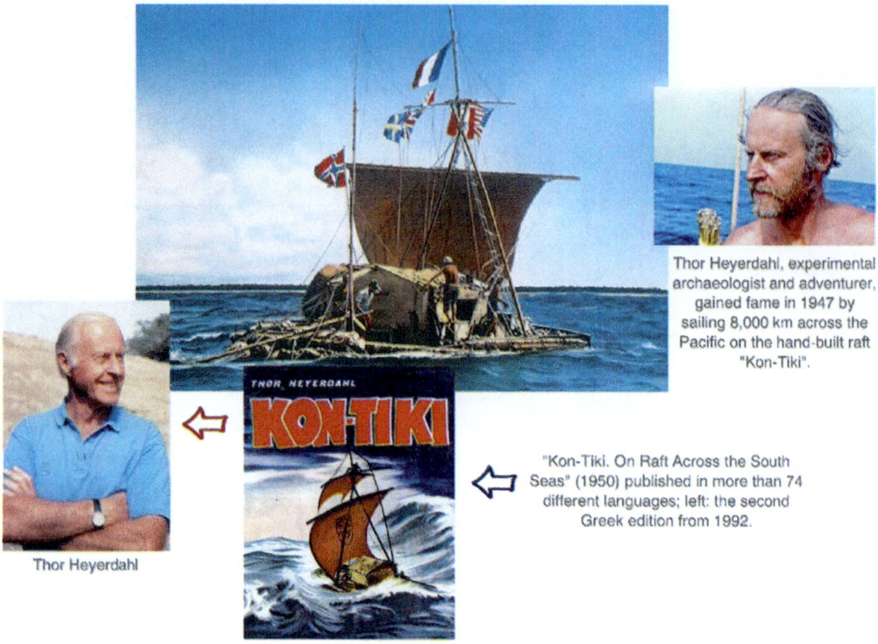

Figure 1.4.10 Thor Heyerdahl and the Kon-Tiki expedition, in which he proved that early intercontinental contacts were possible using rafts and ships made from bio-based materials. Courtesy of The Kon-Tiki Museum, Oslo, Norway (https://www.kon-tiki.no/).

peoples of South America to travel to Polynesia on rafts to colonise it. The floating body of the raft consisted of nine up to 14 m long balsa logs connected with 30 mm thick hemp ropes (Kon-Tiki museet, 2024; Wikipedia, 2024). Heyerdahl published the expedition in his book Kon-Tiki (Heyerdahl, 1950). Thor Heyerdahl had written the script for the documentary film, edited an earlier version for lectures, and assisted with the film's production[1] (edited by Olle Nordemar) (see Fig. 1.4.10).

Clarke and Merlin examined the historical significance of hemp fibre, highlighting, in particular, its use in boat construction (Clarke and Merlin, 2013). In Europe, hemp was essential for marine equipment such as cordage, nets, flags, and the sealing material known as "oakum". This consisted of hemp fibres impregnated with tar and was traditionally used to make the seams and joints between wooden planks watertight—a crucial contribution to the safety of sailing ships. These uses of fibres for waterproofing are in line with the uses of flax described above, and the resulting parts can be considered as composite materials in their own right, with the hemp fibres reinforcing the tar matrix. As Clarke and Merlin explored, the durable fabric 'canvas' owes its name to the original use of hemp ("cannabis") and has etymological roots in Latin and Greek, as Benet also notes (Benet, 1975; Clarke and Merlin, 2013). A

[1] Personal communication with Reidar Solsvik, Curator, The Kon-Tiki Museum, Oslo, Norway.

notable example of the use of hemp in historic shipbuilding is the French frigate L'Hermione, brought to Boston by the Marquis de Lafayette in 1780. During the reconstruction of this iconic warship at Rochefort in France, hemp fibres were used as filler and sealant between the planks (Hermione-La Fayette Association, 2011; Sciolino, 2007). The reconstructed L'Hermione, which played an essential role in the American Revolution, was launched in Rochefort in 2012. These examples illustrate the historical importance of hemp for marine equipment and boat construction (Clarke and Merlin, 2013).

Over recent decades, research has increasingly explored natural fibre composites for marine applications. Ansell uses material maps to assess natural fibres against conventional materials, examining properties such as Young's modulus, density, price, embodied energy, and carbon footprint (Ansell, 2014). He concludes that certain natural fibres hold significant potential for marine use, provided their stiffness and strength are adequate for composite reinforcement. However, addressing moisture absorption remains critical. Ansell proposes solutions such as fibre surface treatments, matrix modifications, and optimised composite designs to mitigate moisture issues, making natural fibre composites more viable for boatbuilding and other marine applications.

Composite materials are today ubiquitous in our everyday lives, but their development was very significant in the 20th century, particularly because of the needs of the aerospace and military industries during the Second World War (de Bruyne, 1939), which stimulated research into new materials that were lighter and stronger than traditional metals. It was against this backdrop that the first glass fibre-reinforced composites appeared (Verpoest, 2020). These materials, composed of a fibre-reinforced polymer matrix, offer a unique combination of mechanical properties.

After the war, the development of synthetic polymers and the emergence of new fibres, such as carbon, opened up new prospects for composites. Carbon fibres, in particular, offer exceptional tensile strength and high stiffness at very low mass. These properties make them the material of choice for cutting-edge applications such as aeronautics, Formula 1 racing, musical instruments, or the space industry (Verpoest, 2020). Polymer matrix composites represent the vast majority of composites used today. The matrix, generally a thermoplastic or thermosetting resin, plays an essential role in transmitting loads between the fibres and ensuring the cohesion of the material. The fibres give the composite its mechanical strength. Over the last 25 years, plant fibres have made significant inroads into composites, even though their use had already been explored in the 20th century, particularly for aeronautical and automotive applications (Bourmaud et al., 2018; Müssig, 2010; Verpoest, 2018).

While natural fibres have the advantage of being renewable and biodegradable, their use in composites brings a number of challenges. Their heterogeneity, their sensitivity to humidity, their individualisation in composites, and their production cost are all limiting factors. Current research is focussing on their use in new and more efficient processes, on their durability and ageing, and on the quality of the available preforms.

Despite these possible scientific and technical issues, there is growing interest in natural fibre composites for marine applications, as highlighted by reviews such as

Yang et al. (2018) and El Hawary et al. (2023). Yang et al. emphasise the need for innovations like chemical modifications, adhesion promoters, and optimised manufacturing to address processability and mechanical performance challenges. Continued research and market expansion are crucial for improving properties, reducing costs, and enhancing the economic viability of these bio-based composites. El Hawary et al. illustrate current applications, including surfboard, canoe, and boat construction (El Hawary et al., 2023). Examples range from a rowing boat using hybrid glass and sugar palm fibre with polyester resin to a 6-metre bamboo composite boat hull and a 4.6-metre racing skiff with sandwich panels of epoxy resin and cork core. Notable developments include the Baltic 68 Café Racer Pink Gin Verde with flax-based ampliTex composites and the TaraTari, a 9-metre boat (Gold of Bengal) made from jute fibre-reinforced polyester resin, showcasing the versatility and potential of these sustainable materials in marine industries.

To provide an overview of current trends in the marine application of natural fibres, Fig. 1.4.11 shows various products made from natural fibre composites, including canoes, kiteboards, motor boats, and sailing yachts, to illustrate their versatility and sustainable potential.

One example of the performance of flax composites that deserves special mention in 2024 is the use of components in high-performance racing yachts for the Vendée Globe, a nonstop regatta for single-handed sailors considered the toughest

Figure 1.4.11 Examples of natural fibre composites in marine applications include kiteboards, motor boats, and sailing yachts.
Courtesy of GREENBOATS, Bremen, Germany.

single-handed race in the world. The story of the use of flax in one of these high-performance boats, Team Malizia's Seaexplorer, also links two people. Friedrich Deimann, who laid the foundation for the GreenBoats company with his masterpiece of a canoe made of flax composites, and the high-performance sailor Boris Herrmann, who relies on components made of flax composites during the Vendée Globe, which leads eastwards around the globe along the Southern Ocean in the "Roaring Forties" (see Fig. 1.4.12).

5. End of life and reusing of plant fibre materials

This section addresses plant fibre composites' end-of-life and recycling potential, highlighting the benefits of sustainability and life cycle assessment. Historical background, such as the reuse of textiles in ancient Egypt and modern examples of recycling plant fibre composites for new applications, is considered. The challenges and opportunities in developing recyclable products, including the complexity of composites and their particular properties and issues to consider, such as water absorption,

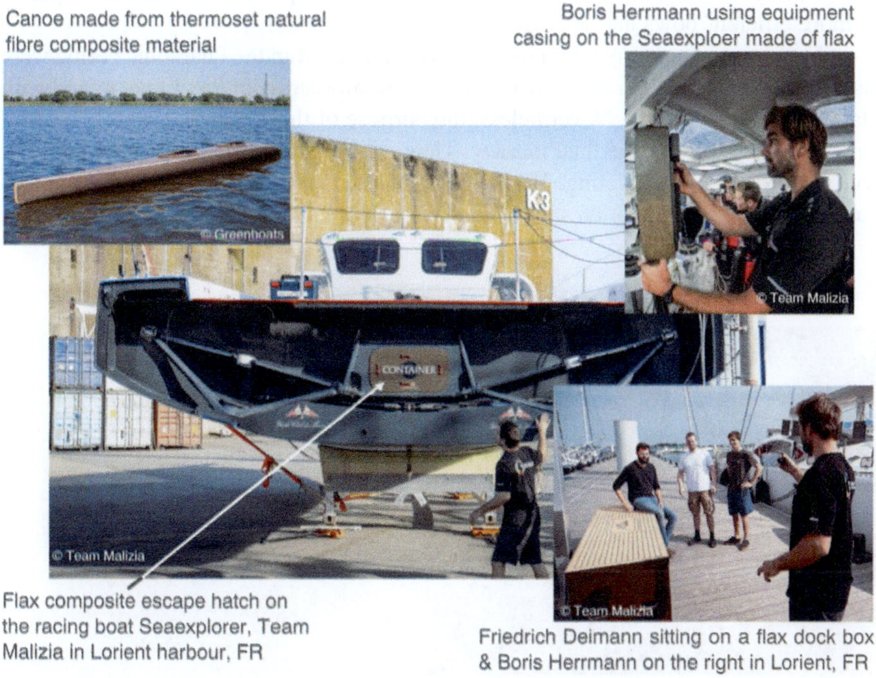

Figure 1.4.12 Marine applications of natural fibre composites, from Friedrich Deimann's flax composite canoe, the foundation of GreenBoats, to high-performance racing yachts like Team Malizia's Seaexplorer, used in the 2024 Vendée Globe, the world's most challenging single-handed regatta.

will be discussed to emphasise the need for sustainable innovation in bio-based composites.

Recycling will become increasingly important as it provides a sustainable source of renewable carbon, helps to replace fossil carbon, and reduces greenhouse gas emissions. Integrating recycling with other renewable carbon sources, such as biomass and CO_2 utilisation, is crucial in mitigating climate change and transitioning to a circular economy (RCI, 2024). Different studies have demonstrated the interest in recycling among the different end-of-life options, including landfill, incineration, and composting. For example, Beigbeder et al. (2019) proposed an LCA (Life Cycle Analysis) of PP wood composites and highlighted a strong interest in the recycling option, especially for terrestrial acidification, photochemical oxidant formation, climate change, and freshwater eutrophication impacts, compared with other end-of-life options. Similar results were obtained on PP-flax by Bensadoun et al. (2016), particularly for terrestrial acidification, but also on the overall LCA single score.

One of the first plant fibre composites was probably the linothorax armour part, used in the Mediterranean area from the VIIth century BCE to the beginning of the Roman Empire; its main ambassador was Alexander the Great (Lush, 2007). This protection tool was made of flax reinforcement mixed with animal or vegetal glue; it was approximately 12 mm thick, providing impact protection, similar to a 2 mm thick bronze plate. This defence accessory progressively disappeared with the development of more efficient bows and arrowheads (Aldrete et al., 2013).

In Egypt, some mummy wrapping and hypocephali from the Pharaonic period were also composite materials, some of which were reinforced with flax fibres mixed with plant resins (Fig. 1.4.13). Nevertheless, the purpose of these materials was to accompany the deceased on their final journey. They were designed for optimum durability, with no prospect of recycling. While the recyclability of such materials was not a question, the same cannot be said for flax fibres.

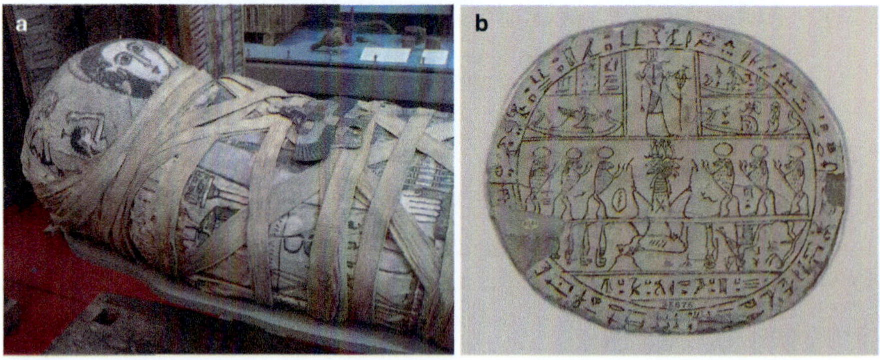

Figure 1.4.13 Mummified remains of a young woman named Cleopatra, a thick layer of bandages wraps the body, with resinous-like material poured in between some of the layers, Roman period (a), Hypocephalus made of stuccoed linen, Ptolemaic period, 305-30 BCE–(b) Author's personal photos, British Museum, London, UK.

Flax has always been considered a precious product; from Ancient Egypt to the Middle Ages, flax textiles were used in trade exchange. For example, the title of "director of the king's linen office" is mentioned as early as the third Dynasty, demonstrating the importance of flax, also used as tax; each producer was required to bring 1/10 of their harvest to the royal administration (Baer, 1962). Linen textiles were also used as currency, for example, to buy copper or metal objects and even houses or enslaved people (Vogelsang-Eastwood and Haeringen, 1992). The King also gave them as privileges or gifts, representing a form of retribution. Linen cloth was classified according to quality, size, and colour; some were assimilated into a luxury production to a privileged social category.

Due to the high value of linen, there was a great interest in recycling or reusing the textiles. In case of huge damage, clothes and textiles were reused or retransformed in another object (Vogelsang-Eastwood, 2000). Investigations on a mummy linen shroud have also demonstrated that this textile was possibly used before as a sail (Rougé, 1987), being one of the first examples of flax textile recycling. The work of Garmi et al. (Garmi and Meunier, 2020) also suggests that the textiles used to waterproof boats could, after modifications be used for other purposes. The same authors also highlighted the presence of different textiles and types of stitching on the same boat sail, showing that this sail could have been made from 12 different fabrics, again demonstrating recycling practices in ancient Egypt. This reusing practice was also applied for producing some cords and lamp wicks by artisans, who received textiles from the royal administration (Valbelle, 1977). This recycling was thus used in many domains, not only for mummy wrapping.

The recycling concepts already realised at these ancient times represent the highest level of the recycling pyramid in the form of reuse according to today's standards (see Fig. 1.4.14).

This should be a model for new developments in the field of natural fibre composites because, in the history of modern composites, we often find ourselves at the lowest levels of the inverted pyramid of disposal, recovery, and, in a few cases, recycling.

Today, with the decrease in fossil-fuel-based resources, the recycling of composite materials is an important research sector. Bio-based composite materials are also included, especially due to the high cost of plant fibres induced by high demand and the decrease in the yields caused by climate change, and especially drought (Melelli et al., 2022). In their review of the recycling of natural fibre composites,

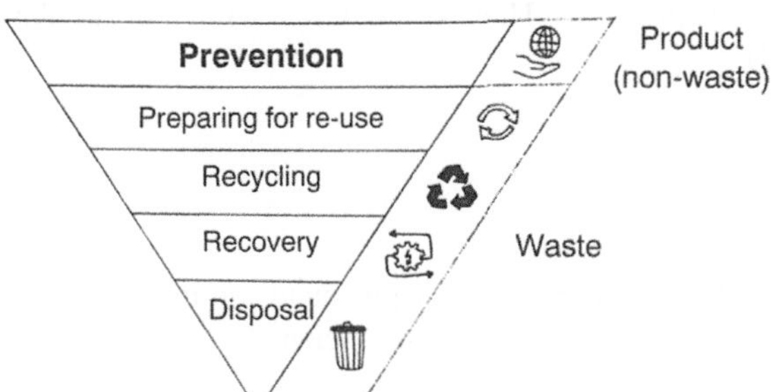

Figure 1.4.14 Waste hierarchy—waste prevention is the preferred option, and landfilling waste should be a last resort (EU Waste Framework Directive, 2023).

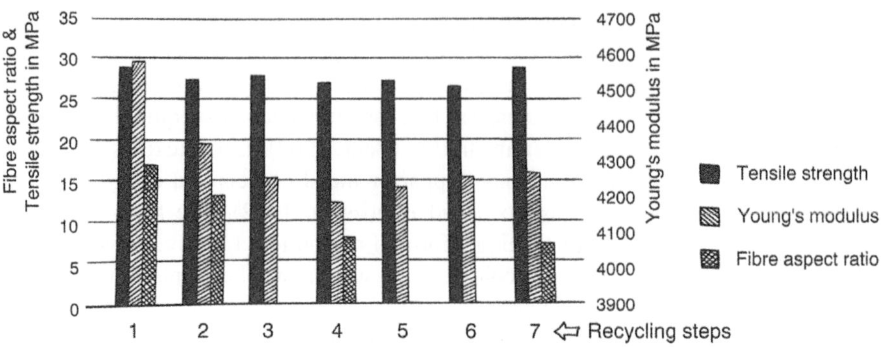

Figure 1.4.15 Recycling of hemp fibre-reinforced Polypropylene composites. Inspired from (Bourmaud and Baley, 2007)

Zhao et al. (2022) focus on the challenges and opportunities that arise primarily for thermoplastic natural fibre-reinforced composites. They generally describe good recyclability for these thermoplastic composites, meaning they can usually be recycled about 4–6 times before their thermo-mechanical properties change.

This good resistance in mechanical properties is mainly induced by the relative consistency of the fibre aspect ratio; indeed, successive processing cycles favour length decreases but also diameter drop due to the progressive separation of the fibre bundles (Fig. 1.4.15) (Albrecht et al., 2018; Bourmaud and Baley, 2007). Adhesion promoters, compatibilizers, and chain extenders should be considered to improve the interface between natural fibres and polymer matrices and reduce degradation after recycling. The recycling behaviour of polyolefin-based plant fibre composites is generally good due to the good behaviour of PP, even after thermo-mechanical cycles. However, this is different for sensitive matrices such as PLA with a pronounced decrease in strength after several cycles (Le Duigou et al., 2008).

The recycling of large parts, such as boats or panels, described in Section 4. (Fig. 1.4.9), is highly dependent on the matrix. When thermoset resins are used, crushing with reincorporation of low-cost materials or incineration are generally the only options, with plant fibres being too sensitive for resin extraction or solvolysis treatment. That's why future developments need to consider using thermoplastic matrices as much as possible. In most cases, it is not enough; an efficient recycling also needs the implementation of a complete value chain and the design of easily deconstructable parts by considering recycling and end-of-life at the design stage of industrial parts. In line with the ancient Egyptian concept, a more desirable approach would be to repair and reuse whole products or parts of them. This would align with the circular economy's aims (see Fig. 1.4.14).

6. Conclusion

The history of natural fibres, exemplified by flax, reveals a profound continuity in their use, adaptability, and relevance over the centuries. From ancient Egyptian sails to contemporary marine composites, natural fibres have demonstrated unparalleled versatility, reflecting humanity's ability to innovate while preserving essential techniques. Early civilisations used fibres as crucial tools and showed remarkable foresight in recycling practices, a necessity driven by material scarcity and high value, mirroring today's sustainability efforts.

As a case study, flax highlights the important intrinsic qualities of plant fibres: strength, durability, and environmental friendliness. The evolution from manual retting and spinning in ancient times to advanced composite manufacturing highlights the seamless integration of traditional knowledge with modern science. Today, flax and other natural fibres contribute to high-performance materials in the automotive and marine industries, addressing modern challenges such as resource scarcity and climate impact.

The enduring lesson of history is the cyclical value of natural fibres. Ancient recycling methods, such as turning sails into mummy wraps, resonate with modern circular economy principles. As we face environmental challenges, the past offers a blueprint for integrating sustainability into material design and lifecycle management.

By combining historical insight with technical innovation, natural fibres can be positioned as key components in transitioning to a sustainable future. Their journey from ancient ropes and textiles to cutting-edge composites exemplifies a harmonious balance between tradition and progress, highlighting the timeless relevance of bio-based materials.

Acknowledgement

Our special thank goes to Johanna Malika Müssig for her creative work with the hand-drawn illustrations.

References

Albrecht, K., Osswald, T., Baur, E., Meier, T., Wartzack, S., Müssig, J., 2018. Fibre length reduction in natural fibre-reinforced polymers during compounding and injection moulding—experiments versus numerical prediction of fibre breakage. J. Compos. Sci. 2, 20–37. https://doi.org/10.3390/jcs2020020.

Aldrete, G.S., Bartell, S.M., Aldrete, A., 2013. Reconstructing Ancient Linen Body Armor: Unraveling the Linothorax Mystery. Johns Hopkins University press, Baltimore, USA.

Andersson Strand, E., 2012. The textile chaîne opératoire: using a multidisciplinary approach to textile archaeology with a focus on the Ancient Near East. Paleorient 38, 21–40.

André, J., 1964. Nature du lin et horticulture. In: Budé (Ed.), Pline l'Ancien, Histoire Naturelle. Paris.

Andresen, S.T., Karg, S., 2011. Retting pits for textile fibre plants at Danish prehistoric sites dated between 800 b.c. and a.d. 1050. Veg. Hist. Archaeobot. 20, 517–526. https://doi.org/10.1007/s00334-011-0324-0.

Ansell, M.P., 2014. 14 - natural fibre composites in a marine environment. In: Hodzic, A., Shanks, R. (Eds.), Natural Fibre Composites. Woodhead Publishing, pp. 365–374. https://doi.org/10.1533/9780857099228.3.365.

Baer, K., 1962. The low price of land in ancient Egypt. J. Am. Res. Cent. Egypt (JARCE) 1, 25–45. https://doi.org/10.2307/40000856.

Bats, A., 2023. Flax in the economy of ancient Egypt. In: Camilleri, A. (Ed.), Flax-Linen – Fibre of Civilisations. Alliance for European Flax-Linen and Hemp, Paris, France, pp. 84–97.

Beigbeder, J., Soccalingame, L., Perrin, D., Bénézet, J.-C., Bergeret, A., 2019. How to manage biocomposites wastes end of life? A life cycle assessment approach (LCA) focused on polypropylene (PP)/wood flour and polylactic acid (PLA)/flax fibres biocomposites. Waste Manag. 83, 184–193. https://doi.org/10.1016/j.wasman.2018.11.012.

Benet, S., 1975. In: Rubin, V. (Ed.), Early Diffusion and Folk Uses of Hemp. De Gruyter Mouton, pp. 39–50. https://doi.org/10.1515/9783110812060.39.

Bensadoun, F., Vanderfeesten, B., Verpoest, I., Van Vuure, A.W., Van Acker, K., 2016. Environmental impact assessment of end of life options for flax-MAPP composites. Ind Crops Prod 94, 327–341. https://doi.org/10.1016/j.indcrop.2016.09.006.

Beresford-Jones, D., Pullen, A., Chauca, G., Cadwallader, L., García, M., Salvatierra, I., Whaley, O., Vásquez, V., Arce, S., Lane, K., French, C., 2018. Refining the maritime foundations of Andean civilization: how plant fiber technology drove social complexity during the preceramic period. J Archaeol Method Theory 25, 393–425. https://doi.org/10.1007/s10816-017-9341-3.

Blue, L., 2002. Myos Hormos/Quṣeir al-Qadīm. A roman and Islamic port on the red sea coast of Egypt — a maritime perspective. In: Proceedings of the Seminar for Arabian Studies, 32, pp. 139–150.

Bourmaud, A., Baley, C., 2007. Investigations on the recycling of hemp and sisal fibre reinforced polypropylene composites. Polym. Degrad. Stabil. 92, 1034–1045. https://doi.org/10.1016/j.polymdegradstab.2007.02.018.

Bourmaud, A., Beaugrand, J., Shah, D.U., Placet, V., Baley, C., 2018. Towards the design of high-performance plant fibre composites. Prog. Mater. Sci. 97, 347–408. https://doi.org/10.1016/j.pmatsci.2018.05.005.

Bourmaud, A., Konschak, K., Buffet, C., Calatraba, M., Rudolph, A.L., Kervoëlen, A., Gautherot, B., Bonnin, E., Beaugrand, J., 2023. A circular approach for the valorization of tomato by-product in biodegradable injected materials for horticulture sector. Polymers 15. https://doi.org/10.3390/polym15040820.

CELC, 2021. Flax/Linen Barometer. https://allianceflaxlinenhemp.eu/en/european-flax-linen-hemp-media-library/9/2021-european-flax-linen-barometer. (Accessed 08 January 2026).

Clarke, R.C., Merlin, M.D., 2013. Evolution and Ethnobotany, First ed. University of California Press. https://doi.org/10.1525/j.ctt3fh2f8.

de Bruyne, N.A., 1939. Plastic progress - some further developments in the manufacture and use of synthetic materials for aircraft construction. Flight 77–79.

Desch, H.E., Dinwoodie, J.M., 1996. Timber: Structure, Properties, Conversion, and Use, seventh ed. Taylor and Francis. https://doi.org/10.1007/978-1-349-13427-4.

D Diderot, JLR d'Alembert, 1765. Ju - Mamira. Encyclopédie ou Dictionnaire raisonné des sciences, des arts et des métiers.

Dimova, B., Harris, S., Gleba, M., 2021. Naval power and textile technology: sail production in ancient Greece. World Archaeol. 53, 762–778. https://doi.org/10.1080/00438243.2021.2015428.

Eder, M.B.I., 2010. Natural fibres–function in nature. In: Müssig, J. (Ed.), Industrial Applications of Natural Fibres: Structure, Properties and Technical Applications. John Wiley & Sons, Ltd., Chichester, United Kingdom, pp. 22–39.

Ejstrud, B., Andresen, S., Appel, A., Gjerlevsen, S., Thomsen, B., 2011. From Flax to Linen: Experiments with Flax at Ribe Viking Centre. Ribe Viking Centre, University of Southern Denmark.

El Hawary, O., Boccarusso, L., Ansell, M.P., Durante, M., Pinto, F., 2023. An overview of natural fiber composites for marine applications. J. Mar. Sci. Eng. 11. https://doi.org/10.3390/jmse11051076.

Endress, M.E., Liede-Schumann, S., Meve, U., 2014. An updated classification for Apocynaeae. Phylotaxa 159, 175–194. https://doi.org/10.11646/phytotaxa.159.3.2.

Garmi, D., Meunier, L., 2020. Les textiles archéologiques d'époque romaine en contexte nautique en France (Ier-IIIe siècle après J.-C.). Gallia 77, 187–211. https://doi.org/10.4000/gallia.5513.

Goudenhooft, C., Bourmaud, A., Baley, C., 2019. Flax (Linum usitatissimum L.) fibers for composite reinforcement: exploring the link between plant growth, cell walls development, and fiber properties. Front. Plant Sci. 10, 411. https://doi.org/10.3389/fpls.2019.00411.

Graupner, N., Labonte, D., Humburg, H., Buzkan, T., Dörgens, A., Kelterer, W., Müssig, J., 2017. Functional gradients in the pericarp of the green coconut inspire asymmetric fibre-composites with improved impact strength, and preserved flexural and tensile properties. Bioinspir. Biomim. 12, 026009. https://doi.org/10.1088/1748-3190/aa5262.

Hendricks, K., 2018. Biosourcing: what are the available reinforcements on the market? In: Verpoest, I., Baley, C. (Eds.), Flax and Hemp Fiber Composites, a Market Reality : The Biobased Solutions for the Industry. JEC Group, Paris, pp. 33–45.

Hermione-La Fayette Association, 2011. End of the caulking, representing a total length of 12 km of sealing and 255 kg of hemp used. https://fregate-hermione.com/en/lhermione/ships-reconstruction/. (Accessed 08 January 2026).

Herrero-Otal, M., Romero-Brugués, S., Piqué Huerta, R., Homs, A., De Diego, M., Palomo, A., 2023. Describing the neolithic cord production process: raw materials, techniques and experimental archaeology in La Draga (Girona, Spain; 5207–4862 cal BC). J. Archaeol. Sci. Rep. 50, 104092. https://doi.org/10.1016/j.jasrep.2023.104092.

Heyerdahl, T., 1950. The Kon-Tiki Expedition: By Raft across the South Seas., London, England. George Allen and Unwin Ltd., London translated by F.H. Lyon.

Hopf, M., 2008. Plant remains and early farming in Jericho 1. In: Dimbleby, G. (Ed.), The Domestication and Exploitation of Plants and Animals. New York, USA.

Jenkins, N., 1983. La Barque royale de Chéops. France-Empire, Paris.

Kon-Tiki museet, 2024. The Kon-Tiki Museum Shows Thor Heyerdahl's Famous Life's Work. Kon-Tiki museet, Oslo. https://www.kon-tiki.no/.

Kvavadze, E., Bar-Yosef, O., Belfer-Cohen, A., Boaretto, E., Jakeli, N., Matskevich, Z., Meshveliani, T., 2009. 30,000-year-old wild flax fibers. Science (1979) 325, 1359. https://doi.org/10.1126/science.1175404.

Le Duigou, A., Pillin, I., Bourmaud, A., Davies, P., Baley, C., 2008. Effect of recycling on mechanical behaviour of biocompostable flax/poly(l-lactide) composites. Composite Part A 39, 1471–1478. https://doi.org/10.1016/j.compositesa.2008.05.008.

Leuzinger, U., Rast-Eicher, A., 2011. Flax processing in the Neolithic and Bronze Age pile-dwelling settlements of eastern Switzerland. Veg. Hist. Archaeobot. 20, 535–542. https://doi.org/10.1007/s00334-011-0286-2.

Lipke, P., 1984. The Royal Ship of Cheops. BAR International Series.

Lukešová, H., Palau, A.S., Holst, B., 2017. Identifying plant fibre textiles from Norwegian Merovingian period and Viking age graves: the late iron age collection of the University Museum of Bergen. J. Archaeol. Sci. Rep. 13, 281–285. https://doi.org/10.1016/j.jasrep.2017.03.051.

Lush, D., 2007. Body Armor in the phalanx of Alexander's army. Ancient World 38, 15–37.

Maier, U., Schlichtherle, H., 2011. Flax cultivation and textile production in Neolithic wetland settlements on Lake Constance and in Upper Swabia (south-west Germany). Veg. Hist. Archaeobot. 20, 567–578. https://doi.org/10.1007/s00334-011-0300-8.

Malézieux, J., 1963. le lin en France. Inf. Geogr. 27, 47–59.

Meeks, D., 1996. Navigation maritime et navires égyptiens: Les éléments d'une controverse. In: Techniques et Économies Antiques et Médiévales. Le Temps de l'innovation. Aix En Provence, pp. 175–194.

Melelli, A., Durand, S., Alvarado, C., Kervoëlen, A., Foucat, L., Grégoire, M., Arnould, O., Falourd, X., Callebert, F., Ouagne, P., Geairon, A., Daniel, S., Jamme, F., Mauve, C., Gakière, B., Bourmaud, A., Beaugrand, J., 2022. Anticipating global warming effects: a comprehensive study of drought impact of both flax plants and fibres. Ind Crops Prod 184. https://doi.org/10.1016/j.indcrop.2022.115011.

Melelli, A., Roselli, G., Proietti, N., Bourmaud, A., Arnould, O., Jamme, F., Beaugrand, J., Migliori, A., Girolami, G. Di, Cinaglia, P., Santulli, C., 2021. Chemical, morphological and mechanical study of the ageing of textile flax fibers from 17th/18th-century paintings on canvas. J. Cult. Herit. 52, 202–214. https://doi.org/10.1016/j.culher.2021.10.003.

Moseley, M., 1975. In: Cummings, A., Merlo, P. (Eds.), The Maritime Foundations of Andean Civilization.

Musselman, L., 2012. A Dictionary of Bible Plants, first ed. Cambridge University Press. https://doi.org/10.1017/CBO9781139035323.

Müssig, J., 2010. Industrial Applications of Natural Fibres: Structure, Properties and Technical Applications. John Wiley & Sons, Ltd, Chichester, United Kingdom. https://onlinelibrary.wiley.com/doi/book/10.1002/9780470660324.

Müssig, J., Hughes, M., 2012. Reinforcements: fibres. In: Reux, F., Verpoest, I. (Eds.), Flax and Hemp Fibres: A Natural Solution for the Composite Industry. JEC, Paris, pp. 39–60.

Nicholas, D., 1992. Medieval Flanders, second ed. Routeledge.

Nour, M.Z., Iskander, Z., Osman, M.S., Moustafa, A.Y., 1950. In: Cairo (Ed.), The Cheops Boats, Part I.

Ollerenshaw, P., 1999. Textile business in Europe during the First World War: the linen industry, 1914–18. Bus. Hist. 41, 63–87. https://doi.org/10.1080/00076799900000202.

Pomey, P., Poveda, P., 2018. Gyptis: sailing replica of a 6th-century-BC archaic Greek sewn boat. Int. J. Naut. Archaeol. 47, 45–56. https://doi.org/10.1111/1095-9270.12294.

Rast-Eicher, A., Thijsse, S., 2001. Anbau und Verarbeitung von Lein: experiment und archäologisches material. Z. für Schweiz. Archaol. Kunstgesch. 58, 47–56.

RCI, 2024. The Renewable Carbon Initiative. nova Institute, Hürth, Germany. https://renewable-carbon-initiative.com/.

Rougé, J., 1987. La momie contenait-elle les fragments d'une voile? Nouv. Arch. Mus. Hist. Nat. Lyon 25, 91–99.

Schenkel, W., 1994. Les systèmes d'irrigation dans l'Egypte ancienne et leurs genèse. Archéo-Nil 4, 27–35.

Sciolino, E., 2007. Following in Lafayette's Footstep, or rather, His Wake. New York Times.

Searle, G.O., 1946. Research in the flax industry. Ann. Appl. Biol. 33, 326–331.

Soriga, E., 2017. A diachronic view on fulling technology in the Mediterranean and the ancient near east: tools, raw materials and natural resources for the finishing of textiles. In:

Gaspa, S., Michel, C., Nosch, M.L. (Eds.), Textile Terminologies from the Orient to the Mediterranean and Europe, 1000 BC to 1000 AD. Zea Books, Lincoln, pp. 24–46. https://doi.org/10.13014/K2CJ8BNP.

Tallet, P., Marouard, G., Laisney, D., 2012. Un port de la IVe dynastie au Ouadi al-Jarf (mer Rouge). Bull. Inst. Fr. Archeol. Orient. 112, 399–446.

Thygesen, L.G., Eder, Æ.M., 2007. Dislocations in single hemp fibres — investigations into the relationship of structural distortions and tensile properties at the cell wall level. J. Mater. Sci. 42 (2), 558–564. https://doi.org/10.1007/s10853-006-1113-5.

Tricomi, A., 2014. Textile archaeology in roman Venetia. Archaeol. Text. Rev. 56, 78–81.

Valamoti, S.M., 2011. Flax in Neolithic and bronze age Greece: archaeobotanical evidence. Veg. Hist. Archaeobot. 20, 549–560. https://doi.org/10.1007/s00334-011-0304-4.

Valbelle, D., 1977. Catalogue des poids à inscriptions hiéroglyphiques et hiératiques de Deir el-Médineh. Documents de fouilles, t. XVI, IFAO 5001–5423, pp. 5–21.

Van Zeist, W., Bakker-Heeres, J.A.H., 1975. Evidence for linseed cultivation before 6000 bc. J. Archaeol. Sci. 2, 215–219. https://doi.org/10.1016/0305-4403(75)90059-X.

Verpoest, I., 2020. 50 Years of Composites. JEC Group, Paris. https://preview.jec-world.events/catalog/50-years-of-composites/50-years-of-composites/n1-2020. (Accessed 08 January 2026).

Verpoest, I., 2018. What is the current status of flax and hemp fibre composites? In: Flax and Hemp Fiber Composites, a market reality. JEC Group, Paris.

Viklund, K., 2011. Flax in Sweden: the archaeobotanical, archaeological and historical evidence. Veg. Hist. Archaeobot. 20, 509–515. https://doi.org/10.1007/s00334-011-0325-z.

Vogelsang-Eastwood, G., 2000. Textiles. In: Nicholson, P.T., Shaw, I. (Eds.), Ancient Egyptian Materials and Technology. Cambridge University Press Cambridge, Cambridge SE - xviii, p. 702. Illustrations; 28 cm. https://worldcat.org/title/38542531.

Vogelsang-Eastwood, G.M., Haeringen, J., 1992. The so-called boy spinner of Beni Hassan. In: Eyre, C.J. (Ed.), The Market Women of Pharaonic Egypt. IFAO, Cairo, pp. 95–96.

Whitewright, J., 2007. Roman rigging material from the red sea port of Myos Hormos. Int. J. Naut. Archaeol. 36, 282–292. https://doi.org/10.1111/j.1095-9270.2007.00150.x.

Whiting, A.G., 1943. A Summary of the Literature on Milkweeds (Asclepias Spp.) and Their Utilization. https://doi.org/10.22004/ag.econ.330497. Miscellaneous 330497.

Wikipedia, 2024. Thor Heyerdahl & Kon-Tiki Expedition. Wikipedia. https://en.wikipedia.org/wiki/Kon-Tiki_expedition.

Yang, M.F.M., Hamid, H., Abdullah, A.M., 2018. Potential use of cellulose fibre composites in marine environment—a review. In: Öchsner, A. (Ed.), Engineering Applications for New Materials and Technologies. Springer International Publishing, Cham, pp. 25–55. https://doi.org/10.1007/978-3-319-72697-7_3.

Zhao, X., Copenhaver, K., Wang, L., Korey, M., Gardner, D.J., Li, K., Lamm, M.E., Kishore, V., Bhagia, S., Tajvidi, M., Tekinalp, H., Oyedeji, O., Wasti, S., Webb, E., Ragauskas, A.J., Zhu, H., Peter, W.H., Ozcan, S., 2022. Recycling of natural fiber composites: challenges and opportunities. Resour. Conserv. Recycl. 177, 105962. https://doi.org/10.1016/j.resconrec.2021.105962.

Nano and nanoengineered composites reinforced with carbon nanotubes

1.5

Jae-Moon Jeong[1] *and Brian L. Wardle*[1, 2]
[1]Department of Aeronautics and Astronautics, Massachusetts Institute of Technology, Cambridge, MA, United States; [2]Department of Mechanical Engineering, Massachusetts Institute of Technology, Cambridge, MA, United States

1. Introduction

At the time of this writing, more than three decades of major investment and progress in nanoscience and nanotechnology have followed the initial interest sparked by Iijima's popular identification of the useful and unique properties of carbon nanotubes (CNTs) in 1991, along with others (Iijima, 1991; Monthioux and Kuznetsov, 2006; Bethune et al., 1993). Nanomaterial innovations have shown remarkable achievements, particularly in the aerospace field, which has been an important driving force for nanomaterial research since its early stages, including carbon fibre composites. Such research in the aerospace sector, particularly toward composite applications, has led to the advancement of high-end applications such as nanomodified, enhanced, and nanoengineered materials and their composites (Balasubramanian and Burghard, 2005; Iijima, 1991; Grzybowski and Huck, 2016; Kreupl et al., 2002; Koo, 2006, 2016; Ajayan et al., 2006; Chou et al., 2010; Spitalsky et al., 2010; Green et al., 2009; Thostenson et al., 2001, 2005; Coleman et al., 2006b; Ajayan and Tour, 2007; Schulte and Windle, 2007; Vaia and Wagner, 2004; Kinloch et al., 2018; Rao et al., 2018; Natarajan, 2022; Gorbatikh et al., 2016; Wardle et al., 2018; Rathinavel et al., 2021; Soni et al., 2020; Kumar et al., 2020; Mohd Nurazzi et al., 2021), particularly polymer composites comprised CNTs. CNTs are used broadly here, and we leave details of their structure, including chiral angles and various special forms (such as double-wall and pea-pod CNTs) to textbooks, for example (Dresselhaus et al., 1996). Two broadly accepted types of CNTs are single-(SWCNTs) and multi-wall (MWCNTs) CNTs, as shown in Fig. 1.5.1 (left). A pictorial comparison of CNTs to carbon fibres and their composites is provided in Fig. 1.5.1 (right). These may also be compared to graphene and other allotropes of carbon (see Fig. 6.1 in Chapter 6 of Part II in this book). SWCNTs exhibit a helical structure of a single crystalline carbon layer, now described as a sheet of graphene. MWCNTs comprise multiple concentric graphene layers, typically of larger diameter than SWCNTs (Sun et al., 1998). The MWCNT structure also enables strong interactions with polymer matrices, facilitating the achievement of enhanced mechanical properties in

The History of Composites. https://doi.org/10.1016/B978-0-443-34119-9.00007-3

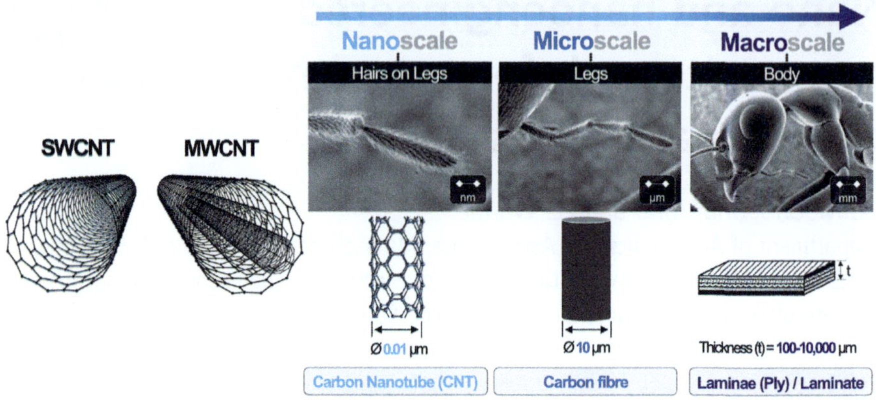

Figure 1.5.1 CNTs and the scale of things from composite and nature perspectives: (*left*) single- and multi-wall CNTs illustrated, and (*right*) scale and organisation in CNT and carbon fibre composites with reference to ant anatomy (right based on original concept by Dr. Enrique J. Garcia, MIT).

nanocomposite materials. The diameter of SWCNTs typically ranges from 0.4 to 3 nm (Balasubramanian and Burghard, 2005), while MWCNTs can reach diameters up to 100 nm. Key generalised properties of CNTs are summarised in Table 1.5.1 (Yu et al., 2000; Arnold et al., 2005; Ma et al., 2010; Yang et al., 2010; Cao et al., 2013; Laurent et al., 2010; Hexcel, 2023).

From a composite practitioner's perspective, CNTs are a rational candidate to go beyond glass and carbon microfibres as a superior reinforcing fibre to create composites with advantaged mass-specific properties, moving existing GFRP and CFRP composites towards higher specific stiffness and strength, as shown in Fig. 1.5.2. CNTs may be thought of as the ultimate version of an aerospace carbon fibre, with the crystalline carbon plane oriented along the fibre axis, giving maximum strength and stiffness in that direction. Of course, these mechanical properties are just a subset of advantaged intrinsic and scale-dependent properties of CNTs, many of which have

Table 1.5.1 Physical properties of carbon materials at room temperature: SWCNTs (Yu et al., 2000; Arnold et al., 2005; Ma et al., 2010), MWCNTs (Yang et al., 2010; Cao et al., 2013; Laurent et al., 2010), Carbon Fibre (Hexcel, 2023).

Properties	SWCNTs	MWCNTs	Carbon fibre (IM7, aerospace)
Young modulus (GPa)	320–1470	27–950	276
Tensile strength (GPa)	13–52	11–63	5–6
Bulk density (g/cm^3)	1.1–1.2	1.0–2.0	1.78
Electrical conductivity (S/cm)	10^2–10^6	10^3–10^5	10^3

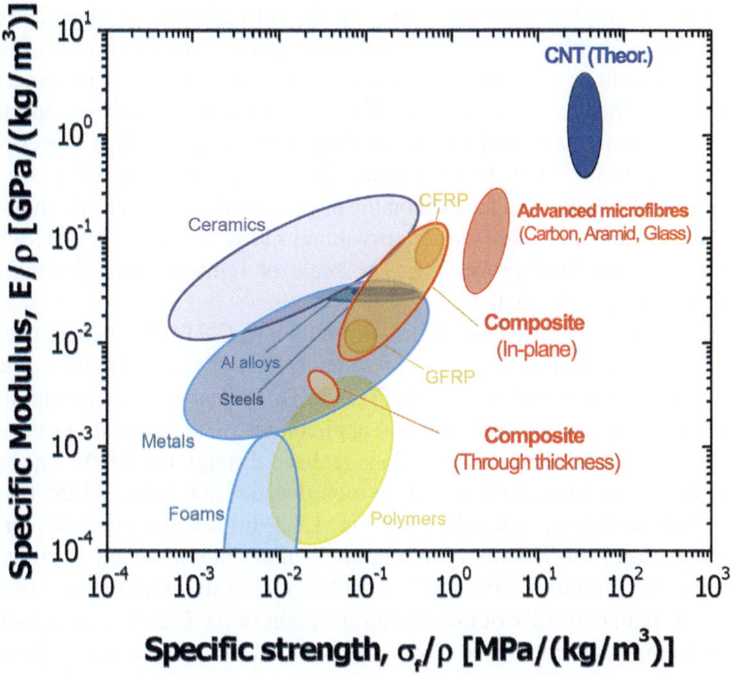

Figure 1.5.2 Specific stiffness and strength for comparing lightweight structural materials, including metal alloys, composites, and CNTs, adapted after original concept by Dr. Roberto Guzman de Villoria, MIT.

been leveraged in other industries, such as nanoelectronics, energy storage (Landi et al., 2009; Zhu et al., 2021), sensing, and human health (Akhtar and Chang, 2020; Kumar et al., 2017; Nishimoto et al., 2022; Jang et al., 2021). In the context of composites, non-mechanical properties of CNTs have been proposed to impart multifunctional attributes to composites beyond their primary mechanical function (Qu et al., 2020; Gibson, 2010; Sung et al., 2016; Vilatela et al., 2012; Kim et al., 2019), as will be discussed and highlighted briefly later in the chapter.

The specific stiffness versus strength Ashby chart (Ashby, 1992) in Fig. 1.5.2 shows why many applications have transitioned from metals to composites and why CNT-based nanocomposites may be even further advantaged. In a log-log chart inspired by Ashby, the materials in the upper right corner of Fig. 1.5.2 are the "better" materials. Advanced composites formed by combining continuous and aligned (mi-cro)fibres such as glass and carbon, with polymers as a matrix, perform better than metallic alloys in terms of mass-specific properties. If the combination of polymers and high-performance fibres delivers advanced composites, it is suspected that combining CNTs with polymers results in even better composites. It should be noted that efforts have also been made to improve out-of-plane properties as well as in-plane properties through nanomaterial addition. Matrix-dominated properties, such as inter-laminar strength and toughness and through-thickness stiffness, fall several orders of magnitude below in-plane properties for even the most advanced composites (see the "Composite through thickness" vs. "Composite in-plane" labels in Fig. 1.5.2) and can significantly benefit from nanoscale reinforcement. Employing nanomaterials for structural benefit has received considerable attention from researchers and industry since the 1990s, with many challenges identified and a general lack of success in early attempts (Chou et al., 2010; Kinloch et al., 2018; Van Noorden, 2011; De Volder et al., 2013). Early challenges focused on the dispersion of nanomaterials in polymers, fighting the fundamental high surface-to-volume ratios of nanomaterials, which increase viscosity and limit processing. The scale of reinforcement (having enough CNTs) was also an early challenge.

With many similarities between carbon fibres and CNTs, the experience gained from the development of carbon fibres provides valuable insight into the current state of CNT-comprised composites (refer to Fig. 1.5.3). The early development of carbon fibres focused on aerospace (and defence) applications, with many military and space projects as well as commercial applications realised through the 1970s. The adoption of carbon fibres in commercial aerospace took decades and followed the S-curve of aerospace structural composites shown in Fig. 1.5.3. It took many decades for carbon fibres to transition from laboratory to a large % of structural weight fraction, and as one example, the initial S-curve for U.S. military aircraft started in the 1970s, and by 1990, approximately 40% of the structural weight of the F-22 was composite (Harris, 2000). Composites comprised CNTs are expected to follow a similar process and trend (see notional blue S-curve in Fig. 1.5.3), but perhaps with faster adoption due to the experience of carbon fibre composites, or slower as CFRP is a dominant structural technology, as discussed later in the chapter.

While Fig. 1.5.3 shows the carbon fibre and carbon nanotube S-curves for aerospace composites, CNTs are much older than their identification in 1991. The popular identification of CNTs as a very interesting and novel material certainly belongs to Iijima (1991), but CNTs were observed in electron microscopy and other studies dating to at least the 1950s (Wikipedia, 2024b). Further, it is likely that human-driven processes knowingly or unknowingly created CNTs. A documented example is that CNTs have been found in 17th-century Damascus steel, demonstrating the unconscious utilisation of such materials by humans (Smith, 1960; Reibold et al., 2006). It has been suggested that the superior strength and durability of Damascus steel may derive in part to the presence of CNTs, or CNTs may just be a by-product of the steel process. Natural processes may also create CNTs, for instance, recent studies have suggested the possibility that carbon nanostructures similar to CNTs could form naturally form in space (Abbe et al., 2019; Chen and Li, 2019). Thus, CNTs certainly predate the 1990s, but that early history is still being clarified, and the purposeful use of CNTs in composites began, as shown in Fig. 1.5.3, in the 1990s. To better understand the developmental stages of CNTs, the Gartner hype cycle concept seems relevant. The hype cycle represents the technology development process from the initial recognition of new technology to the eventual productivity stage, and similar patterns exist in the research and development of CNTs, as noted by others and shown in Fig. 1.5.4 focussing on composites. While we will return to some of the milestones in Fig. 1.5.4 later in the chapter, it is useful here to identify that polymer nanocomposites

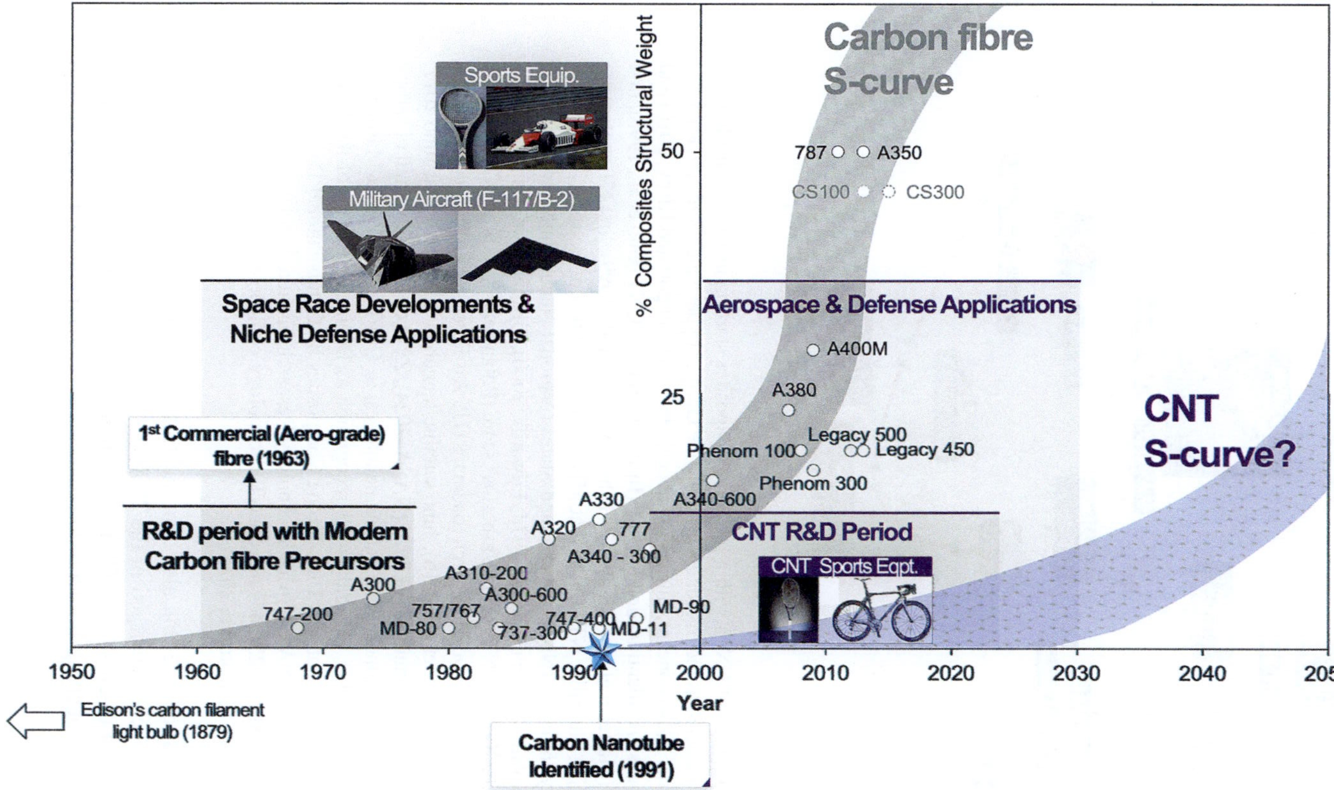

Figure 1.5.3 Development of carbon fibres and CNTs in commercial aerospace transport aircraft establishing technology S-curves (Rogers and Williams, 1983), after (Gorbatikh et al., 2016; Wardle et al., 2018) using industry sources and data (Harris, 2000; Black, 2011; Croft, 2008; Gerzanics, 2009; HEXAGON, 2024) note that Bombardier's CS100 and CS300 (now Airbus A220 variants) have been reported as 46% composite, but the media source is somewhat unclear, so these values are included but greyed out.

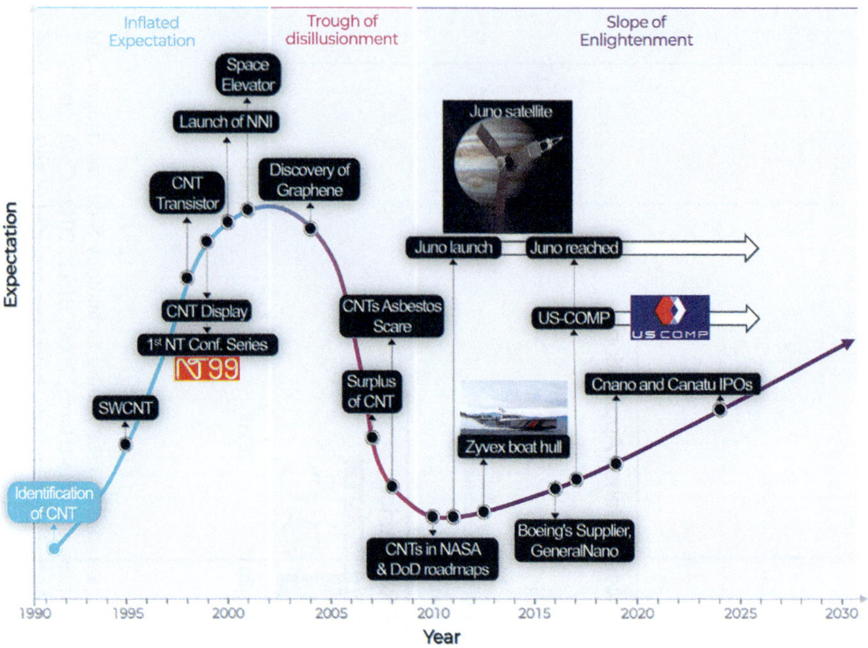

Figure 1.5.4 Timeline of CNTs with a focus on composites starting in 1991 with Iijima's seminal paper that acted as the trigger event, using the Gartner hype cycle (Fenn and Raskino, 2008; Gartner Inc, 2024) concept, Illustration concept from (Davenport and Kirby, 2016) and timeline references are: Identification of CNT (Iijima, 1991), SWCNT (Guo et al., 1995), first CNT transistor (Martel et al., 1998), first CNT display (Choi et al., 1999), first NT conference in series, (ongoing annually; NT24 hosted at MIT in 2024), launch of National Nanotechnology Initiative (NNI) in the United States (National Nanotechnology Initiative, 2000), space elevator concept (Clarke, 1978; Edwards, 2000), Discovery of graphene (Novoselov et al., 2004), "CNTs are like asbestos" scare (Poland et al., 2008), CNTs in NASA & Department of Defence (DoD) research roadmaps (US Air Force, 2010; National Research Council et al., 2012; Betser et al., 2016), Juno satellite launch in 2012, arrival at Juno 2017, and extended mission continuing as of 2025 (NASA, 2024), Zyvex Marine boat hull of carbon fibre reinforced plastic enhanced with carbon nanotubes, or CFRP-CNT (Zyvex Marine, 2012; NASA Spinoff, 2017), GeneralNano's recognition as Boeing's Supplier-of-the-Year (Duvelius, 2016), The Institute for Ultra-Strong Composites by Computational Design (US-COMP) was awarded as a NASA Space Technology Research Institute,[1] and continues as of this writing (US-COMP, 2025), advanced CNT Technology companies Cnano (Bloomberg, 2025) and Canatu Plc (CanatuPlc, 2024).

dominated the first decade (but also continue today), that yarns and sheets and other hierarchical approaches emerged in the 2000s (and still continues today).

We start with a discussion of CNT-based nanocomposites focussing on polymer matrices but also touching on ceramic-, and metal-matrix nanocomposites as well,

[1] N.B. author BLW was part of this Institute

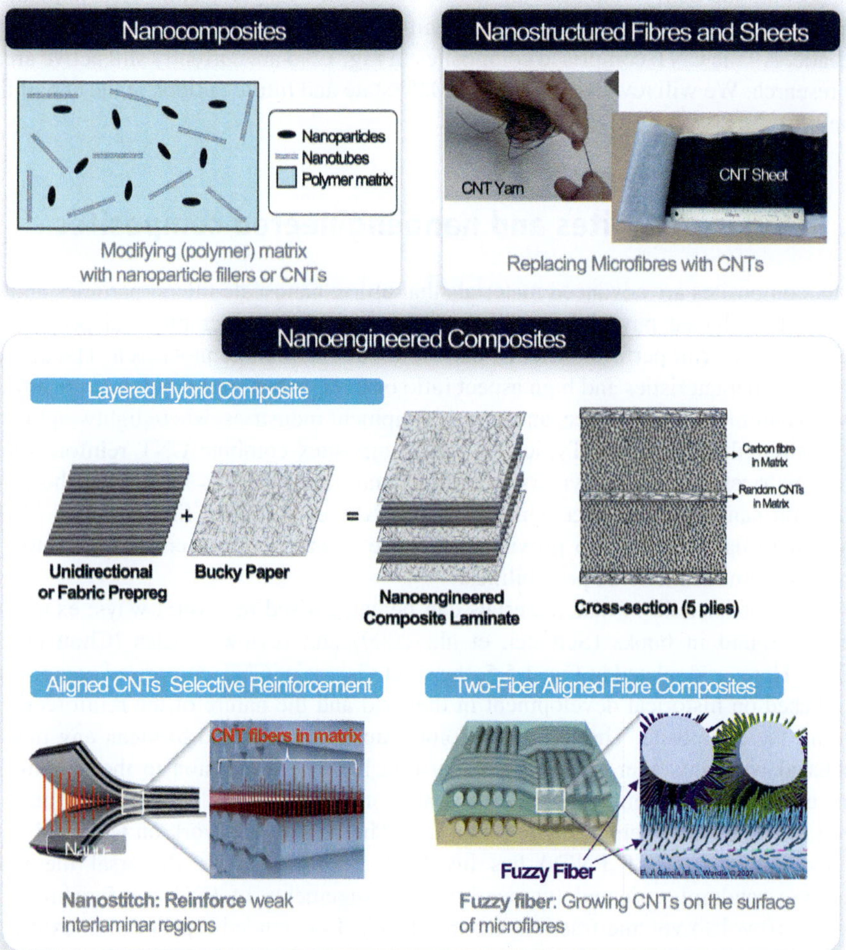

Figure 1.5.5 Broad categories of composites comprised of carbon nanotubes: Nanocomposites, CNT yarns and sheets, and hybrid or nanoengineered composites.

organisation of CNTs into yarns and sheets for reinforcement, and finally hierarchical composites comprised of CNTs within nanoengineered composites as shown in Fig. 1.5.5. Organisation of CNTs to create yarns or sheets from (sometimes) aligned CNTs as a reinforcement architecture as shown in Fig. 1.5.5 is a principal effort in hierarchy. However, for these to be effective as composites, the yarns and sheets need to be bound together with a matrix, or they remain a yarn or dry fabric incapable of taking compression and shear as required by structural composites. Hence, sheets and yarns of CNTs supported by a matrix are a nanocomposite with some order or arrangement of CNTs, not dissimilar to woven or collimated (carbon) fibre architectures. Figs. 1.5.4 and 1.5.5 can be linked in the following way: the first decade and into the second had heavy emphasis on nanocomposites, with yarns and sheets and nano-engineered composites having more significant emphasis beginning in the second decade. All the CNT-comprised composites in Fig. 1.5.5 are currently still active areas of research. We will revisit the current (2025) state and future outlook at the end of the chapter.

2. Nanocomposites and nanoengineered composites

Nanocomposites are advanced materials that utilise nanoscale fillers (particles and/or fibres) to enhance physical, mechanical, electrical, and other physical properties, achieving superior performance that conventional materials cannot reach. The unique physical characteristics and high aspect ratio of CNT "fibre" reinforcement are attractive to automotive, aerospace, and sports equipment industries, where lightweight and high strength are crucial. Typically, nanocomposites combine CNT reinforcement with matrices, such as polymers, ceramics, and metals. CNTs have been the most common nanoscale reinforcement/filler (Spitalsky et al., 2010; Sun and Sue, 2010) studied to date. CNTs also provide metal-like electrical and thermal conductivity, providing multifunctional possibilities.

The broader field of nanomaterials can be categorised in several ways; examples can be found in books (Schodek et al., 2009) and review articles (Chou et al., 2010). Here, as depicted in Fig. 1.5.5, the subdivisions of CNT-comprised composites are based on historical development in the field and the nature of the reinforcement within the composite. The term nanocomposites have emerged to mean any matrix material with any filler or reinforcement that has one dimension in the nanometre range. As we focus here on CNTs, then nanocomposites have CNTs in a host matrix, with the focus here being polymer matrices. The bulk of the work on CNT polymer nanocomposites (CNT PNCs) has involved mixing or other dispersal methods, yielding randomly-oriented (or nearly-so) arrangements of CNT reinforcement at low (<10 vol%) volume fraction. Giving the CNTs extended aspect ratio, texturing, or aligning them in some fashion is logical from an engineering perspective of controlling and maximising contribution to bulk properties. Several groups and companies (such as Huntsman Corporation and DexMat among others) have focused on texturing CNTs by creating yarns and sheets of CNTs, sometimes with alignment of the CNTs, as dry-fibre preforms for creating composites. The yarn fibres and sheets would then have a matrix to surround the CNT "fibres" much like carbon microfibres reinforce polymer matrices in CFRP. However, it has proved challenging to introduce matrices or matrix precursors, including polymers, into the CNT yarns and sheets. Nonetheless, since the idea is to introduce a binder or matrix into the yarns and sheets, these would then be textured or aligned-CNT nanocomposites. Thus, CNT reinforcement can be classified (see Fig. 1.5.5) as nanocomposites, nanostructured fibres, and sheets that are seen as replacements for carbon fibres or hybrid or nanoengineered composites that combine both or introduce novel concepts and hierarchical architectures.

Section 2.1 will focus on polymer-matrix nanocomposites, Section 2.2 will consider fibres and sheets using CNT nanofibres assemblies, and Section 2.3 will discuss nanoengineered (or hybrid CNT and microfibre) composite systems, which create hierarchical and sometimes multi-layered structures by combining CNTs with existing materials (mainly fibre-reinforced plastic (FRP) composites). Most of the extant CNT composites work has been toward structural properties, but of course, other physical properties are also of interest, such as electrical and magnetic properties for EMI shielding (Shaver et al., 2009; Camponeschi et al., 2007; Oliva-Avilés et al., 2011). These non-structural physical properties, primarily studies on electrical and thermal properties (mostly electrical conductivity, see early review (Bauhofer and Kovacs, 2009)) are not of primary focus here, but are broadly in the space of "multi-functional properties/applications" of CNT-comprised composites. An early review article on this topic is instructive (Gibson, 2010), and many useful composites with advantaged attributes have been identified, such as ionic and electrical conductivity for actuators, supercapacitors, and battery electrodes, for example, (Kinloch et al., 2018; Pandey and Thostenson, 2012; Zhang and Peijs, 2020; Veedu et al., 2006; Behabtu et al., 2013), as well as EMI shielding and sensing capabilities (Wu et al., 2022; Vargas-Bernal and Tecpoyotl-Torres, 2021; Zhong, 2012; Shin et al., 2021).

2.1 Nanocomposites reinforced with CNTs

We focus in this section on polymer nanocomposites, with brief mention at the end on metal- and ceramic-matrix nanocomposites. CNTs in a polymer nanocomposite (PNC) may be considered as a filler, not dissimilar to a microparticle or nanoparticle, or as reinforcing fibres more akin to unidirectional and woven microfibre CFRP composites. While there are many nanoparticles that may be considered, including Buckyballs and graphene, CNTs in polymers to date have garnered great interest. While much of the early work (and it continues) focused on mixing and dispersing approaches that treat the CNTs like particulate fillers, the high aspect ratio (AR) can be 10^6 or more (even a short 10 μm long 10 nm diameter CNT give AR = 1000) of CNTs suggests they should be treated as long-fibre composites, such as GFRP and CFRP, rather than as fillers. Taking the CNT then as a fibre, we make several observations based on comparisons to long-fibre advanced CFRP composites:

(i) CNTs as fibres have low density and high crystallinity (orientation of graphite crystal planes), relative to carbon microfibres. Carbon fibres have a graphitic alignment of varying degrees and crystal sizes, but CNTs (ignoring defects) may be considered ideal or pristine carbon fibres. CNTs are also hollow fibres, so the centre of the CNT does not contribute to mechanical properties, but of course, this makes the bulk density of CNTs lower, improving specific properties as shown in Fig. 1.5.2. Interestingly, hollow carbon microfibres have been developed (e.g., see recent contribution here (Shirolkar et al., 2021))

(ii) The scale of CNTs gives these fibres ~1000X more surface-to-volume than micro-scale carbon fibres. This means there is much more surface area for a given loading (composite vol%) for CNTs relative to carbon fibres, with implications for surface-dominated properties, processing, and structure. On the last, structure, it has been established experimentally and theoretically, that hard nanostructures like CNTs can alter the polymer, leading to "interphase effects." Thus, unlike CFRP, it is expected that some polymer alteration local to the CNTs could occur, dependent on the surface of the CNT and the polymer, for example, (Coleman et al., 2006a, 2006b; Gohn et al., 2020; Yue et al., 2015; Taghizadeh and Favis, 2013). Some early work with PNCs, particularly thermoplastics, interpreted the reinforcement effect as being solely due to nanomaterial reinforcement. However, in many cases, the nanomaterials actually served as nucleation sites, crystallising a significant fraction of the polymer, thereby increasing the PNC modulus. The extent and degree of polymer alteration remain active areas of research and depend on details of the nanostructured reinforcement (e.g., molecules adhered to the CNT surface), the specific polymer (broadly, thermoplastic or thermoset) (Schadler et al., 2007; Winey and Vaia, 2007; Winey et al., 2007), and processing parameters of the PNC. Thermoplastics as a class appear to be more sensitive than thermosets to interphase effects, with nanostructures potentially initiating significant crystallinity in some thermoplastics (Ryan et al., 2006; Coleman et al., 2006a)

(iii) Fibrer-matrix (here fibre-polymer) interactions are usually characterised in two ways:

- **i.** The strength of the bond between the CNT and polymer is of critical concern, as in carbon fibre composites this strength, usually quantified as interfacial shear strength (IFSS), controls the toughness of the CFRP. IFSS in microfibre CFRP is engineered to be of intermediate strength so that cracks in the polymer or composite propagate at this interface rather than through the carbon microfibres, giving the composite toughness (Lachman et al., 2012; National Research Council et al., 2005). IFSS is typically ~50 MPa (for epoxy) (Wang et al., 2018; Wu et al., 2020; Montes-Morán et al., 2001) and IFSS for CNTs and epoxy is of similar magnitude or higher (Zhandarov and Mäder, 2015; Mani and Sharma, 2022; Kinloch et al., 2018; Wan et al., 2005; Wagner et al., 1998), again indicating good compatibility of CNTs with traditional polymers.
- **ii.** The interaction between a reinforcement (CNT "fibre" here) and the polymer matrix is further quantified via a shear lag parameter that allows one to understand how quickly load is transferred around a broken fibre and then back into that fibre by the surrounding matrix and/or composite. Conversely, it allows one to determine a critical length of a short fibre for full load transfer into that fibre. In typical microfibre FRP composites, the critical length is on the order of 10 fibre diameters, or 50–100 microns given 5–10 micron diameter carbon fibres. Several studies, both experimental and modelling, have indicated that the critical length scaling for CNTs is similar to carbon microfibres, indicating efficient load transfer, with sub-micron critical lengths of ~10 CNT diameters, or sometimes slightly larger (Papageorgiou et al., 2020; Wan et al., 2005; Wagner et al., 1998). This compares with a normalised shear lag distance of ~10,000 for graphene, taking graphene thicknesses of 0.34 nm (Gong et al., 2010; Papageorgiou et al., 2020).

Both IFSS and shear lag will be a function of the bonding between the polymer and the CNT, especially at CNT defect sites. Such bonding in CFRP is controlled by fibre sizing, but typically CNTs are not sized, although their surfaces can be modified in various ways to alter polymer-CNT interactions, for example, recent work (Park et al., 2024; Liu, 2005; Khare et al., 2014; Zhang et al., 2008) has selectively damaged CNTs to increase the binding sites available to the polymer and thus tailored CNT-polymer interaction.

(iv) Composites take advantage of fibre properties by including high vol% of fibres, in unidirectional CFRP ~60 vol% and woven CFRP ~50 vol%. Achieving such high "loadings" (or packing) or simply vol% with CNTs in a PNC has generally been quite challenging (Kinloch et al., 2018; Rao et al., 2018; De Volder et al., 2013; Bulmer et al., 2021; Islam et al., 2022). High vol% is desirable certainly for mechanical properties such as stiffness and strength, but also for electrical and thermal conductivity and likely other properties. The 50–60 vol% of existing advanced CFRP composites maximise the CF vol% that can be achieved in bulk manufacturing. Rule of mixtures (RoM, see discussion in Chapter 6) calculations show the dependence of many properties on vol%. Mixing techniques dominant in the first decades of CNT research generally found viscosity to be too high to process PNCs above 5 vol%. Further, achieving high vol% requires extremely small (nm-scale) inter-CNT spacings that are on the order of the polymer chain characteristic lengths, such as the radius of gyration (Schadler et al., 2007; Winey and Vaia, 2007; Winey et al., 2007), implying that any interphase effects could dominate the matrix at high CNT vol%.

(v) CNTs are of course not straight fibres, and indeed they are not collimated at all relative to UD CFRP (or even woven CFRP). The mechanics of woven and wavy microfibre composites (historically called micromechanics, see Chapter 4.3 in Part II of this book) have been studied in great detail to modify standard RoM calculations that presume a straight reinforcing fibre (see, e.g., (Mital, 1996; Miravete, 1999)). The impact is significant, for example, typical woven CFRP has ~50% of the stiffness of equivalent UD CFRP).

Thus, polymer nanocomposites show many characteristics of microfibre CFRP, with positive polymer-CNT interactions, but difficulties in processing to obtain high vol% of CNTs, achieving alignment (with minimal waviness), and controlling polymer process-structure interactions. Work in the first decade focused on the synthesis of CNTs and studying physical properties, with some applications of CNTs to composites primarily through mixing approaches in the first two decades of Fig. 1.5.4, although it continues somewhat still today (De Villoria and Miravete, 2007). This led to the understanding that processing is a significant challenge (points ii, iv, and v above) in CNT-based nanocomposites, and has been extensively covered in research articles and textbooks, for example, (Koo, 2006). Numerous CNT processing approaches have been introduced, and continue to be, for CNT-polymer composites, including blends or masterbatches (e.g. (Cheaptubes, 2024),) and electrospinning (e.g. (Ko et al., 2006; Dzenis, 2004; Haggenmueller et al., 2000; Liu et al., 2007),), among others. These approaches are applied to many types of nanoparticles and nanofibres, including CNTs. Chemical processing, including solution intercalation, in situ polymerisation, and emulsion polymerisation, aims to control CNT distribution, and solution intercalation allows polymers to infiltrate between CNTs to enhance interaction (Li and Ishida, 2003; Shen et al., 2002), and in situ polymerisation, including emulsion polymerisation, involves the insertion and polymerisation of monomers in the presence of CNTs (Maneshi et al., 2011; Zhang et al., 2009). However, mechanical methods dominate the literature, such as roll milling, melt intercalation, and high-shear mixing, which leverage heat and shear forces to improve nanoparticle dispersion and polymer matrix homogeneity (Paiva and Covas, 2016; Rosca and Hoa, 2009; Seyhan et al., 2009; Li and Shimizu, 2009; Zunjarrao et al., 2006; Yuan et al., 2014). These mechanical methods enhance polymer matrix homogeneity and achieve dispersion, albeit at low vol% for CNTs.

As a general finding, CNTs (as high aspect ratio fillers) are more difficult to process via these methods than particles. It is interesting to note that many of these methods, and even traditional polymer fibre production methods, have been revisited with CNTs (Parnian, 2022; Hisham et al., 2024; Hendrixson, 2021) as additive manufacturing has received renewed focus in this century. Fused-filament fabrication (FFF) with thermoplastic polymers has been adapted by many groups to work with CNT-comprised nanocomposite feedstocks.

Last, we consider non-polymer matrices for our nanocomposites just briefly. Much less work has been done in this area, and given the typical processing routes of metals and ceramics, significant processing challenges exist even beyond those for polymer matrices. Fortunately, review articles have appeared summarising advances and challenges for both matrices (Sundaram et al., 2018; Chan et al., 2021; Bogue, 2011; Tjong, 2009; Ahmad et al., 2015; Ramachandran et al., 2022a; Bakshi et al., 2010). CNT-reinforced metal nanocomposites leverage the superior mechanical and thermal properties of CNTs to significantly enhance the strength, hardness, and wear resistance of metal matrices, making these composites widely applicable in fields where durability is crucial (Bakshi et al., 2010). The powder metallurgy method, for instance, involves mixing metal powders with CNTs followed by sintering, which it markedly improves strength and hardness (Kim et al., 2007; Ghorbani et al., 2018; Goh et al., 2005). Melting and solidification processes, suited for low-melting metals like magnesium, produce metal-CNT composites with enhanced heat resistance and mechanical strength, despite potential CNT damage during solidification (Li et al., 2009; Han et al., 2015). Thermal spray techniques form CNT-reinforced coatings with improved wear resistance and thermal conductivity (Fauchais et al., 2001; Laha et al., 2007; Cho et al., 2012). Electrochemical deposition enables the co-deposition of CNTs with metals, significantly increasing hardness and wear resistance, yielding improved mechanical properties (Tao et al., 2017; Arai, 2021; Praveen and Venkatesha, 2009; Arai et al., 2008). Such CNT-reinforced metal nanocomposites show substantial promise for applications in aerospace, electronic packaging, and automotive industries. By contrast, CNT-reinforced ceramic nanocomposites remain relatively underexplored, yet show substantial promise (Ramachandran et al., 2022a). CNTs may address the inherent brittleness of ceramics and significantly improve mechanical and thermal properties, vital for sustaining strength and durability under extreme conditions such as high temperatures (Estili et al., 2010; Ramachandran et al., 2021a, 2022b). Key fabrication methods include sol-gel, colloidal routes, powder metallurgy, and in-situ growth (Almeida et al., 2013; Silva et al., 2012; Sharma et al., 2019; Arunkumar et al., 2021; Gojny et al., 2003; Ramachandran et al., 2021b). The in situ growth technique directly grows CNTs on ceramic matrices, enhancing mechanical properties by reducing dispersion steps (Chen et al., 2015; Sun et al., 2012). These reinforced ceramic nanocomposites hold potential for applications in aerospace, automotive, and high-durability electronics (Gohardani et al., 2014).

2.2 CNT yarns and sheets

In the second decade in Fig. 1.5.4, macroscopic assemblies of CNTs (upper right in Fig. 1.5.5) began to emerge for various applications, including composites. Note that creating nanocomposites via mixing etc. continues through this period as well and to this day. CNT-comprised yarns and sheets are a logical extension from existing microfibre composites, where tows are woven into dry-fabric preforms and prepreg sheets are formed from tows of individual microfibres. These yarn and sheet efforts were towards the basic idea that long CNTs, organised in an aligned geometry along the fibre or sheet axis, could outperform traditional composites with the CNT yarns and sheets leveraging individual CNT properties to improve on advanced microfibres and their composites, that is, better "fibres" would make better advanced composites (see Fig. 1.5.2). Through the 2000s and continuing till now, assemblies of discontinuous (up to millimetre-scale) CNTs in yarns and sheets are created via several techniques. One of the earliest to emerge commercially was floating-catalyst CNT growth and collection notably from NanoComp Technologies in the U.S. and Prof. Alan Windle's group at Univ. of Cambridge in the UK (Li et al., 2004; Lashmore et al., 2009). These methods allow for bulk scale production of CNTs and formation into yarns and sheets via collection methods and various post-processing techniques, e.g (Tran et al., 2016a, 2016b; Liu et al., 2016). An early and well-publicised application success involved NanoComp Technologies and Lockheed Martin (see Fig. 1.5.6 left and middle) collaborating to put CNT sheet material (Rawal et al., 2013) on the Juno exploration mission. NASA's Juno mission, which was launched in 2011 (Meador et al., 2010), is now conducting an extended (beyond-original-mission) exploration of Jupiter and its moons (NASA, 2024; Wikipedia, 2024a). The CNT sheets provide electrostatic discharge protection for the spacecraft, a function usually provided by metal foils for the spacecraft.

Dispersing CNTs in a solution and depositing them, for example, (Rinzler et al., 1998), represents an alternative and effective method for obtaining sheet-like CNT

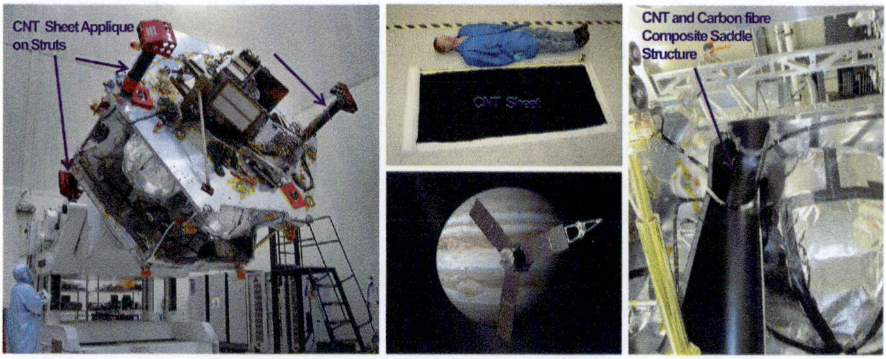

Figure 1.5.6 CNT sheet and hybrid CNT carbon fibre composites on Juno satellite: (*left*) CNT sheet applique application areas on Juno spacecraft, (*middle*) scale of CNT sheet and artist's rendering of Juno satellite, (*right*) saddle structure installed on Juno (Rawal, 2018).

assemblies. CNT sheets are sometimes referred to as Buckypaper although the original use of the term referred to solution deposition techniques. An alternative technique for creating CNT yarns is by spinning and electrospinning; exemplary is vertically-aligned CNTs grown on a substrate followed by CNT yarn is collection via different spinning and collection processes, for example, (Zhang et al., 2004; Dzenis, 2004). Recently a specialised conference series, the *International Workshop on Multi-Functional Nanocarbon Fibres,* has emerged with two meetings thus far (MNFConference, 2018). As noted in the introduction, the dry CNT yarns and sheets need to be combined with a polymer matrix to form a nanocomposite, and such has proved difficult. An example is a recent effort by NASA towards creating better composites from CNT yarns, through computational design (US-COMP) (Odegard, 2017) and/or experiments, for example, (Odegard et al., 2018; Kim et al., 2023). In the next section, we will see examples of yarns and sheets as elements in hybrid, or nanoengineered, composites.

2.3 CNTs in nanoengineered composites

Nanoengineered composites are created in various ways and hierarchical architectures by integrating nanoscale elements into traditional macroscale materials, forming hybrid composites that enhance performance and add new functionalities (see some examples in Fig. 1.5.7). Such composites, nearly always combined with existing microfibre-based fibre-reinforced plastics (FRP) such as CFRP, leverage the unique properties of nanomaterials such as carbon nanotubes (CNTs) to improve mechanical strength, toughness, and durability, especially for high-load applications in the aerospace sector. Work in this area emerged in the 2000s with a key review article on hierarchical composites from 2010 (Qian et al., 2010b). As depicted in Fig. 1.5.5 (bottom), nanoengineered composites combine CNTs and traditional composites in different, hybrid, ways. Some examples are CNT-doped polymer matrices (such as masterbatches mentioned earlier) that are infused into traditional dry microfibre weaves, for example, (Aravand et al., 2014, 2016; Ma et al., 2021; Kim et al., 2009; Yourdkhani et al., 2018), replacing carbon fibres with hybrid CNT-comprised carbon fibres (through incorporating CNTs into polymer precursor fibre before pyrolysis (Andrews et al., 1999; Chae et al., 2005)), and two-fibre (CNT and microfibres) hierarchical systems such as nanostitching, fuzzy fibres and other architectures. Such hierarchical composites began to be realised experimentally in the 2000s, and work on them continues today. Some work had been bio-inspired and biomimetic, drawing from natural systems that have many levels of hierarchy, such as nacre and bone. An important milestone in nanoengineered composites is also achieved in the Juno mission (see Fig. 1.5.6 right), where Lockheed's Applied NanoStructured Solutions (ANS) group provided a hybrid carbon fibre and CNT composite saddle structure on the Juno satellite as well (Rawal, 2018), an accomplishment much less well known than the CNT sheet appliques. Here, we highlight some non-exhaustive examples of hybrid nanoengineered composites comprised of CNTs that have some organisation/alignment that parallels and pairs with aligned microfibres in traditional advanced composites (Wardle et al., 2018; Gorbatikh et al., 2016), that is, we do not focus on facile approaches that give random distributions of

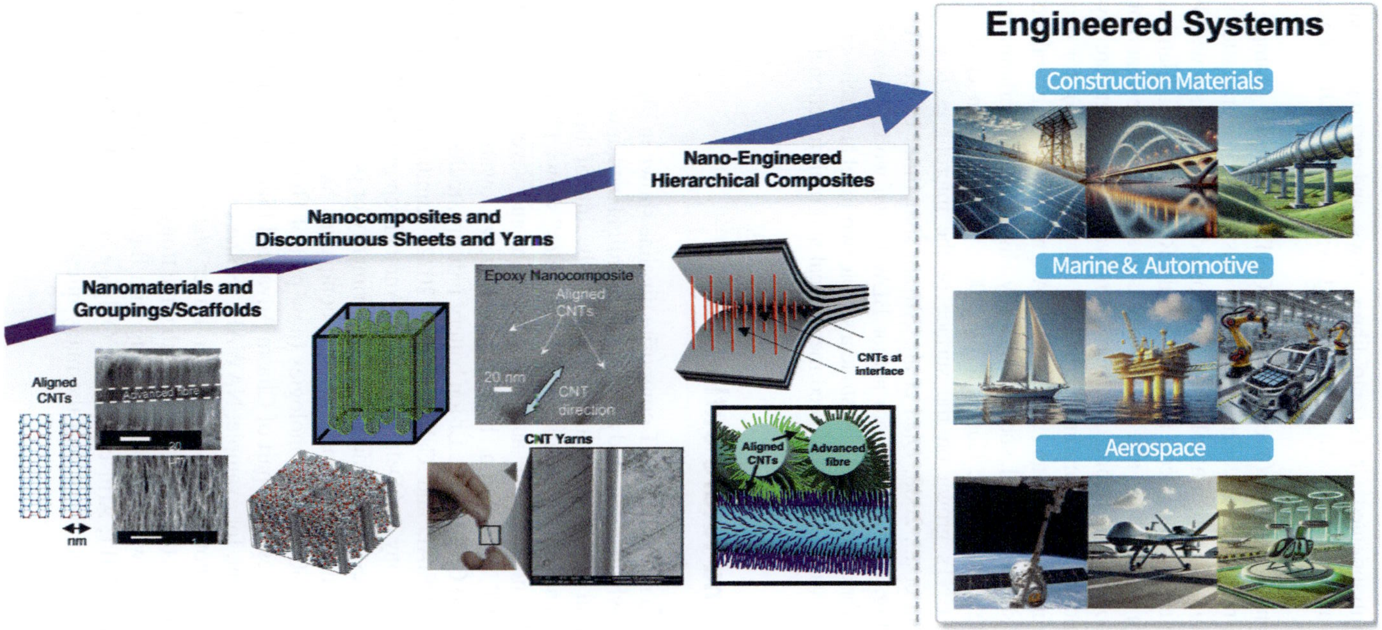

Figure 1.5.7 From nanostructure to engineered systems: Scaling and scaffolding approach to create improved engineered systems, after concept from (Wardle et al., 2018).

CNTs, such as mixtures forming masterbatches mentioned above, or deposition or spraying randomly-oriented CNTs onto fibres, tows, and prepregs (Thakre et al., 2011; Bekyarova et al., 2007; An et al., 2018; Lavagna et al., 2018; Yousefi et al., 2022).

The first aligned two-fibre (microfibres and CNTs) composite example combines impregnated yarns/sheets (see section 2.2) and existing FRP composites simply by layering or laminating, essentially having CNT-based plies laminated with FRP plies, similar to combining metal sheets and traditional FRP in fibre metal laminate (FML) composites such as GLARE and TiGr. In this way, the (somewhat horizontally-aligned) CNT polymer nanocomposite plies using yarns and sheets are layered with traditional composites plies (unidirectional and woven fibre plies), for example, (Li et al., 2018; Xu et al., 2016; Wang et al., 2015). Aligned CNTs in several architectures are more obviously hierarchical, such as vertically-aligned CNT interlaminar reinforcement (e.g., Zeng et al., 2010; Garcia et al., 2008; Kalfon-Cohen et al., 2018) and radially-aligned CNTs grown on the surface of the microfibres (e.g., Yamamoto et al., 2009; Qian et al., 2008; Qian et al., 2010a), sometimes called "fuzzy fibres," both shown in Fig. 1.5.5 lower from author BLW's own work. These aligned-CNT and microfibre hybrid composites illustrate a simple two-level hierarchy of fibre scales, and have been shown to have enhanced mechanical and multifunctional performance.

3. Trends and future outlook

It is worthwhile to revisit the CNT composite timeline shown in Fig. 1.5.4 and assess the current state, which is depicted as being on an enlightened upward slope, but much less than the initial hyped period in the 1990s. At the time of this writing, an internet search of "carbon nanotube manufacturing companies" shows 13 companies (Misaki, 2024), with ~10 known to the authors in addition, and several large corporations which have CNT production businesses, largely towards providing powder (randomly-aligned) CNTs to the Li-ion battery industry, including LG chem, Kumho Petrochemical, OCSiAl, and Cabot (Chem, 2022; Kumho Petrochemical, 2025; OCSiAl, 2020; Anthony and Delahunt, 2020). Another report lists 68 total companies (Tracxn Technologies, 2024), many working primarily or partially in the composites market. Several of those composite-aligned companies have been around since the 2000s, notably NanoComp Technologies (now part of Huntsman), Nanocyl, Southwest Nanotechnologies (now part of Chasm Advanced Materials), and Tortech, while others are more recent, for example, NAWAH.[2] Some succumbed during the "trough of disillusionment," notably Baytubes from Bayer Materials Science. The demise of Baytubes was perhaps the strongest signal that the trough was real, as many academic and corporate groups were actively working with Baytubes at the time, for example, (Loos and Schulte, 2011; Armstrong, n.d.; Quigley et al., 2014; Müller et al., 2011). As well, the field remembers well the many questions related to CNTs during the

[2] N.B. author BLW has licenced technology and research partnerships with NAWAH

"CNTs are like asbestos" scare in 2008 (Poland et al., 2008). Fortunately, exposure and toxicity work relatively quickly to address the scare from a proper environmental health and safety perspective. Many positives happened to recover from the trough, including the continuation of academic meetings on the science and application of nanotubes (NT Conference Series) which continues today (NTConference, 2025), GeneralNano being named Boeing's Supplier-of-the-Year in 2016 (Duvelius, 2016), the launch and mission completion of the Juno satellite with CNTs (see Fig. 1.5.6), CNTs being identified in all major research plans in the U.S. (PCAST, AFOSR, NASA, etc. (US Air Force, 2010; Roco et al., 2011; National Research Council et al., 2012)), an entire MRS Bulletin special issue dedicated to nanomaterials in space applications (Arepalli and Moloney, 2015) with many CNT examples, and a survey of the 100 most influential materials scientists from the first decade of this century showed ~80% cited their primary research area as nanotechnology (Anon, 2011). Furthermore, all the leading journals in composite materials during the 2000s strongly featured works with CNT reinforcement, and review articles on CNT-comprised composites began to appear. The application field was further buoyed by companies producing CNTs commercially and two stock market IPOs for Cnano in China and Canatu in Finland (Bloomberg, 2025; CanatuPlc, 2024). In aerospace, major OEM programs were running in the early 2000s as the trough was reversed, including Airbus/EADS UTOPIUM project, Boeing's Atoms to Airplanes initiative, and Lockheed Martin's Applied NanoStructured Solutions (ANS) entity (see contribution on Juno spacecraft above), now part of Cabot Corp (Caron, 2019). Given all of this, it can be argued we are on the slope of enlightenment for CNT-comprised composites in Fig. 1.5.4.

We have seen CNTs in space on NASA's Juno mission (see Fig. 1.5.6 and related discussion), and they have been tested on NASA MISSES missions (De Groh and Banks, 2017, 2021) in space-environment exposure studies and appeared as CNT sensors for air quality monitoring and trace gas detection on the International Space Station (ISS) (Li and Lu, 2009) as well. So what about the notional S-curve in Fig. 1.5.3 for CNT-composites in aircraft, as compared to the established S-curve for traditional composites? If we consider an S-curve on the same pace as carbon fibre composites, as drawn in Fig. 1.5.3, we would be in the beginning stages with no weight% of CNTs reported in commercial aircraft (as was true in the 1950s for carbon fibres). Yet we now know that aerospace and defence was utilising carbon fibres during the space race and early public awareness was built from sporting goods and racing introductions of (lower-grade than aerospace) carbon fibre composites. Similarly, for CNT composites, we see many research programs (e.g. Carter et al., 2010; McSweeney, 2025; Shah et al., 2012; Pirtle et al., 2019), toward flight and space applications that underly perceived programs with CNT composites in the engineering community. We know of a number of specific CNT composites that have flown, that is, they are flying: (i) Lockheed Martin's F-35 fighter jet has been widely reported to have CNT composites in a fairing (Trimble, 2011), (ii) CNT composite overwrapped pressure vessel (COPV) that was flown on a sounding rocket test in 2017 (Biaggi-labiosa et al., 2018), (iii) CNT-nanocomposite crack sensors, and (iv) ice protection systems (IPS) have recently undergone flight testing (Radulski, 2023; Garcia et al., 2008; Hein et al., 2023),[3] and additional aerospace CNT-composite technologies are on the horizon (Hadhazy, 2020). CNTs are creating their S-curve in the aircraft space, potentially on pace to equal the tremendous impact that micro-scale carbon fibres and their composites have had in that same industry.

References

Abbe, E., Renger, T., Sznajder, M., Klemmed, B., Sachse, E., Hübner, R., Schüler, T., Bärtling, Y., Muchow, B., Tajmar, M., 2019. A material experiment for small satellites to characterise the behaviour of carbon nanotubes in space–development and ground validation. Adv. Space Res. 63, 2312–2321.

Ahmad, I., Yazdani, B., Zhu, Y., 2015. Recent advances on carbon nanotubes and graphene reinforced ceramics nanocomposites. Nanomaterials 5, 90–114.

Ajayan, P.M., Tour, J.M., 2007. Nanotube composites. Nature 447, 1066–1068.

Ajayan, P.M., Schadler, L.S., Braun, P.V., 2006. Nanocomposite Science and Technology. John Wiley & Sons.

Akhtar, I., Chang, S.-H., 2020. Highly aligned carbon nanotubes and their sensor applications. Nanoscale 12, 21447–21458.

Almeida, V., Balzaretti, N., Costa, T., Machado, G., Gallas, M., 2013. Surfactants for CNTs dispersion in zirconia-based ceramic matrix by sol–gel method. J. Sol. Gel Sci. Technol. 65, 143–149.

An, Q., Tamrakar, S., Gillespie Jr, J.W., Rider, A.N., Thostenson, E.T., 2018. Tailored glass fibre interphases via electrophoretic deposition of carbon nanotubes: fiber and interphase characterization. Compos. Sci. Technol. 166, 131–139.

Andrews, R., Jacques, D., Rao, A.M., Rantell, T., Derbyshire, F., Chen, Y., Chen, J., Haddon, R., 1999. Nanotube composite carbon fibers. Appl. Phys. Lett. 75, 1329–1331.

Anon., 2011. Top Materials Scientists of the Past Decade. Times Higher Education.

Anthony, E., Delahunt, S., 2020. Cabot Corporation Completes Acquisition of Shenzhen Sanshun Nano New Materials Co., Ltd. (SUSN). Businesswire. https://www.businesswire.com/news/home/20200401005752/en/Cabot-Corporation-Completes-Acquisition-of-Shenzhen-Sanshun-Nano-New-Materials-Co.-Ltd-SUSN.

Arai, S., 2021. Fabrication of metal/carbon nanotube composites by electrochemical deposition. Electrochemistry 2, 563–589.

Arai, S., Fujimori, A., Murai, M., Endo, M., 2008. Excellent solid lubrication of electrodeposited nickel-multiwalled carbon nanotube composite films. Mater. Lett. 62, 3545–3548.

Aravand, M., Lomov, S.V., Verpoest, I., Gorbatikh, L., 2014. Evolution of carbon nanotube dispersion in preparation of epoxy-based composites: from a masterbatch to a nanocomposite. Express Polym. Lett. 8, 596–608.

Aravand, M.A., Shishkina, O., Straumit, I., Liotta, A.H., Wicks, S.S., Wardle, B.L., Lomov, S. V., Gorbatikh, L., 2016. Internal geometry of woven composite laminates with "fuzzy" carbon nanotube grafted fibers. Compos. Appl. Sci. Manuf. 88, 295–304.

Arepalli, S., Moloney, P., 2015. Engineered nanomaterials in aerospace. MRS Bull. 40, 804–811.

[3] N.B. author BLW was involved in (iii) and (iv)

Armstrong, D., Carbon Nanotubes Enhance Properties of Aluminum. RP News Wires. https://www.reliableplant.com/Read/18066/carbon-nanotubes-enhance-properties-of-aluminum. [Accessed 31 Oct. 2024].

Arnold, M.S., Stupp, S.I., Hersam, M.C., 2005. Enrichment of single-walled carbon nanotubes by diameter in density gradients. Nano Lett. 5, 713–718.

Arunkumar, T., Anand, G., Subbiah, R., Karthikeyan, R., Jeevahan, J., 2021. Effect of multi-walled carbon nanotubes on improvement of fracture toughness of spark-plasma-sintered yttria-stabilized zirconia nanocomposites. J. Mater. Eng. Perform. 30, 3925–3933.

Ashby, M., 1992. Materials Selection in Mechanical Design. Pergamon Press, Oxford.

Bakshi, S.R., Lahiri, D., Agarwal, A., 2010. Carbon nanotube reinforced metal matrix composites-a review. Int. Mater. Rev. 55, 41–64.

Balasubramanian, K., Burghard, M., 2005. Chemically functionalized carbon nanotubes. Small 1, 180–192.

Bauhofer, W., Kovacs, J.Z., 2009. A review and analysis of electrical percolation in carbon nanotube polymer composites. Compos. Sci. Technol. 69, 1486–1498.

Behabtu, N., Young, C.C., Tsentalovich, D.E., Kleinerman, O., Wang, X., Ma, A.W., Bengio, E.A., Ter Waarbeek, R.F., De Jong, J.J., Hoogerwerf, R.E., 2013. Strong, light, multifunctional fibers of carbon nanotubes with ultrahigh conductivity. Science 339, 182–186.

Bekyarova, E., Thostenson, E.T., Yu, A., Itkis, M.E., Fakhrutdinov, D., Chou, T.-W., Haddon, R.C., 2007. Functionalized single-walled carbon nanotubes for carbon fiber–epoxy composites. J. Phys. Chem. C 111, 17865–17871.

Bethune, D.S., Kiang, C.H., De Vries, M., Gorman, G., Savoy, R., Vazquez, J., Beyers, R., 1993. Cobalt-catalysed growth of carbon nanotubes with single-atomic-layer walls. Nature 363, 605–607.

Betser, J., Ewart, R., Chandler, F., 2016. Science and Technology (S&T) Roadmap Collaboration between SMC, NASA, and Government Partners. AIAA SPACE 2016.

Biaggi-labiosa, A., J Werkheiser, M., F Moore, D., 2018. Carbon Nano-Tube (CNT) Reinforced COPV. NASA. https://techport.nasa.gov/projects/13663.

Black, S., 2011. Embraer's Marco Túlio Pellegrini. CompositesWorld. https://www.compositesworld.com/columns/qa-embraers-marco-tlio-pellegrini.

Bloomberg, 2025. Jiangsu Cnano Technology Co. Ltd. Bloomberg. Available: https://www.bloomberg.com/profile/company/1705895D=3ACH[Accessed Jan. 7 2025].

Bogue, R., 2011. Nanocomposites: a review of technology and applications. Assem. Autom. 31, 106–112.

Bulmer, J.S., Kaniyoor, A., Elliott, J.A., 2021. A meta-analysis of conductive and strong carbon nanotube materials. Adv. Mater. 33, 2008432.

Camponeschi, E., Vance, R., Al-Haik, M., Garmestani, H., Tannenbaum, R., 2007. Properties of carbon nanotube–polymer composites aligned in a magnetic field. Carbon 45, 2037–2046.

CanatuPlc, 2024. Combination of Lifeline SPAC I and Canatu. Canatu Plc. https://canatu.com/investors/listing/combination-of-lifeline-spac-i-and-canatu/.

Cao, Q., Yu, Q., Connell, D.W., Yu, G., 2013. Titania/carbon nanotube composite (TiO_2/CNT) and its application for removal of organic pollutants. Clean Technol. Environ. Policy 15, 871–880.

Caron, E., 2019. Cabot Corporation Expands Advanced Carbons Portfolio with New ATHLOS™ Carbon Nanostructures for Engineered Polymers. Cabot. https://investor.cabot-corp.com/news-releases/news-release-details/cabot-corporation-expands-advanced-carbons-portfolio-new.

Carter, J., Michael;, G., Elizabeth, N., 2010. National Nanotechnology Initiative Final Report. Strategic Planning Skakeholder Workshop. https://www.nano.gov/sites/default/files/pub_resource/nni_strategic_plan_stakeholder_rpt.pdf.

Chae, H.G., Sreekumar, T., Uchida, T., Kumar, S., 2005. A comparison of reinforcement efficiency of various types of carbon nanotubes in polyacrylonitrile fiber. Polymer 46, 10925–10935.

Chan, K.F., Zaid, M.H.M., Mamat, M.S., Liza, S., Tanemura, M., Yaakob, Y., 2021. Recent developments in carbon nanotubes-reinforced ceramic matrix composites: a review on dispersion and densification techniques. Crystals 11, 457.

Cheaptubes, 2024. Carbon Nanotube Masterbatches. https://www.cheaptubes.com/product-category/industrial-grade-carbon-nanotubes/carbon-nanotube-masterbatches/?srsltid=AfmBOop-0nJUBvKOhC4AlPmaltDJGWHJMMY9opJ7k87_MySTjwfBtpbF.

Chem, L.G., 2022. LG Chem to Build Fourth Carbon Nanotube Plant. LG Chem. https://www.lgcorp.com/media/release/25247.

Chen, T., Li, A., 2019. Synthesizing carbon nanotubes in space. Astron. Astrophys. 631, A54.

Chen, M., Yin, X., Li, M., Chen, L., Cheng, L., Zhang, L., 2015. Electromagnetic interference shielding properties of silicon nitride ceramics reinforced by in situ grown carbon nanotubes. Ceram. Int. 41, 2467–2475.

Cho, S., Takagi, K., Kwon, H., Seo, D., Ogawa, K., Kikuchi, K., Kawasaki, A., 2012. Multi-walled carbon nanotube-reinforced copper nanocomposite coating fabricated by low-pressure cold spray process. Surf. Coating. Technol. 206, 3488–3494.

Choi, W., Chung, D., Kang, J., Kim, H., Jin, Y., Han, I., Lee, Y., Jung, J., Lee, N., Park, G., 1999. Fully sealed, high-brightness carbon-nanotube field-emission display. Appl. Phys. Lett. 75, 3129–3131.

Chou, T.-W., Gao, L., Thostenson, E.T., Zhang, Z., Byun, J.-H., 2010. An assessment of the science and technology of carbon nanotube-based fibers and composites. Compos. Sci. Technol. 70, 1–19.

Clarke, A.C., 1978. The Fountains of Paradise. Hachette UK.

Coleman, J.N., Cadek, M., Ryan, K.P., Fonseca, A., Nagy, J.B., Blau, W.J., Ferreira, M.S., 2006a. Reinforcement of polymers with carbon nanotubes. The role of an ordered polymer interfacial region. Experiment and modeling. Polymer 47, 8556–8561.

Coleman, J.N., Khan, U., Blau, W.J., Gun'ko, Y.K., 2006b. Small but strong: a review of the mechanical properties of carbon nanotube–polymer composites. Carbon 44, 1624–1652.

Croft, J., 2008. Embraer Phenom 300: Bolder Big Brother. FlightGlobal. https://www.flightglobal.com/embraer-phenom-300-bolder-big-brother/80192.article.

Davenport, T.H., Kirby, J., 2016. Only Humans Need Apply: Winners and Losers in the Age of Smart Machines. Harper Business, New York.

De Groh, K.K., Banks, B.A., 2017. The MISSE-9 polymers and composites experiment being flown on the MISSE-flight facility. In: International Space Station Research and Development (ISSR&D) Conference 2017, Venice, Italy.

De Groh, K.K., Banks, B.A., 2021. MISSE-flight Facility Polymers and Composites Experiment 1-4 (PCE 1-4).

De Villoria, R.G., Miravete, A., 2007. Mechanical model to evaluate the effect of the dispersion in nanocomposites. Acta Mater. 55, 3025–3031.

De Volder, M.F., Tawfick, S.H., Baughman, R.H., Hart, A.J., 2013. Carbon nanotubes: present and future commercial applications. Science 339, 535–539.

Dresselhaus, M.S., Dresselhaus, G., Eklund, P.C., 1996. Science of Fullerenes and Carbon Nanotubes: Their Properties and Applications. Elsevier.

Duvelius, A., 2016. General Nano, Receives Boeing 'Supplier of the Year, Technology Category' Award. UC Spin-Off Company. https://www.uc.edu/news/articles/legacy/enews/2016/05/uc-spin-off-company-general-nano-receives-boeing-supplier-of-the-year-technology-category-award.html.

Dzenis, Y., 2004. Spinning continuous fibers for nanotechnology. Science 304, 1917–1919.

Edwards, B.C., 2000. Design and deployment of a space elevator. Acta Astronaut. 47, 735–744.

Estili, M., Kwon, H., Kawasaki, A., Cho, S., Takagi, K., Kikuchi, K., Kawai, M., 2010. Multiwalled carbon nanotube-reinforced ceramic matrix composites as a promising structural material. J. Nucl. Mater. 398, 244–245.

Fauchais, P., Vardelle, A., Dussoubs, B., 2001. Quo vadis thermal spraying? J. Therm. Spray Technol. 10, 44–66.

Fenn, J., Raskino, M., 2008. Mastering the Hype Cycle: How to Choose the Right Innovation at the Right Time. Harvard Business Press.

Garcia, E.J., Wardle, B.L., Hart, A.J., 2008. Joining prepreg composite interfaces with aligned carbon nanotubes. Compos. Appl. Sci. Manuf. 39, 1065–1070.

Gartner Inc, 2024. Hype Cycle for Emerging Technologies 2024. Gartner, Inc. https://www.gartner.com

Gerzanics, M., 2009. FLIGHT TEST: Phenom 100 - Building on a Legacy. FlightGlobal.

Ghorbani, A., Sheibani, S., Ataie, A., 2018. Microstructure and mechanical properties of consolidated Cu-Cr-CNT nanocomposite prepared via powder metallurgy. J. Alloys Compd. 732, 818–827.

Gibson, R.F., 2010. A review of recent research on mechanics of multifunctional composite materials and structures. Compos. Struct. 92, 2793–2810.

Goh, C., Wei, J., Lee, L., Gupta, M., 2005. Development of novel carbon nanotube reinforced magnesium nanocomposites using the powder metallurgy technique. Nanotechnology 17, 7.

Gohardani, O., Elola, M.C., Elizetxea, C., 2014. Potential and prospective implementation of carbon nanotubes on next generation aircraft and space vehicles: a review of current and expected applications in aerospace sciences. Prog. Aero. Sci. 70, 42–68.

Gohn, A.M., Seo, J., Colby, R.H., Schaake, R.P., Androsch, R., Rhoades, A.M., 2020. Crystal nucleation in poly (ether ether ketone)/carbon nanotube nanocomposites at high and low supercooling of the melt. Polymer 199, 122548.

Gojny, F.H., Nastalczyk, J., Roslaniec, Z., Schulte, K., 2003. Surface modified multi-walled carbon nanotubes in CNT/epoxy-composites. Chem. Phys. Lett. 370, 820–824.

Gong, L., Kinloch, I.A., Young, R.J., Riaz, I., Jalil, R., Novoselov, K.S., 2010. Interfacial Stress Transfer in a Graphene Monolayer Nanocomposite arXiv:1007.1953.

Gorbatikh, L., Wardle, B.L., Lomov, S.V., 2016. Hierarchical lightweight composite materials for structural applications. MRS Bull. 41, 672–677.

Green, M.J., Behabtu, N., Pasquali, M., Adams, W.W., 2009. Nanotubes as polymers. Polymer 50, 4979–4997.

Grzybowski, B.A., Huck, W.T., 2016. The nanotechnology of life-inspired systems. Nat. Nanotechnol. 11, 585–592.

Guo, T., Nikolaev, P., Thess, A., Colbert, D.T., Smalley, R.E., 1995. Catalytic growth of single-walled manotubes by laser vaporization. Chem. Phys. Lett. 243, 49–54.

Hadhazy, A., 2020. Resetting Expectations for Carbon Nanotubes. https://aerospaceamerica.aiaa.org/features/resetting-expectations-for-carbon-nanotubes/.

Haggenmueller, R., Gommans, H.H., Rinzler, A.G., Fischer, J.E., Winey, K.I., 2000. Aligned single-wall carbon nanotubes in composites by melt processing methods. Chem. Phys. Lett. 330, 219–225.

Han, G., Wang, Z., Liu, K., Li, S., Du, X., Du, W., 2015. Synthesis of CNT-reinforced AZ31 magnesium alloy composites with uniformly distributed CNTs. Mater. Sci. Eng., A 628, 350–357.

Harris, C.E., 2000. Assessment of the state-of-the-art in the design and manufacturing of large composite structure. In: 4th Conference on Aerospace Materials: NASA Technical Reports Server. https://ntrs.nasa.gov/api/citations/20010067276/downloads/20010067276.pdf.

Hein, B., Botura, G., Hamman, M., Slane, C., 2023. Carbon Nanotube (CNT) Based Electrothermal Ice Protection System Flight Tests.

Hendrixson, S., 2021. Untangling the Promise of Carbon Nanotubes for 3D Printing. https://www.additivemanufacturing.media/articles/untangling-the-promise-of-carbon-nanotubes-for-3d-printing.

HEXAGON, 2024. Composite Materials in Aerospace. HEXAGON. https://hexagon.com/resources/resource-library/composite-materials-aerospace.

Hexcel, 2023. HexTow® IM7 Carbon Fiber Datasheet. Hexcel Corporation. https://www.hexcel.com/user_area/content_media/raw/IM7_HexTow_DataSheet.pdf.

Hisham, M., Dileep, C., Jacob, L., Butt, H., 2024. Additive manufacturing of carbon nanocomposites for structural applications. J. Mater. Res. Technol. 28, 4674–4693.

Iijima, S., 1991. Helical microtubules of graphitic carbon. Nature 354, 56–58.

Islam, M.H., Afroj, S., Uddin, M.A., Andreeva, D.V., Novoselov, K.S., Karim, N., 2022. Graphene and CNT-based smart fiber-reinforced composites: a review. Adv. Funct. Mater. 32, 2205723.

Jang, D., Farooq, S., Yoon, H., Khalid, H., 2021. Design of a highly flexible and sensitive multifunctional polymeric sensor incorporating CNTs and carbonyl iron powder. Compos. Sci. Technol. 207, 108725.

Kalfon-Cohen, E., Kopp, R., Furtado, C., Ni, X., Arteiro, A., Borstnar, G., Mavrogordato, M. N., Sinclair, I., Spearing, S.M., Camanho, P.P., 2018. Synergetic effects of thin plies and aligned carbon nanotube interlaminar reinforcement in composite laminates. Compos. Sci. Technol. 166, 160–168.

Khare, K.S., Khabaz, F., Khare, R., 2014. Effect of carbon nanotube functionalization on mechanical and thermal properties of cross-linked epoxy–carbon nanotube nanocomposites: role of strengthening the interfacial interactions. ACS Appl. Mater. Interfaces 6, 6098–6110.

Kim, K.T., Cha, S.I., Hong, S.H., 2007. Hardness and wear resistance of carbon nanotube reinforced Cu matrix nanocomposites. Mater. Sci. Eng., A 449, 46–50.

Kim, M., Park, Y.-B., Okoli, O.I., Zhang, C., 2009. Processing, characterization, and modeling of carbon nanotube-reinforced multiscale composites. Compos. Sci. Technol. 69, 335–342.

Kim, H., Wilburn, B.R., Castro, E., Garcia Rosales, C.A., Chavez, L.A., Tseng, T.-L.B., Lin, Y., 2019. Multifunctional SENSING using 3D printed CNTs/$BaTiO_3$/PVDF nanocomposites. J. Compos. Mater. 53, 1319–1328.

Kim, J.-W., Gardner, J.M., Sauti, G., Wincheski, R.A., Jensen, B.D., Wise, K.E., Siochi, E.J., 2023. Multi-scale hierarchical carbon nanotube fiber reinforced composites towards enhancement of axial/transverse strength and fracture toughness. Compos. Appl. Sci. Manuf. 167, 107449.

Kinloch, I.A., Suhr, J., Lou, J., Young, R.J., Ajayan, P.M., 2018. Composites with carbon nanotubes and graphene: an outlook. Science 362, 547–553.

Ko, F.K., Lam, H., Titchenal, N., Ye, H., Gogotsi, Y., 2006. Coelectrospinning of Carbon Nanotube Reinforced Nanocomposite Fibrils. ACS Publications.

Koo, J.H., 2006. Polymer Nanocomposites. McGraw-Hill Professional Pub, New York, NY, USA.

Koo, J.H., 2016. Fundamentals, Properties, and Applications of Polymer Nanocomposites. Cambridge University Press.

Kreupl, F., Graham, A.P., Duesberg, G., Steinhögl, W., Liebau, M., Unger, E., Hönlein, W., 2002. Carbon nanotubes in interconnect applications. Microelectron. Eng. 64, 399–408.

Kumar, S., Rani, R., Dilbaghi, N., Tankeshwar, K., Kim, K.-H., 2017. Carbon nanotubes: a novel material for multifaceted applications in human healthcare. Chem. Soc. Rev. 46, 158–196.

Kumar, A., Sharma, K., Dixit, A.R., 2020. Carbon nanotube-and graphene-reinforced multiphase polymeric composites: review on their properties and applications. J. Mater. Sci. 55, 2682–2724.

Kumho Petrochemical, 2025. Carbon Nanotube. Kumho Petrochemical.

Lachman, N., Wiesel, E., de Villoria, R.G., Wardle, B.L., Wagner, H.D., 2012. Interfacial load transfer in carbon nanotube/ceramic microfiber hybrid polymer composites. Compos. Sci. Technol. 72, 1416–1422.

Laha, T., Liu, Y., Agarwal, A., 2007. Carbon nanotube reinforced aluminum nanocomposite via plasma and high velocity oxy-fuel spray forming. J. Nanosci. Nanotechnol. 7, 515–524.

Landi, B.J., Ganter, M.J., Cress, C.D., DiLeo, R.A., Raffaelle, R.P., 2009. Carbon nanotubes for lithium ion batteries. Energy Environ. Sci. 2, 638–654.

Lashmore, D., Brown, J.J., Dean Jr, R.C., Antoinette, P.L., 2009. Systems and Methods for Synthesis of Extended Length Nanostructures. Google Patents.

Laurent, C., Flahaut, E., Peigney, A., 2010. The weight and density of carbon nanotubes versus the number of walls and diameter. Carbon 48, 2994–2996.

Lavagna, L., Massella, D., Pantano, M.F., Bosia, F., Pugno, N.M., Pavese, M., 2018. Grafting carbon nanotubes onto carbon fibres doubles their effective strength and the toughness of the composite. Compos. Sci. Technol. 166, 140–149.

Li, Y., Ishida, H., 2003. Solution intercalation of polystyrene and the comparison with poly (ethyl methacrylate). Polymer 44, 6571–6577.

Li, J., Lu, Y., 2009. Carbon nanotube based chemical sensors for space and terrestrial applications. ECS Trans. 19, 7.

Li, Y., Shimizu, H., 2009. Toward a stretchable, elastic, and electrically conductive nanocomposite: morphology and properties of poly [styrene-b-(ethylene-co-butylene)-b-styrene]/multiwalled carbon nanotube composites fabricated by high-shear processing. Macromolecules 42, 2587–2593.

Li, Y.-L., Kinloch, I.A., Windle, A.H., 2004. Direct spinning of carbon nanotube fibres from chemical vapor deposition synthesis. Science 304, 276–278.

Li, Q., Viereckl, A., Rottmair, C.A., Singer, R.F., 2009. Improved processing of carbon nanotube/magnesium alloy composites. Compos. Sci. Technol. 69, 1193–1199.

Li, T., Li, M., Gu, Y., Wang, S., Li, Q., Zhang, Z., 2018. Mechanical enhancement effect of the interlayer hybrid CNT film/carbon fiber/epoxy composite. Compos. Sci. Technol. 166, 176–182.

Liu, P., 2005. Modifications of carbon nanotubes with polymers. Eur. Polym. J. 41, 2693–2703.

Liu, L.Q., Tasis, D., Prato, M., Wagner, H.D., 2007. Tensile mechanics of electrospun multiwalled nanotube/poly (methyl methacrylate) nanofibers. Adv. Mater. 19, 1228–1233.

Liu, P., Fan, Z., Mikhalchan, A., Tran, T.Q., Jewell, D., Duong, H.M., Marconnet, A.M., 2016. Continuous carbon nanotube-based fibers and films for applications requiring enhanced heat dissipation. ACS Appl. Mater. Interfaces 8, 17461–17471.

Loos, M.R., Schulte, K., 2011. Is it worth the effort to reinforce polymers with carbon nanotubes? Macromol. Theory Simul. 20, 350–362.

Ma, P.-C., Siddiqui, N.A., Marom, G., Kim, J.-K., 2010. Dispersion and functionalization of carbon nanotubes for polymer-based nanocomposites: a review. Compos. Appl. Sci. Manuf. 41, 1345–1367.

Ma, H., Aravand, M.A., Falzon, B.G., 2021. Synergistic enhancement of fracture toughness in multiphase epoxy matrices modified by thermoplastic and carbon nanotubes. Compos. Sci. Technol. 201, 108523.

Maneshi, A., Soares, J.B., Simon, L.C., 2011. An efficient in situ polymerization method for the production of polyethylene/clay nanocomposites: effect of polymerization conditions on particle morphology. Macromol. Chem. Phys. 212, 2017–2028.

Mani, A., Sharma, S., 2022. Interfacial shear strength of carbon nanotube reinforced polymer composites: a review. Mater. Today Proc. 50, 1774–1780.

Martel, R., Schmidt, T., Shea, H., Hertel, T., Avouris, P., 1998. Single-and multi-wall carbon nanotube field-effect transistors. Appl. Phys. Lett. 73, 2447–2449.

McSweeney, K., 2025. Surviving Space with Carbon Nanotubes and 3D Printing. Northrop Grumman. https://www.northropgrumman.com/space/surviving-space-with-carbon-nanotubes-and-3d-printing.

Meador, M.A., Files, B., Li, J., Manohara, H., Powell, D., Siochi, E.J., 2010. Draft nanotechnology roadmap technology area 10. In: National Aeronautics and Space Administration (NASA), p. 5.

Miravete, A., 1999. 3-D Textile Reinforcements in Composite Materials. Woodhead Publishing.

Misaki, A., 2024. 13 Nanotube Manufacturers in 2024. https://us.metoree.com/categories/2704/ .

Mital, S.K., 1996. Simplified Micromechanics of Plain Weave Composites. National Aeronautics and Space Administration, Office of Management.

MNFConference, 2018. International Workshop on Multi-Functional Nanocarbon Fibres. https://www.mnfworkshop.net.

Mohd Nurazzi, N., Asyraf, M.M., Khalina, A., Abdullah, N., Sabaruddin, F.A., Kamarudin, S. H., Ahmad, S.b., Mahat, A.M., Lee, C.L., Aisyah, H., 2021. Fabrication, functionalization, and application of carbon nanotube-reinforced polymer composite: an overview. Polymers 13, 1047.

Montes-Morán, M., Martınez-Alonso, A., Tascón, J., Young, R., 2001. Effects of plasma oxidation on the surface and interfacial properties of ultra-high modulus carbon fibres. Compos. Appl. Sci. Manuf. 32, 361–371.

Monthioux, M., Kuznetsov, V.L., 2006. Who should be given the credit for the discovery of carbon nanotubes? Carbon 44, 1621–1623.

Müller, M.T., Krause, B., Kretzschmar, B., Pötschke, P., 2011. Influence of feeding conditions in twin-screw extrusion of PP/MWCNT composites on electrical and mechanical properties. Compos. Sci. Technol. 71, 1535–1542.

NASA, 2024. Mission to Jupiter, Juno. https://www.jpl.nasa.gov/missions/juno/.

NASA Spinoff, 2017. Carbon Nanotube Resin Shores up Boats, Bikes. https://spinoff.nasa.gov/Spinoff2017/cg_7.html.

Natarajan, B., 2022. Processing-structure-mechanical property relationships in direct formed carbon nanotube articles and their composites: a review. Compos. Sci. Technol. 225, 109501.

National Nanotechnology Initiative, 2000. National Nanotechnology Initiative. https://www.nano.gov/national-nanotechnology-initiative.

National Research Council, Division on Engineering, Physical Sciences, Aeronautics, Space Engineering Board, & Steering Committee for NASA Technology Roadmaps, 2012. NASA Space Technology Roadmaps and Priorities: Restoring NASA's Technological Edge and Paving the Way for a New Era in Space.

National Research Council, Division on Engineering, Physical Sciences, National Materials Advisory Board, & Committee on High-Performance Structural Fibers for Advanced Polymer Matrix Composites, 2005. High-Performance Structural Fibers for Advanced Polymer Matrix Composites. National Academies Press.

Nishimoto, R., Sato, Y., Wu, J., Saizaki, T., Kubo, M., Wang, M., Abe, H., Richard, I., Yoshinobu, T., Sorin, F., 2022. Thermally drawn CNT-based hybrid nanocomposite fiber for electrochemical sensing. Biosensors 12, 559.

Novoselov, K.S., Geim, A.K., Morozov, S.V., Jiang, D.-e., Zhang, Y., Dubonos, S.V., Grigorieva, I.V., Firsov, A.A., 2004. Electric field effect in atomically thin carbon films. Science 306, 666–669.

NTConference, 2025. Previous and Coming NT Conferences. https://www.photon.t.u-tokyo.ac.jp/nt/.

OCSiAl, 2020. World's Largest Graphene Nanotube Manufacturer OCSiAl Achieves Production Capacity of 75 Tonnes Per Year. https://ocsial.com/news/-worlds-largest-graphene-nanotube-manufacturer-ocsial-achieves-production-capacity-of-75-tonnes-per-year-/.

Odegard, G., 2017. The Institute for Ultra-Strong Composites by Computational Design (US-COMP). https://www.nasa.gov/directorates/stmd/space-tech-research-grants/the-institute-for-ultra-strong-composites-by-computational-design-us-comp/.

Odegard, G.M., Liang, R., Wise, K.E., 2018. Special issue on carbon nanotube composites. Compos. Sci. Technol. 166.

Oliva-Avilés, A., Avilés, F., Sosa, V., 2011. Electrical and piezoresistive properties of multi-walled carbon nanotube/polymer composite films aligned by an electric field. Carbon 49, 2989–2997.

Paiva, M.C., Covas, J.A., 2016. Carbon nanofibres and nanotubes for composite applications. In: Fibrous and Textile Materials for Composite Applications, pp. 231–260.

Pandey, G., Thostenson, E.T., 2012. Carbon nanotube-based multifunctional polymer nanocomposites. Polym. Rev. 52, 355–416.

Papageorgiou, D.G., Li, Z., Liu, M., Kinloch, I.A., Young, R.J., 2020. Mechanisms of mechanical reinforcement by graphene and carbon nanotubes in polymer nanocomposites. Nanoscale 12, 2228–2267.

Park, J.G., Evers, C., Jolowsky, C., Vondrasek, B., Thagard, K., Czabaj, M., Ku, B., Wang, Y., Dumitrică, T., Odegard, G.M., 2024. Gamma-ray irradiation to achieve high tensile performance of unidirectional CNT yarn laminates. Carbon 216, 118530.

Parnian, P., 2022. A short review on: recent advances in the use of carbon nanotubes in additive manufacturing of polymer matrix composites. In: Macromolecular Symposia. Wiley Online Library, p. 2100339.

Pirtle, T.J., E Baker, C., Danielle, D.M., 2019. Carbon Nanotube Infused Launch Vehicle Structures. TechPort. https://techport.nasa.gov/projects/94199.

Poland, C.A., Duffin, R., Kinloch, I., Maynard, A., Wallace, W.A., Seaton, A., Stone, V., Brown, S., MacNee, W., Donaldson, K., 2008. Carbon nanotubes introduced into the abdominal cavity of mice show asbestos-like pathogenicity in a pilot study. Nat. Nanotechnol. 3, 423–428.

Praveen, B., Venkatesha, T., 2009. Electrodeposition and properties of Zn–Ni–CNT composite coatings. J. Alloys Compd. 482, 53–57.

Qian, H., Bismarck, A., Greenhalgh, E.S., Kalinka, G., Shaffer, M.S., 2008. Hierarchical composites reinforced with carbon nanotube grafted fibers: the potential assessed at the single fiber level. Chem. Mater. 20, 1862–1869.

Qian, H., Bismarck, A., Greenhalgh, E.S., Shaffer, M.S., 2010a. Carbon nanotube grafted carbon fibres: a study of wetting and fibre fragmentation. Compos. Appl. Sci. Manuf. 41, 1107–1114.

Qian, H., Greenhalgh, E.S., Shaffer, M.S., Bismarck, A., 2010b. Carbon nanotube-based hierarchical composites: a review. J. Mater. Chem. 20, 4751–4762.

Qu, S., Dai, Y., Zhang, D., Li, Q., Chou, T.-W., Lyu, W., 2020. Carbon nanotube film based multifunctional composite materials: an overview. Funct. Compos. Struct. 2, 022002.

Quigley, J.P., Herrington, K., Baird, D.G., 2014. Enhanced electrical properties of polycarbonate/carbon nanotube nanocomposites prepared by a supercritical carbon dioxide aided melt blending method. Polymer 55, 6167–6175.

Radulski, E., 2023. Metis Design's WISP Technology Named Finalist in DoD Maintenance Innovation Challenge. http://www.navysbir.com/success/docs/SPT-Metis_Design_CO-23.pdf.

Ramachandran, K., Carmine, Z., Yoshida, K., Tsunoura, T., Jayaseelan, D.D., 2021a. Experimental investigation and mathematical modelling of water vapour corrosion of Ti_3SiC_2 and Ti_2AlC ceramics and their mechanical behaviour. J. Eur. Ceram. Soc. 41, 4761–4773.

Ramachandran, K., Subramani, R.R., Arunkumar, T., Boopalan, V., 2021b. Mechanical and thermal properties of spark plasma sintered Alumina-MWCNTs nanocomposites prepared via improvised colloidal route. Mater. Chem. Phys. 272, 125034.

Ramachandran, K., Boopalan, V., Bear, J.C., Subramani, R., 2022a. Multi-walled carbon nanotubes (MWCNTs)-reinforced ceramic nanocomposites for aerospace applications: a review. J. Mater. Sci. 57, 3923–3953.

Ramachandran, K., Leelavinodhan, S., Antao, C., Copti, A., Mauricio, C., Jyothi, Y.L., Jayaseelan, D.D., 2022b. Analysis of failure mechanisms of oxide-oxide ceramic matrix composites. J. Eur. Ceram. Soc. 42, 1626–1634.

Rao, R., Pint, C.L., Islam, A.E., Weatherup, R.S., Hofmann, S., Meshot, E.R., Wu, F., Zhou, C., Dee, N., Amama, P.B., 2018. Carbon nanotubes and related nanomaterials: critical advances and challenges for synthesis toward mainstream commercial applications. ACS Nano 12, 11756–11784.

Rathinavel, S., Priyadharshini, K., Panda, D., 2021. A review on carbon nanotube: an overview of synthesis, properties, functionalization, characterization, and the application. Mater. Sci. Eng., B 268, 115095.

Rawal, S., 2018. Applied functional materials and structures for spacecraft systems. In: 1st International Symposium on Mechanics. Aberdeen, Scotland, United Kingdom.

Rawal, S., Brantley, J., Karabudak, N., 2013. Development of carbon nanotube-based composite for spacecraft components. In: 2013 6th International Conference on Recent Advances in Space Technologies (RAST). IEEE, pp. 13–19.

Reibold, M., Paufler, P., Levin, A.A., Kochmann, W., Pätzke, N., Meyer, D., 2006. Carbon nanotubes in an ancient Damascus sabre. Nature 444, 286.

Rinzler, A., Liu, J., Dai, H., Nikolaev, P., Huffman, C., Rodriguez-Macias, F.J., Boul, P., Lu, A. H., Heymann, D., Colbert, D.T., 1998. Large-scale purification of single-wall carbon nanotubes: process, product, and characterization. Appl. Phys. Mater. Sci. Process 67.

Roco, M.C., Mirkin, C.A., Hersam, M.C., 2011. Nanotechnology research directions for societal needs in 2020: retrospective and outlook.,

Rogers, E.M., Williams, D., 1983. Diffusion of. Innovations. The Free Press, Glencoe, IL.

Rosca, I.D., Hoa, S.V., 2009. Highly conductive multiwall carbon nanotube and epoxy composites produced by three-roll milling. Carbon 47, 1958–1968.

Ryan, K., Cadek, M., Nicolosi, V., Walker, S., Ruether, M., Fonseca, A., Nagy, J., Blau, W., Coleman, J., 2006. Multiwalled carbon nanotube nucleated crystallization and reinforcement in poly (vinyl alcohol) composites. Synth. Met. 156, 332–335.

Schadler, L., Brinson, L., Sawyer, W., 2007. Polymer nanocomposites: a small part of the story. Jom 59, 53–60.

Schodek, D.L., Ferreira, P., Ashby, M.F., 2009. Nanomaterials, Nanotechnologies and Design: An Introduction for Engineers and Architects. Butterworth-Heinemann.

Schulte, K., Windle, A.H., 2007. Carbon nanotube (CNT)–polymer composites. Compos. Sci. Technol. 67, 777.

Seyhan, A.T., Tanoğlu, M., Schulte, K., 2009. Tensile mechanical behavior and fracture toughness of MWCNT and DWCNT modified vinyl-ester/polyester hybrid nanocomposites produced by 3-roll milling. Mater. Sci. Eng., A 523, 85–92.

Shah, T., Jones, M., Alberding, M., Laszewski, M., 2012. Carbon Nanostructures for electromagnetic shielding and lightning strike protection applications in aircraft. In: ESA Workshop on Aerospace. EMC, pp. 1–5.

Sharma, N., Syed, A.N., Ray, B.C., Yadav, S., Biswas, K., 2019. Alumina–MWCNT composites: microstructural characterization and mechanical properties. J. Asian Ceram. Soc. 7, 1–19.

Shaver, J., Parra-Vasquez, A.N.G., Hansel, S., Portugall, O., Mielke, C.H., Von Ortenberg, M., Hauge, R.H., Pasquali, M., Kono, J., 2009. Alignment dynamics of single-walled carbon nanotubes in pulsed ultrahigh magnetic fields. ACS Nano 3, 131–138.

Shen, Z., Simon, G.P., Cheng, Y.-B., 2002. Comparison of solution intercalation and melt intercalation of polymer–clay nanocomposites. Polymer 43, 4251–4260.

Shin, B., Mondal, S., Lee, M., Kim, S., Huh, Y.-I., Nah, C., 2021. Flexible thermoplastic polyurethane-carbon nanotube composites for electromagnetic interference shielding and thermal management. Chem. Eng. J. 418, 129282.

Shirolkar, N., Maffe, A., DiLoreto, E., Arias-Monje, P.J., Lu, M., Ramachandran, J., Gulgunje, P., Gupta, K., Park, J.G., Shih, K.-C., 2021. Multichannel hollow carbon fibers: processing, structure, and properties. Carbon 174, 730–740.

Silva, P.R., Almeida, V.O., Machado, G.B., Benvenutti, E.V., Costa, T.M., Gallas, M.r.R., 2012. Surfactant-based dispersant for multiwall carbon nanotubes to prepare ceramic composites by a sol–gel method. Langmuir 28, 1447–1452.

Smith, C.S., 1960. A history of metallography: the development of ideas on the structure of metals before 1890. University of Chicago Press, Chicago.

Soni, S.K., Thomas, B., Kar, V.R., 2020. A comprehensive review on CNTs and CNT-reinforced composites: syntheses, characteristics and applications. Mater. Today Commun. 25, 101546.

Spitalsky, Z., Tasis, D., Papagelis, K., Galiotis, C., 2010. Carbon nanotube–polymer composites: chemistry, processing, mechanical and electrical properties. Prog. Polym. Sci. 35, 357–401.

Sun, L., Sue, H.J., 2010. Epoxy/carbon nanotube nanocomposites. In: Epoxy Polymers: New Materials and Innovations, pp. 185–211.

Sun, X., Yu, R., Xu, G., Hor, T., Ji, W., 1998. Broadband optical limiting with multiwalled carbon nanotubes. Appl. Phys. Lett. 73, 3632–3634.

Sun, K., Yu, J., Zhang, C., Zhou, X., 2012. In situ growth carbon nanotube reinforced SiCf/SiC composite. Mater. Lett. 66, 92–95.

Sundaram, R.M., Sekiguchi, A., Sekiya, M., Yamada, T., Hata, K., 2018. Copper/carbon nanotube composites: research trends and outlook. R. Soc. Open Sci. 5, 180814.

Sung, D.H., Kang, G.-H., Kong, K., Kim, M., Park, H.W., Park, Y.-B., 2016. Characterization of thermoelectric properties of multifunctional multiscale composites and fiber-reinforced composites for thermal energy harvesting. Compos. B Eng. 92, 202–209.

Taghizadeh, A., Favis, B.D., 2013. Carbon nanotubes in blends of polycaprolactone/thermoplastic starch. Carbohydr. Polym. 98, 189–198.

Tao, J., Chen, X., Hong, P., Yi, J., 2017. Microstructure and electrical conductivity of laminated Cu/CNT/Cu composites prepared by electrodeposition. J. Alloys Compd. 717, 232–239.

Thakre, P.R., Lagoudas, D.C., Riddick, J.C., Gates, T.S., Frankland, S.-J.V., Ratcliffe, J.G., Zhu, J., Barrera, E.V., 2011. Investigation of the effect of single wall carbon nanotubes on interlaminar fracture toughness of woven carbon fiber—epoxy composites. J. Compos. Mater. 45, 1091–1107.

Thostenson, E.T., Ren, Z., Chou, T.-W., 2001. Advances in the science and technology of carbon nanotubes and their composites: a review. Compos. Sci. Technol. 61, 1899–1912.

Thostenson, E.T., Li, C., Chou, T.-W., 2005. Nanocomposites in context. Compos. Sci. Technol. 65, 491–516.

Tjong, S.C., 2009. Carbon Nanotube Reinforced Composites: Metal and Ceramic Matrices. John Wiley & Sons.

Tracxn Technologies, 2024. Carbon Nanotubes Overview.

Tran, T.Q., Fan, Z., Liu, P., Myint, S.M., Duong, H.M., 2016a. Super-strong and highly conductive carbon nanotube ribbons from post-treatment methods. Carbon 99, 407–415.

Tran, T.Q., Fan, Z., Mikhalchan, A., Liu, P., Duong, H.M., 2016b. Post-treatments for multifunctional property enhancement of carbon nanotube fibres from the floating catalyst method. ACS Appl. Mater. Interfaces 8, 7948–7956.

Trimble, S., 2011. Lockheed Martin Reveals F-35 to Feature Nanocomposite Structures. https://www.flightglobal.com/lockheed-martin-reveals-f-35-to-feature-nanocomposite-structures/100174.article.

US Air Force, 2010. Technology Horizons: A Vision for Air Force Science & Technology during 2010–2030. AF/ST-TR-10-02, vol 2.

US-COMP, 2025. In: Computational Development of Materials for Space Applications. https://us-comp.com/#:~:text=The%20Institute%20for%20Ultra-Strong,Research%20Institute%20awarded%20in%2020.

Vaia, R.A., Wagner, H.D., 2004. Framework for nanocomposites. Mater. Today 7, 32–37.

Van Noorden, R., 2011. The trials of new carbon. Nature 469, 14.

Vargas-Bernal, R., Tecpoyotl-Torres, M., 2021. Nanocomposites for space applications: a review. In: Research Anthology on Synthesis, Characterization, and Applications of Nanomaterials, pp. 1681–1705.

Veedu, V.P., Cao, A., Li, X., Ma, K., Soldano, C., Kar, S., Ajayan, P.M., Ghasemi-Nejhad, M. N., 2006. Multifunctional composites using reinforced laminae with carbon-nanotube forests. Nat. Mater. 5, 457–462.

Vilatela, J.J., Khare, R., Windle, A.H., 2012. The hierarchical structure and properties of multifunctional carbon nanotube fibre composites. Carbon 50, 1227–1234.

Wagner, H.D., Lourie, O., Feldman, Y., Tenne, R., 1998. Stress-induced fragmentation of multiwall carbon nanotubes in a polymer matrix. Appl. Phys. Lett. 72, 188–190.

Wan, H., Delale, F., Shen, L., 2005. Effect of CNT length and CNT-matrix interphase in carbon nanotube (CNT) reinforced composites. Mech. Res. Commun. 32, 481–489.

Wang, S., Downes, R., Young, C., Haldane, D., Hao, A., Liang, R., Wang, B., Zhang, C., Maskell, R., 2015. Carbon fiber/carbon nanotube buckypaper interply hybrid composites: manufacturing process and tensile properties. Adv. Eng. Mater. 17, 1442–1453.

Wang, H., Zhang, X., Duan, Y., Meng, L., 2018. Experimental and numerical study of the interfacial shear strength in carbon fiber/epoxy resin composite under thermal loads. Int. J. Polym. Sci. 2018, 3206817.

Wardle, B.L., Koo, J.H., Odegard, G.M., Seidel, G.D., 2018. Advanced nanoengineered materials. In: Bhat, B.N. (Ed.), Aerospace Materials and Applications.

Wikipedia, 2024a. Juno (Spacecraft). https://en.wikipedia.org/wiki/Juno_(spacecraft.

Wikipedia, 2024b. Timeline of Carbon Nanotubes. https://en.wikipedia.org/wiki/Timeline_of_carbon_nanotubes.

Winey, K.I., Vaia, R.A., 2007. Polymer nanocomposites. MRS Bull. 32, 314–322.

Winey, K.I., Kashiwagi, T., Mu, M., 2007. Improving electrical conductivity and thermal properties of polymers by the addition of carbon nanotubes as fillers. MRS Bull. 32, 348–353.

Wu, Q., Zhao, R., Zhu, J., Wang, F., 2020. Interfacial improvement of carbon fiber reinforced epoxy composites by tuning the content of curing agent in sizing agent. Appl. Surf. Sci. 504, 144384.

Wu, H., Zhu, C., Li, X., Hu, X., Xie, H., Lu, X., Qu, J.P., 2022. Layer-by-Layer assembly of multifunctional NR/MXene/CNTs composite films with exceptional electromagnetic interference shielding performances and excellent mechanical properties. Macromol. Rapid Commun. 43, 2200387.

Xu, H., Tong, X., Zhang, Y., Li, Q., Lu, W., 2016. Mechanical and electrical properties of laminated composites containing continuous carbon nanotube film interleaves. Compos. Sci. Technol. 127, 113–118.

Yamamoto, N., Hart, A.J., Garcia, E.J., Wicks, S.S., Duong, H.M., Slocum, A.H., Wardle, B.L., 2009. High-yield growth and morphology control of aligned carbon nanotubes on ceramic fibers for multifunctional enhancement of structural composites. Carbon 47, 551–560.

Yang, K., He, J., Su, Z., Reppert, J.B., Skove, M.J., Tritt, T.M., Rao, A.M., 2010. Inter-tube bonding, graphene formation and anisotropic transport properties in spark plasma sintered multi-wall carbon nanotube arrays. Carbon 48, 756–762.

Yourdkhani, M., Liu, W., Baril-Gosselin, S., Robitaille, F., Hubert, P., 2018. Carbon nanotube-reinforced carbon fibre-epoxy composites manufactured by resin film infusion. Compos. Sci. Technol. 166, 169–175.

Yousefi, N., Fisher, S.J., Burgstaller, C., Shaffer, M.S., Bismarck, A., 2022. Hierarchical carbon fibre composites incorporating high loadings of carbon nanotubes. Compos. Sci. Technol. 222, 109369.

Yu, M.-F., Files, B.S., Arepalli, S., Ruoff, R.S., 2000. Tensile loading of ropes of single wall carbon nanotubes and their mechanical properties. Phys. Rev. Lett. 84, 5552.

Yuan, M., Johnson, B., Koo, J.H., Bourell, D., 2014. Polyamide 11-MWNT nanocomposites: thermal and electrical conductivity measurements. J. Compos. Mater. 48, 1833–1841.

Yue, H., Monreal-Bernal, A., Fernández-Blázquez, J.P., Llorca, J., Vilatela, J.J., 2015. Macroscopic CNT fibres inducing non-epitaxial nucleation and orientation of semi-crystalline polymers. Sci. Rep. 5, 16729.

Zeng, Y., Ci, L., Carey, B.J., Vajtai, R., Ajayan, P.M., 2010. Design and reinforcement: vertically aligned carbon nanotube-based sandwich composites. ACS Nano 4, 6798–6804.

Zhandarov, S., Mäder, E., 2015. Estimation of the local interfacial strength parameters of carbon nanotube fibers in an epoxy matrix from a microbond test data. Carbon 86, 54–57.

Zhang, H., Peijs, T., 2020. Multifunctional composites based on hierarchical micro-–nanostructures: design, manufacturing, properties, and applications. In: Fiber-Reinforced Nanocomposites: Fundamentals and Applications. Elsevier.

Zhang, M., Atkinson, K.R., Baughman, R.H., 2004. Multifunctional carbon nanotube yarns by downsizing an ancient technology. Science 306, 1358–1361.

Zhang, Z., Liu, B., Chen, Y., Jiang, H., Hwang, K., Huang, Y., 2008. Mechanical properties of functionalized carbon nanotubes. Nanotechnology 19, 395702.

Zhang, K., Wu, W., Meng, H., Guo, K., Chen, J.-F., 2009. Pickering emulsion polymerization: preparation of polystyrene/nano-SiO_2 composite microspheres with core-shell structure. Powder Technol. 190, 393–400.

Zhong, W.-H., 2012. Nanoscience and Nanomaterials: Synthesis, Manufacturing and Industry Impacts. DEStech Publications, Inc.

Zhu, S., Sheng, J., Chen, Y., Ni, J., Li, Y., 2021. Carbon nanotubes for flexible batteries: recent progress and future perspective. Natl. Sci. Rev. 8, nwaa261.

Zunjarrao, S., Sriraman, R., Singh, R., 2006. Effect of processing parameters and clay volume fraction on the mechanical properties of epoxy-clay nanocomposites. J. Mater. Sci. 41, 2219–2228.

Zyvex Marine, 2012. Zyvex Marine Launches LRV-17 Long Range Vessel as the First Nano-Composite Manned Boat. PR Newswire. https://www.prnewswire.com/news-releases/zyvex-marine-launches-lrv-17-long-range-vessel-as-the-first.html.

Graphene Composites

1.6

Dimitrios G. Papageorgiou[1] *and Robert J. Young*[2]
[1]School of Engineering and Materials Science, Queen Mary University of London, London, United Kingdom; [2]National Graphene Institute and Department of Materials, University of Manchester, Manchester, United Kingdom

1. Graphene

1.1 The quest for graphene

One of the earliest reports on the preparation of graphene dates back to the early 1960s when Boehm and coworkers (Boehm et al., 1962) described the preparation of "extremely thin carbon foils," through the reduction of graphite oxide (GO), that they identified by transmission electron microscopy (TEM). This led to the proposal in 1994 that "graphene" be recognised formally as the terminology for single-layer sheets of graphite (Boehm et al., 1994). The study of graphite oxide dates back more than 150 years ago. In his attempt to determine the "atomic weight" of graphite, Brodie reported the synthesis of graphite oxide, often now called graphene oxide, though the oxidation of graphite (Brodie, 1859). This process was refined by a number of workers (Hummers and Offeman, 1958) and the state-of-the-art in the preparation of graphene oxide in 2010 was reviewed and summarised by Ruoff and co-workers (Dreyer et al., 2010). It was thought originally that it was impossible to prepare pristine graphene from the reduction of graphene oxide although a study (Voiry et al., 2016) has shown that material with electronic properties similar to those of pristine graphene can be produced from the reduction of graphene oxide through the use of pulsed microwave heating.

Wallace undertook a theoretical study upon the electronic structure of graphite in 1947 using the tight-binding approximation (Wallace, 1947). This involved determining the zone structure of a "single hexagonal layer" of carbon atoms, that is, monolayer graphene, and pointed to some interesting electronic properties (Katsnelson, 2020). Investigations into the mechanical properties of graphite, both theoretically (Gillis, 1984) and experimentally, showed impressive mechanical properties in directions parallel to the covalently-bonded planes, with an in-plane elastic constant, c_{11}, of more than 1000 GPa being measured (Blakslee et al., 1970). This was consistent with the very high values of Young's modulus (>400 MPa) being reported as early as 1970 for high-modulus carbon fibres consisting of aligned graphite crystals (Watt, 1970).

The potential and promise of graphene with its attractive attributes led in the last century to a number of attempts to isolate monolayer material in spite of eminent scientists, such as Peierls, predicting that monolayer 2D crystals could not be prepared as

The History of Composites. https://doi.org/10.1016/B978-0-443-34119-9.00008-5

they would be thermodynamically unstable (Peierls, 1935). Nevertheless, it was known that single layer graphite could be produced on substrates such as by epitaxial growth on the surface of TiC (Itoh et al., 1991) or by the decomposition of hydrocarbons on the surface of platinum (Land et al., 1992). A number of attempts were subsequently made to prepare graphene using techniques such as physical force by the manipulation of an AFM (atomic force microscope) tip on graphite islands (Lu et al., 1999) and intercalation (Shioyama, 2001), although it was found in the latter case that the thin graphite layers tended to form scrolls.

1.2 The isolation and identification of graphene

By the start of the 21st century, based on the exciting developments that were taking place in the field of carbon nanotubes, as outlined in the chapter on carbon nanotube composites. It became clear that both single-layer graphene and graphene-based composites could also have interesting physical properties. There was therefore enormous excitement following the report in 2004 of the preparation and isolation of individual graphene layers in Manchester by Novoselov, Geim, and coworkers through exfoliation using adhesive tape (Novoselov et al., 2004). They identified the material using AFM, found that the material was stable and, most significantly, showed that it was a semi-metal with spectacular electronic properties. This prompted a rapid upsurge in interest in the study of graphene culminating in Novoselov and Geim being awarded the 2010 Nobel prize in physics (Novoselov, 2011; Geim, 2011).

It is interesting to speculate why it took so long to isolate graphene. Very thin specimens are needed for transmission electron microscopy (TEM) and adhesive tape exfoliation was developed in the 1960s for the TEM analysis of layered materials, such as graphite, when electron microscopes became available commercially. In 1964, Baker and Kelly investigated the differences between irradiated and unirradiated graphite using TEM and stated that they prepared their specimens by "repeated cleavage using Sellotape" (Baker and Kelly, 1964). They were interested mainly in interfacial dislocations involving the twisting of the graphene layers that necessitated the graphene to be at least bilayer material. Kelly reported many years later (Kelly, 2013) that any monolayer material that they prepared was of no interest at the time, as the electron microscopes available in the 1960s did not have sufficient resolution to analyse the material. This should be contrasted with the current situation whereby atomic resolution can be undertaken routinely on graphene using state-of-the-art TEM (Robertson and Warner, 2013). What is perhaps of more interest is that following his early studies upon graphite, Kelly went on to have a distinguished career in the field of composites, such that he is sometimes referred to as "the father of composite materials" (Beaumont, 2014).

1.3 The superlative properties of graphene

Once the isolation of graphene had been reported, the research floodgates opened up and it was recognised that graphene was the foundation for all studies upon graphite and nanocarbon materials as shown schematically in Fig. 1.6.1 (Geim and Novoselov,

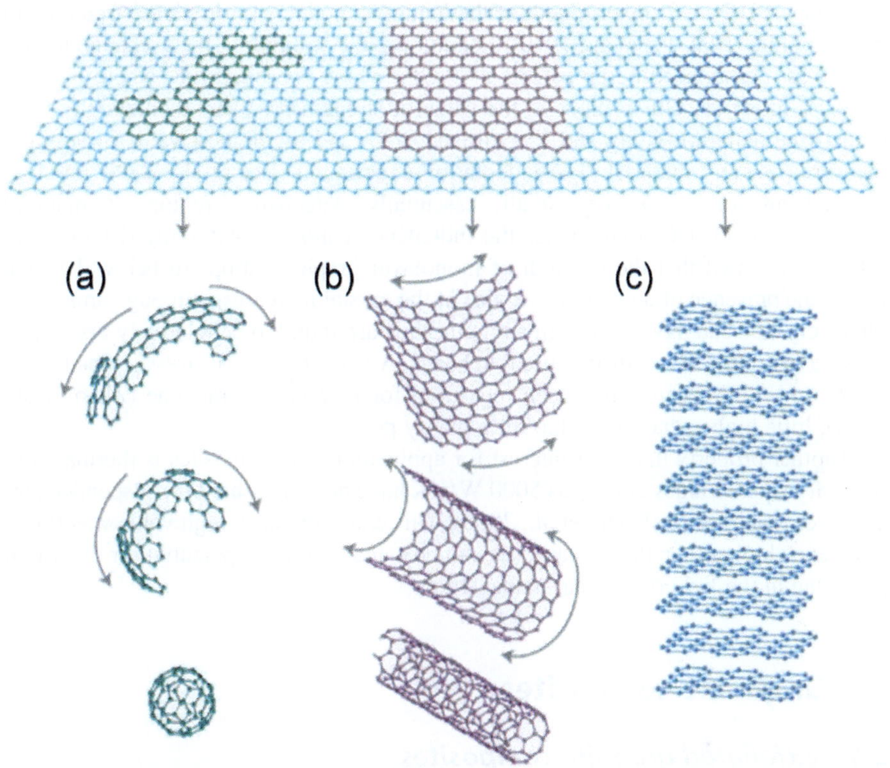

Figure 1.6.1 Graphene—the mother of all sp^2 carbon materials: (a) Fullerenes, (b) nanotubes and (c) graphite (Geim and Novoselov, 2007).
Reproduced with permission of Springer Nature.

2007). Initial interest was in the field of electronics whereby it was demonstrated that graphene could be obtained as high-quality crystals and behaves as a semi-metal, with the carriers acting as massless Dirac fermions (Geim and Novoselov, 2007) leading to very high levels of electrical conductivity in defect-free crystals (Castro Neto et al., 2009).

Of more interest to the field of composites was the 2008 study of Lee et al. (2008). They characterised the mechanical properties of monolayer graphene in an atomic force microscope (AFM) through the nanoindentation of graphene membranes, suspended over small holes on a silicon substrate. Stress-strain curves for the graphene were derived from the indentation experiments and it was found that failure occurred at strains of over 20%. An earlier study (Liu et al., 2007) had undertaken an *ab initio* calculation of the stress strain curve of graphene using density functional theory and close agreement was found between the predicted and measured curves (Lee et al., 2008). A value of Young's modulus of 1000 ± 100 GPa was determined from the indentation experiment which compares very well with the theoretical estimate (Liu et al., 2007) and contrasts with the value of 210 GPa for steel. It is also similar to the value of 1020 GPa determined for the Young's modulus for in-plane stretching in bulk graphite (Blakslee et al., 1970). In addition, the strength of the graphene monolayer was measured to be up to 130 ± 10 GPa which is around *E*/8, where *E* is the Young's modulus. This is close to the value of strength predicted theoretically for a defect-free material (Kelly and Macmillan, 1986) and the highest value of strength measured for any material. The indentation experiment (Lee et al., 2008) involves the deformation of a very small, essentially defect-free, region of material (~300 nm^2 or 11,000 atoms under the indenter). A more recent study (Zhao et al., 2019) has shown that the strength of monolayer graphene drops to below 10 GPa, due to the presence of defects, as the size of the crystal increases. Although small graphene crystals can have very high strength, they are found to be relatively brittle and only have limited levels of fracture toughness. A value of critical stress intensity factor K_c of ~4.0 MPa $m^{1/2}$ has been measured for monolayer graphene (Zhao et al., 2022), little higher than the value for a glassy polymer.

Another property that is of interest for applications in composites is thermal conductivity where values as high as 5000 W/mK have been measured for suspended single layer graphene (Balandin et al., 2008). This extraordinarily high value was found to be even higher than that of carbon nanotubes and pointed to potential for the use of graphene in the field of thermal management.

2. Graphene composites

2.1 Exfoliated graphite composites

The use of exfoliated graphite in polymer composites can be traced back to the patent literature from more than 100 years ago (Aylesworth, 1915, 1916) when Aylesworth reported the preparation of composites by incorporating expanded graphite (termed "fluffy graphite") into a composite with a thermosetting phenolic resin binder. He suggested that the fluffy graphite, exfoliated by acid-intercalation and high-temperature treatment, is "composed of a large number, probably many hundreds, of microscopically thin laminae superposed on each other like the leaves in a book" (Aylesworth, 1915). A number of applications of this "new composition of matter" were proposed including valve seats, gaskets or frictionless bearing blocks.

Work continued on the exfoliation of graphite using strong acids (Carr, 1970) and in 1987 Chung (1987) reviewed the state-of-the-art in both the reversible and irreversible exfoliation of graphite, in view of the many applications of the exfoliated material in gaskets, seals and packing. She pointed out that carbon fibres could also be exfoliated and composites could also be produced with thermoplastic polymers. It was also recognised that composites consisting of exfoliated graphite could be made to be electrically conductive (Wang et al., 1986) and by 2002 the technology was well established (Celzard et al., 2002) along with the development of composites with graphite nanosheets (Chen et al., 2003).

An important step forward in the development of nano-graphite reinforcements (nanoplatelets) for nanocomposites was the 2003 thesis of Fukushima (2003) followed by his patent with Drzal (Drzal and Fukushima, 2013). This study was inspired by earlier studies upon nano-clay reinforced polymers (Usuki et al., 1993; Pinnavaia and Beall, 2001). It was demonstrated that nano-sized graphite materials, suitable for incorporation into nanocomposites, could be produced through a combination of acid intercalation and microwave heating, followed by pulverisation. Nanoplatelets were produced initially with a diameter of the order of 1 μm and a thickness of around 10 nm, giving a specific surface area of 100 m^2/g (Fukushima, 2003). It was found that addition of the nanoplatelets to a variety of different polymers led to nanocomposites with impressive properties, presented initially in a Ph.D. thesis (Fukushima, 2003) and subsequently in a series of research papers by Drzal, Fukushima and others (Cho et al., 2005; Kalaitzidou et al., 2007, 2008). It was demonstrated that the addition of the original nanoplatelets to a polyimide resin led to an increase in the storage modulus and a reduction in thermal expansion coefficient (Cho et al., 2005). A subsequent study using 10 nm thick nanoplatelets with different lateral dimensions (1 and 15 μm) in polypropylene showed an improvement in flexural stiffness and electrical percolation at a loading as low as 0.3 vol% for the larger nanoplatelets (Kalaitzidou et al., 2007). A further study looked crystal nucleation and electrical percolation on the same materials (Kalaitzidou et al., 2008).

2.2 Graphite oxide composites

An alternative route to the preparation of graphite-based composites is through the use of graphite oxide which is readily obtainable in large quantities, and relatively easy to exfoliate and disperse in a polymer matrix. It also has built-in functional groups available for bonding to form strong interfaces with polymer matrices. Graphite oxide also has a major advantage for processing in that it can be exfoliated in water so that nanocomposites can be readily prepared with the use of water-soluble polymers. This was recognised as early as 1997 using polymers such as poly(ethylene oxide) (Matsuo et al., 1997) and in 2004 in poly(vinyl alcohol) (Hirata et al., 2004) both with graphite oxide.

Following the isolation and identification of graphene in 2004 (Novoselov et al., 2004) the terminology for "graphite oxide" was replaced by "graphene oxide" although it seems that differences in these materials are essentially a matter of semantics. Important papers then appeared in 2006 (Stankovich et al., 2006) and 2008 (Ramanathan et al., 2008) on "graphene-based composites," respectively, using reduced or functionalized graphene oxide. It was later found that nanocomposites could also be prepared using solution-based methods with non-water-soluble polymers such as poly (methyl methacrylate) (Das et al., 2009) and polyurethanes (Kim et al., 2010b) by modifying the graphene oxide chemically.

It was demonstrated in 2011 that graphene oxide produced using Hummers' method is composed of functionalized graphene sheets decorated by strongly-bound oxidative debris that acts as a surfactant to stabilise the aqueous suspensions (Rourke et al., 2011). The accepted model proposed for the structure of graphene oxide is presented in Fig. 1.6.2. The oxidative debris can be removed from the graphene-like

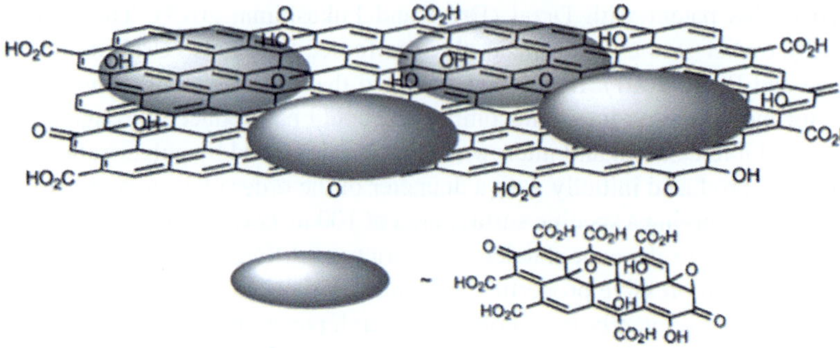

Figure 1.6.2 The structure of graphene oxide produced by Hummers' method showing the oxidative debris that can be removed by base washing (Rourke et al., 2011). Reproduced with permission of John Wiley and Sons.

sheets under basic conditions and they becomes electrically conducting. Moreover, this base-washed graphene oxide cannot easily be resuspended in water.

The vast majority of these early studies upon graphene nanocomposites were undertaken using graphene oxide, often in its reduced form (Kim et al., 2010a; Young et al., 2012). The mechanical properties of graphene oxide are, however, inferior to those of graphene due to the disruption of the structure through oxidation and the presence of sp^3 rather than sp^2 bonding. This also leads to a reduction in the electrical conductivity. The electrical and mechanical properties of graphene oxide can be improved greatly through reduction, either chemically or thermally (Dreyer et al., 2010). High-resolution TEM has been used to study the structure of both graphene oxide (Young et al., 2012) and reduced graphene oxide (Erickson et al., 2010). The structure of graphene oxide is remarkably similar to that of graphene at low magnification and the electron diffraction patterns are virtually identical. This means that the sheets of graphene oxide are not completely amorphous, and that the structure has both short- and long-range order (Wilson et al., 2009). Suk and coworkers undertook a study of the AFM indentation of a suspended film of graphene oxide that had not been reduced (Suk et al., 2010). They measured a Young's modulus of 208 ± 23 GPa when 0.7 nm was used for the effective thickness of the graphene. This is around 20% of the Young's modulus of graphene due to both damage to the graphene lattice and the higher effective thickness of the GO sheets as the result of the presence of the functional groups.

2.3 Nomenclature and terminology

We have shown that the development of graphene and graphene-based composites spans back over many years and the terminology employed has changed over that period of time. Moreover, different groups of workers have used their own terminology, abbreviations and description for their materials that can cause confusion. Attempts to clarify the situation and categorise the graphite and graphene family of material have been made by Bianco et al. (2013) and Wick et al. (2014). Three fundamental characteristics were identified for carbon-based two-dimensional materials; the number of layers, the C/O ratio and the lateral dimensions as shown in Fig. 1.6.2. The classification of graphene-based materials can be undertaken through consideration of these three properties. Monolayer graphene is the one-atom-thick material in which sp^2-bonded carbon atoms are arranged hexagonally. Few-layer graphene is made up of 2-5 sheets of graphene, and multi-layer graphene consists of 5–10 graphene layers. Graphite nanoplatelets consist of graphene sheets but their lateral dimensions/thickness is larger than those of many-layer graphene (more than 10 layers, thickness of less than 100 nm). Exfoliated graphite is the multi-layer material that is prepared by the partial exfoliation of graphite and its 3D crystal stacking is retained.

Graphite oxide, produced from the oxidation of graphite using strong oxidising agents, is termed graphene oxide following the exfoliation of graphite oxide using techniques such as sonication. The C/O ratio of GO is normally between two and three and the same terminology can be used for GO that is used for graphene in terms of the different numbers of layers. Finally, the term "reduced graphene oxide" (rGO) is used for GO material following reduction, whereby the oxygen content is reduced significantly (Fig. 1.6.3).

In 2017 a good practice guide was published by the UK National Physical Laboratory on the characterisation and nomenclature of graphene (Pollard et al. (2017)).

2.4 Model graphene composites

2.4.1 Raman spectroscopy and deformation

One of the many interesting aspects of the history of graphene and related materials has been the way in which Raman spectroscopy has been employed in its characterisation (Li et al., 2023). The Raman effect was first discovered from the inelastic scattering of light by liquids in the 1920s in India and Chandrasekhara Venkata Raman was awarded the 1930 Nobel Prize in Physics. There was soon a rapid upsurge in interest in the investigations of Raman scattering from wide range of liquids and crystals with the Raman spectrum of diamond first being reported in 1930 (Ramaswamy, 1930; Robertson and Fox, 1930). Graphitic carbons absorb light strongly and it appears that attempts to obtain their Raman spectra were unsuccessful until lasers became available in the 1960s (Li et al., 2023). An important development was the reports of the Raman spectra of graphite (Tuinstra and Koenig, 1970b) and carbon fibres (Tuinstra and Koenig, 1970a), met initially with considerable scepticism (Li et al., 2023), showing well-defined spectra from sp^2 carbon materials.

In the following years, interest in using the Raman technique in the analysis of carbon materials developed slowly. The technique came to the forefront when it was found that it was extremely useful for the characterisation of new-discovered nanocarbon materials such as C_{60} fullerenes (Cooper and Young, 1999) and carbon nanotubes (Dresselhaus et al., 2005). In particular, it was found to be invaluable for the characterisation of graphene (Ferrari et al., 2006). The lack of a band gap in graphene means that all wavelengths of the incident laser radiation are resonant such that the Raman spectrum of graphene is massively resonantly enhanced and so very strong.

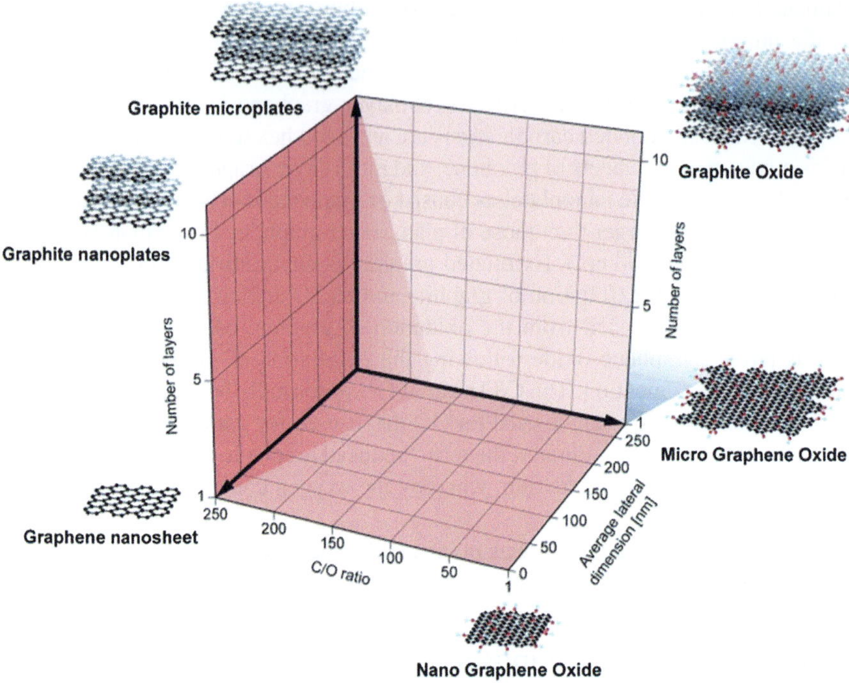

Figure 1.6.3 Classification for different graphene-type materials based on the number of layers, the C/O ratio, and the lateral dimensions (Wick et al., 2014).
Reproduced with permission of John Wiley and Sons.

Remarkably, this enables well-defined Raman spectra to be obtained from single graphene monolayers that are only one atom thick, and moreover, the ability to "count" the number of layers in multilayer graphene materials (Ferrari et al., 2006; Gupta et al., 2006; Malard et al., 2009) as shown in Fig. 1.6.4.

In addition to its use for the characterisation of the structure of graphitic materials, an important finding in the field of composite mechanics was the discovery that the position of the Raman bands in such materials is sensitive to the application of stress or strain. This was first demonstrated for the deformation of carbon fibres in the 1980s (Robinson et al., 1987; Sakata et al., 1988). In combination with the development of the Raman microscope (Williams et al., 1994), this led to the ability to obtain spectra from regions as small as 1 μm in individual carbon fibres in composites, giving a completely new insight into the micromechanics of deformation of carbon fibres and nanotubes in composites (Li et al., 2023). Another significant finding for carbon fibres was that the rate of band shift (in cm^{-1}/% strain) was found to be proportional to the Young's modulus of the fibres (Cooper et al., 2001). Following the original report of the isolation and identification of monolayer graphene (Novoselov et al., 2004), there were a number of attempts to measure Raman band shifts during the deformation of monolayer graphene. Unfortunately, a variety of shift rates (Raman band position

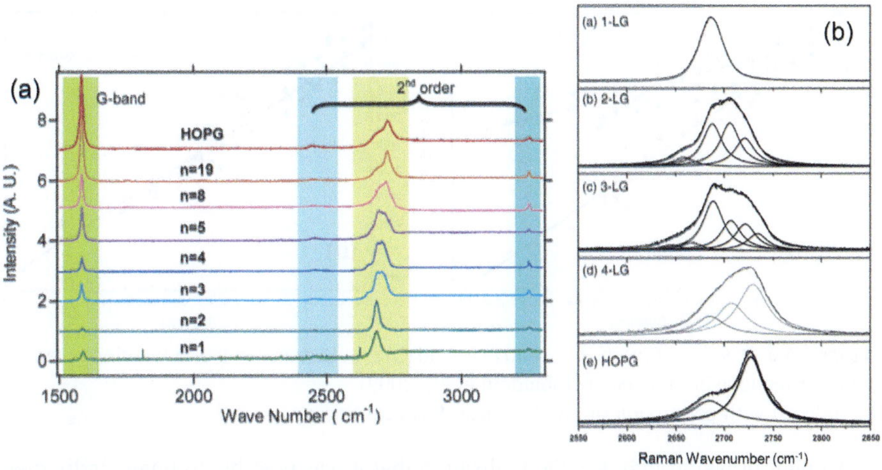

Figure 1.6.4 (a) Raman spectra of graphene with different numbers of layers, and HOPG. (b) Measured Raman 2D band of graphene with different number of layers and HOPG, with the corresponding peak fitting.
(a). Reprinted (adapted) with permission from (Gupta et al., 2006). (Reproduced with Permission of the American Chemical Society). (b). Reprinted from (Malard et al., 2009). (Reproduced with permission of Elsevier).

as a function of strain) were reported as such measurements often suffered from slippage of the graphene on the substrate (Li et al., 2023). A careful study of the deformation of monolayer graphene combined with theoretical analysis (Mohiuddin et al., 2009) showed that the G and 2D bands shifted approximately linearly with strain and that there was splitting of the G bands as the result of the loss of symmetry from the distortion on the hexagonal graphene lattice (Fig. 1.6.5). Subsequently, it was confirmed that there was a universal relationship for the stress or strain induced Raman band shifts for all sp^2 carbon materials including carbon fibres, nanotubes, and graphene (Frank et al., 2011; Li et al., 2023).

2.4.2 Single graphene flake composites

It was not clear in the early studies of the mechanical properties of graphene nanocomposites how the mechanics of reinforcement might be elucidated. Only modest improvement in stiffness were found on the addition of graphene to polymer matrices (Kim et al., 2010b) and it was suggested (Kim et al., 2010a) that the Mori-Tanaka theory (Mori and Tanaka, 1973) might be used to analyse the behaviour. An alternative approach investigated (Young et al., 2012) was to use a self-consistent micromechanics method developed by Hill (Halpin and Kardos, 1976). Unfortunately, neither of these approaches enabled the detailed micromechanics of reinforcement to be elucidated.

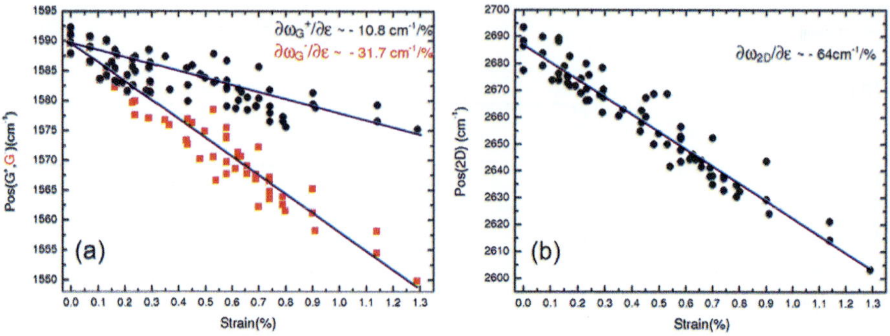

Figure 1.6.5 Position of (a) the ω_{G-} and ω_{G+} and (b) ω_{2D} Raman bands as a function of tensile strain for monolayer graphene (Mohiuddin et al., 2009).
Reproduced with permission of the American Physical Society.

A major breakthrough was the realisation that it was possible to image individual graphene flakes in polymer matrix in an optical microscope and use Raman spectroscopy to follow their deformation (Gong et al., 2010). This involved preparing a model composite consisting of monolayer graphene sandwiched between thin polymer layers as shown in Fig. 1.6.6. Graphene monolayer absorbs around 2.3% of visible light enabling it to be imaged even though it is only one atom thick (Blake et al., 2007). The model composite can be deformed by bending the PMMA substrate and Fig. 1.6.6(c) shows the reversible shift of the 2D Raman band obtained from a spot of the order of 1 μm diameter on the monolayer flake. Since it is known that the 2D band undergoes a well-defined shift with applied strain it is then possible to map the strain from point to point in the monolayer using the Raman microscope translation stage with Fig. 1.6.6(b) as the strain calibration. The distribution of strain

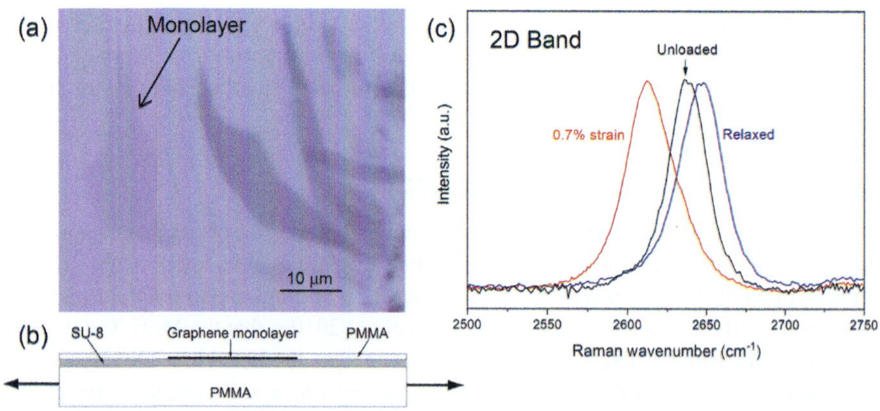

Figure 1.6.6 Single monolayer graphene model composite. (a) Optical micrograph showing the monolayer graphene flake investigated. (b) Schematic diagram (not to scale) of a section through the composite. (c) Change in the position of the Raman 2D band during loading and unloading of the graphene composite (Gong et al., 2010).
Reproduced with permission of John Wiley and Sons.

along the monolayer was found to build up from zero at the end of the flakes and then plateau in the middle, following the predictions of shear lag theory (Kelly and Macmillan, 1986; Young et al., 2012; Papageorgiou et al., 2017). Moreover, the strength of the interface between the graphene and polymer matrix was found to be of the order of 2 MPa. This confirmed that conventional composite micromechanics could be used to model deformation and stress transfer in the model graphene nanocomposite. Subsequently, one of the authors of this study (Gong et al., 2010), Kostya Novoselov, was awarded the 2010 Nobel Prize in Physics.

The original study upon model composites was developed and extended by the authors (Young et al., 2011; Gong et al., 2012, 2013) and others (Wang et al., 2016). Investigations of the deformation micromechanics were undertaken using few-layer flakes (Gong et al., 2012) as shown in Fig. 1.6.7(a) and it was found that shear lag theory was still obeyed for monolayer and bilayer as seen in Fig. 1.6.7(b). It was also demonstrated (Young et al., 2011) that the *x-y* stage of the Raman microscope could be employed to map the distribution of strain in a single monolayer flake (Fig. 1.6.7(c)) and detect cracking at high strain as shown in Fig. 1.6.7(d). Subtle changes that were found in the shape of the 2D Raman band during deformation of few-layer graphene also showed that easy shear could occur between the graphene layers leading to a reversible loss of Bernal stacking (Gong et al., 2013).

Fig. 1.6.8 shows the results of study upon effect of graphene functionalisation (using a O_3/H_2O gaseous mixture) on the interfacial stress transfer between a monolayer and PMMA polymer matrix (Wang et al., 2016). The distribution of strain in monolayer, unfunctionalized graphene was mapped and fitted successfully with the shear-lag model (Fig. 1.6.8(a) with interfacial sliding being observed at 0.7% strain. The gaseous mixture successfully functionalized the graphene and it formed hydrogen bonds with the carbonyl groups of the PMMA. This resulted in better stress transfer than with the pristine graphene and an interfacial shear strength some 4× higher than of the unfunctionalized material (Fig. 1.6.8(b)).

2.5 The mechanics of graphene nanocomposites

Once the study of model composites demonstrated that conventional micromechanics could be employed to elucidate how graphene interacts with the matrix, it was then possible to develop micromechanics to model and predict the overall properties of bulk graphene composites (Papageorgiou et al., 2017, 2020). Several key micromechanical parameters had be considered, including the intrinsic graphene properties, the volume fraction of graphene, its aspect ratio, dispersion/orientation within the matrix, and the strength of the interface between graphene and the matrix. A fundamental concept in micromechanics is the rule of mixtures (ROM), which provides a first approximation of the effective mechanical properties of a composite's, such as its Young's modulus. This is based upon the properties and volume fractions of the individual components (i.e., the graphene and matrix material):

$$E_c = E_f V_f + E_m (1 - V_f) \tag{1.6.1}$$

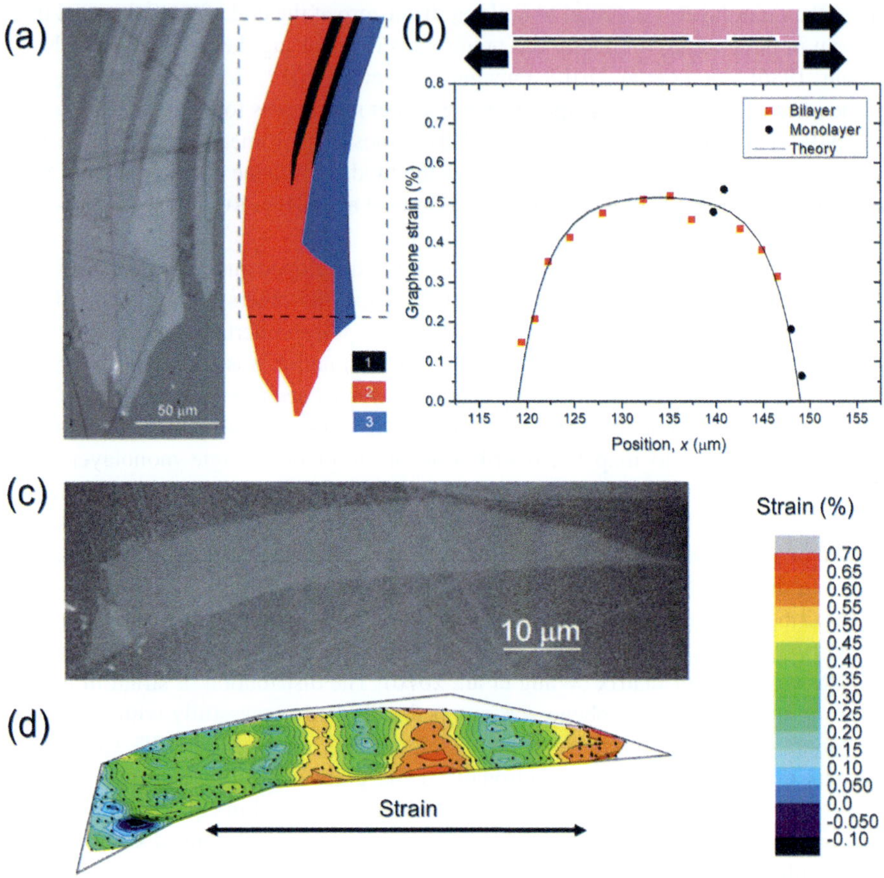

Figure 1.6.7 (a) Optical micrograph and schematic diagram of a graphene flake on a PMMA beam showing monolayer, bilayer, and trilayer regions. (b) Variation of strain in the monolayer and bilayer regions of graphene at an applied strain of 0.6% (Gong et al., 2012). (c) Optical micrograph of a graphene monolayer. (d) Contour maps of strain over the monolayer at 0.8% strain. The low strain regions are due to the presence of cracks (Young et al., 2011). Reproduced with permission of the American Chemical Society.

where E_c, E_f and E_m are the Young's modulus of the composite, filler, and matrix, and V_f is the volume fraction of the filler. Early studies (Kim et al., 2010a; Young et al., 2012) demonstrated that the addition of graphene to a polymer matrix led to a monotonic increase in E_c with filler volume fraction (Fig. 1.6.9(a)) as predicted by Eq. (1.6.1). However, the value of E_f need to fit the data was well below the 1000 GPa measured on isolated graphene flakes (Lee et al., 2008), typically of the order of a few GPa for rigid polymers (Guest et al., 2023) and 10s of MPa for rubbers (Li, 2018). This demonstrates that although the ROM is a useful starting point, it oversimplifies the complexities of the deformation of actual composite systems including filler orientation and stress transfer, necessitating the use of more refined models for accurate predictions.

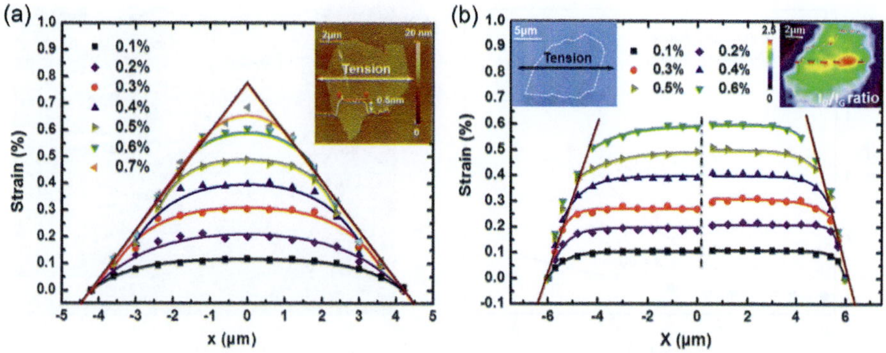

Figure 1.6.8 (a) Strain distribution of monolayer graphene in the direction of tensile axis (the inset shows an AFM image of the monolayer material), (b) strain distribution of oxidised graphene at different strain levels (the left inset shows the optical image of graphene sheets and the right inset represents the Raman map of the I_D/I_G ratio). The solid lines in the graph represent fits with the nonlinear shear-lag theory.
Reprinted (adapted) with permission from (Wang et al., 2016). (Reproduced with permission of the American Chemical Society).

As well as leading to an increase in stiffness, it was found that the addition of graphene to polymer matrices also gave rise to an increase in toughness. One of the main mechanisms was found to be the pull-out of the graphene from the polymer matrix (Liu et al., 2021), as shown in Fig. 1.6.9(b) leading to an increase in fracture energy. Fig. 1.6.9(c) and (d) show that there is an approximately linear increase in both tearing and excess fracture energies with the volume fraction of GNPs for an FKM rubber (Liu et al., 2021) and an epoxy resin (Chong et al., 2016), respectively.

A modification of the ROM was proposed to interpret the dependence of the nanocomposites Young's modulus on the volume fraction of filler:

$$E_c = \underbrace{E_f \eta_o \eta_l}_{E_{\text{eff}}} V_f + E_m (1 - V_f) \tag{1.6.2}$$

where η_o and η_l are the Krenchel orientation factor and length factor of the filler and E_{eff} is the effective Young's modulus of the filler; that is the modulus that contributes towards reinforcement. The length factor is given by the following equation: $\eta_l = 1 - \frac{\tanh\left(\frac{ns}{2}\right)}{\frac{ns}{2}}$ where s is the filler aspect ratio (width/thickness) and $n = \sqrt{\frac{2G_m}{E_f}\left(\frac{t}{T}\right)}$. The parameter G_m is the shear modulus of the matrix, t is the thickness of the flake and T is the thickness of the resin surrounding the flake.

The length factor η_l depends on the geometry of the individual graphene flakes, particularly their aspect ratio, s. For flakes with finite dimensions $\eta_l < 1$ and η_l only approaches unity as $s \rightarrow \infty$, pointing to the importance of using well-exfoliated

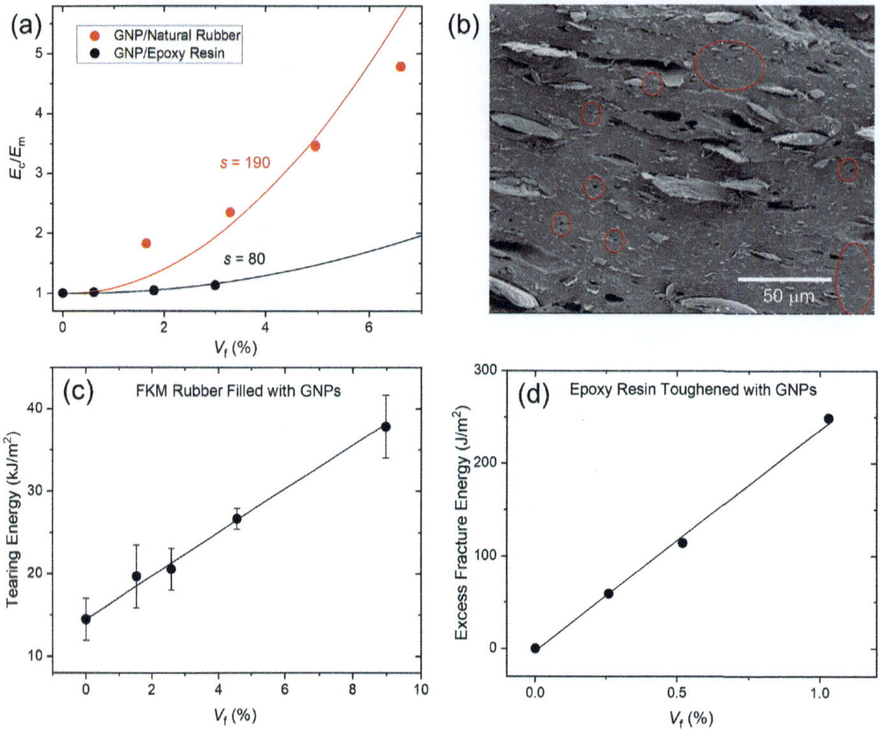

Figure 1.6.9 (a) Dependence of E_c/E_m upon V_f for natural rubber (Li, 2018) and an epoxy resin (Guest et al., 2023) reinforced with GNPs. Values of s were chosen to fit the data using Eq. (1.6.6). (b) SEM micrograph of the surface of an FKM elastomer nanocomposite reinforced with GNPs after failure by tearing (Liu et al., 2021). (Reproduced with the permission of Elsevier) (c) Dependence of the tearing energy of an FKM rubber upon the V_f of GNPs (Liu et al., 2021) (d) Dependence of the excess fracture energy of an epoxy resin on the V_f of GNPs (Chong et al., 2016).

graphene flakes with high aspect ratios (Coleman et al., 2011). Moreover, for $\eta_l = 1$ the theory requires that both the matrix and the filler behave elastically during deformation, with perfect interfacial bonding.

The orientation of the fillers, characterised by η_o, is critical in controlling E_c. According to Krenchel's original analysis (Krenchel, 1964), random orientation reduces the modulus of fibre- or nanotube-reinforced materials by a factor of five ($\eta_o = 1/5$). The Krenchel factor for a material with oriented reinforcement is equal to 1, whereas for a material reinforced with randomly-aligned nanoplatelets, it was shown to be 8/15 (Li et al., 2024). Random orientation of graphene-like (2D) fillers, therefore, reduces the modulus by just over one half compared to a fully-aligned material. Polarised Raman spectroscopy was found to be a useful tool for quantifying the spatial orientation of fillers in graphene-based nanocomposites (Fig. 1.6.10(a)) (Li et al., 2016). In addition, this study confirmed that the reinforcing efficiency of nanoplatelets is

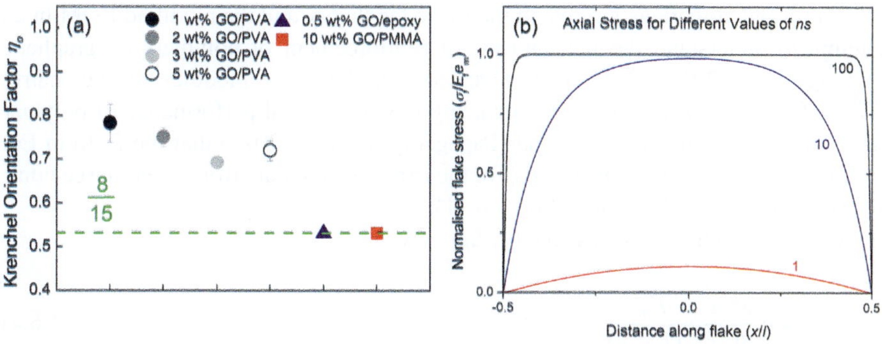

Figure 1.6.10 (a) The Krenchel factors of the GO-reinforced composites. The dashed line represents the Krenchel value of 8/15 for materials reinforced with randomly oriented nanoplatelets (Li et al., 2016). (b) Predicted variation of normalised axial stress along the length of a short nanoplatelet embedded in a matrix. The values for the product *ns* are provided for reference (Young et al., 2018).
Reproduced with permission from Elsevier.

significantly higher than that of fibre- or nanotube-based materials, when both are oriented randomly.

The efficiency of load transfer in graphene nanocomposites is determined primarily by the interfacial interactions between graphene and the matrix. The high aspect ratio of graphene facilitates efficient stress transfer from the matrix to the graphene sheets. Several mechanisms contribute to this process. The shear-lag model (Cox, 1952), developed originally for fibre-reinforced polymers, describes the transfer of stress from the matrix to the graphene through shear stresses at the interface, with the effectiveness of this mechanism depending on the strength of the interface and the aspect ratio of the graphene sheets. For a given matrix strain e_m, the variation of strain in the flake e_f, with position x, across the monolayer, the shear-lag theory takes the form:

$$e_f = e_m \left[1 - \frac{\cosh\left(ns\frac{x}{l}\right)}{\cosh\left(\frac{ns}{2}\right)} \right] \tag{1.6.3}$$

where, l is the length of graphene flake in the x direction and s is its aspect ratio. According to the theory, the shear stress at the ends of the filler is at its maximum and decreases to zero after half of the critical length (Fig. 1.6.10(b)). In addition, the tensile stress at the ends of the filler starts at zero and increases to its maximum value after reaching half of the "critical length" of reinforcement (l_c). From Fig. 1.6.8, it can be seen that the strain increases to approximately 90% of the plateau within about 1.5 μm from the flake's edge, indicating a critical reinforcement length of around 3 μm. Based on the common assumption that effective reinforcement requires flakes or fibres to be 8–10 times longer than the critical length, flakes in the range of 25–30 μm are necessary to achieve highly efficient reinforcement in composites.

Young and coworkers combined the shear-lag theory and the modified ROM in an attempt to understand the mechanisms of reinforcement of polymers by graphene (Young et al., 2018). This work revealed that the shear modulus of the matrix ($G_m = E_m/2(1+\nu)$) is crucial in enhancing the mechanical performance of polymer nanocomposites since it was found (Papageorgiou et al., 2017) that the E_f from Eq. (1.6.1) is directly proportional to E_m for the experimental data from a very large number of studies in the literature (Fig. 1.6.11).

According to the shear-lag theory, E_f is given by

$$E_f \approx \eta_o \frac{s^2}{6} \frac{t}{T} \frac{E_m}{2(1+\nu)} \tag{1.6.4}$$

where s is the aspect ratio of the filler, and t/T is related to the proximity of neighbouring particles and can be assumed that $t/T \backsim V_f$. Combining Eqs. (1.6.1) and (1.6.4), if $E_f >> E_m$, one can obtain

$$E_c \approx E_m \left[1 - V_f + \frac{s^2}{12} \frac{t}{T} \frac{\eta_o}{2(1+\nu)} V_f \right] \tag{1.6.5}$$

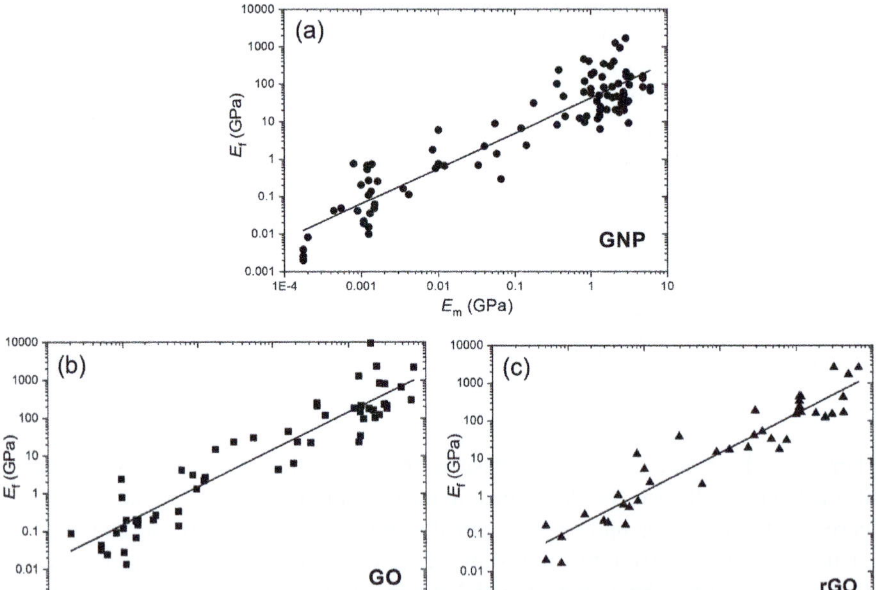

Figure 1.6.11 The modulus of graphene-related materials as a function of the modulus of the matrix for (a) GNPs, (b) GO and (c) rGO. The continuous lines represent linear fits to the data, with a slope of 1.
Adapted from (Papageorgiou et al., 2017) (Reproduced with permission from Elsevier).

by substituting t/T with V_f then

$$E_c \approx E_m\left[1 - V_f + \frac{s^2}{12}\frac{\eta_o}{2(1+\nu)}V_f^2\right] \tag{1.6.6}$$

Notably, this equation highlights that, for most polymers, the aspect ratio and orientation of the fillers alongside the modulus of the matrix, are more critical towards reinforcement than the inherent modulus of the fillers. It is shown in Fig. 1.6.9(a) that this equation can be used to fit experimental modulus data for nanocomposite with matrix materials ranging from flexible polymers such as natural rubber to rigid polymer such as an epoxy resin.

Earlier studies (Young et al., 2012) used semi-empirical micromechanical equations to model the reinforcement of graphene-based composites, such as the ones proposed by Halpin and Tsai (Halpin and Kardos, 1976) which can be expressed as:

$$E_c = E_m\left(\frac{1+\xi\eta V_f}{1-\eta V_f}\right)$$

where $\eta = \frac{(E_f/E_m)-1}{(E_f/E_m)+\xi}$ and ξ is related to the filler aspect ratio l/d (length/diameter equivalent to s), typically expressed as $\xi = \frac{2\left(\frac{l}{d}\right)}{1+\frac{l}{d}}$. This parameter was intended to capture the influence of filler shape on the modulus of the composite, making it a critical component of the model. In the context of nanocomposites, however, the use of the shape parameter ξ introduces several significant limitations about filler shape, orientation, homogeneity and interfacial bonding; this parameter therefore ended up being used simply as a fitting parameter.

Despite the significant progress made in understanding the micromechanics of graphene nanocomposites, several manufacturing challenges still remain today. Achieving uniform dispersion of graphene continues to be a major challenge, as aggregation and agglomeration of the flakes was shown to impair the properties of the composites significantly (Li et al., 2018). In addition, developing methods to enhance the graphene-matrix interface without compromising other properties (for example, functionalisation without the loss of conductivity) remains another research challenge.

The understanding of micromechanics is not only essential for structural applications, where graphene nanocomposites are designed with tailored properties for specific load-bearing applications, but also for functional applications (Tomes et al., 2024). In functional applications, such as electronics, sensors, and energy storage devices, the mechanical properties of graphene nanocomposites are these days often coupled with their electrical and thermal properties to create multifunctional materials.

2.6 *Multifunctionality in graphene-based nanocomposites*

While the extraordinary mechanical properties of graphene originally garnered significant attention for nanocomposites, its influence extends far beyond structural reinforcement. Multifunctional graphene nanocomposites now represent a new frontier in material science, offering the possibility of integrating multiple desirable properties into a single material system. Unlike traditional composites, that are often designed with a singular focus, such as mechanical reinforcement, graphene nanocomposites were found to simultaneously provide enhancements in electrical conductivity, thermal management, electromagnetic interference (EMI) shielding, and chemical resistance, among others. This multifunctionality has the potential to open up a vast array of applications across industries ranging from electronics and energy storage to environmental protection and aerospace engineering. It should be stated clearly at this point though, that "graphene-based" nanocomposites that were reported in the vast majority of publications, referred to materials multiple layers (a few nm) thick rather than a single layer of material. As a 2D material, graphene has a high surface area, which is advantageous for applications that require significant interaction with the environment, such as sensing or catalysis. Its high electrical conductivity makes it an excellent candidate for electronic applications, while its thermal conductivity allows it to efficiently manage heat in high-performance devices. Moreover, the chemical stability and impermeability of graphene make it an ideal candidate for barrier applications, protecting underlying any materials from corrosion, oxidation, or chemical attack.

2.6.1 *Enhanced conductivity in graphene nanocomposites*

Monolayer graphene is renowned for its exceptional electrical conductivity, with an intrinsic carrier mobility of 250,000 $cm^2\ V^{-1}s^{-1}$ at room temperature (Novoselov et al., 2005). When incorporated into polymer matrices, graphene was found to transform an insulating material into one with significant electrical conductivity. The ability to achieve percolation at relatively low graphene loading levels was shown to be one of its key advantages (Marsden et al., 2018) even though it was found that the high aspect ratio and 1D geometry of single- or multi-walled carbon nanotubes could lead to the formation of a percolation network at lower filler loadings (Fig. 1.6.12(a)) (Du et al., 2011). Graphene, however, was found to offer an advantage over CNTs in terms of processability, as graphene composites exhibit significantly lower viscosity than CNT composites for the same filler loadings. This allows for a greater amount of graphene to be incorporated into the matrix, resulting in improved conductivity values.

Potential applications of electrically-conductive graphene nanocomposites are predicted to be vast, ranging from flexible electronics (Han et al., 2017) and conductive inks (Tran et al., 2018) to antistatic coatings (Cui et al., 2019). In particular, the use of graphene in flexible electronics offers promising prospects for wearable devices, where the combination of flexibility, lightweight, and conductivity is crucial. Furthermore, graphene's tuneable conductivity, that can be adjusted by altering its functionalisation or careful dispersion/orientation within the matrix, allows for precise control

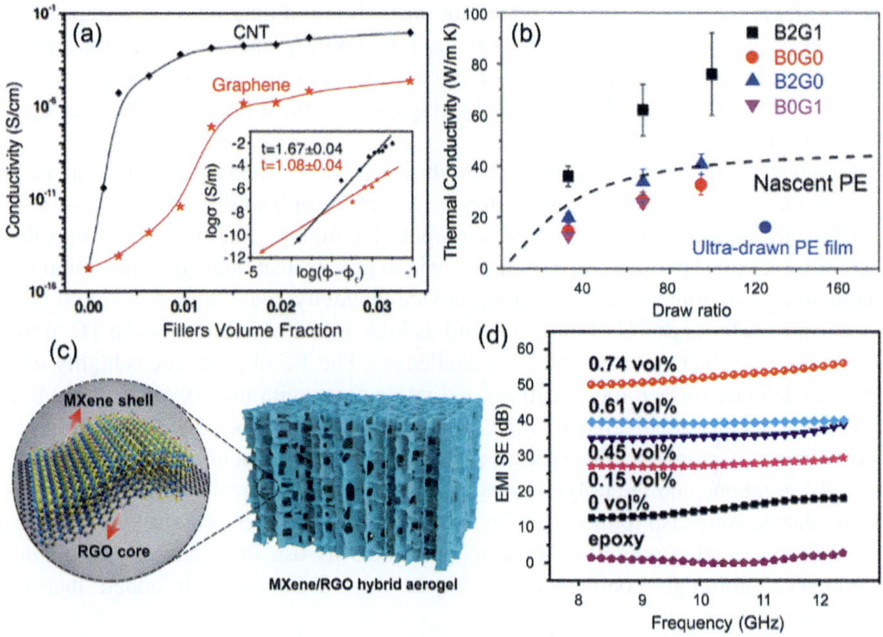

Figure 1.6.12 (a) An example of the percolation threshold in high-density polyethylene (HDPE) composites is demonstrated using carbon nanotubes (CNTs, black) and graphene nanosheets (red). (b) Comparison of thermal conductivity of graphene/UHMWPE samples at different draw ratios (Pan et al., 2019). (c,d) $Ti_3C_2T_x$ MXene/rGO/epoxy hybrid aerogel and the EMI SE as a function of MXene content (Zhao et al., 2018).
(a). Adapted from (Du et al., 2011), (b). (Reproduced with permission from John Wiley and Sons) (Reproduced with permission from Elsevier), (c,d). (Reproduced with permission from the American Chemical Society).

over the electrical properties of the composite, making it suitable for a variety of electronic applications.

2.6.2 Thermal conductivity enhancement in graphene nanocomposites

The intrinsic thermal conductivity (TC) of monolayer graphene is remarkably high, being measured to be up to 5000 W/mK in pristine monolayers (Balandin et al., 2008). Graphene was found to enhance the thermal conductivity of a matrix material but the thermal conductivity of many of the graphene nanoplatelets that are used in polymer nanocomposites is significantly lower than that of monolayer graphene. In polymer matrices such as epoxy resins, the TC values were found to increase only from ~0.2–0.3 W/mK (typical for polymers) to ~1 W/mK with the addition of ~10 wt% graphene. Nevertheless, smart engineering techniques have been developed to achieve considerably higher thermal conductivity enhancements with graphene-based nanocomposites, including fabrication of a 3D interconnected network whereby values of 4.5 W/mK have been reported for the addition of only 0.96 vol% doped GO materials (Yuan et al., 2019), or by drawing UHMWPE/graphene/2-(2*H*-benzontriazol-2-yl)-4,6-ditertpentylphenol nanocomposites (Pan et al., 2019), which achieved specific TC values of ~75 W/mKρ, surpassing those of most metals and nanocomposites (Fig. 1.6.12(b)).

Graphene nanocomposites have been shown to be especially useful as thermal interface materials (TIMs) and heat sinks, where efficient heat dissipation is essential (Shahil and Balandin, 2012). They were proposed to highly effective in electronic devices such as smartphones and computers, where proper heat management is vital for maintaining performance and extending device longevity. Nevertheless, even though the exceptionally-high TC of graphene and its high aspect ratio enhance the TC of the nanocomposite, there were found to be challenges. The TC of graphene is highly sensitive to defects, surface modifications and interactions with the polymer matrix that can reduce its TC drastically. Its high aspect ratio leads to poor dispersion, folding, and twisting of graphene sheets, further lowering TC. Interfacial thermal resistance between graphene and the polymer matrix, as well as contact resistance between graphene sheets, were also found to hinder its effective thermal performance. In addition, the high electrical conductivity of graphene limits its use in electrically-insulating composites, although coatings can mitigate this, although with added thermal resistance.

2.6.3 Graphene-based sensors

The high surface area and chemical sensitivity of graphene make it an excellent candidate for sensing applications. Graphene-based nanocomposites have been developed for chemical, biological, and environmental sensors, where the detection of minute changes in electrical conductivity can signal the presence of specific analytes (Cui et al., 2019). For instance, it was found that the sensitivity of graphene to changes in its environment could be harnessed to detect gases, biomolecules and temperature changes with high precision by following specific measurable properties of graphene such as electrical resistance, capacitance, or optical characteristics. In gas sensors, graphene nanocomposites were shown to be able to detect trace amounts of gases such as ammonia, nitrogen dioxide, or volatile organic compounds at very low concentrations, often in the parts per billion range (Varghese et al., 2015). The responsiveness of graphene to these gases was found to be enhanced when combined with other materials, such as metal oxides, that could increase the sensitivity and selectivity of the sensor further.

2.6.4 Electromagnetic interference (EMI) shielding effectiveness of graphene nanocomposites

Graphene nanocomposites were also found to exhibit excellent electromagnetic interference (EMI) shielding properties, making them suitable for applications in electronics and telecommunications. The high electrical conductivity of pristine graphene and its large surface area enable it to absorb and reflect electromagnetic waves effectively. Nevertheless, the shielding effectiveness of graphene-based composites was found to be determined largely by the connectivity of the conductive network and the electromagnetic properties of the conductive fillers.

In polymer-graphene nanocomposites, the EMI shielding effectiveness (SE) has been shown to reach 20–30 dB with as little as 10 wt% graphene loading, sufficient to block 99.99% of incident electromagnetic radiation (Xia et al., 2022). By introducing higher graphene loadings, or tuning the filler orientation, the shielding effectiveness can increase to over 50 dB, which is typically required in high-performance applications such as military and aerospace equipment. The orientation of graphene-related materials in polymers leads to high electrical conductivity, which is favourable for reflection and absorption of electromagnetic waves; as a result, a number of studies have focused on creating oriented graphene composites or graphene-based composites with 3D interconnected porous structures. Wei and colleagues demonstrated the fabrication of an ultrathin, aligned pristine graphene nanosheet films using a scalable scanning centrifugal casting (SCC) method (Wei et al., 2020). This highly-aligned laminated structure exhibited excellent electrical conductivity and an impressive electromagnetic (EM) wave shielding effectiveness (SE) of 93 dB at a thickness of approximately 100 μm (Wei et al., 2020). Furthermore, dual filler 3D threshold networks have been shown to improve shielding performance due to the synergistic effects of multiple conductive fillers. Zhao and colleagues developed $Ti_3C_2T_x$ MXene/reduced graphene oxide (rGO) hybrid aerogels featuring double 3D threshold networks using hydrothermal and directional freeze-drying techniques (Zhao et al., 2018). These composites formed an aligned honeycomb microstructure (Fig. 1.6.12 (c)), with graphene sheets serving as the inner skeleton and $Ti_3C_2T_x$ sheets as the outer shell. Thanks to the synergistic effects of the dual 3D networks created by the hybrid fillers, the composites were found to achieve a conductivity of 695.9 S/m and a shielding effectiveness of 50 dB with a $Ti_3C_2T_x$ content of only 0.74 vol% (Fig. 1.6.12(d)).

Compared to conventional materials such as metals (e.g., copper and aluminium), graphene nanocomposites have been found to offer similar EMI shielding performance at significantly lower weights and with added flexibility. For example, while metals can achieve shielding effectiveness of over 50 dB, they are heavier and less adaptable to applications requiring lightweight, flexible shielding. In contrast, graphene-based composites have been shown to provide excellent shielding with minimal impact on weight, making them ideal for modern electronic devices.

3. The road to market adoption

Following its isolation in 2004, initial research into graphene focused on understanding its fundamental properties and demonstrating proof-of-concept applications. Early studies established the remarkable features of graphene, leading to widespread interest from both academia and industry. During the early years of graphene, however, the production of high-quality graphene remained limited to laboratory-scale techniques such as mechanical exfoliation and chemical vapour deposition (CVD). These methods, while effective for research purposes, were not fully scalable for mass production, posing a significant barrier to commercial adoption. An analysis of data from the Chemical Abstracts Service (CAS) upon the trends in the research and development of graphene up to 2016 (Zou et al., 2018) sheds some light upon early activities in the field. They found that over 78,000 research papers and more than 23,000 patents had been published up to this date.

Despite the excitement surrounding the potential of graphene, early research revealed a critical challenge in graphene-based nanocomposites, translating these exceptional properties from the nanoscale to macroscopic applications. The intrinsic properties of monolayer graphene were difficult to be reproduced on the performance of bulk composites, since it is impractical and not cost-effective to reinforce polymers with monolayer graphene. For this reason, a number of multilayer graphene-related materials appeared in the market, but without the presence of trustworthy and verified standards, the adoption of these materials from industry was challenging (Kovtun et al., 2019). Commercial graphene-related materials (GRMs) often exhibit high levels of defects, with many samples having sp^2 fractions below 95%, indicating significant disruptions in the graphene lattice. Achieving high exfoliation grades, such as specific surface areas (SSA) comparable to ideal monolayer graphene (2600 m^2/g), remains unattainable at the industrial scale without introducing substantial damage to the sp^2 network. It was found that there was substantial variability in exfoliation, thickness, and lateral size among commercial GRMs, further compounded by inconsistent measurement techniques across producers (Kovtun et al., 2019). Fig. 1.6.13 illustrates how commercial graphene-related materials fall short of ideal graphene across key properties such as defect density, lateral size, SSA, and bulk density. These inconsistencies, driven by varying production methods and a lack of standardised characterisation frameworks, underscore the need for pragmatic benchmarking approaches. Despite ongoing efforts by organisations such as ISO to address these issues, the study of Kovtun et al. (2019) confirms that producing high-quality, consistent graphene remains a major industrial challenge.

The next phase in the journey of graphene to market adoption focused on scaling up production and improving the quality of graphene-related materials. Several methods emerged as potential solutions, each with its own advantages and drawbacks: 1) chemical vapour deposition can produce high-quality monolayer graphene with excellent electrical properties, but the process is limited mainly from the challenges in transferring graphene onto different substrates, 2) liquid-phase exfoliation that involves the exfoliation of graphite in solvents offers higher yields but mainly produces graphene with defects and multilayer structures, affecting its electrical and mechanical properties, and 3) reduced graphene oxide (rGO) can be produced at a large scale and then reduced chemically to form rGO. Unfortunately, rGO contains a number of defects that reduce its thermal and electrical conductivity compared to pristine graphene. The presence of functional groups on its surface can, however, lead to the formation of bonds with polymer matrices and as a result, to stronger interactions and better mechanical properties.

Over the last decade, many graphene companies emerged, attempting to industrialise graphene production (Kovtun et al., 2019). Large-scale manufacturing facilities

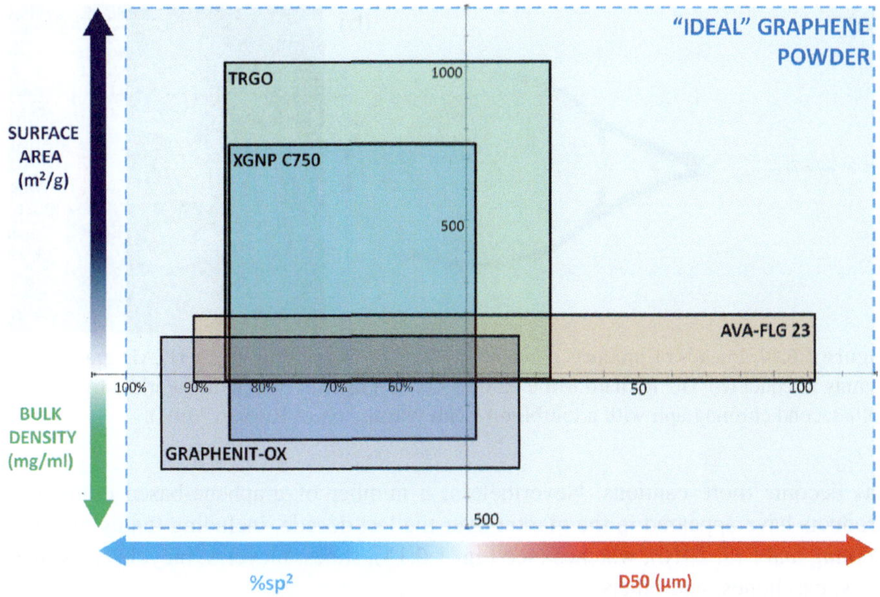

Figure 1.6.13 A four-axis representation to compare key characteristics of selected GRM samples. The X-axis displays lateral size on the positive side and the percentage of sp^2 bonds on the negative side, while the Y-axis shows specific surface area positively and bulk density negatively. Only samples with notable performance in at least one parameter were included for simplicity. The light-blue rectangle represents the theoretical attributes of "ideal" graphene.
Adapted from (Kovtun et al., 2019) with permission of IOP Publishing).

were established, particularly in Europe, North America, and Asia, aimed at producing graphene in kilogram or ton quantities. Despite these advances, the quality and consistency of graphene remained significant challenges, with no single production method emerging as the dominant approach. Graphene can vary significantly in terms of layer thickness, lateral size, defect density, and purity, depending on the production method. These variations make it difficult for manufacturers to integrate graphene reliably into products and for researchers to obtain meaningful results without batch-to-batch variation. The development of industry-wide standards for defining and categorising graphene materials has complicated efforts to develop commercial applications and the ISO/TS 21356–1:2021 that specifies the sequence of methods for characterising the structural properties of graphene is a step in the right direction. Finally, the intense media attention surrounding the discovery of graphene and early research created high expectations that have proved to be difficult to meet. As a result, many companies and investors became cautious as the anticipated graphene revolution unfolded more gradually than expected. Over time, some businesses faced difficulties in sustaining operations, despite significant initial investments and efforts to commercialise the technology. This "hype cycle" has made it more challenging for new graphene-based technologies to attract investment and interest, as the market

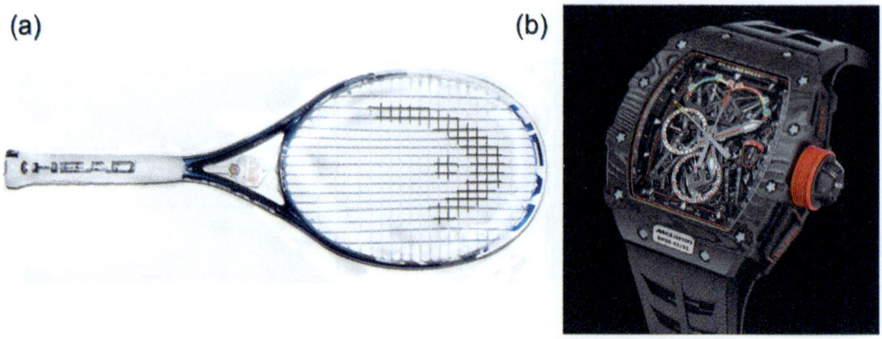

Figure 1.6.14 Images of products containing graphene composites. (a) A HEAD graphene Tennis Racquet (b) The Richard Mille RM 50–03 McLaren F1 watch, the world's lightest split-second chronograph with a tourbillon (with permission of Richard Mille).

has become more cautious. Nevertheless, a number of graphene-based composite products have appeared in the market over the last decade, including tennis racquets (Young and Liu, 2016), watches (see Fig. 1.6.14), shoes, bicycles, bicycle tyres, helmets, earphones, and others.

4. Conclusions

The journey of graphene from its initial isolation to potential market adoption has been marked by both remarkable breakthroughs and persistent challenges. The exceptional properties of graphene, particularly its mechanical strength, electrical conductivity and thermal conductivity, have opened up a vast array of possibilities for multifunctional nanocomposites. The critical role of the mechanics of graphene-based nanocomposites has been highlighted as a way to understand the complex interactions between graphene and the matrix material and the parameters that need to be carefully tuned to allow further improvement. While significant progress has been made in scaling up graphene production and improving its quality, challenges such as achieving uniform dispersion, enhancing graphene-matrix interfaces, and establishing industry-wide standards persist. We emphasise that the full potential of graphene composites hinges on overcoming these challenges and fostering collaboration between academia and industry. Nevertheless, with continued research and development, graphene composites are poised to revolutionise various industries, offering solutions for lightweight, high-performance materials with tailored functionalities.

References

Aylesworth, J. W. (1915). "Expanded graphite and composites thereof". USA Patent 1137373.
Aylesworth, J. W. (1916). "Expanded graphite". USA Patent 1191383.

Baker, C., Kelly, A., 1964. The effect of neutron irradiation on the elastic moduli of graphite single crystals. Philos. Mag. 9, 927–951. https://doi.org/10.1080/14786436408211906.

Balandin, A.A., Ghosh, S., Bao, W., Calizo, I., Teweldebrhan, D., Miao, F., Lau, C.N., 2008. Superior thermal conductivity of single-layer graphene. Nano Lett. 8, 902–907. https://doi.org/10.1021/nl0731872.

Beaumont, P.W.R., 2014. A celebration of the life of professor Tony Kelly ScD FRS FREng PhD CBE DL. Appl. Compos. Mater. 21, 391–395. https://doi.org/10.1007/s10443-014-9406-3.

Bianco, A., Cheng, H.-M., Enoki, T., Gogotsi, Y., Hurt, R.H., Koratkar, N., Kyotani, T., Monthioux, M., Park, C.R., Tascon, J.M., 2013. All in the graphene family – a recommended nomenclature for two-dimensional carbon materials. Carbon 65, 1–6. https://doi.org/10.1016/j.carbon.2013.08.038.

Blake, P., Hill, E.W., Castro Neto, A.H., Novoselov, K.S., Jiang, D., Yang, R., Booth, T.J., Geim, A.K., 2007. Making graphene visible. Appl. Phys. Lett. 91. https://doi.org/10.1063/1.2768624.

Blakslee, O.L., Proctor, D.G., Seldin, E.J., Spence, G.B., Weng, T., 1970. Elastic constants of compression-annealed pyrolytic graphite. J. Appl. Phys. 41, 3373–3382. https://doi.org/10.1063/1.1659428.

Boehm, H.P., Clauss, A., Fischer, G.O., Hofmann, U., 1962. Das adsorptionsverhalten sehr dünner Kohlenstoff-Folien. Z. Anorg. Allg. Chem. 316, 119–127. https://doi.org/10.1002/zaac.19623160303.

Boehm, H.P., Setton, R., Stumpp, E., 1994. Nomenclature and terminology of graphite intercalation compounds (IUPAC Recommendations 1994). Pure Appl. Chem. 66, 1893–1901. https://doi.org/10.1351/pac199466091893.

Brodie, B.C., 1859. XIII. On the atomic weight of graphite. Phil. Trans. Roy. Soc. Lond. 149, 249–259. https://doi.org/10.1098/rstl.1859.0013.

Carr, K.E., 1970. Intercalation and oxidation effects on graphite of a mixture of sulphuric and nitric acids. Carbon 8, 155–166. https://doi.org/10.1016/0008-6223(70)90110-7.

Castro Neto, A.H., Guinea, F., Peres, N.M.R., Novoselov, K.S., Geim, A.K., 2009. The electronic properties of graphene. Rev. Mod. Phys. 81, 109–162. https://doi.org/10.1103/RevModPhys.81.109.

Celzard, A., Krzesińska, M., Bégin, D., Marêché, J.F., Puricelli, S., Furdin, G., 2002. Preparation, electrical and elastic properties of new anisotropic expanded graphite-based composites. Carbon 40, 557–566. https://doi.org/10.1016/S0008-6223(01)00140-3.

Chen, G., Wu, D., Weng, W., Wu, C., 2003. Exfoliation of graphite flake and its nanocomposites. Carbon 41, 619–621. https://doi.org/10.1016/S0008-6223(02)00409-8.

Cho, D., Lee, S., Yang, G.M., Fukushima, H., Drzal, L.T., 2005. Dynamic mechanical and thermal properties of phenylethynyl-terminated polyimide composites reinforced with expanded graphite nanoplatelets. Macromol. Mater. Eng. 290, 179–187. https://doi.org/10.1002/mame.200400281.

Chong, H.M., Hinder, S.J., Taylor, A.C., 2016. Graphene nanoplatelet-modified epoxy: effect of aspect ratio and surface functionality on mechanical properties and toughening mechanisms. J. Mater. Sci. 51, 8764–8790. https://doi.org/10.1007/s10853-016-0160-9.

Chung, D.D.L., 1987. Exfoliation of graphite. J. Mater. Sci. 22, 4190–4198. https://doi.org/10.1007/BF01132008.

Coleman, J.N., Lotya, M., O'neill, A., Bergin, S.D., King, P.J., Khan, U., Young, K., Gaucher, A., De, S., Smith, R.J., Shvets, I.V., Arora, S.K., Stanton, G., Kim, H.-Y., Lee, K., Kim, G.T., Duesberg, G.S., Hallam, T., Boland, J.J., Wang, J.J., Donegan, J.F., Grunlan, J.C., Moriarty, G., Shmeliov, A., Nicholls, R.J., Perkins, J.M., Grieveson, E.M.,

Theuwissen, K., Mccomb, D.W., Nellist, P.D., Nicolosi, V., 2011. Two-dimensional nanosheets produced by liquid exfoliation of layered materials. Science 331, 568–571. https://doi.org/10.1126/science.1194975.

Cooper, C.A., Young, R.J., 1999. Investigation of structure/property relationships in particulate composites through the use of Raman spectroscopy. J. Raman Spectrosc. 30, 929–938. https://doi.org/10.1002/(SICI)1097-4555(199910)30:10<929::AID-JRS462>3.0.CO;2-7.

Cooper, C.A., Young, R.J., Halsall, M., 2001. Investigation into the deformation of carbon nanotubes and their composites through the use of Raman spectroscopy. Compos. Appl. Sci. Manuf. 32, 401–411. https://doi.org/10.1016/S1359-835X(00)00107-X.

Cox, H.L., 1952. The elasticity and strength of paper and other fibrous materials. Br. J. Appl. Phys. 3, 72. https://doi.org/10.1088/0508-3443/3/3/302.

Cui, G., Bi, Z., Zhang, R., Liu, J., Yu, X., Li, Z., 2019. A comprehensive review on graphene-based anti-corrosive coatings. Chem. Eng. J. 373, 104–121. https://doi.org/10.1016/j.cej.2019.05.034.

Das, B., Eswar Prasad, K., Ramamurty, U., Rao, C.N.R., 2009. Nano-indentation studies on polymer matrix composites reinforced by few-layer graphene. Nanotechnology 20, 125705. https://doi.org/10.1088/0957-4484/20/12/125705.

Dresselhaus, M.S., Dresselhaus, G., Saito, R., Jorio, A., 2005. Raman spectroscopy of carbon nanotubes. Phys. Rep. 409, 47–99. https://doi.org/10.1016/j.physrep.2004.10.006.

Dreyer, D.R., Park, S., Bielawski, C.W., Ruoff, R.S., 2010. The chemistry of graphene oxide. Chem. Soc. Rev. 39, 228–240. https://doi.org/10.1039/B917103G.

Drzal, L.T., Fukushima, H., 2013. Expanded graphite and products produced Therefrom. USA Patent US8501858 B2.

Du, J., Zhao, L., Zeng, Y., Zhang, L., Li, F., Liu, P., Liu, C., 2011. Comparison of electrical properties between multi-walled carbon nanotube and graphene nanosheet/high density polyethylene composites with a segregated network structure. Carbon 49, 1094–1100. https://doi.org/10.1016/j.carbon.2010.11.013.

Erickson, K., Erni, R., Lee, Z., Alem, N., Gannett, W., Zettl, A., 2010. Determination of the local chemical structure of graphene oxide and reduced graphene oxide. Adv. Mater. 22, 4467–4472. https://doi.org/10.1002/adma.201000732.

Ferrari, A.C., Meyer, J.C., Scardaci, V., Casiraghi, C., Lazzeri, M., Mauri, F., Piscanec, S., Jiang, D., Novoselov, K.S., Roth, S., Geim, A.K., 2006. Raman spectrum of graphene and graphene layers. Phys. Rev. Lett. 97, 187401. https://doi.org/10.1103/PhysRevLett.97.187401.

Frank, O., Tsoukleri, G., Riaz, I., Papagelis, K., Parthenios, J., Ferrari, A.C., Geim, A.K., Novoselov, K.S., Galiotis, C., 2011. Development of a universal stress sensor for graphene and carbon fibres. Nat. Commun. 2, 255. https://doi.org/10.1038/ncomms1247.

Fukushima, H., 2003. Graphite Nanoreinforcements in Polymer Nanocomposites. Michigan State University, USA. PhD Thesis.

Geim, A.K., 2011. Nobel lecture: random walk to graphene. Rev. Mod. Phys. 83, 851–862. https://doi.org/10.1103/RevModPhys.83.851.

Geim, A.K., Novoselov, K.S., 2007. The rise of graphene. Nat. Mater. 6, 183–191. https://doi.org/10.1038/nmat1849.

Gillis, P.P., 1984. Calculating the elastic constants of graphite. Carbon 22, 387–391.

Gong, L., Kinloch, I.A., Young, R.J., Riaz, I., Jalil, R., Novoselov, K.S., 2010. Interfacial stress transfer in a graphene monolayer nanocomposite. Adv. Mater. 22, 2694–2697. https://doi.org/10.1002/adma.200904264.

Gong, L., Young, R.J., Kinloch, I.A., Haigh, S.J., Warner, J.H., Hinks, J.A., Xu, Z., Li, L., Ding, F., Riaz, I., Jalil, R., Novoselov, K.S., 2013. Reversible loss of Bernal stacking during the deformation of few-layer graphene in nanocomposites. ACS Nano 7, 7287–7294. https://doi.org/10.1021/nn402830f.

Gong, L., Young, R.J., Kinloch, I.A., Riaz, I., Jalil, R., Novoselov, K.S., 2012. Optimizing the reinforcement of polymer-based nanocomposites by graphene. ACS Nano 6, 2086–2095. https://doi.org/10.1021/nn203917d.

Guest, J., Kinloch, I.A., Young, R.J., 2023. The role of filler aspect ratio in the reinforcement of an epoxy resin with graphene nanoplatelets. J. Mater. Sci. 58, 9473–9485. https://doi.org/10.1007/s10853-023-08603-3.

Gupta, A., Chen, G., Joshi, P., Tadigadapa, S., Eklund, 2006. Raman scattering from high-frequency phonons in supported n-graphene layer films. Nano Lett. 6, 2667–2673. https://doi.org/10.1021/nl061420a.

Halpin, J.C., Kardos, J.L., 1976. The Halpin-Tsai equations: a review. Polym. Eng. Sci. 16, 344–352. https://doi.org/10.1002/pen.760160512.

Han, T.-H., Kim, H., Kwon, S.-J., Lee, T.-W., 2017. Graphene-based flexible electronic devices. Mater. Sci. Eng. R Rep. 118, 1–43. https://doi.org/10.1016/j.mser.2017.05.001.

Hirata, M., Gotou, T., Horiuchi, S., Fujiwara, M., Ohba, M., 2004. Thin-film particles of graphite oxide 1:: high-yield synthesis and flexibility of the particles. Carbon 42, 2929–2937. https://doi.org/10.1016/j.carbon.2004.07.003.

Hummers, W.S., Offeman, R.E., 1958. Preparation of graphitic oxide. J. Am. Chem. Soc. 80, 1339. https://doi.org/10.1021/ja01539a017.

Itoh, H., Ichinose, T., Oshima, C., Ichinokawa, T., Aizawa, T., 1991. Scanning tunneling microscopy of monolayer graphite epitaxially grown on a TiC(111) surface. Surf. Sci. 254, L437–L442. https://doi.org/10.1016/0039-6028(91)90620-8.

Kalaitzidou, K., Fukushima, H., Askeland, P., Drzal, L.T., 2008. The nucleating effect of exfoliated graphite nanoplatelets and their influence on the crystal structure and electrical conductivity of polypropylene nanocomposites. J. Mater. Sci. 43, 2895–2907. https://doi.org/10.1007/s10853-007-1876-3.

Kalaitzidou, K., Fukushima, H., Drzal, L.T., 2007. A new compounding method for exfoliated graphite–polypropylene nanocomposites with enhanced flexural properties and lower percolation threshold. Compos. Sci. Technol. 67, 2045–2051. https://doi.org/10.1016/j.compscitech.2006.11.014.

Katsnelson, M.I., 2020. The electronic structure of ideal graphene. In: Katsnelson, M.I. (Ed.), The Physics of Graphene, 2 ed. Cambridge University Press, Cambridge.

Kelly, A., 2013. Personal Communication to Young. R. J., Queen's College, Cambridge.

Kelly, A., Macmillan, N.H., 1986. Strong Solids. Clarendon Press, Oxford.

Kim, H., Abdala, A.A., Macosko, C.W., 2010a. Graphene/polymer nanocomposites. Macromolecules 43, 6515–6530. https://doi.org/10.1021/ma100572e.

Kim, H., Miura, Y., Macosko, C.W., 2010b. Graphene/polyurethane nanocomposites for improved gas barrier and electrical conductivity. Chem. Mater. 22, 3441–3450. https://doi.org/10.1021/cm100477v.

Kovtun, A., Treossi, E., Mirotta, N., Scidà, A., Liscio, A., Christian, M., Valorosi, F., Boschi, A., Young, R.J., Galiotis, C., Kinloch, I.A., Morandi, V., Palermo, V., 2019. Benchmarking of graphene-based materials: real commercial products versus ideal graphene. 2D Mater. 6, 025006. https://doi.org/10.1088/2053-1583/aafc6e.

Krenchel, H. (1964). "Fibre reinforcement: theoretical and practical investigations of the elasticity and strength of fibre-reinforced materials", Copenhagen, Akademisk Forlag.

Land, T.A., Michely, T., Behm, R.J., Hemminger, J.C., Comsa, G., 1992. STM investigation of single layer graphite structures produced on Pt(111) by hydrocarbon decomposition. Surf. Sci. 264, 261–270. https://doi.org/10.1016/0039-6028(92)90183-7.

Lee, C., Wei, X., Kysar, J.W., Hone, J., 2008. Measurement of the elastic properties and intrinsic strength of monolayer graphene. Science 321, 385–388. https://doi.org/10.1126/science.1157996.

Li, J., Guan, T., Zhang, Z., Fu, Y.-T., Guo, F.-L., Huang, P., Li, Z., Li, Y.-Q., Fu, S.-Y., 2024. Orientation of discontinuous fillers in polymer composites: modelling, characterization, control and applications. Prog. Mater. Sci., 101360 https://doi.org/10.1016/j.pmatsci.2024.101360.

Li, S., 2018. Structure/Property Relationships in Elastomers Filled with Graphite Nanoplatelets. University of Manchester, UK. PhD Thesis.

Li, Z., Chu, J., Yang, C., Hao, S., Bissett, M.A., Kinloch, I.A., Young, R.J., 2018. Effect of functional groups on the agglomeration of graphene in nanocomposites. Compos. Sci. Technol. 163, 116–122. https://doi.org/10.1016/j.compscitech.2018.05.016.

Li, Z., Deng, L., Kinloch, I.A., Young, R.J., 2023. Raman spectroscopy of carbon materials and their composites: graphene, nanotubes and fibres. Prog. Mater. Sci. 135. https://doi.org/10.1016/j.pmatsci.2023.101089.

Li, Z., Young, R.J., Wilson, N.R., Kinloch, I.A., Vallés, C., Li, Z., 2016. Effect of the orientation of graphene-based nanoplatelets upon the Young's modulus of nanocomposites. Compos. Sci. Technol. 123, 125–133. https://doi.org/10.1016/j.compscitech.2015.12.005.

Liu, F., Ming, P., Li, J., 2007. Ab initio calculation of ideal strength and phonon instability of graphene under tension. Phys. Rev. B 76, 064120. https://doi.org/10.1103/PhysRevB.76.064120.

Liu, M., Hui, J.H., Kinloch, I.A., Young, R.J., Papageorgiou, D.G., 2021. Deformation and tearing of graphene-reinforced elastomer nanocomposites. Compos. Commun. 25, 100764. https://doi.org/10.1016/j.coco.2021.100764.

Lu, X.K., Yu, M.F., Huang, H., Ruoff, R.S., 1999. Tailoring graphite with the goal of achieving single sheets. Nanotechnology 10, 269–272. https://doi.org/10.1088/0957-4484/10/3/308.

Malard, L.M., Pimenta, M.A., Dresselhaus, G., Dresselhaus, M.S., 2009. Raman spectroscopy in graphene. Phys. Rep. 473, 51–87. https://doi.org/10.1016/j.physrep.2009.02.003.

Marsden, A.J., Papageorgiou, D., Valles, C., Liscio, A., Palermo, V., Bissett, M., Young, R., Kinloch, I., 2018. Electrical percolation in graphene–polymer composites. 2D Mater. 5, 032003. https://doi.org/10.1088/2053-1583/aac055.

Matsuo, Y., Tahara, K., Sugie, Y., 1997. Structure and thermal properties of poly(ethylene oxide)-intercalated graphite oxide. Carbon 35, 113–120. https://doi.org/10.1016/S0008-6223(96)00123-6.

Mohiuddin, T.M.G., Lombardo, A., Nair, R.R., Bonetti, A., Savini, G., Jalil, R., Bonini, N., Basko, D.M., Galiotis, C., Marzari, N., Novoselov, K.S., Geim, A.K., Ferrari, A.C., 2009. Uniaxial strain in graphene by Raman spectroscopy: G peak splitting, Grüneisen parameters, and sample orientation. Phys. Rev. B 79, 205433. https://doi.org/10.1103/PhysRevB.79.205433.

Mori, T., Tanaka, K., 1973. Average stress in matrix and average elastic energy of materials with misfitting inclusions. Acta Metall. 21, 571–574. https://doi.org/10.1016/0001-6160(73)90064-3.

Novoselov, K.S., 2011. Nobel lecture: graphene: materials in the flatland. Rev. Mod. Phys. 83, 837–849. https://doi.org/10.1103/RevModPhys.83.837.

Novoselov, K.S., Geim, A.K., Morozov, S.V., Jiang, D., Katsnelson, M.I., Grigorieva, I.V., Dubonos, S.V., Firsov, A.A., 2005. Two-dimensional gas of massless Dirac fermions in graphene. Nature 438, 197–200. https://doi.org/10.1038/nature04233.

Novoselov, K.S., Geim, A.K., Morozov, S.V., Jiang, D., Zhang, Y., Dubonos, S.V., Grigorieva, I.V., Firsov, A.A., 2004. Electric field effect in atomically thin carbon films. Science 306, 666–669. https://doi.org/10.1126/science.1102896.

Pan, X., Shen, L., Schenning, A.P.H.J., Bastiaansen, C.W.M., 2019. Transparent, high-thermal-conductivity ultradrawn polyethylene/graphene nanocomposite films. Adv. Mater. 31, 1904348. https://doi.org/10.1002/adma.201904348.

Papageorgiou, D.G., Kinloch, I.A., Young, R.J., 2017. Mechanical properties of graphene and graphene-based nanocomposites. Prog. Mater. Sci. 90, 75–127. https://doi.org/10.1016/j.pmatsci.2017.07.004.

Papageorgiou, D.G., Li, Z., Liu, M., Kinloch, I.A., Young, R.J., 2020. Mechanisms of mechanical reinforcement by graphene and carbon nanotubes in polymer nanocomposites. Nanoscale 12, 2228–2267. https://doi.org/10.1039/C9NR06952F.

Peierls, R., 1935. Quelques propriétés typiques des corps solides. Ann. Inst. Henri Poincaré Phys. Théor. 5, 177–222.

Pinnavaia, T.J., Beall, G.W. (Eds.), 2001. Polymer-Clay Nanocomposites. John Wiley and Sons, Chichester.

Pollard, A.J., Paton, K.R., Clifford, C.A., Legge, E., 2017. Characterisation of the Structure of Graphene". Good Practice Guide No.145. National Physical Laboratory, Teddington, UK.

Ramanathan, T., Abdala, A.A., Stankovich, S., Dikin, D.A., Herrera-Alonso, M., Piner, R.D., Adamson, D.H., Schniepp, H.C., Chen, X., Ruoff, R.S., Nguyen, S.T., Aksay, I.A., Prud'homme, R.K., Brinson, L.C., 2008. Functionalized graphene sheets for polymer nanocomposites. Nat. Nanotechnol. 3, 327–331. https://doi.org/10.1038/nnano.2008.96.

Ramaswamy, C., 1930. Raman effect in diamond. Nature 125, 704. https://doi.org/10.1038/125704b0.

Robertson, A.W., Warner, J.H., 2013. Atomic resolution imaging of graphene by transmission electron microscopy. Nanoscale 5, 4079–4093. https://doi.org/10.1039/C3NR00934C.

Robertson, R., Fox, J.J., 1930. Infra-red spectrum of diamond by infra-red spectrometer and Raman methods. Nature 125, 704. https://doi.org/10.1038/125704a0.

Robinson, I.M., Zakikhani, M., Day, R.J., Young, R.J., Galiotis, C., 1987. Strain dependence of the Raman frequencies for different types of carbon fibres. J. Mater. Sci. Lett. 6, 1212–1214. https://doi.org/10.1007/BF01729187.

Rourke, J.P., Pandey, P.A., Moore, J.J., Bates, M., Kinloch, I.A., Young, R.J., Wilson, N.R., 2011. The real graphene oxide revealed: stripping the oxidative debris from the graphene-like sheets. Angew. Chem. Int. Ed. 50, 3173–3177. https://doi.org/10.1002/anie.201007520.

Sakata, H., Dresselhaus, G., Dresselhaus, M.S., Endo, M., 1988. Effect of uniaxial stress on the Raman spectra of graphite fibres. J. Appl. Phys. 63, 2769–2772. https://doi.org/10.1063/1.340975.

Shahil, K.M.F., Balandin, A.A., 2012. Thermal properties of graphene and multilayer graphene: applications in thermal interface materials. Solid State Commun. 152, 1331–1340. https://doi.org/10.1016/j.ssc.2012.04.034.

Shioyama, H., 2001. Cleavage of graphite to graphene. J. Mater. Sci. Lett. 20, 499–500. https://doi.org/10.1023/A:1010907928709.

Stankovich, S., Dikin, D.A., Dommett, G.H.B., Kohlhaas, K.M., Zimney, E.J., Stach, E.A., Piner, R.D., Nguyen, S.T., Ruoff, R.S., 2006. Graphene-based composite materials. Nature 442, 282–286. https://doi.org/10.1038/nature04969.

Suk, J.W., Piner, R.D., An, J., Ruoff, R.S., 2010. Mechanical properties of monolayer graphene oxide. ACS Nano 4, 6557–6564. https://doi.org/10.1021/nn101781v.

Tomes, O., Soul, A., Zhang, H., Bilotti, E., Papageorgiou, D.G., 2024. A novel electrical depercolation model for stretchable nanocomposite strain sensors. Nanoscale 16, 6163–6175. https://doi.org/10.1039/D3NR05234F.

Tran, T.S., Dutta, N.K., Choudhury, N.R., 2018. Graphene inks for printed flexible electronics: graphene dispersions, ink formulations, printing techniques and applications. Adv. Colloid Interface Sci. 261, 41–61. https://doi.org/10.1016/j.cis.2018.09.003.

Tuinstra, F., Koenig, J.L., 1970a. Characterization of graphite fiber surfaces with Raman spectroscopy. J. Compos. Mater. 4, 492–499. https://doi.org/10.1177/002199837000400405.

Tuinstra, F., Koenig, J.L., 1970b. Raman spectrum of graphite. J. Chem. Phys. 53, 1126–1130. https://doi.org/10.1063/1.1674108.

Usuki, A., Kojima, Y., Kawasumi, M., Okada, A., Fukushima, Y., Kurauchi, T., Kamigaito, O., 1993. Synthesis of nylon 6-clay hybrid. J. Mater. Res. 8, 1179–1184. https://doi.org/10.1557/JMR.1993.1179.

Varghese, S.S., Lonkar, S., Singh, K., Swaminathan, S., Abdala, A., 2015. Recent advances in graphene based gas sensors. Sensor. Actuator. B Chem. 218, 160–183. https://doi.org/10.1016/j.snb.2015.04.062.

Voiry, D., Yang, J., Kupferberg, J., Fullon, R., Lee, C., Jeong, H.Y., Shin, H.S., Chhowalla, M., 2016. High-quality graphene via microwave reduction of solution-exfoliated graphene oxide. Science 353, 1413–1416. https://doi.org/10.1126/science.aah3398.

Wallace, P.R., 1947. The band theory of graphite. Phys. Rev. 71, 622–634. https://doi.org/10.1103/PhysRev.71.622.

Wang, G., Dai, Z., Liu, L., Hu, H., Dai, Q., Zhang, Z., 2016. Tuning the interfacial mechanical behaviors of monolayer graphene/PMMA nanocomposites. ACS Appl. Mater. Interfaces 8, 22554–22562. https://doi.org/10.1021/acsami.6b03069.

Wang, Y.S., O'gurkis, M.A., Lindt, J.T., 1986. Electrical properties of exfoliated-graphite filled polyethylene composites. Polym. Compos. 7, 349–354. https://doi.org/10.1002/pc.750070512.

Watt, W., 1970. Production and properties of high modulus carbon fibres. Proceedings of the Royal Society A 319, 5–15. https://doi.org/10.1098/rspa.1970.0161.

Wei, Q., Pei, S., Qian, X., Liu, H., Liu, Z., Zhang, W., Zhou, T., Zhang, Z., Zhang, X., Cheng, H.-M., Ren, W., 2020. Superhigh electromagnetic interference shielding of ultrathin aligned pristine graphene nanosheets film. Adv. Mater. 32, 1907411. https://doi.org/10.1002/adma.201907411.

Wick, P., Louw-Gaume, A.E., Kucki, M., Krug, H.F., Kostarelos, K., Fadeel, B., Dawson, K. A., Salvati, A., Vázquez, E., Ballerini, L., 2014. Classification framework for graphene-based materials. Angew. Chem. Int. Ed. 53, 7714–7718. https://doi.org/10.1002/anie.201403335.

Williams, K.P.J., Pitt, G.D., Batchelder, D.N., Kip, B.J., 1994. Confocal Raman microspectroscopy using a stigmatic spectrograph and CCD detector. Appl. Spectrosc. 48, 232–235. https://doi.org/10.1366/0003702944028407.

Wilson, N.R., Pandey, P.A., Beanland, R., Young, R.J., Kinloch, I.A., Gong, L., Liu, Z., Suenaga, K., Rourke, J.P., York, S.J., Sloan, J., 2009. Graphene oxide: structural analysis and application as a highly transparent support for electron microscopy. ACS Nano 3, 2547–2556. https://doi.org/10.1021/nn900694t.

Xia, Y., Gao, W., Gao, C., 2022. A review on graphene-based electromagnetic functional materials: electromagnetic wave shielding and absorption. Adv. Funct. Mater. 32, 2204591. https://doi.org/10.1002/adfm.202204591.

Young, R.J., Gong, L., Kinloch, I.A., Riaz, I., Jalil, R., Novoselov, K.S., 2011. Strain mapping in a graphene monolayer nanocomposite. ACS Nano 5, 3079–3084. https://doi.org/10.1021/nn2002079.

Young, R.J., Kinloch, I.A., Gong, L., Novoselov, K.S., 2012. The mechanics of graphene nanocomposites: a review. Compos. Sci. Technol. 72, 1459–1476. https://doi.org/10.1016/j.compscitech.2012.05.005.

Young, R.J., Liu, M., 2016. The microstructure of a graphene-reinforced tennis racquet. J. Mater. Sci. 51, 3861–3867. https://doi.org/10.1007/s10853-015-9705-6.

Young, R.J., Liu, M., Kinloch, I.A., Li, S., Zhao, X., Vallés, C., Papageorgiou, D.G., 2018. The mechanics of reinforcement of polymers by graphene nanoplatelets. Compos. Sci. Technol. 154, 110–116. https://doi.org/10.1016/j.compscitech.2017.11.007.

Yuan, H., Wang, Y., Li, T., Wang, Y., Ma, P., Zhang, H., Yang, W., Chen, M., Dong, W., 2019. Fabrication of thermally conductive and electrically insulating polymer composites with isotropic thermal conductivity by constructing a three-dimensional interconnected network. Nanoscale 11, 11360–11368. https://doi.org/10.1039/C9NR02491C.

Zhao, S., Zhang, H.-B., Luo, J.-Q., Wang, Q.-W., Xu, B., Hong, S., Yu, Z.-Z., 2018. Highly electrically conductive three-dimensional $Ti_3C_2T_x$ MXene/reduced graphene oxide hybrid aerogels with excellent electromagnetic interference shielding performances. ACS Nano 12, 11193–11202. https://doi.org/10.1021/acsnano.8b05739.

Zhao, X., Mao, B., Liu, M., Cao, J., Haigh, S.J., Papageorgiou, D.G., Li, Z., Young, R.J., 2022. Controlling and monitoring crack propagation in monolayer graphene single crystals. Adv. Funct. Mater. 32, 2202373. https://doi.org/10.1002/adfm.202202373.

Zhao, X., Papageorgiou, D.G., Zhu, L., Ding, F., Young, R.J., 2019. The strength of mechanically-exfoliated monolayer graphene deformed on a rigid polymer substrate. Nanoscale 11, 14339–14353. https://doi.org/10.1039/c9nr04720d.

Zou, L., Wang, L., Wu, Y., Ma, C., Yu, S., Liu, X., 2018. Trends analysis of graphene research and development. J. Data Inform. Sci. 3, 82–100. https://doi.org/10.2478/jdis-2018-0005.

Fibre-hybrid composites: A history of discrepancies between industry and academia

1.7

Thanasis Chatziathanasiou [1] *and Yentl Swolfs* [2]
[1]Department of Materials Engineering, KU Leuven, Leuven, Belgium; [2]Composite Materials Group, Department of Materials Engineering, Katholieke Universiteit Leuven (KU Leuven), Leuven, Belgium

1. Introduction

Hybrid composites are generally defined as composites with two reinforcement types in a single matrix. The reinforcement type with the lowest failure strain is generally referred to as the low-elongation component, whereas the one with the highest failure strain is the high-elongation component. Some authors also use the terminology brittle and ductile, but this can be misleading because the high-elongation component is often still a brittle fibre, like glass fibre.

Hybrid composites are a very broad term that also applies to fibre metal laminates (FMLs). Early developments were driven by the need for materials combining the high strength and stiffness of metals with the lightweight and fatigue-resistance of fibres. The concept gained momentum in the seventies and eighties with research into ARALL (aramid reinforced aluminium laminate) and later GLARE (glass laminate aluminium reinforced epoxy). These laminates were created by stacking thin metal layers, such as aluminium, with high-performance fibres embedded in polymer matrices. The synergy of these components resulted in materials with exceptional resistance to crack propagation, improved impact tolerance, and superior fatigue life compared with traditional aluminium alloys. This early work laid the foundation for the widespread adoption of FMLs in modern aerospace structures, including the Airbus A380, where weight reduction and durability are critical.

A second class of hybrid composites would be nano-hybrids, which incorporate either a combination of fibre, matrix, and nanoparticles or a matrix with two or more types of nanoparticles. Since nano-hybrids are comprehensively discussed in chapter 1.5 in Part II of this book, we will not delve deeper into them here. Instead, this historical overview will limit itself to a third class of hybrid composites, namely fibre-hybrid composites, which contain at least two fibrous reinforcements in a single matrix. These composites present a unique opportunity to alleviate some of the intrinsic drawbacks associated with individual fibre types. For instance, by strategically placing carbon fibres on the outer layers of a laminate and glass fibres on the inner layers, the flexural stiffness of the carbon fibre composites can almost be achieved, but at a significantly

The History of Composites. https://doi.org/10.1016/B978-0-443-34119-9.00009-7

lower cost. This ability to combine the properties of different fibres allows fibre-hybrid composites to be tailored to specific application requirements, enhancing their performance while balancing cost and sustainability. In addition, the strategic placement of fibres within the composite allows engineers to optimise material behaviour in local, critical regions without significantly affecting the total cost or weight.

In some cases, synergetic effects arise. These synergies can imply two things. First, features arise that are not present in the reference composites, which are the composites with only one fibre type. Common examples are damage sensing and pseudo-ductility. Second, some properties are better than expected based on simple rule-of-mixtures or models. A notable example is the increased failure strain of brittle fibres when well-dispersed ductile fibres are introduced into the composite, which is often referred to as the classic hybrid effect.

Synergetic effects are usually stronger when the dispersion, which is the degree of mixing of the fibre types, is higher. The type of hybrid configuration plays a crucial role in this regard, as this governs the dispersion. Usually, three main configurations (see Fig. 1.7.1) are defined:

- Interlayer, sometimes also called layer-by-layer
- Intralayer, sometimes also called yarn-by-yarn or co-woven
- Intrayarn, sometimes also called fibre-by-fibre or comingled

These configurations can be expanded to include local reinforcements and macroscale hybridisation. Local reinforcement is the strategic strengthening of a specific area or component to address localised stresses, loads, or weaknesses. Macroscale hybridisation is the combination of multiple parts into a larger structure, where not all parts contain the same fibre types. This is common in large wind turbine blades, where the load-carrying beams are made of carbon fibre profiles, but the aerofoil is made of glass fibre composites.

2. The early years of fibre-hybridisation

2.1 Early adoption in the industry

The origins of fibre-hybrid composites in industrial applications can be traced back to the sixties, with one of the earliest and most iconic examples being the Ford GT40's

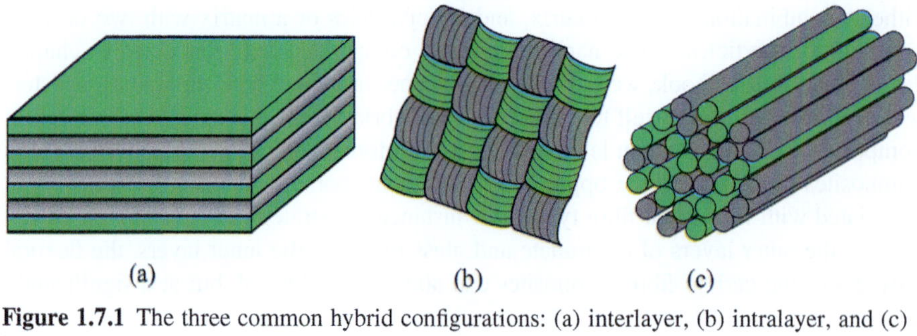

(a) (b) (c)

Figure 1.7.1 The three common hybrid configurations: (a) interlayer, (b) intralayer, and (c) intrayarn.
Reprinted from Swolfs et al. (2014b), with permission from Elsevier.

victory (see Fig. 1.7.2a) at the 1968 Le Mans endurance race (Phillips, 1969). At that time, glass fibre composites were already widely used, but carbon fibres were only recently invented and still very expensive. Making the full car body out of carbon fibre composite was dismissed purely based on price. Instead, the glass fibre composite car body of the Ford GT40 was locally reinforced with unidirectional carbon fibre strips. Carbon fibre technology was still in its infancy, presenting significant manufacturing challenges such as fraying and fluffing of the fibres during handling. Despite these difficulties, even the modest addition of 1.4 kg of carbon fibre strips reduced the car's body weight from 69 to 42 kg (Phillips, 1969). These weight-savings were instrumental in the Ford GT40's consecutive wins of the 1968 and 1969 Le Mans endurance races. This approach also rapidly became common practice in Formula 1 cars (see Fig. 1.7.2b).

The success of carbon-glass fibre hybrids in the Ford GT40 paved the way for a strong expansion of carbon fibre-stiffened sports cars (Phillips, 1969). predicted that the expansion would go hand in hand with reducing the carbon fibre price, but he likely did not anticipate full car bodies made of carbon fibre composites. In addition, the GT40 experience clearly established that the fatigue performance of hybrid composites was comparable to, if not better than, traditional metal-based components.

The seventies saw rapid expansion of fibre-hybrid composites into other industries, including aerospace and marine applications. Helicopter blades (Anonymous, 1974; Salkind, 1975) and aeroplane components (Summerscales and Short, 1978) began incorporating hybrid designs to take advantage of their enhanced mechanical properties. Boats also benefited from carbon/glass hybrids, as evidenced by their role in an Olympic gold medal for the West German four-person sculling team (Anonymous, 1973b) and a powerboat world record (Anonymous, 1973a).

While mechanical improvements drove much of the early adoption, functional benefits of hybridisation also began to emerge. It was, for example, suggested to add carbon fibres to glass fibre composites to accelerate the cure by resistive heating (McCann, 1972) or restore some of the electric properties lost by the switch from

(a) (b)

Figure 1.7.2 The Ford GT40 Mk I (Papisc, 2012), and (b) detail of a Formula 1 car with half of the part unpainted, clearly showing the black carbon fibre reinforcement strips.
Reprinted from Phillips (1969) with permission from Elsevier.

metal to glass fibre composites (Phillips, 1976b). Improved vibration behaviour was also claimed as a benefit in yacht design (Anonymous, 1973c).

2.2 Early academic research

The early industrial successes in the late sixties and early seventies triggered significant academic research into fibre-hybrid composites (Hayashi, 1972). was a pivotal paper, as it was the first report of the classic hybrid effect, which is the improvement in failure strain of a low-elongation fibre when combined with a high-elongation fibre. This discovery triggered significant research into a broad range of hybrid effects in the seventies and eighties (Phillips, 1976a; Marom et al., 1978; Fischer et al., 1979; Xing et al., 1981; Ji, 1982; Yao and Chou, 1989). Most of this research focused on carbon/glass combinations, with some exceptions for aramid fibres. Other fibre types, such as basalt fibres, polyethene fibres, and natural fibres, had not yet gained widespread attention.

A critical challenge in these early studies was establishing a correct baseline for the nonhybrid reference composites. The classic hybrid effect is the relative increase in failure strain of the low-elongation plies or bundles inside a fibre-hybrid composite compared with the failure strain they would have in a composite with only the low-elongation fibres. As reviewed by Kretsis (1987), some of the reported hybrid effects were very high, such as 116% (Aveston and Sillwood, 1976) and 84% (Bunsell and Harris, 1974). Even the original report (Hayashi, 1972) of 45% was later pointed out to be likely a result of inaccuracies in measuring the reference failure strain of unidirectional carbon fibre composites (Swolfs et al., 2014b). The addition of glass fibre layers, especially when added to the outer layers of the laminate, helps to shield stress concentrations at the grips (Czél et al., 2016; Fazlali et al., 2023). Therefore, the tensile strength of the carbon/glass hybrid is easier to measure accurately, whereas the reference tensile strength of the carbon fibre composite tends to be underestimated. This discrepancy likely explains many of the exaggerated hybrid effects reported during the seventies and eighties (Kretsis, 1987). It only became clear in later years that the dispersion of the low- and high-elongation fibres in these early reports was too low to warrant any significant hybrid effect (Swolfs et al., 2014b, 2019).

When Kretsis conducted a comprehensive review of fibre-hybrid composites in 1987, the synergetic effects of hybridisation in compression, flexure, and shear were still poorly understood (Kretsis, 1987). Although not directly reviewed by him at the time, hybridisation results in terms of impact resistance also remained unclear.

3. The silent era followed by an improved understanding

After the initial boom in the seventies, the academic interest in fibre-hybrid composites gradually slowed down in the eighties and only picked up significant steam again in the late nineties. This lull in activity can be attributed to gaps in understanding and the lack of control over critical aspects of fibre-hybrid composite behaviour. The focus at that time shifted more toward improving composite manufacturing processes and developing a better understanding of composite mechanics and damage development. Despite the slowdown, several influential studies were published in the eighties with a focus on predicting the tensile behaviour of fibre-hybrid composites, building on the early work of Zweben (1977). Zweben's model assumed a 1D packing of alternating low- and high-elongation fibres, and used a simple shear-lag model. In the 1980s, this was expanded to dynamic effects (Xing et al., 1981), Monte Carlo simulations (Fukuda and Chou, 1982), and more complex packings (Fukunaga et al., 1984a, 1989).

In parallel, experimental work on fibre-hybrid composites continued to develop. This gradually created a better understanding of the synergetic effects in fibre-hybrid composites. The main parameters controlling the classic hybrid effect in longitudinal tension are now well understood (Swolfs et al., 2014b). The flexural behaviour was also intensely studied, demonstrating significant advantages of fibre-hybridization. However, the findings in this area were not always consistent. Some authors even reported the flexural strengths of fibre-hybrids to be up to 40% higher than the flexural strengths of both reference composites (Dong et al., 2012). This can be partially explained by the fact that the compression strength of a composite is typically different from its tensile strength, and this can matter significantly in the asymmetric laminates tested by (Dong et al., 2012).

These inconsistencies were even more pronounced in terms of impact resistance. Many studies were published where penetration impact tests were performed on asymmetric laminates with carbon on one side and aramid, glass, or other fibre types on the other side. The results, however, varied widely. For instance, placing the carbon fibre layers toward the impacted side improved penetration impact resistance in some studies (Park and Jang, 2001; Sayer et al., 2010), and deteriorated it in other studies (Jang et al., 1989; Onal and Adanur, 2002). The inconsistencies highlight a lack of understanding of the complex material behaviour and a lack of support from advanced models. These challenges underscored the need for a more systematic approach to studying fibre-hybrid composites, paving the way for a renewed focus in the late nineties.

4. Recent developments

4.1 Recycling and re-use

The increased drive toward the recycling of composites has both created new opportunities and underscored the importance of renewed research into fibre-hybrid composites. This is because recycling processes can inherently produce hybrid composites, either due to contamination of the carbon fibre input material with glass fibres, or due to using a mixture of carbon fibre types. This is perhaps not the case yet because most commercial recycling plants still mainly use manufacturing waste as input material. Such waste is relatively easy to track, and hence, its composition is relatively homogeneous and well-known. In the future, however, the expected shift toward processing end-of-life composite waste will introduce greater heterogeneity and uncertainty in input materials. Research into hybrid composites will be vital to help manufacturers (1) better understand which waste streams should ideally be combined or avoided and (2) predict the performance of the recyclate.

Recycling processes for carbon fibre composites can achieve nearly the same fibre stiffness and strength as virgin carbon fibres (Oliveux et al., 2015). The major remaining limitation is that most recycling processes inevitably reduce fibre lengths to a few millimetres, which diminishes composite stiffness and strength. To counteract this, academic research and the industry have gradually refocused on aligning discontinuous fibres. Alignment processes can lead to most fibres being aligned within 3 degrees of the nominal ply angle and to fibre volume fractions in the range of 41%–55% (Yu et al., 2014; Heider et al., 2019; Such et al., 2014) provides an excellent overview of the historical developments of alignment processes.

Aligned discontinuous hybrids were introduced when alignment processes were first developed around 1980 (Richter, 1980; Parratt and Potter, 1980). After dispersing the discontinuous fibres in a liquid (usually water), the short fibres are aligned by exploiting laminar flow, followed by vacuum extraction to remove the liquid (Richter, 1980). Both studies have already achieved excellent dispersion of carbon and glass fibres, together with the tunability of the hybrid volume fraction (Parratt and Potter, 1980). In addition showed that pseudo-ductility was possible with this type of fibre-hybrids, although they used the term "yielding" instead of pseudo-ductility.

After a standstill of 3 decades, the last decade has shown a renewed interest in alignment processes due to the increased need for the recycling of carbon fibres. Several processes have been developed to align discontinuous fibres after first bringing them into an aqueous solution: the HiPerDiF process (Longana et al., 2017; Yu et al., 2014, 2015, 2016) (University of Bristol, now commercialised by Lineat Composites in the UK), the TuFF process (Heider et al., 2019; Legenstein et al., 2024) (University of Delaware) and the University of Nottingham process (Wong et al., 2009; Liu et al., 2021). All these processes leverage the simplicity of mixing multiple fibres into an aqueous solution, enabling excellent dispersion of fibres and straightforward fibre-hybridisation, as illustrated in Fig. 1.7.3. Discontinuous fibres can also be aligned via carding processes, but these usually require longer fibre, result in poorer alignment, and may increase fibre damage (Such et al., 2014).

4.2 New preforms

The early developments of fibre-hybrid composites followed shortly after the initial developments of carbon fibres. Since then, however, many improvements in textile and processing technology (see chapter 2.3 in Part II of this book) have significantly expanded the possibilities for fibre-hybrid research. These new technologies have enabled the development of new fibre-hybrid architectures, like intralayer weaves (Swolfs et al., 2014a; Wu et al., 2018; Hine et al., 2014), comingled yarns (Selver et al., 2016), thin ply hybrids (Wisnom et al., 2016), and aligned discontinuous

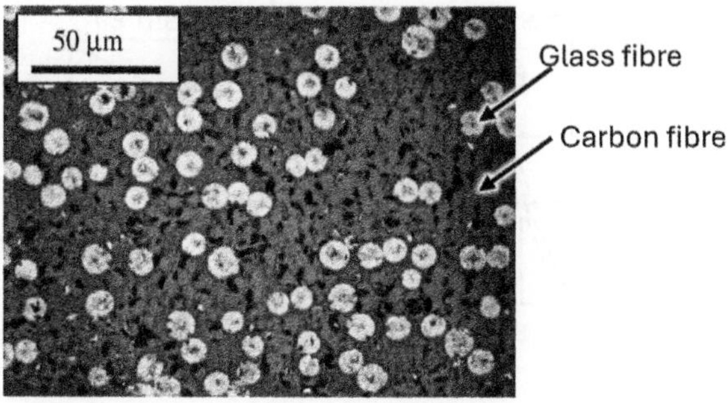

Figure 1.7.3 Microstructure of an aligned discontinuous fibre-hybrid composite with glass fibres in light grey and carbon fibres in dark grey.
Adapted from Yu et al. (2015).

preforms. These new preforms achieve a very high dispersion, which has extensively been proven to maximise the synergetic effects (Swolfs et al., 2014b, 2015, 2019; Fukunaga et al., 1984b). Most of these studies, however, did not delve into measuring the classic hybrid effect apart from (Wisnom et al., 2016), who measured a hybrid effect of up to 20%.

These new preforms remain scarce in the industry mainly due to their higher cost. Instead, most industrial applications still focus on local reinforcements and, to a lesser extent, interlayer hybrids (Swolfs et al., 2019).

4.3 Pseudo-ductility

Pseudo-ductility has received extensive attention over the last decade, as it offers a way to create some ductility by combining components that are brittle on their own. It is defined as "*the ability to deform significantly with a softening response but without complete fracture, while maintaining load carrying capability*" (Wisnom et al., 2024). An important parameter to monitor is the pseudo-ductile strain, as it provides a way to compare different pseudoductility strategies (Wisnom et al., 2024). defined this parameter as "*the difference between the final failure strain, and the elastic strain at the same stress based on the initial modulus E_0*" (see Fig. 1.7.4).

Among various strategies to achieve pseudo-ductility, fibre-hybridisation has proven to be one of the most effective. Much of the research into pseudo-ductile hybrids was triggered by a large UK project called HiPerDuCT (Wisnom et al., 2024). However, the original idea already originated in the early days of fibre-hybrid composites, as it was first reported in the early seventies when only a few percent of carbon fibre was added (Dixon et al., 1972; Dukes and Griffiths, 1971). They referred to it as "pseudo-yield behaviour" (Summerscales and Short, 1978). Some early works on fibre-hybrids, such as (Bunsell and Harris, 1974; Aveston and Sillwood, 1976), can also be seen as precursors to pseudo-ductile hybrid designs.

A key strategy for creating pseudo-ductile fibre-hybrids is using thin plies. Thin plies release less energy when they break, limiting or preventing delamination. This is key to avoiding catastrophic failure as well as significant load drops when the brittle component fails. This enables the higher-elongation component to continue carrying the load, creating a pseudo-ductile regime. The development of damage mode maps

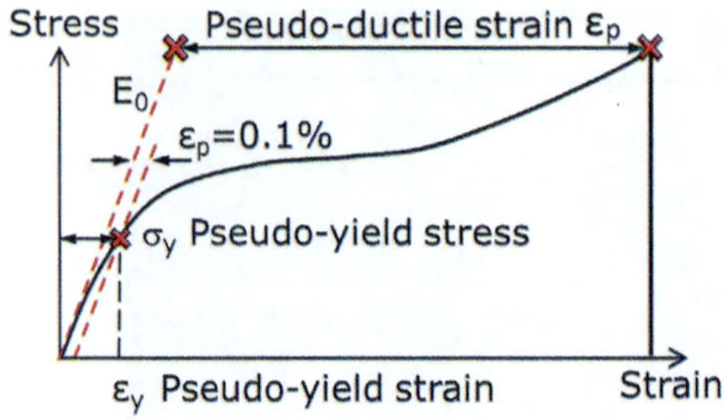

Figure 1.7.4 Definition of the pseudo-ductile strain.
Reprinted from Wisnom et al. (2024).

by Jalalvand et al. (2014) created a convenient method to predict the tensile behaviour of interlayer hybrids (see Fig. 1.7.5).

A second method to create pseudo-ductile fibre-hybrids is using aligned discontinuous preforms. This approach exploits fibre fragmentation instead of the layer fragmentation used by thin-ply hybrids. As explained in section "4.1 Recycling and re-use" and shown in Fig. 1.7.3, aligned discontinuous fibre-hybrids can lead to excellent dispersion. This dispersion creates opportunities for strong synergies (Yu et al., 2015). achieved a pseudo-ductile strain of 1.1% and a tensile modulus of 110 GPa by combining a high-modulus carbon with an E-glass fibre. Although they did not quantify it, their aligned discontinuous fibre-hybrid clearly increased the failure strain of the reference carbon fibre, demonstrating significant hybrid effects. Similarly, Longana et al. (2017) combined reclaimed and virgin carbon fibres to create fibre-hybrids with notable pseudo-ductility.

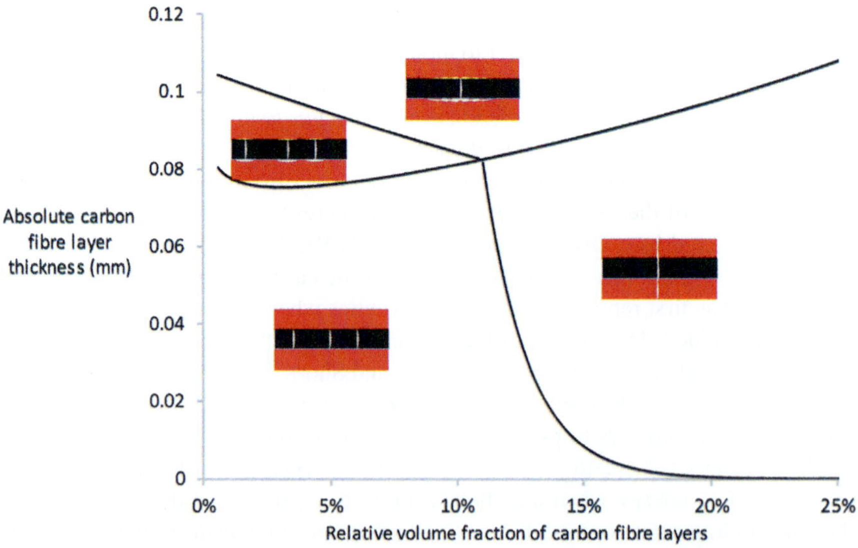

Figure 1.7.5 Example of a damage mode map for a carbon/glass fibre-hybrid composite.
Reprinted from Swolfs et al. (2019) with permission from SAGE.

4.4 New fibre types and combinations

In the early days (see section "2 The early years of fibre-hybridisation"), the focus was on combining carbon and glass fibres with some studies focussing on carbon/aramid (Zweben, 1977; Adam et al., 1989) or glass/aramid (Perry and Adams, 1975). Since then, many new fibre types have been introduced, and the range of carbon and glass fibres has significantly expanded. It is now relatively common to combine a high-strength with a high-modulus carbon fibre (Sapozhnikov et al., 2020; Czél et al., 2017). Because high-modulus carbon fibres can be 3–4 times stiffer than high-strength carbon fibres, they can strongly increase the stiffness relative to a high-strength carbon fibre composite.

In addition to synthetic fibres, using natural fibres in hybrid composites has gained considerable traction. The combination of two types of natural fibres is increasingly common. This is mainly because natural fibres offer a wide range of stiffnesses, strengths, failure strains, impact resistance, and other properties. Fibre-hybridization offers a way to balance the good and the bad properties better.

Despite the promising potential of these new fibre combinations, their direct industry application has been somewhat limited. Instead, most industrial applications have remained focused on fibre-hybrids with low dispersion. Typical examples remain local reinforcements of the carbon fibre laminates. This is, for example, commonly done in surfboards, bike frames, and masts (Swolfs et al., 2019; Nehls, 2024).

5. Outlook

An important drawback of most fibre-hybrid composites is the additional difficulties in recycling. When a fibre-hybrid composite enters the recycling process, its behaviour can vary significantly depending on the fibre types used. For instance, fibres that are not glass or carbon-based will likely not withstand the harsh conditions required to remove the matrix during pyrolysis and solvolysis. Glass fibres, while more likely to survive the recycling process, face considerable challenges in maintaining their strength during recycling. Some progress has been made in terms of upgrading recycled glass fibre performance (Thomason et al., 2016), but questions about the economic value of recycled glass fibres remain. Recycling processes will, therefore, remain focused on recovering carbon fibres, with glass fibres often viewed as a contaminant. While glass fibres can be separated after recycling by using the difference in density or electrostaticity (Thirunavukkarasu et al., 2021), they do require an extra step in the process. The industrial implementation of these separation processes still requires substantial development and investment.

At the moment, there is still a significant mismatch: industry focuses on local reinforcements and interlayer hybrids, whereas academia focuses on high-dispersion routes. This gap is gradually closing though, with more and more industrial examples appearing, such as the A/C pillar of a BMW i3 car, which is made of a carbon/glass intralayer hybrid braid (Swolfs et al., 2019). Further research should help to bridge this gap further, for example, by providing better predictions of fibre-hybrid mechanical performance (see next paragraph). Moreover, reducing the costs of these high-dispersion preforms will likely be critical in facilitating their broader industrial adoption.

Models for the tensile properties of fibre-hybrid composites were already developed in the seventies, but they were still relatively simple. Many advanced models to predict the tensile behaviour of fibre-hybrids have been developed in academia in the last decade (Tavares et al., 2016; Swolfs et al., 2016; Mesquita et al., 2019; Mishnaevsky Jr and Dai, 2014). These models lean on the developments of fibre break models for nonhybrids, starting with (Rosen, 1964; Zweben and Rosen, 1970) and continued by the works of Curtin, Okabe, and Takeda (Zhou and Curtin, 1995; Curtin and Takeda, 1998; Okabe et al., 2005), but tend to focus more on Monte Carlo-based methods and random packings. The models for fibre-hybrids tend to focus on well-dispersed hybrids, with the main exception being (Mesquita et al., 2019), who focused on interlayer hybrids. All these models, however, focus on longitudinal tension. Models for fibre-hybrid composites do exist to cover other loading conditions, such as flexure (Dong and Davies, 2012) and impact (Zhao et al., 2020). However, they are scarce and usually focus on meso- or macroscale. Expanding modelling efforts to cover a broader range of mechanical properties and loading conditions will be crucial for aligning academic research with industrial needs.

We highlighted the discrepancy between the industry focussing on low-dispersion fibre-hybrids and academia focussing more on high-dispersion routes. To propel the sustainability of composites, a greater synergy is required between these two sectors. Industry should target to improve dispersion, and academia should emphasise local stiffening strategies that align with current industrial needs. This shift will help ensure that the technical advances made in academia can be more readily applied to commercial products.

In addition to the technical benefits of fibre-hybrid composites, we cannot underestimate the importance of commercial or advertising benefits. These arguments are often crucial in consumer goods, like many sporting articles. Some fibre-hybrid composites were introduced based on marketing arguments before substantial scientific evidence existed. A good example is the addition of flax fibres to carbon fibre bike frames in the late 2000s based on the arguments of improved damping. Academic research to confirm that flax-carbon fibre-hybrids indeed improved damping in bike frames proved difficult (Vanwalleghem, 2010). Instead, most evidence of improved damping is based on coupon-level tests (Le Guen et al., 2016). We expect that the gap between commercial and technical arguments will narrow down in the future as our fundamental understanding of the behaviour of fibre-hybrid composites improves further.

Overall, fibre-hybrid composites are an excellent way to embrace the full design freedom of composites. As advances in preform technology, mechanical understanding, modelling, and recycling processes are steadily improving, we can expect to see fibre-hybrids gain greater traction in both industrial applications and consumer goods.

References

Adam, T., Fernando, G., Dickson, R.F., Reiter, H., Harris, B., 1989. Fatigue life prediction for hybrid composites. Int. J. Fatig. 11, 233–237.

Anonymous, 1973a. Carbon fiber helps create powerboat records. Reinforc Plast 17, 7.

Anonymous, 1973b. Carbon fiber helps gain gold for West German rowers. Reinforc Plast 17, 69.

Anonymous, 1973c. Carbon fibres chosen for round-the-world yacht. Composites 4, 195–196.

Anonymous, 1974. Carbon fibres in helicopter blades. Composites 5, 144.

Aveston, J., Sillwood, J.M., 1976. Synergistic fiber strengthening in hybrid composites. J. Mater. Sci. 11, 1877–1883.

Bunsell, A.R., Harris, B., 1974. Hybrid carbon and glass fiber composites. Composites 5, 157–164.

Curtin, W.A., Takeda, N., 1998. Tensile strength of fiber-reinforced composites: I. Model and effects of local fiber geometry. J. Compos. Mater. 32, 2042–2059.

Czél, G., Jalalvand, M., Wisnom, M.R., 2016. Hybrid specimens eliminating stress concentrations in tensile and compressive testing of unidirectional composites. Compos. Appl. Sci. Manuf. 91, 436–447.

Czél, G., Jalalvand, M., Wisnom, M.R., Czigány, T., 2017. Design and characterisation of high performance, pseudo-ductile all-carbon/epoxy unidirectional hybrid composites. Compos. B Eng. 111, 348–356.

Dixon, R., Ramsey, B., Usher, P., 1972. Design and build of the GRP hull of HMS Wilton. In: Proceeding of the Symposium on GRP Ship Construction. RINA, London, p. 32.

Dong, C., Davies, I.J., 2012. Optimal design for the flexural behaviour of glass and carbon fiber reinforced polymer hybrid composites. Mater. Des. 37, 450–457.

Dong, C.S., Duong, J., Davies, I.J., 2012. Flexural properties of S-2 glass and TR30S carbon fiber-reinforced epoxy hybrid composites. Polym. Compos. 33, 773–781.

Dukes, R., Griffiths, D., 1971. Marine aspects of carbon fiber and glass fiber/carbon fiber composites. In: Proc., Int. Conf. On Carbon Fibers, Their Composites, and Applications.

Fazlali, B., Upadhyay, S., Ashokbhai Ashodia, S., Mesquita, F., Lomov, S.V., Carvelli, V., Swolfs, Y., 2023. Specimen designs for accurate tensile testing of unidirectional composite laminates. Compos. Appl. Sci. Manuf. 175, 107799.

Fischer, S., Marom, G., Tuler, F.R., 1979. Hybrid effects in composites: comparison between interlaminar and translaminar configurations. J. Mater. Sci. 14, 863–868.

Fukuda, H., Chou, T.W., 1982. Monte Carlo simulation of the strength of hybrid composites. J. Compos. Mater. 16, 371–385.

Fukunaga, H., Chou, T.W., Fukuda, H., 1984a. Strength of intermingled hybrid composites. J. Reinforc. Plast. Compos. 3, 145–160.

Fukunaga, H., Chou, T.W., Fukuda, H., 1989. Probabilistic strength analyses of interlaminated hybrid composites. Compos. Sci. Technol. 35, 331–345.

Fukunaga, H., Chou, T.W., Schulte, K., Peters, P.W.M., 1984b. Probabilistic initial failure strength of hybrid and non-hybrid laminates. J. Mater. Sci. 19, 3546–3553.

Hayashi, T., 1972. On the improvement of mechanical properties of composites by hybrid composition. In: Proceedings of the 8th International Reinforced Plastics Conference, 10–12 October 1972, Brighton, UK, pp. 149–152.

Heider, D., Tierney, J., Henchir, M.A., Gargitter, V., Yarlaggada, S., Gillespie Jr, J.W., Sun, J., Sietins, J.M., Knorr, D., 2019. Microstructural evaluation of aligned, short fiber TUFF material. In: SAMPE 2019. SAMPE, Charlotte, USA.

Hine, P.J., Bonner, M., Ward, I.M., Swolfs, Y., Verpoest, I., Mierzwa, A., 2014. Hybrid carbon fiber/nylon 12 single polymer composites. Compos. Appl. Sci. Manuf. 65, 19–26.

Jalalvand, M., Czél, G., Wisnom, M.R., 2014. Numerical modelling of the damage modes in UD thin carbon/glass hybrid laminates. Compos. Sci. Technol. 94, 39–47.

Jang, B.Z., Chen, L.C., Wang, C.Z., Lin, H.T., Zee, R.H., 1989. Impact resistance and energy-absorption mechanisms in hybrid composites. Compos. Sci. Technol. 34, 305–335.

Ji, X., 1982. On the hybrid effect and fracture mode of interlaminated hybrid composites. In: Hayashi, T., Kawata, K., Umekawa, S. (Eds.), International Conference on Composite Materials IV, Tokyo, Japan.

Kretsis, G., 1987. A review of the tensile, compressive, flexural and shear properties of hybrid fiber-reinforced plastics. Composites 18, 13–23.

Le Guen, M.J., Newman, R.H., Fernyhough, A., Emms, G.W., Staiger, M.P., 2016. The damping–modulus relationship in flax–carbon fiber hybrid composites. Compos. B Eng. 89, 27–33.

Legenstein, A., Füssel, L., Heider, D., Gillespie, J.W., Cender, T.A., 2024. Stretch-steering of highly aligned discontinuous fiber tape with automated fiber placement. Compos. B Eng. 287, 111801.

Liu, Z., Turner, T.A., Wong, K.H., Pickering, S.J., 2021. Development of high performance recycled carbon fiber composites with an advanced hydrodynamic fiber alignment process. J. Clean. Prod. 278, 123785.

Longana, M.L., Yu, H., Jalavand, M., Wisnom, M.R., Potter, K.D., 2017. Aligned discontinuous intermingled reclaimed/virgin carbon fiber composites for high performance and pseudo-ductile behaviour in interlaminated carbon-glass hybrids. Compos. Sci. Technol. 143, 13–21.

Marom, G., Fischer, S., Tuler, F.R., Wagner, H.D., 1978. Hybrid effects in composites: conditions for positive or negative effects versus rule-of-mixtures behavior. J. Mater. Sci. 13, 1419–1426.

Mccann, J., 1972. Moulding Carbon Fibre Propellers. Aeromodeller, pp. 284–287.

Mesquita, F., Swolfs, Y., Lomov, S.V., Gorbatikh, L., 2019. Ply fragmentation in unidirectional hybrid composites linked to stochastic fiber behaviour: a dual-scale model. Compos. Sci. Technol. 181, 107702.

Mishnaevsky Jr., L., Dai, G., 2014. Hybrid carbon/glass fiber composites: micromechanical analysis of structure–damage resistance relationships. Comput. Mater. Sci. 81, 630–640.

Nehls, G., 2024. Rein4ced, Dieffenbacher Automate Large-Scale Production of CFRP Bike Frames. https://www.compositesworld.com/news/rein4ced-dieffenbacher-automate-large-scale-production-of-cfrp-bike-frames.

Okabe, T., Sekine, H., Ishii, K., Nishikawa, M., Takeda, N., 2005. Numerical method for failure simulation of unidirectional fiber-reinforced composites with spring element model. Compos. Sci. Technol. 65, 921–933.

Oliveux, G., Dandy, L.O., Leeke, G.A., 2015. Current status of recycling of fiber reinforced polymers: review of technologies, reuse and resulting properties. Prog. Mater. Sci. 72, 61–99.

Onal, L., Adanur, S., 2002. Effect of Stacking Sequence on the mechanical properties of glass-carbon hybrid composites before and after impact. J. Ind. Textil. 31, 255–271.

Papisc, 2012. A Gulf-Liveried Ford GT40. https://en.wikipedia.org/wiki/John_Wyer.

Park, R., Jang, J.S., 2001. Impact behavior of aramid fiber glass fiber hybrid composite: evaluation of four-layer hybrid composites. J. Mater. Sci. 36, 2359–2367.

Parratt, N.J., Potter, K.D., 1980. Mechanical behaviour of intimately-mixed hybrid composites. In: Bunsell, A.R., Bathias, C., Martrenchar, A., Menkes, D., Verchery, G. (Eds.), 3rd International Conference on Composite Materials, Paris, France. Pergamon Press SARL.

Perry, J.L., Adams, D.F., 1975. Charpy impact experiments on graphite/epoxy hybrid composites. Composites 6, 166–172.

Phillips, L.N., 1976a. The hybrid effect - does it exist? Composites 7, 7–8.

Phillips, L.N., 1976b. On the usefulness of glass-fiber-carbon hybrids. In: British Plastics Federation Congress, Brighton, United Kingdom.

Phillips, N.L., 1969. Improving racing-car bodies. Composites 1, 50–51.

Richter, H., 1980. Single fiber and hybrid composites with aligned discontinuous fibres in polymer matrix. In: Bunsell, A.R., Bathias, C., Martrenchar, A., Menkes, D., Verchery, G. (Eds.), 3rd International Conference on Composite Materials, Paris, France. Pergamon Press SARL.

Rosen, B.W., 1964. Tensile failure of fibrous composites. AIAA J. 2, 1985–1991.

Salkind, M.J., 1975. The twin beam composite rotor blade. Fibre Sci. Technol. 8, 91–102.

Sapozhnikov, S.B., Swolfs, Y., Lomov, S.V., 2020. Mode I and II interlaminar critical energy release rates in all-carbon interlayer unidirectional fiber-hybrids based on ultrahigh-modulus and high-strength fibres. Compos. Struct. 236, 111886.

Sayer, M., Bektasc, N.B., Sayman, O., 2010. An experimental investigation on the impact behavior of hybrid composite plates. Compos. Struct. 1256–1262.

Selver, E., Potluri, P., Hogg, P., Soutis, C., 2016. Impact damage tolerance of thermoset composites reinforced with hybrid commingled yarns. Compos. B Eng. 91, 522–538.

Such, M., Ward, C., Potter, K., 2014. Aligned discontinuous fiber composites: a short history. J. Multifunc. Compos. 2, 155–168.

Summerscales, J., Short, D., 1978. Carbon fiber and glass fiber hybrid reinforced plastics. Composites 9, 157–166.

Swolfs, Y., Crauwels, L., Van Breda, E., Gorbatikh, L., Hine, P., Ward, I., Verpoest, I., 2014a. Tensile behaviour of intralayer hybrid composites of carbon fiber and self-reinforced polypropylene. Compos. Appl. Sci. Manuf. 59, 78–84.

Swolfs, Y., Gorbatikh, L., Verpoest, I., 2014b. Fibre hybridisation in polymer composites: a review. Compos. Appl. Sci. Manuf. 67, 181–200.

Swolfs, Y., Mcmeeking, R.M., Gorbatikh, L., Verpoest, I., 2015. The effect of fiber dispersion on initial failure strain and cluster development in unidirectional carbon/glass hybrid composites. Compos. Appl. Sci. Manuf. 69, 279–287.

Swolfs, Y., Verpoest, I., Gorbatikh, L., 2016. Maximising the hybrid effect in unidirectional hybrid composites. Mater. Des. 93, 39–45.

Swolfs, Y., Verpoest, I., Gorbatikh, L., 2019. Recent advances in fiber-hybrid composites: materials selection, opportunities and applications. Int. Mater. Rev. 64, 181–215.

Tavares, R.P., Melro, A.R., Bessa, M.A., Turon, A., Liu, W.K., Camanho, P.P., 2016. Mechanics of hybrid polymer composites: analytical and computational study. Comput. Mech. 57, 405–421.

Thirunavukkarasu, J., Turner, T.A., Pickering, S.J., 2021. separation-of-composites-for-recycling-measurement-of-electrostatic-charge-of-carbon-and-glass-fiber-particles. Int. J. Electr. Comput. Eng. 15.

Thomason, J.L., Nagel, U., Yang, L., Sáez, E., 2016. Regenerating the strength of thermally recycled glass fibres using hot sodium hydroxide. Compos. Appl. Sci. Manuf. 87, 220–227.

Vanwalleghem, J., 2010. Study of the damping and vibration behaviour of flax-carbon composite bicycle racing frames. Master thesis, Ghent University..

Wisnom, M.R., Czél, G., Swolfs, Y., Jalalvand, M., Gorbatikh, L., Verpoest, I., 2016. Hybrid effects in thin ply carbon/glass unidirectional laminates: accurate experimental determination and prediction. Compos. Appl. Sci. Manuf. 88, 131–139.

Wisnom, M.R., Pimenta, S., Shaffer, M.S.P., Robinson, P., Potter, K.D., Hamerton, I., Czél, G., Jalalvand, M., Fotouhi, M., Anthony, D.B., Yu, H., Longana, M.L., Wu, X., Bismarck, A.,

2024. High performance ductile and pseudo-ductile polymer matrix composites: a review. Compos. Appl. Sci. Manuf. 181, 108029.

Wong, K.H., Turner, T.A., Pickering, S.J., Warrior, N.A., 2009. The potential for fiber alignment in the manufacture of polymer composites from recycled carbon fiber. SAE Int. J. Aerosp. 2, 225–231.

Wu, W., Wang, Q., Li, W., 2018. Comparison of tensile and compressive properties of carbon/glass interlayer and intralayer hybrid composites. Materials 11, 1105.

Xing, J., Hsiao, G.C., Chou, T.W., 1981. A dynamic explanation of the hybrid effect. J. Compos. Mater. 15, 443–461.

Yao, L.N., Chou, T.W., 1989. Analysis of hybrid effect in unidirectional composites under longitudinal compression. Compos. Struct. 12, 27–37.

Yu, H., Longana, M.L., Jalalvand, M., Wisnom, M.R., Potter, K.D., 2015. Pseudo-ductility in intermingled carbon/glass hybrid composites with highly aligned discontinuous fibres. Compos. Appl. Sci. Manuf. 73, 35–44.

Yu, H., Longana, M.L., Swolfs, Y., Wisnom, M.R., Potter, K.D., 2016. Hybrid effect of carbon/glass composites as a function of the strength distribution of aligned short carbon fibres. In: 17th European Conference on Composite Materials, Munich, Germany.

Yu, H., Potter, K.D., Wisnom, M.R., 2014. A novel manufacturing method for aligned discontinuous fiber composites (High Performance-Discontinuous Fibre method). Compos. Appl. Sci. Manuf. 65, 175–185.

Zhao, Y., Cao, M., Tan, H.X., Ridha, M., Tay, T.E., 2020. Hybrid woven carbon-Dyneema composites under drop-weight and steel ball impact. Compos. Struct. 236, 111811.

Zhou, S.J., Curtin, W.A., 1995. Failure of fiber composites: a lattice green function model. Acta Metall. Mater. 43, 3093–3104.

Zweben, C., 1977. Tensile strength of hybrid composites. J. Mater. Sci. 12, 1325–1337.

Zweben, C., Rosen, B.W., 1970. A statistical theory of material strength with application to composite materials. J. Mech. Phys. Solid. 18, 189–206.

The history of advanced composites in the United States – A personal story

1.8

R. Byron Pipes
Purdue University, West Lafayette, IN, United States

1. Birth of advanced composites in the United States – 1967–87

The author recalls his first contact in 1967 with what would become one of the major developments in advanced materials – carbon fibre composites. As a graduate student at Princeton University, the author attended a graduate seminar presented by an engineer from Union Carbide who passed around a sample of carbon fibre-reinforced polymer composites. Most attendees did not know what to make of this new fledgling technology, nor did the presenter seem to have a good explanation or vision of the role that this new material form might play in American society. The next encounter with this emerging technology occurred in 1969 when the author interviewed for an engineering position at a major aerospace conglomerate – General Dynamics. Here he encountered an engineering team, entitled the Composites Structures Group, who were challenged to develop the materials, design, and manufacturing technology sufficient to fly a composite aircraft structure. Indeed, by 1970, several aerospace companies had already begun to fly prototype, nonflight critical elements aboard their military aircraft.

Multiple visionaries led and guided this new technology development from the very beginning: The late Dr. Alan M. Lovelace began federal service with the United States Air Force in 1954, serving at the Air Force Materials Laboratory, Wright–Patterson Air Force Base, Ohio, and was named Director in 1967. From 1972 to 1973, he served as Director of Science and Technology with the Air Force Systems Command, Andrews Air Force Base, Washington, D.C., and during 1973–74 was Deputy Assistant Secretary of the Air Force for Research and Development. The late General Bernard A. Schriever began his distinguished 35-year military career with the United States Air Force in 1931. During his service, he headed all Air Force ballistic missile and space development programs, including the development and deployment of the Thor, Atlas, Titan, and Minuteman missile systems, from 1954 to 1966. Finally, the "hands-on leader," Mr. George P. Peterson, had a distinguished 35-year career in the aeronautics field, which he began in 1951 by joining the United States Air Force Materials Laboratory at Wright–Patterson Air Force Base in the Nonmetallic Materials Division, and culminated in his retirement in

The History of Composites. https://doi.org/10.1016/B978-0-443-34119-9.00010-3

1985 as Director of the Materials Laboratory, Air Force Wright Aeronautical Laboratories, Aeronautical Systems Division, Air Force Systems Command, Wright–Patterson Air Force Base, Ohio.

The program structure developed by these three individuals consisted of the assignment of specific airframe structural elements to each of the major aircraft original equipment manufacturers (OEMs) with the goal of developing design and manufacturing technology sufficient to yield improved structural performance and weight savings over metallic structure. General Dynamics, where the author was employed, was assigned the fuselage. The remaining structural components, such as the wing, empennage, and landing gear, were assigned to other aircraft OEMs.

The reinforcing fibre in favour at the time was a high-performance fibre produced in the United States by Hamilton Standard. However, it was not actually a fibre in the classical sense. Rather, it was a filament produced by chemical vapour deposition of boron onto the surface of a 12-micron tungsten filament. The boron fibre diameter, shown in Fig. 1.8.1, was in the range of 100–200 microns, clearly an order of magnitude greater than the conventional carbon and glass fibres. In the mid-1960s, the U.S. Air Force had invested in the development of boron/epoxy composites fabricated in the form of continuous, collimated fibre prepreg. In this form, prepreg tape consisted of collimated boron fibres with prepreg thicknesses modestly larger than the single boron fibre, 127 microns. To date, most prepreg thicknesses have remained virtually equal to this value for both thermoset and thermoplastic polymer prepregs. One of the most interesting characteristics of this early composite material form was that it was produced in the form of a 3-inch wide tape product with the goal of automated

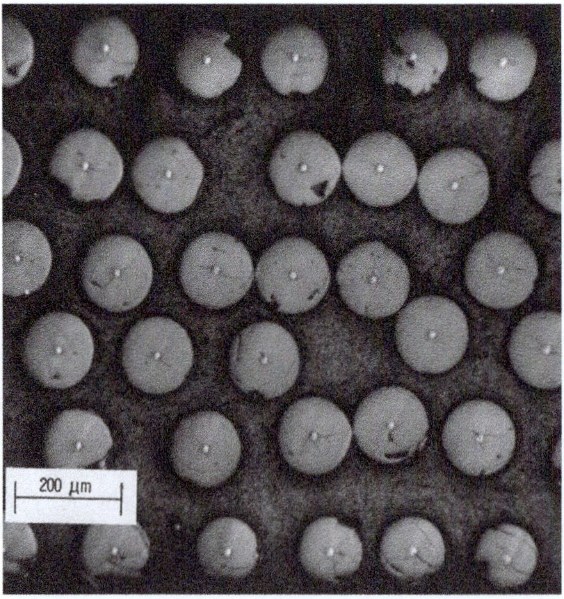

Figure 1.8.1 Boron-epoxy composite, c.1970. Pipes et al. (1977).

tape laying for manufacture. Yet tape laying remained restricted to tool geometries of only modest curvature.

The first high-performance composite to fly was constructed of boron-epoxy prepreg, and the structural element was the horizontal stabiliser of the General Dynamics F-111 (c.1970) shown in Fig. 1.8.2.

Boron-epoxy laminates were also employed to reinforce the wing carry-through structure of the F-111 when crack propagation emanating from fastener holes limited the flight hours of the vehicle. Tapered thickness, angle-ply laminates of $[+30/-30]_{ns}$ fibre orientation were bonded to the surface of the carry-through structure and thereby reduced the stress level sufficiently to achieve enhanced life of the aircraft. Boron fibre was adopted by both Grumman and McDonnell Douglas in the early 1970s to produce empennage components for the F-14A shown in Fig. 1.8.3 and the F-15 aircraft shown in Fig. 1.8.4. Thereby, the first application of high-performance composites in aircraft production programs was established.

In contrast, General Dynamics replaced boron fibre in its fuselage program with polyacrylonitride (PAN)-based carbon fibre. Discovered in 1960 by both Japanese and English national research laboratories, the PAN-based carbon fibre had excellent structural properties, but at that time, this was not the primary reason that boron was replaced. Rather, the Air Force sponsor, under the leadership of George Peterson, purchased the entire Hamilton Standard production capacity in order to make it available for this time-sensitive research and development program. This prudent decision was, however, in direct conflict with the scale and rate of the boron fibre required by the very programs led by the Peterson team. General Dynamics learnt that it would be

Figure 1.8.2 The first boron-epoxy horizontal stabiliser flew on the General Dynamics F-111 in 1970.
Reproduced from https://www.militarymedia.net/2021/10/general-dynamics-f-111-aardvark.html

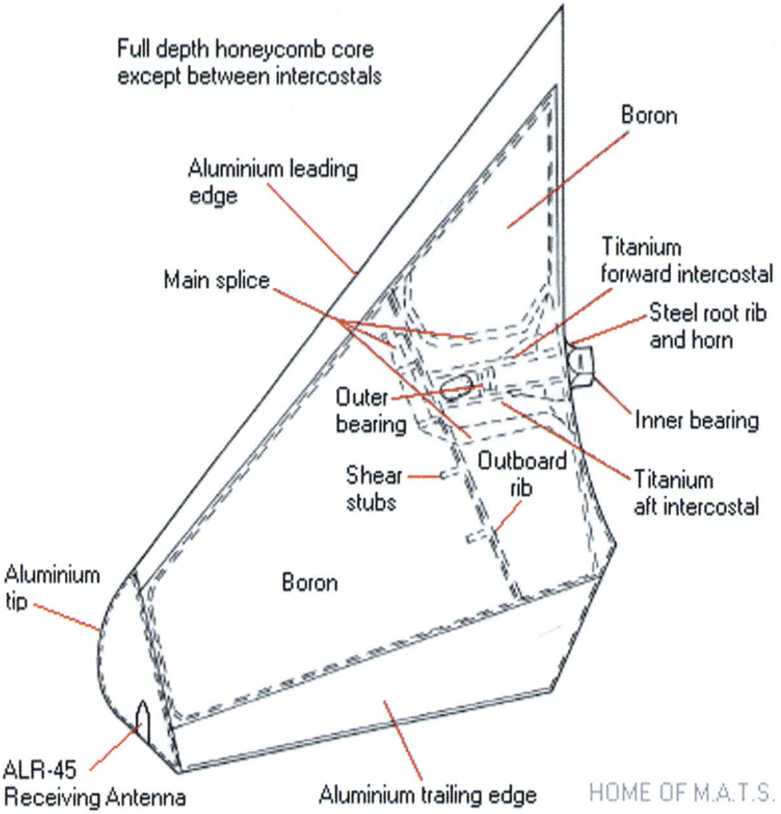

Figure 1.8.3 First production of the boron-epoxy horizontal stabiliser for the Grumman F-14A, c.1971.
Reproduced from http://www.anft.net/f-14/f14-detail-horizstab.htm

unable to meet its contractual delivery of fuselage technology when the disposition program for the scarce boron fibre was announced.

PAN-based carbon was the answer to this dilemma. Earlier, Rolls-Royce had invested heavily in PAN carbon fibre development for fan blade applications in its new engine technology. Epoxy-based prepregs of continuous PAN-based carbon fibre were available from Fothergill and Harvey in the United Kingdom in metre-length form. General Dynamics made the transition to carbon fibre in 1970–71 and received carbon fibre/epoxy prepreg in metre lengths by daily air shipments. With this new material in hand, General Dynamics designed, manufactured, and tested to failure the fuselage of the Northrop F-5, and the structure met both the performance and economic requirements of the Air Force fuselage program. Thus, the advanced composites era was launched.

Figure 1.8.4 Boron-epoxy empennage of the McDonnell Douglas F-15, c.1972. Reproduced from https://www.flight-manuals-online.com/product/mcdonnell-douglas-f-15-eagle/

2. Analysis of composites

The analysis of highly anisotropic structural systems was an entirely new subject to the aerospace engineering community of the United States in the 1960s. Much of the required knowledge had been codified by Russian scholars such as Lekhnitskii and Ambartsumyan. However, it was the researchers of the Forest Products Laboratory that first introduced early concepts of the anisotropy of wood to the American engineering world, and this was the basis for many of the early methods of analysis adopted. Prediction of effective or volume average micromechanical properties of composites was based largely on closed-form analyses and energy methods for the establishment of property bounds. A major development of this period was laminated plate theory, and much of the early work utilised composite laminate theory to predict stresses within multiaxial composite laminates and effective laminate properties. Biaxial failure theories were developed for the prediction of the onset of failure of individual lamina of the laminate and thereby laminate failure. Great diversity existed in approaches for the determination of biaxial laminate strength, with experimental test data and the choice of those with responsibility for developing design allowables for aerospace.

Classical fracture mechanics approaches for multiaxial laminates were found to only be suitable for interlaminar crack propagation, and the development of methods for characterisation of mode I and mode II interlaminar fracture produced the double-cantilever and edge-notched flexure tests that are now American Society for Testing Materials (ASTM) standards. Damage mechanics and cohesive elements within finite-element methods filled a need for damage propagation predictions, and this approach allowed for diffused damage to be modelled.

Only recently have unified, multiscale approaches to composites analysis begun to be developed. Earlier, two-dimensional stress fields received the greatest attention, while plate and shell approaches provided efficient calculations. Furthermore, the wide array of micromechanical models for the prediction of effective composite properties was largely disconnected from macro-scale composite structural performance.

3. The free-edge problem

In the late 1960s, it was noticed that laminate theory, based on the plane stress assumption, was unable to treat the issue of traction-free conditions at free edges, and discussions in the U.S. composites community raised a challenge to develop an analysis to treat this problem. Several scholars developed approximate approaches, but finite-difference solutions to the three-dimensional equations of elasticity provided the first foundational analysis of what became known as the "free-edge problem." It should be noted that three-dimensional, anisotropic finite-element methods were not generally available at the time. Studies of the free-edge problem initially focused on symmetric, angle-ply laminates, $[+\theta/-\theta]_{ns}$, as shown in Fig. 1.8.5. The solutions revealed the expected interlaminar shearing stress singularity at the free edge and showed the phenomenon to be restricted to a boundary layer at the free

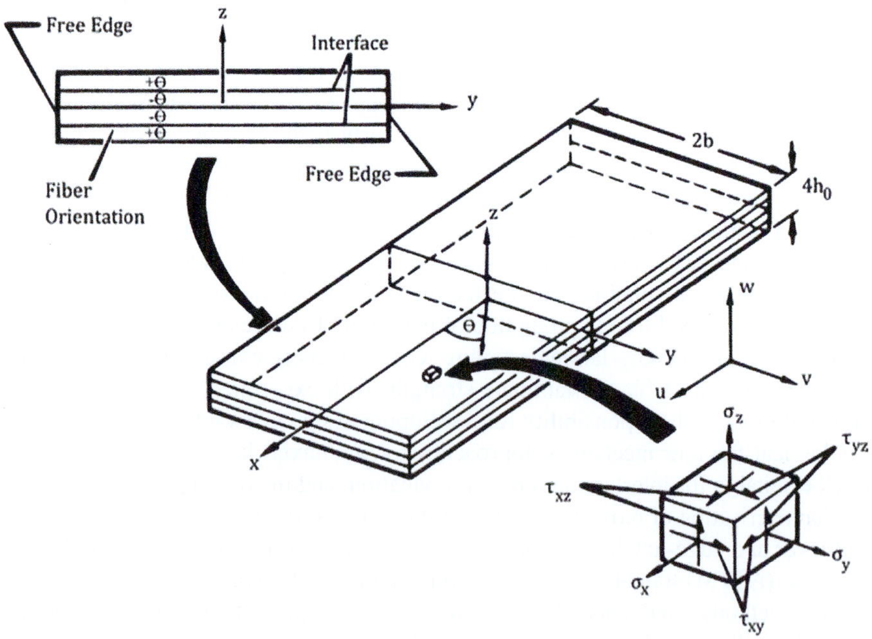

Figure 1.8.5 The free-edge problem solution, c.1970 (Pipes and Pagano, 1970). Reproduced from Pipes and Pagano (1970).

edge approximately equal to the laminate thickness. Next, the solution to the bidirectional laminate, $[0/90]_{ns}$, revealed the presence of an interlaminar normal stress singularity at the free edge, the sign of which was determined by the laminate stacking sequence. Previous data developed for the influence of stacking sequence on laminate tensile strength was found to show close agreement with the developed predictions where the interlaminar normal stress was tensile.

The story of just how this problem came to the attention of the composites community follows: James Ashton was and is an extraordinary individual who would go on to gain an MBA from Harvard to match his Massachusetts Institute of Technology (MIT) PhD and to lead the manufacturing of General Dynamics' next plane, the F-16. To kick-start the author's work in composites, he decided to offer the author an unsolved problem in composite mechanics that he said would attract financial support for the Air Force. Described as the "free-edge problem," Ashton taught that certain in-plane stresses in laminates must vanish at free edges, but that laminated plate theory had no way to accommodate this requirement. Further, he stated that he would sponsor the author to visit the Air Force Materials Laboratory in Dayton, Ohio, to seek Air Force funding. It seems that he said that Dr. Nicholas J. Pagano would be happy to receive the author and likely arrange for the funding. After a flight from Dallas Love Field to Dayton, the author was admitted to Wright–Patterson Air Force Base, where the Materials Laboratory was located, and made his way to Dr. Pagano's office. The author was greeted warmly, and the desired solution of the free-edge problem was discussed. However, in the discussions, it was revealed that Dr. Pagano had no intention of providing funding for a young engineer to solve this important problem. Rather, he planned to solve it himself. During the exchange, he stated several times that a numerical solution of the equations of elasticity for the prediction of the 3D free-edge stress field would be impossible due to singularities in the solution and that only an analytic solution was possible.

Solutions of large numbers of algebraic equations in matrix form were common at the time, but real computing power lay in the future. Consider that at the time, the IBM 360 computer had orders of magnitude less computing power than the smartphone of today. The formulation of a three-dimensional problem in anisotropic elasticity for a finite-width laminate, subjected to an axial loading, required an approximation of a specified form of the axial displacement that reflected the independence of the solution of the coordinate in the loading direction. This approach minimised the size of the matrix required in the finite-difference formulation and, thereby, the required computing power. This deformation state was termed "uniform axial extension" and expressed the axial displacement as the sum of two terms. The first term was the applied axial strain multiplied by the axial coordinate, and the second term was the unknown function in terms of the two coordinates in the plane perpendicular to the loading direction: $u = \varepsilon_x x + U(y,z)$, for an x-y-z coordinate system. Pagano questioned how this relationship was developed and then wrote 10 pages of equations, starting with the assumption of independence of the stress field on the coordinate in the loading direction, and proved that the assumed strain–displacement relationship was exact!

The first problem chosen for study was the symmetric, angle-ply laminate of $\pm\theta$ orientation. Limitations on the number of unknowns in the finite-difference formulation due to computer power allowed only four lamina to be modelled. Just as Pagano had said, a singularity did appear to occur at the free edge and at the interface between the $+\theta$ and $-\theta$ layers.

However, the solution did not suffer from the singularity as Pagano had predicted; it simply predicted finite values where later convergence studies would validate the existence of the singularity. The offer to share this work with Nick Pagano as coauthor was an act that changed the author's life.

Perhaps the most important contribution that Pagano brought to this partnership was his ability to write beautifully with clarity and elegance. His rewrite of the draft paper produced an elegant and seminal document (Pipes and Pagano, 1970).

Of course, the partnership began to grow with the success of that first paper, but the full study of the solution obtained had not yet been completed. The next laminate studied was the bidirectional or [0/90]s symmetric laminate. Unlike the angle-ply laminate, the normal stress in the thickness direction, sigma z, was found to be non-zero and to change sign with the stacking sequence of the [0/90]s laminate. That is, for a tensile coupon of finite width, the peak value of the interlaminar stress was tensile for the [0/90/90/0] laminate and compressive for the [90/0/0/90] laminate. Pagano took on the task of developing what would become known as the "stacking sequence" paper. Furthermore, he was also familiar with experimental data that showed that stacking sequence strongly influenced the fatigue strength of yet another finite-width laminate, the [±15/±45]s symmetric laminate. By now, it was known that the interlaminar normal stress was proportional to the mismatch in the lamina Poisson's ratios, and the difference for the [±15/±45] was of the same order of magnitude as that of the [0/90] laminate. These observations, coupled with the results for the [0/90]s laminate, provided the foundation for one of the most important papers of that decade (1971) to show the importance of stacking sequence on the tensile properties of finite-width laminates. This work was extended in two papers with a primary focus on the influence of the free edge on laminate tensile strength (1971–73) (Pagano and Pipes, 1971, 1973; Pipes et al., 1973).

Pagano's challenge to obtain an analytic solution for the free-edge problem haunted the author. With the numerical solution in hand, insights were sought that might provide the basis for the desired analytic solution. Earlier, it had been noted that the values of three of the six stress components in the finite-difference solution results for the angle-ply laminates were very small when compared to the other three. Thus, the first step in approaching an analytic solution was to assume these minor stress components to be identically zero. What followed were multiple pages of algebra needed to reduce the three governing displacement-equilibrium equations to a single equation. The coefficients in the resulting equation were extremely complex combinations of the stiffness and compliance terms in the anisotropic constitutive relationship, and what resulted was a single partial differential equation with a single dependent variable. What was even more miraculous was that the equation was separable and thereby could be solved by the method of separation of variables. Programmed in Fortran, the small deck of punch cards containing the solution was only one inch in thickness as opposed to the full box (18 inches) of cards required for the finite-difference solution. Expressing boundary conditions in the form of a Fourier series produced a solution for the axial displacement as a function of the rectangular coordinates in the plane normal to the loading direction. Convergence studies undertaken by comparing results as a function of the number of terms in the Fourier series solution revealed the same singularity location as the finite-difference solution had suggested. Further, the results of the two approaches were in excellent agreement, except in regions of high gradients where the analytic solution provided superior results. Without Pagano's challenge to develop an analytic solution for the free-edge problem, this achievement would never have occurred (Pipes and Pagano, 1974).

It is noteworthy that this free-edge analytic solution, represented in Fig. 1.8.5, lay dormant for almost 40 years before graduate students Johnathan Goodsell, Andrew Ritchey, and Alex Kravchenko developed solutions for thermal residual stresses, anticlastic bending, and twisting of finite-width, angle-ply laminates, shown in Fig. 1.8.6, and the resulting publication 43 years after the first paper was published (Pipes et al., 2010; Goodsell et al., 2013; Goodsell and Pipes, 2016; Pagano, 1974; Goodsell and Pipes, 2016)!

Concurrently, Wenbin Yu and a graduate student, B. Peng, joined in this reemergence and published a treatment of the same problem with his Structural Genome concepts (Peng et al., 2016).

There have been many treatments of the free-edge problem over the past 50 years, and while the author has followed much of this work, no attempt has been made to present a comprehensive treatment of this important subject. Rather, the author chose

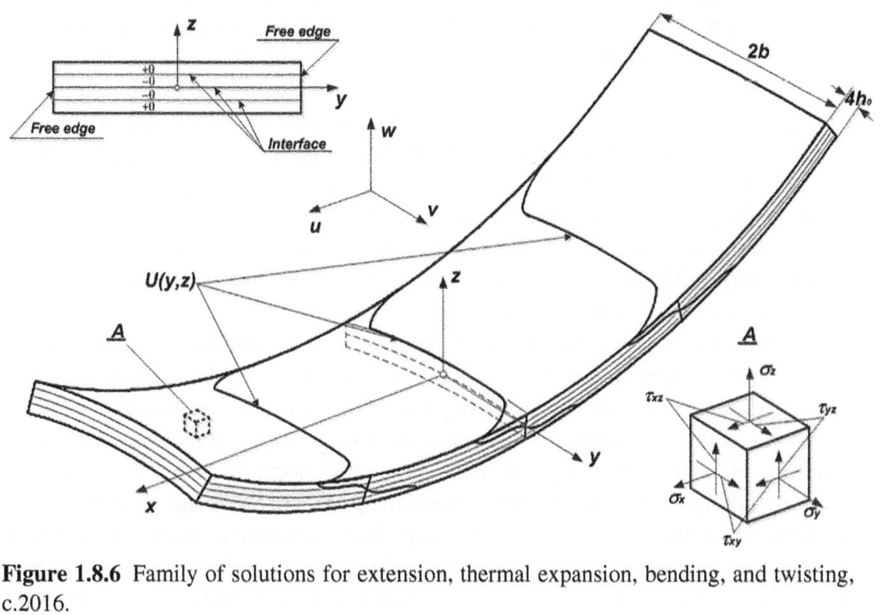

Figure 1.8.6 Family of solutions for extension, thermal expansion, bending, and twisting, c.2016.
Reproduced from Goodsell and Pipes (2016).

a personal experience to represent the untold number of discoveries made by scholars of the field during this period.

4. University research organisations in the United States

The competitive environment for academic leadership of the field of composite materials began to develop in the early 1970s. The competition began with the establishment of a composites research centre at the University of Washington – St. Louis under the leadership of Stephen Tsai. When Tsai moved to the Air Force Materials Laboratory as Chief Scientist, the academic centre he had established languished and was then adopted by the National Science Foundation (NSF). At the time, the NSF was a single-investigator-dominated organisation, and the adoption of a mission-oriented research centre was all too soon abandoned.

In 1974, the Centre for Composite Materials (CCM) of the University of Delaware began as a shared goal of its four faculty founders to make it the "Mecca" for composite materials research and education in the world, and indeed, this goal was accomplished in its first 2 decades (1974–94).

In the same period, the National Aeronautics and Space Administration (NASA) recognised the importance of composites to the field of aeronautics and established a centre of excellence at Virginia Polytechnic Institute and State University in 1974, the same year as the founding of CCM. Funding for the new Virginia Tech centre was reported to be in the range of $1M/year. The NASA Langley Research Centre of Hampton, Virginia, assumed responsibility for the new centre and provided funding for research and internships for students in the program. Thus, an unusual competitive environment existed with the impending decline of the Washington University – St. Louis program and the emergence of a substantial program at Virginia Tech. Of course, the multiple programs in composites throughout the academic community had begun to develop, and these programs would also participate in the competition for preeminence. What was unclear to CCM was how it could establish a funding base in this competitive environment wherein conventional funding remained dominated by single-investigator programs, while interdisciplinary research demanded multi-investigator support.

It was in this context that the industrial experience of the founders of CCM guided the evolution of the new centre. This choice was more driven by the absence of industry participation in competitive university programs than by the wisdom of CCM founders, but as history would prove, the choice was shown to be the correct strategy for success. However, like any transformation, the enormous task could only be achieved by taking the first small steps toward the goal. This took the form of a small advertisement in the *Chemical Engineering News* magazine stating that CCM welcomed opportunities to work with industry. After several visits by interested companies seeking to understand the new emerging composites technology, one small company, the Rogers Corporation of Rogers, CN, made a commitment to establish a sponsored program with CCM. Rogers supplied the automotive industry with brake pistons manufactured of asbestos-reinforced phenolics. The understanding needed to replace asbestos with glass fibres was the subject of the Rogers program, and it would go on for a decade and lead to the successful development of multiple new product innovations by what is now a speciality engineered materials company headquartered in Chandler, Arizona.

By 1978, the success of the Rogers program led to the establishment of an industrial consortium entitled "The Application of Composite Materials to Industrial Products." The success of the Rogers program introduced CCM to the automotive industry giants of Ford and General Motors. Both organisations recognised the potential for weight savings in future products, and they introduced CCM to members of their supply chains. So it was that the small Rogers program led to the development of a consortium in which the OEMs brought the industry together in support of the research and development required to establish composites technology for the automotive industry. By 1984, the consortium had grown to 40 industrial sponsors worldwide, and annual funding exceeded \$1.6M (\$4.825M in 2024 dollars).

It is important to point out that the concept of the industrial consortium was not new in 1978, although there were a few notable examples in existence at the time. The university leadership had decided by then that programs that supported industry to form consortia would be supported by the university by foregoing indirect costs. Thus, the total funding provided by consortia members would go to the support of faculty and student researchers and facilities. This commitment by the university expanded the purchasing power of these funds by almost one-third and thereby accelerated the progress that could be made in the production of research and graduates. In terms of spendable budget, the \$1.6M/year consortium funding provided the equivalent of a \$2.1M/year federal award (\$6.33M in 2024 dollars).

Another important aspect of the development of the CCM was that of the roles played by both faculty affiliates and members of the staff. Always envisioned as an organisation that brought faculty and students of multiple disciplines together for the pursuit of research and education, many functions of CCM were conducted by its staff. Several of the affiliate faculty served in positions of associate directors with responsibility for guiding the research in specified areas. Each faculty member received support for one or more graduate students and summer support.

Both professional and administrative staff were developed by CCM to conduct programs necessary for the enterprise to function. Renewal of consortium membership by industrial sponsors required the annual delivery of products of the centre that justified continued investment. These included two industry symposia per year, multiple research reports delivered, and updates to the *Delaware Composites Design Guide*, a compendium of evolving technology. When asked for the greatest need to promote composites technology within General Motors, the answer was for a design guide like that available for conventional metallic materials, and focused on composites. CCM produced a seven-volume set over the following 7 years with contributions from noted external experts including Dr. James Whitney (retired) of the Air Force Research Laboratory, Professor Keith Kedward (deceased), Professor Peter Beaumont of Imperial College, Professor Michael Bader (retired) of University of Surrey, Dr. Alan Isham (retired) of Owens Corning, Professor Thomas Hahn of University of California Los Angeles (UCLA) and Dr. Carl Zweben of General Electric. Professor Jack Gillespie and Leif Carlsson of Florida Atlantic University were Review Editors. A retired DuPont executive, Henry Cadot, served in the role of monthly newsletter author and consortium fee collector. The role of Deputy Director was held by William Dick, who later joined the University of Illinois–Urbana Champaign as Assistant Dean for Large Programs and abetted numerous research programs of scale there.

Early in the life of CCM, a number of international scholars affiliated with the centre to help deliver the programs of education and research. These scholars often coadvised graduate students and participated in the two annual symposia as speakers. Several contributed to the *Delaware Composites Design Guide*. For these services, the international scholars received both remuneration and recognition. These individuals included Professor Klaus Friedrich (deceased) of Kaiserslautern, Professor and Rector (retired) Marcel Crochet of the Catholic University of Louvain, Professor Pat Mallon of the University of Ireland – Galway, and Vice Chancellor (deceased) Anthony Kelley of the University of Surrey. All these individuals shared stories of the successes of CCM with their colleagues in Europe and served to introduce CCM to the industries and universities of Europe. The first United States – Japan Conference on Composite Materials (1981) was planned at CCM by visitors from Japan. Professor Tsuyoshi Hayashi, NAE (deceased) of the University of Tokyo, hosted the American delegation in Japan. Several members of the delegation were CCM faculty.

During the 1980s, the interest in composite materials research and education grew substantially in Europe and Japan. CCM's reputation had grown rapidly in these two regions of the world through the interactions described previously. In 1988, a celebration of the 350th anniversary of Sweden's first arrival to North America at Newcastle, Delaware, brought a Swedish delegation to the region and to CCM. Professor Jan Backlund (retired) of the Royal Institute of Technology and Dr. Leif Carlsson of Chalmers University were then affiliated with CCM.

This period also brought about a self-study to determine the technological barriers to the more widespread adoption of composite technology, and the study called for a change in direction in CCM's programs of research and education to embrace manufacturing science and engineering. At the same time, the National Academy of Engineering undertook a study of engineering education in the United States, led by George M. Low, former NASA executive and president of Rensselaer Polytechnic Institute. The committee recommended that single-investigator research programs be complemented by "convergent research, education, and technology translation at U.S. universities that will lead to strong societal impacts." The NSF responded to the study by announcing the creation of the Engineering Research Centres (ERC) program under the leadership of NSF Director Erich Bloch (formerly IBM) and Assistant Director for Engineering Professor Nam Suh (MIT). The first six recipients of ERC awards included the University of Delaware's CCM with the centre title, National Centre for Composites Manufacturing Science and Engineering, and an initial 3-year award of $6M ($18.1M in 2024 dollars). The NSF ERC was organised around five thrust areas: mechanics and design science, manufacturing and processing science, computation software and information transfer, materials design, and materials durability.

The principal investigators (PIs) for the proposal were Professors R. Byron Pipes, PI, with co-PIs: Roy L. McCullough, Tsu-Wei Chou, Selcuk Guceri, Arthur Metzner, and Jerold Schultz.

The ERC award to CCM came in large part because CCM's industrial consortium program had already put into practice many of the elements described in the ERC program call for proposals. While several of the other five recipients of ERC awards in the initial award year were just beginning the transformation to large integrated programs, the CCM was already well on its way. Furthermore, the NSF award in dollar value was a relatively small addition to the comprehensive program already in place at CCM. Another major development was the award of an Army Science Center of Excellence to CCM with the recognition that the U.S. Army composites research programs are moving from Watertown, Massachusetts, to Aberdeen, Maryland, just a short distance from the Delaware campus. This fortuitous move by the U.S. Army gave CCM the advantage of proximity in developing a long-term relationship with a major funding source for composites research and education.

One of the major accomplishments of the Delaware CCM was the establishment of the Medal of Excellence in Composite Materials in 1984, mentioned earlier. The medal celebrates excellence in visionary impact, novel theories, discovery of patented processes, and/or exceptional publications. To date, a total of 41 recipients worldwide have been awarded the CCM Medal of Excellence. The first four recipients were all elected to the National Academy of Engineering of the United States and were chosen to represent the leadership of the field globally. Appropriately, their likenesses are engraved on the face of the medal. The eldest of the group was Tsuyoshi Hayashi. The late Dr. Tsuyoshi Hayashi was Professor Emeritus of the University of Tokyo, Japan, and the President of the Japan Plastic Inspection Association, Tokyo. Born in Iijima, Nagano Prefecture, Japan, in 1911, Dr. Hayashi graduated from the Department of Aeronautics, Tokyo Imperial University, in 1935. He earned a Doctor of Engineering at Tokyo Imperial University in 1947 for his work describing the "Theory of Elasticity of Orthogonal Anisotropic Solids." The late Anthony Kelly was an Emeritus Professor and a Distinguished Research Fellow at Cambridge. He was a Fellow of the Royal Society and a Fellow of the Royal Academy of Engineering. He served as Vice Chancellor of the University of Surrey. The late Zvi Hashin was the Nathan Cummings Chair of Mechanics and Solids in the Department of Solid Mechanics, Materials and Structures, Tel Aviv University, Tel Aviv, Israel. Finally, Stephen W. Tsai was Chief of the Mechanics and Surface Interactions Branch, Nonmetallic Materials Division, Materials Laboratory, Air Force Wright Aeronautical Laboratories, Aeronautical Systems Division, Air Force Systems Command, at Wright–Patterson Air Force Base, Ohio. He is currently Professor Emeritus at the Structures and Composites Laboratory of Stanford University. A list of recipients by year is shown in Table 1.8.1. An image of the Medal of Excellence in Composite Materials is shown in Fig. 1.8.7.

Table 1.8.1 List of recipients of the composite medal of excellence.

Recipient	Affiliation	Year
Zvi Hashin, NAE	Tel Aviv University	1984
Tsuyoshi Hayash, NAE	Tokyo University	1984
Anthony Kelly, FRS, NAE	Surrey University	1984
Stephen Tsai, NAE	Stanford University	1984
Walter Rosen	Materials Sciences Corp.	1985
Alan M. Lovelace	United States Air Force	1986
George Peterson, NAE	United States Air Force	1986
Bernard A. Schriever	United States Air Force	1986
Roger Bacon	Union Carbide Corporation	1987
Akio Shindo	Toa Nenryo, Co., Ltd.	1987
William Watt	UK National Physical Lab.	1987
Karl M. Prewo	United Technologies Research Centre	1988
Richard M. Christensen, NAE	Lawrence Livermore Lab.	1989
Derek Hull, FREng FRS	University of Cambridge	1990
Roger R. Naslain	Laboratory for Thermostructural Composites	1991
Herbert Blades	E.I. du Pont de Nemours & Co.	1992
Stephanie J. Kwolek, NMT	E.I. du Pont de Nemours & Co.	1992
Paul Morgan	E.I. du Pont de Nemours & Co.	1992
Georg Menges	University of Aachen	1993
R. Byron Pipes, NAE	Rensselaer Polytechnic Institute	1994
H. Thomas Hahn	UCLA	1996
C. T. Sun	Purdue University	1997
George S. Springer, NAE	Stanford University	1999
Roy L. McCullough	University of Delaware	2002
John C. Halpin	United States Air Force	2005
Nicholas J. Pagano	United States Air Force	2005
James M. Whitney	United States Air Force	2005
Takahashi Ishikawa	Japan Exploratory Aerospace	2008
Alan G. Miller	The Boeing Company	2008
Tsu-Wei Chou	University of Delaware	2009
Anoush Poursartip	University of British Columbia	2009
Isaac Daniel	Northwestern University	2014
Karl Schulte	Hamburg University of Technology	2014
Ignaas Verpoest	Katholieke Universiteit Leuven	2014
Lawrence T. Drzal	Michigan State University	2016
Daniel Wagner	Hebrew University of Jerusalem	2016
Michael R. Wisnom, FREng	Bristol University	2019
Leslie Jay Cohen	HITCO	2019
Klaus Friedrich	Leibniz-Institut für Verbundwerkstoffe	2021
Jack R. Vinson	University of Delaware	2021

Figure 1.8.7 The medal of excellence in composite materials, 1984–2021. Reproduced from https://www.ccm.udel.edu/about-us/medal-of-excellence/

5. A scholar in composite materials and structures

The success of advanced composites during the past 5 decades is in no small part due to the scholars who contributed their knowledge to this field as it emerged. While it is impossible to capture the contributions of all the scholars engaged in this work, the author has selected the individual who most influenced his career, Nicholas J. Pagano, as an example of the many major contributors to this emerging field. An image of Pagano is shown in Fig. 1.8.8.

Pagano was born in Torrington, Connecticut, on January 11, 1933, where in high school, he demonstrated a keen ability in mathematics. Nick graduated from high school in 1950 and was admitted to the undergraduate program at Drexel Institute of Technology (later to become Drexel University). He completed his bachelor's in civil engineering in 1955. At the time, all Drexel students participated in undergraduate programs that required work internships spread over the course of the 5 years required to complete the degree. Salaries earned in the internships paid the required tuition that private institutions like Drexel required. Land surveying in the civil engineering curriculum was an important and economically advantageous skill. He developed knowledge in surveying and, while a student, cofounded a company to provide surveying services to customers in Philadelphia, where city property boundaries were of significant economic importance. He joined the Drexel faculty upon graduation in

Figure 1.8.8 Nicholas J. Pagano, c.1974.
Reproduced from Private photo.

1955 with a bachelor's degree, as was common at the time, since the degree was considered that of a professional engineer with only registration required for practice.

While enrolled in a course in elasticity, taught by Professor Pei Chi Chou, his excellent set of notes from the lectures formed the basis of a textbook that he coauthored with Professor Chou, "Elasticity – Tensor, Dyadic and Engineering Approaches," in 1967. The 1992 edition is published by Dover Publications. A description on the back cover of the book describes its uniqueness: "Written for advanced undergraduates and beginning graduate students, this exceptionally clear text treats both the engineering and mathematical aspects of elasticity."

Pagano completed his master's degree at Drexel while on the Faculty of Civil Engineering in 1962. After Sputnik was launched in 1957, American higher education challenged its faculty to achieve doctoral status. Pagano entered Lehigh University in Bethlehem, Pennsylvania, and studied engineering mechanics with Professor George C. Sih, where he completed the PhD in 1967. In reading the history of Lehigh, I learnt that Nick was appointed to the Lehigh faculty in mechanics in 1963–64. At that time, George Sih was an associate professor in mechanics, but the Department was Mechanical Engineering and Mechanics. Paul Paris was also an associate professor at the time, and Jim Rice was a graduate student along with Pagano.

According to Pagano, the Director of the Air Force Materials Laboratory at the time, Alan Lovelace (mentioned earlier for his role in developing the Air Force program in advanced composites), began a search for expertise in elasticity, and Nick was hired in 1966. He and his family made the journey from Bethlehem to Dayton, Ohio, where he began an extraordinary career. He would go on to make major contributions to the field of anisotropic elasticity and to the solution of seminal problems of importance in the emerging field of fibre-reinforced composite materials. He completed the PhD at Lehigh in 1967, just 1 year after joining Air Force Materials Laboratory (AFML). While Pagano never served in the military, he told me that he learnt belatedly, after arriving at Drexel, that he had been admitted to the Naval Academy in Annapolis.

Pagano was strongly influenced by two major academic advisors. First at Drexel, he studied under the supervision of Pei Chi Chou (deceased 2014). Chou was born on December 1, 1924, in Ichang, Hubei, China. He was the son of Hung Lieh and Shiu Lan (Kao) Chou. He received a bachelor of science degree from National Central University, China, in 1946. Master of Science, Harvard University, 1949. Doctor of Science, New York University, 1951. Professor Chou taught at Drexel from 1953 until 1995.

Professor George C. Sih served on the Lehigh faculty between 1958 and 1995, when he became professor emeritus of mechanical engineering and mechanics. His educational degrees began at the University of Portland, where he received the bachelor of science degree in 1953. In 1958, he received the master of science degree from New York University, and he completed the PhD at Lehigh University in 1960, just 4 years before Nick joined the Lehigh faculty. Professor Sih authored or coauthored over 10 archival volumes treating advanced topics in fracture mechanics. His active scholarship spanned from 1960 to 2017.

Thus, Pagano stood on the shoulders of at least two academic giants as he began his career at the Air Force Materials Laboratory. Professor J. N. Reddy edited a selection of Nick Pagano's work in 2010, entitled "Mechanics of Composite Materials – Selected Works of Nicholas J. Pagano," Kluwer Academic Publishers, and much of the following is drawn from that source (Reddy, 2010).

Pagano's earliest archival publications were coauthored with John Halpin and James Whitney, although he had several collaborators throughout his career. Many of his publications treated subjects that had never been investigated earlier in the United States and that made lasting contributions to the understanding of anisotropic systems. Some of his investigations led to corrections to earlier, less rigorous solutions. It is noteworthy that his solution for the through-thickness coefficient of thermal expansion for symmetric composite laminates (1974) (Pagano, 1974) was one such paper where he corrected an earlier incorrect solution. Halpin and Pagano were among the first to show that the behaviour of a tensile coupon with off-axis fibre orientation was strongly dependent on the specimen length-to-width ratio. Their tyre cord rubber samples illustrated the phenomena for all to see. Halpin completed his doctorate at the University of Akron, where Goodyear Tire and Rubber likely provided a natural source for the test samples. Composite cylindrical geometries were considered of great significance in the early stages of composites' biaxial failure modes since the geometry appeared to offer the potential for application of arbitrary states of biaxial stress needed to measure biaxial failure. However, as learnt in later developments, the cylindrical test specimen required deeper understanding before its potential in composites testing could be assessed. In one such study, Pagano, Halpin, and Whitney showed that composite cylindrical geometries could exhibit tensile buckling instabilities typically associated with compressive phenomena (1968) (Pagano and Halpin,

1968; Pagano et al., 1968; Pagano and Whitney, 1970; Pagano, 1974). Pagano and Whitney next demonstrated an approach to the geometric design of cylindrical geometry specimens (1970).

Many of the papers that Pagano authored during the late 1960s and early 1970s were published in the first journal dedicated to composite materials, the *Journal of Composite Materials*. It is noteworthy that the Editor-in-Chief of the *Journal of Composite Materials* was Stephen Tsai, the Chief Scientist of the Air Force Materials Laboratory and later Chief of the Composite Mechanics Branch at the Materials Laboratory, where Nick held his position. Small world!

Nicholas J. Pagano was intellectual integrity *in persona*. While a kind and sensitive human being and father of eight children, he seldom suffered fools in intellectual discussions. On one occasion at a major technical meeting, a participant took the microphone to ask a question of the speaker, but instead of a question, he offered his own discourse on the subject of the speaker. Nick took the microphone next and asked if there were others in the room who wished to present a dissertation on the subject in the disguise of a question! He often asked the question that everyone in the meeting knew should be asked, but all were hesitant to ask. To be embarrassed by one of Nick's questions was an experience most strived to avoid, but as is the case in technical discourse, speakers whose comments violated the foundations of anisotropic mechanics were publicly chastised for doing so.

6. The future of advanced composites

I believe that the future of advanced composites is a bright one because the engineering world has just begun to exercise the anisotropic nature of these material systems to achieve optimum products. The evolution of living systems has exercised the same anisotropic "designs" for millions of years to achieve our optimum lifeforms. Current composite applications have focused more on quasi-isotropic designs because the penetration of composite knowledge into the engineering workforce has been retarded by the lack of treatment of the subject in typical engineering education. However, with the successful development of commercial aircraft by Boeing and Airbus, major gains have been achieved in the creation of a much larger composites workforce and accompanying supply chain.

Perhaps the most exciting technological development on the horizon is the digital revolution in composites manufacturing. Largely an empirical process in the first 50 years, composites manufacturing is being impacted by the digital revolution in all engineering processes. The virtual twin now offers digital representations and sources of truth for all these phenomena and can provide for decision-making independent of and in advance of asset deployment. Computational power is at the foundation of these advances, and increases in computing power will soon reach beyond zettascale ($>10^{21}$). Computers the size of large buildings, requiring nuclear power plants to provide their power, are under construction. Artificial intelligence is at the foundation of the author's optimism for the future of advanced composites. Coupled with Large

Language Methods, computers of the future will provide the engineer with direct and immediate access to simulation tools without the typical nine to twelve months required to gain training for their use today. The global expertise for advanced composites design, manufacturing, and performance will be assembled into composites "experts" wherein rapid advances in technology and innovation can receive guidance with conventional language as the communication vehicle. There is no doubt that composite engineering of the future will also be extraordinarily exciting and beneficial to the prosperity of our society.

In the foregoing, the author has attempted to "paint a picture" of the extraordinary processes and individuals that have been responsible for the creation of the field of advanced composites. Given that no observer could have witnessed this entire undertaking, the author chose to use examples from his personal experience to illustrate the enormous contributions and innovations that, when brought together, resulted in significant scientific advances, human talent, and organisational structure development and the resulting enterprises, now termed "advanced composites."

References

Goodsell, J., Pipes, R.B., 2016. Free-edge interlaminar stresses in angle-ply laminates: a family of analytic solutions. J. Appl. Mech. 83, 051010-1.

Goodsell, J., Pagano, N.J., Kravchenko, O., Pipes, R.B., 2013. Interlaminar stresses in composite laminates subjected to Anticlastic bending deformation. J. Appl. Mech. 80 (4), 041020.

Pagano, N.J., 1974. Thickness expansion coefficients of composite laminates. J. Compos. Mater. 8, 310–312.

Pagano, N.J., Halpin, J.C., 1968. Influence of end constraint in the testing of anisotropic bodies. J. Compos. Mater. 2, 18–31.

Pagano, N.J., Pipes, R.B., 1971. The influence of stacking sequence on laminate strength. J. Compos. Mater. 5, 50–57.

Pagano, N.J., Pipes, R.B., 1973. Some observations of the interlaminar strength of composite laminates. Int. J. Mech. Sci. 15, 679–688.

Pagano, N.J., Whitney, J.M., 1970. Geometric design of composite cylindrical characterization specimens. J. Compos. Mater. 4, 360–378.

Pagano, N.J., Halpin, J.C., Whitney, J.M., 1968. Tension buckling of anisotropic cylinders. J. Compos. Mater. 2, 154–167.

Peng, B., Goodsell, J., Pipes, R.B., Yu, W., 2016. Generalized free-edge stress analysis using mechanics of structure genome. J. Appl. Mech. 83, 101013-1.

Pipes, R.B., Pagano, N.J., 1970. Interlaminar stresses in composite laminates under uniform axial extension. J. Compos. Mater. 4, 538–548.

Pipes, R.B., Pagano, N.J., 1974. Interlaminar stresses in composite laminates – an approximate elasticity solution. J. Appl. Mech. 41 (3), 668–672.

Pipes, R.B., Kaminski, B.E., Pagano, N.J., 1973. The influence of the free-edge upon strength of angle-ply laminates. In: Analysis of Test Methods for High Modulus Fibers and Composites, ASTM STP-521, pp. 218–228.

Pipes, R.B., Ballintyn, N.J., Scott, W.R., Carlyle, J.M., 1977. Acoustic emission response of metal matrix composites. In: Composite Materials: Testing and Design, ASTM STP-617, pp. 153–169.

Pipes, R.B., Goodsell, J., Ritchey, A., Dustin, J., Gosse, J., 2010. Interlaminar stresses in composite laminates: thermoelastic deformation. Compos. Sci. Technol. 70, 1605–1611.

Reddy, J.N., 2010. Mechanics of Composite Materials – Selected Works of Nicholas J. Pagano. Kluwer Academic Publishers.

History of discontinuous fibre composites

1.9

Paula Hohoff and Tim A. Osswald
Polymer Engineering Center, Department of Mechanical Engineering, University of Wisconsin–Madison, Madison, WI, United States

1. Introduction

Discontinuous fibre composite materials have been utilised in construction for thousands of years. One example is the casted *adobe*[1] brick as illustrated in Fig. 1.9.1, which was used to build a house in Panajachel, Guatemala. In certain parts of the world, this type of discontinuous fibre composite building material is still in use, making it one of the oldest construction methods. Clay, sand, or stones are combined with natural fibres such as straw, small wooden twigs, or *caña brava*, a type of wild cane, to create adobe bricks or casted walls. The brick or wall's fibre components strengthen the clay, keep it cohesive, and lessen the likelihood of cracks and crumbling.

More than 5000 years ago, in ancient Egypt, plywood—a laminated composite made of numerous thin layers of wood—was utilised as a building material for the first time. Using a collagen glue prepared by boiling animal hide and bones, the thin wood layers were joined by the ancient Egyptian method, which also employed the weight of hot sandbags to force the layers together (Dung et al., 2003). Modern plywood is credited to Immanuel Nobel, who along with his son Alfred Nobel invented dynamite in the 1860s. He created a laminated hardwood plate that was stronger than the original wood by adhering crossing veneers together under pressure and steam (Day and McNeil, 1996; Dung et al., 2003). This strength was created in the direction of each layer's fibre orientation.

Greek designs of the screw press, more than two millennia ago, replaced heated sandbags in the production of plywood and other composite materials. These devices were the norm for producing high pressure between two platens for the following two millennia. Steam was eventually routed through channels to heat these press platens. During the second part of the 1800s, early plastics and plastic composite goods were compression moulded using steam-heated screw presses. Around that same period, shellac—an amber-like polymer obtained from Asian insects and reinforced with wood fibre—was moulded into decorative pieces using steam presses (Davis et al., 2003). Eventually, a steam-heated platen hydraulic press (Fig. 1.9.2), developed in 1915 by F. B. Northrup, took the place of the steam-heated screw press (Northrup,

[1] The word "adobe" has been around for at least 4000 years, and most languages pronounce it roughly the same way

The History of Composites. https://doi.org/10.1016/B978-0-443-34119-9.00011-5

Figure 1.9.1 19th century house in Panajachel, Guatemala, that uses cast bricks reinforced with straw fibre for building.
Image courtesy of Tim Osswald © 2018.

1915). The basis for this creation was a four-column hydraulic press that British inventor Joseph Bramah created in the late 1700s. Compression moulding machines of today are unmistakably descended from Northrup's hydraulic press.

When François Charles Lepage patented a plastic substance made of wood fibre and albumen in France and England in 1856, he created one of the earliest discontinuous fibre plastic composites. He named the material bois durci (hardened wood). Sawdust, typically from a hardwood such as rosewood or ebony, combined with albumen from eggs or blood was used to produce bois durci. The albumen was combined with a water-diluted gelatinous material, mineral or metallic additives, and vegetable oils, to give a metallic look. In a steel mould housed in a steam-heated screw

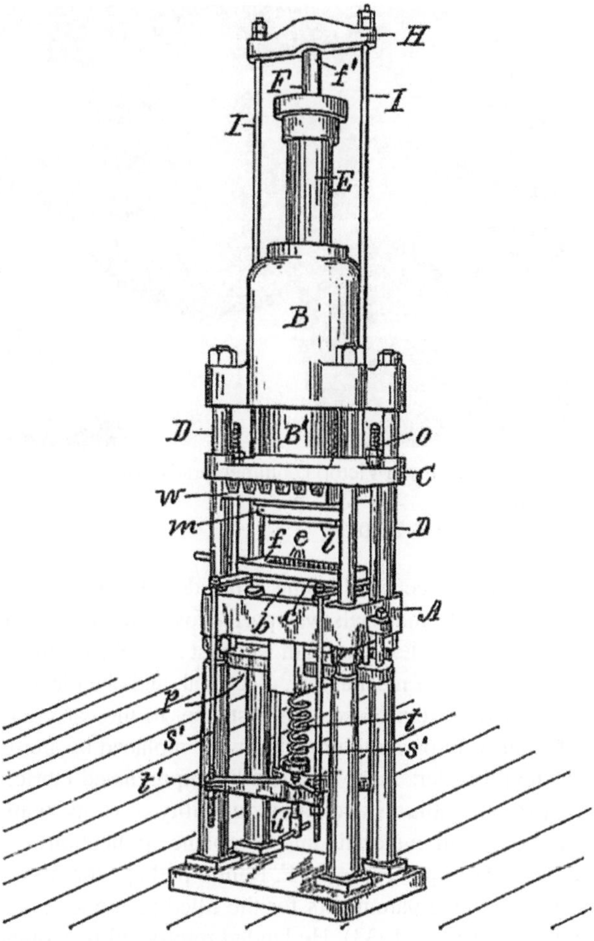

Figure 1.9.2 Northrup's hydraulic press.

press, the dry and wet ingredients were combined and compacted into completed parts. Two ornamental medallions made of this early plastic composite material compression moulded in a screw press are shown in Fig. 1.9.3.

Leo Baekeland created the first synthetic plastic, phenol-formaldehyde, at the start of the 20th century. This polymer is still in use today and can be used as both an unreinforced and reinforced plastic. This development marked a significant advancement in the field of modern polymer composites. Leo Baekeland used asbestos fibres and wood to fill and reinforce phenol-formaldehyde polymers in his Heat and Pressure Patent. He also mentions other reinforcing materials some of which are other fibrous or cellular materials, rubber, mineral powders, and graphite (Baekeland, 1907).

In addition to contributing to the streamlined pre-World War II era, Bakelite offered a material that allowed for the mass production of goods that improved living

Figure 1.9.3 Royal Albert and Queen Victoria commemorative Bois durci medallions from the 1860s (The Kölsch Collection).
Image courtesy of Tim Osswald © 2010.

in the workplace, factory, street, and home. The idea that plastics can be cheaply substituted for higher-quality materials was partly fostered by the ability to produce plastic products in large quantities. The truth was that phenolics outperformed the materials they replaced in terms of mechanical, electrical, and chemical resistance, especially when they were fibre reinforced. For instance, in the early days of aviation, wooden propellers that were prone to cracking were replaced by sturdy, lightweight, and strong composite propellers made of phenolic impregnated fabric. These propellers prevented catastrophic failures and had a durability of many years. This marked the introduction of composite materials into aeroplane applications.

The Princeton University professor, L. F. Rahm, perhaps the first person to teach polymer processing, provided parameters for the filled phenolic resins of that era in his textbook from 1933 (Rahm, 1933). He lauded reinforced phenolics' increased rigidity and strength as well as their noticeably reduced shrinkage when processing it with asbestos fibre. Saw dust filled Bakelite replaced wood as the material of choice for radio housings (Fig. 1.9.4), not necessarily because it was less expensive, but rather because it allowed for the mass manufacturing of the finished goods that were more durable than the handcrafted, lower production rate alternatives. In addition, the designer was able to create parts with a wider range of colour options and more intricate shapes thanks to these new materials.

During the 1930s, the German inventor Max Himmelheber created particle board to make use of the leftover wood waste from sawmills and lumberyards, which made up 60% of the trees that were felled. Phenolic resin was combined with milled wood waste to create this densely packed composite material, which was then pressed into plates. This composite material was unique from others in that very little phenolic was required because its sole purpose was to glue the timber chips and residues together. Particle boards are still widely used in construction today, for instance as the backing for vinyl flooring.

Figure 1.9.4 1945 Fada Bakelite radio (The Kölsch Collection).
Image courtesy of Tim Osswald © 2010.

In 1938, glass was successfully spun into fibreglass and used in flat panels for military aircraft during World War II, using unsaturated polyester resin as a bonding matrix. By the end of the conflict, the military industry used more than seven million pounds of fibreglass materials. In the 1940s, numerous attempts were undertaken to bring fibreglass into the automotive industry to capitalise on the low cost of this lightweight opportunity. Among these were Ford Motor Company car body panels (Fig. 1.9.5). The Ford body panels' warpage[2] prevented them from being commercially released, as did many other plastic composites ventures that came subsequently. Among the designs of early composite vehicles, Project Y, or 1946 Stout Scarab, is the most notable. This was the outcome of Owens Corning and aircraft engineer William B. Stout's demonstration of the utilisation of fibre glass material to construct a vehicle body. This car, which was constructed using the Scarab's existing design as a model, is credited as being the first to use fibre-reinforced composites for all of its body panels and framework.

When General Motors Corporation unveiled the Corvette in 1953, it was a major advancement since it had fibreglass body panels on a steel frame (Fig. 1.9.6). In 1964, GM suppliers produced the body panels of the Corvette for the first time using compression moulding of sheet moulding compound (SMC), which is currently a commonly used discontinuous fibre-reinforced composite material. Still, it was not until the 1970s that the automotive industry became the number one industry in the composites market, passing the marine industry, which had occupied that place since the 1950s.

But from the start, the composites industry has been heavily reliant on the aircraft manufacturing sector. Although the early use of composites in aeroplanes during

[2] The photo in Fig. 1.9.5 demonstrates how the seam around the circumference of the vehicle changes between the trunk lid and the body of the car

Figure 1.9.5 Henry Ford, demonstrating the durability of his new body panels in the 1940s, struck the fibreglass trunk lid with an ax.
Source: From the Collections of The Henry Ford. Gift of Ford Motor Company.

Figure 1.9.6 The 1953 Corvette, first fibre composite production car.
Source: GM Media Archives.

World War II was intended to address a shortage of aluminium, the real benefit of composite materials for aircraft manufacturing is their low weight. Actually, the field that researches composite materials and how to incorporate them into final applications is known as "Leichtbau" in German, which literally translates to "light construction." As a result, composite materials are constantly increasing in the construction of aeroplanes due to the desire to make them lighter. The increase in the last 40 years in the use of composite structural components in aeroplanes is depicted in Fig. 1.9.7. The graph illustrates the industry's growth from the first Boeing 747 and Airbus 300 in 1970, which had less than 5% of composites by weight, to the Boeing 787, which had 50% of composites by weight, and the Airbus 350, which had 52%. Most of the composite materials used in aerospace applications involve continuous carbon fibre reinforced composites. However, some applications in the aerospace industry also use discontinuous fibres.

2. Microstructure development of discontinuous fibre-reinforced composites

Discontinuous fibres are preferred for automotive applications because of the ease of automating the manufacturing process and allowing the mass production of composite

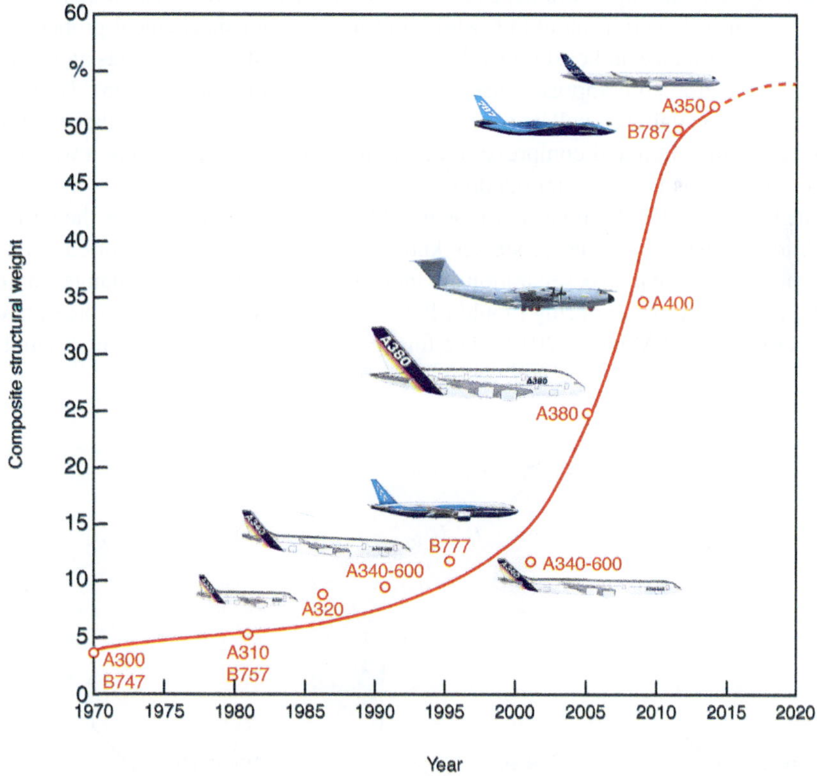

Figure 1.9.7 Composite structural components are becoming more prevalent in aeroplanes design.
Source: Boeing and Airbus.

parts. The two most popular processes for creating components out of these materials are injection moulding and compression moulding, such as the Corvette since 1964. Discontinuous fibres offer superior material qualities, reduced production costs, and the possibility of high-volume production. In high-volume applications in particular, these fibre-reinforced materials are increasingly used in place of metal to reduce the overall weight of the parts. One distinguishing feature of discontinuous fibres is that they are cut and undergo configuration changes in direct reaction to the material's deformation during processing.

Short fibre-reinforced thermoplastics (SFT) and long fibre-reinforced thermoplastics (LFT) are two further classifications for discontinuous fibre-reinforced composites. In general, fibres are referred to as short fibres if their length causes them to be straight and interact with one another less; on the other hand, if their length causes them to curve and interact with one another more, they are referred to as long fibres. These thermoplastic materials are considered short fibre composites when the glass fibres average aspect ratio is less than 80–100. On the other hand, the material performs like a long fibre composite when the aspect ratio is higher than 100. It should be noted that the definition of short and long fibres is arbitrary and dependent on the material system and fibre volume. Because LFT can be processed economically using compression or injection moulding while offering superior mechanical properties to SFT, its performance and cost typically place it between SFT compounds and continuous fibre-reinforced composites used for high-performance applications. Both SFT and LFT are available as pellets as shown in Fig. 1.9.8. The long fibre thermoplastic material for injection and compression moulding is also produced using a well-liked process known as in-line compounding.

The local and global properties of the moulded item are determined by the ultimate condition of the fibres. The processes known as fibre attrition, fibre orientation or alignment, and fibre matrix separation are indicative of the substantial changes in fibre configuration that occur during mould filling of discontinuous fibre-reinforced materials (Osswald and Menges, 2012). The final component has a significant amount of

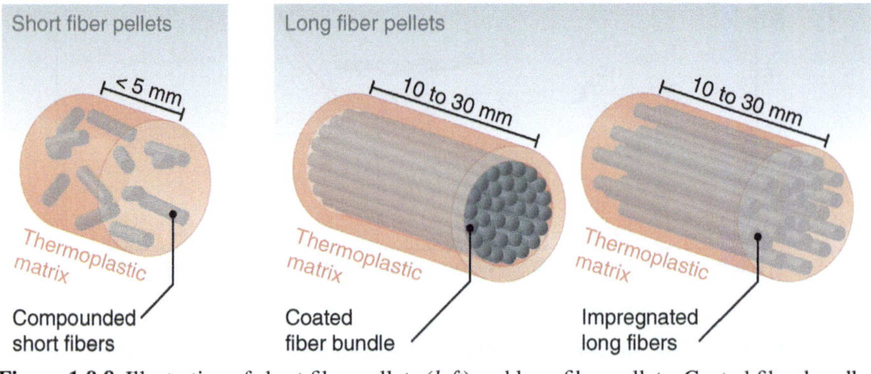

Figure 1.9.8 Illustration of short fibre pellets (*left*) and long fibre pellets: Coated fibre bundles (*centre*) and pultruded fibres (*right*) (Gandhi et al., 2020).

variability due to the process-induced modification of the fibre microstructure. Within a part, there may be variations in the fibre's concentration, length, and alignment.

One of the key elements influencing a moulded part's mechanical characteristics and dimensional stability is the anisotropy's development. Because fibres exhibit a preferential alignment depending on the flow conditions during mould filling, the moulded part's anisotropic qualities result (Goris and Osswald, 2017). Fibre attrition, also known as fibre breakage, is the term used to describe the shortening of the fibres during processing. Injection moulding involves forcing the material through a gate while it is injected, altering its configuration. The fibres experience extreme pressures when the material is forced to flow, which leads to their deformation, buckling, bending, and breaking. This fibre breakage drastically lessens the fibres' reinforcing function (Dung et al., 2003). Last but not least, migration and separation effects during processing can cause variations in the fibre concentration inside a part. On the other hand, a part's mechanical qualities are directly correlated with its fibre concentration. Processing conditions and deformation rates during mould filling, can lead to localised modifications of fibre concentration, which will alter the mechanical performance of a product.

Due to the increased and successful use of fibre-reinforced plastics over time, different types and combinations of plastics and fibres have been developed and tested to make the material even more superior. Discontinuous fibre-reinforced plastics have been constrained by the ability of being able to process them using commercially available plastics processing methods such as injection moulding. This has limited the fibre length to stay in a regime, in which an easy flow through nozzles and gate designs is assured.

The material for fibres has developed over time—from the early asbestos and wood named in Baekeland's "Heat and Pressure" patent to mostly glass fibres, 40 years later. Glass fibres show a superior combination of very high tensile strength compared to their density, isotropy, availability, and cost-effectiveness. Baekeland's initial matrix material phenol formaldehyde was first replaced by unsaturated polyester during the World War II.

After World War II, new materials were available which occurred as big breakthroughs came across in the form of new synthetic thermoplastics. Polyamide, which was first developed by Wallace Humane Carothers at DuPont under the tradename of Nylon (Polyamide 66) in the United States, as well as Perlon (Polyamide 6), developed by Paul Schlack at IG Farben in Germany, marked the first materials used for fibre-reinforced thermoplastics. Today, a versatility of thermoplastics is used as matrix material, which makes discontinuous fibre-reinforced materials an adaptable and customised material for a variety of applications. This adaptivity is further enhanced by the rising use of additives, which strengthens certain properties needed for demanding applications. As an example, most recent developments show the addition of fire retardants to a polypropylene matrix to allow the use of LFT for battery enclosures in the electric vehicle production.

3. The history of fibre modeling

As discontinuous fibre reinforced composites have become more prevalent in automotive applications, the need to be able to predict the properties of such parts has increased. The early work in the 1980s, at the University of Illinois at Urbana–Champaign, led by Professor C.L. Tucker III marks the beginning of the research that ended in today's commercial software programs that are used in the design and optimisation of discontinuous fibre reinforced polymer composite parts. The modeling of fibre microstructure development during processing includes fibre orientation, fibre attrition and fibre-matrix separation. In this chapter, we concentrate on the first two. The modeling of fibre-matrix separation is still in its infancy[3] and will not be covered in this section on the history of modeling. While the other two, fibre orientation and fibre attrition are discussed below in an historical context, starting with the basics of fibre rotation during deformation, derived by Jeffery in 1922 (Jeffery, 1922).

3.1 Modeling fibre orientation

Modeling fibre orientation in discontinuous fibre-reinforced thermoplastics is an important aspect in predicting the mechanical properties of lightweight parts. However, finding an appropriate model is not trivial, which is why many approaches and improvements occurred over time. This chapter summarises the historic development and the key contributions to macroscopic fibre orientation modeling.

3.1.1 1922: Jeffery's model for fibre orientation in dilute suspensions (Jeffery, 1922)

One of the most fundamental models for fibre orientation was developed by G. B. Jeffery in 1922. Jeffery's fibre orientation model describes the motion of a single ellipsoidal particle in a dilute viscous suspension. Thus, the interaction of fibres is not considered in this modeling approach. Fibres are assumed to be perfectly rigid (no bending or breaking) and to not impact the flow field of the fluid. Jeffery solved the equations of motion for a flow field of a Newtonian fluid without turbulences (rot $\boldsymbol{u} = 0$). A single fibre in elongational flow will find a position in stable equilibrium in the direction of flow at an angle of $\phi = 0$ (compare Fig. 1.9.9). For simple shear flows, Jeffery found that the single fibre will not find this stable equilibrium but will rotate in a periodic pattern (compare Fig. 1.9.10). The closer the fibre is to an angle of $\phi = \pi/2$, the more rapid the rotation (Jeffery, 1922; Folgar and Tucker, 1984; Kugler et al., 2020).

This periodic rotational motion, known as "Jeffery orbits," is a function of the type of flow and the aspect ratio of the fibre, where $\boldsymbol{p}$ represents the orientation vector, $\boldsymbol{W} = \frac{1}{2}\left((\nabla \boldsymbol{u})^T - \nabla \boldsymbol{u}\right)$ is the vorticity tensor, ξ is a particle shape function describing the

[3] While commercial software include predictions of fibre migration, they are based on shear driven migration models such as the model developed by J.F. Morris and F. Boulay. Their model is based on normal stresses that arise is curvilinear shear flows of particle suspensions. While their model performs well in simplified flows, such as in tube flow and Couette flow, it does not predict well fibre density distributions that develop during mould filling operations in compression and injection moulding (Morris and Boulay, 1999).

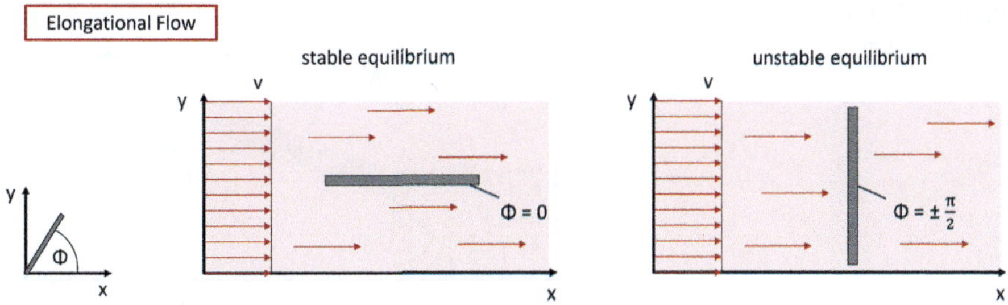

Figure 1.9.9 Stable and unstable equilibrium for a single fibre in elongational flow.

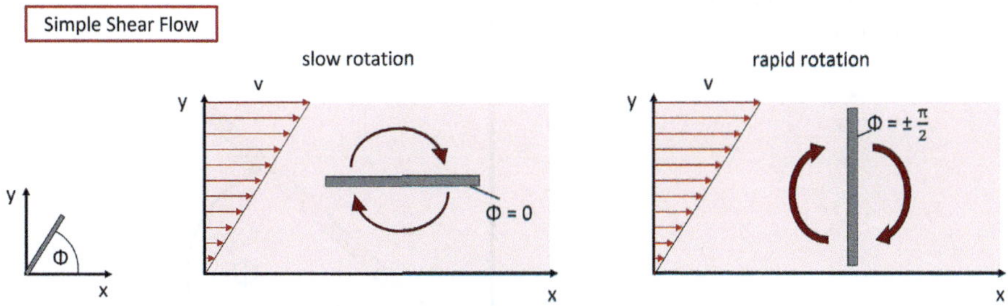

Figure 1.9.10 Continuous rotation of a single fibre in simple shear flow.

aspect ratio of the fibre, and $\boldsymbol{D} = \frac{1}{2}\left((\nabla \boldsymbol{u})^T + \nabla \boldsymbol{u}\right)$ is the rate of strain tensor (compare Fig. 1.9.11) (Kugler et al., 2020):

$$\dot{\boldsymbol{p}} = \boldsymbol{W} \cdot \boldsymbol{p} + \xi\,(\boldsymbol{D} \cdot \boldsymbol{p} - \boldsymbol{D} : \boldsymbol{ppp})$$

Jeffery's model worked as a ground base for the modeling of many dilute suspensions. However, since the processing of discontinuous fibre-reinforced composites usually involves a higher fibre content as well as a non-Newtonian matrix polymer, the prediction accuracy of Jeffery's model for these applications was limited. Therefore, Folgar and Tucker used Jeffery's model as a baseline for a fibre orientation model for concentrated suspensions that also accounted for fibre-fibre interactions (Gandhi et al., 2020; Kugler et al., 2020; Tseng, 2022).

3.1.2 1984: Folgar and Tucker's model for fibre orientation in concentrated suspensions (Folgar and Tucker, 1984)

Since the previous model by Jeffery is lacking fibre-fibre interactions, but these are significantly impacting fibre orientation, Fransisco Folgar and Charles L. Tucker III extended Jeffery's model by a diffusion term to model fibre orientation in concentrated suspensions or semi-dilute suspensions. Their modeling approach focused on a two-dimensional space with a single fibre oriented in a plane oriented in angle ϕ (compare Fig. 1.9.12). (Gandhi et al., 2020; Kugler et al., 2020; Tseng, 2022)

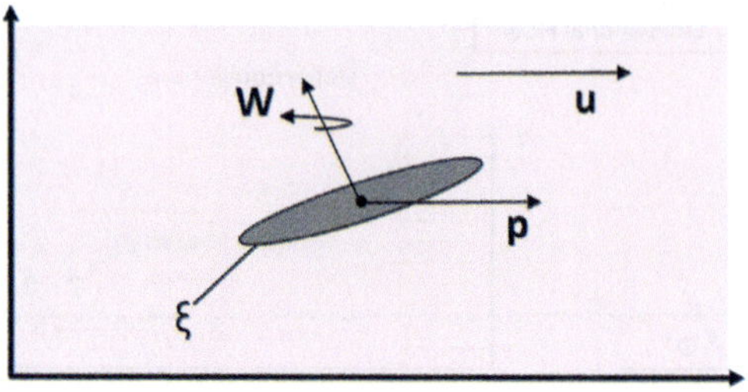

Figure 1.9.11 Ellipsoidal particle in dilute solution.

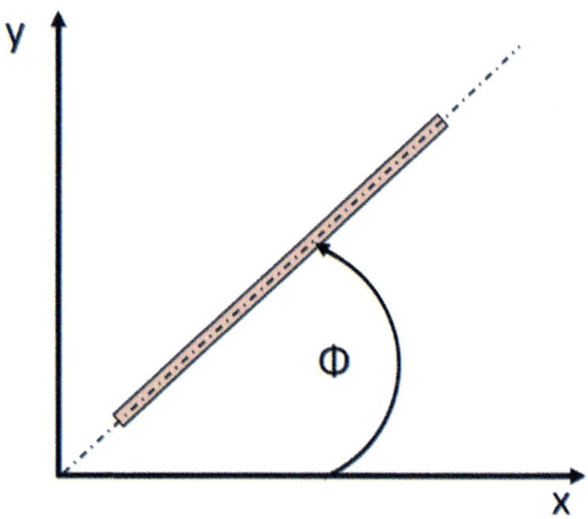

Figure 1.9.12 Single fibre in a two-dimensional plane oriented in angle ϕ, after (Folgar and Tucker, 1984).

The fibre orientation model developed by Folgar and Tucker assumes the fibres to be rigid cylinders (no breaking or bending), uniform in length and diameter, and large enough to neglect Brownian motion. All fibres are perfectly dispersed. A creeping flow (particle inertia is negligible) of a highly viscous fluid (buoyancy is negligible) is assumed for the model. The fluid is further assumed to be incompressible with no external forces or torques acting on the suspension.

Based on experimental observations in concentrated suspensions, Folgar and Tucker found that the flow field and the fibre orientation are not solely impacting the behaviour of a single fibre, but a statistical pattern shows instead. Fibres in concentrated suspensions rotate until they interact with another fibre, reorient due to fibre-fibre interaction, and continue to rotate afterwards. These reorientations of the fibres cause fibre orientation to be random. Thus, Folgar and Tucker used a statistical approach to model fibre orientation predicting a probability distribution function. This mainly started the use of the distribution function ψ_ϕ to model fibre orientation. Looking at Fig. 1.9.12 identifies some boundaries for the statistical description. The single fibre needs to be oriented in an angle ϕ with the characteristics (Folgar and Tucker, 1984; Gandhi et al., 2020; Kugler et al., 2020; Tseng, 2022):

$$\phi_1 = -\frac{\pi}{2} < \phi < \phi_2 = \frac{\pi}{2}$$

Flipping the fibre by 90 degrees must return the same result, which concludes in a periodic distribution function:

$$\psi_\phi\,(\phi+\pi)=\psi_\phi\,(\phi)$$

Furthermore, the function changes with time as the fibres are rotating within a small angular interval between ϕ and $\phi + d\,\phi$. The rotation of the fibres is represented by the average angular velocity of the fibres $\dot{\Phi}$ (Kugler et al., 2020).

All in all, Folgar and Tucker implemented these findings in a diffusion term that was added to the previously described model by Jefferey, resulting in the following model for the fibre motion of a single fibre in concentrated suspensions. $\dot{\gamma}$ represents the scalar magnitude of the rate of strain tensor and C_1 is the interaction coefficient and describes the rate of interaction. In case there is no fibre-fibre interaction, C_1 equals zero and the model returns to Jeffery's initial representation (Folgar and Tucker, 1984; Kugler et al., 2020).

$$\dot{\boldsymbol{p}}=\boldsymbol{W}\cdot\boldsymbol{p}+\xi\,(\boldsymbol{D}\cdot\boldsymbol{p}-\boldsymbol{D}:\boldsymbol{ppp})+C_1\dot{\gamma}\cdot\nabla_s\psi$$

Fibre-fibre interactions are described to occur as the centre of mass of one fibre passes another fibre's centre within the length of the fibre. Both fibres are reoriented due to this interaction in a certain direction. However, the directional bias of reorientation is not accounted for in this model, but reorientation is modelled to be random instead. The frequency of fibre-fibre interactions is proportionally related to the rate of strain $\dot{\gamma}$ as well as other impact factors. These impact factors are captured by the interaction coefficient C_1, which needs to be determined experimentally. It is a function of the fibre aspect ratio l/d, the mean square angle of change of $(\Delta\phi)^2$ and a coefficient c (Folgar and Tucker, 1984; Kugler et al., 2020):

$$C_1=\frac{1}{2}\,\overline{(\Delta\phi)^2}\left(\frac{l}{d}\right)^2 c$$

In contrast to Jeffery, who modelled fibre orientation in terms of a continuous rotation of fibres, Folgar and Tucker found that for steady flows, like simple shear flow or elongational flow, fibre orientation reaches a steady state. This steady state does not equal a state of perfect fibre alignment, but rather a certain distribution of fibre orientation that remains steady, as depictured in Fig. 1.9.13 (Folgar and Tucker, 1984; Kugler et al., 2020; Tseng et al., 2020).

The steady state fibre orientation is dependent on the interaction coefficient C_1 and the type of flow, with elongational flows leading to greater alignment. Furthermore, Folgar and Tucker also concluded that fibre orientation is kinematically not reversible, meaning that fibre orientation will not be reversed when the deformation is reversed (Folgar and Tucker, 1984; Kugler et al., 2020).

Folgar and Tucker's model received great recognition and offers key benefits to modelling fibre orientation. However, it is difficult to solve their model numerically,

Figure 1.9.13 Exemplary steady state fibre orientation in simple shear flow, after (Folgar and Tucker, 1984).

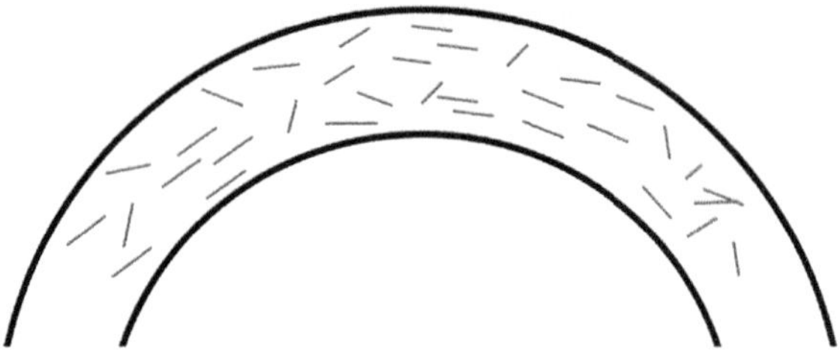

which is why multiple adaption approaches followed (Gandhi et al., 2020; Tseng et al., 2020).

3.1.3 1987: Advani and Tucker's fibre orientation (Advani and Tucker, 1987)

The previous representation of fibre orientation in form of the probability distribution function is too expansive to be solved numerically and to be applied for the prediction of three-dimensional fibre orientation in complex geometries. To solve more complex fibre orientation problems, a similarly general but more concise method was needed. Thus, Suresh G. Advani and Charles L. Tucker III developed a method that uses fibre orientation tensors to describe the fibre orientation state. Orientation tensors were already used previously as intermediate results in the rheology of fibre suspensions or solutions of rigid macromolecules. However, the tensors were mainly used to represents the influence of the orientation on the material's stress tensor. Instead, Advani and Tucker developed the orientation tensor for constitutive equations that could model the instantaneous structure of the material and therefore replace the previous probability distribution function. This leads to a constitutive equation including a second order tensor that can predict the orientational structure of composite materials (Advani and Tucker, 1987; Gandhi et al., 2020; Kugler et al., 2020; Tseng et al., 2020; Tseng, 2022).

To find an appropriate tensor representation, Advani and Tucker first advanced the probability distribution function for orientation ψ to represent the three-dimensional space. This led to the orientation of a single fibre being a function of two angles: $\psi\ (\theta, \phi)$ (compare Fig. 1.9.14). (Advani and Tucker, 1987)

The assumptions for this new modelling approach resemble the previous assumptions for a single fibre in concentrated suspensions. Fibres are assumed to be rigid cylinders, uniform in length and diameter, and perfectly dispersed. This also leads to similar characteristics for the PDF ψ. Again, the function needs to be periodic (as fibre orientation is indistinguishable when fibres are flipped by 90 degrees), normalised (as each fibre has a certain orientation), and meet the continuity condition (ψ changes with time as fibres change orientation and move with bulk motion of fluid).

As an alternative to a representation by the two angles θ and ϕ, Advani and Tucker showed that the probability distribution function can also be described by a unit vector $\boldsymbol{p}$ and thus $\psi = \psi\ (\boldsymbol{p})$, which is also shown in Fig. 1.9.14. The components of this unit

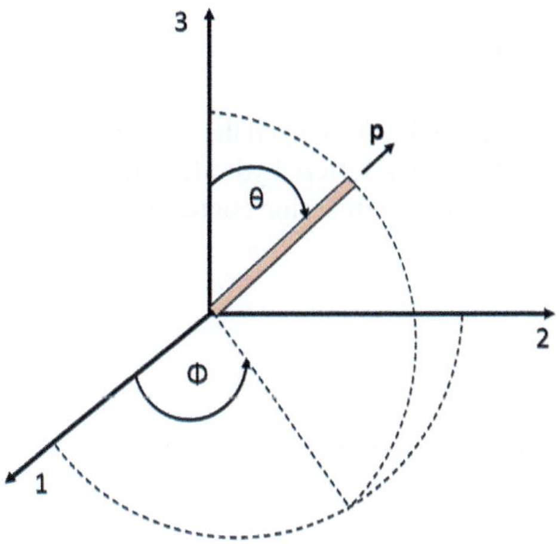

Figure 1.9.14 Coordinate system and definition of θ, ϕ, and $\boldsymbol{p}$ after (Advani and Tucker, 1987).

vector $\boldsymbol{p}$ are interrelated as the length of the fibres is fixed ($p_i\, p_i = 1$) and related to both angles as follows (Advani and Tucker, 1987; Kugler et al., 2020):

$$p_1 = \sin(\theta)\cos(\phi)$$

$$p_2 = \sin(\theta)\sin(\phi)$$

$$p_3 = \cos(\theta)$$

It becomes clear that the unit sphere thus covers all directions of the vector $\boldsymbol{p}$ and the integral over its surface results to

$$\oint d\boldsymbol{p} = \int_{\phi=0}^{2\pi} \int_{\theta=0}^{\pi} sin(\theta) d\theta d\phi$$

Using the probability distribution function is a complete representation of fibre orientation, however too complex and therefore, Advani and Tucker used it as a base for the development of fibre orientation tensors. These tensors are formed by integrating the probability distribution function $\psi(\boldsymbol{p})$ with dyadic products of the unit vector $\boldsymbol{p}$ over all possible directions. Thus, the second order fibre orientation tensor $\boldsymbol{A}$ is formed as follows, where $< >$ represents the integral over all $\boldsymbol{p}$ weighted by $\psi(\boldsymbol{p})$ (Advani and Tucker, 1987; Kugler et al., 2020).

$$\boldsymbol{A} = a_{ij} = \oint p_i p_j\, \psi(\boldsymbol{p})\, d\boldsymbol{p} = < \boldsymbol{pp} >$$

The formation of the fourth order fibre orientation tensor $\mathbb{A}$ follows the same principle, respectively (Advani and Tucker, 1987; Kugler et al., 2020).

$$\mathbb{A} = a_{ijkl} = \oint p_i p_j p_k p_l \, \psi(\boldsymbol{p}) \, d\boldsymbol{p} = <\boldsymbol{pppp}>$$

With these fibre orientation tensors, Advani and Tucker developed the new expression for fibre orientation modelling, where HD indicates the Jeffery hydrodynamic portion of the equation, and IRD represents the isotropic rotary diffusion contributor (Advani and Tucker, 1987; Kugler et al., 2020).

$$\frac{DA}{Dt} = \dot{A} = \dot{A}^{HD} + \dot{A}^{IRD}$$

$$\dot{A}^{HD} = (\boldsymbol{W} \cdot \boldsymbol{A} - \boldsymbol{A} \cdot \boldsymbol{W}) + \xi(\boldsymbol{D} \cdot \boldsymbol{A} + \boldsymbol{A} \cdot \boldsymbol{D} - 2\,\mathbb{A} : \mathbf{D})$$

$$\dot{A}^{IRD} = 2C_1 \dot{\gamma}(\boldsymbol{I} - 3\boldsymbol{A})$$

An infinite series of evolution equations is needed to describe the fourth order fibre orientation tensor. Thus, the series is usually terminated at the stage of the second order fibre orientation tensor and the fourth order tensor is not explicitly computed. This is where the problem of appropriate closure approximations arises that are needed to complete the calculation. Different closure approximations exist and are still being developed but will not be focus of this review (Advani and Tucker, 1987; Kugler et al., 2020; Tseng et al., 2020).

Very simplified, using the second order orientation tensor, the orientation can be described by the three diagonal tensor components A_{11}, A_{22}, and A_{33} which describe the degree of orientation with respect to a defined coordinate system. Usually, the reference coordinates are defined such that the 1-direction represents the flow direction, the 2-direction represents the cross-flow direction, and the 3-direction represents the thickness of the part direction. A completely random orientation (Fig. 1.9.15, A) is given by $A_{11} = A_{22} = A_{33} = 1/3$. A value of $A_{22} = 1$ means that all fibres are oriented in the two direction (Fig. 1.9.15, B). A value of $A_{22} = 0$ means that no fibres are oriented in direction 2. A random planar orientation, as seen in SMC charges is described by $A_{11} = A_{22} = 1/2$ and $A_{33} = 0$ (Fig. 1.9.15, C). It should be noted that the sum of the three diagonal terms of the orientation tensor always equals 1.

3.1.4 2000s: Improvements to Folgar and Tucker's fibre orientation model (Wang et al., 2008; Tseng et al., 2013)

The before mentioned fibre orientation model by Folgar and Tucker using fibre orientation tensors represents a widely applied approach to predict fibre orientation. However, it highly over-predicts the initial fibre orientation rate and is thus critically inaccurate. Especially in simple shear flows, the predicted orientation rate is up to 10 times faster than experimental rates. Therefore, multiple improvements to the model of Folgar and Tucker were developed (Gandhi et al., 2020; Tseng et al., 2020; Tseng, 2022).

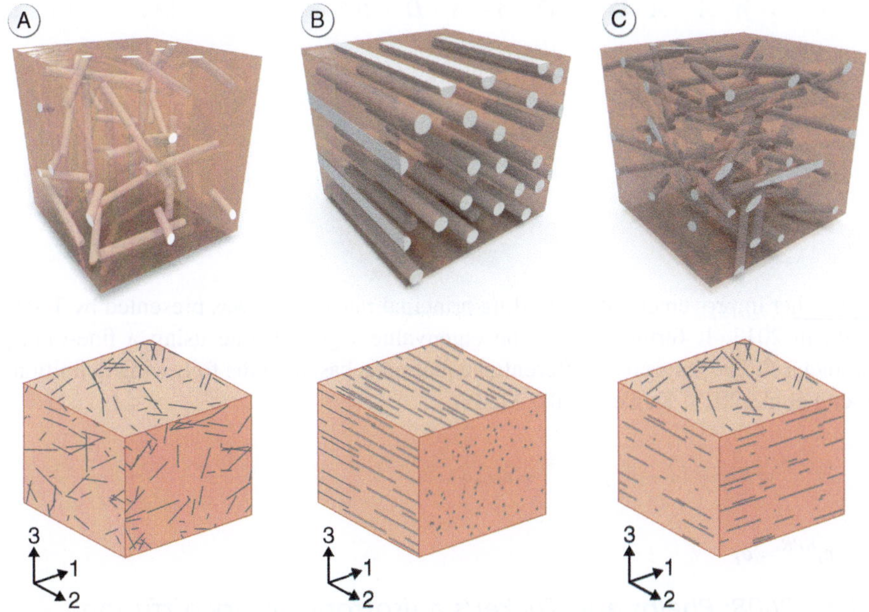

Figure 1.9.15 Orientation of a fibre population within a volume: Randomly oriented fibres (A), 2-directionally oriented fibres (B), and random in-plane fibre orientation (C). The images in the top row illustrate the 3D fibre configuration and the bottom row shows the 2D projection onto the corresponding plane (Gandhi et al., 2020).

In 2001, Huynh et al. introduced a strain reduction factor (SRF) 1/α to slow down the orientation kinetics, with $\alpha \in [0, 1]$. This model adaption decreases the orientation rate due to the fact that concentrations of fibres into clusters decreases strain (Wang et al., 2008; Kugler et al., 2020):

$$\dot{\mathbf{A}} = \alpha \left(\dot{\mathbf{A}}^{HD} + \dot{\mathbf{A}}^{IRD} \right)$$

Following in 2008, Wang et al. developed an objective fibre orientation model that includes slow orientation kinetics, known as the reduced strain closure (RSC) model. It contains of a spectral decomposition to describe the fibre orientation tensor with λ_i representing the eigenvalues and $\mathbf{e}_i$ representing the eigenvectors and a constant $\kappa \in [0, 1]$ (Wang et al., 2008; Kugler et al., 2020):

$$\mathbf{A} = \sum_{i=1}^{3} \lambda_i \boldsymbol{e}_i \boldsymbol{e}_i$$

$$\dot{\mathbf{A}} = \dot{\mathbf{A}}^{RSC} + \kappa \dot{\mathbf{A}}^{IRD}$$

$$\dot{A}^{RSC} = W \cdot A - A \cdot W + \xi[D \cdot A + A \cdot D - 2(\mathbb{A} + (1-\kappa)(\mathbb{L} - \mathbb{M} : \mathbb{A})) : \mathbf{D}]$$

$$\mathbb{L} = \sum_{i=1}^{3} \lambda_i e_i e_i e_i e_i$$

$$\mathbb{M} = \sum_{i=1}^{3} e_i e_i e_i e_i$$

Another improvement, the retarding principal rate (RPR), was presented by Tseng et al., in 2013. It further adapts the eigenvalue's growth rate using a fine-tuning parameter β. The respective differential equation is based on the following definitions (Tseng et al., 2013; Kugler et al., 2020):

$$\dot{\lambda}_i^{RPR} = \kappa\dot{\lambda}_i + (1-k)\,\beta\left(\dot{\lambda}_i^{2} + \dot{\lambda}_j\dot{\lambda}_k\right)$$

$$\dot{e}_i^{RPR} = \dot{e}_i$$

3.1.5 2008: Phelps and Tucker's anisotropic rotary diffusion model (ARD) (Phelps and Tucker, 2009)

Jay H. Phelps and Charles L. Tucker III worked out the anisotropic rotary diffusion fibre orientation model that allows for spatially non-uniform rotary diffusion. The interaction coefficient C_1 was replaced by a rotary diffusion tensor $\boldsymbol{C}$ and therefore the previously described isotropic rotary diffusion contributor (IRD) transforms to the anisotropic rotary diffusion (ARD) contributor to define the fibre orientation (Phelps and Tucker, 2009; Tseng et al., 2013; Gandhi et al., 2020; Kugler et al., 2020):

$$\dot{A} = \dot{A}^{HD} + \dot{A}^{ARD}$$

$$\dot{A}^{ARD} = \dot{\gamma}\,[2C - 2tr(C)A - 5(C \cdot A + A \cdot C) + 10\,\mathbb{A} : C]$$

The effect of rotary diffusion depends on the orientation state but the anisotropic rotary tensor $\boldsymbol{C}$ remains unknown. Phelps and Tucker phenomenologically define the tensor as a general second-order polynomial tensor-valued function form. It is a function of the before mentioned orientation tensor $\boldsymbol{A}$ and the rate of strain tensor $\boldsymbol{D}$ and contains parameters b_1 to b_5 (Phelps and Tucker, 2009; Tseng et al. 2013, 2020; Kugler et al., 2020):

$$C = C\,(D, A) = b_1 I + b_2 A + b_3 A^2 + \frac{b_4}{\dot{\gamma}} D + \frac{b_5}{\dot{\gamma}^2} D^2$$

3.1.6 2013: Improved anisotropic rotary diffusion model (iARD) by Tseng et al. (Tseng et al. (2013)

Huan-Chang Tseng et al. introduced a new dimensionless fibre-rotary-resistance (FRR) tensor in 2013 to improve the existing anisotropic diffusion model, resulting in the improved anisotropic diffusion model (iARD). This tensor $\widetilde{\boldsymbol{L}}$ has the same matrix properties as the fibre orientation tensor $\boldsymbol{A}$, is of second order with a trace tr $(\widetilde{\boldsymbol{L}}) = 1$ (Tseng et al., 2013):

$$\widetilde{\boldsymbol{L}} = \frac{\boldsymbol{L}^T \cdot \boldsymbol{L}}{\boldsymbol{L}^T : \boldsymbol{L}}$$

Its physical meaning is related to mobility obstacles of long fibres, that are described to occur in a gradient direction to the flow fields with respect to the velocity gradient tensor. The improved model thus consists of the previously described isotropic rotary diffusion (IRD) contributor and a fibre-rotary-resistance (FRR) contributor (Tseng et al., 2013):

$$\dot{\boldsymbol{A}}^{iARD} = \dot{\boldsymbol{A}}^{IRD} + \dot{\boldsymbol{A}}^{FRR}$$

$$\dot{\boldsymbol{A}}^{FRR} = 2\dot{\gamma} C_1 C_M \left\{ (\boldsymbol{A} - \widetilde{\boldsymbol{L}}) + 5 \left[(\boldsymbol{A} \cdot \widetilde{\boldsymbol{L}})_{\text{symm}} - (\mathbb{A} : \widetilde{\boldsymbol{L}}) \right] \right\}$$

Both, the IRD and the FRR contributor are simultaneously affected by the shear rate $\dot{\gamma}$. The IRD portion of the term remains to be a function of the dimensionless fibre-fibre interaction coefficient C_1, whereas the FRR portion depends on C_1 as well as the dimensionless parameter for fibre-rotary resistance, C_M (Tseng et al., 2013).

3.1.7 2017: Principal anisotropic rotary diffusion (pARD) by Tseng et al. (Tseng et al. (2018)

Tseng et al. continued to study model improvements, as they found it challenging to find the spatial tensor $\boldsymbol{C}$ for use in the ARD model. Seeking for an alternative to Phelps and Tucker's previous approach of a second-order polynomial tensor-valued function form Tseng et al. designed new principal form of a spatial tensor $\boldsymbol{C}$, that is coaxial with the orientation tensor. The principal directions of $\boldsymbol{C}$ are ordered from the largest to the smallest respective principal value. Thus, the direction in which most of the fibres align equals the first principal direction, and the fewest number of fibres is aligned in the third principal direction, respectively. Furthermore, the larger the principal value, the stronger the alignment of the fibres in this direction. The principal spatial tensor $\widehat{\boldsymbol{C}}$ is characterised by its diagonal with $0 \le \widehat{C}_3 \le \widehat{C}_2 \le \widehat{C}_1 \le 1$. The rotation matrix $\boldsymbol{R}$ consists of the eigenvectors of the orientation tensor $\boldsymbol{A}$ (Tseng et al., 2018).

$$\boldsymbol{C} = C_1 \boldsymbol{R} \cdot \widehat{\boldsymbol{C}} \cdot \boldsymbol{R}^T = C_1 \boldsymbol{R} \cdot \begin{bmatrix} C1 & 0 & 0 \\ 0 & \mathrm{C2} & 0 \\ 0 & 0 & \mathrm{C3} \end{bmatrix} \cdot \boldsymbol{R}^T$$

$$\boldsymbol{R} = [\boldsymbol{e}_1, \boldsymbol{e}_2, \boldsymbol{e}_3]$$

Physically speaking, the principal spatial components $\widehat{C}_1, \widehat{C}_2$, and $\widehat{C}_3$ indicate the ratio of probability for the fibres to lay in the direction of the principal vectors of the fibre orientation tensor. Thus, if all components equal 1 ($\widehat{C}_1 = \widehat{C}_2 = \widehat{C}_3 = 1$), the probability for the fibres to be oriented in a certain direction is equal for all directions and $\boldsymbol{C}$ is isotropic, respectively.

Tseng et al. used the interdependencies of the principal spatial components to simplify $\boldsymbol{C}$ using the parameter Ω, with $0.5 \leq \Omega \leq 1$ (Tseng et al., 2018):

$$\widehat{\boldsymbol{C}} = \begin{bmatrix} 1 & 0 & 0 \\ 0 & \Omega & 0 \\ 0 & 0 & 1 - \Omega \end{bmatrix}$$

Finally, Tseng et al. combined their pARD model and their previously developed RPR model to result to the following fibre orientation model, consisting of Jeffery's hydrodynamic contributor (HD), the principal anisotropic rotary diffusion contributor (pARD), and the retarding principal rate contributor (RPR) (Tseng et al., 2018):

$$\dot{\boldsymbol{A}} = \dot{\boldsymbol{A}}^{HD} + \dot{\boldsymbol{A}}^{pARD} + \dot{\boldsymbol{A}}^{RPR}$$

3.1.8 2018: Mouldflow asymmetrical rotational diffusion model (mARD) by Bakharev et al. (Bakharev et al. (2018)

Alexander Bakharev et al. presented the Mouldflow asymmetrical rotational diffusion model (mARD), a reduced form of the ARD model in 2018. The model replaces the ARD portion of the fibre orientation model with their developed mARD contributor (Bakharev et al., 2018; Favaloro and Tucker, 2019; Kugler et al., 2020):

$$\dot{\boldsymbol{A}} = \dot{\boldsymbol{A}}^{HD} + \dot{\boldsymbol{A}}^{pARD} + \dot{\boldsymbol{A}}^{mARD}$$

$$\dot{\boldsymbol{A}}^{mARD} = 2\dot{\gamma}(\boldsymbol{C} - tr(\boldsymbol{C})\boldsymbol{A})$$

3.2 Modelling fibre breakage

The length of the fibres within the final part also plays a critical role in the mechanical behaviour of parts made from discontinuous fibre-reinforced plastics parts. Similar to fibre orientation, fibre length also evolves throughout the processing of the material.

Fibres start with an initial fibre length and decrease in length during the process. Modelling this fibre breakage, or fibre attrition, helps in predicting the final fibre length and thus the properties of the final product. Therefore, approaches to model fibre attrition have been developed and these efforts continue until today. The evolvements of some models are presented in this chapter.

3.2.1 1938: Burgers' theory (Burgers, 1938)

As early as 1938, Johannes Martinus Burgers from the Netherlands uses Euler's classical buckling theory to model small deformations of rods in suspensions. He developed an approximate theory that described the forces acting on a thin rod in simpler shear flow. Burgers wanted to elucidate irregularities that were found in the streaming birefringence when analysing macromolecules in solutions. He assumed the particles to be rigid and ideally mixed within the solution. Brownian motion and inertial effects are negligible, the particles are thin enough to not have an impact on the flow pattern of the medium, and no slip at the interface occurs (Burgers, 1938; Forgacs and Mason, 1959).

3.2.2 2005: Fibre breakage model by Shon et al. (Shon et al. (2005)

A long time after Burgers' theory, Shon et al. continued the modelling efforts and similarly assumed that glass fibres in polymer flows would break up due to Euler's buckling. The rate at which the fibres break within a continuous mixing process is proportional to the hydrodynamic forces that act on the fibres. They modelled the breakage behaviour and assumed a constant fibre diameter throughout the process. The resulting fibre breakup kinetic model includes a fibre breakup rate constant k_f and a fibre length at which no more buckling occurs L_∞ (Shon et al., 2005):

$$\frac{dL}{dt} = -k_f(L - L_\infty) \approx -k_f L$$

Even though Shon et al. mentioned the dependency of k_f on the geometry of the flow, the stress, the initial fibre diameter, and the fibre's tensile strength, a clear definition of k_f, L_∞, and L were left open (Shon et al., 2005).

3.2.3 2013: Breakage probability model by Phelps et al. (Phelps et al. (2013)

To make fibre breakage analysis easier and thus promote modelling approaches, Jay H. Phelps et al. defined fibre length as an average fibre length distribution. It can be differentiated between the number-average fibre length L_n and the weight-average fibre length L_w. Both lengths depend on the total number of single fibres N_i and each length of one fibre l_i (Phelps et al., 2013):

$$L_n = \frac{\sum N_i l_i}{\sum N_i}$$

$$L_w = \frac{\sum N_i l_i^2}{\sum N_i l_i}$$

It becomes clear that both averages are equal when all fibres have the same length.

Similarly to the previous approaches, Phelps et al. also assume the fibres to break due to the classical Euler buckling theory. Buckling appears under certain compressive forces that are hydrodynamically induced and appear particularly in some critical orientations relative to the flow. The fraction of fibres that are oriented in this critical direction in which they can buckle f_i is a function of a buckling parameter B_i, and the maximum fraction at this certain orientation state $f_{\max}$ (Phelps et al., 2013):

$$f_i = f_{\max}[1 - \exp(1 - B_i)]$$

Many phenomena that impact fibre breakage are unknown until today, which is why Phelps et al. decided to keep it simple and to not overfit their model. Thus, they developed the following model for breakage probability P_i that includes the shear rate $\dot{\gamma}$ and a breakage coefficient C_B, that incorporates the beforementioned $f_{\max}$ together with a constant of proportionality (Phelps et al., 2013):

$$P_i = \begin{cases} 0, B_i < 1 \\ C_B\dot{\gamma}[1 - \exp(1 - B_i)], B_i \geq 1 \end{cases}$$

3.2.4 2021: Novel modelling approach by Bechara et al. (Bechara et al. (2021)

Previous modelling approaches showed inadequate results in predicting fibre breakage. Therefore, Abrahan Bechara et al. developed a new model approach that is similar to previous ideas but based on different assumptions. The model is based on the beam theory and thus acknowledges a distributed hydrodynamically-induced load that is acting on a fibre. Bechara et al. conducted a more in-depth analysis of the key impact factors on fibre breakage and were thus capable of defining the equilibrium length L_∞ and the breakage rate k_f, that were introduced by Shon et al. They found that the equilibrium length is dependent on the material and fibre interaction coefficient λ, the stress acting on the fibre σ_u, the fibre diameter d_f, the matrix viscosity η_m, and the shear rate $\dot{\gamma}$ (Bechara et al., 2021):

$$L_\infty = \lambda\sqrt{\frac{\sigma_u d_f^{\,2}}{\eta_m \dot{\gamma}}}$$

The breakage rate was modelled to be proportional to a scale factor ξ and the shear rate (Bechara et al., 2021):

$$k_f = \xi\dot{\gamma}$$

Bechara et al. assume the material and interaction coefficient λ to be dependent on the fibre concentration and the scale factor ξ to be a function of fibre dispersion. However, more work is needed to understand the dependencies of these parameters (Bechara et al., 2021).

References

Advani, S.G., Tucker, C.L., 1987. The use of tensors to describe and predict fibre orientation in short fibre composites. J. Rheol. 31 (8), 751–784. https://doi.org/10.1122/1.549945.

Baekeland, L.H., 1907. “Method of Making Insoluble Products of Phenol and Formaldehyde”. United States Patent number 942699.

Bakharev, A., Yu, H., Ray, S., Speight, R., 2018. Using New Anisotropic Rotational Diffusion Model to Improve Prediction of Short Fibres in Thermoplastic Injection Molding. In: Proceedings of the 2018 Society of Plastics Engineers (SPE) ANTEC, Orlando, Florida, USA, p. 1835–1839.

Bechara, A., Goris, S., Yanev, A., Brands, D., Osswald, T., 2021. Novel modeling approach for fibre breakage during molding of long fibre-reinforced thermoplastics. Phys. Fluids 33 (7), 073318. https://doi.org/10.1063/5.0058693.

Burgers, J.M., 1938. Second Report on Viscosity and Plasticity. Verhandelingen der Koninklijke Nederlandse Akademie van Wetenschappen, Noord-Holland.

Davis, B.A., Gramann, P.J., Osswald, T.A., Rios, A.C., 2003. Compression molding. Hanser, Munich.

Day, L., McNeil, I., 1996. Biographical Dictionary of the History of Technology. Routledge, London.

Dung, N., Pfeiffer, E., 2003. Bent ply: The art of plywood furniture. Princeton Architectural Press, New York.

Favaloro, A.J., Tucker, C.L., 2019. Analysis of anisotropic rotary diffusion models for fibre orientation. Compos. Appl. Sci. Manuf. 126, 105605. https://doi.org/10.1016/j.compositesa.2019.105605.

Folgar, F., Tucker, C.L., 1984. Orientation behavior of fibres in concentrated suspensions. J. Reinforc. Plast. Compos. 3 (2), 98–119. https://doi.org/10.1177/073168448400300201.

Forgacs, O.L., Mason, S.G., 1959. Particle motions in sheared suspensions. J. Colloid Sci. 14 (5), 457–472. https://doi.org/10.1016/0095-8522(59)90012-1.

Gandhi, U., Osswald, T.A., Goris, S., Song, Y.-Y., Diaz Luque, J., 2020. Discontinuous Fibre-Reinforced Composites: Fundamentals and Applications, first ed. Hanser Publications, Cincinnati.

Goris, S., Osswald, T.A., 2017. Progress on the characterization of the process-induced fibre microstructure of long glass fibre-reinforced thermoplastics. Plast. Eng. 73 (1), 46–47. https://doi.org/10.1002/j.1941-9635.2017.tb01641.x.

Jeffery, G.B., 1922. The motion of ellipsoidal particles immersed in a viscous fluid. Proc. Roy. Soc. Lond. A 102 (715), 161–179. https://doi.org/10.1098/rspa.1922.0078.

Kugler, S.K., Kech, A., Cruz, C., Osswald, T., 2020. Fibre orientation predictions—a review of existing models. J. Compos. Sci. 4 (2), 69. https://doi.org/10.3390/jcs4020069.

Morris, J.F., Boulay, F., 1999. Curvilinear flows of noncolloidal suspensions: the role of normal stresses. J. Rheol. 43, 1213–1237. https://doi.org/10.1122/1.551022.

Northrup, F. B., 1915. U.S. Patent vol. 1,158,830. https://patents.google.com/patent/US1158830A/en?oq=US1158830

Osswald, T.A., Menges, G., 2012. Materials Science of Polymers for Engineers. Hanser, Munich. https://doi.org/10.3139/9781569905241.

Phelps, J.H., Abd El-Rahman, A.I., Kunc, V., Tucker, C.L., 2013. A model for fibre length attrition in injection-molded long-fibre composites. Compos. Appl. Sci. Manuf. 51, 11–21. https://doi.org/10.1016/j.compositesa.2013.04.002.

Phelps, J.H., Tucker, C.L., 2009. An anisotropic rotary diffusion model for fibre orientation in short- and long-fibre thermoplastics. J. Non-Newtonian Fluid Mech. 156 (3), 165–176. https://doi.org/10.1016/j.jnnfm.2008.08.002.

Rahm, L.F., 1933. Plastic Molding. McGraw Hill Company, New York.

Shon, K., Liu, D., White, J.L., 2005. Experimental studies and modeling of development of dispersion and fibre damage in continuous compounding. Int. Polym. Process. 20 (3), 322–331. https://doi.org/10.3139/217.1894.

Tseng, H.-C., Chang, R.-Y., Hsu, C.-H., 2020. Comparison of recent fibre orientation models in injection molding simulation of fibre-reinforced composites. J. Thermoplast. Compos. Mater. 33 (1), 35–52. https://doi.org/10.1177/0892705718804599.

Tseng, H.-C., Chang, R.-Y., Hsu, C.-H., 2013. Phenomenological improvements to predictive models of fibre orientation in concentrated suspensions. J. Rheol. 57 (6), 1597–1631. https://doi.org/10.1122/1.4821038.

Tseng, H.-C., Chang, R.-Y., Hsu, C.-H., 2018. The use of principal spatial tensor to predict anisotropic fibre orientation in concentrated fibre suspensions. J. Rheol. 62 (1), 313–320. https://doi.org/10.1122/1.4998520.

Tseng, H.-C., 2022. Jeffery's orbit leading to the foundation of flow-induced orientation in modern fibre composite materials. J. Non-Newtonian Fluid Mech. 309, 104926. https://doi.org/10.1016/j.jnnfm.2022.104926.

Wang, J., O'Gara, J.F., Tucker, C.L., 2008. An objective model for slow orientation kinetics in concentrated fibre suspensions: theory and rheological evidence. J. Rheol. 52 (5), 1179–1200. https://doi.org/10.1122/1.2946437.

Section 2

Processes

A short history of composites manufacture

2.1

Kevin Potter
Bristol Composites Institute, University of Bristol, Bristol, United Kingdom

1. A note about sources

The academic literature relating to composites manufacture started to pick up a very long time after what I will aim to show was the "golden age" of composites manufacturing developments in the 1950s, so relying on peer-reviewed articles is really a non-starter. The patent databases are a bit better, but as anyone with a few patents to their name will tell you, patenting something does not necessarily turn it into a product, let alone a successful one. Equally, reading older patents with the benefit of hindsight often reveals that what has been patented would not ever have worked in practice. With apologies to those people who would like to put an incontrovertible inventor's name to each advance that has usually not been possible, so the focus here is on the methods and not the inventors.

The development of composites before 1970 seems to has been the preserve of government and corporate labs with relatively little academic research input or publication. For example, the Editorial Board of the first volume of the "Composites" journal in 1969 had 14 members and only three of them were academics. It is by no means unique to the development of composites, but over time the proportion of research activities in universities rose, in common with most areas of science and engineering, to be the dominating influence, at least in the published literature. Moving forward, the editorial board for the first volume of the "Composites Manufacturing" journal in 1990 had 26 members, 14 of them academics. The current editorial board of "Composites Part A" (formed by merging the other two titles) has 36 members only one of them not an academic. This might well be viewed as a natural result of the expansion of the universities world-wide in the past 50 years and the relative shrinking of government and corporate labs, but I feel the consequences of this have yet to be fully explored, although this is not really the place to do it.

2. Introduction to manufacturing with composites

Before moving on to composites-specific manufacturing processes, I would like to say a few words about the wider sweep of manufacturing for the sake of context. Most of the manufactured goods that we use on a daily basis are assemblies of individual components. Assembly processes are essentially generic and universal, even though they

The History of Composites. https://doi.org/10.1016/B978-0-443-34119-9.00012-7

do need to be customised for specific materials and will not be covered here. For the components, we can define four (rather broad and diffuse) manufacturing approaches. We have subtractive manufacture—from flint knapping to modern machining centres. We have casting—including plastics injection moulding. We have forming, from forging to sheet steel pressing and the English wheel, and we have additive manufacture, where the thickness of the component is built up from layers of thinner materials—from papyrus cartonnage to papier-mâché to 3D printing. The great bulk of composites are essentially processed by additive, layer by layer, methods although sheet and bulk forming processes are becoming more common, and for any component of complex geometry, each individual layer must also be formed to the required surface geometry.

The words "complex geometry" can have a very wide range of meanings. In academic papers on forming carbon fibre prepreg, any double curvature seems to be regarded as complex. For Glass Reinforced Plastics (GRP), the ability to generate mouldings of extreme complexity at low cost drove a great deal of the materials and manufacturing development. Fig. 2.1.1 shows a children's fairground ride where all the very complex panels are GRP. It is easy for people building aircraft structures to be dismissive of such things, but it is actually very difficult to see any other material and manufacturing route offering a robust and cost-effective solution here. At a more structural level, it is not at all obvious how a non-composite 100m wind turbine blade could be cost-effectively produced. We need to get away from a hierarchical approach with aerospace carbon fibre primary structures at the top and fairground rides at the bottom and focus more on effective solutions to particular design problems. Anyone wishing to delve further in these matters would do well to start with developing an understanding of product costing, focussing primarily on the distinction between

Figure 2.1.1 A very complex set of glass fibre reinforced plastic mouldings make up this children's fairground ride.

recurring and non-recurring costs, as this largely provides the constraint set for everything else, but does not fit within the specification for this chapter.

Full disclosure: I made my first composite part in a university vacation job over 50 years ago; it was an ultra-high fibre volume fraction tungsten wire reinforced copper matrix rod, about as niche as it is possible to be. I then spent a decade in a government lab, and then another decade first working in a corporate lab, then on to product and process development, technology transfer, factory and SME support environments (in multiple roles for a single employer) before joining the University of Bristol until I "retired." I have worked with glass, carbon, aramid, asbestos, cellulose, tungsten, and boron fibres, with phenolic, furane, epoxy, bismaleimide, and polysialate thermoset resins, with PP, PMMA, Nylon, and PEEK thermoplastic resins in short fibre and continuous fibre formats, plus aluminium and carbon matrices; and too many "alphabet soup" processes to mention. I say this not to try to establish myself as having some unique perspective so much as to say "if I am finding it confusing here, then it is fine if you are too."

Trying to make sense of the wide range of composites manufacturing processes is rather difficult, and it is very tempting to try to build a taxonomy or family tree of processes based on common features such as the split between primarily dry fibre processes and those using prepreg. However, one ends up with lots of exceptions and marginal cases, and the industry has always had a tendency to develop component-specific process variants, limiting the utility of the taxonomic approach if one is trying to cover the whole of the composites industry as it has developed. The approach to be used here will start with a look at the timeline from the start of the synthetic composites industry to the development, commercialisation and early years of carbon fibre, and then focus on the development of the main branches of composites manufacture.

It also needs to be pointed out that the names we give to composites manufacturing processes often only encompass one element of the full manufacturing process chain for a composites part, and in some cases not even that. The process chain for autoclave moulding of a simple composite flat laminate in an aerospace environment might go as follows:

1. Take prepreg out of freezer.
2. Note batch numbers on production records, defrost.
3. Transfer prepreg to cutting table.
4. Cut the required number of sheets of prepreg, mark in accordance with the manufacturing instruction sheet (MIS).
5. Return unused prepreg to freezer, update freezer records.
6. Acquire the relevant tool, inspect for damage.
7. Prepare tool and apply release agent.
8. Remove first backing film from ply 1.
9. Apply ply one to tool in compliance to the MIS, remove second backing film, debulking if required.
10. Repeat until the lay-up is complete.
11. Apply a layer of peel ply over the lay-up.
12. Apply a release film over the lay-up.
13. Apply a breather pack over the lay-up.

14. Apply a suitable mastic line to the tool surface.
15. Apply a vacuum film with vacuum breach units, ensuring that the seal to the mastic is complete.
16. Carry out a vacuum leak-back check to guarantee bag integrity.
17. Insert tool into the autoclave, attach vacuum lines, and thermocouples.
18. Close autoclave, start autoclave cycle.
19. Open autoclave at end of cycle.
20. Take out tool, remove and discard all the bagging materials.
21. Remove flash from laminate edges.
22. Route to inspection.
23. Finalise paperwork.

A very significant number of additional steps might be needed for a more complex or quality critical structure.

The headline manufacturing process here is the Autoclave, but most of these individual steps are common to multiple "processes," and as an added ambiguity if we use an Automated fibre Placement (AFP) machine to lay up the reinforcement it is common to refer to the component as being made by AFP rather than in the autoclave. It is also important to recognise that the bulk of these links in the chain require a significant hands-on manual input, so moving from manual lay-up to automated lay-up does not equate to automated manufacture, just the automation of a single link in the chain of manufacture.

Before going to the history, it is worth spending a little more time on the common factors.

In (almost) all cases once the reinforcement, matrix and an appropriate tool set have been selected and acquired;

1. The reinforcements must be formed to conform to the design intent on/in the tooling.
2. Matrix and reinforcement must be brought together.
3. A combination of energy, pressure and time must be applied to convert the separate matrix and reinforcement phases into a composite material.

Note this is a list of what needs to be done and not a list in chronological order. The chronological order of events for these three steps could be used as another way to categorise different approaches to manufacture. For example:

Resin infusion moulding processes (Group 1) Steps order: 1, 2, 3.

Autoclave/Bag moulding processes (Group 2) Steps order: 2, 1, 3.

Thermoplastic sheet stamping (Group 3) Steps order: 2, 3, 1.

With all the caveats considered earlier, we can probably classify all composites manufacturing processes into one or other of these three groups depending on the order of the principal activities.

Groups 1 and 2 date all the way back to the origins of the composites industry, Group 3 really emerged with the development of PEEK and similar very high-performance thermoplastic resins in the early 1980s, in response to perceived manufacturing and performance limitations with epoxy matrices. It is only slightly cynical to suggest that these materials and the associated manufacturing processes have been "the next big thing" in composites for the past 40 years.

3. The history of composites manufacture

3.1 Prior to 1960

It is undeniable that the story of manufacturing with fibre reinforced materials has a history that can be traced back to human prehistory and far beyond, with wasps possibly being the first to build large complex structures from spit and repurposed wood fibres. However, to provide a clear starting point, I am going to limit discussions to a point after the introduction of synthetic resin systems that could be used as matrices.

The first modern commercial composite material is generally thought to be Gordon Aerolite, a phenolic matrix reinforced with unidirectionally aligned flax fibres first produced in the late 1930s (McMullen, 1984), and also an early example of using wonky spelling in the pursuit of trade names. From a manufacturing perspective, the material may have been historic, but it also had some severe limitations. A combination of poorly collimated fibres, a need for a high fibre volume fraction and a matrix with difficult processing properties meant that very high-pressure press-moulding was the only realistic route forward, requiring very heavy duty and costly presses (i.e., high non-recurring cost). The material was primarily proposed as a like-for-like replacement for aluminium that could be machined into aircraft parts, but adding high machining costs to high processing costs in a material that never had a property suite that could effectively compete with aluminium probably meant it had no significant future outside of name checks for first modern composite. Having said that, I have always shown my students my own old slab of Gordon Aerolite. To be fair, later versions of flax and cotton fibre reinforced phenolic resins such as Tufnol have had significant success as a strong and reliable electrical insulator; absolute mechanical performance is by no means always a critical parameter for success in composite materials.

Gordon Aerolite also represents the first step toward aircraft composites, which from the earliest days could be characterised as prioritising the mechanical performance of the resulting composite structures over the ease with which they could be manufactured.

The first synthetic structurally important fibre to be commercialised was glass fibre (Li et al., 2014). Its centuries long history as a curiosity is irrelevant here, but by 1936 it was in small scale commercial production in the USA. Early applications had the same difficulties with matrices with poor manufacturing properties, but everything really changed in 1943 with the commercial introduction of a room-temperature cure polyester resin (McIntyre, 2003). The combination of formable sheets of chopped glass-fibres with relatively low viscosity resin that readily wet out the reinforcement was game-changing from a manufacturing perspective—even if the mechanical performance was limited, leading toward processes prioritising ease of manufacture over mechanical performance.

Until the end of WW2 almost all glass fibre production was utilised in military applications, especially the radomes that were needed on aircraft mounted radars, although a number of early applications outside the military were reported. Probably the most significant of these was the use of glass reinforcements with polyester resins to mould small boat hulls (Marsh, 2006). Before GRP, building the typical clinker-built wooden hull was a very labour and skills-intensive activity, but required little infrastructure other than a big shed and some tools. With GRP, the high-level skills required to work wood were replaced with a simpler set of skills. Moreover, the hull could be built up as a whole rather than a kit of relatively small intricate parts, leading to a very large reduction in the labour content and hence the costs (once the cost of the tool was amortised over multiple hulls). In this market, GRP was totally disruptive, more or less completely displacing the traditional materials and manufacturing routes over a few years.

The process most commonly used was to become the backbone of the GRP industry for decades, still very often being used today. Starting from a tool surface, first a release agent is applied, then a layer of pigmented resin, or gel coat, is painted on, followed by a layer of glass reinforcement onto which liquid catalysed resin is poured or painted, then worked with rollers, brushes, or other tools to ensure that the glass is saturated (or wet out) with resin and no air is trapped in the layer. Glass fibre has a huge advantage in this process as the wet-out reinforcement is transparent and air pockets are easy to see. This stage is then repeated until the required thickness is achieved; thickness can be varied from point to point as required. For structures needing higher performance, woven glass fabric can be used. Where the geometry is more complex, sheets of short glass fibre or randomly swirled continuous fibre can be used as these reinforcements can be much more easily formed to shape. Now, while this wet layer by wet layer approach was faster and much less labour-intensive than traditional boat-building techniques, it was still very slow compared to the sort of high-volume sheet metal forming processes used in mass market automotive or white goods industries, and it produced mouldings with very poor thickness tolerances and only one smooth surface.

We can get an insight into the state of the US composites industry from a patent applied for by Marco Chemicals Inc. in 1945 and granted in 1950 (Muscat 1950). The patent drawings are reproduced as Fig. 2.1.2 and show what are the earliest recognisable depictions I can find of the resin transfer moulding, vacuum infusion and liquid resin compression moulding processes, from almost 80 years ago. The patent is the third in a series filed in 1945 by Irving Muskat for Marco, but I can find no later patents, so it is not easy to trace further developments.

In some ways, even more interesting are the description of the reasons that the invention was needed. Several difficulties in mouldings are identified, such as entrapped air, wrinkles, high scrap rates, wastage of fibre and resin, surface imperfections, irregular impregnation and so on. Many of these difficulties are particularly associated with the wet lamination process, but as a whole they are quite recognisable even today as ongoing challenges for all types of composites manufacturing processes.

With the success of GRP boat hulls and the improved availability of both glass fibres and polyester resins at the start of the 1950s, there was a surge in attempts to move GRP into the automotive sector, up to and including the demonstration of all GRP car bodies. As noted above, the difficulty in ramping up manufacturing rates for GRP greatly limited the potential for success in the mass market. However, as the costs of tooling up for a car body in pressed steel can run into hundreds of millions

Jan. 24, 1950 I. E. MUSKAT 2,495,640

METHOD OF MOLDING

Filed June 8, 1945

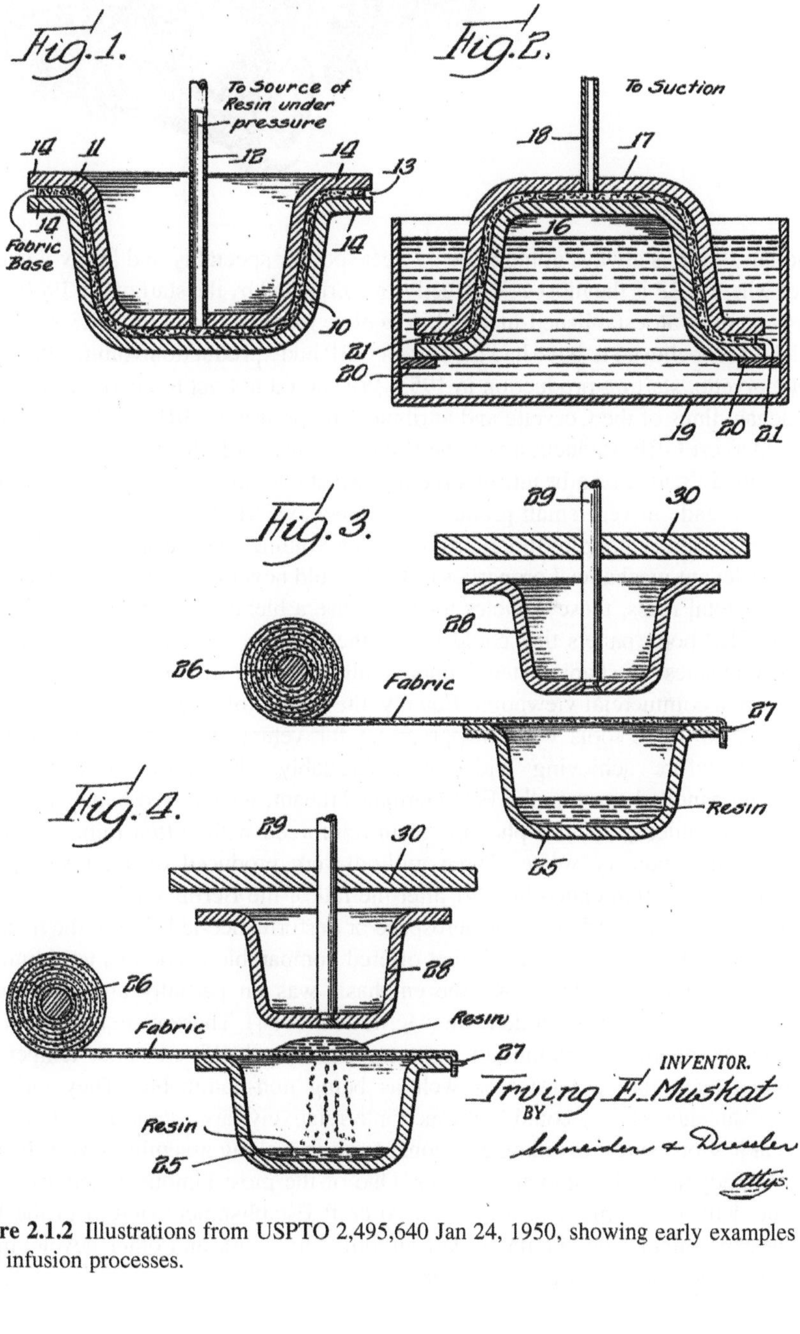

Figure 2.1.2 Illustrations from USPTO 2,495,640 Jan 24, 1950, showing early examples of resin infusion processes.

Figure 2.1.3 GRP body shell of the Chevrolet Corvette.

of dollars, there is always going to be a sweet spot for speciality and low-volume vehicles where GRP is the most cost-effective material and by the start of the 1960s GRP was becoming an established material in vehicle manufacture. Fig. 2.1.3 shows the model that is believed to have had the largest GRP body production volume, the Chevrolet Corvette, starting production in 1953. We should not get too carried away with the elegant lines of the Corvette and attribute it to the use of GRP as the car with the second largest GRP production volume is the three-wheel Reliant Robin made in the UK—nobody's idea of a beautiful structure. Most cars made with GRP body shells have been made in very small production numbers where the production economics make sense, and scaling up to larger production volumes has been very difficult.

The relevant mechanical properties of GRP could never match those of steel, so to minimise total mass, these vehicles tended to use a blend of simple steel structures clad in GRP body panels that could reflect the designer's aesthetic rather than the practical business of shaping steel, although all-GRP cars were produced unsuccessfully from a commercial viewpoint. Equally, the poorly toleranced simple shells produced on the open tools were inappropriate for vehicle assembly and had to be improved before achieving wider use. Arguably, the most successful early composite-panelled car was the East German Trabant, which used better toleranced hot-press moulded panels of phenolic resin reinforced with cotton fibres recovered from garment industry waste. Three million were produced over 34 years from 1957 until production ended in 1991 after the fall of the Berlin Wall.

At the start of the 1950s on the aerospace or performance-led side of the industry, there were still no fibres available that offered comparable properties to aluminium alloys, and, at least in the U.K., the emphasis was on partially aligned asbestos mats impregnated with phenolic resin (McMullen, 1984). These were lacking in performance compared to aluminium but were stiffer and offered better temperature resistance than glass/polyester as well as being non-flammable. They did have some advantages as they could be formed into relatively large structures of complex shape at low pressures, avoiding additional mass in making assemblies as well as offering aerodynamically smooth surfaces. One of the most famous examples of this was the delta wings made at the Royal Aircraft Establishment (RAE) in the U.K. for the experimental de Haviland Dart aircraft. This work and other developments at RAE are outlined in (McMullen, 1984).

These were made in a vacuum only pressure bag moulding process as shown in the patent drawing in Fig. 2.1.4 (Phillips, 1960). This mould was fabricated from a stainless-steel face sheet, backed up by heating elements and encased in a concrete and steel shell with an inner vacuum bag, in what would today be called an integrated tool (note the inventor's name. L.N. Phillips would later play an important role in the development of carbon fibres in the UK). The project demonstrated three important outcomes. The first was that a moulded composite wing skin could meet the load

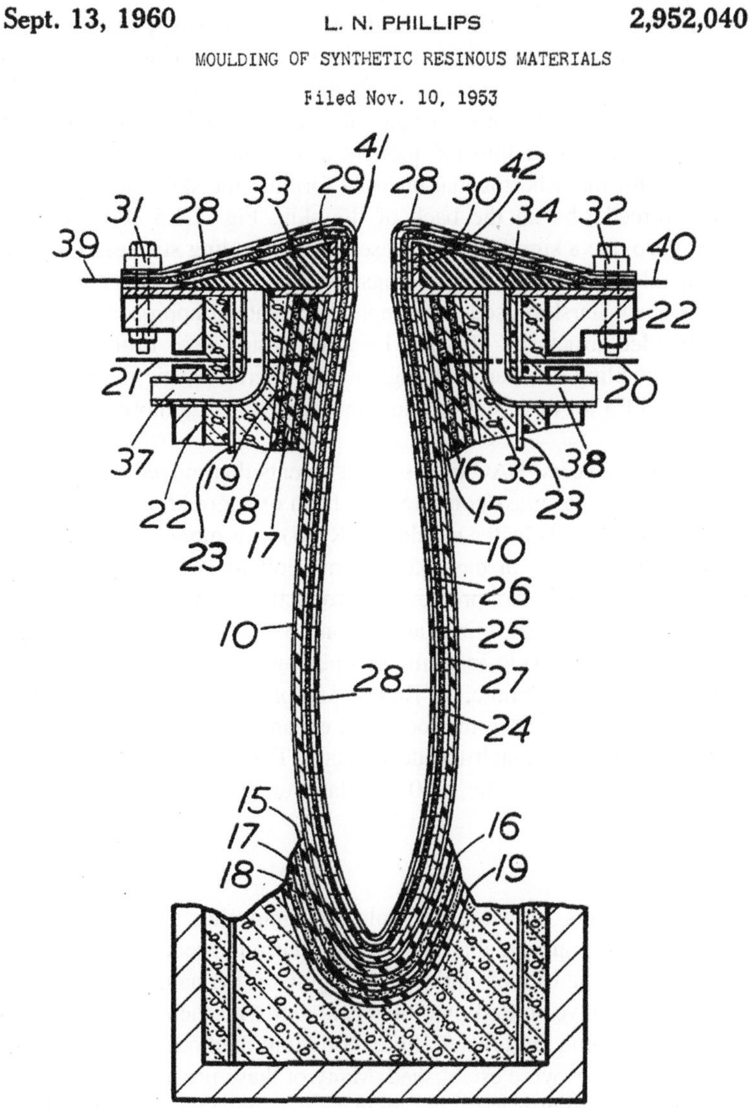

Figure 2.1.4 Mould tool used for asbestos/phenolic wing skin.

and mass targets set at the program outcome. The second was that integrating multiple metal components into a single composite moulding was a valuable approach to weight saving at the system level. The third was that there would be resistance to transitioning from metal to composite, the load requirements were increased after passing the successful first test set, to a level that could not be met within the weight target. In retrospect, avoiding building large asbestos reinforced structures was probably a good idea, but for many years a section of the moulding sat in the corner of the lab I worked in, and asbestos/phenolic military aircraft wing-mounted fuel drop tanks were still in production in the U.K. in the 1980s.

At the same time on the other side of the U.K., another way to make integrated mouldings was being developed by the Bristol Aeroplane Company (BAC). The concept here relied on the same asbestos/phenolic prepregs being used at the RAE to provide the wing skin, but used a randomly reinforced asbestos fibre-reinforced moulding compound forced into the tool by a hydraulic ram to both provide the driving pressure for the asbestos mat material forming the skin and raise a series of randomly reinforced ribs on the back of the skin, Fig. 2.1.5 (Pollard and Rees, 1958a). These provide a significant increase in skin bending stiffness for little additional mass. In addition, two such mouldings could then be bonded together to deliver a wing. The Beech Starship used a conceptually identical approach to build up the wings from bonded structures. 60 years later, we would refer to the process in the BAC 1952 patent filing as overmoulding, and the thermoplastic matrix version has been the subject of a lot of recent development targeting automotive uses of carbon fibre.

Over the next few years, BAC filed several patents that moved the technology on from asbestos/phenolic to glass/polyester and from overmoulding to what we would describe today as Resin Transfer moulding (RTM) at a more developed level than shown in the Marco patent in the 1955 patent (Pollard and Rees, 1958b). In this case, there was clear evidence of problems in production being overcome, with further developments being patented. For example, one patent is specifically focussed on automation of the RTM cycle in terms of opening and closing resin flow valves without human input. At that time, the Bristol Aeroplane Company owned the Bristol Car Company and it is clear that the focus had expanded from aircraft to cars for the developing technology, although the developed RTM technology was used on the Bristol Bloodhound missile in the 1950s and later on the Concorde radome.

3.2 1960–1980

By the end of the 1950s, the basic principles of composites manufacture were well established, at least in conceptual terms, and most current processes had been at least described. There is some lack of clarity or ambiguity as the names given to processes and process variants have shifted over the years, but with the exception of AFP and some recently emerging automated fibre collation methods, all the composites manufacturing processes were established. Many of these processes are recognisably closely related to current processes, but often seem to have been optimised for a specific application. As equipment such as autoclaves became more industrialised, the

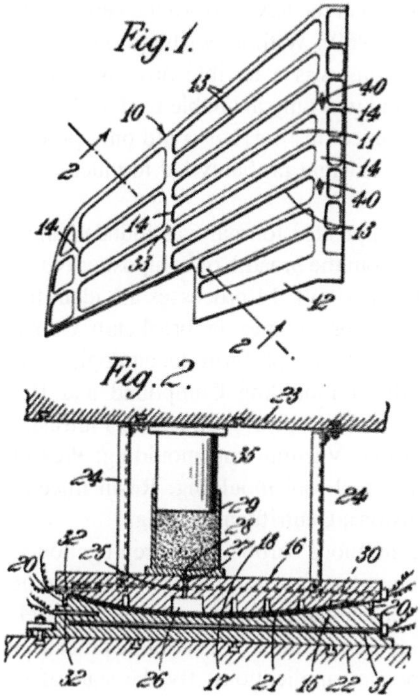

Application date 1952
Compression/overmoulding
of complex geometry

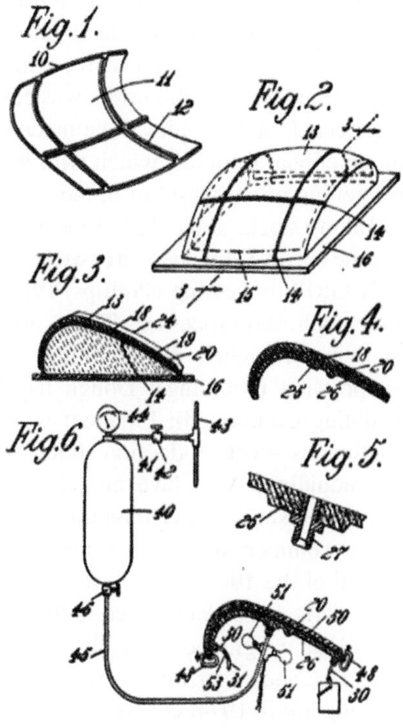

Application date 1955
Resin Transfer Moulding of
complex geometry

Figure 2.1.5 Patent drawings from the Bristol Aeroplane Company in the 1950s.

processes tended to settle toward more standardised approaches. Probably the last two processes to enter the fray were filament winding and pultrusion; although both processes could be regarded as having their roots in the 1940s, they were some years behind the other processes in becoming established. Both processes will be more fully explored in other chapters so will not be covered here.

At that time, modern processes of technology development and analysis such as the use of Technology or Manufacturing Readiness Levels had yet to be produced, and it is probably fair to suggest that none of the processes available then could realistically be defined as being fully developed to TRL9. Even today, composites manufacturing processes struggle to hit right-first-time and freedom from defects targets that would be expected in a fully realised technology operating at a Six Sigma performance level.

From the available literature, it is hard to detect very significant improvements in composites manufacturing processes during the 1960s. I am certainly aware that they happened, but within organisations that had little interest in advertising how they improved things to their competitors. The only process that I can be fairly sure was novel, was the "leaky mould technique" developed at the RAE to turn the newly available carbon fibre tows into laminates (Phillips, 1969). Without wishing to disparage some of the giants of the carbon fibre story, this was a terrible process and was only needed when carbon fibre was emerging and was only available in short lengths. It did enable flat samples to be made for mechanical testing, but could only produce unidirectional samples, which with the gift of hindsight probably led to much of the overselling of carbon fibre in the 1970s.

Taking stock: at the end of the 1960s the Composites Journal was founded and in volume one in 1969, there are two papers setting out the breadth of composites (essentially GRP) processing (Phillips, 1969; Wildman, 1969). 14 processes are identified, some with sub-processes. These are (more modern names in brackets); Contact moulding; Matched die moulding using a variety of preform technologies (wet compression moulding); Dough moulding (Sheet moulding Compound and Bulk moulding Compound); Filament winding; Continuous pipe production; Pultrusion; Continuous sheet production; Injection moulding; Vacuum bag moulding; Pressure bag moulding; Autoclave moulding; Cold matched tool moulding; Resin injection process (Resin Transfer moulding, Resin Infusion); Centrifugal casting.

The commercialisation of unidirectional continuous carbon fibre prepregs toward the end of the 1960s and especially the introduction of woven carbon fibre prepregs in the early 1970s largely led to the adoption of manual lay-up of prepreg on to relatively light weight tooling, followed by autoclave moulding as the baseline process for aerospace and other high performance composites applications. By the start of my working life in 1974, everything seemed to be in place for the new carbon fibre composites to achieve the promise seen by the first researchers aiming to develop new aircraft structural materials 40 years earlier. It is probably fair to suggest that the early carbon fibre prepreg-based materials had multiple limitations in terms of properties such as impact toughness, alongside wafer-thin process windows (Childs, 1980) in processes of limited reliability that were only just starting to attract any academic attention or understanding, so early problems were probably inevitable. The introduction of tougher, flow-controlled matrices and woven carbon fibre fabrics greatly improved matters by the mid-1970s, but with hindsight it is hardly surprising that carbon fibre composites got off to a rocky start. However, by the late 1970s it was already clear that we needed much more productive processes; manual lay-up of prepreg was identified as the major problem and automation was identified as the answer. As a target, that proved to be much easier to state than to achieve, partly because as noted earlier, automating lay-up does not equate to automated manufacture and partly in my view because there was minimal study of how manual lay-up was actually carried out.

50 years on, carbon composite structures have had a major impact on aircraft manufacture, but much less penetration of other markets such as automotive or consumer goods. Today, the world's annual carbon fibre production capacity is still expanding rapidly but is currently still sitting at a few % of the production of glass fibres.

By 1980, almost everything in today's composites manufacturing world had been invented, up to and including AFP, even if very clearly at low TRL, and software was still fairly basic. For example, simple drape models had just been reported (Potter, 1979), but without graphical interfaces were at best very limited.

3.3 Developments post 1980

By 1980, two major process styles had consolidated their predominance for the manufacture of high-performance composite structures of arbitrary geometry—Autoclave moulding and Resin Transfer moulding; the focus in this section will just be on these processes and their variants.

3.3.1 Autoclave moulding

By 1980, the functional process chain for composite structures made by autoclave moulding of thermoset matrix prepreg was well established, as laid out earlier. At that point, most if not all links in the process chain were manual operations, and it is the gradual replacement of these manual operations by at least partially automated operations that characterises the more recent developments in the autoclave moulding process. These developments will be considered, working through the process chain. The focus has generally been on increasing speed and reducing errors compared to the manual baseline. In some cases, such as ply cutting, technology has been adopted from existing industries such as garment manufacture, whereas other composites-specific technology has been adopted such as laser templating.

Developments and improvements can be identified at every stage in the process chain, but whether they have been adopted by specific manufacturers will depend on multiple factors such as product mix and production volumes. Not all of the developments are due to in-factory automation, and some relate more to changes in the supply chain and make or buy decisions on reinforcement cutting or core machining. Even the requirement for on-site refrigerated prepreg storage can be replaced by just in time delivery of prepreg to the shop floor. None of this, of course, eliminates those functions - simply moves them to the supply chain.

In many ways, a typical factory producing composite aircraft secondary structures is an exercise in logistics, producing thousands of cut plies each day that all need to be recorded, tracked for out-time, matched to mould tools, laid up in specific ways and so on. Similar tracking is needed for the mouldings produced, recording material and process details, inspections, and quality outcomes. Initially, this large volume of data was collected in manual analogue systems, with separate systems often used for different elements of the data handling requirements, and often, in my experience, with more focus on meeting regulatory requirements than on capturing and identifying valuable understanding from the mass of data. Replacing separate manual analogue systems with an integrated and more automated digital system is as much an important development as the more obvious elements of process automation and allows that data to be treated as a value rather than just a cost.

Automated ply cutting was probably the first aspect of autoclave manufacture to be widely adopted, with automated ply nesting software helping to minimise in-process waste. Picking the cut plies and assembling them into ply kits was initially a manual task, but currently commercial ply cutters offer automation of this function if the cost/benefit calculation is positive for the specific production environment.

Ensuring that prepreg plies are applied to the mould tool in the correct position is of critical importance; this can be achieved using a series of physical templates datumed to the tool surface to control the ply edge positions. For large or complex structures, this generated major bottlenecks in workflow, slowing production by manual lay-up very significantly. The first laser templating solutions became commercially available in 1989, leading to significant reductions in delays and non-productive labour (Blake and Bird, 1997). These systems became more widely accepted in the following decades with multiple vendors offering their own solutions. Over the decades, since their first introduction the capabilities of these lay-up support systems have expanded to offer ply verification, edge checking, defect, and foreign object damage detection, and aspects of in-process inspection, especially as automated lay-up by AFP has increased its market penetration.

There have been multiple attempts to develop automated lay-up systems that can replace manual lay-up, as this process consumes the bulk of the touch labour in delivering a moulded part. These attempts have not to date delivered a Commercial Off The Shelf (COTS) solution that can deliver the automation of lay-up of all the features of a typical composite secondary or tertiary aircraft structure. Fig. 2.1.6 shows how complex an autoclave moulded part can be, an extremely challenging target for any sort of automation.

However, it is possible to develop a system such as Automated Tape Laying (ATL), that can routinely deliver cut prepreg plies into a simple geometry such as a plate, stringer or single curvature panel, but the great mass of composite parts is simply not deliverable unless the automation solution can handle complex double curvature. It is, of course, a perfectly sensible response to redesign structures to be manufacturable from assemblies of simple or formable, automatically manufacturable geometries (and as we will see this has been key to developments in RTM), but that option falls outside the remit here.

The AFP technology was developed to be able to offer a solution to replacing manual lay-up with full automation on somewhat more complex geometries

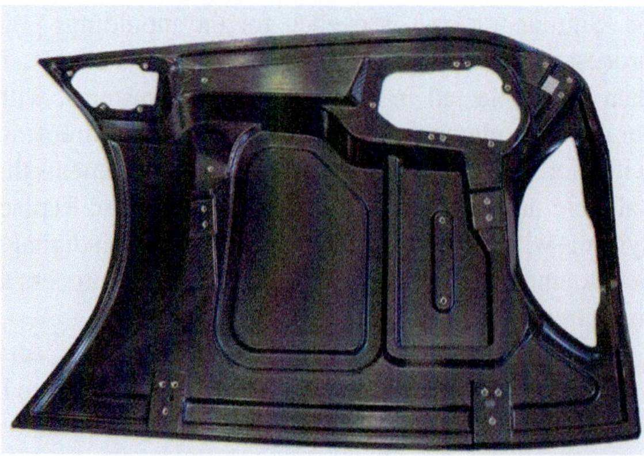

Figure 2.1.6 Autoclave moulded panel of extreme complexity.

(Potter, 2023). The key understanding is that narrow (typically 6 mm wide) tapes of prepreg can be made to follow a doubly curved surface (at least on a geodesic path) without necessarily developing defects due to buckling or wrinkling of the tapes. Handling one tape at a time is clearly inefficient from a lay-up rate perspective and most AFPs will handle multiple tapes, with 8–16 being common. Each tape must be individually controlled and tensioned and needs to be cut and restarted as required. Heating is needed to bring the tapes to the correct conditions for lay-up, and a roller or pressure shoe is used to provide a consolidating force at the point of lay-up. The resulting AFP head must then be held against the tool surface and must be able to be moved across the tool surface as required.

This set of requirements leads to quite a large, heavy and cumbersome piece of equipment in the AFP head, and a much larger piece of equipment to manipulate it at speed across the mould tool, using gantries or tracked heavy-duty industrial robots. AFP has been proven in service to work very well for large primary structures in commercial aircraft, as these structures are really much too large to handle manually and have relatively low levels of double curvature or other geometrically complex features. However, the smaller and more complex geometry parts that make up the great bulk of aircraft composites by number fall outside the capabilities of current AFP technology, and it is hard to see how that gap can be filled. The ATL/AFP history actually dates back to the start of the composites story (Lukaszewicz et al., 2012) and continues today with novel designs of AFP being developed to relieve some of the limitations (Zympeloudis et al., 2015). Today, the tension between attempting full lay-up automation for all structures and focussing on delivering design and production support tools for manual lay-up of very complex parts still needs to be resolved.

Howsoever the lay-up has been achieved; the next link in the process chain is to prepare the tool and lay-up to go into the autoclave. Application of a sequence of release film and breather/bleeder layers to the completed lay-up ensures that vacuum can be applied across the whole part. This is then followed by the application of an impermeable vacuum film, sealed to the tool to allow the external pressure and internal vacuum to be separated without leaks. This is a quality critical process step but has probably been the subject of less research or development than any other link in the process chain and is often still a fully manual process. One development has been the introduction of leak-finding technologies to speed up the process of eliminating all leak paths prior to loading the assembly into the autoclave.

Whilst large primary structures may be cured as a single autoclave load, it is very common for smaller mouldings to be batched together on racks to make up a full autoclave load. Each tool will normally be monitored by thermocouples so that the leading and lagging thermocouples can be used to fine-tune the autoclave control cycle. Connecting up the thermocouples and preparing the autoclave load is another time-consuming task that can be tackled by improved data management (Stewart et al., 2020). Stripping out the mouldings at the end of the cure cycle is another largely manual task that, to date, seems to have resisted automation.

For each of the developments noted above, the primary motivation has been the reduction of the touch labour content and the associated costs or, to be more accurate, the replacement of touch labour inputs by capital expenditure. Small trade moulders outside the aerospace industry are much less likely to adopt these practices than larger organisations in the commercial aerospace supply chain. Even for many companies in that supply chain, the most common response to the need to cut the costs of touch labour has been to out-source production to low labour cost economies.

In contrast, the last major development in autoclave moulding is, paradoxically, to eliminate the autoclave and instead use vacuum only processing in relatively low-cost ovens. In this case the aim is simply to reduce capital expenditures rather than touch labour contents. There is, however, a significant potential for the degradation of part quality, especially in terms of consolidation and voidage, in removing the increased pressure in the autoclave. Special purpose prepregs have been developed to improve the evacuation of air which helps somewhat. Overall, out-of-autoclave processing probably makes most sense for large-scale relatively simple structures where the costs of a new autoclave are unaffordable.

3.3.2 Resin transfer moulding (RTM)

All the steps in the process chain from a roll of reinforcement to the start of lay-up in Resin Transfer moulding (RTM) are the same as in autoclave moulding, although the materials are different and will not be repeated here.

In contrast to autoclave moulding which really developed in the aerospace supply chain, RTM, whilst in part developed for the manufacture of radomes, remained for many years a relatively niche process in the general industry sector, used when more tightly toleranced mouldings were required than could be achieved with single sided mould tools. There were developments reported in the late 1970s using a high-pressure version of RTM with a bismaleimide resin to produce gas turbine blades (Jones and Johnson, 1980), but this work seems to have come to an end without reaching production. The major exception to this is probably the use of RTM to manufacture aircraft propellers (McCarthy et al., 1994), where reliable and tight dimensional and mass tolerances were critical.

It is very difficult to trace developments in the trade moulding community as publication is very infrequent, but we can pick up the trail when interest in the U.K. increased in the early 1980s. An example of this was the collaboration between BP Research and Moore Plastics Ltd., a small trade moulder, in the development of a flywheel energy storage system for regenerative braking in urban buses. The system consisted of a flywheel and a set of five GRP casing components that maintained the flywheel under vacuum to limit losses. As might be imagined, each component has to be very tightly toleranced for reasons of dynamic balance or vacuum tightness, and RTM was the only realistic option. The system was a technical success but did not make it into service, in part, due to deregulation of U.K. buses. The potential of RTM for more advanced applications was recognised, and further developments were carried out focussed on components of great complexity, many of which might be regarded as "black aluminium" designs.

Lay-up of dry reinforcements is actually much more difficult in many ways than laying up prepreg, and some form of tackifier is often needed to hold the plies in place. Early variants of RTM used mould designs that clamped the edges of the dry preforms, which can simplify lay-up, but as preform complexity increases, another approach is needed. The development of using press-formed preforms, which can be trimmed to size off the tool, allowed the complexity of mouldings to be very greatly increased. Autoclave moulding simply cannot compete with the complexity or tolerances available in rigid tool RTM using multiple press-formed preforms.

The availability of accurately trimmed to size preforms allowed another development to be achieved, the routine production of net-shape moulding (Fig. 2.1.7), something that still eludes autoclave moulders. This also supported the transition from unsealed tools where the injection of the resin drove out the air in the tool, to sealed and evacuated tools where high-quality mouldings could be made even with complex geometries where full air removal could not be assured. During this period, the art, rather than the science, of tool design was further developed using techniques such as internal runners to distribute resin and integrated ejectors.

By the end of the 1980s, RTM was attracting interest from both aerospace and automotive companies, and flow and process models started to be developed to support the process. The structural and manufacturing requirements of aerospace and automotive industries are very different, so much of the technology diverged from this point. RTM historically used mix and metre resin delivery machinery dosing volumetrically. The potential for variation in resin composition is unacceptable in aerospace, so single part resins were developed for use in RTM often driven by air pressure for simplicity of control. Mass market automotive production requires a minimal takt time, generating a need for fast cure resins and high-capacity mix and metre machinery; this generates a high back pressure creating a demand for powerful and fast acting presses.

In the low volume automotive market, in 1987, Lotus introduced the VARI (Vacuum Assisted Resin Infusion) process into their vehicle body production (Vaccarella et al., 1991). This used one relatively rigid tool half and a second part with more flexibility, with a partial vacuum pulled between them, to balance the needs for ease and speed of injection and dimensional fidelity in the mouldings produced. Mould loading was entirely manual using glass mats and rovings, and a great deal of machining was needed to turn a moulding into a floor pan or body.

Figure 2.1.7 Complex net-shape moulding made by RTM from more than 30 carbon fibre reinforcement preforms.

The introduction by BMW of the i3 electric vehicle in 2013, with a passenger compartment structure assembled from carbon fibre composites made by high pressure RTM, possibly represents the most advanced use of automotive composites to date. The largest components are the side frames, made up from nine individual press-formed preforms. These were produced in a highly automated factory production line, but it is notable that insertion of the multiple preforms into the mould tool remained a manual task.

In both industries, the ability to inject resin into large components within the resin's cure window led to the development of Compression RTM - where the mould is initially not fully closed, and resin can flood into the tool cavity, followed by a compression step where the resin is forced through the thickness of the preform. This approach can also be scaled up very greatly by adopting single sided tools and injecting resin through a vacuum bag into an infusion network of essentially any size. However, there is a balancing loss of the ability to generate structures of extreme complexity as the process used moves away from fixed and rigid tooling. At present, there are a large number of related processes that can be optimised for specific applications, but certainly no one size fits all solution.

There is a substantial academic literature on RTM and its variants, for example, (Michaud, 2016; Bodaghi et al., 2019; Correia et al., 2004), although only a very small minority relates to the critical issues of part design for RTM, preform design and manufacture, tool and production line design and production line integration. The bulk of the academic work relates to flow prediction and simulation. There is a tendency to assume that only by the use of flow modelling software can the RTM process be understood, and that without the software, good parts cannot be made. It should be remembered that RTM was invented in the 1950s and had its first major uses for very critical aircraft parts in the 1980s —well before process modelling was an established capability.

4. Final remarks

Given that the manufacturing of composite components and structures under factory conditions predates significant academic interest in the understanding and simulation of composites manufacturing processes by decades, it is no surprise that the great progress to date owes more to industrialists than to academics. As the capabilities of the academic simulations improve and become embedded in commercial software and hardware offerings; the impact of the academic outputs should be felt more strongly in improved methodologies for the design and processing of composite structures. Nevertheless, there are still gaps, for example in understanding of some important manual processes, and whilst the models are good at predicting the outcomes of design decisions, they are currently less effective at supporting those decisions. However, it is not clear how far improvements in understanding processing can go unless it is joined by a much better understanding of Design for Manufacture principles as applied to composites and the commercial, industrial, and factory systems in which the design and processing is embedded. I would hope that the next sets of academic research targets take on a much broader, holistic, and systems level approach to push the composites manufacturing into the future leaving behind the innovations from the 1950s.

References

Blake, S., Bird, N.J., 1997. Laser Guidance for Hand Laid Composites: Past, Present and Future. Society of Manufacturing Engineers.

Bodaghi, M., Lomov, S.V., Simacek, P., Correia, N.C., Advani, S.G., 2019. On the variability of permeability induced by reinforcement distortions and dual scale flow in liquid composite moulding: a review. Compos. Appl. Sci. Manuf. 120, 188–210. https://doi.org/10.1016/j.compositesa.2019.03.004.

Childs, R., 1980. Some aspects of the autoclave fabrication process. In: Proceeding of the Symposium on Fabrication Techniques for Advanced Reinforced Plastics,Salford, pp. 1–9.

Correia, N.C., Robitaille, F., Long, A.C., Rudd, C.D., Simacek, P., Advani, S.G., 2004. Use of resin transfer molding simulation to predict flow, saturation, and compaction in the VARTM process. J. Fluids Eng. 126 (2), 210–215.

Jones, W., Johnson, J.W., 1980. A resin injection technique for the fabrication of aero-engine composite components. In: Proceeding of the Symposium: Fabrication Techniques for Advanced Reinforced Plastics, Salford. IPC Science and Technology.

Li, H., Richards, C., Watson, J., 2014. High-performance glass fiber development for composite applications. Int. J. Appl. Glass Sci. 5 (1), 65–81. https://doi.org/10.1111/ijag.12053.

Lukaszewicz, D.H.J.A., Ward, C., Potter, K.D., 2012. The engineering aspects of automated prepreg layup: history, present and future. Compos. B Eng. 43 (3), 997–1009. https://doi.org/10.1016/j.compositesb.2011.12.003.

Marsh, G., 2006. 50 years of reinforced plastic boats. Reinforc Plast 50 (9), 16–19. https://doi.org/10.1016/S0034-3617(06)71125-0.

McCarthy, R., Haines, G., Newley, R., 1994. Polymer composite applications to aerospace equipment. Compos. Manufact. 5 (2), 83±93. https://doi.org/10.1016/0956-7143(94)90059-0.

McIntyre, J.E., 2003. The historical development of polyesters. Mod. Poly.: Chem. Technol. Poly. Copoly. 2, 1.

McMullen, P., 1984. Fibre/Resin composites for aircraft primary structures: a short history, 1936-1984. Composites 15 (3), 222–230. https://doi.org/10.1016/0010-4361(84)90279-9.

Michaud, V., 2016. A review of non-saturated resin flow in liquid composite moulding processes. Trans. Porous Media 115 (3), 581–601. https://doi.org/10.1007/s11242-016-0629-7.

Muscat, I.E. (1950). USPat 2495640A.

Phillips, L.N. (1960). USPat 2952040A.

Phillips, L.N., 1969. Forming processes for carbon fibre and resin. Composites 1 (2), 101–103. https://doi.org/10.1016/0010-4361(69)90008-1.

Pollard H.J., Rees, J. (1958a.1) USPat 2847712A.

Pollard, H.J., Rees, J., 1958b. British Patent Specification, GB790639A, vol. 2.

Potter, K.D., 1979. The influence of accurate stretch data for reinforcements on the production of complex structural mouldings: part 1. Deformation of aligned sheets and fabrics. Composites 10 (3), 161–167. https://doi.org/10.1016/0010-4361(79)90291-X.

Potter, K., 2023. Automated fibre placement. In: Design and Manufacture of Structural Composites. Woodhead Publishing, pp. 125–143. https://doi.org/10.1016/B978-0-12-819160-6.00015-9.

Stewart, A., Fabris, J., Terpstra, C., Shead, M. and Fernlund, G., (2020). Thermal analysis of historical autoclave data using science-based data analytics methods. SAMPE 2020| Virtual Series.

Vaccarella, P.W., Jensen, J.C., Adams, A.A., 1991. Lotus Elan-An RTM Composite Success. https://doi.org/10.33599/nasampe/s.20.0362. SAE Technical Paper.

Wildman, D., 1969. Forming processes for glass fibre and resin – contact moulding. Composites 1 (Issue 1), 35–40. https://doi.org/10.1016/S0010-4361(69)80010-8.

Zympeloudis, E.D., Weaver, P., Potter, K., Kim, B.C., 2015. Manufacturing characteristics of the continuous multi-tow shearing for high-volume manufacturing of tow-steered composites. In: 20th International Conference on Composite Materials (ICCM20). International Conference on Composite Materials, ICCM, p. 4217.

The story of pultrusion

2.2

Eric Moussiaux [1, 2, 3]

[1]Exel Composites Free University of Brussels (VUB); [2]European Composites Industry Association (EuCIA); [3]European Pultrusion Technology Association (EPTA)

1. Introduction

1.1 The pultrusion process

Pultrusion is a very simple one-step continuous process to make composite profiles: pullers pull reinforcements through a bath of resin into a heated tool where the resin is cured. Finished profiles are cut at length by a moving saw at the end. On paper it really is easy (Fig. 2.2.1).

The pultrusion process produces profiles with a constant cross-sectional area, very similar to extrusion of aluminium or plastic profiles. In extrusion, the raw materials are heated above their melting point and pushed through an extrusion die into a cooling bath to fix their new shape. The making of an extrusion profile is a physical transformation. Pultrusion is different in two key aspects. Fibres can't be pushed through the process, hence the need for a pulling mechanism. Most matrices used are thermosetting resins; the profile is shaped in the die by a polymerisation reaction. The making of a thermoset pultrusion profile is a chemical transformation (Meyer, 1985; Starr, 2000).

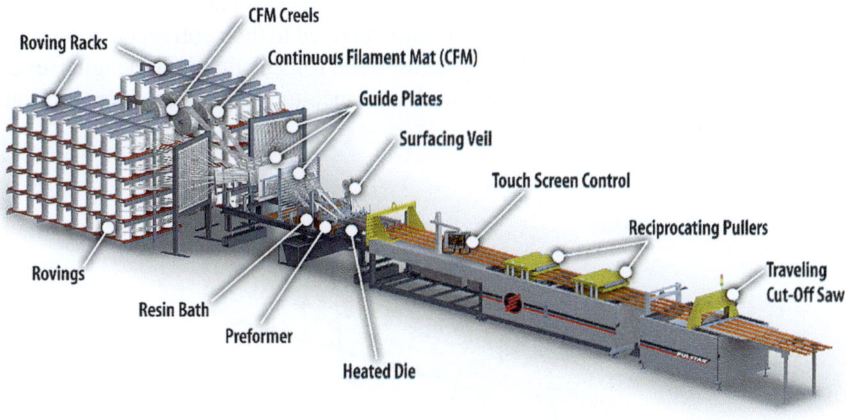

Figure 2.2.1 Schematic view of pultrusion process.
Courtesy Strongwell.

The History of Composites. https://doi.org/10.1016/B978-0-443-34119-9.00013-9

Two types of reinforcements are used in pultrusion: uni-directional and multidirectional. The uni-directional fibres, commonly called rovings, are the basis for the very good mechanical properties along the length of the profile and provide the strength to overcome the sometimes very important pultrusion pulling loads. Multidirectional reinforcements provide the mechanical performance in the transverse or crosswise direction of the profile, or improve the shear performance. Continuous fibre mats, either bonded or stitched together, are mainly used together with rovings to form the traditional mat/roving pultrusion layup (Fig. 2.2.2). Optionally, synthetic veils can be added to create resin-rich layers on the profile surface to protect the fibres from erosion or chemical attack, or simply to improve aesthetics.

Glass is by far the most used fibre; carbon comes in as a good second. Aramids and basalts are very distant thirds. Natural fibres are being investigated everywhere for obvious reasons. Together with filament winding, the pultrusion technology allows for the highest reinforcement contents, up to 80% per volume for example in electrical insulator rods.

Since the very beginning, and still today, unsaturated polyester is the workhorse resin in pultrusion. A wide range of other thermoset resins find their way into specific needs or applications. Vinylesters have better chemical resistance and improved mechanical performances. Epoxies are the primary choice for carbon fibre composites and electrical applications. Polyurethanes excel in toughness. Low viscosity acrylics can be loaded to reach very high fire resistances. Phenolics by far have the best fire properties, provided one can live with their many drawbacks.

Further tailoring of the matrix is achieved by adding fillers (calcium carbonates, kaolin clays, ...), additives for improved fire resistance (aluminium trihydroxides, phosphates, ...), better surface quality (low profiles and low shrinks) ... The "peppers and salts" finish off the resin mix: catalysts, release agents, pigments, ...

Fibres need to be fed into the tool in a smooth continuous way. Bobbins of rovings are stacked on high racks, called creels. The rovings are guided to their right position in the profile in all possible ways: eyelets, multiple holes in plates, tubes, ... Multidirectional reinforcements on rolls are carefully directed to their correct position in the profile via three-dimensional guiding systems. Such rolls having finite lengths, every new layer needs to be connected without disturbance to the process or creating unwanted weak areas in the profiles.

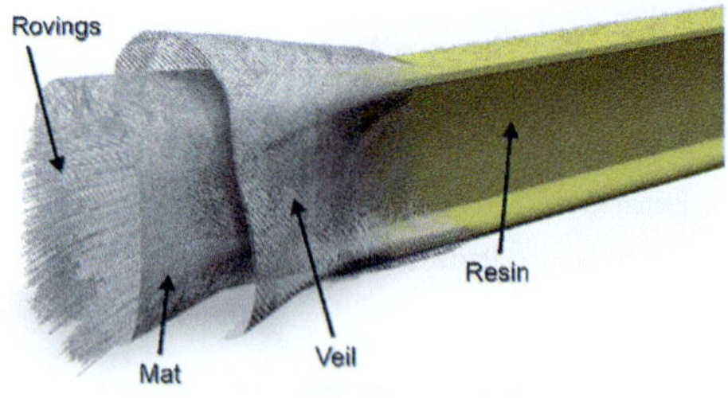

Figure 2.2.2 Reinforcement structure in a pultruded grating bar. Courtesy Strongwell.

Proper fibre guiding is the most critical step in a good continuous pultrusion process. The real technology is behind the scenes, inside experienced brains and CAD computers, but still good ideas can easily be caught by the pultrusion expert's eye. Pultruders are typically very proud and secretive about their guiding technologies.

Reinforcements are impregnated via run-through in a resin bath, via straight injection into the pultrusion die or via any combination thereof (Fig. 2.2.3).

Polymerisation takes place in heated pultrusion dies made of steel (Fig. 2.2.4). As pultruded profiles are normally highly reinforced, friction wear on the tool surfaces is important. Depending on the expected lifetime of a profile, a proper combination of steel type and surface treatment is selected. Traditional chrome-plating is gradually being replaced by better and more ecological solutions. Inside the tool, by a combination of the resin reactivity, the catalysts used in the mix, and the applied heat, resin cure is fast and complete in one step. Pultrusion speeds vary from centimetres per minute for heavy thick-walled parts to metres per minute for small thin-walled profiles.

Figure 2.2.3 Fibre guidings and bath-impregnation.
Courtesy Exel composites.

Figure 2.2.4 Steel pultrusion die.
Courtesy LUM.

Downstream, after exiting the die, fully cured but still hot, the profile must survive the combined pressure and tensile loads of the puller mechanism. Pultrusion machines are either belt pullers or reciprocation (hand-over-hand) clamp systems. At the end of the line, a travelling saw cuts the profile to exact lengths.

Pultrusion is widely acknowledged being the most efficient and cost-effective composites manufacturing technique. The process is indeed continuous and runs 24/7. Speeds can be optimised. Size allowing, several profiles can run side-by-side on one pultrusion line. Double or triple cavity production is very common. Small little profiles can be pulled like spaghetti in tens of cavities. However, pultrusion remains a one-step process: everything needs to be right first time. Optimising "speed & multi-cavity" requires excellent process controls every step of the way (Fig. 2.2.5).

1.2 Key features of composite profiles

Pultruded composites are profiles, linear, two-dimensional elements. Missing the 3D component is the main, and arguably the only real drawback, of the pultrusion process.

Being composites, pultruded profiles share the inherent material properties of their brothers and sisters produced in other ways.

- Good mechanical performance, at the highest reinforcement contents possible
- Light weight
- Excellent thermal properties: insulating, stable at high and low temperatures
- Electrical insulation
- Good weatherability and chemical resistance

Properties can be tailored into successful combinations for a great many of applications. Hybrid reinforcements optimise the weight-stiffness-cost balance. Local conductivity is added to inherently insulating materials. Oil-based matrices are improved to railway fire safety standards.

Figure 2.2.5 Early multicavity production of ladder profiles.
Courtesy Strongwell.

Although pultrusion bears an unjustified reputation of being a simple shape technology, its shape flexibility is surprisingly high. From minute micro pultrusions in medical tools (Fig. 2.2.6), to 2.5m wide Kenworth truck floors with balsawood cores. From large thin-walled bus profiles with excellent surface quality and integrated details (Fig. 2.2.7) to huge multihollow carbon fibre reinforced structural bridge profiles with varying wall thicknesses (Fig. 2.2.8). Moreover, these shapes can be made to surprisingly tight tolerances. And they are easier to handle than competing heavier or more vulnerable materials.

Compared with 3D technologies, pultrusion has the flexibility of length, no need for extra tooling when a different length of a product is needed. Transportation and logistics are the limiting factors when extra long lengths are needed. Size permitting, rods, tubes, or flats can also be spooled over several kilometres of length.

Jeff Martin, one of pultrusion's first marketeers, used to say "to be successful in pultrusion you always need at least three distinctive advantages: mechanical performance, light weight and a corollary[1] benefit" (Goldsworthy and Martin, 1985; Martin and Sumerak, 1984).

2. Technology development

2.1 The inventor (50ies)

Brandt Goldsworthy (1915–2003) is generally recognised as the father of pultrusion, if not just composites all together. As a native Californian, of course Brandt must be involved in developing the first composite surfboard; as a native American, of course he has to build the first composite car parts in the Ford Mustang. Brandt is a great pioneer, a great visionary, a great inventor (with some 50 US patents) ... and a great storyteller.

Brandt remembers an Owens Corning salesperson asking him if he saw some use for this bobbin of glass fibres that were too thin to use in their traditional glass wool insulation: "Hell, we didn't have resin to put around the fibre."

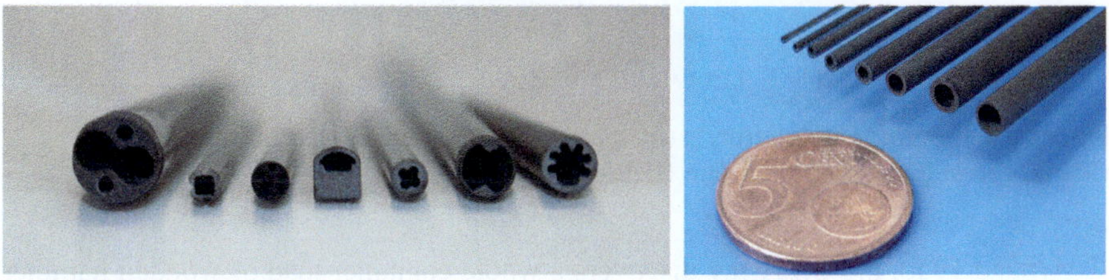

Figure 2.2.6 Micropultrusion, left Polygon (1960s), right DPP (2020s).
Courtesy Polygon, courtesy DPP.

[1] Jeff Martin always used this rather uncommon word meaning "additional."

Figure 2.2.7 Thin-walled (2.5 mm) complex bus profiles.
Courtesy Exel composites.

Figure 2.2.8 Extren carbon and glass fibre reinforced double-webbed beam—36 in x18 in.
Courtesy Strongwell.

Figure 2.2.9 Brandt Goldsworthy's (on the right) first pultrusion machine.
Courtesy Brandt Goldsworthy/Clem Hiel.

Around 1950, the US Army discovers a whole container load of wooden hammer handles has been eaten by termites. They call Brandt Goldsworthy … (Fig. 2.2.9)

On January 13th' 1953, Brandt files his first two patents for what will later be known as "pultrusion" machines (Goldsworthy and Landgraf, 1959; Goldsworthy and Landgraf, 1961; Goldsworthy, 1971). Some years later, inventing radio frequency preheating for pultrusion, he manages to disturb air traffic control over LAX for a couple of days, before being caught red-handed by a mobile radar van.

In 1954, in his laboratory at Swiss electrical company Brown Boveri, and independent of Brandt Goldsworthy, Ernst Kühne develops his version of pultrusion to produce glass fibre-epoxy insulator profiles. Both inventors only meet for the first time in the early seventies.

2.2 The pioneers (60ies–70ies)

Quickly after Goldsworthy's invention, a number of other pioneers enter the American pultrusion scene.

The company now known as Strongwell starts pultruding in Bristol (Virginia) in 1956. Very early on, ladder rail production starts and still continues today. Strongwell produces heavy structural profiles vertically downwards to benefit from gravity (Fig. 2.2.10). Under the John Tickle (owner-president) and Keith Liskey (COO) leadership, Strongwell becomes pultrusion market leader in the eighties and nineties, and is still going strong.

Sam Shobert founds Polygon in 1949 in Indiana to make composites for electrical applications. In 1954 he files a patent application to produce small rods and tubes, pultruding vertically upwards (Shobert, 1961) (Fig. 2.2.11). Factory space is optimized, and the rumour goes that sometimes profiles grow through the roof …

Deep in the Pennsylvanian heartland, Bob Sweet pulls Creative Pultrusions' first profile with a tow-winch on a pick-up truck. It is a triple hollow rectangular box profile for a General Motors storage rack system. The year is 1973.

Ernst Kühne Kunststoffwerk is the first European pultrusion company, founded in 1967. Early seventies, three other still relevant players start in pultrusion: Virgilio Branca with Top Glass in Italy (1972), David Shaw-Steward with machine-maker Pultrex in the UK (1973), and Yrjö Aho with Exel Composites in Finland (1973). Exel Composites, today the most global pultrusion company, starts its ski-pole manufacturing and is the first in Europe to pultrude carbon fibre profiles.

Figure 2.2.10 Vertically downwards pultrusion.
Courtesy Exel composites.

Figure 2.2.11 Vertically upwards pultrusion.
Courtesy Polygon.

2.3 The scientists (80ies–etc.)

Joe Sumerak brings science to the art of pultrusion.

He is the first to analyse and to understand what really goes on inside a pultrusion die. Fibres impregnated with liquid resin enter the cold front end of the pultrusion die. The combination of applied temperature, reactivity, and catalyst action, makes the resin first gel and then cure roughly halfway the die (left side of Fig. 2.2.12). The polymerisation of thermoset resins is an exothermic reaction: once the reaction has started, the profile releases heat toward the die. The right-hand side of Fig. 2.2.12 represents the temperature evolution inside the profile as it travels through the pultrusion die, this is the "pultrusion exotherm." Conjointly on the same graph, the temperature along the length of the die is plotted. The process exotherm lags the die temperature in the first half of the process when the resin is being heated by the tool. The exotherm reaches a peak at the polymerisation, which has to be rapid in pultrusion, and slowly goes down in the remainder of the time in the tool and after its exit.

Joe Sumerak creates the first practical tool to measure the process exotherm, feeding a thermocouple into the reinforcements. It is a very powerful method to diagnose processing problems, to optimize production parameters, to test new resin formulations, heating and cooling technique, etc. Every pultruder should have one. Good knowledge of the process exotherm is as vital to the pultruder, as is blood pressure and pulse to the medical doctor, or a cardiogram to the cardiologist. Joe's early papers are must-reads for every new pultrusion technologist (Sumerak, 1984; Sumerak, 1997).

For 4 decades, Sumerak has stayed at the forefront of development in pultrusion technology, building more powerful simulation tools with increasing computer calculation power as it becomes available (Sumerak, 1994; Sumerak, 2001) (Fig. 2.2.13).

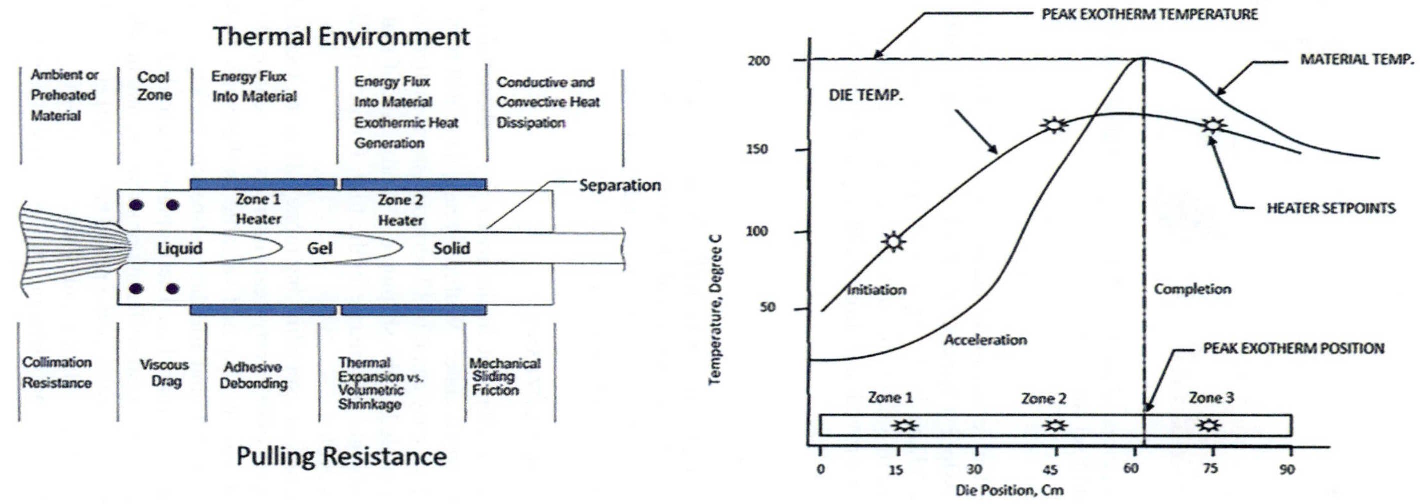

Figure 2.2.12 Pultrusion die dynamics and process exotherm. Courtesy Joe Sumerak.

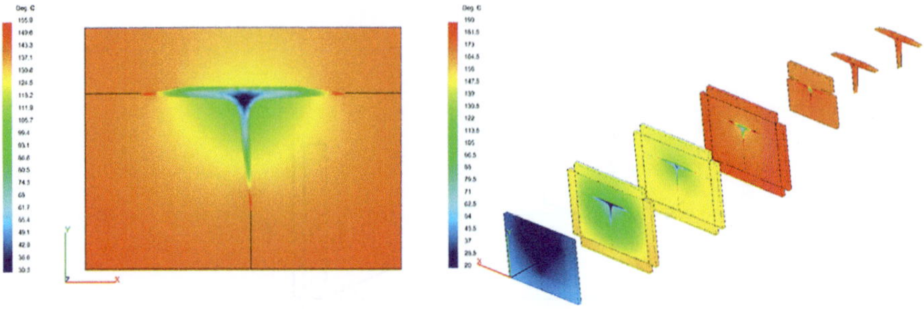

Figure 2.2.13 Numerical process simulation of an unequal-walled T-section profile. Courtesy Joe Sumerak and Exel composites.

Gibson Batch at the University of Minnesota (Batch, 1990) and James Vaughn at the University of Mississippi (Charad et al., 1993; Raper et al., 1998), have been pioneering theoretical modelling in the eighties. In more recent years, the renowned German Fraunhofer Institute has pultrusion research projects going on in several of their labs related to improved injection technology, polyurethanes, … (Strauss, 2019; Behnisch et al., 2024).

2.4 The diversifiers

Whereas the core technology remains basically the same—pulling impregnated fibres through a shaping tool—development and diversification have been happening in every detail aspect of the process: in the materials being pultruded, tooling being used, preprocess technology and postprocess add-ons.

2.4.1 Thermoplastic pultrusion

On the materials side, the elephant in the room is the thermoplastic matrix. Late eighties some industry experts state: "Within 5 years, there will be no thermoset pultrusion left." After all, how easy can it be? Thermoplastic is melted, fibres are impregnated, thermoplastic is cooled in the shaping die, ready. And if that wouldn't work, just mix the components and polymerise in the tool. Profiles could be postformed, over-moulded, easily recycled, …

Big chemical companies are spending big money: Dupont, General Electric, Dow Chemical, … (Moussiaux, 1989). Forty years later, an abundance of technologies has been tried, none has proven as easy and versatile as thermoset pultrusion. What is worse, no market has been found for the cost-benefit package offered. But there may be light at the end of the tunnel … Thermoplastic pultruded strips or rods could be integrated as very effective local stiffening or strengthening in other thermoplastic moulding technologies. Automotive is the prime target market for this new development (Behnisch et al., 2024). Also, Frank Chauzu at CQFD in France should be credited for his relentless thermoplastic pultrusion pioneering work (Chauzu, 2016).

2.4.2 Alternative heating methods

A well-controlled, gradual, and even polymerisation over the whole cross-section is critical for pultruding a quality product at an economical speed. In thick solid parts, the central fibres are not easily heated via the outer surface contact with the die, causing low speeds, internal fibre cracking, or undercured centres. Preheating the impregnated fibres from within is a good solution, provided the right safety measures are implemented of course. For historical reasons, the US pultruders use radio-frequency waves, whereas the European colleagues prefer microwaves.

One niche application, optical fibre cable cores, uses UV-light to cure the small diameter UD rods. No need to pass the impregnated fibres through a heated steel tool. Speeds of over 100 m/min are achieved, one or two orders of magnitude faster than with traditional pultrusion.

2.4.3 Combined technologies

Pultrusion and filament winding are the two composites manufacturing technologies reaching the highest reinforcement contents. Both allow to put the fibres where the stresses are. Both have opposing natural fibre directions: in the longitudinal direction for pultrusion, in the radial direction for filament winding. Pull-winding is a unique technique to combine the best of both worlds. A series of winders with multiple roving bobbins is introduced in the guiding section before entering the die. Pull-wound tubes can be very thin walled, produced to tight tolerances and have spectacular fibre patterns on the surface. Exel Composites is the world's leading expert in this technology (Fig. 2.2.14).

In-line coating methods use the exotherm heat after exiting tool, to solidly cure a very robust layer of paint onto the profile, while avoiding a costly secondary handling. Prime examples are Creative Pultrusion's utility poles and Marvin Composites window profiles (Fig. 2.2.15). Similarly, a thermoplastic external layer can be co-extruded over the outside perimeter.

Adding the third dimension to a linear process technology, has intrigued many pultruder. Everyone can make curved profiles, often not really wanted, adjusting the alignment of fibre guidings, dies, and puller blocks. Doing it consistently and with smaller radii requires more serious changes to the process. German Thomas Technik has pushed those boundaries furthest with its radius-pultrusion (Jansen, 2016).

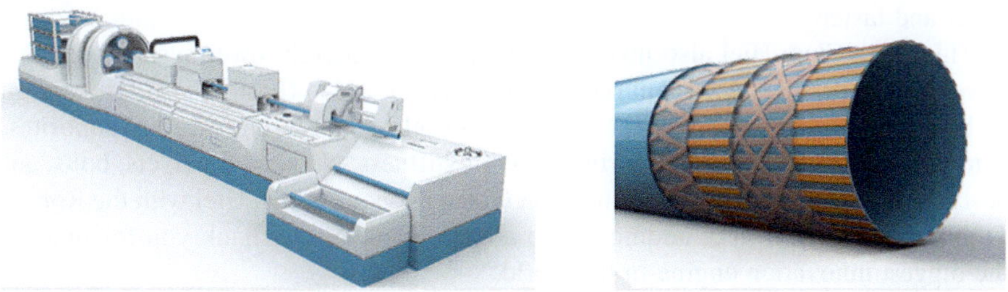

Figure 2.2.14 Pull-winding machine and tube fibre structure.
Courtesy Exel composites.

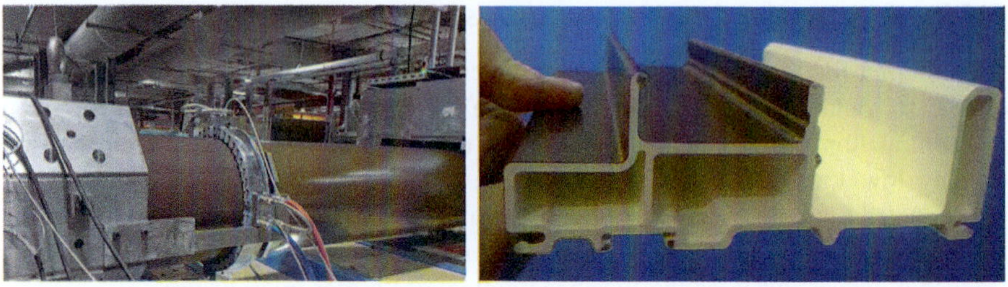

Figure 2.2.15 In-line coating of utility poles (courtesy Creative composites group) and in-line coated window profile.
Courtesy Tecton-Marvin.

3. Historical overview of application developments

3.1 Composite profiles in electrical applications

This chapter describes the author's very personal selection of landmark applications in pultrusion.

Fibreglass ladder rail is the one single product that has brought pultrusion to real industrial scale. The first pultruded fibreglass ladder rail is made as early as 1959 in Bristol (Virginia) by Koppers Company, a Strongwell predecessor, for Louisville Ladders. The breakthrough comes when industry pioneers join forces to write a safety standard for portable ladders. In 1974 the ANSI A14.5 "Code for reinforced plastic portable ladders," is published. It creates a new industry simply stating that to be electrically safe, ladders should be made from nonconductive materials. In the US, these yellow and orange ladders are just everywhere.

In 2003 Belgian professor and innovator Clem Hiel, and his colleague George Korzienowski, file a patent for an "aluminium conductor composite core reinforced cable" for high-voltage power lines (Hiel and Korzeniowski, 2006; Hiel, 2020) (Fig. 2.2.16). In a high-voltage power line, the electricity is transmitted via the aluminium conductors around a central load-bearing core, typically in steel. The very low coefficient of thermal expansion of the carbon fibre reinforced pultruded rod in the core, allows for very stable power lines that don't sag under the heat of ever more loaded power lines. The rods are spooled, multiple kilometres long, and in use all over the world. Total length now surpasses several times the circumference of our planet and is counting faster and faster.

This same Clem Hiel also invents, and realises, a snap-fit tower system with just three, ingeniously designed, pultruded profiles (Hiel, 2017) (Fig. 2.2.17). The design is based on the unique elastic properties of glass fibre reinforced composites, combining high strength and the right level of stiffness. No need for rivets, bolts, adhesives bonds. In January 2017, President Obama honours Clem Hiel with the Ronald Tibbett innovation award, adding: "You are a Dreammaker. Thank you for fuelling the biggest innovation engine in the world."

Figure 2.2.16 traditional steel (*left*) and composite (*right*) cores in aluminium conductors.
Courtesy Clem Hiel.

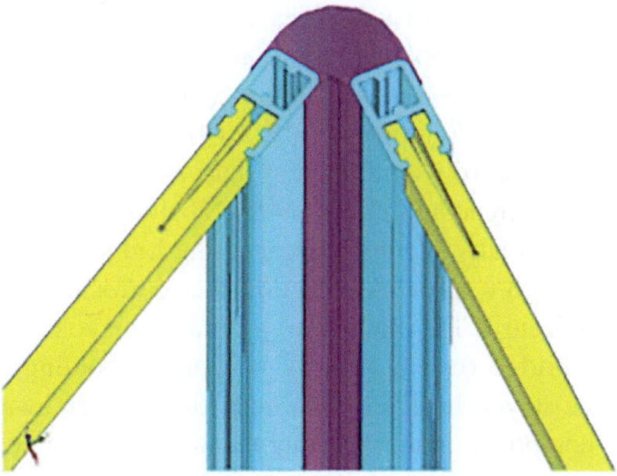

Figure 2.2.17 40 m high pultruded tower and snap-fit principle.
Courtesy Clem Hiel.

3.2 Composite profiles in building and infrastructure

To build something, you need building blocks. To build beam and column structures, you need a range of profiles. Strongwell, again, is the first to develop one in the 1960's: the Extren range of structural pultruded profiles (Fig. 2.2.18). Quickly, guidelines on how to design with them, and especially how to make the connections, are developed. Access structures in chemically aggressive environments are the first

Figure 2.2.18 Original range of Extren structural profiles.
Courtesy Strongwell.

markets, including ladders, handrails, stairs, and pultruded gratings. These applications grow larger and larger, to full structures for cooling towers.

What better way to appeal to a civil engineer than to make a bridge? Thousands of traditional looking truss bridges are installed in remote areas in North America (Winey, 2023). Europe has two real landmark composite bridges in pultrusion. The Aberfeldy Bridge is built in 1992, connecting two halves of a golf course separated by a small river in Scotland (Fig. 2.2.19). The main structure is a triple hollow caisson box, itself made from a modular system with a seven-hollow plank profile as the main element (Burgoyne and Head, 1993). That system is originally developed by UK engineering company Maunsel Engineering Ltd as an enclosure for a large viaduct near Middlesbrough to protect it from further corrosion.

Owner Henrik Thorning rides a tandem bicycle with the local mayor to open his Fibreline bridge over a railway in Danish Kolding in 1997. The 40m long, two-span cable-stayed bridge receives multiple design awards (Fig. 2.2.20).

Two composite manufacturing technologies compete for the growing foot and bicycle bridge market: modular systems with pultruded profiles and one-piece infused

Figure 2.2.19 Aberfeldy bridge and multihollow structural profile.
Courtesy Exel composites.

Figure 2.2.20 Kolding bridge.
Courtesy Fibreline.

bridges. Their very light weights help to save on transport and installation costs, while also reducing the need for heavy foundations. The city of Rotterdam in the Netherlands already has over 100 composite footbridges installed and praises their longevity with minimum maintenance needs (Schutte, 2017).

Pultrusion's wet dream of concrete reinforcements materialises today on the North American continent. For decades the industry has worked together with the American Concrete Institute (ACI), developing all the standards needed for the safe use of composite rebar in heavy-duty structural applications. World-leading glass fibre manufacturer Owens Corning has entered the scene to provide confidence to the road authorities for large-scale application. The leading authority in the field, Professor Antonio Nanni of University of Miami, even becomes the 100th president of the ACI.

The Harkers Island road bridge replacement in hurricane prone North Carolina is the most impressive showcase for composite rebar. The original bridge lasted 60 years in this very harsh environment. The North Carolina Department of Transportation and its contractor Balfour Beatty are determined to make the new bridge a "nonsteel design." The 975 m long bridge is built in 2023–24 and the numbers are dazzling. All concrete piles and girders are pretensioned with Carbon-fibre reinforced polymer strands. The piles alone consume over 100 km of these strands. The two-lane decks are reinforced with more than 200 km of glass fibre rebar produced by nearby Owens Corning (Distefano, 2024) (Fig. 2.2.21).

3.3 Composite profiles in transportation

Automotive is the largest market for composites in transportation but it is difficult for a technology making straight parts to find its spot in a vehicle where virtually nothing is straight. The Dana driveshaft, is the exception proving the point.

Late eighties, Dana Corporation and Strongwell develop a unique product for the General Motors GMT-400 pickup truck. The traditional two-piece steel driveshaft for the rear-wheel drive vehicle, is replaced by a one-piece fibreglass-graphite-aluminium composite, saving 60% on weight, reducing noise and vibrations and saving on maintenance costs. The Spicer Graph Lite driveshaft is pultruded with carbon and glass fibres over an aluminium tube (Strongwell, 1988). The products' life ends when the pick-up trucks changed to front wheel drive (Fig. 2.2.22).

Figure 2.2.21 Harkers Island road bridge replacement, carbon fibre polymer strands and glass fibre rebar.
Courtesy Balfour Beatty.

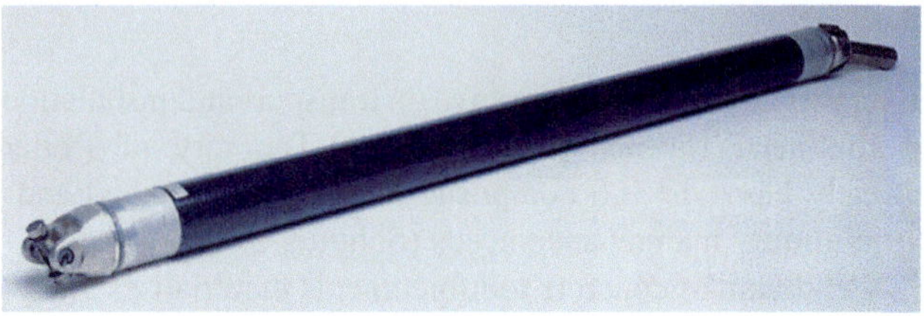

Figure 2.2.22 Dana truck driveshaft.
Courtesy Strongwell.

During the nineties, several attempts are made at producing full containers or trailers with pultruded composites. Other than cars, these vehicles are straight and continuously strive for weight reductions. In 1991 Brandt Goldsworthy wins the grand price of composites, the prestigious Counterpoise Award, for his SMARTcar (Secured Modular Automotive Rail Transport) (Goldsworthy, 1991). The SMARTcar is a container for railway transportation of cars, fully constructed from 11 custom-made pultruded profiles (Fig. 2.2.23). The glassfibre composite sidewalls even allow some light into the otherwise pitch-dark containers. Later Stoughton Trailers applies the same concept for noncorroding marine containers

In Europe, Belgian entrepreneur Jan Verhaeghe first developed his Composittrailer and later his Acrosoma panels. The Composittrailer is fully made from pultruded profiles, saving over 20% on weight. The Acrosoma panel is a 3 m wide foam-cored pultruded panel with the faces stitched together before entering the die, making it the strongest composite panel on the market (Verhaeghe, 2019).

They don't make it to large-scale volumes. The transportation market isn't ready to pay the extra price for the extra benefits. Today, climate change and global emission reduction targets provide a new drive to weight saving in transportation. New kid on the block on the pultrusion scene is Gerret Kalkoffen of German Carbon TT with several developments for large OEM's as Volkswagen and Mercedes (Kalkoffen, 2022).

Bus and train are a different story. Nonstructural, but lightweight and highly functional inner and outer body parts, are commonplace for all European OEM's. Exel Composites develops a European first very complex thin-walled profile for Belgian bus manufacturer Van Hool as early as 1989 (Fig. 2.2.7). It replaces an assembly of seven aluminium profiles with one single pultruded profile. The company adds another first with the full outer body panels for the Bombardier Talent commuter

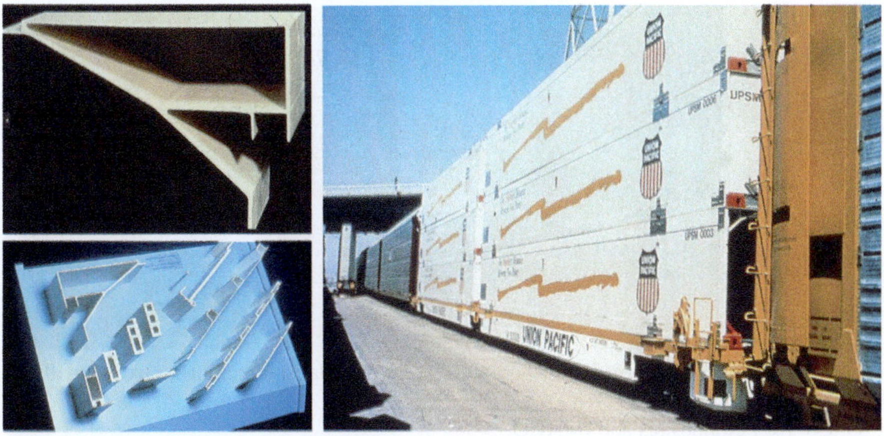

Figure 2.2.23 SMART car, all pultruded railway container for car transportation. Courtesy Brandt Goldsworthy/Clem Hiel.

Figure 2.2.24 Talent trains (2024) and original top corner prototype (1997). Courtesy Exel composites.

trains in 1997. Extreme fire safety requirements make train profiles much more difficult to pultrude. Twenty years later the trains still look as good as new (Moussiaux, 2024) (Fig. 2.2.24).

3.4 Composite profiles at the cutting edge ...

Sports at the Olympic level is the natural place for developments at the cutting edge of technology. What better showcase than winning an Olympic gold medal with your innovative product? This is what Nikolai Bazhukov does in the 1976 Innsbruck Winter Games. He wins the gold medal cross-country skiing with Exel Composites carbon fibre poles (Fig. 2.2.25). Four years later, in Lake Placid (1980), 82% of cross-country skiing medals are won with Exel's poles. As the conical tube is the most efficient shape for a ski-pole, Exel Composites develops a continuous pultrusion-like process with removable conical mandrels. They call it co-winding.

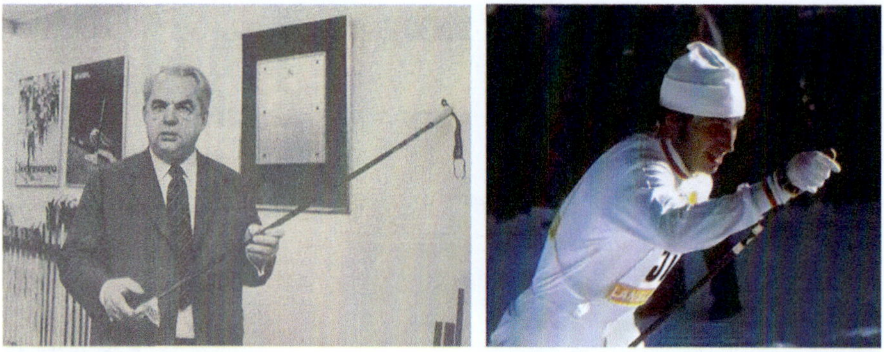

Figure 2.2.25 Exel president Urjö Aho and 1976 gold medalist Nikolai Bashukov© with carbon fibre skipoles.
Courtesy Exel composites and IMAGO.

In 1997, Exel Composites launches a new sport using the same technology: they call it "Nordic walking."

Unidirectional carbon fibre composite is the only material that can carry its own weight into outer space. Carbon fibre composites have the best fatigue properties. What better material for elevator ropes in high buildings? This is what Finnish world-leading elevator manufacturer Kone thinks when developing its Ultrarope together with fellow countrymen Exel Composites (Fig. 2.2.26). The specification is sky-high: keep 2000 MPa strength after five million of combined tensile-bending fatigue cycles. The Ultrarope is used to reach the top of the world's highest building: the Burj Khalifa in Dubai, 828 m high.

Large animals need a strong backbone to carry their weight. So do large wind turbine blades. Wind turbine blades are the single largest market for composites, their structural backbone is the single largest market for pultruded composites worldwide. Wind turbine blades undergo a complex combination of dynamic loads over a very long lifetime of over 25 years. Gravity loading typically acts on the blade in the stiffer direction of the wing shape reaching its maximum in horizontal position, wind loading

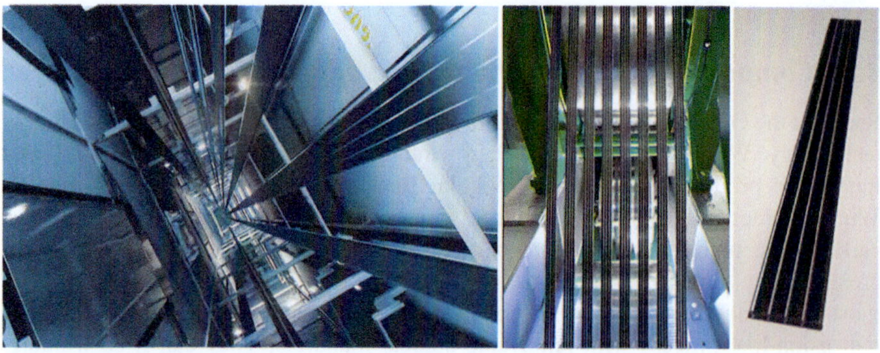

Figure 2.2.26 Kone Ultrarope elevator rope.
Courtesy Kone & Exel composites.

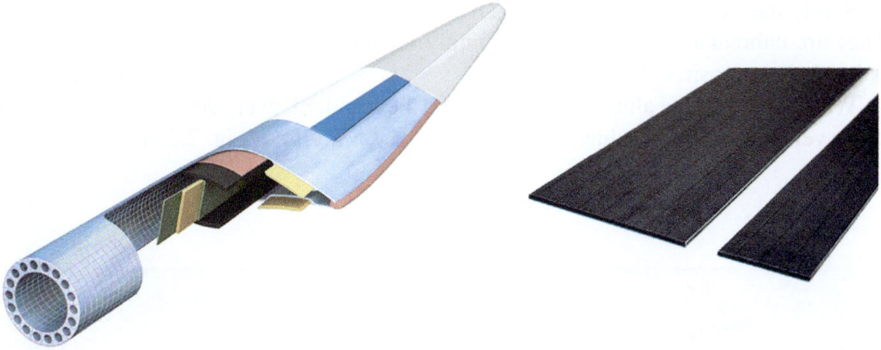

Figure 2.2.27 Wind blade structure (courtesy Gurit) and carbon fibre flats. Courtesy Exel composites.

acts in the weaker direction where the moment of inertia is much smaller. Composites come to the rescue: missing moment of inertia is compensated with local higher modulus. Unidirectionally reinforced carbon fibre flat plates are integrated in the spar cap area where they contribute most to the overall bending stiffness of the blade in the wind direction. Either side of the spar, a double backbone is formed bonding together the flat strips in a three-dimensional shape along the full length of the blade (black zones on the exploded windmill blade view in Fig. 2.2.27). For shorter blades, glass fibre reinforced flats can be selected as a less costly option. Pioneering work started as early as 2000 when Exel Composites' Dr John Hartley developed the wood-carbon fibre composite hybrid spar cap with Vestas UK.

3.5 ... and in every household

Although we don't realise it, we all have pultruded composites at home.

A hammer handle was the first pultruded profile 75 years ago, it still is there now. The most robust working tools for firemen, wood loggers, construction workers ... have a pultruded core inside their plastic handles (Fig. 2.2.28).

Figure 2.2.28 Fibreglass hammer handle & Kim Sjödahl's sledgehammer. Courtesy Exel composites.

Safety flags on children's bicycles stick out on colourful pultruded glassfibre rods. They are unbreakable. A pop-up tent simply wouldn't work without the resilience of pultruded composites (Fig. 2.2.29).

Whereas in Europe aluminium ladders are the standard, every American household has a yellow fibreglass ladder at home. They last a lifetime (Fig. 2.2.30).

Figure 2.2.29 Bicycle safety flags and pop-up tents.

Figure 2.2.30 Fibreglass portable ladders.
Courtesy Strongwell.

4. A sustainable future

4.1 Harmonised solutions via industry cooperation

"Trust your industry associations."

Prague 2002. Keynote speech at the EPTA World Pultrusion Conference. Speaker is Ove Sörensen, ex-CEO of Reynolds Aluminum company. Topic: "what can we learn from the aluminium industry?" (Sörensen, 2002). Associations are needed to safeguard the long-term interests of their industry. They are the unified voice of their members, for industry-wide promotion and education, participation in evolving regulation, development of standards. Companies should be members of their associations and support them with effort and action, if not in kind only. They should have confidence in the long-term benefit and not ask short term return on every euro spent. Sörensen ends his inspiring speech stressing the need for a common identity. Stop with all the different names for our materials and especially all the acronyms (FRP, CFRP, GRP, …). We all make composites !

In Europe, the European Pultrusion Technology Association EPTA is at the service of the pultrusion industry (www.pultruders.com). EPTA itself is a member of the European Composites Industry Association EuCIA (www.eucia.eu). Sister organisations in the US are the Pultrusion Industry Council PIC and the American Composite Manufacturers Association ACMA (www.acmanet.org).

The aforementioned US examples of portable ladders and concrete rebars show the great value of having industry standards toward market penetration. They provide the sound basis for the manufacturers to produce safe solutions in a harmonized way. They provide the confidence the users need to adopt these solutions.

Europe already had the EN 13706, Specifications for pultruded profiles (parts 1–3), published in 2002 after very good industry cooperation in the pultruders workgroup of EuCIA's predecessor GPRMC (European Standard, 2002). That standard included lists of tolerances, acceptable defects, rules of good workmanship. Its main value, however, is the definition of two grades of structural profiles with minimum levels of mechanical performance on the laminate level. These 2 grades, E23 and E17, referring to the overall elastic modulus, have become thé industry standards for specifying pultruded profiles at the worldwide level (except the US) and for every industry.

Standardising a laminate structure conflicts with the very nature of the composites engineer wanting to compose the optimum solution for every application. A construction engineer needs to have reliable values to calculate with for standard profiles he can find on the market, often in smaller quantities. If the volumes for an application are large enough, it will always be possible to raise the bar higher for a customised solution.

Even the aeroplane industry sees the need to standardise on a limited amount of laminate structures. Here, endless possibilities conflict with statistical confidence. It is just too costly to certify new solutions for a product where both lightweight and safety need to be maximised together. The godfather of composites design, Professor Steve Tsai, has responded to this with his new double-double theory in which he proposes a harmonisation to a limited number of different laminate structures (Tsai, 2021).

In the very large construction market, composites only have an estimated share of 1%. The almost simultaneous publication in 2022–23 of two design codes, one on each side of the Atlantic, should be a great lever for deeper penetration. The CEN Technical Specification 19101, "Design of Fibre-Polymer Composites Structures," will become the future Eurocode 11 for composites (European, 2022). The Eurocodes is the suite of codes to which all buildings in Europe need to be designed. The mere existence of a code allowing to design a structure in composites for a 100-year lifetime, establishes composites as a viable construction material for any application. The TS 19101 has a more holistic nature and is applicable to all composite manufacturing techniques. The American Society of Civil Engineers finally approves the ASCE 74-23, "Load and Resistance Factor Design for Pultruded Reinforced Polymer Structures," in May 2023 (American Society of Civil Engineers, 2023) (Fig. 2.2.31). That code is dedicated to pultrusion only and much more specific and detailed.

4.2 *Composites are sustainable and circular …*

The European Composites Industry Association EuCIA summarises the sustainability and circularity of composites perfectly well in Fig. 2.2.32 below.

First of all, composites have a very long use phase, in which they should help to decarbonise the world by being lighter, noncorrosive, thermally and electrically insulating,… proven by Life Cycle Analysis. The very best case is the wind turbine. The large blades can only be made with composites. They pay back their own carbon footprint in less than 1 year for a lifetime of over 25 years. Pultrusion helps providing their backbones via the spar caps as seen earlier.

At the end of the service life, composites can be recycled either in closed or open loops. Recycling is considered closed loop if the inherent material properties are maintained. For composites, this means that the recyclate needs to have a reinforcing effect in the next life: "composites into composites." An abundance of techniques and

Figure 2.2.31 Eric Moussiaux (Exel composites/EuCIA/EPTA) and John Busel (ACMA) celebrating landmark publications at the North American pultrusion Conference (Chicago, 2023).
Courtesy American Society of civil engineers.

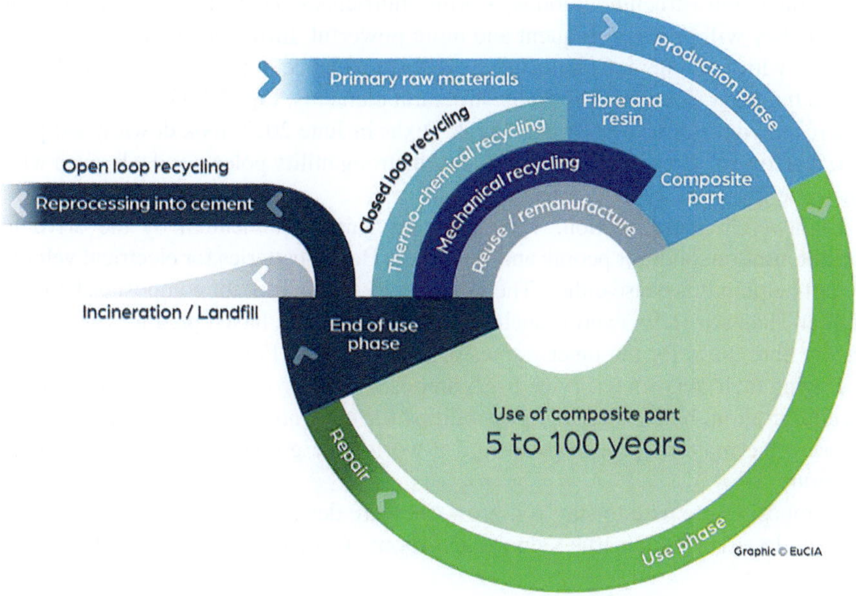

Figure 2.2.32 Composites circularity.
Courtesy EuCIA.

technologies, for thermoplastics and thermosets, are in various stages of development, some are already in production stages. Time will tell which ones will prevail, economy and ecology will decide hand-in-hand, business and green needs will converge.

The main open loop solution for glass fibre reinforced composites is recycling them into cement. At the very high temperatures in the cement ovens, fibres and fillers become very good quality raw materials for cement, replacing virgin minerals. The resin content of the composite helps to fuel the process, reducing the need for primary energy sources (EuCIA and Cembureau, 2023). The LCA for the cement recycling shows that for a typical glass fibre composite about 10% of its own carbon footprint need is returned, reducing the carbon footprint of cement production with some 330 kg per ton of recycled composite (EuCIA, 2024).

4.3 ... and have a strong offer for the green world

The green transition offers new and different opportunities for all materials. Here are just three for pultruded composites.

Energy-efficient buildings. Everywhere a hole needs to be closed in a building, is an opportunity for pultrusion. These holes normally have straight edges. The closings elements need to withstand decades without too much maintenance, and they must be thermally insulating. Plenty of opportunities: windows, doors, curtain walls, greenhouses, skylights, roof products, ...

Resilient infrastructure. Floods, storms, hurricanes, heatwaves, ... with climate change they will be more frequent and more powerful. Infrastructure needs the resilience to withstand them. Composites have this resilience in their very nature. Pultrusion is the main technology for long structural elements. Fig. 2.2.33 shows the point. A derecho windstorm over Bridge Creek Texas in June 2024, took down wood poles and steel towers. Creative Composites' Stormstrong utility poles stood tall, even when hurricane Beryl hit 2 weeks later.

Lightweight transportation. That need has gained momentum by the drive for decarbonisation, both for people and for goods. Heavy batteries for electrical vehicles are not helping this cause either. The sky literally is the limit for composites here. US Lighter Than Air (LTA) and French Flying Whales, both have opted for pull-wound carbon fibre tubes for the inner frames of their airships (Fig. 2.2.34).

During their very long service life, composites are strong enablers for even faster decarbonisation, by their unique possibilities for innovative design of shapes and by tailoring the material properties toward a worldwide growing need for lightweight solutions.

Composites are here to stay and grow, they are designed to perform, and they are made to last. This is our true value proposition: "Composites for sustainable lasting performance."

Figure 2.2.33 Composite utility poles survive a derecho windstorm.
Courtesy RS technologies & Creative composites group.

Figure 2.2.34 Pull-wound carbon fibre tubes for airships.
Courtesy LTA – Flying Whales – Exel composites.

References

American Society of Civil Engineers. 2023. ASCE/SEI 74-23: load and resistance factor design (LRFD) for pultruded fiber reinforced polymer (FRP) structures.

Batch, G., 1990. Analysis and optimization of heater placement in a pultrusion die. Reinforced Plastics/Composites Institute.

Behnisch, F., Janssen, M., Wilhelm, M., 2024. Cost-effective, high-volume and recyclable design for highly loaded components. In: EPTA World Pultrusion Conference.

Burgoyne, C., Head, P., 1993. Aberfeldy bridge, an advanced textile reinforced footbridge. In: Textile Symposium. Frankfurt.

Chauzu, F., 2016. Composites pultruded front crash beam for hyundai. Industrial Process News. Issue 648.

Charad, Y., Roux, J., Vaughn, J., Lackey, E., Hassouneh, A., 1993. Temperature and degree of cure modeling for the pultrusion of graphite-epoxy composites. In: 48th Annual Conference. Reinforced Plastics/Composites Institute.

Distefano, P., 2024. Harkers island bridge replacement project. In: JEC Summit Building and Infrastructure. JEC World.

EuCIA, 2024. Cement Co-processing of end-of-life composites. Press release, European Composites Industry Association.

EuCIA and Cembureau, 2023. Joint contribution of CEMBUREAU and EuCIA to the JRC "recycling" definition project with regard to Co-processing of composite end of life/use material specific to the cement industry. European Composites Industry Association.

European Standard. (2002). EN 13706 – 2002: specifications for pultruded profiles (parts 1 – 3).

European Standard. 2022. Cen TS 19101 - 2022: design of fiber-polymer composite structures.

Goldsworthy B., and Landgraf F. 1961. Apparatus for feeding plastic bars, tubes and like elongated stock. United States Patent US2990091.

Goldsworthy B., and Landgraf F. 1959. Apparatus for producing elongated articles from fiber-reinforced plastic material. United States Patent US2871911.

Goldsworthy B. 1971. "Pultrusion machine and method." United States Patent US3556888.

Goldsworthy, B., 1991. Pultruded composite automobile rail transport system. In: 46th Annual Conference, Reinforced Plastics/Composites Institute. Reinforced Plastics/Composites Institute.

Goldsworthy, B., Martin, J., 1985. Pultruded composites - a blueprint for market penetration: parts A and B. In: 40th Annual Conference, Reinforced Plastics/Composites Institute. Reinforced Plastics/Composites Institute.

Hiel C., Korzeniowski G. 2006. "Aluminum conductor composite core reinforced cable and method of manufacture." United States Patent US 7060326.

Hiel, C., 2017. Two pultruded products and process innovations. In: North American Pultrusion Conference (ACMA).

Hiel, C., 2020. Electric power delivery infrastructure with pultruded composite power cables for increased reliability and reduced carbon emissions. In: EPTA World Pultrusion Conference.

Jansen, K., 2016. Machines for radius, curves and frames. Continuous processing beyond putrusion. In: EPTA World Pultrusion Conference.

Kalkoffen, G., 2022. Lightweight battery cover and chassis. Multiaxial pultrusion for structural automotive composites. In: EPTA World Pultrusion Conference.

Martin, J., Sumerak, J., 1984. Pultruded composites - the case against aluminium extrusions. In: 39th Annual Conference, Reinforced Plastics/Composites Institute. Reinforced Plastics/ Composites Institute.

Meyer, R., 1985. Handbook of Pultrusion Technology. Chapman and Hall.

Moussiaux E. 1989. Thermoplastic pultrusion. internal report Bekaert composites. (Dutch language).

Moussiaux, E., 2024. How composites changed the game for the talent train. Exel Composites press release.

Raper, K., Roux, J., McCarty, T., Vaughn, J., 1998. SPI - vaughn (Univ Mississippi) – "investigation of pressure behavior in a pultrusion die for fiberglass composites.". In: 53th Annual Conference, Reinforced Plastics/Composites Institute.

Shobert S. 1961. Reinforced plastic rods and method of fabricating the same. United States Patent US3007497.

Schutte, W., 2017. FRP composite bridges provide optimal solutions in urban areas. IABSE J. 2017.

Strongwell (1988). Graphite Driveshaft Puts Aerospace Technology on the Road. Strongwell Application Profile., https://www.strongwell.com (accessed 18/11/2025)

Starr., T., 2000. Pultrusion for Engineers. Woodhead Publishing Series in Composites Science and Engineering.

Sörensen, Ö., 2002. Learning from the aluminium industry and your own colleagues. In: In: EPTA World Pultrusion Conference.

Sumerak, J., 1984. Pultrusion process variables and their effect upon manufacturing capability. In: 39th Annual Conference, Reinforced Plastics/Composites Institute. Reinforced Plastics/Composites Institute.

Sumerak J. (1997). Pultrusion process optimization using On-Line and off-line techniques" Composites in Manufacturing Association of the Society of Manufacturing Engineers, , Second Quarter 1997. Vol. 23. No.2.

Sumerak, J., 1994. Pultrusion Die Design Optimization Opportunities Using Thermal Finite Element Analysis Techniques. In: 49th Annual Conference, Reinforced Plastics/Composites Institute. Reinforced Plastics/Composites Institute.

Strauss, S., 2019. Flexibility in PU injection molding. In: North American Pultrusion Conference. ACMA.

Sumerak J. (2001). Influence of Real Die Design Features on Temperature and Cure Predictions from Pultrusion Process Models. Composites Fabricators Association, COMPOSITES. 2001.

Tsai, S., 2021. Double-double : new family of composite laminates. AIAA J. 59.

Verhaeghe J. (2019). Composites in trailers and containers - a personal approach. Composites in Sustainable Mobility Conference, Agoria.

Winey, D., 2023. Discovering the world largest pultruded bridge in Bermuda. In: In: North American Pultrusion Conference (ACMA).

Textile technologies contribution to composites

2.3

Stepan V. Lomov
Composite Materials Group, Department of Materials Engineering, Katholieke Universiteit Leuven (KU Leuven), Leuven, Belgium

1. Introduction

The decades around the turn of the century have seen a fast development of textile composites, in textile technology, composite manufacturing and applications and in design methods, strongly connected to advancements in textile composites mechanics. Started by "borrowing" existing textile technologies of weaving, braiding and knitting, the composites' needs have initiated development of new, highly specialised fibrous structures, such as 3D woven and braided fabrics, non-crimp fabrics, net-shape composite preforms, stitched preforms. These developments created a "pay-back" in textile industry. It received a boost through new knowledge-, value- and investment-intensive requests—an example of a "de-maturation" process (Abernathy and Clark, 1985). The term is used for an innovation-driven revival, which starts new growth from a new, "non-mature" direction, emerging as an answer to new technical developments or customers' demands. It has strongly influenced recovery of the textile industry in Europe and North America in the last decades of the XX century from the deep crisis of the 1960s–1980s (Noor et al., 2008; Noor-Evans et al., 2012).

The present chapter gives a historic overview, on one hand, of the developments of these specific textile products and composites reinforced with them, and on the other hand, of the advancement of textile composites mechanics and design tools for them.

2. The research landscape

2.1 Starting points

In his 1907 patent Baekeland describes "compounding" the phenolic resin with a suitable material such as asbestos fibre, wood fibre, other fibrous material, in this way presenting the composite concept, without mentioning explicitly any reinforcing effect of adding fibres (see Part I of the present book). The textile form of fibrous material was used in the patent application of 1913 by O'Connor (1918), and developed by him together with H. Faber (van Dulken 2002). It is described in the patent as "any suitable fabric, such as paper, muslin or other cloth" impregnated with phenolic resin ("bakelite," invented by L. Baekeland). The invention purpose was claimed to be electric

The History of Composites. https://doi.org/10.1016/B978-0-443-34119-9.00014-0

insulating material. The mechanical applications followed in years 1920, notably the use of phenolic laminate fabric for tough and quiet gears for automotive industry.

However, in 1989 "Handbook of Composites" the chapter on "Woven fabric aerospace structures" (Bailie et al., 1989) stated: "*Any searching review on the literature on modern composite aerospace structures shows it to almost exclusively devoted to unidirectional tape*" and "*fabric composite structures are usually selected [...] on the basis of rather pragmatic reasons such of manufacturing economics, ease of laying up complex geometries and damage containment.*" In spite of these pessimism Bailie further presents an impressive selection of woven glass and carbon reinforced aeronautic parts and concludes that *[fabric composites] "are widely used and there is no doubt that they will continue to be employed in the future, in an expanding range of applications."* These colliding statements catch the moment of the textile reinforcements advent. In the same year 1989 a book "Textile Structured Composites," edited by Chou and Ko (1989), was published, which for several decades stays as a starting point for any research and technology development in the field. The book was a joint endeavour of textile and composite materials scientists, heralding a "marriage" between the centuries-old textile and the few decades-old composites technologies. Its recent namesake (Pastore et al., 2023) revisits the field 30 years later.

In the field of textiles, the preceding decades saw the maturing of "mechanics of flexible fibre assemblies" (Hearle et al., 1969). It connects the mechanical descriptions of fibres, yarns and fabrics, linking scales of microns (fibres), sub-millimetres (yarns) and centimetres (fabrics). Complex regular geometry of twisted/spun yarns and woven/knitted/braided/ ... fabrics and random structure of non-wovens mediates this scale transition. The mechanical behaviour on all scales is highly non-linear and non-reversible, and is governed by principles of minimisation of energy of the fibrous assembly. The ideas of textile mechanics are often based on heuristics and not-too-rigorous mechanics and mathematics, involve specific terminology of textiles and were often looked upon with scorn by hard-boiled specialists in mechanics of solids.[1]

The ultimate aim of the theory for apparel textiles was the translation of a qualitative characterisation of the fabric "handle" into measurable mechanical quantities. These developments started in 1930s with works of Peirce (1930), (1937), (1947) and Womersley (1937), followed after the war in England (J.W.S. Hearle, P. Grosberg, J. Amirbayat, M. Konopasek, G.A.V. Leaf), Australia (R. Postle), and Japan (S. Kawabata), among others, and culminating in the monographs "Structural Mechanics of fibres, Yarns and Fabrics" (Hearle et al., 1969) and "Mechanics of Flexible fibre Assemblies" (Hearle et al., 1980). The textile part of "Textile Structured Composites" is written by authors from this circle.

Textile mechanics provides a description of the reinforcement geometry which is sine *qua non* the initial point for any mechanical modelling. It models mechanical response of dry textiles, but does not proceed to homogenised properties of the solid

[1] The author recalls an episode in early 1990s, when he presented his work on woven fabric mechanics to a well-reputed specialist in theoretical mechanics, and had heard: "An utter nonsense! A thread cannot have bending rigidity by definition!" The expert knew no other threads apart from threads "in Euler's sense".

impregnated material, nor to details of fibre-matrix interactions. In Chou and Ko (1989) the analytical analysis, "bridging" and "mosaic" models, were presented by T.-W. Chou and T. Ishikawa for two-dimensional fabric composites. These models applied a classical laminate analysis to a woven composite locally for elements of warp and weft positioned one above another, and adding to this an estimation of stiffness depreciation due to crimp in places where warp and weft yarns changed their position on face and back of the fabric.

In the book "Mechanics of Textiles and Laminating Composites" (Bogdanovich and Pastore, 1996a) A.E. Bogdanovich and C.M. Pastore presented a systematic treatment of the problem of homogenisation for impregnated textile composites. It summarises the developments of analytical models for woven and braided composites in the previous decade and proposed a framework of micro-meso-macro scale models, which involve homogenisation of unidirectionally (UD) reinforced yarns, discretisation of the woven or braided interlaced yarn structure in a composite volume and calculations of the mechanical response of the composite.

2.2 Growth and maturity

Fig. 2.3.1 shows a classical "S-curve" of growth and maturity of the field on textile composites, expressed as research intensity within the vast composite materials field. It is measured by number of publications indexed in the Science Citation Index (Web of Science, WoS). The number of publications on textile composites (totalling to 40,000) is normalised year-by-year by the number of publications on fibre reinforced

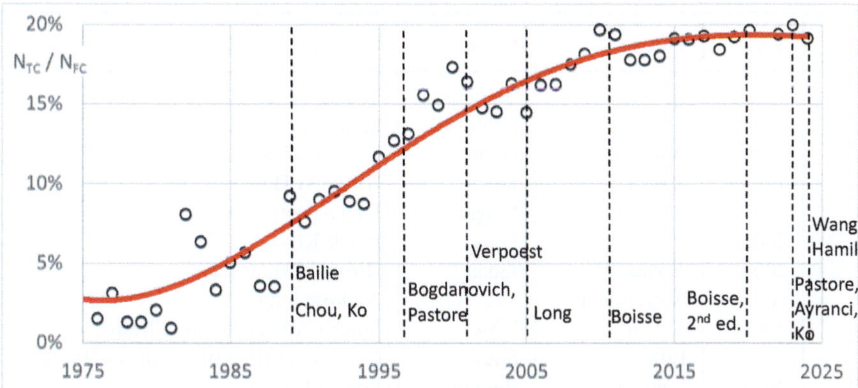

Figure 2.3.1 Publications history on textile composites. The data points are a proportion of papers on textile composites (N_{TC}) in the total number of papers on fibre reinforced composites (N_{FC}), published in the given year. N_{TC} is number of entries in Web of Science, search with composit* and (textile or woven or braid* or fabric or yarn), N_{FC} is number of entries in Web of Science, search with composit* and fib*. Indications of the book publications refer to (Bailie et al., 1989; Chou and Ko, 1989; Bogdanovich and Pastore, 1996a; Verpoest et al., 2000; Long, 2005; Boisse, 2011, 2021; Pastore et al., 2023; Wang et al., 2024).

composites (totalling 220,000). In 2019–2023, each year ~3200 papers were registered by WoS on textile composites and ~16,500 of fibre reinforced composites in general. Publication of important summarising works – books and encyclopedia chapters – is also marked in the graph. The graph reflects the steady growth of the research on textile composites for 3 decades. The growth has stabilised in years 2020, indicating maturity of the field.

The academic and industrial research was supported by a strong international collaboration. One of the vehicles of it is the TexComp series of conferences (International Conference on Textile Composites), started in 1992 in Leuven by I. Verpoest and since then held bi-annually, in alternating locations in Europe, the USA and Japan (Table 2.3.1). The locations of the TexComp conferences give a good idea of the international spread of the research; one should add to the list in Table 2.3.1 more universities in UK, EU and USA, but also in Australia, Canada, China, Korea and New Zealand. Apart from TexComp, the topic was represented in these years in special sessions in every International/European Conference on Composite Materials (ICCM and ECCM); textile composites are in the centre of attention in annual ESAFORM (European Conference on Materials Forming) and bi-annual FPCM (Flow Processes in Composite Materials) series.

Table 2.3.1 TexComp conferences.

#	Year	Place	Country	Chairs
1	1992	Leuven	Belgium	Ignaas Verpoest
2	1994	Leuven	Belgium	Ignaas Verpoest, Frank Ko
3	1996	Aachen	Germany	Thomas Gries
4	1998	Kyoto	Japan	Hiroyuki Hamada
5	2000	Leuven	Belgium	Ignaas Verpoest
6	2002	Philadelphia	USA	Frank Ko
7	2004	Yonezawa	Japan	Hiroyuki Hamada
8	2006	Nottingham	UK	Andrew long
9	2008	Newark	USA	Suresh Advani, John Gillespie
10	2010	Lille	France	Christophe Binetruy, Francois Boussu
11	2013	Leuven	Belgium	Stepan Lomov
12	2015	Raleigh	USA	Alex Bogdanovich
13	2018	Milano	Italy	Valter Carvelli
14	2022	Kyoto	Japan	Asami Nakai, Tetsusei Kurashiki
15	2024	Leuven	Belgium	Yentl Swolfs, Mahoor Mehdikhani, Stepan Lomov, Steven Hallett
16[a]	2026	Bristol	UK	Steven Hallett, Yasmine Abdin
17[a]	2028	Vancouver	Canada	Yasmine Abdin, Louis Lebel

[a]as planned.

3. Textile reinforcements

Textile composite reinforcements are only a small part in a broad scope of technical textiles. A non-exhaustive classification, used by Techtextil (bi-annual Messe Frankfurt exhibition), includes 12 application areas, nicknamed Agrotech, Buildtech, Clothtech, Geotech, Hometech, Indutech, Medtech, Mobiltech, Oekotech, Packtech, Protech, and Sporttech (Rasheed et al., 2020). Composites reinforcements do not constitute a separate category, but are included in the corresponding application "niche," for example, reinforced concrete in Buildtech, carbon fibre composites for aviation and automotive in Mobitech etc. Specific needs of composite materials gave rise to several specific textile technologies.

3.1 Non-crimp fabrics

A non-crimp fabric (NCF) is a textile industry answer to a long-standing challenge faced by designers of composite parts: how one combines a perfect placement of the reinforcing fibres with easy, inexpensive, automated manufacturing of the part. In other words, how to create a reinforcement which combines straightness of the fibres with integrity, ease of handling, and drapability of textile fabrics.

Describing NCFs, two basically different textile processes have to be differentiated:

- The weft insertion is synchronised with the mesh-forming warp-system, as is applied in the MALIMO, 0°/90 degrees rachel machine configuration, and in the multiaxial carousel machine by Karl Mayer), and
- The feeding of the reinforcement and weft insertion is not synchronised with the mesh-forming stitching warp system, so the needles stich into the roving's (multi-axial warp knitting).

The initial invention, which has led to advent of NCF reinforcements, served, however, the needs of apparel, household and other non-technical fabrics production in after-the-war Germany. In 1949 a textile dyeing engineer Heinrich Mauersberger (1909–1982, Fig. 2.3.2a) in Limbach-Oberfrohna, not far from Chemnitz, has invented *Malimo*, or *stitch-bonded fabrics* (*Nähgewirkte variable Gelege*) (Mauersberger and Kemter, 1961). The name *Malimo* combines *Ma-* from the inventor's name, and *-LimO* from the name of his home town. As a legend goes, his wife gave him an idea. She was using a sewing machine to mend a piece of underwear by sewing rows of stitches up and down and across the damaged section – a process that could be used to produce textile substrates at the speed of a sewing or knitting machine. The basic idea of Malimo is that chain-stitch seams (or knitting loops) are used for the production of fabrics by interlinking loose filling threads or drawn parallel weft threads. Intersecting weft threads are routed with a guide rail and connected with chain-stitch seams. Threads that are fed in the process direction to the stitching unit can be fixed through a "zig-zag" chain stitch (trikot knitting pattern).

Mauersberger was granted a DDR patent 8194, February 03, 1949 (U S. Patent 2,890,579, June 16, 1954). Series production of Malimo machines began in 1957 at

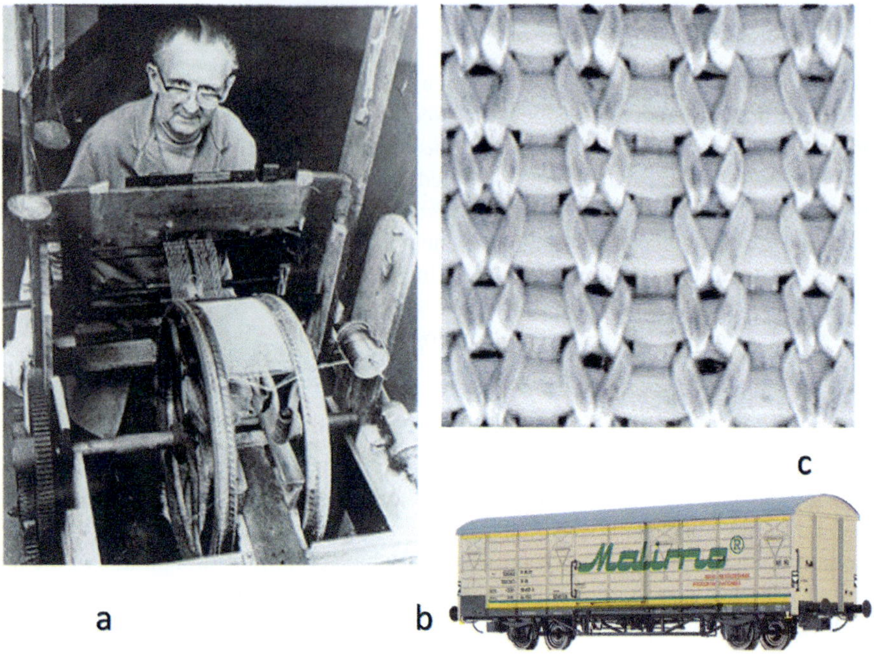

Figure 2.3.2 Malimo: (a) Heinrich Mauersberger and his invention, 1950; (b) Malimo-toy; (c) stitch-bonded biaxial glass fibre fabric.
(a) Reproduced with permission Reindl (2011).

VEB Tüllmaschinenbau Karl-Marx-Stadt (Chemnitz), which was renamed VEB Nähwirkmaschinenbau MALIMO in 1963. In 1992 it was taken over by Karl Mayer to became KARL MAYER MALIMO Textilmaschinenfabrik.

The Malimo technology, producing fabrics faster than weaving by orders of magnitude, became a showpiece of the DDR industry in the 1960s–1980s and part of its mass culture (Fig. 2.3.2b). Malimo products could be found in almost any private household and public space in the DDR. They soon also became used for producing technical textiles, such as backing substrates for PVC coatings, car headliner and insulating materials. The 1000th stitch-bonding machine was delivered in 1972. Mauersberger was awarded *Hero of Labor* title of DDR (1963), and was made an honorary member of the Textile Institute of Manchester (1979).

Uniaxial, biaxial and multiaxial Malimo fabrics with glass, carbon, aramid insertions were widely used as composite reinforcements. In the 1990s they become a standard material for production of wind turbine blades. Often in the description of the material just "glass UD" is mentioned instead of the actually used stitch-bonded fabric (for example, in fatigue databases (Vallons et al., 2015)).

In a Malimo fabric (Fig. 2.3.2c) the position of the needle puncture should be between two weft threads. This prevents inflicting damage to the load-carrying tows by the stitching needles, but creates resin-rich areas in the composite, which limits the overall fibre volume fraction and favours local damage development. The same is true for UD, bi- and multi-axial Malimo fabrics.

The concept of "non-course wise weft insertion" (Schnabel et al., 2011) uses fibrous multiaxial plies, created by parallel weft insertion of fibre tows, which are placed continuously and stitch-bonded (warp-knitted) together, with needles piercing the plies at positions which are not related to the tow boundaries. Standard CSN EN 13,473-1 "Reinforcement - Specifications for multiaxial multiply fabrics" (ENStandards 2000) defines a "multiaxial multiply warp knitted fabric" as

> *"a textile structure constructed out of one or more laid parallel non-crimped not-woven thread plies with the possibility of different orientations, different thread densities of the single thread plies and possible integration of fibre fleeces, films, foams or other materials, fixed by loop systems or by chemical binding systems. The threads can be oriented parallel or alternating crosswise. These products can be produced on machines with insertion devices (parallel-weft or cross-weft) and warp knitting machines or chemical binding systems.*

The scheme of the production process is shown in Fig. 2.3.3a and b. The multiaxial layup of UD layers is created with a weft insertion system and stitch-bonded by a warp-knitting needle bed. The pattern of the warp-knitting (Fig. 2.3.3c and d) is chosen according to requirements related to the drapability of the fabric.

When an NCF is produced, the needles pierce the layup, creating distortions of the parallel fibres, in a form of "channels" (Fig. 2.3.3c) or "cracks" (Lomov et al., 2002), also called "fish eyes" (Colin et al., 2020). The distortions constitute deviations from the ideal fibre direction and cause a "knock-down" of stiffness and strength of NCF composites in comparison with UD laminates by 10%–15% and earlier damage initiation (Truong Chi et al., 2005; Mesquita et al., 2022).

The technology was developed in LIBA Maschinenfabrik in Naila (Bavaria), Germany. LIBA was a family-owned company, founded by Karlheinz Liebrandt in 1945, just after the war, which developed and produced knitting machines. In 1980s it started work on the multi-axial multiply warp knitting equipment, the development which has led after 2 decades to a technology, used for production of NCF reinforcements till now. The technology was nicknamed PARAMAX (Parallel [weft insertion] Multiaxial), Fig. 2.3.3a, and was promoted through the International PARAMAX Committee, based in LIBA and presided during the peak years of development by Christian Wienands. By mid 1990s the technology was mature enough (Hoersting 1998). It was finalised with the key patent "Multiaxial machine with portal structure" (Hoersting et al., 1999), Fig. 2.3.3b, authored by Karlheinz Hoersting, Christian Wienands, and Michael Unglaub.

The multiaxial multiply warp knitting process was implemented in commercially available machines of LIBA as from 1987 (Hoersting, 1998) and was fully realised in early 2000s in the COPCENTRA MAX 3 machine (LIBA, 2012a), suitable for processing glass fibre roving's and reaching an impressive production speed of 100 m/h. Freedom offered by the portal and the layer carriage systems makes possible realisation of layer angles from ±20 degrees–90 degrees relative to the production direction,

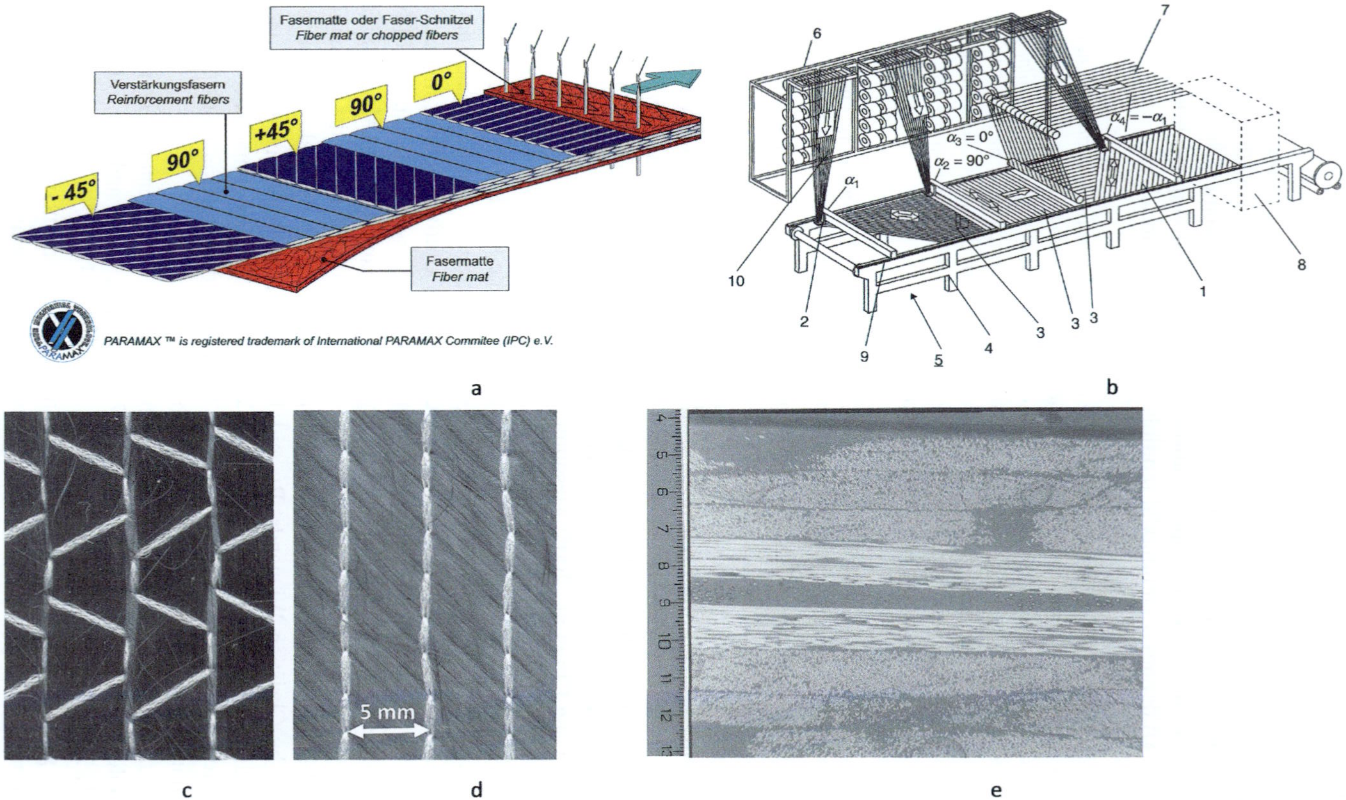

Figure 2.3.3 Multiaxial multiply warp-knitted fabrics: (a) a scheme of the production process; (b) weft-insertion system, patented in DE719726831 C2 (Hoersting et al., 1999); (c and d) face (c) and back (d) of a carbon fibre fabric [C, 156, 0/C, 156, +45/C, 156, 90/C, 156, −45,//PES, 12], as described in Lomov et al. (2002); (e) cross-section of the same fabric.
(a) Reproduced with permission from Elsevier Schnabel et al. (2011). (c and d) reproduced with permission from Elsevier Lomov et al. (2002).

together with 0° warp. Further improvement, realised in the COPCENTRA MAX 5 machine (LIBA, 2012b) follows the LIBA patent DE 10,214,140 (2003) and allows working with brittle carbon fibres. Before being placed, the carbon tows are spread into unidirectional tapes. The tapes are pulled out from the spreading unit or coils, then cut to a predefined length and placed into the transport chain with a clamping device. This principle is also utilised in a Hexcel patent (Bombard et al., 2000). An important element of warp-knitting is the "walking needle." The puncture of the fabric through the compound needles and the simultaneous movement of the fabric can lead to damage and deflection of the fibres. In the walking needle system, an additional horizontal movement—in the direction of production—during the vertical stroke, is carried out in order to reduce the fibre distortion by the needles.

In 2014 Karl Mayer has acquired LIBA Maschinenfabrik, changing its name to Karl Mayer LIBA Textilmaschinenfabrik GmbH, which was further merged with Karl Mayer MALIMO, to be called Karl Mayer Technische Textilien GmbH.

The production of NCFs, their microstructure (Fig. 2.3.3e), behaviour during composites manufacturing (formability, permeability) and mechanical performance was intensively studied in the first decade of XXI century. Among other institutions, the research concentrated in:

Belgium: KU Leuven (S.V. Lomov, I. Verpoest);

Denmark: DTU (L. Mikkelsen);

France: INSA Lyon (Ph. Boisse)

Germany: ITA RWTH Aachen (T. Gries), TU Dresden (C. Cherif, M. Gude), University of Stuttgart (A. Pickett), IVW (P. Mitschang, N. Himmel);

Sweden: TU Lulea (T. Lundstrom, M. Nordlund, J. Varna), Swerea SICOMP (L. Asp, E. Marklund);

UK: University of Nottingham (A. Endruweit, P. Harrison, A. Long, N. Warrior).

The results of the years 2000s were summarised in the book (Lomov, 2011). In the last decade steady 50–80 papers on NCF composites per year are indexed in Web of Science.

LIBA/Karl Mayer machines are used for the production of NCFs by a number of companies, with the NCF market size valued at USD 10.5 Billion in 2023 (Verified-MarketReports, 2024). The main NCF producers are HENGSHEN, NMG Composites, PGTEX, Sinoma Jinjing fibreglass (China), Sigmatex, Vitrulan Group (Finland), Karl Mayer, R&G, SAERTEX, SGL Carbon (Germany), Chomarat, Hexcel, Owens Corning (France), Teijin, Toray (Japan), UMATEX (Russia), Formosa Taffeta (Taiwan), SKAPS Industries, Vectorply (USA).

Applications of NCF composites in aeronautic industry started in the years 2000 with large parts as the pressure bulkhead of Airbus A380 (Middendorf et al., 2011), first flight 2005 (Fig. 2.3.4a), a cargo door of the A400M (2009), the wing of the A220 (2013), wing skins and spars for the Learjet 85 business jet (2014), and smaller parts such as diaphragms, side shells and straps. For helicopter composite airframes smaller components made of carbon NCF are also in production (Dumont et al., 2011; Middendorf et al., 2011).

The development of NCF applications in automotive industry started in early 2000s with such collaborative projects as TECABS in Europe (Fig. 2.3.4b) and

a b

c d

Figure 2.3.4 Applications of NCF: (a) Draping of an NCF sheet over a mould for A380 pressure bulkhead; (b) a team of European TECABS project (2005) demonstrates the floor pan for Volkswagen Lupo (inset), 2005; (c and d) draping of glass fibre UD NCF over a mould for a wind turbine blade.
(a) Reproduced with permission from Elsevier Middendorf et al. (2011); (c and d) Courtesy of T.A. Andersen

NEDO in Japan (Sköck-Hartmann et al., 2011). NCFs are applied in batch production by BMW in the M3-car roof (2020), the M6-roof carline (2012), and the BIW-monocoque of the electric vehicle i3 (2013). In such a body-in-white design, a multi-material concept is applied and NCF composites are used alongside with other types of composite materials and metals.

Glass fibre NCF is a standard reinforcement type for wind turbines, with well-automated production of gigantic blades by resin infusion (Adolphs et al., 2011; Andersen 2018), Fig. 2.3.4c and d. Rotor blade technology for these turbines has evolved in the 40 years since the "energy crisis" of the 1970s and now can be associated more with lightweight aeronautical engineering than with conventional mechanical engineering. NCF composites are used both in blades and in the nacelle. In the blade, a substantial part of the laminate is build-up by UD and triaxial NCF fabrics. In common designs, the UD NCF portion is 50%–60% of the NCF fabrics. Triaxials are reduced in favour of UD and biaxial ±45 degrees fabrics in bigger blades; 0/90 biaxial NCFs are used in nacelles.

More applications of NCF composites include ground transportation (trucks, trains), marine, civil and road construction, sport goods and others.

3.2 Braided reinforcements

Handmade braids have been discovered at Peruvian burial sites dating as far back as 8600–5780 BCE (ropes, bags). Braids were used in the Egyptian, Huari, Chavin and Japanese cultures for both functional and decorative uses; the Peruvian and Japanese are the most widely studied and mimicked (Rana and Fanqueiro, 2015).

Textile Terms and Definitions (Oxenham and Price, 2012) defines braiding as "*The process of interlacing three or more threads in such a way that they cross one another in diagonal formation. Flat, tubular or solid constructions may be formed in this way.*" Braids can be linear products (ropes), curved or plane shell or solid structures with constant or variable cross-section. In their use as composites reinforcements, braids are classified as 2D or 3D, according to the local fabric structure rather than dimensionality of the product. Interestingly, in the early literature and patents term "weaving" may be used instead of "braiding" (for example (Stover et al., 1971; Florentine 1982)).

2D braids are produced as tubular or flat fabrics. The tubular braid can be used as such or pressed to a two layer flat product. The machines for producing braids are called often *Maypole braiders*, as the movement of the carriers (holding bobbins with yarn) is organised in two groups, which move in opposite direction. Their trajectories when two oncoming bobbins meet resemble a merry dance around a maypole with coloured ribbons wrapped around the pole as the dance proceeds. The bobbin movement (Fig. 2.3.5a) is determined by a set of horn gears: rotating discs (hence the machine is also called rotary braider) with notches cut in the periphery. These notches hold and push the carrier during the motion around the track. When a carrier reaches a track intersection, it is forced by the shape of the track to transfer from the notch of one horn gear to the notch of the next one. The horn gears are often arranged on a ring. The periodically repeated transfer process is in the heart of any rotary braiding process. The yarns, taken from the bobbins, interlace directly or are laid on a mandrel, which defines a shape of the produced braid.

The reader is referred to a concise overview of braiding technology in (Branscomb et al., 2013), an overview of applications in (Rana and Fanqueiro, 2015; Carey 2017), and fundamental descriptions of braiding technology in (Ko, 1987; Ko et al., 1990; Potluri et al., 2012a; Kyosev, 2014, 2016; Bogdanovich and Kyosev, 2016). A historical overview of the pre-1950s braiding technology can be found in (Brunnschweiler 1953).

The development of 2D and 3D braided composites, together with textile composites of complex, 3D architecture, was boosted by governmental funding in the 1980s–1990s in the USA, Europe, and Japan.

Examples in the USA are the NASA ACT program (Dow and Dexter, 1997) and the ATCAS (Advanced Technology Composite Aircraft Structures) of Boeing (Fedro and Willden, 1992). In 2D braiding, an important development was the three-axial carbon braided fabric with yarns in [0°, ±60°] orientation, which produces a

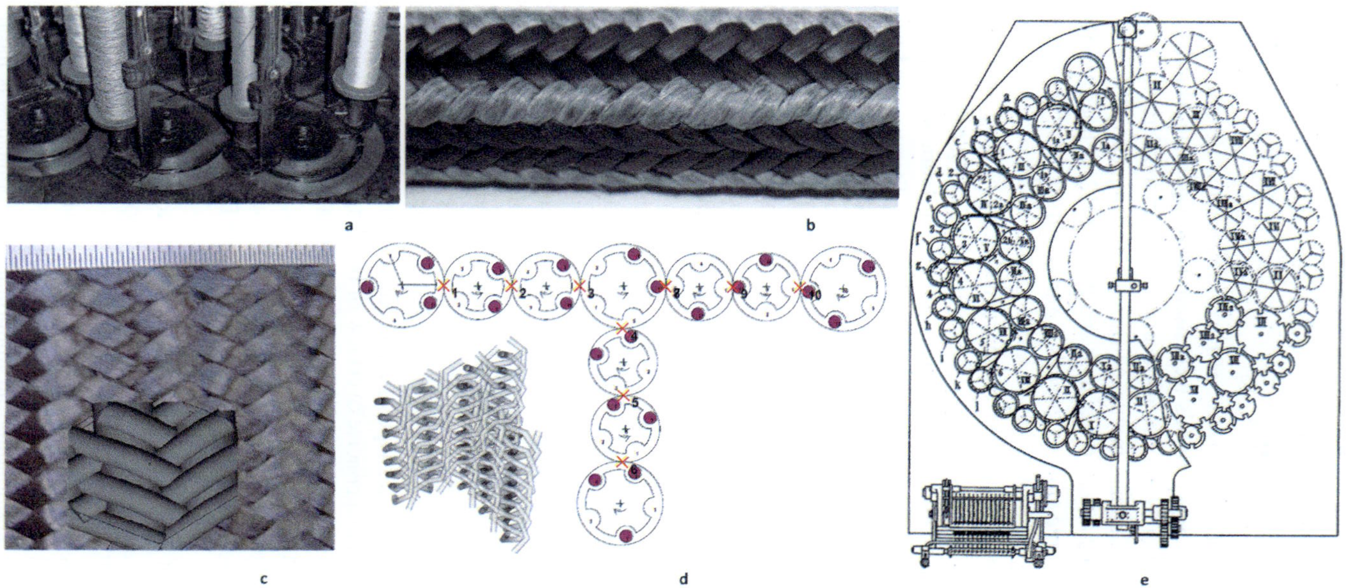

Figure 2.3.5 Rotary braiding and braided composites: (a) Historical machine with sliding carriers: The feet of the carriers slide over the plate following the track; (b) a solid 3D braid; (c) three-axial carbon fibre braid with an overlaid WiseTex geometrical model (Lomov and Boisse, 2021); (d) 3D braid with a scheme of the horngears and a TexMind model; (e) maypole braider for 3D braiding, patent US894022 (Lepperhof 1908). (a) Courtesy Y. Kyosev, adapted from Kyosev (2014); (b) Courtesy Y. Kyosev; (d) Courtesy Y. Kyosev, adapted from Kyosevet al. (2018).

quasi-isotropic composite (Fig. 2.3.5c). The immediately identified drawback of existing 3D textile technologies was the low in-plane strength of these high damage tolerance materials. These early investigations of aircraft fuselage and wing structures by Boeing, Lockheed Martin, Northrop Grumman and McDonnell Douglas Aerospace showed that braids are extremely useful in special applications such as window belts, curved fuselage frames and wing stiffening elements, where flexibility and damage tolerance are essential. The first open publications appeared in 1984 (as indexed in the Science Citation Index); the author of these papers, Frank Ko (Drexel University) was, in the following years, instrumental in the advent of braided composites to the aerospace industry in the USA and worldwide (Ko, 1987, 1990; Chou and Ko, 1989).

In EU Framework Program for Research and Innovation several large projects we dedicated to development of textile and multiaxial composites technologies, design, manufacturing and applications, which included braiding. Examples are MultexComp (1996–99) "Multiaxial textile preforms for complex structural composite parts" and ITOOL (2005–08) "Integrated tool for simulation of textile composites." Important research centres were located in Germany (K. Drechsler, P. Middendorf, Daimler-Crysler, EADS, Stuttgart University, TU Munich; P. Mitschang, IVW, Kaiserslautern; T. Gries, ITA, Aachen; Herzog since 2003), the Netherlands (A. Voscamp, S. Voskamp, Eurocarbon, since 1992; R. Akkerman, U Twente) and UK (P. Potluri, Manchester University; National Composites Center, Bristol), among others.

In Japan the research, starting in 1990s, was centred in Kyoto Institute of Technology (H. Hamada, A. Nakai). In China, also starting in 1990s, the most intensive research in braiding composites was carried on in Donghua University (Baozhong Sun, Bo-Hong Gu).

The maypole machines were further modified for creation of thick tubular braids with varying diameter and cross-section shape. This includes:

(1) Multi-layer interlocked braids: generalisations of conventional 2D Maypole braiding by adjoining two or more "belts" of horngears with a track switch mechanisms used to transfer carriers from one belt to another, creating quasi-laminar braided structures in which distinct parallel planes of 2D braided architectures are interconnected (Brookstein, 1991; Bigaud et al., 2005; Bogdanovich and Kyosev, 2016) reviews a number of specific patented realisations of this principle over the XX-th century, which can be traced down to 1914 (Cobb 1914a,b).
(2) Multistep 2D overbraiding: a pre-selected number of layers is braided over and over the mandrel to obtain the desired wall thickness/areal weight (Grave et al., 2009), pioneered by Eurocarbon since 1994 (Eurocarbon 2024);
(3) Formation of complex hollow profiles with varying cross-section shapes through 2D braiding process with controlled braiding angle.

If the bobbins in a rotary braiding machine move in complex interlacing trajectories rather than circularly, and horn gears are arranged on a plane rather than along a ring, a 3D braid can be produced (Fig. 2.3.5d and e).

Another type of machines can also be used for 3D braiding: "row-and-column," or "Cartesian" braiding. In this case the bobbins are placed in a rectangular grid of positions and get moved in columns or rows with characteristic four-step or two-step movements of either magnetic-driven actuators or planks oriented in two orthogonal positions. The reader is referred to a detailed historical discussion of 3D braiding in the review of Bogdanovich and Kyosev (2016), who worked in 3Tex Inc and North–Carolina State University and was in the centre of the developments of 3D reinforcements in general and 3D braiding in particular during the 1990s–2010s.

Examples of braided composites application in the 1980s–1990s[2] include:

(1) Aerospace: high temperature resistant structures – re-entry vehicles, high temperature heat seals (Ko et al., 1994), turbine rotors (Tan et al., 1991);
(2) Aviation: Boeing 787 aircraft structures, braided composite stiffener for fuselage (Fedro and Willden, 1992), NH-80 helicopter braiding gear (Eurocarbon 2021), net-shape preforms and reinforcements (Weimer et al., 2000a,b);
(3) Light weight electric vehicles: Solectria SunriseTM EV (Ko and Shishoo, 2008);
(4) Sporting goods: 2D braided composite mountain bike frame, China Bicycle, Shenzhen, China, 1992;
(5) Medical devices: artificial ligaments (Laurencin et al., 2001), bio-glass/ceramic braided bone prosthetic implants (Marcolongo et al., 1993).

Fig. 2.3.6 summarises the timeline of 3D braiding according to (Bogdanovich and Kyosev, 2016). Two time-clusters for development of Cartesian and rotary braiding technologies can be seen: 1970s–1990s and 1990s–2000s, respectively. The 3D braiding technology development was largely finished in the years 2000. Newer research in universities and research centres deals with applications and optimisation of braided structures and composite properties (Table 2.3.2).

Forty years after the introduction of braided reinforcements in composites, in the mid-2020s, braiders for composite reinforcement can be purchased off-shelf or ordered on demand at companies such as A&P Technology (USA), Eurocarbon (The Netherlands), Herzog (Germany), Spiraltex (France), Steeger (USA), Muratec, Kokubun (Japan) and T4L (Korea).

The main players in the global braided composites market are identified (MaximizeMarketResearch, 2024) as Exelis Inc., Airbus, Boeing, BMW Group, GKN Aerospace Services Ltd, Highland Industries Inc., Munich Composites GmbH, Revolution Composites LLC, Sigma Precision Components Ltd. (EU and UK), GE Aviation, Albany International Corp, Bally Ribbon Mills (USA) with the majority of applications in the automotive sector (~50%), followed by aerospace, defence and sporting goods. The producers of braided composites in China include Nanjing fibreglass Research & Design Institute Co. Ltd; Beijing Glass Research Institute Co., Ltd; Tianniao High Technology, Tianjin Gongda Aerotech Composite Materials Co.,Ltd, Yunlu Composite Materials (Shanghai) Co., Ltd, Yixing Huaheng High Performance textile CO, Ltd with the main users China Aerospace Science and Technology Corporation (CASC); Aviation Industry Corporation of China, (AVIC), Aero Engine Corporation of China (AECC).

[2] Frank Ko in a letter to the author: "1980s and 1990s, the golden age of textile structural composites".

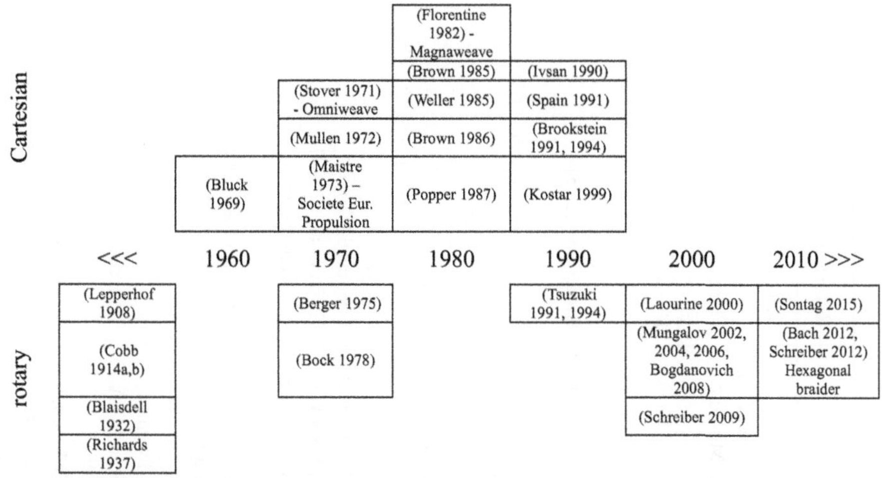

Figure 2.3.6 Timeline of the 3D braiding developments.
Data from Bogdanovich and Kyosev (2016), with inclusion of later sources. The sources cited in the figure: (Bach et al., 2012; Berger, 1975; Blaisdell, 1932; Bluck, 1969; Bock and Flohr, 1978; Bogdanovich et al., 2008; Brookstein et al., 1994; Brown, 1985; Brown and Ratliff, 1986; Maistre, 1973; Cobb, 1914a, 1914b; Florentine, 1982; Ivsan et al., 1990; Kostar and Chou, 1999; Laourine et al., 2000; Lomov and Truevtzev, 1995; Mungalov and Bogdanovich, 2004; Mullen and Roy, 1972; Mungalov and Bogdanovich, 2002; Mungalov et al., 2006; Popper and McConnell, 1987; Richards, 1937; Schreiber et al., 2009, 2012; Sontag et al., 2015; Spain and Bailey, 1991; Stover et al., 1971; Tsuzuki, 1994; Tsuzuki et al., 1991; Weller, 1985).

3.3 3D woven reinforcements

Multilayered woven fabrics were well known to weavers long before the composites era. Heavy apparel fabrics, narrow fabrics, technical fabrics as belts, transporters, paper making machine clothing are examples of multilayered weaves. The name "3D fabrics" starts being applied to composites reinforcements to stress the most important feature of them: through-the-thickness load-carrying connections in the weave structure. The limits of the present book and chapter do not allow discussion of terminological subtleties of the term "3D fabrics"; the detailed analysis of possible logics of distinctions between 2D and 3D fabrics, 3D woven and non-woven fabrics, etc., can be found in (Bogdanovich and Chen, 2015). This paper contains a detailed analysis of the principles, achievements and history of 3D weaving. Other sources include (Chou and Ko, 1989, 1992; Bogdanovich and Pastore, 1996a; Bogdanovich and Mohamed, 2009; Miravete 1999; Tong et al., 2002; Coupé abd Boisse 2011; Chen 2015).

Following Bogdanovich and Chen (2015), 3D woven reinforcements can be classified into three categories: (1) 3D multi-layer, (2) 3D non-crimp orthogonal, and (3) 3D dual interlaced woven fabrics.

Table 2.3.2 Examples of innovations in braiding technology and applications, 2000–2024.

Group	Innovation	Year	Authors	References
Software and optimisation	CAE for 3D-braided textiles	2000	Schneider, Pickett, Wulfhorst, ESI, ITA	Schneider et al. (2000)
	Simulations of crush behaviour of braided composites	2012	Fiendler, TU Munich	Feindler (2012)
	Software for braiding and the braid geometry simulations, configuration of a braider	2012	Kyosev, TexMind	Kyosev (2015), (2024)
	Quick optimisation tool chain for braided preforms design	2013	Birkefeld, U Stuttgart	Birkefeld (2013)
	Extension of the theory of the equivalent horn gears for 3D braiding machines	2018	Kyosev, Glessner, Hochsch Niederrhein	Kyosev and Gleβner (2018)
	Multiscale non-linear meso-FE damage models of braided composites	2020 2022	Zhang et al., NW poly U He, Ge et al., Beijing IT	Dang et al. (2020), He et al. (2022), Hu et al. (2022), Cai et al. (2024)
Braiding technology	Damage in carbon tows during braiding	2013	Ebel et al., TU Munich Soulat et al., ENSAIT	Ebel et al. (2016) Calba et al. (2023)
	Fully electronically controlled carriers	2015	von Reden, composites United	van Reden (2015)
	Overbraiding on non-circular mandrels	2017	Monnot et al., Polytechnique Montreal Vu et al., U Twente	Monnot et al. (2017) Vu et al. (2023)
	Feedback control for fibre orientation	2021	Vollbrecht et al., ITA	Vollbrecht et al. (2021)
	System for quality control	2022	Maidl, LLC München	Maidl (2022)
	High speed magnetic switches and multiple rings, rotary braiding	2024	Brauener, Budillon, ADMEDES	Budillon and Brauner (2024)

Aircraft components	UD braids for stiffeners	2014	Middendorf et al., EADS	Metzner et al. (2014)
	Overview of airframe applications	2016	Carey et al., U Alberta	Carey et al. (2016)
	Oxide ceramic matrix composites	2021	Kolloch et al., ITA	Kolloch et al. (2021)
Rockets and space	High temperature structural seals	2002	Newquist et al., Boeing	Newquist et al. (2002)
Automotive, bicycles	Mercedes SLR crashcone	1999–2009	Eurocarbon	Eurocarbon (2024)
	Tubes for mountain bikes	2010	Thouin et all, McGill U	Thouin et al. (2010)
	Industrialisation of the braiding process to the automotive scale	2015	BMW	BMW (2015)
	Spikes for mountain bike wheels	2020	Gulo composites	Wolverine (2023)
Medical applications	Stents for blood vessels	2008	Becker, Ficker, Miksch, Fraunhoffer TFK	Becker et al. (2008), Emonts et al. (2021)
	Artificial ligaments, biodegradable	2005 2013 2021	Cooper et al., Drexel U Barber et al., U Wisconsin Mengsteab et al., U Connecticut	Cooper et al. (2005), Barber et al. (2013), Emonts et al. (2021), Mengsteab et al. (2021)

3.3.1 3D multi-layer woven fabrics

3D multi-layer woven fabrics can be made on conventional 2D weaving looms; they are further distinguished as "warp-" and "weft interlock." Warp interlock 3D woven fabrics are much more common: it is possible to weave them on simple (dobby) looms, as well as on Jacquard's. Many specific types of these fabrics were patented in the XXth century for different technical and apparel applications. Jacquard looms with electronic control are required for weft interlock fabrics.

The weft layers in a warp interlock fabric, also called "angle interlock," are interwoven with warp, which migrates from one weft layer to another (Fig. 2.3.7a). The trajectories of warps are oriented at a certain angle to the fabric surface. 3D multi-layer warp interlock fabrics found their way in composites in the 1970s and were intensively studied from the 1990s on. Most of the earlier applications were high-temperature ones, required using carbon and ceramic matrices, and it was not the primary design requirement to achieve high fibre volume fractions, keep in-plane yarns as straight and aligned as possible and maximise the in-plane stiffness and strength of the resultant composites. Hence, highly crimped, wavy and irregular 3D interlock weaves were usually accepted (Bogdanovich and Chen, 2015).

Mouritz et al. (1999) states that work on 3D woven composites remained at a low level until the mid-1980s, when interest was renewed because of problems with the traditional 2D laminates in aircraft structures: it was found to be expensive producing complex components from laminates and, secondly, laminates were highly susceptible to Barely Visible Damage (dropped tools, etc.), which is a reflection of a general disadvantage of 2D laminates: susceptibility to delamination. These production and maintenance problems were the main forces behind the effort in advancing 3D woven composites in aircraft structures and components. However, this review (Mouritz

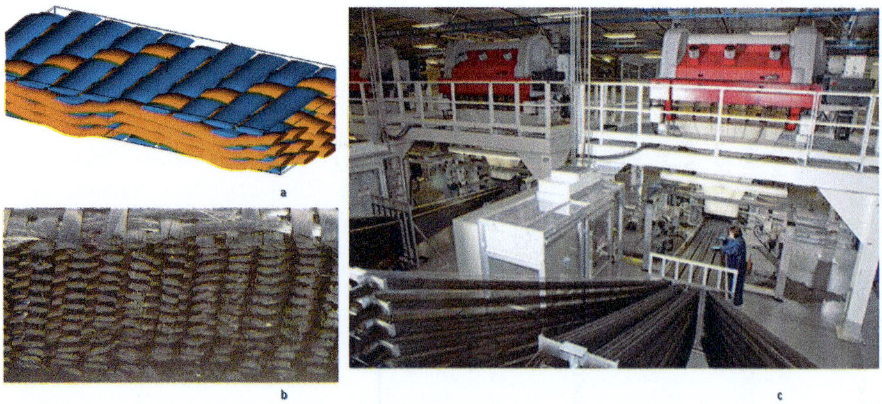

Figure 2.3.7 3D warp angle interlock fabrics: (a) a typical weave structure; (b) a photo of a warp angle interlock carbon fibre fabric; (c) Albany engineered composites (Rochester, NH, USA) loom for weaving of angle interlock fabrics for fan blades.
(a) Image produced with WiseTex software Verpoest and Lomov (2005), Lomov et al. (2022); (c) Reproduced with permission Albany (2024b).

et al., 1999) concludes: *[by the end of the XXth century] "Despite the advantages and potential benefits of 3D woven composites, these materials have failed to find many commercial applications."*

In the early 1980s the technology for 3D weaving was developed in the Leningrad Institute of Textile Industry (now St. Petersburg State University of Technology and Design) by V.A. Gordeev and his coworkers. The original name used by them for their 3D fabrics was "layered & carcassed fabrics"; this technology, still essentially warp interlock, but including vertically positioned woven layers, was used for electric assemblage, composite reinforcements, ballistic protection, hollow architectural structures and other applications. Unfortunately, Gordeev's work was not published in open sources and was known only in a limited circle of the textile community in Russia (Gordeev et al., 1989; Lomov, 1995; Lomov and Gusakov, 1995).

Based on the facilities for 3D weaving, P. Potlury in Manchester University (UK) since the 2010s carried on extensive investigations into the influence of the 3D fabric architecture on processibility and composite mechanical properties (Potluri et al., 2001, 2012a,b; Potluri and Manan, 2007; Dhiman et al., 2015). X. Chen, also in Manchester, since end the 1990s, developed a CAD/CAM software package TexEng for 3D fabrics, and studied their mechanical performance, notably ballistic. In 2008 he started, supported by J.W. Hearle, a series of World conferences on 3D fabrics and their applications (yearly from 2008 in Manchester to 2021 in Zhengzhou).[3]

In ENSAIT (Roubaix, France) F. Boussu established a centre for 3D weaving, conducting since the 2010s broad research on 3D warp interlock fabrics and composites, especially for personal armour applications for ballistic and stab protection (for example (Boussu, 2011; Boussu et al., 2015, 2022; Abtew et al., 2018, 2020; Li et al., 2021),). Multiaxial woven 3D preforms were developed by K. Bilisik in Leeds and later in Erciyes University (Türkiye) (Bilisik, 2009, 2012).

The first ply to ply interlock 3D pattern published was 2.5D Aerotiss, designed by Georges Cahuzac from Aerospatiale, Issac plant (which became part of Airbus Group and now ArianeGroup) at the end of the 1980s (Cahuzak and Monget, 1987, 1988). Aerospatiale had its own weaving loom at that time in the Issac site. High temperature applications were targeted, hence the use of ceramic matrix with these reinforcements.

At the same time activities started on 3D weaving in Société Europeenne de Propulsion – SEP (then Snecma Moteurs, then Snecma Propulsion Solide, then Herakles, then Safran) at Le Haillan, also for CMC reinforcement. As this entity did not have a weaving loom, this work was done in cooperation with technical weaving companies such as Porcher and Hexcel (Bruno Bompard and Jean Aucagne). This 3D weaving development started with multilayer, plain weave based patterns first due to the use of a dobby loom, then multisatin and eventually ply to ply interlock were tested when a Jacquard loom became available to weave carbon and SiC fibres. The first

[3] A remarkable statement was made in the first 3D fabrics conference in Manchester (2008) by Alan Prichard of Boeing, who addressed an audience with all major 3D reinforcement producers present with words: "*we need kilometers of stiffeners and like structural elements in each airplane, which can be easily 3D woven or braided – but who has capabilities of producing these amounts?*" These restrictions given by productivity of 3D textile processes are still largely in place in the present days

serial production of a composite part with a 3D woven reinforcement started in the early 90s with M88 CMC nozzle seals. This part is still in production as this engine powers the Dassault Rafale aircraft.

Probably the most impressive development of 3D angle interlock fabrics is a fan blade for an aeroplane engine (Dambrine et al., 2008). Attempts of development of fan blades using carbon fibre laminates in the late 1960s by Rolls Royce failed because of problems with delaminations and low impact resistance, including bird impact. In the 1990s General Electric successfully developed a 2D composite fan blade for the GE90 engine powering Boeing 777.

In mid-1990s Snecma started work on 3D woven fan blade (polymer matrix), which resulted in the development of a 3D woven angle interlock fabric, produced by Albany Engineered Composites and Safran in a plant, specially built in 2013 in Rochester, NH, USA (Fig. 2.3.7c). Two more plants are in Commercy, France (2014) and Queretaro, Mexico (2018). The developments at Albany were based on their experience in the production of multilayered "clothing" for paper machines (Adanur 1997; Albany 2024a) and complex carbon 3D weaving for preforms; they purchased Techniweave, NH (USA) in 1998, and merged with a Texas-based composite manufacturer in 2006. The combined companies formed the basis of Albany Engineered Composites.

The reinforcement of the fan blade is woven as a single preform by the weaving loom, only a forming step, epoxy matrix infusion and consolidation are necessary to get the blank fan blade (Fig. 2.3.8a and b). Dominique Coupé and Bruno Dambrine at Snecma were instrumental, among others, in the initiation and carrying on 20 years of development in cooperation with John Goering from Albany. A broad research network was created by Safran in France and abroad for work on scientific aspects of the fabric and the part modelling, design and production. This included a MAIA (*Methodes Avancees en Ingénierie mécAnique*) agreement of cooperation between Snecma, ONERA, CNRS, with universities as INSA Lyon, University de Compiegne, LMT Cachan (France), KU Leuven (Belgium), Politechnique Montreal (Canada) and others.

The fan blades were used in the LEAP engine ("Leading Edge Aviation Propulsion"), launched in 2008, produced by CFM International, a 50–50 joint venture between GE Aerospace (USA) and Safran Aircraft Engines (France). LEAP flies on Airbus 320neo, Boeing 737 MAX and the Chinese Comac C919.

Another prominent application of 3D warp interlock woven fabrics is Adaptive Deployable Entry Placement Technology (ADEPT) for planet probes (NASA), developed since 2012, flight test in 2018 (Dutta et al., 2021; Wercinski 2024). The main element of the thermal protection system are foldable screens made of a 3D warp interlock carbon fibre fabric, produced by Bally Ribbon (Bally Ribbon Mills, 2012), and studied in (Vanaerschot et al., 2017) - Fig. 2.3.8c and d.

The main producers of 3D woven composites are Axis, MKWright, Sigmatex, 3D Weaving (EU and UK), Albany International Corp, Bally Ribbon Mills, Boeing, fibre Materials Innovations, GE Aviation, Textum Weaving, TEAM (USA), Texonics (Canada), 3D Wovens (Turkiye).

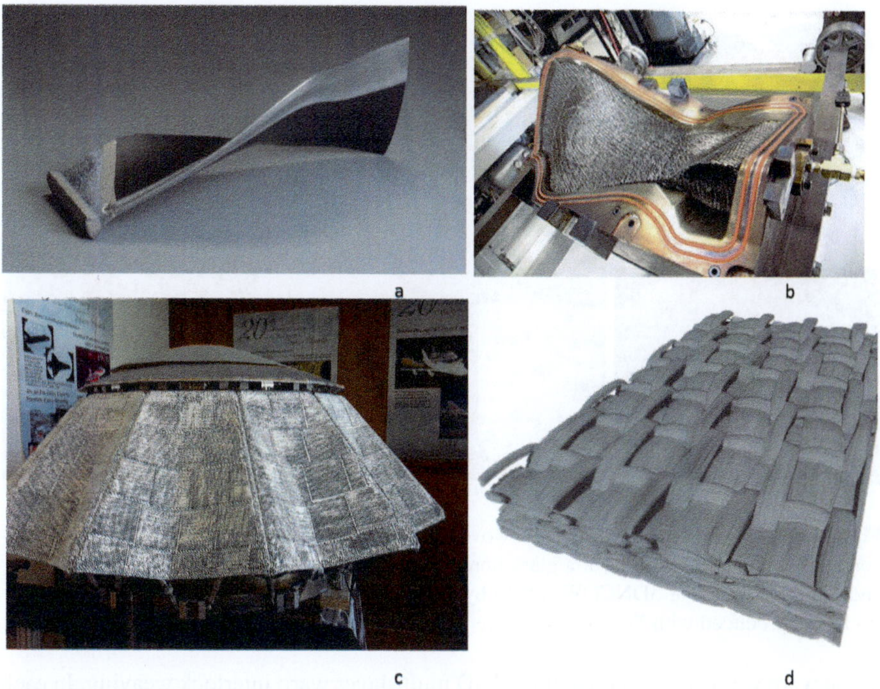

Figure 2.3.8 Application of 3D warp interlock carbon fibre fabrics: (a,b) LEAP engine fan blades, blade shape (a) and its forming (b); (c and d) thermal protection system for adaptive deployable entry placement technology (ADEPT, NASA): a view of the thermal protection system (c) and a computed tomography image of the structure of 6-layer warp interlock 3D fabric, Bally ribbon Mills (d).
(a) Reproduced with permission Albany (2024b); (d) reproduced with permission from Elsevier Vanaerschot et al., (2017)

3.3.2 3D non-crimp orthogonal woven fabrics (3DNCOW)

The concept has been initially introduced by M.H. Mohamed in North Carolina State University College of Textiles in 1989 and elaborated during the 1990s (Mohamed, 1990; Mohamed et al., 1989, 2001; Mohamed and Zhang, 1992). In 1996 M. Mohamed, together with A.E. Bogdanovich and B. Lienhart, have founded 3Tex Inc., a company in Cary, North Carolina, which started operations in 1998 (3Tex, 2024).

The 3DNCOW process can be realised either on the speciality 3D "orthogonal" weaving machines (Fig. 2.3.9c) or on the specially modified 2D weaving machines (for example, rapier Dornier looms). In 3DNCOW two sets of "Z-yarn" are supplied from the same type creels or beams as the warp yarns. None of the warp yarns are functionally used as "binders"; they are not drawn through the harness heddles. Therefore, they cannot be interlaced with the weft yarns. Each end of the two Z-yarn sets is first pulled through the heddle eyelet and then drawn through the reed, forming a binder for the weave. The Z-yarns in the 3DNCOW approach serve the same purpose

Figure 2.3.9 3D non-crimp orthogonal woven fabrics: (a) a typical weave structure; (b) micro-computed tomography image of a glass fibre 3DNCOW; (c) North Carolina state University and 3Tex Inc loom for 3DNCOW (photo by the author, 2008).
(a) Image produced with WiseTex software Verpoest and Lomov (2005), Lomov et al. (2022).

as the warp weavers in a conventional 3D multi-layer warp interlock weaving. In each weaving cycle the entire vertical column of filling (or weft) yarns is simultaneously inserted between the horizontal layers of warp and Z yarns in one step.

The scheme of a typical 3DNCOW is shown in Fig. 2.3.9a and the actual weave geometry – in Fig. 2.3.9b. Z-yarns form a direct connection through the fabric, from its one face surface to the other. Ideally, this connection should be orthogonal to the fabric surface (hence, the term "orthogonal interlock"), while in reality there is always deviation from the orthogonality. The weaving process defines absence of crimp of the warp yarns and a very low crimp (below 0.1%) of the weft yarns (Lomov et al., 2009; Karahan et al., 2010). This leads to high in-plane mechanical properties of 3DNCOW. Comparative experimental studies proving this statement have been conducted for glass and carbon fibre 3DNCOW composites in (Brandt et al., 1996) for the cases of tension, compression, CAI, in (Ivanov et al., 2009b; Lomov et al., 2009; Bogdanovich et al., 2013) for quasi-static tension, including damage initiation and development, and in (Carvelli et al., 2010a; Karahan et al., 2011; Topal et al., 2015) for fatigue.

Applications of 3DNCOW composites include large E-glass segments of recreational boat and yacht hulls; protective personnel armour backings and armoured vehicles' body parts (S-2 glass); glass and carbon fibre structures for civil engineering; automotive brakes and suspension; lids for low pressure storage tanks and medical centrifuges; joints of wind energy generators (3Tex, 2024). In 2016 Sonaca (Belgium) and Bally Ribbon Mills received a JEC Award for the development of a connector ring for a Future Launchers Preparatory Program rocket (ESA), diameter 4.3 m, which is made from 3DNCOW preforms.

3.3.3 3D dual interlaced woven fabrics

3D warp interlock and 3DNCOW fabrics have many similarities, the main being that the through-the-thickness direction of yarns has mainly the weave binding function, rather than creation of a load-carrying function in the third direction. In late the 1990s Khokar (1996) recognised that a "true" 3D weaving must bring about the interlacement of all three orthogonal sets of yarns. N. Khokar coined the term "noobing" (NOOB = "Non-interlaced Oriented Orthogonally and Bind") for the corresponding weaving process (Khokar, 1999, 2001; Khokar and Peterson, 1999); the term "3D dual interlaced" was applied to this type of 3D fabrics in (Bogdanovich and Chen, 2015), following terminology of N. Khokar "dual-directional shedding system" (Khokar 2001). In (Khokar 2013) "noobing" is put in perspective of other dual interlaced weaving processes, patented in 1942–1991. The ideal structure of a 3D dual interlaced fabric is shown in Fig. 2.3.10.

(Bogdanovich and Chen, 2015) states that there is a similarity between this approach and the Cartesian row and column 3D braiding, especially in the final structure of the fabric.

Stig and Hallstrom (Stig and Hallström, 2009) studied experimentally carbon fibre 3D dual woven composites. It was found that the real geometry of the reinforcement is far from the non-crimp idealisation of Fig. 2.3.10 and the reported mechanical tensile and flexural properties are significantly lower than those of their UD counterparts. Judging by the citation analysis of N. Khokar's works, the developments of this type of 3D reinforcements gained interest in 2018–2020, which dropped sharply in the 2020s, with citations only in review papers.

4. Experimental and modelling studies of textile reinforcements

The behaviour of textile reinforcements during composite manufacturing and the mechanical performance of a textile composite are strongly integrated. Fig. 2.3.11a

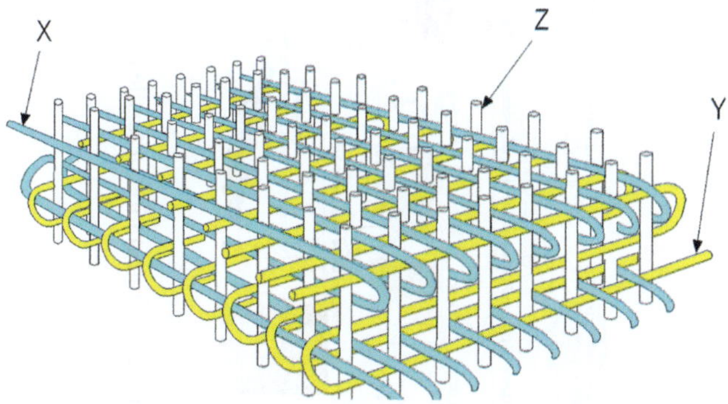

Figure 2.3.10 3D dual interlaced woven fabric (NOOBing) structure, patent WO9803712 (Khokar and Peterson, 1998).

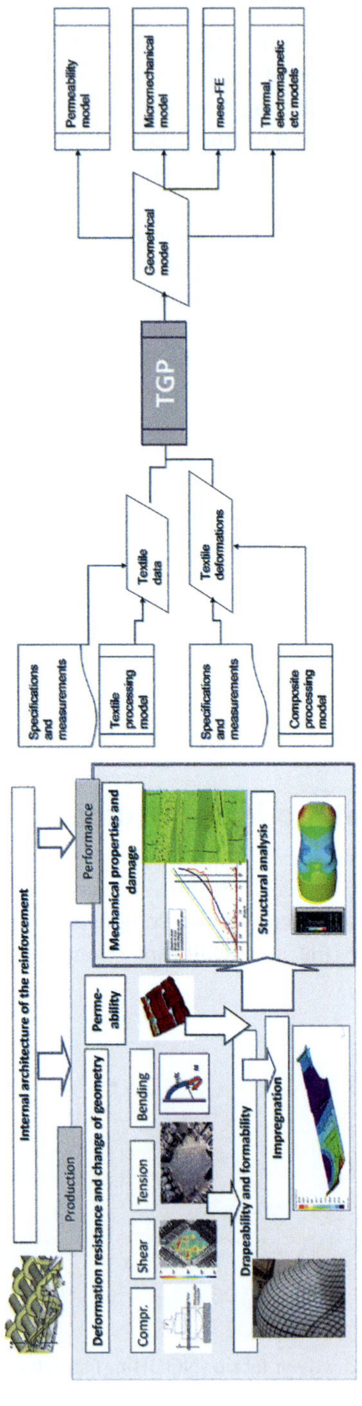

Figure 2.3.11 Integration of textile composites modeling: (a) Elements of the integration; (b) data flow for a Textile geometry preprocessor (TGP).

shows a scheme for this integration, proposed by S.V. Lomov during the TexComp-5 conference (Lomov et al., 2000a,b; 2001a,b). The "keystone" of the integration is the internal structure of the reinforcement. In modelling, this led to the concept of a "textile geometry preprocessor" (Fig. 2.3.11b), which takes data on the textile as manufactured, adds the data on composite processing and creates a geometrical model of the reinforcement local geometry in a composite part, which is further used for analysis of the composite performance (Lomov et al., 2000a,b, 2014).

4.1 Internal geometry of textile reinforcements and textile mechanics

In the first period of textile composite studies, 1980s, there were two directions in understanding and describing the internal geometry of textile reinforcements. One was related to *dry* textiles, not consolidated into a composite, and followed the textile mechanics approach, which is based on the ideas of minimum energy of the equilibrium configuration, due to J. Hearle and R. Postle (de Jong and Postle, 1978; Hearle and Shanahan, 1978), with the primary role of yarn bending. To quote J. Hearle (Hearle et al., 1980), "*[...] the glorious link [of the developed theory] is with Euler and the Bernoulli family, with their theories of the mechanics of flexible slender rods. While mathematicians have admired the beauty of this work, the invention of elliptic integrals, and the grace of the different classes of planar elastica, it is in the technology of textile materials, composed of flexible fibres and yarns, that the subject has found its more direct application.*"

The textile mechanics may be nicknamed "Manchester – Leeds theory," with developments led by J. Hearle in University of anchester Institute of Science and Technology (UMIST) and P. Grosberg, G.A.V. Leaf in Leeds University in the 1960's – 1980's (and also by other researchers in the UK, Australia, New Zealand and the USA). The Textile Institute (Manchester), with its Proceedings, Transaction and Journal, together with the Textile Research Journal (Princeton, USA) promoted these studies. An important point was the NATO Advanced Study Institute meeting "Mechanics of Flexible fibre Assemblies" in 1979, with proceedings published under this title in 1980 (Hearle et al., 1980).

The alternative to the textile mechanics way of description of the reinforcement geometry in impregnated composites is the approximation of this geometry by a set of averaged directions of fibres, representing their primary orientations in the composite. This includes different formulations of the "bridging model" for woven composites (Ishikawa and Chou, 1982; Naik 1994; Naik et al., 2001; Naik and Kuchibhota, 2002; Bogdanovich and Pastore, 1996b), models of braided geometry (Ko et al., 1987; Byun and Chou, 1996; Sun and Qiao, 1997) and the "binary" model of Cox et al. (1994), Cox and Dadkhan (1995). Both routes are presented in (Chou and Ko, 1989).

The textile mechanics approach was generalised for complex 3D woven fabrics and implemented in the WiseTex software by S.V. Lomov and I. Verpoest in KU Leuven (Fig. 2.3.12a). WiseTex was introduced in the early 2000s (Lomov et al.,

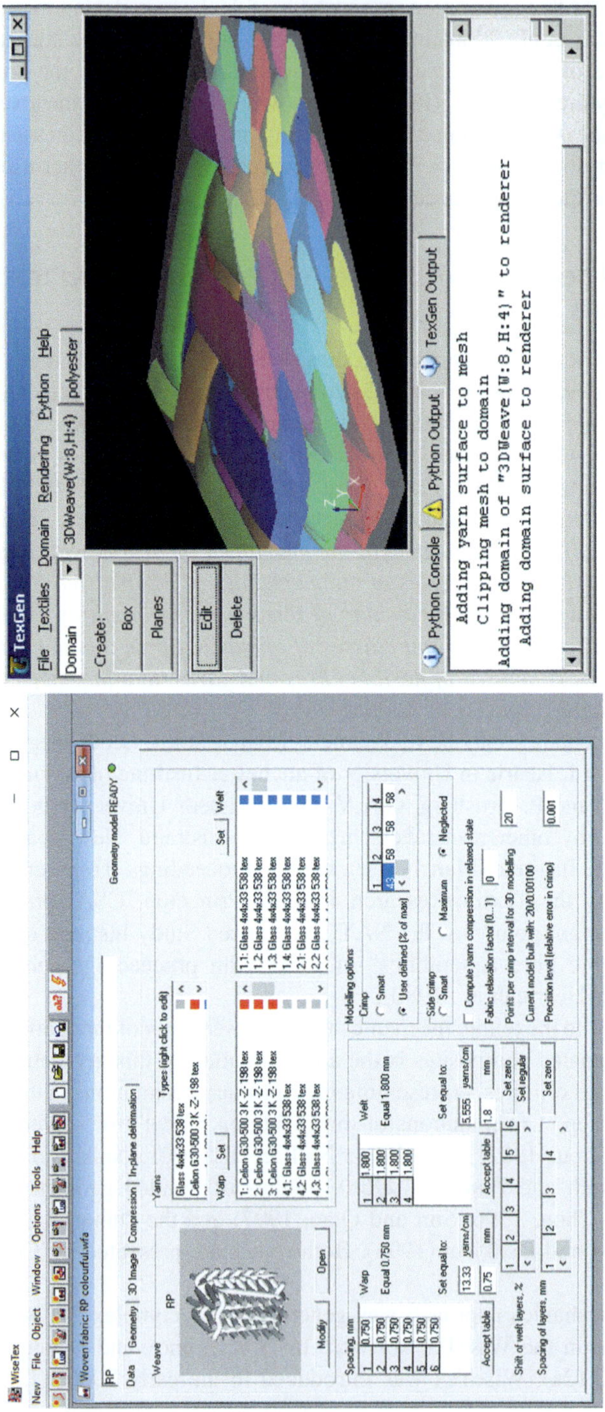

Figure 2.3.12 Screenshots of WiseTex (a) and TexGen (b) software.

2000a,b, 2001a,b, 2005; Verpoest and Lomov, 2005) and was in continuous development since then (Lomov and Aliabadi, 2015; Lomov and Boisse, 2021; Lomov et al., 2014, 2022; Vanaerschot et al., 2016a,b). The software has its origin in the previous work of S.V. Lomov in St-Petersburg in 1990–95, resulting in the CETKA software (Lomov and Primachenko, 1992; Lomov, 1995; Lomov and Gusakov, 1995; Lomov et al., 1995; Lomov and Truevtzev, 1995). The application of this software was 3D paper-making machines clothing; the methodology was extended to composite reinforcements in WiseTex.

WiseTex creates geometrical models of 2D and 3D woven, 2D braided (two-and three-axial) and non-crimp (multiaxial multiply warp knitted) fabrics. It is a textile geometry preprocessor (see Fig. 2.3.11b), which is integrated in a software suite with modules TexComp (composites micromechanics), FlowTex (permeability) and generates the output in XML format, allowing building custom processing like export to finite element models. It is widely used by industry and academy around the globe (~100 licences), and is integrated with various commercial composites modelling software (Digimat, Simcenter, Composites Dream) (Lomov and Boisse, 2021; Lomov et al., 2022).

If WiseTex geometrical models are based on textile mechanics, another widely used textile geometry preprocessor, TexGen (University of Manchester), implements CAD-type of models and is focused on geometry pre-processing for FE simulations (Fig. 2.3.12b). The developments originated from the work of F. Robitaille (Robitaille et al., 1998, 1999a,b). The first TexGen version was developed by B.J. Souter (Robitaille et al., 1998), and further by M. Sherburn and A.C. Long (Sherburn 2007). Version 2 and 3 (2007) are a basis for the subsequent advancement by A.C. Long and L.P. Brown. The description of TexGen can be found in (Brown et al., 2021, 2022).

TexGen is an open-source project (Brown, 2019), freely downloadable:

- executables (https://sourceforge.net/projects/texgen/)
- source code (https://github.com/louisepb/TexGen).

As described in (Brown et al., 2021), a modular approach, particularly with the use of third-party libraries for functions such as rendering and file export, was used. This gives great flexibility to the system, enabling the TexGen functions to be called from Python scripts. The Python interface allows integration with other packages which have a Python capability, for example, the ABAQUS Finite Element Analysis package, and also allows specific functionality to be developed separately from the main development. TexGen has been downloaded more than 50,000 times since its release in 2006–07 and is widely used for applications such as textile and textile composites mechanics, permeability and design optimisation of 3D textiles.

Geometrical models of textiles suffer from artifacts related to the approximate representation of yarn shapes. Several authors, as Hivet and Boisse (2005), (2008) and others (Grail et al., 2013; Sevenois et al., 2016), increased the complexity of the yarn geometry to avoid interpenetration; TexGen contains a module, which identifies penetrating sites and then moved yarn surfaces to avoid them. Another idea is more generic and efficient: use the models as (indeed) pre-processors for "sketching" the geometry, and apply finite element (FE), or other discrete modelling to resolve yarn contacts and yarn shape distortion due to contact interactions, creating a more accurate representation of the geometry. This approach was first followed with ad hoc volumes modification in Lomov et al. (2007), Ivanov et al. (2009a) and by Durville (2007), Durville et al. (2018), with his Multifil software. The similar to Multifil representation of yarns as bundles of elastic beams was used by S.D. Green, S.R. Hallett, B. El Said et al. (Mahadik and Hallett, 2010; Green et al., 2014; El Said et al., 2018), and by the "digital chain" method by Zhou et al. (2004). E. Iarve, E. Zhou et al. (Iarve et al., 2009; Adluru et al., 2023) implemented these algorithms in the software Virtual Textile Morphology Suite. Th. Massart et al. developed a level set-based algorithm for resolving the interpenetrations (Wintiba et al., 2017; Pierreux et al., 2020). P. Badel, Ph. Boisse et al. used a hypoelastic representation of the fibrous yarn's homogenised behaviour in such simulations (Badel et al., 2008).

The advent of X-ray computed tomography (XCT) has established this method as a "golden standard" for the investigation of the internal geometry of textiles and fibre reinforced composites (see examples in Figs. 2.3.8 and 2.3.9). The work in this direction started in the 1990s and flourished in the XXIth century. The search in Science Citation Index on *(tomograph* or CT) and composite* and (fibre* or fibre* or textile* or woven)* returns 20–30 papers per year in 2005–08. Then the number of publication rockets to ~250 papers per year in 2023.

The availability of instruments for detailed quantification of the internal geometry allowed in-depth investigation of the reinforcement variability, which is introduced during textile production and during the impregnation and consolidation. Research on variability started in the mid-2000's, with works in University of Nottingham, University of Cambridge, KU Leuven (Endruweit et al., 2004, 2006; Desplentere et al., 2005; Skordos and Sutcliffe, 2006, 2008), and continued through the 2010s, resulting in detailed characterisation of the variability of tow dimensions and fibre orientations in 2D and 3D woven and NCF composites (Karahan et al., 2010; Olave et al., 2012; Sutcliffe et al., 2012; Mesogitis et al., 2014; Gommer et al., 2016). In the 1990s S.P. Yushanov and A.E. Bogdanovich have developed a stochastic description of woven reinforcements (Yushanov and Bogdanovich 1998a,b, 2000). With the availability of XCT quantification of the 3D geometry, B. Cox and co-workers (Bale et al., 2012) have advanced it using a random variation of the Markov chain representation of the yarn centre lines. This approach was further generalised by Vanaerschot et al. (2013), 2016a,b), and applied to complex 3D reinforcements (Vanaerschot et al., 2017).

4.2 Meso-FE modelling

Predecessor of the meso-level (structure and scale level of a textile unit cell/repeat) FE modelling is "the method of cells," mapping the geometric description of the textile meso-structure into a 3D grid of elements. Developed in the 1990s by A.E. Bogdanovich et al. (Bogdanovich, 1993; Bogdanovich et al., 1992, 2017; Bogdanovich and Pastore, 1996a,b), who used the term "mosaic method" and J. Aboudi, who introduced the term "method of cells" (Aboudi, 1989), it was further developed by Vandeurzen et al. (1998) and A. Prodromou (Prodromou et al., 1998, 2011). The "method of cells" was reincarnated later as voxel meshes for FE.

Meso-level FE modelling of woven composites emerged in the mid-1990s – mid-2000's. The concentrated developments on this early stage were done in Texas A&M University by Whitcomb et al. (1995), (2000); Whitcomb and Srirengan (1996), Whitcomb and Tang (2001), Woo and Whitcomb (2000), Tang and Whitcomb (2003a,b); Goyal et al. (2005), and in Osaka University by M. Zako et al. (Takano et al., 1995; 2001a,b, 2002; Kurashiki et al., 2002; Uetsuji et al., 2002; Zako et al., 2003), Dayton University and Air Force Research Laboratory by E. Iarve et al. (Sihn 2004a,b), University of Nottingham by Crookston et al. (2005a,b) and others. FE models of dry fabrics with complex mechanical behaviour of the yarns (hypoelastic and other models) were built in INSA Lyon by Ph. Boisse et al. (Gasser et al., 2000; Boisse, 2006; Boisse et al., 2005, Hagège et al., 2005; Hivet and Boisse, 2005).

The work in this period was summarised by Lomov et al. (2007), who formulated a "road map" for meso-level modelling of textile composites, and in proceedings of the specialised symposium "Finite element modelling of textiles and textile composites," which met in St.-Petersburg in 2007.

Two main problems were identified at that time as the most seriously impeding "on the fly" creation of the meso-FE models: interpenetration of the yarn volumes constructed using approximate geometrical models and efficient meshing of the inter-yarn volume. The interpenetration problem is solved by modification of the yarn volumes in a geometrical sketch, as described in the previous section. The difficulty with meshing is the need to preserve the real geometry of the yarn and their dimensions, which define local fibre volume fraction, hence local mechanical properties.

The book chapter by Ivanov and Lomov (2020) contains a detailed overview of the contemporary status of meso-FE modelling of textile composites as well as an extensive bibliography. The authors conclude "Despite vast efforts from the research community, a realistic description of the fabric remains a challenge in many practically relevant cases."

A general solution is the use of superimposed meshes for yarns and matrix, which interact through kinematic constraints. For meso-FE of textile composites such methods (under different names) were first investigated by M. Zako et al. (Takano et al., 2001a,b, 2003), who called their method "M^3," and implemented it in the SACOM software (Zako et al., 2005) and in Composites Dream of CTC/Itochu (Sakakibara et al., 2018). Jiang et al. (2008) used "domain superposition" of ANSYS. Iarve et al. (2009) developed an "independent mesh" method, combining it with Rx-FEM based discrete damage modelling (Adluru et al., 2023). Tabatabaei et al. (2014), (2022); Tabatabaei and Lomov (2015); Vorobiov et al. (2017) investigated the application of Abaqus "embedded elements" to meso-FE modelling.

An alternative to the precise meshing of the matrix and yarn volumes (conformal meshing) is voxel meshing, where "voxel" means an element of regular discretisation of the volume into cuboidal elements. TexGen (Brown et al., 2021, 2022) automatically creates voxel meshes for its geometrical model; the same is being done based on "digital chain" geometrical models (Isart et al., 2015). Examples of applications of voxel meshing to textile reinforcements can be found, for example, in Crookston et al. (2005a), (2005b), Green et al. (2014), Liu et al. (2017). The comparison of these two types of meshes is discussed in Doitrand et al. (2015), Fang et al. (2016), Ewert et al. (2020), Zheng et al. (2021), Nagaraju et al. (2022), Sinchuk et al. (2022), Wintiba et al. (2020), Rassineux (2022), Fourrier et al. (2023) have developed methods for construction of robust conformal meshes starting from a voxel mesh.

The voxel mesh is a natural way of creation of an FE model based on an XCT image of textile composites (Naouar et al., 2020). Table 2.3.3 gives an overview of the published works, which started appearing in the second half of the 2010s.

Table 2.3.3 Meso-level XCT-based FE modeling of textile composites.

Properties	Reinforcement type	Voxel	Conformal
Mechanical, impregnated	Twisted bundle	Straumit et al. (2017)	
	NCF	Ilin et al. (2018)	Auenhammer et al. (2020), (2021), Auenhammer et al., (2022a,b)
	2D and 3D woven	Naouar et al. (2014), (2020), Liu et al. (2017), Shishkina et al. (2018), Song et al. (2023)	Mazars et al. (2017), Wintiba et al. (2020), Ai et al. (2021), Yu et al. (2021), Sinchuk et al. (2022), Hernández et al. (2024)
	Braid		Laurent et al. (2014), Jing et al. (2020), Liu et al. (2020), (2024), Melenka and Gholami (2021), Zahid et al. (2021), Gholami and Melenka (2022)
Mechanical, dry	3D woven		Badel et al. (2008), Naouar et al. (2014), (2015), Boisse et al. (2024)
Thermal	2D and 3D woven	Gao et al. (2017); Gigliotti et al. (2018)	Foster et al. (2023)
Permeability *	NCF Woven Braid	Straumit et al. (2016) Syerko et al. (2023a,b) Lomov et al. (2023)	 Syerko et al. (2023a,b)

4.3 Textiles in composite manufacturing

4.3.1 Deformability and drape

Draping of a textile preform over a 3D shaped mould is the key process, defining the quality of the composite part (see Fig. 2.3.4a) – in the same way as fabric drape over a body (with extremely complex 3D shape), and fabric "handle" in general are key for the quality of clothing. In the 1930s the notion of "handle as a measurable quantity" was introduced by Peirce (1930); his colleague J.R. Womersley in the Shirley Institute was first to attempt a rigorous mathematical approach to modelling of drape (Womersley 1937).

The ideas of Peirce were fully developed in the 1970s by S. Kawabata. He has elaborated the "objective evaluation of fabrics" (Kawabata 1975): complexes of mechanical properties of fabric under tension, shear, bending, compression and friction, which define different descriptors of fabric handle (thickness, fullness, weight, firmness, crispness ... - all having Japanese-based terms in the original Kawabata's nomenclature: *sofutusa, kishimi, shinayakasa* ...). S. Kawabata and M. Niwa have developed a special suite of measurement instruments, "Kawabata Evaluation System for Fabrics" (KES-F) for measurements of these properties. In the 1980s J. Amirbayat has linked the basic mechanical properties of fabrics (shear, tension and bending stiffness) to draping and wrinkling of the fabric (Hearle and Amirbayat, 1986; Amirbayat and Hearle, 1989, 1995). Ko et al. (Cai et al., 1994; Yu et al., 1994) applied KES-F characterisation to the reinforcement formability during composites manufacturing.

It soon became clear that the characterisation instruments, used for apparel fabrics, are not suitable for heavy, thick and stiff glass and carbon reinforcement fabrics. Recognising these requirements, a group of international researchers (R. Akkerman, Ph. Boisse, J. Cao, J. Chen, A. Long, J. Sherwood and others) gathered at the University of Massachusetts Lowell for the NSF Workshop on Composite Sheet Forming in September 2001. One direct outcome of the workshop was the establishment of a web-based forum exclusively for research on forming of woven composites and start of the international benchmark; since 2002, the group has taken the ESAFORM conference and society as a platform to exchange and coordinate research on composite material and forming characterisation. The Composites Forming micro-symposium in the annular ESAFORM conferences was led by Ph. Boisse in 2002–2022 and by P. Carlone since then; the summary of 10 (Boisse et al., 2007) and 25 years (Boisse et al., 2022) of ESAFORM give a good historical overview of the field.

Fig. 2.3.13 shows devices for testing fabric deformability. Especially for shear testing, the applicability of devices used for apparel fabrics was questionable. The device for fabric shear testing, a picture frame, was not standardised, and many different variants were in use (Fig. 2.3.13a). The equivalency of the picture frame test and the alternative method, bias extension (Fig. 2.3.13b), was not clear. The international benchmark (Cao et al., 2008) has identified requirements for such tests and the best practice rules. The picture frame, bias extension and biaxial tension tests (Fig. 2.3.13c) were instrumented in the years 2000 with full field optical strain measurement (Lomov and Verpoest, 2006; Lomov et al., 2008c; Zouari et al., 2006;

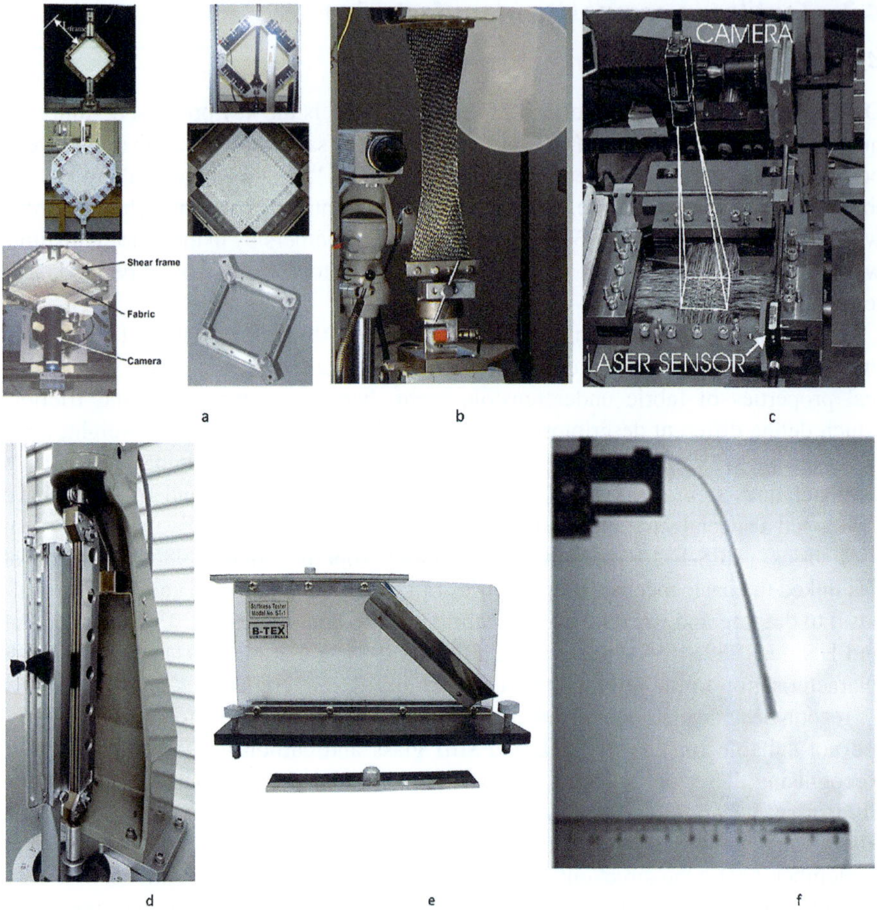

Figure 2.3.13 Fabric deformability measurements: (a and b) shear: (a) Different picture frames used by participants in the benchmark, and (b) bias extension test; (c) biaxial tension test; (d–f) bending, KES-F (d), cantilever (e and f).
(a) Reproduced with permission from Elsevier (Cao et al., 2008), (b) Reproduced with permission from Elsevier (Lomov et al., 2008a,b,c), (c) Reproduced with permission from Elsevier (Willems et al., 2008), (e and f) photos by the author and Cornelissen and Akkerman (2009), courtesy of R. Akkerman.

Willems et al., 2008), allowing assessment of the true strain state of the sample. The bending test for composite reinforcements has evolved from the complexity of the KES-F device (Fig. 2.3.13c) to the simplicity of a (standardised) cantilever test (Fig. 2.3.13e), but instrumented with optical registration of the curved shape of the sample (Fig. 2.3.13f) and identification of the curvature-dependent bending rigidity (Cornelissen and Akkerman, 2009; Liang et al., 2017; Boisse et al., 2018a). Friction of fibres, yarns and fabrics was studied by J. Sherwood et al. (Gorczyca et al., 2004) and Cornelissen et al. (2012), (2013). The fabric compression benchmark has identified big scatter of the compression test results (Yong et al., 2020); its second stage is running at the time of writing.

Meso-level modelling of textile deformation has two goals: first, calculate the force needed for a given deformation and second, calculate the new unit cell geometry after deformation. In textile mechanics models this is done as the collective action of different modes of yarn deformation resistance, namely bending, compression, tension and friction, which was introduced by S. Kawabata (Kawabata et al., 1970a,b, 1973a,b,c; Niwa et al., 1970). In the WiseTex software such approximate calculations are included (Lomov et al., 2006a,b, 2022; Kyosev et al., 2016).

Modelling of fabric drape started from membrane elastic models (Shanahan et al., 1978; Lloyd 1980), and evolved in a general approach of representing fabrics as two-dimensional continua and using the differential geometry of surfaces to describe the shape of the draped fabric and to deduce deformation measures, following the paradigm previously used to establish computational models of elastica theory (Lloyd et al., 1996). Earlier references can be found in Hearle et al. (1980). The end-of-the-century state-of-the-art was summarised in a special issue of Journal of Clothing Science and Theory (vol 8, part 3, 1996).

Finite element models for deformations of dry fabrics with non-linear descriptions of multiple deformation modes are created since the years 2000. The most intensive work in this direction was carried on in INSA Lyon by Ph. Boisse, E. Vidal-Salle and their co-workers (Boisse et al., 1997, 2005, 2006; Boisse, 2006). The breakthrough in these developments was made when hyper-elastic material models for fibrous yarns were introduced (Charmetant et al., 2011). This allowed for a physics-based description of the yarn deformation, with material law parameters identifiable in a well-defined set of experiments. The approach was applied to complex woven structures (Nguyen et al., 2013; Pazmino et al., 2014) and FE models based on XCT images (Naouar et al., 2014, 2015) and coupled meso-macro simulations (Iwata et al., 2019; Wang et al., 2020; Li et al., 2024).

As discussed in Boisse et al. (2024), the development of the theory of the macroscopic deformation of textiles in general and of textile reinforcements in particular, can be presented as the development of continuum mechanics of textiles. Chronologically, textile structures were at first represented by bar systems representing the tensile stiffness of the yarns, with rotational springs for shear (Ben Boubaker et al., 2005). Thereafter came the so-called semi-discrete methods (Hamila et al., 2009) based on the expression of the deformation energy for each representative elementary cell. By describing the kinematics, the deformation energy is measured by well-defined tests. Following this, a mechanical model was proposed, providing the dual quantities of each of these deformation modes. Simple tests decomposing these modes of deformation could then be set up. Since the 2010s, continuous models have been developed. Initially, they were models of purely hyper-elastic behaviour based on geometric invariants (Aimène et al., 2009; Charmetant et al., 2012; Gao et al., 2022). Hyper-elastic models were extended with introduction of the viscosity in large deformations for reinforcements impregnated with a molten matrix (Guzman-Maldonado et al., 2016) and of the hysteretic behaviour of dry reinforcements, accounting for friction of the fibrous structures at fine scales (Denis et al., 2020).

The discussed continuum models are models of Cauchy, based on the first gradient of displacements. In the case of fibrous reinforcements this classical approach has limitations (Ferretti et al., 2014; Boisse et al., 2018b), which arise from the inability of a first gradient model to capture the very low transverse shear and simultaneously to account for the bending stiffness of each fibre. To overcome this difficulty, hyper-elastic second gradient models have been introduced (dell'Isola and Steigmann, 2015; Madeo et al., 2015). An alternative to the second gradient terms is the addition to first gradient potentials of strain energy related to the curvature of the fibres (Mathieu et al., 2015; Boisse et al., 2018b). This theory has been extended to account for the bending stiffness in the finite element plane of shells, not considered in standard shell elements (Steer et al., 2021).

The research direction of mechanics of woven fabrics is summarised in a chapter in "Comprehensive Mechanics of Materials" (Boisse et al., 2024); Chapter 4.2 "No composites without computers: the key role of software tools" of the present book contains a history of forming/draping macro-scale software and modelling methods.

4.3.2 Permeability

In the mid-1980s, works of Adams et al. (1986, 1988); Adams and Rebenfeld (1987, 1991a,b), have introduced the systematic permeability determination of composite reinforcements for Liquid Composites Moulding (LCM), applying the Darcy's theory of flow in porous media. Darcy's equation is the basis for simulation methods for resin transfer moulding, which were developed in the early 1990s (Bruschke and Advani, 1990; Trochu et al., 1994) (see also Chapter 4.2 "No composites without computers: the key role of software tools" of the present book) fuelling the need for reliable methods of permeability determination.

The permeability research, starting from 1990s, was reported and to some extent organised through the international conference "Flow Processes in Composite Materials" started in 1988 in Uxbidge, UK. The 17th FPCM convened in Abu Dhabi in January 2025. FPCM proceedings are available through (Summerscales, 2023). The developments were strongly steered by the 20-year long sequence of international benchmark exercises.

The first regional benchmark study on in-plane permeability (Parnas et al., 1995), was followed by Lundström et al. (2000), Grössing et al. (2015), Fauster et al. (2018) and by initial efforts to establish databases (Parnas et al., 1997). It became clear that differences in test equipment and methodology led to an orders of magnitude variation in results (Hoes et al., 2004; Liu and Parnas, 2007).

In 2006, the research community around FPCM organised the first international benchmark study. It was initiated by ONERA and KU Leuven, presented at FPCM-10 in 2010 (Laine et al., 2010), and results were published a year later (Arbter et al., 2011). 11 participants investigated the same fabric, but no specifications were defined regarding the measurement method and the test parameters. An advantage of this strategy was that, in the context of technology screening, it provided an overview of all the methods used and distinguished good practices from the usual pitfalls. One main difference was between unsaturated and saturated flow tests, another between linear and radial tests. A scatter of the measured permeability values of more than one order of magnitude was the result of this benchmark.

Based on this experience, a second benchmark was organised, led by Ecole Polytechnique Montreal, this time with a mandatory predefined measurement procedure, prepared by the core team of the exercise (Alms et al., 2010). The results published in 2014 (Vernet et al., 2014) showed that by defining the minimum requirements for the equipment, and the measurement procedure and analysis, satisfactory reproducibility of data (scatter below 25%) using different systems could be achieved.

In 2016, the United Kingdom's National Physical Laboratory (NPL) and the National Composites Center (NCC) published a survey of procedures in use for permeability measurements in LCM (Aktas et al., 2016). This review was prompted by discussions between S.V. Lomov (KU Leuven) and M. Hinton (NCC) during ICCM-15 (Copenhagen) on the possibilities of continuing the work toward standardisation of permeability measurements. At the 2016 conference in Kyoto, Japan, the FPCM community launched a third and final international benchmark on in-plane permeability characterisation. The benchmark, led by D. May of the Leibniz-Institut für Verbundwerkstoffe (IVW, Kaiserslautern), with 19 participants, focused only on radial methods, defining even stricter guidelines and included an extensive questionnaire on the applied methods, test equipment and data evaluation (May et al., 2019).

After three consecutive international benchmark exercises, it became possible to advance to standardisation. The ISO project for the standard development, inside the ISO committee ISO/TC61/SC13/WG1, which is in charge of "Reinforcements and reinforcement products," started in 2020 and the standard was issued in 2023 as ISO 4410 "Experimental characterisation of in-plane permeability of fibrous reinforcements for liquid composite molding" (ISO 2023).

Apart from these benchmarks, parallel experimental studies were initiated on 3D permeability measurements (Yong et al., 2021) and compressibility of fibrous reinforcements (Yong et al., 2020). At the time of writing, the second phase of this exercise is running.

Another related benchmark is on virtual permeability, led by E. Syerko and Ch. Binetruy (Nantes). The participants calculate the permeability of a fibrous reinforcement, using as input an XCT image of (a) unidirectional fibrous ply and (b) twill weave glass fibre fabric (Syerko et al., 2023a,b, 2024). The second stage of this benchmark is also running at the time of writing.

4.4 Mechanical performance of textile composites

Elastic properties of textile composites can be calculated without complex FE discretisation, but still using yarn crimped trajectories. Homogenisation of the stiffness is performed with orientation averaging (iso-strain assumption) (Chou and Ko, 1989; Bogdanovich and Pastore, 1996a) or with the Mori-Tanaka algorithm; in the latter case the impregnated yarns are represented by a set of ellipsoidal inclusions (Huysmans et al., 1998, 2001; Lomov et al., 2000a,b). Tsai (1992) gives practical guidelines for knockdown factors of stiffness of 2D woven carbon and glass fibre composite in comparison with cross-ply UD laminates.

Damage in textile composites is a complex phenomenon, with many types of damage and developing through several stages (Fig. 2.3.14). This complexity was studied and described in the 1990s in the paradigm of damage types in cross-ply laminates, for example Tsai and Hahn (1980), Naik (1994), Talreja and Singh (2012), Talreja and Varna (2016). The complexity, depicted in Fig. 2.3.14, was fully formulated in the 2010s (Wisnom 2012; Gorbatikh et al., 2016, 2018, 2023); it was extensively discussed in conjunction with damage processes in UD ply multiaxial laminates in annual Composite Damage Workshops, started in 2009 by F. Paris, S. Pinho, M. Quaresimin, K. Schulte, B. Sorensen, R. Talreja, J. Varna, M. Wisnom and others, and running till the time of writing (2024).

Such a complex interaction of damage events and processes has to be studied with multi-instrument methodology, which combines sample pre-test investigation using NDT techniques, in-situ registration of damage initiation and development during the test and post-mortem analysis of the sample (Lomov et al., 2008b; Oz et al., 2018). The most popular methods for in-situ damage observation during lab-scale are digital image correlation (DIC) and acoustic emission.

For textile composites full-field strain measurements were introduced in the years 2000 (Lee et al., 2004; Lomov et al., 2008a), capitalising on the intensive developments of full-field strain measurement equipment and inverse identification methods (Chalal et al., 2004; Grediac 2004). DIC was extensively used afterwards in textile composite studies, becoming a “must” instrument for any composites mechanical testing lab. An important role in proliferation of the optical full-field measurement methods and their use for parameter identification for composites was played by the CompTest series of conferences on Composite Testing and Model Identification, initiated by F. Pierron and M. Wisnom in 2003 (Chalon-en-Champagne); the 12th CompTest took place in Riga in 2025.

Acoustic emission (AE) registration was applied to textile composites starting from the mid-2000s (Huguet et al., 2002; Sause 2016), following work from the 1970s on, dedicated to laminates and filament wound vessels (Bunsell 1977; Williams and Lee, 1978). It reveals damage initiation thresholds (see Fig. 2.3.14b), seen in the graph of AE energy or peak amplitude (Gorbatikh et al., 2016, 2018, 2023). These thresholds were studied first in (Lomov et al., 2008b; Truong Chi et al., 2008) for NCF and braided composites, and then for 2D and 3D woven (Lomov et al., 2009; Bogdanovich et al., 2013), stitched (Carvelli et al., 2010b) and other textile composites. The AE events can be classified according to the AE signal parameters in the time and frequency domain and connected to the type of damage (Fig. 2.3.14a and c) (Huguet et al., 2002; Gutkin et al., 2011; Sause et al., 2012; Li et al., 2014, 2015, 2016). AE-based damage characterisation is often questioned because of uncertainties of the damage identification (Oz et al., 2017). Multiple instrument damage monitoring is most likely the way to resolve these difficulties (Lomov et al., 2018).

Strength and damage FE models of textile composites are still in the development phase, in spite of the fact damage models were already created together with the very first meso-level FE models in the 1990s, for example, the works of Zako et al. (1994), Takano et al. (1999). They followed a damage mechanics scheme, with a multi-axial, multi-damage-mode criterion for onset of damage in impregnated

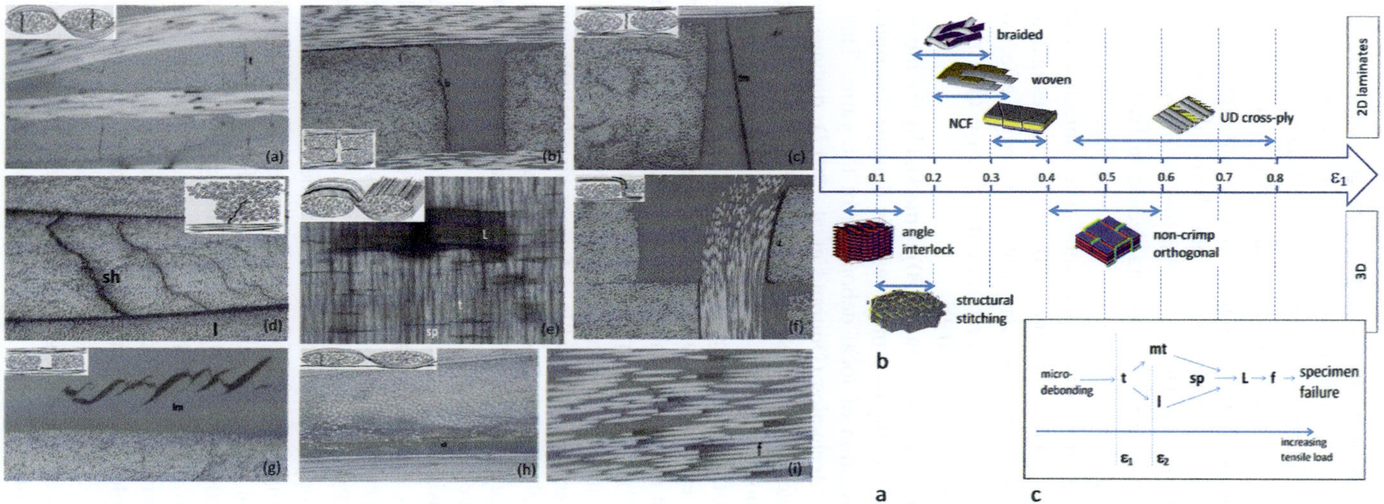

Figure 2.3.14 Damage in textile composites (Gorbatikh et al., 2016, 2018, 2023): (a) damage types: (a) Transverse matrix cracks inside yarns "t"; (b) transverse cracks at yarn boundaries "b" and local delaminations between yarns "l"; (c) transverse cracks in the matrix pockets "mt"; (d) shear matrix cracks "sh"; (e) longitudinal cracks in the yarn "sp"; (f) cracks in Z-yarns; (g) longitudinal cracks in the matrix pockets; (h) interplay cracks – delaminations "L"; (i) fibre breakage "f"; the loading and horizontal directions coincide.

(a) detailed description of the experiments depicted in the images is in Ivanov et al. (2009b); Lomov et al. (2009); Bogdanovich et al. (2013), reproduced with permission from Elsevier; (b) damage thresholds for different types of textile reinforcements (initiation of transverse cracking) reproduced with permission from Elsevier; (c) a typical damage sequence in textile composites under tension. Reprinted with permission.

yarns (considered as UD composites) and a stiffness degradation scheme, also multi-axial and multi-mode. The popular damage onset criteria were those allowing segregation of fibre breaks, transverse and shear cracking of the impregnated yarns (Hoffmann, Puck …). The degradation algorithm in the very early work of the 1990s was a 1/100 reduction, attributed to Blackletter (Blackketter et al., 1993). Later, in the years 2000, the Murakami scheme with damage tensor was widely used (Murakami 1988). The restrictions of the chapter size do not allow a detailed presentation of the history of these developments. The reader can find an overview of the early work in Lomov et al. (2007) and for the following decade in Ivanov et al. (2017).

Application of periodic boundary conditions is a pre-requisite for meso-level FE modelling of transversely periodic textile composites. The subject was introduced by Tang and Whitcomb (Tang and Whitcomb, 2003a,b) and investigated in detail in Li and Sitnikova (2020). W. Van Paepegem et al. proposed an effective scheme for the definition of periodic boundary conditions in the case of a non-periodic mesh (Jacques et al., 2014; Garoz et al., 2019).

The last decade (end of the years 2010 and 2020) is marked by the introduction in meso-FE textile composite modelling of a more realistic damage description, which includes initiation and propagation of local debonding of the yarns. This includes methods of Finite Fracture Mechanics (Doitrand et al., 2017), damage evolution laws accounting for the material toughness (Liu et al., 2017), use of regularised extended finite element method (Rx-FEM) (Adluru et al., 2023) and allows a direct representation in the model of different damage modes.

As summarised in Ivanov et al. (2017): geometrical modelling tools just started to approach automated creation of the internal geometry, suitable for quality meshing and preserving a realistic fibre volume fraction inside the yarns. There are still pending fundamental questions on multiscale modelling of non-periodic interaction of plies of non-matching orientations, modelling of processing features and damage at larger structural scales. Solutions to these challenges begin to appear. There is a clear perspective that an adequate automated FE tool for meso-level analysis of woven composites will emerge in the near future.

Fatigue of textile composites did not catch much attention till the years 2000. In Harris' book on composites fatigue (Harris 2003) only a short chapter by Naik (2003) is dedicated to woven composites. A systematic study of textile composites fatigue was performed by Van Paepegem and Degrieck (2001a,b), (2002), (2004). Apart from developing a methodology for experimental studies, he has introduced a "cycle jump" method of modelling high cycle fatigue of textile composites. This method follows the stiffness degradation concept (Kachanov, Schulte, Reifsneider, Sidoroff and others), developed in the 1980s. The "cycle jump" method was presented in Van Paepegem and Degrieck (2001a,b), and is described in detail in Van Paepegem et al. (2015), and is used in all modelling works for high-cycle fatigue. It can be applied also to variable fatigue load, for example, as was done for 3D braided composites in Ge et al. (2024).

In the years 2000 the research into fatigue of textile composites was intensified because of practical needs of wind energy and automotive industries. A number of 2D and 3D woven and braided, NCF and structurally stitched composites were studied, both with carbon and glass fibres, mostly with thermosetting matrices. The results of these works were summarised in the book "Fatigue of Textile Composites," edited by Carvelli and Lomov (2015) and in Carvelli et al. (2017).

5. Conclusion

Textile composites after 50 years of their development are a mature structural material with an industry of high production volume and large capital investments. Important composite products, windmill blades for example, are just impossible to produce without off-shelf availability of kilometres of metres-wide fabrics with a well-organised stable fibrous structure. Reciprocally, composite reinforcements are one of the *raisons d'etre* of the knowledge-intensive, high-end product textile industry.

An important promise of textile composites, mentioned on the very first pages of any publication discussing perspectives of them, is adaptability of the fibrous structure to the expected loads and hence optimisation of this structure. Critically assessing the validity of this claim, one has to admit that the expectations are exaggerated. Textile technologies, versatile as they are, are versatile within their own boundaries. It is rather a variety of choices of fibre arrangement technology, which creates a wide range of choices for the designer of a new material. There are forays towards optimisation of 3D woven or braided structures (Cox et al., 2016; Wang and Hamila, 2024), but robotic automated fibre placement offers much more freedom in the realisation of a topologically optimised fibre architecture, in comparison with traditional textile technologies.

Looking from the bright side, textile composites give to a clever designer wide possibilities for a manufacture – performance – cost integration, allowing operating with square metres of fabric instead of meticulous positioning of centimetres of tapes, and offering a rich "Zoo" of prefabricated fibrous structures, parameters of which can be adjusted in the textile production. A commonplace statement "lack of reliable design methods and tools hinders the wide use of < ... *here the name of a material type is inserted* ... >" is often used on textile composites. The material of this chapter, as well as Chapter 4.2 "No composites without computers: the key role of software tools" by A. Pickett, shows that this pessimistic statement is outdated, with mechanics of textiles and textile composites developed extensively and implemented in off-shelf software tools. Wide applications range, some part of it presented in this chapter, and half a century of developments ensure a bright future for textile composites as a structural material.

Acknowledgements

The author is grateful to a number of people, who helped collecting information for this chapter:

NCF composites: Thomas Gries, Christoph Greb (ITA, RTWH, Germany).

Braided composites: Dominik Granich (ITA, RTWH, Germany), Frank Ko (University of British Columbia, Canada), Yordan Kyosev (Dresden University, Germany), Peter Middendorf (University of Stuttgart, Germany), Qiang Liu, Man Zhu (Nanjing University of Aeronautics and Astronautics, China), Damien Soulat (ENSAIT, Roubaix, France), Chao Zhang (Northwestern Polytechnical University, China).

3D woven composites: Francois Boussu (ENSAIT, France), Francois Charleux, Dominique Coupé (Safran, France), Keith Sharp (USA).

Textile mechanics: David Lloyd (Manchester and Leeds Universities, UK).

References

A World Leader in Research, Development and Manufacturing of 3-Dimensional Engineered Fiber Preforms for Composite Products. Retrieved 14/08/2024, from https://3tex.com/

Abernathy, W.J., Clark, K.B., 1985. Innovation: mapping the winds of creative destruction. Res. Pol. 14 (1), 3–22. https://doi.org/10.1016/0048-7333(85)90021-6.

Aboudi, J., 1989. Micromechanical analysis of composites by the method of cells. Appl. Mech. Rev. 42, 193–221. https://doi.org/10.1115/1.3152428.

Abtew, M.A., Bruniaux, P., Boussu, F., Loghin, C., Cristian, I., Chen, Y., Wang, L.C., 2018. Female seamless soft body armor pattern design system with innovative reverse engineering approaches. Int. J. Adv. Manuf. Technol. 98 (9–12), 2271–2285. https://doi.org/10.1007/s00170-018-2386-y.

Abtew, M.A., Boussu, F., Bruniaux, P., Loghin, C., Cristian, I., 2020. Effect of structural parameters on the deformational behaviors of multiply 3D layer-by-layer angle-interlock para-aramid fabric for fiber-reinforcement composite. J. Compos. Sci. 4 (4). https://doi.org/10.3390/jcs4040145.

Adams, K.L., Rebenfeld, L., 1987. In-plane flow of fluids in fabrics: structure/flow characterization. Textil. Res. J. 57 (11), 647–654. https://doi.org/10.1177/004051758705701104.

Adams, K.L., Rebenfeld, L., 1991a. Permeability characteristics of multilayer fiber reinforcements. Part I: experimental observations. Polym. Compos. 12 (3), 179–185. https://doi.org/10.1002/pc.750120307.

Adams, K.L., Rebenfeld, L., 1991b. Permeability characteristics of multilayer fiber reinforcements. Part II: theoretical model. Polym. Compos. 12 (3), 186–190. https://doi.org/10.1002/pc.750120308.

Adams, K.L., Miller, B., Rebenfeld, L., 1986. Forced in-plane flow of an epoxy resin in fibrous networks. Polym. Eng. Sci. 26 (20), 1434–1441. https://doi.org/10.1002/pen.760262013.

Adams, K.L., Russel, W.B., Rebenfeld, L., 1988. Radial penetration of a viscous liquid into a planar anisotropic porous medium. Int. J. Multiphas. Flow 14 (2), 203–215. https://doi.org/10.1016/0301-9322(88)90006-7.

Adanur, S., 1997. Paper Machine Clothing. CRC Press.

Adluru, H.K., Zhou, E., Hoos, K.H., Popelar, C., Ballard, M.K., Mollenhauer, D., Riha, D., Selvarathinam, A.S., Iarve, E.V., 2023. Performance prediction of interlock woven composites by independent mesh method. Compos. Appl. Sci. Manuf. 165, 107317. https://doi.org/10.1016/j.compositesa.2022.107317.

Adolphs, G., Skinner, C., 2011. 21 - non-crimp fabric composites in wind turbines. In: Lomov, S.V. (Ed.), Non-Crimp Fabric Composites. Woodhead Publishing, pp. 481–495e. https://doi.org/10.1533/9780857092533.4.481.

Ai, S.G., Song, W.L., Chen, Y.F., 2021. Stress field and damage evolution in C/SiC woven composites: image-based finite element analysis and in situ X-ray computed tomography tests. J. Eur. Ceram. Soc. 41 (4), 2323–2334. https://doi.org/10.1016/j.jeurceramsoc.2020.12.026.

Aimène, Y., Vidal-Sallé, E., Hagège, B., Sidoroff, F., Boisse, P., 2009. A hyperelastic approach for composite reinforcement large deformation analysis. J. Compos. Mater. 44 (1), 5–26. https://doi.org/10.1177/0021998309345348.

Aktas, A., Sims, G., Lira, C., Stojkovic, M., 2016. Survey of Procedures in Use for Permeability Measurements in Liquid Composite Moulding Processes, NPL Report Mat 82. Teddington. National Physical Laboratory. https://eprintspublications.npl.co.uk/7091/.

Albany, 2024a. Paper Machine Clothing. Retrieved 14/08/2024, from https://www.albint.com/business/mc/en-us/pmc/pages/default.aspx.

Albany, 2024b. 3D Weaving. Retrieved 13/08/2024, from https://www.albint.com/engineered-composites/capabilities/process-capabilities/3d-weaving/

Alms, J., Correia, N., Advani, S., Ruiz, E., 2010. Experimental Procedures to Run Longitudinal Injections to Measure Unsaturated Permeability of LCM Reinforcements. Retrieved 23/08/2024, from. https://www.researchgate.net/publication/273327662_Experimental_procedures_to_run_longitudinal_injections_to_measure_unsaturated_permeability_of_LCM_reinforcements.

Amirbayat, J., 1995. Dimensionless groups in textile mechanics and the stories they tell. J. Textil. Inst. 86 (2), 251–256. https://doi.org/10.1080/00405009508631331.

Amirbayat, J., Hearle, J.W.S., 1989. The anatomy of buckling of textile fabrics: drape and conformability. J. Textil. Inst. 80 (1), 51–69. https://doi.org/10.1080/00405008908659185.

Andersen, T.A., 2018. Here Is the World's Largest Wind Turbine Blade. Retrieved 03/08/2024, from. https://www.dtu.dk/english/news/all-news/nyhed?id=00d429c3-eb5b-4e1a-8102-8016d0f15699.

Arbter, R., Binetruy, C., Bizet, L., Bréard, J., Beraud, J.M., Comas-Cardona, S., Demaria, C., Endruweit, A., Ermanni, P., Gommer, F., Hasanovic, S., Klunker, F., Lavanchy, S., Laine, B., Long, A., Lomov, S.V., Michaud, V., Morren, G., Henrat, P., Ruiz, E., Sol, H., Trochu, F., Verleye, B., Wietgrefe, M., Wu, W., Ziegmann, G., 2011. Experimental determination of the permeability of textiles: a benchmark exercise. Composites Part A 42 (9), 1157–1168. https://doi.org/10.1016/j.compositesa.2011.04.021.

Auenhammer, R.M., Mikkelsen, L.P., Asp, L.E., Blinzler, B.J., 2020. Dataset of non-crimp fabric reinforced composites for an X-ray computer tomography aided engineering process. Data Brief 33. https://doi.org/10.1016/j.dib.2020.106518.

Auenhammer, R.M., Mikkelsen, L.P., Asp, L.E., Blinzler, B.J., 2021. Automated X-ray computer tomography segmentation method for finite element analysis of non-crimp fabric reinforced composites. Compos. Struct. 256. https://doi.org/10.1016/j.compstruct.2020.113136.

Auenhammer, R.M., Jeppesen, N., Dahl, V.A., Mikkelsen, L.P., 2022a. X-ray computed tomography data structure tensor orientation mapping for finite element models – STXAE. Software Imp. 11, 100216. https://doi.org/10.1016/j.simpa.2021.100216.

Auenhammer, R.M., Jeppesen, N., Mikkelsen, L.P., Dahl, V.A., Blinzler, B.J., Asp, L.E., 2022b. Robust numerical analysis of fibrous composites from X-ray computed tomography image data enabling low resolutions. Compos. Sci. Technol. 109458. https://doi.org/10.1016/j.compscitech.2022.109458.

Bach, C., Schreiber, F., Theelen, K., Schulte Südhoff, E., Lee, H.-Y., Ko, F.K., Gries, T., 2012. 3D hexagonal braiding: opportunities for manufacturing complex braided structures. Mitteilungen/Institut für Textiltechnik 59, 99.

Badel, P., Vidal-Salle, E., Maire, E., Boisse, P., 2008. Simulation and tomography analysis of textile composite reinforcement deformation at the mesoscopic scale. Compos. Sci. Technol. 68 (12), 2433–2440. https://doi.org/10.1016/j.compscitech.2008.04.038.

Bailie, J. (1989). Woven fabric aerospace structures. Handbook of composites. C. T. Herakohich and Y. M. Tarnopol'skii. 2: 353-391.

Bale, H., Blacklock, M., Begley, M.R., Marshall, D.B., Cox, B.N., Ritchie, R.O., 2012. Characterizing three-dimensional textile ceramic composites using synchrotron x-ray micro-computed-tomography. J. Am. Ceram. Soc. 95 (1), 392–402. https://doi.org/10.1111/j.1551-2916.2011.04802.x.

Bally Ribbon Mills, 2012. 3-D Woven Materials for Space Flight Applications. BRM. Retrieved 14/08/2024, from https://www.ballyribbon.com/projects/nasa/.

Barber, J.G., Handorf, A.M., Allee, T.J., Li, W.J., 2013. Braided nanofibrous scaffold for tendon and ligament tissue engineering. Tissue Eng Part A 19 (11–12), 1265–1274. https://doi.org/10.1089/ten.tea.2010.0538.

Becker, M., F. Ficker and R. Miksch (2008). Variation braiding technology by the example of novel stent structures. STRUTEX 2018. Liberec, Czech Republic. 1: 11–13.

Ben Boubaker, B., Haussy, B., Ganghoffer, J.-F., 2005. Discrete models of fabric accounting for yarn interactions. Rev. Eur. Des. Eléments Finis 14 (6–7), 653–675. https://doi.org/10.3166/reef.14.653-675.

Berger, H., 1975. Apparatus for Braiding Knotless Netting. USA. US3866512.

Bigaud, D., Dréano, L., Hamelin, P., 2005. Models of interactions between process, microstructure and mechanical properties of composite materials—a study of the interlock layer-to-layer braiding technique. Compos. Struct. 67 (1), 99–114. https://doi.org/10.1016/j.compstruct.2004.01.009.

Bilisik, K., 2009. Multiaxis 3D weaving: comparison of developed two weaving methods (tube-rapier weaving versus tube-carrier weaving) and effects of bias yarn path to the preform properties. Fibers Polym. 11 (1), 104–114. https://doi.org/10.1007/s12221-010-0104-y.

Bilisik, K., 2012. Multiaxis three-dimensional weaving for composites: a review. Textil. Res. J. 82 (7), 725–743. https://doi.org/10.1177/0040517511435013.

Birkefeld, K., 2013. Virtual Optimisation of Braided Preforms Considering Production Aspects. PhD, University of Stuttgart. https://doi.org/10.18419/opus-3926.

Blackketter, D., Walrath, D., Hansen, A., 1993. Modeling damage in a plain weave fabric-reinforced composite material. J. Compos. Technol. Res. 15, 136–142. https://doi.org/10.1520/CTR10364J.

Blaisdell, S.B., 1932. Braiding Brake Lining and Machine for Making Same. USA. US1885676.

Bluck, B.M., 1969. High Speed Bias Weaving and Braiding. USA. US3426804.

BMW, 2015. BMW 7er 2015: Video Erklärt Carbon Core Leichtbau-Konzept. Retrieved 25/08/ 2024, from https://www.bimmertoday.de/2015/06/09/bmw-7er-2015-video-erklart-carbon-core-leichtbau-konzept/.

Bock, K., Flohr, G., 1978. Process and Machine for Production of Braided Packing. USA. US4096781.

Bogdanovich, A.E., 1993. Three-dimensional analysis of anisotropic spatially reinforced structures. Compos. Manufact. 4 (4), 173–186. https://doi.org/10.1016/0956-7143(93)90002-P.

Bogdanovich, A. (2015). 2 - Solid Three-Dimensional Woven Textiles. Advances in 3D Textiles. X. Chen, Woodhead Publishing: 21-52. https://doi.org/10.1016/B978-1-78242-214-3.00002-4.

Bogdanovich, A. E. (2016). An Overview of Three-Dimensional Braiding Technologies. Advances in Braiding Technology. Y. Kyosev, Woodhead Publishing: 3-78. https://doi.org/10.1016/C2014-0-03190-5.

Bogdanovich, A.E., Mohamed, M., 2009. Three-dimensional reinforcements for composites. SAMPE J. 45 (6), 8–28.

Bogdanovich, A.E., Mungalov, D., Duke, P.W., 2008. New 3-D rotary braiding technology for high-speed manufacturing of unitary, complex shape preforms and composites. In: First World Conf. On 3D Fabrics and Their Applications. Manchester.

Bogdanovich, A.E., Pastore, C.M., 1996a. Mechanics of Textile and Laminated Composites. Chapman and Hall, London.

Bogdanovich, A.E., Pastore, C.M., 1996b. Material-smart analysis of textile-reinforced structures. Compos. Sci. Technol. 56, 291–309. https://doi.org/10.1016/0266-3538(95)00109-3.

Bogdanovich, A. E. and C. M. Pastore (1992). On the Structural Analysis of Textile Composites. Topics in Composite Materials and Structures. V. Birman and A. Nagar. New York, ASME: 1-19.

Bogdanovich, A.E., Karahan, M., Lomov, S.V., Verpoest, I., 2013. Quasi-static tensile behavior and progressive damage in carbon/epoxy composite reinforced with 3D non-crimp orthogonal woven fabric. Mech. Mater. 62, 14–31. https://doi.org/10.1016/j.mechmat.2013.03.005.

Bogdanovich, A. E. (2017). Multi-scale Progressive Failure Modeling: From Nano-Structured Carbon Fibers to Textile Composites. The Structural Integrity of Carbon Fiber Composites: Fifty Years of Progress and Achievement of the Science, Development, and Applications. P. W. R. Beaumont, C. Soutis and A. Hodzic. Cham, Springer International Publishing: 767-793. https://doi.org/10.1007/978-3-319-46120-5_26.

Boisse, P., 2006. Meso-macro approach for composites forming simulation. J. Mater. Sci. 41 (20), 6591–6598. https://doi.org/10.1007/s10853-006-0198-1.

Boisse, P. (Ed.), 2011. Composite Reinforcements for Optimum Performance. Woodhead Publishing, Oxford.

Boisse, P. (Ed.), 2021. Composite Reinforcements for Optimum Performance, second ed. Woodhead Publishing, Oxford.

Boisse, P., Buet, K., Gasser, A., Hanklar, S., Launay, J., 1997. Biaxial mechanical behaviour of textile composite reinforcements. Mecanique Industrielle Et Materiaux 50 (3), 140–144.

Boisse, P., Zouari, B., Gasser, A., 2005. A mesoscopic approach for the simulation of woven fibre composite forming. Compos. Sci. Technol. 65, 429–436. https://doi.org/10.1016/j.compscitech.2004.09.024.

Boisse, P., Zouari, B., Daniel, J.L., 2006. Importance of in-plane shear rigidity in finite element analyses of woven fabric composite preforming. Compos. Appl. Sci. Manuf. 37 (12), 2201–2212. https://doi.org/10.1016/j.compositesa.2005.09.018.

Boisse, P., R. Akkerman, J. Cao, J. Chen, S. V. Lomov and A. Long. (2007). Composites Forming. Advances in Material Forming. Esaform 10 Years on. F. Chinesta and E. Cueto, Springer: 61-79. https://doi.org/10.1007/978-2-287-72143-4_5.

Boisse, P., Colmar, J., Hamila, N., Naouar, N., Steer, Q., 2018a. Bending and wrinkling of composite fiber preforms and prepregs. A review and new developments. Comp. Part B 141, 234–249. https://doi.org/10.1016/j.compositesb.2017.12.061.

Boisse, P., Hamila, N., Madeo, A., 2018b. The difficulties in modeling the mechanical behavior of textile composite reinforcements with standard continuum mechanics of Cauchy. Some possible remedies. Int. J. Solid Struct. 154, 55–65. https://doi.org/10.1016/j.ijsolstr.2016.12.019.

Boisse, P., Akkerman, R., Carlone, P., Karger, L., Lomov, S.V., Sherwood, J., 2022. Advances in composite forming through 25 years of ESAFORM. Int. J. Material Form. 15, 39. https://doi.org/10.1007/s12289-022-01682-8.

Boisse, P., Hamila, N., Lomov, S.V., Naouar, N., Vidal-Sallé, E., 2024. Mechanics of woven fabrics. Comprehens. Mech. Materials 4, 113–149. https://doi.org/10.1016/B978-0-323-90646-3.00017-4.

Bombard, B., P. Olry, R. Duval, A. Bruyere, D. Coupe and J. Aucagne (2000). Method for Producing Multiaxial Fibrous Webs. EU, Safran Aircraft Engines SAS; Hexcel Reinforcements SAS; Safran Ceramics SA; Lexvall SA. EP0972102B1.

Boussu, F., 2011. The use of warp interlock fabric inside textile composite protection against ballistic impact. Textil. Res. J. 81 (4), 344–354. https://doi.org/10.1177/0040517510385170.

Boussu, F., Cristian, I., Nauman, S., 2015. General definition of 3D warp interlock fabric architecture. Compos. B Eng. 81, 171–188. https://doi.org/10.1016/j.compositesb.2015.07.013.

Boussu, F., Abtew, M.A., Bruniaux, P., 2022. 3D warp interlock fabric structure and their applications in soft and hard armour protections. Appl. Compos. Mater. 29 (1), 65–82. https://doi.org/10.1007/s10443-021-09955-2.

Brandt, J., Drechsler, K., Arendts, P.J., 1996. Mechanical performance of composites based on various three-dimensional woven-fibre preforms. Compos. Sci. Technol. 56, 381–386. https://doi.org/10.1016/0266-3538(95)00135-2.

Branscomb, D., Beale, D., Broughton, R., 2013. New directions in braiding. J. Eng. Fib. Fabr. 8 (2). https://doi.org/10.1177/155892501300800202.

Brookstein, D.S., 1991. A comparison of multilayer interlocked braided composites with other 3D braided composites. In: 36th International SAMPE Symposium. San Diego, CA, vol. 1, pp. 141–150.

Brookstein, D.S., Skelton, J., Dent, J.R., Dent, R.W., Rose, D.J., 1994. Solid Braid Structure. US5357839.

Brown, R.T., 1985. Through-the-thickness braiding technology. In: 30th National SAMPE Symposium. Anaheim, CA. pp. 1509–1518.

Brown, R.T., Ratliff, E.D., 1986. Method of Sequenced Braider Motion for Multiply Braiding Apparatus. USA. US4621560.

Brown, L.P., 2019. Louisepb/TexGen: TexGen v3.11.0. Retrieved 18/08/2024, from. https://zenodo.org/records/3241493.

Brown, L.P., Long, A.C., 2021. 8 - Modeling the Geometry of Textile Reinforcements for Composites: TexGen. Composite Reinforcements for Optimum Performance, second ed. Woodhead Publishing, pp. 237–265. https://doi.org/10.1016/B978-0-12-819005-0.00008-3. P. Boisse.

Brown, L. P. (2022). TexGen. Advanced Weaving Technology. Y. Kyosev and F. Boussu. Cham, Springer International Publishing: 253-291. https://doi.org/10.1007/978-3-030-91515-5_6.

Brunnschweiler, D., 1953. Braids and braiding. J. Textil. Inst. 44, P666–P686. https://doi.org/10.1080/19447015308687874.

Bruschke, M.V., Advani, S.G., 1990. A finite element/control volume approach to mold filling in anisotropic porous media. Polym. Compos. 11 (6), 398–405. https://doi.org/10.1002/pc.750110613.

Budillon, F. and M. Brauner (2024). Highly Flexible Production Process for Economical Manufacturing of Patient Specific Braided Implantats. International Week of Narrow Textiles. Dresden.

Bunsell, A.R., 1977. Acoustic emission for proof testing of carbon fibre-reinforced plastics. NDT E Int. 10 (1), 21–25. https://doi.org/10.1016/0308-9126(77)90108-0.

Byun, J.-H., Chou, T.-W., 1996. Process-microstructure relationship of 2-step and 4-step braided composites. Compos. Sci. Technol. 56, 235–251. https://doi.org/10.1016/0266-3538(95)00112-3.

Cahuzak, G. (1988). Woven Reinforcement for a Composite Material. France. FR19870002012.

Cahuzak, G. and F. Monget (1987). Method for Manufacturing Three-Dimension Woven Composite Reinforcing Elements, Machine for Implementing Same and Product Obtained. France. FR2612950B1.

Cai, Z., Yu, J.Z., Ko, F.K., 1994. Formability of textile preforms for composite applications. Part 2: evaluation experiments and modelling. Compos. Manuf. 5 (2), 123–132. https://doi.org/10.1016/0956-7143(94)90063-9.

Cai, Y., Zhao, Z., Liu, P., Xing, J., Zhang, C., 2024. Revealing the failure mechanism of 2D triaxially braided composites under off-axial tension through mesoscale simulations. Compos. Sci. Technol. 257, 110838. https://doi.org/10.1016/j.compscitech.2024.110838.

Calba, J., Soulat, D., Legrand, X., Renauld, S., 2023. Damage investigation on the carbon tows during rewinding and braiding processes. Fibers 11 (3). https://doi.org/10.3390/fib11030030.

Cao, J., Akkerman, R., Boisse, P., Chen, J., Cheng, H.S., Graaf, E.F.d., Gorczyca, J.L., Harrison, P., Hivet, G., Launay, J., Lee, W., Liu, L., Lomov, S.V., Long, A., Luycker, E.d., Morestin, F., Padvoiskis, J., Peng, X.Q., Sherwood, J., Stoilova, T., Tao, X.M., Verpoest, I., Wiggers, J., Willems, A., Yu, T.X., Zhu, B., 2008. Characterization of mechanical behavior of woven fabrics: experimental methods and benchmark results. Compos. Part A 39, 1037–1053. https://doi.org/10.1016/j.compositesa.2008.02.016.

Carey, J.P., 2017. Handbook of Advances in Braided Composite Materials, Theory, Production, Testing and Applications. Woodhead Publishing - Elsevier, Amsterdam. https://doi.org/10.1016/C2014-0-03943-3.

Carey, J. P., G. W. Melenka, A. Hunt, B. Cheung, M. Ivey and C. Ayranci (2016). 6 - Braided Composites in Aerospace Engineering. Advanced Composite Materials for Aerospace Engineering. S. Rana and R. Fangueiro, Woodhead Publishing: 175-212. https://doi.org/10.1016/B978-0-08-100037-3.00006-7.

Carvelli, V., Lomov, S.V. (Eds.), 2015. Fatigue of Textile Composites. Elsevier - Woodhead, Cambridge.

Carvelli, V., Gramellini, G., Lomov, S.V., Bogdanovich, A.E., Mungalov, D., Verpoest, I., 2010a. Fatigue behaviour of non-crimp 3D orthogonal weave and multi-layer plain weave E-glass reinforced composites. Compos. Sci. Technol. 70 (14), 2068–2076. https://doi.org/10.1016/j.compscitech.2010.08.002.

Carvelli, V., Tomaselli, V.N., Lomov, S.V., Verpoest, I., Witzel, V., Van Den Broucke, B., 2010b. Fatigue and post-fatigue tensile behaviour of non-crimp stitched and unstitched carbon/epoxy composites. Compos. Sci. Technol. 70, 2216–2224. https://doi.org/10.1016/j.compscitech.2010.09.004.

Carvelli, V., Jain, A., Lomov, S.V., 2017. Fatigue of Textile and Short Fiber Reinforced Composites. Wiley - ISTE, London. https://doi.org/10.1002/9781119413592.

Chalal, H., Meraghni, F., Pierron, F., Grediac, A., 2004. Direct identification of the damage behaviour of composite materials using the virtual fields method. Compos. Appl. Sci. Manuf. 35 (7–8), 841–848. https://doi.org/10.1016/j.compositesa.2004.01.011.

Charmetant, A., Vidal-Salle, E., Boisse, P., 2011. Hyperelastic modelling for mesoscopic analyses of composite reinforcements. Compos. Sci. Technol. 71, 1623–1631. https://doi.org/10.1016/j.compscitech.2011.07.004.

Charmetant, A., Orliac, J.G., Vidal-Salle, E., Boisse, P., 2012. Hyperelastic model for large deformation analyses of 3D interlock composite preforms. Compos. Sci. Technol. 72 (12), 1352–1360. https://doi.org/10.1016/j.compscitech.2012.05.006.

Chen, X. (Ed.), 2015. Advances in 3D Textiles. Woodhead Publishing.

Chou, T.-W., 1992. Microstructural Design of Fiber Composites. Cambridge University Press, Cambridge. https://doi.org/10.1017/CBO9780511600272.

Chou, T.-W., Ko, F.K., 1989. Textile Structural Composites. Elsevier, Amsterdam.

Cobb, H.Z., 1914a. Braided Fabric for Hose. US1104777.

Cobb, H.Z., 1914b. Braided Fabric for Hose. US1104778.

Colin, D., Bel, S., Hans, T., Hartmann, M., Drechsler, K., 2020. Virtual description of non-crimp fabrics at the scale of filaments including orientation variability in the fibrous layers. Appl. Compos. Mater. 27 (4), 337–355. https://doi.org/10.1007/s10443-020-09819-1.

Cooper, J.A., Lu, H.H., Ko, F.K., Freeman, J.W., Laurencin, C.T., 2005. Fiber-based tissue-engineered scaffold for ligament replacement: design considerations and in vitro evaluation. Biomaterials 26 (13), 1523–1532. https://doi.org/10.1016/j.biomaterials.2004.05.014.

Cornelissen, B., Akkerman, R., 2009. Analysis of yarn bending behaviour. In: 17th International Conference on Composite Materials (ICCM 17). Edinburgh. https://research.utwente.nl/en/publications/analysis-of-yarn-bending-behaviour.

Cornelissen, B., Rietman, B., de Rooij, M.B., Akkerman, R., 2012. Tow Mechanics: A Contact Mechanics Approach of Friction in Fibrous Tows during Forming. Material Forming - Esaform 2012, Pts 1 & 2, pp. 504–506. https://doi.org/10.4028/www.scientific.net/KEM.504-506.325, 325-330.

Cornelissen, B., Rietman, B., Akkerman, R., 2013. Frictional behaviour of high performance fibrous tows: friction experiments. Compos. Appl. Sci. Manuf. 44, 95–104. https://doi.org/10.1016/j.compositesa.2012.08.024.

Coupé, D. (2011). 4 - Woven Reinforcements for Composites. Composite Reinforcements for Optimum Performance. P. Boisse, Woodhead Publishing: 89-115. https://doi.org/10.1533/9780857093714.2.89.

Cox, B.N., Dadkhan, M.S., 1995. The macroscopic elasticity of 3D woven composites. J. Compos. Mater. 29 (6), 785–819. https://doi.org/10.1177/002199839502900606.

Cox, B.N., Carter, W.C., Fleck, N.A., 1994. A binary model of textile composites: I Formulation. Acta Mater. 44 (10), 3463–3479. https://doi.org/10.1016/0956-7151(94)90479-0.

Cox, B.N., Nilakantan, G., Sudre, O., Marshall, D.B., 2016. Generating virtual specimens for complex non-periodic woven structures by converting machine instructions into topological ordering rules. Compos. Struct. 141, 63–78. https://doi.org/10.1016/j.compstruct.2016.01.009.

Crookston, J.J., Long, A.C., Jones, I.A., 2005a. A summary review of mechanical properties prediction methods for textile reinforced polymer composites. Proc. IME J. Mater. Des. Appl. 219 (L2), 91–109. https://doi.org/10.1243/146442005X10319.

Crookston, J. J., M. N. Sherburn, L. G. Zhao, J. Ooi, A. C. Long and I. A. Jones (2005b). Finite Element Analysis of Textile Composite Unit Cells Using Conventional and Novel Techniques. Proceedings ICCM-15. Durban: CD Edition.

Dambrine, B., J. N. Mahieu, J. Goering and K. Oliellete (2008). Development of 3D Woven, Resin Transfer Molded Fan Blades. Recent Advances in Textile Composites. S. G. Advani and J. W. Gillespie. Lancaster, Destech Publications, Inc: 19-24.

Dang, H., Zhao, Z., Liu, P., Zhang, C., Tong, L., Li, Y., 2020. A new analytical method for progressive failure analysis of two-dimensional triaxially braided composites. Compos. Sci. Technol. 186, 107936. https://doi.org/10.1016/j.compscitech.2019.107936.

de Jong, S., Postle, R., 1978. A general energy analysis in fabric mechanics using optimal control theory. Textil. Res. J. 48 (3), 127–135. https://doi.org/10.1177/004051757804800.

dell'Isola, F., Steigmann, D., 2015. A two-dimensional gradient-elasticity theory for woven fabrics. J. Elasticity 118 (1), 113–125. https://doi.org/10.1007/s10659-014-9478-1.

Denis, Y., Morestin, F., Hamila, N., 2020. A hysteretic model for fiber-reinforced composites at finite strains: fractional derivatives, computational aspects and analysis. Comput. Mater. Sci. 181, 109716. https://doi.org/10.1016/j.commatsci.2020.109716.

Desplentere, F., Lomov, S.V., Woerdeman, D.L., Verpoest, I., Wevers, M., Bogdanovich, A., 2005. Micro-CT Characterization of variability in 3D textile architecture. Compos. Sci. Technol. 65, 1920–1930. https://doi.org/10.1016/j.compscitech.2005.04.008.

Dhiman, S., Potluri, P., Silva, C., 2015. Influence of binder configuration on 3D woven composites. Compos. Struct. 134, 862–868. https://doi.org/10.1016/j.compstruct.2015.08.126.

Doitrand, A., Fagiano, C., Irisarri, F.X., Hirsekorn, M., 2015. Comparison between voxel and consistent meso-scale models of woven composites. Compos. Appl. Sci. Manuf. 73, 143–154. https://doi.org/10.1016/j.compositesa.2015.02.022.

Doitrand, A., Fagiano, C., Hild, F., Chiaruttini, V., Mavel, A., Hirsekorn, M., 2017. Mesoscale analysis of damage growth in woven composites. Compos. Appl. Sci. Manuf. 96, 77–88. https://doi.org/10.1016/j.compositesa.2017.02.018.

Dow, M.B., Dexter, H.B., 1997. Development of Stitched, Braided and Woven Composite Structures in the ACT Program and at Langley Research Center (1985 to 1997). NASA. https://ntrs.nasa.gov/api/citations/19980000063/downloads/19980000063.pdf.

Dumont, F. and C. Weimer (2011). 19 - Non-crimp Fabric: Preforming Analysis for Helicopter Applications. Non-crimp Fabric Composites. S. V. Lomov, Woodhead Publishing: 449-460. https://doi.org/10.1533/9780857092533.4.449.

Durville, D. (2007). Finite Element Simulation of Textile Materials at Mesoscopic Scale. Proceedings of Symposium "Finite Element Modelling of Textiles and Textile Composites", St.-Petersburg.

Durville, D., Baydoun, I., Moustacas, H., Perie, G., Wielhorski, Y., 2018. Determining the initial configuration and characterizing the mechanical properties of 3D angle-interlock fabrics using finite element simulation. Int. J. Solid Struct. 154, 97–103. https://doi.org/10.1016/j.ijsolstr.2017.06.026.

Dutta, S., Karlgaard, C.D., Korzun, A.M., Green, J.S., Tynis, J.A., Williams, J.D., Yount, B., Cassell, A.M., Wercinski, P.F., 2021. Adaptable deployable entry and placement technology sounding rocket one modeling and reconstruction. J. Spacecraft Rockets 59 (1), 236–259. https://doi.org/10.2514/1.A35090.

Ebel, C., A. Mierzwa and K. Kind (2016). 13 - Yarn Damage during Braiding of Reinforcement Fibers for Composites. Advances in Braiding Technology. Y. Kyosev, Woodhead Publishing: 319-354. https://doi.org/10.1016/B978-0-08-100407-4.00013-2.

El Said, B., Daghia, F., Ivanov, D., Hallett, S.R., 2018. An iterative multiscale modelling approach for nonlinear analysis of 3D composites. Int. J. Solid Struct. 132–133, 42–58. https://doi.org/10.1016/j.ijsolstr.2017.08.017.

Emonts, C., Grigat, N., Merkord, F., Vollbrecht, B., Idrissi, A., Sackmann, J., Gries, T., 2021. Innovation in 3D braiding technology and its applications. Textiles 1 (2), 185–205. https://doi.org/10.3390/textiles1020009.

Endruweit, A., Long, A.C., Robitaille, F., Rudd, C.D., 2004. Dependence of Permeability Variations on the Textile Structure. Proceedings ECCM-11: CD Edition.

Endruweit, A., McGregor, P., Long, A.C., Johnson, M.S., 2006. Influence of the fabric architecture on the variations in experimentally determined in-plane permeability values. Compos. Sci. Technol. 66 (11–12), 1778–1792. https://doi.org/10.1016/j.compscitech.2005.10.031.

ENStandards (2000). CSN EN 13473-1 Reinforcement - Specifications for Multiaxial Multiply Fabrics. https://www.en-standard.eu/csn-en-13473-1-reinforcement-specifications-for-multi-axial-multi-ply-fabrics-part-1-designation/.

Eurocarbon, 2021. NH-90 Helicopter Landing Gear. Retrieved 29/11/2024, from https://www.eurocarbon.com/nh-90-helicopter-landing-gear/.

Eurocarbon, 2024. About Eurocarbon. Retrieved 29/11/2024, from https://www.eurocarbon.com/about/.

Ewert, A., Drach, B., Vasylevskyi, K., Tsukrov, I., 2020. Predicting the overall response of an orthogonal 3D woven composite using simulated and tomography-derived geometry. Compos. Struct. 243, 112169. https://doi.org/10.1016/j.compstruct.2020.112169.

Fang, G., El Said, B., Ivanov, D., Hallett, S.R., 2016. Smoothing artificial stress concentrations in voxel-based models of textile composites. Compos. Appl. Sci. Manuf. 80, 270–284. https://doi.org/10.1016/j.compositesa.2015.10.025.

Fauster, E., Berg, D.C., Abliz, D., Grössing, H., Meiners, D., Ziegmann, G., Schledjewski, R., 2018. Image processing and data evaluation algorithms for reproducible optical in-plane permeability characterization by radial flow experiments. J. Compos. Mater. 53 (1), 45–63. https://doi.org/10.1177/0021998318780209.

Fedro, M. J. and K. Willden (1992). Characterization and manufacture of braided composites for large commercial aircraft structures. Ninth DOD (NASA) FAA Conference on Fibrous Composites in Structural Design, vol. 2 https://ntrs.nasa.gov/citations/19950022057.

Feindler, N. (2012). Charakterisierungs- und Simulationsmethodik zum Versagensverhalten energieabsorbierender Faserverbundstrukturen. München, Verlag Dr. Hut.

Ferretti, M., Madeo, A., dell'Isola, F., Boisse, P., 2014. Modeling the onset of shear boundary layers in fibrous composite reinforcements by second-gradient theory. Z. Angew. Math. Phys. 65 (3), 587–612. https://doi.org/10.1007/s00033-013-0347-8.

Florentine, R.A., 1982. Apparatus for Weaving a Three-Dimensional Article. USA. US4312261.

Foster, C.W., Collins, L.N., Panerai, F., Roberts, S.A., 2023. Assessing thermophysical properties of parameterized woven composite models using image-based simulations. Compos. Sci. Technol. 241. https://doi.org/10.1016/j.compscitech.2023.110136.

Fourrier, G., Rassineux, A., Leroy, F.-H., Hirsekorn, M., Fagiano, C., Baranger, E., 2023. Automated conformal mesh generation chain for woven composites based on CT-scan images with low contrasts. Compos. Struct. 308, 116673. https://doi.org/10.1016/j.compstruct.2023.116673.

Gao, X.G., Han, X., Song, Y.D., 2017. X-ray computed tomography based microstructure reconstruction and numerical estimation of thermal conductivity of 2.5D ceramic matrix composite. Ceram. Int. 43 (13), 9790–9797. https://doi.org/10.1016/j.ceramint.2017.04.157.

Gao, S., Qu, Z., Zhang, Y., Zhao, Z., Liang, B., 2022. An efficient hyper-elastic model for the preforming simulation of Carbon-Kevlar hybrid woven reinforcement. Chin. J. Aeronaut. 35 (12), 321–335. https://doi.org/10.1016/j.cja.2022.05.007.

Garoz, D., Gilabert, F.A., Sevenois, R.D.B., Spronk, S.W.F., Van Paepegem, W., 2019. Consistent application of periodic boundary conditions in implicit and explicit finite element simulations of damage in composites. Compos. B Eng. 168, 254–266. https://doi.org/10.1016/j.compositesb.2018.12.023.

Gasser, A., Boisse, P., Hanklar, S., 2000. Analysis of the mechanical behaviour of dry fabric reinforcements. 3D simulations versus biaxial tests. Comput. Mater. Sci. 17, 7–20. https://doi.org/10.1016/S0927-0256(99)00086-5.

Ge, J., Liu, Z., Hu, X., Liu, X., Li, B., He, C., Liang, J., 2024. A new progressive fatigue damage model for the three-dimensional braided composites subjected to locally-variable-amplitude loading. Int. J. Fatig. 184, 108313. https://doi.org/10.1016/j.ijfatigue.2024.108313.

Gholami, A., Melenka, G.W., 2022. Finite element analysis of 2-D tubular braided composite based on geometrical models to study mechanical performances. Mech. Adv. Mater. Struct. 29 (28), 7542–7558. https://doi.org/10.1080/15376494.2021.2001879.

Gigliotti, M., Y. Pannier, Y. Sinchuk, R. Antoranz-Gonzalez, M. C. Lafarie-Frenot and S. V. Lomov (2018). "X-ray μCT based assessment of thermal cycling induced cracks in non-crimp 3D orthogonal woven composite materials with porosity." IOP Conf. Series: Materials Science and Engineering (13th International Conference on Textile Composites (TEXCOMP-13)), vol. 406: 012008.

Gommer, F., Endruweit, A., Long, A.C., 2016. Quantification of micro-scale variability in fibre bundles. Compos. Appl. Sci. Manuf. 87, 131–137. https://doi.org/10.1016/j.compositesa.2016.04.019.

Gorbatikh, L., Lomov, S.V., 2016. Damage Accumulation in Textile Composites. Modeling Damage, Fatigue and Failure of Composite Materials. R. Talreja and J. Varna. Elsevier (Woodhead Publishers), Cambridge, pp. 41–60. https://doi.org/10.1016/B978-1-78242-286-0.00003-0.

Gorbatikh, L., Lomov, S.V., 2018. 2.15 Damage in Architectured Composites. Comprehensive Composite Materials II. Vol. 2. P. W. R. Beaumont and C. H. Zweben. Academic Press, Oxford, pp. 291–306. https://doi.org/10.1016/B978-0-12-803581-8.09919-7.

Gorbatikh, L., Lomov, S.V., 2023. The process of cracking in textile composites subjected to quasistatic tensile and tension-tension fatigue loadings. In: Talreja, R., Varna, J. (Eds.), Modeling Damage, Fatigue and Failure of Composite Materials, second ed. Elsevier (Woodhead Publishers), Cambridge, pp. 37–57. https://doi.org/10.1016/B978-0-443-18489-5.00021-7.

Gorczyca, J.L., Sherwood, J.A., Chen, J., 2004. Friction at the tool-fabric interface during the thermostamping of woven commingled glass-polypropylene composite fabrics. In: Proceedings ESAFORM-2004. Trondheim, pp. 337–340.

Gordeev, V.A., Dudko, D.A., Mokeev, M.N., 1989. Woven Electric Assemblage (In Russian). Moscow, Legprombytizdat.

Goyal, D., Tang, X.D., Whitcomb, J.D., Kelkar, A.D., 2005. Effect of various parameters on effective engineering properties of 2 x 2 braided composites. Mech. Adv. Mater. Struct. 12 (2), 113–128. https://doi.org/10.1080/15376490490493998.

Grail, G., Hirsekorn, M., Wendling, A., Hivet, G., Hambli, R., 2013. Consistent Finite Element mesh generation for meso-scale modeling of textile composites with preformed and compacted reinforcements. Compos. Appl. Sci. Manuf. 55, 143–151. https://doi.org/10.1016/j.compositesa.2013.09.001.

Grave, G., K. Birkefeld, T. Reden, K. Drechsler, Y. Kyosev and A. Rathjens (2009). Simulation of 3D Overbraiding – Solutions and Challenges. Second World Conference on 3D Fabrics and Their Applications. Greenfield SC. https://doi.org/10.13140/2.1.4459.1680.

Grediac, M., 2004. The use of full-field measurement methods in composite material characterisation: interest and limitations. Composites Part A 35, 751–761. https://doi.org/10.1016/j.compositesa.2004.01.019.

Green, S.D., Matveev, M.Y., Long, A.C., Ivanov, D., Hallett, S.R., 2014. Mechanical modelling of 3D woven composites considering realistic unit cell geometry. Compos. Struct. 118, 284–293. https://doi.org/10.1016/j.compstruct.2014.07.005.

Grössing, H., May, D., Kaufmann, S., Schledjewski, R., Mitschang, P., 2015. An evaluation of the reproducibility of capacitive sensor based in-plane permeability measurements: a benchmarking study. Express Polym. Lett. 9, 129–142. https://doi.org/10.3144/expresspolymlett.2015.14.

Gutkin, R., Green, C.J., Vangrattanachai, A., Pinho, S.T., Robinson, P., Curtis, P.T., 2011. On acoustic emission for failure investigation in CFRP: pattern recognition and peak frequency analyses. Mech. Syst. Signal Process. 25, 1393–1407. https://doi.org/10.1016/j.ymssp.2010.11.014.

Guzman-Maldonado, E., Hamila, N., Naouar, N., Moulin, G., Boisse, P., 2016. Simulation of thermoplastic prepreg thermoforming based on a visco-hyperelastic model and a thermal homogenization. Mater. Des. 93, 431–442. https://doi.org/10.1016/j.matdes.2015.12.166.

Hagège, B., Boisse, P., Billoët, J.-L., 2005. Finite element analyses of knitted composite reinforcement at large strain. Rev. Eur. Des. Eléments Finis 14 (6–7), 767–776. https://doi.org/10.3166/reef.14.767-776.

Hamila, N., Boisse, P., Chatel, S., 2009. Semi-discrete shell finite elements for textile composite forming simulation. Int. J. Material Form. 2 (1), 169–172. https://doi.org/10.1007/s12289-009-0518-5.

Harris, B., 2003. Fatigue in Composites. Woodhead Publishing.

He, C., Ge, J., Lian, Y., Geng, L., Chen, Y., Fang, D., 2022. A concurrent three-scale scheme FE-SCA2 for the nonlinear mechanical behavior of braided composites. Comput. Methods Appl. Mech. Eng. 393, 114827. https://doi.org/10.1016/j.cma.2022.114827.

Hearle, J.W.S., Amirbayat, J., 1986. Analysis of drape by means of dimensionless groups. Textil. Res. J. 56 (12), 727–733. https://doi.org/10.1177/004051758605601203.

Hearle, J.W.S., Shanahan, W.J., 1978. An energy method for calculations in fabric mechanics. J. Textil. Inst. 69 (4), 81–110. https://doi.org/10.1080/00405007808631425.

Hearle, J.W.S., Grosberg, P., Baker, S., 1969. Structural Mechanics of Fibres, Yarns and Fabrics. Wiley Interscience, N.Y.

Hearle, J.W.S., Amirbayat, J., Twaites, J.J., 1980. Mechanics of Flexible Fibre Assemblies. Alphen Aan Den Rijn, Sijthoff and Nordhof.

Hernández, L.F.G., Wintiba, B., Li, A.Q., Berke, P.Z., Massart, T.J., 2024. Incorporation of compaction effects in the automated generation of 3D woven composites representative volume elements by geometrical modelling. J. Compos. Mater. https://doi.org/10.1177/00219983241226793.

Hivet, G., Boisse, P., 2005. Consistent 3D geometrical model of fabric elementary cell. Application to a meshing preprocessor for 3D finite element analysis. Finite Elem. Anal. Des. 42, 25–49. https://doi.org/10.1016/j.finel.2005.05.001.

Hivet, G., Boisse, P., 2008. Consistent geometrical model of interlock fabrics. Int. J. Material Form. 1 (1), 875–878. https://doi.org/10.1007/s12289-008-0275-x.

Hoersting, K., 1998. Rationalising the Production of Long Fibre Reinforced Composite Materials Using Multiaxial Multiply Fabrics. Aachen, Shaker Verlag.

Hoersting, K., Wienands, C., Unglaub, M., 1999. Multiaxial Machine with Portal Structure. Liba Maschinenfabrik Gmbh, Germany. DE19726831C2.

Hoes, K., Sol, H., Dinescu, D., Lomov, S.V., Parnas, R., 2004. Study of nesting induced scatter of permeability values in layered reinforcement fabrics. Composites Part A 35 (12), 1407–1418. https://doi.org/10.1016/j.compositesa.2004.05.004.

Hu, X., He, C., Ge, J., Zhang, Q., Liang, J., 2022. Progressive damage and failure analysis of three-dimensional braided composites under multiaxial loadings. Int. J. Solid Struct. 254–255. https://doi.org/10.1016/j.ijsolstr.2022.111853, 111853.

Huguet, S., Godin, N., Gaertner, R., Salmon, L., Villard, D., 2002. Use of acoustic emission to identify damage modes in glass fibre reinforced polyester. Compos. Sci. Technol. 62 (10), 1433–1444. https://doi.org/10.1016/S0266-3538(02)00087-8.

Huysmans, G., Verpoest, I., Van Houtte, P., 1998. A poly-inclusion approach for the elastic modelling of knitted fabric composites. Acta Mater. 46 (9), 3003–3013. https://doi.org/10.1016/S1359-6454(98)00021-4.

Huysmans, G., Verpoest, I., Van Houtte, P., 2001. A damage model for knitted fabric composites. Composites Part A 32 (10), 1465–1475. https://doi.org/10.1016/S1359-835X(01)00045-8.

Iarve, E.V., Mollenhauer, D.H., Zhou, E.G., Breitzman, T., Whitney, T.J., 2009. Independent mesh method-based prediction of local and volume average fields in textile composites. Compos. Appl. Sci. Manuf. 40 (12), 1880–1890. https://doi.org/10.1016/j.compositesa.2009.04.034.

Ilin, K., Lomov, S.V., Ivens, J., 2018. FE modeling of inhomogeneities in laminated fibrous composites based on micro-computed tomography. In: ECCM18 - 18th European Conference on Composite Materials Athens: 3-11.04.

Isart, N., El Said, B., Ivanov, D.S., Hallett, S.R., Mayugo, J.A., Blanco, N., 2015. Internal geometric modelling of 3D woven composites: a comparison between different approaches. Compos. Struct. 132, 1219–1230. https://doi.org/10.1016/j.compstruct.2015.07.007.

Ishikawa, T., Chou, T.W., 1982. Stiffness and strength behaviour of woven fabric composites. J. Mater. Sci. 17 (11), 3211–3220. https://doi.org/10.1007/BF01203485.

ISO, 2023. 4410 Experimental Characterization of In-Plane Permeability of Fibrous Reinforcements for Liquid Composite Moulding. https://www.iso.org/standard/79944.html.

Ivanov, D.S., Lomov, S.V., Baudry, F., Xie, H., Van Den Broucke, B., Verpoest, I., 2009a. Failure analysis of triaxial braided composite. Compos. Sci. Technol. 69, 1372–1380. https://doi.org/10.1016/j.compscitech.2008.09.013.

Ivanov, D.S., Lomov, S.V., Bogdanovich, A.E., Karahan, M., Verpoest, I., 2009b. A comparative study of tensile properties of non-crimp 3D orthogonal weave and multi-layer plain weave E-glass composites. Part 2: comprehensive experimental results. Compos. Part A 40, 1144–1157. https://doi.org/10.1016/j.compositesa.2009.04.032.

Ivanov, S.G., Beyens, D., Gorbatikh, L., Lomov, S.V., 2017. Damage development in woven carbon fibre thermoplastic laminates with PPS and PEEK matrices: a comparative study. J. Compos. Mater. 51, 637–647. https://doi.org/10.1177/0021998316653460.

Ivanov, D.S., Lomov, S.V., 2020. Modeling of 2D and 3D woven composites. In: Irving, P.E., Soutis, C. (Eds.), Polymer Composites in the Aerospace Industry, second ed., pp. 23–57. https://doi.org/10.1016/b978-0-08-102679-3.00002-2.

Ivsan, T.J., Bailey, C., Llewell, J., 1990. Apparatus and Method for Braiding Fiber Strands. US4922798, USA.

Iwata, A., Inoue, T., Naouar, N., Boisse, P., Lomov, S.V., 2019. Coupled meso-macro simulation of woven fabric local deformation during draping. Compos. Part A 118, 267–280. https://doi.org/10.1016/j.compositesa.2019.01.004.

Jacques, S., De Baere, I., Van Paepegem, W., 2014. Application of periodic boundary conditions on multiple part finite element meshes for the meso-scale homogenization of textile fabric composites. Compos. Sci. Technol. 92 (0), 41–54. https://doi.org/10.1016/j.compscitech.2013.11.023.

Jiang, W.-G., Hallett, S.R., Wisnom, M., 2008. Development of domain superposition technique for the modeling of woven fabric composites. In: Camanho, P.P. (Ed.), Mechanical Response of Composites. Springer. https://doi.org/10.1007/978-1-4020-8584-0_14.

Jing, X., Cheng, Z., Teng, X.F., Yang, X.G., Shi, D.Q., 2020. Reconstruction of meso-structure and numerical simulations of the mechanical behavior of three-dimensional four-directional braided ceramic matrix composites. Ceram. Int. 46 (18), 29309–29320. https://doi.org/10.1016/j.ceramint.2020.08.107.

Karahan, M., Lomov, S.V., Bogdanovich, A.E., Mungalov, D., Verpoest, I., 2010. Internal geometry evaluation of non-crimp 3D orthogonal woven carbon fabric composite. Compos. Part A 41, 1301–1311. https://doi.org/10.1016/j.compositesa.2010.05.014.

Karahan, M., Lomov, S.V., Bogdanovich, A.E., Verpoest, I., 2011. Fatigue tensile behavior of carbon/epoxy composite reinforced with non-crimp 3D orthogonal woven fabric. Compos. Sci. Technol. 71, 1961–1972. https://doi.org/10.1016/j.compscitech.2011.09.015.

Kawabata, S., 1975. The Standardisation and Analysis of Hand Evaluation. Textile Machinery Society of Japan, Osaka.

Kawabata, S., Niwa, M., Nanashima, Y., Kawai, H., 1970a. The theoretical investigation on the biaxial tensile properties of plain knitted fabrics. Part I. Theory. J. Text. Mach. Soc. Jpn. 23 (5), 1–14. https://doi.org/10.4188/transjtmsj1965b.23.T95.

Kawabata, S., Niwa, M., Nanashima, Y., Kawai, H., 1970b. The theoretical investigation on the biaxial tensile properties of plain knitted fabrics. Part III. Imrovements in the theory. J. Text. Mach. Soc. Jpn. 23 (10), 13–27. https://doi.org/10.4188/transjtmsj1965b.23.T223.

Kawabata, S., Niwa, M., Kawai, H., 1973a. The finite-deformation theory of plain weave fabrics. Part I. The biaxial-deformation theory. J. Textil. Inst. 64 (1), 21–46. https://doi.org/10.1080/00405007308630416.

Kawabata, S., Niwa, M., Kawai, H., 1973b. The finite-deformation theory of plain weave fabrics. Part II. The uniaxial-deformation theory. J. Textil. Inst. 64 (2), 47–61. https://doi.org/10.1080/00405007308630417.

Kawabata, S., Niwa, M., Kawai, H., 1973c. The finite-deformation theory of plain weave fabrics. Part III. The shear-deformation theory. J. Textil. Inst. 64 (2), 42–85. https://doi.org/10.1080/00405007308630418.

Khokar, N., 1996. 3D fabric-forming processes: distinguishing between 2D weaving, 3D weaving and an unspecified non-interlacing process. J. Textil. Inst. 87 (1), 97–106. https://doi.org/10.1080/00405009608659059. Part 1.

Khokar, N., 1999. A classification of shedding methods. J. Textil. Inst. 90 (4), 570–579. https://doi.org/10.1080/00405000.1999.10750054 part 1.

Khokar, N., 2001. 3D-weaving: theory and practice. J. Textil. Inst. 92 (1), 193–207. https://doi.org/10.1080/00405000108659570.

Khokar, N., 2013. Making the uniaxial noobing process industrially relevant. In: 5th World Conference on 3D Fabrics and Their Applications. Dehli. https://www.diva-portal.org/smash/get/diva2:887876/FULLTEXT01.pdf.

Khokar, N., Peterson, E., 1998. A Device for Producing Integrated Nonwoven Three Dimensional Fabric. Sweden. Patent WO9803712.

Khokar, N., Peterson, E., 1999. An experimental uniaxial noobing device: construction, method of operation, and related aspects. J. Textil. Inst. 90 (2), 225–242. https://doi.org/10.1080/00405009908690624.

Ko, F.K., 1987. Braiding, Engineered Materials Handbook.

Ko, F.K., 2008. 15 - textile composites for automotive structural components. In: Shishoo, R. (Ed.), Textile Advances in the Automotive Industry. Woodhead Publishing, pp. 320–343. https://doi.org/10.1533/9781845695040.4.320.

Ko, F.K., Pastore, C.M., Lei, C., Whyte, D.W., 1987. A fabric geometry model for 3D braid reinforced FP/Al-Li composites. In: International Sampe Metals Conference: Competitive Advancements in Metals/metals Processing.

Ko, F.K., Pastore, C.M., Head, A.A., 1990. Atkins & Pearce Handbook of Industrial Braiding. Atkins & Pearce.

Ko, F. K., Z. Cai and R. Mutharasan (1994). Development of Braided Rope Engine Seals, NASA TM-105902, NASA. https://ntrs.nasa.gov/api/citations/19940024335/downloads/19940024335.pdf.

Kolloch, M., Puchas, G., Grigat, N., Vollbrecht, B., Krenkel, W., Gries, T., 2021. Process chain development for the fabrication of three-dimensional braided oxide ceramic matrix composites. Materials 14 (21). https://doi.org/10.3390/ma14216338.

Kostar, T.D., Chou, T.-W., 1999. Braided structures (chapter 7). In: Miravete, A. (Ed.), 3-D Textile Reinforcements in Composite Materials. Woodhead Publ. Ltd, Cambridge, pp. 217–240.

Kurashiki, T., Zako, M., Verpoest, I., 2002. Damage development of woven fabric composites considering an effect of mismatch of lay-up. Composites for the Future. In: Sol, H., Degrieck, J. (Eds.), Proceedings 10th European Conference on Composite Materials (ECCM-10) (Brugge: CD edition).

Kyosev, Y., 2014. Braiding Technology for Textiles. Elsevier - Woodhead Publishing, Cambridge. https://doi.org/10.1016/B978-0-85709-135-2.50018-0.

Kyosev, Y., 2015. 13 - Computer Assisted Design (CAD) Software for the Design of Braided Structures. Braiding Technology for Textiles. Woodhead Publishing, pp. 315–336. https://doi.org/10.1533/9780857099211.4.315.

Kyosev, Y., 2016. In: Advances in Braiding Technology. Woodhead Publishing Series in Textiles.

Kyosev, Y., 2024. TexMind. Retrieved 15/08/2024, from. https://texmind.com/wp/.

Kyosev, Y., Gleßner, P., 2018. Extended horn gears in 3D maypole braiding: theoretical analysis, gear arrangement and prediction of the floating length. J. Text. Fibr. Mater. 1. https://doi.org/10.1177/2515221118786741, 2515221118786741.

Kyosev, Y., Lomov, S.V., Kuster, K., 2016. The numerical prediction of the tensile behaviour of multilayer woven tapes made by multifilament yarns. In: Kyosev, Y. (Ed.), Recent Developments in Braiding and Narrow Weaving. Springer, pp. 69–82. https://doi.org/10.1007/978-3-319-29932-7_7.

Kyosev, Y., Hübner, M., Cherif, C., 2018. Virtual development and numerical simulation of 3D braids for composites. IOP Conf. Ser. Mater. Sci. Eng. 406 (1), 012025. https://doi.org/10.1088/1757-899X/406/1/012025.

Laine, B., Arbter, R., Binetruy, C., Bizet, L., Bréard, J., Beraud, J.M., Demaria, C., Endruweit, A., Ermanni, P., Gommer, F., Hasanovic, S., Klunker, F., Lavachy, S., Long, A., Lomov, S.V., Michaud, V., Morren, G., Henrat, P., Ruiz, E., Sol, H., Trochu, F., Verleye, B., Wietgrefe, M., Ziegmann, G., Vautrin, A., Wu, W., Beauchene, P., 2010. Experimental determination of textile permeability: a benchmark exercise. In: 10th

International Conference on Flow Processes in Composite Materials (FPCM-10). Ascona, Switzerland: CD Edition.

Laourine, E., Schneider, M., Wulfhorst, B., 2000. Production and analysis of 3D braided textile preforms for composites. In: 5th Int. Conf. On Textile Composites (TexComp-5). In: .

Laurencin, C.T., Cooper, J.A., Ko, F.K., Lu, H.H., Attawia, M.A., 2001. Ligament Replacement Constructs and Methods for Production and Use Thereof. Germany. WO2001072241A1.

Laurent, C.P., Latil, P., Durville, D., Rahouadj, R., Geindreau, C., Orgeas, L., Ganghoffer, J.F., 2014. Mechanical behaviour of a fibrous scaffold for ligament tissue engineering: finite elements analysis vs. X-ray tomography imaging. J. Mech. Behav. Biomed. Mater. 40, 222–233. https://doi.org/10.1016/j.jmbbm.2014.09.003.

Lee, J.R., Molimard, J., Vautrin, A., Surrel, Y., 2004. Application of grating shearography and speckle shearography to mechanical analysis of composite material. Composites Part A 35, 965–976. https://doi.org/10.1016/j.compositesa.2004.01.023.

Lepperhof, B., 1908. Braiding Machine. USA. US894022.

Li, S., Sitnikova, E., 2020. Representative Volume Elements and Unit Cells. Woodhead Publishing - Elsevier. https://doi.org/10.1016/C2017-0-03559-0.

Li, L., Lomov, S.V., Yan, X., Carvelli, V., 2014. Cluster analysis of acoustic emission signals for 2D and 3D woven glass/epoxy composites. Compos. Struct. 116, 286–299. https://doi.org/10.1016/j.compstruct.2014.05.023.

Li, L., Lomov, S.V., Yan, X., 2015. Correlation of acoustic emission with optically observed damage in a glass/epoxy woven laminate under tensile loading. Compos. Struct. 123, 45–53. https://doi.org/10.1016/j.compstruct.2014.12.029.

Li, L., Swolfs, Y., Straumit, I., Yan, X., Lomov, S.V., 2016. Cluster analysis of acoustic emission signals for 2D and 3D woven carbon fiber/epoxy composites. J. Compos. Mater. 50, 1421–1435. https://doi.org/10.1177/0021998315597742.

Li, M.R., Wang, P., Boussu, F., Soulat, D., 2021. Investigation of impact performance of 3-dimensional interlock polymer fabrics in double and multi-angle pass stabbing. Mater. Des. 206. https://doi.org/10.1016/j.matdes.2021.109775.

Li, H., Jin, Q.R., Liu, Z.F., Ma, C., Ge, J.R., Zhu, Y.C., Liang, J., 2024. A mapping-based method capturing the mesoscopic morphological characteristics of 3D woven fabric torsion structures. Compos. Sci. Technol. 252. https://doi.org/10.1016/j.compscitech.2024.110630.

Liang, B., Chaudet, P., Boisse, P., 2017. Curvature determination in the bending test of continuous fibre reinforcements. Strain 53 (1), e12213. https://doi.org/10.1111/str.12213.

LIBA, 2012a. Copcentra Max 3 CNC. Retrieved 01/08/2024, from. https://interjacm.wordpress.com/wp-content/uploads/2012/09/liba.pdf.

LIBA, 2012b. Copcentra Max 5 CNC. Retrieved 01/08/2024, from. https://interjacm.wordpress.com/wp-content/uploads/2012/09/liba.pdf.

Liu, Q., Parnas, R., 2007. New setup for in-plane permeability measurement. Composites Part A 38, 954–962. https://doi.org/10.1016/j.compositesa.2006.06.024.

Liu, Y., Straumit, I., Vasiukov, D., Lomov, S.V., Panier, S., 2017. Prediction of linear and nonlinear behavior of 3D woven composite using mesoscopic voxel models reconstructed from X-ray micro-tomography. Compos. Struct. 179, 568–579. https://doi.org/10.1016/j.compstruct.2017.07.066.

Liu, X.D., Zhang, D.T., Sun, J., Yu, S., Dai, Y.F., Zhang, Z.W., Sun, J., Li, G., Qian, K., 2020. Refine reconstruction and verification of meso-scale modeling of three- dimensional five -directional braided composites from X-ray computed tomography data. Compos. Struct. 245. https://doi.org/10.1016/j.compstruct.2020.112347.

Liu, X., Liu, C., Ge, J., Zhang, D., Liang, J., 2024. Deep learning and integrated approach to reconstruct meshes from tomograms of 3D braided composites. Compos. Sci. Technol. 255, 110737. https://doi.org/10.1016/j.compscitech.2024.110737.

Lloyd, D.W., 1980. The Analysis of Complex Fabric Deformation. Mechanics of Flexible Fibre Assembly, pp. 311–342.

Lloyd, D.W., Mete, F., Hussain, K., 1996. An approach to the theoretical mechanics of static drape. Int. J. Cloth. Sci. Technol. 8 (3), 43–58. https://doi.org/10.1108/09556229610120363.

Lomov, S.V., 1995. Prediction of Geometry and Mechanical Properties of Woven Technical Fabrics with Mathematical Modelling. Thesis: Doctor of Technical Sciences (Dr Habil.), SPbSUTD. https://dlib.rsl.ru/01000145410.

Lomov, S.V. (Ed.), 2011. Non-crimp Fabric Composites: Manufacturing, Properties and Applications. Woodhead Publisher Ltd, Cambridge.

Lomov, S.V., 2015. From a virtual textile to a virtual woven composite. In: Aliabadi, M.H. (Ed.), Woven Composites (Computational and Experimental Methods in Structures, vol. 6. Imperial College Press, London, pp. 109–140. https://doi.org/10.1142/9781783266180_0003.

Lomov, S.V., 2021. Modeling the geometry of textile composite reinforcements: WiseTex - chapter 7. In: Boisse, P. (Ed.), Composite Reinforcements for Optimum Performance, second ed., pp. 199–236. https://doi.org/10.1016/B978-0-12-819005-0.00007-1.

Lomov, S.V., Gusakov, A.V., 1995. Modellirung von drei-dimensionalen gewebe Strukturen. Technische Textilen 38, 20–21.

Lomov, S.V., Primachenko, B.M., 1992. Mathematical modelling of two-layered woven fabric under tension, accounting for non-linear deformation of the yarns. Technologia Tekstilnoy Promyshlennosty (1), 49–53 (in Russian).

Lomov, S.V., Truevtzev, N.N., 1995. A software package for the prediction of woven fabrics geometrical and mechanical properties. Fibres Text. East. Eur. 3 (2), 49–52.

Lomov, S.V., Verpoest, I., 2006. Model of shear of woven fabric and parametric description of shear resistance of glass woven reinforcements. Compos. Sci. Technol. 66, 919–933. https://doi.org/10.1016/j.compscitech.2005.08.010.

Lomov, S.V., Gusakov, A.V., Cassidy, C., 1995. 3D fabrics: technology, structure properties and mathematical simulation. In: EUROMECH-334, Textile Composites and Textile Structures. Lyon, pp. 187–200.

Lomov, S.V., Gusakov, A.V., Huysmans, G., Prodromou, A., Verpoest, I., 2000a. Textile geometry preprocessor for meso-mechanical models of woven composites. Compos. Sci. Technol. 60, 2083–2095. https://doi.org/10.1016/S0266-3538(00)00121-4.

Lomov, S.V., Huysmans, G., Luo, Y., Prodromou, A., Verpoest, I., 2000b. Textile composites: modelling strategies. In: TexComp 5: The 5th Internatiomal Conference on Textile Composites. Leuven, K.U.Leuven: s.P.

Lomov, S.V., Huysmans, G., Luo, Y., Parnas, R., Prodromou, A., Verpoest, I., Phelan, F.R., 2001a. Textile composites: modelling strategies. Compos. Part A 32 (10), 1379–1394. https://doi.org/10.1016/S1359-835X(01)00038-0.

Lomov, S.V., Huysmans, G., Verpoest, I., 2001b. Hierarchy of textile structures and architecture of fabric geometric models. Textil. Res. J. 71 (6), 534–543. https://doi.org/10.1177/004051750107100611.

Lomov, S.V., Belov, E.B., Bischoff, T., Ghosh, S.B., Truong Chi, T., Verpoest, I., 2002. Carbon composites based on multiaxial multiply stitched preforms. Part 1: geometry of the preform. Compos. Part A 33 (9), 1171–1183. https://doi.org/10.1016/S1359-835X(02)00090-8.

Lomov, S.V., Verpoest, I., Robitaille, F., 2005. Manufacturing and internal geometry of textiles. In: Long, A.C. (Ed.), Design and Manufacture of Textile Composites. Woodhead, pp. 1–60. https://doi.org/10.1533/9781845690823.1.

Lomov, S.V., Willems, A., Vandepitte, D., Verpoest, I., 2006a. Simulations of shear and tension of glass/PP woven fabrics. In: Juster, N., Rosochowski, A. (Eds.), The 9th International Conference on Material Forming ESAFORM, April 26-28, 2006, pp. 783–786. Glasgow.

Lomov, S.V., Willems, A., Verpoest, I., Zhu, Y., Barburski, M., Stoilova, T., 2006b. Picture frame test of woven fabrics with a full-field strain registration. Textil. Res. J. 76 (3), 243–252. https://doi.org/10.1177/0040517506061032.

Lomov, S.V., Ivanov, D.S., Verpoest, I., Zako, M., Kurashiki, T., Nakai, H., Hirosawa, S., 2007. Meso-FE modelling of textile composites: road map, data flow and algorithms. Compos. Sci. Technol. 67, 1870–1891. https://doi.org/10.1016/j.compscitech.2006.10.017.

Lomov, S.V., Boisse, P., Deluycker, E., Morestin, F., Vanclooster, K., Vandepitte, D., Verpoest, I., Willems, A., 2008a. Full field strain measurements in textile deformability studies. Compos. Part A 39, 1232–1244. https://doi.org/10.1016/j.compositesa.2007.09.014.

Lomov, S.V., Ivanov, D.S., Truong Chi, T., Verpoest, I., Baudry, F., Vanden Bosche, K., Xie, H., 2008b. Experimental methodology of study of damage initiation and development in textile composites in uniaxial tensile test. Compos. Sci. Technol. 68, 2340–2349. https://doi.org/10.1016/j.compscitech.2007.07.005.

Lomov, S.V., Ivanov, D.S., Verpoest, I., Zako, M., Kurashiki, T., Nakai, H., Molimard, J., Vautrin, A., 2008c. Full field strain measurements for validation of meso-FE analysis of textile composites. Compos. Part A 39, 1218–1231. https://doi.org/10.1016/j.compositesa.2007.09.011.

Lomov, S.V., Bogdanovich, A.E., Ivanov, D.S., Mungalov, D., Karahan, M., Verpoest, I., 2009. A comparative study of tensile properties of non-crimp 3D orthogonal weave and multilayer plain weave E-glass composites. Part 1: materials, methods and principal results. Compos. Part A 40, 1134–1143. https://doi.org/10.1016/j.compositesa.2009.03.012.

Lomov, S.V., Verpoest, I., Cichosz, J., Hahn, C., Ivanov, D.S., Verleye, B., 2014. Meso-level textile composites simulations: open data exchange and scripting. J. Compos. Mater. 48, 621–637. https://doi.org/10.1177/0021998313476327.

Lomov, S.V., Breite, C., Carrella-Payan, D., Carvelli, V., Ersoy, N., Gigliotti, M., Antoranz-Gonzalez, R., Ivanov, S., Kersani, M., Lafarie-Frenot, M.-C., Mavrogordato, M., Mehdikhani, M., Melnikov, A., Mesquita, F., Oz, F., Pannier, Y., Pyl, L., Schöberl, E., Sinclair, I., Spearing, S.M., Swolfs, Y., Van Hemelrijk, D., Zhu, M., Gorbatikh, L., 2018. Multi-instrument multi-scale experimental damage mechanics for fibre reinforced composites. In: IOP Conf. Series: Materials Science and Engineering (13th International Conference on Textile Composites (TEXCOMP-13)), vol 406, p. 012057.

Lomov, S.V., 2022. WiseTex — a virtual textile composites software. In: Kyosev, Y., Boussu, F. (Eds.), Advanced Weaving Technology. Springer Nature, pp. 293–318. https://doi.org/10.1007/978-3-030-91515-5_7.

Lomov, S.V., Abaimov, S.G., Kandinskii, R., Mahato, B., Swolfs, Y., 2023. Micro-CT based permeability calculations with a limited size of the model volume. In: 15th Conference on Flow Processes in Composite Materials (FPCM-15). West Lafayette, IND, USA, p. 3833.

Long, A.C. (Ed.), 2005. Design and Manufacture of Textile Composites. Woodhead Publishing Ltd, Cambridge.

Lundström, T.S., Stenberg, R., Bergström, R., Partanen, H., Birkeland, P.A., 2000. In-plane permeability measurements: a Nordic round-robin study. Compos. Part A 31, 29–43. https://doi.org/10.1016/S1359-835X(99)00058-5.

Madeo, A., Ferretti, M., dell'Isola, F., Boisse, P., 2015. Thick fibrous composite reinforcements behave as special second-gradient materials: three-point bending of 3D interlocks. Z. Angew. Math. Phys. 66 (4), 2041–2060. https://doi.org/10.1007/s00033-015-0496-z.

Mahadik, Y., Hallett, S.R., 2010. Finite element modelling of tow geometry in 3D woven fabrics. Compos. Part A 41, 1192–1200. https://doi.org/10.1016/j.compositesa.2010.05.001.

Maidl, S.J., 2022. Integrated Sensor System for Monitoring the Braiding Process as a Use Case for Digitalization in Composite Production Technology. PhD, Technische Universität München. https://mediatum.ub.tum.de/doc/1660997/document.pdf.

Maistre, M.A., 1973. Construction of a Three Dimensional Structure. Germany. P2301696.8.

Marcolongo, M., Ducheyne, P., Ko, F.K., LaCourse, W., 1993. Composite Materials Using Bone Bioactive Glass and Ceramic Fibers. USA. US5468544A.

Mathieu, S., Hamila, N., Bouillon, F., Boisse, P., 2015. Enhanced modeling of 3D composite preform deformations taking into account local fiber bending stiffness. Compos. Sci. Technol. 117, 322–333. https://doi.org/10.1016/j.compscitech.2015.07.005.

Mauersberger, H., Kemter, H., 1961. Neue Textiltechnologien. Malimo. Berlin, VEB Verl. Volk U. Wissen.

MaximizeMarketResearch, 2024. Braided Composites Market – Global Industry Analysis and Forecast (2021-2027). Retrieved 08/08/2024, from. https://www.maximizemarketresearch.com/market-report/global-braided-composites-market/91379/.

May, D., Aktas, A., Advani, S.G., Berg, D.C., Endruweit, A., Fauster, E., Lomov, S.V., Long, A., Mitschang, P., Abaimov, S., Abliz, D., Akhatov, I., Ali, A.M., Allen, T.D., Bickerton, S., Bodaghi, M., Caglar, B., Caglar, H., Chiminelli, A., Correia, N., Cosson, B., Danzi, M., Dittmann, J., Ermanni, P., Francucci, G., George, A., Grishaev, V., Hancioglu, M., Kabachi, M.A., Kind, K., Deléglise-Lagardère, M., Laspalas, M., Lebedev, O.V., Lizaranzu, M., Liotiert, P.-J., Middendorf, P., Morán, J., Park, C.-H., Pipes, R.B., Pucci, M.F., Raynal, J., Rodriguez, E.S., Schledjewski, R., Schubnel, R., Sharp, N., Sims, G., Sozer, E.M., Sousa, P., Thomas, J., Umer, R., Wijaya, W., Willenbacher, B., Yong, A., Zaremba, S., Ziegmann, G., 2019. In-plane permeability characterization of engineering textiles based on radial flow experiments: a benchmark exercise. Compos. Part A 121, 100–114. https://doi.org/10.1016/j.compositesa.2019.03.006.

Mazars, V., Caty, O., Couegnat, G., Bouterf, A., Roux, S., Denneulin, S., Pailhes, J., Vignoles, G.L., 2017. Damage investigation and modeling of 3D woven ceramic matrix composites from X-ray tomography in-situ tensile tests. Acta Mater. 140, 130–139. https://doi.org/10.1016/j.actamat.2017.08.034.

Melenka, G.W., Gholami, A., 2021. Fiber identification of braided composites using micro-computed tomography. Compos. Commun. 27. https://doi.org/10.1016/j.coco.2021.100813.

Mengsteab, P.Y., Freeman, J., Barajaa, M.A., Nair, L.S., Laurencin, C.T., 2021. Ligament regenerative engineering: braiding scalable and tunable bioengineered ligaments using a bench-top braiding machine. Regen Eng Transl Med 7 (4), 524–532. https://doi.org/10.1007/s40883-020-00178-8.

Mesogitis, T.S., Skordos, A.A., Long, A.C., 2014. Uncertainty in the manufacturing of fibrous thermosetting composites: a review. Compos. Appl. Sci. Manuf. 57, 67–75. https://doi.org/10.1016/j.compositesa.2013.11.004.

Mesquita, F., Melnikov, A., Rajpurohit, A., Singery, V., Sanial, P., Lomov, S.V., Swolfs, Y., 2022. Tensile failure strain and microstructure of unidirectional carbon fibre non-crimp fabric composites. Compos. Part B 243, 110123. https://doi.org/10.1016/j.compositesb.2022.110123.

Metzner, C., Gessler, A., Weimer, C., Beier, U., Middendorf, P., 2014. Performance assessments on unidirectional braided CFRP materials. In: Proceedings of the Low Cost Composite Processing, from Aerospace OOA to Automotive Thermoplastics. Paris, SAMPE Europe.

Middendorf, P., Metzner, C., 2011. 18 - aerospace applications of non-crimp fabric composites. In: Lomov, S.V. (Ed.), Non-Crimp Fabric Composites. Woodhead Publishing, pp. 441–449. https://doi.org/10.1533/9780857092533.4.441.

Miravete, A. (Ed.), 1999. 3-D Reinforcements in Composite Materials. Woodhead Publishing, Cambridge.

Mohamed, M.H., 1990. Three-dimensional textiles. Am. Sci. 78 (6), 530–541.

Mohamed, M.H., Zhang, Z.-H., 1992. Method of Forming Variable Cross-Sectional Shaped Three-Dimensional Fabrics, U.S. Patent 5,085,252, Issued to NCSU.

Mohamed, M.H., Zhang, Z., Dickinson, L., 1989. 3-D Weaving of Net Shapes. 1st Japan Int. SAMPE Symposium.

Mohamed, M.H., Bogdanovich, A.E., Dickinson, L.C., Singletary, J.N., Lienhart, R.B., 2001. A new generation of 3D woven fabric preforms and composites. SAMPE J. 37 (3), 8–28.

Monnot, P., Lévesque, J., Laberge Lebel, L., 2017. Automated braiding of a complex aircraft fuselage frame using a non-circular braiding model. Compos. Appl. Sci. Manuf. 102, 48–63. https://doi.org/10.1016/j.compositesa.2017.07.011.

Mouritz, A.P., Bannister, M.K., Falson, P.J., Leong, K.H., 1999. Review of applications for advanced three-dimensional fibre textile composites. Compos. Part A 30, 1445–1461. https://doi.org/10.1016/S1359-835X(99)00034-2.

Mullen, C.K., Roy, P.J., 1972. Fabrication and Properties Description of Avco 3D Carbon-Carbon Cylindrical Composites. In: National SAMPE Symp. Los Angeles: III-A-Two-1-8.

Mungalov, D., Bogdanovich, A., 2002. Automated 3-D Braiding Machine and Method. USA. US6439096.

Mungalov, D., Bogdanovich, A., 2004. Complex shape 3D braided composite preforms: structural shapes for marine and aerospace. SAMPE J. 7 (3), 7–21.

Mungalov, D., Duke, P., Bogdanovich, A., 2006. Advancements in design and manufacture of 3-D braided preforms for complex composite structures. In: 38th SAMPE Fall Techn. Conf. Dallas, TX.

Murakami, S., 1988. Mechanical modeling of material damage. J. Appl. Mech. 55, 280. https://doi.org/10.1115/1.3173673.

Nagaraju, H.T., Sankar, B.V., Kim, N.H., Subhash, G., 2022. An extrapolation method to remove spurious stress concentration in micromechanical analyses of composites using pixel-based meshes. Compos. Struct. 290, 115522. https://doi.org/10.1016/j.compstruct.2022.115522.

Naik, N.K., 1994. Woven Fabric Composites. Lancaster-Basel, Technomic.

Naik, N. (2003). Woven-fibre thermoset composites. Fatigue in composites: science and technology of the fatigue response of fibre-reinforced plastics. B. Harris: 296-313. https://doi.org/10.1533/9781855738577.3.296.

Naik, N.K., Kuchibhota, R., 2002. Analytical study of strength and failure behaviour of plain weave fabric composites made of twisted yarns. Compos. Part A 33, 697–708. https://doi.org/10.1016/S1359-835X(02)00012-X.

Naik, N.K., Azad, S.N.M., Durga Prasad, P., Thuruthimattam, B.J., 2001. Stress and failure analysis of 3D orthogonal interlock woven composites. J. Reinforc. Plast. Compos. 20 (17), 1485–1523. https://doi.org/10.1177/073168401772679110.

Naouar, N., Vidal-Salle, E., Schneider, J., Maire, E., Boisse, P., 2014. Meso-scale FE analyses of textile composite reinforcement deformation based on X-ray computed tomography. Compos. Struct. 116 (0), 165–176. https://doi.org/10.1016/j.compstruct.2014.04.026.

Naouar, N., Vidal-Salle, E., Schneider, J., Maire, E., Boisse, P., 2015. 3D composite reinforcement meso F.E. analyses based on X-ray computed tomography. Compos. Struct. 132, 1094–1104. https://doi.org/10.1016/j.compstruct.2015.07.005.

Naouar, N., Vasiukov, D., Park, C.H., Lomov, S.V., Boisse, P., 2020. Meso-FE modelling of textile composites and X-ray tomography. J. Mater. Sci. 55, 16969–16989. https://doi.org/10.1007/s10853-020-05225-x.

Newquist, C.W., Verzemnieks, J., Keller, P.C., Rorabaugh, M., 2002. Advanced High Temperature Structural Seals, NASA/CR—2002-211973. NASA. https://ntrs.nasa.gov/api/citations/20020090799/downloads/20020090799.pdf.

Nguyen, Q.T., Vidal-Sallé, E., Boisse, P., Park, C.H., Saouab, A., Bréard, J., Hivet, G., 2013. Mesoscopic scale analyses of textile composite reinforcement compaction. Compos. B Eng. 44 (1), 231–241. https://doi.org/10.1016/j.compositesb.2012.05.028.

Niwa, M., Kawabata, S., Nanashima, Y., Kawai, H., 1970. The theoretical investigation on the biaxial tensile properties of plain knitted fabrics. Part II. Comparison between theoretical and experimental results. J. Text. Mach. Soc. Jpn. 23 (6), 1–14. https://doi.org/10.4188/transjtmsj1965b.23.T120.

Noor, F., Peters, S., Stingelin-Stutzmann, N., Smith, P., 2008. Going against the Grain: The Dematurity of the European Textile Industry. https://repository.gatech.edu/bitstreams/635e8763-95ba-4568-9900-3d2676753d71/download.

Noor-Evans, F., Peters, S., Stingelin, N., 2012. 7 - nanotechnology innovation for future development in the textile industry. In: Horne, L. (Ed.), New Product Development in Textiles. Woodhead Publishing, pp. 109–131. https://doi.org/10.1533/9780857095190.2.109.

O'Connor, D.J., 1918. Process of Making Composite Material. U. P. Office. USA, 1384432; 3.

Olave, M., Vanaerschot, A., Lomov, S.V., Vandepitte, D., 2012. Internal geometry variability of two woven composites and related variability of the stiffness. Polym. Compos. 33, 1335–1350. https://doi.org/10.1002/pc.22260.

Oxenham, W., Price, R. (Eds.), 2012. Textile Terms and Definitions Manchester. The Textile Institute.

Oz, F., Ersoy, N., Lomov, S.V., 2017. Do high frequency acoustic emission events always represent fibre failure in CFRP laminates? Compos. Part A 103, 230–235. https://doi.org/10.1016/j.compositesa.2017.10.013.

Oz, F., Ersoy, N., Mehdikhani, M., Lomov, S.V., 2018. Multi-instrument in-situ damage monitoring in quasi-isotropic CFRP laminates under tension. Compos. Struct. 196, 163–180. https://doi.org/10.1016/j.compstruct.2018.05.006.

Parnas, R.S., Howard, J.G., Luce, T.L., Adwani, S.G., 1995. Permeability characterisation. Part 1: a proposed standard reference fabric for permeability. Polym. Compos. 16 (6), 429–445. https://doi.org/10.1002/pc.750160602.

Parnas, R.S., Flynn, K.M., Dal-Favero, M.E., 1997. A permeability database for composites materials. Polym. Compos. 18 (5), 623–633. https://doi.org/10.1002/pc.10313.

Pastore, C.M., Ayranci, C., Ko, F.K., 2023. Textile Structural Composites. DESTech Publications.

Pazmino, J., Mathieu, S., Carvelli, V., Boisse, P., Lomov, S.V., 2014. Hyperelastic model for forming simulations of a noncrimp 3d orthogonal weave e-glass composite reinforcement. In: Proceedings of the 16th European Conference on Composite Materials (ECCM-16), Seville: Electronic Edition, s.P.

Peirce, F.T., 1930. A "handle" of cloth as a measurable quantity. J. Textil. Inst. 20 (9), T377–T417. https://doi.org/10.1080/19447023008661529.

Peirce, F.T., 1937. The geometry of cloth structure. J. Textil. Inst. 28 (3), T45–T96. https://doi.org/10.1080/19447023708658809.

Peirce, F.T., 1947. Geometrical principles applicable to the design of functional fabrics. J. Textil. Inst. 17 (3), 123–147. https://doi.org/10.1177/004051754701700301.

Pierreux, G., Van Hemelrijck, D., Massart, T.J., 2020. Automated RVE computations for evaluation of microdamage initiation in structural stitched non-crimp fabric composites. J. Compos. Mater. 54 (30), 4751–4771. https://doi.org/10.1177/0021998320937471.

Popper, P., McConnell, R., 1987. A new 3-D braid for integrated parts manufacturing and improved delamination resistancedthe 2-step process. In: 32nd Int. SAMPE Symp. Covina, CA: 92-103.

Potluri, P., Manan, A., 2007. Mechanics of non-orthogonally interlaced textile composites. Compos. Appl. Sci. Manuf. 38 (4), 1216–1226. https://doi.org/10.1016/j.compositesa.2006.04.008.

Potluri, P., Sharma, S., Ramgulam, R., 2001. Comprehensive drape modelling for moulding 3D textile preforms. Compos. Part A 32 (10), 1415–1424. https://doi.org/10.1016/S1359-835X(01)00040-9.

Potluri, P., 2012a. Braiding. In: Nicolais, L., Lee, S.M. (Eds.), Wiley Encyclopedia of Composites, second ed., vol. 1. Wiley, Hoboken, NY, pp. 138–152.

Potluri, P., Hogg, P., Arshad, M., Jetavat, D., Jamshidi, P., 2012b. Influence of fibre architecture on impact damage tolerance in 3D woven composites. Appl. Compos. Mater. 19 (5), 799–812. https://doi.org/10.1007/s10443-012-9256-9.

Prodromou, A., Vandeurzen, P., Huysmans, G., Ivens, J., Verpoest, I., 1998. Micromechanical modelling of textile composites using variational pribciples. In: Hoa, S., de Wilde, V.,W. P., Bain, W.R. (Eds.), Computer Methods in Composite Materials VI. Computational Mechanics Publications, Southhampton - Boston, pp. 360–370.

Prodromou, A., Lomov, S.V., Verpoest, I., 2011. The method of cells and the mechanical properties of textile composites. Compos. Struct. 93, 1290–1299. https://doi.org/10.1016/j.compstruct.2010.09.022.

Rana, S., Fangueiro, R., 2015. Braided Structures and Composites. Production, Properties, Mechanics, and Technical Applications. CRC Press.

Rasheed, A., 2020. Classification of technical textiles. In: Ahmad, S., Rasheed, A., Nawab, Y. (Eds.), Fibers for Technical Textiles. Springer International Publishing. https://doi.org/10.1007/978-3-030-49224-3_3.

Rassineux, A., 2022. Robust conformal adaptive meshing of complex textile composites unit cells. Compos. Struct. 279. https://doi.org/10.1016/j.compstruct.2021.114740.

Reindl, J., 2011. Historische Aufarbeitung der Textilindustrie in Chemnitz. Retrieved 01/08/2024, from. https://www.openpr.de/news/584394/Historische-Aufarbeitung-der-Textilindustrie-in-Chemnitz.html.

Richards, N.E., 1937. Braiding Machine. USA. US2085334.

Robitaille, F., Souter, B.J., Long, A.C., Rudd, C.D., 1998. A geometrical model for textile preforms. In: 19th SAMPE Europe International Conference. Paris, pp. 151–162.

Robitaille, F., Clayton, B.R., Long, A.C., Souter, B.J., Rudd, C.D., 1999a. Geometric modelling of industrial preforms: warp-knitted textiles. In: Proceedings of the Institute of Mechanical Engineers 214 Part L: 71-90. https://doi.org/10.1177/146442070021400203.

Robitaille, F., Clayton, B.R., Long, A.C., Souter, B.J., Rudd, C.D., 1999b. Geometric modelling of industrial preforms: woven and braided textiles. In: Proceedings of the Institute of Mechanical Engineers 213 Part L: 69-81. https://doi.org/10.1177/146442079921300.

Sakakibara, T., Yamamoto, T., Yamada, G., Imaoku, A., Ohtagaki, R., Zako, M., 2018. Recent developments in "COMPOSITES DREAM" for meso-FE modelling of advanced materials. In: IOP Conf. Series: Materials Science and Engineering (13th International Conference on Textile Composites (TEXCOMP-13) 406: 012027.

Sause, M.G.R., 2016. In: Situ Monitoring of Fiber-Reinforced Composites. Springer. https://doi.org/10.1007/978-3-319-30954-5.

Sause, M.G.R., Gribov, A., Unwin, A.R., Horn, S., 2012. Pattern recognition approach to identify natural clusters of acoustic emission signals. Pattern Recognit. Lett. 33, 17–23. https://doi.org/10.1016/j.patrec.2011.09.018.

Schnabel, A., Gries, T., 2011. Production of non-crimp fabrics for composites. In: Lomov, S.V. (Ed.), Non-crimp Fabric Composites: Manufacturing, Properties and Applications. Woodhead Publishing, Oxford, pp. 3–41. https://doi.org/10.1533/9780857092533.1.3.

Schneider, M., Pickett, A., Wulfhorst, B., 2000. A New Rotary Braiding Machine and CAE Procedures to Produce Efficient 3D-Braided Textiles for Composites. SAMPE USA. SAMPE, Long Beach, CA. https://www.nasampe.org/store/viewproduct.aspx?id=4303308.

Schreiber, F., Ko, F.K., Yang, H.J., Amalric, E., Gries, T., 2009. Novel three-dimensional braiding approach and its products. In: Proc. 17th Int. Conf. On Compos. Mater. (ICCM-17).

Schreiber, F., Theelen, K., Schulte Südhoff, E., Lee, H.-Y., Ko, F.K., Gries, T., 2012. 3D-hexagonal braiding: possibilities in near-net shape preform production for lightweight and medical applications. Jeju Island, Korea.

Sevenois, R.D.B., Garoz, D., Gilabert, F.A., Spronk, S.W.F., Fonteyn, S., Heyndrickx, M., Pyl, L., Van Hemelrijck, D., Degrieck, J., Van Paepegem, W., 2016. Avoiding interpenetrations and the importance of nesting in analytic geometry construction for Representative Unit Cells of woven composite laminates. Compos. Sci. Technol. 136, 119–132. https://doi.org/10.1016/j.compscitech.2016.10.010.

Shanahan, W.J., Lloyd, D.W., Hearle, J.W.S., 1978. Characterising the elastic behaviour of textile fabrics in complex deformation. Textil. Res. J. 48 (9), 495–505. https://doi.org/10.1177/004051757804800901.

Sherburn, M., 2007. Geometric and Mechanical Modelling of Textiles. PhD, University of Nottinghan. https://eprints.nottingham.ac.uk/id/eprint/10303.

Shishkina, O., Matveeva, A., Wiedemann, S., Hoehne, K., Wevers, M., Lomov, S.V., Farkas, L., 2018. X-ray computed tomography-based FE-homogenization of sheared organo sheets. In: ECCM18 - 18th European Conference on Composite Materials Athens: P_3.33.

Sihn, S., Iarve, E., Roy, A.K., 2004a. Three-dimensional stress analysis of textile composites: Part I. Numerical analysis. Int. J. Solid Struct. 41, 1377–1393. https://doi.org/10.1016/j.ijsolstr.2003.10.017.

Sihn, S., Iarve, E., Roy, A.K., 2004b. Three-dimensional stress analysis of textile composites: Part II. Asymptotic analysis. Int. J. Solid Struct. 41, 1395–1410. https://doi.org/10.1016/j.ijsolstr.2003.11.024.

Sinchuk, Y., Shishkina, O., Gueguen, M., Signor, L., Nadot-Martin, C., Trumel, H., Van Paepegem, W., 2022. X-ray CT based multi-layer unit cell modeling of carbon fiber-reinforced textile composites: segmentation, meshing and elastic property homogenization. Compos. Struct. 298, 116003. https://doi.org/10.1016/j.compstruct.2022.116003.

Sköck-Hartmann, B., Gries, T., 2011. 20 - automotive applications of non-crimp fabric composites. In: Lomov, S.V. (Ed.), Non-Crimp Fabric Composites. Woodhead Publishing, pp. 461–480. https://doi.org/10.1533/9780857092533.4.461.

Skordos, A.A., Sutcliffe, M.P.F., 2006. Geometric variability of woven textiles. In: Procedings of the 8th International Conference on Textile Composites (TexComp-8). Nottingham: CD Edition.

Skordos, A.A., Sutcliffe, M.P.F., 2008. Stochastic simulation of woven composites forming. Compos. Sci. Technol. 68 (1), 283–296. https://doi.org/10.1016/j.compscitech.2007.01.035.

Song, Y.Y., Qu, Z.L., Liao, H.T., Ai, S.G., 2023. Material twins generation of woven polymer composites based on ResL-U-Net convolutional neural networks. Compos. Struct. 307. https://doi.org/10.1016/j.compstruct.2023.116672.

Sontag, T., Yang, H., Gries, T., Ko, F., 2015. 7 - recent advances in 3D braiding technology. In: Chen, X. (Ed.), Advances in 3D Textiles. Woodhead Publishing, pp. 153–181. https://doi.org/10.1016/B978-1-78242-214-3.00007-3.

Spain, R.G., Bailey, C., 1991. Apparatus and Method for Braiding Fiber Strands and Stuffer Fiber Strands. USA US4984502.

Steer, Q., Colmars, J., Naouar, N., Boisse, P., 2021. Modeling and analysis of in-plane bending in fibrous reinforcements with rotation-free shell finite elements. Int. J. Solid Struct. 222–223, 111014. https://doi.org/10.1016/j.ijsolstr.2021.03.001.

Stig, F., Hallström, S., 2009. Assessment of the mechanical properties of a new 3D woven fibre composite material. Compos. Sci. Technol. 69 (11–12), 1686–1692. https://doi.org/10.1016/j.compscitech.2008.04.047.

Stover, E. R., W. C. Mark, I. Marforwitz and W. Mueller (1971). Preparation of an Omniweave Reinforced CarboneCarbon Cylinder as a Candidate for Evaluation in the Advanced Heat Shield Screening Program. Wright-Patterson AFB, OH, U.S. Air Force.

Straumit, I., Hahn, C., Winterstein, E., Plank, B., Lomov, S.V., Wevers, M., 2016. Computation of permeability of a non-crimp carbon textile reinforcement based on X-ray computed tomography images. Composites Part A 81, 289–295. https://doi.org/10.1016/j.compositesa.2015.11.025.

Straumit, I., Vandepitte, D., Wevers, M., Lomov, S.V., 2017. Identification of the flax fibre modulus based on an impregnated quasi-unidirectional fibre bundle test and X-ray computed tomography. Compos. Sci. Technol. 151, 124–130. https://doi.org/10.1016/j.compscitech.2017.07.029.

Summerscales, J., 2023. International Conference Series on Flow Processes in Composite Materials. Retrieved 23/08/2024, from. https://ecm-academics.plymouth.ac.uk/jsummerscales/fpcm/.

Sun, H.-Y., Qiao, X., 1997. Prediction of the mechanical properties of three-dimentionally braided composites. Compos. Sci. Technol. 57, 623–629. https://doi.org/10.1016/S0266-3538(96)00154-6.

Sutcliffe, M.P.F., Lemanski, S.L., Scott, A.E., 2012. Measurement of fibre waviness in industrial composite components. Compos. Sci. Technol. 72 (16), 2016–2023. https://doi.org/10.1016/j.compscitech.2012.09.001.

Syerko, E., Schmidt, T., Barisin, T., Schladitz, K., May, D., Binetruy, C., Silva, L., Lomov, S. V., Advani, S., 2023a. Workshop: first insights from the second mesoscale stage of the virtual permeability benchmark on engineered textiles. In: 15th Conference on Flow Processes in Composite Materials (FPCM-15). West Lafayette, IND, USA.

Syerko, E., Schmidt, T., May, D., Binetruy, C., Advani, S.G., Lomov, S.V., Silva, L., Abaimov, S., Aissa, N., Akhatov, I., Ali, M., Asiaban, N., Broggi, G., Bruchon, J., Caglar, B., Digonnet, H., Dittmann, J., Drapier, S., Endruweit, A., Guilloux, A., Kandinskii, R., Leygue, A., Mahato, B., Martinez-Lera, P., Matveev, M., Michaud, V., Middendorf, P., Moulin, N., Orgéas, L., Park, C.H., Rief, S., Rouhi, M., Sergeichev, I., Shakoor, M., Shishkina, O., Swolfs, Y., Tahani, M., Umer, R., Vanclooster, K., Vorobyev, R., 2023b. Benchmark exercise on image-based permeability determination of engineering textiles: microscale predictions. Composites Part A 167, 107397. https://doi.org/10.1016/j.compositesa.2022.107397.

Syerko, E., T. Schmidt, D. May, C. Binetruy, S. V. Lomov, S. Advani and A. F. Silva (2024). Mesoscale results from the virtual permeability benchmark on engineering textiles. 21st European Conference on Composite Materials (ECCM-21). Nantes. vol. 8: 1312-1320 https://doi.org/10.60691/yj56-np80.

Tabatabaei, S.A., Lomov, S.V., 2015. Eliminating the volume redundancy of embedded elements and yarn interpenetrations in meso- finite element modelling of textile composites. Comput. Struct. 152, 142–154. https://doi.org/10.1016/j.compstruc.2015.02.014.

Tabatabaei, S.A., Lomov, S.V., Verpoest, I., 2014. Assessment of embedded element technique in meso-FE modelling of fibre reinforced composites. Compos. Struct. 107, 436–446. https://doi.org/10.1016/j.compstruct.2013.08.020.

Tabatabaei, S., Bedogni, E., Melro, A.R., Ivanov, D.S., Lomov, S.V., 2022. Meso-scale progressive damage simulation of textile composites using mesh superposition. Int. J. Solid Struct. 256, 111987. https://doi.org/10.1016/j.ijsolstr.2022.111987.

Takano, N., Zako, M., 1995. Three-dimensional microstructural design of woven fabric composite material by homogenisation method. In: Bees, W.J., Newas, G.M., Takezono, S., et al. (Eds.), PVP-Vol.302. Composites for the Pressure Vessel Industry. https://doi.org/10.1299/kikaia.61.1038. ASME. H00965: s.p.

Takano, N., Uetsuji, Y., Kashiwagi, Y., Zako, M., 1999. Hierarchical modelling of textile composite materials and structures by the homogenisation method. Model. Simulat. Mater. Sci. Eng. 7, 207–231. https://doi.org/10.1088/0965-0393/7/2/006.

Takano, N., Zako, M., Okazaki, T., 2001a. Efficient modelling of microscopic heterogenity and local crack in composite materials by finite element mesh superposition method. JSME International Journal, Series A 44 (4), 602–609. https://doi.org/10.1299/jsmea.44.602.

Takano, N., Zako, M., Okazaki, T., 2001b. Local and microscopic analysis of woven fabric composite material under bending by finite element mesh superposition method. In: Proceedings of 7th Japan International SAMPE Symposium & Exhibition, November 13-16. Tokyo, pp. 725–728.

Takano, N., Zako, M., Okazaki, T., Terada, K., 2002. Microstructure-based evaluation of the influence of woven architecture on permeability by asymptotic homogenisation theory. Compos. Sci. Technol. 62, 1347–1356. https://doi.org/10.1016/S0266-3538(02)00076-3.

Takano, N., Zako, M., Okuno, Y., 2003. Multi-scale finite element analysis of porous materials and components by asymptotic homogenisation theory and enchanced mesh superposition method. Model. Simulat. Mater. Sci. Eng. 11, 137–156.

Talreja, R., Singh, C.V., 2012. Damage and Failure of Composite Materials. Cambridge University Press, Cambridge. https://doi.org/10.1017/CBO9781139016063.

Talreja, R., Varna, J. (Eds.), 2016. Modeling Damage, Fatigue and Failure of Composite Materials. Elsevier (Woodhead Publishers), Cambridge.

Tan, T.M., Pastore, C.M., Ko, F.K., 1991. Engineering design of tough ceramic matrix composites for turbine components. J. Eng. Gas Turbines Power 113 (2), 312–317. https://doi.org/10.1115/1.2906564.

Tang, X., Whitcomb, J.D., 2003a. Progressive failure behaviour of 2D woven composites. J. Compos. Mater. 37 (14), 1239–1259. https://doi.org/10.1177/0021998303037014002.

Tang, X., Whitcomb, J.D., 2003b. General techniques for exploiting periodicity and symmetries in micromechanics analysis of textile composites. J. Compos. Mater. 37 (13), 1167–1189. https://doi.org/10.1177/00219983030370130.

Thouin, M., Ghiasi, H., Lessard, L., 2010. Design of a carbon fiber bicycle stem using a novel internal bladder resin transfer molding technique. Adv. Compos. Lett. 19 (1). https://doi.org/10.1177/096369351001900105, 096369351001900105.

Tong, L., Mouritz, A.P., Bannister, M.K., 2002. 3D Fibre Reinforced Polymer Composites. Elsevier.

Topal, S., Baiocchi, L., Crocombe, A.D., Ogin, S.L., Potluri, P., Withers, P.J., Quaresimin, M., Smith, P.A., Poole, M.C., Bogdanovich, A.E., 2015. Late-stage fatigue damage in a 3D orthogonal non-crimp woven composite: an experimental and numerical study. Compos. Appl. Sci. Manuf. 79, 155–163. https://doi.org/10.1016/j.compositesa.2015.08.020.

Trochu, F., Gauvin, R., Dong Ming, G., Boudreault, J.-F., 1994. RTMFLOT - an integrated software environment for the computer simulation of the resin transfer molding process. J. Reinforc. Plast. Compos. 13 (3), 262–270. https://doi.org/10.1177/073168449401300306.

Truong Chi, T., Vettori, M., Lomov, S.V., Verpoest, I., 2005. Carbon composites based on multiaxial multiply stitched preforms. Part 4: mechanical properties of composites and damage observation. Compos. Part A 36, 1207–1221. https://doi.org/10.1016/j.compositesa.2005.02.004.

Truong Chi, T., Ivanov, D.S., Klimshin, D.V., Lomov, S.V., Verpoest, I., 2008. Carbon composites based on multiaxial multiply stitched preforms. Part 7: mechanical properties and damage observations in composite with sheared reinforcement. Compos. Part A 39, 1380–1393. https://doi.org/10.1016/j.compositesa.2008.05.004.

Tsai, S.W., 1992. Theory of Composites Design. Think Composites, Dayton.

Tsai, S.W., Hahn, H.T., 1980. Introduction to Composite Materials.

Tsuzuki, M., Kimbara, M., Fukuta, K., Machii, A., 1991. Three-dimensional Fabric Woven by Interlacing Threads with Rotor Driven Carriers. USA. US5067525.

Tsuzuki, M., 1994. Three-dimensional Woven Fabric with Varied Thread Orientations. USA. US5348056.

Uetsuji, Y., Zako, M., Nishiyabu, K., 2002. Numerical analysis and in-situ SEM observation of damage development for woven fabric composite materials. J. Soc. Mat. Sci. - Japan 51 (10), 1147–1153. https://doi.org/10.2472/jsms.51.1147.

Vallons, K., 2015. Databases for fatigue analysis in composite materials. In: Carvelli, V., Lomov, S.V. (Eds.), Fatigue of Textile Composites. Elasevier - Woodhead Publishing, Amasterdam - London, pp. 75–82. https://doi.org/10.1016/B978-1-78242-281-5.00004-3.

van Dulken, S., 2002. Inventing the 20th Century: 100 Inventions that Shaped the World from the Airplane to the Zipper. NYU Press.

Van Paepegem, W., Degrieck, J., 2001a. Fatigue degradation modelling of plain woven glass/epoxy composites. Compos. Part A 32 (10), 1433–1441. https://doi.org/10.1016/S1359-835X(01)00042-2.

Van Paepegem, W., Degrieck, J., 2001b. Experimental set-up for and numerical modelling of bending fatigue experiments on plain woven galss/epoxy composites. Compos. Struct. 51, 1–8. https://doi.org/10.1016/S0263-8223(00)00092-1.

Van Paepegem, W., Degrieck, J., 2002. Coupled residual stiffness and strength model for fatigue of fibre reinforced composites. Compos. Sci. Technol. 62, 687–696. https://doi.org/10.1016/S0266-3538(01)00226-3.

Van Paepegem, W., Degrieck, J., 2004. Simulating in-plane fatigue damage in woven glass fibre reinforced composites subject to fully reversed cyclic loading. Fatig. Fract. Eng. Mater. Struct. 27, 1197–1208. https://doi.org/10.1111/j.1460-2695.2004.00851.x.

Van Paepegem, W., 2015. 2 - the cycle jump concept for modelling high-cycle fatigue in composite materials. In: Carvelli, V., Lomov, S.V. (Eds.), Fatigue of Textile Composites. Woodhead Publishing, pp. 29–55. https://doi.org/10.1016/B978-1-78242-281-5.00002-X.

van Reden, T., 2015. Erweiterung der Systemgrenzen der Flechttechnik durch elektronisch gesteuerte Klöppel. Verlag Dr. Hut, München.

Vanaerschot, A., Cox, B.N., Blacklock, M., Kerckhofs, G., Wevers, M., Lomov, S.V., Vandepitte, D., 2013. Stochastic framework for quantifying the geometrical variability of laminated textile composites using micro-computed tomography. Compos. Part A 44, 122–131. https://doi.org/10.1016/j.compositesa.2012.08.020.

Vanaerschot, A., Cox, B.N., Lomov, S.V., Vandepitte, D., 2016a. Multi-scale modelling strategy for textile composites based on stochastic reinforcement geometry. Comput. Methods Appl. Mech. Eng. 310, 906–934. https://doi.org/10.1016/j.cma.2016.08.007.

Vanaerschot, A., Cox, B.N., Lomov, S.V., Vandepitte, D., 2016b. Experimentally validated stochastic geometry description for textile composite reinforcements. Compos. Sci. Technol. 122, 122–129. https://doi.org/10.1016/j.compscitech.2015.11.023.

Vanaerschot, A., Panerai, F., Cassell, A., Lomov, S.V., Vandepitte, D., Mansour, N.N., 2017. Stochastic characterisation methodology for 3-D textiles based on micro-tomography. Compos. Struct. 173, 44–52. https://doi.org/10.1016/j.compstruct.2017.03.107.

Vandeurzen, P., Ivens, J., Verpoest, I., 1998. Micro-stress analysis of woven fabric composites by multilevel decomposition. J. Compos. Mater. 32 (7), 623–651. https://doi.org/10.1177/002199839803200702.

VerifiedMarketReports, 2024. Global Non Crimp Fabric Market by Type (Unidirectional, Multi Directional), by Application (Aerospace, Medical), by Geographic Scope and Forecast. Retrieved 02/08/2024, from. https://www.verifiedmarketreports.com/product/non-crimp-fabric-market/.

Vernet, N., Ruiz, E., Advani, S., Alms, J.B., Aubert, M., Barburski, M., Barari, B., Beraud, J. M., Berg, D.C., Correia, N., Danzi, M., Delaviare, T., Dickert, M., Di Fratta, C., Endruweit, A., Ermanni, P., Francucci, G., Garcia, J.A., George, A., Hahn, C., Klunker, F., Lomov, S.V., Long, A., Louis, B., Maldonado, J., Meier, R., Michaud, V., Perrin, H., Pillai, K., Rodriguez, E., Trochu, F., Verheyden, S., Wietgrefe, M., Xiong, W., Zaremba, S., Ziegmann, G., 2014. Experimental determination of the permeability of engineering textiles: benchmark II. Compos. Appl. Sci. Manuf. 61, 172–184. https://doi.org/10.1016/j.compositesa.2014.02.010.

Verpoest, I., Lomov, S.V., 2005. Virtual textile composites software WiseTex: integration with micro-mechanical, permeability and structural analysis. Compos. Sci. Technol. 65 (15–16), 2563–2574. https://doi.org/10.1016/j.compscitech.2005.05.031.

Verpoest, I., 2000. 2.18 - composite preforming techniques. In: Kelly, A., Zweben, C. (Eds.), Comprehensive Composite Materials. Pergamon, Oxford, pp. 623–669. https://doi.org/10.1016/B0-08-042993-9/00168-6.

Vollbrecht, B., Kohler, C., Kolloch, M., Jung, F., Grigat, N., Gries, T., 2021. Developing a camera-based measuring system to feedback control the fibre orientation for the braiding process of CFRP. Adv. Indus. Manufact. Eng. 3, 100059. https://doi.org/10.1016/j.aime.2021.100059.

Vorobiov, O., Tabatabaei, S.A., Lomov, S.V., 2017. Mesh superposition applied to meso-FE modelling of fibre reinforced composites: cross-comparison of implementations. Int. J. Numer. Methods Eng. 111, 1003–1024. https://doi.org/10.1002/nme.5487.

Vu, A.N., Grouve, W.J.B., Akkerman, R., 2023. Modeling of yarn interactions for non-axisymmetric biaxial overbraiding simulations. Compos. Appl. Sci. Manuf. 167, 107421. https://doi.org/10.1016/j.compositesa.2022.107421.

Wang, P., Hamila, N. (Eds.), 2024. Advanced Structural Textile Composites Forming. Woodhead Publishing - Elsevier.

Wang, J., Wang, P., Hamila, N., Boisse, P., 2020. Mesoscopic analyses of the draping of 3D woven composite reinforcements based on macroscopic simulations. Compos. Struct. 250. https://doi.org/10.1016/j.compstruct.2020.112602.

Wang, Z.-P., Cox, B.N., Kuehsamy, S.J., Jhon, M.H., Sudre, O., Sridhar, N., Conduit, G.J., 2024. A compact yet flexible design space for large-scale nonperiodic 3D woven composites based on a weighted game for generating candidate tow architectures. Comput. Aided Des. 167, 103637. https://doi.org/10.1016/j.cad.2023.103637.

Weimer, C., Preller, T., Mitschang, P., Drechsler, K., 2000a. Approach to net-shape preforming using textile technologies. Part I: edges. Compos. Appl. Sci. Manuf. 31 (11), 1261–1268. https://doi.org/10.1016/s1359-835x(00)00073-7.

Weimer, C., Preller, T., Mitschang, P., Drechsler, K., 2000b. Approach to net-shape preforming using textile technologies. Part II: holes. Compos. Appl. Sci. Manuf. 31 (11), 1269–1277. https://doi.org/10.1016/s1359-835x(00)00074-9.

Weller, R., 1985. AYPEX: a new method of composite reinforcement braiding. In: 3-D Composite Materials, NASA Conf, pp. 229–249. Annapolis, MD.

Wercinski, P., 2024. Adaptable Deployable Entry and Placement Technology (ADEPT). Retrieved 14/08/2024, from. https://flightopportunities.ndc.nasa.gov//technologies/139/.

Whitcomb, J., Srirengan, K., 1996. Effect of various approximations on predicted progressive failure in plain weave composites. Compos. Struct. 34, 13–20. https://doi.org/10.1016/0263-8223(95)00125-5.

Whitcomb, J., Tang, X., 2001. Effective moduli of woven composites. J. Compos. Mater. 35 (23), 2137–2144. https://doi.org/10.1177/002199801772661380.

Whitcomb, J., Srirengan, K., Chapman, C., 1995. Evaluation of homogenization for global/local stress analysis of textile composites. Compos. Struct. 31, 137–149. https://doi.org/10.1016/0263-8223(95)00011-9.

Whitcomb, J., Chapman, C.D., Tang, X., 2000. Derivation of boundary conditions for micromechanics analyses of plain and satin woven composites. J. Compos. Mater. 34 (9), 724–747. https://doi.org/10.1177/0021998300034009.

Willems, A., Lomov, S.V., Verpoest, I., Vandepitte, D., 2008. Optical strain fields in shear and tensile testing of textile reinforcements. Compos. Sci. Technol. 68, 807–819. https://doi.org/10.1016/j.compscitech.2007.08.018.

Williams, J.H., Lee, S.S., 1978. Acoustic emission monitoring of fiber composite materials and structures. J. Compos. Mater. 12 (4), 348–370. https://doi.org/10.1177/002199837801200402.

Wintiba, B., Sonon, B., Ehab Moustafa Kamel, K., Massart, T.J., 2017. An automated procedure for the generation and conformal discretization of 3D woven composites RVEs. Compos. Struct. 180, 955–971. https://doi.org/10.1016/j.compstruct.2017.08.010.

Wintiba, B., Vasiukov, D., Panier, S., Lomov, S.V., Kamel, K.E.M., Massart, T.J., 2020. Automated reconstruction and conformal discretization of 3D woven composite CT scans with local fiber volume fraction control. Compos. Struct. 248, 112438. https://doi.org/10.1016/j.compstruct.2020.112438.

Wisnom, M.R., 2012. The role of delamination in failure of fibre-reinforced composites. Philos. Trans. R. Soc. A Math. Phys. Eng. Sci. 370 (1965), 1850. https://doi.org/10.1098/rsta.2011.0441.

Wolverine, T., 2023. The Difference between Triaxially Braided and UD Carbon Spokes. Retrieved 28/08/2024, from. https://gulocomposites.com/journal/the-difference-between-triaxially-braided-and-ud-carbon-spokes.

Womersley, J.R., 1937. The application of differential geometry to the study of the deformation of cloth under stress. J. Text. Inst. Trans. 28 (3), T97–T113. https://doi.org/10.1080/19447023708658810.

Woo, K., Whitcomb, J.D., 2000. A post-processor approach for stress analysis of woven textile composites. Compos. Sci. Technol. 60, 693–704. https://doi.org/10.1016/S0266-3538(99)00165-7.

Yong, A.X.A., Aktas, A., May, D., Endruweit, A., Lomov, S.V., Advani, S., Hubert, P., Abaimov, S.G., Abliz, D., Akhatov, I., Ali, M.A., Allaoui, A., Allen, T., Berg, D.C., Bickerton, S., Caglar, B., Causse, P., Chiminelli, A., Comas-Cardona, S., Danzi, M., Dittmann, J., Dransfeld, C., Ermanni, P., Fauster, E., George, A., Gillibert, J., Govignon, Q., Graupner, R., Grishaev, V., Guilloux, A., Kabachi, M.A., Keller, A., Kind, K., Large, D., Laspalas, M., Lebedev, O.V., Lizaranzu, M., Long, A.C., Lopez, C., Masania, K., Michaud, V., Middendorf, P., Mitschang, P., van Oosterom, S., Schubnel, R., Sharp, N., Sousa, P., Trochu, F., Umer, R., Valette, J., Wang, J.H., 2020. Experimental characterisation of textile compaction response: a benchmark exercise. Compos. Part A 142, 106243. https://doi.org/10.1016/j.compositesa.2020.106243.

Yong, A.X.H., Aktas, A., May, D., Endruweit, A., Advani, S., Hubert, P., Abaimov, S.G., Abliz, D., Akhatov, I., Ali, M.A., Allen, T., Berg, D.C., Bickerton, S., Brauner, C., Brütsch, D., Caglar, B., Caglar, H., Causse, P., Chiminelli, A., Cohades, A., Comas-Cardona, S., Danzi, M., Dittmann, J., Dransfeld, C., Ermanni, P., Fauster, E., Garcia-Manrique, J.A., George, A., Graupner, R., Grishaev, V., Guilloux, A., Hancioglu, M., Harizi, W., Herman, T., Huang, W., Kabachi, M.A., Keller, A., Kind, K., Laspalas, M., Lebedev, O.V., Lizaranzu, M., Long, A.C., Masania, K., Michaud, V., Middendorf, P., Salvatori, D., Schubnel, R., Sharp, N., Sozer, M., Thomas, J., Trochu, F., Umer, R., Valette, J., Wang, J.H., Willenbacher, B., 2021. Out-of-plane permeability measurement for reinforcement textiles: a benchmark exercise. Compos. Appl. Sci. Manuf. 148, 106480. https://doi.org/10.1016/j.compositesa.2021.106480.

Yu, J.Z., Cai, Z., Ko, F.K., 1994. Formability of textile preforms for composite applications. Part 1: characterisation experiments. Compos. Manuf. 5 (2), 113–122. https://doi.org/10.1016/0956-7143(94)90062-0.

Yu, G.Q., Jia, Y.F., Dong, C.Q., Gao, X.G., Wang, F., Song, Y.D., 2021. Geometric parameters characterization of minicomposite and modulus prediction of 2D composite based on X-ray computed tomography. Ceram. Int. 47 (21), 30391–30398. https://doi.org/10.1016/j.ceramint.2021.07.219.

Yushanov, S.P., Bogdanovich, A.E., 1998a. Analytical probabilistic modelling of initial failure and reliability of laminated composite structures. Int. J. Solid Struct. 33 (7–8), 665–685. https://doi.org/10.1016/S0020-7683(97)00081-4.

Yushanov, S.P., Bogdanovich, A.E., 1998b. Stochastic theory of composite materials with random waviness of the reinforcements. Int. J. Solid Struct. 35 (22), 2901–2930. https://doi.org/10.1016/S0020-7683(97)00351-X.

Yushanov, S.P., Bogdanovich, A.E., 2000. Fibre waviness in textile composites and its stochastic modelling. Mech. Compos. Mater. 36 (4), 501–532. https://doi.org/10.1007/BF02262808.

Zahid, M., Sharma, R., Bhagat, A.R., Karra, P., 2021. Image-based numerical modeling for the effective thermo-elastic property of 4D carbon/carbon composite at high temperatures. Compos. Struct. 267. https://doi.org/10.1016/j.compstruct.2021.113881.

Zako, M., Tsujikami, T., Tsumura, T., 1994. Development of finite element method considered an anisotropic damaged state for composite materials. In: The 37th Japan Congress on Materials Research. Kyoto, Society of Materials Science, Japan, pp. 174–179.

Zako, M., Uetsuji, Y., Kurashiki, T., 2003. Finite element analysis of damaged woven fabric composite materials. Compos. Sci. Technol. 63, 507–516. https://doi.org/10.1016/S0266-3538(02)00211-7.

Zako, M., Kurashiki, T., Kubo, F., Imura, M., 2005. A multi-scale analysis for structural design of fibrous composites - M3 method. In: 15th International Conference on Composite Materials (ICCM-15). Durban: CD Edition.

Zheng, T., Guo, L.C., Tang, Z.H., Wang, T.T., Li, Z.X., 2021. Comparison of progressive damage simulation of 3D woven composites between voxel and conformal discretization models. Mech. Mater. 158, 103860. https://doi.org/10.1016/j.mechmat.2021.103860.

Zhou, G., Sun, X., Wang, Y., 2004. Multi-chain digital element analysis in textile mechanics. Compos. Sci. Technol. 64, 239–244. https://doi.org/10.1016/S0266-3538(03)00258-6.

Zouari, B., Daniel, J.L., Boisse, P., 2006. A woven reinforcement forming simulation method. Influence of the shear stiffness. Comput. Struct. 84, 651. https://doi.org/10.1016/j.compstruc.2005.09.031.

The era of automation and robotisation of composite manufacturing

2.4

Klaus Drechsler
Technical University of Munich, Munich, Germany

1. Introduction

The cradle of composite technology lies in military aircraft and niche vehicles. The main focus here was on good mechanical properties and low mould costs for small quantities. In the following years, the focus of developments was primarily on optimising these mechanical properties. Higher stiffness and strength of the carbon fibre and higher damage tolerance and temperature resistance of the matrix systems were the goals. The optimisation led to more and more application areas becoming interested in the lightweight construction potential of composites. Manual production based on prepregs and fabrics was the predominant process technology. However, cycle times of several hours for complex and large components made it difficult to use composites in civil aircraft construction or even more in high-volume vehicle construction or the sporting goods industry, especially in high-wage countries. As a result, the production of composite tennis rackets or bicycle frames moved almost entirely from Western countries to Asia. It was not until the 1980s, that research and development projects were launched to automate the manufacturing process and thus making it faster, more robust and more cost-effective, thereby opening up new areas of application, in which the high lightweight construction potential and good functional properties could be utilised. The first successful step was the use of various textile technologies to process the fibres into complex preforms. Braiding technology was considered as well as embroidery and sewing techniques. At the same time, injection technologies, such as the high-pressure RTM process, were introduced as a second processing step. The development of robot-based fibre placement technology began in the 1990s, particularly for aircraft construction, initially for thermosets and later for thermoplastic composites. This enabled the automated production of large-area components with outstanding mechanical properties. In addition to optimising the lightweight construction potential, the focus was always on improving sustainability by minimising waste and offcuts. On basis of these and many other developments, the foundation has been laid for the use of composites in large-scale automotive construction, in the energy industry and for future applications, for example, in the field of urban air mobility.

In the following, some of the most important developments are described in detail.

The History of Composites. https://doi.org/10.1016/B978-0-443-34119-9.00015-2

2. Automated textile preforms technologies

Textile technology and fibre composite technology are very closely linked. It makes sense to use this partly ancient technology to process fibres into semi-finished products. Since the beginning of composite technology, fabrics have been used in various weave types to drape fibre structures with a 0/90-degree fibre orientation onto a tool in a hand lay-up process using dry fabrics and later in prepreg technology. Woven fabrics are also very popular for visible carbon applications in particular. Weaving technology also offers many possibilities for creating multi-axis or three-dimensionally shaped fibre structures, and also a fibre reinforcement in the thickness direction is possible. However, due to the low productivity, these semi-finished products are only used in special niche applications.

Multiaxial fabrics, which are produced on the basis of warp knitting technology, offer greater flexibility in terms of fibre orientation. Upto nine individual layers are laid on top of each other with high productivity and then fixed together using a multi-needle sewing system. Due to low fibre undulation, composites made from multiaxial fabrics tend to have higher strength and stiffness, while the damage tolerance is somewhat lower than that of woven fabrics.

Braided hoses, from which complex hollow structures such as tennis rackets can be produced, have found a wide range of applications, particularly in the sporting goods industry. However, this process is also strongly characterised by manual activities.

Another interesting textile technology for fibre composites is embroidery (textile fibre placement, TFP) (Coppola et al., 2023). It is also an ancient technology which has been adapted to carbon fibre processing in the '80s. This process enables the automated production of load-appropriate fibre structures. The reinforcing fibres are sewn onto a thin base textile in curved strips. This creates near-net-shape structures with minimal waste and high lightweight construction potential for complex loads such as hole reinforcements. Productivity can be increased with multi-head embroidery machines. Some typical embroidery structures are shown in Fig. 2.4.1, and a multi-head embroidery machine offering high productivity is shown in Fig. 2.4.2.

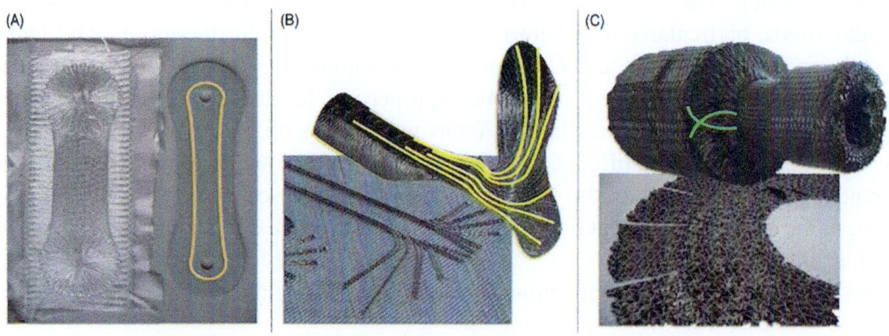

Figure 2.4.1 Typical textile fibre placement preforms (Uni Stuttgart, IFB).

Figure 2.4.2 Typical multi head embroidery machine (BIONTEC).

These four technologies exemplify the high potential of textile technology for processing fibres for composite technology, but they also illustrate the shortcomings. More details can be found in the chapter of Stepan Lomov.

In particular, the manual work steps for further processing and the high level of waste lead to inefficient processes in some cases. This has resulted in many research and development projects around the world with the aim of achieving greater automation combined with higher productivity and more robust processes, greater flexibility in terms of fibre architecture and less waste. One key to this is the use of industrial robots in conjunction with conventional textile technologies such as braiding or sewing, as well as the development of completely new processes such as fibre placement or fibre patch placement.

In robot-based braiding, braiding cores are guided by an industrial robot and moved through the braiding point (Ebel, 2017). This can take place on very complex paths and the fibre orientation can be adjusted by coupling the control of the braiding machine speed and the robot movement. In this way, complex, hollow fibre structures with biaxial or tri-axial fibre reinforcement are produced with a high degree of automation. Typical applications are drive shafts or hollow profiles for automotive engineering or the sporting goods industry.

Fig. 2.4.3 shows a robot-assisted braiding machine, and Fig. 2.4.4 some typical braided structures.

Both, structural cores and lost cores, such as blow-moulded parts, waxes, or low-melting alloys, can be used as core material. After removing the core, the spar structures can be reshaped and further processed into complex, even curved profiles with optimal fibre orientation. Examples of applications could be stiffeners of bulkheads for aircraft fuselages.

The degree of automation and productivity of the manufacturing process are very high with the appropriate machine size. Braiding machines with up to 196 bobbins have been realised, which also enable large component diameters in the range of

Figure 2.4.3 Robot-assisted braiding machine (TU Munich, LCC).

30 cm. The corresponding wall thicknesses are achieved by repeated over-braiding of the cores. It is also interesting to note that the waste is typically less than 10%. The latest developments are moving in the direction of intelligent bobbins, which, for example, detect fibre breakage in advance and enable flexible control of the yarn tension. This allows fibre placement to be further optimised and the process to become

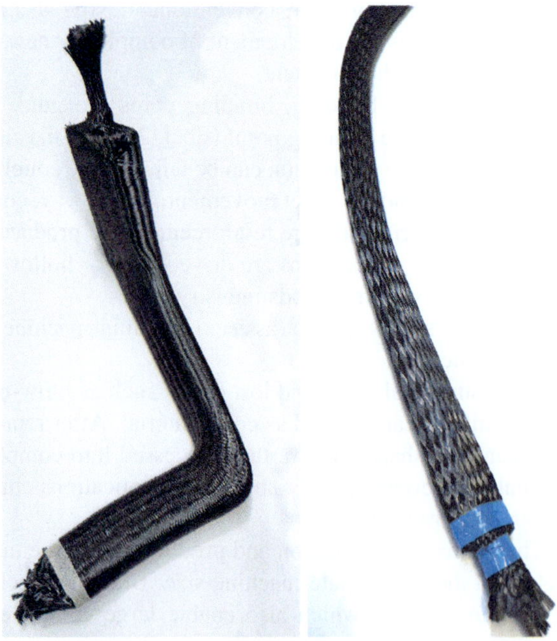

Figure 2.4.4 Typical braided preforms (TU Munich, LCC).

more robust. This is coupled with appropriate process monitoring, foor example, for the early detection of fibre damage.

It is important that the component design is adapted to the possibilities of braiding. In bicycle construction, for example, this "design-to-manufacturing" can lead to completely new frame concepts with a reduced number of parts, which can reduce the assembly effort and thus the costs. Fig. 2.4.5 shows a typical example of a braided CFRP bicycle frame.

Serial applications in automotive engineering have also been realised on the basis of robot-assisted braiding technology. Examples include energy-absorbing A-pillars or sills.

Another interesting application of automated braiding is the production of closed ring structures that can be used for the manufacture of bicycle rims, for example. For this purpose, braiding machines with a separable braiding body and corresponding fusible core concepts have been developed. More than 10.000 parts per year can be produced with one manufacturing cell. Fig. 2.4.6 shows a typical CFRP bicycle rim.

Another interesting robot-based process is 3D sewing. Various sewing heads have been developed for this, which are manipulated by a robot and only require one-sided accessibility. Sewing can be applied for two different tasks. First, the sewing threads can be used to combine different base textiles with each other to facilitate automated handling of large and complex preforms. Second, load carrying fibres can be inserted in the thickness direction to achieve a corresponding three-dimensional fibre reinforcement that improves the interlaminar properties. The result is improved damage tolerance or a higher energy absorption capacity by avoiding delamination.

A one-sided sewing technology, the so-called blind stitch, was originally developed for surgery. Here, a curved needle penetrates the semi-finished textile product and the sewing thread is knotted as it emerges before the curved needle is retracted

Figure 2.4.5 Braided bicycle frame (TUM LCC).

Figure 2.4.6 Braided bicycle rim (TUM LCC).

again. The advantage of this technology is that sewing can be carried out directly in a sewing tool, as the needle does not emerge from the back of the preform.

This is the case with the two-needle sewing process, in which two needles enter the material from one side, and the knotting takes place on the reverse side.

Tufting is particularly interesting for optimising the interlaminar properties. Fibres are inserted into a preform in the z-direction using a tufting head. The distance and direction of insertion of the fibres can be controlled by the robot control. This allows composites with a high damage tolerance and a high energy absorption capacity to be realised as no delamination occurs. Sewing threads made of glass, carbon or aramid fibres can be used. The aim is to optimally adapt the proportion of z-threads to the requirements, as they reduce the in-plane properties by causing a certain amount of damage to the base textile structure.

A typical stitching cell is shown in Fig. 2.4.7.

Based on these robot-based technologies, complex production cells have been developed that combine braiding and tufting, for example, and thus enable the fully automated production of complex, 3D-reinforced preforms. One application example is crash cones with a high energy absorption capacity for automotive applications (see Fig. 2.4.8) or frame structures for aircraft construction. A corresponding production cell is shown in Fig. 2.4.9.

Various injection and infusion processes have been developed to impregnate the textile semi-finished products, which are used depending on the required cycle time. These processes range from vacuum-assisted infusion for aircraft construction to high-pressure RTM for automotive construction. In interlinked processes and on the basis of robot-based handling systems, production times in the range of 5 minutes are possible from the fibre to the finished component. An RTM production cell with

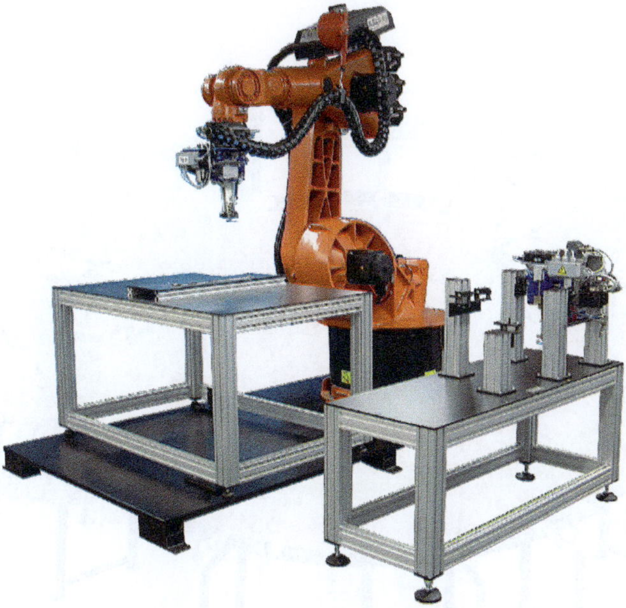

Figure 2.4.7 Typical robot stitching cell (Uni Stuttgart, IFB).

the interlinked process steps is shown in Fig. 2.4.10. Detailed information about the basic composite processes can be found in the chapter of Kevin Potter.

Increasing automation can also be seen in the area of composite recycling. Although there is currently still a pronounced development dynamic in this field of work, some major industrial business models have already been established (Oliveux

Figure 2.4.8 Energy absorbing crash cone based on automated preforming technologies (TUM LCC).

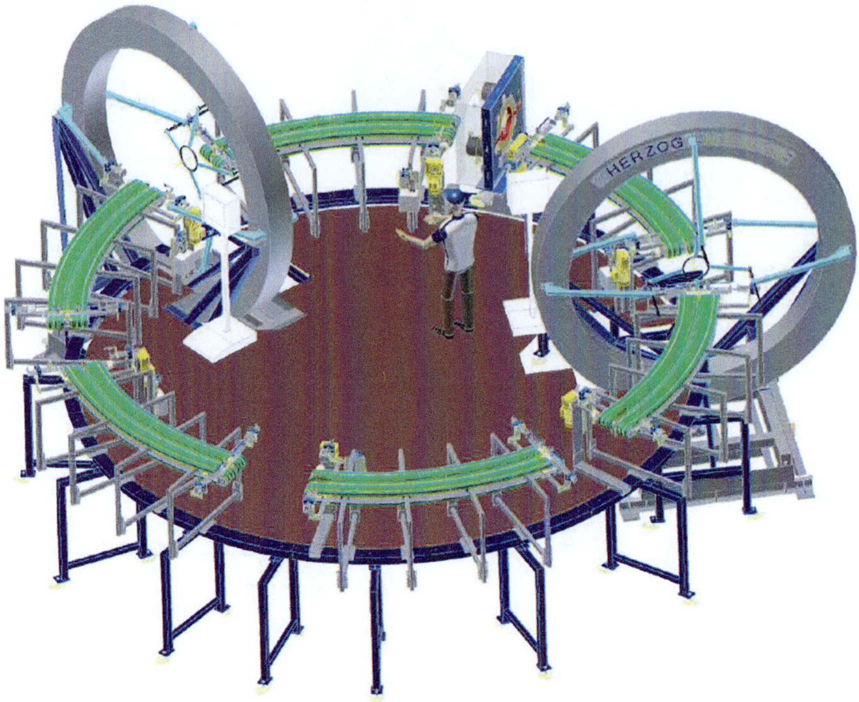

Figure 2.4.9 Preforming cell for automated fuselage frame production.

et al., 2015). One example is the processing of reclaimed carbon fibres using adapted textile manufacturing methods and paper technology (Wölling et al., 2017). Automated and digitalised plant equipment is already being used here. The processed recycled carbon fibres represent a major modification compared to the classic application scenarios of such systems (e.g., natural fibres as cellulose), but extensive expertise can be utilised on the machine engineering side. Fig. 2.4.11 shows an example of a semi-industrial-scale wetlaid nonwoven line as a modification of a classical paper machine. This enables the production of a uniform nonwoven semi-finished product despite fluctuating input material parameters and also features a very high material throughput. The semi-finished products can be used as structural components of a nonwoven-reinforced composite as part of a second-life application. Alternatively, attractive functional property profiles can also be created using such technical nonwovens, for example in the field of filtration or as an electrochemical membrane (H_2 fuel cell as well as battery applications). Another advantage is the possibility of creating fibre blends, for example, in the form of so-called hybrid nonwovens. Here, a thermoplastic matrix component is already integrated into the material, so that subsequent thermoplastic composite processing is enabled (Sauer et al., 2019). High-performance in-line quality monitoring and closed loop control technology play an increasingly important role, especially in applications that require a high degree of

Figure 2.4.10 Automated production cell for high-volume composite production (fill).

homogeneity. The monitored material parameters include, for example, a continuous detection of defect patterns, a grammage measurement, a residual-moisture determination, as well as the degree of fibre orientation respectively the degree of isotropy. Here, extensive use can be made of sensor and control technology from the field of industrial paper plant equipment, although decisive adaptation measures are yet necessary in the context of recycled carbon fibres. For comparison, in today's industrial paper machines, a downstream process control is carried out right up to the point of actively adjusting defined units and process control variables to correct any deviations detected with minimal manual interaction.

Figure 2.4.11 Wet laid nonwoven machine for recycled fibres (www.bewegt-content.com |C. Strohmayr).

3. Robot-based fibre placement technologies

Parallel to the automation of textile preform technologies, various laying technologies for unidirectional tapes were initially developed, particularly for prepregs. The aim here was not only to increase the degree of automation and robustness in production but also to enable the more automated production of very large shell structures for aircraft for empennages, wings or fuselage. The Automated Tape Laying (ATL) has been developed since 1970s for shells with lower geometrical complexity and therefore limited need for fibre steering (Brasington et al., 2021). Commercialised in early '80s ATL heads provide positioning, cutting and placement of prepreg tapes of 75 mm up to 600 mm on curing moulds and demonstrated the high potential of automation especially in aircraft structure manufacturing.

Even if the investment costs for the corresponding systems are quite high, they are amortised quite quickly thanks to the high productivity. This results from tapes being laid down by complex laying heads using a gantry system. Fig. 2.4.12 shows a typical Automated Tape-laying machine (ATL).

Moreover, in late '80s hybrid concepts of Filament winding and ATL were initially investigated to provide tape steering capabilities for high complex 3D geometries and Automated fibre Placement (AFP) technology was born. Initially mainly used in the US for military applications for fighter aircraft or space launch systems, AFP became baseline technology for the new full CFRP commercial aircraft fuselages of Boeing 787 and Airbus A350 XWB in early 2000s.

The disadvantage of ATL is that (1) only relatively simple geometries can be realised because the wide tapes cannot be guided on curved paths (known as fibre steering) and (2) large geometric curvatures cannot be manufactured with the large laying heads. For these applications, fibre placement systems have been developed in which

Figure 2.4.12 Typical Tape-laying machine (Dirk et al., 2012).

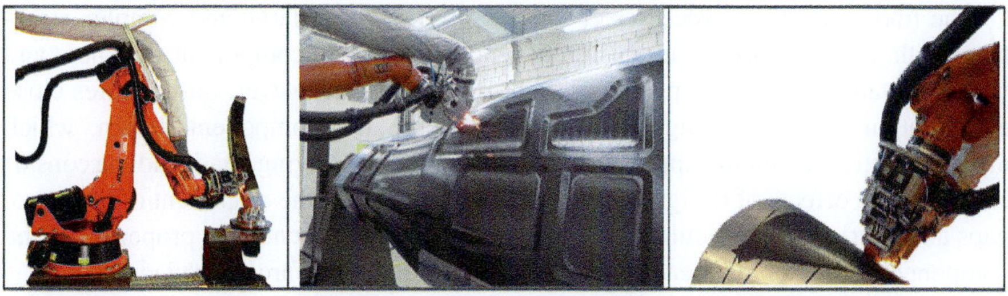

Figure 2.4.13 Robot-based fibre placement on complex parts (Fraunhofer IGCV).

a relatively small laying head can be guided by a robot on complex, three-dimensional paths. This is illustrated by the Advanced Fibre Placement (AFP) system in Fig. 2.4.13, where the narrow tapes can also be placed in a curved position, and several tapes can be placed in parallel to increase productivity.

The laying head must fulfil various functions. The temperature is used to adjust the stickiness of the tape (tack), the pressure of the pressure roller ensures adhesion or pre-compacting, and a cutting unit makes it possible to stop or restart the tape at any point.

The basic principle of an AFP system is shown in Fig. 2.4.14.

Various basic principles can be used for temperature control, from open gas flames to UV lamps and lasers, which differ in their energy efficiency, controllability, and maximum temperature.

The basic principle can also be applied to thermoplastic tapes. Here, even online consolidation is possible, which means that no subsequent pressing or autoclave process is required after depositing. The prerequisite is a high process temperature above the melting temperature of the matrix, which requires laser heating, as well as high pressure when depositing the tapes.

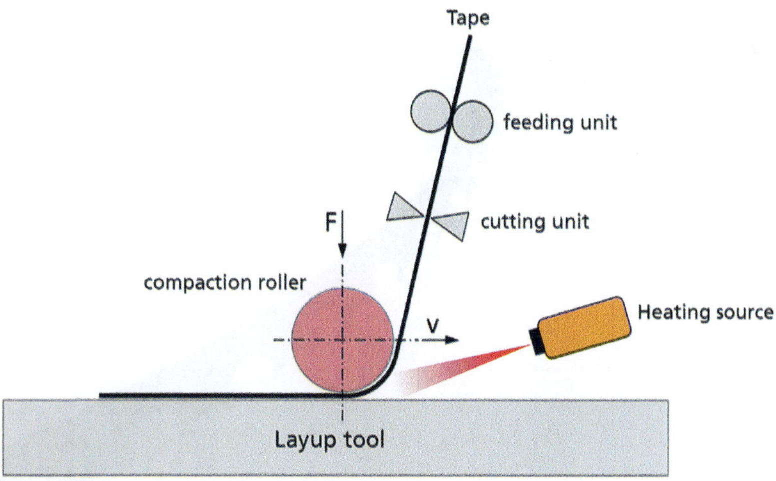

Figure 2.4.14 Principal of the fibre-placement process (TUM LCC).

The fibre placement processes enable the production of large and complex structures with a high lightweight construction potential as the component thickness and fibre orientation can be adapted very well to the loads. However, this requires close coordination with the design. This concerns both the component design, which must take into account the appropriate accessibility of the laying head, and the consideration of the effects of design in the layout. In addition to the alignment of the tapes, gaps and overlaps in particular are quality features that influence the properties of the components and can also lead to rejections if certain limits are exceeded.

Digitalisation is also increasingly finding its way into the AFP process (Online-Prozessmonitoring im ATL und AFP et al., 2023). Cameras or laser cutting sensors are integrated into the laying heads to monitor the quality of the deposit online. In particular, the aim is to reduce the effort required for subsequent non-destructive testing and to minimise rejections.

Fig. 2.4.15 shows a typical fuselage frame structure that was manufactured using an AFP process. Compared to the conventional titanium structure, both the cost and the weight of this component were reduced.

A further example is the helicopter shell in Fig. 2.4.16. Sandwich elements integrated into the CFRP structure were used to reinforce this shell component. In order to realised the sandwich structure, step ramps with adjusted fibre steering must be realised without damaging the core structure by high roller pressure. Therefore, adapted machine programming and layup strategies for the laying heads are required.

The productivity of AFP technology depends on the system technology but, above all, on the complexity of the components. For simple geometries and large-area structures, up to 100 kg per hour are possible, but for small components with a complex laminate structure, the layup rate is significantly reduced.

Figure 2.4.15 Fuselage frame structure based on the AFP-process (Fraunhofer IGCV).

Figure 2.4.16 Helicopter sandwich shell structure manufactured by robot-based AFP (Airbus Helicopters Deutschland GmbH).

4. Fibre patch placement

Fibre patch placement has been developed in the frame of public funded projects by Airbus based on pick-and-place facilities from the electronic part production (Fischer et al., 2015). In further developments by the spin-off company CEVOTEC of the Technical University of Munich, industrial robots have been used to increase the productivity.

The process enables a new degree of freedom in automated preforming technologies. Typical FPP parts are too complex in terms of curvature to be efficiently manufactured by ATL or AFP systems. FPP therefore enables the automated production of parts that could previously only be produced using manual lay-up. Key differentiators to robotic AFP are:

F09E Ability to handle a wide variety of materials (in parallel, in one system)

F09E Ability to lay up on more complex and smaller shapes/surfaces (compared to AFP capabilities)

F09E Ability to design one-layer laminates with defined overlaps

F09E Opportunity to significantly reduce debulking steps through defined deposition pressure

A patch can be made from any material that is sufficiently flexible and generates enough tack during heating to create a stable preform. The FPP production systems enable the automated lay-up of carbon or glass fibre prepreg, dry fibre material, adhesive film and other technical (fibre) materials on complex 3D geometries.

The patch size can be adjusted to the size and complexity of the component. It ranges from around 12.7 × 60 mm to 200 × 300 mm and above (Horn et al., 2017). Defined patches are automatically cut from a tape and precisely placed by one or multiple robots and a flexible patch gripper. The process is implemented as a series of individually controlled patch placements, enabling a superior level of process control for the entire laminate lay-up cycle. Patches, when cut in-line, are usually rectangular or trapezoidal. Free shapes are possible through offline cutting on a separate cutting system and subsequent pick-up from a table.

The schematic principle of the FPP-process is shown in Fig. 2.4.17. Fig. 2.4.18 shows the realisation of an actual FPP device; the processing steps 1–5 are correlating for both figures.

The key to a fast and automated lay-up process for complex shapes is a form-flexible silicone-foam-based patch gripper. The activation of the patch material can be provided through gripper-integrated heating or by external heating sources. The pick-up of patches is based on vacuum generation within the gripper base plate by using compressed air. The physical principle is known as the Coanda effect and eliminates the need for a centralised vacuum supply on the shop floor. A quick switchover from suction to blow-off supports the placement of the patch on the tool, regardless of its geometric shape.

The gripper is available in tailored sizes to match product requirements. It adapts to highly complex surfaces. Even across 90 degrees edges and double-curved surfaces, patches are placed precisely and without detrimental draping effects. The provided compression is sufficiently high to reduce or fully eliminate the need for time-consuming intermediate debulking steps. The gripper can be equipped with a compaction-force sensor.

FPP production systems are modular by design, and the modules can be combined flexibly to create a production set-up optimised to client and part requirements (Dr. et al.). This enables either the highly efficient production of a specific product category or increased flexibility across multiple products.

Fig. 2.4.19 shows two exemplary FPP components.

5. Additive manufacturing

3D printing supports the development and production of components since the '80s in many areas. The integration of short or continuous fibres into the printing process has

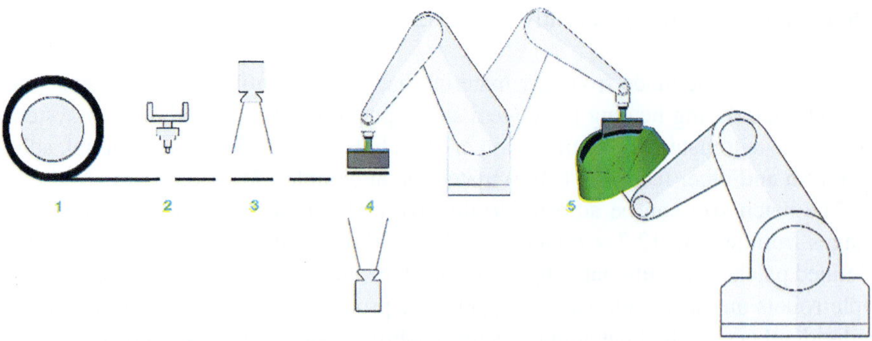

Figure 2.4.17 The schematic process steps of FPP (CEVOTEC).

1 | Feed fiber tape 2 | Cut tape into patches 3 | Inspect quality 4 | Pick up, check position 5 | Place fiber patch

Figure 2.4.18 FPP-device and the subsequent processing steps (CEVOTEC).

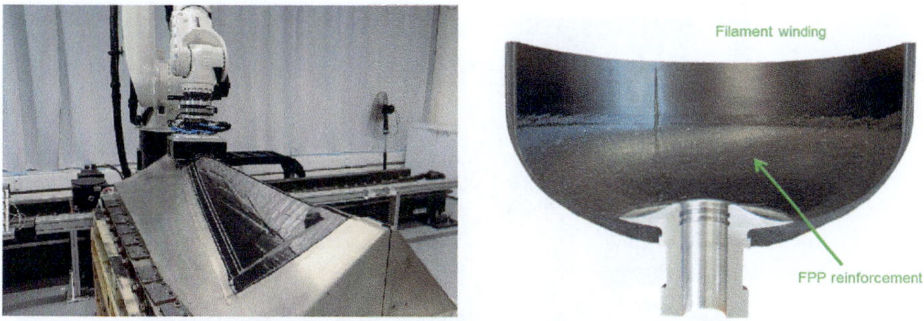

(Left) Fairing demonstrator, public funded project: ACoSaLUS, (right) Cut-out of an FPP dome reinforcement

Figure 2.4.19 Typical FPP components (CEVOTEC).
(*Left*) Fairing demonstrator, public funded project: ACoSaLUS, (*right*) Cut-out of an FPP dome reinforcement.

created an interesting process for the production of complex structural components with interesting mechanical properties, which implements the vision of tool-free production.

Fig. 2.4.20 schematically shows two processes for short and endless fibre integration.

On this basis, many systems have been developed in recent years, from small laboratory systems to large robot-based production cells, which enable the automated production of complex components, similar to the AFP process. Fig. 2.4.21 shows a typical system.

Typically, thermoplastics are used as the matrix system, but there are also systems that work with thermosets and UV-curing systems.

Compared to fibre placement, however, the properties of the manufactured components are significantly worse. This is due to the less perfect fibre placement of the rovings compared to the tapes, the lower fibre volume content and, in particular, the

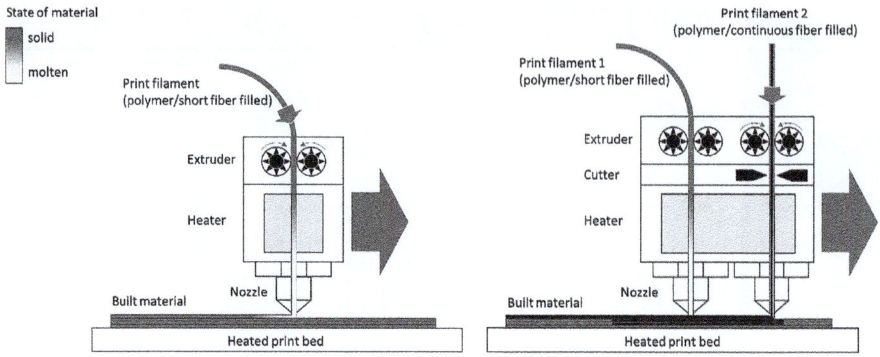

Figure 2.4.20 Principal of short and endless fibre integration in the 3D printing process (TUM, LCC).

Figure 2.4.21 Typical robot-assisted large-scale 3D-printing process (TUM, LCC).

higher void content due to the missing compaction of the material. The application possibilities for highly stressed structural components are therefore limited, although a combination of additively manufactured components with localised reinforcement using AFP is very interesting from a structural-mechanical point of view as well as in terms of production technology. An exemplary component is shown in Fig. 2.4.22.

Fibre-reinforced 3D printing is particularly interesting for tool making (Consul et al., 2020). On the one hand, the production time is much shorter than with metal tools and on the other hand, the weight can be significantly reduced, and the thermal expansion can be adapted to the fibre composite component. Even the integration of heating and cooling channels is possible. Fig. 2.4.23 shows a typical mould. This process is called indirect additive manufacturing.

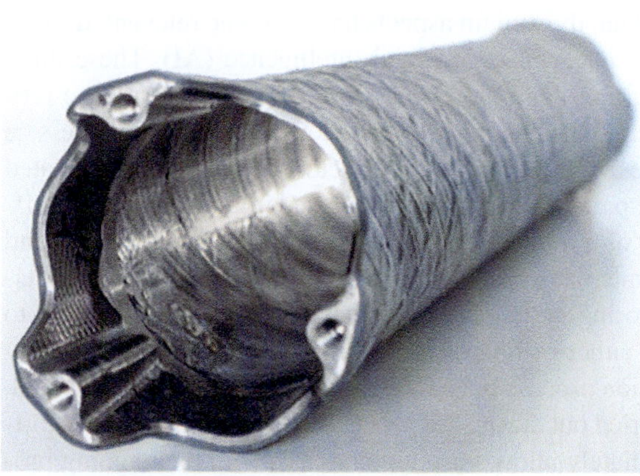

Figure 2.4.22 Driveshaft with combined 3D printing and fibre-placement (TUM, LCC).

Figure 2.4.23 3D-printed mould for composite manufacturing (Consul et al., 2021), (TUM, LCC).

6. Digitalisation and artificial intelligence

Automation is inherently connected to methods of digitalisation. As automation requires (partial) autonomy of the machines in manufacturing processes, a digital representation of real-world assets is key in modern-day composite manufacturing. The trend clearly goes toward an increasing amount of data gathered during the manufacturing process with the goal of increasing process stability, speed and quality.

With increasing amounts of data, three main aspects have become relevant: digitalisation, process monitoring and, lately also, artificial intelligence (AI). These three areas are additionally becoming more and more interconnected: to increase the process stability, speed and/or quality, an increased knowledge about the process is necessary. While the knowledge about the process characteristics has been generated for a long time by taking sample parts from the production and investigating them offline, nowadays, the availability of a multitude of sensor and measuring systems and adequate data processing equipment at a reasonable price nourished the implementation of advanced process monitoring systems. These systems can be implemented to increase process understanding either offline or online. While offline monitoring results in insights into the production process only after the part has been manufactured, online process monitoring is carried out during the part production. Consequently, offline monitoring does not immediately allow for countermeasures in the ongoing production. This means that if a faulty process leading to a damaged part is recognised, usually, the entire batch of produced parts must be scrapped. On the contrary, online monitoring systems are implemented in or near the production machines to process the gathered data immediately. Depending on the system used, a faulty production process can be identified immediately, and the number of scrap parts can be reduced significantly. Ideally, the online process monitoring system does not only allow for the immediate stop of production (reactive) but also for the prevention of process instabilities and interruptions (predictive).

It becomes clear that if a production is to be transferred to a more predictive process monitoring, the "amount of digitalisation" necessary is increasing. In fact, the better the virtual representation of the production process, the better the data basis for implementing predictive algorithms. "Better" in this context does not necessarily mean more data, but high-quality data from multiple sources. This trend culminates in the complete virtual representation of a production plant. If the entirety of processes and processing machines is digitised, this is referred to as "digital twin." In this context, terms like "Industry 4.0" and, recently, with the introduction of AI technologies in the production environment, "Industry 5.0" have been coined.

In the area of composite manufacturing, significant advancements have been achieved in recent years with the integration of novel AI methods in various production technologies. The method used herein is mainly influenced by the type of data that can be acquired from the processing machines. While conventional sensors (like pressure or temperature sensors used in LCM processes) provide only a single data signal at a local position, the implementation of sensors for locally distributed information (like camera systems or laser scanners) is also possible. A combination of multiple data sources (at different locations/with different physical measuring techniques) widens the knowledge about the current state of the process and thus can improve the basis for evaluations using AI algorithms. Generally speaking, if less process data is captured, simpler models like shallow neural networks, decision trees, and others will be utilised. If more data is available, deep learning methods like deep neural networks, together with big data, are used.

Examples include the implementation of stationary (Maidl et al., 2021) and bobbin-carrier-mounted sensor systems measuring the yarn tension in the braiding

Figure 2.4.24 Sensor-integrated radial braiding machine with localisation and indication of process defects (fibrous rings) (Diemar et al., 2024), (TUM, LCC).

process to predict machine downtimes, as can be seen in Fig. 2.4.24. Other examples are the implementation of image-based sensor systems to retrieve the surface profile of tapes laid in the AFP process and assess the suitability for further processing or the optimisation of the draping behaviour in dry preforming or the determination of flow characteristics from sensor data in the LCM process and the prediction of dry spots.

An AFP machine with a mounted laser scanner can be seen in Fig. 2.4.25. The scanner measures the height distribution along a line behind the laying head. Alternatively, conventional camera systems can also be utilised. Information about the surface topology and, thereby, the occurrence of defects such as gaps or overlaps then results from the formation of shadows caused by a light source. Either way, this can serve as an input for an AI algorithm, for example, a convolutional neural network (CNN), to classify the currently placed layer as suitable (OK) or unsuitable (NOK) for further processing. An even more advanced AI task is the semantic segmentation of defect regions in the captured image data. Then, not only a classification for the entire

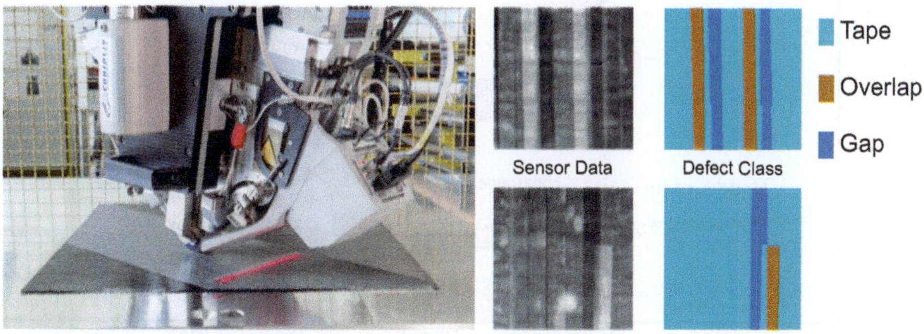

Figure 2.4.25 *Left*: Automated Fibre Placement (AFP) with laser scanner for online process monitoring (TUM, LCC). *Right*: Sensor data and corresponding annotation map as input for a convolutional neural network (TUM, LCC).

image (OK/NOK) but a classification of every pixel in the image to a predefined defect class (gaps/overlaps/etc.) is possible (Diemar et al., 2024).

7. Summary and outlook

The production technologies for composite materials can be categorised into different decades. Until the 1970s, hand lay-up dominated, especially with textile fabrics. Typical areas of application were glider aircraft and niche vehicles. In the 1980s, prepreg construction became increasingly popular, initially using manual processes before ATL processes were used to manufacture large structures for commercial aircraft. In the 1990s, the variety of technologies became ever greater. Automated preform technologies in combination with RTM processes were developed primarily for the automotive industry, while the range of components used in aircraft construction was constantly expanded, particularly through AFP processes in 2000–2010s for Airbus A350 and Boeing 787.

The various processes are continuously being developed and linked and supplemented by new technologies in 2020s such as 3D printing, especially to meet the increasing volumes in commercial aircraft production or for new types of air vehicles in the field of urban air mobility.

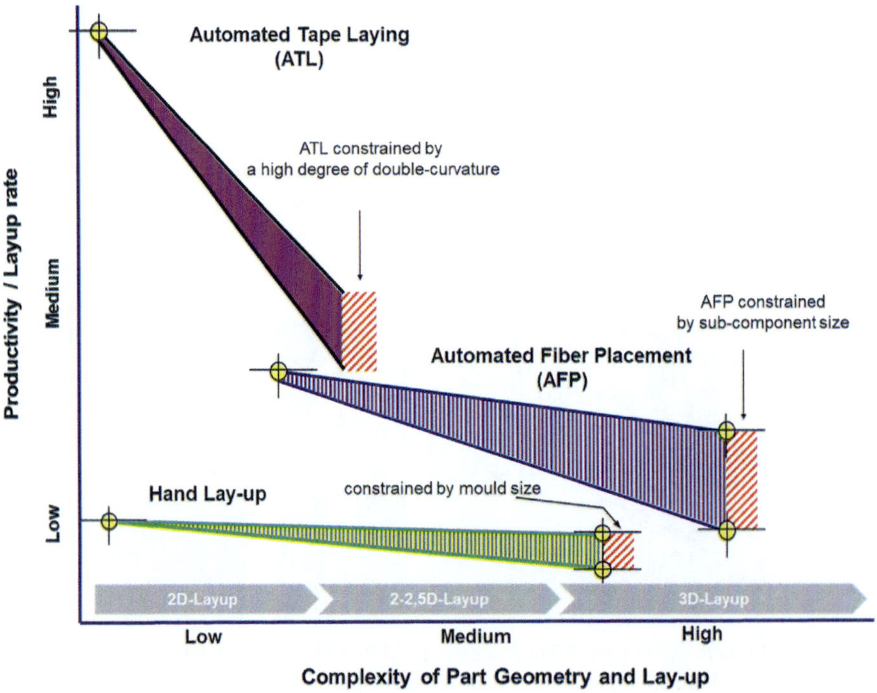

Figure 2.4.26 Influence of part complexity on the productivity (Fraunhofer IGCV).

The new technologies also play an important role in the energy transition. Hydrogen tanks, battery boxes, or blades for wind turbines are inconceivable without the use of composites, but they require cost-effective and large-scale production processes. The automated and robot-based processes described fulfil the corresponding requirements.

The optimal technology must be selected depending on the requirements. Fig. 2.4.26 illustrates this using the example of component complexity.

Digitalisation and AI will play an increasingly important role. By monitoring all relevant data online and processing it using artificial intelligence methods, not only can processes be further optimised, but it will also be possible to consistently link design, layout, tool making, and production through to recycling and the circular economy.

References

Brasington, A., Sacco, C., Halbritter, J., Wehbe, R., Harik, R., 2021. Automated fiber placement: a review of history, current technologies, and future paths forward. Composites Part C: Open Access 6, 100182.

Consul, P., Chaplin, A., Drechsler, K., 2020. Big area additive manufacturing of a thermoplastic tooling for a large thermoset composite structure. In: SAMPE Europe Conference 2020.

Consul, P., Chaplin, A., Tagscherer, N., Zaremba, S., Drechsler, K., 2021. Interlaminar strength in large-scale additive manufacturing of slow crystallizing polyaryletherketone carbon composites. Polym. Int. 70, 1099–1108. https://doi.org/10.1002/pi.6168.

Coppola, A.M., Huelskamp, S.R., Tanner, C., Rapking, D., Ricchi, R.D., 2023. Application of tailored fiber placement to fabricate automotive composite components with complex geometries. Compos. Struct. 313, 116855.

Diemar, F., Trappmann, T., Drechsler, K., 2024. Machine learning-based classification of lay-up defects in the automated fiber placement process with artificial image data. In: SAMPE Europe Conference Belfast 2024.

Dirk, H.-J.A.L., Ward, C., Potter, K.D., 2012. The engineering aspects of automated prepreg layup: history, present and future. Composites, Part B 43, 997–1009. https://doi.org/10.1016/j.compositesb.2011.12.003.

Dr D. Sikoutris, Dr F. Lenz, H. Weidinger, H. Morgenstern, Prof Dr N. Majić: Fiber patch placement in aerostructures: Investigation on material equivalency. https://www.cevotec.com/fiber-patch-placement-in-aerostructures-investigation-on-material-equivalency/. Accessed on 22. January.2025,

Ebel, C., 2017. High Speed Braiding for Manufacturing Composite Parts. Dissertation. https://mediatum.ub.tum.de/?id=1335638.

Fischer, B., Horn, B., Bartelt, C., Blößl, Y., 2015. Method for an automated optimization of fiber patch placement layup designs. Int. J. Compos. Mater. 5 (2), 37–46. https://doi.org/10.5923/j.cmaterials.20150502.03.

Horn, B., Neumayer, J., Drechsler, K., 2017. Influence of patch length and thickness on strength and stiffness of patched laminates. J. Compos. Mater. 52, 2199–2212. https://doi.org/10.1177/0021998317740413.

Maidl, S., Fernández Villalba, Á., Kind, K., Drechsler, K., 2021. Development of sensor integrated braiding rings for the automated detection of braiding defects. Mater. Today Proc. 34, Part 1, 74–81.

Oliveux, G., Dandy, L.O., Leeke, G.A., 2015. Current status of recycling of fibres reinforced polymers: review of technologies, reuse and resulting properties. Prog. Mater. Sci. 72, 61–99.

Online-Prozessmonitoring im ATL und AFP, Yadav, N., Schledjewski, R., 2023. Review of in-process defect monitoring for automated tape laying. Composites Part A 173, 107654.

Sauer, M., Feil, J., Manis, F., Betz, T., Drechsler, K., 2019. Thermoplastic multi-material nonwovens from recycled carbon fibres using wet-laying technology. Key Eng. Mater. 809, 210–216, 22nd Symposium on Composites.

Wölling, J., Schmieg, M., Manis, F., Drechsler, K., 2017. Nonwovens from recycled carbon fibres – comparison of processing technologies. Proced. CIRP 6, 271–276, 1st CIRP Conference on Composite Material Parts Manufacturing CIRP CCMPM.

The emergence of high-performance thermoplastic composites

2.5

David Leach
Composite Material Solutions, Walnut Creek, CA, United States

1. Introduction

The first modern composites were developed in the early 1900s using thermosetting resins such as phenolics and polyesters. Epoxy resins followed in the 1930s and provided a significant increase in temperature resistance, strength, and stiffness (see Part I). The first use of long fibre composites was with flax fibres in the 1930s and these composites were used in aircraft in World War 2 (see Part I). Thermoset resins were readily available, simple to formulate and more importantly had low viscosities and therefore were easy to impregnate into high fibre volume composites. Glass fibres (GF) were developed in the 1930s and carbon fibres (CF) in the 1960s, and the demand for lightweight, high strength materials grew considerably with their availability, particularly in sporting goods and aircraft applications. Epoxy resins provided high mechanical strength and service temperature for aerospace but were brittle. Through the 1970s, epoxy/GF composites were used increasingly in commercial aircraft secondary structures, such as fairings and control surfaces, and epoxy/CF materials in defence applications.

Thermoplastic polymers were developed from the 1930s onward with increasing thermal and mechanical performance. A wide range of thermoplastic polymers were formulated, but most had relatively low glass-rubber transition temperature (T_g) and, hence, low service temperature, along with low strength, and stiffness compared to the thermosetting resins. These aspects were offset by high toughness and the ability to mould parts quickly because there is no chemical reaction during part fabrication (see Part I). As a result, unreinforced and short fibre reinforced thermoplastics became widely used in commodity and durable goods applications. Higher performance thermoplastic polymers were developed in the 1980s, including amorphous polymers such as the poly(aryl-ether-sulphone) (PAES) family, poly(ether-imide) (PEI), and semicrystalline polymers of poly(phenyl-sulfide) (PPS) and the poly(aryl-ether-ketone) family (PAEKs). The semicrystalline polymers have glass transition temperatures in excess of 90°C, and amorphous polymers have glass transition temperatures over 200°C. In the early 1980s poly(ether-ether-ketone) (PEEK), a member of the PAEK family, was commercialised. PEEK had the combined benefits of good environmental resistance, with increased service temperature (T_g of 143°C), and high toughness. The commercial availability of PEEK opened the opportunity for

The History of Composites. https://doi.org/10.1016/B978-0-443-34119-9.00016-4

much higher performance thermoplastic polymer composites in multiple forms which could meet aerospace requirements.

2. Thermoplastic composite materials

2.1 Polymers

A wide range of thermoplastic polymers are available, and these are generally categorised into Commodity, Engineering, and High Performance Polymers, Fig. 2.5.1. The main characteristics of moving up the pyramid are increasing service temperature and increasing mechanical performance. The polymers suitable for aerospace structural applications are in the High Performance Polymer bracket and are highlighted in Fig. 2.5.1. The representative thermoplastic polymers and their thermal properties are summarised in Table 2.5.1.

2.2 Composite product forms

Unreinforced polymers and short fibre or particle reinforced plastics were widely available by the 1970s but thermoplastic composites (TPCs) with high loadings of long or continuous fibre reinforcement were not considered practical. At the SAMPE Conference in 1980, Prof Anthony Kelly FRS of Surrey University said: '*I believe that reinforced thermoplastics have a brighter future in the long run because the physical properties of the plastic arise from a more stringent chemistry, namely that necessary to produce linear polymerization, and hence should lead to better controlled and more reproducible physical properties.*' (Cogswell, 1992).

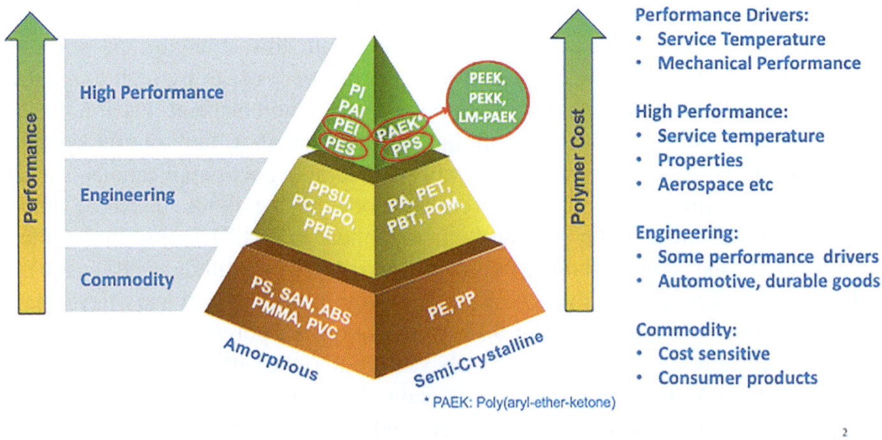

Figure 2.5.1 Classification of thermoplastic polymers.
(Image credit: D Leach).

Table 2.5.1 Thermoplastic polymers for high-performance composites (Leach and Pratte, 2024).

Polymer	Morphology	Polymer structure	T_g (°C)	T_m (°C)	T_p (°C)
PESU Poly(ether-sulfone)	Amorphous	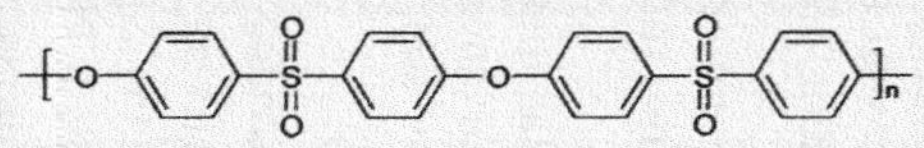	220	–	360
PEI Poly(ether-imide)	Amorphous	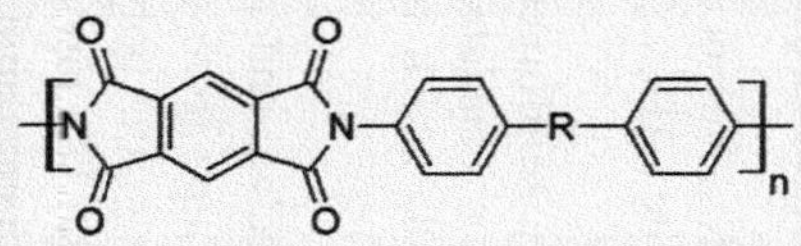	217	–	332
Poly(phenylene-sulfide)	Semicrystalline	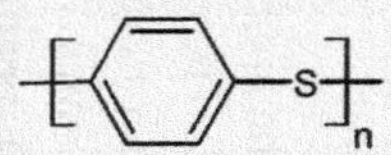	93	280	329
PEEK Poly(ether-ether-ketone)	Semicrystalline	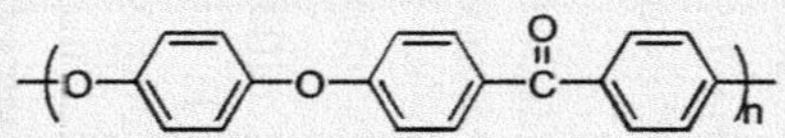	143	343	390
PEKK Poly(ether-ketone-ketone)	Semicrystalline	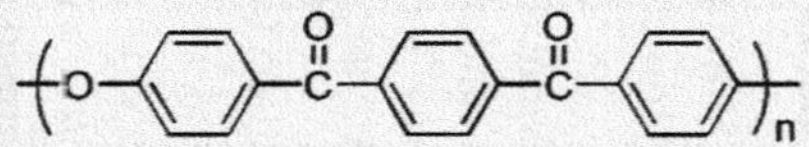	159	337	380
Low melt PAEK LM-PAEK	Semicrystalline	PEEK–PEDEK Copolymer	147	305	350

T_g: Glass-rubber transition temperature, T_m: Melt temperature, T_p: Typical processing temperature.

Thermoset resins have low viscosities and could be readily impregnated into fibres. Thermoplastic polymers have much higher viscosities and could only be combined with fibres in compounding operations using high shear, which limited the length and quantity of the reinforcing fibres. For example, typical short fibre compounds contained up to ~30% by weight of reinforcement. It was recognised that the ability to incorporate much higher loadings of long or continuous fibres would open up many new opportunities. The polymers of interest for these more structural materials all have high process temperatures, generally 300°C–400°C, further complicating the impregnation process. In the previous material iterations, thermoplastic polymers had essentially displaced thermosets in unreinforced and short fibre reinforced materials, so it was anticipated that the same would happen with highly reinforced composites.

The viscosity and high process temperatures meant that the impregnation processes would be more difficult and therefore higher cost, and so the focus was on demanding applications where the performance premium was high. Several materials suppliers followed this path, and two emerged with commercially viable products:

- Imperial Chemical Industries (ICI) in the UK developed a unidirectional (UD) tape material called Aromatic Polymer Composite (APC). This product is now part of Syensqo.
- TenCate in The Netherlands developed fabric reinforced laminates, also referred to as 'organosheets' under the trade name Cetex. TenCate Advanced Composites was acquired by Toray Industries Inc. in 2018, which now supplies the materials through Toray Advanced Composites.

The two companies took different approaches and provided different material forms as summarized in Table 2.5.2. They came from different backgrounds with ICI being a large polymer manufacturer ranging from poly(ethylene) to PEEK, while TenCate was primarily a specialist weaver. The UD tapes with PEEK matrix developed by ICI were aimed at very high-performance applications, in particular aerospace structures. The combination of GF with engineering polymers from TenCate were more suitable for aircraft interiors and secondary structures. While the unidirectional tapes gave higher performance and more design freedom, they required more involved processing to fabricate a finished part. The fully consolidated organosheets from TenCate were a semi-finished product which was easier to fabricate into components. TenCate also operated within an ecosystem in the Netherlands, which included an end-user in Fokker Aircraft (now GKN), and applied research through the Technical University Delft. By the mid-1980s, both material forms were available and were being evaluated for aircraft applications (Cogswell and Leach, 1984; van Dreumel and de Groot, 1987). The interest in the potential of TPCs was recognised quickly, leading to an in-depth assessment by the US National Research Council (US NRC, 1987). Many more manufacturers entered the market later, and a wide range of UD tapes, fabrics and organosheets are available now from Teijin, Hexcel, Porcher, Mitsubishi, Victrex, Barrday, Gurit, Concordia, Lingol, and others, in addition to Syensqo and Toray.

Table 2.5.2 Initial continuous fibre thermoplastic composite suppliers and product forms.

Manufacturer	Location	Primary business function	Product form	Main polymer	Main reinforcement
Imperial Chemical Industries (ICI)	United Kingdom	Polymer manufacturer	Unidirectional tape	PEEK	Carbon fibre
TenCate	Netherlands	Fabric weaver	Fabric laminates	PEI, PPS	Glass fabric

3. Development and characterisation of thermoplastic composites

Two of the main benefits of TPCs are improved performance (particularly toughness) and rapid part fabrication. Both aspects were investigated extensively on the newly available TPCs through the 1980s, and this section focuses on the early assessments that set the scene for production use.

3.1 Performance

The performance of TPCs, and particularly their toughness, was one of the main areas of interest. The epoxy resins used in aerospace applications were formulated for high service temperature, high stiffness, and high strength. The high resin T_g was needed to ensure adequate service temperature even after hot-wet conditioning, which causes plasticisation of epoxy resins. These highly cross-linked formulations were brittle, and this proved to be a limitation due to potential delamination of the composites, particularly in the presence of damage. Toughening mechanisms were known in the 1980s, mainly through the addition of elastomers, but this led to reduced T_g and hence service temperature, particularly after hot-wet conditioning. The toughness and consequent damage tolerance were seen as a major design limitation of epoxy matrix composites because impact damage that may not be visible on the surface of a component could cause internal delamination, which can propagate under compression loading. Extensive characterisations of damage tolerance were undertaken through 'compression after impact' tests, and fundamental interlaminar fracture toughness measurements such as double cantilever beam for mode I, and end-notched flexure for mode II. The effects of matrix toughness on composite interlaminar toughness showed the benefits of tough, thermoplastic matrix composites (Hunston et al., 1987). This led to many assessments of impact damage and interlaminar toughness, for example (Leach et al., 1987). In further studies, the residual compression strength after impact of a PEEK/CF composite has been shown to be up to three times that of a moderately toughened epoxy matrix composite, as shown in Fig. 2.5.2 (Levin et al., 2013).

Other performance benefits of the high-performance thermoplastic polymers include very low flammability (Leach et al., 1985), insensitivity to moisture (Ma and Yur, 1989), and high resistance to aggressive fluids (Ma et al., 1992; Pritchard and Randles, 1988). All of these evaluations confirmed that thermoplastic composites, particularly based on PPS and PEEK, met aerospace requirements. By 1990, several of the TPCs had been evaluated extensively for aerospace applications (Carlile et al., 1989) with entire conferences devoted to highly reinforced thermoplastic composites (Newaz, 1989).

3.2 Processing

One aspect which received much attention was the processing of composites containing semicrystalline polymer matrices. The polymer industry was very familiar with

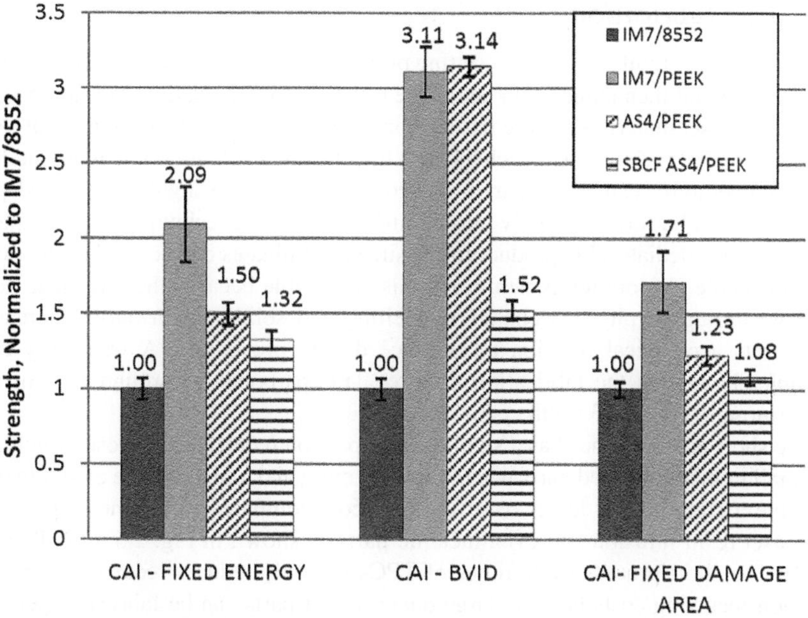

Figure 2.5.2 Residual compression strength after impact (CAI) for thermoplastic (PEEK) and toughened thermoset epoxy (8552) composites (Levin et al., 2013). *BVID*: Barely visible impact damage; *SBCF*: Stretch broken carbon fibre.
(Image credit: S Levin).

processing semicrystalline polymers going back to the 1930s with the development of polyethylene, but crystallinity was new to the aerospace industry. There are some important differences between processing TPCs and thermosets aside from the higher viscosity and higher processing temperatures of TPCs. For thermoset resins, the critical part of the process cycle is the heat-up phase due to the reduction in viscosity and gelation of the resin. When processing semicrystalline thermoplastics, the cooling phase is critical due to the nucleation and crystal growth of the polymer. There were some confusing studies evaluating model materials such as single CF in a polymer on a hot stage microscope, and more thorough assessments relating process conditions (in particular cooling rate), to crystallinity and properties (Blundell et al., 1985). There were also investigations regarding spherulite nucleation from the surface of CF known as transcrystallinity. One of the main areas of study was the relationship of processing conditions and crystallinity to toughness (for example, Davies et al., 1991). In general, these studies showed that the development of crystallinity does not start from the fibre surface (Crick et al., 1987) and that PEEK is insensitive to cooling rates up to very high rates of around 100°C/minute (Davies et al., 1991).

3.3 Part fabrication

Continuous fibre reinforced thermosetting prepregs are typically laid-up ply-by-ply on the part tool and then cured in an autoclave or oven, which is a time-consuming process. Thermoset materials require frozen storage and have a short working life at room temperature due to their reactive chemistry. In contrast, TPCs are stable indefinitely at room temperature, can be prepared and consolidated using highly automated processes and can be formed rapidly into a finished part using techniques such as stamp forming. Intermediate TPC product forms such as laminates can be used as the materials are stable at room temperature. All this is possible because thermoplastic polymers go through a physical change of melting and solidification only, whereas in thermosets a chemical reaction takes place during processing. A frequently used example shows that the fabrication time for TPC parts can be around 10% of that of thermosets, Fig. 2.5.3 (Wilkerson, 2021).

The short process times has a significant impact on recurring part cost particularly for small to medium sized parts where rapid forming and a high degree of automation can give a significantly lower recurring part cost compared to thermoset prepregs, thermoset resin infusion and even metallic parts as shown in Fig. 2.5.4 (Wilkerson, 2020). The short process cycle times of TPCs not only reduces recurring costs, but also non-recurring costs because large quantities of parts can be fabricated per year on a single piece of process equipment and tooling.

Fabrication cycle times of TPCs can be very short, but high process temperatures are required, generally 300°C–400°C, compared to thermosets at 120°C–180°C. Many different methods and equipment are available to fabricate TPC parts compared to thermoset composites, although different equipment is needed to realise the economic benefits, for example presses instead of autoclaves. This can present a

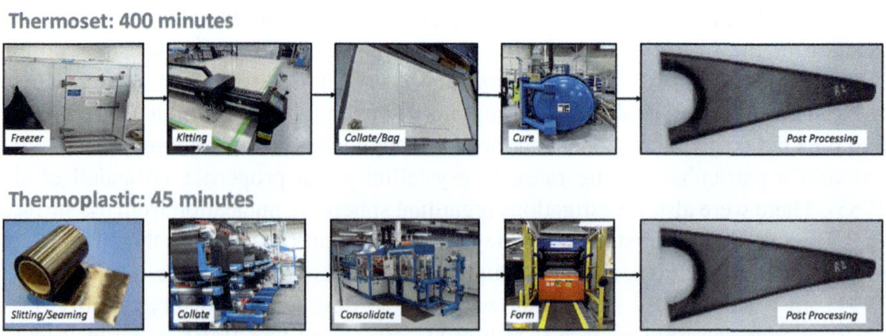

Figure 2.5.3 Part processing times for thermoset and thermoplastic composite parts (Wilkerson, 2021).
(Image credit: The Boeing Company).

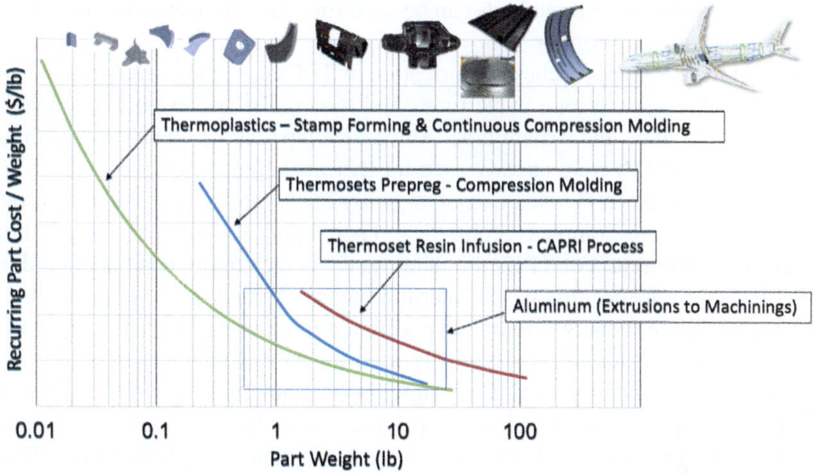

Figure 2.5.4 Aircraft recurring part cost by component size (Wilkerson, 2020). (Image credit: The Boeing Company).

significant barrier to change due to the need to invest in new equipment and apply different fabrication techniques.

Another aspect of TPC fabrication that can be a benefit and a challenge is selecting from the wide range of methods that can be used. There are three main fabrication steps of material preparation, consolidation and forming. A summary of the main techniques is shown in Table 2.5.3. TPC prepregs have no tack and therefore it was necessary to develop material preparation methods that would both position plies in the correct orientations and secure them in place, for example using local heating or ultrasonic tacking. In many cases, a partially or fully consolidated blank is prepared which is then formed rapidly into the final geometry, for example, using stamp

Table 2.5.3 TPC fabrication techniques.

Fabrication method	Material preparation	Consolidation	Forming
Pick & place	✓		
Low energy ATL/AFP	✓		
High energy ATL/AFP	✓	✓	
Oven/VBO		✓	✓
Autoclave		✓	✓
Continuous moulding		✓	✓
Static press: Single/dual		✓	✓
Stamp form			✓

ATL: Automated tape laying; *AFP*: Automated fibre placement; *VBO*: Vacuum-bag only.

forming. The same process methods can be used for different operations, in particular for consolidation and forming.

Much of the early approaches to fabrication of TPCs used the available equipment, such as autoclaves, with long cycle times and required single-use process materials such as bagging films and sealant tape. Automated tape laying (ATL) and automated fibre placement (AFP) were already used for thermosetting composites in the 1980s, with wider bandwidth ATL used for flat or gently contoured lay-up, and narrow bandwidth AFP for more complex geometries. Their applicability to thermoplastics is different to thermosets. For TPCs the main factor is the energy level applied during ply lay-up, and whether this is low thermal energy, to tack plies to each other, or much higher thermal energy, to achieve a high degree or full consolidation. In the early developments, hot gas was typically used for AFP, which heated both the substrate and the incoming material. Over time, laser techniques were developed which could provide highly focused and controlled heat to the nip area. Another difference is that ATL and AFP are typically not used to produce a final part geometry in TPCs but rather a lay-up or blank that will be fully consolidated or formed in a subsequent operation to produce the final part. Automated and rapid fabrication techniques were developed through the 1980s and matured to production readiness.

3.4 Assembly

TPCs can be assembled using conventional mechanical attachment and adhesive bonding, but the major opportunity is the ability to re-melt thermoplastics in welding or co-consolidation processes. Various types of welding were already in use for thermoplastic polymers in the 1980s and it was a logical extension to apply these approaches to highly reinforced TPCs. Multiple welding methods are possible including resistance, ultrasonic, induction and thermal, for example with laser heating, as shown in Fig. 2.5.5 (Gardiner, 2018). These methods were all assessed in the 1980s through many evaluation and development programs, including collaborative initiatives to compare the different welding methods, for example (Taylor, 1987). To avoid the problem of the substrates melting and deconsolidating, techniques were developed using a lower processing temperature thermoplastic polymer film at the weld interface, which is co-consolidated on the surface of the parts to be joined (Cogswell et al., 1989). One example uses PEI polymer film, with process temperature of 330°C, co-consolidated onto the surface of PEEK matrix parts, with a melt temperature of 343°C. This approach is effective when the film has a lower process temperature than the substrate matrix and is miscible with the substrate matrix.

3.5 Evaluation status

The highly reinforced TPCs discussed here were developed in the early 1980s and by 1990 a large body of development work and evaluations had been completed by a wide range of organisations including material suppliers, end users, equipment manufacturers, government and academic groups. These extensive characterisations of this new class of materials in a short time period established that TPCs had properties

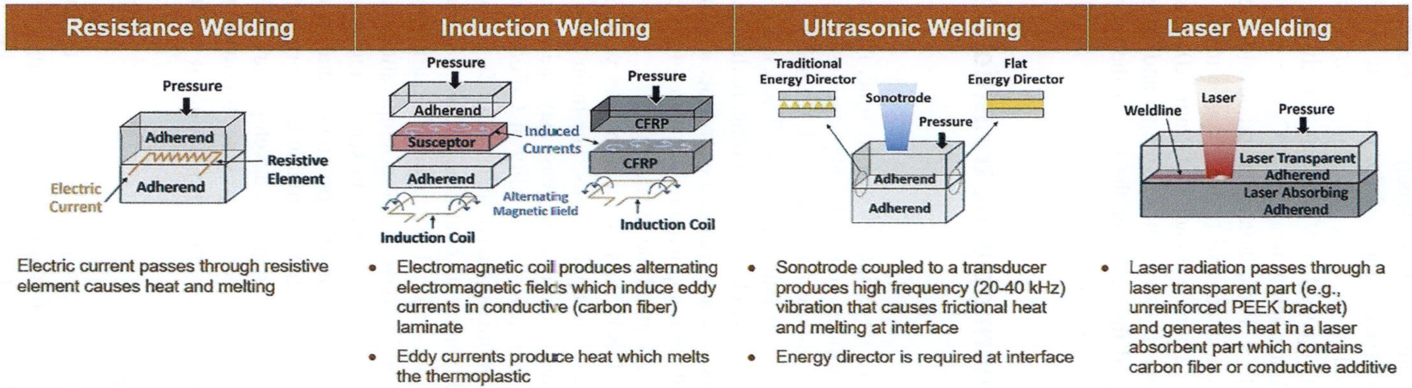

Figure 2.5.5 Thermoplastic composite welding methods (Gardiner, 2018). (Image credit: CompositesWorld).

and processing that could bring significant benefits to a wide range of demanding applications, in particular aerospace. Much of the material on PEEK UD tapes was brought together by Cogswell in the book 'Thermoplastic Aromatic Polymer Composites' published in 1992 (Cogswell, 1992). Even though this was written over 30 years ago, it highlights the fundamental work on TPCs over this period and still serves as an excellent primer on TPCs. The evaluations and development of processing and fabrication methods have continued, and there is now a much larger body of work available in the literature, for example (Leach and Pratte, 2024) and a recent extensive literature review of thermoplastic composites (Kreuger and Bergan, 2024). The evaluations discussed here formed the basis for the use of thermoplastic composites in aerospace applications by the early 1990s.

4. Thermoset to thermoplastic transition experience

4.1 Overview

With the extensive experience developed in a short time, it was expected that TPCs would rapidly take over advanced composite applications, but the implementation has taken a more varied path, and even now in 2024, the usage of thermoplastic composites in aerospace is much lower than thermosets. There is considerable enthusiasm for TPCs, although the drivers have changed over time. In this section, we will explore the path of TPC use and how this has changed over time, as different priorities have emerged, focussing on production applications but also referencing some significant development programs that advanced the technologies.

Moving back to the early 1980s, thermoset composites were becoming established in defence and commercial aircraft applications. The main drivers were the ability to reduce the weight of structures, and thus increase payload or fuel economy, and an added benefit in defence is the low observability of CF composites. But composite structures were significantly more expensive than metallics, not only the raw materials but also in the handling and fabrication operations. If we were to start over today and propose that a new lightweight material was available but it would need frozen storage, have a working life of a few days at room temperature, require cure cycles of many hours and have a significant reduction in service temperature in the presence of moisture, it would be difficult to convince program managers of the benefits. However, when the reinforcing fibres became available the thermosetting resins were the only viable option, and a whole infrastructure developed around the use of these materials. The adoption of TPCs has been progressive following the parallel developments in materials and fabrication technologies, as shown in Fig. 2.5.6.

4.2 Initial applications

There were many economic and logistical reasons to use thermoplastic composites, but the use of TPCs required a different manufacturing approach. Several initial applications were driven entirely by performance, where the defence industry was

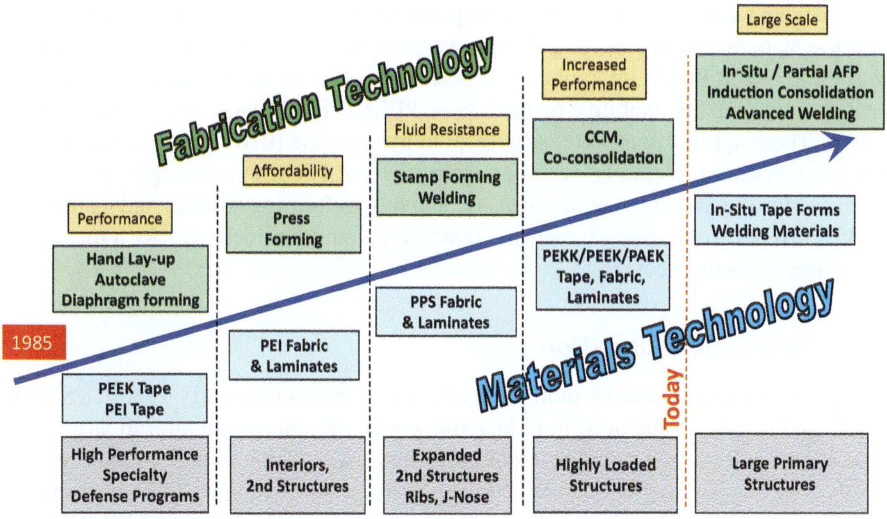

Figure 2.5.6 Thermoplastic composites adoption curve.
(Image credit: D Leach).

willing to pay a high premium to reach performance goals using PEEK/CF UD tapes. Two early production applications were on the Lockheed Martin F117A (Vosteen and Hadcock, 1994) and the Lockheed-Martin F-22 (Black, 2016) shown in Fig. 2.5.7. The F-22 forward fuselage and weapons bay doors were manufactured using PEEK/CF UD tapes (Black, 2016). The F-22 applications were driven by the need for high toughness in these parts, which could suffer significant damage from runways and due to opening the weapons bay doors during flight. The much higher toughness

Figure 2.5.7 Lockheed-Martin F-22 with PEEK/CF forward fuselage and weapons bay doors (Black, 2016).
(Image credit: D Ramey Logan, Wikimedia Commons, CC BY-SA 4.0).

and damage tolerance of PEEK/CF compared to the available thermoset composites was the main driver for selecting thermoplastic composites. Another early application was the Dassault Rafale internal central fuselage section which was driven by the high temperature and flammability resistance of PEEK/CF UD tape, Fig. 2.5.8 (Dellus, 1992). These applications were successful in meeting the performance requirements, but the fabrication methods were not optimised for TPCs. For example, parts were fabricated using autoclave or superplastic diaphragm forming (Barnes and Cattanach, 1985) methods, often using expensive material preparation methods. As a result, the structures were high cost.

4.3 Economic fabrication

The price of oil increased by more than a factor of three in the early 1980s (US EIA, 2024) and this drove the need to reduce the weight of commercial aircraft structures. The existing thermoset composite solutions had low toughness and were relatively costly, and this gave a great incentive to apply the newly developed TPCs in commercial aircraft. High volume applications would give the greatest benefit, and as the commercial aircraft industry is risk-averse, the initial applications were mainly for aircraft interiors or lightly loaded structures. The PEI/glass fabric laminates developed by TenCate (now Toray Advanced Composites) eliminated the need for raw material assembly and consolidation operations, and they could be formed into parts rapidly using a thermal folding or stamp forming processes. PEI is an amorphous polymer and therefore the cooling rate could be rapid and did not need to be controlled. Fokker, now part of GKN, based in the Netherlands was instrumental in developing many of the early applications in the late 1980s, and as an aircraft manufacturer, they had the ability to move parts into production. The early applications included stamp-formed interior support brackets and ice protection plates for the Fokker 50, and

Figure 2.5.8 Dassault Rafale with PEEK/CF central fuselage structure (Dellus, 1992). (Image credit: Tim Felce (Airwolfhound) Creative Commons CC BY-SA 2.0).

deep-drawn hinge covers for the Fokker 100 cargo compartment (Offringa, 2024). One of the early high-volume applications was sandwich floor panels using glass fabric/PEI with thermofolded edges for the Fokker 100 cargo compartment, which led to the development of floor panels for Gulfstream aircraft. These are still in production, and many thousands of panels are supplied per year. Several of these early TPC applications are shown in Fig. 2.5.9 (Gardiner, 2011). In addition to the ability to form parts quickly and economically, many of the applications utilised the high toughness and low fire, smoke, and toxicity of PEI composites.

Another application that emerged in the 1990s was due to the need to install smoke detectors in the cargo holds of commercial aircraft. Many detectors are required even in a narrow-body aircraft and therefore the housings are a high-volume application. The smoke detector housings have a deep draw and must be lightweight, low cost, damage tolerant, and flammability resistant. The parts could have been fabricated using thermosets such as flame retarded epoxy or phenolic resins, but this would require considerable touch labour for the lay-up process. Stamp formed PEI/glass fabric proved to be the solution with production cycle time reduced by 75% compared to thermosets (Sloan, 2013). 'Design for manufacturing' principles were applied to optimise the part design for rapid fabrication. An example of the installed smoke detector housing is shown in Fig. 2.5.10.

4.4 Expanded secondary structures

The success of the smaller and less critical fabric reinforced applications opened the next phase of larger scale and more structurally demanding applications. This was also made possible by the expansion of the materials available, notably with CF fabrics and the use of PPS polymer, which has much higher environmental resistance than PEI, and the expansion of fabrication and assembly methods. Fokker and Airbus were instrumental in moving applications forward, including the landing gear door of the Fokker 50 regional passenger aircraft using carbon fabric/PPS, the A340 and later the A380 wing leading edges both in glass fabric/PPS, shown in Fig. 2.5.9. These combined many of the TPC fabrication advantages including stamp forming of ribs, and welding of the ribs to the aerodynamic skins (Offringa, 2005). Assembly using resistance welding was qualified and validated on the Fokker F50 main landing gear door which flew for 3.5 years and was then inspected with no deterioration of the weld quality. This development paved the way to use thermoplastic composites on the A340 inboard wing fixed leading edge, or J-nose, which was developed and manufactured by Fokker for Airbus UK. The skin and spar were autoclave moulded using partially impregnated glass fabric/PPS 'semi-preg' and the ribs were stamp formed. The leading edge was assembled using a resistance welding approach (Offringa, 2005). The business case was based on a 20% weight reduction relative to the heredity aluminium structure and a cost reduction versus thermosets. The success of the A340 leading edge led to a similar technology for the A380 outboard leading edge, Fig. 2.5.11. The same materials were used but with a change in the welding methodology. A ship set of A380 J-noses consists of 16 assemblies, each with a length

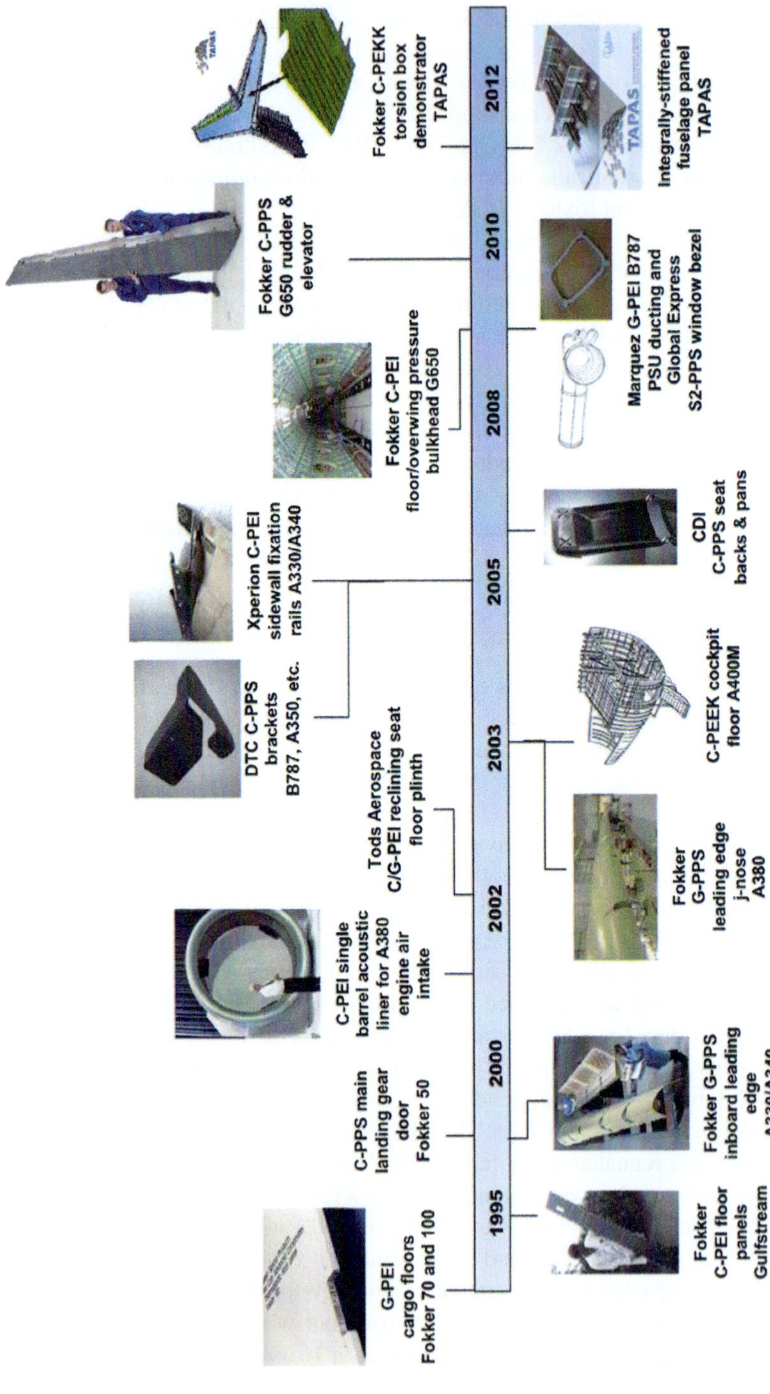

Figure 2.5.9 Evolution of commercial aircraft TPC applications (Gardiner, 2011). (Image credit: CompositesWorld).

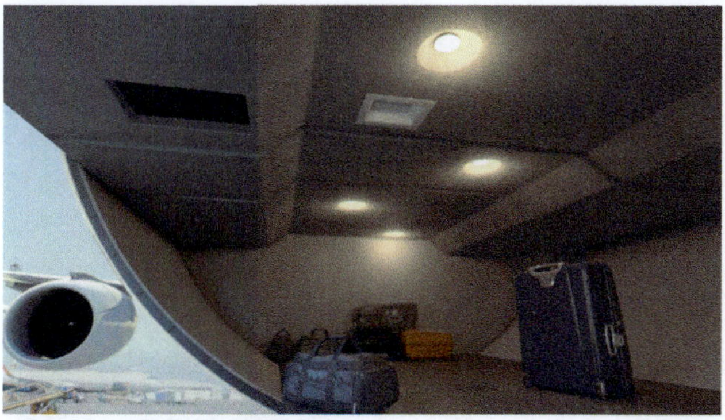

Figure 2.5.10 High-volume smoke detector housings for commercial aircraft cargo hold stamp-formed using PEI/glass fabric.
(Image credit: Toray Advanced Composites).

of about 4 m. The Airbus A340 and A380 aircraft with thermoplastic leading edges are shown in Fig. 2.5.12.

These production applications showed successfully that larger scale TPC assemblies were feasible and demonstrated both weight and cost saving using a combination of fabrication and assembly methods. As such, they were important production applications in the transition of aircraft structures to TPCs.

4.5 Structural applications

The initial applications of TPC UD tapes were based on performance, in particular toughness. During the late 1980s and 1990s the toughness of epoxy-based composites was increased considerably through mechanisms such as thermoplastic toughening, the development of semi-interpenetrating approaches and interleaving (Almen et al., 1988) which negated one of the benefits of TPCs. In addition, thermoset fabrication methods and infrastructure were already well established. Some TPC applications emerged, for example, the Agusta-Westland (now Leonardo) EH101 floor panels, which required structural performance, high damage tolerance and fluid resistance, but widespread adoption of high-performance TPCs was not realised and thermosets remained the baseline technology. Other fabrication methods were being developed for TPCs, and one was the use of vacuum-only (VBO) consolidation. It had been thought that high pressure was required to consolidate and form TPCs, but Airbus developed a VBO process to manufacture the A400M cockpit floor, which was later transitioned to Daher (Sloan, 2014a), Fig. 2.5.13. This large structure is manufactured using PEEK/CF UD tape and it carries significant structural loads and therefore very high quality was required on the VBO components.

Other constraints on the use of UD tapes were the challenges of rapid forming and the limited products available. The weave construction of fabric reinforcement provides control of fibre deformations during fabrication. On the other hand, there are no constraints in forming UD tapes, which can result in the creation of unacceptable

Figure 2.5.11 Airbus A380 leading edge J-nose manufactured by Fokker using glass fabric/PPS (Offringa, 2005).
(Image credit: GKN Fokker).

Figure 2.5.12 Airbus A340 and A380 TPC leading edges in flight.
(Image credit: GKN Fokker).

features, such as wrinkles (Chamberlain et al., 2020). For an extended period, essentially the only UD tape available was from ICI (later Fiberite, Cytec, Solvay, and now Syensqo) using PEEK matrix. This material was recognised to have excellent performance but was costly, and the high process temperature limited some fabrication methods. Cytec made PEKK UD tapes available at lower cost and lower processing

Figure 2.5.13 Airbus A400M cockpit floor made with PEEK/CF unidirectional tape using vacuum only consolidation (Sloan, 2014a).
(Image Credit: J Sloan).

temperature with minimal reduction in performance versus PEEK (Bai et al., 2003). Other suppliers also started providing UD tapes and with other matrices, including PPS and PEI.

Fabrication methods expanded and matured, including ATL, AFP, and continuous compression moulding (CCM), which is a very efficient process for making profiles and laminates (Spelz, 2008). This was also a phase of some significant large-scale TPC development programs, including DRAPCOCOT (Coiffier-Colas et al., 2005), the Survivable, Affordable, Repairable Aircraft Program (SARAP) (Carstensen et al., 2008), and Thermoplastic Affordable Primary Aircraft Structure (TAPAS) (Sloan, 2014b). The DRAPCOCOT and SARAP programs both included thermoplastic AFP by Automated Dynamics Corporation, now Trelleborg Sealing Solutions Albany (TSSA). As well as developing the AFP process for TPCs, the programs included integrated assembly, for example, by placing pre-formed stiffeners in the AFP tool and then fibre placing over them to consolidate the skin and fuse it to the stiffeners in one operation. The SARAP program developed a Sikorsky UH60 Blackhawk helicopter fuselage section with fully integrated floor, frames, and skins. The overall assembly demonstrated a 23% weight reduction, 40% lower recurring cost, and 47% lower nonrecurring cost versus the baseline (Carstensen et al., 2008). A similar integrated TPC helicopter tail-boom assembly manufactured by TSSA is shown in Fig. 2.5.14. TAPAS was a Dutch-Airbus program that ran from 2009 to 2017. It developed several TPC fabrication and assembly technologies for representative demonstrator components, including a stiffened fuselage panel, torsion box, and an engine pylon upper spar. The combination of a broader range of materials and fabrication processes incrementally opened further applications.

Figure 2.5.14 Integrated TPC H-1 helicopter tail boom demonstrator (Hauber, 2023). (Image credit: Trelleborg Sealing Solutions Albany).

4.6 TPCs come of age

This positioned TPCs for two major composites-intensive commercial aircraft programs in the Boeing 787 and Airbus A350. Both these aircraft make widespread use of thermoset composite structures, and this drove the desire to use composites in detailed parts which would have previously been made using aluminium. If these parts had remained aluminium, then considerable mitigation would have been needed to avoid galvanic corrosion between the aluminium and CF composites. TPCs also offered cost savings versus thermosets in smaller detailed parts, and weight savings compared to metallic solutions. Both programs make extensive use of TPCs in smaller structural components using PPS, PEEK, and PEKK polymers with CF reinforcement in UD tape and fabric forms. This provided a platform for larger-scale implementation of cost-effective fabrication methods such as continuous compression moulding and stamp forming, as well as the maturation of material databases and processing methodology. The A350 uses 5000 clips in more than 2000 designs, manufactured using carbon fabric with PPS or PEEK in a stamp-forming process (Miaris et al., 2014), and typical applications on A350 are shown in Fig. 2.5.15. A range of structural parts for various aircraft programs using TPCs with CF UD tapes and fabrics are shown in Fig. 2.5.16.

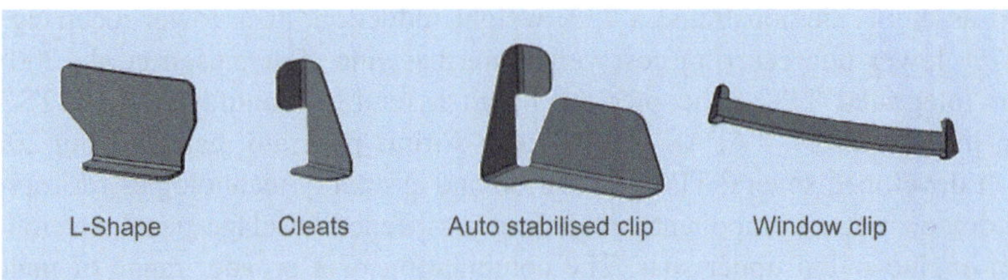

Figure 2.5.15 Airbus A350 TPC components (Miaris et al., 2014). (Image credit: Airbus).

Figure 2.5.16 TPC structural applications using CF UD tape and fabric reinforcement. (Image credit: ATC Manufacturing).

A further step in the adoption of TPCs came in the late 2000s with the development of welded assemblies for the Gulfstream G650 control surfaces fabricated by GKN Fokker using PPS/CF fabric materials shown in Fig. 2.5.17 (Ingen et al., 2010). These

Figure 2.5.17 Gulfstream G650 rudder fabricated by GKN Fokker using PPS/CF fabric (Ingen et al., 2010). (Image credit: GKN Fokker.).

primary structure parts use a combination of fabrication technologies, including autoclave consolidation of the skins and spar, stamp forming of the ribs and induction welding of the ribs to the skins. There is now more than 12 years of in-service experience for these structures, which provide 20% cost reduction and 10% weight reduction compared to the previous thermoset composite solution (Ingen et al., 2010).

TPC use was assessed recently, and the typical commercial aircraft applications are shown in Fig. 2.5.18 (Stratview, 2024). TPCs are now used in a wide range of components, including in primary structures, although these parts are still limited in size, typically to less than 1 m in length.

4.7 Supporting collateral

Along with developments in materials and fabrication processes there have been significant advances in some of the supporting information and methods enabling broader use of TPCs.

One of these is the availability of design data, material, and process specifications. Most of the applications have been implemented by major airframers such as Airbus, Boeing, and Lockheed Martin, and these companies have developed their own proprietary specifications and databases. The cost and time to develop this information is considerable and this is a constraint for smaller companies using new materials. The FAA and NASA developed the Advanced General Aviation Transport Experiments (AGATE) and later the National Center for Advanced Materials Performance (NCAMP) systems to establish standardised and publicly available specifications and databases to democratise the use of composite materials (NCAMP, 2024). The initial materials in this system were thermoset composites but in recent years, several thermoplastic composites have been added, with more in progress (NCAMP, 2024).

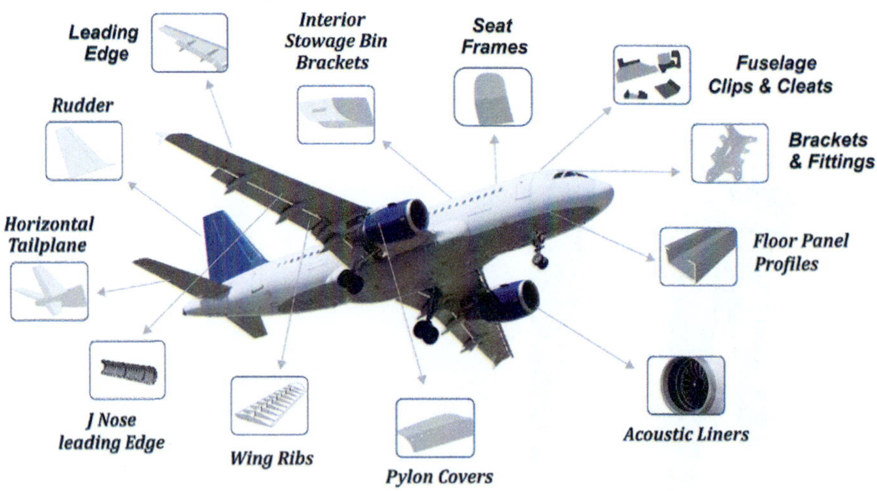

Figure 2.5.18 Typical thermoplastic composite aircraft applications (Stratview, 2024). (Image credit: Stratview Research).

Another important aspect is the simulation and modelling of composite processing. In the early years of TPC use, there was a lot of 'trial and error' in developing forming techniques and in the design of tooling to meet the final part dimensions. As we have discussed, process temperatures for TPCs are significantly higher than for thermosets and therefore, the thermal stresses will be higher in TPC components. This will lead to residual stresses and dimensional changes, often referred to as 'spring–in' which affect the ability to meet dimensional tolerances. For thermosets, the 'stress-free temperature' is generally the cure temperature, but for TPCs it is the T_g for amorphous matrix materials or the crystallisation temperature (T_c) for semicrystalline materials. This is further complicated by the fact that the T_c depends on the cooling conditions. Several simulation programs are available now which have been successful in predicting the dimensional changes and hence tool design for different materials and process conditions (Convergent, 2024).

Many TPC fabrication methods rely on forming a flat lay-up or laminate into a shaped part. This will cause reorientation of the plies and can result in unacceptable features such as wrinkles. Predictive simulation methods are now available which can successfully model the ply and fibre orientation movement (Postera and Leach, 2023). The ability to simulate and predict material behaviour during forming operations considerably reduces the time and cost of developing tools and parts.

5. Future direction

Moving into the 2020s, there is great interest in TPCs for structural aircraft applications. Transportation is committed to achieve net-zero carbon emissions by 2050 (ICAO, 2022), which requires lightweight materials, and these must be fabricated economically. TPCs are sustainable, as the materials and parts can be recycled or downcycled, and the applications demand high volumes and high-rate manufacturing. The interest extends across commercial aircraft, advanced air mobility, defence, unmanned air vehicles, and electric vehicles. This is driving the need for high-rate manufacturing for large-scale parts and assemblies, which is being advanced through several collaborative programs such as Clean Aviation in Europe (Cleansky, 2024) and the NASA Hi-Rate Composites Manufacturing (HiCAM) program in the USA, which includes TPCs (NASA, 2024). These programs address large-scale and high-rate manufacturing of TPC structures, including the evaluation of different manufacturing and assembly techniques. For example, the Cleansky program has manufactured an 8.5 m long, 4 m diameter lower fuselage skin using AFP, shown in Fig. 2.5.19 (Nehls, 2021), which has been incorporated into the 'multi-functional fuselage demonstrator' (MFFD) manufactured using TPCs, Fig. 2.5.20 (Airbus, 2024). The NASA HiCAM program will advance manufacturing readiness through the manufacture of fuselage and wing demonstrators (NASA, 2024).

Figure 2.5.19 Fuselage section made by the STUNNING project. An 8.5-metre-long skin made of TPC as part of the Clean Sky 2 multifunctional fuselage demonstrator (MFFD) manufactured using an AFP process (Nehls, 2021).
(Image credit: Royal NLR).

Figure 2.5.20 Clean Sky multi-functional TPC fuselage demonstrator (MFFD) (Airbus, 2024).
(Image credit: Airbus).

6. Conclusions

The adoption of TPCs provides a lesson in the introduction of a new technology in a highly regulated industry. TPCs became available in the 1980s and it seemed that the disruptive benefits of their performance and processing were compelling reasons for the rapid displacement of thermoset composites.

Referring back to Fig. 2.5.6, the progressive implementation of TPCs has been dependent on several factors, particularly the parallel developments of materials and fabrication technologies. The initial applications were driven by performance, particularly toughness and flammability. While there was potential for rapid fabrication, the available equipment and process technologies were based on thermoset methods, and were not economical for TPCs. Implementing appropriate fabrication processes required new equipment and methods, which was a major barrier to acceptance. The development of toughened thermosets, and the fact that the whole supply chain and infrastructure for thermoset composites were already established, created a major delay in the acceptance of thermoplastics.

The next wave of adoption was for higher volume but lower performance applications with glass fabric reinforcement, where the rapid fabrication of large quantities of small parts by stamp forming gave a considerable economic advantage. The higher toughness and flammability resistance were additional benefits. This approach gradually progressed to more structural components through the use of carbon fabric TPCs and then to UD tapes, along with the application of other rapid forming techniques which had been matured in the interim. The ability to assemble parts using welding has a multiplicative benefit on cost reduction.

The market drivers for TPCs have also changed over time. The initial need was performance, then later cost and rate, progressing into assembly, and a major factor now is sustainability. A further driver was the desire to move away from aluminium components, not only for weight reduction but to ensure compatibility with large, thermoset composite structures such as skins.

The use of TPCs in aerospace structures has been incremental and additive, and the adoption has taken much longer than originally anticipated. While the benefits of thermoplastics may appear to be compelling, there were many barriers to change. Aerospace structures are highly regulated and considerable supporting information is required, even for non-structural applications. A transition is needed in capital equipment and manufacturing approaches for the adoption of a new technology, such as thermoplastics. This only occurs when there is a new program where investments can be justified, but new programs are infrequent in aerospace.

Despite this long arc of implementation, there are once again compelling reasons for thermoplastic composites, based on lightweight, high-rate manufacturing, cost-effective assembly, and sustainability. The next generation of aircraft will make much more use of composites, and thermoplastic composites will be key to achieving the goals.

References

Airbus, 2024. Fantastic Thermoplastics. https://www.airbus.com/en/newsroom/stories/2025-01-fantastic-thermoplastics, 17 January 2025, Accessed 24 December 2025.

Almen, G., MacKenzie, P., Malhotra, V., Maskell, R., McGrail, P.T., Sefton, M., 1988. Thermoset/thermoplastic alloys: An approach to tough, high performance resin matrices.

In: Proceedings of the 20th International SAMPE Technical Conference. Society for the Advancement of Material and Process Engineering, Covina, California USA, pp. 46–59.

Bai, J.M., Leach, D.C., Pratte, J.F., 2003. PEKK composites - opening new opportunities in properties and processing. In: CETMA THEPLAC Workshop on Thermoplastic Matrix Composites, Italy.

Barnes, A.J., Cattanach, J.B., 1985. Advances in thermoplastic composites fabrication technology. In: Materials Engineering Conference, Leeds.

Black, S., 2016. Thermoplastic Composite Wings on the Horizon. CompositesWorld. https://www.compositesworld.com/articles/thermoplastic-composite-wings-on-the-horizon.

Blundell, D.J., Chalmers, J.M., Mackenzie, M.W., Gaskin, W.F., 1985. Crystalline morphology of the matrix of PEEK – carbon fiber aromatic polymer composite – I Assessment of crystallinity. SAMPE Q. 16 (4), 22–30.

Carlile, D.R., Leach, D.C., Moore, D.R., Zahlan, N., 1989. Mechanical properties of the carbon fiber/PEEK composite APC-2/AS4 for structural applications. In: Newaz, G.M. (Ed.), Advances in Thermoplastic Matrix Composite Materials, ASTM STP 1044. American Society for Testing and Materials, Philadelphia, pp. 199–212.

Carstensen, T., Townsend, W., Goodworth, A., 2008. Development and validation of a virtual prototype airframe design as part of the survivable affordable repairable airframe program. In: American Helicopter Society 64th Annual Forum, Montreal, Canada, April 29 – May 1, 2008.

Chamberlain, C., Leach, D.C., McCrea, T., 2020. Cost effective thermoplastic composites in aerostructures. SAMPE J.

Cleansky, 2024. EU Cleansky Program. https://www.clean-aviation.eu/clean-sky-2/key-demonstrators/multi-functional-fuselage-demonstrator.

Cogswell, F.N., Leach, D.C., 1984. Continuous fibre reinforced thermoplastics; a change in the rules for composite technology. Plast. Rubber Process. Appl. 4, 271–276.

Cogswell, F.N., 1992. Thermoplastic Aromatic Polymer Composites. Butterworth Heinemann, Oxford UK.

Cogswell, F.N., Meakin, P.J., Smiley, A.J., Harvey, M.T., 1989. Thermoplastic interlayer bonding of aromatic polymer composites. In: 34th International SAMPE Symposium, pp. 2315–2325. Society for the Advancement of Materials and Process Engineering, 1989.

Coiffier-Colas, C., Sibois, H., Lefebure, P., 2005. Future for high performance thermoplastic composites and aircraft helicopters. In: Proceedings of JEC Composites, Paris, pp. 14–19.

Convergent, 2024. Convergent Manufacturing Technologies Process Simulation Software. Convergent Technologies. https://www.convergent.ca/.

Crick, R.A., Leach, D.C., Meakin, P.J., Moore, D.R., 1987. Interlaminar fracture morphology of carbon fiber/PEEK composites. J. Mater. Sci. 22 (6), 2094–2104.

Davies, P., Cantwell, W.J., Jar, P.-Y., Richard, H., Neville, D.J., Kausch, H.H., 1991. Cooling rate effects in carbon fiber/PEEK Composites (Third Volume), ASTM STP 1110. In: O'Brien, T.K. (Ed.), Proceedings of the Composite Materials: Fatigue and Fracture. American Society for Testing and Materials, Philadelphia, pp. 70–88.

Dellus, S., 1992. Thermoplastic developments at Dassault aviation. In: Proceedings of the Composites in Manufacturing '92. Society of Manufacturing Engineers, Danbury, Connecticut USA. Paper EM92-107, 7-8 January 1992, Society of Plastics Engineers.

Gardiner, G., 2011. Thermoplastic Composites: Primary Structure? CompositesWorld. https://www.compositesworld.com/articles/thermoplastic-composites-primary-structure.

Gardiner, G., 2018. Welding Thermoplastic Composites. CompositesWorld. https://www.compositesworld.com/articles/welding-thermoplastic-composites.

Hauber, D., 2023. Additive manufacturing of advanced composite structures. In: SAMPE Conference 2023, Society for the Advancement of Materials and Process Engineering, Seattle WA, pp. 17–20.

Hunston, D.L., Moulton, R.J., Johnston, N.J., Bascom, W.D., 1987. Matrix resin effects in composite delamination: mode I fracture aspects. In: Johnston, N.J. (Ed.), Toughened Composites. American Society for Testing and Materials, Philadelphia, pp. 74–94. ASTM STP 937.

ICAO, 2022. States Adopt Net-Zero 2050 Global Aspirational Goals for International Flight Operations. International Civil Aviation Organization. https://www.icao.int/environmental-protection/long-term-global-aspirational-goal-ltag-international-aviation.

Ingen, J., Buitenhuis, A., van Wijngaarden, M., Simmons, I.I.I.F., 2010. Development of the Gulfstream G650 induction welded thermoplastic elevators and rudder. In: SAMPE Conference, Seattle, WA, 2010, Society for the Advancement of Material and Process Engineering, Covina, California USA.

Kreuger, R., Bergan, A., 2024. Advances in Thermoplastic Composites Over Three Decades – A Literature Review NASA Technical Report TM-20240005376. National Aeronautics and Space Administration.

Leach, D.C., Briggs, P.J., Carlile, D.R., 1985. Mechanical and fire properties of aromatic polymer composites. In: 3rd European Symposium on Spacecraft Materials in Space Environment. European Space Agency ESA-SP 232, The Netherlands, pp. 153–160.

Leach, D.C., Curtis, D.C., Tamblin, D.R., 1987. Delamination behavior of carbon fiber/poly (etheretherketone) (PEEK) composites. In: Johnston, N.J. (Ed.), Toughened Composites. American Society for Testing and Materials, Philadelphia, pp. 358–380. ASTM STP 937.

Leach, D., Pratte, J., 2024. Overview of high performance thermoplastic composites. In: ASM Handbook, Volume 21, Composites. ASM International, Materials Park, Ohio USA.

Levin, S., Dequine, D., Lamm, D., Crocco, J., 2013. Damage tolerance of thermoplastic stretch-broken carbon fiber prepreg vs continuous carbon fiber pre-preg. In: Proceedings of the SAMPE Conference 2013, Society for the Advancement of Materials and Process Engineering, Long Beach CA, 6-9 June, 2013.

Ma, C.C., Lee, C.-L., Tai, N.-H., 1992. Chemical resistance of carbon fiber-reinforced poly (ether-ether-ketone) and poly(phenylene-sulfide) composites. Polym. Compos. 13 (6), 435–440.

Ma, C.C., Yur, S.W., 1989. Environmental effect on the water absorption and mechanical properties of carbon fiber reinforced PPS and PEEK composites. In: 47th SPE Annual Technical Conference (ANTEC). Society of Plastics Engineers, pp. 1496–1500.

Miaris, A., Edelmann, K., von Hayek Boelingen, M., 2014. A350 WXB: thousands of thermoplastic composite parts in an FRP aircraft - the breakthrough to a highly automated composite manufacturing. In: SAMPE Conference Proceedings. Seattle, WA, 2-5 June 2014. Society for the Advancement of Material and Process Engineering, Covina, California USA.

NASA, 2024. NASA Hi-Rate Manufacturing (HiCAM) Program. https://www.nasa.gov/directorates/armd/hi-rate-composite-aircraft-manufacturing-hicam-project/.

NCAMP, 2024. National Center for Advanced Materials Performance. https://www.wichita.edu/industry_and_defense/NIAR/Research/ncamp.php.

Nehls, G., 2021. STUNNING Project Successfully Develops 8.5-Meter Thermoplastic Fuselage Skin. CompositesWorld. https://www.compositesworld.com/news/stunning-project-successfully-develops-85-meter-thermoplastic-fuselage-skin.

ASTM STP 1044. In: Newaz, G.M. (Ed.), 1989. Advances in Thermoplastic Matrix Composite Materials. American Society for Testing and Materials, Philadelphia.

Offringa, A., 2005. Thermoplastics, from press forming to co-consolidation. In: Proceedings of the SAMPE Technical Conference, Seattle WA, 31 Oct – 3 Nov 2005. Society for the Advancement of Material and Process Engineering, Covina, California USA.

Offringa, A., 2024. GKN Fokker, Private Communication. Hoogeveen Netherlands.

Postera, R., Leach, D.C., 2023. Influence of material form on fabrication of thermoplastic composite window frame. In: Proceedings of the SAMPE Conference 2023, Seattle WA, 18-20 April 2023. Society for the Advancement of Material and Process Engineering, Covina, California USA.

Pritchard, G., Randles, S.J., 1988. Long fibre thermoplastics for aerospace applications. In: Paper 15, 16th British Plastics Federation Reinforced Plastics Congress. Blackpool.

Sloan, J., 2013. Compression Molding Mass Out of Aircraft Interiors. CompositesWorld. https://www.compositesworld.com/articles/compression-molding-mass-out-of-aircraft-interiors.

Sloan, J., 2014a. Inside a Thermoplastic Composites Hotbed. https://www.compositesworld.com/articles/inside-a-thermoplastic-composites-hotbed.

Sloan, J., 2014b. TAPAS 2: Next steps in Thermoplastic Aerostructures. CompositesWorld.

Stratview, 2024. Aerospace & Defense Thermoplastic Composites Market: Market Size, Share & Forecast Analysis (2024-2030). Report SRAD156, Stratview, stratviewresearch.com.

Spelz, U., 2008. Serial production of CFRP thermoplastic profiles with continuous compression molding technique. In: Proceedings of the SAMPE Europe Conference 2008. Society for the Advancement of Material and Process Engineering, Covina, California USA.

Taylor, N.S., 1987. The feasibility of welding APC-2 thermoplastic composite material. In: The Welding Institute Research Bulletin. The Welding Institute, Abington, Cambridge UK.

US Energy Information Administration, 2024. US Crude Oil First Purchase Price. https://www.eia.gov/dnav/pet/hist/LeafHandler.ashx?n=pet&s=f000000__3&f=a.

US National Research Council (NRC), 1987. The Place for Thermoplastic Composites in Structural Components. In: National Materials Advisory Board. NMAB-434. National Academy Press, Washington, D.C.

van Dreumel, W.H.M., de Groot, M., 1987. An experimental nose-wheel door of continuous fibre-reinforced thermoplastic. Composites 18 (5), 405–408.

Vosteen, L., Hadcock, R., 1994. Composite chronicles: a study of the lessons learned in the development, production, and service of composite structures. In: NASA Contractor Report 4620. National Aeronautics and Space Administration.

Wilkerson, R., 2020. Aerospace applications for high performance thermoplastic composites. In: ACMA Thermoplastic Composites Conference, Virtual, March 2020. American Composite Manufacturers Association, Arlington, VA, USA.

Wilkerson, R., 2021. Thermoplastics Manufacturing Technologies for Aerospace Applications. Aerodef Satellite Session, 20 January 2021. Society of Manufacturing Engineers, Southfield, Michigan USA.

Historical perspective on modelling of processing of polymer composites

2.6

Suresh G. Advani[1] *and Christophe Binetruy*[2]
[1]Department of Mechanical Engineering, University of Delaware, Neward, DE, United States; [2]Nantes Université, Ecole Centrale de Nantes, CNRS, GeM, UMR 6183, Nantes, France

1. Introduction

The journey of the development of polymer composite process models has been marked by significant breakthroughs, driven by the growing demand for lightweight, high-performance cost-effective materials. The modelling of polymer composite processing has evolved significantly over the decades, reflecting advancements in materials characterisation and science, engineering, and computational techniques. Early development of models in 1970s relied heavily on empirical approaches which were a natural offshoot of trial-and-error approaches being practiced, relying on experimental data to correlate process parameters such as temperature and pressure with material properties. Also, rule of thumb approaches such as simple guidelines and heuristics was used to guide manufacturing processes, often based on experience and intuition.

Composite manufacturing processes can be divided into two very different ways of processing: Short fibre composite processing and continuous fibre composites processing. The early methods for short fibre composite processing organically evolved from polymer processing methods such as extrusion, injection moulding and compression moulding. To extend the application of polymer products and make them more durable, short fibres were added to create a suspension and polymer production methods were used to make high volume composite parts. Hence, the modelling of these processes stemmed from modelling of such polymer processes.

We can follow the history of these processes separately. In the case of continuous fibre composites, modelling was first promoted to improve properties by getting rid of defects such as fibre waviness and wrinkling and voids and was mainly driven by the space race between US and USSR in the 1960s to manufacture light weight materials. The early models developed there were mainly empirical characterisation of resin cure kinetics (Kamal, 1974) and fibre compaction to tune the pressure and temperature cycle during Autoclave processing. In addition to Autoclave processing, we address the history of development in Filament winding, pultrusion, Sheet Forming, Liquid Composite moulding, and Tow and Tape Placement.

The History of Composites. https://doi.org/10.1016/B978-0-443-34119-9.00017-6

2. Discontinuous fibre composites

2.1 Early history

Discontinuous fibre composites can be divided into two categories. Aligned long discontinuous fibres (ALDF) and short fibres. ALDF materials were developed for forming methods with a desire to manufacture high performance structural properties and they will be discussed under advanced composites history.

The modelling of short fibre suspensions in composite processes such as extrusion, injection and compression moulding arose from models developed for polymer processing. The father of polymer processing was Robert Byron Bird at the University of Wisconsin at Madison. He not only developed mathematical models based on fluid mechanics principles for polymers that formed the fundamental underpinnings to model polymer processes but also wrote a comprehensive text on Dynamics of Polymeric Fluids (Bird et al., 1987) which many modellers of composite processes based on polymer processes reference and compare their numerical solutions to analytic solutions provided in that text.

There were others such as Stanley Middleman who went on to publish a textbook on Fundamentals of Polymer processing through McGraw Hill in 1977 (Middleman, 1977), and Tadmor and Gogos (2013) who provided mathematical treatment to processes such as Extrusion, Injection moulding, and Compression moulding in a textbook called Principles of Polymer Processing.

2.2 Injection moulding simulations

National Science Foundation (NSF) in early 1970s recognised the importance of process modelling for composites manufacturing processes such as injection and compression moulding as these composites' parts were replacing metal parts for automobile secondary structural components such as fenders, dashboards, and door interiors. NSF established an injection moulding program at Cornell University in 1974 with Professor K.K. Wang as the director. K.K. Wang had graduated from University of Wisconsin program where the polymer program was very strong.

The first major breakthrough in modelling and simulation of polymer suspension flow which applied to thin cavities was brought about by Hieber and Shen (1995) who used Hele-Shaw approximation to reduce Navier Stokes equations from four nonlinear equations to one equation that required just to solve for pressure and then extract the velocities from the pressure field. This was a game changer in late 1970s as it allowed one to examine fibre suspension flow in complex moulds as the computational needs were reduced by three–four orders of magnitude making it possible for 1970 computers to address these issues. As workstations were introduced in the early 1980s by IBM and Sun, the computations ventured into addressing viscoelastic nature of polymeric flows and also fibre orientation of the short fibres in these flows.

On the other side of the Atlantic Ocean, this flurry of activities was summarised by JRA Pearson and Richardson from Imperial College in their edited volume on

Computational Analysis of Polymer Processing in Elsevier Science Publishing Co., in 1983 (Richardson and Pearson, 1983). It provided comprehensive modelling efforts on physics of polymer processes in Extrusion, Injection moulding, and Coating flows. The volume mathematically described the polymer mechanics by JRA Pearson followed by Roger Tanner from University of Sydney with a chapter on how to address the viscoelasticity of flows and Die Swell in Extrusion. Ken Walters from Wales and Marcel Crochet from Belgium laid the groundwork on how to simulate these flows using finite difference and finite element methods. This later on was packaged into a commercial software by Crochet called Polyflow. Other important contributors that led to development of models were Vincent and Agassant (1986) from France who first published experimental and theoretical studies on fibre orientation tracking and analysis in 1985. Menges at IKV (Menges and Wübken, 1973) in Germany was generating data on fibre orientation in injection moulding in early 1980s which was useful for the development of models.

In 1983 Byron Pipes and Roy McCullough from University of Delaware recognised the importance of modelling and describing fibre orientation during processing of short fibre suspensions when manufacturing short fibre composites by injection and compression moulding. Their student Rick Givler with the help of numerical methods expert Professor Marcel Crochet, who was on Sabbatical at University of Delaware, implemented Jeffery's derivation (Jeffery, 1922) of how a single ellipsoid particle orients in flows in a two-dimensional geometry (Givler et al., 1983).

Experimental results showed although qualitatively the fibre orientation prediction in short fibre composites was encouraging, one could hardly represent the motion of concentrated fibre suspensions to be represented by rotation of a single fibre. In 1984 C.L. Tucker with his student Folgar and Tucker (1984) developed a phenomenological model for fibre orientation rotation in concentrated suspensions through experimental observations. They retained Jeffery's equation but added a diffusive term that randomised the fibre orientation due to fibre-fibre interactions and provided a method to characterise this from simple shear flow experiments in Couette flow apparatus.

However, one needed to evaluate the entire distribution function of fibre orientation as the fibres rotate and convect with the fluid in the mould. This means that even in two-dimensional flows to predict final orientation at each spatial location it would be necessary to specify the initial fibre orientation distributions. This would require one to solve for at least 30–60 initial fibre orientation angles to represent the initial fibre orientation distribution in 2D and even more in 3D to find the final orientation distribution function at a given location as a function of time. One would need to repeat this for each node in the mesh which would require one to solve 30–60 variables at each node of the mesh making the problem highly computationally expensive. In 1980s this was also a tall order for computations.

To address this issue in 1986 (Advani and Tucker, 1987) at Illinois realised that even if you evaluated the entire distribution function during flow and found the final fibre orientation state at each location in the mould you only needed the fourth moments of the distribution function at each location to predict the physical and mechanical properties. So they came up with the idea of using second order tensorial description which required one to solve for only two variables to describe 2D fibre orientation state and five variables for 3D orientation state at each node making it possible to track the orientation of fibres as the suspension filled the mould.

However, this development caused a closure issue for orientation tensor equations as the second-order orientation evolution equation contained a fourth-order orientation tensor term. This was addressed by approximating the fourth order terms with second-order terms by using a hybrid closure approximation (Advani and Tucker, 1990). Over the years, many closure approximations have been introduced which continue to make the predictions more accurate. Tucker has written an authoritative book on this topic which provides a great summary of the state of fibre orientation predictions and measurement methods today (Tucker, 2022)

The birth of industrial simulations of injection moulding can be traced back to the southern hemisphere, specifically to Colin Austin from Australia, who launched Mouldflow in 1978. Colin Austin worked for an injection moulding manufacturer and he was interested in predicting the pressure requirements for the mould. He used one-dimensional finite differences strip geometries to predict pressures and cooling rates. He did a market survey and found that injection moulders were interested so he launched Mouldflow in 1978 in which he stitched one-dimensional solutions for linear and radial flows to approximately predict two-dimensional flows. Mouldflow's early versions focused primarily on 2D polymer flow simulations, allowing manufacturers to predict flow front patterns and identify defects such as weld lines. They also created a polymer database of over 1000 materials that made it easier for moulders to use.

Around the same time, the Cornell Injection moulding Program (CIMP) was interested in prediction of flow of short fibre suspension and fibre orientation. They used finite elements to model the complex geometries and implemented the Heiber and Shen model to predict the velocity field which allowed them to predict fibre orientation more accurately. Noticing the industrial interest in Mouldflow, CIMP launched their company called C-mould in 1986 which was based on finite elements. In 2000, Mouldflow acquired C-mould and included the fibre orientation tensors to predict the orientation state in their simulations which was very useful to moulders and designers who were trying to tailor the mechanical properties. In 2008, Mouldflow was acquired by Autodesk for 297M dollars. They integrated the software with their computer-aided design and engineering tools which allowed designers to couple mould design with manufacturing simulation within the same environment streamlining the workflow. Sophisticated 3D injection moulding simulation tools that modelled not just the flow but also temperature, cooling rates, material shrinkage, and mechanical stresses are now available for designers of the injection moulders who use them routinely.

NSF recognised the importance of training manufacturing engineers to use these modelling and simulation tools. In 1994, NSF awarded an Education and Training Program in Composite Materials for Durable Goods Industries jointly to Michigan State University and the University of Delaware to develop new courseware and software to advance low-cost composite manufacturing. Education delivery mechanisms used included computer-based simulations and interactive learning, broadcasting of classes over satellite networks, and workshops on design and processing of composites. Workshops were usually conducted through case studies to familiarise the industry folks with the application of such tools. Figure 2.6.1 shows a case study of the simulation of an injection moulded radiator cap which displays the final fibre orientation.

2.3 SMC and compression moulding

Sheet moulding Compound (SMC) containing unsaturated polyester resin and short glass fibres were formed in sheets. The sheets were cut and stacked in a mould which was compressed in a press with heat to form and cure the composite part. Tiffin (1967) and Kerr (1972) were the first to report the use of compression moulding. GM's Corvette was the first one to debut SMC in their car in 1973 paving the way for the rest of the industry. The modelling of this process was not reported until 1980s. The main contributors to these early models for compression moulding were James Lee from Ohio State University, Barone and Caulk from GM Research Centre, and Charles Tucker from University of Illinois. James Lee (Lee, 1981) predicted the temperature profiles inside plate sections of SMC parts during moulding by a mathematical model in which addition polymerisation was coupled with heat transfer. He also used the model to predict the cure time of different part thicknesses, mould temperatures, and initiator concentrations. Glass fibres played a role as a heat sink as addition of glass fibres to SMC not only lowered the maximum exotherm but also reduced the cure time.

Barone and Caulk (1981) who worked for the GM research Centre also published on the heat transfer between the mould and SMC during its flow and cure with the intent to optimise the mould design. However, it was Charles Tucker (1984) who was the first to simulate the compression moulding process using finite element methods and explore the fibre orientation during compression moulding in 1983 with his students C.C. Lee and Francisco Folgar. He developed a constitutive expression to model fibre-fibre interaction which even today is used by injection and compression moulding industry. Tucker's next student Tim Osswald worked with Barone and Osswald (1987) on use of boundary elements to model the charge shape as it flowed inside a mould. Tucker's group was one of the first ones that worked with GM to predict the movement of the SMC in compression moulding of Navistar's truck

Figure 2.6.1 Mould filling simulation of a radiator cap showing final fibre orientation state.

wind deflector and was validated with GM incomplete filling trials during the moulding process (Jackson et al., 1986).

In late 1980s, Jose Castro (Castro and Lee, 1987) focused on addressing the non-isothermal behaviour of SMC as usually the mould was hot and the SMC charge was cold and this created a temperature gradient through the thickness of the charge which he discovered was quite significant. As computation power became cheaper it became easier for researchers to build three-dimensional simulations and address other issues such as compressibility, the anisotropy, fibre bundle formation, fibre interactions, and void formation. However, the interest in modelling this process was waning as the moulders had figured out with early development in the 1980s of how to make SMC parts without defects.

Around 2000s recycling was becoming important and GE plastics introduced a new thermoplastic material called GMT which contained polypropylene and glass fibres to manufacture compression-moulded automotive parts. Wakeman et al. (1999) explored GMT materials for compression moulding in 1999. Other researchers who worked with this material were Andy Hrymak in Canada, and Jan-Anders Mansson at EPFL for compression moulding but the process never quite took off and the interest in dwelling deep into modelling of this process diminished. However numerical advancement continued to simulate this process in more details which led to incorporation of such simulations in softwares such as Mouldex3D (Tseng et al., 2017).

The recent introduction of new SMC grades using carbon or glass fibres with volume fractions exceeding 30% has necessitated the development of new models to account for the pronounced mechanical interactions between the fibre bundles and

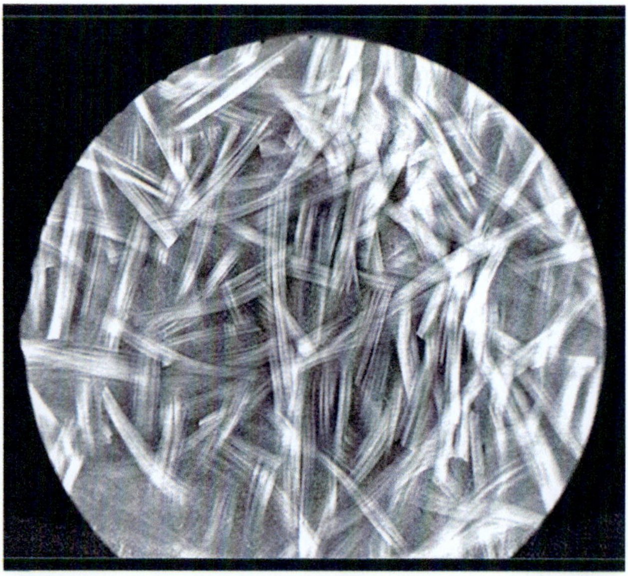

Figure 2.6.2 Micrograph showing fibre bundles and filaments in of new SMC grades using carbon fibres with volume fractions exceeding 30%.

between the filaments within the bundle (Mahe et al., 2023). Fig. 2.6.2 shows the approach adopted to address this dual scale evolution.

3. Evolution of process models for continuous fibre composites

The 1960s saw the advent of lightweight structure composites in the race for the space age. All manufacturing of composites was done by trial and error and manufacturing personnel either did not believe that there was any science in composites manufacturing or thought it was too complex to learn anything useful from it. The academic community in the 70s who wanted to explore the composites research area focused on property predictions of these continuous fibre composites. Stephen Tsai at Stanford was among the first to develop property prediction models followed by Jack Vinson at Delaware. Composite Centre was established at Delaware first in 1974 followed by one at Stanford and later one in Imperial College, London.

In 1980s understanding of composite science and engineering principles was becoming important mainly to address the issue of cost, quality, and product development time. George Springer was one of the early scientists who recognised this and started developing process models for composite manufacturing processes such as Autoclave processing and Filament Winding. Byron Pipes and Roy McCullough also recognised the importance of manufacturing science of composites and established the first National Science Foundation Engineering Center at the University of Delaware in 1985.

Rapid progress was being made in this field but it was all scattered across different parts of USA and hence Tim Gutowski of MIT (who was Nam Suh's student) put together a manufacturing conference in Atlanta in 1988 (Gutowski, 1988). It brought all the early founders of process modelling together who were applying scientific principles and/or developing some general principles prevalent in various composite manufacturing processes. They were George Springer from Stanford, John Kardos from Washington University, John Muzzy from Georgia Tech, K.K. Wang from Cornell, C.L.Tucker from Illinois, and John Muzzy from Georgia Tech. This group and others were invited the same year for a workshop organised by James Seferis of University of Washington (who was Roy McCullough student) on the island of Capri in Italy. At this workshop many composite manufacturing engineers and scientist from companies such as Boeing and McDonnell Douglas were present and they took notice of the importance of process modelling in composites manufacturing.

3.1 Autoclave and out-of-autoclave processing

In autoclave processing, once the specified number of prepreg plies have been placed (either manually or with the assistance of a machine) on the tool, the component is bagged and sealed prior to undergoing autoclave curing. The bagging process involves the placement of various materials, including peel ply, release fabric, bleeder ply, breather ply, a caul plate, and a flexible heat-resistant membrane, over the tool and component. Subsequently, pressure and curing cycles are applied to the laminate stack. This serves two purposes: firstly, to remove any voids that may have formed, and secondly, to ensure that the stacked prepreg plies are sufficiently compacted.

This process is distinguished by a pronounced interplay between the thermokinetics and rheo-kinetics of resin and the flow of the latter within the consolidating fibrous network. Consequently, the development of models is essential for defining optimal autoclave parameters such as the pressure and temperature. To observe the flow pattern during the compression of viscous uncured prepregs, Springer in 1982 designed an experiment in which resin and fibre substitutes were used (Springer, 1982). The results of this experiment suggested that the top ply moves downward until it touches the next ply, at which point the two plies move together toward the next ply. All other plies remain motionless and the plies move consecutively, resulting in two distinct regions of moving plies and stationary plies.

These observations resulted in the formulation of an initial model by Loos and Springer (1983). Subsequently, it was demonstrated that the thickness distribution of the layers is continuous, as it is a consequence of hydrodynamic compaction, which is itself scaled by the continuous pressure gradient. The primary challenges in addition to flow and compaction were to control the curing of the resin and facilitate the removal of excess resin present within the laminate prior to its gelation, to achieve the desired fibre volume fraction. This requires the integration of cure kinetics modelling with the flow of resin in a deformable fibre bed.

Halpin et al. (1983) at Washington University at St. Louis was the first to explore the processing science of thermoset prepreg systems. With his students, Dave and Dudukovic in 1986 he described the resin flow and cure in autoclave processing (Dave et al., 1986). His group solved the mystery that had baffled McDonnell Douglas in St. Louis (which later became part of Boeing) as to why they were not able to get rid of voids in the composites they fabricated in the autoclave even after using very high pressures. Kardos was able to explain that the applied pressure was being shouldered by the fibres and as resin was not under pressure, the dissolved volatiles were free to grow. He conducted an experiment with pressure transducers embedded within prepreg layers which clearly showed how the resin pressure was different across the thickness of the composite although the applied autoclave pressure was uniform and constant.

This was a seminal finding for the aerospace industry and for the first time showed them that understanding of processing science was useful for an industry that was marred in trial and error culture. Since then process models and processing science for autoclaves have advanced addressing the cure kinetics and resin flow to develop optimised temperature and pressure cycles for the processing of the composites in the autoclave.

Dielectric and other type of sensors were introduced to monitor cure kinetics of the composite in conjunction with the optimised cycle to reduce manufacturing time for autoclave processed composites. Over time autoclave processing made void free reliable composite parts, however as the cost of autoclaves was prohibitive, modelling showed that one could obtain adequate consolidation with an applied vacuum without the use of an autoclave. This led to development of out-of-autoclave or Oven-cured composites.

Player et al. (2000) at MIT first showed that one could cure a composite out of the autoclave. He used E-Beam curing but it was not a difficult leap to then resort to Oven based curing. This process was also called Vacuum Bag Only (VBO) cure. The patent for this process was issued in 2002 (Xu et al., 2002). The process modelling for Oven based curing could be done only under one atmospheric pressure by the application of vacuum and it was Michaud et al. (2009) who through their modelling showed that prepregs used in out of autoclave processing needed to have air evacuation pathways to reduce or eliminate voids as the pressure was not sufficient to dissolve the entrapped air into the prepreg resin. Hence prompted by this understanding through process modelling, the material suppliers started to manufacture prepregs specifically for Out of Autoclave processing with designed pathways for air evacuation.

Modellers (Arafath et al., 2009; Cender et al., 2013) were then able to estimate time needed for air evacuation by characterising air permeability. The equations, based on Darcy's Law and the ideal gas law predicted the time required to evacuate a given mass fraction of gas which was directly proportional to permeability.

3.2 Filament winding

In the process of wet filament winding, dry fibre bundles are pulled through a resin bath (the wetting stage) and then immediately wound onto a rotating mandrel (the tool). These impregnated layers are then cured in an oven or at room temperature. The relative velocity of the rotating mandrel and the motion of the incoming material determine the winding angles. It is also possible to use a prepreg roving in this process, which eliminates the need for the resin bath.

The origins of winding technology can be traced back to ancient Egypt, where the practice of wrapping mummies in tapes soaked in resin can be seen as the precursor of modern filament winding with synthetic materials. Filament winding began in the early 1940s and is one of the oldest methods of processing composites. Because of its simplicity, it was one of the first to be automated and therefore one of the first to guarantee consistent part results. As a result of this simplicity and ease of automation, it remains one of the most competitive processes for producing high quality lightweight parts. The initial models were developed for the purpose of simulating the filament winding of thin and thick thermosetting matrix cylinders, and subsequently, thermoplastic cylinders. In a pioneering study, Springer and colleagues developed in the late 1980s a generic model that integrated thermomechanical, void growth, and flow submodels (Calius and Springer, 1990; Lee and Springer, 1990).

While the first two models were already established as a result of their use in autoclave consolidation modelling, the third was a novel contribution. During the consolidation phase, the tensile stress within the fibres can result in their movement through the resin, specifically during the initial stages of the winding process when the resin viscosity is relatively low. As a result, this mechanism gives rise to alterations in fibre tension and the radial position of the deposited layers. Subsequently, consolidation and bonding submodels were incorporated to address the specific case of thermoplastic tapes (Mantell and Springer, 1992).

Although these initial models addressed the case of tubes with circumferential (hoop), polar, and helical windings, other technological advancements during the same period demonstrated that filament winding could be employed to fabricate more intricate structures of revolution by incorporating additional axes of motion. Filament winding can be used to produce a variety of complex shapes, including nozzles, aeroplane inlets and vanes, elbows, and T-pieces to name a few. In addition, it can be employed to create toroidal tanks and other intricate structures (Munro, 1988). New modelling techniques pertaining to the management of fibre trajectories have emerged (Carrino et al., 2004).

The study of fibre trajectories on the mandrel surface has become a crucial aspect in the production of parts with the desired mechanical properties, reduction in manufacturing times, and cost savings. Two approaches have been studied and modelled, namely geodesic and nongeodesic trajectories. The importance of geodesic trajectories for filament winding is important because in geodesic trajectories no lateral forces of any kind are experienced, thus fibre slippage is avoided and fibre stability is ensured. In the initial stages of composite manufacturing, geodesic trajectories were the predominant method employed for filament winding (Pidgainyi et al., 1967). However, with the advent of winding on complex geometries, nongeodesic paths have become a common place feature of the process (Menges et al., 1978).

Nongeodesic paths can also be used to increase design freedom and optimise mechanical properties. They use frictional force to prevent fibres from slipping off the mandrel surface and maintain fibre stability (Tarnopol'skii et al., 1998). It is noteworthy that the concept of fibre trajectory description using differential geometry emerged at the same time in the modelling of filament winding (Wells and McAnulty, 1987) and the draping of fibre reinforcements (Tam and Gutowski, 1990), although it was not developed by the same researchers. With the advancement of winding machines and computers, the capabilities of the filament winding process were further enhanced with the introduction of sophisticated simulation and control programs. These programs enabled the prediction of roving paths on the mandrel and the determination of required machine movements, thus facilitating the optimisation of the winding process.

3.3 Pultrusion

In thermoset pultrusion process, a continuous dry fibre bundle is saturated with resin as it is pulled through a resin bath and fed into a die where it is subjected to compaction and heating. The cross-sectional shape of the pulled fibre bundles is formed by these dies. Within the die, the composite is compacted and cured. Since the advent of the pultrusion process in the mid-1950s to make fishing rods, it has undergone continuous modification with the objective of obtaining products in a shorter time frame, with greater complexity, and of improved quality. modelling of the thermoset process evolved a couple of decades later to shed light on the relationship between the pulling speed and tension, residence time, curing, shape of the die, pressure, and temperature development within the die and impregnation and cure quality.

The early work in 1970s discussed the pultruded products and the process from the viewpoint of process engineering and curing of the resin. Price (1979) in his thesis in 1979 was the first one to propose a more fundamental approach. He suggested three dimensionless variables that accounted for temperature, residence time, and volume change due to the cure reaction and put forth empirical relationships among these three dimensionless variables by conducting experiments with a laboratory-scale pultruder. The pultrusion process is limited to straight parts with a constant cross-section (beams, frame sections). The early impetus for modelling was to achieve a uniform degree of cure in the cross section of a pultruded product, hence the temperature within the pultrusion die was considered to be the main processing variable which therefore required modelling of the cure kinetics within the die. A mathematical model to simulate this was proposed by Han et al. (1986) in 1986. The initial modelling focused on temperature profile and cure kinetics. It was Jeffery Roux in 1994 who first published a model to describe the resin flow in the die and its relation to the pulling force (Gorthala et al., 1994). In early 1990s instead of using a resin bath, pressurised resin injection was proposed as a slight modification of the process. Kardos was one of the first one to propose a numerical model to simulate this process (Kommu et al., 1998).

Thermoplastic pultrusion was attempted later without the resin bath but with thermoplastic resins introduced in continuous tape forms or with thermoplastic powder interdispersed in between and within the fibre tows. The initial models for thermoplastics were developed in early 1990s by Astrom, Hahn, and coworkers (Astrom, 1992; Larock et al., 1989). Lee and Springer developed the first full model for thermoplastics in 1991. The general model incorporates three fundamental submodels: a heat transfer submodel, a resin flow submodel for consolidation within the taper, and a pulling force submodel.

The early contributors to the modelling of pultrusion process were Roux, Babu, Kardos, Astrom, and James Lee. In the recent years, we have seen other researchers such as Baran, Sandberg, and Spangenberg (Sandberg et al., 2021) who have been refining the models for the pultrusion process which is making it possible to pultrude complex sections which were not possible in early 2000 and also improve the throughput.

3.4 Forming

The mechanical modelling of the forming of fibrous reinforcements and composite prepregs has been informed by the advances made in metal sheet forming. From the 1970s to the 1980s, significant advancements were made in the fields of sheet metal forming mechanics, numerical modelling, and computational methods. The research topics were concentrated on numerical formulations, including membrane versus shell, static implicit versus dynamic explicit, tool/sheet contact, and material modelling (yield functions, work hardening rules, and failure criteria).

Forming was an attractive method if one could use metal forming methods to make net shape composites. The initial contributions to drape modelling focused on the dry fibrous reinforcements. The fundamental concept of woven cloth deformation using the Pin-Joined idealisation of the shear mode was initially proposed in 1956 by Mack and Taylor (1956). The earliest documented reference to a computer model for fabric drape is from 1979 by Potter (Potter, 1979). A more comprehensive analysis was subsequently published by Robertson et al. (1981) and in 1984 (Robertson et al., 1984). These two papers likely mark the beginning of the kinematic drape modelling activity. Similarly, Bergsma and Huisman (1988); Van West et al. (1990) employed a comparable methodology to simulate, respectively, the deep drawing of fabric-reinforced composite sheets and the draping of bidirectional fabrics over arbitrary surfaces.

The use of finite element models allows for the generation of more accurate and detailed simulation results for forming processes. Among the early research devoted to the FE modelling of the draping behaviour of fabrics was the work of Collier et al. (1991). In this work, nonlinear shell elements were applied in combination with orthotropic material behaviour. In the field of composites, the FEM approach has been employed in the numerical modelling of draping processes in 1993 by Boisse et al. (1993) and 1994 by Long and Rudd. (1994). An early application of finite element techniques to the modelling of cloth can be found in the work of Terzopoulos and Fleischer (1988). In this work, the modelling was employed primarily to achieve more realistic computer graphics.

Early modelling methods for composites were explored as processing engineers were trying to form mildly curved sections of continuous fibre composites in an autoclave using a diaphragm. Gutowski (1985) reported on this in 1985. Neil Cogswell who was working for ICI with their continuous fibre prepregs along with T C Rogers realised that the rheology of the composite was very important to address the shearing and slipping between the prepregs for the forming process in the early 1980s. Pat Mallon and Conchur O'Bradaigh in 1988 and later Conchur O'Bradaigh with Byron Pipes at Delaware (O'Bradaigh and Pipes, 1991; Mallon et al., 1989) were the first to model the forming process using implicit finite element methods with the ideal fibre reinforced fluid model developed by Rogers (1989). This new fast processing methods in which resin flow and rheology played an important role was hailed by inaugurating a new international conference on Flow Processing in Composite Materials (FPCM) in Limerick, Ireland in 1991 with Pat Mallon and Neil Cogswell as the initial chairs (Cogswell, 1991).

In modelling prepreg as a transverse isotropic fluid, the huge challenge was that as a prepreg with continuous fibres had an infinite viscosity in the fibre direction, the composite did not flow in that direction. This anisotropy in material property resulted in numerical instability and was very difficult to model. When forming with a blank of continuous fibre prepregs using a rigid or complaint tool or with pressure into the shape of the mould resulted in wrinkles and other defects due to the anisotropy or slippage between layers for large, sharp, or compound curvature parts as shown in Fig. 2.6.3.

Other researchers have instead looked at prepregs as deformable solids, and have drawn on knowledge gained in the field of metal sheet forming to propose approaches suitable for composites. Such is the case of De Luca and Pickett who developed in

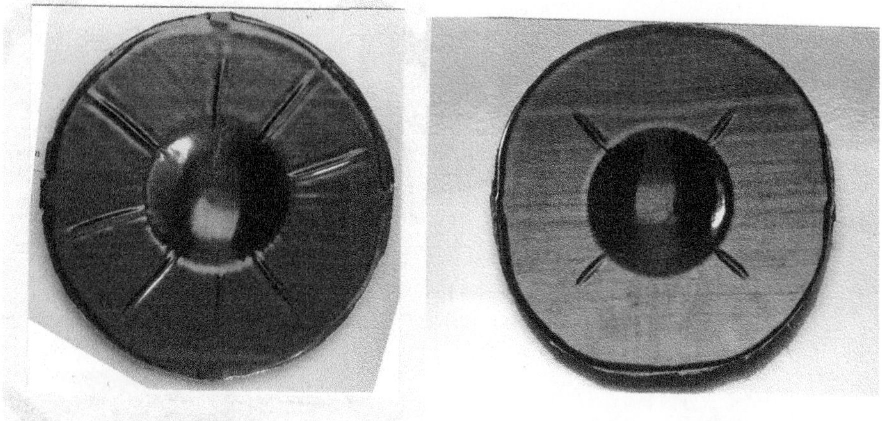

Figure 2.6.3 Wrinkle formation during forming with APC (PEEK with continuous carbon fibres).

1995 the ESI PAM-FORM commercial software package for comprehensive analysis of forming using an explicit FE solver derived from ESI PAM STAMP software (Pickett et al., 1995; De Luca et al., 1998).

The group led by Akkerman at the University of Twente have developed a software called Aniform (Haanappel et al., 2014) which models the process for highly anisotropic materials and have developed numerical methods to address the singularity due to infinite viscosity. However, due to the challenges associated with infinite viscosity of the material in the fibre direction which restricted the geometries that could be formed with continuous fibres, the possibility of using discontinuous fibres was entertained for forming. Okine (1989) who worked with Nam Suh at MIT first addressed the forming process during his Ph.D. in 1982 and then went on to Dupont to develop Long Discontinuous fibre material (LDF) specifically designed for the forming process as the material could now deform in the fibre direction as the fibres were not continuous. Micromechanics models were developed to predict the anisotropic viscosity of this material but before more sophisticated models could be developed, LDF material was abandoned as its mechanical properties did not come close to continuous fibre composite properties because all the fibre ends in the LDF material were at the same location within the LDF sheet and this was the weakest link in the composite as the failure was governed by the properties of the resin rich region with the absence of fibres.

It was not until 2013 that Kevin Potter and his team came up with a novel method to align these long discontinuous fibres in a prepreg where the fibre ends were dispersed randomly (Yu et al., 2014; Yu and Potter, 2013). Under a DARPA program, University of Delaware developed a better version of Potter's system to get such fibre alignment of 57% in prepregs called Tailorable Universal Feedstock for Forming (TuFF) in 2019 (Yarlagadda et al., 2019a, 2019b), and modelling of this material is being pursued using Aniform to address its viscoelastic rheology and stretchability for forming

Figure 2.6.4 Creating short fibre aligned material (TuFF) for forming [CCM].

like sheet metal components. This approach to forming is also very conducive to forming and is being pursued at UD-CCM as shown in Fig. 2.6.4.

In response to the emergence of novel challenges associated with environmental constraints, the formation of stacks of discontinuous plies, or patches, has also emerged as a promising approach. The forming of these preforms entails intricate mechanisms of formation, viscous sliding over considerable distances, and compression flow of the patches. This necessitates the adaptation of previously developed solutions. This issue has given rise to a number of recent developments in the field. To reduce the computational costs associated with the formation of anisotropic multilayer media and the compression flow possibilities of these layers, a new pinching shell element has been developed by Mulye et al. (2023). This approach unifies the aforementioned processes, thereby reducing the overall computational costs (Pittman et al., 1982).

3.5 Liquid Composite moulding

"Liquid Composite moulding" (LCM) is an umbrella term used to describe a class of processes that involve the infiltration of a dry fibrous preform with a low-viscosity polymeric resin. When the need to manufacture net shape structural composite parts became evident, a family of processes was invented in which a dry assembly of woven or stitched fabric could be conformed to the mould geometry and placed in it and resin injected under pressure to fill all the empty spaces between the fibres. There were many ways to accomplish this, and four processes emerged with slight variations over time and continued to be practiced today.

The first one was Resin Transfer moulding (RTM) in which both mould surfaces were rigid in which the dry preform was enclosed before resin injection, next to reduce cost there was emergence of RTMLight in which the top mould surface was replaced by a Caul plate but this limited the complexity one could introduce in the part geometry. To reduce the cost further and to make large structures such as wind blades, one sided tool was used with a vacuum bag on top and the resin injected under atmospheric pressure and the process was called Vacuum Assisted Resin Transfer moulding (VARTM). This process was still very slow so SCRIMP (Seemann, 1994) was invented by addition of a distribution media on the top surface to speed up the flow. Automotive companies were more interested in reducing the cycle time, so they came up with an injection compression moulding also known as Compression Resin Transfer moulding (CRTM) process with expensive rigid tooling. The primary distinction between these processes lies in the boundary conditions and the stationary nature of the fibrous preform.

The important modelling need in all these processes was to accurately predict the flow pattern, impregnation quality, and the associated pressure field, particularly in the context of local saturation of fibre bundles so that one could couple the mould design with the manufacturing process. Fortunately, the modelling of these family of processes was similar and over time has addressed the nuances of each process which are discussed below.

3.6 Resin Transfer moulding

In 1985 Carl Johnson, a Ford employee, who was on the industrial board for first Engineering Research Centre in Manufacturing Science of Composites awarded to the Center for Composite materials at the University of Delaware revealed that RTM was something automotive industry was exploring as it had the potential to manufacture net shape structural components. He added that the industry would gain tremendously if the moulders could understand how the resin was flowing in the mould as they were getting many dry spots in the components they were manufacturing.

Suresh Advani in 1987 had just joined the University of Delaware and saw the opportunity to simulate RTM by modifying his code that he had worked on for simulation of compression moulding while doing his Ph.D. with Charles Tucker at the University of Illinois at Urbana–Champaign. He realised that his code with its Finite Element/Control Volume (FE/CV) approach could simulate moving boundary of the resin front with minor modifications in the governing equation and the boundary conditions. This was made possible by use of the Sun workstation 3/160 proudly showing it off in Fig. 2.6.5. With his Ph.D. student Michael Bruschke, they published the first paper (Bruchke and Advani, 1989; Bruschke and Advani, 1990) that used FE/CV method to simulate the isothermal flow of resin in complex geometries in 1989 which they later called Liquid Injection moulding Simulation (LIMS) (Šimáček and Advani, 1998).

Figure 2.6.5 Suresh Advani with his Sun 3/160 workstation where he first simulated the resin transfer moulding process.

In 1990 Francois Trochu at École Polytechnique de Montréal, Canada working in Raymond Gauvin's group used boundary fitted finite difference methods to simulate RTM. After many discussions with Advani, he realised that finite difference methods were not conducive to model complex geometries (Tucker, 1989; Trochu and Gauvin, 1992) and were quite slow due to the process of remeshing at each time step and switched to using a finite element-based simulation which they called RTMFLOT in 1994. Their code was later purchased by ESI in 2001, a French company, and called it PAM-RTM. Nonconforming finite elements were employed to approximate the pressure field and to guarantee the conservation of flow rate at the interelement boundaries (Soukane and Trochu, 2006).

There was a lot of activity in early 1990s to address flow simulation in RTM.

The other groups that joined Advani from Delaware and Gauvin and Trochu from École Polytechnique de Montréal in advancing the modelling were Lee et al. (1994) from Ohio State, Rudd et al. (1997) from Nottingham, and Loos and MacRae (1996) from Virginia State. These groups focused more on addressing the nonisothermal aspect of the process by coupling the temperature and cure kinetics of the resin. Due to computation limitations in 1990s, they opted to solve coupled complex equations in simple planar moulds instead of simpler equations in complex geometries.

Lin et al. (1991) addressed the first nonisothermal simulation of the process in 1991 where they modelled the temperature and cure but the simulation time was days and made it prohibitively expensive to simulate large complex parts. In later years many other researchers addressed the three-dimensional flow with changing temperature and cure but the nonlinear and coupled nature required many hours to solve for the final temperature and cure field. Some researchers also tried to couple this with residual stress calculations but this work has still not fully transitioned from academia to industry as this would require one to characterise the thermal properties of the resin and cure kinetic parameters along with viscosity dependence on temperature and cure (Zobeiry et al., 2016). Most industrial parts were complex and were performed under isothermal conditions, and the filling process was usually complete before the heat transfer effects significantly affected the process.

Two other phenomena in RTM were identified that influence the flow and needed to be modelled. They were the role of capillary action and race tracking of the resin. James Lee's group in Ohio State (Patel et al., 1995) were the first ones to postulate in 1995 the role of capillary action that could trap voids either within the fibre tows or in between the fibre tows which was later modelled by the Montreal group. Advani's group (Bickerton and Advani, 1997) continued to identify and model important features in the RTM process such as (1) race-tracking of the resin along corners and inserts due to imperfect fit of the fabric which changed its permeability and (2) recognition and modelling of the dual scale flow in woven and stitched fabrics due to different impregnation rates within the tows compared with in-between the tows.

In late 1980s there was a lot of discussion if flow in fabrics was non-Darcian as many researchers were measuring nonlinear pressure increase when resin was injected under constant flow rate in the mould. However, in early 1990s it was recognised that this was caused due to dual-scale architecture of the fabric and modelling of this phenomena conclusively showed that Darcy's law could describe the physics observed in dual-scale fabrics (Pillai and Advani, 1995).

In addition to the FE/CV and pure FEM approaches previously discussed, alternative methods were developed in the 2000s to model the complex phenomena such as the coupling between the fluid flow and the displacement and/or deformation of fibrous preforms. These include the level-set method for tracking moving boundaries in both fluid and solid domains as shown by Soukane and Trochu (2006) and meshless methods such as the SPH method proposed by Comas-Cardona et al. (2005a, 2005b).

Another breakthrough in simulation of RTM came in 1996 when Maier et al. (1996) came up with an algorithm that could solve for pressure in LIMS at each time step by matrix multiplication rather than matrix inversion which resulted in one to two orders of magnitude speed up in the solution. The speed up was exponentially higher for large meshes which made it possible to simulate resin flow in large complex structures such as a 40-foot-long keel beam of a helicopter within hours with the power of 1990 PCs as shown in Fig. 2.6.6. This also made it possible to optimise the flow by running hundreds of simulations within a few hours as the speed up was from hours to seconds to run one simulation for moderate meshes of complex parts with 5000 nodes. This paved the way for optimisation of the process for best gate and vent locations even in the presence of race tracking and for multiple permutations of race tracking along edges as race-tracking strength could vary during manufacturing from one part to the next.

As computer power became cheaper and sensors that could detect resin arrival could be implemented, one could forecast race tracking within each production part and this made simulations even more useful so they could design offline a control

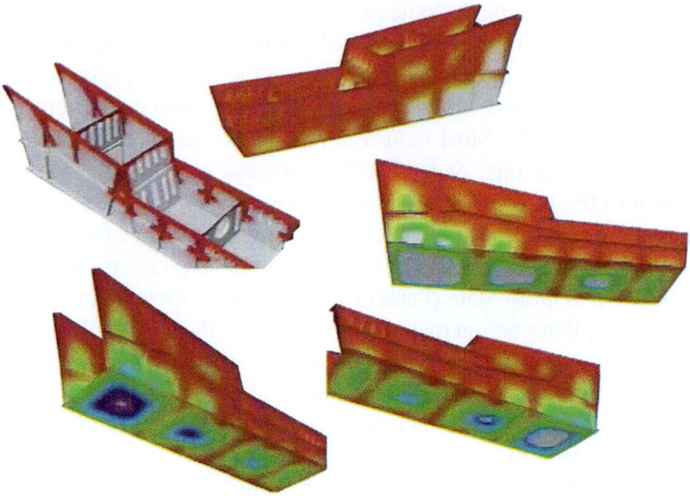

Figure 2.6.6 Mould filling simulation of keel beam showing time history of resin flow.

action for that scenario and correct the flow pattern in situ to avoid formation of a dry spot. Advani's group demonstrated this with experimental validation in 2002 on a planar mould containing a triangular insert (Bickerton et al., 2001) funded by Office of Naval Research under AMIPC grant (Advanced Materials Intelligent Processing Centre). They assumed that race-tracking was either weak or strong along the edges and addressed a limited number of possible permutations. Today a digital twin could be easily developed from the simulation process using reinforced learning as most of the real physics has been incorporated in the simulation and the data obtained from the manufacturing floor could be used to update the race-tracking conditions along the boundaries to sync the manufacturing and the simulation process.

All groups who were simulating the flow and heat transfer in RTM realised the importance of preform permeability as it was an input to their simulation and found experimental approaches to measure permeability based on the early work done by Adams and Rebenfeld (1987) on directional in-plane permeabilities of geotextiles in 1987. Richard Parnas, who at that time had established a Composites group at National Institute of Science and Technology(NIST) in Maryland developed a standard for measuring permeability of a fabric with the intent to create a database for various fabrics and to their surprise found that small changes in the setup led to permeability values that differed in magnitude by orders of magnitude (Parnas et al., 1995; Luce et al., 1995).

Another manufacturing step that changed the permeability of bidirectional or stitched fabrics was the act of draping the preform on the curved surfaces of the mould as this would cause shearing of the fibre tows and also tow slippage changing the local fibre volume fraction and fibre orientation which would change the permeability locally of the draped fabric. Simacek developed a crude version of this called DRAPE in early 1995 which only took into account the shearing of the fabric to predict permeability of draped fabrics (Bickerton et al., 1997). Later on, Vistagy included this feature in their software fibre SIM which was developed in the 1990s to help Lockheed Martin in the design process of composite parts for fighter jets by generating manufacturing information to interface with automated cutters, laser projectors, and fibre placement machines. However, one still needed to experimentally measure permeability for a planar fabric and this has its own history which is summarised below that eventually led to an ISO standard (May et al., 2024).

3.7 Permeability

Once it was established that Darcy's law could be used to model the Liquid Composite moulding processes, fabric permeability determination became a centrepiece for the modelling and simulation of the family of these injection processes. It was in 1991 that Richard Parnas who was the leader of a composites group at National Institute for Standards and Technology (NIST) in Maryland, USA did a very early round robin experiment on measuring the permeability of the same fabric at three different locations and found that small changes in the set-up could lead to large differences in permeability values and proposed the first permeability characterisation standard in 1995 (Parnas et al., 1995; Luce et al., 1995). That standard was not formalised as NIST lost interest in this endeavour and Parnas left NIST. As a result, researchers continued to innovate with various methods to measure permeability in 1D, 2D, and 3D directions with flow visualisation, pressure transducers, and point and linear resin arrival sensors.

However, there was no reference standard to validate or verify the method's accuracy or efficacy. As permeability architecture was also dual scale and one needed to stack multiple layers to extract a bulk permeability value, hence results varied from one group to the next by more than 100%. The permeability also depended on fibre volume fraction and slight variations could strongly influence the measured value. Researchers started to analytically and numerically model permeability on a unit cell level to address the issue of change in permeability with fibre volume fraction. All these efforts were discussed at the biennial FPCM conference (Flow Processes in Composite Materials) in 2004 which lasted over 3 hours after the conference was over. This clearly revealed a need for a universally accepted standard. A first international permeability benchmark exercise, initiated by ONERA (Office National d'Études et de Recherches Aérospatiales, France) and KU Leuven, was conducted and published in 2011 for this purpose (Arbter et al., 2011). The study showed that without proper guidelines and guardrails, the permeability values for the same fabrics could differ by orders of magnitude.

Justin Alms, a student at Delaware developed a guidelines document to conduct permeability experiments in 2011 delineating all the major pitfalls, and based on a common desire of researchers to standardise the determination of permeability, hence a second benchmark was conducted with Hexcel fabrics. A total of 12 participants from all around the world carried out the in-plane permeability measurements of a carbon fabric using their respective setups following the guidelines (Vernet et al., 2014). The results were promising, and the inaccuracies and inconsistencies were identified and validated. At this point David May from Technische Universität Kaiserslautern and Ana Yong of NPL (National Physical Laboratory) in summer of 2016 took over the reins of developing permeability measurement standards for 2D and through-the-thickness permeability and also the compression force measurement of the fabric (May et al., 2019; Yong et al., 2021a, 2021b).

An ISO standard has been finally developed (May et al., 2024), for permeability measurement but there is still a lack of database which could provide permeability of fabrics at different volume fractions. To introduce online permeability measurement of preforms, models of transient air flow in fibre reinforcements have been developed. These models account for the compressibility of the gas and the rarefaction of the air molecules when the flow is induced by the creation of a vacuum in the preforms prior to resin injection (Hou et al., 2013; Cender et al., 2016). Binetruy's group (Syerko et al., 2023) initiated a benchmark on permeability prediction based on the microstructure obtained from a CT scan and its comparison to the measurement benchmark which will prove helpful in developing a database. The lack of this has prevented the integration of modelling and simulation of the Liquid moulding processes by industry. However, with AI and digital twin emergence, the incorporation of this in actual manufacturing process is still an attractive possibility.

3.8 VARTM

As new variants of RTM process were invented such as VARTM and CRTM, RTM simulations were tweaked to address the additional physics that played an important role. VARTM became a viable process once SCRIMP (an acronym which stands for Seemann Composites Resin Infusion Moulding Process) was patented by Bill Seemann in 1990 (Seemann, 1990; Seemann, 1998) in which flow along the top face was accelerated by addition of a very high permeability distribution media which reduced the filling time by orders of magnitude and made it possible to manufacture large parts such as wind blades. This required 3D simulations of resin flow as there was a velocity variation even through the thickness of the preform due to the high permeability distribution media. modelling of this distribution media was a challenge as it was thin and its permeability was much higher.

PAM-RTM opted to solve the flow in three-dimensional mesh with the top layer having very high permeability to represent the flow media. However, this resulted in inaccurate solutions as the anisotropy in permeability required one to scale the element aspect ratio accordingly to avoid distorted elements. As the permeability of the fabric was usually one to two orders lower in the thickness direction compared with the in-plane directions, it was fine to have the element dimension in the thickness direction to be much smaller than in plane directions and still maintain a decent aspect ratio and not cause any distortion to maintain accuracy. This was not true for the distribution media as its permeability in the thickness direction was two to three orders of magnitude higher than the in-plane directions.

Simacek and Advani (2004) solved this problem by attaching a two-dimensional permeability media to a three-dimensional mesh of the fabric. This was mathematically correct as in the governing equation one was only solving for the pressure which is a scalar. Similar approach was also used to model race tracking by attaching one-dimensional elements along the edges of the mesh. They used the same modelling approach to represent fibre tow filling by attaching one-dimensional element at each node of the mesh which simulated filling of the tow by assigning a lower permeability for the tow filling. Their results showed that the tows filled at a much slower rate validating the nonlinear pressure at the inlet with a constant filling rate which was observed in the experiments. The faster solution also made it possible to explore the best distribution media design in the VARTM/SCRIMP process (Sas et al., 2015).

In VARTM/SCRIMP process the other modelling challenge was that as the top surface was a bag instead of a rigid mould, the pressure distribution during the flow of resin was coupled with the compaction and hence the permeability of the fabric making the problem nonlinear (Correia et al., 2005). Researchers addressed this issue and found that if an equivalent permeability was used for the fabric in VARTM, the resin flow pattern was not that different as compared to when the coupled solution was employed (Simacek and Advani, 2016).

3.9 CRTM

In the classical Compression RTM (CRTM) a small gap between the mould surface and the fibre preform is created by partially closing the mould. The injection gate is closed after the required amount of resin is injected into the gap which flows easily and may partially impregnate the preform. The mould is closed to its final thickness and fibre volume fraction by applying high compression pressure to squeeze the resin into the preform. Other variants of this process were also being explored such as closed mould CRTM, Wet Compression moulding (WCM), HP-RIM (High Pressure Resin Injection moulding), and HP-RTM.

This offered new challenges to modellers as high injection rates were causing fibre washout as the force from the compression equipment needed to be higher than the drag force of resin to avoid fibre displacement. A further challenge inherent to the CRTM with injection gap, when compared to the RTM, is the modelling of the flow into the aforementioned gap. This may be approached in two ways: firstly, as a Stokes flow; secondly, if the gap is sufficiently thin for the lubrication approximation to be valid, by a Poiseuille flow. When the resin flows in a direction that is parallel to that of the fibrous reinforcement deformation, the pressure gradient can cause the fibrous material to deform nonuniformly through the thickness (Lee et al., 2023). In addition, this phenomenon can result in in-plane fibre washing (Bodaghi et al., 2020; Hautefeuille et al., 2020).

In 2004 an Automotive Composites Consortium (ACC) was formed with Ford, Chrysler, and GM as the main partners to develop the technology to build a carbon composite- B-pillar with minimum of 60% weight savings over steel at a cost close to that of steel, with rapid and optimal manufacturing processes. CRTM was identified as the process and LIMS was used to simulate and design this process with multiple injection locations to manufacture for the first time a structural component for a regular passenger vehicle (Simacek et al., 2008). The program was later abandoned due to cost issues and reluctant of US companies to invest in this switch in early 2000s and it was not until 2011 when BMW took the plunge and designed i3 with fully integrated carbon composite frame using Compression RTM process and reduced the cycle time down to 8 min. Although i3 and i8 were discontinued later, they showed the world that a composite passenger car was possible to make at high production rates and today all EVs have benefited from this technology.

3.10 Automated tape and fibre placement

Automated Tape Placement (ATP) and Automated fibre Placement (AFP) are processes by which multiple strips of composite material or powder-coated fibre tows are laid down with pressure on a tool surface in their proper position by an automated machine (usually robotic system). The dispensing head is equipped with a heating source and a compaction roller, which are used to debulk the tows and consolidate the layers while promoting adhesion between successive layers. The adhesion and void removal between the incoming materials and the deposited substrate is governed by processing conditions such as heating, compaction, and tensioning systems. Despite the deployment of sophisticated robotic systems, numerous challenges remain that necessitate the utilisation of numerical tools for the simulation of material deposition to ensure good quality of composites made by this process.

Automated fibre placement is a process that bears resemblance to filament winding. In this process, the fibres are placed onto the surface of the mandrel in a sequential manner, with each strip being placed one at a time. Nevertheless, AFP diverges from filament winding in several key respects.

- The fibres are driven toward the surface of the mould. Accordingly, flexible tows are unsuitable, strips exhibiting a certain bending stiffness are employed.
- Filament winding can be conducted using either wet or dry winding techniques. In contrast, fibre placement requires the use of prepregs or powdered fibre strips, precluding the possibility of employing a wet fibre placement process. This explains why the initial developments concentrated on thermoplastic tapes and why it was not until the late 1990s that the advent of an appropriate powder-coated fibre tow became a possibility.
- In the fibre placement process, a roller is employed to exert pressure on the fibres as they are positioned on the surface of the mould or on the previously deposited material. The pressure serves to consolidate the fibres as they are being laid down, thereby enabling the use of surfaces other than convex ones (which are required for filament winding).
- In the fibre placement process, heat is typically applied at the nip point, which is the point where the fibre strip meets the surface of the mould. The application of heat reduces the polymer viscosity. The combination of heat and pressure provides the driving force for flow and consolidation.

Over the years, process modeller recognised the role of mass, momentum, and heat transfer in this process coupled with adhesion between layers and the fibre displacement and compaction. The APC (Aromatic Polymer Composite) thermoplastic composite tape process was first developed in 1980 by Imperial Chemical Inc (currently AkzoNobel). APC tape which was a carbon fibre prepreg impregnated with PEEK resin was suitable for tape placement processes due to its resin-rich surfaces and easier transverse flow of the resin. The initial models for automated thermoplastic tape placement were developed inlate 1980s to address their conformability and their adhesion ability. Gutowski conducted comprehensive studies about the mechanics of tapes and their change in geometry during the layup process (Gutowski and Bonhomme, 1988). During the same period, Grove conducted further modelling work, employing a finite element model of heat conduction to investigate the temperature distribution and thermal cycling associated with the fabrication of continuous carbon fibre-reinforced PEEK (APC2) laminates by tape laying with a laser heat source (Grove, 1988).

Also, in the late 1980s, Loos conducted experimental investigations and developed a model to quantify the extent to which intimate contact and bond formation can occur at successive ply interfaces in thermoplastic tows. This model comprised submodels that defined the degree of nonuniformity of tow heights across the width of a prepreg. In addition, a viscoelastic mechanics model was developed to simulate the response of tows to a compressive load (Loos and Dara, 1987).

Later in the 1990s the role of consolidation and squeeze flow was recognised and modelled by Ranganathan et al. (1995). This consolidation modelling work was crucial to address microvoids that were noticeable in the final part fabricated by ATP. It was also found through nonlocal void filling modelling approach that initial void content in the tape played a major role on the void content after consolidation (Simacek et al., 2013). Other aspects such as fibre waviness due to overlaps between tapes being placed adjacent to each other and resin-rich regions due to gaps between tapes when placed next to each other were being formed and process modelling in later years tried to address this issue as well. This modelling work is still ongoing today addressing the issue of adhesion between layers, void content in the final part, resin-rich regions, and fibre waviness (Simacek et al., 2022).

4. Conclusions

The inception of the modelling work in manufacturing of short fibre composite processes was logically a continuation of the work done in polymer processing, where the presence of short fibres influenced the rheology of the fibre suspension, and fibres in the suspension changed their orientation in the flowing suspension. The final orientation of the fibres induced by the flow was important as it influenced the physical and mechanical properties of the composite. The continuous fibre composites process modelling benefited from other research areas such as textile draping in the context of clothing, sheet metal forming, and the mechanics of porous media for wet impregnation processes.

The field of polymer composite process modelling has undergone a significant evolution, progressing from simple, empirical approaches to more sophisticated methods which address more detailed phenomena that have been experimentally observed and have a general consensus that they play a vital role in the manufacturing processes. The initial research on continuous fibre based composite manufacturing processes focused on modelling of the thermokinetic and rheokinetic coupling of thermoset resins used in such processes. Subsequently, the influence of fibres was also incorporated by accounting for stresses experienced by the fibres during the process. The dynamics of viscous behaviour of the resin with the elastic behaviour of the fibres was coupled to estimate pressures, forces, and flow rates through process models to evaluate the role of various process, material, and geometric parameters.

Process modelling as it evolved over the years has also contributed to development of new processes and new materials over the years. For example SCRIMP process which evolved from VARTM as process modelling showed how one could reduce the fill time dramatically with high permeability flow enhancement additions to the process. Another example is how injection pultrusion process evolved due to calculation of resin pressure by modelling resin flow in a porous domain which eliminated the use of a resin bath in the pultrusion process.

Process modelling has also helped development of new materials. For example, from process modelling, it was clear that low viscosity of resin is preferred to ensure good impregnation and if the viscosity is high then the resin should travel only short distances. This led to the development of powder-impregnated prepregs and commingled carbon or glass fibres with thermoplastic resin fibres so thermoplastic resin did not have to travel large distances. When Out-ofAutoclave processes were discovered, new prepregs with air pathways had to be developed to evacuate the air from the prepregs before consolidation to prevent voids as only one atmospheric pressure was available to manufacture them and if there was air present, 1 atmosphere was not sufficient to compress them.

The understanding of the fibre orientation flow physics played a key role in the development of High Performance-Discontinuous Fibre Method (HiPerDiF) at Bristol and Tailorable Universal Feedstock for Forming (TuFF) at University of Delaware in which all the short fibres were aligned in one direction during the formation of the preform which could be then converted into a prepreg with any resin and used for stamp forming or automated tow placement processes which can pave the path for rapid manufacturing with composites.

Material characterisation has gone hand in hand with process modelling as one needs material parameters such as viscosity and cure kinetic parameters of the resin and compressibility, conformability, volume fraction, and permeability of the fibres for any process model. Also as we try to build in more and more observed phenomena into the model, more parameters need to be characterised. This usually has been the bottleneck on the use of process models by industrial engineers as values of these parameters are usually not available and would require a lot of effort and time to characterise them.

There may be a strong inclination to consistently use the same raw materials to minimise characterisation costs and development times. However, this propensity is contrary to the fundamental principle of composites, which is to provide a tailored solution for each individual component. However with the emergence of digital twins and machine learning tools, process modelling is bound to play a key role in the advancement and automation of composite manufacturing processes as less material parameter characterisation will be necessary because process models and sensors will provide the needed online feedback for continuous improvement of the process.

References

Adams, K.L., Rebenfeld, L., 1987. In-plane flow of fluids in fabrics: structure/flow characterization. Textil. Res. J. 57 (11), 647–654.

Advani, S.G., Tucker, I.I.I.,C.L., 1987. The use of tensors to describe and predict fibre orientation in short fibre composites. J. Rheol. 31 (8), 751–784.

Advani, S.G., Tucker III, C.L., 1990. Closure approximations for three-dimensional structure tensors. J. Rheol. 34 (3), 367–386.

Arafath, A.R.A., Fernlund, G., Poursartip, A., 2009. Gas transport in prepregs: model and permeability experiments. In: Proceedings of International Conference on Composite Materials, p. 19.

Arbter, R., Béraud, J.M., Binetruy, C., Bizet, L., Bréard, J., Comas-Cardona, S., Demaria, C., Endruweit, A., Ermanni, P., Gommer, F., Hasanovic, S., 2011. Experimental determination of the permeability of textiles: a benchmark exercise. Compos. Appl. Sci. Manuf. 42 (9), 1157–1168.

Åström, B.T., 1992. Development and application of a process model for thermoplastic pultrusion. Compos. Manuf. 3 (3), 189–197.

Barone, M.R., Caulk, D.A., 1981. Optimal thermal design of compression moulds for chopped-fibre composites. Polym. Eng. Sci. 21 (17), 1139–1148.

Barone, M.R., Osswald, T.A., 1987. Boundary integral equations for analyzing the flow of a chopped fibre reinforced polymer compound in compression moulding. J. Non-Newtonian Fluid Mech. 26 (2), 185–206.

Bergsma, O.K., Huisman, J., 1988. Deep drawing of fabric reinforced thermoplastics. In: Computer Aided Design in Composite Material Technology.

Bickerton, S., Šimáček, P., Guglielmi, S.E., Advani, S.G., 1997. Investigation of draping and its effects on the mould filling process during manufacturing of a compound curved composite part. Compos. Appl. Sci. Manuf. 28 (9–10), 801–816.

Bickerton, S., Advani, S.G., 1997. Experimental investigation and flow visualizationoftheresin-transfermould-fillingprocessinanon-planargeometry. Compos. Sci. Technol. 57 (1), 23–33.

Bickerton, S., Stadtfeld, H.C., Steiner, K.V., Advani, S.G., 2001. Design and application of actively controlled injection schemes for resin-transfer moulding. Compos. Sci. Technol. 61 (11), 1625–1637.

Bird, R.B., Armstrong, R.C., Hassager, O., 1987. Dynamics of Polymeric Liquids. In: Fluid Mechanics, 1.

Bodaghi, M., Simacek, P., Correia, N., Advani, S.G., 2020. Experimental parametric study of flow-induced fibre washout during high-injection-pressure resin transfer moulding. Polym. Compos. 41 (3), 1053–1065.

Boisse, P., Gelin, J.C., Sabhi, H., 1993. Experimental study and finite element simulation of glass fibre fabric shaping process. ASME Mater. Div. Publ. 46, 587–608.

Bruschke, M., Advani, S.G., 1989. Flow analysis in resin transfer moulding: experiments and predictions. SPE Tech. Pap. 35, 1769.

Bruschke, M.V., Advani, S.G., 1990. A finite element/control volume approach to mould filling in anisotropic porous media. Polym. Compos. 11 (6), 398–405.

Calius, E.P., Springer, G.S., 1990. A model of filament-wound thin cylinders. Int. J. Solid Struct. 26 (3), 271–297.

Cogswell, F.N., 1991. Flow processes in composite materials' 91: University of Limerick, Ireland, 4–5 July 1991. Composites 22 (5), 402–404.

Carrino, L., Polini, W., Sorrentino, L., 2004. Method to evaluate winding trajectories in robotized filament winding. J. Compos. Mater. 38 (1), 41–56.

Castro, J.M., Lee, C.C., 1987. Thermal and cure analysis in sheet moulding compound compression moulds. Polym. Eng. Sci. 27 (3), 218224. https://www.ccm.udel.edu/research/program-highlights/tuff/.

Cender, T.A., Simacek, P., Advani, S.G., 2013. Resin film impregnation in fabric prepregs with dual length scale permeability. Compos. Appl. Sci. Manuf. 53, 118–128.

Cender, T.A., Simacek, P., Davis, S., Advani, S.G., 2016. Gas evacuation from partially saturated woven fibre laminates. Transport Porous Media 115, 541–562.

Collier, J.R., Collier, B.J., O'Toole, G., Sargand, S.M., 1991. Drape prediction by means of finite-element analysis. J. Textil. Inst. 82 (1), 96–107.

Comas-Cardona, S., Groenenboom, P., Binetruy, C., Krawczak, P., 2005a. A generic mixed FE-SPH method to address hydro-mechanical coupling in liquid composite moulding processes. Compos. Appl. Sci. Manuf. 36 (7), 1004–1010.

Comas-Cardona, S., Groenenboom, P.H., Binétruy, C., Krawczak, P., 2005b. Simulation of liquid composite moulding processes using a generic mixed FESPH method. Revue Européenne des Eléments 14 (6–7), 867–883.

Correia, N.C., Robitaille, F., Long, A.C., Rudd, C.D., Šimáček, P., Advani, S.G., 2005. Analysis of the vacuum infusion moulding process: I. Analytical formulation. Compos. Appl. Sci. Manuf. 36 (12), 1645–1656.

Dave, R., Kardos, J.L., Dudukovic, M.P., 1986, October. Process modelling of thermosetting matrix composites: a guide for autoclave cure cycle selection. In: Proceedings of the American Society for Composites, 1st Technical Conference, pp. 137–153.

De Luca, P.L.A.P.P., Lefébure, P., Pickett, A.K., 1998. Numerical and experimental investigation of some press forming parameters of two fibre reinforced thermoplastics: APC2-AS4 and PEI-CETEX. Compos. Appl. Sci. Manuf. 29 (1–2), 101–110.

Folgar, F., Tucker, I.I.I.,C.L., 1984. Orientation behaviour of fibres in concentrated suspensions. J. Reinforc. Plast. Compos. 3 (2), 98–119.

Givler, R.C., Crochet, M.J., Pipes, R.B., 1983. Numerical prediction of fibre orientation in dilute suspensions. J. Compos. Mater. 17 (4), 330–343.

Gorthala, R., Roux, J.A., Vaughan, J.G., 1994. A model to predict resin pressure/back flow in the tapered inlet of a pultrusion die. J. Reinforc. Plast. Compos. 13 (6), 484–497.

Grove, S.M., 1988. Thermal modelling of tape laying with continuous carbon fibre-reinforced thermoplastic. Composites 19 (5), 367–375.

Gutowski, T.G., 1985. Resin flow/fibre deformation model for composites. SAMPE Q. 16 (4).

Gutowski, T.G., 1988. The manufacturing science of composites. Proc. Manufacturing International. IV.

Gutowski, T.G., Bonhomme, L., 1988. The mechanics of prepreg conformance. J. Compos. Mater. 22 (3), 204–223.

Halpin, J.C., Kardos, J.L., Dudukovic, M.P., 1983. Processing science: an approach for prepreg composite systems. Pure Appl. Chem. 55 (5), 893–906.

Han, C.D., Lee, D.S., Chin, H.B., 1986. Development of a mathematical model for the pultrusion process. Polym. Eng. Sci. 26 (6), 393404.

Haanappel, S.P., tenThije, R.H., Sachs, U., Rietman, B., andAkkerman, R., 2014. Formability analyses of uni-directional and textile reinforced thermoplastics. Compos. Appl. Sci. Manuf. 56, 80–92.

Hautefeuille, A., Comas-Cardona, S., Binetruy, C., 2020. Consolidation and compression of deformable impregnated fibrous reinforcements: experimental study and modelling of flow-induced deformations. Compos. Appl. Sci. Manuf. 131, 105768.

Hieber, C.A., Shen, S.F., 1995. A finite-element/finite-difference simulation of the injection moulding filling process. J. Non-Newtonian Fluid Mech. 56 (3), 361.

Hou, Y., Comas-Cardona, S., Binetruy, C., Drapier, S., 2013. Gastransportin fibrous media: application to in-plane permeability measurement using transient flow. J. Compos. Mater. 47 (18), 2237–2247.

Yu, H., Potter, K.D., 2013. Method and apparatus for aligning discontinuous fibres. In: UK patent, Patent application, (1306762.4).

Yu, H., Potter, K.D., Wisnom, M.R., 2014. A novel manufacturing method for aligned discontinuous fibre composites (high performance-discontinuous fibre method). Compos. Appl. Sci. Manuf. 65, 175–185.

Test Methods for the Experimental Characterization of in-plane Permeability of Fibrous Reinforcements for Liquid Composite Moulding, 441, 2023. ISO.

Jackson, W.C., Advani, S.G., Tucker, C.L., 1986. Predicting the orientation of short fibres in thin compression mouldings. J. Compos. Mater. 20 (6), 539–557.

Jeffery, G.B., 1922. The motion of ellipsoidal particles immersed in a viscous fluid. Proc. R. Soc. Lond. - Ser. A Contain. Pap. a Math. Phys. Character 102 (715), 161–179.

Kamal, M.R., 1974. Thermoset characterization for mouldability analysis. Polym. Eng. Sci. 14 (3), 231–239.

Kerr, R.C., 1972. SMC Compression moulding. Modern Plastics 49 (4), 68.

Kommu, S., Khomami, B., Kardos, J.L., 1998. Modelling of injected pultrusion processes: a numerical approach. Polym. Compos. 19 (4), 335–346.

Larock, J.A., Hahn, H.T., Evans, D.J., 1989. Pultrusion processes for thermoplastic composites. J. Thermoplast. Compos. Mater. 2 (3), 216–229.

Lee, L.J., 1981. Curing of compression moulded sheet moulding compound. Polym. Eng. Sci. 21 (8), 483–492.

Lee, C.C., Folgar, F., Tucker, C.L., 1984. Simulation of Compression Moulding for Fibre-Reinforced Thermosetting Polymers.

Lee, S.Y., Springer, G.S., 1990. Filament winding cylinders: I. Process model. J. Compos. Mater. 24 (12), 1270–1298.

Lee, L.J., Young, W.B., Lin, R.J., 1994. Mould filling and cure modelling of RTM and SRIM processes. Compos. Struct. 27 (1–2), 109–120.

Lee, J., Duhovic, M., Allen, T., May, D., Kelly, P., 2023. Computational modelling and analysis of transverse liquid composite moulding processes. Compos. Appl. Sci. Manuf. 167, 107433.

Lin, R., Lee, L.J., Liou, M., 1991. Non-isothermal mould filling and curing simulation in thin cavities with preplaced fibre mats. Int. Polym. Process. 6 (4), 356–369.

Long, A.C., Rudd, C.D., 1994. A simulation of reinforcement deformation during the production of preforms for liquid moulding processes. Proc. IME B J. Eng. Manuf. 208 (4), 269–278.

Loos, A.C., Dara, P.H., 1987. Processingofthermoplasticmatrixcomposites. In: Review of Progress in Quantitative Nondestructive Evaluation. Springer US, Boston, MA, pp. 1257–1265.

Loos, A.C., MacRae, J.D., 1996. A process simulation model for the manufacture of a blade-stiffened panel by the resin film infusion process. Compos. Sci. Technol. 56 (3), 273–289.
Loos, A.C., Springer, G.S., 1983. Curing of epoxy matrix composites. J. Compos. Mater. 17 (2), 135–169.
Luce, T.L., Advani, S.G., Howard, J.G., Parnas, R.S., 1995. Permeability characterization. Part 2: flow behaviour in multiple-layer preforms. Polym. Compos. US6391436B1.
Mack, C., Taylor, H.M., 1956. 39—the fitting of woven cloth to surfaces. Journal of the Textile institute Transactions 47 (9), T477–T488.
Mahé, F., Binetruy, C., Advani, S., Férec, J., Eck, B., 2023. A revisited compression flowmodel for concentrated suspensions with fibre-friction interactions. J. Non-Newtonian Fluid Mech. 319, 105090.
Maier, R.S., Rohaly, T.F., Advani, S.G., Fickie, K.D., 1996. A fast numerical method for isothermal resin transfer mould filling. Int. J. Numer. Methods Eng. 39 (8), 1405–1417.
Mallon, P.J., O'bradáigh, C.M., Pipes, R.B., 1989. Polymeric diaphragm forming of complex-curvature thermoplastic composite parts. Composites 20 (1), 48–56.
Mantell, S.C., Springer, G.S., 1992. Manufacturing process models for thermoplastic composites. J. Compos. Mater. 26 (16), 23482377.
May, D., Aktas, A., Advani, S.G., Berg, D.C., Endruweit, A., Fauster, E., Lomov, S.V., Long, A., Mitschang, P., Abaimov, S., Abliz, D., 2019. Inplane permeability characterization of engineering textiles based on radial flow experiments: a benchmark exercise. Compos. Appl. Sci. Manuf. 121, 100–114.
May, D., Advani, S.G., Duhovic, M., Endruweit, A., Fauster, E., George, A., Liotier, P.J., Lira, C., Lomov, S.V., Pucci, M.F., Schmeer, S., 2024. A new ISO standard for the experimental characterization of in-plane permeability of fibrousreinforcements. Compos. A Appl. Sci. Manuf., 108592
Menges, G., Wübken, G., 1973. In: 31st Annual Technical Conference of Society of Plastics Engineers (SPE ANTEC) Montreal, Canada, 19, pp. 519–522.
Menges, G., Wodicka, R., Hl, B., 1978. Non-geodesic Winding on a Surface of Revolution.
Middleman, S., 1977. Fundamentals of Polymer Processing. McGraw-Hill, New York, p. 468.
Mulye, P.D., Morançay, L., Binetruy, C., Comas-Cardona, S., Leygue, A., Guillon, D., 2023. Unified numerical process modelling of forming and consolidation for thermoplastic composites with prepreg patches. Front. Mater. 10, 1176482.
Munro, M., 1988. Review of manufacturing of fibre composite components by filament winding. Polym. Compos. 9 (5), 352–359.
Okine, R.K., 1989. Analysis of forming parts from advanced thermoplastic composite sheet materials. J. Thermoplast. Compos. Mater. 2 (1), 50–76.
O'Bradaigh, C.M., Pipes, R.B., 1991. Finite element analysis of composite sheet-forming process. Compos. Manuf. 2 (3–4), 161–170.
Parnas, R.S., Howard, J.G., Luce, T.L., Advani, S.G., 1995. Permeability characterization. Part 1: a proposed standard reference fabric for permeability. Polym. Compos. 16 (6), 429–445.
Patel, N., Rohatgi, V., Lee, L.J., 1995. Micro scale flow behaviour and void formation mechanism during impregnation through a unidirectional stitched fibreglass mat. Polym. Eng. Sci. 35 (10), 837–851.
Pickett, A.K., Queckbörner, T., De Luca, P., Haug, E., 1995. An explicit finite element solution for the forming prediction of continuous fibre-reinforced thermoplastic sheets. Compos. Manuf. 6 (3–4), 237–243.
Pidgainyi, Y.M., Morozova, V.M., Dudko, V.A., 1967. Method of calculating the geodesic winding characteristics of shells of revolution. Polym. Mech. 3 (6), 721–725.

Pillai, K.M., Advani, S.G., 1995. Numerical and analytical study to estimate the effect of two length scales upon the permeability of a fibrous porous medium. Transport Porous Media 21, 1–17.

Pittman, J.F., Wood, R.D., Alexander, J.M., Zienkiewicz, O.C., 1982. Proceedings Numiform, p. 82.

Player, J., Roylance, M., Zukas, W., Roylance, D.K., 2000. UTL consolidation and out-of-autoclave curing of thick composite structures. In: 32nd International SAMPE Technical Conference, pp. 757–767.

Potter, K.D., 1979. The influence of accurate stretch data for reinforcements on the production of complex structural mouldings: part 1. Deformation of aligned sheets and fabrics. Composites 10 (3), 161–167.

Price Jr, H.L., 1979. Curing and Flow of Thermosetting Resins for Composite Material Pultrusion. Old Dominion University.

Ranganathan, S., Advani, S.G., Lamontia, M.A., 1995. A non-isothermal process model for consolidation and void reduction during in-situ tow placement of thermoplastic composites. J. Compos. Mater. 29 (8), 10401062.

Richardson, S.M., Pearson, J.R.A. (Eds.), 1983. Computational Analysis of Polymer Processing. Applied Science Publishers, New York, pp. 138–177.

Robertson, R.E., Hsiue, E.S., Sickafus, E.N., Yeh, G.S.Y., 1981. Fibre rearrangements during the moulding of continuous fibre composites. I. Flat cloth to a hemisphere. Polym. Compos. 2 (3), 126–131.

Robertson, R.E., Hsiue, E.S., Yeh, G.S.Y., 1984. Continuous fibre rearrangements during the moulding of fibre composites. II. Flat cloth to a rounded cone. Polym. Compos. 5 (3), 191–197.

Rogers, T.G., 1989. Rheological characterization of anisotropic materials. Composites 20 (1), 21–27.

Rudd, C.D., Long, A.C., Kendall, K.N., Mangin, C., 1997. Liquid Moulding Technologies: Resin Transfer Moulding, Structural Reaction Injection Moulding and Related Processing Techniques. Elsevier.

Sandberg, M., Yuksel, O., Baran, I., Hattel, J.H., Spangenberg, J., 2021. Numerical and experimental analysis of resin-flow, heat-transfer, and cure in a resin-injection pultrusion process. Compos. Appl. Sci. Manuf. 143, 106231.

Sas, H.S., Šimáček, P., Advani, S.G., 2015. A methodology to reduce variability during vacuum infusion with optimized design of distribution media. Compos. Appl. Sci. Manuf. 78, 223–233.

Seemann, W.H., 1990. Plastic transfer moulding techniques for the production of fibre reinforced plastic structures. U.S. Patent 4 (902), 215.

Seemann, W., 1994. Unitary vacuum bag for forming fibre reinforced composite articles. U.S. Patent 5 (316), 462.

Seemann, W.H., Tunis III, G.C., Perrella, A.P., Haraldsson, R.K., Everitt, W.E., Pearson, E.A., 1998. U.S. Patent No. 5,721,034.

Šimáček, P. and Advani, S. G., 1998. https://sites.udel.edu/lims/.

Šimáček, P., Advani, S.G., 2004. Desirable features in mould filling simulations for liquid composite moulding processes. Polym. Compos. 25 (4), 355–367.

Simacek, P., Advani, S.G., 2016. "equivalent" permeability and flow in compliant porous media. Compos. Appl. Sci. Manuf. 80, 107–110.

Simacek, P., Advani, S.G., Iobst, S.A., 2008. Modelling flow in compression resin transfer moulding for manufacturing of complex lightweight high-performance automotive parts. J. Compos. Mater. 42 (23), 2523–2545.

Simacek, P., Advani, S.G., Gruber, M., Jensen, B., 2013. A non-local void filling model to describe its dynamics during processing thermoplastic composites. Compos. Appl. Sci. Manuf. 46, 154–165.

Simacek, P., Kermani, N.N., Gargitter, V., Advani, S.G., 2022. Role of resin percolation in gap filling mechanisms during the thin ply thermosetting automated tape placement process. Compos. Appl. Sci. Manuf. 152, 106677.

Soukane, S., Trochu, F., 2006. Application of the level set method to the simulation of resin transfer moulding. Compos. Sci. Technol. 66 (7–8), 1067–1080.

Springer, G.S., 1982. Resin flow during the cure of fibre reinforced composites. J. Compos. Mater. 16 (5), 400–410.

Syerko, E., Schmidt, T., May, D., Binetruy, C., Advani, S.G., Lomov, S., Silva, L., Abaimov, S., Aissa, N., Akhatov, I., Ali, M., 2023. Benchmark exercise on image-based permeability determination of engineering textiles: microscale predictions. Compos. Appl. Sci. Manuf. 167, 107397.

Tadmor, Z., Gogos, C.G., 2013. Principles of Polymer Processing. John Wiley Sons.

Tam, A.S., Gutowski, T.G., 1990. The kinematics for forming ideal aligned fibre composites into complex shapes. Compos. Manuf. 1 (4), 219228.

Tarnopol'skii, Y.M., Peters, S.T., Beil', A.I., 1998. Filament winding. In: Handbook of Composites. Springer US, Boston, MA, pp. 456–475.

Tavares, S.S., Michaud, V., Månson, J.A., 2009. Through thickness air permeability of prepregs during cure. Compos. Appl. Sci. Manuf. 40 (10), 1587–1596.

Terzopoulos, D., Fleischer, K., 1988. Deformable models. Vis. Comput. 4 (6), 306–331.

Tiffin, A.J., 1967. Moulding techniques for sheet moulding compounds. SPE J. 23 (2), 21.

Trochu, F., Gauvin, R., 1992. Limitations of a boundary-fitted finite difference method for the simulation of the resin transfer moulding process. J. Reinforc. Plast. Compos. 11 (7), 772–786.

Tseng, H.C., Chang, R.Y., Hsu, C.H., 2017. Numerical predictions of fibre orientation and mechanical properties for injection-moulded long-glass-fibre thermoplastic composites. Compos. Sci. Technol. 150, 181186.

Tucker, C.L., 1989. Fundamentals of Computer Modelling for Polymer Processing. Carl Hanser Verlag, Munich.

Tucker III, C.L., 2022. Fundamentals of Fibre Orientation: Description, Measurement and Prediction. Carl Hanser Verlag GmbH Co KG.

van West, B.P., Pipes, R.B., Keefe, M., 1990. A simulation of the draping of bidirectional fabrics over arbitrary surfaces. J. Text. Inst. 81 (4), 448–460.

Vernet, N., Ruiz, E., Advani, S., Alms, J.B., Aubert, M., Barburski, M., Barari, B., Beraud, J. M., Berg, D.C., Correia, N., Danzi, M., 2014. Experimental determination of the permeability of engineering textiles: benchmark II. Compos. Appl. Sci. Manuf. 61, 172–184.

Vincent, M., Agassant, J.F., 1986. Experimental study and calculations of short glass fibre orientation in a center gatedmoulded disc. Polym. Compos. 7 (2), 76–83.

Wakeman, M.D., Cain, T.A., Rudd, C.D., Brooks, R., Long, A.C., 1999. Compression moulding of glass and polypropylene composites for optimised macro-and micro-mechanical properties II. Glass-mat-reinforced thermoplastics. Compos. Sci. Technol. 59 (5), 709–726.

Wells, G.M., McAnulty, K.F., 1987. Computer Aided Filament Winding Using Non-geodesic Trajectories. ICCM-VI, pp. 1–161.

Xu, G.F., Repecka, L., Mortimer, S., Peake, S., Boyd, J., 2002. Manufacture of void-free laminates and use thereof. U.S. Patent US6391436B1.

Yarlagadda, S., Deitzel, J., Heider, D., Tierney, J., Gillespie Jr, J.W., 2019a. Tailorable universal feedstock for forming (TUFF): overview and performance. In: SAMPE 2019-Charlotte, NC, May 2019.

Yarlagadda, S., Advani, S., Deitzel, J., Heider, D., Molligan, D., Roseman, D., Simacek, P., Tierney, J., Gillespie Jr, J.W., 2019b. Formability of TuFF composite blanks. In: SAMPE 2019-Charlotte, NC, May 2019.

Yong, A.X., Aktas, A., May, D., Endruweit, A., Lomov, S.V., Advani, S., Hubert, P., Abaimov, S.G., Abliz, D., Akhatov, I., Ali, M.A., 2021. Experimental characterisation of textile compaction response: a benchmark exercise. Compos. Appl. Sci. Manuf. 142, 106243.

Yong, A.X.H., Aktas, A., May, D., Endruweit, A., Advani, S., Hubert, P., Abaimov, S.G., Abliz, D., Akhatov, I., Ali, M.A., Allen, T., 2021. Out-ofplane permeability measurement for reinforcement textiles: a benchmark exercise. Compos. Appl. Sci. Manuf. 148, 106480.

Zobeiry, N., Forghani, A., Li, C., Gordnian, K., Thorpe, R., Vaziri, R., Fernlund, G., Poursartip, A., 2016. Multiscale characterization and representation of composite materials during processing. Philos. Trans. R. Soc. A Math. Phys. Eng. Sci. 374 (2071), 20150278.

Section 3

Applications

The incredible, triumphant advance of carbon composites to airframe structures in Europe

3.1

Axel Herrmann [1], *Marc Fette* [2] *and Arnt Offringa* [3]
[1]Airbus Operation GmbH, Stade, Germany; [2]Composite Technology Centre/CTC GmbH - An Airbus Company, Stade, Germany; [3]GKN Aerospace, Hoogeveen, the Netherlands

1. Introduction

The development and use of carbon fibre reinforced plastics (CFRP) for primary structures in commercial aviation is a fascinating story of innovation, technical challenges, and advances in materials science. This chapter provides a chronological account of the key events, developments, and innovations in the use of carbon fibre reinforced composites in commercial aviation in Europe from the 1970s to the present day (Fig. 3.1.1).

2. 1970s: the first research and applications

The use of carbon composites in civil aviation in Europe began in the 1970s, albeit on a rather experimental and limited scale. These materials were new at the time and not

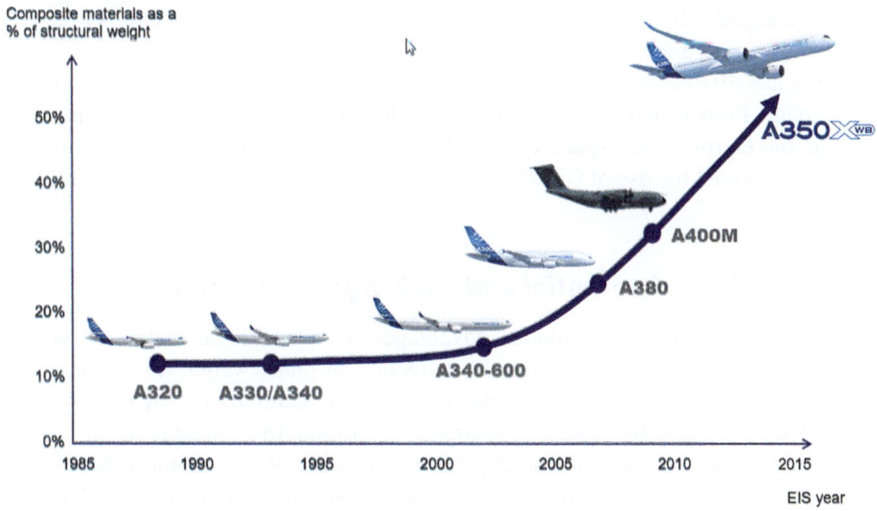

Figure 3.1.1 Portion of carbon composite materials used in Airbus aircrafts. (Source: Airbus).

The History of Composites. https://doi.org/10.1016/B978-0-443-34119-9.00018-8

yet widely used. Nevertheless, they laid the foundation for today's extensive use in modern aviation.

2.1 Background and developments

Carbon composite materials were created from a mixture of carbon fibres and resins (e.g., epoxy resins) as a matrix, resulting in lightweight but strong and rigid structures. These materials had already been used in the military and aerospace industries by the 1970s, due to their high strength, low density and excellent fatigue resistance. In the 1970s, technological advancements, with the focus on more efficient and powerful planes, characterised civil aviation in Europe. Large European manufacturers, such as Airbus, began to gain initial experience with the new materials, as they offered the potential to reduce weight and thus fuel consumption.

2.2 First applications

The first applications of fibre-reinforced composites in civil aviation involved smaller structural components that were less critical to the overall plane structure. Examples include flaps and covers for systems such as hydraulic or electrical components.

There were also many test programs in the 1970s. European aerospace companies, such as Airbus and BAE Systems, carried out tests with carbon composites on components fitted to planes of the time to analyse their performance in real-life conditions. In particular, the Airbus consortium, which was founded in 1970, investigated the possibilities of using these new materials. The Airbus consortium's first plane model, the Airbus A300, was introduced in 1972. These wide-body planes used aluminium as the main material for the structure, but the first CFRP components for nonstructural parts were also tested. This was a milestone, as Airbus already recognised at this time that CFRP could play a greater role in the future.

In the late 1970s, more intensive research programs began in Europe, particularly in Germany, France, and the UK, to further explore the potential of CFRP. Institutions such as the German Aerospace Center (DLR) and the French ONERA carried out research to extend the use of CFRP to larger structural parts such as wings or fuselage sections.

2.3 The view of potential and challenges at the time

As today, CFRP offered significant advantages over traditional materials such as aluminium. These materials promised a reduction in plane weight due to their low density, high strength and greater rigidity compared to aluminium, which led to lower fuel consumption and therefore cost savings in operation. In contrast to metal components, CFRP materials were not susceptible to corrosion, which should reduce maintenance costs. Fibre composites are also more resistant to fatigue caused by the cyclical stresses of plane operation. This extends maintenance intervals and reduces overall costs for the life of the plane, which was already recognised as potential at the time but needed to be investigated further.

However, there were also major challenges, as the production of CFRP materials was still very expensive in the 1970s, particularly due to the complex production and processing of carbon fibres. There was also a lack of knowledge, as engineers and manufacturing companies still had little experience of working with the new materials, which led to longer cycles of development. In addition, the inspection and quality assurance of CFRP components was complex, as cracks and other defects were often difficult to detect and the corresponding nondestructive testing methods were not yet sufficiently developed, if at all.

In the 1970s, the applications of CFRP in civil aviation were limited, but it was clear that the potential of these materials in the coming decades was considerable. This was to be confirmed in the 1980s and 1990s, when the use of carbon composites by plane manufacturers such as Airbus increased exponentially. The long-term goal of building lighter, more efficient planes remained a driving factor.

To summarise, the 1970s in Europe were a time of transition and the first trials of carbon composites in civil aviation. While the application was initially limited to smaller components, the research and testing carried out during this decade laid the foundations for the later widespread use of these advanced materials throughout the aviation industry.

3. The 1980s: the breakthrough in aviation

The first research steps in the field of CFRP in the commercial aviation sector were achieved in Europe in the 1980s. Airbus was one of the first companies to use carbon composite materials on a large scale. The Airbus A310, which entered service in 1983, used CFRP for components such as control surfaces. Carbon composites were a modern material that plays a key role, especially in the aerospace industry, due to its excellent properties in terms of strength and weight savings. In the Airbus A310, a twin-engine commercial plane developed in the 1980s based on the Airbus A300, the use of CFRP played an important role in improving structural efficiency and reducing overall weight.

3.1 The use of CFRP on the Airbus A310

The Airbus A310, which was launched in the 1980s as a further development of the Airbus A300, was a pioneer in the use of fibre-reinforced composites in modern aviation. The Airbus A310 was one of the first commercial planes to use CFRP materials on a large scale for primary and secondary structures. The main advantages of these materials include their lightness, strength, corrosion resistance and fatigue resistance, which helps to improve fuel efficiency and the longevity of plane components (Fig. 3.1.2).

The most important areas in which CFRP was used on the Airbus A310 were the tail components. CFRP was used for various components of the tail unit. One of the most important applications of fibre-reinforced composites in the Airbus A310

Figure 3.1.2 First flight of the Airbus A310 from Toulouse, France on April 3, 1982. (Source: Airbus).

is the vertical tail plane (VTP) or vertical stabiliser, which has been made of CFRP in every civil plane program at Airbus from that time until today. This enabled a considerable weight reduction by 22% compared to metallic alternatives without compromising on stability or strength.

Components such as ailerons, elevators, and landing flaps, which are used to control the plane and are crucial for the stability of the plane in operation, are also partly made of fibre composites. These parts are subject to high loads and must be particularly light and strong, which makes the use of CFRP sensible. The use of CFRP helped to reduce the weight of the moving parts, which not only lowered the overall weight but also increased the reaction speed and efficiency of these components.

Secondary structures such as various covers, access panels, and internal reinforcements were also made from fibre composites. These components are not directly load-bearing but receive help from the lightweight yet robust properties of composite materials.

The production of fibre composite components requires specialised processes to achieve the desired material properties. For the Airbus A310, the "prepreg" autoclave process with pre-impregnated carbon fibre materials, known as "prepregs", which have already been impregnated with a controlled amount of resin, was primarily used for structural components, such as the vertical stabiliser. These materials are inserted into a mould in several semi-finished product layers, whereby the fibres can be oriented in different directions to improve the strength as needed. The prepregs are then cured in an autoclave under heat and pressure, which hardens the resin and creates a rigid, stable structure.

3.2 Further development of CFRP applications for the Airbus A320

The Airbus A320, which was launched in 1988, used CFRP for other structural components such as ailerons. The Airbus A320, one of the most successful and best-selling commercial planes in the world, first entered service in 1988 and marked a significant advance in aviation technology. The use of CFRP played a key role in the development of the Airbus A320, with Airbus increasingly focussing on the use of composite materials from this model onwards. CFRP was a key part in reducing weight, increasing strength and improving overall performance.

The Airbus A320 uses CFRP in several key plane parts, especially in areas where weight savings are critical and high strength is needed, such as the vertical stabiliser. The vertical stabiliser of the A320 is made of CFRP. This part handles the directional stability of the plane and must withstand high aerodynamic forces. The use of CFRP makes it possible to create a lighter, yet rigid and resilient structure (Fig. 3.1.3).

Carbon composites are also used for control surfaces, such as the ailerons, elevators, and landing flaps. These components must be particularly light, as they are movable and must be able to react quickly to control commands. CFRP makes this possible thanks to its high strength and rigidity combined with low weight. The Airbus A320 also features secondary structural components, such as side members, which are made of CFRP. These contribute to the overall structure and help to further reduce the weight of the plane. Secondary structures such as various covers, flaps and access panels are made of fibre-reinforced composites. These parts are less stressed but benefit from the weight and corrosion advantages of the materials. The use of CFRP in the A320 resulted in significant weight saving compared to the traditional use of aluminium alloys. This led to a reduction in fuel consumption, which is particularly important for the economic operation of the planes and the reduction of emissions. With the Airbus A320neo, the wingtips, known as sharklets, have also been made from CFRP in recent years (Fig. 3.1.4).

The prepreg autoclave process is one of the most frequently used manufacturing processes for the CFRP components and structures of the Airbus A320. Automated prepreg lay-up is now carried out for large structures like wing-covers and fuselage

Figure 3.1.3 Airbus A320 vertical tail plane made of CFRP in Airbus plant in Stade. (Source: C. Brinkmann, Airbus).

Figure 3.1.4 Airbus A321XLR in the final assembly line (FAL) at the Airbus plant in Hamburg.
(Source: Airbus).

shells. Other structural parts like the VTP-and HTP-shells, and Section 19 of the rear fuselage, are using the Automated Tape Laying (ATL) process.

In addition, the Resin Transfer Moulding (RTM) process has been used since the mid-2000s for smaller components, such as the side guide vans as load introductions into the vertical stabilisers. In the RTM process, dry and usually pre-formed semi-finished fibre products are placed in a closed mould and the thermosetting resin is then injected into the mould. The resin penetrates the semi-finished fibre product before it is cured under pressure and heat. This process enables precise control of the resin quantity and distribution, which results in high part quality, higher part complexity and faster processing times.

The Airbus A320 uses fibre composites in many areas of primary and secondary structure. The biggest challenges included developing new manufacturing processes and ensuring the long-term reliability of the material under operating conditions. The processing of CFRP required special manufacturing techniques, such as autoclave processes, which were complex and expensive.

3.3 The use of carbon composites for regional planes at Fokker

Fokker was one of the first to use carbon composites in regional planes, and it was a pioneer in this field. Fokker has been at the forefront of using composites in planes since the 1980s, with the legendary Fokker 50 and Fokker 100 regional planes leading the way. And from the early 1990s, the Fokker 70 joined the party. The use of composites was a notable change. It made it possible to reduce the weight of the plane, increase structural strength and reduce operating costs. These incredible innovations had a massive impact on the aviation industry, helping Fokker to establish itself as a major player in the regional plane market. Fokker was one of the first plane manufacturers to use fibre composites in various load-bearing components, including the flaps and rudder. These components are crucial for the stability and control of the plane. The use of CFRP enabled a significant reduction in weight while improving structural integrity. These parts are highly loaded and need to be lightweight to enable smooth control and fast manoeuvres. The ailerons of the Fokker planes were also made of CFRP. Ailerons are critical to the lateral control of the plane, and the low weight of fibre composites allowed for better controllability and efficiency.

The prepreg autoclave process was one of the most popular manufacturing processes at Fokker. Another great process for manufacturing fibre composites is the vacuum infusion process. The dry semi-finished fibre products and layers are placed in a mould, which is then placed under vacuum. The resin is injected into the mould, evenly impregnating the semi-finished fibre products. The vacuum is the key to success here. It ensures that the resin penetrates all areas and prevents any air pockets from forming. The vacuum infusion process was the perfect choice for larger components such as wing covers and panelling made of GRP, where precise resin distribution and high strength are needed. The RTM process was also used on occasion for some complex, smaller components.

The advantages of using composites in Fokker planes are numerous. They include significant weight reduction, increased corrosion resistance, higher fatigue strength and improved maintainability. These benefits were a huge boon to Fokker planes, making them more economical and increasing their market acceptance. Fokker used some of the most advanced manufacturing processes out there, including prepreg, RTM, and vacuum infusion, to produce fibre composite components in a more efficient way. These processes ensured that the planes were lightweight and durable, giving them a huge advantage in the highly competitive regional plane market.

In the early 1990s, Fokker started the development of thermoplastic composite technologies and applications. Initial applications were glass and carbon fibre reinforced PEI stamp-formed brackets and covers for the Fokker 100, as well as carbon/PEI outer wing panels for the Fokker 50. A breakthrough was the development of innovative glass fibre reinforced thermoplastic cargo floor panels, with thermofolded edges for increased durability in use. These became standard on all Fokker 70 and Fokker 100 planes, and a modified version of these floors formed the basis of the Airbus Beluga floors (Fig. 3.1.5).

Later, carbon fibre reinforced thermoplastic floors were developed for Gulfstream business jets, which are in production in large quantities to this day.

In 1997, a prototype thermoplastic landing gear door was built and flown successfully for several years on an NLM Fokker 50. It gained important in-service experience for welded thermoplastic composite structures. This experience formed the basis of welded thermoplastic wing fixed leading edges of the Airbus A340-500/600 and a few years later also for the A380. This innovative Fokker designed and built product was a breakthrough for Airbus, being a large, welded structure. Weight reduction was a driver, and the product is a 'post-buckled' design, meaning that the skin is allowed to buckle when the wing bends upward in flight. The A340 application was for four products and on the later A380 this number was increased four-fold to 16 large assemblies per plane. Resistance welding is used throughout for assembly. Smaller parts such as ribs and stringers are stamp-formed (Figs. 3.1.6 and 3.1.7).

Figure 3.1.5 Airbus Beluga with thermoplastic floor, featuring thermo-folding of panel and panel edges, as well as ultrasonic welding of thermoplastic inserts.
(Picture source GKN Fokker).

Today, GKN Fokker manufactures large numbers of induction welded rudders and elevators for Gulfstream and Dassault business jets. The basic induction welding technology is on licence from KVE Composites Group and was merged with GKN Fokker's weld tooling and automation ability (Fig. 3.1.8).

Figure 3.1.6 Experimental welded thermoplastic landing gear door Fokker 50 (Willem van Dreumel, then Ten Cate's CTO, on the left).
(picture source GKN Fokker).

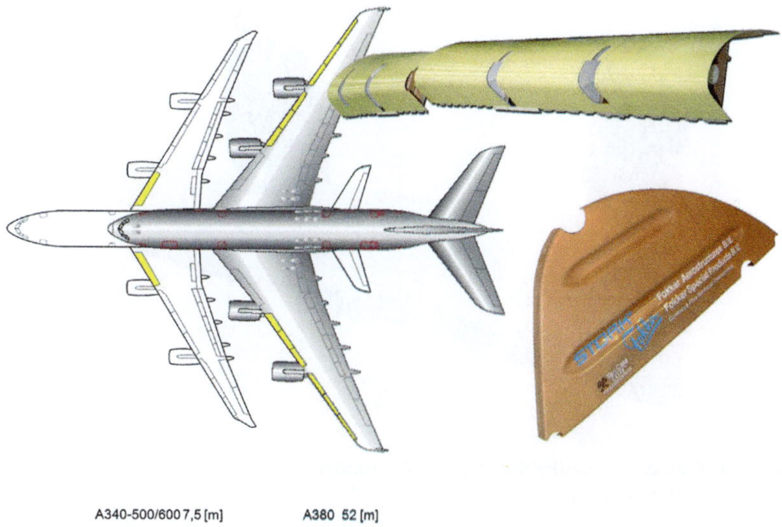

Figure 3.1.7 Welded glass/PPS thermoplastic composite wing fixed leading edges (so-called 'J-noses') on A340-500/600 and A380, with stamp-formed ribs (right bottom). (Picture source GKN Fokker).

4. The 1990s: the expansion of the usage of CFRP

In the 1990s, Airbus increased its use of CFRP in civil aviation, especially in the Airbus A340 and Airbus A330 long-haul planes. Here, carbon composites were used not only for control surfaces, but also for primary structural elements such as the rear pressure bulkheads and wing panelling. Other European plane manufacturers such as Fokker and Dornier are also increasingly focussing on CFRP in structural applications. The 1990s saw significant advances in CFRP materials and manufacturing processes, including improved resin systems and automated manufacturing processes. This led to higher production rates and lower costs, which increased the cost-effectiveness of CFRP parts.

4.1 The use of CFRP on the Airbus A340 and Airbus A330

The Airbus A330 and Airbus A340 are long-haul planes that were developed in the 1990s and share many similarities in their design, particularly on the use of fibre composites. These materials were used to reduce plane weight, increase structural strength and reduce maintenance costs. As Airbus built on its experience with earlier models such as the Airbus A310 and Airbus A320, the proportion of composites in both plane models was significantly increased.

In both the A330 and the A340, the vertical stabiliser is made of carbon fibre-reinforced plastic. This part handles directional stability and control and must withstand high aerodynamic forces. The control surfaces are also crucial for flight control

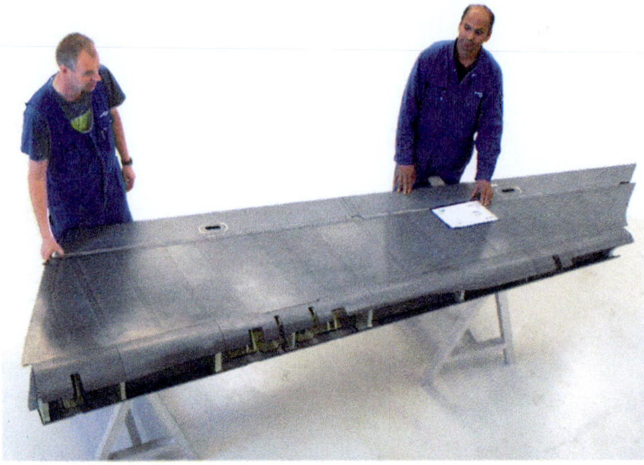

Figure 3.1.8 Welded carbon/PPS thermoplastic rudder.
(picture source GKN Fokker).

and are made of CFRP. They must be lightweight to ensure a fast response to control commands and at the same time sufficiently robust to withstand aerodynamic loads. In addition, landing flaps are made of CFRP to minimise weight and maximise aerodynamic efficiency. As these parts are moved often and need to be particularly light and resilient, CFRP offers the greatest properties.

In addition, the pressure bulkhead, which provides an airtight seal between the passenger area and the rear of the plane and is therefore a safety-critical part, was manufactured from CFRP for the first time on the Airbus A340. Some areas of the leading edges of the wings, particularly in the transitions to the engines, were made from composites. These parts must be able to withstand aerodynamic loads, but at the same time be light enough to improve the overall performance of the wings. The prepreg autoclave process was primarily used for the structural components of the Airbus A340 and is still used for the Airbus A330.

By using CFRP and other fibre composite materials, the Airbus A330 and Airbus A340 were made more economical, efficient and durable, which had a decisive influence on their success in international aviation (Fig. 3.1.9).

4.2 The use of CFRP at Dornier

The use of CFRP at the German company Dornier began more than 50 years ago with the development and manufacture of the first thermoset parts for the AEROS A/B satellite and the Dornier/Dassault "Alpha Jet" military trainer programs. In the following 2 decades, this was also extended to thermoplastic matrix materials for the Dornier Do 228 and 328 light commuter aircrafts.

From a technological point of view, Dornier plays a special role in the history of CFRP. Apart from North American Rockwell in the USA, Dornier was the only aerospace company in the world that had the knowledge to build its own textile machines

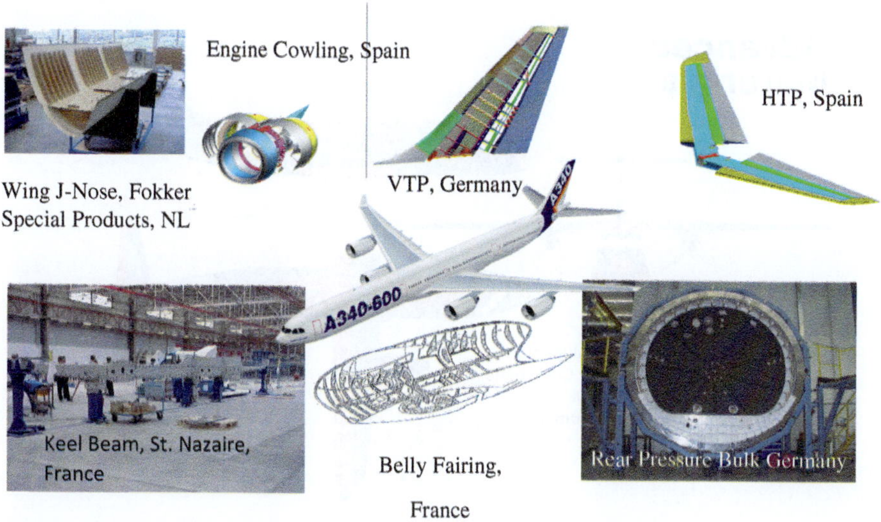

Figure 3.1.9 Major composite applications at the Airbus A330-600. (Source: Airbus).

for processing carbon fibre yarns into fabrics. By working closely with pioneers of glass and carbon fibre weaving in the USA, France, Germany and the Netherlands since 1970, Dornier was not only able to define and develop the technical specifications and layout of the FRP parts for its programs.—It also played a unique role in ensuring the immaculate quality of the dry fabrics needed for pre-impregnation, which were preferably produced on Dornier weaving machines. The latter developed into the world's leading production tools for carbon fibre fabrics and are today built by Lindauer DORNIER GmbH as a branch of Dornier, which is still family-owned even after the sale of the company to Mercedes-Benz in 1985. The Do 328 - still the most modern commercial planes in active service today - was conceived from the very beginning, in the late 1980s, as a showcase for Dornier's technological prowess and was far ahead of its time thanks to the use of highly developed fibre composite materials.

The Dornier Do 328, which took off on its maiden flight in 1991, was a marvel of engineering with movable control surfaces made of innovative fibre composite material, including carbon (CFRP) and glass fibre (GFRP) reinforced plastic.

These parts must be not only light to ensure the efficiency and manoeuvrability of the plane, but also able to withstand high loads. The structure of the pressure dome at the aft of the cabin was for the first time realised in aramid FRP (AFRP) (Fig. 3.1.10).

Dornier also relied heavily on the prepreg autoclave process for structural components made of CFRP; they enable the precise and efficient production of high-quality plane parts. This led to some amazing improvements! The Dornier Do 328 became more fuel-efficient, had lower operating costs, and a longer service life. As recent projects to stretch the fuselage and mount SAF capable or electric powerplants show the

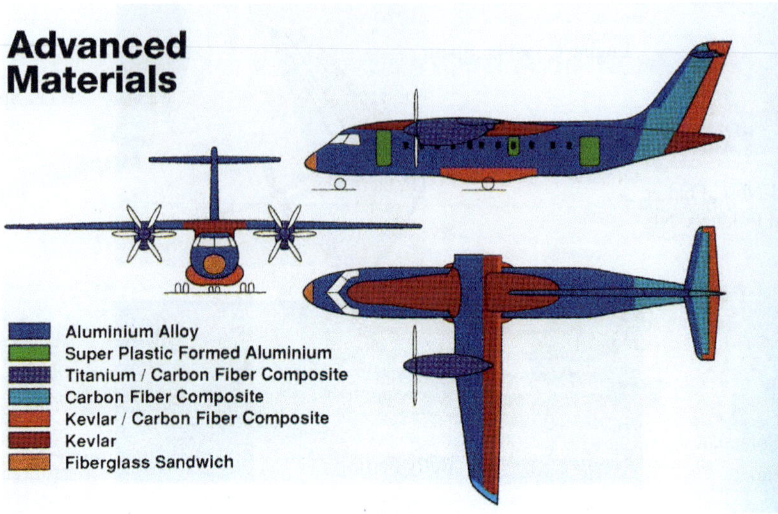

Figure 3.1.10 Composite parts used in the Dornier Do 328.
(Source: Airbus Corporate Heritage).

Do 328 is still today a competitive design for future developments in regional air transport.

5. The years 2000: the breakthrough in primary structures

The Airbus A380, which entered service in 2007, was the first wide-body plane to make extensive use of CFRP for primary structures. Parts of the fuselage structure, the wing boxes and the central wing box were made of carbon composites. This significantly reduced the overall weight of the plane and contributed to improved fuel efficiency.

The Airbus A380, the world's largest passenger plane, makes extensive use of fibre composites in its construction to save weight and maximise structural efficiency despite its size. The use of advanced materials such as CFRP, glass fibre reinforced plastics (GFRP) and other composite materials such as Glass Laminate Aluminium Reinforced Epoxy (GLARE) plays a key role in improving the aircraft's fuel efficiency, strength and durability. As the A380 is designed for long haul, high-capacity flights, reducing weight using lightweight but robust materials is particularly important (Fig. 3.1.11).

The Airbus A380 uses CFRP in many load-bearing wing components, particularly in the wing box and rib structures. The wing box, a critical part of the wing structure, is largely made of CFRP. Parts of the fuselage are also made of CFRP, particularly in the upper areas and the tapered tail sections. Horizontal stabilisers are also largely

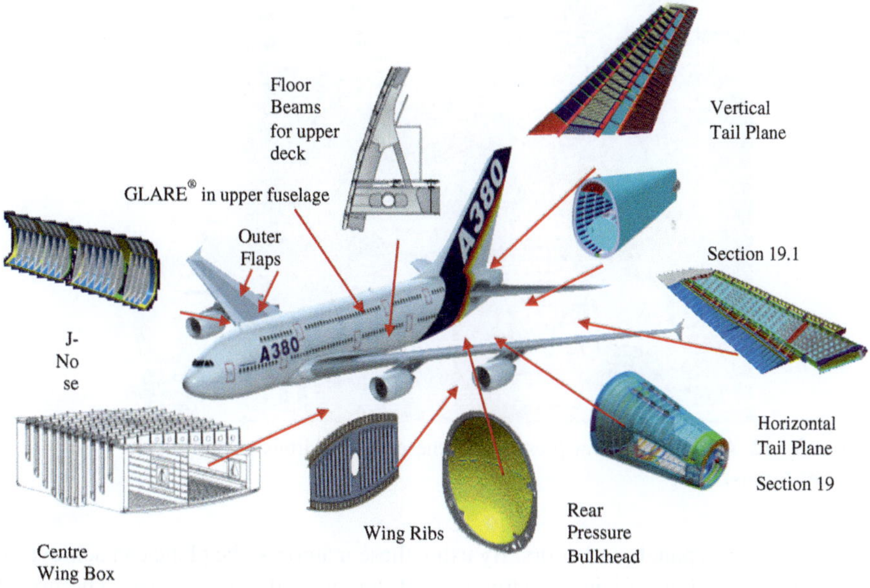

Figure 3.1.11 CFRP structures in the A 380.
(Source: Airbus).

made of CFRP. These components are particularly important for the control and stability of the plane. Thanks to the use of CFRP, these huge structures are lighter and yet extremely stable, which improves the handling of the plane and at the same time, reduces its weight. Furthermore, the control surfaces of the Airbus A380, such as the ailerons, elevators and rudders, are made of CFRP. These moving parts are particularly important for controlling the plane and benefit from the lightweight yet robust properties of fibre-reinforced composites.

The prepreg autoclave process was the dominant manufacturing process for primary CFRP structures in this plane. However, the geometrically complex CFRP rear pressure bulkhead was manufactured using the so-called film infusion process (Fig. 3.1.12).

As a speciality, the hybrid composite material GLARE was used by Airbus for the first time in the Airbus A380. It is mainly used in the upper fuselage segments of the A380, as these parts of the plane are subject to high compressive loads during flight. GLARE offers the advantage of improved fatigue resistance and better behaviour under compressive loads compared to pure aluminium. By using GLARE in these critical areas, Airbus was able to reduce the overall weight of the plane without jeopardising its structural integrity. The corrosion resistance and increased damage tolerance also help to extend the life of the plane and reduce maintenance costs. However, the production of these GLARE structures proved to be expensive.

The Airbus A380 uses fibre composite materials such as CFRP, GFRP and GLARE in many areas, from primary structures such as the wings and empennage to secondary

Figure 3.1.12 Airbus A380 rear pressure bulkhead in the Airbus plant in Stade. (Source: Airbus).

structures such as panels and doors. By using these materials, the plane can achieve an elevated level of fuel efficiency, strength and durability despite its size. The use of carbon composites in such large structures presented engineers with new challenges, particularly in terms of material fatigue, repair methods and inspection techniques. The long-term durability of the CFRP components had to be proven through extensive testing and far-reaching simulations.

6. The years 2010: the maturation and expansion of CFRP technologies

The Airbus A350 XWB, whose first commercial flight took place in 2015, marks a new milestone in the use of CFRP in aviation. Around 53% of the plane is made of carbon composites, including the fuselage and wings. Alongside the Boeing 787 Dreamliner, the Airbus A350 is one of the first commercial planes in which CFRP is used on such a large scale for primary structures. The Airbus A350 is one of the most modern and technologically advanced commercial planes in the world, relying extensively on the use of composites. These materials play a crucial role in the structure of the plane and contribute significantly to its high efficiency, lightness and strength. The A350 uses fibre composites in various areas of the plane, including primary structures, secondary structures and control surfaces. The extensive use of composites significantly reduces the aircraft's weight and fuel consumption without compromising on structural integrity.

In the years 2010, new CFRP processes such as RTM and out-of-autoclave (OoA) technologies were developed, further reducing production costs and increasing efficiency. The recyclability of CFRP also became a prominent issue as the material was increasingly used in aviation. The industrial scalability of production and the

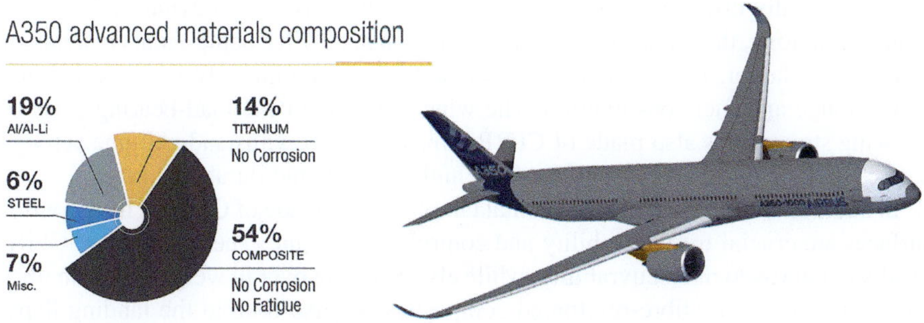

Figure 3.1.13 Usage of advanced materials in the Airbus A350. (Source: Airbus).

reduction of cycle times were major challenges, but these were successfully addressed through automation and advanced production lines (Fig. 3.1.13).

The primary structures of the A350, which bear the main loads, are largely made of CFRP. This is a truly innovative approach, as traditional commercial planes have so far mainly relied on aluminium alloys in these areas.

The fuselage of the Airbus A350 is largely made of CFRP, as are the complex keel beam and the centre wing box. This lightweight yet strong structure enables a significant weight reduction, which lowers fuel consumption. Using carbon fibre composites in the fuselage also provides increased resistance to corrosion and fatigue, reducing maintenance costs over the life of the plane. CFRP allows Airbus to keep the cabin pressure at a more comfortable level for passengers, increasing comfort on long-haul flights. The structural advantages of CFRP also make it possible to incorporate larger windows and a more spacious cabin design (Fig. 3.1.14).

The wings of the Airbus A350 also consist to a significant extent of CFRP. The use of fibre-reinforced composites makes it possible to make the wings more

Figure 3.1.14 Airbus A350 fuselage section assembly. (Source: Airbus).

aerodynamically efficient, as they can react more flexibly to aerodynamic loads. At around 32m long, the wing shells are among the largest CFRP components in aviation history. The larger, lighter, and more efficient wings significantly improve the aircraft's range and fuel consumption. The wing box, one of the load-bearing parts of the wing structure, is also made of CFRP. This area must withstand extreme aerodynamic forces, and the use of CFRP offers high strength and durability here.

In addition, the vertical and horizontal tail planes are made of CFRP. These control surfaces are crucial for the stability and control of the plane. The lightness of CFRP results in improved manoeuvrability, while also reducing overall weight. Carbon fibre composites and other fibre-reinforced composites are also used in the landing flaps, ailerons, and other movable control surfaces. These parts must be light and robust, as they are subject to aerodynamic loads when used for plane control and manoeuvrability (Fig. 3.1.15).

The tried and tested prepreg autoclave process is used primarily in the CFRP production of the primary structures of the Airbus A350. This is combined with the automated fibre placement (AFP) process, for example for fuselage and wing shells, with a high degree of automation. The RTM process is used for complex components such as window frames, which are being produced for the first time at Airbus from a tailored fibre placement (TFP) semi-finished product as a reinforcing semi-finished product. The complex door structures are also manufactured using the RTM process.

For the first time at Airbus, thermoplastic composites made of carbon fibre-reinforced organo-sheets with a matrix of polyphenylensulfide (PPS) were also implemented in a hot forming process for the so-called cleats and clips as structural connecting elements in the fuselage sections. For the first time at Airbus, production residues have been reused as a gusset filler in the stringers of the Airbus A350 wing shells for several years.

The use of advanced manufacturing processes and materials enables Airbus to produce high-quality, lightweight yet high-performance components. These processes make it possible to exploit the full potential of fibre-reinforced composites and make the Airbus A350 one of the most efficient and advanced commercial planes in the world.

Figure 3.1.15 The process of manufacturing the upper wing cover of the Airbus A350 at the Airbus plant in Stade.
(Source: Airbus).

7. The years 2020s: future developments and outlook—what is next?

Research and development are currently focused on developing carbon composite materials with enhanced properties such as increased damage tolerance, better environmental friendliness and higher temperature resistance. Overall, multi-physics simulations considering material, structure, function, manufacturing process and industrial system have an increasingly significant role in research and development. These advances are expected to further expand the use of CFRPs in new plane generations. Combining carbon composites with other advanced materials such as metal matrix composites and integrating CFRP into hybrid structures are topics of current interest. These developments aim to further increase the performance of planes.

Rate capability and production flexibility, while simultaneously reducing costs and increasing reproducibility, also play a decisive role in research and development of new materials, processes and construction methods. A holistic view of the product and the industrial system in the context of automation, digitalisation and ecological efficiency throughout the life cycle plays an increasingly significant role.

Highly efficient OoA technologies, such as RTM, vacuum infusion, and SMC compression moulding, are coming into focus. In this context, thermoplastic fibre-reinforced composites are also becoming increasingly important in aviation, as they offer many advantages over conventional thermoset fibre-reinforced composites. They are expected to lead to improved structural efficiency and functional integration through new joining options, such as welding. The recyclability of these materials is also promising. In view of the growing importance of sustainability, the development of recycling processes for CFRP is also being intensified. The recycling of CFRP waste and design for recyclability are important research priorities for reducing the ecological footprint of aviation.

Figure 3.1.16 Future aircraft concepts of Airbus: this is just a draft to give you an idea of what is planned.
(Source: Airbus).

Since the 1970s, CFRP has come a long way from being a new, poorly understood material to becoming a key technology in commercial aviation. Europe, in particular Airbus, has played a leading role in the research, development and application of carbon composites. With continuous innovation and overcoming technical challenges, CFRP has the potential to continue to shape the aviation industry in the decades to come (Fig. 3.1.16).

Further reading

A350 XWB Family. Online: https://www.airbus.com/en/products-services/commercial-aircraft/passenger-aircraft/a350-family, accessed 29.04.2025.

Airbus Commercial, 2024. Global Market Forecast - GMF 2024-2043. https://www.airbus.com/sites/g/files/jlcbta136/files/2024-07/GMF%202024-2043%20Presentation_4DTS.pdf. (Accessed 29 April 2025).

Breuer, U.P., 2016. Commercial Aircraft Composite Technology. Springer/Basel, Schweiz.

Edelmann, K., 2012. Automatisiert CFK-Thermoplast Fertigung für den A350 XWB. Lightweight Design 2, 42–47.

Fecker, A., 2014. Technik im Flugzeugbau: Von den fliegenden Kisten zur A350 XWB. MotorBuch Verlag, Stuttgart.

Fette, M., 2019. Die Airbus A350-XWB-Familie im Überblick. In: ATZ/MTZ Sonderausgabe. Springer, pp. 40–41. Nr. 24.

Fette, M., Herrmann, A.S., 2017. Zivile Luftfahrtindustrie im Wandel der Digitalisierung. In: ATZ/MTZ Sonderausgabe, 29. Springer, Nr, pp. 38–41.

International Civil Aviation Organisation (ICAO). Trends in Emissions that Affect Climate Change. Online: https://www.icao.int/environmental-protection/Pages/ClimateChange_Trends.aspx, accessed 29.04.2025.

International Air Transport Association (IATA) - IATA. Climate Change Factsheet. Online: https://www.iata.org/contentassets/d13875e9ed784f75bac90f000760e998/fact_sheet_on_climate_change.pdf, accessed 29.04.2025.

Klein, B., Gänsicke, T., 2019. Leichtbau-Konstruktion – Dimensionierung, Strukturen, Werkstoffe und Gestaltung. Springer, Vieweg/Wiesbaden.

Lengsfeld, H., et al., 2014. Faserverbundwerkstoffe: Prepregs und ihre Verarbeitung. Hanser, Fachbuch München.

Composite applications and technology at Boeing

3.2

Gwen M. Gross, John Keune, Patrick Enjuto, Holly Thomas, John Griffith and Tia Benson Tolle
The Boeing Co., Arlington, VA, United States

1. Introduction

The rise of the modern composite material and aerospace is integrally entwined with each other, and thus, Boeing's history of the use of composite materials begins almost from the day the company opened its doors. The fabric wings built in the old shipyard building where William Boeing started his company (Fig. 3.2.1) are a great example of how, right from the earliest days of the aerospace industry, the demanding design requirements to achieve flight were reliant on innovative material choices and thoughtful development. All images in this chapter are courtesy of the Boeing Co.

While these earliest structures used wood, canvas (cloth), and resin (glue) as a composite structure, today's modern Boeing aircraft use a large variety of carbon, glass, ceramic, and resin chemistries across commercial, defence, and space applications. As design challenges continued to evolve over the last 100 years, advances in materials, including composites, became a key enabler for the aerospace industry to achieve milestones that many thought intangible at the turn of the 20th century. The moon landing and International Space Station are obvious feats of wonder, but just the daily number of flights for civil transport and commerce would seem unreal to someone watching the first test flight of the B-1 Seaplane in December of 1919, Boeing's first civil transport plane (The Boeing Co, 2024b). Over its lifetime, this plane flew 350,000 miles (563,000 km), an early demonstration of the potential of composite materials. Thus, while modern composite applications may have started with military applications for performance, it was natural that they would then expand to commercial applications for life cycle cost reductions, which were enabled by composites (lighter weight structures, for example). In this way, the applications of a variety of materials that fall within the general family of "composites" by the military helped to reduce the risk for commercial applications.

2. Early history (pre-1950s—1960s)

2.1 *Materials and technologies—pre-1950s*

As mentioned previously, the earliest applications of composites at Boeing utilised mostly birch plywood for frames upon which canvas (cloth) was stretched and then

The History of Composites. https://doi.org/10.1016/B978-0-443-34119-9.00019-X

Figure 3.2.1 Cloth wing fabrication in the red barn. A former shipyard was the Boeing company's first home. Affectionately called the red barn, it is now part of the Museum of Flight in Seattle. In the red barn, engineers worked upstairs while builders and accountants worked downstairs.

saturated with a resinous system. This was known as doped fabric, and the dope was typically made of cellulose-based resins such as cellulose acetate or nitrate applied over the cotton canvas material. This made the aero structure waterproof while allowing the canvas to remain flexible and overall lightweight. This type of technology was common on components such as rudders and ailerons.

Later, during the Second World War, glass fibres were incorporated into resins to make new composite structures in support of the war effort. Phenolic resins such as Bakelite had demonstrated this utility previously with other fibres, and with the availability of cheap glass filaments in the late 1930s, there were a variety of resin chemistries and glass fibre combinations that were utilised. The most notable were epoxy resins, which showed an increased level of robustness and performance as compared to the phenolics, and polyester resins, which could be cured at room temperature, allowing for flexibility in the manufacture of glass fibre reinforced composites.

This ease of manufacturing a composite part, often called producibility within Boeing, was important given the high demand for aircraft during the Second World War. At this time, the advances in composite technology were being utilised exclusively for military applications. Like the cloth composite structures, these parts were made using a hand lay-up process where the resin was introduced into the reinforcement (glass fibre) at the point of part manufacture. However, unlike dope, resins could be introduced to the glass fibre reinforcement prior to being incorporated into the part structure. One could think of this as a version of aeroplane papier-mâché.

Also of note during this time was the use of just wood and resin to make a composite laminate. While development of this type of technology occurred prior to the

First World War for use on some aerostructures, it really was not until the development of the waterproof Duramold that usage took off significantly in response to the shortage of metal alloys during the Second World War. The large difference in this product from earlier versions was the use of phenolic resin to laminate the wood sheets rather than bio-derived albumin glue. The durability of this product is what led to industry adoption (Schlazberg, 1994).

It should also be noted that the first steps into modern honeycomb core laminates occurred at this time. The 1938 patent by Norman de Bruyne for a honeycomb-compatible adhesive overcame most of the technical challenges that previous trials for honeycomb core had encountered—how to keep large face sheets in intimate contact with the core structure. Just a few years later, Reinforced Plastics, which later became Hexcel, began working to produce laminated honeycomb materials specifically for use in aerospace (EconHP Holding GmbH, 2024).

2.2 Applications—pre-1950s

Boeing's application of these technologies was typical of other aeroplane manufacturers in the United States at that time. As mentioned, the use of composite technologies for wing, fuselage and control surface skins was very common with the doped canvas materials. Laminated woods were also common, with the most famous example by the H4 Hercules prototype, affectionately known as the Spruce Goose (The Boeing Co, 2024b). While not a Boeing aircraft at the time, this aeroplane captured the American public's attention and is the most referred to example of the wood composite laminates used.

An early example of doped fabric composites could be seen on the Douglas (a Boeing heritage company) biplane, the Cloudster (Fig. 3.2.2). This was the first Douglas product and the first aeroplane in history to airlift a useful load exceeding its own weight. The strut-braced wood, wire and doped-cloth biplane was powered by the First World War surplus 400-horsepower Liberty engine. The goal of development for the Cloudster was to achieve the first cross-continental flight, a feat that it failed to achieve in 1921. The plane did go on to become the flagship plane of one of the first scheduled passenger airlines, routinely carrying up to 12 passengers up and down the California coast. In 1926, the Cloudster was forced to make a landing on a beach during a chartered flight to Mexico, where it was caught in the high tide and destroyed (The Boeing Co, 2024b).

Other technologies developed during the Second World War influenced the application space of composite materials on Boeing aircraft. While radar was developed prior to the start of the war, initially, only limited aircraft were fitted with the technology due to its overall size and bulk. As the war effort continued and other advances came along, on-aeroplane radar became seen as a critical technology for the Allies in the war effort. The use of epoxy-glass fibre composites for military aeroplanes became an area of expanded application of composite technology due to the inherent radar transparency of the glass reinforcing fibres and inherently waterproof nature of the cured epoxy laminate.

Figure 3.2.2 The Cloudster on the ground after conversion to a passenger aeroplane capable of carrying 12 persons. An early example of a commercial aircraft utilising doped fabric composites.

2.3 Material and technology—1950s through 1960s

Coming out of the Second World War, most composite structures were utilising glass fibre-based materials. The corrosion resistance and lightweight nature of these materials made them very attractive in many industries. After the war, automotive and naval applications became the dominant industries. For this reason, there began to emerge more variety in the material forms available, the resin types, and even the glass types used for the fibre. Considerations such as the drape and tack of the material, often related to the weave of the glass, as well as the variety of differing filament diameters, and to a lesser extent, the filament glass type, all meant an expansion of the design space for which glass fibre composites could prove beneficial for structural design. This same expansion in variety and choice, however, meant that additional scrutiny was beginning to be placed on the consistency and repeatability of composite materials. Of note during this time is the use of phenolic-glass face sheets in aeroplane interiors coupled with Nomex honeycomb core for helping to meet some of the newly identified regulations around passenger safety.

As mentioned previously, honeycomb core and foam structures had caught attention for aerospace applications due to the lightweight and strong designs that could, in theory, be realised. A few different materials, including metal and paper, had been explored for honeycomb core, but the real breakthrough came in the mid-1960s with the production of Nomex-based honeycomb core (EconHP Holding GmbH, 2024). The inherent flame-resistant properties of the meta-aramid chemistry of this material made it attractive for more than just structural composite applications. It quickly found adoption for interior panels in commercial aircraft, where it remains a critical technology to this day. When integrated into a sandwich structure, the combination of honeycomb and face sheets functions as an I-beam to achieve bending stiffness at a significantly lower cost and weight than a solid structure. For this reason, when launched in 1969, the 747–100 was reported to have 1500 m^2 of honeycomb core in interior parts and 2000 m^2 for exterior parts. Interior applications include floor panels, sidewalls, ceilings, stowage bins, class dividers, bulkheads, lavatories and galleys.

During this same period, additional supporting fibres were also under development. For example, the graphitic material form was developed in 1956 by Union Carbide, but as discussed previously in this book, additional development was needed before the form of the modern carbon fibre we recognise today was fully commercialised (Gorss, 2003). It was not until 1969 that consistent carbon filaments were available for use in aerospace-grade materials. Note that when looking at historical papers from this time, it was more common to refer to these as graphite fibres than carbon. The earliest carbon fibres were typically produced from rayon or pitch. For aerospace applications in the United States, the higher tensile properties achieved with rayon fibres, which were commercially available in 1959, were the focus of development in the 1960s for graphite-based composites (Gorss, 2003). In later years, the US aligned with international efforts around the polyacrylonitrile (PAN) precursors for aerospace carbon fibre applications.

Boron fibres had found their way to market in 1966, a few years before carbon fibre, despite the demonstration of continuous filament boron fibre occurring later in 1959 (Schjelderup and Purdy, 1971). In these early offerings, it was initially thought that boron would become the primary basis for advanced composites. As noted by June and Lager, "*Boron has come to the fore primarily because of some early deficiencies of graphite composites, namely, low interlaminar shear and compressive strength caused by the low transverse strength of the fibre and the difficulty of achieving a good bond at the fibre matrix interface* ..." (June and Lager, 1973). So, while the potential tensile strength of carbon fibre was attractive, as compared to the compression strength of boron fibre, the inconsistent performance of carbon fibre composites limited the applications for the first years they were readily available. In time, the overall lack of flexibility and large diameter of boron fibres would limit their design space except in cases where the compression properties proved advantageous.

2.4 Applications—1950s through 1960s

Initial applications of "modern" composites, i.e., fibres in a resin matrix, for both defence and commercial aircraft, understandably started with secondary structure. As noted by Pinckney, "*The first, and probably most important* [limit on use of composites], *is a lack of historically accumulated service experience* ..." (Pinckney, 1964). Despite this, the application of glass fibre composites continued to expand through the 1950s and 1960s. For example, the Boeing B-52 Stratofortress bomber utilised glass fibre skins and spars for the vertical fin. These were both adhesively bonded and mechanically fastened to electrically isolate the upper portion of the fin from the lower portion (Pinckney, 1964). This aeroplane underwent its first flight test in 1952 after a rapid redesign when the US Air Force command directed Boeing to deliver a jet-powered bomber instead of the propeller-driven version originally proposed (The Boeing Co, 2024b). Another early structural application of glass fibre composites to improve performance compared to metal was the rocket case on the Minuteman missile system (Hackman, 1973).

Besides a shift from propeller to jet propulsion, rapid changes in composite materials also occurred in these decades. The defence industry began looking at other fibre reinforced materials beyond just glass fibre composite secondary structure (e.g., radome on the B-29 Superfortress, Fig. 3.2.3). As mentioned in the previous section, early focus was on boron fibres due to the extremely high specific strength and modulus as well as consistency of the available reinforcement. Multiple programs and structures made this initial foray into the advanced composite application space to start building that in-service experience with these new composite materials.

For example, McDonnell Douglas (a Boeing heritage company) substituted boron/epoxy face sheets for the original aluminium for the honeycomb sandwich structure on the F-4 Phantom rudder. Forty-five were flown in service beginning in 1969 (Hackman, 1973). This aeroplane was originally introduced in 1960. Similarly, North American Aviation (later a part of Rockwell International and then Boeing) made a substitution for the aluminium wing skins on the F-100 Super Saber. These were replaced with solid laminate boron/epoxy, where the inner surface of the thick composite laminate was machined to meet the existing interfaces and fastened to the existing metal substructure (Kiersarsky and Ong, 1970). The structure underwent a full-scale static test and flight test (Peterson, 1971).

Additional development and testing included the successful flight tests for both boron/epoxy (aft) and glass fibre/epoxy (forward) rotor blades on the CH-47. The flight test did not occur until 1972, but development was kicked off in 1966 (Salkind, 1976). A picture showing the typical blade configuration is shown in Fig. 3.2.4.

Blades were of particular interest for composite applications due to the difference in fatigue performance with time. The US Army Heavy Lift Helicopter (HLH) program investigated multiple composite blade options. Under this program, Boeing

Figure 3.2.3 The Boeing B-29 Superfortress, the most technologically advanced aeroplane produced during the Second World War, with multiple updates and derivatives made in the decades following the war.

Figure 3.2.4 Boeing Vertol CH-47 Chinook helicopter. Circa 1968.

designed a hybrid structure primarily consisting of glass fibre with local application of carbon for stiffening and tuning (Lewis and Gorum, 1976). Another aeroplane that did some early work with carbon fibre was the Douglas A-4 Skyhawk. This single-seat subsonic aircraft was originally manufactured starting in 1954. In the late 1960s, a full-scale experimental horizontal stabiliser with carbon/epoxy solid laminate skins bonded and fastened to carbon/epoxy sandwich shear webs was fabricated and tested (Salkind, 1976).

Boeing commercial products started including sizable amounts of glass fibre/epoxy, both laminate and sandwich construction, in the late 1950s and early 1960s. Structures included nose radomes, antenna enclosures, wing-to-body fairing panels, access hole close-out panels, control surfaces, and interior structures. To give an idea of scale, the 727 (first flight 1963), the 737 (first flight 1967), and the 747 (first flight 1969) had approximately 170, 280, and 930 m^2 (1800, 3000, and 10,000 ft^2) of composites, respectively (Bauman and Swedlow, 1981). One unique application of glass fibre composite was the solid laminate used for the leading-edge variable camber flap (VCF) on the 747. This solution to an aerodynamic high lift problem necessitated a flexible skin that could change shape between the retracted and deployed position while still meeting all the strength, durability and stiffness requirements (Cole, 1980). An interesting fact is that these composites have also found their way into the development tools used for new aeroplanes. In 1958, Boeing integrated foam-filled glass fibre composite blades in the Boeing transonic wind tunnel. These successfully replaced spruce blades that failed due to high temperatures, which in turn had replaced the original steel blades that failed due to impacts from failed test articles (Pinckney, 1964).

With the experience and success of boron and carbon reinforced plastics in defence applications, it was not long until the commercial programs of Boeing began to explore the same spaces. The first Boeing-produced composite part was a boron composite wing flap that was tested on a 707–320 Intercontinental starting in 1969 (The

Boeing Co, 2024c). It was reported that testing confirmed the in-service reliability of the new high-strength, low-weight, structural material, with design engineers reporting that the boron foreflap was 25% lighter than an aluminium foreflap (Fig. 3.2.5).

2.5 *Reaching forward*

It should be noted that toward the end of the 1960s, there was a large push to increase fundamental understanding of composite materials and how they behaved within the broader American aerospace community. The power behind this came from the Air Force Materials Laboratory. A personal accounting of this history by Stephen Tsai was published in 2005 (Tsai, 2005). The research at that time was focused on what was dubbed advanced composites, with a large focus on boron fibres, closely followed by carbon fibres. Not only was the work focused on the development of new materials, but also on the foundational science of how these materials functioned and how best to design for them. This initiative helped build the underlying understanding of composites within the industry that allowed for the expansion of their use in the decades to come.

3. Building confidence (1970s—1990s)

In the following decades, the application of composite materials, most notably carbon fibre and glass fibre based, expanded from small niche applications at the beginning of the 1970s through military primary structural parts in the 1980s and on to commercial aircraft adoption of composites for large-scale flight-critical parts by the end of the 1990s. This timeframe represents a huge period of growth, not just in the application of the materials but also in the maturation of material properties and designs that take advantage of said properties.

Figure 3.2.5 In 1969, Boeing began to test the first Boeing-built composite component, the boron composite wing fore flap, on a 707–320 Intercontinental.

3.1 Materials and technologies—1970s

One of the first shifts in composite material usage was away from boron fibres to carbon fibres for reinforcement. The larger diameter boron fibres created difficulty in manufacturing parts with higher curvature and proved to be more expensive in the long run (Schjelderup and Purdy, 1971). There was still a lot of work being done looking at different combinations of resins and reinforcements at this time, including the application of aramid fibres such as DuPont's Kevlar fibres. While the chemistry of Kevlar is aramid-based, such as Nomex, it did not reach commercialisation until the early 1970s. Glass fibre materials were still the most common composite forms for applications at this time.

In terms of resin development, epoxy was dominating the landscape by the 1970s. Modern epoxy resins were initially commercialised in the 1950s, but it was not until the 1970s that they achieved the performance and success that we associate with them today. The epoxy resins at this time were what are commonly referred to as first-generation resins—brittle, untoughened resins. However, these resins were often easier to work with in the production environment (less solvent as compared to polyester resins, for example), demonstrated long working life (as compared to isocyanate-based systems), and when cured, demonstrated good chemical and thermal resistance while being corrosion resistant. By the 1970s, a variety of backbone structures had been synthesised with the epoxide cross-linking group. Coupled with a variety of curing agent options, it was now possible for formulation chemists to more precisely tailor the desired performance attributes of the advanced composite materials. This extended beyond basic structural composite materials to include adhesives, paints and speciality coatings that can be used to protect metallic aerostructures.

Another critical change at this time was the form composite materials were procured in– a preimpregnated material form known as prepreg, where the resin and fibre were already present in a set ratio. The first prepreg machine was developed in the late 1960s to support NASA (Sloan, 2013). This material form was significantly easier for production shops to work with as compared to the wet lay-up processes. Some processes, such as resin transfer moulding and filament winding, still require the part producer to combine the resin and fibre reinforcement of interest in-house. This is still the case today, where most composite parts are produced utilising prepregs.

Prepregs at this time were produced by drawing the reinforcement through a bath of resin, typically solvated, and then sent through a drying oven prior to being rolled onto a core. This near continuous production process helped to industrialise prepreg materials to an extent where volumes and cost were attractive to other industries beyond aerospace, including automotive and construction. Examples of the composite systems from this period include Narmco's Rigidite 5208 (a legacy Syensqo product now known as Cycom 5208) as well as Hexcel's HexPly 3501–6 and 3502 systems (previously listed under Hercules). These resins are still available in the market as of the time of this publication and have been offered on a variety of reinforcements through the years, with carbon fibre being the most common.

The processing technology that came to prominence during this time, for Boeing applications as well as within the industry, was the consistent use of autoclave curing.

Autoclaves had been in use for years for things like rubber vulcanisation, but it was not until the addition of gas pressurisation post-Second World War that the composites industry started to see the benefits of this technology. As mentioned previously, many of the resins utilised at this time solvents. In addition, epoxy resins are inherently hydrophilic and will contain water even without other volatiles being present. The precise control of pressure and temperature in an autoclave, along with vacuum application, meant that it was now possible to produce composite laminates with minimal levels of porosity. This resulted in more predictable and controlled properties for the parts produced. This not only helped to build confidence in composites in general but also allowed for continued advancement of the work NASA, AFRL and others had undertaken in characterising and understanding the fundamentals of how these advanced composite materials behaved.

3.2 Applications—1970s

The 1970s saw the defence industry transition the use of composites to more and more primary structures, while the civil aircraft industry continued with secondary structure demonstrators of increasing complexity, as predicted by Schjelderup and Purdy (1971).

Initial demonstrator applications in the late 1960s led to McDonnell Douglas (a Boeing heritage company) selecting boron composites for the F-15 horizontal and vertical stabilisers that are still in production today (Fig. 3.2.6) (Peterson, 1971). These are boron/epoxy skins over aluminium honeycomb core sandwich structures, very similar to the F-4 rudders used to develop service experience. The F-15 also implemented composites for the speed brake in the form of carbon epoxy sandwich starting at the 60th production aircraft, and was the first carbon/epoxy component on a military aircraft (Bauman and Swedlow, 1981).

Figure 3.2.6 The F-15 "Eagle" painted for the United States bicentennial celebration in 1976. The aeroplane utilises boron-epoxy composite skins on the vertical and horizontal tail surfaces.

The Rockwell International (a Boeing heritage company) B-1 used composites for a variety of structures. In addition to the now almost common glass fibre composite applications, the B-1B aft equipment bay doors, weapon bay doors, flaps, and some fairings are all carbon/epoxy face sheets over aluminium honeycomb sandwich structure (Bauman and Swedlow, 1981). The dorsal longeron is a boron/epoxy laminate bonded to titanium for additional stiffness. Interestingly, composite vertical stabilisers were designed and went through full-scale testing for the B-1A but did not make it into production on the B-1B (Salkind, 1976; Bauman and Swedlow, 1981).

Beginning in the mid-1970s, the Naval Air Systems Command (NAVAIR) began procurement of two new aircraft from McDonnell Douglas that would have much greater usage of composites than anything previous. While both the F/A-18 (first flight 1978) and AV-8B Harrier (first flight 1981) have carbon/epoxy wings, the implementation is notably different. It should be noted that the first prototype of this aeroplane, the YAV-8B that flew in 1978, did not contain all of the composite technologies that were present on the final flight aircraft, which had a composite wing, fuselage and control surfaces, compromising 27% of the structural weight (Campbell, 2006). The AV-8B wing is continuous from tip to tip (~9m (30 ft)) and includes composite substructure (spars and ribs) in addition to the skins. Wings were produced utilising a fabrication technique known as drape forming. Fuselage and control surfaces were co-cured to unitise the structure for piece part count reduction (Campbell, 2004). Conversely, the F-18 wing is composite skins over a metallic torque box, and these skins are bonded to titanium fittings at the root for attachment to the fuselage. About 10% of the F-18 and 15% of the AV-8B by weight are specifically carbon epoxy (Weinberger et al., 1977).

The NASA Aircraft Energy Efficiency (ACEE) Composite Primary Aircraft Structures Program was the major driver behind expanded use of composites in commercial aircraft applications. It ran for 11 years from 1975 to 1986 and was referred to as "*the golden age of composites research in the United States*" (Dow, 1987). Boeing, McDonnell Douglas and Lockheed each had two of the six major technology development efforts (Vosteen, 1978).

McDonnell Douglas applications were a rudder and a vertical stabiliser, both on the DC-10 aircraft. The rudder is an all-laminate rib-stiffened two-spar carbon epoxy box structure cured in a single operation on expanding silicone rubber mandrels without using an autoclave. It was certified in 1976, and 13 rudders flew on the aircraft of six different airlines (Rawdon, 1983). The vertical stabiliser is a laminate rib-stiffened four-spar carbon epoxy box structure with honeycomb sandwich skins. Skins were mechanically fastened to the webs and spar substructure that was secondarily bonded together in a complex operation to save weight, given the constraints imposed by replacing an existing metal structure (Palmer et al., 1984). The structure was certified in 1986 after an extensive test program including full-scale static, fatigue and flight test articles, entering service in 1987 (Dow, 1987).

Prior to ACEE, NASA initiated a program with Boeing to investigate the effects of long-term ground and flight exposure on carbon/epoxy materials. In addition to panels placed at select locations around the world, Boeing certified an aluminium honeycomb core with carbon fibre epoxy facesheet 737 spoiler in 1973. A total of 114 composite spoilers were fabricated using one of three materials (two fibres, three resins) for the program. Of these, 108 flew on 27 aircraft with five major airlines. Monitored for 15 years, they accumulated more than 2.5M flight-hours and almost 3.5M landings (Coggeshall, 1989). Boeing ACEE applications were the 727 elevator and 737 horizontal stabiliser. The 727 elevator is a two-spar design that uses honeycomb skins mechanically fastened to honeycomb ribs and solid laminate spars. It was certified in 1979, and five shipsets (10 elevators) entered service with a single airline (Dow, 1987). The 737 horizontal stabiliser is also a two-spar design but uses I-section integrally stiffened skins mechanically fastened to honeycomb ribs and solid laminate spars (Vosteen, 1978). It was certified in 1982 and the five shipsets that entered revenue service in 1984 were the first commercial primary structure to do so (Dow, 1987).

3.3 Materials and technologies—1980s

During the 1980s, a large shift in the focus on materials development began to coalesce around making resin systems that were less prone to damage without sacrificing the strength and modulus of the composite systems. Multiple approaches to achieving this were explored by the different material suppliers, some of which are still utilised today, including rubber adduct epoxy resins, interpenetrating thermoplastic networks, as well as interlayer thermoplastic particle tougheners, which include a variety of particle types, including core shell rubbers (Mi et al., 2022). The reason behind this interest can be seen in the images shown below (Fig. 3.2.7). The top figure is typical of an early epoxy-carbon fibre composite laminate when impacted. The lower image, however, shows how the overall damage area and laminate structure changes are minimised with the application of a toughened epoxy resin system. The qualification and development of multiple toughened epoxy resin systems occurred in the 1980s and extended into the early 1990s for use in Boeing structural composite parts. For early

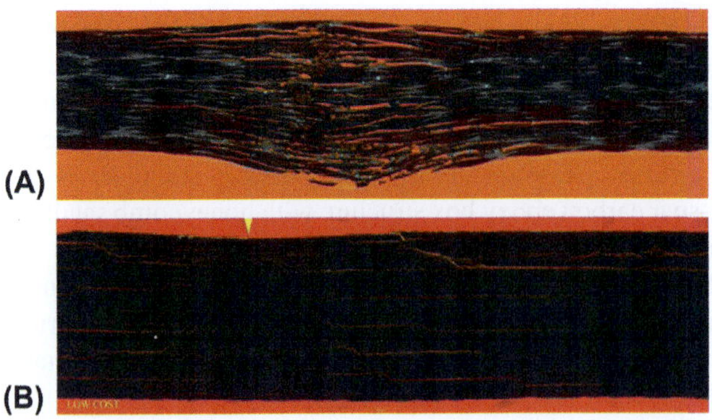

Figure 3.2.7 Cross-section of different carbon fibre composite materials showing the difference in impact damage due to resin toughness. (A) Representative image of early untoughened epoxy-carbon fibre composite system (B) Representative image of toughened epoxy-carbon fibre composite system. Boeing generated images.

toughened systems, there was always a trade-off with decreased hot/wet compression strength of varying levels.

Other advances in resin chemistry at this time included the introduction of two new types of resins specific for high temperature applications—polyimide (PI) and bismaleimide (BMI). These resin systems were commonly utilised with carbon fibre. Both new chemistries provided increased design space but came with some challenges for production utilisation. Both require higher temperature cure cycles and extended time at temperature to achieve the final cross-linking state necessary for the thermal, chemical and electrical properties desired. They are also less forgiving in terms of producibility and handling characteristics, requiring a higher attention to detail for both design and fabrication as compared to typical epoxy systems.

Resin development and process development for resin transfer moulding accelerated during this time. The closed mould system was introduced in the late 1970s, but Boeing's research in this area occurred in the 1980s. It was not until the end of the decade that the technology was mature enough for consideration for use on secondary structures for commercial aircraft (non-flight critical parts). While the process could result in laminates with controlled resin content and good laminate quality, the process required resins that were very low in viscosity, did not contain volatiles (or were easy to degas), and had good pot life (not too reactive) to support the infusion process.

Other important advances in the production of composite parts took advantage of the increasing prevalence of prepreg material forms. Automated tape and fibre placement equipment was first documented in the 1970s but began to see application in production spaces in the 1980s (Brasington et al., 2021; Dirk et al., 2011). The use of automation led to highly consistent quality and cost-effective manufacture of advanced composite parts as compared to the manual methods that were common to this point. Larger parts and more complex shapes were possible while also reducing the ergonomic impact on workers. These technologies continue to advance and are integral to fourth-generation manufacturing occurring today in aerospace (Brasington et al., 2021). An example of the scale of composite parts made with this technology is shown in Fig. 3.2.8.

3.4 Applications—1980s

In 1982, the U.S. Air Force initiated a program to investigate the damage tolerance of composite aircraft structures. A significant part of this program was the fabrication and testing of two full-scale wing boxes. Boeing was responsible for the transport wing architecture and designed a multi-rib structure with low modulus skins (10/80/10) and high modulus stringers (60/30/10) and additional 0-deg "planking plies" under the stringers. The box was impacted at multiple locations, load cycled for two lifetimes, and residual strength tested in compression (Whitehead and Horton, 1988). Results successfully met all requirements and became a basis for damage tolerance of the composite structure going forward. This was part of the overall maturation of certification of civil composite structure that was also occuring at this time (Soderquist, 1981).

Figure 3.2.8 Automated lay-up of carbon fibre composite materials fabrication for the B-2 Stealth Bomber, including some of the largest structural components made from composite materials at the time.

Building off the work done for the ACEE program and the damage tolerance box, Boeing designed and fabricated large composite primary wing structures for two separate military programs. For the A-6E, this was a replacement of the original metal wing with a carbon/epoxy structure to address fatigue problems that were discovered in service (National Air and Space Museum, 2024). Conversely, the B-2 made use of a significant quantity of composites in the baseline design for stealth considerations, including some of the largest structural components at the time (Deo et al., 2001). Both programs relied heavily on automated tape laying (ATL) equipment for manufacturing.

During this time, Boeing commercial aircraft continued to limit the use of composites in production applications to secondary structure. For the 757 and 767, this was still extensive (>1500 kg) and included control surfaces, wing-to-body fairings, landing gear doors, nacelle components, and fixed trailing edges of the wing and empennage. Note these are in addition to the glass fibre composite structure in use since the 1960s and described earlier. Of note was the integration of aramid fibre/epoxy material by itself and in hybrids with carbon and glass fibre composites (Boeing Commercial Airplane Company, 1984). This, in turn, led to discovering the sensitivity of some materials to thermal cycling and the potential for moisture ingression into honeycomb sandwich structure (Akay et al., 1997).

3.5 Materials and technologies—1990s

By the late 1980s, epoxy resins comprised over 90% of aerospace composites (Hamerton, 1994). Many of the toughened epoxy-carbon fibre composites developed in the 1980s were qualified and implemented on both defence and civil aerostructures during this time at Boeing. This includes the second-generation toughened matrix systems such as Toray's 3900–2 interlayer particle toughened resin, as well as the homogenous toughened resins, Cytec's 977–3 and Hexcel's 8552. These resins are available on a variety of carbon fibre reinforcements, most often intermediate modulus (IM), as well as supporting glass fibre versions. At the turn of the century, the utilisation of interlayer toughened material would become dominant within the primary structure for Boeing commercial aircraft, while defence applications would rely more on the homogenous toughened systems. The increase in damage tolerance with the interlayer method was important for commercial aircraft, where resistance to incidental impacts from things such as luggage belts and passenger ramps (for example) was important to customer airlines. However, the quantity of these chemistries is inversely proportional to the maximum service temperature of the composite material, an important consideration for defence aircraft, which often require higher-performance windows. Thus the different approach between commercial and defence applications for toughened epoxy composites. This is reflected in the multitude of university research papers that began to appear at this time.

Speciality materials continued to appear in the market this decade. In epoxy resins, the use of carbon-fibre composite tooling that was compatible with the 177°C toughened epoxy prepregs was of interest. Of these epoxy resin tooling materials, systems such as Cytec's LTM10 tooling resin, which could be cured without the use of an autoclave, were important to help reduce the cost of tooling, in particular, low-cost tooling for rapid prototype development efforts. However, for manufacturing runs needing consistent composite tools for many parts, these epoxy resins were often not stable enough. Depending upon the epoxy resin, dimensional stability could be a concern or embrittlement of the resin, leading to leaks through the tooling with continued use. In this way, BMI became the choice for durable composite tools where the need to produce hundreds of parts using the same tool. Both structural and tooling applications were dominated by the Cycom 5250–4 BMI resin system.

Cyanate ester resins gained attention during this time, with carbon fibre and quartz fibre reinforcing fibres. The general strength and toughness of this resin, as compared to untoughened epoxies, along with low moisture absorption and good thermal resistance, made it attractive for many applications, but it was the unique electrical properties of this resin that carved a niche for it in applications such as radomes (Hamerton, 1994). One factor that has limited the use of cyanate esters is the sensitivity to moisture during fabrication; facilities must adequately control humidity and out-of-time conditions to achieve consistent cure and part quality. Some variants utilise a copolymer where the cyanate ester is doped with an epoxy resin, but there are trade-offs in thermal and electrical properties. In addition, the quartz is expensive compared to other composite materials. However, it offers unique properties and thus Boeing continues to utilise this composite system.

It should be noted that the rapidly expanding electronics industry of the 1990s drove extensive research into thermomechanical stable resin systems. Advances in BMI's, PI's and cyanate esters all came about through this industry. Areas of cross-industry interest also included flammability properties and better producibility characteristics. However, aerospace adoption of advances was limited to specific systems due to the extended service life needed for many composite systems (Liaw et al., 2012; Hamerton, 1994).

Like materials, fabrication processes initially utilised in defence applications in the 1970s and 1980s were transitioned into Boeing's commercial division during this decade. The use of automated lay-up was integral in the composite empennage built by Boeing for the 777 program, where the horizontal and vertical empennage parts were made entirely from toughened epoxy-carbon fibre composite material (Quinlivan, 1993). More sophisticated techniques for forming complex shapes utilising automation were also developed during this time, including automated stiffener forming, which was applied on the 777.

Building on the resin development work in the 1980s, out-of-autoclave technologies also began to make some inroads into production. Resin transfer moulding produced parts became more routine in production. Some of these applications were outside of Boeing, for some defence contracts, but the advances made were closely monitored as Boeing applications, both defence and commercial, began to look for ways to make composite part production more efficient and cost-effective. Even for traditional fabrication methods such as hand lay-up, the use of laser project systems to ensure ply alignment instead of physical templates began to make their way into production shops in the 1990s, with this focus on cost and efficiency.

A major research and development effort was undertaken in the 1990s under the NASA Advanced Composite Technology (ACT) initiative "… *to develop materials, structural mechanics methodology, design concepts and fabrication procedures that offer the potential to make composite structures cost-effective compared to aluminum structure. Goals for the ACT program included 30–50% weight reduction, 20–25% acquisition cost reduction, and provided the scientific basis for predicting materials and structures performance*" (Davis, 1992).

Boeing's Advanced Technology Composite Aircraft Structures (ATCAS), under the ACT initiative, had the primary objective "… *to develop an integrated technology and demonstrate a confidence level that permits the cost- and weight-effective use of advanced composite materials in primary structures of future aircraft with the emphasis on pressurized fuselages*" (Ilcewicz et al., 1997). The focus of the ATCAS program was on design, analysis, fabrication and testing of transport fuselage concepts that exploit the Automated Fibre Placement (AFP) process but included the evaluation of multiple structural configurations.

3.6 Applications—1990s

In addition to taking advantage of some of the newer toughened materials, the V-22 is also one of the first major implementations of AFP fabrication. Composites comprise 50% of the structural weight of the aircraft, and the use of AFP on the fuselage reduced manufacturing costs as much as 53% via unitised structure (Deo et al., 2001). In this same vein, redesign of the F/A-18 to the Super Hornet configuration nearly doubled the use of composites to 20% of the structural weight. The major structural material was upgraded to a new IM carbon-fibre toughened epoxy, and advanced manufacturing techniques were used, such as AFP on the engine inlet ducts (Deo et al., 2001).

Continuing the trend of increased composites usage, the F-22 is comprised of about 26%. The majority of this is an approximately even split (12% each) of toughened epoxy and the first major use of bismaleimide (BMI) resin in locations where higher temperature capability is needed. The F-22 also includes a small amount (~1%) of reinforced thermoplastics. This included the use of composites for engine components or as protective wraps for engine components and ducting. On the manufacturing side, resin transfer moulding (RTM) makes an appearance alongside the now almost common ATL and AFP processes (Renstrom, 1997).

The rapid maturation of composites during this time led to the incorporation on new and previously launched aircraft. The C-17 is a good example of this. Original development of the C-17 was done in the 1980s, with the first of these McDonald Douglas-built aircraft occurring in 1991 (The Boeing Co, 2024b). However, after the first 50 aeroplane were built, a new composite horizontal stabiliser torque box was incorporated. It was reported by Northrop Grumman (Vought), who designed and built the box as a subtier to Boeing, that the new box had 2000 fewer parts and 42,000 fewer fasteners than the original metal unit, weighed about 250 kg less (~20% weight savings), and was shown to be very durable by achieving more than 200% of the design limit load during predeliver testing (Mathews and Engle, 1999). At the end of the day, the C-17 was purported to have more than 6500 kg of composite materials, ~6% of the total aeroplane weight. This included composite landing gear pods with sections that were fibre-placed sandwich structures (Fig. 3.2.9).

Similarly, at 9% of the structural weight, the 777 used three times more composites than previous transport aircraft. This includes all the aforementioned applications, plus the first production implementation by Boeing of primary structure on a

Figure 3.2.9 C-17 Globemaster III landing on wet tarmac. Note the size of the tail where the composite horizontal stabiliser torque box was incorporated.

commercial transport. Floor beams are solid laminate carbon fibre epoxy C-channels fabricated utilising advanced forming processes. The vertical and horizontal stabiliser are two-spar, multi-rib designs manufactured with carbon fibre/epoxy (Quinlivan, 1993). They include a simplified design without "planking plies" below the stringers made possible by the advancements in toughened materials (Harradine and Quinlivan, 1989). A picture of these stringers being prepared for cure is shown in Fig. 3.2.10. It should be noted that the 777 and other commercial aircraft, such as the MD-80/MD-90, continued to make increasing use of composites for applications noted previously, such as interiors, control surfaces, radomes and others.

Other sectors that saw significant growth in the use of composites during this decade were space and satellite applications. The rapid evolution of telecommunications that began in the 1990s meant that the need for both launch vehicles (i.e., rockets) as well as satellites created a growing market. While Boeing (and heritage companies) had been offering satellite devices since the 1980s, this increasing demand meant that multiple derivatives and new satellite structures were brought to market in the 1990s (The Boeing Co, 2024b). While similar composite materials are utilised in space applications, there are additional stringent requirements for hot and cold temperature performance, low moisture absorption, and out-gassing from the laminate resin, which are unique to space applications (Toor, 2018). Typical applications of carbon fibre composites are strap elements, boom, housing and solar panels, where designers can take advantage of the anisotropic, tailorable properties of the composites using different orientations to achieve high stiffness, specific strength, low thermal expansion/conductivity and dimensional stability in this challenging environment. Kevlar, glass and quartz fibre composites are applied in the fabrication of satellite antennas due to their low transmission loss and electrical conductivity, similar to defence and commercial radome applications.

Figure 3.2.10 Autoclave where the 777 empennage part is produced at Boeing's Frederickson facility in Washington.

3.7 Gaining momentum

By the end of the 1990s, both commercial and military aircraft were utilising composite materials in flight-critical parts, showing how the industry's confidence in the technology had progressed. By the end of this period, the composite materials industry could no longer be thought of as new. Development efforts such as the Joint Strike Fighter program and the Bird of Prey YF-118G research plane helped demonstrate new technology. Development efforts were no longer focused on fundamental understanding but rather on challenges in design and production efficiency. For example, the YF-118G utilised out-of-autoclave oven-cured materials and disposable tooling to very rapidly prototype large, unitised part fabrication. Many first-generation materials began to fall out of favour, while those with a proven track record of performance and robust production usage were being utilised for a wider variety of parts. Efforts to protect the supply of materials for defence led to different partnering efforts within the supply base. With confidence high in the fundamental technology, the cost competitiveness of carbon-fibre materials and parts relative to metal drove development in technology as the 1990s ended, and the industry raced toward even grander goals.

4. Going big

By the turn of the century, composites were accepted as a standard material of choice in aerospace. The industry at this time was looking toward expanded applications and focused on production and cost efficiency so that those new applications made sense not only from a performance perspective but also from a business standpoint.

4.1 Materials and technologies—2000s to present day

The higher volumes of composite materials needed to support the larger structures and higher build rates of this time drove some of the advances in composite material prepreg production in the early part of the 2000s. Most epoxy-based material systems that had been produced utilising solvent impregnation were switched to hot-melt impregnation based on the success of hot-melt systems such as the Hexcel (HexPly) 3501–6, which had been demonstrated to successfully produce quality laminates for large unitised structures as part of the ACT initiative (continued under the NASA/Boeing Advanced Subsonic Technology Composite Wing Program) (Karal, 2001). In addition, some prepregs that had historically been 36-inch or 48-inch (0.91 or 1.22 m) widths were scaled up to 60-inch (1.52 m) widths, enabling bigger shapes for forming and hand lay-up operations. Many material producers in this period also diversified the locations of their prepreg material production to more efficiently support the expanding globalisation of composite part production. Having the same material produced in multiple locations helps stabilise the supply chain and protect against both geopolitical and natural disruptions.

New materials being developed in this period were focused on enabling production efficiencies. Generically, epoxy resin systems that could cure more rapidly and be utilised in press or stamp forming operations began to gain more traction in aerospace, leveraging lessons from automotive and consumer goods. Out-of-autoclave materials continued to gain a lot of attention. Cycom 5215–1, brought to market by Cytec (now Syensqo), was most likely a first-generation brittle resin system, but it was not long until they introduced a toughened epoxy-carbon fibre system, Cycom 5320–1. This system can be handled like a normal autoclave prepreg system, including automated laydown and forming, but can then be cured with just a vacuum bag, achieving consistent enough laminate quality to be utilised for structural aerospace parts.

The other major material advancement in this time frame was advanced thermoplastic carbon fibre composite materials utilising polyaromatic chemistries. While other thermoplastic composite chemistries had been researched and even implemented during the 1990s, structural composite applications were still limited. Advances in the underlying thermoplastic prepreg quality and the use of highly tough, thermally and chemically resistant chemistries such as polyetheretherketone (PEEK) changed this. The promise of more rapid part production while still being compatible with AFP made the transition from thermoset to thermoplastic usage more attractive in some cases. The application space for this technology is mostly medium to small parts due to the high cost of the material and the high processing temperatures required.

4.2 Applications—2000s to present day

In the 2000s, Boeing revolutionised the large transport commercial aeroplane industry by launching the 787 program (Stansell, 2024). With an airframe comprised of 50% carbon fibre reinforced plastic and other composites, the 787 introduced the first all-composite wing in a twin-aisle aeroplane and the first all-composite fuselage, unlocking the design flexibility and corrosion/fatigue resistance advantages of composite materials. While the ATCAS program had concentrated on fuselage sections comprised of four segments, the 787 program developed the technology that enabled the manufacturing of full one-piece barrels that were then integrated with composite wing and tail sections (Fig. 3.2.11). These large fuselage sections are comprised of a co-cured skin-stringer architecture where the fuselage skin is manufactured using the AFP process. The development and integration of complex and novel manufacturing and design technologies at scale were supported by comprehensive evaluations at all levels of the building block. Such a development benefited from a highly integrated team that met the complex challenges associated with engineering a revolutionary product. In its first decade in service, the 787 program had made more than 4.4 million flights and served over 900 million passengers, and, as of 2024, the 787 has avoided more than 77 billion kilogrammes of carbon emissions. Building on the experience of the 787, the 777X wing skins and spars are some of the largest composite structures being fabricated today.

Increased interest in ceramic matrix composites occurred in this period, with Boeing commercial looking at an acoustic engine nozzle that offers the potential for lower weight, longer life, and greater noise abatement than metallic engine nozzles (Karadimas and Salonitis, 2023). The nozzle underwent ground tests on a full-scale engine

Figure 3.2.11 Boeing final assembly line is in Charleston, South Carolina. Note each of the individual composite parts comprising the major structural elements of the 787.

and flight tests in 2013 time frame. An example of this part is shown in Fig. 3.2.12. The work is in support of the Federal Aviation Administration's Continuous Lower Energy Emissions and Noise (CLEEN) Program. The CLEEN Program is a cost-share program with the FAA to conduct ground and flight demonstrations of airframe and engine technologies with the potential of reducing greenhouse gas emissions and community noise (Federal Aviation Administration, 2024).

Figure 3.2.12 Test of ceramic matrix composite acoustic engine nozzle.

5. Looking forward

Composites are a part of Boeing's fundamental materials toolbox and will continue to be going forward. Benefits such as expanded performance windows (both hot and cold), continued focus on lightweight structures for efficient fuel burn, resistance to chemical attack, and in-service durability are just a few of the reasons as seen in the historical examples discussed herein. Looking to the future, material systems that enable more sustainable manufacturing practices, in-service efficiency, increased customer comfort, as well as materials that can support recycling and overall circularity, will help push and inspire Boeing to continue developing the next generation of composite materials and parts (The Boeing Co., 2024a).

References

Akay, M., Mun, K.A., Stanley, A., 1997. Influence of moisture on the thermal and mechanical properties of autoclave and oven-cured Kevlar-49/epoxy laminates. Compos. Sci. Technol. 57 (5), 565–571.

Bauman, K., Swedlow, J., 1981. Research Priorities for Advanced Fibrous Composites (NASA-CR-165414). National Technical Information Service, Springfield, VA.

Boeing Commercial Airplane Company, 1984. ACEE Composite Structures Technology (NASA-CR-172358). National Aeronautics and Space Adminstration, Hampton, VA.

Brasington, A., et al., 2021. Automated fiber placement: A review of history, current technology and future paths forward. Composties Part C: Open Access 6, 100182.

Campbell, F., 2004. Manufacturing Processes for Advanced Composites. Elsevier Ltd., s.l.

Campbell, F., 2006. Manufacturing Technology for Aerospace Structural Materials. Elsevier Ltd, s.l.

Coggeshall, R.L., 1989. Boeing/NASA Composite Components Flight Service Evaluation (NS-CR-181898). National Aeronautics and Space Adminstration, Hampton, VA.

Cole, J.B., 1980. Airplane Wing Leading Edge Variable Camber Flap. NASA Langley Research Center, s.l., pp. 225–235

Davis, J.G., 1992. Overview of the ACT Program, NASA Report N95-28463. National Aeronautics and Space Administration, Hampton, VA.

Deo, R.B., Starnes, J.H., Holzwarth, R.C., 2001. Low-Cost Composite Materials and Structures for Aircraft Applications (RTO-MP-069(II)). St. Joseph Print Group, InC., Loen, Norway, pp. 23–34.

Dirk, H., Lukaszewicz, J., Ward, C., Potter, K.D., 2011. The engineering aspects of automated prepreg layup: History, present and future. Composites, Part B 43, 997–1009.

Dow, M.B., 1987. The ACEE Program and Basic Composites Research at Langley Research Center (1975 to 1986). NASA Refence Publications, Hampton, VA.

EconHP Holding GmbH, 2024. The History of Sandwich Construction and Honeycombs. https://www.econhp.com/history/.

Federal Aviation Administration, 2024. Continuous Lower Energy, Emissions, and Noise (Cleen) Program. https://www.faa.gov/sites/faa.gov/files/2022-07/eeSC-Mar2020-CLEENBenefitsSummary.pdf.

Gorss, J., 2003. High Performance Carbon Fibers - A National Historic Chemical Landmark. The American Chemical Society, Washington, DC.

Hackman, L., 1973. Military aircraft. In: Applications of Composites Materials, ASTM STP 524. American Society for Testing and Materials, Philadelphia, PA, pp. 43–75.

Hamerton, I., 1994. Introduction to cyanate ester resins. In: Chemistry and Technology of Cyanate Ester Resins. Springer Dordrecht, Tunbridge Wells, pp. 1–6.

Harradine, P.J., Quinlivan, J.T., 1989. Composites and the Commercial Jet - A Boeing Viewpoint (AIAA 89-2126). American Institute of Aeronautics and Astronautics, Seattle, WA, pp. 1–4.

Ilcewicz, L., et al., 1997. Advanced Technology Composites Fuselage - Program Overview, NASA Report NS-CR-4734. National Aeronautics and Space Administration, Hampton, VA.

June, R., Lager, J., 1973. Commercial aircraft. In: Applications of Composite Materials, ASTM STP 524. American Society for Testing and Materials, Philadelphia, PA, pp. 1–42.

Karadimas, G., Salonitis, K., 2023. Ceramic matrix composites for aero engine applications - a review. Applied Sciences 13 (3017), 1–42.

Karal, M., 2001. AST Composite Wing Program - Executive Summary, NASA Report NASA-CR-2001-210650. National Aeronautics and Space Administration, Hampton, VA.

Kiersarsky, S., Ong, C., 1970. Advanced Composites for Aerospace, NASA Report NASA-CR-110322. Belcomm, Inc., Washington, DC.

Lewis, R., Gorum, A., 1976. Army Applications of Advanced Composites. Metallurgical Society of AIME, Boston, MA, pp. 1226–1249.

Liaw, D.-J., et al., 2012. Advanced polyimide materials: Syntheses, physical properties and applications. Prog. Polym. Sci. 37 (7), 907–974.

Mathews, M., Engle, G., 1999. News release - deliveries begin on Boeing C-17 composite horizontal tail. https://boeing.mediaroom.com/1999-01-28-Deliveries-Begin-on-Boeing-C-17-Composite-Horizontal-Tail.

Mi, X., et al., 2022. Toughness and its mechanism in epoxy resins. Prog. Mater. Sci. 130, 100977.

National Air and Space Museum, 2024. Grumman A-6E Intruder. https://airandspace.si.edu/collection-objects/grumman-6e-intruder/nasm_A19940152000.

Palmer, J., Stephens, C., Sutton, J., 1984. McDonnell-Douglas Report: ACEE Composites Structure Technology (NASA-CR-172359). National Aeronautics and Space Administration, Hampton, VA.

Peterson, G.P., 1971. Development and Performance of Advanced Composite Structures. American Institute of Aeronautics and Astronautics, Anaheim, CA, pp. 71–367.

Pinckney, R.L., 1964. Recent Applications of Composite Structure in Modern Commercial Aircraft. American Institute of Aeronaturics and Astronaturics, Wichita, KS, pp. 64–164.

Quinlivan, J., 1993. Challenges and Payoff of Composites in Transport Aircraft: 777 Empennage and Future Applications, NASA Report N95-29031. National Aeronautics and Space Administration, Hampton, VA, pp. 25–47.

Rawdon, B.K., 1983. DC-10 Composite Upper Aft Rudder Service Experience. Society of Manufacturing Engineers, Anaheim, CA.

Renstrom, R., 1997. It Ain't No Junkyard Dogfighter: Air Force's F-22 Makes Extensive Use of Composites. https://www.plasticsnews.com/article/19970127/NEWS/301279998/it-ain-t-no-junkyard-dogfighter-air-force-s-f-22-makes-extensive-use-of-composites.

Salkind, M.J., 1976. Fibers Composite Structures. Metallurgical Society of AIME, Boston, MA, pp. 5–30.

Schjelderup, H., Purdy, D., 1971. Advanced Composites - The Aircraft Material of the Future. AIAA 3rd Aircraft Design and Operations Meeting, Seattle, WA. American Institute of Aeronautics and Astronautics, pp. 71–779.

Schlazberg, E., 1994. Ideology and technical choice: The decline of the woodern airplane in the United States, 1920-1945. Technol. Cult. 35 (1), 34–69.

Sloan, J., 2013. In: In-house prepregging: Cost/benefit calculus. https://www.compositesworld.com/articles/(597).

Soderquist, J.R., 1981. Certification of Civil Composite Aircraft Structure. SAE, Anaheim, CA.

Stansell, K., 2024. The 787 dreamliner difference. IQ Innovation Quarterly 8 (29), 4–11.

The Boeing Co, 2024a. Accelerating innovation for a sustainable future. https://www.boeing.com/sustainability/environment/ecodemonstrator#overview.

The Boeing Co, 2024b. Select Products in Boeing History. The Boeing Co, Crystal City, VA.

The Boeing Co, 2024c. Tech that connects. IQ Innovation Quarterly 8 (29), 10–13.

Toor, Z.S., 2018. Space applications of composite materials. J. Space Technol. 8 (1), 65–70.

Tsai, S.W., 2005. Three decades of composites activities at US Air Force materials laboratory. Compos. Sci. Technol. 65 (15–16), 2295–2299.

Vosteen, L.F., 1978. Composite Structures for Commercial Transport Aircraft, NASA Report TM-78730. NASA Langley Research Center, Hampton, VA.

Weinberger, R.A., Somoroff, A.R., Riley, B., 1977. U.S. Navy Certification of Composite Wings for the F-18 and Advanced Harrier Aircraft. American Institute of Aeronautics and Astronautics, Reston, VA.

Whitehead, R., Horton, R., 1988. Damage Tolerance of Composites - Development of Requirements and Compiance Demonstration, Report AFWAL-TR-87-3030-1. Air Force Wright Aeronautical Laboratories, Wright Patterson, OH.

No wind energy without composites: The early years as seen from Denmark

3.3

Hans Lilholt and Bo Madsen
Wind Energy Materials and Components Division, Department of Wind and Energy Systems, Technical University of Denmark, Risø Campus, Roskilde, Denmark

1. Wind and its energy content

The wind is part of the atmospheric system surrounding the earth as a thin layer, up to about 1000 km above the earth surface. The wind is generated from pressure differences generated by varying temperatures, typically high pressures and low pressures. The wind will in general move from the high-pressure region to the low-pressure region, and the rotation and movement of the earth cause the wind direction to be a circulation around the high or the low pressure. These wind patterns develop and die away on a long-term basis and thus cause the wind and its kinetic energy content to be very variable over time, both on a daily basis and on longer time scales, such as weeks, months, and the year. The energy content in the wind is very large. The obvious exploitation of wind energy is complicated, both the extraction of the energy to make it accessible to practical human uses, and the long-time stability of this energy production.

The practical human exploitation of the energy in the wind requires an extraction machine of some sort, to convert the kinetic energy of the moving air masses to some other energy form, for example, electricity. A theoretical ideal extraction machine can at maximum capture 16/27, or about 59%, of the kinetic energy in the wind. This was demonstrated in 1920 by the German physicist Betz (1920), who derived the equation for this maximum, based on the physical laws of conservation of mass and energy (see Box 3.3.1). His analysis does not imply the form of this ideal extraction machine. Actual practical extraction machines, called wind turbines, were developed, based on a rotor of normally three aerodynamically shaped blades, with the rotor axis mounted horizontally on the top of a tower. The wind speed, and thus the kinetic energy, increases with the height over the ground, so a tall tower would potentially extract more energy. This is, of course, counteracted by the practical problems and costs of building a (very) high tower, so a compromise has been established, that the height of the tower typically is about equal to the rotor diameter.

The local wind speed and wind direction are, often strongly, influenced by the shape of the ground or its roughness as seen from the wind. These roughness elements are hills and mountains, trees and other vegetation, as well as buildings and other

The History of Composites. https://doi.org/10.1016/B978-0-443-34119-9.00020-6

Box 3.3.1 Wind and its energy content

Wind is a mass m of air moving at a certain speed v_1 (Betz, 1920). The kinetic energy U is

$$U = ½ * m * v_1^2$$

The air approaching the rotor is contained in a tubular cylinder. The rotor slows down the wind, which thus delivers energy to the rotor. The mass of a slice Δx of air of density ρ at the rotor with area A_t is

$$m = \rho * A_t * \Delta x$$

and the mass flow for time Δt and wind speed v_t at the rotor is

$$m/\Delta t = \rho * A_t * \Delta x/\Delta t = \rho * A_t * v_t$$

The wind speed upwind of the rotor is v_1 and the lower wind speed downwind of the rotor is v_2. Thus, the kinetic energy delivered to the rotor U_t is

$$U_t = ½ * m * (v_1^2 - v_2^2)$$

and the power of the rotor P_t, in unit energy per time, is

$$P_t = U_t/\Delta t = ½ * \rho * A_t * v_t * (v_1^2 - v_2^2)$$

Assuming that the wind speed changes continuously from far upwind to far downwind of the rotor, the wind speed at the rotor is

$$v_t = ½ * (v_1 + v_2)$$

The only unknown parameter is v_2, and it can be found by maximising P_t by setting $dP_t/dv_2 = 0$ and solving for v_2. This corresponds to an ideal wind extraction machine. The result is

$$v_2 = 1/3 * v_1 \text{ and } v_t = 2/3 * v_1$$

Inserting these into the equation for P_t gives.

$$P_t = ½ * \rho * A_t * v_1^3 * 16/27.$$

The power in the undisturbed wind is $P_w = ½ * \rho * A_t * v_1^3$ and thus the ideal machine can extract the fraction 16/27 or 59% of the wind energy. The factor 16/27 is called the Betz limit (Betz, 1920).

human constructions. The ideal terrain would be flat with no obstructions, but because wind turbines will normally be placed in nonideal regions of the natural landscape, knowledge of the terrain and its wind pattern is therefore of interest for the best

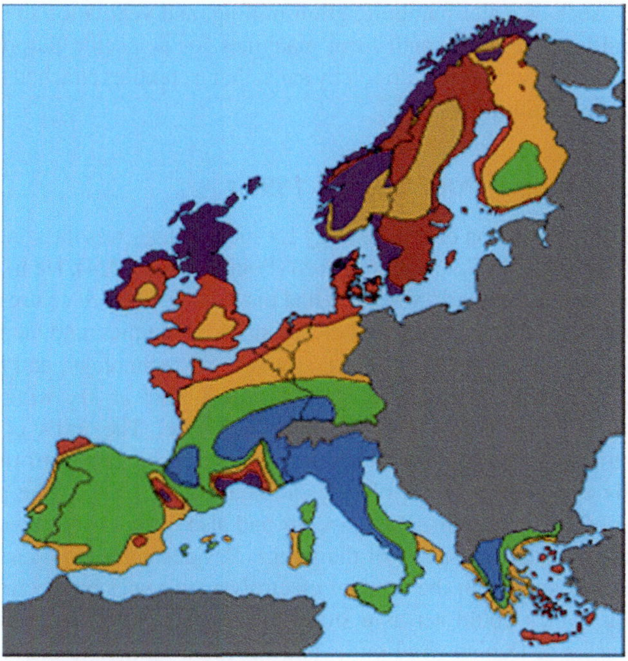

Figure 3.3.1 Wind atlas for Europe, strong winds in north-western Europe (Nielsen et al., 1998).

placement of wind turbines. These aspects of the wind in various regions with different terrain characteristics were recognised in the early years of wind energy exploitation. The Meteorology Section of the (then) Risø National Laboratory in Denmark had already in 1980 developed a chart of the wind energy content in Denmark, and later extended these wind atlases to the rest of Europe (in 1989, see Fig. 3.3.1) and to the rest of the world (Nielsen et al., 1998). These wind atlases have become important in selecting optimal sites for wind turbines and for wind parks composed of many wind turbines placed together.

The wind atlas for Europe shows that the best wind energy potential is located in the northwestern Europe, especially Scotland and Denmark, as well as along the North Sea and the English Channel. These best locations are all land based, because in the first about 30 years of wind energy wind turbines were, out of necessity, placed on land to be close to the energy distribution systems like the grid systems. Only later, in the 1990′s, wind parks were built offshore, where the wind generally is stronger than on land and also is unobstructed, so the wind energy potential is greater, but so are the problems and costs of building the wind turbines and connecting them to the land-based distribution systems.

2. Pre year zero

Since start of humanity, the wind has been exploited locally by the people living in the different regions, because it supplied them with energy for various activities in the local societies, such as water pumping, grain milling, and ventilation in hot climates. These wind mills were often small local machines for a specific purpose, made as wooden constructions with rotors driven by sails, that is, textiles attached to some supporting structures of the wings.

2.1 Danish pioneer Poul la Cour 1890′s

An early case of exploitation of wind energy for modern uses was the studies made by the Danish pioneer Poul la Cour (1846–1908) (Poul la Cour, 2024). He lived in Askov in southern Jutland of Denmark, where he had his private research centre at his home. He was interested in several technical topics, like aerodynamics, and the practical exploitations of them. In the 1890′s he studied wind mill technology, and built experimental wind mills with sturdy towers, rotors with four or six wings, and with wooden plate-elements attached to the wing structure. The mills converted the wind energy to electricity to be used in his house and farm. He performed tests on the wind mills and made a range of measurements to investigate the performance and efficiency of the wind mills. Poul la Cour had also built a wind tunnel for closer studies of the aerodynamic aspects of the wings. Besides energy production, he also focused on energy storage in the form of electrolysis of water by electricity, to produce hydrogen, which could act as a storage medium. Poul la Cour also acted as consultant to users of Danish wind mills, and educated farmers to handle their electrical installations.

2.2 The Gedser wind mill 1957 to 1967

One of Poul la Cour's students was Johannes Juul (1887–1969) (Juul, 2024), who maintained interest in wind energy and wind mills. Johannes Juul was employed at a Danish local electricity utility called SEAS. During the years around the second World War, there were activities to develop wind energy in several countries like Denmark, Germany, Holland, Russia, and the United States. Johannes Juul was leading a Danish development program for wind energy, which, with support from the post-World War II Marshall Plan, included the construction and building of the Gedser wind mill, named after its location on the southern tip of one of Denmark's southern islands. The Gedser wind mill was ready in 1957 and was part of a testing and measurement program up to 1967, and over the years it was the world largest wind mill.

The Gedser wind mill, in Fig. 3.3.2, had a tower of height 24 m made of steel and wood, and a rotor of diameter 24 m with three blades. The rotor was placed up-wind on a box (later called nacelle) on top of the tower. Because of the up-wind design, the box needed an active yawing control mechanism (electromechanical) to keep the rotor facing into the wind at all times. Each blade was a steel structure carrying a wooden wing of aerodynamic shape of constant width. The wooden wing covered about 8.5 m of the total wing length of 12 m, with the inner part of each blade left bare. The three blades were not self-supported but were stabilised by struts consisting of steel rods between the blades and connected to the hub.

Figure 3.3.2 The Gedser wind mill (Jensen, 2018).

As a further safety measure, Johannes Juul invented aerodynamic tip brakes, where the outer part of the wing, about 1 metre length, would be released and turn at right angle to the wing to act as a brake, when wind speed increased beyond the accepted limit. The rotor of the Gedser wind mill had a constant rotational speed, independent of wind speed, driving an induction generator of nominal power 200 kW. The electricity produced by the generator was connected to the grid system. The Gedser wind mill operated for the 10 years of the Danish development program without maintenance. Its rather simple and sturdy design, safety aspects and low cost were key elements in making it known worldwide as the "Danish design."

The Gedser wind mill was intact, but not operating, when the interest in wind energy was renewed in the early 1970′s, and it became an important starting point for the modern development of wind energy. The Gedser wind mill has sometimes been called "the mother of the modern wind turbine."

2.3 New sources of energy

Already in 1961 the United Nations had organised a 10-day conference on "New Sources of Energy" held in Rome (United Nations, 1961). Within the regime of wind energy, a large number of topics were discussed. One topic was the fundamental design of the wind turbine related to the orientation of the rotation axis for the rotor, that is, horizontal axis designs, as in the "Danish design," or vertical axis designs. Further to the horizontal axis designs, there was a discussion of the up-wind positioning of the rotor, which requires an active yawing control, and the down-wind positioning of the rotor, which was assumed to be self-aligning by the direct action of the wind itself.

For the rotor, the discussion was about the number of blades. The important parameter for the rotor is the area it sweeps in the wind, as seen from the equation developed by Betz (see Box 3.3.1). Thus, the number of blades could be one blade, two blades, three blades and upward toward "many" blades. Few, like one or two, blades would reduce the amount of material to be used for the blades, and keep weight and costs down, while on the contrary few blades might make the rotor unstable during rotation. Many blades would work in the opposite direction. A theoretical analysis on this compromise would give the optimum number as 2.7 blades (!), so this is increased to three blades in practice. This was probably the conclusion also reached by Johannes Juul in his practical approach to the design of the Gedser wind mill.

With respect to the rotor blades, sometimes called wings, discussions were on the aerodynamic shape of the blades, their structural design, and the materials used. In particular, the interaction between these aspects was of interest to obtain the optimum aerodynamic shape and the lowest weight for strong, reliable, and safe blades. In the first period of development of wind energy, the only available aerodynamic shapes or profiles for blades were those developed for wings of aircrafts over the previous 50 years. These profiles were called NACA profiles, where NACA is the National Advisory Committee for Aeronautics, which was founded in 1915 and changed name to NASA in 1958. The NACA aircraft profiles were used for rotor blades, before detailed studies developed improved aerodynamic shapes.

The materials for the blades were discussed in relation to properties like stiffness, strength, and density. Both existing materials like metals and wood, and potential, new materials like fibre reinforced composites were considered. Tests of materials and components were discussed, in particular accelerated tests to study the expected very long life times of materials for rotor blades. This is of special concern with the many rotations experienced by the blades, and thus the large number of cycles in fatigue of the materials.

At the conference, Johannes Juul argued for wind energy in its own right, and not just as a replacement for other fuels, and he estimated and hoped that about 20% of Denmark's electricity demand could be supplied by wind energy.

Many of the topics discussed at the conference were to become of great concern during development of wind energy. Little realistic experience existed, and little action was taken at the time.

3. Year zero = 1973

In 1973 the war of Israel against Egypt and Syria, with support from United States and other countries, made the Arab nations in October 1973 start an embargo on production and export of oil, leading to dramatic increases in oil prices on the world market. This oil crisis became an epoch-making event, which started the thoughts and activities about energy production and energy consumption.

3.1 The early 1970s

In the early 1970′s many power plants produced electricity from coal and oil. Some countries had in addition electricity supply from nuclear power plants. In Denmark, nuclear energy for the benefit of society had been studied at Risø National Laboratory since the late 1950′s (Nielsen et al., 1998). Several concepts and separate parts of the nuclear exploitation were under investigation, such as the production of nuclear fuel in the form of U–Al alloys for plate fuel elements, and UO_2 pellets for tubular fuel elements. These studies of nuclear energy had developed several techniques and accumulated experience on materials which later became of interest and found use in the development of wind energy.

The Danish electricity utilities were, during the 1960′s, not particularly interested in nuclear energy, claiming that they had efficient power stations running on oil of low price on the world market. However, in the early 1970′s the expected price for nuclear energy was slowly decreasing, making the electricity utilities show increasing interest in nuclear power plants. In Denmark, the electricity utilities were local, private companies, and they expected that a decision to build a nuclear power plant was an in-house matter. At the same time, in the early 1970′s, there was a growing resistance against nuclear power in developed countries. In Denmark, this resistance gained momentum among the left-wing population, and an organisation was established by these people. Their activities made the politicians and the Parliament concerned, and in a period of few weeks, the public feeling changed, so that the question of building a nuclear power plant was to be decided by Parliament and not by the electricity companies themselves. The public opposition increased in the following years of the 1970′s, and this effectively blocked for nuclear power in Denmark. As nuclear energy was not an option in Denmark, other production methods for useful forms of energy, such as electricity, were taken up by many individuals, with a strong focus on wind energy.

3.2 The first modern rotor blades

An early pioneer on wind energy produced by small wind mills, or wind turbines as they were to be called, was Erik Grove living in Sparkær in northern Jutland of Denmark (Winds of Change, 2003). In August 1977 he founded a company, Økær Vind Energi, for production of wings for wind mills. He borrowed a mould from the Tvind school (see below). The wings, or blades, were 4.5 m in length, and built of composite materials. He used glass fibre random mats and polyester, as was common for, for example, sailing boats. The two composite constituents were placed in

Box 3.3.2 Wind loads and gravity loads on rotor and blades

The wind will exert a load on the rotor, caused by the pressure drop from the upwind side to the downwind side as the wind passes through the rotor plane, while activating the blades to cause rotation (Mikkelsen, 2016). The Bernoulli equation relating the pressure and the speed of an air stream gives the pressure drop Δp across the rotor plane as

$$\Delta p = ½ * \rho * (v_1^2 - v_2^2)$$

The total load acting on the rotor of area A_t is

$$Q_t = \Delta p * A_t = ½ * \rho * A_t * (v_1^2 - v_2^2)$$

At the maximum power P_t we have $v_2 = 1/3\ v_1$ (see Box 3.3.1), and thus

$$Q_t = 4/9 * \rho * A_t * v_1^2$$

This total load on the rotor plane is distributed equally on the number of blades, and with the normal three-bladed rotor, this gives the total load for one blade

$$Q_{bl} = 4/27 * \rho * A_t * v_1^2$$

The total load Q_t on the rotor plane is uniformly distributed on the rotor plane, and thus the load per area k is

$$k = Q_t/A_t = 4/9 * \rho * v_1^2$$

The load on a circular ring-shaped area of the rotor plane of width dx is

$$dq_{ring} = k * A_{ring} = k * 2\pi * x * dx$$

With the normal three-bladed rotor, this gives the load intensity on one blade

$$dq_{bl}/dx = 1/3 * dq_{ring}/dx = 8/27 * \pi * \rho * v_1^2 * x$$

If dq_{bl}/dx is integrated along the blade length with x from 0 to L, we recover the total load Q_{bl} for one blade, given above, where $A_t = \pi * L^2$.

The linear relation for dq_{bl}/dx means that, when testing rotor blades, the applied load must increase linearly from root to tip of the blade, as shown by the increasing number of sand bags in Fig. 3.3.5.

For short blades loads from gravity are not important. On the contrary, modern large and very large blades have significant loads from gravity, when the blades are in the horizontal positions during rotation. For one blade with a material cross-sectional area A_{mat} and a blade material with mass density ρ_{mat}, the total gravity load is

$$Q_{gr} = \rho_{mat} * A_{mat} * L * g$$

with the gravitational constant $g = 9.81\ m/s^2$.

The wind loads and the gravity loads acting on a blade are illustrated schematically in Fig. 3.3.3 (Mikkelsen, 2020).

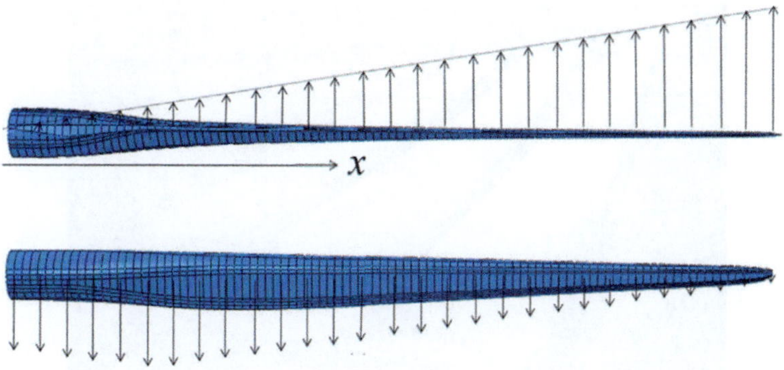

Figure 3.3.3 Distribution of flapwise wind load and edgewise gravity load along a rotor blade (Mikkelsen, 2020).

the mould with the aerodynamic shape, and cured to produce one of the first composite rotor blades in Denmark, and the first set of three wings was delivered to the customer in late November 1977. These early wings were a shell construction as the load-bearing part, and the internal cavity was filled with PU foam (Grove, 2024). Erik Grove also had made a test stand, in which the wing could be loaded in bending to test and record its deflections and shape changes during simulated wind loading. In the test stand the wing was mounted horizontally, and the load consisted of sand bags, which were stacked on the surface of the wing.

The very first tests of the wings could not be completed because the rather flexible wing would touch the floor, and no further loading and bending could be performed. This caused Erik Grove to ask for support from the Risø National Laboratory, and discussions led to improvements in the composition of the composite material, by placing as many fibres as possible along the length of the wing, thus improving the stiffness of the wing through the use of a near unidirectional fibre composite. Stiffness was further improved by ensuring a high fibre content, by controlling and minimising the amount of matrix polyester used during fabrication.

In January 1978, Erik Grove was asked to produce improved wings of 5 m length. He made a model (plug) for the new wing and the corresponding mould, and he also improved the composite material by using glass fibre filaments in bundles, which were wetted with polyester matrix in a special machine. These measures allowed production of one wing per day, and the first set of three wings were delivered in May 1978, see Fig. 3.3.4

An important decision for Erik Grove regarding the new blades was the direction of rotation for the rotor, clockwise or anticlockwise, because this determined the way the blade should be twisted to optimise its efficiency (see Box 3.3.4). The old wind mills of Danish farmers were traditionally rotating anticlockwise, and so were some of the early small electricity-producing wind mills. It was believed

Figure 3.3.4 Year 1978: The first three wings of 5 m length manufactured by Erik Grove's company Økær Vind Energi. The weight is 80 kg, the circular ring at the root has diameter 0.31 m and 24 bolts; safety steel wires are sticking out of each wing, the wires were attached to the tower and supposed to hold the wings in case of wing failure (Winds of Change, 2003).

that the Tvind school (see below) was following this trend. Erik Grove and his wife did not support this leftist-oriented school, and discussions at the kitchen table let to the decision that their blades and rotors should rotate clockwise. Because some of the then small companies, buying blades from Erik Grove, later became the large dominating wind turbine producers, the clockwise rotation became the Danish standard, and then the world standard.

Because a number of failures of rotors and blades had occurred, various brake systems were included in the blades, typically tip brakes with spring-release. Around 1980, still longer blades were asked for by customers who were building larger wind turbines with rotor diameter of 15 m and power of about 50 kW. Erik Grove produced the blades of 7.5 m length, and the blades were also tested to control their bending characteristics. The original small test stand had become larger, and was also improved by tilting the fixture plate for the root of the wing, so that in the start position the wing was pointing upward, as seen in Fig. 3.3.5. The wind load increases linearly from root to tip of the wing (see Box 3.3.2), which means that few sand bags were used at the root and many sand bags were needed at the tip; this posed a problem of stability of the stack of sand bags, as seen in Fig. 3.3.5. It is interesting, that the stiffness of the composite wing was such, that the quite heavy steel plate for the test fixture was (slightly) bent into a planar S-shape by the loading of the wing. This required recording of and correction for bending of the steel plate to establish the true bending characteristics of the wing.

3.3 Other materials were attempted

Enthusiasts looked at other materials for practical wind energy. A Danish metallurgically trained engineer attempted to produce wings in aluminium, by welding thin Al-plates into an aerodynamic shape. These wings could only operate for very short times

Figure 3.3.5 May 1982: Wing/rotor blade of length 7.5 m being tested in bending under simulated wind load, with sand bags as loads, notice that the load is largest near the blade tip. The wing was manufactured by company Coronet/Aerostar/Alternergy in Slagelse west of Copenhagen, the design was by Erik Grove, Aeroform (Winds of Change, 2003).

before they failed or were permanently bent. This is caused by the short fatigue life of aluminium and its low yield strength.

In the USA, a wind energy program was initiated, and also here some first attempts were focused on metals especially aluminium. A fairly large wind turbine was designed and constructed with rotor blades in aluminium, and with the rotor placed down-wind of the tower. The wind turbine was placed on the ridge of a mountain range, with the wind normally blowing across the ridge, under the assumption the strongest wind would be on top of the mountains. The large topographic scales of the mountains caused the wind to blow up one side of the mountain range and "overshoot" with some turbulent wind on the opposite side, causing the rather small-scale wind turbine, probably with height about 50 m, to stand in the lee on the ridge. But even with these unexpected light winds the rotor blades failed after very short operating times. Again this was due to the low fatigue strength of aluminium.

As presented above, in these early years of wind energy exploitation, several materials were considered for the rotor blades. Naturally, the initial focus was on the well-known metals, steel and aluminium. These turned out not to have sufficient fatigue strength and life times for the surprisingly large number, typically 100 to 200 million (see Box 3.3.3), of cycles to be experienced by the rotor blades over the nominal 20 years life time, which was set in these early years. Therefore, alternative materials were considered. Wood offered a possibility, especially with its low density, but concern was raised about its uniformity in structure with the risk of knots. Rather early, the focus was on glass fibre composites with a matrix of polymer, typically polyester or epoxy. Such composite materials presented attractive property profiles, with low density as well as high stiffness when fibres were oriented along the length of the rotor blades. Fatigue properties of glass fibre composites were, on

Box 3.3.3 Rotation, blade tip speed and number of revolutions

The relation between rotor radius R and linear speed of the blade tip through the air v_{tip} is

$$v_{tip} = R * \omega$$

where ω is the rotational speed in radian per second. This is related to the number of revolutions per minute, rpm, by

$$\omega = rpm * 2\pi / 60$$

This means that, for example, 20 rpm corresponds to a rotational speed ω of about 2 s^{-1} = 2 Hz.

Combination of the equations gives the relation for the rpm

$$rpm = \omega * 60 / 2\pi = v_{tip} / R * 60 / 2\pi$$

The number revolutions N for a design life time of 20 years is

$$N = 20 \text{ years} * 365 * 24 * 60 * rpm$$

In practice the optimum extraction of energy by a wind turbine occurs when the blade tip speed is about 7 times the wind speed. With a reference wind speed of 13–14 m/s this gives a blade tip speed of about 100 m/s.

For a large wind turbine with 100 m blade length, the rotational speed is $rpm = \omega * 60/2\pi = v_{tip}/R * 60/2\pi = 9.5$ revolutions per minute.

For this large turbine with rpm = 10, the number N is about 100 million revolutions.

the contrary, not well known, and fairly large programmes were initiated to understand the fatigue behaviour of composites and establish data for design of rotor blades.

4. Many activities in Denmark

Among other people with an interest in harvesting wind energy were master smiths (Danish: Smedemestre), either as individuals or in groups. The smiths had the practical experience in working with metals and manufacturing of steel structures, which was one, but not the only, part for wind turbines. The smiths needed assistance from civil and electrical engineers, and one of the smiths, Preben Maegaard, established

Nordisk Folkecenter for Vedvarende Energi (Nordic Folk-Center for Renewable Energy) in 1978 in the windy north-western part of Denmark. The smiths and engineers at the Center produced the necessary calculations, workshop drawings, components and prototypes, all of which were provided free of charge to the various individual master smiths.

In 1984, one such wind mill, designed by Preben Maegaard, was placed at the Center. It had a 75 kW generator and a rotor diameter of 17 m, and was named DANmark 17 (Smidt, 2024). The three 8 m long wings in glass fibre composites were manufactured by the company LM Wind Power. The rotor main shaft plus gear plus generator formed a compact box, called nacelle, on top of the tower. The nacelle was given a characteristic blue colour. This wind mill is still working today, after 40 years, having operated for nearly three times as many hours as the expected lifetime. The only component which has been replaced is the electronic control. In the period 1978 to 1992, with help from the Nordic Folk-Center, a series of wind mills of various designs were developed and produced, with the final design of 1992 being a 525 kW wind mill. This wind mill was placed at Hanstholm, the north-western point of Denmark, facing into the windy North Sea, and was, at the time, the largest commercial wind mill in Denmark.

4.1 Private companies

From about 1980, the activities in wind energy increased rapidly in Denmark. Many small companies started up, often based on the technical idea of an individual enthusiastic entrepreneur, and with various degrees of commercial success. Company names like Riisager, Kongsted, Vestas, Giro, Vølund, Danwin, Bonus, Nordtank and others emerged. Fig. 3.3.6 shows some of the many different appearances of wind turbines, which the solution to harvest of wind energy took over the years. Most of these activities disappeared or merged, and eventually only few companies, like Vestas and Siemens, survived and became very large and internationally oriented.

4.2 Danish rotor blades

The Danish wings or rotor blades were all designed as sturdy engineering structures, which could support the expected loads for long times, with a nominal design life time of 20 years, and with conservative safety factors where knowledge was limited, especially regarding fatigue behaviour of composite materials. This design approach meant that direct failures of composite blades were extremely few, and some failures of rotors were caused by loss of control rather than materials failure.

In contrast to Denmark, other countries with an aircraft industry, like England and Germany, took outset in the optimised wings of aircrafts for the design and construction of rotor blades for their wind turbines. These early rotor blades were rather light and flimsy compared with the heavy and sturdy Danish blades, and in the early years some failures were encountered.

Figure 3.3.6 Years 1976–1985. Six small wind mills, all with generator power less than 100 kW, notice the different design solutions selected by these six Danish companies (Lilholt, 2024).

5. Tvind wind turbine

In western Jutland of Denmark a group of school teachers had established an alternative folk high school for young people, based on a more free form of teaching with many practical activities (Tvindkraft, 2010). In 1970 the group bought a farm of the name Tvind in windy western Denmark. The school wanted to be independent and self-sufficient in as many aspects as possible. This should be the case also for their energy supply for the school and its facilities, and in 1975 wind energy was decided unanimously by the teachers in plenum, as this was their principle of decision making. A windmill team was established consisting of teachers, students and volunteers. They decided to build a wind turbine with a rotor diameter of 54 m, mounted down-wind on a tower of height 53 m, and with a generator with a power of 2000 kW. The planned Tvind turbine was at the time by far the largest and was developed and built over the next 3 years.

The three rotor blades of length 27 m for the Tvind turbine were made of glass fibre composites with an epoxy matrix. The size and scale as well as the composite material of these blades were rather advanced at the time. The use of epoxy, rather than the more common polyester, was unique. The glass fibres were applied in the form of unidirectional bundles, which were impregnated with the liquid epoxy before they were placed in the mould, to build up the lower part of the blade shape. The use of glass fibres aligned with the length of the blade was unique. Then, the load supporting structure of ribs and webs were inserted, before the upper part of the blade was completed. This was a near optimal exploitation of the potential of fibre composite materials. The process, shown in Fig. 3.3.7, was entirely manual, relying on the large number of school pupils, who were both available and enthusiastic. After allowing for cure of the epoxy, finally the carry-out of the 5200 kg blade from the production tent was by pure man power under the guidance of a foreman with megaphone. In Fig. 3.3.10 of the completed turbine the twist along the blades is visible. This twist of the aerodynamic profile is needed to optimise the orientation of the profile relative to the effective wind direction (see Box 3.3.4).

Figure 3.3.7 Tvind wind turbine, manufacturing of blades: (a) Applying the epoxy, (b) placing the impregnated glass fibre bundles, (c) installing the ribs and webs, (d) carry-out of the completed blade of weight 5200 kg (Tvindkraft, 2010).

Figure 3.3.8 Tvind wind turbine: (a) Design deliberations in plenum, (b) an enthusiastic member of the Tvind windmill team (Tvindkraft, 2010).

The tower of the Tvind turbine was made of reinforced concrete, and the tower construction was designed with help from professional civil engineers, some of whom gave consultancy free of charge. But still the discussions and decisions were taken in plenum by the windmill team, Fig. 3.3.8. The pictures in Fig. 3.3.9 show the steps in the building of the tower, from the symbolic breaking the ground with spades, spoons and cups by everyone who was at Tvind that day, followed by excavating with heavy machinery, over placing of the steel rods, to the casting of the concrete for the tower. Finally, Fig. 3.3.10, some assistance from large machinery was needed to lift and mount the blades.

The Tvind wind turbine was built nearly entirely by the school itself and with little public knowledge about the work. But there was a certain element of outside assistance from enthusiastic individuals. Besides the professional civil engineers, also researchers from Risø National Laboratory assisted with aerodynamic and structural

Figure 3.3.9 Tvind wind turbine: (a) The ground for the tower being excavated, (b) the steel bars for reinforcement of the concrete being layed out, (c) the casting of the concrete in progress (Tvindkraft, 2010).

Figure 3.3.10 Tvind wind turbine: Completion of the turbine, notice the blade twist, the rotor is downwind of the tower (Tvindkraft, 2010).

Box 3.3.4 Wind relative to blade and blade twist

The aerodynamic shape of the rotor blade (and of aircraft wings) is shown schematically in Fig. 3.3.11. The line connecting the nose of the blade and the trailing edge is called the cord. The attack angle α between the cord and the resulting wind direction should be in the range 10–15° to ensure a smooth, so called laminar, flow of air around the aerodynamic profile. This flow produces the maximum lift and the minimum drag for the profile, and is thus the required situation for maximum efficiency of the blade and thus the energy production. The optimal attack angle has different values for different aerodynamic profiles.

The resulting wind speed is the vector difference between the wind speed and the (large) linear speed of the rotor blade, where the vector is characterised by its size (speed) and its direction, as shown in Fig. 3.3.11. The resulting wind direction, as seen by the blade, is dramatically changed from the real wind direction.

Because the linear blade speed is much larger than the wind speed, the orientation of the blade is close to in-plane with the rotor plane. The turbine runs with a fixed revolution speed, given in rpm, and thus the linear speed of each part of the blade varies from root to tip. The wind speed varies due to weather conditions. Thus, in the vector diagram both the blade speed vector and the wind speed vector can vary for these two separate reasons. Therefore, the *resulting* wind speed and direction vary "all the time," and the blade profile orientation relative to the wind ideally needs to be adjusted continuously to maintain the optimal attack angle for the aerodynamic profile.

The linear blade speed at a given position r along the blade, is given by the relation

$$v_{blade} = r * \omega$$

Box 3.3.4 Wind relative to blade and blade twist—cont'd

where ω is the rotational speed and is related to the revolutions via

$$\omega = \text{rpm} * 2\pi/60\ [s^{-1}]$$

The resulting wind direction β is found from geometry of the vector diagram as

$$\tan\beta = v_{wind} / v_{blade}$$

The change in β is ideally handled by giving the blade profile an equivalent *twist angle* from root to tip. Fig. 3.3.11 shows the twist orientation of the blade profile at the blade tip (small profile) and at the root (large profile). This is an important part of the design process. But this built-in twist along the blade is only optimal at a given wind speed, as seen from the vector diagram, and thus a compromise must be reached, relative to wind speed pattern at the wind turbine site. The blade twist is strongest at the inner part of the blade and moderate closer to the tip.

For a rotor blade with given twist, the angle β will change when the *wind speed changes*. For a decrease in wind speed from $v_{wind,1}$ to $v_{wind,2}$ the angle β decreases, as seen in Fig. 3.3.11, which shows a given aerodynamic profile of the blade. To maintain the attack angle α between the resulting wind direction and the cord, the profile of the blade must be rotated from position 1 and position 2 by an amount β_1 - β_2. This rotation control of the blade along its longitudinal axis is called pitch regulation, and is an important element of the operation of a wind turbine to optimise the energy production under varying wind conditions.

rotor design, and moreover, researchers, together with several students, from Technical University of Denmark manufactured a 400 kW frequency converter, which was the power that could be handled by the public grid. In addition a 600 kW immersion heater capacity was installed. On the contrary, there was no official or formal contact between the Tvind project and the Danish government projects (see below), although the two efforts were separately presenting major efforts into the promotion of wind energy. The Tvind organisation made all their accumulated knowledge and experience freely available.

6. Nibe wind turbines

In the years 1976–1980, there was great activity under the Danish government research program for wind energy, which was initiated by the Ministry for Trade

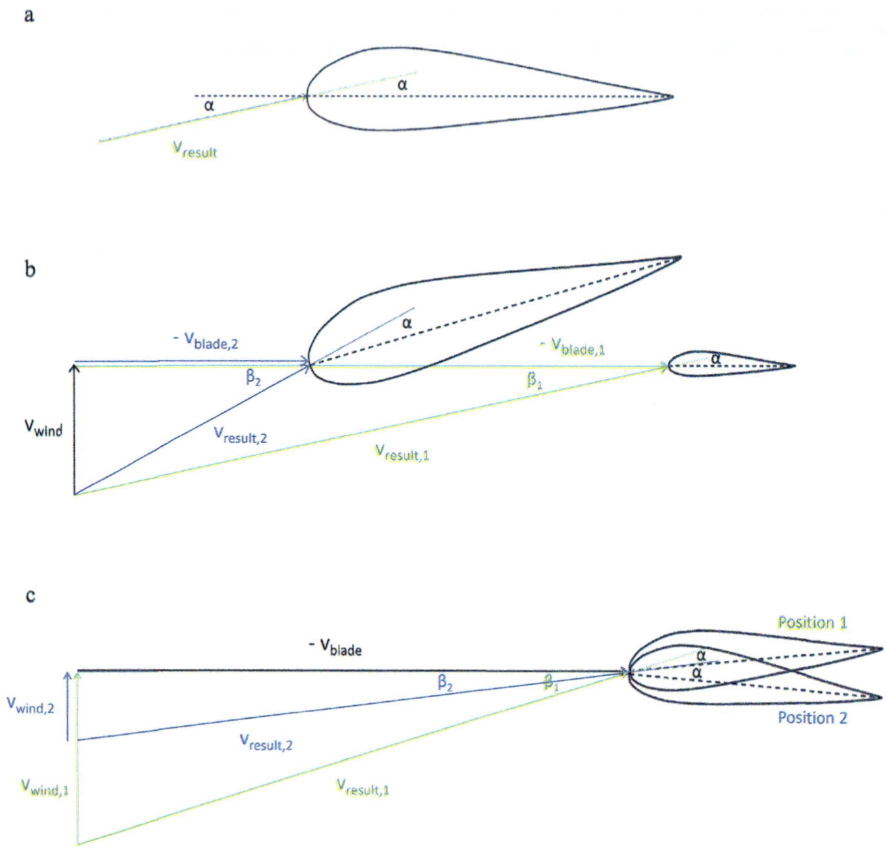

Figure 3.3.11 (a) aerodynamic profile of rotor blade and the attack angle α between the resulting wind direction and the cord of the profile, (b) the vector diagram for wind speed and linear rotor blade speed shows the resulting wind direction and the orientations of the blade profile to maintain the attack angle α, at the blade tip (small profile) and at the blade root (large profile), this change in orientation of the profile constitutes the twist of the rotor blade, (c) the vector diagram for a decrease in wind speed, with the corresponding resulting wind direction, and the change in orientation of the blade to maintain the attack angle α, this change in orientation of the blade constitutes the pitch regulation.

and the Directorate for Technology (Jensen, 2018). The action was in five fields: (1) practical measurements of operation, energy production, and loads and stability of the 200 kW Gedser wind turbine (see also above); (2) development and building of two 630 kW experimental wind turbines; (3) practical measurements of operation, energy production, and loads and stability of these two turbines; (4) measurements and mapping of regional wind energy resources, called wind atlas; (5) establishment of a test station for approval of small wind turbines for government financial support.

The practical measurements on the Gedser wind turbine were possible, because, although its operation had been terminated in 1967, the turbine was still in good shape and could be refurbished. These measurements gave early and important data for the loads on the rotor, and in particular, the energy production at different wind speeds, the so-called power curves showing power as a function of wind speed.

6.1 The Nibe project

The largest effort in the Danish program was the design and development of the two large experimental wind turbines, to be called the Nibe wind turbines after their location in the windy north-western Denmark. The Danish electricity utilities were, somewhat like their noninterest in nuclear energy a few years earlier, not particularly interested in wind energy, but they could not ignore it. So, they became active partners in the Nibe project through their two regional organisations, Elsam for western Denmark and Elkraft for eastern Denmark, with the national electricity organisation DEFU as project leader. They were all professional electrical and civil engineers, and further expertise was given from Risø National Laboratory, on structural rotor blade design (Construction Department) and materials (Metallurgy Department), and from Technical University of Denmark on aerodynamics (Fluid Dynamic Department).

A number of design and materials problems were addressed in the two slightly different Nibe wind turbines. Their common features were: a three-bladed rotor mounted up-wind, a rotor diameter of 40 m, a tower of concrete and height 45 m, and a generator of 630 kW. The Nibe turbine A was the conservative design with the three blades supported with struts, by inspiration from the Gedser wind mill. The Nibe turbine B was the more daring design with the three blades self-supported as cantilever beams.

The leadership of the Nibe project was concerned also about nontechnical aspects, with focus on the potential public acceptance, or otherwise, of wind energy and especially the wind turbines as objects placed in nature. Several architectural designs of the wind turbines were commissioned, and the project group (of mainly technical people) selected the design, which is seen in Fig. 3.3.12. This also shows the geometrical design aspect with the horizontal rotor axle tilted slightly upward, and thus the rotor plane slightly backward, by 6°, to give more clearance for the rotor blades when passing the tower in their down-pointing position. This was done in anticipation of the expected rather large deflection of the blades.

In the initial phase of the Nibe project, selection of the structural load-bearing materials for the rotor blades was addressed, and, for completeness, steel, aluminium, wood and composites were considered, although the general feeling in the project group was that composites would be the optimal, and only realistic, material. Steel was out-ruled on the basis of the need for a large oven to anneal the unavoidable welding seams in a steel construction, while aluminium was out-ruled due its low yield strength. For both metals, a concern was also the nonavailability of fatigue data for the required very long life times of the blades, from 100 to 200 million cycles. Wood had its potentials, but no local expertise on wood was readily available. So, the project steering group received a report advocating composites with glass fibres for the rotor blades.

Figure 3.3.12 Nibe wind turbines: (a) Nibe A in background, Nibe B in foreground, notice window at half height of tower in Nibe B, (b) climbing the internal vertical ladder of 45 m in the tower of Nibe B, (c) Nibe B, initial trials and recording data, 1980. Courtesy: Nibe Project (1977–1988).

The use of composites for components of the required size, shape and function of rotor blades also presented its challenges. The design concept for the blades was a load-bearing beam, so called spar, along the blade and aerodynamic shells to be glued to and around the spar. This spar had a flattened tubular D-shape with the round part of the D following the contour of the aerodynamic blade shape. The D-shaped spar was to be made by winding the glass fibres around a mandrel with shape corresponding to the inner shape of the D-spar. The blades, and thus also the spar, were tapered from a large cross-section at the blade root to a small cross-section at the blade tip, and the wall thickness of the spar was also decreasing from root to tip. The spar had to be straight in the longitudinal direction to allow the completed tubular spar to be pulled off the mandrel.

A further restriction for the Nibe blades was the size of the manufacturing facilities at the company Vølund in Viborg in northern Denmark. This company was the only available Danish location with a filament winding facility of usable size. The company normally produced, by filament winding, composite containers and tanks of some 8–10 m length and with diameters of about 2 m. The blade manufacturing group of three–five persons from Metallurgy Department of Risø National Laboratory were allowed to modify the filament winding machine, and extend it to 13.5 m, permitting the winding of D-spars, which were trimmed to 12 m effective length. This however influenced the construction of the 20 m blades in that the inner 8 m spar for the blade was made as a conventional welded steel structure. A 1 metre steel connecting fixture further was needed to assemble the two parts of the 20 m spar, consisting of steel and composite. All this added to blade weight, but was considered acceptable and not critical, because the main part of the weight was close to the centre of the rotor.

The final design of the blades for Nibe turbine A and B are shown in Fig. 3.3.13. Blade design A was conservative with struts between the blades, being fixed at the 8 m position, and to the centre of the rotor. The use of struts meant that the aerodynamic shells were interrupted at the 8 m position, and that the blades were in fixed orientation relative to the wind during operation. Regulation of power at high wind speeds was obtained by stalling of the blades. Blade design B was more advanced with non-interrupted 20 m long aerodynamic shells, and with regulation of the blades along their longitudinal axis, by rotation at the root, allowing the aerodynamic profile to be turned into optimal position relative to the wind. This control is called pitch regulation (see Box 3.3.4).

6.2 Materials aspects and blade manufacturing

The material selected for the load-bearing spars of the Nibe turbines was composites with glass fibres in a polyester matrix (Johansen et al., 1980). The static properties, like stiffness, were well known at the time, while only few experimental data existed for the fatigue behaviour at very large number of cycles, of the order of 350 million cycles for a 20 years design life time. From the few available fatigue data, the limiting strain for this design life time was estimated, conservatively, at 0.1% strain. A volume fraction of glass fibres of 50 vol% was selected to obtain a rather high stiffness and a good quality of the composite. The optimal orientation of the fibres is along the blade axis. This was not possible with traditional filament winding of bundles, both for

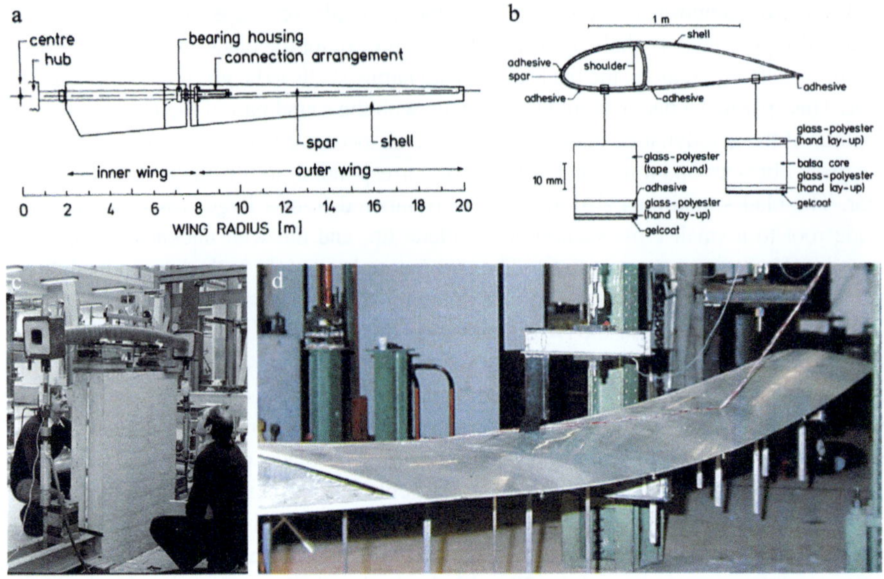

Figure 3.3.13 Nibe rotor blades, structural design and testing of blade: (a) Structural design, Nibe A is supported by stays, Nibe B is free cantilever beam, (b) materials design, load-bearing D-spar of glass fibre composite with fibres aligned along the blade, and aerodynamic shells of sandwich with core of balsa wood, (c) model spar of glass fibre/polyester composite tested for stiffness and strength, (d) outer 6 m of rotor blade tested in bending, deflections are recorded by vertical rulers, observed by binoculars from a safe distance, BKF-Centralen at technical University of Denmark.
(a and b) Johansen et al. (1980), (c and d) Courtesy: Nibe Project (1977–1988).

practical problems of placing the fibres onto the mandrel, and because the D-spar was tapered from root to tip.

The practical solution was to use a so called transverse tape with fibres placed across the tape, so that winding of the tape onto a mandrel would place the fibres along the axis of the mandrel and thus along the axis of the spar (Johansen et al., 1980). This practical manufacturing solution was tried on a small model-spar with constant and square cross-section, which was tested in bending for control of stiffness and strength, Fig. 3.3.13.

The transverse glass fibre tape, used for the Nibe spars, was a tape of 20 cm width, delivering fibres of length 20 cm. Most fibres, 93%, were placed across the tape with the rest (7%) being supporting fibres placed along the tape, Fig. 3.3.14. According to composite theory, the 20 cm fibres are "long" in terms of stress transfer between fibres and matrix, and thus act as continuous fibres. Tape winding requires a certain pitch to advance the winding along the mandrel. To produce a homogeneous and uniform fibre distribution in the composite, tape winding of the Nibe spars was made with a pitch of 10 cm per revolution giving a tape-overlap of 10 cm, and this half width overlap ensured no weak cross-section at fibre ends in the spar.

Figure 3.3.14 Nibe rotor blades, manufacturing of blades: (a) glass fibre transverse tape with 93% of fibres transverse, and 7% longitudinal, the tape is 20 cm wide, and the longitudinal direction is marked by white arrow, (b) tape winding process at Vølund company in Viborg in northern Denmark, inserting the wooden wedges during winding, (c) assembling process of gluing the shells to the spar, at OL boats, Ringsted near Copenhagen, handling the clamps and the lead blocks, (d) centre of gravity and uniformity of weight of three rotor blades are calibrated by brass and wooden rods inserted along the spar.
(a and b and c), (Johansen et al., 1980), (d) Courtesy: Nibe Project (1977–1988).

The pitch of the winding caused the fibres to be not perfectly aligned with the spar axis, but to have a small off-axis angle θ. The winding of the spar is performed from root to tip and back from tip to root, this results in a composite laminate with fibres orientations $\pm\ \theta$. The 12 m spar was tapered from root (at rotor radius 8 m) to tip (at rotor radius 20 m), Fig. 3.3.13.The tapering was designed to give same strain (at max 0.1%) in the spar during bending. This taper had two effects: (i) at the root the off-axis angle was 4° and at the tip the angle was 12°, (ii) during winding the tape was always "too long" along the side facing the spar tip, due to the slightly smaller circumference of the spar closer toward the tip. To maintain the correct placement of the fibres along the curved part of the D-shape the excess tape was "packed" on the vertical part of the D-shape of the spar. This packing was done by inserting wooden wedges temporarily for each revolution, and withdrawing them after the next revolution. This tedious job was done manually, as in Fig. 3.3.14.

The manufacturing process was wet tape winding, meaning that the glass fibre tape was soaked in a polyester bath, before it was running onto the mandrel. The soaking was indirect in the sense that the tape was passing over a rotating steel drum which itself was dipping onto the bath, leaving a thin layer of polyester on the surface of the drum. A steel bar parallel to the drum was mounted in an adjusting fixture to control the layer thickness, so that the correct minimum amount of polyester was impregnated into the tape, to ensure the required high volume fraction of 50 vol% of glass fibres in the final composite. The spools with glass fibre tape and the polyester impregnation bath were mounted on a platform moving along the winding machine at the required speed to follow the winding pattern. Furthermore, the platform could swing around a vertical axis, steered by a guiding rail mounted on the floor, so that the entrance angle for the tape was adjusted to follow the local fibre off-axis angle from 4° at the root to 12° at the tip.

Each single layer of glass fibre tape with its accompanying polyester constituted a layer of the composite laminate and had a nominal thickness of 0.25 mm. The spar wall thickness varied from 24 mm at the root to 6 mm at the tip, so this implied winding more layers at the inner part of the spar, leading to 96 layers at the root and 24 layers at the tip. The layers reaching only a short distance from the root were wound first, ensuring that the final full length (12 m) layers formed the outer smooth surface of the spar. This was important, because the outer surface and shape of the spar needed to fit inside the separately made aerodynamic shells in the completion of the whole rotor blade. The 12 m spar weighed 365 kg, and the whole winding process with its many travels along the spar lasted about 4 h, leaving the production rate at one spar per day.

During the winding, various elements of the production were controlled. The viscosity of the polyester was controlled to ensure optimal wetting of the glass fibres in the tape, and special cure-delaying chemicals were added to the polyester to prevent premature curing during the 4 h winding operation. Fibre orientation and wall thickness of the spar were controlled at intervals.

After the completion of winding, the spar was overwrapped with nylon tape to compress the outer layers, and to contain the low viscosity polyester. After the overnight cure of the polyester, the nylon tape was removed to give a "fresh" spar surface for the subsequent gluing process with the aerodynamic shells. The spar was pulled off the mandrel by hydraulic equipment, not always an easy operation, leaving the mandrel ready for the day's production. A total of seven spars were produced, with six spars being used for the two Nibe wind turbines and one spar was kept as reserve.

The Risø composite team was not particularly popular with the staff of the Vølund company, because they interrupted the normal production routines. The Vølund staff did not follow the details of the spar winding, and the next summer they regretted this, because they now saw a potential new production line for rotor blade spars. The Vølund staff had to pay the Risø team a 2 days consultancy and instruction in the spar winding (!).

As noted, winding with the transverse glass fibre tape gave a small off axis fibre angle θ, due to the necessary pitch of winding. Since the process was alternating between winding from root to tip and back from tip to root, the resulting fibre pattern in the composite laminate was a $\pm\ \theta$ orientation. This meant that the stiffness of the glass/polyester composite varied from 35 GPa at root (with $\theta = 4^{\circ}$) to 32 GPa at tip (with $\theta = 12^{\circ}$).

The D-shape of the spar was not ideal. The load-bearing parts were the upper and lower rather flat parts of the D, and these positions were precisely those where the compaction during winding of the glass fibre tape was minimal, leading to lower fibre volume fraction than the design value. On the contrary, the sharp corners and the "nose" of the D were rather well compacted, leading to higher local fibre volume fractions. This general situation for a winding process for load-bearing spars was very much improved during the following years with further development of rotor blade design, mainly by industrial companies.

Besides the detailed control of the winding process, also the resulting properties of the composite materials in the spars were measured and evaluated: composite samples were cut from the excess length of the spars, and density, fibre volume fraction and porosity fraction were measured. Mechanical testing of the samples delivered stress-strain curves, from which stiffness and strength were calculated. Finally a long-term program for fatigue testing was initiated.

The aerodynamic shells for the Nibe blades were produced by conventional boat building technologies, consisting of sandwich constructions with a central core of balsa wood and outer skins of randomly oriented glass fibre mat/polyester composites. The shell for one blade consisted of three parts, one curved part and two nearly flat parts, which were all glued onto the spar and to each other, by a rather complex manual process, as can be seen in Fig. 3.3.14 from the company OL Boats. A critical aspect here was the fit between the outer spar surface and the inner shell surface, where the unavoidable gap had to be filled by the glue consisting of polyester plus mineral powder. This glue only added weight to the blade and of course had to be minimised, but in practice a couple of 100 kilogrammes of glue were needed for each blade.

An experimental blade was made for testing. The outer 6 m (of the 12 m blade) was loaded in bending to simulate the wind load, see Fig. 3.3.13. For safety reasons the bending was upward, so in case of failure, the blade would slam into the concrete floor in the laboratory. Also for safety reasons, the bending deflections were measured by free hanging rulers, which were observed with binoculars from a distance.

The blades were calibrated for same centre of gravity and same weight by inserting brass and wooden rods in a tube mounted along the "back" of the D-spar during the assembly process for spar and blade, see Fig. 3.3.14. The three blades had to be identical to avoid the risk of vibrations of the rotor during operation. The completed 12 m blade had a weight of 910 kg, including the steel connector, see Fig. 3.3.13. A modern experimental blade of length 12.6 m has a weight of 398 kg (Haselbach et al., 2020).

6.3 The Nibe wind turbines and their operation

When the Nibe wind turbines started operating, turbine A in August 1979 and turbine B in March 1980, testing and measurement programmes were initiated for both energy production and performance and stability of the various parts of the turbines, such as rotor, gears and generator. The energy production was monitored continuously by people from the utilities Elsam and Elkraft, and over a rather long period the energy (electricity) production was stable, and the efficiency of the turbines was evaluated.

Moreover, a materials control program for the Nibe turbines was initiated related to the glass fibre composites in the rotor blades. During the gluing operation in manufacturing of the blades, strain gauges were placed on the outer surface of the D-spar of one blade, at a number of positions along the blade. This allowed a continuous monitoring of the strain history of the blade, which would give information about the potentially increasing deformation of the blades, as time goes by, which might be caused by a continuous reduction in stiffness, due to increasing damage of the composite material. Initially no increase in strain was detected, and the test program was terminated, due to lack of funding, before meaningful data could be collected. The strain gauges and their signals were also used to record the free vibration of the rotor blades, by pulling the blade tip with a rope from the tower window, see Fig. 3.3.12, releasing the rope and letting the blade vibrate freely until it came to stand still. The strain vibration pattern was used to evaluate the resonance frequency of the blade for comparison with the original calculations. Furthermore, simple bending tests of the blades on the turbine were performed (Lilholt, 2024). These tests were made from the platform inside the tower at half height, where the window allowed access to the blade tip in its downward pointing position. The bending tests were made by pulling a rope, attached to the blade tip, and reading off a load monitor. The displacement of the rope along the floor was measured by simple steel rulers. Altogether, these tests for materials control were initiated, but not continued due to lack of funding.

The two Nibe wind turbines and the related measurement program, mainly recording the energy production, were running from 1977 to 1988.

7. Fatigue behaviour of fibre composite materials

At the Risø National Laboratory, work continued in the 1980′s and expanded on materials development, with focus on materials testing, especially fatigue at very long times corresponding to the expected life times for wind turbines. With the Nibe turbines as an example, a design life time of 20 years will bend the blades in fatigue about 350 million times at a rotational speed of 33 rpm, which is equivalent to a cyclic loading frequency of about 3.5 Hz (3.5 s^{-1}). Laboratory testing should accelerate this cyclic loading considerably to achieve results and data for fatigue behaviour within a realistic time. But increasing the frequency is met with the problem of self-heating of the composite material in the test specimens. Due to a low thermal conductivity, a temperature gradient is set up in the specimens with an increased temperature in the inside and the lower, external temperature at the outside. Cooling of the test specimens will keep the outside temperature low but cannot remove the gradient, so the test specimens are on average at an increased temperature. This will give nonrealistic, pessimistic test results of fatigue performance, which will lead to a conservative inefficient design. The challenge of self-heating makes it difficult to perform realistic, accelerated experiments. At Risø National Laboratory it was decided to run fatigue testing at a moderately increased frequency of 5–10 Hz. As 1 day has 86,400 s, it meant that 1 day testing was equivalent to nearly one million cycles at

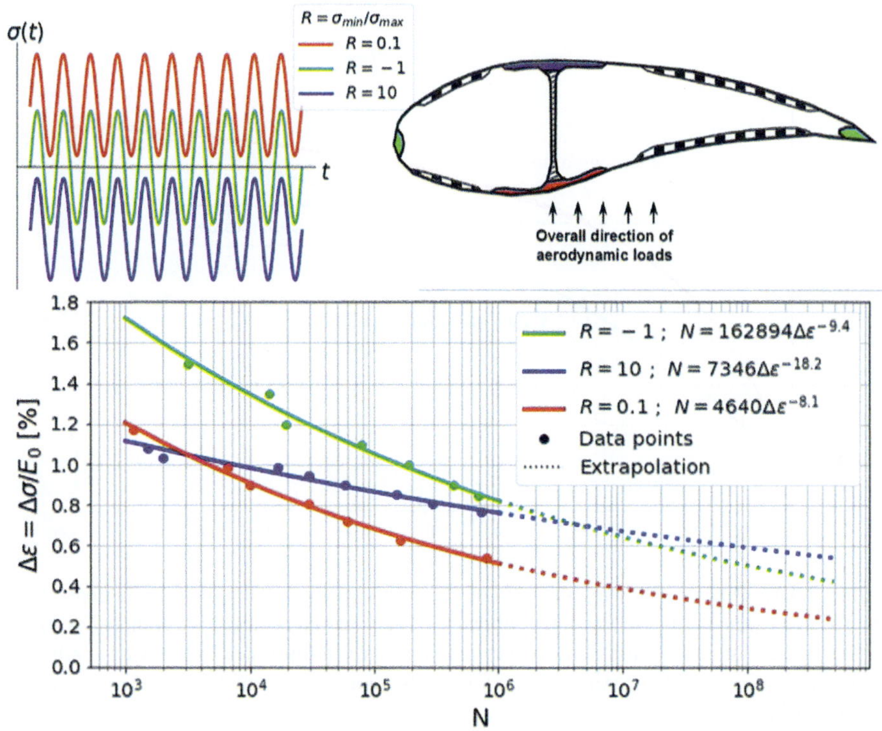

Figure 3.3.15 Fatigue diagram for glass/polyester composites with fibre volume fraction 0.59, showing the life time in number of cycles N, as function of the strain range $\Delta\varepsilon = \Delta\sigma/Eo$. The stress amplitude $\Delta\sigma$ is shown top left as function of time t, for three loading cases: Tension-tension fatigue, R = 0.1, *red*, compression-compression fatigue, R = 10, *blue*, tension-compression fatigue, R = −1, *green*. The R-ratio is defined from the min. stress and the max. stress of the amplitude. The colour codes in the sketch of the blade show the regions of the blade which are loaded in the respective fatigue mode.
Courtesy: Lars P. Mikkelsen, DTU, redrawn from data in Mortensen et al. (2023).

10 Hz. Thus, it needed 100 days of constant, reliable running of the test machines to achieve 100 million cycles. This turned out to be a practical challenge, especially when good results, see Fig. 3.3.15, required a large number of tests and several test machines to have trustworthy data and statistics.

Due to these challenges of fatigue testing, also materials models were developed to describe and understand the degradation of the composites, both damage to their microstructure and reduction in especially stiffness (Sørensen and Goutianos, 2019; Sørensen et al., 2021).These materials models were used to attempt realistic predictions of the load or strain limits for the 100 to 200 million cycles life times of wind rotor blades. During the years Risø National Laboratory was also giving consultancy services to the Danish wind turbine industry, and with several manufacturing projects involving different rotor blade designs and geometries as well as different composite

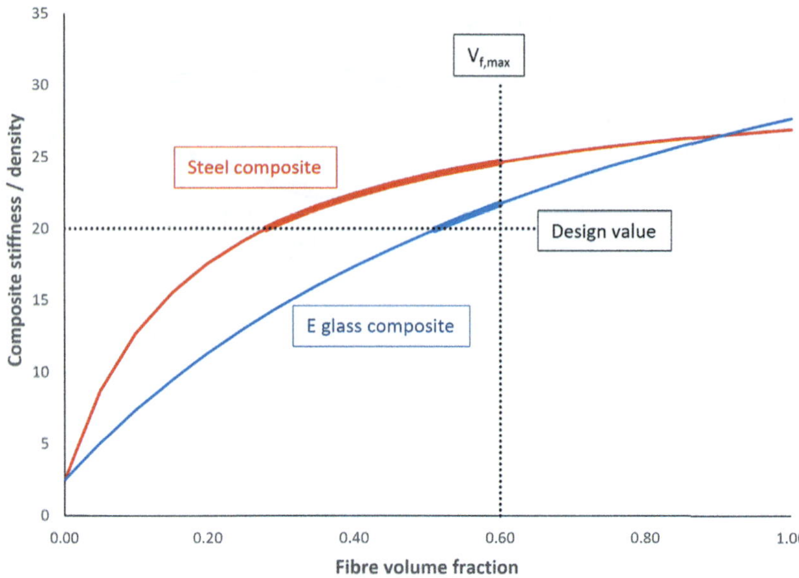

Figure 3.3.16 Diagram showing the ratio of composite stiffness (GPa)/composite density (g/cm^3) as a function of volume fraction of fibres, V_f. The potential of steel fibre composites (*red* curve) over traditional E-glass fibre composite (*blue* curve) is illustrated by the heavy parts of the respective curves. The maximum fibre volume fraction is 0.60, and the design value is 20 GPa/(g/cm^3).

materials, like glass fibres in matrices of thermoplastic polymers, and like steel fibres in a polyester matrix, see Fig. 3.3.16 below.

From 1985 the European Union initiated research programmes, called Frame Work Programmes, for large projects with partners from both industry and universities. Over the period from 1988 to 1995, EU supported three successive projects, JOULE 1, JOULE 2, and JOULE 3, all focussing of the fatigue behaviour of composite materials with special reference to wind energy applications, and with Risø National Laboratory as the project leader.

During the 1980′s and 1990′s, still larger wind energy companies consolidated themselves, and established their own research and development activities, and their own, often large scale, test facilities especially for rotor blades.

8. Materials principles for rotor blades

Steel is generally too heavy relative to its mechanical properties. Long time fatigue behaviour for welded parts is not reliable.

Wood has a structure like composite materials with long cellulose fibres aligned with the growth direction of trunk and branches, and with a low density, giving acceptable mechanical properties relative to weight. Concern about uniformity and the risk of knots have prevented wood from being used. A potential for cut-up and re-glued structures in the form of laminated beams might be a future solution for rotor blades.

Glass fibre composite materials were early candidates for rotor blades. Early use of random fibre orientation in form of glass fibre mats demonstrated clearly too low stiffness, and emphasised the need for a high proportion of long aligned fibres. Early successful uses of unidirectional fibres in glass/polymer composites were the Tvind wind turbine with aligned glass fibre bundles, and the Nibe wind turbines with transversely oriented fibres in tape winding.

A design and selection parameter for materials is the often used ratio of stiffness relative to density, E/ρ. A plot of this parameter as a function of fibre volume fraction, for various composites, in Fig. 3.3.16, gives guide lines for the "best" composite, that is, the highest value for E/ρ. With practical manufacturing limitation on fibre volume fraction of about 60%, and a minimum required stiffness, the curves show that glass fibre/polyester is inferior to *steel fibre/polyester composites* in the actual range of fibre volume fractions. This was the rationale behind a Danish project from the company LM Wind Power with partners from universities, Risø National Laboratory and European industry, which focused on development of steel fibre/polyester composites for rotor blades. After about 2 years it had to be realised that the practical handling of steel fibres, in the form of woven fabrics, was unmanageable for the doubly curved surfaces of rotor blades. The reason was that steel fibres were only available as rather thick "fibres" of diameter 100 μm, which combined with the high stiffness of 200 GPa for steel, made the individual fibres rather nonflexible, and the fabrics not easy to drape over the curved surfaces of the rotor blades.

This project was actually redirected toward *hybrid fibre composites*, based on two different types of fibres in combination. By combining glass fibres and carbon fibres, an improved stiffness could be obtained. These efforts were taken up to allow design and manufacturing of very long rotor blades of 80–100 m, without increasing the blade weight too dramatically. The high cost of carbon fibres was also a parameter, which resulted in an optimal glass/carbon fibre fabric with 1/3 carbon fibres and 2/3 glass fibres. This was successfully used in rotor blades of 88.4 m from LM Wind Power, see Fig. 3.3.17, with an optimised weight of 34 tons. This blade was manufactured in 2016 and was the largest and most advanced blade at the time (LM Wind Power, 2024). Fig. 3.3.18 shows the potential weight saving for hybrid blades, for example, at a length of 100 m the weight saving is 10 tons on a blade weight of about 50 tons.

Pultruded profiles of composites with unidirectionally oriented fibres are common constructional products, with a similarity to wood planks. They have high fibre volume fractions and thus high stiffness. They can be used as "building blocks" for larger structures. LM Wind Power has recently exploited this concept (Korsgaard, 2024), by using pultrusions of carbon fibre composites as load-bearing material in the spar caps of the blades. The spar caps can be seen in Fig. 3.3.15 as the red and blue parts of the blade profile.

Figure 3.3.17 Large rotor blade manufactured by LM Wind Power: This 88.4 m blade was manufactured in 2016 and was the longest and most advanced blade in the world. The blade weight of 34 tons was optimised by using composites with a combination of glass fibres and carbon fibres in a hybrid construction (LM Wind Power, 2024).

9. Design principles for rotor blades

All rotor blades have an outer shape, which generates the aerodynamic airflow with the lift and drag which translate into rotational energy. At the same time the blades must take up the mechanical loads from wind and from gravity. This function has been achieved in various ways, from the beginning in 1970′s up to present days. The early design was a load-bearing D-spar in the Nibe wind turbines, this spar was rather heavy and inefficient, but manufacturing facilities made it necessary. Later designs divided the various sections of the D-spar into separate parts. The LM Wind Power LM 19.1 rotor blade, see Fig. 3.3.19, had the upper and lower load-bearing parts at the thickest sections of the blade, combined with shear webs, corresponding to the vertical sections of the D-spar. The load-bearing parts were made from high performance composites with high fibre volume fractions and aligned fibres. This central box like part of the blades became larger, stronger and heavier, during the developments in the following years, while the rest of the blades remained light and often was made with sandwich constructions. Modern very large rotor blades of about 100 m length have shear webs, which themselves are made as sandwich constructions to stabilise them against buckling. All these design elements are introduced to keep the blade weight down, while still ensuring a stiff and strong blade.

The length wise design of rotor blades developed quickly from the rather clumsy design of the Nibe rotor blades with an inner (8 m) steel part and an outer (12 m) glass fibre composite part. The improved blade design consisted of a circular root of steel for fastening of the blade to the hub on the nacelle. The blade itself had a short transition section from the aerodynamic profile to the circular steel root. The sizes increased from root diameter 0.31 m and 24 bolts (see Fig. 3.3.4) to root diameter 5 m and over 150 bolts (see Fig. 3.3.22).

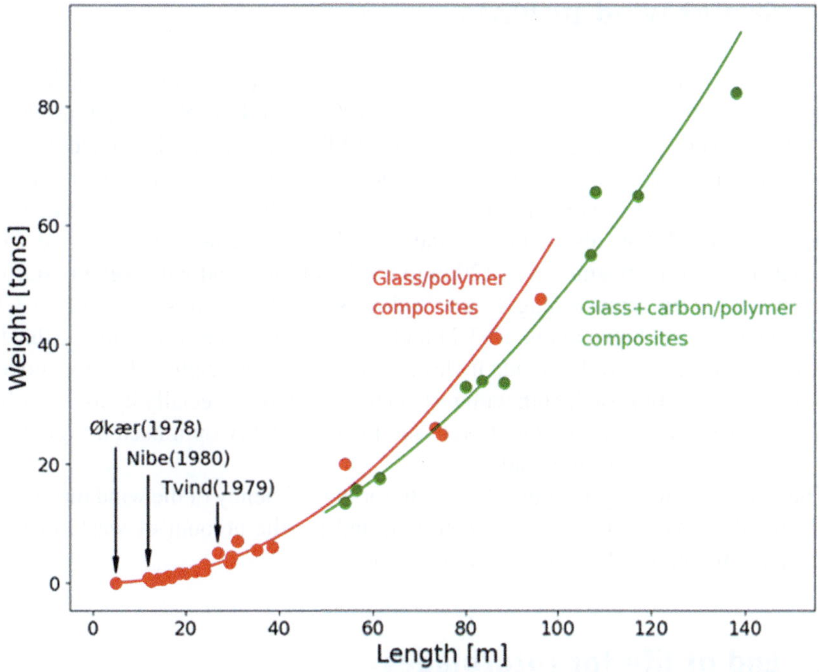

Figure 3.3.18 Diagram showing weight in tons as a function of length in metre for one rotor blade. The red curve and data represent glass/polymer composites, and the green curve and data represent hybrid of glass & carbon/polymer composites. The curves are fitted to the data, and illustrate the weight reduction obtained by using a certain amount of stiffer carbon fibres in the composites.
Courtesy: Lars P. Mikkelsen, DTU.

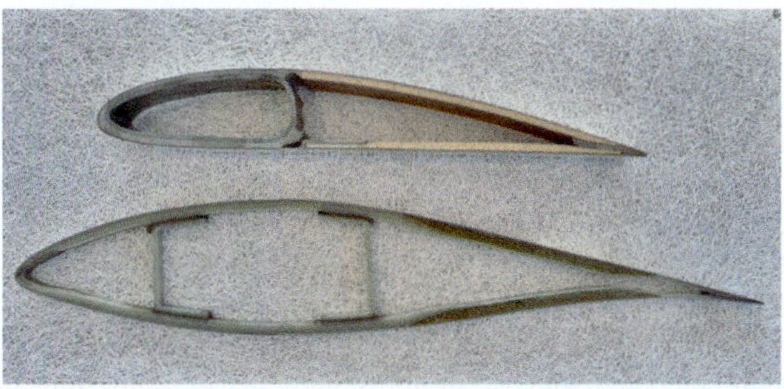

Figure 3.3.19 Section of Nibe rotor blade (top) and of LM Wind Power LM 19.1 rotor blade (below), the design principles can be compared: The heavy and inefficient D-spar in the Nibe blade against the lighter and simpler design with thick parts (spar caps) and vertical shear webs as a box-like configuration in the LM 19.1 blade.
Courtesy: Jakob Bech, DTU.

10. Size of wind turbines

The early wind turbines had a power rating, for the generator, of few 10's of kW, with a large step taken by the Tvind wind turbine at 400 kW and the Nibe wind turbines at 630 kW, to ever increasing power levels of 10 MW and larger. The wind turbines themselves increased correspondingly in dimensions, with rotor diameters reaching 240 m, and the tower of similar heights, meaning that the tip of an upward pointing blade would be 360 m above the ground. The blade lengths thus increased from less than 10 m in the beginning to 120 m in modern times, and with weights starting at about 100 kg and approaching 85 tons today. Short blades are shown in Fig. 3.3.4 and long blades are shown in Figs. 3.3.21 and 3.3.22. It should be remembered that the early rotor blades were rather crude in design and were overweight, while the modern blades are highly optimised, both with respect to materials, especially by using carbon fibres to replace some of the glass fibres, and through highly optimised design of the load-bearing geometry of the blade.

The reduction in weight is important, both for the efficiency of the wind turbine, for the materials consumption in manufacturing, and for the amount of waste material produced at the end of life for the wind turbines.

11. End of life for rotorblades

The increasing focus on wind as an energy production method without emission of CO_2 has over the years led to increasing concern over the handling of the rotor blades when their service life had terminated (Beauson, 2022). Both the number of wind turbines and the size of them has produced an enormous quantity of mainly glass fibre composite waste, estimated to be of the order of 35,000 tons per year (in 2022) (Beauson, 2024). The situation is both one of material quantity and one of material form. The material itself is a combination of glass fibres and polyester, and the design of the composite in terms of content of fibres and the placement and orientation of the fibres is characteristic for the intended application, that is, rotor blades. Therefore, the blades themselves are not directly re-usable.

Ideas and attempts to separate the fibres from the matrix for re-use as "new" fibres and "new" matrices have been made, but with limited success. One idea was to heat the composite to about 500°C, thus letting the polymer matrix burn off to produce energy, and leave the glass fibre free and clean for re-use. The positive element was that the polymer, originally produced from crude oil, was delivering the energy content of the crude oil, which it would have produced if it were burnt off directly. The glass fibres recovered from the process turned out to have reduced strength compared with virgin fibres, and thus were not immediately applicable for re-use. The problems for this idea and other attempts have not been solved. One method, called cement co-processing, is based on the above idea, where the composite is used as fuel in the cement production, and the polymer delivers energy, and the glass fibres eventually act as fillers and reinforcements in the final cement (EuCIA, 2024).

12. Challenges due to size

The ever increasing size of wind turbines, in terms of power, leads to the corresponding larger rotor diameters and thus the increased linear speeds of the blades tips. This causes noise, such that higher blade speed gives higher noise level. The limiting tip speed of about 100 m/s (see Box 3.3.3) results in acceptable noise levels.

These increased and high linear speeds for the rotor blades, in particular for their outer parts, have over the last 10–15 years revealed a partly unexpected phenomenon, namely the erosion of the leading edge of the blades (Mishnaevsky Jr. et al., 2021). This leading edge erosion is caused by the high speed impact of rain drops onto the surface, resulting in a wear process of the composite material, see Fig. 3.3.20. This wear is causing a slow disintegration of the material, as well as resulting in a non-smooth surface, leading to disturbed air flow and thus reduced energy production. Research projects and counter measures are in progress to solve the problems. Improved materials combinations, especially for the polymer matrix, are attempted. Also coatings on the leading edge surface are studied.

Studies to understand the rain drop erosion process are in progress, and the Risø National Laboratory has installed a rain erosion facility allowing detailed experiments to be made. But the problems have not been solved, so repair technologies are an accompanying necessity. Inspection of rotor blades by drones have become wide spread, and in situ repair technologies are developed. As an example, the company Clobotics have developed a repair unit, which can "sit" on the leading edge, when

Figure 3.3.20 Leading edge of rotor blade, showing the erosion caused by impact of rain drops: The surface is very rough, the composite material is heavily attacked, and exposed glass fibres (brownish) are seen in the middle of the picture, the blade comes from the Vindeby wind park, and has been operating for 25 years, at tip speed about 60 m/s.
Courtesy: Jakob Bech, DTU.

Figure 3.3.21 Transport on roads: (a) Three blades of 7.5 m length from Økær Vind Energi, about 1980, (b) one 88.4 m blade from LM Wind Power, 2016.(a) (Winds of Change, 2003), (b) (LM Wind Power, 2024).

the blade is at stand still in horizontal position. The surface is cleaned, prepared, and new material added in a continuous process, operated remotely from the ground.

The size of all components of modern wind turbines has presented a number of logistic problems in transporting them from factory to wind site. In particular the very long, about 100 m, and heavy rotor blades require carefully planned transport on roads, taking note of traffic, corners, roundabouts, bridges and other obstacles, see Fig. 3.3.21. Air lift of blades has been considered, with no clear evaluation of its potential. The increasing use of sea based wind parks requires sea transport, and challenges are related to placing the rotor blades on existing ships.

13. Testing of wind turbines

From the beginning of wind energy times it was realised that a system to ensure the safety of wind turbines was needed. In the Danish wind energy research programmes of the 1970′s a Test Station for Small Wind Turbines was established. It was located on the site of Risø National Laboratory, where a narrow strip of land in the agricultural research fields was allocated as test site for small wind mills. This strip was running north-south, so the turbines were exposed to the generally westerly winds blowing in from across the Roskilde fjord. On another small strip of land nearby four huts were placed to house the staff and facilities for recording measurements on the wind turbines.

The many small companies of those days could have their wind turbine placed on one of five positions on the test strip for a lengthy period of 6–9 months, where both the mechanical and electrical safety of the turbine were recorded. The energy production was also measured, and allowing for seasonal variations in wind speeds and thus energy content, the annual energy production was estimated. These measurements were the basis for an approval system, which allowed the companies both to advertise their particular wind turbine and potentially receive governmental financial support for development. The approval system was instrumental in activating the market for small wind turbines as well as establishing contact between the growing research in wind energy at Risø National Laboratory and the needs of industry.

Figure 3.3.22 The Østerild test centre in windy north-western Denmark: (a) Three large rotor blades of length about 100 m, (b) wind turbines on test site.
Courtesy: Tom L. Andersen, DTU.

In the following years of growth in wind energy, the Test Station became an integral part of the Wind Energy Department. In the 1990′s the wind turbines became so large that a test site was needed with space for the large turbines and with stronger winds. The Høvsøre Test Centre on the west coast of Jutland in Denmark was established in 2002 and had space for five wind turbines with towers of up to 110 m and rotor diameter of 110 m. The facilities were expanded in 2019 to seven spaces and maximum size 133 m. In 2012 an even larger test station was established in north-western Jutland, the most windy region of Denmark. This Østerild Test Centre, Fig. 3.3.22, could accommodate seven turbines with towers of 167 m and rotor diameters of 167 m, and in 2019 expansion gave a total of nine spaces, with five of these allowing sizes up to 220 m. The test centres are used by companies for testing and improving their prototype wind turbines before they are released for production.

Besides the complete wind turbines, in particular the blades themselves are tested for comparison with the design process including both composite materials and geometrical aspects. The Large Scale Facility, situated at Risø National Laboratory, is such a test stand, where rotor blades can be tested in bending under wind load patterns,

Figure 3.3.23 Large scale facility at DTU Risø Campus: (a) Test stand for 45 m blades, 31 m IWES blade ready for testing, (b) test stand for 25 m blades, experimental 12.6 m rotor blade tested to complete failure.
Courtesy: (a) Kim Branner, ReliaBlade project, DTU, (b) Xiao Chen, DiscoverBlade project PI, DTU.

both for stiffness characteristics and fatigue behaviour, see Fig. 3.3.23. The large test base can accommodate blades up to 45 m in length, and due to the flexibility of glass fibre composites the start position of the blade is pointing upward, to permit the large deflection of the blade. This test facility can be compared with the 1982 test stand for the small rotor blades in Fig. 3.3.5. For very large deflections, the floor of the hall can be opened, so that the blade tip will end in a position below floor level. Tests to deliberately overload an experimental blade to failure have been performed, as shown in Fig. 3.3.23.

14. …… and next: offshore wind parks

The wind energy started on land, and from the Wind Atlas it was possible to identify windy regions for the ever larger wind turbines. Over the same period a slowly increasing concern and resistance against the wind turbines as elements in nature built up. The obvious place to look was to the sea and ocean, but initially concerns here were on the practical difficulties and extra costs involved in both erecting the wind turbines off shore and the complexities of inspection, service and repair. The advantages at sea is the stronger and more uniform and unobstructed wind, and disadvantages are the harsher weather with the effect of water and salt spray on the structural parts of the wind turbine.

The world's first off shore wind park was established in southern Denmark on the Vindeby field about 2 km off the north coast of Lolland in shallow waters, (Vindeby Wind Farm, 2024). This important step was taken in 1991 with a park of a total of 5 MW power with 11 wind turbines each of 450 kW power with tower height 35 m and rotor diameter 35 m. After 25 years of service the park was decommissioned in 2017, having produced a total of 240 GWh electricity, corresponding to a capacity factor of 22%. This is much lower than the general value of about 40% for later and modern wind parks. The Vindeby wind park gave valuable information and experience for the further tremendous development in wind at sea.

One of the most recent wind parks with large power and very large wind turbines is the Vineyard Wind Park off the north east coast of USA. It consists of 62 wind turbines each of 13 MW, giving a total of 806 MW, or nearly 1 GW, which is about the size of a nuclear power station. The wind turbines have a tower of 260 m and a rotor diameter of 220 m, and thus a maximum height of 370 m.

The wind turbines are ever increasing in numbers and in size, and the corresponding energy production is enormous and is finding its applications along several routes. The chain of energy conversions, each step with its loss of energy, is from wind to rotational mechanical energy, to electricity, to be used directly in the grid system, or, for example, for cars, or for electrolysis of water to produce hydrogen, or for other ways of energy storage. The hydrogen can enter the chemical processes with CO_2 or CO for production of artificial fuels, called Power-to-X, like car petrol and aircraft fuel. These developments are relying on wind energy—and some day they will be history.

Acknowledgement

The elements of this historical presentation have been collected from a multitude of sources of different origins, spanning the time since the early years of wind energy. We have tried to trace the origins or sources, but not succeeded in all cases, and we apologize to those sources which we have not been able to identify. We should particularly like to thank for the contributions from colleagues of the previous Risø National Laboratory and the present Department of Wind and Energy Systems, Technical University of Denmark: Tom L. Andersen, Justine Beauson, Jakob I. Bech, Kim Branner, Xiao Chen, Peter Hjuler Jensen, Lars P. Mikkelsen, Steffen Rasmussen, Bent F. Sørensen. We furthermore thank the following people who are among the central actors in the development of wind turbines: Erik Grove of Aeroform, Britta Engelbrecht Jensen of Tvindkraft, and John Korsgaard of LM Wind Power.

References

Beauson, J., 2022. End-of-life of wind turbine blades – value chain, recycling and composite materials. PhD thesis. In: Department of Wind and Energy Systems. Technical University of Denmark. https://doi.org/10.11581/dtu.00000221.

Beauson, J., 2024. Personal Communication.

Betz, A., 1920. Das Maximum der theoretisch möglichen Ausnützung des Windes durch Windmotoren. Zeitschrift für das gesamte Turbinenwesen 17 (26), 307–309.

EuCIA, 2024. Report by SGS Intron B. V. "LCA of composite waste processing," 7 pages. In: https://eucia.eu/wp-content/uploads/2024/10/LCA-of-Composite-Waste-through-Cement-Co-Processing.pdf.

Grove, E., 2024. Personal Communication.

Haselbach, P.U., Semenov, S., Berring, P., 2020. DTU's blade research and demonstration platform. IOP Conf. series: Mater. Sci. Eng. 942 (1), 012043. https://doi.org/10.1088/1757-899X/942/1/012043.

Jensen, P.H., 2018. 40 Years Anniversary Presentation. Personal Communication.

Johansen, B.S., Lilholt, H., Lystrup, A., 1980. Wingblades of glass fibre reinforced polyester for a 630 kW wind turbine, Design, fabrication and materials testing. In: Bunsell, A.R., Bathias, C., Martrenchar, A., Menkes, D., Verchery, G. (Eds.), Proceedings of the 1980 International Conference on Composite Materials (ICCM 3). Pergamon Press, Paris, pp. 1355–1367.

Juul, J., 2024. In: Wikipedia. https://en.wikipedia.org/wiki/Johannes_Juul.

Korsgaard, J., 2024. Personal Communication.

Lilholt, H., 2024. Personal Recollections.

Mikkelsen, L.P., 2016. A simplified model predicting the weight of the load carrying beam in a wind turbine blade. IOP Conf. series: Mater. Sci. Eng. 139 (1), 012038. https://doi.org/10.1088/1757-899X/139/1/012038.

Mikkelsen, L.P., 2020. The fatigue damage evolution in the load-carrying composite laminates of wind turbine blades. In: Fatigue Life Prediction of Composites and Composite Structures, second ed. Woodhead Publishing Series in Composites Science and Engineering, pp. 569–603. https://doi.org/10.1016/B978-0-08-102575-8.00016-4 (Chapter 16).

Mishnaevsky Jr., L., Hasager, C.B., Bak, C., Tilg, A.-M., Bech, J.I., Rad, S.D., Fæster, S., 2021. Leading edge erosion of wind turbine blades: understanding, prevention and protection. Renew. Energy 169, 953–969. https://doi.org/10.1016/j.renene.2021.01.044.

Mortensen, U.A., Rasmussen, S., Mikkelsen, L.P., Fraisse, A., Andersen, T.L., 2023. The impact of the fiber volume fraction on the fatigue performance of glass fiber composites. Compos. Appl. Sci. Manuf. 169, 107493. https://doi.org/10.1016/j.compositesa.2023.107493.

Nielsen, H., Nielsen, K., Petersen, F., Jensen, H.S., 1998. Til Samfundets Tarv - Forskningscenter Risøs Historie (To Society's Need – Research Centre Risø's History), Danmarks Tekniske Universitet, Risø Nationallaboratoriet for Bæredygtig Energi, pp. 388–410.

Poul la Cour, 2024. In: Wikipedia. https://en.wikipedia.org/wiki/Poul_la_Cour.

Smidt, H., 2024. Smedemestermøllen Fylder Rundt Og Høster Stadig Vinden. Nordiske Medier.

Sørensen, B.F., Goutianos, S., 2019. Micromechanical model for prediction of the fatigue limit for unidirectional fibre composites. Mech. Mater. 131, 169–187. https://doi.org/10.1016/j.mechmat.2019.01.023.

Sørensen, B.F., Goutianos, S., Mikkelsen, L.P., Fæster, S., 2021. Fatigue damage growth and fatigue life of unidirectional composites. Compos. Sci. Technol. 211, 108656. https://doi.org/10.1016/j.compscitech.2021.108656.

Tvindkraft, 2010. Tvindkraft. Retrieved 9 September 2024, from. www.tvindkraft.dk.

United Nations, August 1961. Proceedings of the United Nations Conference on New Sources of Energy, vol 7, pp. 21–31.

Vindeby Offshore Wind Farm. September 15, 2024. In Wikipedia. https://en.wikipedia.org/wiki/Vindeby_Offshore_Wind_Farm.

LM Wind Power, 2024. https://www.lmwindpower.com/en/products/blade-types/longest-blade-in-the-world.

Winds of Change, 2003. Erik Grove. Retrieved 2 September 2024, from. www.windsofchange.dk.

Citius, altius, fortius: The crucial role of composites in sports

3.4

Helena Teixidó[1], *Véronique Michaud*[1] *and Ignaas Verpoest*[2]
[1]Laboratory for Processing of Advanced Composites (LPAC), Institute of Materials (IMX), Ecole Polytechnique Fédérale de Lausanne (EPFL), Lausanne, Switzerland; [2]Composite Materials Group, Department of Materials Engineering, Katholieke Universiteit Leuven (KU Leuven), Leuven, Belgium

1. Introduction

Apart from very few cases where specific equipment is not required (such as walking or running barefoot, or swimming), the practice of sports, whether for leisure or competition, requires making or acquiring specific parts, such as poles, bats, balls, adapted shoes, rolling, sliding or floating devices. Historically, natural materials, such as wood, plants, stones, leather, natural latex, and beeswax, were employed to create these adapted devices that could be used to play games or enable improved transport. As examples, Mesoamerican ball games, probably the ancestors of racquetball games, were played at least 3700 years ago in South America using natural rubber balls and stone, then ceramic racquets; a 5000-year-old petroglyph of a skier named Rødøy-man, has been found in Norway, and was used as the logo for the 1994 Olympics in Lillehammer (Blomster and Salazar Chávez, 2020).

Competition sports have driven the development of improved devices, which often require to be durable and strong, but also very light, and, depending on the cases, able to dissipate or elastically restitute energy optimally. These objects can be quite expensive and made in small quantities; as athletes are often ready to invest in equipment specifically made for them to improve performance, provided that it is allowed by the rules, sports applications have always represented a very low entry barrier to the use of any novel material combination. As a result, lightweight metallic alloys have been adopted in sport as soon as they were developed for aerospace applications, for example in boat hulls, skis or pole vault. Composite materials were no exception: as they could easily replace wood or plant structures that also show anisotropy, they entered sports applications at about the same time as they were developed for structural applications in aerospace or transport, in the early 40s, as will be discussed further.

In turn, sports applications have constituted a very appreciated test bed for companies developing novel composite material combinations to provide early feedback from manufacturers and users, since the safety regulations are often limited, and the time for development and continuous improvement very short, about 1 year (Mansson, 2020). Composite manufacturing methods span a very wide range in terms of

The History of Composites. https://doi.org/10.1016/B978-0-443-34119-9.00021-8

cost, complexity, and production volumes, from the homemade surfboard prototype by hand layup in a garage, up to aerospace grade autoclave manufacturing of carbon fibre reinforced masts from prepregs, through the large volume manufacturing of skis or tennis racquets by hot press-moulding. It is thus possible to manufacture several prototypes and test them immediately in real conditions, and to quickly adapt to the needs of an athlete or a team.

The field of sports is very varied, with different needs in terms of physical and mechanical part properties, but also in terms of regulations, and diversity of embodiments. Many composite types have thus found either very niche applications, for example, thanks to their electrical conductivity in fencing guards, or widespread use in boats, cycling or racquet games thanks to their high specific stiffness as compared to other alternatives. The large community of sports enthusiasts, willing to invest in advanced equipment to come closer to the professional athletes they so admire, also creates a vibrant market for sports goods made of advanced composites, and encourages part producers to experiment further, to provide new models and innovations each season.

This chapter proposes a historical overview of the use of composites in sports, focussing on the most relevant Olympic sports in alphabetic order in Section 2 and summarising these in Section 3 in light of the various requirements and common features for materials in sport applications.

2. Historical overview of the use of composites in Olympic sports

2.1 Archery

Archery evolved alongside human civilisation, and can probably be traced back to 70.000 BCE, because flint arrowheads from that era have been unearthed in South African caves (Archery, n.d; Archery - Paris 2024, n.d; Sport - History, n.d). Around 10.000 BCE, Egyptian and Nubian cultures used bows and arrows for hunting and warfare. The oldest European bows, found in Holmegaard (Denmark), date back to the Neolithic period (from 12.000 BCE). Their shape has strong similarities with the Assyrian longbow, and later evolved into the medieval European longbow.

One of the technically most outspoken composite innovations might be the composite bow, made from a combination of wood, bamboo, bone, cattle tendons, horn and silk bonded with natural pine resin. It was invented by Mongolians around 1200 BCE (Fig. 3.4.1), or by Parthians, Assyrians, and Tartars, according to Gordon in "Structures" (Gordon, 2003), and well described by Hardy in "Longbow" (Hardy, 1976). According to this author, the composite bow was composed of a wooden core (tough wood such as maple was often used), to which tendons at the tension side and horn bone at the compression side had been adhesively bonded using bitumen, found in wells in the desert.

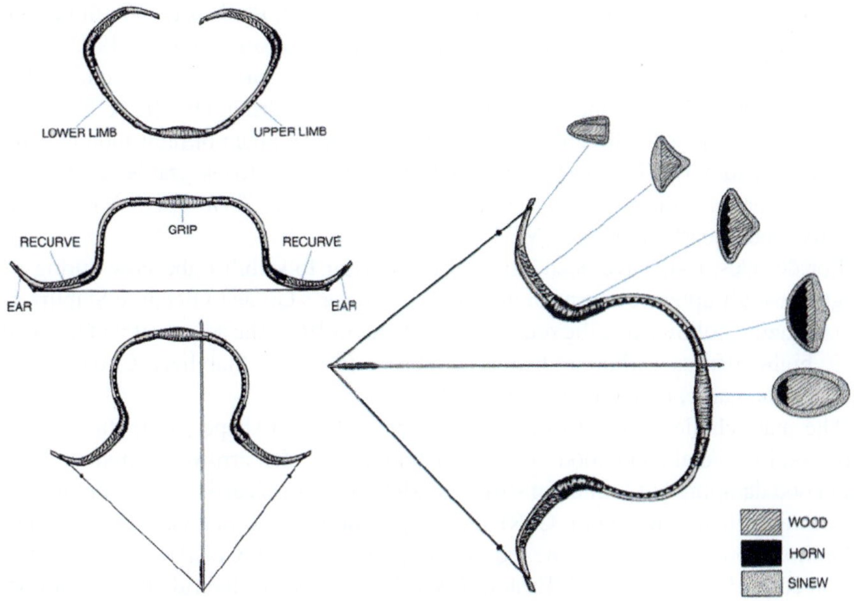

Figure 3.4.1 Classic Mongolian "composite" bow (1200 BCE).
Picture reprinted with permission from https://edspace.american.edu/silkroadjournal/v16_2018_rumschlag/.

Already in 1933, the Fred Bear Co. (USA) first used glass fibre composites (GFRP) in bows (Archery History – The Evolution of the Recurve Bows, n.d). This is remarkable because glass fibres were produced industrially in the same period by Corning (USA). In 1946, the same company filed a patent for a "glass fibre reinforced bow" and in 1950 for a "*synthetic laminated bow*". In the latter, the choice of glass fibres was justified because of their high energy storage capacity. The bow is a sandwich structure, with a rather thick wooden core in the handle area, tapering down toward the tips, and a glass fibre composite multilayered skin.

Nowadays, only the "recurve bow" is used in archery sports and must be used during Olympic competitions. All parts of the modern recurve bow use composites: riser, limbs, stabilisers, visors (Fig. 3.4.2) but also arrows. It is not surprising that the limbs, which have to store a maximum amount of energy, first turned to composites. The first

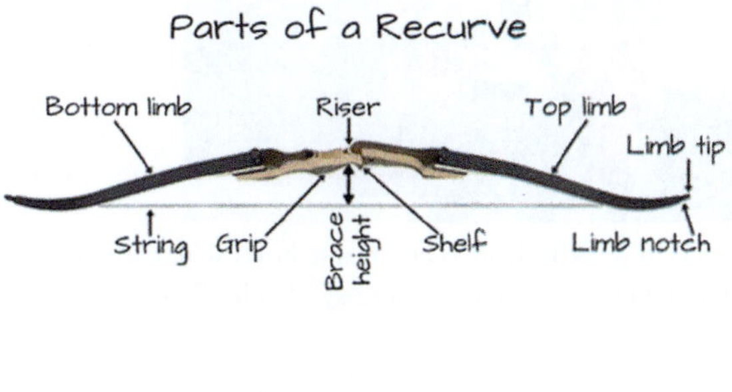

Figure 3.4.2 Parts of a recurve bow.
Picture reprinted with permission from https://www.hunting-bow.com/blogs/news/choosing-a-recurve-bow.

glass fibre composite limbs have been produced in 1969, the first carbon fibre composite stabilisers in 1987 and the first carbon fibre composite arrows in 1982.

The central part, the risers, remained either wood or metal, but to reduce their weight, carbon fibre composites (CFRP) are used in ultralight NIK'A ET9 riser (ET-9 Carbon Fiber ILF Recurve Riser, n.d), a complex layup of high-modulus unidirectional and woven carbon-fibre prepreg. WiaWis claims to use graphene enhanced carbon prepregs in their "*Meta DX Graphene Carbon Nano Tech*" risers, however the benefit is not clearly indicated (META DX, n.d).

For decades, rods have been attached to the riser to stabilise the bow during the lengthy and ultraprecise aiming at the target (A Buyer's Guide to Recurve Stabilizers, n.d). Initially, hollow, metallic rods were used, but in 1987, the advent of widely available, high-stiffness carbon rods for stabilisers (Recurve Stabilizer Choices, n.d) changed the game, as shown in Fig. 3.4.3.

The main challenge for arrows is combining a perfect shape (straightness), high stiffness, low weight and good vibration damping. Wooden arrows are light and provide good damping but limited in stiffness. Metallic arrows can be stiffer, but heavier and vibration damping is poor. CFRP arrows (Recurve Arrow Shafts, n.d) can achieve an optimal compromise of all the requirements. The first full CFRP arrows have been produced in 1982. Since 1984, Easton (USA) has produced a hybrid arrow composed of a high-strength carbon fibre composite outer layer bonded to a 7075 aluminium alloy core (4MM X10 Parallel Pro, n.d).

2.2 *Athletics*

Athletics is certainly one of the oldest sports on record, since names of the champions in ancient competitions in Greece have already been documented as far back as 776 BCE (Athletics - Paris 2024, n.d; Nelson, 2024). Next to several running races, the

Figure 3.4.3 Archers using WIAWIS (Win & Win, Korea) composite bows with carbon fibre risers, limbs, and stabilisers (European Games, 2019; photo courtesy of John Stanley).

ancient Olympic Games also included pentathlon (running, long jump, discus throw, javelin throw, and wrestling). In the second half of the 19th century, modern athletics emerged almost simultaneously in England and the United States, leading eventually to the start of the modern Olympics in 1896 in Athens.

2.2.1 *Pole vault*

In pole vaulting, the kinetic energy of the running athlete must be converted in the most efficient way to "throw" the body as high as possible over the bar. Upon hitting the take-off box, the pole is bent elastically and absorbs about half of the athlete's kinetic energy. As the pole gets straight again, the released elastic energy is used to "throw" the athlete over the bar. An ideal pole hence would combine a high strength with a high resilience or elastic energy storage.

Fig. 3.4.4 illustrates the progression of materials use in pole vaulting. Initially, solid wooden poles have been used, and from 1900 bamboo was in vogue (Linthorne, 2019), because of its hollow structure. During the period 1940–60, some attempts were made to use steel or aluminium. It is in a way surprising that it took till the 1960s before glass fibre composites were employed. The American athlete George Davies broke the pole vault record in 1961 with a "*fibreglass pole*" (risepolevault, 2023). From the 1960s on, GFRP hollow tubes are generally used because they allow for much higher energy storage and a large flexibility (up to 170 degrees bending), combined with high strength. A typical pole consists of three layers (from in to out), wound around a slightly tapered, metallic rod. First, a "spiral layer" is applied, with fibres in the hoop direction to prevent an oval shape during bending. Then, a "full body wrap" textile with most fibres in 0° direction in a textile as long as the pole, is wrapped around, and finally the "sail piece" (a trapezium shaped textile) is added in the central region to control bending. The pole, weighing 1.5–3 kg, is often not perfectly straight but very slightly bent, so that the shock during take-off is less brutal and energy loss in the athlete's body is minimised.

Vaulting poles fully made from CFRP would be too stiff and too brittle, hence limiting the amount of absorbed energy. A trend is, however, observed toward hybrid poles, in which carbon fibres are added at critical places, mostly to the second and third layer, for performance optimisation and weight reduction.

2.2.2 *Javelin throwing*

Lightweight and high stiffness are again the predominant criteria when selecting materials for javelins: the lower the weight, the lower the energy needed to throw it as far as possible. To control the trajectory and to reduce the aerodynamic drag, vibrations, and fluttering, the javelin should be as stiff as possible for a given cross-section.

Initially, javelins were made of wood (mainly Finnish birch), then steel and aluminium were used. Composite javelin production is reported in an early patent from 1980 (Saffire, 1983), in which a "rotary winding" technique is used to apply aramid and glass fibre composite tapes, hybridised with carbon fibre prepregs as an external layer; fibre orientations vary between 0° and 30 degrees. Nemeth (a former

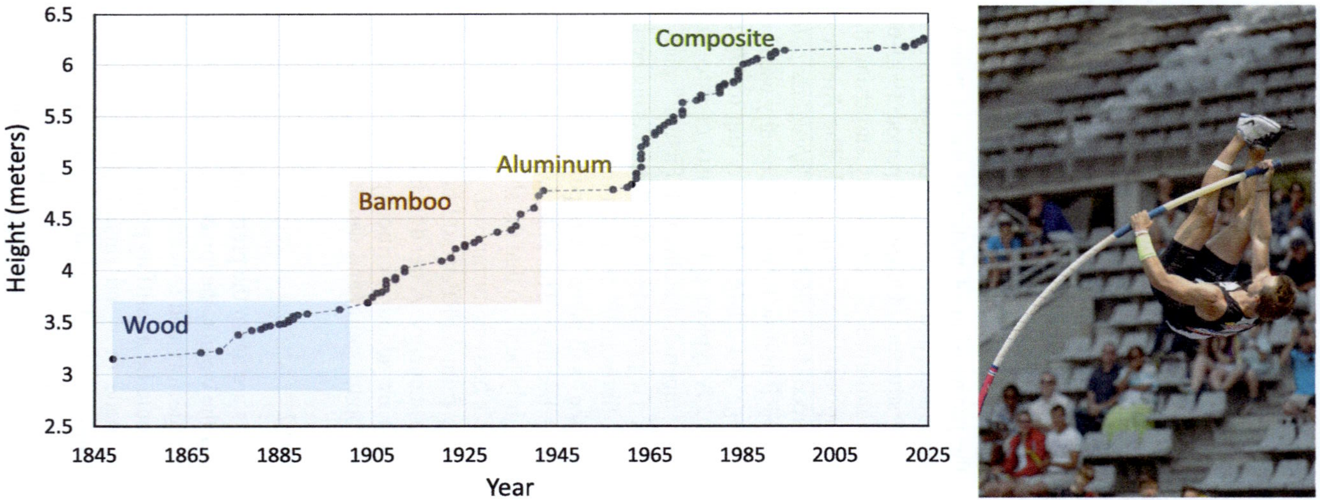

Figure 3.4.4 Left: Men's pole vault world record progression, and material used for the pole; and right: pole vault showing high flexibility (© Marie-Lan Nguyen/Wikimedia Commons, CC BY-SA 3.0.https://commons.wikimedia.org/w/index.php?curid=27478865).

Olympic champion in javelin) produces "hybrid" javelins: an aluminium core with carbon fibre composite skin for "*minimal vibration*" (Nameth Javelins, n.d). Finally, hybrid carbon-glass fibre composite javelins are currently available on the market.

2.2.3 Discus throwing

In traditional discs, the flat or slightly curved sides are made from wood or aluminium, whereas the rim is made from steel, brass or bronze alloys. To realise a steady flight path, it is important to keep the disc rotating/spinning while flying. For this, the gyroscopic properties of the spinning disc are used against aerodynamic forces changing the ideal flight path. To optimise the gyroscopic properties, CFRP sides help keep the weight below the maximum allowed (2 kg for men, 1 kg for women), while concentrating mass as much as possible in the metallic rim (Fig. 3.4.5). Top athletes prefer the highest moment of inertia (80%–92% of the weight in the rim), hence the sides must be as light as possible. However, it is difficult to get the disc rotating because the high moment of inertia requires a complex action of the index or middle finger at the moment of release.

2.2.4 Running

Running shoes should combine minimal weight with appropriate stiffness and excellent support of the athlete's foot. Linthorne (2019) discusses the recent advent of shoes with a "carbon fibre composite plate" to stiffen the shoe under the metatarsal joints. During running, at the moment of "touchdown", the carbon composite plate is bent and stores elastic energy, which is released during the toe-off phase of the movement. Carbon fibre composites have an optimum elastic energy storage capacity per weight. For short distances and for jumping, the mass of the shoe should be as low as possible, hence only a thin foam sole is added. However, for longer distances shock-absorbing foam is indispensable. The best marathon shoes hence are composed of a thick midsole made from a highly compliant and resilient foam material to further increase energy storage, while its low stiffness is compensated by the carbon fibre composite inserted plate.

Figure 3.4.5 Carbon fibre composite throwing disc with galvanised steel rim, by Polanik. Picture, reprinted with permission from https://polanik.shop/en_GB/p/CCD19-1-CARBON-IN-STEEL-Competition-discus/1337.

The very first shoe of this type seems to have been invented by Zoot in 2014 (Zoot Ultra TT 4.0 Running Shoe, n.d), the Ultra TT 4.0 carbon running shoe (Fig. 3.4.6). In 2017, Nike Vaporfly came on the market, and made the concept widely popular. In a paper quoted in (Sahota, 2020), it is stated that "*The (VaporFly) shoe stored more mechanical energy at an 87% energy return, twice as much as other shoes*". There is a lot of debate among scientists on whether such claims are justified, see (Hoogkamer et al., 2018; Patoz et al., 2022) and a recent paper in Nature (Beck et al., 2020). The current marathon world record was achieved by Kelvin Kiptum in October 2023 in Chicago, running the race in 2 h and 35 s, wearing a prototype version of Nike's Alphafly 3.

2.3 Climbing

Indoor climbing is a relatively recent sport, with the inception of artificial climbing walls dating back to 1964, when Don Robinson created the first modern wall by inserting rock pieces into a corridor wall at the University of Leeds (Ward, 2018). Competitive climbing began in 1985 at the SportRoccia event near Turin, Italy, with the first artificial wall competition held in Vaulx-en-Velin, France, the following year. By the early 1990s, international events moved to purpose-built walls to minimise environmental impact. The modern era of sport climbing began with the inaugural World Championships in Frankfurt, Germany, in 1991 (History of Sport Climbing, n.d).

The most basic and widely used climbing walls, affectionately known as 'woodies' in the climbing community, are constructed from plywood. However, composites have emerged as indispensable elements in the construction and design of climbing walls, due to their impact resistance, durability, lightweight design, and mouldability. GFRP has become the leading material for crafting sturdy, innovative climbing structures due to its ability to endure heavy use and challenging conditions while offering longevity and minimal maintenance. The 80-metre tall climbing wall in Copenhagen, constructed from 24 metric tons of glass fibre/polyester resin panels, exemplifies GFRP's exceptional resilience, maintaining its usability in the extreme temperatures, high winds, and corrosive salt air of the Scandinavian climate (Copenhill, 2021).

Furthermore, GFRP is extensively used in crafting high-volume climbing holds, such as volumes and macros, enhancing their strength. In the late 1980s, European climbing wall companies shifted from heavy concrete to crafting three-dimensional additions using wood and GFRP (Jaffe, 2016). Primary materials include nonwoven mats, chosen for their lack of specific fibre orientation, combined with polyester resin. In some cases, an external layer, such as quartz sand PU paint, is applied to the surface of the hollow composite holds to ensure optimal friction.

2.4 Cycling

The first bicycle competition is said to have happened in 1868 at Saint-Cloud Park (near Paris), a 1200-metre race on a heavy wooden, front-wheel driven "vélocipède". In the 1880s, Thomas Humber, John Kemp Starley and others in the United Kingdom invented a completely novel design: the "safety bicycle" had chain rear-drives, giving

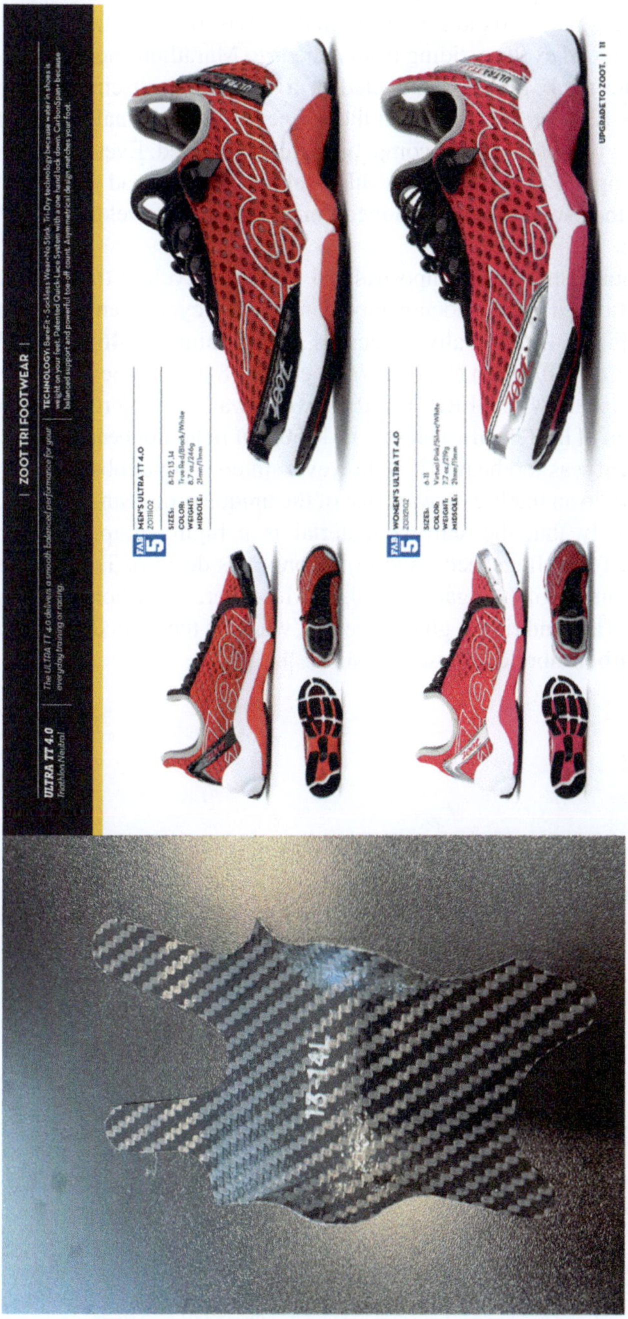

Figure 3.4.6 First running shoe with "a carbon-fibre chunk embedded in the sole": the Zoot Ultra TT 4.0. Left: CFRP part, right: Examples of the model (with permission from Zoot Customer services).

bicycles the shape they still have today. From the very beginning of the "modern" Olympics (1896), cycling has been present (Cycling Road - Paris 2024, n.d; Cycling Track - Paris 2024, n.d). Aristidis Konstantinidis (Aristidis Konstantinidis, n.d) won the gold medal in road cycling, riding from Athens to Marathon and back (86 km) in 3 h 22 min. On the track, athletes competed in five different disciplines, from sprint (over 2 km) to a 12-hour race, in which the winner (Adolf Schmal, Austria) covered 295 km, and only two of the seven competing cyclists arrived. Ever since, cycling has developed as a popular sport with many different disciplines: road racing (from 1-day competitions to the 3-week Tour de France), mountain and gravel biking, cyclocross, track biking, etc.

It might be surprising that composites did not enter cycling through the gate of cycling competitions. The first documented composite bicycle seems to be the Spacelander bicycle (Fig. 3.4.7), initially conceived in aluminium (1946) by the American designer Benjamin G. Bowden (Pil and Verpoest, 2018). As in the 1950s, the knowledge of manufacturing with GFRP spread, Bowden was able to produce a glass-fibre composite version of his (electric) bicycle in 1960, but it did not become a commercial success, because it was too heavy. It is, however, interesting to observe that Bowden already exploited, in an intelligent way, one of the unique processing characteristics of composites, namely that the starting material is a highly drapable fibrous cloth (compared to stiff metallic sheets used in the previous design). This enabled him to create a very rounded-off, "organic" design. Moreover, the whole frame consisted of only two parts (left and right), glued together, which is the main design concept still used today in carbon fibre composite bicycles.

Figure 3.4.7 The **Spacelander** (electric) **bicycle** by the American designer Benjamin G. Bowden (1960).
Picture from https://www.brooklynmuseum.org/opencollection/objects/2587, Brooklyn Museum, Marie Bernice Bitzer Fund, 2001.36.

As happens all too often when a new material becomes available, the frame of the first carbon composite bicycles was a "black copy" of the existing "tube and lug" metallic frames. Carlton Cycles (UK, later part of Raleigh) already in 1969 started designing and producing their "Carlton Flyer" (Mossberg – first US producer of CF frames, 2010; Timeline, n.d; Road.cc, 2010), using carbon fibre composite tubes glued into aluminium lugs. Similar concepts were later used by Assos (Switzerland) in their Track Bike (1976), using for the first time tubes with an oval cross-section, and by Allan (Italy, in collaboration with Toray, 1976). The latter is important, because all these innovations started only a few years after the first industrial production of carbon fibres by Toray in Japan (1970)—the UK's Carlton Flyer bike used carbon fibres provided by the Royal Aircraft Establishment (RAE). Vitus (France) produced its Carbon 3 in the early 1980s, and Look followed in 1986 with its KG86, which was used by Greg Lemond to win the 1986 Tour de France.

The Italian company Colnago (1986) seems to have been the first to develop a "full carbon" bicycle frame, where not only the tubes but also the lugs and all other connections were made of CFRP. Almost simultaneously in the United States (Nelson, 2001), the companies Huffy and Kestrel filed patents describing a technology where lugs and tubes would be cocured into a monocoque frame, using inflatable bladders. Trek further developed this technology, first alone to develop the Trek 5000 (which had a too high failure rate), then together with Ron Nelson and Dimitrije Milovich at Radius Eng. (US Patent 5,264, 519), resulting in the Trek 5500, resulting in the improved OCLV technology (*Optimum Compaction, Low Void*), as shown Fig. 3.4.8. Their production efficiency being higher, Trek outperformed Kestrel and Lance Armstrong won the 1999 Tour de France—a victory which was taken away from him because of doping. OCLV technology developed during the 1990's, by Radius and TREK, gathered Tour de France successes in 1998–2005, and was further produced and commercialised by TREK into the 2000's and beyond.

The first CFRP monocoque mountain bike was designed by Chris Hornzee-Jones (initially in UK, moved to California, USA) in 1992 (Darlow, 2021). This drastically different frame concept deviated fundamentally from the classical "diamond" (parallelogram with one diagonal) frame and was later sold to Lotus, producing 400 frames in 1993–94. They were "real" monocoques, where the whole frame was produced as a hollow structure (using inflatable bladders) in one step.

In an excellent review paper (Malizia and Blocken, 2020), Malizia and Blocken describe how the "freedom of design" offered by composites, compared to metals, has enabled the development of bicycles with drastically lower aerodynamic resistance, culminating in the world hour record of Chris Boardman (1996), riding a Lotus 110. This bike was a further development of the "airfoil" Lotus 108, designed by Mike Burrows in 1992 (Fig. 3.4.9). Boardman covered 56.375 km, 7 km more than Eddy Merckx in 1972 on a traditional, steel diamond frame bicycle.

In the following 25 years, the monocoque frame manufacturing concepts "two adhesively bonded halves" and the "one-piece monocoque with inflatable bladder" have been predominant and have been further optimised. However, the large number of manual handlings involved has always been a major drawback: one frame often requires a few hundred individual prepreg carbon pieces to be laid down precisely into

Figure 3.4.8 Trek 5500 bicycle, with details on CFRP lugs, used by Lance Armstrong during his first Tour de France win (1999). In the background, some sketch drawings of the OCLV patent are presented (with permission from Dimitrije Milovich, Radius engineering, Inc.).

the moulds, leading to high labour costs. Hence, carbon composite bicycle manufacturing gradually moved to lower-wage countries such as China and Taiwan, and, more recently, to Vietnam and Indonesia.

This not only creates logistic problems because the main markets for carbon composite bicycles are the European Union and United States, resulting in expensive and time-consuming transportation. It also often occurs that production quality or design issues are only discovered during the use of bicycles in the European Union or United States, while in the meantime, production continues in China, Taiwan, or Vietnam. To address this problem, "reshoring" has become a major trend, namely bringing back the production to the countries where the markets are. The Belgian company Rein4ced (REIN4CED, n.d) realised a triple innovation in 2019: use of thermoplastic matrix and hybrid carbon-steel fibre composites (for improved impact resistance), combined with a press-forming process of tailored blanks, which are laid up robotically from UD prepreg tapes, followed by a proprietary assembly step. This allowed them to reshore bicycle production to Europe; hence they started a plant in Belgium (Leuven) with an annual production capacity of 20,000 frames. Rein4ced produces amongst others one type of electric mountain bike frames for the European mountain bike brand Kellys (Fig. 3.4.10).

Figure 3.4.9 The Lotus 108 track bike, a monocoque design with an inflatable bladder (1992), by Paul Hudson from United Kingdom—Design Museum, CC BY 2.0, https://commons.wikimedia.org/w/index.php?curid=56854232.

It is expected that more reshoring will happen, driven by the need for increased design flexibility and optimised manufacturing agility. Local production will also reduce the environmental footprint, a target addressed as well using bio-based materials. Replacing (partially) carbon fibres by natural fibres such as flax has been realised already in 2008 by Museeuw Bikes (Belgium) (Atkins, 2008), a few years later followed by Schwinn in their Vestige bike (2011, 80% flax, 20% carbon) and by the Look 765 CarboFlax (2017). The additional and important advantage of using natural fibres is the increased vibration damping.

Figure 3.4.10 The KELLYS THEOS F90 mountain bike of Kellys (Slovakia), with a frame made by Rein4ced (Belgium) using carbon fibre/steel fibre hybrid thermoplastic composites in a robotised process.
Picture reprinted with permission from https://kellysbike.com/en/.

Innovations in the use of composites in bicycles are too many to be covered completely! The racing car company Lotus, always looking for extreme performances, has recently astonished carbon bike fanatics with the Look HB.T track bike (Arthurs-Brennan, 2021) designed and produced in collaboration with Hope Technology (UK), allowing the Great Britain Cycling Team to top the track cycling medal table during the Tokyo Olympics in 2020.

2.5 *Equestrian*

Equestrian sports boast ancient origins, with evidence of horse-riding dating back to 3500 BCE. In ancient Greece, the excitement of chariot racing gripped Olympic audiences, while the art of dressage was immortalised by Greek historian Xenophon in his seminal work, "On Horsemanship". Show jumping emerged in the 19th century in England, born from the practical challenges of fox hunting, and gained widespread acclaim at events like the 1869 Dublin Horse Show. Eventing, initially a rigorous test for cavalry horses, debuted as a competitive sport in 1902, combining dressage, cross-country, and stadium jumping into a thrilling showcase of equestrian skill and partnership (History of Equestrian, n.d).

The development of materials in equestrian sports has always prioritised the well-being of the horse and enhancing the interaction between rider and animal. Composites were first introduced in the 2000s, starting with saddles, where lightweight, flexible materials such as GFRP and CFRP replaced traditional wood and metal (Mackechnie-Guire et al., 2018; Murray et al., 2023; Mutel et al., 2016). These materials improved comfort, adaptability, and shock absorption, reducing stress on the horse's back and optimising rider control. Composites were soon applied to other equestrian equipment, including horseshoes and protective gear. Carbon and aramid fibres in horseshoes provide better durability and reduce strain on the horse's joints, while composite helmets and leg protectors offer superior safety and impact resistance.

2.6 *Golf*

Although golf was one of the first sports to codify its rules in 1754 in St. Andrews (Scotland), the origins probably go back to similar sports in China in the eighth century ("Chuiwan", meaning *"hitting a small ball"*) and The Netherlands, where "*kolven*" was played at the end of the 14th century. Modern golf, however, originated in Scotland in the mid-15th century.

Golf club shafts were initially made from hickory wood, until, in 1893, a Scottish blacksmith Thomas Horsburgh made the first steel shaft (History of the Golf Shaft, n. d). Being a solid shaft, it was very heavy, hence soon hollow shafts appeared. During the following decades, all kinds of steel alloys were evaluated and used, but it took another 80 years until, in 1969, Frank Thomas made the first "graphite" shaft (Takamatsu, 2000) in the Shakespeare Sporting Goods Co. (USA). Since carbon fibres were still in their infancy at that time, and only very small quantities of (noncontinuous) carbon fibre were produced, this can be considered a ground-breaking application of these new fibres.

Around the same time, aircraft engineer James Flood developed his version of a carbon-fibre composite golf club, which was the start of the Aldila company (1972, San Diego, USA). One year later, 1973, Gay Bowers won the Pacific Ocean Masters with a composite shaft, but it is not clear which brand of golf club he used. Anyway, this remarkable event suddenly increased dramatically the demand for carbon fibres, and a posteriori justified Toray's audacious decision to build, already in 1971, their first industrial plant, producing 300 kg per month of continuous Torayca T200 and T300 fibres.

Since these early times, the manufacturing method for carbon fibre composite golf club shafts has hardly changed, as can be observed in numerous videos (Golf Shaft, n. d; Noll, 2022). Basically, carbon fibre composite prepreg layers with different fibre orientations are wrapped around a slightly conical steel rod, the latter to ease demoulding (Fig. 3.4.11). Local reinforcements with specific fibre orientations can be added as well. Several methods to apply compression during curing are possible (a.o. shrinking plastic tapes), but one of the most crucial steps is grinding the shaft to its exact dimensions, straightness, weight and weight distribution. The first scientific papers only appeared in the 1990s, mainly addressing design optimisation, dynamic behavior and damping (Howell, 1992; Swider et al., 1994).

Although the basic design of golf shafts has not changed over the past decades, further optimisation of performance is being pursued through various innovations. In the "*GOST*" (Graphite On Steel Technology), "multi-material shafts" from Nippon Steel, a patented special adhesive technology is used, so that the graphite layers properly adhere to the steel shaft interior (Nippon N.S. PRO Modus3 GOST Hybrid Shafts, n.d). Mitsubishi Chemicals, active in producing carbon fibres since the 1960s and involved in golf shaft production ever since, is now using superlow resin volume fraction prepregs (down to 20%) in their Vanquish golf shafts (MEET DIAMANATM

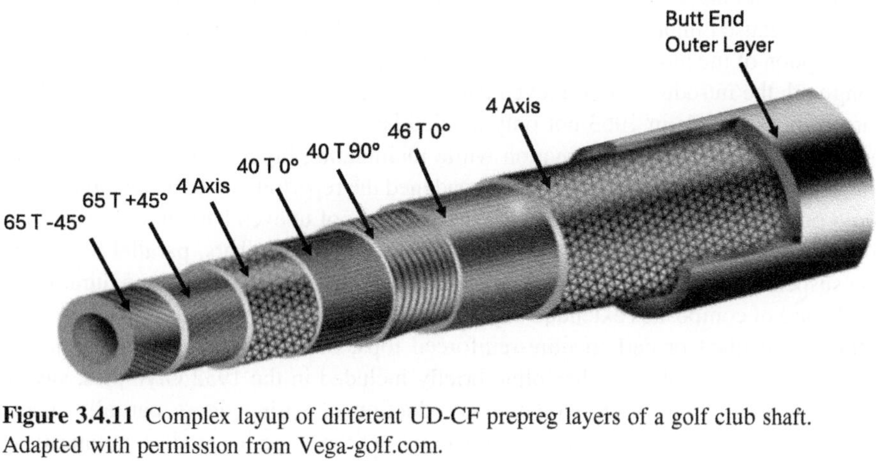

Figure 3.4.11 Complex layup of different UD-CF prepreg layers of a golf club shaft. Adapted with permission from Vega-golf.com.

BB, n.d) for *"lighter weight while maintaining strength"*. The weight is down to 50–65 g, but even 29 g has been reached with the Mitsubishi Grand Bassara. Further optimisation is achieved by hybridisation. Enigma Golf Shafts (Germany) is hybridizing carbon weaves with aramid fibres (The Enigma Driver Shaft, n.d). Using very thin carbon-fibre UD plies made by NTPT (Switzerland), for reaching the same thickness, the number of layers can be increased and hence the laminate layup can be tuned with higher precision (TPT Golf Shafts, n.d). In the new millennium, some scientific papers on golf clubs have been addressing new issues: for instance, Zako (Zako et al., 2002) studied the advantages of replacing UD prepreg-based laminates by triaxially woven preforms.

Even golf club heads use carbon-fibre composites instead of steel or titanium. In the TaylorMade Stealth (Horaczek, 2022), the club face is made up of 60 layers of carbon fibre prepregs, making it 40% lighter than its titanium equivalent. Foldable golf trolleys also appeared on the market, reducing drastically the weight: the Go-Carbon+ (The Go-Carbon+ Electric Golf Trolley, n.d) electric golf trolley by Athena (Switzerland) weighs only 4.1 kg *(including battery).*

2.7 Gymnastics

Gymnastics, originating in ancient Greece, was first recorded by Philostratus in "Gymnastics: The Ethics of an Athletic Esthetic". It involved physical exercises, such as running, jumping, and wrestling, aimed at harmonising body and mind, a concept championed by Plato, Aristotle, and Homer. Derived from the Greek verb gumnázō ("to train naked"), gymnastics was initially practised for overall fitness by both genders but shifted to military training for men after Rome's conquest of Greece in 146 BCE. The term "artistic gymnastics" emerged in the early 1800s, distinguishing it from military training, and it gained momentum with competitions flourishing in European schools and athletic clubs (History of Acrobatic Gymnastics, n.d).

In early modern Olympics, gymnasts struggled with inconsistent apparatuses, such as parallel bars made from either flexible wood or rigid hollow iron rails, which often broke and caused injuries. This led to efforts to standardise construction, resulting in the adoption of the modern "drop" parallel bar design by the 1952 Helsinki Olympics, along with the introduction of uneven bars. The game-changing introduction of glass fibre composite rails in 1963 not only tackled issues such as pole breakage but also increased overall elastic deformation while maintaining high strength. This innovation substantially improved safety and broadened the repertoire of skills that gymnasts could perform (History of parallel bars, n.d; History of uneven bars, n.d). Over time, GFRP rods coated in wood became the standard for uneven bars, parallel bars, and ring suspension bars, providing a balance between strength, flexibility, and durability.

The use of composites extended beyond bars; springboards and vaulting tables also incorporated glass or carbon fibre-reinforced tops, offering improved performance and resilience. Tumbling, a discipline briefly included in the 1932 Olympics, saw a key evolution in the 1960s with the introduction of spring floors partially made from glass fibre composites. These innovations set the stage for modern gymnastics, where composite materials continue to shape the development of safer, more advanced equipment.

2.8 Hockey

The oldest historical records on hockey go back to Egypt and Persia (2000 BCE) and to Ethiopia (1000 BCE), but also Greeks, Romans and Aztecs would have played a similar game. In China, it appeared around 1000 (AD) in Inner Mongolia, and later during the Ming Dynasty (14th to 17th century), when it was also played in Punjab (India). In Europe, already in the Early Middle Ages, a kind of hockey was played, initially in Ireland and Iceland, but quickly spreading out as entertainment for the local clergy and aristocracy. From the beginning of the 19th century, the game developed in public schools in England. From there, it conquered Western Europe, the British colonies and the United States.

According to ancient paintings in graves, about 2000 years ago Egyptians used bamboo to make their sticks. Later, hockey sticks were traditionally made from ash wood, but Indians preferred the tougher mulberry wood. The shape changed drastically due to the Dutch inventor Toon Coolen (1982), so that nowadays all hockey sticks have a "hook-like" head (Field Hockey, n.d; How-to-Choose-a-Field-Hockey-Stick, 2020).

In the 1970s, the first wooden field hockey sticks were reinforced with external layers of glass fibre composites (Cains, 2015). The development of full glass fibre and/or carbon fibre composite field hockey sticks only started in the 1980s, one of the very early producers (1983) being the Belgian tennis racquet company Snauwaert and its designer Joris Van Raemdonck.

A combination of reinforcing fibres is often used. In such hybrids, according to commercial descriptions, carbon fibres add stiffness and "power", glass fibres are used for durability and strength, whereas aramid fibres enhance shock absorption. Manufacturing is mostly done by wrapping tapes around a foam core, and curing the stick in a metal mould, as can be discovered in an Adidas video (Total Hockey, 2020) about their Adidas 2023 Shosha Kromaskin 100% carbon.

According to Al Rashid (Al Rashid et al., 2020) "*around 90% of hockey equipment in the world is produced in Sialkot, Pakistan*". Different fibre orientations are being used, mainly 0° and ±45 degrees, and next to sticks with a foam core, hollow sticks are also produced using the well-known bladder moulding principle. Wrinkling of the reinforcing tapes or layers during draping and curing is a frequent source of strength and durability reduction. To address this problem, Crown (Modern Manufacturing, n. d) uses an expanding foam inside the stick, to reduce prepreg wrinkling. Moreover, they added a longitudinal rib, connecting the front and back face of the stick, to prevent buckling and reinforce the stick (like stringers in the wing of an aeroplane).

Ice hockey emanated from hockey and was developed in Canada, and became widespread toward the end of the 19th century. Ice hockey sticks were initially made of wood, from one piece, then laminated, until the end of 1990s, when composite materials took over. First some GFRP layers were laminated together with wood layers, then CFRP was used, with sticks made in one or two pieces (shaft and a slightly curved blade) for better handling of spare parts in case of failure. The first company to successfully develop "one piece" carbon fibre composite sticks was Composite Busch SA (Switzerland) in 1992. The main challenge beyond lightweight is to preserve a high flexibility of the stick, and a high modulus so as to maximise strain energy restitution when shooting the puck (Laliberte, 2009). The manufacturing is quite similar to that of field hockey, with hollow tubes in the shaft, and a fully reinforced blade.

2.9 Rowing and canoe

Canoes, kayaks, and rowing boats, originally developed worldwide primarily for transportation, reflect diverse cultural adaptations. Rowing, with its ancient roots in Egypt, Greece, and Rome, transitioned into a competitive sport, notably through events such as the Oxford–Cambridge boat race initiated in 1828. Meanwhile, kayaks, stemming from traditional Inuit hunting and fishing vessels derived their name from the Inuit term for "man boat", and evolved into recreational vessels, epitomised by the establishment of the Royal Canoe Club in 1866. Canoe sprint, the original form of the sport, emphasises speed and reaching the finish line first, while canoe slalom emerged in the 1930s in Switzerland, as an alternative to slalom skiing and was practised initially on flatwater rather than whitewater courses (History of Canoe Slalom, n.d; History of Canoe Sprint, n.d; History of Rowing, n.d).

Rowing and canoeing, such as sailing, experienced a major transformation with the introduction of composite materials. Early boats, made from wood or metal, were heavy and limited in performance. However, the 1960s saw the introduction of glass and nylon fibre composites, which, while still relatively heavy, offered improved durability and performance (Walsh, 2016). This sparked innovation among rowers, canoeists, and kayakers, paving the way for advancements in the 1970s when carbon and aramid fibres reduced weight and further enhanced performance across water sports (Mckenzie and Berglund, 2018; Secher and Volianitis, 2007).

At the 1972 Munich Olympics, Luigi Colani designed a ground-breaking prototype for the German Olympic eight team, featuring a titanium frame covered with an epoxy, glass, and carbon fibre composite skin (Colani, n.d). Despite initial promise in speed tests, the boat faced challenges during rowing trials, earning the nickname "Glass Casket". Empacher (Germany) modified the boat, leading to the final "Bodensee-Vierer" for the Munich competition, securing a gold medal and marking a historic moment as the first sandwich-built plastic rowing boat (1923–2023, 100 years of Empacher boatyard, n.d). This shift to full composite boats culminated at the 1996 Atlanta Games, where wooden boats saw their last victories in kayak races.

Ted Van Dusen of Composite Engineering (USA) significantly advanced composite rowing and kayaking shells in the 1970s. He developed a rowing shell featuring an aramid fibre composite hull with wood stringers and boron in the laminate, leading to Olympic success in 1976. With the advent of carbon fibres in the mid-80s, Van Dusen seamlessly incorporated aerospace composites into their designs, marking a significant leap forward in rowing and kayaking shell construction, including the introduction of aerodynamic wing riggers in rowing boats (Horney, 2016). The 1988 Seoul Olympics saw the debut of the "Eagle" kayak, a vacuum-infused composite boat that helped American Greg Barton win two gold medals and became a foundational design in kayak racing (Robinson et al., 2002).

Modern rowing shells, along with canoe and kayak designs, are engineered for precision, emphasising lightweight, stiff, and durable hulls for competitive performance. Meeting demands for higher strength-to-weight ratios, sandwich composite technology, led by carbon fibre/epoxy and sometimes reinforced with aramid fibres, is widely used (Fig. 3.4.12). Structural cores, typically honeycomb or resilient foam, enhance integrity, while manufacturing methods such as wet layup, vacuum infusion, and autoclave-cured prepreg technology ensure quality, with prepreg standing out for precision and efficiency (Filter, 2019). New models aiming at improving sustainability for recreational use were also recently produced with natural fibres, such as the Melker sea kayak from Värmdö, which was produced with flax fibres and cork.

Composite paddles and oars also evolved. GFRP paddles debuted at the 1984 Los Angeles Games, but early versions faced limitations. In 1972, Guest, Keen and Nettlefolds, a British company introduced carbon fibre oars, but high costs limited their use. By 1977, the Dreissigacker brothers, founders of Concept2, created a more affordable carbon fibre oar, which became the standard. They introduced the asymmetrical "hatched" blade in 1991, which revolutionised rowing. For canoeing, the Swedish "wing" paddle, introduced in 1986, enhanced efficiency and was widely adopted by the 1988 Seoul Olympics (Secher and Volianitis, 2007).

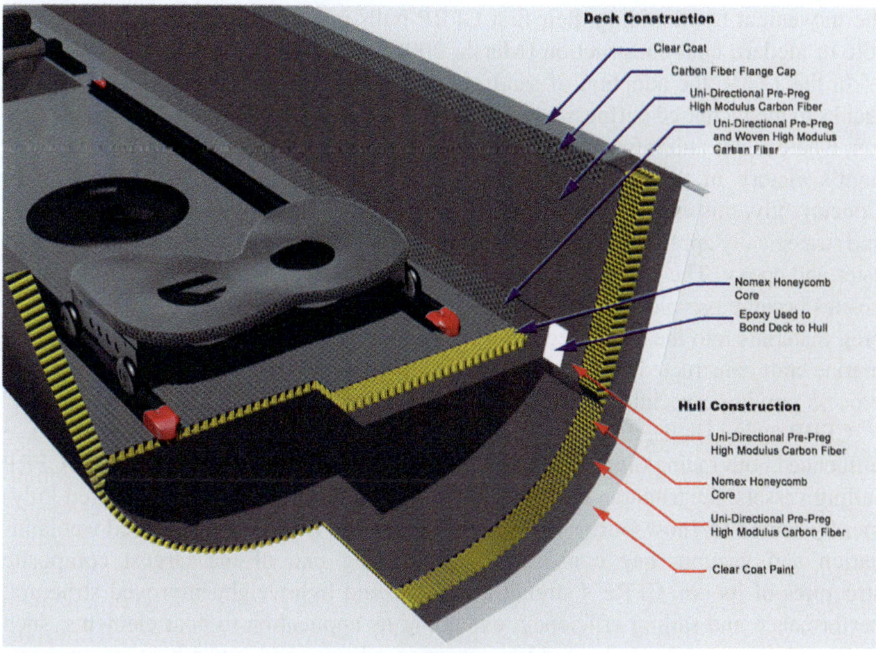

Figure 3.4.12 Rowing hull composition from Wintech (USA).
Picture reprinted with permission from https://www.wintechracing.com.

2.10 Sailing

Sailing, rooted in ancient civilisations such as Mesopotamia and Egypt, has played a pivotal role in the evolution of society and global commerce. Over time, it evolved from transportation and trade to recreation and competitive sport. International sailing competitions began in 1851, initiated by members of the New York Yacht Club who sought to engage in a race held in England. To participate, they constructed a 31-m-long schooner named America, voyaged to England and emerged victorious, securing a prize known as the Hundred Guineas Cup. In commemoration of this inaugural international sailing contest, the trophy was subsequently renamed The America's Cup (History of Sailing, n.d).

The use of composites in sailing began during WWII when the U.S. Army, recognising the corrosion-resistant qualities of GFRP, built the first GFRP boats for military use, some reportedly used in the Normandy landings (Verpoest, 2020). In 1942, Ray Greene (USA) is credited with crafting the first sailing dinghy from glass fibres and polyester resin. The commercial production market flourished in the 1950s, featuring the introduction of the GFRP yacht Arion in 1951. By 1957, Ray Greene achieved the production of 175 units of a Sparkman & Stephens design boat (Fig. 3.4.13), and in 1959, Pearson Corp fuelled the trend further with the Triton sailboat (Spurr, 2012). Similar developments unfolded in Europe, with W&J Tod's Tod dinghy and the Flying 20 sailboat produced in 1951 and 1952, respectively. Halmatic company claimed the construction of the world's first large GFRP boat, the Perpetua, a 15-m-long motor cruiser, in 1954. French companies Jeanneau and Beneteau joined the movement by moulding their first GFRP hulls in the 1960s, solidifying GFRP's role in modern boat construction (Marsh, 2006; Neşer, 2017).

In the 1970s, the adoption of sandwich constructions saw a notable breakthrough thanks to the increased stiffness/weight ratio, exemplified by the yacht Sir Thomas Lipton, where a glass fibre foam sandwich laminate played a crucial role in Geoffrey William's victory in the 1968 Single-Handed Transatlantic Race (The Transat, n.d). Concurrently, this era saw the emergence of CFRP, often combined with other fibres, and strategically applied in various boat components and alternative resins such as vinyl ester and epoxy. The Munich Olympics in 1972 marked a turning point as fibre composites became prevalent in various competition boats. After the incorporation of prepreg materials into the aerospace industry by companies such as Hexcel in the 1970s, marine craft constructed with prepreg materials began appearing around 1980, with the process proving especially well-suited for carbon fibre composite structures.

CFRP racing shells, developed by companies such as Carbocraft in England, have influenced both racing dinghies and high-performance yachts. One of the early CFRP sailing vessels, the trimaran Colt Cars GB, was designed in 1980 by Structured Polymer Systems Ltd. (now Gurit). Made from carbon and aramid fibres, it used wet lamination and vacuum bag consolidation, becoming one of the largest composite structures of its era. CFRP's strength, rigidity, and lightweight improved structural performance and sailing efficiency, extending its application to boat elements, such as masts, spars, sails, foils, rudders, steering wheels, and other boat components (Marsh, 2006).

Figure 3.4.13 A flier created by Greene's Company in 1959 advertising their GFRP boats. Picture from https://www.fiberglassics.com/library/index.php?title=Ray_Greene.

Sail evolution also progressed. Nylon, used since the 1950s for its quick-drying qualities, struggled with shape retention. In the 1980s, aramid, polyethylene, polyester, and carbon fibres improved sail efficiency and rigging longevity. In the 1990s, North Thin Ply Technology (Switzerland) developed a new method for creating moulded sails, later enhanced by North Sails (USA) with the 3Di technology. The process involves layering carbon fibre, polyethylene, or polyester yarn between two polyester films, which are then heat-cured, resulting in sails 30% lighter than traditional ones (Pearson, 2016).

The integration of foils revolutionised 21st-century sailing by enhancing lift and reducing drag. Although hydrofoil technology originated with Emmanuel Denis Fargot's 1869 patent for a rowing boat, Eric Tabarly further advanced the concept in 1980 with the Trimaran Paul Ricard (Hydroptère), which used aluminium foils and set a transatlantic record despite its inability to fully lift out of the water (Rubinstein, 2023; Sheahan, 2021). The adoption of CFRP transformed hydrofoils due to its stiffness and lightweight, allowing the production of monolithic CFRP parts and paving the way for today's high-speed sailing technology (Fig. 3.4.14).

2.11 Skateboarding

Skateboarding emerged in the early 1950s in California, USA, as a pastime for surfers seeking to recreate the sensation of riding waves to the street during flat wave periods. Initially, enthusiasts improvized by affixing roller skates to wooden boards, but by 1959, the introduction of the first commercially produced skateboards, the Roller Derby, signalled the sport's evolution into a distinct activity (Hawk, 2024; History of Skateboarding, n.d).

The 7-ply maple skateboard, first introduced in the 1970s, set the industry standard with its strength, durability, and flexibility, with veneers hot-pressed and glued using epoxy. Veneers are typically stacked with a $[0_2/90/0/90/0_2]$ layup, with the highest quality veneer plies positioned at the top and bottom face sheets to withstand higher levels of tension and compression. Initially, the grassroots nature of skateboarding prioritised accessible materials such as wood, but as the sport progressed and performance became crucial, manufacturers began exploring fibre-reinforced composites (Prentiss et al., 2011).

This shift to advanced materials came later in the 2000s, reflecting the sport's transition from casual riding to technical street skating. Notable brands such as Santa Cruz and Powell Peralta embraced this evolution, integrating layers of glass and carbon fibres into their decks. For instance, the Santa Cruz VX model features patented Quad X Technology in 2013, while Powell Peralta Flight decks utilise triaxial glass fibre layers (Santa Cruz, 2020). These innovations allow riders to push the limits of performance, significantly enhancing strength and flexibility and marking a new era in skateboard technology.

2.12 Ski

Skiing has ancient roots, born from the necessity to traverse snowy, mountainous terrain. The word "ski" itself comes from the Norwegian term "skid," meaning a split length of wood. Archeological evidence, including images and wooden ski fragments found in peat bogs across Russia, Finland, Sweden, and Norway, shows that skiing dates back 5000 years. Initially used for transportation, skiing evolved into a

Figure 3.4.14 AC45f sailing on hydrofoils with one daggerboard raised above the water. Picture by Nick Dimbleby - https://www.flickr.com/photos/landrovermena/27917479043/, CC BY 2.0, https://commons.wikimedia.org/w/index.php?curid=50441776.

competitive sport by the 19th century. Modern downhill skiing began in the 1850s, thanks to Norwegian Sondre Norheim's innovations. Ski jumping emerged in Norway, with Ole Rye's 9.5-m jump in 1808, while ski mountaineering began in Europe in the late 19th century, with Wilhelm Paulcke's 1897 traverse of the Bernese Oberland. Freestyle skiing, or "hotdogging," took off in the 1960s in America, driven by advances in equipment and a spirit of expression (History of Alpine Skiing, n.d; History of Freestyle Skiing, n.d; History of Ski Jumping, n.d; History of Ski Mountaineering, n.d).

Advanced ski designs began in 1926 when Rudolph Lettner introduced the first steel edge into traditional wooden skis for better grip (Hogen, 2021). By 1932, laminated "Splitkein" skis, combining hardwood with a lighter core, and the introduction of a formaldehyde-based glue, Aerolite in 1935, paved the way for the transition to engineered ski structures (Fangel, n.d). In the late 1940s, Paul Michal revolutionised ski design with the Dynamic K race ski, featuring 24 layers of hickory, a hidden low-drag continuous L-section edge, and the pioneering "Cellolix" base—an insulating plastic armour plate made from cellulose fibre-reinforced plastic. The solid celluloid base provided a tough, waterproof surface that glided better than heat-conducting materials such as aluminium and steel, making it particularly promising for ski racing (Dynamic hier et aujourd'hui, n.d; Masia, 2022).

In the 1950–60s, the sport's growing popularity led to experimentation with composite materials to enhance ski performance. In 1950, aeronautical engineer Howard Head introduced the Head Standard Ski, featuring a plywood core sandwiched between aluminium layers and a phenol formaldehyde resin finish. In 1954, Kofler in Austria introduced the first polyethylene base, which ski factories instantly adopted, as it was easily repaired and could slide well, meaning wax was no longer needed.

There had been several attempts to build all-GFRP sandwich skis starting as early as 1952, but none made it into production until 1959, when Montreal ski designers Fred Langendorf and Art Molnar invented the first successful GFRP ski featuring a foam core, the Toni Sailer ski, manufactured by G.M. Plastics in Granby, Quebec (Soden, 2013). This innovative construction quickly replaced both wood and aluminium for most recreational skis, becoming the industry standard within 10 years. In 1963, Molnar designed a later model with a ribbed GFRP core where the ribs were separated by air channels. This latter design made the ski extremely light but still strong. Around the same time, Bill Kirschner invented the GFRP ski using a wet-wrap process in a spruce core, leading to the production of the K2 Holiday ski, which contributed to the decline of metal skis (Masia, n.d).

Dynamic in France invented the "torsion box" in 1964, using sandwich beam principles to develop the legendary VR7 ski (V for "verre," R for "resin," and seven for its years of development). The model evolved into the VR17 in 1967, introducing the first antivibration system. Made with epoxy resin, which is tougher than polyester initially used, the new ski reduced the amount of steel used in the edge, and used thicker glass layers that improved torsional stiffness and vibration damping (Masia, 2022). The VR17 became a slalom benchmark, winning Olympic medals in 1968, 1972, and 1976.

In parallel, in the 1960s, Sherman Poppen invented the "snurfer," a snow toy that combined two skis and a rope, laying the foundation for modern snowboarding (Macarthur, 2008). Since the 1970s, the ski and snowboard industry has advanced ski construction by using composites such as GFRP and adding materials such as aramid, carbon, and ceramic fibres, enhancing strength, flexibility, and impact resistance for tailored performance (Van De Velde, 2005).

Carbon fibres have historically been utilised in skis and snowboards in small percentages, often incorporated into braided ski designs with longitudinal carbon tows within a glass sock. However, as the prices of carbon fibre have decreased and ski designs have become wider, manufacturers are increasingly exploring its use to reduce weight and modify the flexural behaviour in flexion and shear. Previously, the reliance on heavier aluminium skis, which provided stability through "mass damping" by remaining steady on snow, delayed the shift toward lighter carbon fibre constructions during the 1980 and 1990s, paving the way for modern carving skis with wider tips and narrower waists.

Nowadays, skis are highly engineering sandwich structures, combining a high-molecular-weight high-density polyethylene (HMW-HDPE) base for gliding, layers of composite materials and sometimes aluminium arranged symmetrically around a wooden or foam core, and a polymer top-sheet for design and protection, as well as steel edges, as illustrated in Fig. 3.4.15. Constructions differ depending on the uses, from lightweight and very stiff, honeycomb-cored, jumping skis, to lightweight touring skis, to geometries adapted to powder snow or carved turns in prepared slopes, as well as snowboards with adapted shear/longitudinal stiffness. Novel developments have included modifications to improve the vibrational behavior and stability of the ski while keeping the weight low and tune bending and twisting properties.

1. Polyamid | 2. Non woven mat / Epoxy prepreg | 3. GFRP | 4. Rubber | 5. Titanal® (aluminium alloy) | 6. Non woven mat / Epoxy prepreg | 7. Wood Core | 8. Melamine based sidewall | 9. Titanal® | 10. GFRP | 11. Steel edges | 12. UHMWPE ski base

Figure 3.4.15 Example of ski construction.
Picture made by André Habermacher from Stöckli Swiss Sports AG, reprinted with permission.

As for all sports equipment, recent trends increasingly consider sustainability aspects, by reducing the number of materials (going back to full wood), making use of recycled materials, introducing more bio-based elements (flax, bamboo fibres) and exploring bio-based/recyclable/debondable novel adhesive systems (Essential ski, n.d; R3 skis, n.d), as well as lower energy manufacturing methods. Improved designs to confer nonlinear or nonsymmetrical properties to the skis in bending and torsion are also progressing, with the example of the carbon steering control (Technology Adaptative Contact Length, n.d), or the design of snowboards making use of coupling effect in nonsymmetrical laminates (Kaufmann et al., 2010).

Given the advancements in ski technology, boot innovation also became necessary. Early efforts focused on improving stiffness and control by using thicker leathers and stiffer soles, though this made boots uncomfortable (Colonna et al., 2013). In 1948, Robert Lange pioneered the use of glass fibre-reinforced polyester resin, combining laminated plastic with leather for better waterproofing and durability (Brand history Lange, n.d). In 1958, he developed the first injection-moulded ABS plastic ski boot, later replaced by Dupont's Adiprene in 1965 for improved durability, leading to the first modern plastic ski boots, which achieved mass production by 1966. In parallel, Rosemount, who was competing with Lange, introduced to the market the first ski boots completely made of composite materials, using glass fibres and epoxy resin. Ski boots were strong, lightweight, and weatherproof, featuring metal hinges riveted to them. Split into two hinged halves, they allowed easy entry and secure closure with buckles and elastic seals to manage GFRP rigidity (Masia, n.d).

Composite materials later revolutionised other ski equipment, like poles. In 1990, David P. Goode popularised carbon fibre poles, which had existed for over a decade, replacing the lightweight aluminium standard that had dominated since the 1960s. These composite poles, favoured by experts for their exceptional strength-to-weight ratio, offer enhanced performance and agility on the slopes (Wanglee, 2022).

2.13 Sleigh sports

Sleigh sports—bobsleigh, luge, and skeleton racing—trace their origins to Switzerland, where they evolved from early forms of winter transportation. Bobsleigh was invented in the late 1860s when Swiss innovators combined two skeleton sleds and added a steering mechanism, leading to the formation of the first bobsleigh club in St. Moritz in 1897. Luge, a high-speed sport where athletes race feet-first on a specialised sled, began in the Swiss Alps, with the first international race held in Davos in 1883. Skeleton racing, the world's first sliding sport, also developed in St. Moritz, initially as a pastime for wealthy tourists, with the Cresta Run becoming its iconic location. Switzerland's influence on these thrilling winter sports is undeniable, as the country played a key role in their development and popularisation (History of Bobsleigh, n.d; History of Luge, n.d; History of Skeleton, n.d).

Sleigh sports have undergone a dramatic transformation driven by innovations in materials and design. Early bobsleighs were built from wood, later replaced by heavy steel sleds to maximise speed through mass. Until 1952, the lack of weight limits for bobsleigh crews and sleds encouraged a strategy based on sheer weight, favouring bulkier, heavier setups to generate momentum on the track. This approach changed fundamentally with the introduction of bobsleigh weight rule in 1952, which capped the combined mass of the sled and crew. The rule redefined the sport, shifting the focus from weight to agility, precision, and optimised engineering (Robinson, 2010).

GFRP, introduced in the mid-20th century, emerged as a key material to meet these new demands. Its lightweight, durable, and versatile properties allowed sleds to achieve superior aerodynamics and enhanced stability. Designers could now create streamlined shapes that reduced drag, significantly improving speed and control. GFRP also absorbed impacts better than steel, offering improved safety for athletes navigating icy tracks at high velocities. By the 1970s, GFRP sleds had become a standard in Olympic bobsleigh and luge competitions, marking a turning point that redefined the technical and athletic nature of these high-speed winter sports (Dabnichki, 2015).

2.14 Surfing

The origins of surfboarding trace back to ancient fishermen in Peru, and it later flourished in Polynesia during the sixth century, when it became a beloved activity among all social classes. Early surfboards, simple and finless, marked the beginning of the sport. Modern surfing gained attention through James King's 18th-century accounts of Hawaiian surfers and was later revolutionised by the lifeguard George Freeth, who developed lighter, more practical boards, and Duke Kahanamoku, a Hawaiian Olympic swimmer whose demonstrations and advocacy introduced surfing to audiences worldwide, including in Australia (History of Surfing, n.d).

Surfboard construction underwent a significant transformation with the early development of composite materials, as recently reviewed by Zimmermann et al. (2024). Originally made from solid wood, early surfboards were heavy and lacked fins, limiting their manoeuvrability. A turning point came in 1929 when Tom Blake introduced the first hollow surfboard, nicknamed "cigar boxes." This design greatly reduced the board's weight, and by 1935, the addition of fins further enhanced its stability and control (Feldmann, n.d).

World War II-driven materials technology advancements spurred experimentation, leading to Pete Peterson's 1946 creation of the first hollow glass fibre composite (GFRP) board reinforced with a stringer. Following this, in 1947, Joe Quigg and Bob Simmons introduced the first polyurethane foam-core sandwich surfboards, sheathed with a thin plywood layer. They had solid balsa rails and a glass fibre composite covering and experimented with all-GFRP fins (Fig. 3.4.16). Simmons also experimented with polystyrene foam, but its reaction with polyester resin led him to laminate the foam within solid wood structures (1948–1953: The Birth of the Malibu, n.d).

In 1950, George Downing's "Rocket" surfboard, constructed with balsa, glass fibres and resin, marked a breakthrough by introducing the first removable fin, pushing the boundaries of big-wave surfing. GFRP, initially a waterproof coating, played a crucial role in enhancing manoeuvrability with deeper fins. Foam cores revolutionised the weight-to-buoyancy balance, leading surfboard makers to adopt alternative plastics and include a stringer as a method to reinforce foam blanks. Gordon Clark's 1961

Figure 3.4.16 First glass fibre composite fin made by Bob Simons in the 1950s. Photograph by Guy Motil, reprinted with permission from the California Surf Museum https://surfmuseum.org/stories/bob-simmons-foam-sandwich-surfboard-1950/.

development of polyurethane foam surfboards established Clark Foam as a major producer.

The 1967 "Shortboard Revolution" marked a shift to smaller, lighter and more agile boards, with all-GFRP construction and hard plastic fins, allowing for new wave-riding techniques. In the late 1960s, the introduction of the round-tail represented a significant advancement in the shortboard revolution, renowned for its ease of turning. In 1979, Michel Barland pioneered the computerised shaping machine, enabling precise surfboard designs. In the 1980s, epoxy resins were first applied in timber laminated surfboards. Simon Anderson revolutionised board design in 1981 with the "thruster," featuring a three-fin setup, offering greater control and responsiveness, shaping the contemporary shortboard design prevalent today. Throughout the 1980s, the focus in design centred on refining shortboards, resulting in thinner, lighter, and narrower boards, marking a moment of calm in design evolution (Correia et al., 2022).

2.15 Tennis and badminton

2.15.1 Tennis

The very early history of tennis goes back to the "jeu de paume" in France in the 11th century, where players were hitting a small ball with bare hands against a wall. Later, gloves were used and finally these evolved into racquets, a kind of wooden extension of the hand (History of Tennis, n.d; Tennis - Paris 2024, n.d). The word "tennis" originates from the French word "tenez" *(hold! take! receive!)*, and the word racquet derives from the Arabic "rakhat" *(palm of the hand)*.

From the 16th century on, "real tennis" or "royal tennis" was played in an enclosed area, and complex rules were established. By the second half of the 19th century, "lawn tennis" emerged, first in England on cricket courses, with rules similar to the current ones.

Initially made from solid wood, tennis racquets got a "composites flavor" when Dunlop introduced the Dunlop MaxPly (Dunlop Max-Fort Racquet, n.d; Mann, 2023) to the market in 1932: solid wood was replaced by a laminate of thin wooden plies, allowing for optimised bending and torsional stiffness at minimum weight. It was still used by John McEnroe, winning Wimbledon in 1981, playing against Bjorn Borg.

In the meantime, experiments with steel (Wilson, end of 1960s) and aluminium tennis racquets did not result in a satisfactory performance/weight balance, having a *"too hard feeling"* while being too heavy.

It is not evident to find out who has built the very first composite tennis racquet. In a recent, extensive historical-technical study (Taraborrelli et al., 2019), mentioned that the first composites were used as an external reinforcing ply on wooden tennis racquets (Slazenger, 1961; Wilson, 1966). The first full carbon fibre composite racquets would have been the Phantom by Slazenger (1974) and the TR10 by Rossignol (1977). According to Lammer & Kotze (Lammer and Kotze, 2003) the world's first complete glass fibre composite racquet was introduced by the PDP Sports Company around 1975.

The early 1980s was the era when most tennis racquet brands changed to carbon fibre composites, some of the best-known ones being the racquets of Donnay (Björn Borg), Dunlop (John McEnroe, Steffi Graf), Prince (André Agassi, Monica Seles), Wilson (Pete Sampras, Roger Federer) and Head (Novak Djokovic).

The manufacturing method is similar for most brands *(an excellent video can be found on YouTube) (Tennis Warehouse, 2020)*). More than a dozen unidirectional prepreg pieces, with different ply orientations, are laid on a flat surface upon each other in a specific sequence. The stack is rolled up on a metal rod, which has been covered with a nylon tube, and will later act as a bladder in the mould. After slight heating, the metal rod is removed, some air pressure is applied inside the bladder, and the tubular preform is draped into the mould. Some local reinforcements and other items such as small balancing weights or damping items are finally added, before the mould is closed and full pressure is applied to the bladder. The mould is then heated and curing of the (mostly epoxy) resin is completed in the appropriate time duration. After demoulding, the racquet is deburred, holes are drilled, and grommets are fixed before the final finishing (painting, etc.). All this requires a lot of manual labour, and even more: highly experienced personnel.

From the very beginning, alternatives for this labour-intensive manufacturing process have been explored. Already in 1983, Wilson used (hybrid) braids and foam cores (Origins of the ProStaff Original, n.d). Braids allow to reduce the layup time, foams avoid the use of bladders, whereas the combination of carbon with aramid fibres *"is effective in damping frame shock and vibration. Additionally, they are foam-filled to help achieve uniform weight/balance"*. On the ITF (International Tennis Federation) "history page" (Mann, 2023), all important tennis racquet-related patents are listed.

All professional tennis racquets nowadays are made from carbon fibre composites, even using the most advanced high strength and high modulus carbon fibres (Rackets, n.d) such as TORAYCA T1100G and M40X.

Because vibrations are harmful to the elbow and tiring for the muscles, continuous efforts have been made to optimise damping by hybridisation. Next to aramid fibres, natural fibres have a promising future, because they combine this property with an ecological advantage. It was first applied by Artengo (2010, Fig. 3.4.17), and recently in 2023 by Babolat in their newest Babolat Pure Aero (Babolat in Partnership, n.d).

The time-consuming and costly manufacturing process with thermoset prepregs could be drastically reduced by using thermoplastic injection moulding. According again to Lammer & Kotze, Dunlop developed an injection moulding process in 1980 for their popular Max 200G, a hollow frame of carbon fibre-reinforced polyamide 66. A removable, low-melting temperature bismuth-tin alloy was used as the core of the frame.

2.15.2 Badminton

The old game of "battledore and shuttlecock" was played in ancient times in Greece, India, and China, and it used a tool made of parchment or gut stretched across a

Figure 3.4.17 Artengo flax fibre tennis racquet (2010) with permission from Decathlon, company owning Artengo.

wooden frame (Badminton, 2024; Badminton - Paris 2024, n.d). It is not known exactly when this ancient sport transformed into its actual "badminton" form, but it was brought by British officers to India in the 1860s. It gradually spread out over other British colonies, and from there to Europe and East Asia, where it is highly popular. Most probably, the sport was named after a place where it was often played, namely Badminton, the county estate of the dukes of Beaufort in Gloucestershire.

Initially made from solid wood, badminton racquets followed the trend in tennis racquet technology: from the 1960s on, they turned to aluminium or steel, the first mainly for the head, the second for the shaft. In 1978, the first "carbon composite racquets" were produced by Yonex in Japan: the Carbonex 8–8500B (Yonex Carbonex 8-8500 B Worth, n.d), weighing less than 100 g (Fig. 3.4.18). Since then, most competition badminton racquets use carbon fibre composites, whereas (cheaper) racquets for leisure still stick to aluminium and steel.

The manufacturing method of carbon fibre composite (CFRP) racquets is quite different from tennis racquets. The shaft is hollow and made in a similar way as the tennis racquet frame, but it is straight and has a much smaller (circular) cross-section. The head is not hollow but solid as it is made from folded and compressed carbon fibre prepregs, which are then draped into the mould. The head and (precured) shaft are then connected with a CFRP T-joint, sliding the two horizontal tips into the two ends of the folded prepregs and the vertical tip into the hollow shaft. After closing the mould, curing takes place without internal bladders. Several videos can be found online that describe the manufacturing process in detail (How Is a Badminton Racket Made, 2022).

The performance is optimised by applying many types of hybridisations: the Yonex "Nanoflare 800 Pro" contains high-stiffness Toray M40X carbon fibres and

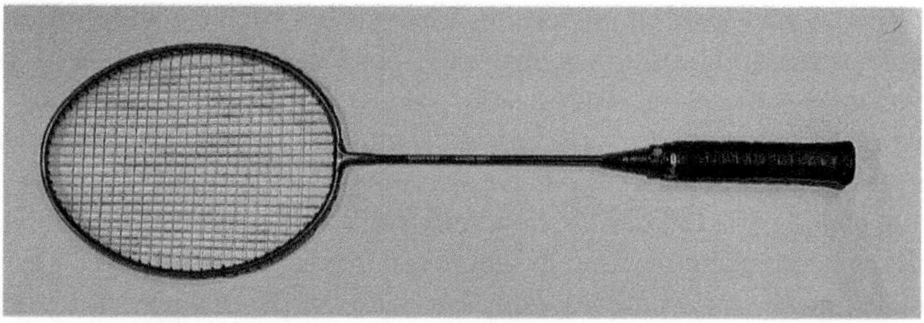

Figure 3.4.18 Original Yonex Carbonex 8–8500B CFRP badminton racquet (1978). Picture reprinted with permission from https://www.badmintoncentral.com/forums/index.php?threads/yonex-carbonex-8-b-8500-b-worth.188415/.

UHMW (ultrahigh molecular weight) polyethylene fibres for shock absorbance. The Yonex Carbonex 8000 head is hybridised with a titanium mesh for a "*more stable shot*" or even with carbon nanotubes.

3. Concluding remarks

As reported above, a large variety of sports applications led to different developments, with a widespread application of composites in several forms. Nonetheless, several common trends arise for most sports, which will be presented in this concluding Section.

The first common point and reason for use, as clearly reported above, is that composites help improve performance. In general, performance is related to an increase in specific properties (i.e., properties over density), such as specific stiffness and strength. In some cases, limited deformation under stress is required, as in boat hulls and masts, racquets, or bicycles, requiring high specific stiffness, whereas in other cases, a sweet spot in bending stiffness is required, which composite structures (or sandwich structures) can be tailored to reach, making use of the adequate combination of fibre types and orientations.

Many sports applications require high energy restitution after deformation, thus high stiffness and large strains, archery and pole vault being the most obvious cases. Yet, others require high damping and energy absorption, such as protective equipment, helmets, and flooring materials, which are often related to the improvement of safety for athletes beyond simple performance. The reproducibility in properties has also helped composite replace wood in many examples, such as the parallel bars in gymnastics, to ensure improved safety. Additional requirements beyond these can lead to a multiphysics approach, such as designing composites with the needed electrical conductivity, wear, and sliding properties, among others. As a result, many composite applications are found in sports applications under the form of hybrid structures, either with a combination of fibres (glass/carbon/aramid/flax) but also as a combination of composites with wood or metal to truly optimise the performance as a function of a precise set of requirements, which cannot be achieved with one material alone.

In addition, a main asset of composites in sports has been the wide variety of materials to choose from (fibres, fillers, matrices), and the wide adaptability of these materials to a range of processes. They enable producing prototypes or uniquely tailored parts, that are fully adapted to the needs of the athletes, and to the specific needs of a given sport, for example moulding composites in precisely tuned shapes to improve the hydrodynamics of a foil design, or aerodynamic shells. As a result, athletes often have their own equipment made to fit their exact needs (this is often the case for skis and tennis racquets, for example). This has allowed para-athletes to benefit from advanced composite parts that are fully adapted to their requirements, whether these are prostheses, adapted grips, or wheelchairs (Verpoest et al., 2024). The possibilities are endless, and without composites, most paralympic sports would not be accessible.

In parallel, sports equipment manufacturers are always in search of a competitive edge to increase their customer base and attract new users. This goes hand in hand with marketing, and the need for a good story to tell. As a result, composites were also introduced as they convey an image of high-tech and performance, which relates well with sports, as observed with increased reference to the use of graphene in skis or tennis racquets (Young and Liu, 2016)

Finally, the need to consider the environmental impact beyond the pure performance of the sports equipment is increasing both in high-level competition and for everyday athletes. Composite materials also follow this evolution toward lower impact materials, from bio-based renewable resources to the use of recycled materials and better designs aiming at circularity. As a result, the adoption of new forms of composite materials and processes will probably continue to benefit from the fast pace and early adopting business approach of sports for many years to come.

References

1923-2023, 100 Years of Empacher Boatyard. Available at: https://www.empacher.com/en/about-us (accessed 1 January 2025).

1948-1953: The Birth of the Malibu: Bob Simmons, Joe Quigg, Matt Kilvin, and George Downing. Available at: https://www.surfresearch.com.au/1948_Simmons.html (accessed 1 January 2025).

4MM X10 Parallel Pro. (n.d.). , available at: https://eastonarchery.com/arrows_/x10-parallel-pro/ (accessed 1 January 2025).

A Buyer's Guide to Recurve Stabilizers. (n.d.). Available at: https://archery360.com/2020/05/14/a-buyers-guide-to-recurve-stabilizers/ (accessed 1 January 2025).

Al Rashid, A., Khalid, M.Y., Imran, R., Ali, U., Koc, M., 2020. Utilization of Banana fiber-reinforced hybrid composites in the sports industry. Materials 13 (14), 3167. https://doi.org/10.3390/ma13143167.

Archery. (n.d.), available at: https://en.wikipedia.org/wiki/Archery (accessed 1 January 2025).

Archery - Paris 2024. (n.d.). Available at: https://olympics.com/en/paris-2024/sports/archery (accessed 1 January 2025).

Archery History – The Evolution of the Recurve Bows. (n.d.). Available at: https://www.elongoutdoor.com/en/new/Archery-History-The-Evolution-of-the-Recurve-Bows.html (accessed 1 January 2025).

Aristidis Konstantinidis. (n.d.). Available at: https://en.wikipedia.org/wiki/Aristidis_Konstantinidis (accessed 1 January 2025).

Arthurs-Brennan, M., 2021. Lotus x Hope HB.T: Team GB's Track Bike in Detail available at: https://www.cyclingweekly.com/news/racing/olympics/bike-future-hope-lotus-british-cycling-went-back-drawing-board-446431. (Accessed 1 January 2025).

Athletics - Paris 2024. (n.d.). Available at: https://olympics.com/en/paris-2024/sports/athletics (accessed 1 January 2025).

Atkins, B., 2008. Museeuw Flax Bike Launch available at: https://www.bikeradar.com/news/museeuw-flax-bike-launch. (Accessed 1 January 2025).

Babolat in Partnership with Eco-Technilin Introduce Flax Fibres in Pure Aero Tennis Racquets. (n.d.). Available at: https://www.jeccomposites.com/news/spotted-by-jec/babolat-in-partnership-with-eco-technilin-introduce-flax-fibres-in-pure-aero-tennis-racquets/(accessed 1 January 2025).

Badminton - Paris 2024. (n.d.). Available at: https://olympics.com/en/paris-2024 (accessed 1 January 2025).

Badminton, 2024. Britannica available at: https://www.britannica.com/sports/badminton. (Accessed 1 January 2025).

Beck, O.N., Golyski, P.R., Sawicki, G.S., 2020. Adding carbon fiber to shoe soles may not improve running economy: a muscle-level explanation. Sci. Rep. 10 (1), 17154. https://doi.org/10.1038/s41598-020-74097-7.

Blomster, J.P., Salazar Chávez, V.E., 2020. Origins of the mesoamerican ballgame: earliest ballcourt from the highlands found at Etlatongo, Oaxaca, Mexico. Sci. Adv. 6 (11). https://doi.org/10.1126/sciadv.aay6964.

Brand history Lange. (n.d.). Available at: https://www.dynastar-lange.com/fi-en/History-LANGE.html (accessed 1 January 2025).

Cains, T., 2015. The Materials and Technology used in the Design of Hockey Sticks. available at: https://www.azom.com/article.aspx?ArticleID=11710. (Accessed 1 January 2025).

L. Colani. (n.d.). River & Rowing Museum Collection, Available at: https://collection.rrm.co.uk/persons/252 (accessed 1 January 2025).

Colonna, M., Nicotra, M., Moncalero, M., 2013. Materials, designs and standards used in ski-boots for alpine skiing. Sports 1 (4), 78–113. https://doi.org/10.3390/sports1040078.

Copenhill, 2021. : the tallest climbing wall in the world. In: JEC available at: www.jeccomposites.com/news/spotted-by-jec/copenhill-the-tallest-climbing-wall-in-the-world/?news_type=applications,business&end_use_application=building-civil-engineering, design-furniture-and-home,sports-leisure-recreation&tax_product=glass-fiber#:~:text=At%2080m%20high%2C%20with%20a,climbing%20wall%20in%20the%20world. (Accessed 1 January 2025).

Correia, J.M.D., Serra, G.F., Alves de Sousa, R.J., Pereira, A.B., Fernandes, F.A.O., 2022. Expanded (black) cork for the development of an eco-friendly surfboard: environmental impact and mechanical properties. Sustainability 14 (2), 668. https://doi.org/10.3390/su14020668.

Cycling Road - Paris 2024. (n.d.). Available at: https://olympics.com/en/paris-2024/sports/cycling-road (accessed 1 January 2025).

Cycling Track - Paris 2024. (n.d.). Available at: https://olympics.com/en/paris-2024/sports/cycling-track (accessed 1 January 2025).

Dabnichki, P., 2015. Bobsleigh performance characteristics for winning design. Procedia Eng. 112, 436–442. https://doi.org/10.1016/j.proeng.2015.07.221.

Darlow, J., 2021. Mountain bike history: world's first carbon monocoque mountain bike. available at: https://www.mbr.co.uk/news/bike_news/lotus-sport-336454. (Accessed 1 January 2025).

Dunlop Max-Fort Racquet. (n.d.). Available at: https://artsandculture.google.com/asset/dunlop-max-fort-racquet-dunlop/zQEcbEL_qaf9GA (accessed 1 January 2025).
Dynamic hier et aujourd'hui. (n.d.). available at: https://dynamic.ski/hier-et-aujourdhui/ (accessed 1 January 2025).
Essential ski. (n.d.). available at: https://www.rossignol.com/ch-fr/essential-ski.html (accessed 1 January 2025).
ET-9 Carbon Fiber ILF Recurve Riser. (n.d.). available at: https://www.elongoutdoor.com/en/product/product-36-258.html (accessed 1 January 2025).
Fangel, G. (n.d.). "History of laminated skis", available at: https://woodenskis.com/SplitkeinHistory.htm (accessed 1 January 2025).
Feldmann, K. (n.d.). "The history of surfboard design: Tom Blake's 'cigar box'", available at: https://surfsimply.com/magazine/the-history-of-surfboard-design-tom-blakes-cigar-box (accessed 1 January 2025).
Field Hockey. (n.d.). available at: https://en.wikipedia.org/wiki/Field_hockey (accessed 1 January 2025).
Filter, B.K., 2019. Chapter 13 - design and materials in rowing. In: Subic, A. (Ed.), Materials in Sports Equipment, second ed. Woodhead Publishing, pp. 375–401. https://doi.org/10.1016/B978-0-08-102582-6.00013-7.
Golf Shaft. (n.d.). available at: https://www.m-chemical.co.jp/carbon-fiber/en/product/golf/ (accessed 1 January 2025).
Gordon, J.E., 2003. Structures. Ingram Publisher Services, pp. 80–84.
Hardy, R., 1976. In: Stephens, P. (Ed.), Longbow: A Social and Miliatry History.
Hawk, T., 2024. Skateboarding. Britannica available at: https://www.britannica.com/sports/skateboarding. (Accessed 1 January 2025).
History of Acrobatic Gymnastics. (n.d.). available at: https://olympics.com/en/sports/acrobatic-gymnastics/ (accessed 1 January 2025).
History of Alpine Skiing. (n.d.). available at: https://olympics.com/en/sports/alpine-skiing/ (accessed 1 January 2025).
History of Bobsleigh. (n.d.). available at: https://olympics.com/en/sports/bobsleigh/ (accessed 1 January 2025).
History of Canoe Slalom. (n.d.). available at: https://olympics.com/en/sports/canoe-kayak-slalom/ (accessed 1 January 2025).
History of Canoe Sprint. (n.d.). available at: https://olympics.com/en/sports/canoe-kayak-flat-water/ (accessed 1 January 2025).
History of Equestrian. (n.d.). available at: https://olympics.com/en/sports/equestrian/ (accessed 1 January 2025).
History of Freestyle Skiing. (n.d.). available at: https://olympics.com/en/sports/freestyle-skiing/ (accessed 1 January 2025).
History of Luge. (n.d.). available at: https://olympics.com/en/sports/luge/ (accessed 1 January 2025).
History of parallel bars. (n.d.). available at: https://www.gymmedia.com/ghent2001/appa/bars/history_ba.htm (accessed 1 January 2025).
History of Rowing. (n.d.). available at: https://olympics.com/en/sports/rowing/ (accessed 1 January 2025).
History of Sailing. (n.d.). available at: https://olympics.com/en/sports/sailing/ (accessed 1 January 2025).
History of Skateboarding. (n.d.). available at: https://olympics.com/en/sports/skateboarding/ (accessed 1 January 2025).

History of Skeleton. (n.d.). available at: https://olympics.com/en/sports/skeleton/ (accessed 1 January 2025).

History of Ski Jumping. (n.d.). available at: https://olympics.com/en/sports/ski-jumping/ (accessed 1 January 2025).

History of Ski Mountaineering. (n.d.). , available at: https://olympics.com/en/sports/ski-mountaineering/ (accessed 1 January 2025).

History of Sport Climbing. (n.d.). available at: https://olympics.com/en/sports/sport-climbing/ (accessed 1 January 2025).

History of Surfing. (n.d.). available at: https://olympics.com/en/sports/surfing/. (accessed 1 January 2025).

History of Tennis. (n.d.). available at: https://en.wikipedia.org/wiki/History_of_tennis (accessed 1 January 2025).

History of the Golf Shaft. (n.d.). available at: http://www.golfclubshaftreview.com/history-of-the-golf-shaft.html (accessed 1 January 2025).

History of uneven bars. (n.d.). available at: https://www.gymmedia.com/Rotterdam10/ag/appa/unevenbars/history_ub_e.htm (accessed 1 January 2025).

Hogen, J., 2021. The skiers who changed skiing, 14 December, available at: https://realskiers.com/revelations/the-skiers-who-changed-skiing/. (Accessed 1 January 2025).

Hoogkamer, W., Kipp, S., Frank, J.H., Farina, E.M., Luo, G., Kram, R., 2018. A comparison of the energetic cost of running in marathon racing shoes. Sports Med. 48 (4), 1009–1019. https://doi.org/10.1007/s40279-017-0811-2.

Horaczek, S., 2022. How TaylorMade replaced titanium with carbon fiber in its new driver face. available at: https://www.popsci.com/technology/taylormade-stealth-carbon-fiber-driver/. (Accessed 1 January 2025).

Horney, S., 2016. Builder profile: Van Dusen racing boats. Canoe News available at: https://uscanoe.org/. (Accessed 1 January 2025).

How Is a Badminton Racket Made, 2022. available at: https://badmintonbites.com/how-is-a-badminton-racket-made/. (Accessed 1 January 2025).

How-to-Choose-a-Field-Hockey-Stick, 2020. available at: https://www.khelmart.com/Resource/How-to-Choose-a-Field-Hockey-Stick_1. (Accessed 1 January 2025).

Howell, D.D., 1992. The design of filament wound graphite/epoxy golf shafts. In: 37th International SAMPE Symposium, pp. 1392–1405.

Jaffe, M., 2016. Innovations in Climbing: Volumes. Momentum Climbing available at: https://momentumclimbing.com/innovations-in-climbing-volumes/. (Accessed 1 January 2025).

Kaufmann, J., Kroll, L., Odenwald, S., 2010. Application-specific design of sports equipment from anisotropic fiber-reinforced plastics with stiffness related coupling effect. Procedia Eng. 2 (2), 2569–2574. https://doi.org/10.1016/j.proeng.2010.04.033.

Laliberte, D.J., 2009. Biomechanics of ice hockey slap shots: which stick is best? Sport J. 12 (1). https://web.archive.org/web/20090806023625/http://www.thesportjournal.org/article/biomechanics-ice-hockey-slap-shots-which-stick-best.

Lammer, H., Kotze, J., 2003. Materials and tennis rackets. Mat. Sports Equipm. 1, 222–248. https://doi.org/10.1016/B978-1-85573-599.

Linthorne, N., 2019. Chapter 14 - design and materials in athletics. In: Subic, A. (Ed.), Materials in Sports Equipment, second ed. Woodhead Publishing, pp. 405–411. https://doi.org/10.1016/B978-0-08-102582-6.00014-9.

Macarthur, C.J., 2008. Sherman Poppen: snowboarding's Godfather. In: Skiing Heritage available at: https://www.summitdaily.com/news/snowboardings-predecessor-the-snurfer/. (Accessed 1 January 2025).

Mackechnie-Guire, R., Mackechnie-Guire, E., Fisher, M., Mathie, H., Bush, R., Pfau, T., Weller, R., 2018. Relationship between saddle and rider kinematics, horse locomotion, and thoracolumbar pressures in sound horses. J. Equine Vet. Sci. 69, 43–52. https://doi.org/10.1016/j.jevs.2018.06.003.

Malizia, F., Blocken, B., 2020. Bicycle aerodynamics: history, state-of-the-art and future perspectives. J. Wind Eng. Ind. Aerod. 200, 104134. https://doi.org/10.1016/j.jweia.2020.104134.

Mann, A., 2023. The history of tennis racquets. available at: https://www.tennisnerd.net/history/the-history-of-tennis-racquets/28502. (Accessed 1 January 2025).

Mansson, J.-A., 2020. Driving innovation with sports. MRS Bull. 45 (6), 405–407. https://doi.org/10.1557/mrs.2020.151.

Marsh, G., 2006. 50 years of reinforced plastic boats. Reinforc Plast 16–19. https://doi.org/10.1016/S0034-3617(06)71125-0.

Masia, S., 2022. Dynamic reinvents the slalom ski. International Skiing History Association available at: https://www.skiinghistory.org/online-magazine/dynamic-reinvents-slalom-ski. (Accessed 1 January 2025).

Masia, S. (n.d.). Bill Kirschner - William Kirschner, K2 founder. International Skiing History Association. Available at: https://www.skiinghistory.org/lives/bill-kirschner (accessed 1 January 2025).

Masia, S. (n.d.). "A short, colourful history of ski boots. International Skiing History Association. Available at: https://www.skiinghistory.org/news/short-colorful-history-ski-boots (accessed 1 January 2025).

Mckenzie, D., Berglund, B., 2018. Handbook of Sports Medicine and Science: Canoeing. Wiley-Blackwell. https://doi.org/10.1002/9781119097198.

MEET DIAMANATM BB. (n.d.). , available at: https://mitsubishigolf.com (accessed 1 January 2025).

META DX. (n.d.). available at: https://www.wiawis.com/us/archery-handle-meta (accessed 1 January 2025).

Modern Manufacturing. (n.d.). available at: https://www.crownhockey.com/Technology/index.html (accessed 1 January 2025).

Mossberg – first US producer of CF frames (1974), 2010. available at: https://www.bikeforums.net/classic-vintage/1180482-mossberg-first-us-producer-cf-frames-1974-a.html. (Accessed 1 January 2025).

Murray, R., Fisher, M., Fairfax, V., MacKechnie-Guire, R., 2023. Saddle thigh block design can influence rider and horse biomechanics. Animals 13 (13). https://doi.org/10.3390/ani13132127.

Mutel, F., Reux, F., Glowacz, F., Lagoutte, C., Morel, A., Laparra, A., Saint-Martin, T., et al., 2016. Cantilevered saddle tree: combining a lightweight thermoplastic composite material with an innovative design. In: JEC Composites Magazine No103, p. 56.

Nameth Javelins. (n.d.). available at: https://nemethjavelins.hu/javelins/ (accessed 1 January 2025).

Nelson, R., 2001. All-composite bicycle frames -past, present, and future. In: Utah SAMPE Chapter available at: https://www.slideshare.net/RonNelson5/allcomposite-bicycle-frames-past-present-and-future?from_action=save. (Accessed 1 January 2025).

Nelson, B., 2024. Athletics. Britannica available at: https://www.britannica.com/sports/athletics. (Accessed 1 January 2025).

Neşer, G., 2017. Polymer based composites in marine use: history and future trends. Procedia Eng. 194, 19–24. https://doi.org/10.1016/j.proeng.2017.08.111.

Nippon N.S. PRO Modus3 GOST Hybrid Shafts. (n.d.). , available at: https://www.nordicagolf.nl/nippon-n-s-pro-modus3-gost-hybrid-shafts (accessed 1 January 2025).

Noll, R., 2022. Shaft Education: what goes into making a graphite shaft? available at: https://golf.com/gear/golf-accessories/what-goes-into-making-graphite-shaft-fujikura/. (Accessed 1 January 2025).

Origins of the ProStaff Original. (n.d.). available at: https://www.tennis-warehouse.com/Reviews/W6085/ProstaffOrigins.html (accessed 1 January 2025).

Patoz, A., Lussiana, T., Breine, B., Gindre, C., 2022. The Nike Vaporfly 4%: a game changer to improve performance without biomechanical explanation yet. Footwear Sci. 14 (3), 147–150. https://doi.org/10.1080/19424280.2022.2077844.

Pearson, W.E., 2016. Textiles to composites: 3D moulding and automated fibre placement for flexible membranes. In: Marine Applications of Advanced Fibre-Reinforced Composites. Elsevier Inc., pp. 305–334. https://doi.org/10.1016/B978-1-78242-250-1.00014-4

Pil, L., Verpoest, I., 2018. Fibre-Fixed Composites in Design. Stichting Kunstboek.

Prentiss, A.M., Skelton, R.R., Eldredge, N., Quinn, C., 2011. Get rad! The evolution of the skateboard deck. Evol. Educ Outreach 4 (3), 379–389. https://doi.org/10.1007/s12052-011-0347-0.

R3 skis. (n.d.). available at: https://genuineguidegear.com/pages/r3-skis/ (accessed 1 January 2025).

Rackets. (n.d.). available at: https://www.cf-composites.toray/markets/qol/sporting/rackets.html (accessed 1 January 2025).

Recurve Arrow Shafts. (n.d.). available at: https://www.canadaarcheryonline.com/collections/recurve-arrow-shafts (accessed 1 January 2025).

Recurve Stabilizer Choices – narrowing the possibilities. (n.d.). , available at: https://eastonarchery.com/2019/02/recurve-stabilizer-choices-narrowing-the-possibilities/ (accessed 1 January 2025).

REIN4CED develops more impact resistant composite than carbon. (n.d.). , available at: https://rein4ced.com (accessed 1 January 2025).

risepolevault. (2023), "The first fiberglass poles", available at: https://www.tiktok.com/@risepolevault/video/7250487252344802606 (accessed 1 January 2025).

Road.cc, 2010. Carlton Flyer – The First Carbon Bike. Youtube available at: https://www.youtube.com/watch?v=ugCR53fksjI. (Accessed 1 January 2025).

Robinson, L., 2010. Advanced materials are a game changer in the winter Olympics. J. Occup. Med. 62 (2), 17–19. https://doi.org/10.1007/s11837-010-0024-8.

Robinson, M., Holt, L., Pelham, T., 2002. The technology of sprint racing canoe and kayak hull and paddle designs. Int. Sports J. 6, 68–85.

Rubinstein, B., 2023. Tabarly : la foil histoire du trimaran Paul Ricard. Voile & Moteur available at: https://www.voileetmoteur.com/voiliers/actualite-voile/regates-courses/tabarly-la-foil-histoire-de-paul-ricard/72662. (Accessed 1 January 2025).

Saffire, V. (1983), "Method of making a javelin". Patent number 4404053, granted Sep 13, 1983, https://patents.justia.com/patent/4404053 (accessed 1 January 2025).

Sahota, R., 2020. Carbon-plated running shoes: Fad or Science-based Style? available at: https://www.engineering.com/carbon-plated-running-shoes-fad-or-science-based-style/. (Accessed 1 January 2025).

Santa Cruz, V.X., 2020. Skateboard deck skate test. available at: https://www.skatedeluxe.com/blog/en/santa-cruz-vx-skateboard-deck-skate-test-review/. (Accessed 1 January 2025).

Secher, N.H., Volianitis, S., 2007. Handbook of Sports Medicine and Science: Rowing. Wiley. https://doi.org/10.1002/9780470757116.

Sheahan, M., 2021. Foiling: the history of the hydrofoiler. Yachting World available at: https://www.yachtingworld.com/features/foiling-the-history-of-the-hydrofoiler-135741. (Accessed 1 January 2025).

Soden, B., 2013. Made in Canada. Skiing History, pp. 30–33.

Sport - History. (n.d.). available at: https://www.worldarchery.sport/sport/history (accessed 1 January 2025).

Spurr, D., 2012. Wood to glass. Professional Boatbuilder Magazine available at: https://www.proboat.com/2012/01/wood-to-glass/. (Accessed 1 January 2025).

Swider, P., Ferraris, G., Vincent, B., 1994. Theoretical and experimental dynamic behavior of a golf club made of composite-material. Modal Anal. Int. J. Anal. Exp. Modal Anal. 9 (1), 57–69.

Takamatsu, T., 2000. Development of PAN-based carbon fibre. In: Paper in Japanese in Technology and Civilization, vol 12, p. 1.

Taraborrelli, L., Grant, R., Sullivan, M., Choppin, S., Spurr, J., Haake, S., Allen, T., 2019. Materials have driven the historical development of the tennis Racket. Appl. Sci. 9 (20). https://doi.org/10.3390/app9204352.

Technology Adaptative Contact Length. (n.d.). , available at: https://www.stoeckli.ch/inten/ski/technology (accessed 1 January 2025).

Tennis - Paris 2024. (n.d.). available at: https://olympics.com/en/paris-2024 (accessed 1 January 2025).

Tennis Warehouse, 2020. How a HEAD Tennis Racquet is Made – Behind the Scenes in Kennelbach, Austria. available at: https://www.youtube.com/watch?v=n62fqfBUuIY. (Accessed 1 January 2025).

The Enigma Driver Shaft. (n.d.). available at: https://shop.enigmagolfshafts.com/ (accessed 1 January 2025).

The Go-Carbon+ Electric Golf Trolley – A Noble Designer Throw. (n.d.). , available at: https://at-hena.ch/en/produkt/go-carbon/ (accessed 1 January 2025).

The Transat CIC. (n.d.). “L’histoire de la course”, available at: https://www.thetransat.com/history (accessed 1 January 2025).

Timeline. (n.d.). available at: http://carltoncycles.uk/history/timeline/(accessed 1 January 2025).

Total Hockey, 2020. Adidas Kromaskin Hockey Sticks by Hypetex. available at: https://www.youtube.com/watch?v=RInztV97Vfk. (Accessed 1 January 2025).

TPT Golf Shafts. (n.d.). available at: https://tptgolf.com (accessed 1 January 2025).

Van De Velde, K., 2005. Textile composites in sports products. In: Textiles in Sport. Elsevier Ltd., pp. 309–322. https://doi.org/10.1533/9781845690885.5.309

Verpoest, I., 2020. 50 years of Composites: Past, Present and Future. JEC. https://preview.jec-world.events/catalog/50-years-of-composites/50-years-of-composites/n1-2020. (Accessed 1 January 2025).

Verpoest, I., Michaud, V., Teixido, H., 2024. Composites in Olympic and Paralympic sports. JEC. https://preview.jec-world.events/catalog/composites-in-olympic-paralympic-sports. (Accessed 1 January 2025).

Walsh, L., 2016. History of Canoeing in the Olympics. available at: https://mpora.com/kayaking/canoeing-kayaking-olympics-complete-history-sport. (Accessed 1 January 2025).

Wanglee, W., 2022. How Ski Poles Industrialized and Diversified, Becoming More Efficient and Safe (3623 BC to 2019 AD) available at: https://scholarblogs.emory.edu/historyofskiing/2022/05/05/how-ski-poles-industrialized-and-diversified-becoming-more-efficient-and-safe-3623-bc-to-2019-ad/. (Accessed 1 January 2025).

Ward, M., 2018. How the Leeds wall changed climbing history. UKC available at: https://www.ukclimbing.com/articles/features/how_the_leeds_wall_changed_climbing_history-10122. (Accessed 1 January 2025).

Yonex Carbonex 8 B-8500 B Worth. available at: https://www.badmintoncentral.com/forums/index.php?threads/yonex-carbonex-8-b-8500-b-worth.188415/. (Accessed 1 January 2025).

Young, R.J., Liu, M., 2016. The microstructure of a graphene-reinforced tennis racquet. J. Mater. Sci. 51, 3861–3867. https://doi.org/10.1007/s10853-015-9705-6.

Zako, M., Nakanishi, Y., Matsumoto, K., Matsumoto, N., 2002. Identification of equivalent elastic parameters of triaxial woven fabric composites for golf club shafts. In: Fourth Canada-Japan Workshop on Composites, first ed. CRC Press.

Zimmermann, M.V.G., de Almeida, M.K., Ponsoni, L.V., dos Santos Salazar, M., Romanzini, D., Borsoi, C., Ornaghi, J.,H., et al., 2024. Advances in composite materials for surfboard manufacturing—a review. J. Appl. Polym. Sci., e56488 https://doi.org/10.1002/app.56488.

Zoot Ultra TT 4.0 Running Shoe. (n.d.). available at: https://www.zafiri.com/review-zoot-ultra-tt-4-running-shoe/ (accessed 1 January 2025).

3.5 Light weighting for automotive with composite technologies and its limitations in mass production vehicles

Nobuya Kawamura [1,2,3]
[1]Toyota Motor Corporation, Toyota, Aichi, Japan; [2]Scalable Advanced Composite Technology, Nisshin, Aichi, Japan; [3]Joby Aviation, Santa Cruz, CA, United States

1. Introduction

I have been very fortunate to be involved in lightweighting development since the beginning of my career in the automotive industry, starting as a Material Development and Engineering staff. My experience spans not only mass production applications but also, in my mid-career, advanced carbon composite developments for Formula One race car and a super sports car. In the later stage of my career, I transitioned into a corporate strategy management role, focusing on enhancement of lightweight technology applications. This progression has provided me with comprehensive understanding of material and process technology in a wider view, corporate strategic planning, and the essential decision making framework required by global corporation to balance environmental responsibilities with customer satisfaction.

So, why lightweighting? The benifits are evident when observing Formula One racing, where agility and performance are direct results of extreme lightweighting and precision-weight-balance. However, in the context of mass production, where cars are built every minute across multiple production facilities around the globe, you are required to consider the environmental impact each car will have once they are delivered to the market. Indeed, the transportation sector, in terms of the overall ratio of the greenhouse gas (GHG) emission, holds a significant amount, and lightweighting of vehicles, together with fuel efficiency improvement technologies of the power plants, became a focused topic since the early year 2000 as one of the solutions to increase fuel efficiency and, therefore, to reduce the GHG emission.

Extensive analysis of lightweighting potential were conducted by U.S. government agencies such as the EPA (Environmental Protection Agency) and NHTSA (National Highway Traffic Safety Administration). They were contracted to engineering organisations such as Lotus Engineering (Phase 1), FEV (Forschungsgesellschaft für Energietechnik und Verbrennungsmotoren GmbH), and EDAG GmbH (Phase 2) on light-duty vehicles and light-duty trucks to determine the GHG emission reduction and CAFE (corporate average fuel economy) requirements on new vehicles to be

The History of Composites. https://doi.org/10.1016/B978-0-443-34119-9.00022-X

launched around the year 2020. Toyota Venza and Honda Accord were chosen for the light-duty vehicles, and full-size pickup trucks from major U.S. manufacturers for light-duty truck evaluations. Each vehicle was completely disassembled into individual components, and a bill of materials were generated with weight and cost estimates. Furthermore, the engineering companies made suggestions for alternative technologies on each component or subassembly in three distinct categories: readily available technologies, near-future technologies, and advanced developments. The most common approach for the readily available technologies would be to reduce the thickness of the metal on the monocoque, replace steel components with lighter metallic materials such as aluminium or magnesium, or reduce plastic material density by foam expansions, etc.

By such method, in the phase 1 study by Lotus on the Toyota Venza, the conclusion on the light weighting opportunity was reported to be 38% in mass reduction at only a 2% cost increase, achievable by the year 2020. Major original equipment manufacturers (OEMs) were asked by the government agencies to provide their feedback on these reports. In the case of Toyota, as one of the vehicles studied was a Toyota brand, detailed engineering analysis was possible. During my time in the corporate strategic weight reduction group, we have conducted a detailed analysis on this report with individual design departments, such as the influence of reducing the guages of the monocoque on its structural integrity in terms of crashworthiness, noise vibration and harshness (NVH), and regulation conformities, or concerns on quality, durability, and cost of the components by design or material changes. The reply to the agencies on the analysis concluded that half of the amount of weight reduction could be realised with minimal cost or quality impact to the consumer. Finally, the U.S. government determined a fuel economy standard of 54.5 mpg (miles per gallon) or 4.3 liter per 100 km for vehicle model years (MY) 2022–2025 as a requirement.

Similar harsh requirements from other global countries have motivated the automotive industries to focus not just on the fuel efficiency improvement of the powertrains but also on reducing the running resistance of vehicles. Fig. 3.5.1 shows the comparison of simulated total energy loss calculation, under the same combined running mode, between old versus new models for an ICE (internal combustion engine) and a hybrid motor, where the acceleration resistance and inclination resistance are expressed as moment of inertia. A couple of points to be noted here are that the relative proportion of energy loss by running resistances within total energy loss (red box in Fig. 3.5.1) increases as the efficiency of the powertrains increases. Moreover, the moment of inertia has not improved as much as the other resistance factors throughout the model iterations, indicating that there may be more opportunities for fuel efficiency improvements by lightweighting. Examples of such lightweight developments that have or may have a positive impact are given in the following section.

[1] All figures mentioning "Toyota" or "Toyota internal document": courtesy of Toyota Motor Company.

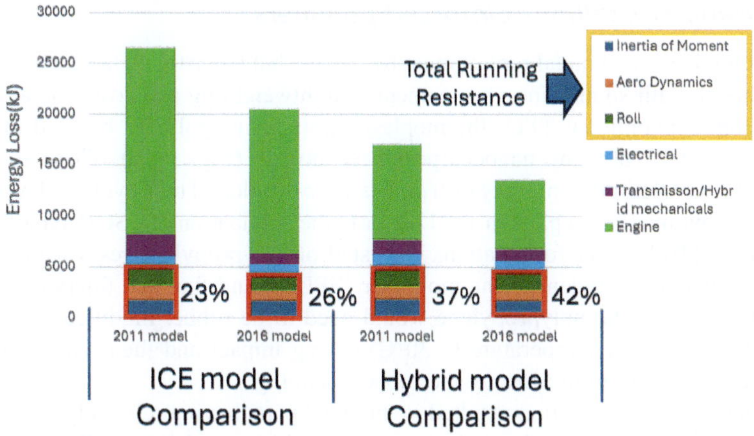

Figure 3.5.1 Energy loss reduction achieved for ICE and hybrids by newer model changes. Toyota internal document[1]

2. Conventional lightweighting of automotives

Long before the higher demand for fuel economy improvements began, replacements of metallic components by injection moulding of fibre-reinforced thermoplastics became popular, more for the cost improvements by reduction of post-machining, processing time, and functional integration, but they have also helped to reduce the weight significantly. The following are some of the representative results that had been applied by a majority of OEMs by the late 20th century.

2.1 Injection-moulded fibre-reinforced engineering plastics for metallic replacement

The first application of plastic intake manifolds started in the early 90s by Porsche, replacing heavy aluminium casting, using low-temperature melting metal as the removable core to injection mould glass-reinforced polyamides. This one-piece design and process was quickly replaced by a multipiece friction vibration-welded design and process due to its cost advantage. Material-wise, polyamide 6 is typically used because of the relatively lower temperature environment, lower material cost, and lower warpage by postshrinkage, which is important in achieving quality weld strength during vibration welding. Similarly, aluminium cast engine valve covers were also replaced by thermoplastic injection moulding. In this case, polyamide 66 was also used to cope with the temperature of the engine oil, which it needs to contain, and possible heat radiation from the exhaust pipes. Other engine components, such as the throttle body and cooling water inlet and outlet pipes, were replaced by injection-moulded glass-reinforced polyphenylene sulfide (PPS), due to high dimensional tolerances or heat resistance requirements.

2.2 Nano-crystalline-reinforced polymers

This example is not a metal replacement by plastic, but a plastic replacement by plastic, although with significant advancement in lightweighting and cost. The advancement from conventional TPO (thermoplastic polyolefin) polymer blended material was achieved by utilising nanocomposite technology (Fig. 3.5.2). These so-called Toyota Super Olefin Polymer (TSOP) grades were developed by Toyota and were first applied to bumper fascia parts in 1991. Prior to the application of TSOP, typical polypropylene (PP) bumper formulations consisted of polypropylene resin with rubbers such as ethylene propylene diene monomer (EPDM) and fine talc fillers in roughly a 60:30:10 ratio. The polypropylene would need to be rubber-modified to improve its brittleness at low temperature (−30°C) during impact and then reinforced with talc to recover the rigidity lost by the rubber blending.

Such material formulation resulted in moulded parts having relatively low surface hardness, requiring parts to be grain-patterned to hide scratch marks. With the nanocomposite technology, where the newly developed nano-crystalline structure enhanced polypropylene and polyolefin elastomers reinforce each other on the nanoscale, the TSOP material was able to increase rigidity and surface hardness at a higher flow rate (lower molecular weight) without reducing low-temperature impact strength, preventing the bumper from cracking or shattering by impact during accidents.

Such improvements in the material property enabled larger-sized bumpers to be injection moulded at a thinner wall thickness, with a faster cycle time, and without a grained surface due to better scratch resistance, therefore enabling them to replace even the reaction injection moulding (RIM) urethane bumper facias used for high-end vehicles to meet the 5 mph damage tolerance requirement. This had a significant impact on cost and weight reduction throughout the entire fleet of vehicles. Although this development was initiated by Toyota, eventually the entire industry followed the path, and gradually the RIM bumpers or XENOYTM, a polycarbonate/polybutylene

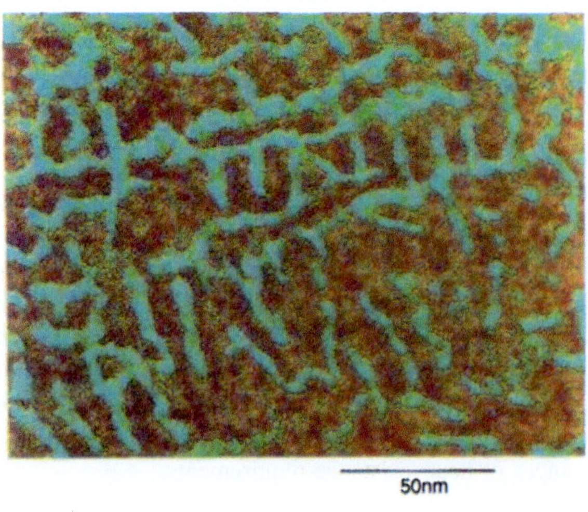

Figure 3.5.2 TSOP: Image analysed high-order structure of polypropylene (PP) phase (amorphous, crystalline). Toyota Motor Corporation internal document

terephthalate (PC/PBT) bumpers were replaced by polypropylene material like Toyota's Super Olefin Polymer (TSOP, Fig. 3.5.3).

3. Exploring lightweighting of automotives with composite technologies

The above-mentioned composites for injection moulding did provide lightweighting opportunities, but due to the relatively short fibre composition, the bigger mass reduction opportunities in the primary and secondary structural applications were limited. No need to mention it, but for the material to perform structurally, the fibres need to be utilised in a longer length. Such long fibre-reinforced composite technologies for structural applications started with glass fibre-reinforced sheet moulding compound (SMC) or bulk moulding compound (BMC) in the mid-1960s. For the glass fibre-reinforced SMC applications in large volume production, applications were still limited to exterior panels, and the aim was more on design freedom with integrated functions (see Fig. 3.5.4) or detachable retrofitting of truck bed roofs, also called canopies (see Fig. 3.5.5). BMC was used more for replacing steel or aluminium casing or housings for mechanical units to enhance anticorrosion performance in harsh environments.

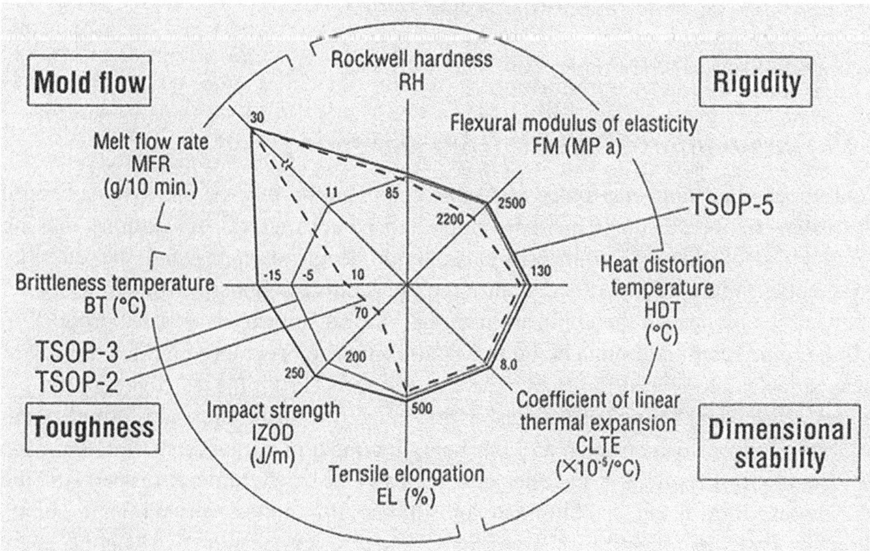

Figure 3.5.3 Property comparison of conventional TPO versus TSOP (Toyota, 1995).

Figure 3.5.4 Engine hood with moulded-in fins (Toyota, 1988).

Figure 3.5.5 Toyota Land Cruiser SMC canopy (Toyota, 1984).

3.1 Fibre reinforced plastic (FRP) leaf spring

Continuous glass fibre-reinforced leaf spring application may be the first successful glass fibre-reinforced composite to be industrialised in structural applications, making optimal use of the load transferring capability (high strength) and the ductility (reasonable failure strain) of the continuous glass fibres. General Motors' Corvette has been the pioneer in the application of the FRP leaf spring on a series production vehicle (Fig. 3.5.8). Although being in a sports car category, more than 1.6 million of these vehicles have been built over the last 60 years. From the third-generation C3 model, an FRP leaf spring has been used, first in its front suspension and from the generation to follow, also on the rear, both being mounted in a transverse direction. The FRP leaf spring replaces a metallic coil spring. Due to the unique property of the continuous fibres, it can also eliminate antiroll bars if mounted on two pivotal points, allowing a weight reduction of more than 15 kg. A similar approach was taken by Volvo on their C90 and CX90 rear suspension in early 2000.

On the commercial vehicle side, the Mercedes-Benz Sprinter also adopted the transversely mounted FRP leaf suspension. For the conversion of longitudinally mounted steel leaf suspension to FRP, the advantages gained are limited to just a simple weight reduction by material change and do not provide any functional integration. Therefore, it becomes difficult to justify the cost increase of converting from steel to FRP, although there is a gain in corrosion resistance and better fatigue resistance at an additional cost.

In terms of material and process, the application of high-pressure resin transfere moulding (RTM), together with fast curing and a relatively ductile polyurethane resin, has enabled shorter cure times and reduced the overall process time of the parts, which allowed the application to expand to higher production volume vehicles.

3.2 SMC truck beds

In the early 2000s, the application of glass fibre-reinforced SMC for the boxes of the pickup truck bed became popular. Until then, the heavier steel or aluminium truck beds had to be further covered with high density polyethlene (HDPE) liners, which were vacuum moulded from HDPE sheets, adding even more weight. So the weight savings on these high-volume-selling light-duty trucks, which suffer from high average fuel consumption, are significant. Application of SMC to the truck beds also resulted in better dent resistance, corrosion durability, and multifunctional design capability, such as integrated storage boxes.

Now, an even more enhanced material technology is being explored, where the conventional glass fibre-reinforced SMCs are combined with braided glass fibre laminates. Especially by using a QISO®, a 0 ± 60-degree triaxial braid provided by A&P Technology, a balanced laminate can be achieved at the lowest ply amount (Fig. 3.5.6). The innovative approach is said to enable a 20% reduction in the thickness of the part with enhanced impact damage tolerance compared to the conventional SMC structures.

Figure 3.5.6 Triaxial braid-reinforced SMC.
Courtesy of ID Composite International

3.3 Metal overmoulding technologies

Overmoulded metallic components with injection-moulded engineering plastics are a combined metal/plastic composite technology. The technology can be performed with or without the use of an adhesive at the interface. In case of not utilising adhesives at the interface, the injection-moulded part is secured by mechanical interlocks, provided by punched holes in the metallic part, but they lack a tight and secure fit. A successful example of the adhesive interface approach would be cross-car beams. The design approach of a crossbeam differs per OEM due to the differences in the structural philosophy of the monocoque architecture or the existing manufacturing infrastructure of your component supply chain. This can result in a large reduction of weight, if the metal stamping and welding process elimination balances with the investment in injection moulding machines for the overmoulding, and if the mounting interface design can incorporate an efficient assembly process.

3.4 Compression-moulded glass fibre-reinforced mats

General Motors' Corvette in 1968 was the first car to incorporate AZDEL, a continuous strand, random glass fibre-reinforced polypropylene mat for stamp forming, to replace metallic bumper beams. The application expanded as the material evolved to be reinforced with unidirectional glass fibres and with clever collapsable body mount crash boxes (Fig. 3.5.7). These compression-moulded beams were able to withstand a 5 mph collision without the heavier shock absorber systems used at the beginning, enabling a reduction of more than 30% of the weight compared to the steel system. Since these beams were compression moulded, the cross sections are either C-channel or W-channel shaped, which, compared to a closed box cross section, are not as efficient in deformation resistance. Later, some attempts to inject overmold cross ribs in the open channels were made, but eventually they were replaced by aluminium extrusion beams with reinforcement ribs to form a multicell cross section, which was postpressed to the needed curved shapes to fit the exterior styling.

3.5 Compression-moulded continuous fibre reinforcements

For some of the high-performance cars, such as the BMW M3, enhancement of the driving performance by reduction of the front overhang mass is important. The use of continuous fibre reinforcements was explored by Jacob Plastics GmbH for the front and rear bumper beams (Fig. 3.5.8), replacing the aforementioned AZDEL-like material types. Also in this case, woven glass fibre textiles were impregnated with a polyamide matrix instead of polypropylene to make full use of the mechanical properties gained by the continuous fibres.

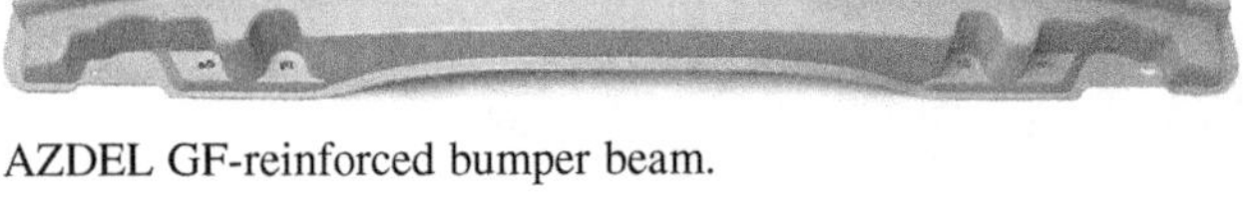

Figure 3.5.7 AZDEL GF-reinforced bumper beam.
own picture

Figure 3.5.8 Compression-moulded textile laminate beam (Mitschang, 2012).

Typically, the textile prepregs were laminated to their designed orientations and ply thickness and were then consolidated into a flat sheet as a semiproduct. The main reason to go through this extra process was to gain fully consolidated laminates with minimum void contents by static heat pressing. Conventionally, the textile prepreg process for mass production is done by a continuous film melting process, laminating the resin to the textile through compression rollers. In such process, it can be difficult to fully impregnate without any voids. These sheets were cut to their required shapes, preheated to just above the melt temperature of the resin matrix in a heating zone or in an oven, shuttled to the press, and moulded into their final shapes. Reinforcement rib features can be made either by secondary injection overmoulding or by adding patches of random fibre-reinforced charges, similarly preheated and compressed at the same time the beam is compression moulded, so the random fibre material would flow into the rib cavity of the tool to acquire a final part in one step. This can be an economic solution, but usually it has a weight penalty due to the excessive amount of random fibre material covering the surface of the part.

Just as a note, to enhance the driving performance of a sports car, a unique attempt was made on the Lexus RC-F models, where the aluminium profile beam was reinforced with a thermoset carbon fibre (CF) unidirectional (UD) prepreg, adhesively cocured onto the back side of the aluminium beam. In this way, the crash energy absorption of the beam is enhanced by having the carbon fibres loaded in tension (Fig. 3.5.9).

3.6 Random fibre moulding with in-situ continuous fibre reinforcements

For relatively more complex shapes compared to bumper beams, injection moulding has an advantage. Modular front-end structures for body frames, integrating radiator, cooling fan and headlamp mounts, etc., were developed using glass-reinforced polyamide injection moulding with continuous fibre textile reinforcements across the top width to eliminate conventional steel body frames. This approach has become more popular in Europe due to the larger role the system suppliers play in the subassembly of components for the OEMs. OEMs also benefit from being able to access the engine

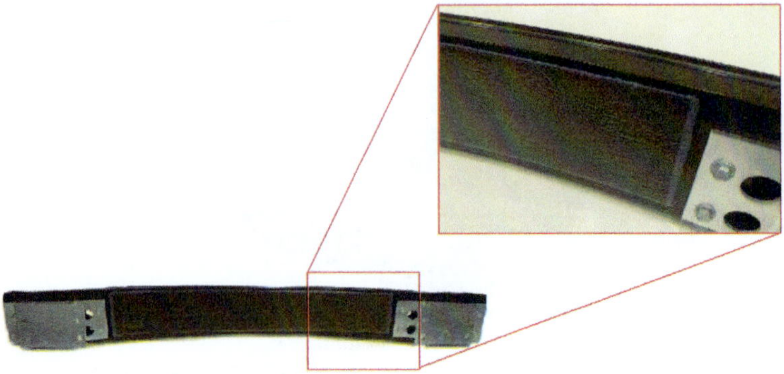

Figure 3.5.9 CF-reinforced Al bumper beam.
own picture

compartment directly from the front side during assembly to ensure a better ergonomic working environment for the workers.

Personally, I have found these cases to be interesting. For the OEMs, it can be difficult to successfully benefit from such approach as the design functions of the components to be integrated lie across unique design sections throughout the company.

Alternatively, an injection-moulded polypropylene matrix with long glass fibre reinforcement has been successfully implemented by the development of high-flow, high-crystalline polypropylene grades to avoid applying excess shear to the glass fibre during injection moulding. This was achieved by Mazda and applied to their Mazda 6 model in 2001.

Another clever way of utilising the advantages of continuous fibre properties can be seen in underbody components, such as the spare wheel well (SWW) for passenger cars. Here, the composite materials started to replace steel equivalents for their advantages in rust protection, NVH, parts, and functional integrations, as well as weight reduction. Initially, a thermoset SMC was used, as it can be put through the e-coat and paint baking process together with the metallic body. Later, it was replaced by press-moulded glass mat thermoplastic (GMT) materials, which are randomly distributed long glass fibre strands tufted to a mat and impregnated with polypropylene resin films. This GMT material was additionally reinforced in the longitudinal and transverse direction of the SWW with Twintex, a UD glass fibre comingled with polypropylene fibres, to cradle the spare wheel for better fracture resistance during rear collisions, restricting the spare wheel from falling off the vehicle and potentially causing secondary accidents. Typically, 30%–40% weight saving was achievable.

3.7 Long fibre thermoplastic-direct (LFT-D) moulding

Reduction in process steps from raw material to component moulding can be achieved with the LFT-D process. Since there is no need for a preimpregnation process of the fibres with resin matrix, laminating the prepregs, or compounding the resin with fibres and pelletising for injection moulding, the LFT-D process becomes attractive in terms of cost reduction for the part manufacturer, because he has control over the raw material purchases.

LFT-D can be done using either a compression moulding or an injection moulding method. The injection moulding process would allow for the reduction of labour costs and would orient the fibres by directly injecting them into the tool cavity at high velocity. On the other hand, compression moulding would require the compounded composite slabs to be carried into the tool; it also requires postmachine holes, as the material flow within the cavity is not as controllable as injection moulding in terms of heat and pressure distribution, making it difficult to flow around big openings in the part design. The advantage of compression moulding would be that it is easier to locally reinforce with continuous fibre patches, whereas the injection process will have too much pressure in the plane direction, which will cause distortion of the continuous fibre charges. One of the first attempts to take advantage of the LFT-D process was to strengthen the instrument panel. Other components, such as door panel modules and front-end modules, were also manufactured with LFT-D.

Some extensive developments have been conducted by institutes and consortia such as the Japan National Composite Center in Nagoya University, using the LFT-D process with carbon fibre reinforcements. The aim was to offset the cost of carbon fibre material by enabling cost-efficient processing. An automotive monocoque based on a carbon fibre-reinforced polyamide 6 matrix with the LFT-D process was prototyped in 2017 (Fig. 3.5.10), replacing the original aluminium chassis of an existing commercial model and achieving a 10% weight reduction while maintaining the same strength and rigidity requirements.

3.8 Carbon fibre-reinforced sheet moulding compound (C-SMC)

C-SMC applications in the structural body panels have been explored. This expanded SMC usage from fairings and cosmetic panels, typical for glass fibre-reinforced SMC, to primary structural components. An extreme example is the Lamborghini Sesto

Figure 3.5.10 LFT-D underbody.

Elemento chassis in 2009 (Fig. 3.5.11), where the entire lower body structure was compression moulded with C-SMC (Lamborghini brand name: Forged-Composite), enabling a V10 supercar to achieve a kerb weight of under 1 ton. In 2010, the Lexus LFA was marketed using a similar C-SMC for its entire upper back body structure (Figs. 3.5.12 and 3.5.14). This experience at Toyota has led to the CF SMC liftgate application for the 2017 Prius PHV, achieving a 40% weight reduction compared to an equivalently sized aluminium counterpart (Fig. 3.5.13).

The much higher production rate of 5000 vehicles per month posed significant challenges to the OEMs as well as to the material suppliers in terms of cost efficiency, quality consistency, and process repetitiveness. At the same time, this technology was applied to the Lexus LC model for the door inner panel and the trunk inner panel, saving 47% weight. The significant achievement here was that, although the parts were exposed in the cosmetic zones, the painting of the parts was avoided by the fine-tuning by the OEM of the formulation and quality control on the material supplier side, as well as that of the compression moulding parameters. No need to mention, but avoiding having to paint has a positive impact on the GHG emissions.

Figure 3.5.11 Lamborghini Sesto Elemento tub.
Own picture at the Paris Motor Show, 2010

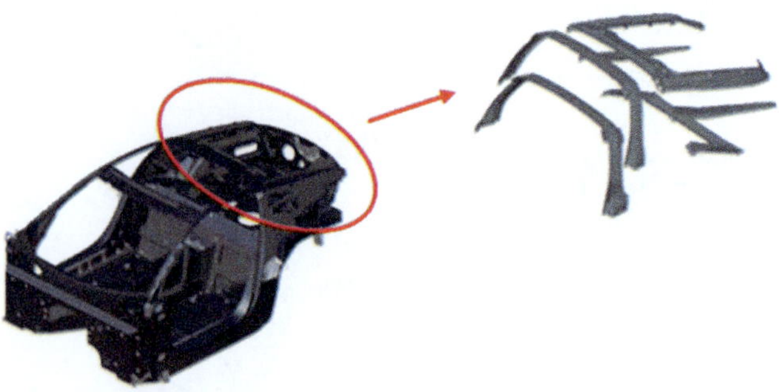

Figure 3.5.12 Lexus LFA upper back body.
Toyota Motor Corporation internal document

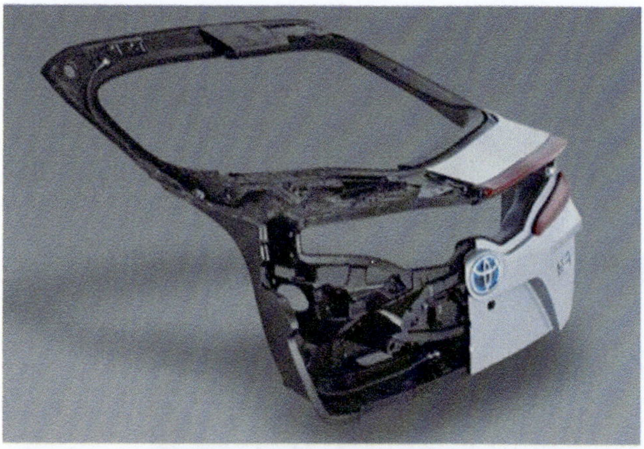

Figure 3.5.13 Prius C-SMC liftgate (Toyota, n.d.a.).

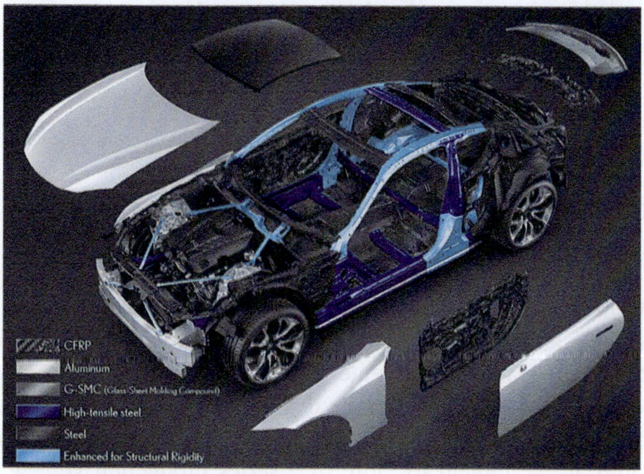

Figure 3.5.14 Lexus LC multimaterial body.
Toyota Motor Corporation internal document

Another example to follow was the Toyota GR Corolla roof panel introduced in 2022 (Fig. 3.5.15), which aimed at decreasing, for this entry-level sports car, the centre of gravity at an affordable cost in comparison to conventional RTM panels. The marble-like appearance with uneven roughness of compression-moulded random chopped fibre material has been accepted by the market ever since Lamborghini started to apply their Forged-Composite technology for not just the aforementioned structural components but also to aesthetic parts in the interior trims and to their midship-mounted engine hood. These exterior components were painted with a clear coat for UV protection.

Figure 3.5.15 GR Corolla C-SMC roof (Toyota, 2024).

3.9 Carbon fibre-reinforced RTM

RTM moulding is a standard method of producing carbon fibre exterior aesthetic panels, like parts of the front engine hood, roof panel, and rear trunk lid. It is applied in high-end vehicles for better weight distribution to enhance drivability. Vacuum-assisted RTM has been a popular solution for low-volume production cars due to the lower tooling and process equipment cost. However, the arrival of high-pressure RTM, together with advancements in the resin mixing and injection equipment, new resin formulations with internal mold release, and snap cure cycles for enhanced production rate from hours to minutes, all together have enabled the RTM technology to be applied to more of a series production vehicle. BMW was an early adopter of these applications to their M3 E46 CSL sports models since 2003 and continued to expand to other M series with their in-house production cell.

Other RTM applications are the monocoques for low-volume supercars starting from the 2003 Mercedes-Benz SLR McLaren, the 2010 Lexus LFA (Fig. 3.5.16), and the 2011 Lamborghini Aventador. McLaren MP4 (Fig. 3.5.17) took the composite monocoque manufacturing one step further and developed a one-shot RTM process in 2011 (Autoevolution, 2015).

3.10 The Lexus LFA development

As for the development of the Lexus LFA's carbon monocoque, although not everything is done by RTM, a little more detail can be shared, as I have personally led the composite material development from 2004, when the original vehicle development

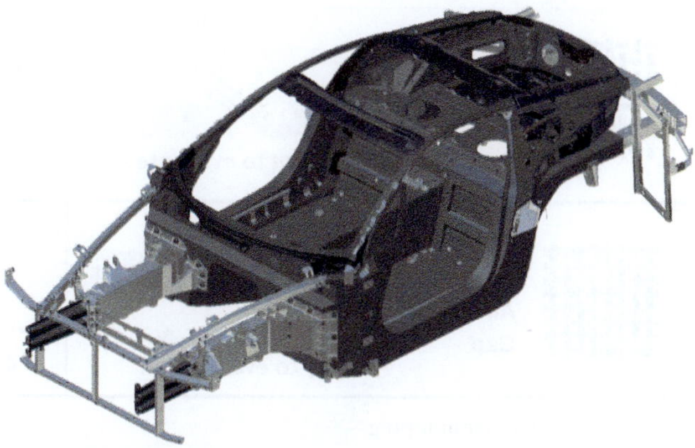

Figure 3.5.16 Lexus LFA monocoque.
Toyota Motor Corporation internal document

Figure 3.5.17 McLaren MP4-12C monocoque.
Tokumeigakarinoaoshima, CC0, via Wikimedia Commons

was still carried out with an aluminium structure. Faced with a task to enable production of five monocoques per day, the production rate of the LFA was determined to be only one per day for 2 years at 500 units total. This had to be done at a reasonable cost in comparison to the aluminium counterpart. Hence, the composite R&D team sought for an out-of-autoclave (OOA) solution (= minimum capital investment), with heavier ply laminates to reduce hand labour and reduce curing time, in order to win the competition.

The solution chosen was to develop a 450 g/m^2, biaxial NCF (noncrimp fabric)-based semiprepreg (Figs. 3.5.18 and 3.5.19), which can be vacuum bag cured. The

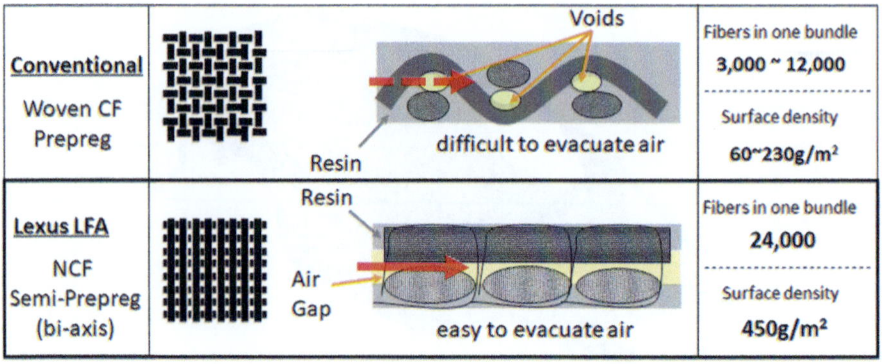

Figure 3.5.18 Lexus LFA NCF semiprepreg.
Toyota Motor Corporation internal document

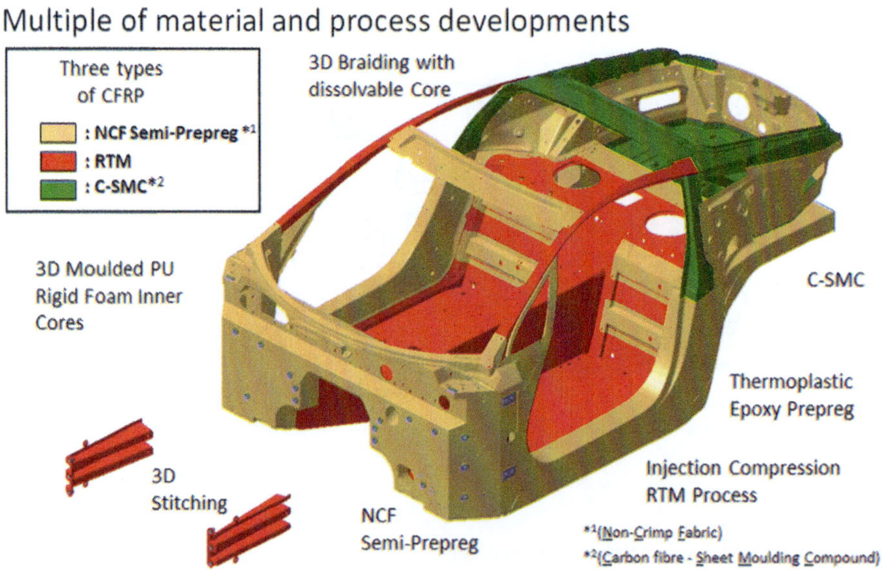

Figure 3.5.19 Lexus LFA Monocoque's multimaterial and process.
Toyota Motor Corporation internal document

original design and manufacturing approach, suggested by my team, consisted of large sandwich panel preforms in much simpler shapes, mostly for cost reasons, but that was not how the final product turned out to be. During the conceptual design stage of the monocoque, extra emphasis was placed on the body stiffness requirement due to the high handling performance of a super sports car.

Finally, the side member structures required multiple inner rib reinforcements in their closed box section to stabilise deformation from front to rear suspension loads without interfering with surrounding components that already existed and were more mature in design (Fig. 3.5.20). To achieve this, a 3D moulded rigid polyurethane core foaming process was developed to avoid using expensive material and the numerical control machining process. This foam core multirib structure was also used for the firewall structure, as LFA was a front V10 4.8L engine mounted configuration with a short crash stroke, meaning that the composite firewall needed to withstand the intrusion of the engine unit during crash impact at a relatively wide lateral distance. Consequently, the NCF semiprepreg with the OOA process was only used for panels, hat sections, and simple single-cell box crossbeams. In contrast, the highly loaded structures were all autoclave cured due to concerns about insufficient pressure to negate bulk factors related to complicated cross-section design in combination with multiple foam cores and bulky semiprepreg layups. Regardless, the semiprepregs were still used.

Although the monocoque resulted in a much more complicated design, I believe the overall cost advantage of the material and laminating process with the heavyweight NCF semiprepreg compared to that of the standard prepregs did allow us to keep our cost down. Even though it was still far more expensive than the aluminium version, the project was eventually granted to the composite solution. The increased body rigidity at minimum weight increase has complemented the performance of the car significantly, as shown in Fig. 3.5.23.

Another innovative development was the thermoplastic carbon fibre reinforced plastic (CFRP) composite for impact energy absorption. The tight packaging design of the LFA required an efficient solution for the energy absorption in pole collision to the side sill at a relatively short stroke. The ductility of the newly developed epoxy-based thermoplastic matrix, in comparison to that of the thermoset, offered clear advantages, and a sinusoidal wave-shaped energy absorption component (Fig. 3.5.21) was implemented as the side sill inner structure.

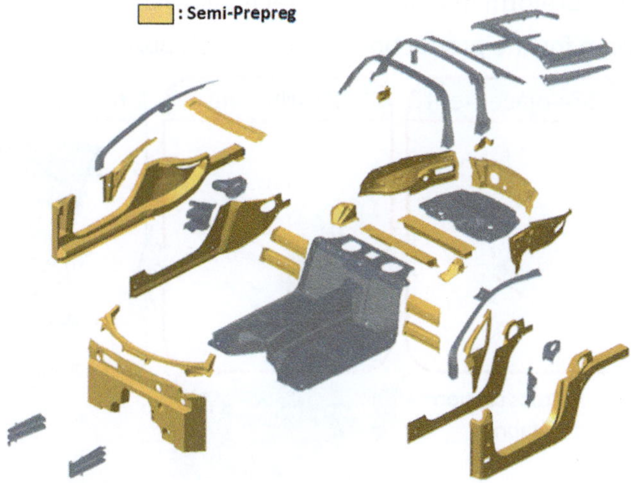

Figure 3.5.20 Application of the NCF semiprepreg technology.
Toyota Motor Corporation internal document

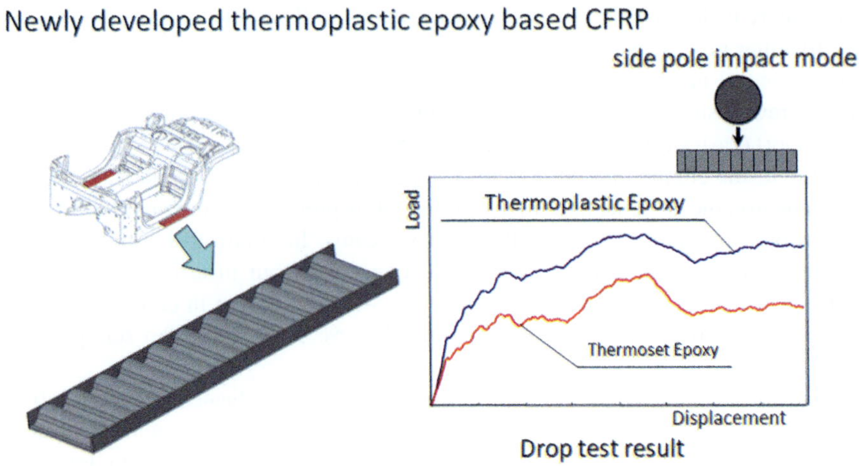

Figure 3.5.21 Thermoplastic impact energy absorber.
Toyota Motor Corporation internal document

There are many more composite developments implemented in the LFA, as can be seen in Fig. 3.5.19, but the last one I would like to mention is the 3D carbon fibre stitched impact energy absorber for the front crash structure (Fig. 3.5.22). This could be an interesting development, due to its uniqueness and the existing stories behind it. It was originally developed by Toyota Industry, a subsidiary of Toyota Motor Corporation, for space applications in the late 1990s and was used to manufacture the roll hoop structure for Toyota Formula One cars. Actually, a weaving machine manufacturing company established in 1926, is at the origin of Toyota Motor Corporation. The front crash structure could also have been done with a thermoplastic

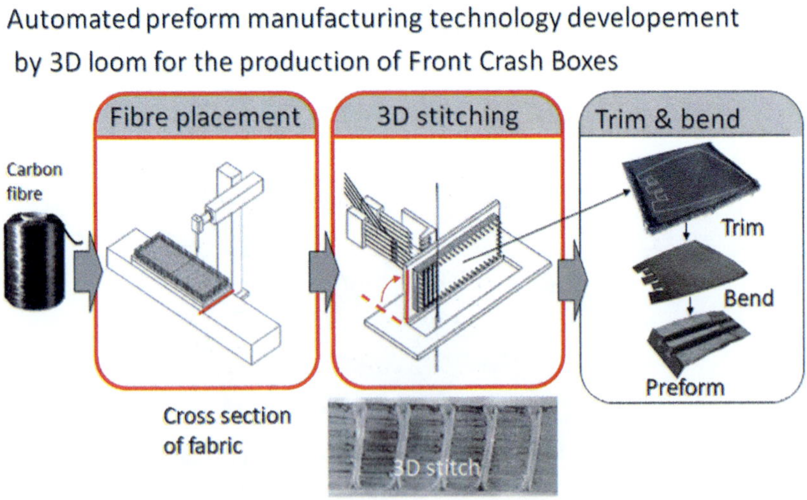

Figure 3.5.22 Carbon fibre 3D stitching technology for the front crash structure of the LFA.
Toyota Motor Corporation internal document

composite as well, but for the long-desired super sports car, marketed ever since the Toyota 2000GT (1967–1970), it was nice to have a story to tell, connecting the Lexus LFA to the original weaving machine company and the Formula One technology.

The significance of the LFA development (Fig. 3.5.23) was that it was led by Toyota from scratch with help from many group companies and was manufactured in-house. Personally, if the car had been designed with the intention of a composite monocoque from the very beginning, then the entire design concept and packaging could have been done very differently, allowing much simpler technology and process developments. Many of the facilities introduced to manufacture this multitechnology monocoque and the manufacturing team still remain operational today, which adds internal capability to respond to future needs for lightweight solutions. One example of that is the C-SMC technology that was carried on to mass production models such as the Prius PHV (Fig. 3.5.13), Lexus LC (Fig. 3.5.14), and GR Corolla (Fig. 3.5.15), as explained in Section 3.8.

As for the relatively higher volume RTM production, the BMW i3's upper body was designed as subassemblies, produced by RTM, and finally robotically assembled by adhesive bonding. This upper body was matched with an aluminium underbody, carrying the motor and batteries. Such modular approach enabled BMW to manufacture a total of 250,000 vehicles in the time span of 2013–2022 (Fig. 3.5.24).

4. Hopes and limitations of composite technologies as a solution for the industry

The composite technology developments in the automotive industry have advanced significantly in the last 40 years or so of my career as an automotive engineer, and

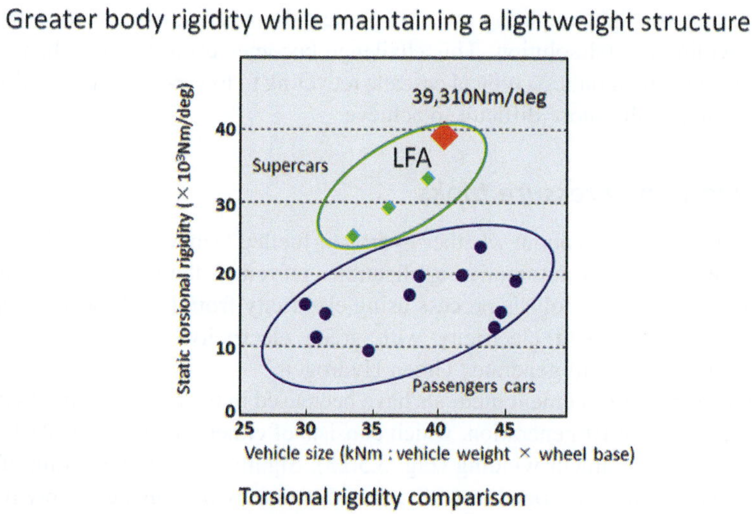

Figure 3.5.23 Body rigidity comparison.
Toyota Motor Corporation internal document

Figure 3.5.24 BMW i3 upper (CFRP) and lower (aluminium) body (Edgar Jiménez from Milton Keynes, UK, CC BY-SA 2.0, via Wikimedia Commons).

I was extremely fortunate to witness and to experience the evolution. The hopes certainly were raised as we explored continuous fibre reinforcements and carbon fibre reinforcements together with thermoplastic matrices, but the balance between the cost, including processability, and performance limitations, due to the limited choice of affordable resin systems, has hampered the expected growth of composite applications at an industrial scale.

On the other hand, there are applications where a composite technology makes the most sense, not just from the lightweighting point of view but also due to its capability to meet or exceed requirements deriving from the new technologies. Such example might be the alternative propulsion systems as shown in the examples to follow.

At the same time, a successful recycling scenario remains to be an issue. Certainly, if the reasons for lightweighting are environmental concerns, one cannot turn away from providing a total solution. This challenge becomes even greater when we use more expensive materials, as typical cascade recycling to lower-value applications becomes economically more difficult to achieve.

4.1 Hydrogen pressure tanks

While there are debates about whether hydrogen-fuelled technology would become a sustainable solution for automotive applications, currently, the majority of hydrogen is produced by the electrolysis process using electricity from fossil fuels. A replacement of fossil-generated electricity with green electricity will enable a carbon emission-free process to generate "Green Hydrogen."

Carbon fibre reinforcement methods have been used in the high-pressure hydrogen tanks since the type III generation, which consists of either a metallic or plastic liner with carbon fibre filament winding (Fig. 3.5.25). Significant lightweighting of these tanks has been achieved, together with cost reduction, by optimising the ply balance between the dome and the hoop area of the cylindrical tanks. Eliminating unnecessary

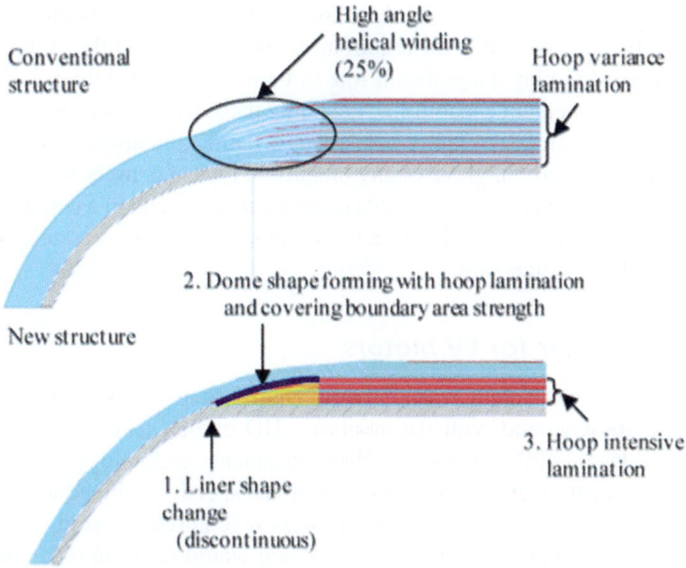

Figure 3.5.25 Hydrogen tank cross-section comparison.
Toyota Motor Corporation internal document

helical direction winding at the dome area and increasing the ratio of hoop winding at the cylinder area where strength is needed allowed for a reduction of carbon fibre usage by 40% on the second-generation Toyota Mirai, introduced into the market in 2014 (Fig. 3.5.26).

4.2 Electric vehicle (EV) battery cases

As the number of Li-ion battery-powered system applications increases, improvements in the containment of flames during battery runaway accidents will become

Figure 3.5.26 Toyota Mirai (Toyota, 2019).

increasingly important. This is also the case for the eVTOL technology that I am currently involved in, and while the requirements may differ slightly, composite technology is expected to play a significant role in both cost and weight efficiency, with enhanced heat management capabilities remaining the same. Choices in the resin type may differ due to regulatory requirements. But if flame containment is necessary, it seems that whatever carbon, glass, or any other fibre type, a densely woven UD laminate or chopped UD layers in a preferable quasi-isotropic layup to a certain minimum thickness will be the key factor in successfully preventing high-pressure flame from bursting through the composite.

4.3 Magnet rotor for EV motors

For the last few years, Trelleborg (Albany, United States) has been displaying a concept for a rotor wound with thermoplastic UD carbon tape using their unique in-situ consolidated AFP (automated fibre placement) technology. The aim is to secure the permanent magnets during high rotational speed, combined with weight efficiency. At the same time, a minimum thickness is required to avoid increasing the distance between the magnet and the stator, which would result in reducing the efficiency of the motor power. In this application, carbon fibre composites provide benefits other than weight saving, such as the relatively low conductivity to reduce the interference with the magnet compared to a steel casing, better fatigue properties, and protection of the brittle magnets from impacts (Figs. 3.5.27 and 3.5.28).

Figure 3.5.27 In-situ AFP composite rotor mock-up.
own picture taken at JEC-World, 2021

Figure 3.5.28 In-situ AFP composite sleeve mock-up.
own picture taken at JEC, 2021

Currently, the resin matrix material used for this application is polyether ether ketone (PEEK), which has a relatively high melt temperature of 380°C, exceeding the roughly 130°C demagnetisation temperature of permanent magnets. But by using a laser heating-assisted AFP system, the temperature experienced by the magnet is instantaneous and causes no harm to the magnet properties. If the availability of this technology expands and hence the cost reductions are achieved, it may not be so hard to justify moving away from conventional metallic housings with adhesively bonded solutions.

5. Summary

There are many other examples of successful lightweighting or product enhancements by clever use of composite technologies that I have not touched on in this chapter, such as seat pans and seat backs, or other examples that a tier one supplier, together with an OEM, has successfully introduced by justifying its benefits against their previous design. As mentioned in the beginning, the goal of this chapter is not to list a complete history of composite applications in automotive but to highlight those that had or may have enough impact on the entire industry to enable our advancement toward the reduction of environmental stress caused by the transportation sector.

References

Autoevolution, 2015. https://www.autoevolution.com/news/mclaren-reveals-sports-series-carbon-fibre-monocoque-tub-91116.html.

Mitschang, P.K.H., 2012. Polymer and composite moulding technologies for automotive applications. In: Advanced Materials in Automotive Engineering, 210–229.

Toyota, n.d.a. https://global.toyota/en/newsroom/toyota/22701478.html.

Toyota, (2024). Downloadable assets. https://global.toyota/jp/newsroom/toyota/37061800.html.

Toyota, (1984). https://global.toyota/en/album/images/26333704/#.

Toyota, (1988). https://oldtownclassics.com/wp-content/uploads/2021/07/IMG_8240-1200x900.jpg.

Toyota, (1995). Toyota develops new high-performance resin material. Accessed Toyota Times. https://global.toyota/en/detail/7907490.

Composites in structural engineering and architecture

3.6

Thomas Keller[1], João R. Correia[2], Joaquim Barros[3], Seyed Arman Hosseini[4] and Brahim Benmokrane[4]
[1]CCLab, École Polytechnique Fédérale de Lausanne, EPFL ENAC IIC CCLAB, Lausanne, Switzerland; [2]CERIS, Instituto Superior Técnico, Universidade de Lisboa, Lisboa, Portugal; [3]ISISE/ARISE, Univ. of Minho, Campus de Azurém, Guimarães, Portugal; [4]Department of Civil & Building Engineering, University of Sherbrooke, Sherbrooke, QC, Canada

1. Introduction

1.1 Main reasons for using composites in structural engineering and architecture

Fibre-polymer composites, also referred to as fibre-reinforced polymer (FRP) composites, are generally comprised of glass, carbon, aramid, or basalt fibres embedded in a thermoset polymer matrix of isophthalic polyester, vinylester, or epoxy. These composites are being increasingly considered for applications in structural engineering and architecture due to the various advantages they offer compared with conventional materials, namely reinforced concrete, steel, and timber.

One of the key advantages of fibre-polymer composites is the high strength-to-weight ratio and, when using carbon fibres, also the high stiffness-to-weight ratio. For new construction, this allows designing lightweight structures that are easier to handle, transport, and install, requiring light lifting equipment on the construction site. For the strengthening of existing constructions, FRPs are effective in increasing the load-bearing capacity, with negligible weight increase and geometry alteration of the strengthened members, being relatively faster to apply than traditional techniques based on conventional materials.

Another major advantage of fibre-polymer composites over traditional materials is their resistance to environmental agents, which makes them particularly well suited for applications under exposure to marine environments, de-icing salts, moisture, or chemicals. For such harsher exposure conditions, the noncorrodibility and rot-resistance of FRP allow for reducing maintenance requirements, thus providing significant life cycle cost reductions during service life.

Due to their flexibility and free-formability, FRP can be moulded into intricate shapes, enabling complex and aesthetic designs, which may be a relevant feature for form-active facades, roof structures, or decorative panels. Here, the low thermal conductivity of FRP and their potential for function integration can also be an important feature, as they are able to provide thermal insulation and translucency in a single layer, contributing to the energy efficiency and lighting of buildings.

The History of Composites. https://doi.org/10.1016/B978-0-443-34119-9.00023-1

Finally, the electro-magnetic transparency afforded by FRP can also be a relevant advantage in some applications, such as in buildings with medical facilities (e.g., magnetic resonance imaging) or in bridges over railway lines with strict requirements in terms of clearance to the catenary.

1.2 Main applications of composites in structural engineering and architecture

The main applications of fibre-polymer composites have involved two main forms: (i) bending components, or (ii) tensile components, according to their two-dimensional or one-dimensional structural action, respectively (Keller, 2002). Bending components include profiles (Fig. 3.6.1a) and sandwich slab panels (Fig. 3.6.1b), either web-core or foam-core, generally produced by pultrusion or vacuum infusion. Tensile components include laminates (Fig. 3.6.2a) and fabrics (Fig. 3.6.2b), respectively produced by pultrusion and applied (on site) by hand lay-up, as well as cables and (pultruded) bars (Fig. 3.6.3).

Profiles and sandwich panels are typically applied in new construction, either in all-composite structures or in hybrid structures, where they are combined with components made of other structural materials. Strips and sheets are generally applied to strengthen existing structures and increase their flexural, shear, or axial load-bearing capacity, which is already a well-established state-of-the-art technique. Cables are used as external reinforcement or strengthening of new or existing structures, as an alternative to steel tendons or cables. Composite bars are used as internal

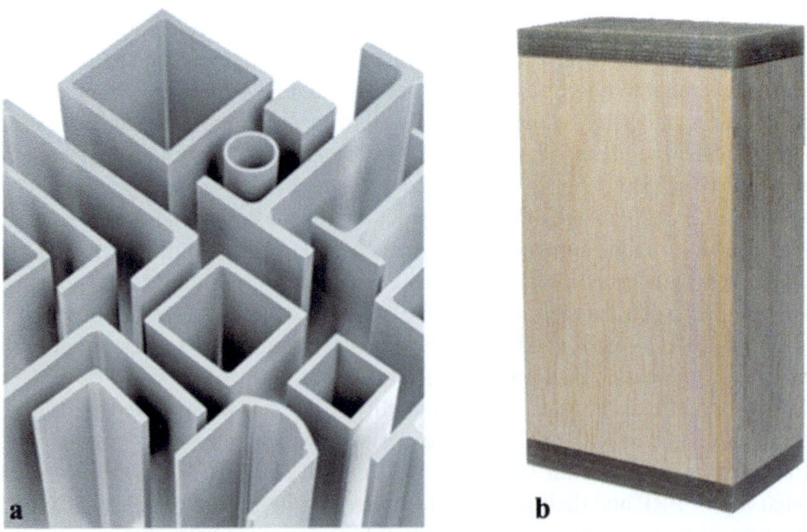

Figure 3.6.1 a) Glass-polyester profiles (©Fiberline, used with permission of Springer Nature, Correia (2015), in "Materials for Construction and Civil Engineering. Science, Processing, and Design, 2015, Springer Nature") and b) sandwich panel composed of glass-vinylester face sheets and balsa core.

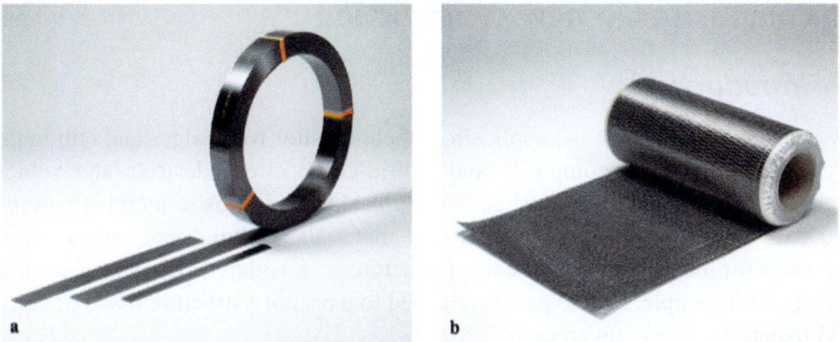

Figure 3.6.2 Carbon fibre: a) laminates (used as strips) and b) fabrics (© S&P Clever Reinforcement Portugal).

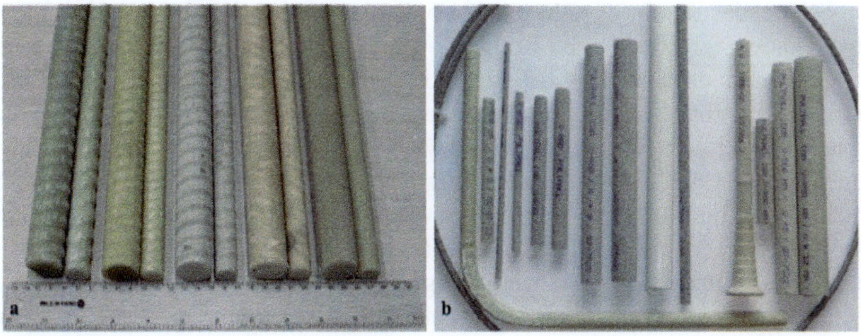

Figure 3.6.3 Fibre-polymer bars: a) different surface conditions and b) various shapes. Authorized reprint from Benmokrane et al. (2021a), ACI Special Publication 346: 93–112, 2021.

reinforcement of new or repaired concrete structures, being an alternative to steel (or stainless steel) rebars.

1.3 Content and structure of the chapter

This chapter provides an overview of the history of fibre-polymer composites in structural engineering and architecture, and it is organised according to the above-mentioned applications. Following the present introduction, Section 3. describes the history of composites in new construction, covering the different phases of development for applications in both bridge and building construction. Subsequently, Section 3. presents the history of composites for strengthening existing structures and composite cables. Next, the history of composite rebars in concrete construction is described in Section 4.. Finally, Section 5. presents some concluding remarks about this chapter and future trends regarding the development of fibre-polymer composites in structural engineering and architecture.

2. Composites in new construction

2.1 Introduction

New construction refers to the application of composites for bridges and buildings. In the case of bridges, all-composite and hybrid-composite pedestrian and vehicular bridges are considered; in hybrid-composite bridges, composite members are combined with members made of conventional materials. In buildings, only the use of composites for the primary load-bearing structure is considered. The use of secondary elements, for example, facade panels attached to a primary structure made of conventional materials, is not covered.

2.2 Composites in bridge construction

2.2.1 Pedestrian bridges

Composite pedestrian bridges have been built since the 1980s. In the USA, the first bridges, known as "Techtonics Bridges", were installed in 1987. The truss bridges were composed of standardised pultruded profiles (Keller, 2003), thus mimicking steel or timber truss bridges; the decks were made of timber in most cases. A more material-tailored bridge concept was developed for the Aberfeldy (cable-stayed) Bridge, built in 1992, see Fig. 3.6.4a (Burgoyne and Head, 1993). The bridge girders and deck were built-up from pultruded multicellular panels, three-way tubular profiles, and toggle connectors. In parallel, since 1986, several pedestrian bridges have been built in China. The girders were designed with multicellular sections composed of hand lay-up composite sandwich panels with honeycomb core. They were suspended to various reinforced concrete supporting systems (pylons, frames, arches) (Keller, 2003).

Since these first applications of composites for pedestrian bridges, numerous other bridges have been built. They can be classified into all-composite and hybrid-composite bridges, and, in the former case, material-substitution and material-tailored concepts can be distinguished. Material-substitution designs are mostly composed of standardised pultruded profiles, as the first bridges were. The profiles are used in typical beam, truss, arch, or cable-stayed systems. Examples are the

Figure 3.6.4 Pedestrian bridges: a) Aberfeldy Bridge, UK, 113 m total length, 1992 (Wikimedia Commons, CC BY-SA 2.0, Andrew Curtis, bottom and right side slightly cropped); b) Pontresina Bridge, Switzerland, 2 × 12.5 m span, 1997 (CC BY NC, Thomas Keller); c) Rhyl/Foryd Bridge, UK, 2 × 32 m span, 2013 (© AM Structures Ltd.).

Pontresina Bridge in Switzerland with two lateral trusses, installed in 1997 (Fig. 3.6.4b), the Kolding Bridge in Denmark with a cable-stayed system, also built in 1997, or the Lleida Bridge in Spain with a top arch system, built in 2011. When looking at these bridges, the novel material can normally not be recognised; they have the appearance of steel bridges. The required material quantity of such bridges is high in most cases, since the cross-sectional dimensions are imposed by the compact joints between components or members. The material quantity can be significantly reduced by developing material-tailored systems, where the shape and fibre architecture are adaptable to the internal forces, and compact joints can be prevented or their number minimised. Such bridges are often composed of large-scale prefabricated sandwich modules, produced by vacuum infusion, which also makes the manufacturing of complex shapes possible. An example is the Rhyl/Foryd Bascule Bridge, built in 2013 in the UK (Gurit, 2024), shown in Fig. 3.6.4c. In hybrid bridges, the main load-bearing system is normally made of conventional materials, and only the decks are composed of composites, using commercial pultruded or vacuum-infused deck systems for pedestrian loads (Keller, 2024).

2.2.2 Vehicular bridges—first phase

The first composite vehicular bridges were installed from the early 1980s onwards, in Bulgaria (Ginzi Bridge, 1981/82), China (Miyun Bridge, 1982), the United Kingdom (Bonds Mill Lift Bridge, 1994), and the United States (No-Name Creek Bridge, 1996) (Hollaway and Head, 2001; Keller, 2003). The bridges built in Bulgaria (11.9 m length) and China (20.7 m span) were composed of beams and sandwich box-girders, respectively. Due to their hand lay-up manufacturing, stiffnesses were low and imperfections significant. The deck of the UK bridge (8.2 m length), composed of the same thin-walled multicellular panels as the Aberfeldy (pedestrian) Bridge, was not able to resist vehicular loads in the long term; the bridge is planned to be replaced (Cotswold Canals in Pictures, 2024). The development of composite vehicular bridges thus stopped in these countries. Development progressed elsewhere, however, and two main phases can be identified: (i) a first phase of rapid growth of composite vehicular bridge installations, but equally rapid slowdown, almost exclusively in the USA, and (ii) a subsequent (current) second phase of moderate but continuous application in the United States, Europe, and other regions. The development thereby focused on bridge decks. Most of the bridges built were hybrid bridges with composite decks on steel girders or composite deck slab bridges (without girders); composite decks on composite girders represented a minority (Keller, 2024). It should also be noted that composite bridge decks do not require a waterproofing layer, as concrete decks do, which simplifies the kerb detailing and associated maintenance significantly (Keller, 2024).

Focussing on the first phase, in the USA, about 40% of the bridges were considered structurally deficient or functionally obsolete in the mid-1990s. A research and development program was therefore launched in 1995, targeting infrastructure renewal, and emphasising the use of composites to reduce life-cycle costs and construction time. In line with this program, composite bridge decks were developed and installed from

1996 onwards. Four proprietary deck systems dominated the deck installations, that is, the decks developed by the companies Hardcore Composites, Kansas Structural Composites, Martin Marietta Composites (MMC), and Creative Pultrusions (CP). The Hardcore deck was a vacuum-infused web-core sandwich deck with variable deck thickness, where an orthogonal web system was formed in the core by assembling fibre-wrapped foam blocks prior to infusion. The Kansas deck was a sandwich deck with a honeycomb core also of variable thickness, manufactured by hand lay-up techniques. The remaining two deck systems were pultruded/bonded web-core decks of constant thickness, called "DuraSpan" (MMC), see Fig. 3.6.5a, and "Superdeck" (CP). Two bridges using the MMC deck system, an all-composite and a hybrid-composite bridge, are shown in Figs. 3.6.6 and 3.6.7, respectively, also depicting different types of crash barrier fixations. The former bridge, named "Tech 21" (Materials Technology for the 21st Century), was developed based on transferring military technologies (Keller, 2024).

Until 2003, about 120 composite vehicular bridge decks or whole composite vehicular superstructures were installed in the USA (Market Development Alliance of the FRP Composites Industry, 2003; Triandafilou and O'Connor, 2009). After 2003, however, the installations abruptly stopped (Potyrała, 2011). This stop is likely related to the appearance of damage in some of the installed decks, already after short periods of use in some cases (Reising et al., 2004; Hong and Hastak, 2006; Triandafilou and O'Connor, 2010). Damages were mainly attributed to the effects of temperature variations and associated through-thickness gradients, insufficient manufacturing quality, and too large dimensional tolerances. The deck installations demonstrated, however,

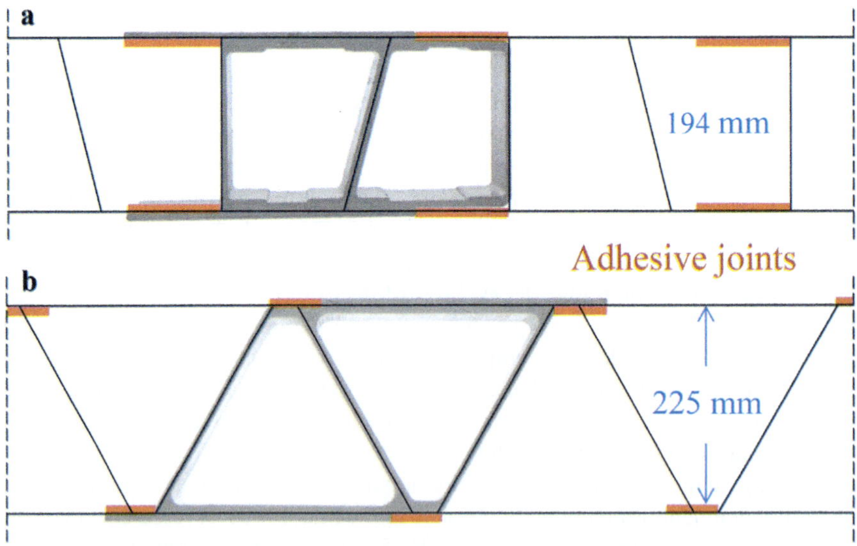

Figure 3.6.5 Adhesively bonded pultruded profiles for bridge decks: a) with double-trapezoidal cell shape; b) double-triangular cell shape, both with outstand flanges for adhesive connections to form web-core sandwich panels (CC BY NC, Thomas Keller).

Figure 3.6.6 a) Smith Road (Tech 21) Bridge, Butler County, Ohio, 10.1 × 7.3 m, 1997; b) two of three U-shaped hand-laminated glass fibre composite box girders; c) connections of crash barriers to deck (CC BY NC, Thomas Keller).

Figure 3.6.7 a) Darke County Bridge, state Route 47, Ohio, 15.2 × 14.0 m, 1999; b) steel longitudinal and transverse girders below deck panels; c) connections of crash barriers to steel girders below deck (CC BY NC, Thomas Keller).

that, in comparison with reinforced concrete decks, installation times could be significantly reduced and lane closures avoided due to prefabrication and the light weight of the composite panels (Reising et al., 2004). The four deck types mentioned above, nevertheless, subsequently disappeared from the market (Keller, 2024).

In Europe, only one relevant composite deck system was developed for vehicular bridges: a pultruded/bonded web-core deck produced by Fiberline, Denmark, named "ASSET", see Fig. 3.6.5b. The deck was installed from 2002 to 2014 on several vehicular bridges (Summerscales, 2024). However, the production of the deck was subsequently stopped.

Reasons for the disappearance of the pultruded/bonded web-core decks might have been (i) too little flexibility in application due to geometrical limitations and orthotropy, (ii) limited fatigue resistance, and (iii) complicated shear-stud connections between decks and beams in the USA (Keller, 2024).

2.2.3 Vehicular bridges—second phase

A second phase of vehicular bridge deck applications began around 2010, mainly with the appearance of three companies on the construction market, that is, Composite Advantage in the USA, and FibreCore and 3A Composites in Europe. These companies developed proprietary sandwich deck systems based on vacuum infusion, namely the "FibreSpan" system in the USA, the "FibreCore/InfraCore" system in the Netherlands, and the "Colevo" system in Switzerland (Keller, 2024).

The FibreSpan and FibreCore systems are composed of web-core panels with narrowly spaced webs integrated into a foam core. The narrowly spaced webs provide

Figure 3.6.8 Avançon Bridge, Switzerland, 11.5 m length, 2012: a) composite-balsa sandwich lay-up before infusion (used with permission of American Society of Civil Engineers, "GFRP-Balsa Sandwich Bridge Deck: Concept, Design, and Experimental Validation," Journal of Composites for Construction, Keller et al., 18(2), 04013043, 2014); b) application of adhesive on steel top flange; c) lifting of bridge superstructure into place.

more continuous support of the top face sheets and thus increase fatigue resistance compared with the decks of the first phase. The infusion moulding also allows more versatility compared with pultruded decks. Features such as camber, slopes, kerbs, drain scuppers, and panel joint details can be integrated into the deck during manufacturing; furthermore, complex layout plans can be designed. The Colevo deck has a homogeneous end-grain balsa core and thus provides fully continuous support to the top face sheet, as shown in Fig. 3.6.8; simple kerb and dewatering details can also be conceived. The bridge decks are adhesively bonded to the main girders (Keller, 2024).

Four application fields emerged in the USA during this second phase, in which composite decks can offer economic benefits. These fields are: (i) movable bridges, (ii) historic steel truss bridges, (iii) bridges with steel gratings, and (iv) lateral pedestrian deck additions to existing vehicular bridges. The light weight of composite decks is the main advantage in all these applications (Reeve, 2010). In Europe, most of the installed decks also involve deck replacements and movable bridges. Bridges can be upgraded by replacing heavy concrete decks with lightweight composite decks. The replacement of existing concrete superstructures by lightweight composite decks on steel girders even allows for the widening of existing bridges, if the additional traffic weight on new lanes can be compensated by the weight reductions from changing the superstructure, as it was possible for the Avançon Bridge in Switzerland, shown in Fig. 3.6.8 (Keller, 2024).

2.3 Composites in building construction

2.3.1 Initiation phase

The appearance of composites in building construction was closely linked to the beginning of the Space Age, which began with the launch of the first Sputnik satellite in 1957. A spirit of optimism and bright future emerged after World War II and design was influenced by the futuristic shapes of satellites and space technology. Composites could respond to the spirit of that time, as their characteristics enabled complex shapes to be manufactured and materials to be made transparent and coloured. Thanks to their physical properties and free formability, composites can fulfil, in addition to structural functions, building physics functions normally attributed to the nonload-bearing building envelope (such as thermal insulation), and architectural functions (such as complex shapes and transparency) (Keller, 2002, 2024).

Further driven by the chemical industry, such as Monsanto in the USA or BASF in Europe (both of which were keen to enter the construction market), numerous designs of housing units emerged, such as (i) the first all-composite house in 1955, influenced by the growth of a snail's shell and exhibiting a moulded-in interior, (ii) the famous Monsanto "House of the Future" in 1957, designed in collaboration with MIT (Fig. 3.6.9a), or a bit later, (iii) the "Tube House" (Fig. 3.6.10a), and (iv) the "Futuro House" (Fig. 3.6.9b); details of these houses are given in (Schwabe and Saechtling, 1959; Doernach, 1974; Quarmby, 1974; Isler, 1977; Voigt, 2007).

All these housing units had some common characteristics in that they (i) were designed modular and targeted for serial production, (ii) were built in glass fibre composite sandwich construction, (iii) had curved shapes mainly to provide geometrical stiffness, and (iv) integrated structural, building physics and architectural functions into the sandwich components, in most cases. Composite laminates and hybrid members could, furthermore, already bridge significant spans and be designed with a certain transparency, as demonstrated in Fig. 3.6.11 (Keller, 2024).

The phase of optimism and economic growth ended abruptly, however, with the first and second oil crises during the Arab-Israeli and Iran–Iraq wars in 1973 and 1979, respectively. Composite housing development also stopped, the reasons for which were, in addition to the economic crisis, (i) a low acceptance due to the impossibility of adaptation to individual needs, associated with the serial production, (ii) the impossibility of building more than one storey, due to limitations in material properties and connection technology, (iii) not yet sufficiently developed automated processes, and (iv) deficiencies in detailing. According to Isler (1998), in Switzerland, the cost doubled due to the oil crises, the quality of manufacturing was low, maintenance was missing, and architects did thus not show any interest due to a negative connotation of composites as a poor material. Furthermore, the houses generally did not respond to building acoustic and fire behaviour requirements defined in the standards (Keller, 2024).

Figure 3.6.9 a) Monsanto "House of the Future" at Disneyland, Anaheim, California, 1957 (Wikimedia Commons, CC BY-2.0, Orange County Archives); b) Futuro House, Matti Suuronen, 1968 (Wikimedia Commons, CC BY-SA 4.0, Henning Schlottmann).

Figure 3.6.10 a) "Tube House," Heinz Isler, St. Gallen, Switzerland, 1969; b) interior view showing coherent formal language of windows and "plastic" furniture; c) manufacturing on rotating steel mandrel, coating in progress (© gta Archiv/ETH Zürich, Heinz Isler).

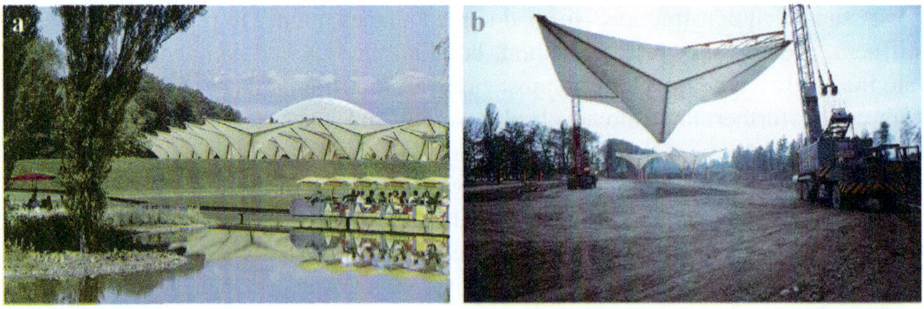

Figure 3.6.11 a) Pavillon "Les échanges" Expo'64, Heinz Hossdorf, Lausanne, 1964 (ETH-Bibliothek Zürich, Bildarchiv/Fotograf: Comet Photo AG (Zürich)/Com_BC25-004-017/CC BY-SA 4.0, top and bottom cropped); b) installation of one translucent umbrella composed of glass fibre composite laminates bonded to a steel frame (© Schnetzer Puskas Ingenieure AG).

2.3.2 Revival in 1990s

After the decline of composite applications in building construction at the end of the 1970s, it took more than a decade before a revival of composites in architecture occurred. In the meantime, the material properties and manufacturing methods, such as vacuum infusion and pultrusion, significantly improved. Sandwich panels and pultruded profiles became available, which exhibited mechanical properties and shape sizes that made possible even vehicular bridge construction (see above), multi-storey buildings, and large-scale roofs, as demonstrated by the Eyecatcher Building composed of pultruded profiles in Fig. 3.6.12, and the function-integrated sandwich roof of the Novartis building in Fig. 3.6.13 (Keller, 2024).

More recent buildings, whose appearance is characterised by composite members, are (i) the Halley mobile research station VI on the Brunt Ice Shelf in Antarctic (2013), (ii) the House of Dior in Seoul (2015) with a composite envelope of complex shape, (iii) the Orthodox Trinity Cathedral in Paris (2016) exhibiting large-scale composites domes, or (iv) the Steve Jobs Theater in Cupertino, USA (2017) covered by a carbon fibre composite roof (Keller, 2024).

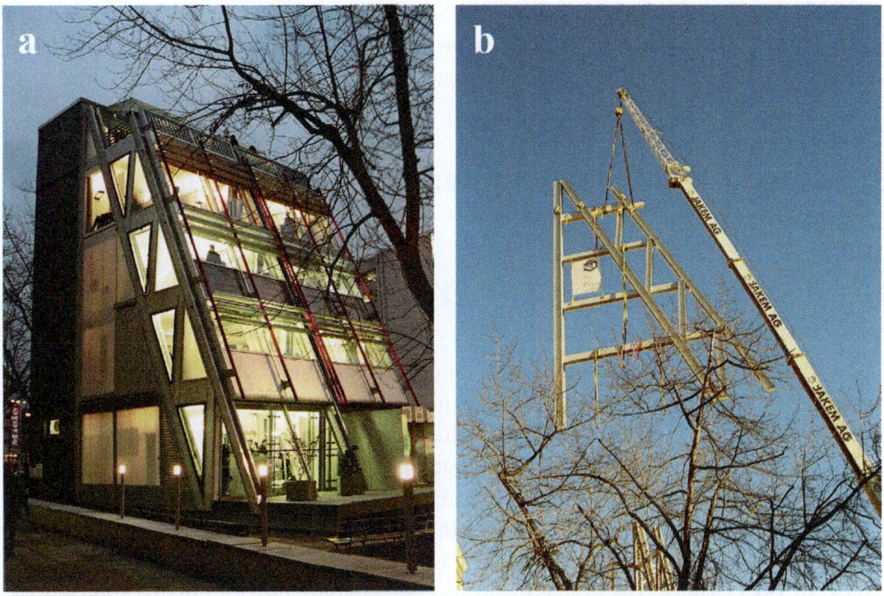

Figure 3.6.12 a) Eyecatcher Building, Basel, Switzerland, 12.7 × 10.0 × 14.6 m (length×width×height), 1998; b) installation of one of three preassembled frames (CC BY NC, Thomas Keller).

Figure 3.6.13 a) Novartis Campus Entrance Building, Basel, Switzerland, roof 21.6 × 18.5 m, 2006; b) installation of glass fibre composite sandwich panel segment (CC BY NC, Thomas Keller).

2.4 Development of design manuals and standards

The first design manual for the design of composite structures was published in the United States, in 1984 (ASCE, 1984). It was followed, in Europe, by a design manual issued by a pultruder in 1995 (Fiberline, 1995), and the EUROCOMP design code and handbook in 1996 (Clarke, 1996). Subsequently, several European countries published design guidelines, that is, the Netherlands, Germany, the United Kingdom, and Italy, until the European Technical Specification CEN/TS 19101 appeared in 2022 (CEN, 2022). The most recent design standard was issued in the United States, in 2023; it is limited to structures composed of pultruded profiles (ASCE, 2023).

3. Composites for strengthening existing structures and cables

3.1 Introduction to composite systems and strengthening techniques

The aim of having lightweight, stiff, and noncorrodible cables for cable-stayed bridges (Rhine Bridge Mannheim, design competition for the Messina Crossing) and large roofs (Olympia Stadium Munich) have, in 1969–1971, triggered a collaboration between Prof. Fritz Leonhardt's team and EMPA researchers for the development of FRP cables (Meier, 2024). During this research activity, the EMPA team found an alternative application for CFRP: using lamellas for the poststrengthening of existing structures such as bridges. The effectiveness of this approach was demonstrated in 1987 (Meier, 1987a), and was implemented in 1991 for the first time worldwide on the Ibach Bridge near Lucerne (Meier and Deuring, 1991). The first successful application of CFRP cables occurred in 1996 in the Stork Bridge in Winterthur (Meier and Meier, 1996).

Since then, FRP composites have been applied to the strengthening of existing structures, mainly using externally bonded reinforcement (EBR), near-surface mounted (NSM), and embedded through section (ETS) techniques. These techniques have been used to increase the axial, flexural, shear, punching resistance, and energy absorption and dissipation capacity, mainly of reinforced concrete (RC) structures (e.g., Motavalli and Czaderski, 2007; Siddika et al., 2020), but applications in timber (e.g., Saad and Lengyel, 2022) and masonry (e.g., Babatunde, 2017) structures have also been relevant.

Carbon-FRPs (CFRPs) have been the most used with these strengthening techniques, but FRPs with glass, aramid, and basalt fibres (GFRP, AFRP, and BFRP, respectively) have also been adopted. These FRPs are bonded to the structures to be strengthened with, in general, polymer-based adhesives (PBA), mainly epoxy (EBA), but cement-based adhesives (CBA) have also been explored. Thin composite plates formed by fibre-reinforced concrete (FRC), including FRP grids, EBR-CFRP sheets, and NSM-CFRP laminates, have also been adopted for several strengthening purposes.

The design of strengthening solutions with these techniques has been assisted with guidelines prepared by several organisations.

3.2 Externally bonded reinforcement (EBR) technique

Figure 3.6.14 schematically illustrates the application of the EBR technique to the flexural and shear strengthening of beams and columns of an RC frame. After the treatment of the concrete substrate where the FRP will be applied, by removing a more deteriorated thin layer of the cement matrix, a polymer-based material of high fluidity is applied on the substrate to increase the concrete properties of this medium. An FRP layer is then bonded to the treated concrete substrate using, in general, an EBA. Further layers of FRP can be applied by their successive bonding with adhesive. Figure 3.6.14 shows a major part of possible EBR applications. Flexural strengthening is assured with EBR strips or sheets applied on the faces of the members subjected to tension and with the fibres aligned with the tensile stresses. The first investigations on structural strengthening with EBR-FRP were conducted in the 1980s (Fardis and Khalili, 1981; Meier 1987a). The first successful application of EBR-FRP occurred at the beginning of the 1990s and corresponded to flexural strengthening (Fig. 3.6.15). This was followed by the formation of companies dedicated to this sector (e.g., S&P Clever Reinforcement, 1995) and sparked the interest of several companies from the construction sector in this new market (e.g., Sika company 1997).

Experimental research has demonstrated the possibility of EBR-FRP to increase significantly the load-carrying capacity of RC beams and slabs at the serviceability and ultimate limit state (SLS and ULS, respectively) design conditions (e.g., Meier

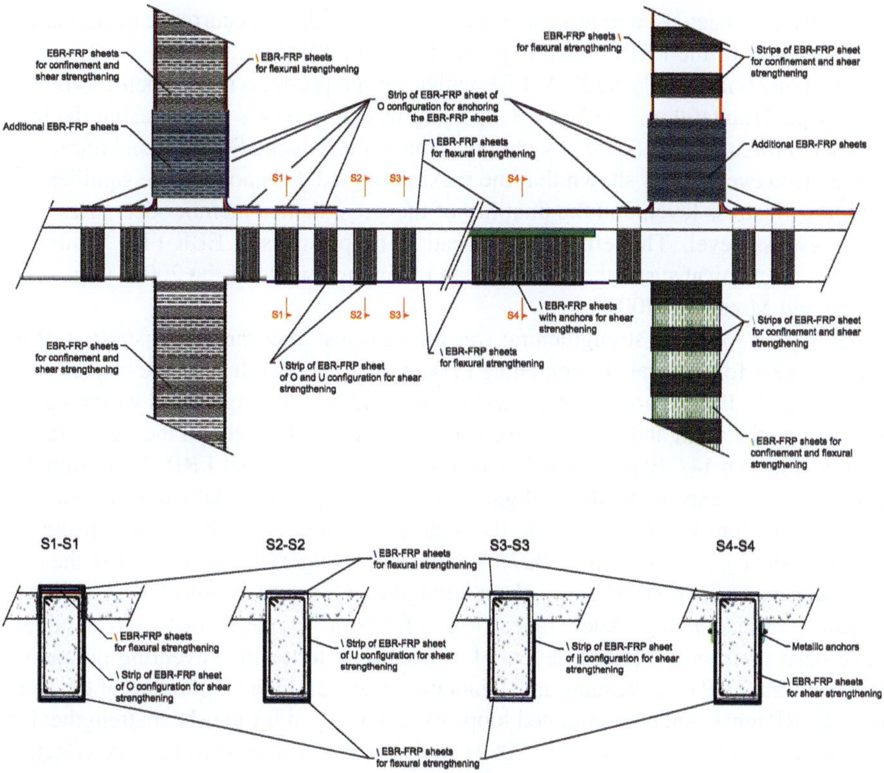

Figure 3.6.14 Schematic representation of the EBR technique for the flexural and shear strengthening of RC beams and columns of an RC frame.

Figure 3.6.15 One of the first applications of EBR-FRP: Ibach Bridge near Lucerne, 1991 (© Prof. Urs Meier).

and Kaiser, 1991; Teng et al., 2002). It has been shown that the strengthening effectiveness increases with the quality of the concrete substrate and the flexural strengthening ratio (e.g., Barros et al., 2007). However, it was also shown that the maximum tensile stress possible to be supported by the FRP decreases with its axial stiffness (e.g., Oehlers and Moran, 1990). Anchorages of several configurations and in different materials have been used to postpone, and even to avoid, the occurrence of premature failure modes of the FRP system (e.g., Kalfat et al., 2018).

The potential of applying EBR-FRP with a certain prestress started being explored later (e.g., Triantafillou, 1992), by evidencing its effectiveness in increasing the load at crack initiation, at SLS and at yield initiation of the existing flexural steel reinforcements. However, it was shown that the maximum load does not change significantly with the prestress level, and the ductility of the strengthened member decreases with the prestress level. The efficient application of prestressed EBR-FRP laminates without mechanical systems was demonstrated in the middle of the 2000s (e.g., Czaderski and Motavalli, 2007).

EBR-FRP for shear strengthening was initiated just after the demonstration of its effectiveness for flexural strengthening (Triantafillou, 1992). It can be executed with continuous or discrete arrangements of the FRP, and by adopting O (full wrapping), U (embracing the soffit and lateral faces) and || (applied exclusively on the lateral faces) shapes (Fig. 3.6.14). Experimental tests have demonstrated that FRP shear strengthening arrangements with fibres aligned as much as possible with the direction of the principal tensile strains provide the highest contribution to the shear capacity of the member (e.g., Barros et al., 2007). While an O configuration can mobilise the tensile capacity of the FRP, U and || shear strengthening configurations tend to fail prematurely by debonding. Anchorages (Fig. 3.6.14) of several formats and materials have been used with different levels of efficiency in terms of preventing premature failure of the FRP (e.g., Koutas and Triantafillou, 2012). In the beginning of this century, CFRP tensioned nonlaminated loops were developed for the shear strengthening of concrete members (Lees et al., 2002) and were demonstrated to be very effective when applied with specific engineered equipment (Fig. 3.6.16).

Figure 3.6.16 Tensioned shear strengthening of a concrete heritage building with nonlaminated loops (© Prof. Urs Meier).

FRP systems have also been used to increase the punching capacity of RC slabs, with a level of efficiency that depends on the geometric and material characteristics of the FRP, and on the concrete quality (e.g., Chen and Li, 2005).

One of the main applications of FRP is the confinement of RC columns for increasing their axial load-carrying capacity and axial deformability by using EBR sheets in the form of strips (partial confinement) or continuous wrapping (full confinement), Fig. 3.6.14. In the first applications, GFRP was used (e.g., Fardis and Khalili, 1981). Later, other types of EBR-FRP were explored, even in the form of prefabricated tubes (e.g., Nanni and Bradford, 1995). FRPs have been applied with full and partial wrapping configurations (Fig. 3.6.17) and using more than one type of FRP (at material and arrangement levels) to obtain competitive solutions from the technical and economic points of view, taking advantage of the different modulus of elasticity and tensile strength of FRP available in the market (e.g., Wu et al., 2008). When confinement arrangements are combined with FRP systems with fibres aligned with the axis of the column, a significant increase has been reported not only in terms of axial load and deformation capacity but also under lateral loads representative of seismic events (e.g., Pantelides et al., 2008).

3.3 Near-surface mounted (NSM) technique

The first investigations on the NSM technique for the flexural strengthening of RC structures occurred at the end of the 1990s (Blaschko and Zilch, 1999), using the first CFRP reinforcements produced by S&P Clever Reinforcement company. Figure 3.6.18 illustrates the application of the NSM technique on the flexural and shear strengthening of RC members. According to this technique, grooves are open

Figure 3.6.17 Applications of EBR-CFRP for the strengthening of RC columns: full (left, © S&P Clever Reinforcement Portugal), discrete (top right), and localised in the extremities (bottom right, © Christoph Czaderski).

on the concrete cover of the member to strengthen, with a geometry that depends on the FRP reinforcement to be installed into these grooves. FRP rods and laminates have been used (e.g., De Lorenzis et al., 2000), mainly in CFRP, with the laminates providing higher strengthening efficiency due to the larger ratio between their contact surface with the adhesive and their cross-sectional area (e.g., Costa and Barros 2013). EBA has been the most used to bond NSM reinforcements to concrete substrates and into the groove. Due to the significant loss of mechanical and bond properties of epoxy at temperatures above its glass transition temperature (e.g., Firmo et al., 2015), CBA has also been explored in an attempt to ensure high levels of strengthening performance of the NSM system in a fire scenario (e.g., Täljsten and Blanksvärd, 2007). Providing cold regions in the extremities of the NSM strengthening arrangements using fire-protective systems (e.g., Azevedo et al., 2023) or adopting FRP reinforcements with inclined extremities anchored in the core cross-section of the member to strengthen (Azevedo et al., 2022) have demonstrated to be efficient strategies for preserving relatively high strengthening levels in a fire scenario. The FRP reinforcements with inclined extremities have the extra advantage of assuring simultaneous flexural and shear strengthening for RC beams, and flexural and punching strengthening for RC slabs (Barros et al., 2017).

The application of NSM with a certain prestress started to be explored in the middle of the first decade of the present century, by using CFRP rods and laminates (e.g., Nordin and Täljsten, 2006). The conclusions already pointed out that the strengthening

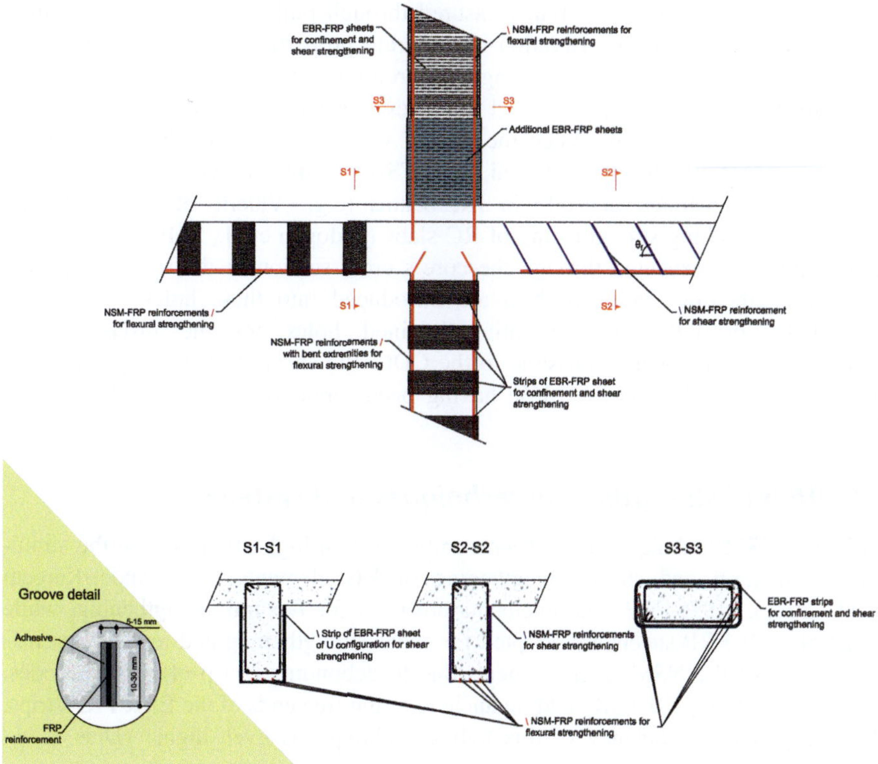

Figure 3.6.18 Schematic representation of the NSM technique for the flexural and shear strengthening of RC beams and columns.

effectiveness of prestressed EBR systems are also applicable to the prestressed NSM reinforcements (e.g., Rezazadeh et al., 2014).

The assessment of the effectiveness of NSM for shear strengthening was made almost in parallel with its application for flexural strengthening (e.g., De Lorenzis and Nanni, 2001). It was shown that the strengthening efficiency increases with the quality of the concrete substrate, with the shear strengthening ratio, by disposing the reinforcements as much as possible orthogonal to the critical diagonal crack, CDC, (e.g., Barros et al., 2007), and as deeper as possible into the grooves (Barros et al., 2016).

The application of NSM-FRP reinforcements with a certain prestress was initiated in the 2020s. It was demonstrated that applying NSM-FRP with a certain prestress assures higher shear-strengthening efficiency than their passive counterparts, but the prestress level should be limited to avoid detrimental effects on the deformation capacity of the shear-strengthened member (Guo et al., 2024).

3.4 Embedded through section (ETS) technique

EBR and NSM systems are not recommended for RC structures with concrete substrate of strength quality below a certain limit, which, according to (ACI

PRC-440.2-23, 2023) is 1.4 MPa, measured through pull-off tests. Besides natural weathering effects, the deterioration of concrete properties can be caused by fire occurrence and corrosion of existing steel reinforcements. In several situations, the stirrups are the most affected by these occurrences, requiring a shear-strengthening intervention. When the quality of the concrete of the core's cross-section was not significantly affected, the ETS technique can be very effective, as was first demonstrated in 2007 in RC beams (e.g., Valerio et al., 2007) and later in the punching strengthening of RC slabs (Erdogan et al., 2010). In this technique, holes are executed through the core's cross-section in the region requiring shear strengthening, and FRP bars are introduced into these holes and bonded with EBA or CBA. By executing inclined holes for the installation of FRP bars as orthogonal as possible to the CDC, the shear strengthening efficiency increases, with values up to 136% having been reported (e.g., Breveglieri et al., 2016).

3.5 Hybrid strengthening techniques and systems

EBR and NSM techniques have been combined with high efficiency on the simultaneous flexural and shear strengthening of RC elements (e.g., Abdel-Kareem et al., 2021). NSM-FRP reinforcements are used for flexural strengthening, while strips of EBR-FRP sheets are adopted for shear strengthening and to avoid premature failure of the NSM reinforcements due to debonding or rip-off failure modes, Fig. 3.6.14 (fib 90, 2019). By adding anchors to the free ends of the EBR-FRP strips, the strengthening efficiency of this hybrid technique is even higher (Dias et al., 2021).

EBR and NSM techniques have also been combined for increasing simultaneously the axial, flexural and shear capacity, and energy dissipation performance of RC columns (e.g., Zeng et al., 2022). The NSM FRP reinforcements are applied in the lateral faces of the column for providing the intended flexural capacity, being anchored in the concrete members connected to the column, while EBR-FRP sheets with the fibres oriented transversally to the axis of the column, arranged in full, localised, or discrete confinement configurations, provide the intended shear strengthening and concrete confinement (Figs. 3.6.17 and 3.6.18).

NSM has also been also combined with textile-reinforced mortars (TRM) with high effectiveness, mainly for the strengthening of RC columns under seismic loading conditions (e.g., Bournas and Triantafillou, 2009).

In the middle of the 2010s, composite plates (CP) were used for flexural (Esmaeeli and Barros 2015), shear (Baghi and Barros, 2016), and seismic strengthening (Esmaeeli et al., 2015). CP are prefabricated or cast-in-place thin plates made by FRC reinforced with FRP sheets or laminates. The FRC is, in general, a tensile strain hardening material (Li et al., 1995) with a tensile deformability compatible with the FRP reinforcements. The tensile strain hardening nature of the FRC allows the CP to be fixed to the substrate, not only with adhesive, but also with metallic anchors without premature rupture of the FRC around the anchors (Baghi and Barros, 2016).

3.6 Cables

Despite having triggered the research on the use of FRP for infrastructures (Meier, 1987b), FRP cables have only been used when the technology for anchoring a bundle of FRP wires was demonstrated to be feasible from both mechanical and economic points of view (Meier and Meier, 1996). Cables, mainly formed by carbon fibres, have been used typically to control the deflection and increase the flexural capacity and fatigue life of bridges and structural members of relatively long span, whose stiffness has degraded at a level that compromises their design performance at SLS and ULS conditions (Liu et al., 2015). FRP cables have also been used for replacing damaged steel cables in bridges (Fig. 3.6.19). The factors that contribute to the success of this strengthening technology are the immunity of the CFRP cables to corrosion, their superior fatigue life over steel strands, and the evolution of anchors that allow proper fixation with relatively small initial and long-term losses of prestress.

3.7 Development of design guidelines

The experimental, numerical and analytical research has contributed to the development of design guidelines by several organisations, such as ACI PRC-440.2-23 (ACI, 2023), CSA-S806-12 (CSA, 2021), CNR-DT 200 (CNR, 2013), fib Bulletin 90 (FIB, 2019), GB 50367-2013 (GB Standards, 2013), TR55 (Concrete Society, 2012). These guidelines have promoted the continuous increase in the use of FRPs as alternative solutions for the strengthening of the built environment at competitive costs and with technical advantages due to the support they give to designers. The predictive performance of these guidelines has been assessed with the support of databases that include relevant information from experimental tests. Machine learning algorithms have been contributing not only to this assessment, but also to reveal the most important variables affecting the phenomena simulated by the design guidelines, and to propose design methodologies of higher accuracy (Mohammadi et al., 2024).

Figure 3.6.19 Application of CFRP cables in the rehabilitation of the Stork Bridge Winterthur (1996), left, and in the Verdasio Bridge (1998), middle and right (© Prof. Urs Meier).

4. Composite rebars in concrete constructions

4.1 *Introduction to composite rebars in concrete*

The development of advanced materials in concrete construction has long aimed to address the durability challenges associated with traditional steel reinforcement. While steel is resilient in strength and widely used, it is prone to corrosion when exposed to aggressive environmental conditions such as marine environments, de-icing salts, and moisture. This corrosion not only compromises the structural integrity of concrete members but also results in significant maintenance costs and frequent repairs over the structure's lifespan (Benmokrane and Rahman, 1998; Benmokrane et al., 2002; Mufti et al., 2005). To mitigate these issues, various corrosion-resistant solutions were explored, including galvanised coatings, electrostatic spray fusion-bonded (powder resin) coatings, and polymer-impregnated concrete epoxy coatings (Plecnik and Ahmad, 1988). Among these, FRP reinforcing bars emerged as a promising alternative to steel (ACI, 2007; Nanni, 1999).

The development of FRP reinforcement for concrete structures began in the 1960s in the United States (Bank, 2006), followed by the 1970s in Europe (Taerwe and Matthys, 1999) and Japan (Fukuyama, 1999). However, it was during the 1980s that substantial progress in research, field applications, and commercialisation was achieved (Bakis et al., 2002). Since the 1980s, significant global efforts have been made to integrate FRP rebars into design codes and guidelines. In the United States, initiatives by the National Science Foundation and the Federal Highway Administration played a critical role in advancing this technology by supporting research at universities and research institutions (Bakis et al., 2002). In 1991, the American Concrete Institute (ACI) established Committee 440 to provide the construction industry with science-based design guidelines, construction specifications, and recommendations for inspection and quality control related to FRP rebars. The committee's work culminated in the publication of the first edition of the "Guide for the Design and Construction of Structural Concrete Reinforced with FRP Bars" in 2001, which significantly accelerated the adoption of FRP rebars (ACI, 2006).

4.2 *Types of FRP rebars*

FRP bars are typically made from glass, carbon, or aramid fibres, with basalt fibres emerging as a recent alternative to glass. These fibres are bonded and protected by thermosetting resins like epoxy, polyester, or vinyl ester, which enable stress transfer between fibres. This material also offers a high strength-to-weight ratio and nonmagnetic properties, broadening its applicability (Benmokrane et al., 2017, 2021a, 2023; Ortiz et al., 2024; Rashid et al., 2005; Fang et al., 2016; Refai et al., 2015; Mehany et al., 2023). Manufacturing techniques for FRP bars include pultrusion, braiding, and weaving, producing mostly solid, round cross-sections, though other shapes are available (Liu et al., 2022). Bending these FRP bars after the curing process is impossible due to the rigid three-dimensional network of thermosetting resins, posing challenges at construction sites. Thermoplastic resins, which soften at high temperatures, offer a solution, enabling easier shaping and reuse. Developments in the field of thermoplastic GFRP reinforcing bars have been reported in recent years (Benmokrane et al., 2021c; Ibrahim et al., 2024a, b). Surface conditioning of FRP bars enhances their bond with surrounding materials like concrete or adhesives. As shown in Fig. 3.6.3a, common methods include sand coating, ribbing, helical wrapping, or grooving to improve mechanical interlock, along with chemical treatments to boost adhesion. Furthermore, FRP bars are available in various shapes to suit different structural applications, including straight and bent bars, stirrups, L and U shape bars, headed bars, and spirals (Fig. 3.6.3b). In addition, multidimensional FRP reinforcements, such as prefabricated grids or on-site fabricated shapes, provide versatile solutions for various structural needs. Many researchers have contributed extensively in this field to improve the understanding of the mechanical properties, durability, and structural behaviour of FRP-reinforced concrete members, confirming their suitability in diverse construction scenarios (Uomoto et al., 2002; Benmokrane et al., 2002; Benmokrane et al., 2021a; Hosseini et al., 2023; ACI 440 Committee, 2022a, b).

4.3 Applications of GFRP rebars in concrete structures

As awareness of GFRP's durability and strength grew, its applications in construction expanded. GFRP rebars are now used in various structures where reducing maintenance and extending service life are essential. This section describes the historical application of GFRP in bridge decks, marine infrastructure, water treatment plants, and more, illustrating each with relevant real-world examples.

4.3.1 Bridge piles and deck slabs

Corrosion of steel reinforcement in concrete bridges is a major issue, limiting infrastructure longevity, especially in North America, where de-icing salts accelerate deterioration. The 2013 ASCE Report Card noted many bridges as structurally deficient, with corrosion a leading cause of failure. To address this, FRP bars, with their noncorroding and durable properties, have been integrated into bridge design since the late 1990s to extend service lives from 75 to 150 years. Successful field applications of FRP bars are evident in projects across Canada and the US. For example, the Nipigon River Bridge (Fig. 3.6.20) in Ontario — 2013 to 2017 — used GFRP bars for the deck slab reinforcement, addressing steel expansion and corrosion issues (Arafa et al., 2016; Mohamed and Benmokrane, 2018). Similarly, the Clyde River Bridge (Fig. 3.6.21) in Prince Edward Island — 2017 to 2020 — is fully reinforced with GFRP bars, including pile caps, abutments, and deck slabs, setting a precedent for corrosion-free bridge design (Benmokrane et al., 2021a). GFRP bars, chosen for their cost-effectiveness compared with carbon and aramid alternatives, are primarily used for internal reinforcement in deck slabs, barriers, and girders. A recent study also investigated precast GFRP-reinforced concrete piles for the Arthur Drive Bridge project (Fig. 3.6.22) in Panama City, FL, USA — 2017. The 61 × 61 cm piles, reinforced with GFRP bars and ties, were successfully installed and monitored, with stresses remaining within allowable limits (Benmokrane et al., 2021b).

Figure 3.6.20 Cable-stayed Nipigon River Bridge, ON, Canada — 2013 to 2017.
Authorized reprint from Benmokrane et al. (2021a), ACI Special Publication 346: 93–112, 2021.

Figure 3.6.21 Worldwide first GFRP-reinforced bridge crossing (Clyde River Bridge, PEI, Canada — 2017–2020).
Authorized reprint from Benmokrane et al. (2021a), ACI Special Publication 346: 93–112, 2021.

4.3.2 Water treatment facilities

GFRP bars are highly effective in water treatment facilities due to their excellent corrosion resistance in chemical and moisture-rich environments. The challenge is to design these tanks with noncorroding FRP reinforcing bars. The world's first concrete chlorination water treatment tank reinforced entirely with FRP bars was built in Thetford Mines, Canada — 2012 (Fig. 3.6.23). The tank, with a 4500 m^3 capacity, is 30 m wide, 30 m long, and 5 m high, supported by a raft foundation. It was designed

A B

Figure 3.6.22 GFRP pile cage and installation (Panama City, FL, USA – 2017).
Authorized reprint from Benmokrane et al. (2021a), ACI Special Publication 346: 93–112, 2021.

A B

Figure 3.6.23 FRP-reinforced concrete water tank (QC, Canada – 2012).
Authorized reprint from Benmokrane et al. (2021a), ACI Special Publication 346: 93–112, 2021.

per CAN/CSA-S806-02 (CSA, 2002), using high-modulus GFRP bars as reinforcement for the foundation, walls, and top slab. The construction, performance under service conditions, and monitoring indicated satisfactory results, meeting serviceability and strength requirements. The GFRP bars effectively addressed steel-corrosion issues and facilitated an efficient construction process. This successful application demonstrated the effective use of GFRP bars in RC water treatment tanks, potentially extending the life of such structures to 100 years, compared with steel-reinforced concrete, which typically requires major restoration after 25 years (Mohamed and Benmokrane, 2014; Benmokrane and Mohamed, 2014).

4.3.3 Tunnel segments and RC soft eyes

GFRP applications extend to tunnel construction, specifically in soft-eye sections for launching tunnel boring machines (TBMs) and placing precast concrete tunnel lining (PCTL) segments. Conventional steel-reinforced walls pose challenges for TBM breakthroughs, necessitating complex preparations. GFRP's anisotropic

Figure 3.6.24 GFRP reinforcement for soft eyes (ON, Canada –2016) (authorised reprint from Benmokrane et al. (2021a), ACI Special Publication 346: 93–112, 2021) and tunnel segments (QC, Canada – 2022) (used with permission of American Society of Civil Engineers, "Long-Term Durability of Nonpultruded Curvilinear GFRP Bars Exposed to an Alkaline Environment: Experimental Studies and Modeling," Journal of Composites for Construction, Hosseini et al., 28(3), 04024005, 2024).

properties—high tensile strength and corrosion resistance—streamline excavation processes and reduce costs. GFRP-reinforced soft eyes, built with bore piles or diaphragm walls, aid in efficient TBM operations. Successful implementations of GFRP reinforcement include several Canadian tunnel projects, such as the South Tunnels and the Eglinton Crosstown LRT in Toronto, Ontario, which began construction in 2016. These projects utilised GFRP reinforcement cages measuring up to 19 m in length and 600–1100 mm in diameter. The field performance of these reinforcement systems was exceptional, with no reported issues, effectively validating their practicality and efficiency in real-world tunnel construction applications (Mohamed et al., 2020). An effective solution to the corrosion issue in PCTL segments is replacing steel reinforcement with noncorroding GFRP bars (Fig. 3.6.24). The results showed that both GFRP- and steel-reinforced specimens failed in flexure, with the GFRP-reinforced specimen demonstrating a higher load-carrying capacity and ultimate load than the steel-reinforced one (Hosseini et al., 2024).

4.3.4 Secant piles (coastal area)

GFRP secant piles provide a corrosion-resistant solution for infrastructure in aggressive coastal environments, helping to reduce maintenance costs associated with traditional steel reinforcement. The Florida Department of Transportation (FDOT) has adopted GFRP reinforcement to enhance durability and lower life-cycle costs. Its light weight makes for easier handling and reduces construction expenses. The Flagler Beach County dune restoration project had a steel sheet pile bulkhead facing severe corrosion. The solution was a buried secant-pile system using 1847 secant piles reinforced with GFRP bars and spirals (Fig. 3.6.25) – 2019. The assembly and installation processes were closely monitored to ensure constructability and gather data on transportation, cage assembly, and placement. This project underscores the benefits of GFRP reinforcement in improving the longevity and durability of coastal infrastructure (Nolan et al., 2019).

Figure 3.6.25 GFRP reinforcement for secant piles in a coastal area (FL, USA – 2019). Authorized reprint from Benmokrane et al. (2021a), ACI Special Publication 346: 93–112, 2021.

4.3.5 Concrete pavement

In addition to bridge decks, GFRP bars have shown significant promise for enhancing the durability of transportation infrastructure, such as railway ties, retaining walls, and continuously reinforced concrete pavements (CRCPs). CRCPs reinforced with conventional steel often face issues like corrosion and wide cracks due to aggregate-interlock degradation and steel-bar rupture, which compromise long-term performance. As a real-world example to address these issues, the Ministry of Transportation of Quebec (MTQ), in partnership with the University of Sherbrooke, began exploring GFRP bars in 2006 with a pilot project on Highway 40 in Montréal, as shown in Fig. 3.6.26. Initial findings demonstrated that GFRP-reinforced sections maintained strain and crack width within ACI 440.1R and AASHTO limits. The MTQ expanded the use of GFRP to a 306 m section of CRCP on Highway 40 West – 2013, using fibre-optic sensors for extensive monitoring over 6 years. Design modifications included an increased reinforcement ratio to accommodate the lower modulus of GFRP. Results showed that the GFRP-reinforced section managed crack widths and spacing effectively, even under severe conditions involving heavy use of de-icing salts. These findings highlight the potential of GFRP bars to reduce maintenance and enhance the lifespan of transportation infrastructure (Benmokrane et al., 2020).

4.4 Development of design codes and guidelines

Several design codes and guidelines have been developed to support the use of GFRP bars in reinforced concrete structures. Key standards include the AASHTO LRFD

Figure 3.6.26 Construction of the CRCP using GFRP reinforcement in Highway 40, Montréal, QC, Canada –2013.
Authorized reprint from Benmokrane et al. (2021a), ACI Special Publication 346: 93–112, 2021

Bridge Design Guide Specifications (AASHTO, 2018) and several ASTM International standards, such as D7205/D7205M (ASTM, 2021a), D7914/D7914M (ASTM, 2021b), D7957/D7957M (ASTM, 2022), and D8505/D8505M (ASTM, 2023). The American Concrete Institute (ACI) has also contributed with ACI 440.11-22 (ACI, 2022a), ACI PRC-440.1-15 (ACI, 2015), and ACI SPEC-440.5-22 (ACI, 2022b).

In Canada, the Canadian Standards Association (CSA) has developed CSA S806-12 (CSA, 2012, reapproved 2022), Chapter 16 of CSA S6-19 (CSA, 2019a), and CSA S807-19 (CSA, 2019b). Internationally, guidelines include the Annex (R) of the recent version of EN 1992-1-1:2023 that focuses on FRP reinforcement (CEN, 2023), Guidelines from the Association Française de Génie Civil (AFGC, 2023) from France, JSCE-97 (JSCE, 1997) from Japan, CNR-DT 203/2006 (CNR, 2006) from Italy, and the Interim Guidelines from the Institution of Structural Engineers (ISE, 1999) from the United Kingdom. These standards collectively ensure the safe, durable, and efficient design and application of GFRP reinforcement in construction projects.

5. Concluding remarks

Fibre-polymer composites started to be adopted in structural engineering and architecture in the mid-20th century, mostly due to their lightness, high strength-to-weight ratio, design flexibility, resistance to environmental agents and minimal maintenance requirements. After approximately 70 years of pioneering applications of composites, they are now considered mainstream construction materials, and their primary applications can be categorised into new construction, structural strengthening, and concrete reinforcement.

In new constructions, FRP composites have been used in bridges since the 1980s, through the application of pultruded profiles and sandwich panels. In all-composite footbridges, different structural systems were adopted (e.g., beam, truss, arch, or cable-stayed), comprising material-substitution and material-tailored designs, typically involving the use of respectively pultruded profiles and vacuum-infused sandwich panels with more complex shapes. In hybrid-composite footbridges, pultruded or vacuum-infused composite decks are connected to other structural materials that constitute the main load-bearing system. As for vehicular bridges, two main development phases can be distinguished. The first phase, which occurred between 1996 and 2003 (mostly in the USA), typically involved the use of proprietary (generally pultruded) sandwich deck systems, in most cases in hybrid bridges with steel girders or in composite deck slab bridges. At some point, due to the identification of damage in some decks after short periods of use, this type of application stopped. In the 2010s, a second phase of development started, slower than the first one, involving the use of proprietary vacuum-infused sandwich deck systems, which proved more versatile and flexible than those used in the first phase. In building construction, there was also an initiation phase, which occurred between the 1950s and the 1970s, when various modular housing units with bold architecture and curved shapes were developed. These projects explored the versatility of composites, enabling free-form designs and the integration of structural, building physics and architectural functions. At some point, due to both high costs and technical limitations, the interest in these first composite buildings decreased. A revival stage started in the 1990s, triggered by improvements in the manufacturing processes, which allowed overcoming the abovementioned limitations.

For structural strengthening, FRP systems applied according to different techniques, such as externally bonded reinforcement (EBR), near-surface mounted (NSM), and embedded-through-section (ETS), presented major developments since the 1980s and are now well-established among practitioners. These systems generally involve the application, by bonding, of FRP strips, sheets, or bars in various configurations tailored to the main purpose of the strengthening intervention. These FRP strengthening systems proved to be very efficient in enhancing the load-bearing capacity of existing structures, addressing deficiencies in terms of axial, flexural, shear, and punching capacities with minimal weight addition, while being very efficient for seismic retrofitting. More recently, the potential of such FRP systems has been further expanded through technological advancements in anchorage methods, prestressed systems, and hybrid techniques and systems, including the combination of FRP composites with textile-reinforced mortars (TRM) and FRC. FRP cables, initially developed to provide a light and noncorrodible solution for cable-stayed bridges and large roofs, proved to be very effective in replacing damaged steel cables in bridges due to their superior environmental and fatigue resistance. For this specific application, recent advancements have been made in anchoring systems of cables, which enabled the reduction in short- and long-term prestress losses.

Composite rebars for concrete reinforcement started to be developed in the 1960s, but the most relevant progress in research, commercialisation, and field applications occurred during and after the 1980s. Mostly driven by their resistance to corrosion, and the possibility they offer in terms of maintenance reduction and service life extension, thermoset GFRP rebars are finding increasing use in marine, de-icing, and chemically aggressive environments, providing a noncorrosive alternative to steel rebars for concrete structures. Key applications of GFRP rebars include concrete bridge decks and piles, tanks for water treatment facilities, tunnel segments and soft eyes, secant piles in coastal areas, and pavements. Recent progress in polymer engineering has enabled the development of thermoplastic GFRP rebars, which can be bent after production and also offer a promising potential for more sustainable reinforcement solutions.

Despite the various advantages of composites for structural engineering and architecture, several challenges remain, which hamper a wider application. This includes (i) the initial costs (which hinders the use in new construction and concrete reinforcement), (ii) the low stiffness of glass fibre composites, (iii) the typically brittle failure modes, (iv) the behaviour at elevated temperature and under fire exposure, posing a major limitation for building applications, and (v) the limited durability design provisions.

Looking forward, the potential of composites for structural engineering and architecture remains high. The way forward to widespread their use in this field may involve, among others, the following developments (i) emerging technologies in constituent materials (e.g., natural fibres, bio-based resins, thermoplastic resins, and cement-based adhesives for strengthening) and manufacturing methods (additive manufacturing, 3D printing, integration of sensors); (ii) advances in quality control and certification of composite products; (iii) development of phenomenological degradation models; (iv) demonstration of improved sustainability compared with alternative structural solutions; (v) reliable and mature design standards for the various types of applications (important advances were recently made in this respect in various regions); and (vi) education of the various construction stakeholders (designers, contractors owners).

References

AASHTO, 2018. AASHTO LRFD Bridge Design Guide Specifications for GFRP-Reinforced Concrete, second ed. American Association of State Highway and Transportation Officials, Washington, DC.

Abdel-Kareem, A.H., Debaiky, A.S., Makhlouf, M.H., Abdel-Baset, M., 2021. Shear strengthening of reinforced concrete beams using NSM/EBR techniques. Adv. Concr. Constr. 12 (1), 63–74. https://doi.org/10.12989/acc.2021.12.1.063.

ACI, 2006. ACI 440.1R-06. Guide for the Design and Construction of Structural Concrete Reinforced with FRP Bars. ACI Committee 440. American Concrete Institute, Farmington Hills, MI.

ACI, 2007. ACI 440R -07. Report on Fiber-Reinforced Polymer (FRP) Reinforcement for Concrete Structures. ACI Committee 440. American Concrete Institute, Farmington Hills, MI.

ACI, 2015. ACI PRC-440.1-15. Guide for the Design and Construction of Structural Concrete Reinforced with Fiber-Reinforced Polymer Bars. American Concrete Institute, Farmington Hills, MI.

ACI, 2022a. ACI CODE-440.11-22. Building Code Requirements for Structural Concrete Reinforced with Glass Fiber-Reinforced Polymer (GFRP) Bars - Code and Commentary, first ed. American Concrete Institute, Farmington Hills, MI.

ACI, 2022b. ACI SPEC-440.5-22. Construction with Glass Fiber-Reinforced Polymer. Reinforcing Bars - Specification. American Concrete Institute, Farmington Hills, MI.

ACI, 2023. ACI PRC-440.2-23. Design and Construction of Externally Bonded Fiber-Reinforced Polymer (FRP) Systems for Strengthening Concrete Structures-Guide. ACI Committee 440. American Concrete Institute, Farmington Hills, MI.

AFGC, 2023. Use of FRP (Fiber Reinforced Polymer) Rebars for Reinforced Concrete Structures. Association Française de Génie Civil, France.

Arafa, A., Farghaly, A., Ahmed, E., Benmokrane, B., 2016. Laboratory testing of GFRP-RC panels with UHPRFC joints of Nipigon River Cable Stayed Bridge in Northwest Ontario, Canada. J. Bridge Eng. 21 (11), 05016006. https://doi.org/10.1061/(ASCE)BE.1943-5592.0000943.

ASCE, 1984. Structural Plastics Design Manual. American Society of Civil Engineers, New York, NY.

ASCE, 2023. ASCE Standard 74-23: Load and Resistance Factor Design (LRFD) for Pultruded Fiber Reinforced Polymer (FRP) Structures. American Society of Civil Engineers, Reston, VA.

ASTM, 2021a. ASTM D7205/D7205M-21. Standard Test Method for Tensile Properties of Fiber Reinforced Polymer Matrix Composite Bars. ASTM International, West Conshohocken, PA. https://doi.org/10.1520/D7205_D7205M-21.

ASTM, 2021b. ASTM D7914/D7914M-21. Standard Test Method for Strength of Fiber Reinforced Polymer (FRP) Bent Bars in Bend Locations. ASTM International, West Conshohocken, PA. https://doi.org/10.1520/D7914_D7914M-21.

ASTM, 2022. ASTM D7957/D7957M-22. Standard Specification for Solid Round Glass Fiber Reinforced Polymer Bars for Concrete Reinforcement. ASTM International, West Conshohocken, PA. https://doi.org/10.1520/D7957_D7957M-22.

ASTM, 2023. ASTM D8505/D8505M-23. Standard Specification for Basalt and Glass Fiber Reinforced Polymer (FRP) Bars for Concrete Reinforcement. ASTM International, West Conshohocken, PA. https://doi.org/10.1520/D8505_D8505M-23.

Azevedo, A.S., Firmo, J.P., Correia, J.R., Chastre, C., Biscaia, H., Franco, N., 2022. Fire behaviour of CFRP-strengthened RC slabs using different techniques – EBR, NSM and CREatE. Compos. B Eng. 230, 109471. https://doi.org/10.1016/j.compositesb.2021.109471.

Azevedo, A.S., Firmo, J.P., Correia, J.R., Firouz, R.M., Barros, J.A.O., 2023. Fire behaviour of reinforced concrete slab strips strengthened with prestressed NSM-CFRP laminates. Eng. Struct. 297, 116982. https://doi.org/10.1016/j.engstruct.2023.116982.

Babatunde, S.A., 2017. Review of strengthening techniques for masonry using fiber reinforced polymers. Compos. Struct. 161, 246–255. https://doi.org/10.1016/j.compstruct.2016.10.132.

Baghi, H., Barros, J.A.O., 2016. Shear strengthening of reinforced concrete T-beams with hybrid composite plate. J. Compos. Construct. 20 (6), 04016036. https://doi.org/10.1061/(ASCE)CC.1943-5614.0000693.

Bakis, C.E., Bank, L.C., Brown, V.L., Cosenza, E., Davalos, J.F., Lesko, J.J., Machida, A., Rizkalla, S.H., Triantafillou, T.C., 2002. Fiber-reinforced polymer composites for construction - state-of-the-art review. J. Compos. Construct. 6 (2), 73–87. https://doi.org/10.1061/(asce)1090-0268(2002)6:2(73.

Bank, L.C., 2006. Composites for Construction: Structural Design with FRP Materials. John Wiley & Sons, New York.

Barros, J.A.O., Rezazadeh, M., Laranjeira, J.P.S., Hosseini, M.R.M., Mastali, M., Ramezansefat, H., 2017. Simultaneous flexural and punching strengthening of RC slabs according to a new hybrid technique using U-shape CFRP laminates. Compos. Struct. 159, 600–614. https://doi.org/10.1016/j.compstruct.2016.10.009.

Barros, J.A.O., Dias, S.J.E., Baghi, H., Ventura-Gouveia, A., 2016. New shear strengthening configurations of NSM CFRP laminates for RC beams. ACI Struct. J. 113 (06), 1275–1287. https://doi.org/10.14359/51689029.

Barros, J.A.O., Dias, S.J.E., Lima, J.L.T., 2007. Efficacy of CFRP-based techniques for the flexural and shear strengthening of concrete beams. Cement Concr. Compos. 29 (3), 203–217. https://doi.org/10.1016/j.cemconcomp.2006.09.001.

Durability of fiber reinforced polymer (FRP) composites for construction. In: Benmokrane, B., Rahman, H. (Eds.), 1998. Proceedings of the First International Conference on Durability of FRP Composites for Construction (CDCC'98). Department of Civil Engineering, University of Sherbrooke, Sherbrooke, QC, Canada, August 5–7, 1998, p. 706.

Benmokrane, B., Wang, P., Ton-That, T.M., 2002. Durability of glass fiber-reinforced polymer reinforcing bars in concrete environment. J. Compos. Construct. 6 (3), 143–153. https://doi.org/10.1061/(ASCE)1090-0268(2002)6:3(143).

Benmokrane, B., Mohamed, M.H., 2014. Extending the service life of water treatment structures: glass fiber-reinforced polymer bar replaces steel in concrete water tanks in Quebec. ACI Concrete Int. Mag. 36 (2), 40–45.

Benmokrane, B., Manalo, A., Bouhet, J.C., Mohamed, K., Robert, M., 2017. Effects of diameter on the durability of GFRP bars conditioned in alkaline solution. J. Compos. Construct. 21 (5), 04017040–04017041. https://doi.org/10.1061/(ASCE)CC.1943-5614.0000814.

Benmokrane, B., Bakouregui, A.S., Mohamed, H.M., Thébeau, D., Abdelkarim, O.I., 2020. Design, construction, and performance of continuously reinforced concrete pavement reinforced with GFRP bars: case study. J. Compos. Construct. 24 (5), 05020004. https://doi.org/10.1061/(ASCE)CC.1943-5614.0001064.

Benmokrane, B., Mohamed, H.M., Mohamed, K., Mousa, S., 2021a. Recent Canadian Developments Related to Unconventional Reinforcing for Concrete Structures, Design Codes, and Applications in Buildings and Bridges, vol 346. ACI Special Publication, pp. 93–112.

Benmokrane, B., Mousa, S., Mohamed, H.M., El-Safty, A., Nolan, S., 2021b. Design, construction, testing, and behavior of driven precast concrete piles reinforced with GFRP bars and spirals. J. Bridge Eng. 26 (8), 04020116. https://doi.org/10.1061/(ASCE)BE.1943-5592.0001755.

Benmokrane, B., Mousa, S., Mohamed, K., Sayed-Ahmed, M., 2021c. Physical, mechanical, and durability characteristics of newly developed thermoplastic GFRP bars for reinforcing concrete structures. Constr. Build. Mater. 276, 122200. https://doi.org/10.1016/j.conbuildmat.2020.122200.

Benmokrane, B., Mehany, S., Shield, C., Nanni, A., Brown, V., 2023. Physical properties, longitudinal tensile properties, and bond strength of the new generation of GFRP bars. J. Compos. Construct. 27 (5), 04023026. https://doi.org/10.1061/JCCOF2.CCENG-4300.

Blaschko, M., Zilch, K., 1999. Rehabilitation of concrete structures with CFRP strips glued into slits. In: Proceedings of the 12th International Conference on Composite Materials, Paris.

Bournas, D.A., Triantafillou, T.C., 2009. Flexural strengthening of RC columns with NSM FRP or stainless steel. ACI Struct. J. 106 (4), 495–505. https://doi.org/10.14359/56615.

Breveglieri, M., Aprile, A., Barros, J.A.O., 2016. RC concrete beams strengthened in shear using the Embedded Through-Section technique: experimental results and analytical formulation. Compos. B Eng. 89, 266–281. https://doi.org/10.1016/j.compositesb.2015.11.023.

Burgoyne, C.J., Head, P.R., 1993. Aberfeldy Bridge - an advanced textile reinforced footbridge. In: TechTextil Symposium, Frankfurt.

CEN, 2022. CEN/TS 19101:2022. Design of Fiber-Polymer Composite Structures. European Committee for Standardization, Brussels.

CEN, 2023. EN 1992-1-1:2023. Eurocode 2 - Design of Concrete Structures - Part 1-1: General Rules and Rules for Buildings, Bridges and Civil Engineering Structures. European Committee for Standardization, Brussels.

Chen, C.-C., Li, C.-Y., 2005. Punching shear strength of reinforced concrete slabs strengthened with glass fiber-reinforced polymer laminates. ACI Struct. J. 102 (4), 535–542. https://doi.org/10.14359/14557.

Clarke, J.L. (Ed.), 1996. EUROCOMP Design Code and Handbook – Structural Design of Polymer Composites. E&FN Spon.

CNR, 2006. CNR-DT 203. Guide for the Design and Construction of Concrete Structures with Fiber-Reinforced Polymer Bars. CNR - Advisory Committee on Technical Recommendations for Construction, Rome.

CNR, 2013. CNR-DT 200 R1. Guide for the Design and Construction of Externally Bonded FRP Systems for Strengthening Existing Structures. CNR - Advisory Committee on Technical Recommendations for Construction, Rome.

Concrete Society, 2012. TR55. Design Guidance for Strengthening Concrete Structures Using Fiber Composite Materials, third ed. The Concrete Society. Technical Report nº 55. Amendment nº1 October 2013.

Correia, J.R., 2015. Fiber-reinforced polymer (FRP) composites. In: Margarido, F., Gonçalves, C. (Eds.), Materials for Construction and Civil Engineering. Science, Processing, and Design. Springer Nature.

Costa, I.G., Barros, J.A.O., 2013. Critical analysis of fiber-reinforced polymer near-surface mounted double-shear pull-out tests. Strain 49, 299–312. https://doi.org/10.1111/str.12038.

Cotswold Canals in Pictures, 2024. Bond's Mill Bridge, Stonehouse. https://www.cotswoldcanals.net/bonds-mill-bridge#.

CSA, 2002. CAN/CSA-S806.02. Design and Construction of Building Components with Fiber-Reinforced Polymers. Canadian Standards Association, Mississauga, ON.

CSA, 2012. CSA S806-12 (R2021). Design and Construction of Building Structures with Fiber-Reinforced Polymers. Canadian Standards Association, Mississauga, ON.

CSA, 2019a. Chapter 16 of CSA S6-19. The Canadian Highway Bridge Design Code. Canadian Standards Association, Mississauga, ON.

CSA, 2019b. CSA S807-19. Specification for Fiber-Reinforced Polymers. Canadian Standards Association, Mississauga, ON.

CSA, 2021. CSA-S806-12. Design and Construction of Building Components with Fiber-Reinforced Polymers. Canadian Standards Association, Toronto, Canada.

Czaderski, C., Motavalli, M., 2007. 40-Year-old full-scale concrete bridge girder strengthened with prestressed CFRP plates anchored using gradient method. Compos. B Eng. 38 (7–8), 878–886. https://doi.org/10.1016/j.compositesb.2006.11.003.

De Lorenzis, L., Nanni, A., 2001. Characterization of FRP rods as near surface mounted reinforcement. J. Compos. Construct. 5 (2), 114–121. https://doi.org/10.1061/(ASCE)1090-0268(2001)5:2(114).

De Lorenzis, L., Nanni, A., La Tegola, A., 2000. Strengthening of reinforced concrete structures with near surface mounted FRP rods. In: International Meeting on Composite Materials, PLAST 2000, Milan, Italy.

Dias, S.J.E., Silva, J.R.M., Barros, J.A.O., 2021. Flexural and. shear strengthening of reinforced concrete beams with a hybrid CFRP solution. Compos. Struct. 256, 113004. https://doi.org/10.1016/j.compstruct.2020.113004.

Doernach, R., 1974. Bausysteme mit Kunststoff. Deutsche Verlags-Anstalt, Stuttgart.

Erdogan, H., Binici, B., Ozcebe, G., 2010. Punching shear strengthening of flat-slabs with CFRP dowels. Mag. Concr. Res. 62, 465–478. https://doi.org/10.1680/macr.2010.62.7.465.

Esmaeeli, E., Barros, J.A.O., 2015. Flexural strengthening of RC beams using hybrid composite plate (HCP): experimental and analytical study. Compos. B Eng. 79, 604–620. https://doi.org/10.1016/j.compositesb.2015.05.003.

Esmaeeli, E., Barros, J.A.O., Sena-Cruz, J.M., Varum, H., Melo, J., 2015. Assessment of the efficiency of prefabricated hybrid composite plates (HCPs) for retrofitting of damaged interior RC beam–column joints. Compos. Struct. 119, 24–37. https://doi.org/10.1016/j.compstruct.2014.08.024.

Fang, H., Zou, F., Liu, W., Wu, C., Bai, Y., Hui, D., 2016. Mechanical performance of concrete pavement reinforced by CFRP grids for bridge deck applications. Compos. Part B Eng. 110, 315–335. https://doi.org/10.1016/j.compositesb.2016.11.015.

Fardis, M.N., Khalili, H., 1981. Concrete encased in fiberglass reinforced plastic. ACI J. Proc. 78 (6), 440–446. https://doi.org/10.14359/10527.

FIB, 2019. Fib Bulletin 90. Externally Applied FRP Reinforcement for Concrete Structures. The International Federation for Structural Concrete, Technical Report. https://doi.org/10.35789/fib.BULL.0090.

Fiberline, 1995. Fiberline Design Manual, Denmark.

Firmo, J.P., Correia, J.R., Bisby, L.A., 2015. Fire behaviour of FRP-strengthened reinforced concrete structural elements: a state-of-the-art review. Compos. B Eng. 80, 198–216. https://doi.org/10.1016/j.compositesb.2015.05.045.

Fukuyama, H., 1999. FRP composites in Japan. Concr. Int. 21 (10), 29–32.

GB Standards, 2013. GB 50367-2013. Code for Design of Strengthening Concrete Structures. China Architecture and Building Press (in Chinese).

Guo, H., Liao, H., Su, M., Zhang, B., Li, S., Peng, H., 2024. Shear strengthening of RC beams with prestressed NSM CFRP: influencing factors and analytical model. Compos. Struct. 342, 118262. https://doi.org/10.1016/j.compstruct.2024.118262.

Gurit, 2024. Rhyl Harbour Bridge, Wales. https://www.gurit.com/rhyl-harbour-bridge-rhyl-wales-uk/.

Hollaway, L.C., Head, P.R., 2001. Advanced Polymer Composites and Polymers in the Civil Infrastructure. Elsevier Science, Oxford, UK.

Hong, T., Hastak, M., 2006. Construction, inspection, and maintenance of FRP deck panels. J. Compos. Construct. 10 (6), 561–572. https://doi.org/10.1061/(asce)1090-0268(2006)10:6(561).

Hosseini, S.A., Farghaly, A.S., Eslami, A., Nanni, A., Benmokrane, B., 2023. Bond behaviour of lap spliced GFRP bars in concrete members: a state-of-the-art review and design recommendations. Constr. Build. Mater. 411, 134714. https://doi.org/10.1016/j.conbuildmat.2023.134714.

Hosseini, S.M., Mousa, S., Mohamed, H., Benmokrane, B., 2024. Long-Term durability of nonpultruded curvilinear GFRP bars exposed to an alkaline environment: experimental studies and modeling. J. Compos. Construct. 28 (3), 04024005. https://doi.org/10.1061/jccof2.cceng-4454.

Ibrahim, B., Iraqy, A., Cousin, P., Benmokrane, B., 2024a. Physical and mechanical properties and durability characteristics of new pultruded GFRP bars with thermoplastic resin. J. Compos. Construct. 28 (4), 06024001. https://doi.org/10.1061/jccof2.cceng-4607.

Ibrahim, B., Nguyen, K.Q., Cousin, P., Robert, M., Benmokrane, B., 2024b. Performance of newly developed thermoplastic resin for FRP bars compared with vinyl-ester resin under laboratory accelerated aging testing. J. Compos. Construct. 28 (5), 04024030. https://doi.org/10.1061/jccof2.cceng-4533.

ISE, 1999. Interim Guidance on the Design of Reinforced Concrete Structures Using Fiber Composite Reinforcement. Institution of Structural Engineers, London, UK.

Isler, H., 1977. Tragende bauteile aus kunststoff: anwendungsbeispiele. Schweizerische Bauzeitung 95 (3), 13–20.

Isler, H., 1998. Personal Communication on 22 January 1998.

JSCE, 1997. Design guidelines of FRP reinforced concrete building structures. In: Concrete Engineering Series, vol 23. Japan Society of Civil Engineering, Tokyo, Japan.

Kalfat, R., Gadd, J., Al-Mahaidi, R., Smith, S.T., 2018. An efficiency framework for anchorage devices used to enhance the performance of FRP strengthened RC members. Constr. Build. Mater. 191, 354–375. https://doi.org/10.1016/j.conbuildmat.2018.10.022.

Keller, T., 2002. Fiber reinforced polymer materials in bridge construction. In: Proceedings of IABSE Symposium, towards a Better Built Environment - Innovation, Sustainability, Information Technology, Melbourne.

Keller, T., 2003. Use of fiber reinforced polymers in bridge construction. In: Structural Engineering Documents, SED 7, International Association for Bridge and Structural Engineering. IABSE, Zurich.

Keller, T., 2024. Composites in Structural Engineering and Architecture. EPFL Press, Lausanne. https://doi.org/10.55430/6225TKVA01.

Koutas, L., Triantafillou, T.C., 2012. Use of anchors in shear strengthening of reinforced concrete T-beams with FRP. J. Compos. Construct. 17 (1), 101–107. https://doi.org/10.1061/(ASCE)CC.1943-5614.0000316.

Lees, J.M., Winistörfer, A.U., Meier, U., 2002. External prestressed CFRP straps for the shear enhancement of concrete. J. Compos. Construct. 6 (4), 249–256. https://doi.org/10.1061/(ASCE)1090-0268(2002)6:4(249).

Li, V.C., Mishra, D.K., Wu, H.C., 1995. Matrix design for pseudo strain-hardening fiber reinforced cementitious composites. Mater. Struct. 28 (183), 586–595. https://doi.org/10.1617/s11527-022-01944-6.

Liu, Y., Zwingmann, B., Schlaich, M., 2015. Carbon fiber reinforced polymer for cable structures - a review. Polymers 7 (10), 2078–2099. https://doi.org/10.3390/polym7101501.

Liu, Y., Zhang, H.-T., Tafsirojjaman, T., Dogar, A.U.R., AlAjarmeh, O., Yue, Q.-R., Manalo, A., 2022. A novel technique to improve the compressive strength and ductility of glass fiber reinforced polymer (GFRP) composite bars. Constr. Build. Mater. 326, 126782. https://doi.org/10.1016/j.conbuildmat.2022.126782.

Market Development Alliance of the FRP Composites Industry, 2003. Global FRP Use for Bridge Applications: Vehicular. Harrison, NY.

Mehany, S., Mohamed, H.M., Benmokrane, B., 2023. Flexural behavior and serviceability performance of lightweight self-consolidating concrete beams reinforced with basalt fiber-reinforced polymer bars. ACI Struct. J. 120 (3), 47–60. https://doi.org/10.14359/51738502.

Meier, U., 1987a. Brückensanierungen mit Hochleistungs-Faserverbundwerkstoffen. Material und Technik 87 (4), 125–128.

Meier, U., 1987b. Proposal for a carbon fiber reinforced composite bridge across the Strait of Gibraltar at its narrowest site. Proc. IME B J. Eng. Manufact. 201 (2), 73–78. https://doi.org/10.1243/PIME_PROC_1987_201_048_02.

Meier, U., 2024. Advancements in carbon fiber reinforced polymer tendons for structural rehabilitation. In: Keynote Lecture. SMAR 2024 Conference, Salerno, Italy.

Meier, U., Deuring, M., 1991. The application of fiber composites in bridge repair. A world premiere at the gates of Lucerne. Str. Verk. 77 (9), 523–525.

Meier, U., Kaiser, H., 1991. Strengthening of structures with CFRP laminates. In: Advanced Composite Materials in Civil Engineering Structures, ASCE Specialty Conference, pp. 224–232.

Meier, U., Meier, H., 1996. CFRP finds use in cable support for bridge. Mod. Plast. 4, 87–98.

Mohamed, H., Ali, A., Hadhood, A., Mousa, S., Abdelazim, W., Benmokrane, B., 2020. Testing, design, and field implementation of GFRP RC soft-eyes for tunnel construction. Tunn. Undergr. Space Technol. 106, 103626. https://doi.org/10.1016/j.tust.2020.103626.

Mohamed, H.M., Benmokrane, B., 2018. Construction of Nipigon River Cable-stayed bridge using precast concrete panels reinforced with glass FRP rebars. In: Proceedings of the 9th International Conference on FRP Composites in Civil Engineering (CICE2018), Paris, France.

Mohamed, H.M., Benmokrane, B., 2014. Design and performance of reinforced concrete water chlorination tank totally reinforced with GFRP bars. J. Compos. Construct. 18 (1), 05013001. https://doi.org/10.1061/(ASCE)CC.1943-5614.0000429.

Mohammadi, A., Barros, J.A.O., Sena-Cruz, J.M., 2024. Multi-objective optimization for an unbiased model predicting the contribution of EBR-CFRP strengthening system to the shear- resistance of RC beams. J. Compos. Construct. 6 (6), 04024066. https://doi.org/10.1061/JCCOF2.CCENG-4814.

Motavalli, M., Czaderski, C., 2007. FRP composites for retrofitting of existing civil structures in Europe: state-of-the-art review. In: International Conference of Composites & Polycon. American Composites Manufacturers Association, Tampa, FL, USA, pp. 17–19.,

Mufti, A., Onofrei, M., Benmokrane, B., Banthia, N., Boulfiza, M., Newhook, J., Bakht, B., Tadros, G., Brett, P., 2005. Durability of GFRP reinforced concrete in field structures. In: Shield, C.K., Busel, J.P., Walkup, S.L., Gremel, D.D. (Eds.), 7th International Symposium on Fiber-Reinforced Polymer (FRP) Reinforcement for Concrete Structures, SP-230. American Concrete Institute (ACI), Farmington Hills, MI, pp. 1361–1377.

Nanni, A., 1999. Composites: coming on strong. Concr. Construct. 44 (1), 120–124.

Nanni, A., Bradford, N.M., 1995. FRP jacketed concrete under uniaxial compression. Constr. Build. Mater. 9 (2), 115–124. https://doi.org/10.1016/0950-0618(95)00004-Y.

Nolan, S., Levine, S., Denty, L., Steputat, C., Nanni, A., 2019. Low-Impact Secant-Pile Bulkhead for Protecting SR-A1A along Flagler Beach. Shoreline. Florida Shore & Beach Preservation Association, p. 26.

Nordin, H., Täljsten, B., 2006. Concrete beams strengthened with prestressed near surface mounted CFRP. J. Compos. Construct. 10 (1), 60–68. https://doi.org/10.1061/(ASCE)1090-0268(2006)10:1(60).

Oehlers, D.J., Moran, J.P., 1990. Premature failure of externally plated reinforced concrete beams. J. Struct. Eng. 116 (4), 978–995. https://doi.org/10.1061/(ASCE)0733-9445(1990)116:4(978).

Ortiz, J.D., Hussain, Z., Hosseini, S., Benmokrane, B., Nanni, A., 2024. Lap splice assessment of GFRP rebars in reinforced concrete beams under flexure. Constr. Build. Mater. 419, 135408. https://doi.org/10.1016/j.conbuildmat.2024.135408.

Pantelides, C.P., Okahashi, Y., Reaveley, L.D., 2008. Seismic rehabilitation of reinforced concrete frame interior beam-column joints with FRP composites. J. Compos. Construct. 12 (4), 435–445. https://doi.org/10.1061/(ASCE)1090-0268(2008)12:4(435).

Plecnik, J., Ahmad, S.H., 1988. Transfer of composite Technology to Design and Construction of Bridges. Final report to US DOT, contract no. DTRS 5683-C000043.

Potyrała, P.B., 2011. Use of Fiber Reinforced Polymer Composites in Bridge Construction - State of the Art in Hybrid and All-Composite Structures. Master dissertation. Polytechnic University of Catalonia, Barcelona, Spain.

Quarmby, A., 1974. The Plastics Architect. Pall Mall Press, London.

Rashid, M.A., Mansur, M.A., Paramasivam, P., 2005. Behavior of aramid fiber-reinforced polymer reinforced high strength concrete beams under bending. J. Compos. Construct. 9 (2), 117–127. https://doi.org/10.1061/(asce)1090-0268(2005)9:2(117).

Reeve, S., 2010. FRP bridge decking – 14 years and counting. Reinforc Plast. https://doi.org/10.1016/S0034-3617(10)70031-X.

Refai, A., Abed, F., Altalmas, A., 2015. Bond durability of basalt fiber-reinforced polymer bars embedded in concrete under direct pullout conditions. J. Compos. Construct. 19, 04014078. https://doi.org/10.1061/(ASCE)CC.1943-5614.0000544.

Reising, R.M.W., Shahrooz, B.M., Hunt, V.J., Neumann, A.R., Helmicki, A.J., Hastak, M., 2004. Close look at construction issues and performance of four fiber-reinforced polymer composite bridge decks. J. Compos. Construct. 8 (1), 33–42. https://doi.org/10.1061/(asce)1090-0268(2004)8:1(33).

Rezazadeh, M., Costa, I.G., Barros, J.A.O., 2014. Influence of prestress level on NSM CFRP laminates for the flexural strengthening of RC beams. Compos. Struct. 116 (1), 489–500. https://doi.org/10.1016/j.compstruct.2014.05.043.

Saad, K., Lengyel, A., 2022. Strengthening timber structural members with CFRP and GFRP: a state-of-the-art review. Polymers 14, 2381. https://doi.org/10.3390/polym14122381.

Schwabe, A., Saechtling, H., 1959. Bauen mit Kunststoffen. Ullstein AG, Berlin.

Siddika, A., Al Mamun, M.A., Ferdous, W., Alyousef, R., 2020. Performances, challenges and opportunities in strengthening reinforced concrete structures by using FRPs - a state-of-the-art review. Eng. Fail. Anal. 111, 104480. https://doi.org/10.1016/j.engfailanal.2020.104480.

Summerscales, J., 2024. Fiber-reinforced Polymer Composite Bridges in Europe. University of Plymouth, UK. https://ecm-academics.plymouth.ac.uk/jsummerscales/composites/bridges.htm.

Taerwe, L.R., Matthys, S., 1999. FRP for concrete construction. Concr. Int. 21 (10), 33–36.

Täljsten, B., Blanksvärd, T., 2007. Mineral-based bonding of carbon FRP to strengthen concrete structures. J. Compos. Construct. 11 (2), 120–128. https://doi.org/10.1061/(ASCE)1090-0268(2007)11:2(120).

Teng, J.G., Chen, J.F., Smith, S.T., Lam, L., 2002. FRP Strengthened RC Structures. John Wiley & Sons, Inc., West Sussex, UK.

Triandafilou, L.N., O'Connor, J.S., 2009. FRP composites for bridge decks and superstructures: state of the practice in the U.S. In: Proceedings of International Conference on Fiber Reinforced Polymer (FRP) Composites for Infrastructure Applications.

Triandafilou, L.N., O'Connor, J.S., 2010. Field issues associated with the use of fiber-reinforced polymer composite bridge decks and superstructures in harsh environments. Struct. Eng. Int. 20 (4), 409–413. https://doi.org/10.2749/101686610793557663.

Triantafillou, T.C., 1992. Shear strengthening of reinforced concrete beams using epoxy-bonded FRP composites. ACI Struct. J. 89 (3), 235–244. https://doi.org/10.14359/531.

Uomoto, T., Mutsuyoshi, H., Katsuki, F., Misra, S., 2002. Use of fiber reinforced polymer composites as reinforcing material for concrete. J. Mater. Civ. Eng. 14 (3), 191–209. https://doi.org/10.1061/(asce)0899-1561(2002)14:3(191).

Valerio, P., Ibell, T.J., Darby, A.P., 2007. Deep embedment of FRP bars for shear strengthening of prestressed concrete bridges. In: Proceedings of Fiber-Reinforced Polymers for RC Structures (FRPRCS-8) (Patras).

Voigt, P., 2007. Die Pionierphase des Bauens mit glasfaserverstärkten Kunststoffen (GFK) 1942 bis 1980. PhD thesis, Bauhaus-Universität Weimar. https://www.db-thueringen.de/receive/dbt_mods_00036183.

Wu, G., Wu, Z.S., Lu, Z.T., Ano, Y.B., 2008. Structural performance of concrete confined with hybrid FRP composites. J. Reinforc. Plast. Compos. 27 (12), 1323–1348. https://doi.org/10.1177/0731684407084989.

Zeng, X., Jiang, S.-F., Deng, K., Huang, H., Cui, E., 2022. Seismic performance of circular RC columns strengthened in flexure using NSM reinforcement and externally bonded BFRP sheets. Eng. Struct. 256, 114033. https://doi.org/10.1016/j.engstruct.2022.114033.

Section 4

The theory behind composites, and how it evolved

On the evolution of the laminated plate theory for composite materials: From trace via quad to double-double

4.1

Stephen W. Tsai
Department of Aeronautics & Astronautics, Stanford University, Stanford, CA, United States

1. History

Scientists around the world did not wait until the advent of carbon fibres to try to understand the behaviour of heterogeneous materials (meaning: composed of two or more distinct materials or phases). In Section 7 of Part I of this book, reference has been made to the interesting review paper by Herakovich (2012), where he describes in quite some detail the main steps in the developments of the mechanics of composites, from micro-over meso-to macro- or structural mechanics.

Laminated plate theory is a special case of theory of plates or theory of plane elasticity well developed by S.P. Timoshenko and his contemporaries in Russia and elsewhere more than 100 years ago. The reader is referred to a historical overview (Timoshenko, 1953). Lekhnitskii's books on Anisotropic Elasticity and Anisotropic Plates, first published in 1947 (Lekhnitskii, 1947; Lekhnitskii et al., 1968) were written with wood in mind when it was used for aircraft in the beginning of aviation. Comparable books on wood were published by the US Forest Products Laboratory, and a comparable one from the United Kingdom was found. A similar report was published by DLVLR/Braunschweig in the same time frame. In the US, one early publication on laminated plate theory was done by Stanley Dong for his thesis at the University of California, Berkely, in the early 1960s (Dong et al., 1962).

Laminated plate theory for composite materials is the critical link between fibres and structures. That link between fibre and the binding matrix material to unidirectional plies is covered by micro-mechanics; while that between laminates and structures is macro-mechanics. The coverage in this chapter may be termed meso-mechanics that bridges between unidirectional plies and laminates where laminates are recognised as the building blocks of laminated structures.

Quad laminates are governed by a set of rules that have made their theory complex and their structures difficult to design, produce and test. It is difficult to challenge these established practices because design tools and property data have been generated through the years. New approached are often turned away from

The History of Composites. https://doi.org/10.1016/B978-0-443-34119-9.00024-3

lack of data, the challenge in learning new tools and concepts, or no known path for government certification. The old Quad remains the accepted practice for 60 years and counting.

The premise of this chapter is based on the recognition that materials and laminates play different and independent roles in the behaviour of composite materials and structures. Not unlike metals, stiffness and strength of composite materials set the level of performance. Like titanium is 40% stiffer than aluminium, different composite materials can be rated in the magnitude for their relative stiffness and strength. That same level applies to all laminates made from the composite materials. Laminates on the other hand are geometric quantities that set their stiffness and strength uncoupled and independent from materials. Thus, if laminate A is 20% stiffer or stronger than laminate B, it is the same for all materials. There are 15 carbon/epoxy (CFRP) materials and for double-double there are 15 laminates. So there are over 200 combinations of material and laminate. If we know the rating of materials and laminates, we need to test only 15 material/laminate combinations. All other combinations can be scaled. Testing is to confirm what is already known. That is different from blind testing that has no expectation what the data would be. This is one example how composite laminates can be more easily learnt and validated experimentally. It is the goal of this chapter to propose a history changing strategy for composite laminate so it can be simplified and better utilised as a competitive structural material.

The chapter is based on the works of the author during the last decade (Tsai, 2019; Tsai et al., 2023; Kappel et al., 2024; Tsai, 2021a,b).

2. Rules for quad

Quad laminates are controlled by a number of rules, some with good reasons but others are arbitrary and subjective. In general they make composite laminates difficult to design, produce and test. They increase the weight and cost of structures. They make optimisation difficult if not impossible. The quality of product is often poor because of internal flaws and discontinuities in the laminates. Specific rules are listed in the following:

1. Mid-plane symmetry; required to prevent warping after laminate consolidation
2. 10% rule: required every ply angles at a minimum of 10% of total number of plies to cover unknown secondary loads
3. Balanced laminate: so there will not have shear coupling in 16 and 26 components
4. Nesting of plies: no more than four plies in any ply group so delamination is less prone
5. Adjacent ply angle difference: no more than 45° to reduce the mismatch between adjacent plies

These rules may look innocent but have imposed conditions that would make the resulting laminates unnecessarily complex to design and costly to produce. In its place, a double-double laminate (DD) is much simpler to design and produce. This will be the intent of this chapter.

3. One parameter for stiffness

Description of stiffness and strength are commonly described by 4 parameters for stiffness (E_x, E_y, ν_x, and E_s); and five or more for strength (X, X′, Y, Y′, and S). It is not possible to compare materials and laminates by their relative stiffness and strength when they are defined by multiple parameters. With one parameter however such comparison can be straightforward. The mass of materials is measured by 1 parameter. Their relative values are easily seen. One parameter for stiffness and one for strength have been found and will be presented herein.

The one parameter for stiffness of materials and laminates is the trace, or also known as the Tsai's modulus. It is the sum of the diagonal components of the stiffness matrix [Q] of a unidirectional ply:

$$\text{Trace} = Q_{xx} + Q_{yy} + 2Q_{ss} = \text{Tsai's modulus} \qquad 4.1.1$$

The derivation of trace for the stiffness of [0] ply for representative CFRP from their engineering constants to plane stress stiffness [Q] follows Eq. (4.1.1):

$$\begin{aligned} &Q_{xx} = E_x/(1-\nu_x\nu_y);\; Q_{yy} = E_y/(1-\nu_x\nu_y);\; Q_{ss} = E_s \\ &Q_{xy} = \nu_y E_x/(1-\nu_x\nu_y) = \nu_x E_y/(1-\nu_x\nu_y) = \nu_y Q_{xx} = \nu_x Q_{yy} \end{aligned} \qquad 4.1.2$$

These constants can be normalised by their trace as [Q*], shown in the last column of Fig. 4.1.1:

$$\text{Trace}^* = Q_{xx}{}^* + Q_{yy}{}^* + 2Q_{ss}{}^* = 1 \qquad 4.1.3$$

One critical master ply value in the bottom row is $Q_{xx}{}^* = 0.885$ with cv = 1.5%. This allows a quick determination of trace from Q_{xx}; i.e.,

$$\text{Trace} = Q_{xx}/0.885 \qquad 4.1.4$$

It can also be shown that for CFRP:

$$\text{Trace} = E_x/0.88 \qquad 4.1.5$$

Similarly, for Hi-modulus CFRP the factor is 0.95; Kevlar, 0.88; E-glass, 0.65; S-glass, 0.66 (as will be discussed later, see Fig. 4.1.9). These relations are easier to find trace, because E_x can be measured directly, or estimated from fibre stiffness multiplied by fibre volume fraction. The accuracy of this simple relation for materials, shown in Fig. 4.1.1, is within 5% error, with an average error for all materials of 1.6%.

Laminate in-plane stiffness [A] is defined by two distinct parts: the material part in trace; the geometric part in the partitioning of trace for each laminate. This is shown in

CFRP	Ex	Ey	Es	nu	Qxx	Qyy	Qxy	Qss	Trace	Qxx*	Qyy*	Qxy*	Qss*	Trace*
IM6/epoxy	203	11.20	8.40	0.32	204	11.26	3.60	8.40	232	0.879	0.049	0.016	0.036	1.00
IM7/977-3	191	9.94	7.79	0.35	192	10.00	3.50	7.79	218	0.883	0.046	0.016	0.036	1.00
T300/5208	181	10.30	7.17	0.28	182	10.35	2.90	7.17	206	0.880	0.050	0.014	0.035	1.00
IM7/MTM45	175	8.20	5.50	0.33	176	8.24	2.72	5.50	195	0.901	0.042	0.014	0.028	1.00
T800/Cytec	162	9.00	5.00	0.40	163	9.08	3.63	5.00	183	0.895	0.050	0.020	0.027	1.00
IM7/8552	159	8.96	5.50	0.32	160	9.01	2.88	5.50	180	0.889	0.050	0.016	0.031	1.00
T800S/3900	151	8.20	4.00	0.33	152	8.25	2.72	4.00	168	0.903	0.049	0.016	0.024	1.00
T300/F934	148	9.65	4.55	0.30	149	9.71	2.91	4.55	168	0.888	0.058	0.017	0.027	1.00
T700 C-Ply 6	141	9.30	5.80	0.30	142	9.36	2.81	5.80	163	0.871	0.057	0.017	0.036	1.00
AS4/H3501	138	8.96	7.10	0.30	139	9.01	2.70	7.10	162	0.857	0.056	0.017	0.044	1.00
T650/epoxy	139	9.40	5.50	0.32	140	9.47	3.03	5.50	160	0.872	0.059	0.019	0.034	1.00
T4708/MR6	142	7.72	3.80	0.34	143	7.77	2.64	3.80	158	0.903	0.049	0.017	0.024	1.00
T700/2510	126	8.40	4.20	0.31	127	8.45	2.62	4.20	144	0.883	0.059	0.018	0.029	1.00
AS4/MTM45	128	7.93	3.65	0.30	129	7.97	2.39	3.65	144	0.894	0.055	0.017	0.025	1.00
T700 C-Ply 5	121	8.00	4.70	0.30	122	8.05	2.41	4.70	139	0.875	0.058	0.017	0.034	1.00
									Average: Master ply	0.885	0.051	0.016	0.032	1.00
									cv %	1.5%	10.4%	9.9%	17.4%	0.00

Figure 4.1.1 Trace values for CFRP, and master normalized stiffness of Q_{xx}* (Tsai, 2021b).

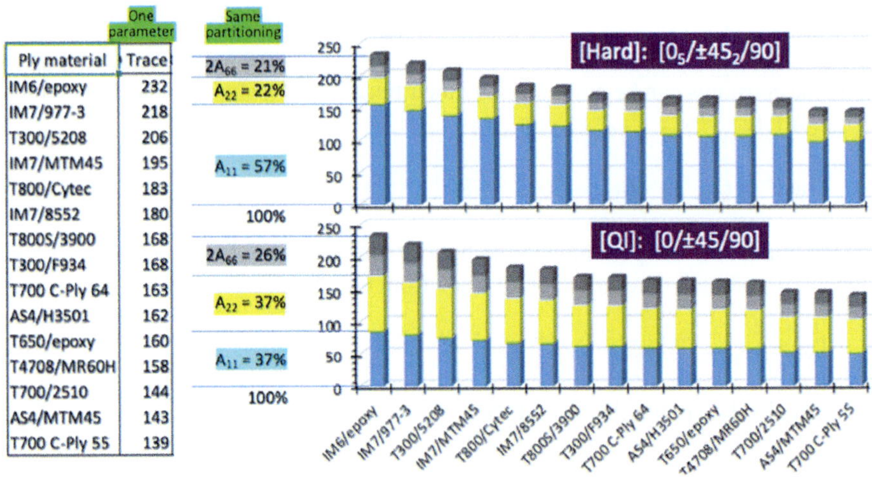

Figure 4.1.2 Trace and two laminates stiffness as partitioned traces (Tsai, 2021b).

Fig. 4.1.2 where the material part in trace is in the column on the left; and the partitioning of trace for two laminates—hard $[0_5/\pm45_2/90]$ and quasi-isotropic $[0/\pm45/90]$—are shown in the middle column. The stiffness of each CFRP is simply the product of trace and the partitioning. Trace determines the magnitude of the stiffness of each material, and the partitioning of trace for each laminate remains the same for all materials.

We have found one stiffness parameter for all CFRP materials. It is easy to rate their relative stiffness. IM6/epoxy is 232 and IM7/977, 218 GPa. IM6 is 232/218 = 1.13, or 13% stiffer than IM7. It is 13% stiffer for every laminate, not just the two laminates shown in this figure.

Conversely, if Hard laminate has A_{11} at 57% of trace and QI at 37%, their ratio 57/37 = 1.8 or 80% stiffer. That ratio remains the same for every material. Laminates like materials have fixed relative stiffnesses. Relative stiffnesses between any two laminates remain the same for all materials. Such laminate rating will be covered later in this paper.

Testing need not be applied to combinations of every material and every laminate. For stiffness of CFRP laminates, we only need to know their trace. If we trust the rule-of-mixtures equation of $E_x = v_f E_f$, we can trust Eq. (4.1.4) to find $Trace = E_x/0.88 = v_f E_f/0.88$. Tests may be waived for preliminary design. The intent here is to show that with the right parameter material and laminate stiffness can be much simpler than what current practice has presumed. Tests of materials and laminates are for confirmation and calibration, not exercises with no expectation.

4. One parameter for strength

It is the intent of this section to show that a similar one-parameter for the strength of CFRP and laminates can be found. While the detail of this single parameter and master failure envelops will be shown in detail later, it is important to show how a single parameter for strength can simplify design and testing. It also increases understanding and confidence in the use of composite materials.

Trace as a single parameter for laminate stiffness is exact while failure envelopes are curve fitting of uniaxial tensile and compressive strength data to include failure in combined strain and stress space. They are not based on any first principles, nor physics. They are needed for design and testing. There are dozens of failure criteria proposed through the years. Some are better than others. We hope to select the simplest so design and testing can be more easily accomplished.

With a single parameter we can rate the strength of materials and laminates. That was done with stiffness with trace. Our goal now is to show this parameter, with detail discussions to be covered later. The parameter that seems to work best is the enclosed area of the failure envelope, or the average strength of a failure envelope in stress space. The radius R of a circle with the same area as the failure envelope turns out to be a good single parameter for strength. This is shown in Fig. 4.1.3 where the last-failure-envelope (LPF) in blue is matched with a red circle having the same area with radius R. The centres of the circle for the two DD in [75/15] and [45/0] are displaced by Δ as indicated by red crosses, along the angle of inclination of 45 and 15°, respectively. The curves on the right are the rectilinear plot of the polar plots on the left.

With this radius of the same area as the LPF envelope, we can use this R factor to rate materials and laminates as we can see in Fig. 4.1.4 for 10- and 8-ply Quad

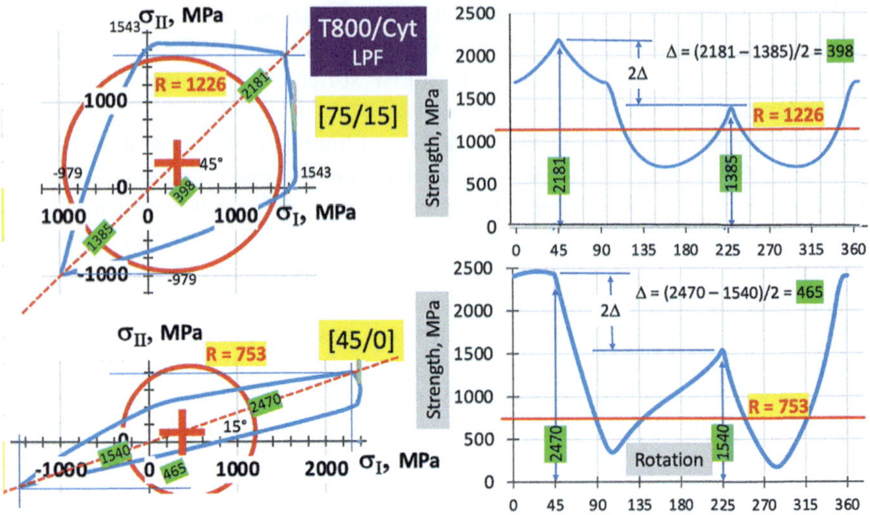

Figure 4.1.3 Polar and rectilinear plots of failure envelopes of two T800/Cyt laminates (Tsai et al., 2023).

laminates. For each envelope, its value of radius R, in MPa, is shown in red numerals. The rating of materials and laminates is shown on the left; while the selected failure envelopes of three materials and four laminates are shown on the right. Note that the size of envelopes among materials follows their rating; that is, T3/F9 is 63% weaker than T4708, and T800/Cyt is 25% stronger than T4708. Those ratios remain the same for all laminates. Conversely, the relative strengths between two laminates remain the same for all materials although their shapes can be totally different. Testing of materials and laminates need not be done for all combinations, say, among 15 CFRP and 15 laminates would lead to 225 combinations. If we believe in material and laminate ratings using radius R as the single parameter, we need to test only one material/laminate combination; or a total of 15. The rest can be estimated by scaling with values shown in arrows. The [90/0] is shown in the bottom row. Strictly speaking it is not a Quad. It is the simplest laminate that can be tested because its Poison's ratio is near zero. The mechanical grip introduced minimal stress concentration, and the load required for failure is one half of the one for [0].

5. Transformation and invariants of [Q]

We now return to the stiffness of composites. It all starts with the transformation of unidirectional ply [0] expressed in the fourth power of sine and cosines. It can be replaced by multiple angles and invariants shown in Fig. 4.1.5. This is made possible from trigonometric identities shown in the lower right box. Stiffness components in [Q] are replaced by U''s with definitions shown in the lower left box. In particular three invariants in U_1, U_4, and U_5 are shown.

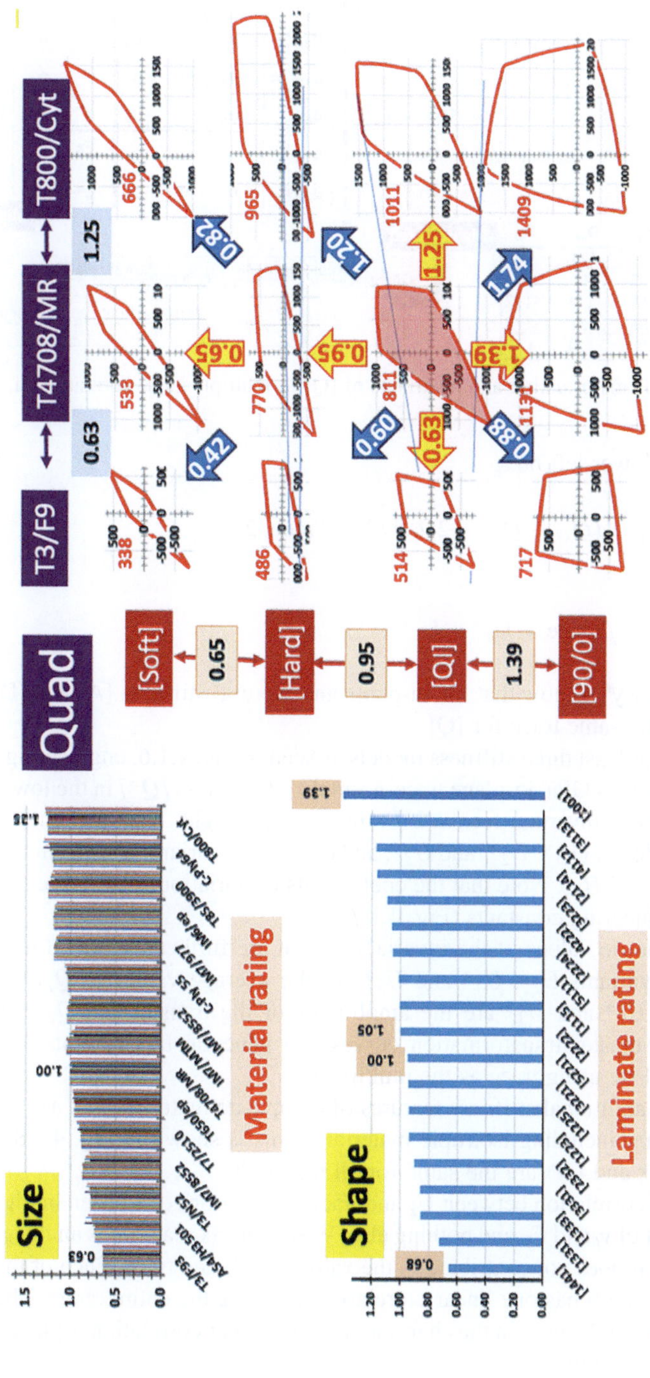

Figure 4.1.4 Rating of CFRP and quad laminates of LPF envelopes and their scaling from one test.

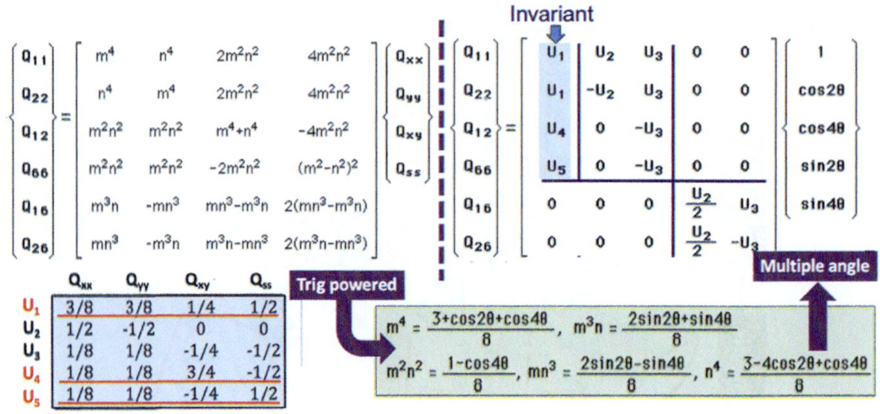

Figure 4.1.5 Transformation and invariants of [Q] in multiple angle format (Tsai, 2021b).

From the lower left box:

$$U_1 + U_5 = (Q_{xx} + Q_{yy} + 2Q_{ss})/2 = \text{Trace}/2;$$

or 4.1.6

$$(\text{U}_1 + \text{U}_5)/\text{Trace} = \text{U}_1{}^* + \text{U}_5{}^* = ½$$

It is also easy to show that the in-plane and flexural stiffness [A] and [D] are also confined to the same trace for [Q].

There are at least three stiffness models as seen in Fig. 4.1.6: engineering constants in the upper left in GPa, in-plane trace-normalized stiffness *[Q*]* in the lower left, and trace-normalized invariant format U*'s in the upper right. Note the similarity in the column profiles of Q_{xx}*, U_2*, and U_3*; and constant invariant values among materials in U_1*, U_4*, and U_5*. Note that the coefficients of variations of U's are smaller than those of engineering constants (Fig. 4.1.7).

The geometric values of the invariant format are further illustrated in Fig. 4.1.7. The three invariants U_1*, U_4* and U_5* are the average values of Q_{11}*, Q_{12}*, and Q_{66}*; while U_2* and U_3* are the amplitudes of the transformed Q_{11}*, Q_{12}*, and Q_{66}*. These master transformation curves can generate transformation curves for each material by using trace as the multiplier.

There are additional stiffness features of composite materials that are no different from any other including isotropic materials. This is shown in Fig. 4.1.8 where the values of U_1* and U_5* are the same for all materials.

There is a correlation between U_2 and trace as functions of E_x seen in Fig. 4.1.9. E_x is the product of v_f and E_f, and nothing else. While matrix material is an integral part of a composite, it does not contribute to the values of U_2 and trace. Each of the first two charts in Fig. 4.1.9 has one linear correlation covering the entire range. The ratios of their slopes are indicated. In the chart on the right, linear correlation is piece-wise; that is, applies to each material.

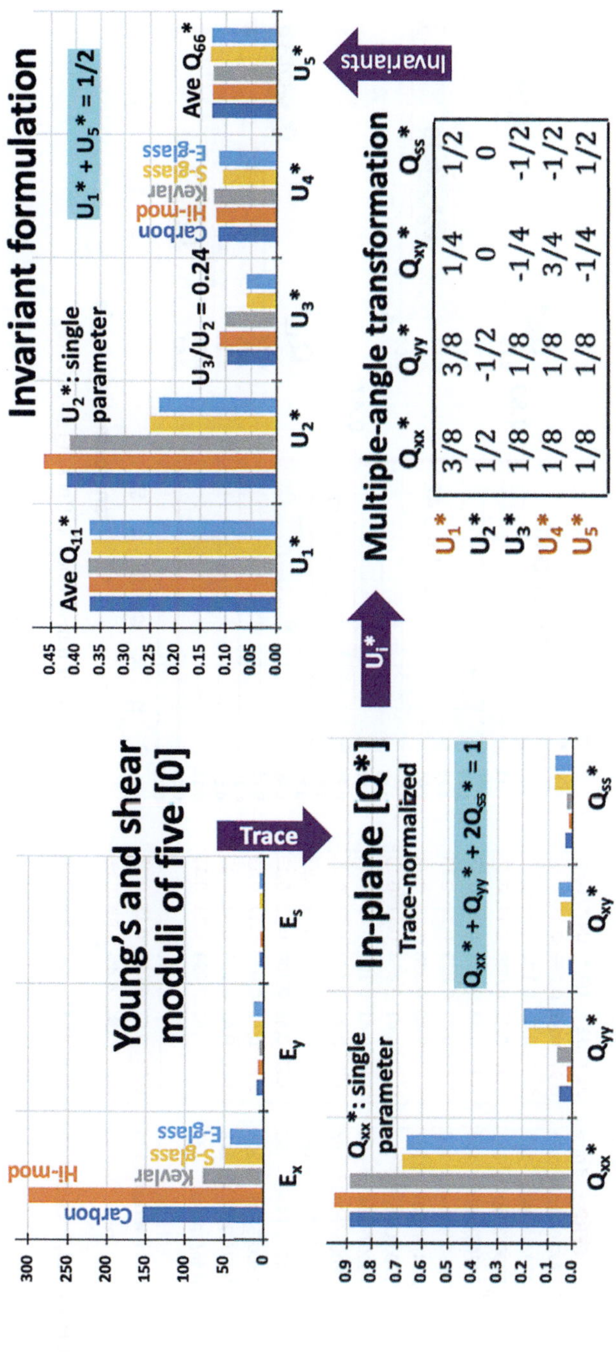

Figure 4.1.6 Three formats in elastic constants of five common composite materials [0].

CFRP	Trace	U1 *	U2 *	U3 *	U4 *	U5 *
T700 C-Ply 55	139	0.37	0.41	0.095	0.11	0.13
AS4/MTM45	143	0.37	0.42	0.102	0.12	0.13
T700/2510	144	0.37	0.41	0.099	0.12	0.13
T4708/MR60	158	0.37	0.43	0.103	0.12	0.13
T650/epoxy	160	0.37	0.41	0.095	0.11	0.13
AS4/H3501	162	0.37	0.40	0.088	0.10	0.13
T700 C-Ply 64	163	0.37	0.41	0.094	0.11	0.13
T300/F934	168	0.37	0.41	0.100	0.12	0.13
T800S/3900	168	0.37	0.43	0.103	0.12	0.13
IM7/8552	180	0.37	0.42	0.098	0.11	0.13
IM7/MTM45	195	0.37	0.43	0.100	0.11	0.13
T300/5208	206	0.37	0.42	0.095	0.11	0.13
IM7/977-3	218	0.37	0.42	0.094	0.11	0.13
IM6/epoxy	232	0.37	0.42	0.094	0.11	0.13
Average	174	0.37	0.42	0.097	0.11	0.13
cv		0.4%	2.0%	4.4%	3.9%	1.1%

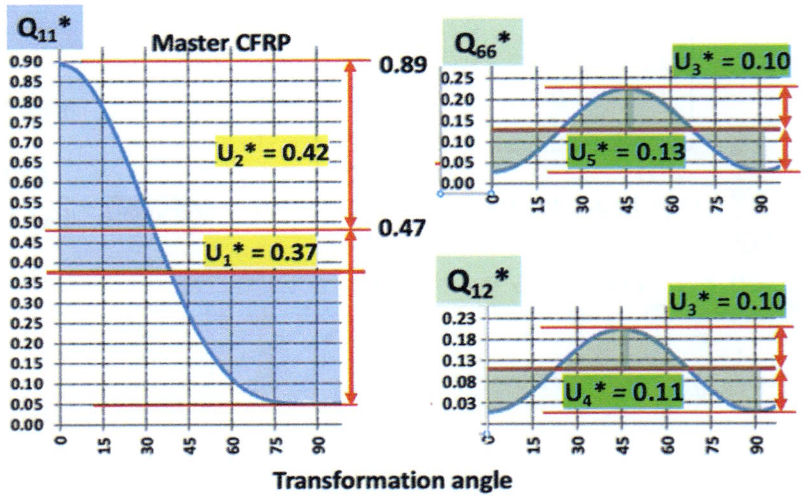

Figure 4.1.7 Geometric values of U's as the average and amplitude of transformed stiffness (Tsai, 2021a).

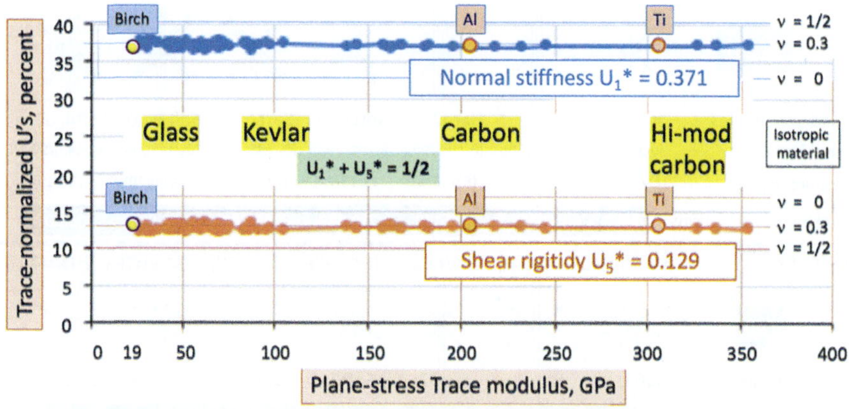

Figure 4.1.8 Universal constants in U_1* and U_5* for all materials (Tsai, 2021a).

6. Double-double (DD) laminates

As mentioned earlier that Quad laminates in 0, ±45, 90 plies were first conceived in 1960s. A new family of laminates in DD is proposed herein. Quad is a collection of laminates while DD is a field-based. The difference is fundamental, not unlike those between particles and waves, DC and AC, and analogue and digital. A comparison between Quad and DD is shown in Fig. 4.1.10 where their respective ply orientations are shown. The number of plies for Quad varies from 8 to 24 among the those shown in the top row. The matching DD (either exactly or nearly equal) has the same four plies for all laminates as seen in the lower row. A more dramatic comparison is seen in the respective stacking sequences that can be thousands for Quad to 1 recommended for DD. Quad cannot have balanced flexural stiffness (orthotropic) because +45 and −45 plies occupy difference z-axis positions and cannot cancel each other

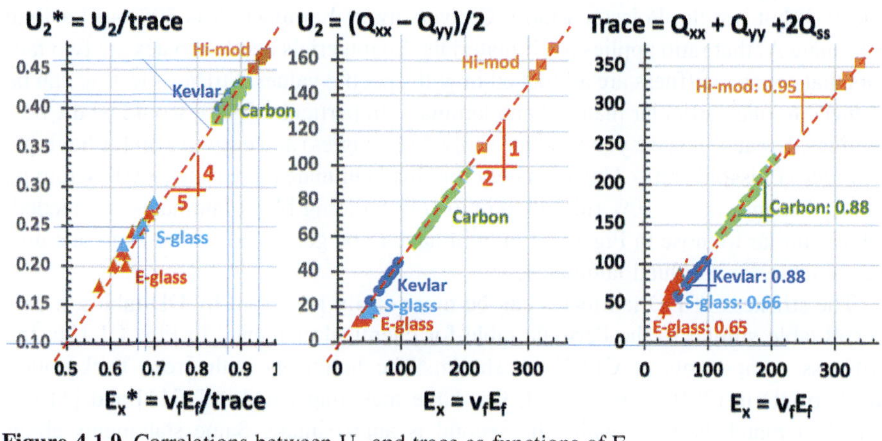

Figure 4.1.9 Correlations between U_2 and trace as functions of E_x

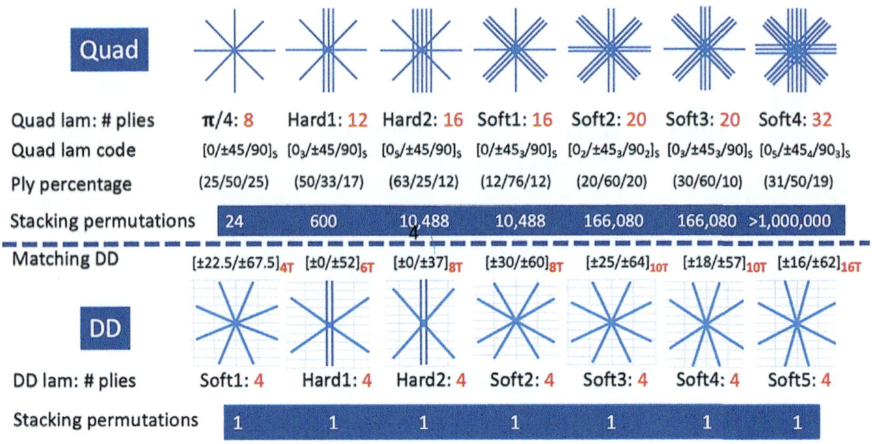

Figure 4.1.10 Visual comparison between quad and DD and their stacking permutations (Tsai, 2021b).

as they can in in-plane stiffness. The non-zero shear coupling components in 61 and 62 are sources of complexity for Quad that cannot be avoided.

There is a fundamental difference between the tabular stiffness data for Quad and DD laminates. Quad laminates depend on the number of plies in the building block. For the 8-ply family the master ply stiffness in percentage of trace is shown in a carpet plot in Fig. 4.1.17. The plot has a horizontal axis in the number of plies in [±45] in 2, 4, and 6, as represented by 11, 22, and 33. The vertical axis is represented by the number of plies in [0] in 1 through 5. So laminate in [5111] is the top left laminate. For A_{11}*, the value is 63; that is, 63% of trace. This percentage is the master value so it applies to all materials: always 63%.

The advantage of having one stiffness parameter in trace makes stiffness of materials and laminates simple. If the trace of Material A is 20% higher than Material B, this ratio applies to all laminates and not only components of [A*] but also E_1 degree and E_2°, but not the Poisson's ratio. Conversely, if Laminate A is 30% stiffer than Laminate A, that ratio applies to all materials. So materials and laminates are two factors that control stiffness are all locked in with specific values. Stiffness testing can be reduced to one particular material and laminate. In particular [90/0] would be the best laminate because it is easy to produce (only 2 ply angles) and easy to grip for having a near-zero Poisson's ratio. Unfortunately for Quad laminate there is no [90/0] within its family because of the 10% rule. If 6- or 10-ply building block laminates are needed, tables similar to those in Fig. 4.1.11a, will have to be generated. There is no one universal table for all Quad laminates.

The stiffness of DD laminates can be more easily presented by DD tables, a universal table similar to the Periodic table for chemical elements. In Fig. 4.1.11b, the stiffness components of CFRP are shown. The laminates with green background are two DD in [75/15] and [90/30] that have matching Quad in [3311] and [1133]. The first match is exact while the second is approximate. Same statements about

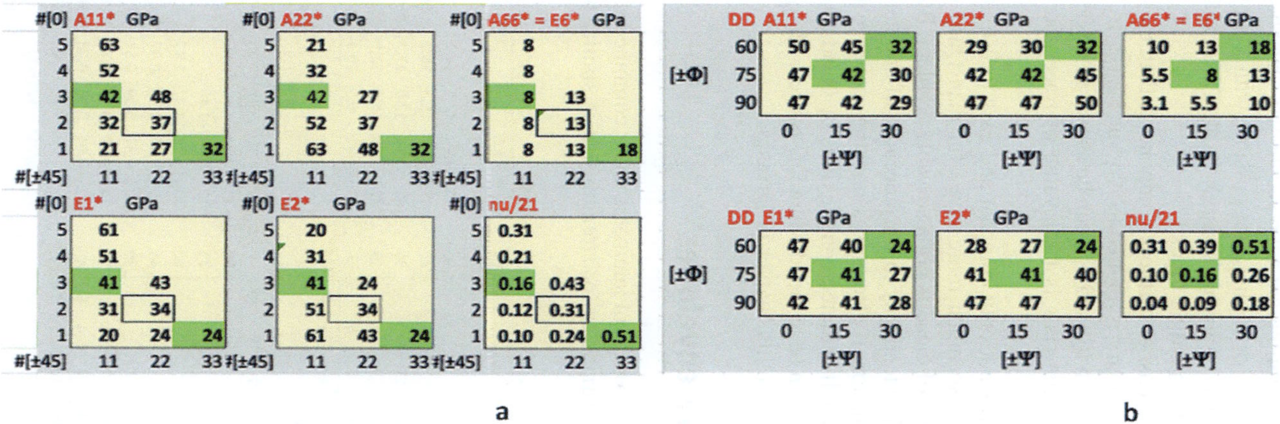

Figure 4.1.11 Carpet plots of master percentage of trace: (a) 8-ply quad CFRP laminates; (b) DD CFRP.

the scaling of materials and laminates cited for Quad would be applicable. They are equivalent. For DD that is field-based, interpolation and extrapolation among laminates are possible. Such operations are not possible with Quad because their laminates are discrete; that is, either 2 or 3 plies, not possible with 2.5 plies. The DD Tables shown in this figure are universal; that is, apply to all CFRP, unlike Quad that varies between 6-, 8-, and 10-ply building blocks.

DD tables for more extensive stiffness components than those in Fig. 4.1.11b are shown in Fig. 4.1.12. Symmetry is applied to A_{11}* and A_{22}*, E_1* and E_2*, and major and minor Poisson's ratios, in ν_{21} and ν_{12}, can be shown in single DD tables. The only difference among CFRP materials is their value of trace, the one and only single parameter.

DD data presentation is not only compact and easy to use; it shows further what field-based laminates are so much more powerful than Quad laminates. The non-balanced flexural stiffness of Quad is not a solvable problem. Buckling and vibration become more difficult. The tendency to warp because of non-zero shear coupling can be one cause for delamination. It also makes stacking sequence and blending without rational solutions. Other self-inflicted complexities by Quad rules such as mid-plane symmetry and 10% rules can be avoided when DD is used. Many other advantages of DD will be brought open as this chapter progresses.

7. Master failure envelopes

Failure envelopes define strength of laminates under combined strains or stresses. Our goal is to find a single parameter for strength that is similar to trace as the single

	A11*	0	15	30	45	60	75	90	
	0	89							90
	15	83	78						75
	30	71	66	53					60
[±Φ]	45	58	53	40	27				45
	60	50	45	32	19	11			30
	75	47	42	30	17	9	6		15
	90	47	42	29	16	8	6	5	0
		90	75	60	45	30	15	0	A22*
					[±Ψ]				

A66*	0	15	30	45	60	75	90
0	3.1						
15	5.5	8.0					
30	10.4	12.9	17.7				
45	12.9	15.3	20.2	22.6			
60	10.4	12.9	17.7	20.2	17.7		
75	5.5	8.0	12.9	15.3	12.9	8.0	
90	3.1	5.5	10.4	12.9	10.4	5.5	3.1
					A12* = A66* - 1.5		
				[±Ψ]			

	E1*	0	15	30	45	60	75	90	
	0	88							90
	15	80	71						75
	30	61	51	30					60
[±Φ]	45	50	41	22	11				45
	60	47	40	24	11	6			30
	75	47	41	27	13	7	5.4		15
	90	47	41	28	14	7	5.4	5.2	0
		90	75	60	45	30	15.0	0.0	E2*
		E1* = 1/a11*			E2* = 1/a22*				
					[±Ψ]				

nu/21*	0	15	30	45	60	75	90	
0	0.32							90
15	0.74	1.10						75
30	1.08	1.32	1.43					60
45	0.70	0.83	0.97	0.77				45
60	0.31	0.39	0.51	0.47	0.31			30
75	0.10	0.16	0.26	0.26	0.17	0.08		15
90	0.04	0.09	0.18	0.20	0.13	0.05	0.02	0
	90	75	60	45	30	15	0	nu/12*
	nu/21* = -a21/a11				nu/12* = -a12/a22			
				[±Ψ]				

Figure 4.1.12 Full-field DD tables showing all master percentage of trace in stiffness components (Tsai, 2021a).

parameter for the stiffness of laminates. As claimed earlier, failure envelopes are empirical estimates of laminate strength under combined strain or stress domains. Only uniaxial tensile and compressive strengths of balanced (orthotropic) coupons can provide the most reliable data. Combined strain or stress tests are not so easily done because the complexity in specimen configuration and dimensions, clamping and loading methods, bi-axial testing machines, and data acquisition and analysis. Ultimate failure as viewed in the last-ply-failure (LPF) is easy because coupons go from one piece to two or more pieces. First-ply-failure (FPF) cannot be tracked as easily. Our strength will be based on LPF in this work.

The first step to simplify failure envelopes of laminates is the omni envelope as seen in Fig. 4.1.13. It is the common core of all ply orientations in strain space. In the in-plane stretching of laminates, ply failures in strain space are independent of the stacking sequence of the laminate. It is, therefore, viewed as a material constant. Fig. 4.1.13 shows the LPF under in-plane loading.

Table on the left shows the controlling ply where failure occurs. They are shown with yellow background. The most important point is that LPF is controlled by either [0] or [90] plies. All intermediate ply angles do not control. In8 tensile failure, it is expected to be fibre failure. In compressive failure, it can be buckling or fibre failure. The omni domain is the most conservative envelope that covers all ply orientations. This envelope is isotropic. Then the general strain can be switched to principal strain, as shown in the lower right polar diagram.

T700/2510
E_m^* = 0.15:
Fiber dominated

Ply orientation →

LPF strain vector ↓

Angle	0	±15	±30	±45	±60	±75	90	Omni	
0	17.1	17.8	20.0	23.7	27.4	28.1	27.4	17.1	X
15	16.9	17.4	18.9	21.4	24.4	26.8	27.7	16.9	
30	17.6	17.9	18.8	20.0	21.5	22.7	23.1	17.6	
45	19.6	19.6	19.6	19.6	19.6	19.6	19.6	19.6	
60	23.1	22.7	21.5	20.0	18.8	17.9	17.6	17.6	
75	27.7	26.8	24.4	21.4	18.9	17.4	16.9	16.9	
90	27.4	28.1	27.4	23.7	20.0	17.8	17.1	17.1	X
105	20.9	23.0	27.7	26.9	22.5	19.4	18.5	18.5	
120	15.9	17.9	24.4	30.4	26.9	22.8	21.5	15.9	
135	13.2	14.7	20.6	32.1	34.3	29.4	27.4	13.2	S
150	11.9	13.1	17.8	30.4	45.1	42.7	39.9	11.9	
165	11.4	12.4	16.3	26.9	51.9	70.2	70.4	11.4	
180	11.6	12.5	15.7	23.6	43.9	85.0	111.7	11.6	X′
195	12.7	13.4	15.9	21.3	31.9	48.8	59.9	12.7	
210	15.0	15.5	17.1	20.0	24.1	28.2	30.1	15.0	
225	19.6	19.6	19.6	19.6	19.6	19.6	19.6	19.6	
240	30.1	28.2	24.1	20.0	17.1	15.5	15.0	15.0	
255	59.9	48.8	31.9	21.3	15.9	13.4	12.7	12.7	
270	111.7	85.0	43.9	23.6	15.7	12.5	11.6	11.6	X′
285	70.4	70.2	51.9	26.9	16.3	12.4	11.4	11.4	
300	39.9	42.7	45.1	30.4	17.8	13.1	11.9	11.9	
315	27.4	29.4	34.3	32.1	20.6	14.7	13.2	13.2	S′
330	21.5	22.8	26.9	30.4	24.4	17.9	15.9	15.9	
345	18.5	19.4	22.5	26.9	27.7	23.0	20.9	18.5	
360	17.1	17.8	20.0	23.7	27.4	28.1	27.4	17.1	X

Controlling plies — Omni

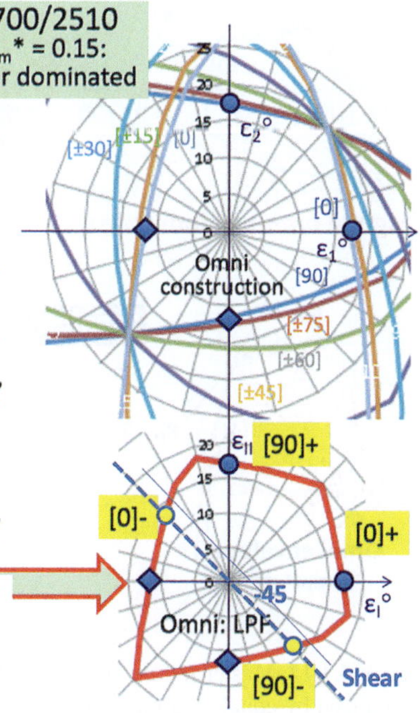

Figure 4.1.13 Omni LPF envelope as controlled by [0] and [90] plies (Tsai et al., 2023).

Since the omni inner core is isotropic, the laminate shear strength can be found by imposing a strain diagonal along −45°. They are shown as blue rim circles with yellow cores. Shear tests are difficult and their data can depend on the test fixtures. The estimated shear in this figure may be as good as what actual measurement can be.

Fourteen envelopes are shown on the top left in Fig. 4.1.14a. When each envelope is normalized by its radius R of a circle with the same area as its LPF envelope, per Fig. 4.1.3, the shapes of all envelopes become similar, as seen in the middle column. Their average is shown on the right with blue rim circles with yellow core; it is the master LPF envelope for all CFRP. This master is for [90/0].

The conversion from principal strain to principal stress envelope is straightforward. This is shown in Fig. 4.1.14b for a DD in [60/0] for T800/Cyt. The key is the degraded stiffness matrix for this material for this [60/0] laminate. The resulting envelope in stress space is shown on the right. There are four axis intercepts, designated in X, X'. Y, Y'. As mentioned previously these uniaxial tensile and compressive strengths are the only reliable and easy-to-measure data. Combined stress tests are difficult and not reliable. It will be explained later in this chapter that the more difficult tests in compression than tension are material insensitive. It will reduce the concern on the lack of reliability of compressive data.

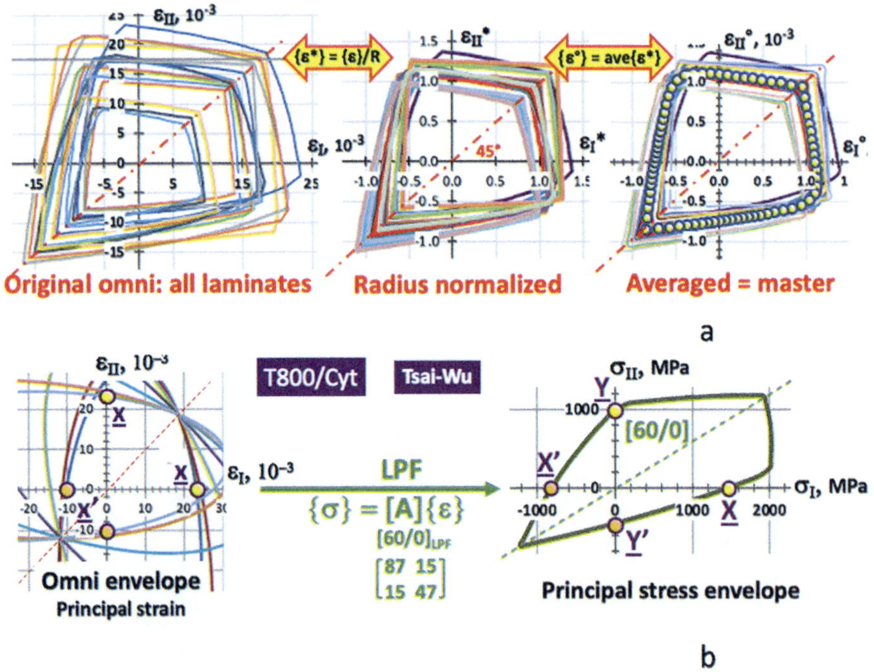

Figure 4.1.14 Master failure envelopes: (a) Finding master [90/0] from original shapes-normalized by equal-area radius R; (b) conversion of strain space envelope to stress space envelope (Tsai et al., 2023).

8. Rating of failure envelopes

Using radius R as the single parameter for failure envelopes it is straightforward to rate materials and laminates. In Fig. 4.1.15, the values of R are listed for 13 CFRP materials in rows and 16 DD laminates in columns. To rate materials, reference materials in T4708 shown in the next to last row will be used. The choice of this material is based on its closeness in value to the master material. To rate laminates, the reference laminate in [90/0]. The choice of this laminate is its highest R value and ease for testing as cited previously. Both are shown with yellow background. Cells with brown background are the chosen 3 CFRP and 3 DD laminates that will be used in an example how scaling would work to be shown in Fig. 4.1.17.

Material rating can be obtained by having the radius R normalized by the reference materials in T4708. Note the unity values for this reference material. In Fig. 4.1.16a, the average values shown in the column on the right are used to rate materials. Their coefficients of variations in the last right column are very small. The cells with brown background are those to be used in a scaling scheme to be presented in Fig. 4.1.17. They represent the master, and the lower and upper bounds.

For each material in Fig. 4.1.16a, its normalized laminate strengths are nearly constant. The material columns on the lower left have flat tops. This means that the normalized laminate strength is constant for each material. In the lower right, the laminate values are expanded in blue columns that have essentially the same value. As we will see presently, the relative strength of laminates as measured by their normalized R's apply to all materials. Such simple relations are analogous the stiffness of materials and laminates when normalized by trace.

The rating of laminates is shown in Fig. 4.1.16b. The reference laminate is [90/0] in the last column on the right. The average normalized R's for 16 laminates are shown in the bottom row. They are plotted in columns in the lower left. They all have "flat tops" meaning that materials and laminates are similarly uncoupled as material rating has claimed. The blue columns in the lower right are normalized material strengths

	CFRP	1	2	3	4	5	6	7	8	9	10	11	12	13	14	15	16		
#	Radius R	75/15	75/0	60/0	60/15	60/30	45/0	45/15	90/0	45/30	±45	30/0	30/15	±30	±15	15/0	0	Average	cv
1	S4/H3501	708	760	619	567	429	434	383	812	256	160	274	235	171	196	207	211	401	56%
2	IM6/ep	1158	1246	1016	929	693	708	623	1331	403	219	427	359	242	279	297	308	640	60%
3	T3/F93	623	670	544	497	373	377	332	717	215	121	230	194	131	155	166	171	345	60%
4	T3/N52	753	809	658	602	453	458	404	864	267	160	284	241	170	197	209	213	421	58%
5	C-Ply 64	1168	1255	1023	936	704	716	632	1341	419	249	447	381	271	315	334	342	658	57%
6	IM7/977	1155	1243	1011	924	691	702	618	1328	400	212	422	353	237	273	291	302	635	61%
7	T800/Cyt	1226	1319	1076	984	735	753	663	1409	430	239	461	390	266	312	331	343	684	59%
8	M7/8552	849	913	742	678	505	513	450	977	288	149	304	253	163	191	206	214	462	62%
9	M7/MTM	1021	1098	891	814	608	614	539	1174	346	184	365	304	197	230	247	255	555	62%
10	T7/2510	861	926	754	690	517	526	464	990	304	175	326	276	193	227	241	248	482	58%
11	T650/ep	934	1004	816	746	560	567	499	1073	326	186	348	294	202	237	260	269	520	59%
12	'4708/MR	984	1058	861	787	588	598	527	1131	341	187	364	306	206	243	260	269	544	60%
13	T8S/3900	1222	1314	1066	974	729	737	648	1405	416	225	444	371	245	290	311	321	670	61%
	Average	974	1047	852	779	583	592	522	1119	339	190	361	304	207	242	258	267	540	
	cv	21%	21%	21%	21%	21%	21%	21%	21%	21%	20%	21%	21%	21%	20%	20%	21%		

Figure 4.1.15 Radius R values for 13 CFRP and 16 DD laminates (Tsai et al., 2023).

#	Mat rating	1 75/15	2 75/0	3 60/0	4 60/15	5 60/30	6 45/0	7 45/15	8 90/0	9 45/30	10 ±45	11 30/0	12 30/15	13 ±30	14 ±15	15 15/0	16 0	Average	cv
3	T3/F93	0.63	0.63	0.63	0.63	0.63	0.63	0.63	0.63	0.63	0.65	0.63	0.63	0.64	0.64	0.64	0.64	0.63	0.8%
1	S4/H3501	0.72	0.72	0.72	0.72	0.73	0.72	0.73	0.72	0.75	0.86	0.75	0.77	0.83	0.81	0.80	0.79	0.76	6.0%
4	T3/N52	0.76	0.76	0.76	0.76	0.77	0.77	0.77	0.76	0.78	0.86	0.78	0.79	0.83	0.81	0.80	0.79	0.79	3.4%
8	M7/8552	0.86	0.86	0.86	0.86	0.86	0.86	0.85	0.86	0.85	0.80	0.84	0.83	0.79	0.79	0.79	0.80	0.83	3.8%
10	T7/2510	0.88	0.88	0.88	0.88	0.88	0.88	0.88	0.87	0.89	0.94	0.90	0.90	0.93	0.93	0.93	0.92	0.90	2.7%
11	T650/ep	0.95	0.95	0.95	0.95	0.95	0.95	0.95	0.95	0.96	1.00	0.96	0.96	0.98	0.97	1.00	1.00	0.96	2.1%
12	'4708/MR	1.00	1.00	1.00	1.00	1.00	1.00	1.00	1.00	1.00	1.00	1.00	1.00	1.00	1.00	1.00	1.00	1.00	0.0%
9	M7/MTM	1.04	1.04	1.03	1.03	1.03	1.03	1.02	1.04	1.01	0.98	1.00	0.99	0.95	0.94	0.95	0.95	1.00	3.6%
6	IM7/977	1.17	1.17	1.17	1.17	1.17	1.17	1.17	1.17	1.17	1.14	1.16	1.16	1.15	1.12	1.12	1.12	1.16	1.8%
2	IM6/ep	1.18	1.18	1.18	1.18	1.18	1.18	1.18	1.18	1.18	1.18	1.17	1.17	1.17	1.14	1.14	1.15	1.17	1.2%
13	T8S/3900	1.24	1.24	1.24	1.24	1.24	1.23	1.23	1.24	1.22	1.21	1.22	1.21	1.19	1.19	1.20	1.20	1.22	1.6%
5	C-Ply 64	1.19	1.19	1.19	1.19	1.20	1.20	1.20	1.19	1.23	1.33	1.23	1.24	1.31	1.29	1.28	1.27	1.23	4.1%
7	T800/Cyt	1.25	1.25	1.25	1.25	1.25	1.26	1.26	1.25	1.26	1.28	1.27	1.28	1.29	1.28	1.27	1.28	1.26	1.2%

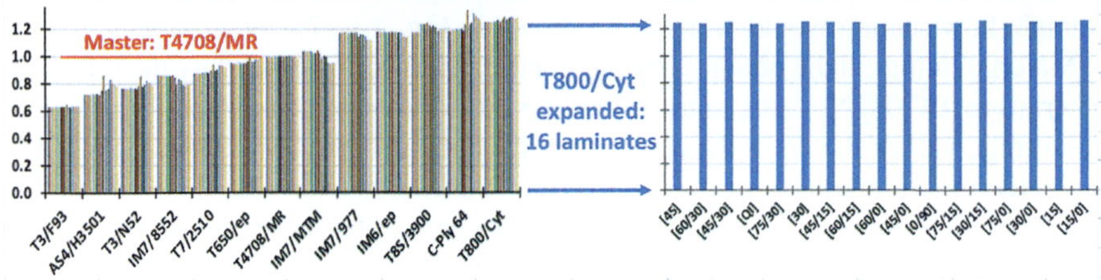

a

#	Lam rating	10 ±45	13 ±30	14 ±15	15 15/0	16 0	12 30/15	9 45/30	11 30/0	7 45/15	5 60/30	6 45/0	4 60/15	3 60/0	1 75/15	2 75/0	8 90/0
1	AS4/H3501	0.20	0.21	0.24	0.26	0.26	0.29	0.32	0.34	0.47	0.53	0.53	0.70	0.76	0.87	0.94	1.00
2	IM6/ep	0.16	0.18	0.21	0.22	0.23	0.27	0.30	0.32	0.47	0.52	0.53	0.70	0.76	0.87	0.94	1.00
3	T3/F93	0.17	0.18	0.22	0.23	0.24	0.27	0.30	0.32	0.46	0.52	0.53	0.69	0.76	0.87	0.94	1.00
4	T3/N52	0.18	0.20	0.23	0.24	0.25	0.28	0.31	0.33	0.47	0.52	0.53	0.70	0.76	0.87	0.94	1.00
5	C-Ply 64	0.19	0.20	0.23	0.25	0.26	0.28	0.31	0.33	0.47	0.53	0.53	0.70	0.76	0.87	0.94	1.00
6	IM7/977	0.16	0.18	0.21	0.22	0.23	0.27	0.30	0.32	0.47	0.52	0.53	0.70	0.76	0.87	0.94	1.00
7	T800/Cyt	0.17	0.19	0.22	0.23	0.24	0.28	0.30	0.33	0.47	0.52	0.53	0.70	0.76	0.87	0.94	1.00
8	IM7/8552	0.15	0.17	0.20	0.21	0.22	0.26	0.30	0.31	0.46	0.52	0.52	0.69	0.76	0.87	0.94	1.00
9	IM7/MTM	0.16	0.17	0.20	0.21	0.22	0.26	0.29	0.31	0.46	0.52	0.52	0.69	0.76	0.87	0.93	1.00
10	T7/2510	0.18	0.19	0.23	0.24	0.25	0.28	0.31	0.33	0.47	0.52	0.53	0.70	0.76	0.87	0.94	1.00
11	T650/ep	0.17	0.19	0.22	0.24	0.25	0.27	0.30	0.32	0.47	0.52	0.53	0.70	0.76	0.87	0.94	1.00
12	T4708/MR	0.16	0.18	0.22	0.23	0.24	0.27	0.30	0.32	0.47	0.52	0.53	0.70	0.76	0.87	0.94	1.00
13	T8S/3900	0.16	0.17	0.21	0.22	0.23	0.26	0.30	0.32	0.46	0.52	0.52	0.69	0.76	0.87	0.94	1.00
	Average	0.17	0.19	0.22	0.23	0.24	0.27	0.30	0.32	0.47	0.52	0.53	0.70	0.76	0.87	0.94	1.00
	cv	7.6%	7.0%	6.5%	6.2%	5.7%	3.4%	2.1%	2.5%	0.9%	0.6%	0.7%	0.3%	0.2%	0.1%	0.0%	0.0%

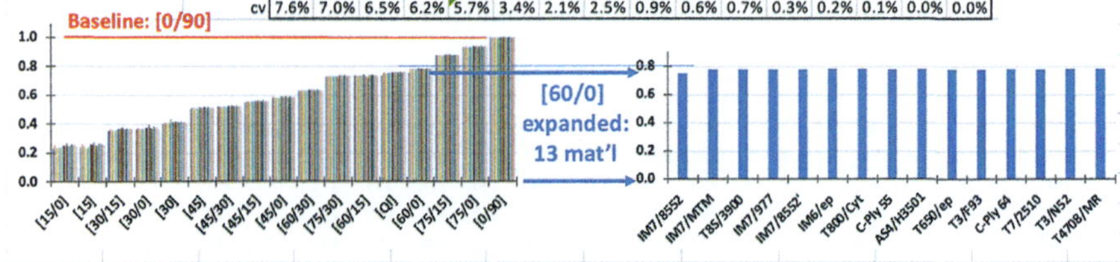

b

Figure 4.1.16 CFRP rating with T4708 as the master material: (a) materials, (b) DD laminates (Tsai et al., 2023).

that are constant for each laminate. The cells with brown background are shown in Fig. 4.1.17 as an example of scaling among materials and laminates. They represent almost the upper and lower bounds and the master of CFRP.

Material and laminate ratings enable the direct scaling of their failure envelopes. In Fig. 4.1.17a, 3 materials with 3 laminates each can have their envelopes in 9

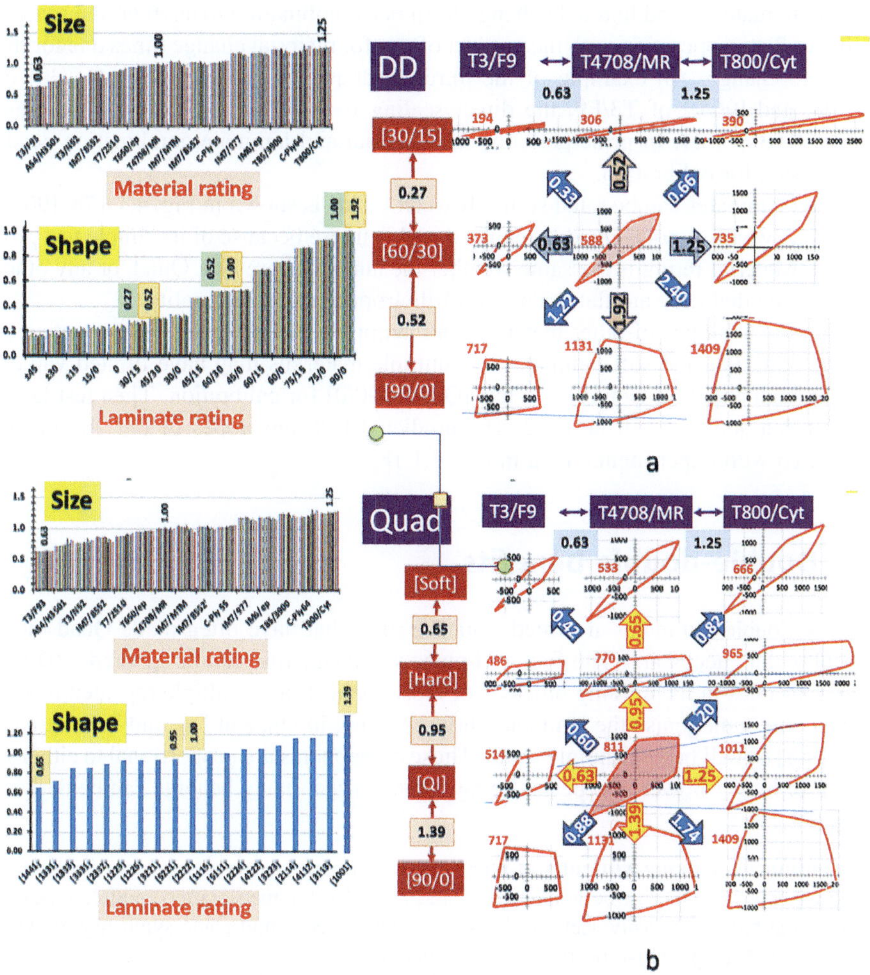

Figure 4.1.17 Materials and laminates rating that facilitate scaling for failure envelopes: (a) DD, (b) quad.

combinations. Material rating specifies the size of the envelopes. Their shapes remain similar but size changes in accordance to the rating of materials. If the rating between two materials is by a factor of 2, it means that the size of one is twice as large as the other. That ratio applies to all laminates between these two materials.

Laminate rating specifies the shapes of envelopes. In Fig. 4.1.17a, [90/0] laminate that is hard is squarish in its shape. For soft laminates such as [30/15] in the top row, the shape is elongated. But their difference can be tracked by radius R as predicted by the laminate rating. Material and laminate rating are independent, uncoupled. The shape change as measured by radius R is applied to all materials. If one laminate has an R value twice as large as another, the ratio remains constant for all materials. Direct scaling is applied directly and their values match what their respect rating are.

If both material and laminate change from one combination to another, the scaling will be a 2-step operation with the product of one for material change times one for the laminate change. An example of the third quadrant laminates between [60/30] of T4708, and [90/0] of T3/F9, the direct scaling is 717/588 = 1.22; 2-step scaling, 1.92 x 0.63 = 1.21. In principle only one material and laminate combination needs to be tested for calibration.

Quad laminates can be rated similarly to DD. This is shown in Fig. 4.1.17b. [90/0] laminates which is not a Quad Is included in this figure because of its simple test procedure. Material rating is the same whether the laminates are DD, Quad, or any other design provided they are flat, balanced (orthotropic) and good quality.

Failure envelopes of other materials and laminates can be scaled. The practice of testing hard, QI, and soft laminates for multiple materials may have limited values. Our recommendation would be to test [QI] or [90/0] for calibration. Then test laminate(s) that actually will be used. The predicted LPF envelopes in Fig. 4.1.17 are compared with experimental data in Fig. 4.1.18.

9. Double-double benefits

Double-double laminates are freed from the rules that have often made Quad more complex than necessary. The first mid-plane symmetry rule can be ignored. DD has only four plies in its building blocks. For most applications, multiple repeated blocks of DD will homogenise the laminate, shear coupling in D_{16} and D_{26} and bend-stretch coupling [B] will be vanishingly small. Those components are proportional to either 1/repeat or higher power in repeats. Being homogenised, many benefits can be derived (Fig. 4.1.19):

1. **Continuous stacking without mid-plane stoppage** is much easier and faster to do, and less prone to error. DD can be homogenised and naturally symmetric after multiple repeated building blocks as easily seen in Fig. 4.1.19a. Quad needs mid-plane symmetry to avoid warping. A 10-ply Quad becomes a 20-ply laminate.
2. **Tapering:** With multiple repeats, the resulting homogeneous DD laminates can be tapered to save weight (41.5%) and cost. Quad laminates are heterogeneous, and cannot be tapered without complex change in laminate properties.
3. **Ply drops for homogenised DD can be in singles and located on the tool or bag si**de leaving the laminate interior without discontinuities and wrinkles. Quad on the other hand must have drops in pairs with mid-plane symmetry, and embedded in the interior of laminates. The resulting interior difference between the two is shown in Fig. 4.1.19b.
4. **One-axis layup** (without cross-plying) of rectangular shapes with various length-to-width ratios are shown in Fig. 4.1.19c and d. It can increase the tape deposition rate 6-fold over a 4-axis layup because the slowdown layup rate at the corners of the rectangle can be avoided. For DD pre-plied tape can best be made with thin plies so the total thickness of 4-ply DD is comparable with regular plies at 125 microns. Such tape thickness will be suitable for ply drops as well. Quad laminates have building blocks too thick (8- to 10-ply thick) even for thin plies. Cannot have 1-axis layup because mid-plane symmetry is required. Ply drops will also be too thick. So will the minimum gage.

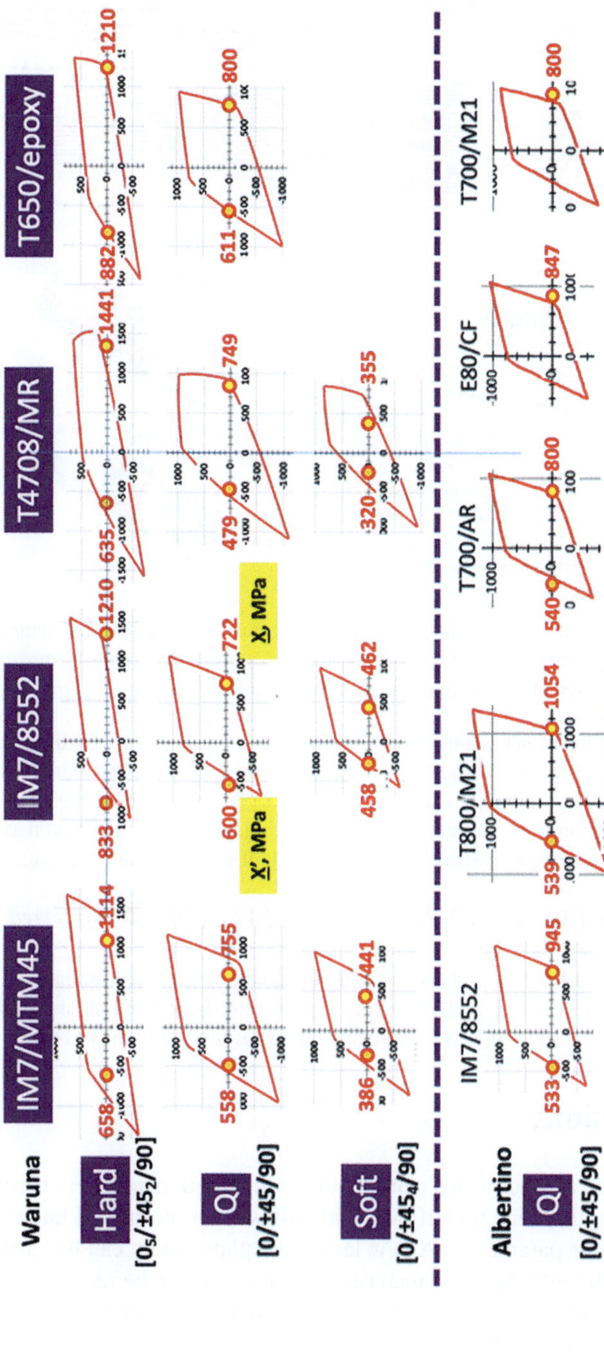

Figure 4.1.18 Comparison between prediction in red lines and data in circle with yellow core. The data provided by Waruna Sereviratne of NIAR (National Institute of Aerospace research, Wichita State University), and Albertino Alteiro of the University of Porto.

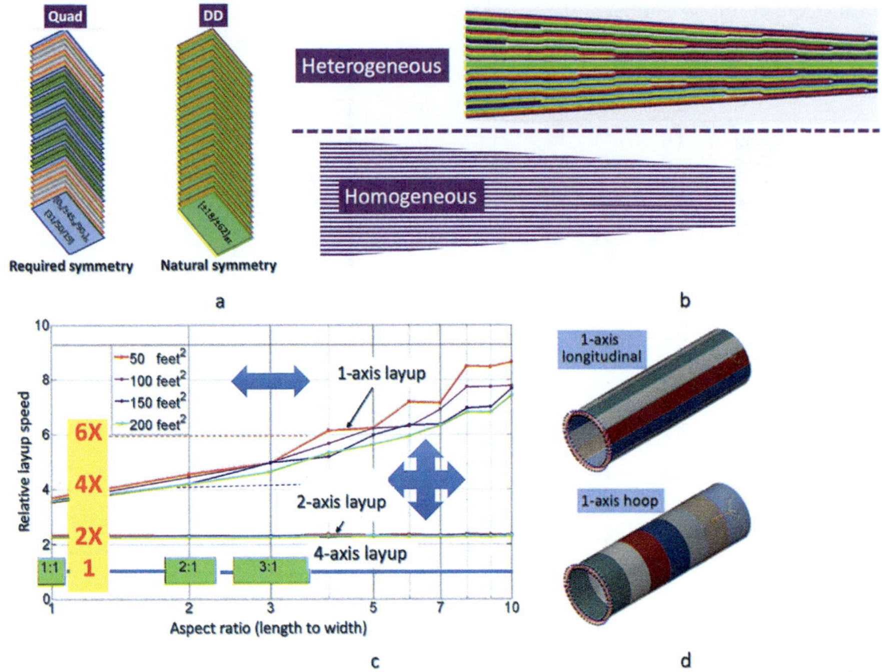

Figure 4.1.19 Double-double benefits: (a) Required versus natural mid-plane symmetries; (b) ply drops in heterogeneous quad and homogenised DD; (c) one-axis layup as compared with 2- and 4-axis; (d) examples of 1-axis layup in all-longitudinal and all-hoop layup (Tsai, 2021a).

5. **Balanced laminates:** Quad cannot have balanced flexural stiffness [D] because +45 and −45 plies cannot cancel each other because they occupy in different position in the laminate. It is an unavoidable complexity and can contribute to delamination failure. There are ways that can reduce or eliminate the shear coupling, making laminates with flexural as well as in-plane balanced (orthotropy). Having repeated DD will reduce shear coupling:

   Stagger 1 in [Φ/−Ψ/−Φ/Ψ]; Stagger 3 in [Φ/−Ψ/Ψ/−Φ] preferred

   Another way to have zero shear coupling is square symmetric laminates, rotated 45°. Rotation of [90/0] to [±45], and vice versa to eliminate shear coupling are obvious. Rotations of [75/15] and [60/30] to reach balance laminates are not intuitive but it true.

10. Conclusions

Rating of materials and laminates can be done with single parameters based on their stiffness and strength. Trace for stiffness and radius R for the area of failure envelope are the critical single parameters. A new laminated plate theory can be based on master stiffness and strength. Specific material and laminate can be recovered from their respective ratings. Property tests are for confirmation. Blind testing with no expectation of the result should see its end.

Homogenisation is the most effective key to simplify design and production. Tapering, ply drops in singles, 1-axis layup, and pre-plied tapes, not possible with Quad, are now feasible. Homogenised laminates are fully balanced or orthotropic; that is, [A*] = [D*]; [B] = 0. SCF of open hole and possibly other stress raisers and impact damage are insensitive to materials. Design and testing can be significantly simplified when the uncoupled material and laminate properties are found and rated.

DD is field-based and offers compact data tables that can zoom for better properties and higher resolution. Quad on the other hand cannot match the versatility and ease of use of DD. Complexity in stacking sequence and blending make optimisation of Quad laminates difficult if not impossible. DD can offer laminate with innovative performances and applications not possible with Quad. It is time to make a choice of DD over Quad, not unlike those historical choices between AC over DC, digital over analogue, and wave over particle. The train is leaving the station. Hope many old and young engineers will get on board.

Nomenclature

[A], [A*]	Laminate in-plane stiffness, absolute GPa, trace-normalized
[B], [B*]	Laminate coupling stiffness, absolute GPa, trace-normalized
[D], [D*]	Laminate flexural stiffness, absolute GPa, trace-normalized
[Q], [Q*]	Ply plane-stress stiffness, absolute GPa, trace-normalized
[QI]	Quasi-isotropic laminates
cv	Coefficient of variation
DD	Double-double laminates in [±Φ/±Ψ]
Quad	Quad laminates in 0, ±45, 90° plies
R, R*, R°	Radius of circle with same area as LPF envelope: MPa, normalized, master
S, S'	Positive and negative shear strengths of laminates, MPa
SCF	Stress concentration factor
Trace	Trace or also known as Tsai's modulus; $Q_{xx} + Q_{yy} + 2Q_{SS}$ $A_{11} + A_{22} + 2A_{66}$
U_i, U_i*	Invariant ply stiffness, absolute GPa and trace-normalized
X, X′	Uniaxial longitudinal ply tensile and compressive strength, MPa
x, x′	Uniaxial ply longitudinal tensile and compressive strain, 10-3
X, X'	Uniaxial laminate longitudinal tensile and compressive strength, MPa
Y, Y′	Uniaxial ply transverse tensile and compressive strength, MPa
Y, Y'	Uniaxial laminate transverse tensile and compressive strength, MPa
Δ	Position of centre of circle with equal area as LPF failure envelope
{σ}, {σ°}	Stress, in-plane components, MPa
{ε}, {ε°}	Strain, in-plane strain components, 10^{-3}

Acknowledgement

The author wishes to acknowledge help from many of his global research team. They include Waruna Sereviratne of the University of Wichita State University; Naresh Sharma of Nashero; Albertino Arteiro of the University of Porto; Daniel Melo Universidade Federal do Rio Grande do Norte (Natal, Brazil); Carlos Cimini, DEMEC-UFMG (Belo Horizonte, Brazil); Dan Wang, High Performance Computing, Singapore; Eric Kappel, German Space and Aeronautics Centre,

Braunsweig; Antonio Miravete, formerly University of Zaragoza, and MIT; Bruno Vermes, formerly Budapest Institute of Technology and Economics; Ignaas Verpoest, professor emeritus, KU Leuven.

References

Dong, S.B., Pister, K.S., Taylor, R.L., 1962. On the theory of laminated anisotropic shells and plates. J. Aero. Sci. 29 (8), 969–975. https://doi.org/10.2514/8.9668.

Herakovich, C.T., 2012. Mechanics of composites: a historical review. Mech. Res. Commun. 41, 1–20. https://doi.org/10.1016/j.mechrescom.2012.01.006.

Kappel, E., Tsai, S.W., 2024. Double-Double Laminates - A Review on the Status Quo in Spring 2024. https://doi.org/10.57965/g2cb-8t21.

Lekhnitskii, S.G., 1947. Anisotropic Plates. OGIZ (ОГИЗ), Moscow - Leningrad, p. 489 (in Russian).

Lekhnitskii, S.G., Tsai, S.W., Cheron, T. 1968. Anisotropic Plates, English translation by. Gordon and Breach. 534.

Timoshenko, S.P., 1953. History of Strength of Materials. McGraw-Hill, p. 480.

Tsai, S.W., 2019. Composite Double-Double and Grid/Skin Structures Low Weight/Low Cost Design and Manufacturing. Stanford University, Stanford.

Tsai, S.W., 2021a. Double–double: new family of composite laminates. AIAA J. 59 (11), 4293–4305. https://doi.org/10.2514/1.J060659.

Tsai, S.W., 2021b. Double-double a New Perspective in the Manufacture and Design of Composites. Stanford University, Stanford, p. 300.

Tsai, S.W., Falzon, B.G., Aravand, A., 2023. Double–Double: Simplifying the Design and Manufacture of Composite Laminates. Stanford University - RMIT University - Belfast University, Belfast, p. 398.

No composites without computers: The key role of software tools

4.2

Anthony Pickett
IFB (Institute of Aircraft Design) University of Stuttgart, Stuttgart, Germany

1. Introduction

Software tools to computational model composites usually fall into one of two types that are either based on an analytical approach or a finite element solution.

Since development of the first FRP composites researchers have proposed many analytical laws based on micro-mechanics that yield simple formulae to characterise mechanical or some other composite constitutive behaviour. In some cases, these solutions have been implemented in analytical software tools to characterise a textile representative volume element (RVE) for properties like stiffness or permeability. From the 1960s one essential analytical tool has been laminate analysis, based on plate theory and classical laminate theory (CLT) which can predict the load-deformation behaviour of a multi-layered laminate composite. For manufacturing some useful analytical based solutions have also been proposed concerning mapping of fabrics for shear deformations and prediction of yarn paths in braiding type processes.

Inevitably analytical approaches suffer limitations, usually related to the assumed geometry and size of composite that can be tackled or the physics represented. FE analysis provides a general-purpose analysis method that can, potentially, overcome these limitations. This technique was first developed in the mid-1950s and applied to composites since the start of the 1960's. Today, most commercial FE packages allow linear composites stress or field analysis. Further, dedicated FE codes have also been developed for resin infusion analysis. And in the past 20 years, more sophisticated explicit FE simulation codes have appeared involving multi-physics to solve a variety of complex composites manufacturing processes such as braiding, draping and forming, and mechanical analysis for problems like impact and crash.

In recent years coupling of software tools has made it possible, in certain cases, to chain stages in composites manufacturing with mechanical performance analysis. For example, effects like fabric shear from draping can be taken into account to modify permeability distribution for an infusion analysis, or modify stiffness for a mechanical analysis. However, this remains a challenging topic that still requires further research and development.

The History of Composites. https://doi.org/10.1016/B978-0-443-34119-9.00025-5

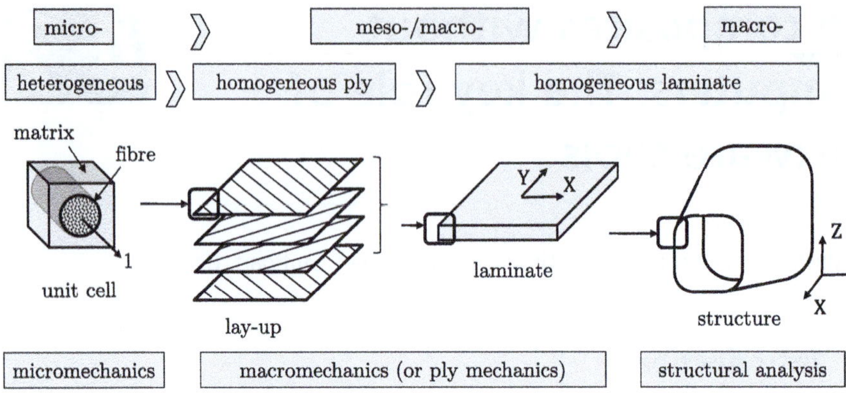

Figure 4.2.1 Length scales used in composites analysis tools.

1.1 Length scales in analysis

Different length scales covering micro-, meso- and macro-, as illustrated in Fig. 4.2.1 can be used to analyse fibre-reinforced composites depending on information available and purpose of the analysis. For instance, micromechanics treats the material as heterogeneous at the individual fibre and matrix scale with typical dimensions $5 \rightarrow \approx 20$ μm, whereas a macromechanics analysis will use a length scale of millimetres for the composite, with plies approximated as an orthotropic homogeneous continua. Fig. 4.2.1 also demonstrates the potential to transfer information from one scale to the next, so properties from a micromechanics analysis could provide ply properties for a macromechanics laminate analysis, which in turn could provide input for a structural applications. This approach is sometimes termed "multi-scale modelling." Also shown in this figure is a unit cell (UC) used in a micromechanics analysis to obtain composite properties, which is the smallest repeating representation of the fibre-matrix composite. For a repeating geometric volume in a textile it is usually referred to as a repetitive volume element (RVE).

1.2 Timeline of composites developments and their analysis tools

Engineering composites first appeared in the mid-1930s by combining glass fibres with polyester resin. Since then new grades of resins, fibre types and fabric forms have been continuously developed, that have been successfully converted to parts using a wide variety of manufacturing processes. Fig. 4.2.2 highlights some important stages in composites development and indicative times, at which different industries have adopted these materials and processes. Also shown are key types of analysis and a guide to their time of introduction.

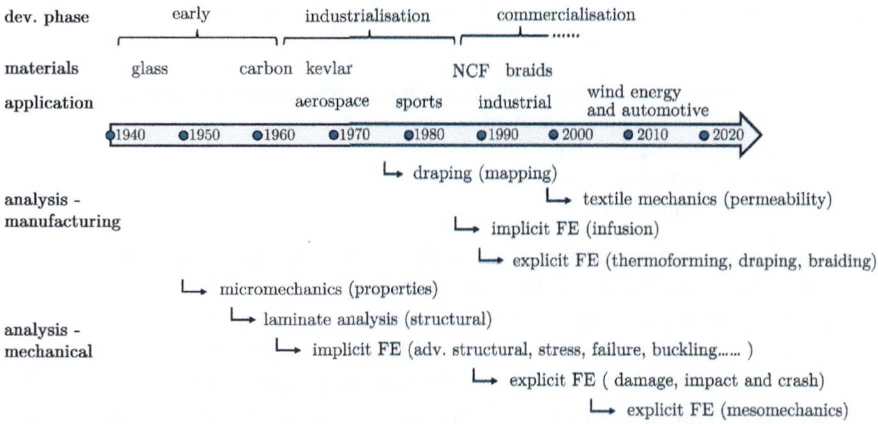

Figure 4.2.2 Timeline of some key composites developments and applications, and their interaction with new analysis tools.

2. Analytical tools for processing

For composites manufacture numerous analytical models have been proposed based on a closed form solution of an RVE. Examples include infusion and consolidation of yarns, and consolidation of plies. These models were largely developed in the 1980s due to great interest at the time in thermoplastics for fast manufacturing. In addition, several analytical models have been proposed to help characterise deformation of yarns in dry fabrics. All these models can be helpful to supplement test measurements, or might be used as a post-processing option, for example, Fig. 4.2.16 uses the consolidation model (Lee and Springer 1987) in an FE code to compute and visualise inter-ply consolidation for a stack of thermoplastic sheets shaped at high temperature.

Concerning yarn processing the key areas are draping, braiding and filament winding. For draping algorithms can predict deformation to identify if a fabric can be shaped over a given geometry without excessive shear. For braiding and winding it is the angles and spacing that are of most interest. These simple tools can help assess feasibility of a manufacturing process and can provide information like yarn angles for use in other analyses such as a CLT or FE stiffness analysis.

2.1 *Fabric drape analysis*

Kinematic draping analysis for fabrics was originally proposed by Mack and Taylor 1956 in the 1950s for applications in the clothing industry. This is basically geometric mapping which appears under various other names including "fitting," "kinematic," "fish-net," "mapping," and the "pin-jointed net" method. These techniques all use a simple kinematic algorithm in which a fabric is idealised as a network of orthogonal fibres, with cross-over points acting as fixed pin-jointed nodes. The main assumptions are:

- Yarn cross-overs act as pin-joints, that is, there is no relative slip between yarns.
- Yarn segments are straight, without crimp, between pin-joints.
- Yarn segments are inextensible.
- Perfect surface contact is assumed.
- Yarn layers are infinitely thin.

To generate a deformed fibre pattern, the method requires information on two initial fibre directions L_1, L_2 of the "net," plus lengths of the arbitrary net segments a, b and location of a starting "seed" point A, as shown in Fig. 4.2.3.

For a solution an algorithm works outward from the seed point A, and uses vectors L_1, L_2 to project points on the surface for the main warp and weft fibre directions. Intermediate points are then found by solving the intersection of each pin-joint node using Eq. (4.2.1a) and enforcing that their intersect lies on the geometry surface given by Eq. (4.2.1b).

$$\begin{aligned} \left(x_{i,j} - x_{i-1,j}\right)^2 + \left(y_{i,j} - y_{i-1,j}\right)^2 + \left(z_{i,j} - z_{i-1,j}\right)^2 = a_i^2 \\ \left(x_{i,j} - x_{i,j-1}\right)^2 + \left(y_{i,j} - y_{i,j-1}\right)^2 + \left(z_{i,j} - z_{i,j-1}\right)^2 = b_j^2 \end{aligned} \qquad (4.2.1a)$$

$$F\left(x_{i,j}, y_{i,j}, z_{i,j}\right) = 0. \qquad (4.2.1b)$$

As these equations suggest, intersection points are computed between two spheres of radius a and b, centred at points $i-1, j$ and $i, j-1$ that lay on the geometry surface. This intersection can be solved directly if the surface is defined as a geometric function, such as a hemisphere of known radius. Or it must be solved numerically if an arbitrary surface has to be fitted. In commercial kinematic drape codes the arbitrary surface is often conveniently defined using a finite element mesh, where each 2D element of the mesh is treated as a facet of the surface geometry to be fitted.

For draping several stand-alone codes do exist and most CAE packages like Catia, Siemens NX and Fibresim include a drape algorithm. Fig. 4.2.4 shows an example

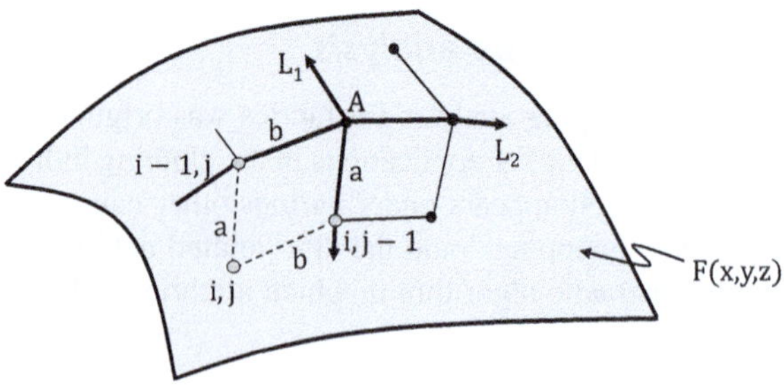

Figure 4.2.3 Geometric mapping to a surface.

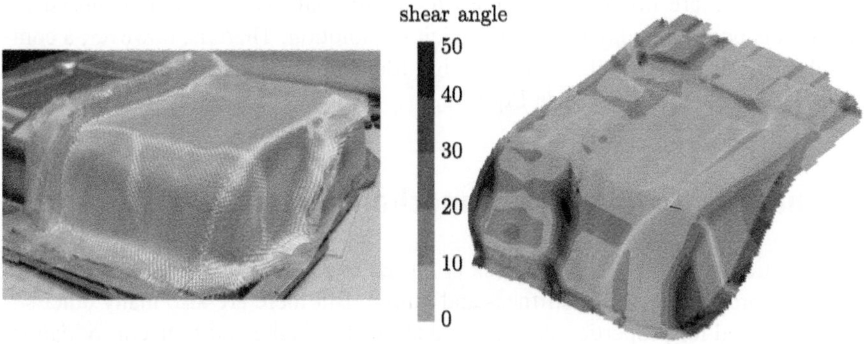

Figure 4.2.4 Example geometric drape analysis.

draping of a biaxial woven fabric over an automotive part using (PAMQUIKFORM 2000). The main limitations of these codes are the aforementioned assumptions and neglect of any fabric stiffness and other process properties such as friction or blank holders. Selection of the correct seed point A is also critical. Nevertheless, these tools are popular as a means to estimate draping, the initial cutting pattern and assess different options to position and orientate a fabric. The limitations mentioned can be overcome with an FE drape analysis.

2.1 Braiding and filament winding analysis

Since the 1980s, these two traditional manufacturing methods have become popular automated processes for certain composites applications. To aid design simple analytical formulae can describe yarn paths depending on mandrel geometry and velocities. As an example, Eq. 4.2.2 gives yarn angle θ and yarn spacing δ that are produced as N_{BX} bi-axial yarns are braided over a mandrel of diameter D at a rotational velocity ω and translational velocity v, as depicted in Fig. 4.2.5.

$$\theta = \tan^{-1}\left(\frac{D\omega}{2v}\right) \text{ and } \delta = \left(\frac{\pi D \cos\theta}{N_{BX}/2}\right) \tag{4.2.2}$$

These formulae are helpful to guide machine setup so a particular yarn architecture can be produced, possibly to help meet certain stiffness or strength criteria of the final part. However, like analytical draping analysis, these formulae ignore many important process effects such as friction, yarn stiffness, and yarn interactions. To the authors

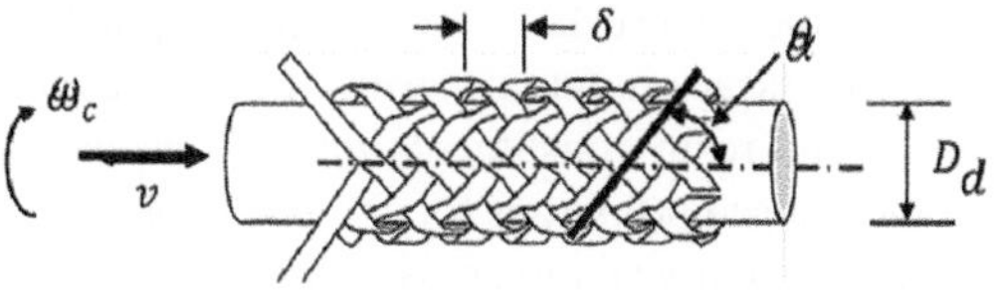

Figure 4.2.5 Notations for an analytical model used for braid yarn architecture.

knowledge there are no dedicated codes for braid analysis, although advanced FE simulation is possible today using an explicit FE solution. There is, however, a commercial package CADWIND to help design filament wound parts that is based on analytical formulae of the type in Eq. (4.2.2).

3. Analytical methods for mechanical analysis

Most analytical methods that have been proposed to calculate composite or laminate mechanical properties are for stiffness and failure. But there are also many other solutions proposed for properties such as conduction, thermal expansion, consolidation, and permeability. These can be computed at different scales depending on the information available and scale at which the composites RVE can be realistically represented. The following gives a brief introduction only for mechanical properties.

3.1 Micromechanical tools

Since widespread acceptance of fibre reinforced polymer composites in the 1950s–60s many approaches have been investigated to try and quantify their orthotropic homogeneous properties from those of the fibre and resin constituents. The starting point for these are the law of mixtures based on the general Voigt theorem (Voigt 1889) which assumes the material is in a state of isostrain. For the unit cell in Fig. 4.2.1 this theory gives $E_1 = E_f V_f + E_m(1 - V_f)$, where E_1 is homogenised stiffness in the fibre direction, E_f is axial fibre stiffness, E_m is resin stiffness and V_f is fibre volume fraction. For homogenised transverse stiffness inverse law of mixtures based on Reuss theorem (Reuss 1929), which assumes a state of isostress applies giving $1/E_2 = V_f/E_f + (1 - V_f)/E_m$. From these laws expressions for Poisson's ratio ν_{12} and shear modulus G_{12} are also possible. Extensive research in this field has led to a plethora of improved theories covering all aspects of stiffness, failure, and other composite properties. Many of these are covered in standard composites textbooks and the dedicated book (Stellbrink 1996).

Micromechanical models are often not especially accurate. Assumptions are needed to enable an analytical solution and required constituent mechanical properties can be difficult, or impossible, to measure experimentally. For failure properties identification of a representative volume to replicate true material damage and ultimate failure behaviour is impossible. Nevertheless, these models can provide useful data and a valuable insight to the mechanisms of load transfer and failure.

Standalone codes for micromechanical analysis are limited with notable ones being MAC/GMC (Arnold et al., 1999), elamX (eLamX 2021), and KU Leuven's code (Verpoest et al., 2007). Furthermore, sometimes, these laws are directly applied in commercial FE codes to help users obtain necessary input properties. In these codes stiffness and failure properties of the fibre and resin constituents must be provided with micromechanical laws providing homogenised properties for composite ply analysis. As an alternative it has been possible in recent years to apply powerful meshing and FE analysis methods to model bundles of fibre at the micro-scale for stiffness and failure properties, a good example is provided by LLorca et al. (2011).

3.2 Mesomechanical tools for textile analysis

Since the 1990s, two academic codes (WiseTex 2000) and (TexGen 2006) have been developed for textile visualisation and analysis. A primary aim of both codes is to graphically model one representative volume of any textile architecture, from which a numerical model computes composite mechanical, or some other property.

These codes approximate the yarn geometry by B-spline type mathematical expressions. Some capabilities are incorporated to try and approximate nesting, but this can only approximate the true preform compaction that occurs during manufacture. Fig. 4.2.6 illustrate some principles used by both codes to build an RVE textile architecture. The basic RVE layout of yarns in one RVE is input and a chequerboard system is used so that yarns laying under or over each other can be set. Graphical views in 3D are then shown to confirm the arrangement. Two simple examples from TexGen are shown in Fig. 4.2.7.

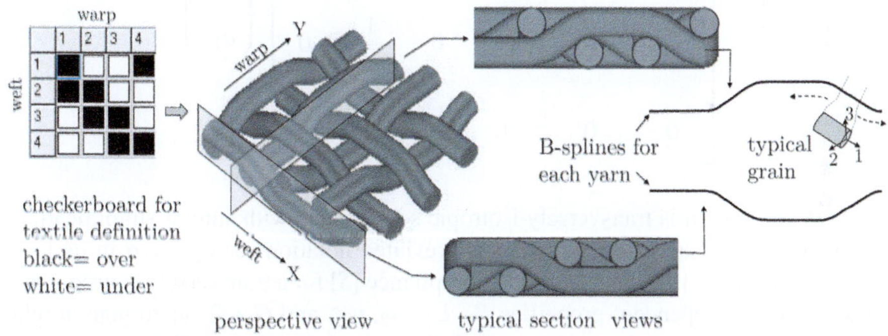

Figure 4.2.6 Example construction of a twill weave.

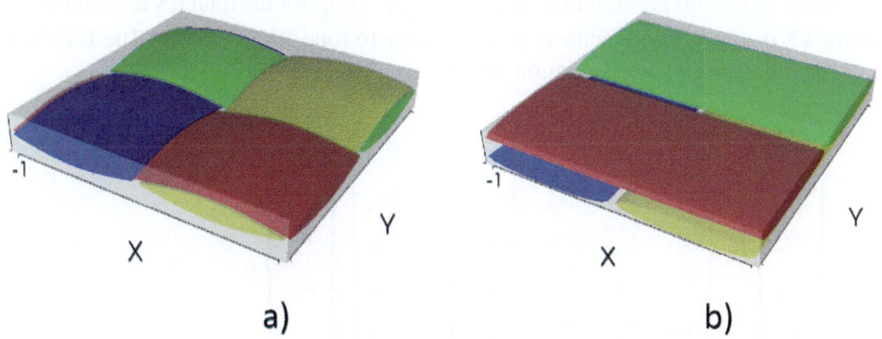

Figure 4.2.7 RVE weaves for analysis. a) Plain weave. b) NCF (or unidirectional).

From this geometric RVE model it is possible to compute mechanical properties. For instance, one method used by WiseTex is the orientation averaging method (Cox and Flanagan 1997), in which each yarn is split into segments, or grains. Either a state of isostrain so that all grains have the same strain state, or a state of isostress so all grains have a constant stress is assumed. For isostrain grain stiffnesses are summed and for isostress grain compliances (1/*E*) are summed. For example, for isostress each grain has the local principle frame compliance matrix relating strains to stresses given by

$$\begin{Bmatrix} \varepsilon_1 \\ \varepsilon_2 \\ \varepsilon_3 \\ \gamma_{23} \\ \gamma_{13} \\ \gamma_{12} \end{Bmatrix} = \begin{bmatrix} \frac{1}{E_1} & -\frac{\nu_{21}}{E_2} & -\frac{\nu_{31}}{E_3} & 0 & 0 & 0 \\ -\frac{\nu_{12}}{E_1} & \frac{1}{E_2} & -\frac{\nu_{32}}{E_3} & 0 & 0 & 0 \\ -\frac{\nu_{13}}{E_1} & -\frac{\nu_{23}}{E_2} & \frac{1}{E_3} & 0 & 0 & 0 \\ 0 & 0 & 0 & \frac{1}{G_{23}} & 0 & 0 \\ 0 & 0 & 0 & 0 & \frac{1}{G_{13}} & 0 \\ 0 & 0 & 0 & 0 & 0 & \frac{1}{G_{12}} \end{bmatrix} \begin{Bmatrix} \sigma_1 \\ \sigma_2 \\ \sigma_3 \\ \sigma_{23} \\ \sigma_{13} \\ \sigma_{12} \end{Bmatrix}. \tag{4.2.3}$$

However, a yarn is transversely isotropic so $E_2 = E_3$, with similar simplifications to Poisson's ratio and shear terms. In abbreviated notation each grain α in its local frame L is $\{\varepsilon^\alpha\}_L = [S^\alpha]_L\,\{\sigma^\alpha\}_L$, where compliance $[S]$ for a transversely isotropic material has five independent properties E_1, E_2, ν_{12}, ν_{21}, and G_{12}. Transformation relations between each grain in the 1, 2, 3 frame and X, Y, and Z frame give the transformed global frame compliance as $[S^\alpha]_G = [T_\epsilon^\alpha]^{-1}[S^\alpha]_L[T_\sigma^\alpha]$.

Finally, all grains are summed $[\overline{S}]_{\text{RVE}} = \sum V^\alpha [S^\alpha]_G$ for the total RVE compliance, where V^α is a weighting factor of grain volume to total RVE volume. The RVE coordinate frame strain-stress relationship is

$$\begin{Bmatrix} \varepsilon_x \\ \varepsilon_y \\ \varepsilon_z \\ \gamma_{yz} \\ \gamma_{xz} \\ \gamma_{xy} \end{Bmatrix} = \begin{bmatrix} \overline{S}_{11} & \overline{S}_{12} & \overline{S}_{13} & \overline{S}_{14} & \overline{S}_{15} & \overline{S}_{16} \\ \overline{S}_{21} & \overline{S}_{22} & \overline{S}_{23} & \overline{S}_{24} & \overline{S}_{25} & \overline{S}_{26} \\ \overline{S}_{31} & \overline{S}_{32} & \overline{S}_{33} & \overline{S}_{34} & \overline{S}_{35} & \overline{S}_{36} \\ \overline{S}_{41} & \overline{S}_{42} & \overline{S}_{43} & \overline{S}_{44} & \overline{S}_{45} & \overline{S}_{46} \\ \overline{S}_{51} & \overline{S}_{52} & \overline{S}_{53} & \overline{S}_{54} & \overline{S}_{55} & \overline{S}_{56} \\ \overline{S}_{61} & \overline{S}_{62} & \overline{S}_{63} & \overline{S}_{64} & \overline{S}_{65} & \overline{S}_{66} \end{bmatrix} \begin{Bmatrix} \sigma_x \\ \sigma_y \\ \sigma_z \\ \sigma_{yz} \\ \sigma_{xz} \\ \sigma_{xz} \end{Bmatrix}.$$

From this coordinate frame properties are retrieved. For instance, if the RVE only has applied stress σ_x, with all other stresses zero, then

$$E_x = \frac{\sigma_x}{\varepsilon_x} = \frac{\sigma_x}{\overline{S}_{11}\sigma_x} = \frac{1}{\overline{S}_{11}} \tag{4.2.4}$$

with similar arguments used for other constants E_y, E_z, G_{yz}, G_{xz}, G_{xy}.

For an isostrain analysis, identical steps are used giving a global frame stiffness $[\overline{E}]$ which is inverted for $[\overline{S}]$ and used in Eq. 4.2.4. These two approached give upper and lower bounds to true stiffness, with neither being particularly good. Consequently, the Voigt-Reuss-Hill model has been proposed (Hill 1952) which advocates taking an average of these two limiting solutions, so stiffness is given by $[\overline{E}]_{\mathrm{VRH}} = \left([\overline{S}]^{-1}_{\mathrm{RVE}} + [\overline{E}]_{\mathrm{RVE}}\right)/2$.

In a similar manner other thermo-mechanical constants can be found, but properties like failure cannot be investigated since there is no possibility to predict stress or strain distribution in the RVE. For this both WiseTex and TexGen allow creation of an FE mesh that can be analysed with commercial FE codes to obtain properties via different applied loads. Furthermore, during the past 20 years RVE textile models have been used to compute fluid flow and estimate permeability. There are several challenges, however, concerning meshing especially at inter-yarn locations and application of periodic boundary is necessary. It is also doubtful if this textile geometry is an accurate representation of the true in-situ compacted architecture, which will lead to inaccurate prediction of stress or flow distribution.

3.3 Laminate analysis based on classical laminate theory (CLT)

CLT is an essential tool to determine stiffness, strength and possibly hygrothermal (thermal and moisture) properties of a stack of plies. It was developed by numerous contributors in the 1960s from earlier work on bonded thin plates by Pister et al. (1959), and is today implemented in many commercial software tools. A good freely available academic code is eLamX (2021), or the simple spreadsheet (Aerospace Engineering 2019).

Essentially the method describes a stack of plies in a laminate of unit size and thickness h, as shown in Fig. 4.2.8. Each ply in the stack has principal orthotropic properties E_1, E_2, ν_{12}, ν_{21}, G_{12} in its fibre direction θ for ply thickness t. CLT performs transformation operations on each ply between fibre frame 1, 2, 3 and Cartesian frame x, y, z and sums their Cartesian frame inplane stiffnesses for matrix $[A]$ and bending for matrix $[D]$ in Eq. (4.2.5). Matrix $[B]$ in Eq. (4.2.5) defines possible coupling terms between inplane loads that may cause bending curvature and moments that may cause inplane stretching. Full derivation of Eqs. (4.2.5) and (4.2.6) is to be found in composites textbooks such as Daniel and Ishai (2005), Gibson (2011), Barbero (2017) and Pickett (2024).

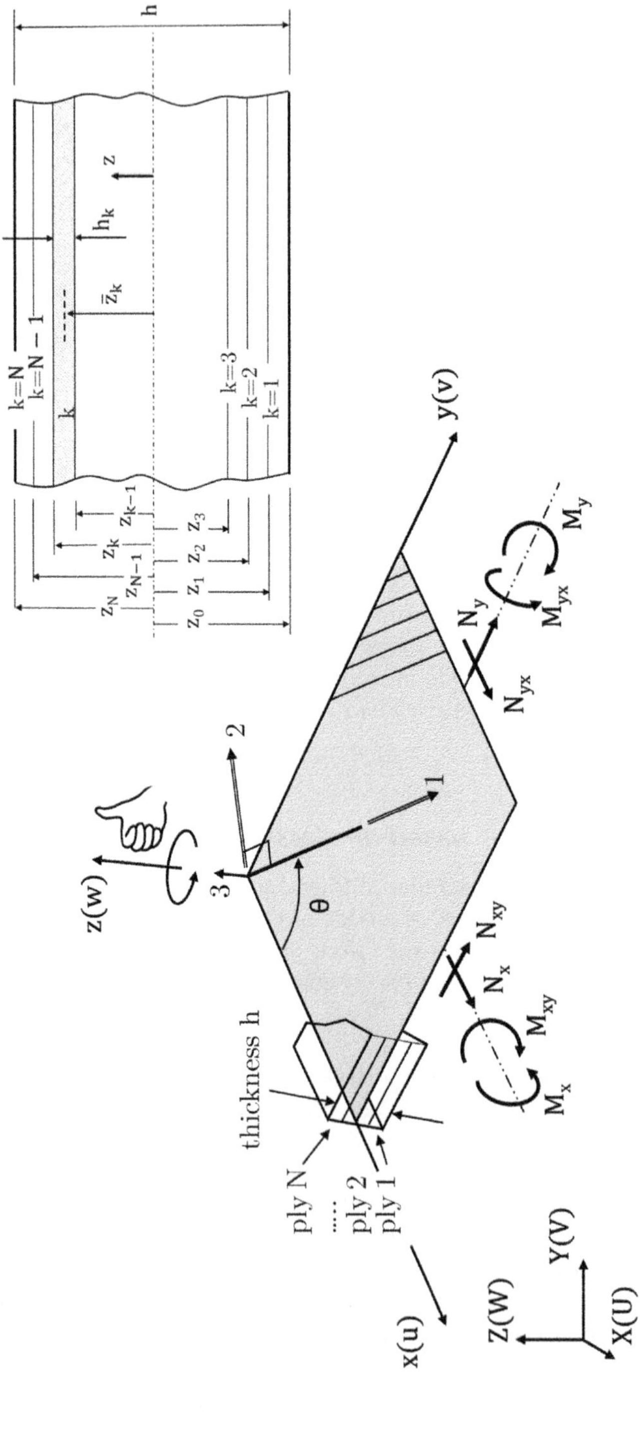

Figure 4.2.8 Laminate conventions for loading, stacking, and fibre orientations.

$$\begin{Bmatrix} N_x \\ N_y \\ N_{xy} \\ M_x \\ M_y \\ M_{xy} \end{Bmatrix} = \begin{bmatrix} A_{11} & A_{12} & A_{16} & B_{11} & B_{12} & B_{16} \\ A_{12} & A_{22} & A_{26} & B_{12} & B_{22} & B_{26} \\ A_{16} & A_{26} & A_{66} & B_{16} & B_{26} & B_{66} \\ B_{11} & B_{12} & B_{16} & D_{11} & D_{12} & D_{16} \\ B_{12} & B_{22} & B_{26} & D_{12} & D_{22} & D_{26} \\ B_{16} & B_{26} & B_{66} & D_{16} & D_{26} & D_{66} \end{bmatrix} \begin{Bmatrix} \varepsilon_x^0 \\ \varepsilon_y^0 \\ \gamma_{xy}^0 \\ \kappa_x \\ \kappa_y \\ \kappa_{xy} \end{Bmatrix} \tag{4.2.5}$$

where

$$A_{ij} = \sum_{k=1}^{N} \left(\overline{Q}_{ij}\right)_k (z_k - z_{k-1}) = \sum_{k=1}^{N} \left(\overline{Q}_{ij}\right)_k h_k \quad i,j = 1,2,6 \tag{4.2.6a}$$

$$B_{ij} = \frac{1}{2} \sum_{k=1}^{N} \left(\overline{Q}_{ij}\right)_k \left(z_k^2 - z_{k-1}^2\right) = \sum_{k=1}^{N} \left(\overline{Q}_{ij}\right)_k h_k \overline{z}_k \quad i,j = 1,2,6 \tag{4.2.6b}$$

$$D_{ij} = \frac{1}{3} \sum_{k=1}^{N} \left(\overline{Q}_{ij}\right)_k \left(z_k^3 - z_{k-1}^3\right) = \sum_{k=1}^{N} \left(\overline{Q}_{ij}\right)_k \left(h_k \overline{z}_k^2 + \frac{h_{k^3}}{12}\right) \quad i,j = 1,2,6. \tag{4.2.6c}$$

In abbreviated notation these equations are expressed as

$$\begin{Bmatrix} \varepsilon^0 \\ k \end{Bmatrix} = \begin{bmatrix} A & B \\ B & D \end{bmatrix}^{-1} \begin{Bmatrix} N \\ M \end{Bmatrix} = \begin{bmatrix} a & b \\ b & d \end{bmatrix} \begin{Bmatrix} N \\ M \end{Bmatrix}.$$

Thus for a given applied loading $\{N, M\}$ the mid-plane laminate Cartesian frame strains $\{\varepsilon^0, k\}$ are found. From this standard bending relations and transformation matrices give fibre frame strains ε_1, ε_2, ε_{12} and material laws yield stresses σ_1, σ_2, σ_{12} for each ply. Furthermore, hygrothermal strains and stresses caused by thermal expansion and/or moisture absorption can be added and failure criteria applied to predict failure, or obtain a scaling factor for applied loads that will cause failure. The more advanced codes include a database of properties that may be at the micromechanical level for a wide range of fibre and resin types, from which micromechanical laws compute homogenised. Alternatively, these properties are directly available in a database or input by the user from test measurements or other sources. Often useful tools are provided based on analytical solutions for buckling and frequency response. And helpful graphics can visualise laminate stress and strain distributions and deformations to help make any necessary design changes.

The main limitation of CLT analysis is that applied loads must be found from hand calculations or some other design procedure for different areas of a component. For simple beams, plates and similar parts this is possible, but becomes far more difficult for complex geometries and loading conditions. In this case, a finite element solution is a more practical approach.

4. Finite element methods

Finite element analysis was originally developed by Turner et al. (1956) in the mid-1950s for aerospace structural analysis. Today, it has evolved to become the most popular general purpose computation method, with applications in many different disciplines ranging from static mechanical and steady state field analysis to transient flow problems such as heat transfer, fluid flow, and impact or crashworthiness.

From the 1960s, this method was quickly adapted for orthotropic composites and laminate analysis using, essentially, the same techniques as classical laminate theory in a finite element framework for thin shell or plate elements. This allowed full structures to be tackled, compared to CLT which is restricted to a loaded laminate. Early commercial FE codes included simple failure criteria for ply or laminate failure. Extension to impact and crash analysis is a highly non-linear problem that involves progressive inter- and intra-ply damage, which has been successfully tackled during the past 25 years using dynamic explicit FE codes.

For manufacturing analysis early work from the 1990s used linear FE field analysis with a control volume (CV) method to predict transient liquid resin infusion of dry fabrics for RTM type processes. And during last 3 decades FE simulation of manufacturing processes for draping, pre-preg forming and thermoforming, and braiding, have also become increasingly important and have been successfully simulated with dedicated developments in explicit FE codes.

These different FE solution methods for "implicit," "explicit," and "implicit CV" analysis are briefly introduced showing application of each to different fields of composites analysis. Opportunity is also taken to introduce some recent work that is ongoing for mesoscopic modelling of these topics.

5. Implicit FE analysis

The most common FE method is implicit in which a stiffness matrix for each element $[k]$ is computed and assembled to give a global structure stiffness $[K]$, so that the system of governing equations are

$$[K]\{u\} = \{P_{ext}\} \tag{4.2.7}$$

where $\{u\}$ and $\{P_{ext}\}$ are vectors of nodal displacements and applied nodal forces. Inversion of $[K]$ multiplied by external applied forces $\{P_{ext}\}$ gives structure nodal displacements $\{u\} = [K]^{-1}\{P_{ext}\}$. Strains in each element are found from nodal displacements and stresses found from these strains and the chosen constitutive law. For a linear problem $[K]$ is constant, so there is a linear relationship between applied loads and displacements, and a solution is therefore straightforward. However, if contact, large deformations or material nonlinearity occur, such as in sheet forming or composites impact damage, then $[K]$ changes with applied load and a more difficult iterative solution scheme or a dynamic explicit FE analysis becomes necessary.

5.1 Implicit FE mechanical analysis

Most composites analyses are undertaken to determine deformations and ply stresses and strains, to predict possible failure conditions. If deformations are relatively small this is performed as a linear FE analysis as described by Eq. (4.2.7). Required laminate input is the same as that for a CLT analysis. Usually shell elements are used for the different laminates, with each ply represented as one integration layer of the element. Sometimes the $[ABD]$ matrix in Eq. 4.2.5 must be computed from a separate CLT analysis and input for stretch-bend stiffness properties of the homogenised shell element; however, using this approach cannot predict individual ply stresses and strains.

Fig. 4.2.9 shows a typical study from an MSc student design project (Pickett 2006) that redesigned a steel automotive part with tape wound CFRP composites. The pre-impregnated tape had UD fibres with known fibre content, so micromechanics was used to estimate homogenised properties that supplement properties from manufacturers data sheets. In principle, a CLT analysis could be used to design laminates to limit deformation for an imposed load, or ensure ply failure does not occur, but even with this simple structure it is difficult to accurately estimate laminate load distribution. An FE analysis overcomes these limitations. If large deformation occurs a geometric non-linear analysis may be needed, or if non-critical plies fail at low load a

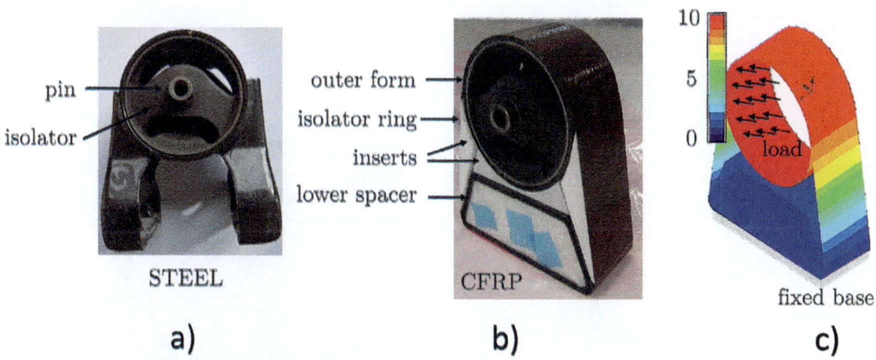

Figure 4.2.9 Engine mount design (a) Steel production version (1.4 kg) (b) Proposed CFRP design (0.49 kg) (c) FE analysis deformations [mm].

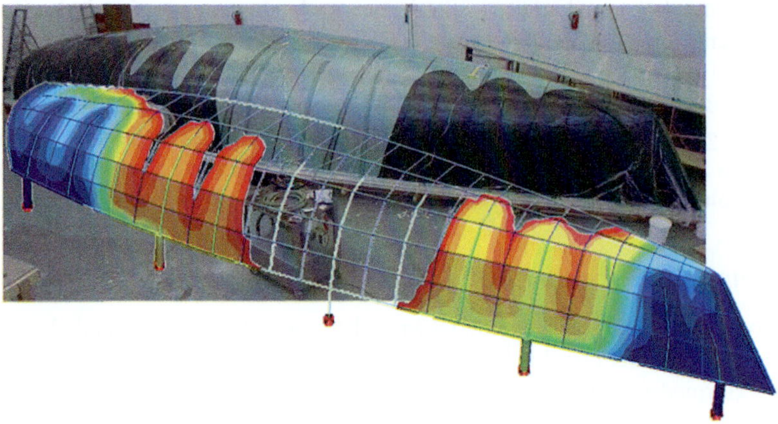

Figure 4.2.10 Example manufacture and infusion analysis of a boat hull (courtesy StandfastYachts and Polyworx, Netherlands).

strategy is needed to damage these and repeat analyses so ultimate load at last ply failure can be identified. In these cases, an explicit FE solution may be a better approach.

6. Control volume FE analysis for infusion

Early manufacturing using resin transfer moulding (RTM) can be traced back to the 1950s, but applications to more advanced structures like those found in the aerospace and other transport industries has only taken place from about the 1980s. Infusion is a notoriously difficult process to control that has greatly benefited from supporting analysis tools. So far three academic FE codes have been developed, from which (LIMS 2019) remains academic, with the other two (PAM-RTM 2005) and

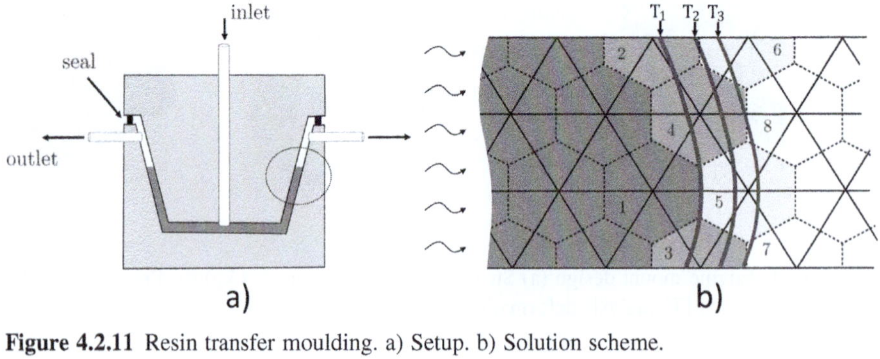

Figure 4.2.11 Resin transfer moulding. a) Setup. b) Solution scheme.

(RTMworx, 2005) now commercialised. An example infusion analysis is shown in Fig. 4.2.10.

Fig. 4.2.11a shows a standard RTM infusion where resin is pressure injected at the inlet and air is released from the outlets, which is stopped once liquid exits the outlet. Tooling holds the dry fabric preform that is separated once the part is sufficiently cured. In Fig. 4.2.11b the FE control volume filling scheme used by these codes is depicted. Briefly, a 2D triangular mesh is shown in solid lines, with dashed lines representing control volumes (CV) that are assigned to each node. Typical flow advance is analysed as a series of linear solutions in a time stepping scheme. This is illustrated in this figure for three successive times T_1, T_2, and T_3 in which the following steps are performed:

1. At time T_1 the algorithm will have just completed filling CV_1, with CV_2, CV_3, and CV_4 only partially filled.
2. A new implicit analysis is made of the updated filled region from which new pressure distribution for the liquid region is computed using $\{P\} = [C]^{-1}\{b\}$. This is analogous to solving Eq. 4.2.7 except $\{b\}$ and $\{P\}$ are nodal flow rates and pressures, and $[C]$ is the assembled global characteristic matrix which is similar to $[K]$, except based on resin viscosity and fabric permeability. Flow rates between CVs are found from pressure gradients between adjacent elements.
3. From flow rates and fill volumes in the CVs, a time interval is determined so that one new CV is completely filled. All other partially filled CVs are topped up and the new filling time T_2 updated. In Fig. 4.2.11b this is CV_2 with further topping up of CV_3, CV_4, and CV_5, using their flow rates and the computed time step.
4. This process repeats with a new FE analysis to fill one new CV, with topping up of others and updating of the filling time T_3. In Fig. 4.2.11, CV_3 is the CV to be completely filled.

7. Explicit FE analysis for dynamic problems

Explicit FE simulation techniques were originally developed for military impact problems and successfully commercialised in the 1980s for automotive crashworthiness. Since the 1990s, these techniques have been applied to composites simulation; first, for progressive damage modelling of impact and crash phenomena and later for manufacturing applications such as thermoforming, draping, braiding, and weaving.

Essentially explicit analysis uses similar meshing and modelling techniques as implicit analysis, except a dynamic solution is sought in the time domain t using the dynamic equations of motion given by

$$\left([\mathrm{M}]\left\{\ddot{u}\right\}\right)_t + ([C]\{\dot{u}\})_t + ([K]\{u\})_t = \{P_{ext}\}_t \tag{4.2.8}$$

where $\left\{\ddot{u}\right\}$ and $\{\dot{u}\}$ are vectors of nodal acceleration and velocities, respectively, and $[M]$ and $[C]$ are structure nodal mass and damping matrices. This equation states that inertia, damping, and internal forces must balance external loads at any time t. Usually damping effects are ignored, so Eq. 4.2.8 is solved directly for $\{\ddot{u}\}$ since both internal nodal forces $[K]\{u\}$ and external nodal forces $\{P_{ext}\}$ are known. From a central finite difference integration scheme forward time predictions for new nodal velocities $\{\dot{u}\}$ and displacements $\{u\}$ are found. For a single dof at a solution step n the central difference equations solved to advance the solution one time step Δt are

$$\begin{aligned} \ddot{u}_n &= (p_n - ku_n)/m = (p_n - p_{int})/m \\ \dot{u}_{n+1/2} &= \dot{u}_{n-1/2} + \Delta t_n \ddot{u}_n \\ u_{n+1} &= u_n + \Delta t_{n+1/2} \dot{u}_{n+1/2}. \end{aligned} \tag{4.2.9}$$

This scheme is conditionally stable and only works if the integration time step is smaller than a critical value that depends on mesh size and material properties. Provided this stability criterion is observed the dynamic equations at each node are uncoupled and a solution can be undertaken on a node-by-node basis, without the need for expensive matrix assembly or inversion. A further advantage is that material nonlinearity, large deformations and contact are easily handled, but the time step stability criterion can lead to high CPU costs if loading is applied over a long duration or is quasi-static.

The explicit simulations presented in this chapter were all undertaken with the commercial code (PAM-CRASH 1985), which was created in the early 1980s by combining an open source 3D solid version of DYNA3D with the Mindlin shell element proposed by Belytschko et al. (1983). Commercialisation for metal crash simulation largely took place through a national project with the German automotive industry and parallel studies (Haug et al. 1986; Chedmail et al., 1986). Since then dedicated composites options for static and impact analysis, and forming in the derivative code (PAM-FORM 1998) have gradually been added. Today, other commercial explicit FE codes LS-DYNA, Abaqus and Radioss will be capable of most of these simulations.

a) b)

Figure 4.2.12 Braiding machines. (a) The prototype 3D braider. (b) A conventional 2D rotary braider.

7.1 Braid and weave simulation

Between 1992 and 2001 two German national research projects (Brandt et al. 1992–1997) and one EU project were run to develop 3D braiding technology, which involved moving bobbins in arbitrary paths over a flat surface via a grid of horngears and mechanical switches as shown in Fig. 4.2.12a. Unfortunately, the technology proved complex and slow, and so far has not been industrially successful. However, the work and advances in computing power and explicit FE codes at the time did enable the author to undertake new simulations of braiding, including the possibility to combine predicted yarn architecture with resin matrix for analysis of composite mechanical properties (Schneider et al. 1999). This helped open new opportunities for the simulation of conventional 2D rotary braiding and other textile weaving processes.

For a simple estimate of yarn paths in a braid the kinematic relations given by Eq. 4.2.2 can be used, which only requires information on mandrel geometry and velocities of the bobbin and mandrel. However, like kinematic methods for draping, these simplified relations ignore many important effects such as yarn stiffness and yarn-to-yarn and yarn-to-tool friction, they also ignore important yarn interactions that take place during their placement. Consequently, results are approximate and it is difficult to identify any potential defects in the preform.

These limitations are largely overcome using FE methods and several publications for rotary braiding have been reported (Birkefeld et al., 2018; Pickett et al., 2009; Pickett and Erber, 2009; Rosso et al., 2019; Swery, 2016) using different commercial FE codes. In these studies explicit integration is necessary to handle the large displacements and extensive contact between yarns that occur. Early work used beam elements, but nowadays shell elements that better represent the oval yarn geometry are also feasible. A typical simulation setup for 2D braiding is shown in Fig. 4.2.13a, where bar elements control pretension and shell elements model yarns,

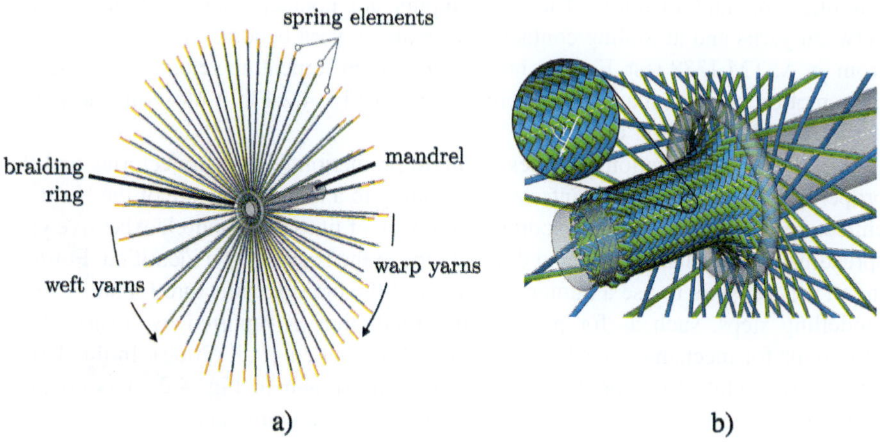

Figure 4.2.13 Two-dimensional braiding simulation. (a) Model setup. (b) Close-up of braiding ring and mandrel, and a partially braided cylindrical preform.

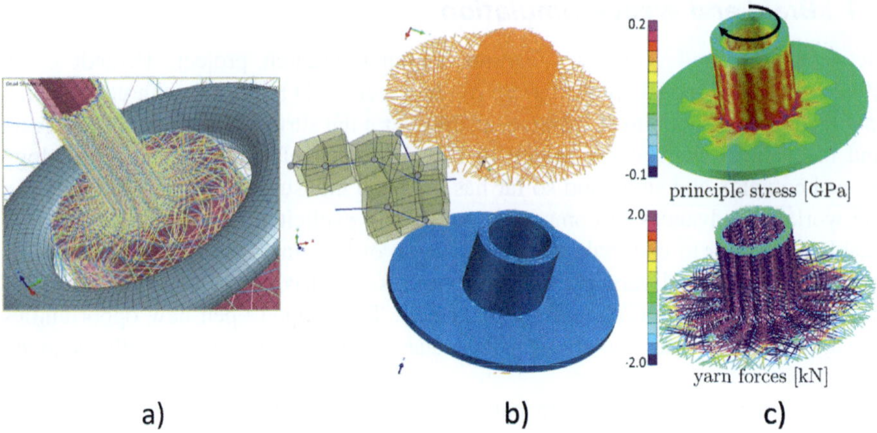

Figure 4.2.14 Simulation of a flange. (a) The braided beam model. (b) Merging of yarn beam and solid matrix models. (c) Results from a static torsion implicit FE analysis.

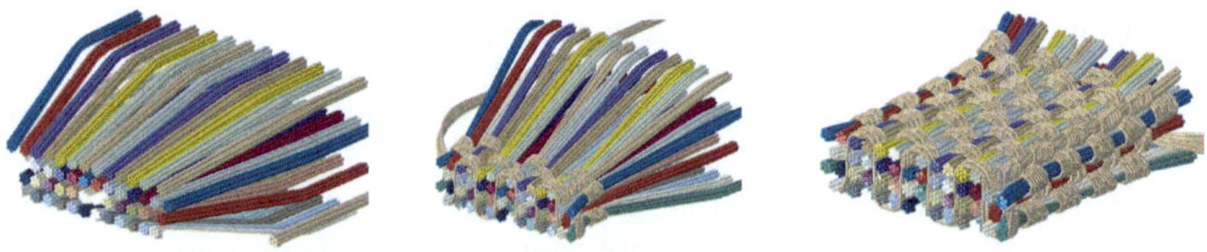

Figure 4.2.15 Weaving simulation of a 3D textile.
Courtesy Reden B.V., Netherlands.

braiding ring and mandrel. The main inputs are contact with Coulomb friction between yarns and at tooling contact, with realistic yarn bending stiffness calibrated from an ASTM 1388 test. Finally, bobbins are given serpentine velocities to describe their motion and the mandrel velocity is set. In Fig. 4.2.13b a close-up view of a partially braided section is shown.

An FE braid simulation provides accurate prediction of yarn trajectories and the proper representation of yarn interlacing, leading to a realistic picture of the braid architecture. Defects such as over-compaction, lack of fit to the mandrel, excessive yarn spreading or poor distribution and defect loops in the braid can be identified. Furthermore, it is possible to use a numerical model of the braid architecture for subsequent modelling steps, such as for permeability prediction as reported by Swery et al. (2016), or for mechanical analysis as reported by Pickett et al. (2009). In this latter, work yarn architecture for the braided flange as shown in Fig. 4.2.14 is coupled with a solid mesh for the resin to analyse stiffness of a component. Alternatively, it might be feasible to use the predicted yarn architecture as input information in a dedicated mesomechanical code like WiseTex to estimate stiffness.

More recently researchers have begun to apply these simulation techniques to weaving with typical results shown in Fig. 4.2.15 for a 3D weave. Again, the main interest is to replicate the weaving process and provide a model for subsequent mechanical analysis that could supplement tests or possibly even replace them. However, it must be emphasised that this remains a research topic today.

7.2 Draping and thermoforming simulation

Simulation of these large deformation highly non-linear processes is only practical using an explicit FE code with suitable material and contact laws. To help achieve this the author initiated a European multi-partner Brite-Euram project in the early 1990s to develop simulation capabilities for composites forming (BE-5092 1992–1996). At the time this linked with ongoing efforts to develop a new explicit FE code PAM-STAMP for metal stamping simulation (Lefebvre et al. 1994). This Brite-Euram project initially focussed on thermoforming of aerospace grade thermoplastic composite plies PEEK and CEI-Cetex for aerospace applications covering pre-heating analysis, forming simulation and validation through industrial processes and parts, which led to the commercial code (PAM-FORM 1998).

Today FE codes for fabric forming can treat dry or pre-impregnated sheets that could have viscous thermoset or thermoplastic resin. These have constitutive laws that model fabric stretch and shear stiffness, with impregnated fabrics also having a superimposed temperature dependent viscous law. Fig. 4.2.16 shows an early study on this topic from the EU project (BE-5092 1992–1996). In this simulation 8 hot plies of PEI-Cetex were stamped between cold metal tools to predict deformations, with a blank holder added to help suppress possible wrinkling. Of special interest was prediction of consolidation, or so-called degree of intimate contact D_{ic}, which varies

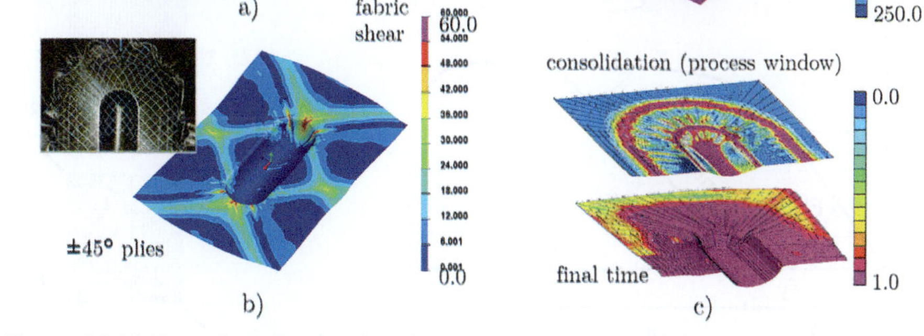

Figure 4.2.16 Example thermoforming simulation. (a) Model setup. (b) Deformation and contour plots of fabric shear. (c) Contours of temperature and consolidation.

$0 \rightarrow 1$ for full consolidation using the inter-ply model (Lee and Springer 1987) with experimental input from Phillips et al. (1996) giving

$$D_{ic} = \frac{a_o b_o}{w_o + b_o} \left\{ \frac{1}{a_o^5} + 5 \frac{(b_o + w_o)}{(a_o b_o)^3} \int_o^t \frac{\psi P_{app(t)}}{\eta(T)} dt \right\}^{\frac{1}{5}}$$

where $a_o = 224.0\ \mu m$, $b_o = 603.0\ \mu m$, $w_o = 1219.0\ \mu m$ and $\psi = 0.73$ for PEI-Cetex. This illustrates one example where an analytical micromechanical model of composites behaviour can be a valuable addition to an FE solution. For this simulation evolution of temperature T and pressure P with time t was necessary to predict inter-ply consolidation. Other results show large fabric shear with some buckling at the ends of the sikken geometry.

7.3 Damage and impact simulation

The underlying idea used to model composites impact is to damage intra- and inter-ply stiffness properties as the material is weakened by excessive loading. For this the concept of continuum damage mechanics (CDM) first proposed by Kachanov (1958) is used. For CFRP materials this involves introducing a new state variable for each failure mode to represent its degree of damage. A good general textbook on the topic is Lemaitre and Chaboche (1994) with theory applied to CFRP plies by Ladevèze and LeDantec (1992) and delamination by Allix and Ladevèze (1992). These concepts were implemented in PAM-CRASH by the author and co-workers during the late 1990s as part of another multipartner EU project HICAS (Johnson

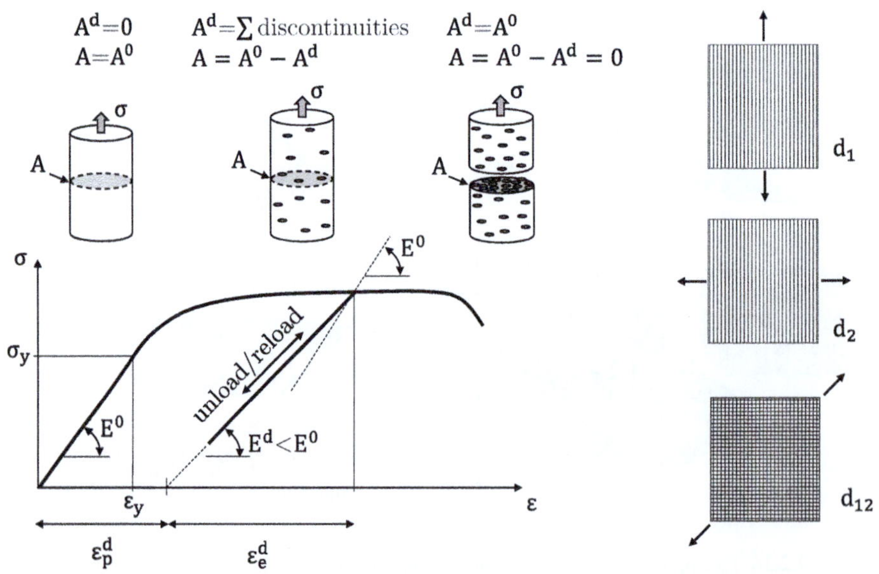

Figure 4.2.17 Inelastic response of an elasto-plastic material with damage.

et al. 2001). Today, other FE crashworthiness codes have implemented similar models and a great deal of work has been independently done by researchers through user interface elements.

Damage causes a permanent loss in strength. For metals, this involves void formation at imperfections (or inclusions) and in composites it is matrix micro-cracks, fibre-matrix debonding and individual fibre failure. Some concepts of damage for a typical elasto-plastic material with damage growth are depicted in Fig. 4.2.17, including schematic diagrams of void (or fracture) growth. Usually damage starts at the same time as plasticity, but it can be at some other loading state. Defining A^d as a measure of surface density of damage discontinuities at any section through the RVE, it will have a value $A^d = 0$ in the elastic region and then grow steadily to full fracture at $A^d = A^0$. For all states the effective material section area resisting applied stress is $A = A^0 - A^d$, and the measure of damage d is given by the scalar variable $d = 1 - A/A^0$, which varies from $0 \rightarrow 1$ representing the transition from undamaged material to full fracture. At any state of loading, it is possible to quantify damage d from a test coupon by measuring the reduced modulus E^d during an unload-reload cycle. State variable d is then found from $E^d = E^0\,(1-d)$. In practice five to six unload-reload cycles are performed to find evolution of d with strain, which is done using unidirectional test coupons loaded in tension, transverse tension, and shear for damage functions d_1, d_2, d_{12}.

The "effective stress" $\tilde{\sigma}$ in a damaged material is applied stress carried by the intact material $\tilde{\sigma} = \sigma/(1-d)$, where $\tilde{\sigma} > \sigma$ due to the reduction of load carrying area $A < A^0$. Using similar arguments strains in the intact material decrease due to relaxation as micro-cracks open, so "effective strain" is $\tilde{\varepsilon} = \varepsilon \times (1-d)$, and for an applied stress σ the effective strain is $\tilde{\varepsilon} = \sigma/E^0(1-d)$. Using Hooke's law for an orthotropic composite the equivalent damage relations are

$$\left\{ \tilde{\varepsilon}_1\ \tilde{\varepsilon}_2\ \tilde{\varepsilon}_{12} \right\} = \begin{bmatrix} \dfrac{1}{E_1^0(1-d_1)} & -\dfrac{\nu_{21}^0}{E_2^0} & 0 \\ -\dfrac{\nu_{12}^0}{E_1^0} & \dfrac{1}{E_2^0(1-d_2)} & 0 \\ 0 & 0 & \dfrac{1}{2G_{12}^0(1-d_{12})} \end{bmatrix} \begin{Bmatrix} \sigma_1 \\ \sigma_2 \\ \sigma_{12} \end{Bmatrix} \tag{4.2.10}$$

where initial values in the symmetry relation $-\nu_{21}^0/E_2^0 = -\nu_{12}^0/E_1^0$ are assumed not to damage.

For delamination between plies the common approach today models this inter-ply failure and separation using an energy criterion with fracture energy G_c obtained from standardised mode I and mode II coupon tests. Usually, either interface solid elements or a contact interface is used, whereby a shaped stress versus displacement curve with area G_c represents the loading and damage behaviour. Once full damage occurs

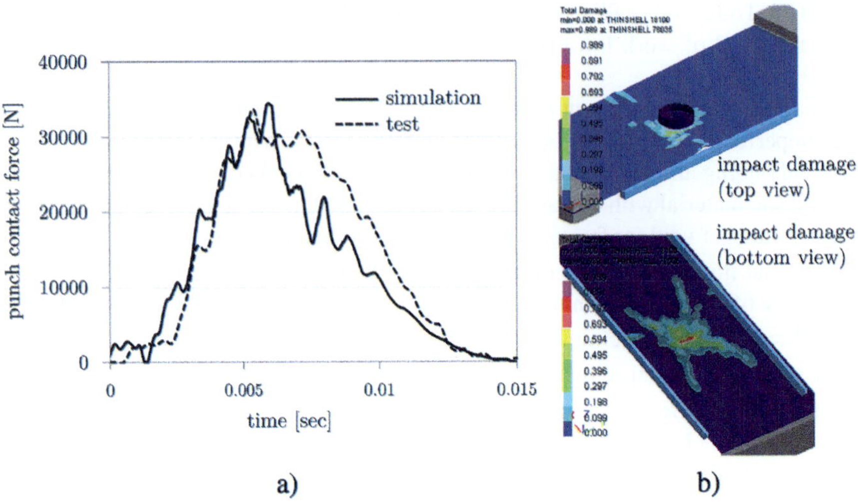

Figure 4.2.18 Example plate impact study. (a) Punch to plate impact force. (b) Contours of ply damage.

fracture energy for crack growth over the area represented by the contact is absorbed and free separation is allowed. Ideally, for a laminate, each ply has its own damage model with each adjoining pairs of plies being tied together with an inter-ply delamination model.

A typical study using these approaches is shown in Fig. 4.2.18. In order to reduce CPU costs the 32 plies were modelled using eight multilayer shell elements each having four plies [0/90/45/-45], so only seven delamination interfaces are modelled. Selected simulation results are shown below, including a comparison of test and simulation punch impact force time histories as well as views of top and bottom ply damage. Further details on this study, including material properties for an alternative bi-axial automotive grade carbon NCF composite are described in Pickett et al. (2009).

8. FE mesomechanics modelling

Most textile fabrics either have intertwined yarns or yarns that are held together by stitching. During forming these fabrics undergo a complex interaction of deformation mechanisms, such as compaction and spreading of individual yarns, inter-yarn slippage at cross-over points and, for stitched fabrics, relative yarn to stitch slippage. These local deformations cannot be represented by a macro-scale model that uses a homogenised constitutive law in continuum finite elements. Consequently, the author and several research students have investigated numerical models to idealise fabric architecture for improved forming simulation that are also able to predict the local imperfections that may occur. Similarly, failure and impact analysis using current state-

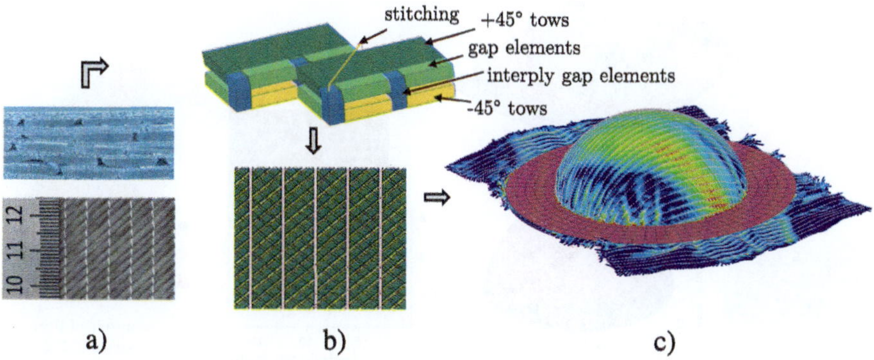

Figure 4.2.19 Mesoscale simulation. (a) Fabric architecture. (b) FE unit cell and fabric idealisation. (c) Mesoscale drape simulation.

of-the-art macro-mechanical methods do have limitations and there is potential for improved modelling of damage mechanisms at the mesoscale.

Meso-mechanical models also open interesting possibilities to couple, or chain results, for use in subsequent analysis steps. However, it should be mentioned there are significant challenges since meshing, model preparation, assignment of valid constitutive laws, computational effort, and even visualisation of results can be difficult. The following demonstrate some selected research in this direction. In all cases a commercial explicit FE code has been used due to the dynamic loading, large deformations and extensive contact that occur. In these studies (PAM-CRASH 1985) has been applied, but it would be possible to use comparable explicit FE codes.

8.1 Mesoscale drape simulation

The challenge to build an FE mesoscale drape model for forming a moderately large geometry is to combine necessary element types so that all key deformation modes can be represented without the model being excessively large, or prohibitively computationally expensive. Further, it must be robust to handle the large deformations and contact that occur. This does necessitate making compromises to the geometric representation and constitutive modelling. For example, Fig. 4.2.19a,b shows a mesoscale fabric model developed by Sirtautas (2017), for a chain stitched ±45 degrees NCF. In this example each yarn is modelled by one row of orthotropic solids, with gap elements for inter-yarn and inter-ply friction/contact, with stitching represented by bar elements. Dedicated testing and calibration is necessary for all constitutive and friction properties. This model was successful to predict yarn spacing and orientations for the moderately large hemisphere in Fig. 4.2.19c. In earlier work similar ideas were applied by Creech (2006); Creech and Pickett (2006) for a tricot stitched NCF.

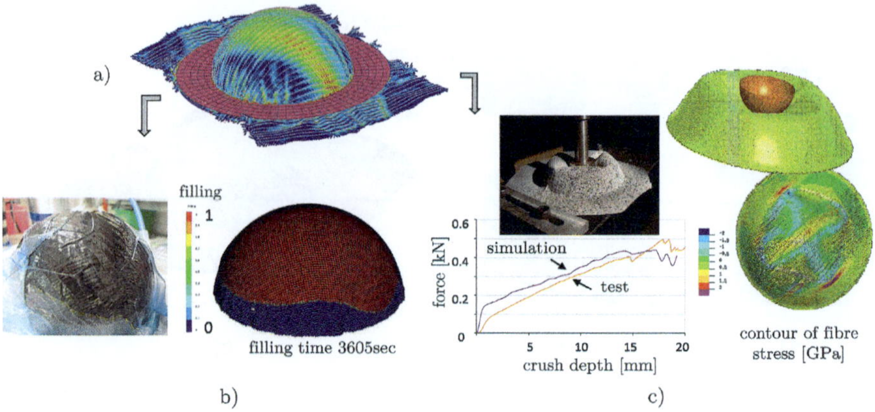

Figure 4.2.20 Chaining of process and mechanical analyses. (a) Drape simulation. (b) Drape-to infusion. (c) Drape-to-impact.

8.2 Chaining drape, infusion, and impact simulations

So far very little work has been published that investigate the effect fabric deformations have on composite properties. Two publications that have addressed this for sheared fabrics are Demaria et al. (2007) for permeability and Hwang et al. (2023) for failure.

An alternative approach could be chaining of results from different mesoscale simulation models. For example, Sirtautas et al. (2013) used the deformed NCF mesh with new permeability and porosity properties for yarn and inter-yarn volumes to perform an infusion analysis with PAM-RTM of the dry fabric, Fig. 4.2.20b. In a similar manner used new mechanical properties for cured yarns and inter-yarn delamination as a basis for in impact analysis, Fig. 4.2.20c. In principle, the advantage of this approach is that thickness, yarn spacings, shear, and other orthotropic effects are taken into account by the deformed architecture and only new yarn and inter-yarn properties are needed for either the infusion or impact analysis.

8.3 Meso-scale impact simulation

Failure and impact analysis using macromechanical methods of the type shown previously in Fig. 4.2.18 are important tools today for industrial applications and design. But these techniques do have some limitations to accurately model the true fracture mechanisms that occur in real laminate composites.

Fig. 4.2.21 illustrates the fracture mechanisms often described for a transversely impacted laminate, which are outlined here for a simple [0/90/0] layup. First, transverse intra-ply cracks occur due to shear loading in the low strength 90 degrees ply, Fig. 4.2.21a. These cracks form at 45 degrees and run parallel to the fibres until they are stopped once they hit the off-axis fibres of an adjacent ply. If sufficient energy is available, a delamination crack will then extend along the interface. These cracks

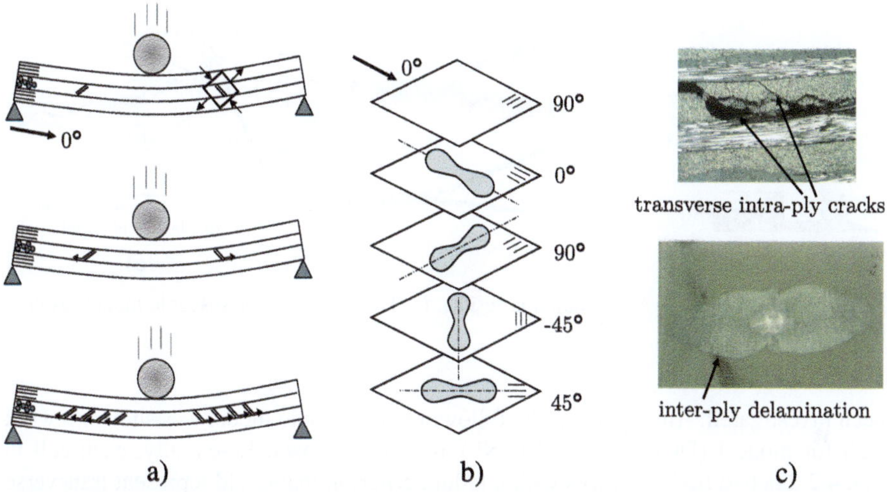

Figure 4.2.21 Formation of fractures in a transversely impacted laminate. (a) Evolution of intra and inter-ply cracks. (b) Laminate delaminations. (c) Test views of intra-ply cracks and delamination.

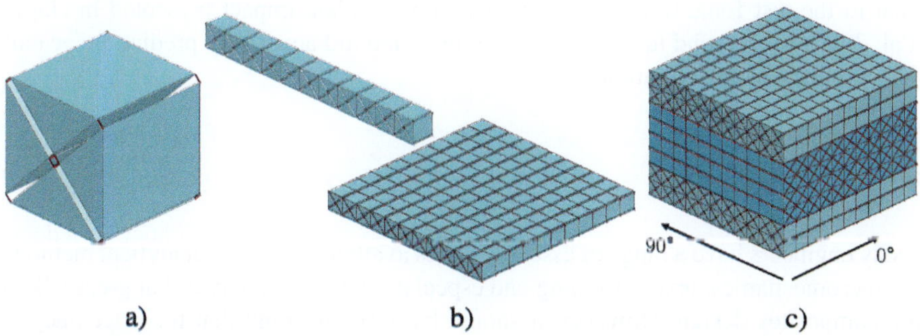

Figure 4.2.22 The mesoscopic laminate model. (a) A single cell, (b) Assembly for a ply, (c) Assembly for a [0/90/90/0] laminate.

form at regular spacings and eventually connect to give a so-called "peanut" shaped delamination. For the multi-layered laminate shown in Fig. 4.2.21b, a series of such delaminations are created, with each orientated along the fibre direction of the adjacent lower ply. Fig. 4.2.21c shows some impact test results that support these arguments and do approximately show the above mentioned "peanut" inter-ply delamination shape.

The widely use macromechanical approach described in Section 7.3 that has independent ply and delamination failure models cannot capture this direct link between intra-ply and inter-ply failure, although it is likely failure in one will promote failure in the other. Furthermore, current ply models cannot model this localised intra-ply failure. In (Jäger 2020; Jäger et al. 2016) several approaches to this problem have

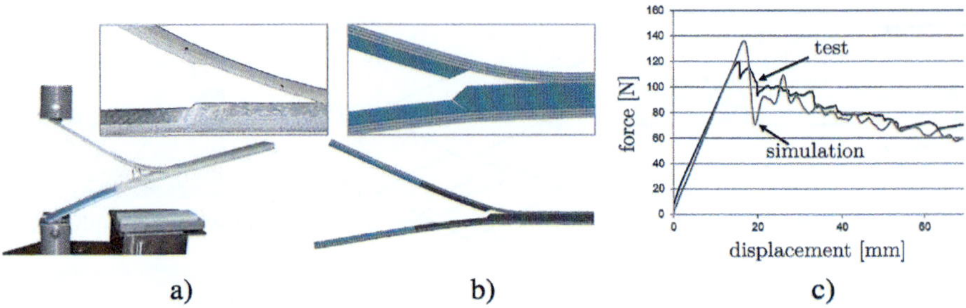

Figure 4.2.23 DCB test and simulation results. (a) DCB test. (b) Mesoscopic model results. (c) Force time histories.

been investigated. Among these, the following mesoscopic model was shown to work well for mode I (DCB), mode II (ENF) and plate impact. Essentially, each cell in Fig. 4.2.22a had tied interfaces with a failure criterion that could represent transverse ply fracture, from this plies were built Fig. 4.2.22b, and assembled with delamination interfaces in Fig. 4.2.22c.

Only the DCB case are presented here in Fig. 4.2.23 for a [0/90/90/0] laminate, demonstrating correct modelling of the fracture sequence and giving a good agreement to the test force time history. Simulations of plate impact presented in (Jäger et al. 2016) were found to agree well with tests and did accurately predict the peanut shaped inter-ply delaminations.

9. Conclusions

Today engineers have a range of essential codes available based on analytical methods for micromechanics, textile forming and especially laminate analyse that greatly help any composites design. However, it should be borne in mind that their accuracy is dependent on material input properties, which is often difficult to obtain for fibres, resins, and plies, and there are assumptions inherent in all these methods. Some care is needed and where possible it is advisable to support results with suitable experimental tests.

Over the past 50 years or more FE codes to analyse metal parts for manufacturing processes, stiffness, failure, crashworthiness, and other loading conditions have been established and are now well validated. Efforts to adapt these methods to composites analysis has proven more challenging, mostly because of the complex nature of fibre reinforced composites. However, today leading commercial FE codes all have extensive capabilities for linear analysis. Failure prediction is an important topic that has advanced over the last 20 years leading to accurate failure criteria for UD composites, but further work is still needed to improve failure for 2D and 3D textile composites and damage modelling for impact and crash, especially if manufacturing effects like preform shearing must be taken into account.

FE codes for manufacture simulation of draping, thermoforming, braiding and infusion are all now available. But one notable difficulty here is the lack of appropriate standardised test procedures for constitutive properties to help calibrate material laws. Finally, in this chapter an opportunity has been taken to highlight some future possibilities for meso-scale analysis. These do offer some unique capabilities for improved accuracy, chaining of manufacturing stages through to parts analysis, and could possibly help reduce materials testing for properties that can be hard to measure.

References

Aerospace Engineering, 2019. CompositeLaminateAnalysisSpreadsheet.

Allix, O., Ladevèze, P., 1992. Interlaminar interface modelling for the prediction of delamination. Compos. Struct. 22.4, 235–242.

Arnold, S.M., et al., 1999. Micromechanics analysis code with generalized method of cells (MAC/GMC): user guide. In: NASA/TM-1999-209070.

Barbero, E.J., 2017. Introduction to Composite Materials Design. CRC Press.

BE-5092, 1992-1996. Brite-Euram Project: Industrial Press Forming of Corltinuous Fibre Reinforced Thermoplastic Sheets and the Development of Numerical Simulation Methods. https://cordis.europa.eu/docs/projects/files/BRE/BRE20143/26839481-6_en.pdf.

Belytschko, T., Lin, J.I., Tsay, C.S., 1983. Explicit algorithms for the nonlinear dynamics of shells. In: Computer Methods in Applied Mechanics and Engineering, 42.

Birkefeld, K., Pickett, A.K., Middendorf, P., 2018. Virtual design and optimisation of braided structures considering production aspects of the preform. In: Beaumont, P.W.R., H Zweben, C. (Eds.), Comprehensive Composite Materials II, 8. Academic Press, Oxford, pp. 85–97.

Brandt, J., Pickett, A.K., Schneider, M., 1992-1997. Rechnergesstütze, automatisierte herstellung von 3D geflechten fur faserverbundbauteile, BMBF project, No. 03N3036.

Chedmail, J.F., et al., 1986. Numerical techniques, experimental validation and industrial applications of structural impact and crashworthiness analysis with supercomputers for automotive industries. In: Marino, C. (Ed.), Conf. Supercomputers for the Automotive Industries. Computational Mechanics Publications, pp. 127–145.

Cox, B.N., Flanagan, G., 1997. Handbook of Analytical Methods for Textile Composites. Langley research center, Hampton, Virginia, US.

Creech, G., 2006. Mesoscopic Finite Element Modelling of Non-crimped Fabrics for Drape and Failure Analysis. School of Applied Sciences, Cranfield University, UK.

Creech, G., Pickett, A.K., 2006. Meso-modelling of non-crimp fabric composites for coupled drape and failure analysis. J. Mater. Sci. 41, 6725–6736.

Daniel, I.M., Ishai, O.I., 2005. Engineering Mechanics of Composite Materials. Oxford University Press.

Demaria, C., Ruiz, E., Trochu, F., 2007. In-plane anisotropic permeability characterization of deformed woven fabrics by unidirectional injection. Part II: prediction model and numerical simulations. Polym. Compos. 28 (6), 812–827.

Gibson, R.F., 2011. Principles of Composite Material Mechanics. CRC Press.

Haug, E., Scharnhorst, T., Du Bois, P., 1986. FEM-Crash, Berechnung eines Fahrzeugfrontalaufpralls. VDI-Ber. 613, 479–505.

Hill, R., 1952. The elastic behavior of a crystalline aggregate. In: Vol. A65. Proceedings of the Physical Society, pp. 349–354.

Hwang, Y., Um, H., Kim, H., 2023. Multi-scale progressive failure analysis of shear deformed woven fabric composites considering its pre-forming process. Compos. Appl. Sci. Manuf. 174, 107713. https://doi.org/10.1016/j. compositesa.2023.107713.

Jäger, S., 2020. Untersuchung Und Simulation Der Schädigenden Wirkung Von Transversalen Stoßlasten Auf Faserverbundkunststoffe. Universität Stuttgart, IFB.

Jäger, S., Pickett, A.K., Middendorf, P., 2016. A discrete model for simulation of composites plate impact including coupled Intra- and interply failure. Appl. Compos. Mater. 23, 179–195.

Johnson, A.F., Pickett, A.K., Rozycki, P., 2001. Computationalmethodsfor predicting impact damage in composite structures. Compos. Sci. Technol. 61 (15), 2183–2192. https://doi.org/10.1016/S0266-3538(01)00111-7.

Kachanov, L.M., 1958. On the creep rupture time (In Russian). In: Izv. AN SSSR, Otd. Tehn. Nauk, 8, pp. 26–31.

Ladevèze, P., LeDantec, E., 1992. Damage modelling of the elementary ply for laminated composites. Compos. Sci. Technol. 43 (3), 257–267. https://doi.org/10.1016/0266-3538(92)90097-M.

eLamX, 2021. Institute of Aerspace Engineering. Dresden. https://tudresden.de/ing/maschinenwesen/ilr/lft/elamx2/elamx.

Lee, W.I., Springer, G.S., 1987. A model of the manufacturing process of thermoplastic matrix composites. J. Compos. Mater. 21 (11), 1017–1055. https://doi.org/10.1177/002199838702101103.eprint.

Lefebvre, D., Haug, E., Hatt, F., 1994. Industrial applications of computer simulation in stamping. J. Mater. Process. Technol. 46 (3), 351–389. https://doi.org/10.1016/0924-0136(94)90121-X.

Lemaitre, J., Chaboche, J.L., 1994. Mechanicsofsolidmaterials. Cambridge University Press.

LIMS, 2019. Liquid Injection Molding Simulation. University of Delaware. https://sites.udel.edu/lims/.

Llorca, J., et al., 2011. Multiscale modeling of composite materials: a roadmap towards virtual testing. Adv. Mater. 23 (44), 5130–5147.

Mack, C., Taylor, H.M., 1956. The fitting of woven cloth to surfaces. J. Textile Inst. 47, 477–488.

PAM-CRASH, 1985. ESI Group, Paris, France.

PAM-FORM, 1998. ESI Group, Paris, France.

PAM-QUIKFORM, 2000. ESI Group, Paris, France.

PAM-RTM, 2005. ESI Group, Paris, France.

Phillips, R., et al., 1996. Processing limits for fibre-reinforced thermoplastic composites. In: 17th Int. SAMPE Europ. Conf, pp. 341–351.

Pickett, A.K., 2006. Mounting tension. In: Materials World.

Pickett, A.K., Erber, A., et al., 2009. Comparison of analytical and finite element simulation of 2D braiding. Plast. Rubber Compos. 38 (910), 387–395. https://doi.org/10.1179/146580109X12540995045769.

Pickett, A.K., Fouinneteau, M.R.C., Middendorf, P., 2009. Test and modelling of impact on pre-loaded composite panels. Appl. Compos. Mater. 16 (4), 225–244. https://doi.org/10.1007/s10443-009-9089-3.

Pickett, A.K., Sirtautas, J., Erber, A., 2009. Braiding simulation and prediction of mechanical properties. Appl. Compos. Mater. 16 (6), 345–364.

Pickett, A.K., 2024. Introduction to Process and Mechanical Modelling of Engineering Composites, third ed.

Pister, K.S., Asce, A.M., Dong, S.B., 1959. Elastic bending of layered plates. In: Journal of the Engineering Mechanics Devision: Proceedings of the American Society of Civil Engineers.

Reuss, A., 1929. Berechnung der Fließgrenze von Mischkristallen auf Grund der Plastizitätsbedingung für Einkristalle. Z. Angew. Math. Mech. 9, 49–58.

Rosso, S. Del, Iannucci, L., Curtis, P., 2019. Finite element simulation of the braiding process. In: *Mech Adv Mater Mod Process* 5.1. Rtmworx (2005). POLYWORX B.V., Nijverdal, Netherlands.

RTMworx, 2005. POLYWORX B.V., Nijverdal, Netherlands.

Schneider, M., Wulfhorst, B., Pickett, A.K., 1999. Micromodelling of yarn achitecture in 3D braids and transfer to macromodelling of composites. In: Texcomp-4.

Sirtautas, J., 2017. Coupled Fabric Draping and Infusion Process Simulation. University, Stuttgart.

Sirtautas, J., Pickett, A.K., Lépicier, P., 2013. A mesoscopic model for coupled drape-infusion simulation of biaxial non-crimp fabric. Compos. B Eng. 47, 48–57. https://doi.org/10.1016/j.compositesb.2012.09.088.

Stellbrink, K.K.U., 1996. Micromechanics of Composites: Composite Properties of Fibre and Matrix Constituents. Hanser.

Swery, E.E., et al., 2016. Towards a complete simulation process chain for the manufacturing of braided composite parts. In: FPCM-13. Kyoto, Japan.

TexGen, 2006. Texgen. School of Mechanical, Materials and Manufacturing Engineering. University of Nottingham, UK.

Turner, M.J., et al., 1956. Siffness and deflection analysis of complex structures. J. Aero.Sci. 23 (9), 805–823.

Verpoest, I., et al., 2007. Department MTM, Katholieke Universiteit Leuven, Belgium. Version: Micromechanics 2.5.

Voigt, W., 1889. Ueber die Beziehung zwischen den beiden Elasticitätsconstanten isotroper Körper. Ann. Phys. 274 (12), 573–587. https://doi.org/10.1002/andp.18892741206.

WiseTex, 2000. An Integrated Tool for Modelling Textiles and Textile Composites. Composite Materials Group, KU Leuven, Belgium.

From micro/meso-mechanics to FEM analysis of aerospace structural details

4.3

Pedro P. Camanho and Albertino Arteiro
INEGI, Faculdade de Engenharia, Universidade do Porto, Rua Dr. Roberto Frias, Porto, Portugal

1. The early days—ply reduction methods and the virtual crack-closure technique

Innovation has always been the lifeblood of the aerospace industry. One key aspect of innovation in aerospace is the introduction of new materials and structural concepts that have the potential to reduce weight and CO_X emissions, without compromising structural integrity and durability, and to promote the integral manufacturing of aerostructures. These drivers partially justify the success of polymer composites in the aerospace industry, despite the higher engineering costs and lead time in terms of design and certification. Simulation is key to mitigating these problems, reducing the number of physical tests in product development, instilling confidence in large-scale structural tests, and enabling certification by analysis supported by testing.

While the analysis tools that predict stiffness, dynamic response, and elastic instabilities of composite structures have matured rather quickly, the prediction of inelastic deformation, in particular damage propagation and ultimate failure, has been challenging the community for the last decades. Efforts have been devoted to predicting intralaminar damage, i.e., cracks that develop in the individual plies and that may take different geometries, and interlaminar cracks, or delamination, that correspond to the separation of the different plies. These damage mechanisms often interact in a complex way, a phenomenon that results in additional challenges for the analysis models.

One of the initial ideas that was put forward in the late 80s and early 90s to simulate intralaminar damage is the concept of the ply reduction method (Chang and Chang, 1987a,b; Chang and Lessard, 1991; Chen and Lee, 1995; Lessard and Shokrieh, 1995; Ochoa and Engblom, 1987). Simply put, the method consists of a stress analysis, typically conducted using the finite element method (FEM); a set of failure criteria that identifies the type of damage that occurs in the ply (matrix cracking and fibre failure); and a set of degradation functions of the respective components of the stiffness tensor (Fig. 4.3.1).

The stiffness degradation rules were either empirical, with a sudden degradation of the components of the stiffness tensor, or, in the case of fibre failure, often based on

The History of Composites. https://doi.org/10.1016/B978-0-443-34119-9.00026-7

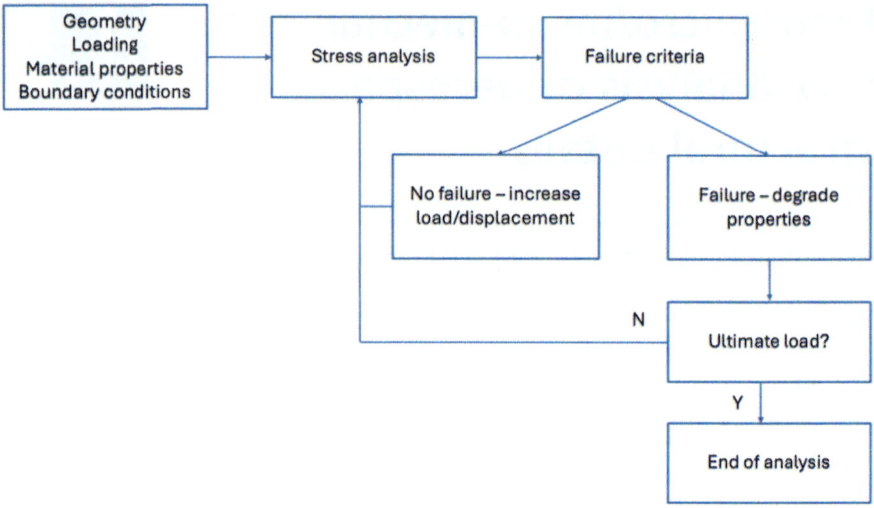

Figure 4.3.1 Overview of the ply reduction method (after Lessard and Shokrieh, 1995).

the Weibull distribution (Chang and Chang, 1987a), an idea that provided a stronger physical background to the models.

While several accurate simulations of failure using the ply reduction method were reported, especially in structural details containing stress concentrations, e.g., open or loaded holes, a fundamental problem still needed to be addressed: mesh sensitivity. In fact, using the ply reduction method, the energy dissipated by damage is a function of the size of the finite element used, and an objective, mesh-independent response is not ensured. Additionally, there was no agreement about the proper failure criteria to use for the identification of the different ply failure mechanisms, an issue that was later addressed in the worldwide failure exercise coordinated by Hinton et al. (2004).

Regarding the simulation of delamination, the early efforts consisted mainly of direct extensions of linear-elastic fracture mechanics (LEFM), taking advantage of the well-defined geometry of the crack: the separation of the composite plies at their interfaces. Notwithstanding the simplicity of applying direct extensions of LEFM, the geometry of the delamination had to be known a priori, e.g., from inspection, or assumed, e.g., taking into account the precision of the inspection methods.

Two analysis models that correspond to extensions of LEFM are worth mentioning: the crack tip element (CTE) and the virtual crack-closure technique (VCCT). The CTE is a simple technique proposed in the 90s by Davidson et al. (1995) to estimate the components of the energy release rate, and of the associated mixed-mode ratio, based on analytical, closed-form solutions established in terms of the forces and moments per unit width, elastic properties of the plies, and geometry of the laminate (including the location of the delamination). The VCCT is an extension of the crack-closure technique (CCT) used to calculate the components of the energy release rate using the work required to close the crack back to its original position. While the CCT requires two separate analyses to calculate nodal forces and the relative displacements associated with crack extension, the VCCT relies on one analysis only, assuming that the crack opening displacement is not significantly affected for a small crack extension; see Fig. 4.3.2.

The implementation of the VCCT requires detailed guidelines concerning its conditions of applicability and the type and size of elements that should be used. All the relevant aspects regarding the appropriate use of the VCCT are thoroughly discussed in the seminal contribution of Krueger (2004).

While the VCCT is fast and highly accurate in determining the components of the energy release rate in scenarios where LEFM applies, its fundamental hypotheses result in limitations when a large fracture process zone develops ahead of the crack tip, and when the structure does not contain a preexisting delamination that can be identified or assumed a priori. It is nevertheless worth mentioning the efforts conducted by Mabson et al. (2007) to generalise the use of the VCCT, introducing the concept of the VCCT element, and its later extensions to simulate delamination under fatigue loading (De Carvalho et al., 2019).

To address the difficulties identified here, the scientific community proposed alternative, more general analysis models that are discussed in the following section.

2. The big leap—continuum damage models and cohesive zone models

Not as often as desirable, there are scientific contributions that have a wide impact in the corresponding community and inspire generations of scientists. Two contributions played a fundamental role, not only in fracture mechanics in general, but also in the models that were subsequently developed specifically for polymer composite materials. These contributions resulted from the work of Kachanov (1958) and Rabotnov (1969), who first proposed the concept of continuum damage, and of Barenblatt

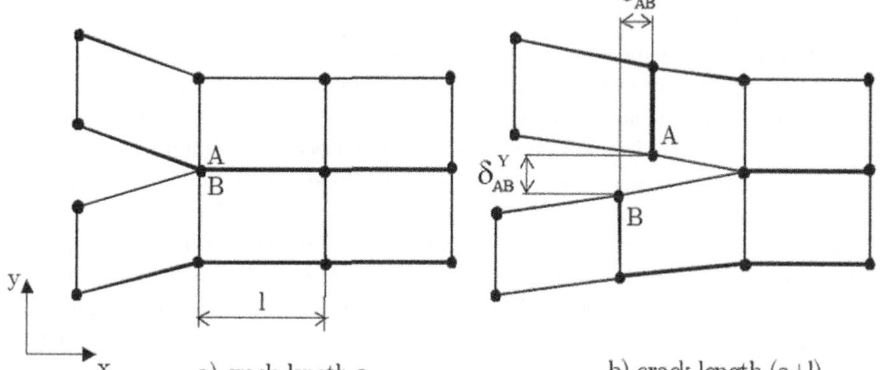

Figure 4.3.2 Virtual crack-closure technique for 2D solid elements.

(1962), who introduced the concept of a cohesive zone ahead of the crack tip, in the spirit of Dugdale's strip zone model (Dugdale, 1960). These were the building blocks that enabled the development of a new generation of analysis models that addressed the main problems identified in the previous chapter regarding the ply reduction methods and the VCCT.

One of the initial contributions to simulate ply damage using continuum damage mechanics and the corresponding internal variables was the work of Talreja (1990), where a damage entity tensor d_{ij} that represents the effect of damage in the strain fields was proposed:

$$d_{ij} = \int_S a_i n_j dS \tag{4.3.1}$$

where $\boldsymbol{n}$ is the unit outward vector normal to the surface S, and $\boldsymbol{a}$ represents the extent and direction of the disturbance of the strain field that results from the presence of damage. The overall mechanical response of the composite material is predicted using a strain energy density function established in terms of the state of damage, e.g., the matrix cracking damage state. Related approaches, also using second-order tensors to represent the kinematics of cracks in composite materials, were proposed by Allen, Harris, and coauthors (Lee et al., 1989; Lo et al., 1991) and applied to the prediction of damage evolution under both quasistatic and fatigue loading. The work proposed by Matzenmiller et al. (1995) should also be emphasised, as it combined a sound thermodynamic basis with the suitability to be implemented in standard Finite Element (FE) codes, as it is based on strain-controlled continuum models. These characteristics were later explored by other researchers.

An alternative approach to simulate one particular damage mechanism, transverse matrix cracking, corresponds to the definition of representative volume elements (RVEs) of composite laminates with discrete representations of matrix cracks and, from the corresponding boundary value, predicts the stiffness reduction of a composite laminate as a function of the applied strain and of the predicted density of matrix cracks (Joffe and Varna, 1999; McCartney, 1998; Nuismer and Tan, 1988; Shahid and Chang, 1995; Tan and Nuismer, 1989). Not only this modelling strategy was strongly founded in the actual mechanics of accumulation of transverse cracks, but it also created the opportunity to formulate continuum damage models implemented using the FEM where the meaning of the damage variable and its evolution kinetics are directly obtained for a discrete representation of the process of accumulation of transverse cracks (Ladevèze et al., 2006; Mayugo et al., 2006, 2010), Figs. 4.3.3 and 4.3.4.

An important landmark in the use of continuum damage mechanics in the simulation of damage in composite materials corresponds to the work of Ladevèze and coauthors, who introduced for the first time the concept of damage meso-models (Ladevèze and Lubineau, 2001, 2002). The original idea put forward by Ladevèze was to define the strain energy density as a function of a set of scalar damage variables associated with the different ply damage mechanisms and to model the composite laminate as a set of plies connected by damageable interfaces. The work conducted

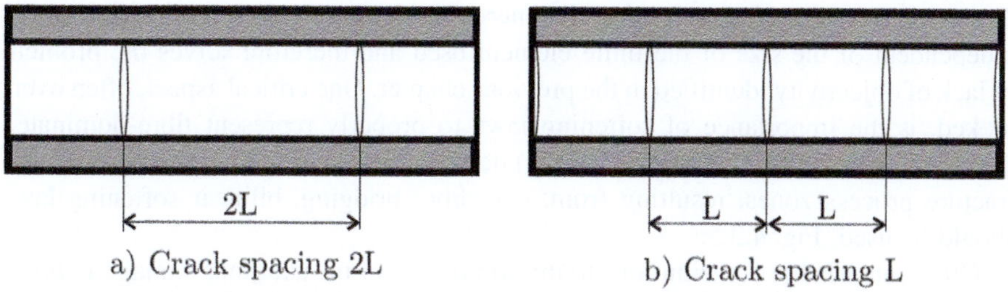

Figure 4.3.3 Discrete representation of transverse cracks (Mayugo et al., 2006).

by Ladevèze and coauthors at ENS-Cachan inspired several researchers who proposed alternative models using the concept of the meso-mechanical model. These include the work of Maimí et al. (2007a,b), which defined a continuum damage model where the complementary free energy density is established in terms of damage variables (d_i) that represent longitudinal and transverse damage:

$$G=\frac{\sigma_{11}^2}{2(1-d_1)E_1}+\frac{\sigma_{22}^2}{2(1-d_2)E_2}+\frac{\sigma_{12}^2}{2(1-d_6)G_{12}}-\frac{\nu_{12}}{E_1}\sigma_{11}\sigma_{22} \quad (4.3.2)$$

Some remarkable features of Maimí's model are the capability to represent crack closure under reverse load cycles and the regularisation of the solution using the concept of the crack band model originally proposed by Bažant and Oh (1983).

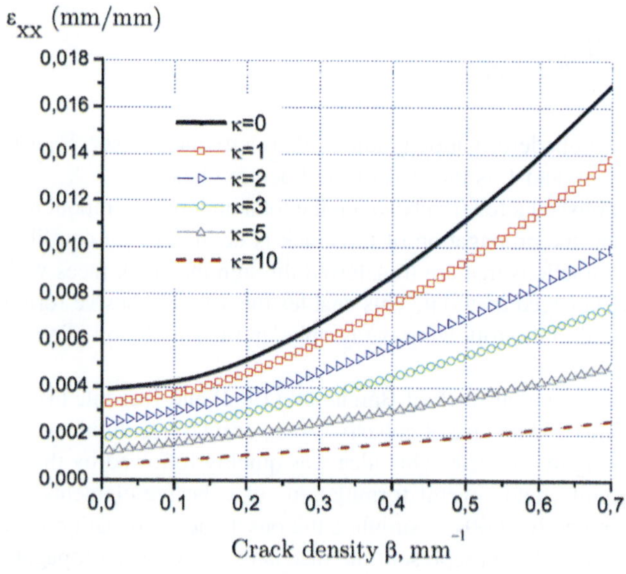

Figure 4.3.4 Predicted accumulation of transverse cracks as a function of the applied axial strain and mode ratio (k) for a ($\pm$25/90)$_3$ laminate (Mayugo et al., 2006).

The crack band model ensures that the energy dissipated by each type of fracture is independent of the size of the finite element used and therefore solves the problem of lack of objectivity identified in the previous chapter. One critical aspect, often overlooked, is the importance of softening laws to properly represent fibre-dominated damage mechanisms. Dávila et al. (2009) demonstrated that in the presence of large fracture process zones, resulting from, e.g., fibre bridging, bilinear softening laws should be used, Fig. 4.3.5.

Other remarkable contributions to the simulation of intralaminar damage in polymer composite laminates using the concepts of continuum damage mechanics include the works of Pineda and Waas (2013), Williams et al. (2003), Laurin et al. (2007), and Pinho et al. (2006a,b) that collectively provided the scientific community with robust analysis models that capture the main physics of the onset and propagation of the different ply failure mechanisms. Notwithstanding all these scientific contributions, there was still room for improvement regarding the kinematic representation of the failure mechanisms, especially in the presence of finite deformations and finite fibre rotations (Tijs et al., 2023). Based on the concept of a composite smeared crack model proposed by Camanho et al. (2013), Leone Jr (2015) proposed the concept of decomposition of the deformation gradient $\boldsymbol{F}$ to improve the kinematic representation of transverse cracks, including the ability to capture the split cracks that blunt the regions of stress concentrations in multidirectional laminates. While in the work of Leone Jr (2015), the contribution of the displacement jump to the total deformation gradient is assumed, Lopes et al. (2023) used homogenisation techniques to calculate the exact form of the contribution of the displacement jump ω (Fig. 4.3.6) to the total deformation gradient, resulting in the following closed-form solution:

$$F = F_B + \frac{|\Gamma|}{V_B} R_\alpha \cdot \omega \otimes e_N \tag{4.3.3}$$

where F_B is the bulk deformation gradient, R_α is a rotation tensor, V_B is the volume of the finite element, and Γ is the area of the fracture plane.

Having established a relation between the total and local components of the deformation gradient, the propagation of the crack is defined by an equilibrium equation that relates the tractions that are transferred through the crack faces with the stresses at the bulk, undamaged material. This enables the use of cohesive formations, which will be described in the following paragraphs, in the simulation of transverse matrix cracking.

Following the seminal work of Barenblatt (1962) and Dugdale (1960), Hillerborg et al. (1976) proposed the use of cohesive zone, or interface, finite element models to simulate cracking in concrete. This idea was quickly followed by the polymer composites community, and several formulations of cohesive elements were proposed in the 1990s and early 2000s to simulate the onset and propagation of delamination. Cohesive zone models can represent the mechanisms of crack propagation under the influence of a fracture process zone that bridges the crack, and a clear relation between

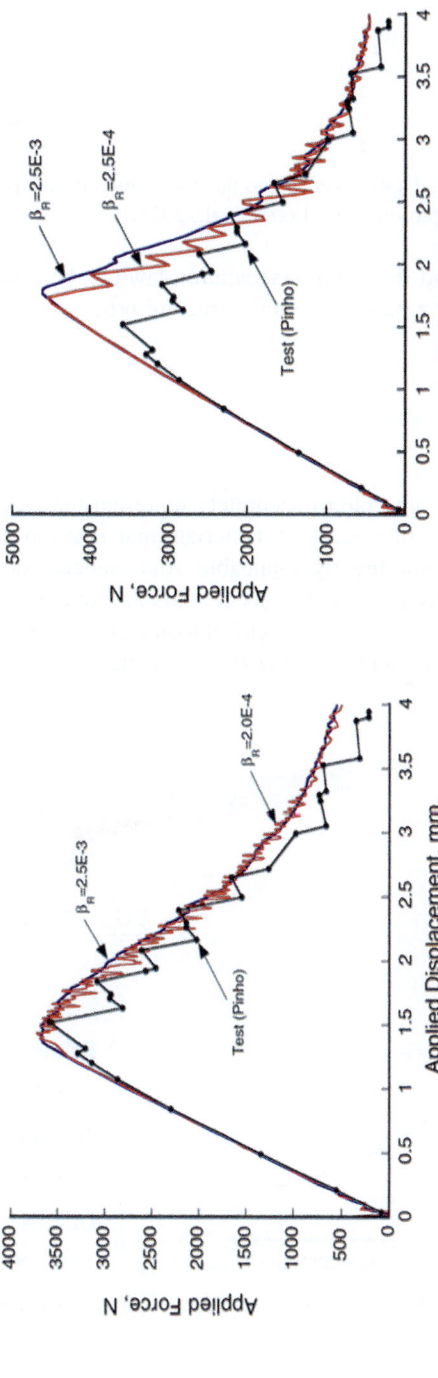

Figure 4.3.5 Comparison between experimental results and predictions using bilinear (left) and linear (right) softening laws (Dávila et al., 2009, reproduced with permission from Springer Nature).

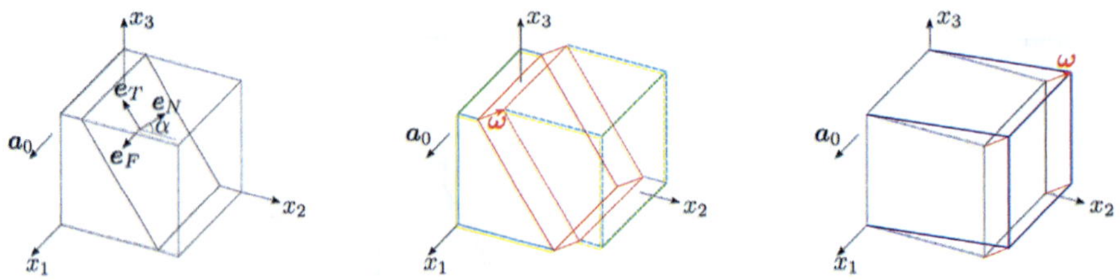

Figure 4.3.6 Fracture plane, displacement jump due to a transverse matrix crack, and the corresponding continuum representation (Lopes et al., 2023).

the energy release rate and the traction-separation law can be established by calculating the J-integral along a path through the fracture process zone:

$$G = \int_{o}^{\omega_f} \sigma(\omega) d\omega \tag{4.3.4}$$

Cohesive zone models are able to accurately represent the mechanisms of crack propagation for different crack sizes in materials that develop a fracture process zone, and when LEFM is not directly applicable. An excellent example of this capability was presented by Dávila's simulations of a double cantilever beam (DCB) with initial cracks of different sizes, where the global load-displacement response changed from ductile, to quasibrittle, and to brittle (for large cracks, where LEFM is applicable), Fig. 4.3.7.

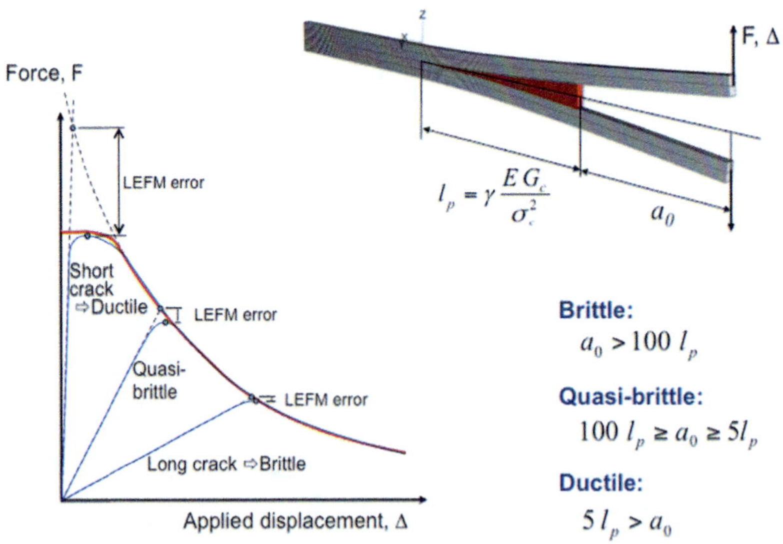

Figure 4.3.7 Simulation of a DCB specimen with different initial crack lengths using cohesive elements.
Figure courtesy of Dr. Carlos Dávila, NASA—LaRC

Three additional developments generalised the applicability of cohesive zone models for the simulation of delamination: the definition of softening laws in the presence of large-scale bridging (Sørensen et al., 2000), the definition of cohesive zone models for delamination using only one scalar damage variable, since the orientation of the fracture plane is known a priori (Allix and Ladevèze, 1992; Turon et al., 2006), and the extension of cohesive zone models to simulate high-cycle fatigue, in particular the seminal work of Turon et al., where a relation between Paris' law and cohesive zone models was proposed for the first time (Turon et al., 2007). It is worth noting that cohesive zone models were also used to represent in a discrete way transverse matrix cracks, which enabled the study and the understanding of the origins of the size effects observed in notched composite laminates (Wisnom, 2010; Wisnom et al., 2010). Discrete representations of transverse matrix cracks were also suggested in the scope of mesh-independent modelling techniques (Chen et al., 2016; Iarve et al., 2006; van der Meer et al., 2010).

With the advances in continuum damage and cohesive zone models described in the previous paragraphs, it became possible to perform simulations of failure in complex composite aircraft structural details, for example, composite laminates with notches and cracks (Furtado et al., 2020), impact and edge impact (Lopes et al., 2016; Arteiro et al., 2020), and bolted joints (Zhuang et al., 2019), Figs. 4.3.8–4.3.11.

3. Recent developments—computational micromechanics and phase-field methods

Well before the appearance of advanced composite materials, the scientific community was already interested in understanding and predicting the properties of

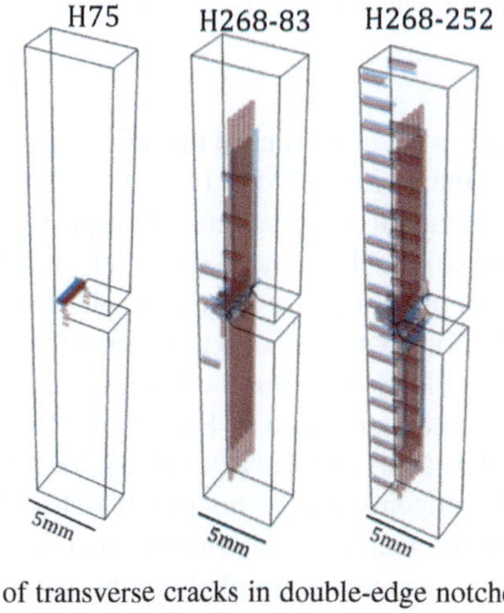

Figure 4.3.8 Simulation of transverse cracks in double-edge notched specimens (reprinted from Furtado et al., 2020, with permission from Elsevier).

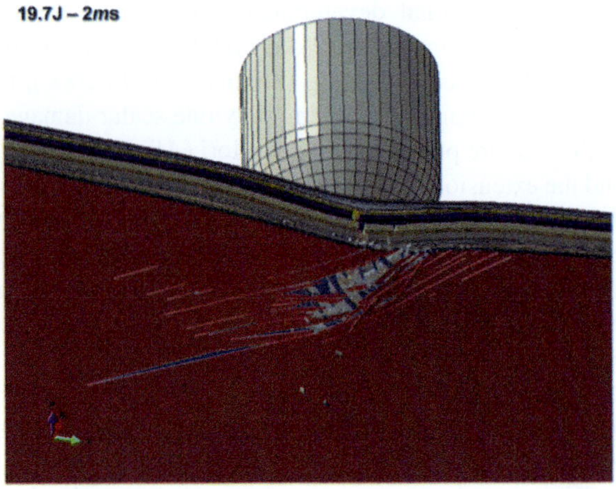

Figure 4.3.9 Simulation of through-the-thickness impact in a composite laminate (reprinted from Lopes et al., 2016, with permission from Elsevier).

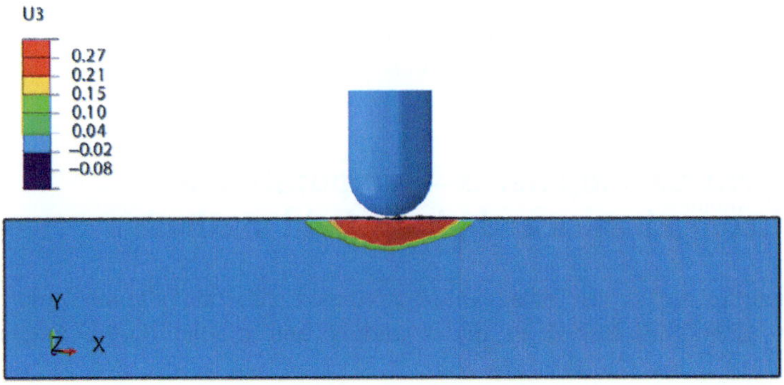

Figure 4.3.10 Simulation of edge impact and compression after edge impact in a carbon fibre-reinforced polymer laminate (reprinted from Arteiro et al., 2020, with permission from Elsevier).

multiphase materials based on the properties of the individual phases. This interest can be traced back to, at least, the works of Voigt (1889) and Reuss (1929) with the proposal of the rule of mixtures and its inverse version, respectively. It was later demonstrated (Paul, 1960) that these models define, respectively, an upper and a lower bound on the elasticity coefficients of multiphase materials, which were further tightened by Hashin and Shtrikman (1963) and extended to transversely isotropic composite materials by Hill (1964) and Hashin (1965). But the first predictive models for the elastic properties of composites can be attributed to Hashin and Rosen (1964), Hill (1965), Budiansky (1965), and later, Mori and Tanaka (1973), the latter marking the advent of predictive mean-field homogenisation techniques (Doghri et al., 2010). Now, while these methods can be very efficient in the determination of the effective elastic properties—and even the inelastic response (provided that no strain softening occurs)—of composites based on the elastic properties—and inelastic response—of the

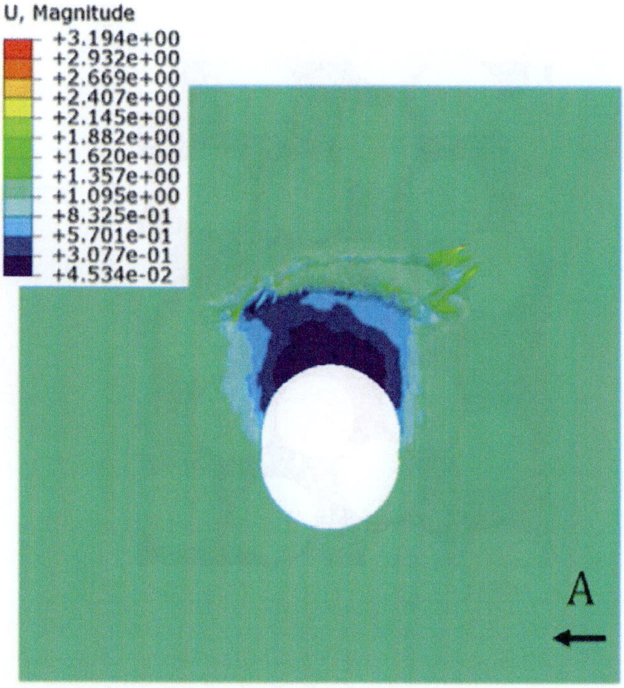

Figure 4.3.11 Simulation of bearing failure in a composite bolted joint (reprinted from Zhuang et al., 2019, with permission from Elsevier).

constituents, the semianalytical prediction of the strength and fracture toughness of composites remains a significant challenge. Benefiting from the ever-increasing speed and reliability of computers, computational micromechanics became an attractive tool to overcome the challenge of determining the effective response of composite materials based on the response of the reinforcing and matrix materials, without major limitations in the constitutive behaviour of the constituents and their interfaces.

But, in addition to the properties of the reinforcing and matrix materials, which are seen as individual homogeneous materials, and their interfaces, computational micromechanics also requires a truthful geometrical representation of the arrangement of the reinforcements. In fact, the efficient definition of materially and statistically RVEs of the microstructure of composite materials is still subject to development (e.g., Tian et al., 2024) by the scientific community, while early studies (Asp et al., 1996a,b; Blackketter et al., 1993; Böhm and Rammerstorfer, 1991; Brockenbrough et al., 1991) based on computational micromechanics employed simple, regular square or hexagonal fibre patterns. Nowadays (Trias et al., 2006a), it is well recognised that, for statistical and reliability analysis or for the simulation of local phenomena and failure, random models must be considered (see Fig. 4.3.12).

An RVE can be defined as the smallest volume of a heterogeneous material whose homogenised response is representative of the global mechanical performance of the

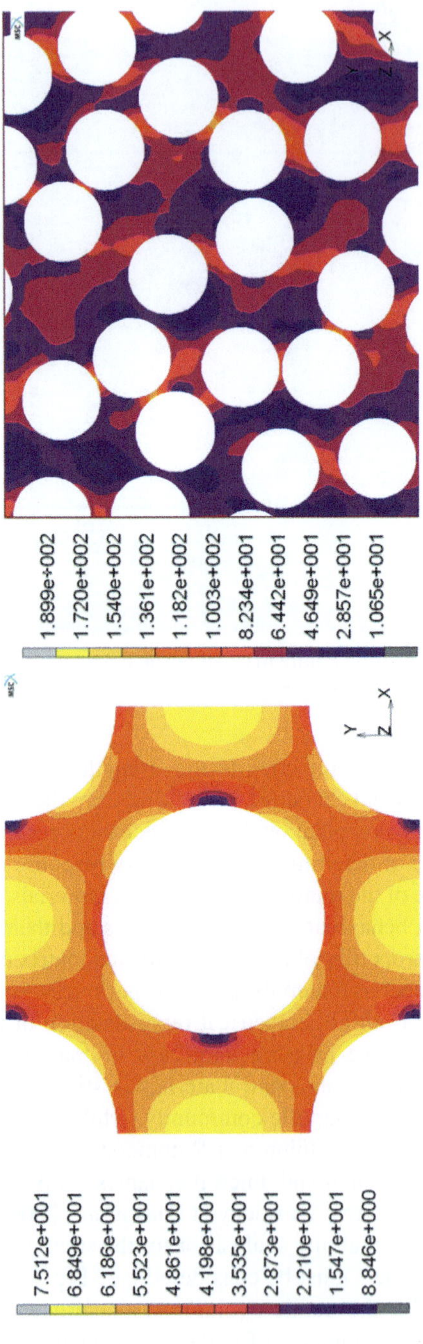

Figure 4.3.12 Von Mises stress in a periodic (left) and random (right) fibre pattern based on computational micromechanics (reprinted from Trias et al., 2006a, with permission from Elsevier).

material system. In the case of composite materials, in addition to the spatial arrangement of reinforcements, other important aspects include the size of the RVE, which must be sufficiently large to become virtually independent of the randomisation of the microstructure, or the volume fraction of the reinforcements. And these two aspects bring two important challenges: (i) the RVE cannot be too large, as it would endanger the possibility of numerically analysing it; (ii) difficulties usually arise in generating distributions with fibre volume fractions greater than 50%. The first challenge often requires either running RVEs of different sizes until RVE size convergence is achieved—defining the minimum size of the RVE (Trias et al., 2006b; Vallmajó et al., 2024)—or running a sufficiently large number of samples of randomly generated (smaller) volume elements (or statistical volume elements, SVEs), whose statistics are representative of the continuum at a higher scale—defining the minimum number of samples to be run (Tallman et al., 2020). The second challenge requires adopting efficient algorithms (Catalanotti, 2016; Ferreira et al., 2022; Melro et al., 2008; Tian et al., 2024; Vila-Chã et al., 2021) to provide materially and statistically representative distributions of fibres.

In the generation of RVEs for computational micromechanics, one important constraint is related to the implementation of the boundary conditions (see Fig. 4.3.13). Periodic boundary conditions (PBCs) are still regarded as the best compromise between accuracy and computational simplicity. However, the usage of PBCs has limitations in the simulation of strain localisation phenomena, such as damage propagation, because they overconstrain the strain field at the boundaries; this has led to new formulations of special boundary conditions (e.g., Coenen et al., 2012) that suppress this problem, which require more complex implementation.

Another key ingredient of computational micromechanics is the definition of the constitutive laws that govern the behaviour of the composite constituents and their interfaces (see Fig. 4.3.13). While the use of an elasto-plastic constitutive model with damage based on paraboloidal yield and failure criteria (Melro et al., 2013a) for the polymer matrix, a nonlinear-elastic continuum damage mechanics model based on a Weibull strength distribution (Tavares et al., 2016) for the reinforcing fibres and a bilinear traction-separation law (Melro et al., 2013b) for the interfaces are the most widely used options, recent studies demonstrate that additional work is required for an accurate representation of the inelastic deformation and fracture of the polymer matrix constrained by the stiff fibres (Chevalier et al., 2019; Klavzer et al., 2024) and of longitudinal failure (Breite et al., 2022) of advanced composite materials. In addition, the characterisation characterisation of the constituents under the conditions that ensure proper representation of their mechanical response in the presence of the different phases remains a significant challenge, with recent contributions relying on in situ characterisation (Naya et al., 2017a,b) and inverse identification of the properties of the matrix, fibres, and interfaces.

Nevertheless, computational micromechanics has found its way into the polymer composites community, with demonstrated applications in the understanding and prediction of complex phenomena such as (see Fig. 4.3.14) ply size effects (Arteiro et al., 2014, 2015; Saito et al., 2014), hybridisation of fibres (Swolfs et al., 2013b; Tavares et al., 2016), intralaminar stress redistributions (Swolfs et al., 2013a, 2015; Tavares

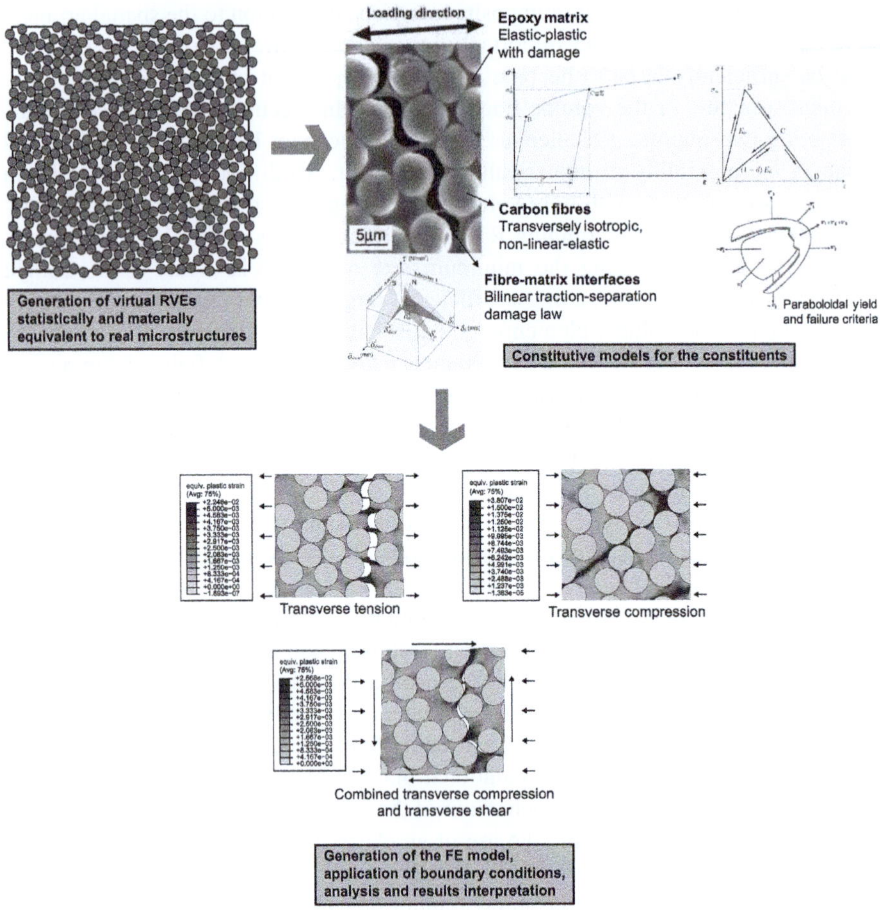

Figure 4.3.13 Main ingredients of computational micromechanics (Arteiro et al., 2019, reproduced with permission from Springer Nature).

et al., 2019) and interlaminar crack friction (Varandas et al., 2017), migration (Varandas et al., 2019), and bridging effects (Naya et al., 2019). Yet, challenges remain with respect to the prediction and representation of complex crack paths that can develop at the microscale (including crack branching and merging), the regularisation of the constitutive softening laws in continuum damage mechanics due to very small elements necessarily used in microscale simulations, or the formulation and solution of coupled, multifield problems at the microscale. Some of these challenges are not exclusive to the microscale, as they can also be found in the implementation of meso- and macroscale models for fibre-reinforced composites.

A recent method that is gaining momentum in the field of computational fracture mechanics is the phase-field method for fracture. It results from the seminal work of Francfort, Marigo, and coauthors (Bourdin et al., 2000; Francfort et al., 2008; Francfort and Marigo, 1998), who expressed Griffith's differential energy balance in a

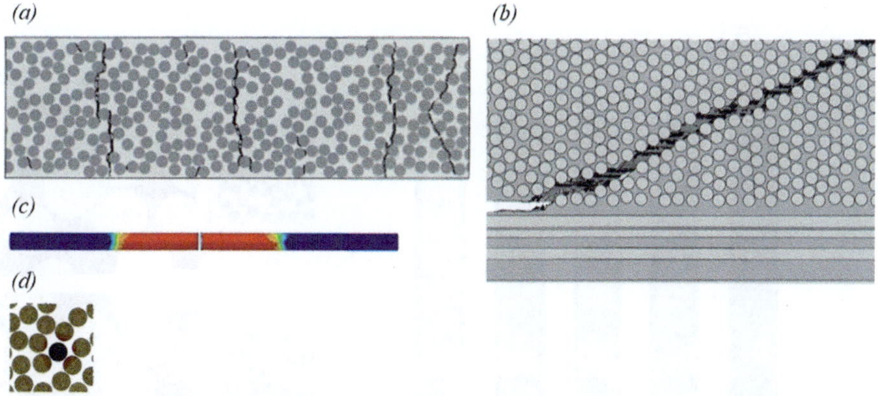

Figure 4.3.14 Examples of application of computational micromechanics. (A) Matrix cracking and in situ effect (reprinted from Arteiro et al., 2014, with permission from Elsevier). (B) Interlaminar crack migration (reprinted from Varandas et al., 2019, with permission from Elsevier). (C) Fiber–matrix decohesion after longitudinal fibre failure. (D) Stress concentrations in the surrounding fibres after fibre failure (Tavares et al., 2016, reproduced with permission from Springer Nature).

variational integral form. Through a damage-like variable, a regularised crack representation can be defined, replacing the surface integral on the unknown crack path by a volume integral within a diffuse damaged zone whose size is given according to a length scale parameter.

Nowadays, we find application of the phase-field method to model fracture of composite materials (see Fig. 4.3.15) at the macroscale (Mitrou et al., 2023; Reinoso et al., 2017), at the mesoscale (Pillai et al., 2020; Quintanas-Corominas et al., 2019), and at the microscale (Espadas Escalante et al., 2019; Guillén-Hernández et al., 2019, 2020, Macías et al., 2023; Tan and Martínez-Pañeda, 2021, 2022). However, full exploitation of the advantages of the phase-field method is still hindered by the need to better understand the effects of the choice of the constitutive law (incl. plasticity), the strain energy decomposition, the phase-field formulation, the length scale, the crack driving forces, the loading conditions, etc., on the (generally anisotropic) behaviour of fibre-reinforced composites. All these aspects will certainly keep the scientific community interested in exploring the application of phase-field to composite materials for some years ahead.

4. Conclusions and the road ahead

This chapter described the quest for robust analysis models for the prediction of the different failure mechanisms that are observed in polymer composites reinforced by unidirectional plies. While the last decades have witnessed significant progress, from the understanding of the failure mechanisms at different length and timescales

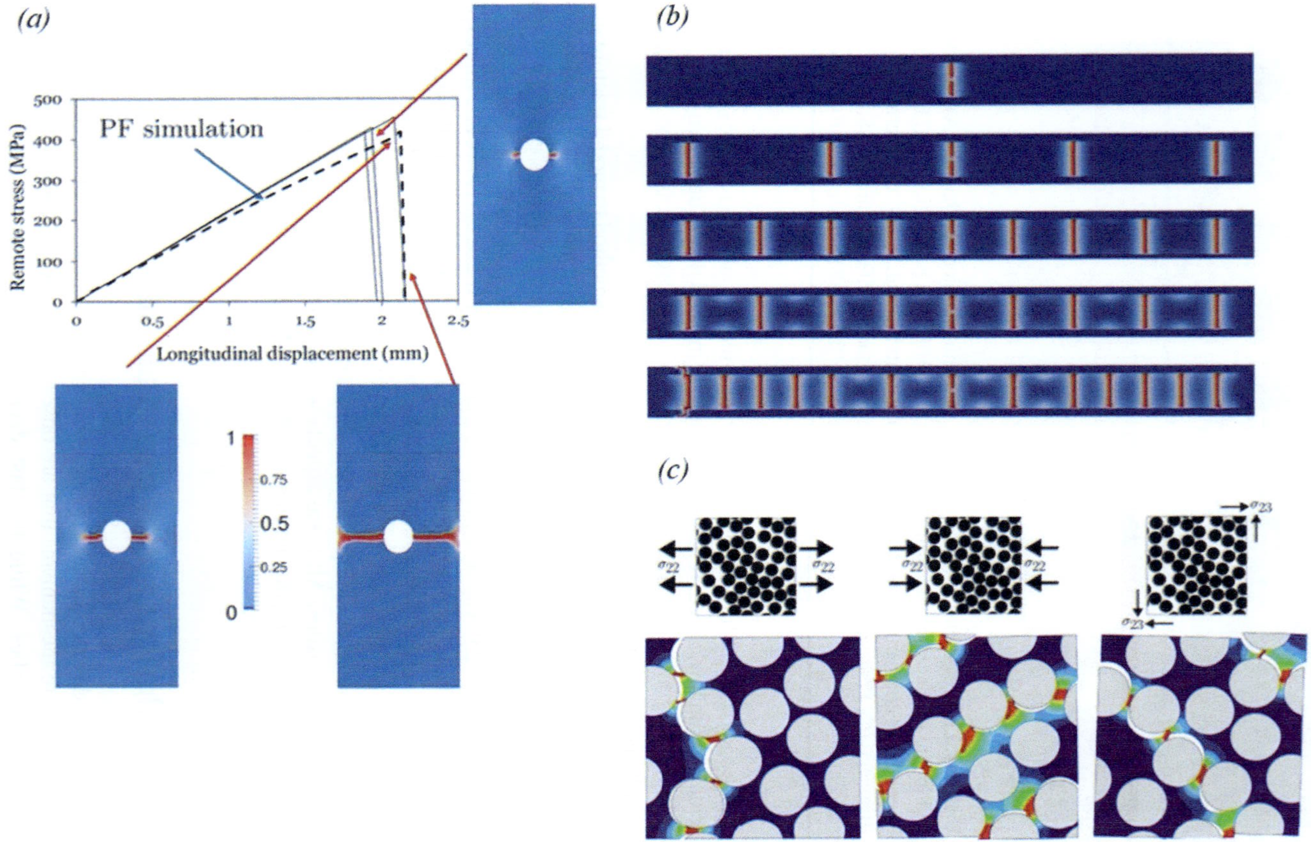

Figure 4.3.15 Applications of the phase-field method to model fracture of fibre-reinforced composites at different length scales. (A) Macroscale (reprinted from Reinoso et al., 2017, with permission from Elsevier). (B) Mesoscale (reprinted from Quintanas-Corominas et al., 2019, with

to the capability to predict failure in complex composite structural details, there is room for improvement in several areas.

One of the most pressing requirements is the use of the analysis models developed by the scientific community in the formal workflow of model validation, verification, and uncertainty quantification (ASME, 2020). The existence of robust models is a necessary, but not sufficient, condition for the change of mindset in the process of certification of composite aerostructures, i.e., certification by analyses supported by experiments rather than certification by experiments supported by analyses. Additional developments in system decomposition methodologies and surrogate models, potentially based on artificial intelligence techniques, are also required to enable this vision.

Another topic where the scientific community should place its efforts is on the development of multiscale models, especially for new materials where there is scarce experimental data about their mechanical behaviour, and when the effects of the microstructure on the structural performance remain elusive. One major hurdle must be overcome—the large computational time that is required to solve the boundary-value problem at the microscale, especially when nonlinear effects are present. There are recent outstanding scientific contributions to solve this problem in the presence of plasticity, namely the self-consistent clustering methods proposed by Bessa et al. (2017), Ferreira (2022), and Liu et al. (2016), and the physically recurrent neural networks proposed by Maia et al. (2023). Notwithstanding these remarkable contributions, where the time to solve the boundary-value problem at the microscale can be drastically reduced, one topic will keep the scientific community busy in the coming years—how to reduce computational time in problems at the microscale that include all the complex features of composite failure?

References

Allix, O., Ladevèze, P., 1992. Interlaminar interface modelling for the prediction of delaminations. Compos. Struct. 22 (4), 235–242. https://doi.org/10.1016/0263-8223(92)90060-P.

Arteiro, A., Catalanotti, G., Melro, A.R., Linde, P., Camanho, P.P., 2014. Micro-mechanical analysis of the *in situ* effect in polymer composite laminates. Compos. Struct. 116, 827–840. https://doi.org/10.1016/j.compstruct.2014.06.014.

Arteiro, A., Catalanotti, G., Melro, A.R., Linde, P., Camanho, P.P., 2015. Micro-mechanical analysis of the effect of ply thickness on the transverse compressive strength of polymer composites. Compos. Appl. Sci. Manuf. 79, 127–137. https://doi.org/10.1016/j.compositesa.2015.09.015.

Arteiro, A., Catalanotti, G., Reinoso, J., Linde, P., Camanho, P.P., 2019. Simulation of the mechanical response of thin-ply composites: from computational micro-mechanics to structural analysis. Arch. Comput. Methods Eng. 26, 1445–1487. https://doi.org/10.1007/s11831-018-9291-2.

Arteiro, A., Gray, P.J., Camanho, P.P., 2020. Simulation of edge impact and compression after edge impact in CFRP laminates. Compos. Struct. 240, 112018. https://doi.org/10.1016/j.compstruct.2020.112018.

ASME, 2020. V V 10 - standard for verification and validation in computational solid mechanics. ISBN: 9780791873168.

Asp, L.E., Berglund, L.A., Talreja, R., 1996a. Effects of fiber and interphase on matrix initiated transverse failure in polymer composites. Compos. Sci. Technol. 56, 651–665. https://doi.org/10.1016/0266-3538(96)00047-4.

Asp, L.E., Berglund, L.A., Talreja, R., 1996b. Prediction of matrix-initiated transverse failure in polymer composites. Compos. Sci. Technol. 56, 1089–1097. https://doi.org/10.1016/0266-3538(96)00074-7.

Barenblatt, G.I., 1962. The mathematical theory of equilibrium cracks in brittle fracture. Adv. Appl. Mech. 7, 55–129. https://doi.org/10.1016/S0065-2156(08)70121-2.

Bažant, Z.P., Oh, B.H., 1983. Crack band theory for fracture of concrete. Matériaux et Construction 16, 155–177. https://doi.org/10.1007/BF02486267.

Bessa, M.A., Bostanabad, R., Liu, Z., Hu, A., Apley, D.W., Brinson, C., Chen, W., Liu, W.K., 2017. A framework for data-driven analysis of materials under uncertainty: Countering the curse of dimensionality. Comput. Methods Appl. Mech. Eng. 317, 633–667. https://doi.org/10.1016/j.cma.2017.03.037.

Blackketter, D.M., Upadhyaya, D., King, T.R., 1993. Micromechanics prediction of the transverse tensilc strength of carobn fiber/epoxy composites: the influence of the matrix and interface. Polym. Compos. 14 (5), 437–446. https://doi.org/10.1002/pc.750140511.

Böhm, H.J., Rammerstorfer, F.G., 1991. Micromechanical investigation of the processing and loading of fibre-reinforced metal matrix composites. Mater. Sci. Eng. A 135, 185–188. https://doi.org/10.1016/0921-5093(91)90558-5.

Bourdin, B., Francfort, G.A., Marigo, J.-J., 2000. Numerical experiments in revisited brittle Fracture. J. Mech. Phys. Solid. 48 (4), 797–826. https://doi.org/10.1016/S0022-5096(99)00028-9.

Breite, C., Melnikov, A., Turon, A., de Morais, A.B., Le Bourlot, C., Maire, E., Schöberl, E., Otero, F., Mesquita, F., Sinclair, I., Costa, J., Mayugo, J.A., Guerrero, J.M., Gorbatikh, L., McCartney, L.N., Hajikazemi, M., Mehdikhani, M., Mavrogordato, M.N., Camanho, P.P., Tavares, R., Spearing, S.M., Lomov, S.V., Pimenta, S., Van Paepegem, W., Swolfs, Y., 2022. Detailed experimental validation and benchmarking of six models for longitudinal tensile failure of unidirectional composites. Compos. Struct. 279, 114828. https://doi.org/10.1016/j.compstruct.2021.114828.

Brockenbrough, J.R., Suresh, S., Wienecke, H.A., 1991. Deformation of metal-matrix composites with continuous fibers: geometrical effects of fiber distribution and shape. Acta Metall. Mater. 39 (5), 735–752. https://doi.org/10.1016/0956-7151(91)90274-5.

Budiansky, B., 1965. On the elastic moduli of some heterogeneous materials. J. Mech. Phys. Solid. 13 (4), 223–227. https://doi.org/10.1016/0022-5096(65)90011-6.

Camanho, P.P., Bessa, M.A., Catalanotti, G., Vogler, M., Rolfes, R., 2013. Modeling the inelastic deformation and fracture of polymer composites – part II: smeared crack model. Mech. Mater. 59, 36–49. https://doi.org/10.1016/j.mechmat.2012.12.001.

Catalanotti, G., 2016. On the generation of RVE-based models of composites reinforced with long fibres or spherical particles. Compos. Struct. 138, 84–95. https://doi.org/10.1016/j.compstruct.2015.11.039.

Chang, F.K., Chang, K.Y., 1987a. Post-failure analysis of bolted composite joints in tension or shear-out mode failure. J. Compos. Mater. 21, 809–833. https://doi.org/10.1177/002199838702100903.

Chang, F.K., Chang, K.Y., 1987b. A progressive damage model for laminated composites containing stress concentrations. J. Compos. Mater. 21, 834–855. https://doi.org/10.1177/002199838702100904.

Chang, F.K., Lessard, L.B., 1991. Damage tolerance of laminated composites containing an open hole and subjected to compressive loadings: part I-analysis. J. Compos. Mater. 25 (1), 2–43. https://doi.org/10.1177/002199839102500101.

Chen, B.Y., Tay, T.E., Pinho, S.T., Tan, V.B.C., 2016. Modelling the tensile failure of composites with the floating node method. Comput. Methods Appl. Mech. Eng. 308, 414–442. https://doi.org/10.1016/j.cma.2016.05.027.

Chen, W.H., Lee, S.S., 1995. Numerical and experimental failure analysis of composite laminates with bolted joints under bending loads. J. Compos. Mater. 29, 15–36. https://doi.org/10.1177/002199839502900102.

Chevalier, J., Camanho, P.P., Lani, F., Pardoen, T., 2019. Multi-scale characterization and modelling of the transverse compression response of unidirectional carbon fiber reinforced epoxy. Compos. Struct. 209, 160–176. https://doi.org/10.1016/j.compstruct.2018.10.076.

Coenen, E.W.C., Kouznetsova, V.G., Geers, M.G.D., 2012. Novel boundary conditions for strain localization analyses in microstructural volume elements. Int. J. Numer. Methods Eng. 90 (1), 1–21. https://doi.org/10.1002/nme.3298.

Davidson, B.D., Hu, H.R., Schapery, R.A., 1995. An analytical crack-tip elemento for layered elastic structures. J. Appl. Mech. Trans. ASME 62 (2), 294305. https://doi.org/10.1115/1.2895931.

Dávila, C.G., Rose, C.A., Camanho, P.P., 2009. A procedure for superposing linear cohesive laws to represent multiple damage mechanisms in the fracture of composites. Int. J. Fract. 158, 211–223. https://doi.org/10.1007/s10704-009-9366-z.

De Carvalho, N.V., Mabson, G.E., Krueger, R., Deobald, L.R., 2019. A new approach to model delamination growth in fatigue using the Virtual Crack Closure Technique without remeshing. Eng. Fract. Mech. 222, 106614. https://doi.org/10.1016/j.engfracmech.2019.106614.

Doghri, I., Adam, L., Bilger, N., 2010. Mean-field homogenization of elasto-viscoplastic composites based on a general incrementally affine linearization method. Int. J. Plast. 26 (2), 219–238. https://doi.org/10.1016/j.ijplas.2009.06.003.

Dugdale, D.S., 1960. Yielding of steel sheets containing slits. J. Mech. Phys. Solid. 8 (2), 100–104. https://doi.org/10.1016/0022-5096(60)90013-2.

Espadas-Escalante, J.J., van Dijk, N.P., Isaksson, P., 2019. A phase-field model for strength and fracture analyses of fiber-reinforced composites. Compos. Sci. Technol. 174, 58–67. https://doi.org/10.1016/j.compscitech.2018.10.031.

Ferreira, B.P., 2022. Towards Data-driven Multi-scale Optimization of Thermoplastic Blends: Microstructural Generation, Constitutive Development and Clustering-based Reduced-Order Modeling (Ph.D. thesis). University of Porto, Porto. https://hdl.handle.net/10216/146900.

Ferreira, B.P., Vila-Chã, J.L.P., Pires, F.M.A., 2022. An adaptive multi-temperature isokinetic method for the RVE generation of particle reinforced heterogeneous materials, part II: numerical assessment and statistical analysis. Mech. Mater. 165, 104068. https://doi.org/10.1016/j.mechmat.2021.104068.

Francfort, G.A., Bourdin, B., Marigo, J.-J., 2008. The variational approach to fracture. J. Elasticity 91 (1), 5–148. https://doi.org/10.1007/s10659-007-9107-3.

Francfort, G.A., Marigo, J.-J., 1998. Revisiting brittle fracture as an energy minimization problem. J. Mech. Phys. Solid. 46 (8), 1319–1342. https://doi.org/10.1016/S0022-5096(98)00034-9.

Furtado, C., Arteiro, A., Linde, P., Wardle, B.L., Camanho, P.P., 2020. Is there a ply thickness effect on the mode I intralaminar fracture toughness of composite laminates? Theor. Appl. Fract. Mech. 107, 102473. https://doi.org/10.1016/j.tafmec.2020.102473.

Guillén-Hernández, T., García, I.G., Reinoso, J., Paggi, M., 2019. A micromechanical analysis of inter-fiber failure in long reinforced composites based on the phase field approach of fracture combined with the cohesive zone model. Int. J. Fract. 220 (2), 181–203. https://doi.org/10.1007/s10704-019-00384-8.

Guillén-Hernández, T., Quintana-Corominas, A., García, I.G., Reinoso, J., Paggi, M., Turon, A., 2020. In-situ strength effects in long fibre reinforced composites: a micro-mechanical analysis using the phase field approach of fracture. Theor. Appl. Fract. Mech. 108, 102621. https://doi.org/10.1016/j.tafmec.2020.102621.

Hashin, Z., 1965. On elastic behaviour of fiber reinforced materials of arbitrary transverse phase geometry. J. Mech. Phys. Solid. 13 (3), 119–134. https://doi.org/10.1016/0022-5096(65)90015-3.

Hashin, Z., Rosen, B.W., 1964. The elastic moduli of fiber reinforced materials. J. Appl. Mech. 31, 223–232. https://doi.org/10.1115/1.3629590.

Hashin, Z., Shtrikman, S., 1963. A variational approach to the theory of the elastic behaviour of multiphase materials. J. Mech. Phys. Solid. 11 (2), 127–140. https://doi.org/10.1016/0022-5096(63)90060-7.

Hill, R., 1964. Theory of mechanical properties of fibre-strengthened materials: I. elastic behaviour. J. Mech. Phys. Solid. 12 (4), 199–212. https://doi.org/10.1016/0022-5096(64)90019-5.

Hill, R., 1965. A self-consistent mechanics of composite materials. J. Mech. Phys. Solid. 13 (4), 213–222. https://doi.org/10.1016/0022-5096(65)90010-4.

Hillerborg, A., Modéer, M., Petersson, P.-E., 1976. Analysis of crack formation and crack growth in concrete by fracture mechanics and finite elements. Cement Concr. Res. 6, 773–782. https://doi.org/10.1016/0008-8846(76)90007-7.

Hinton, M.J., Kaddour, A.S., Soden, P.D., 2004. The world-wide failure exercise: its origin, concept and content. In: Hinton, M.J., Kaddour, A.S., Soden, P.D. (Eds.), Failure Criteria in Fibre-Reinforced-Polymer Composites. Elsevier, pp. 2–28. https://doi.org/10.1016/B978-008044475-8/50002-0. Chapter 1.1.

Iarve, E.V., Mollenhauer, D., Whitney, T.J., Kim, R., 2006. Strength prediction in composites with stress concentrations: classical Weibull and critical failure volume methods with micromechanical considerations. J. Mater. Sci. 41 (20), 6610–6621. https://doi.org/10.1007/s10853-006-0200-y.

Joffe, R., Varna, J., 1999. Analytical modeling of stiffness reduction in symmetric and balanced laminates due to cracks in 90° layers. Compos. Sci. Technol. 59 (11), 1641–1652. https://doi.org/10.1016/S0266-3538(99)00025-1.

Kachanov, L.M., 1958. Time of the rupture process under creep conditions. Izvestiia Akademii Nauk SSSR, Otdelenie Teckhnicheskikh Nauk 8, 26–31.

Klavzer, N., Abatour, M., Chevalier, J., Forest, S., Pardoen, T., 2024. Modelling of size-dependent plasticity in polymer-based composites based on nano- and macroscale experimental results. Compos. Appl. Sci. Manuf. 185, 108309. https://doi.org/10.1016/j.compositesa.2024.108309.

Krueger, R., 2004. Virtual crack closure technique: history, approach, and applications. Appl. Mech. Rev. 57 (2), 109–143. https://doi.org/10.1115/1.1595677.

Ladevèze, P., Lubineau, G., 2001. On a damage mesomodel for laminates: micro–meso relationships, possibilities and limits. Compos. Sci. Technol. 61 (15), 2149–2158. https://doi.org/10.1016/S0266-3538(01)00109-9.

Ladèveze, P., Lubineau, G., 2002. An enhanced mesomodel for laminates based on micromechanics. Compos. Sci. Technol. 62 (4), 533–541. https://doi.org/10.1016/S0266-3538(01)00145-2.

Ladevèze, P., Lubineau, G., Marsal, D., 2006. Towards a bridge between the micro- and mesomechanics of delamination for laminated composites. Compos. Sci. Technol. 66 (6), 698–712. https://doi.org/10.1016/j.compscitech.2004.12.026.

Laurin, F., Carrère, N., Maire, J.-F., 2007. A multiscale progressive failure approach for composite laminates based on thermodynamical viscoelastic and damage models. Compos. Appl. Sci. Manuf. 38 (1), 198–209. https://doi.org/10.1016/j.compositesa.2006.01.018.

Lee, J.W., Allen, D.H., Harris, C.E., 1989. Internal state variable approach for predicting stiffness reduction in fibrous laminated composites with matrix cracks. J. Compos. Mater. 23, 1273–1291. https://doi.org/10.1177/002199838902301205.

Leone Jr, F.A., 2015. Deformation gradient tensor decomposition for representing matrix cracks in fiber-reinforced materials. Compos. Appl. Sci. Manuf. 76, 334–341. https://doi.org/10.1016/j.compositesa.2015.06.014.

Lessard, L.B., Shokrieh, M.M., 1995. Two-dimensional modeling of composite pinned-joint failure. J. Compos. Mater. 29, 671–697. https://doi.org/10.1177/002199839502900507.

Liu, Z., Bessa, M.A., Liu, W.K., 2016. Self-consistent clustering analysis: an efficient multiscale scheme for inelastic heterogeneous materials. Comput. Methods Appl. Mech. Eng. 306, 319–341. https://doi.org/10.1016/j.cma.2016.04.004.

Lopes, C.S., Sádaba, S., González, C., Llorca, J., Camanho, P.P., 2016. Physically-sound simulation of low-velocity impact on fiber reinforced laminates. Int. J. Impact Eng. 92, 3–17. https://doi.org/10.1016/j.ijimpeng.2015.05.014.

Lo, D.C., Allen, D.H., Harris, C.E., 1991. Modeling the progressive failure of composites with continuum damage mechanics. In: Chona, R. (Ed.), Fracture Mechanics 23rd Symposium, 1189. ASTM-STP, pp. 680–695.

Lopes, I.A.R., Danzi, F., Arteiro, A., Pires, F.M.A., Camanho, P.P., 2023. Modelling intralaminar damage mechanisms in fibre reinforced polymers at finite strains. Int. J. Solid Struct. 282, 112449. https://doi.org/10.1016/j.ijsolstr.2023.112449.

Mabson, G.E., Deobald, L.R., Dopker, B., 2007. Fracture interface elements for the implementation of the virtual crack closure technique. In: 48th AIAA/ASME/ASCE/AHS/ADC Structures, Structural Dynamics, and Materials Conference. https://doi.org/10.2514/6.2007-2376.

Macías, J., Arteiro, A., Otero, F., Camanho, P.P., Reinoso, J., 2023. Micro-mechanical analysis of composite materials using Phase-Field models of brittle fracture. Eur. J. Mech. Solid. 102, 105069. https://doi.org/10.1016/j.euromechsol.2023.105069.

Maia, M.A., Rocha, I.B.C.M., Kerfriden, P., van der Meer, F.P., 2023. Physically recurrent neural networks for path-dependent heterogeneous materials: embedding constitutive models in a data-driven surrogate. Comput. Methods Appl. Mech. Eng. 407, 115934. https://doi.org/10.1016/j.cma.2023.115934.

Maimí, P., Camanho, P.P., Mayugo, J.A., Dávila, C.G., 2007a. A continuum damage model for composite laminates: part I – constitutive model. Mech. Mater. 39 (10), 897–908. https://doi.org/10.1016/j.mechmat.2007.03.005.

Maimí, P., Camanho, P.P., Mayugo, J.A., Dávila, C.G., 2007b. A continuum damage model for composite laminates: Part II – computational implementation and validation. Mech. Mater. 39 (10), 909–919. https://doi.org/10.1016/j.mechmat.2007.03.006.

Matzenmiller, A., Lubliner, J., Taylor, R.L., 1995. A constitutive model for anisotropic damage in fiber-composites. Mech. Mater. 20, 125–152. https://doi.org/10.1016/0167-6636(94)00053-0.

Mayugo, J.A., Camanho, P.P., Maimí, P., Dávila, C.G., 2006. A Micromechanics-Based Damage Model for [+/- Theta/90n]s Composite Laminates. NASA/TM-2006-214285.

Mayugo, J.A., Camanho, P.P., Maimí, P., Dávila, C.G., 2010. Analytical modelling of transverse matrix cracking of $\{\pm\theta/90_n\}_s$ composite laminates under multiaxial loading. Mech. Adv. Mater. Struct. 17, 237–245. https://doi.org/10.1080/15376490903056577.

McCartney, L.N., 1998. Predicting transverse crack formation in cross-ply laminates. Compos. Sci. Technol. 58 (7), 1069–1108. https://doi.org/10.1016/S0266-3538(96)00142-X.

Melro, A.R., Camanho, P.P., Pinho, S.T., 2008. Generation of random distribution of fibres in long-fibre reinforced composites. Compos. Sci. Technol. 68 (9), 2092–2102. https://doi.org/10.1016/j.compscitech.2008.03.013.

Melro, A.R., Camanho, P.P., Pires, F.M.A., Pinho, S.T., 2013a. Micromechanical analysis of polymer composites reinforced by unidirectional fibres: part I – constitutive modelling. Int. J. Solid Struct. 50, 1897–1905. https://doi.org/10.1016/j.ijsolstr.2013.02.009.

Melro, A.R., Camanho, P.P., Pires, F.M.A., Pinho, S.T., 2013b. Micromechanical analysis of polymer composites reinforced by unidirectional fibres: part II – micromechanical analyses. Int. J. Solid Struct. 50, 1906–1915. https://doi.org/10.1016/j.ijsolstr.2013.02.007.

Mitrou, A., Arteiro, A., Reinoso, J., Camanho, P.P., 2023. Modeling fracture of multidirectional thin-ply laminates using an anisotropic phase field formulation at the macro-scale. Int. J. Solid Struct. 273, 112221. https://doi.org/10.1016/j.ijsolstr.2023.112221.

Mori, T., Tanaka, K., 1973. Average stress in matrix and average elastic energy of materials with misfitting inclusions. Acta Metall. 21 (5), 571–574. https://doi.org/10.1016/0001-6160(73)90064-3.

Naya, F., González, C., Lopes, C.S., Van der Veen, S., Pons, F., 2017a. Computational micromechanics of the transverse and shear behavior of unidirectional fiber reinforced polymers including environmental effects. Compos. Appl. Sci. Manuf. 92, 146–157. https://doi.org/10.1016/j.compositesa.2016.06.018.

Naya, F., Molina-Aldareguía, J.M., Lopes, C.S., González, C., Llorca, J., 2017b. Interface characterization in fiber-reinforced polymer–matrix composites. JOM 69, 13–21. https://doi.org/10.1007/s11837-016-2128-2.

Naya, F., Pappas, G., Botsis, J., 2019. Micromechanical study on the origin of fiber bridging under interlaminar and intralaminar mode I failure. Compos. Struct. 210, 877–891. https://doi.org/10.1016/j.compstruct.2018.11.064.

Nuismer, R.J., Tan, S.C., 1988. Constitutive relations of a cracked composite lamina. J. Compos. Mater. 22, 306–321. https://doi.org/10.1177/002199838802200401.

Ochoa, O.O., Engblom, J.J., 1987. Analysis of progressive failure in composites. Compos. Sci. Technol. 28, 87–102. https://doi.org/10.1016/0266-3538(87)90092-3.

Paul, B., 1960. Prediction of elastic constants of multiphase materials. Trans. Metall. Soc. AIME 218, 36–41.

Pillai, U., Triantafyllou, S.P., Essa, Y., de la Escalera, F.M., 2020. An anisotropic cohesive phase field model for quasi-brittle fractures in thin fibre-reinforced composites. Compos. Struct. 252, 112635. https://doi.org/10.1016/j.compstruct.2020.112635.

Pineda, E.J., Waas, A.M., 2013. Numerical implementation of a multiple-ISV thermodynamically-based work potential theory for modeling progressive damage and failure in fiber-reinforced laminates. Int. J. Fract. 182, 93–122. https://doi.org/10.1007/s10704-013-9860-1.

Pinho, S.T., Iannucci, L., Robinson, P., 2006a. Physically-based failure models and criteria for laminated fiber-reinforced composites with emphasis on fibre kinking: Part I: development. Compos. Appl. Sci. Manuf. 37 (1), 63–73. https://doi.org/10.1016/j.compositesa.2005.04.016.

Pinho, S.T., Iannucci, L., Robinson, P., 2006b. Physically based failure models and criteria for laminated fibre-reinforced composites with emphasis on fibre kinking. Part II: FE implementation. Compos. Appl. Sci. Manuf. 37 (5), 766–777. https://doi.org/10.1016/j.compositesa.2005.06.008.

Quintanas-Corominas, A., Reinoso, J., Casoni, E., Turon, A., Mayugo, J.A., 2019. A phase field approach to simulate intralaminar and translaminar fracture in long fiber composite materials. Compos. Struct. 220, 899–911. https://doi.org/10.1016/j.compstruct.2019.02.007.

Rabotnov, Y.N., 1969. Creep Problems in Structural Members, North-Holland Series in Applied Mathematics and Mechanics.

Reinoso, J., Arteiro, A., Paggi, M., Camanho, P.P., 2017. Strength prediction of notched thin ply laminates using finite fracture mechanics and the phase field approach. Compos. Sci. Technol. 150, 205–216. https://doi.org/10.1016/j.compscitech.2017.07.020.

Reuss, A., 1929. Berechnung der Fließgrenze von Mischkristallen auf Grund der Plastizitätsbedingung für Einkristalle. ZAMM - Zeitschrift für Angewandte Mathematik und Mechanik 9 (1), 49–58. https://doi.org/10.1002/zamm.19290090104.

Saito, H., Takeuchi, H., Kimpara, I., 2014. A study of crack suppression mechanism of thin-ply carbon-fiber-reinforced polymer laminate with mesoscopic numerical simulation. J. Compos. Mater. 48 (17), 2085–2096. https://doi.org/10.1177/0021998313494430.

Shahid, I.S., Chang, F.K., 1995. An accumulative damage model for tensile and shear failures of laminated composite plates. J. Compos. Mater. 29, 926–981. https://doi.org/10.1177/002199839502900705.

Sørensen, B.F., Gamstedt, E.K., Jacobsen, T.K., 2000. Equivalence of J integral and stress intensity factor approach for large scale bridging problems. Int. J. Fract. 104, 31–36. https://doi.org/10.1023/A:1007689004304.

Swolfs, Y., Gorbatikh, L., Romanov, V., Orlova, S., Lomov, S.V., Verpoest, I., 2013a. Stress concentrations in an impregnated fibre bundle with random fibre packing. Compos. Sci. Technol. 74, 113–120. https://doi.org/10.1016/j.compscitech.2012.10.013.

Swolfs, Y., Gorbatikh, L., Verpoest, I., 2013b. Stress concentrations in hybrid unidirectional fibre-reinforced composites with random fibre packings. Compos. Sci. Technol. 85, 10–16. https://doi.org/10.1016/j.compscitech.2013.05.013.

Swolfs, Y., McMeeking, R.M., Verpoest, I., Gorbatikh, L., 2015. Matrix cracks around fibre breaks and their effect on stress redistribution and failure development in unidirectional composites. Compos. Sci. Technol. 108, 16–22. https://doi.org/10.1016/j.compscitech.2015.01.002.

Tallman, A.E., Swiler, L.P., Wang, Y., McDowell, D.L., 2020. Uncertainty propagation in reduced order models based on crystal plasticity. Comput. Methods Appl. Mech. Eng. 365, 113009. https://doi.org/10.1016/j.cma.2020.113009.

Talreja, R., 1990. Internal variable damage mechanics of composite materials. In: Boheler, J.P. (Ed.), Yielding Damage and Failure of Anisotropic Solids. Mechanical Engineering Publications, pp. 509–533.

Tan, S.C., Nuismer, R.J., 1989. A theory for progressive matrix cracking in composite laminates. J. Compos. Mater. 23, 1029–1047. https://doi.org/10.1177/002199838902301006.

Tan, W., Martínez-Pañeda, E., 2021. Phase field predictions of microscopic fracture and R-curve behaviour of fibre-reinforced composites. Compos. Sci. Technol. 202, 108539. https://doi.org/10.1016/j.compscitech.2020.108539.

Tan, W., Martínez-Pañeda, E., 2022. Phase field fracture predictions of microscopic bridging behaviour of composite materials. Compos. Struct. 286, 115242. https://doi.org/10.1016/j.compstruct.2022.115242.

Tavares, R.P., Melro, A.R., Bessa, M.A., Turon, A., Liu, W.K., Camanho, P.P., 2016. Mechanics of hybrid polymer composites: analytical and computational study. Comput. Mech. 57, 405–421. https://doi.org/10.1007/s00466-015-1252-0.

Tavares, R.P., Guerrero, J.M., Otero, F., Turon, A., Mayugo, J.A., Costa, J., Camanho, P.P., 2019. Effects of local stress fields around broken fibres on the longitudinal failure of composite materials. Int. J. Solid Struct. 156–157, 294–305. https://doi.org/10.1016/j.ijsolstr.2018.08.027.

Tian, W., Zhang, Y., Pei, C., Ge, J., Chao, X., Qi, L., 2024. FE modeling to generate composite RVEs with high volume fractions and various shapes of inclusions. Int. J. Solid Struct. 300, 112933. https://doi.org/10.1016/j.ijsolstr.2024.112933.

Tijs, B.H.A.H., Dávila, C.G., Turon, A., Bisagni, C., 2023. The importance of accounting for large deformation in continuum damage models in predicting matrix failure of composites. Compos. Appl. Sci. Manuf. 164, 107263. https://doi.org/10.1016/j.compositesa.2022.107263.

Trias, D., Costa, J., Mayugo, J.A., Hurtado, J.E., 2006a. Random models versus periodic models for fiber reinforced composites. Comput. Mater. Sci. 38, 316–324. https://doi.org/10.1016/j.commatsci.2006.03.005.

Trias, D., Costa, J., Turon, A., Hurtado, J.E., 2006b. Determination of the critical size of a statistical representative volume element (SRVE) for carbon reinforced polymers. Acta Mater. 54 (13), 3471–3484. https://doi.org/10.1016/j.actamat.2006.03.042.

Turon, A., Camanho, P.P., Costa, J., Dávila, C.G., 2006. A damage model for the simulation of delamination in advanced composites under variable-mode loading. Mech. Mater. 38 (11), 1072–1089. https://doi.org/10.1016/j.mechmat.2005.10.003.

Turon, A., Costa, J., Camanho, P.P., Dávila, C.G., 2007. Simulation of delamination in composites under high-cycle fatigue. Compos. Appl. Sci. Manuf. 38 (11), 2270–2282. https://doi.org/10.1016/j.compositesa.2006.11.009.

Vallmajó, O., Arteiro, A., Guerrero, J.M., Melro, A.R., Pupurs, A., Turon, A., 2024. Micromechanical analysis of composite materials considering material variability and microvoids. Int. J. Mech. Sci. 263, 108781. https://doi.org/10.1016/j.ijmecsci.2023.108781.

van der Meer, F.P., Oliver, C., Sluys, L.J., 2010. Computational analysis of progressive failure in a notched laminate including shear nonlinearity and fiber failure. Compos. Sci. Technol. 70 (4), 692–700. https://doi.org/10.1016/j.compscitech.2010.01.003.

Varandas, L.F., Arteiro, A., Catalanotti, G., Falzon, B.G., 2019. Micromechanical analysis of interlaminar crack propagation between angled plies in mode I tests. Compos. Struct. 220, 827–841. https://doi.org/10.1016/j.compstruct.2019.04.050.

Varandas, L.F., Arteiro, A., Bessa, M.A., Melro, A.R., Catalanotti, G., 2017. The effect of through-thickness compressive stress on mode II interlaminar crack propagation: a computational micromechanics approach. Compos. Struct. 182, 326–334. https://doi.org/10.1016/j.compstruct.2017.09.020.

Vila-Chã, J.L.P., Ferreira, B.P., Pires, F.M.A., 2021. An adaptive multi-temperature isokinetic method for the RVE generation of particle reinforced heterogeneous materials, Part I: Theoretical formulation and computational framework. Mech. Mater. 163, 104069. https://doi.org/10.1016/j.mechmat.2021.104069.

Voigt, W., 1889. Ueber die Beziehung zwischen den beiden Elasticitätsconstanten isotroper Körper. Ann. Phys. 274 (12), 573–587. https://doi.org/10.1002/andp.18892741206.

Williams, K.V., Vaziri, R., Poursartip, A., 2003. A physically based continuum damage mechanics model for thin laminated composite structures. Int. J. Solid Struct. 40 (9), 2267–2300. https://doi.org/10.1016/S0020-7683(03)00016-7.

Wisnom, M.R., 2010. Modelling discrete failures in composites with interface elements. Compos. Appl. Sci. Manuf. 41 (7), 795–805. https://doi.org/10.1016/j.compositesa.2010.02.011.

Wisnom, M.R., Hallett, S.R., Soutis, C., 2010. Scaling effects in notched composites. J. Compos. Mater. 44 (2), 195–210. https://doi.org/10.1177/0021998309339865.

Zhuang, F., Arteiro, A., Furtado, C., Chen, P., Camanho, P.P., 2019. Mesoscale modelling of damage in single- and double-shear composite bolted joints. Compos. Struct. 226, 111210. https://doi.org/10.1016/j.compstruct.2019.111210.

Historical development of damage mechanics for composite materials

4.4

Ramesh Talreja
Texas A&M University, College Station, TX, United States

1. Introduction

The concept of damage and the associated field of damage mechanics have been of important significance in developing methodologies for designing composite structures so that they perform safely under service environments. Like fracture and fracture mechanics, damage and damage mechanics have their roots in metals. Composites such as fibre reinforced polymers respond to mechanical loading differently from metals, and this has posed challenges in appropriately tailoring the modelling approaches for composites. It is therefore useful for today's researchers to know how the historical development of the field of damage mechanics for composite materials has been and thereby appreciate the challenges facing further development of this field.

It is not the intention of this exposition to describe in detail the analyses and approaches involved in damage mechanics of composite materials. For this, the main relevant references will be cited in appropriate places in the text. The purpose here is to narrate the concepts underlying the field as they emerged and developed over the years.

2. The Kachanov and Rabotnov concepts

Kachanov (1958) in treating creep of metals assumed that a stable and gradual process of microcracking took place for most of the creep lifetime and that this process eventually localised to lead to unstable fracture. For the stable process, Kachanov proposed a representation of material degradation (damage) by a certain scalar $0 \le \phi \le 1$ that he called "continuity." This variable has value one in the pristine state and decreases to a small positive value, which Kachanov assumed for simplicity to be 0. Kachanov further assumed that the continuity variable's rate of decrease can be expressed for simplicity by a power law,

$$\frac{d\phi}{dt} = -A\left(\frac{\sigma}{\phi}\right)^{m} \tag{4.4.1}$$

The History of Composites. https://doi.org/10.1016/B978-0-443-34119-9.00027-9

where $A > 0$, and $m \geq 0$ are assumed material constants, and σ is the uniaxial tensile stress under which creep occurs. Kachanov suggested that the actual stress σ divided by ϕ can be interpreted as an "effective stress."

Later, Rabotnov (1969) suggested redefining the internal material state by a damage variable $\omega = 1 - \phi$, and expressing its time rate of increase as

$$\frac{d\omega}{dt} = B\frac{\left(\frac{\sigma}{\sigma_0}\right)^p}{(1-\omega)^q} \tag{4.4.2}$$

where B, σ_0, p, and q are material constants. Thus, Rabotnov did not assume the material degradation to depend on the effective stress proposed by Kachanov (a fact misrepresented in the literature, see Leckie and Hayhurst, 1977). The damage variable ω, subsequently denoted by D, has also been misinterpreted as representing a pristine cross-sectional area A when $D = 0$, and a reduced cross-sectional area A$(1 - D)$ when $0 < D < 1$ (Leckie and Hayhurst, 1977).

2.1 Rice's description of scalar damage variables

In defining (or interpreting) the effective stress in Kachanov's Eq. (4.4.1), an obvious inconsistency exists. While the stress σ is defined at a material point, the continuity variable ϕ is only meaningful as a volume-averaged quantity because of the assumption that it is discrete and distributed within the material volume. In fact, Rice (1971) recognised this in the context of plasticity where he viewed internal changes such as those caused by dislocation motion within a polycrystalline metal as distributed discrete structural rearrangements that determined the constitutive behaviour at a point. He assumed these rearrangements to be described by a finite number n of discrete scalar variables ζ_1, ζ_2, ζ_3, … ζ_n, each characterising a local structural rearrangement. He noted the need for constructing field-like descriptions of these variables for given mechanisms underlying the irreversible response changes. He also argued, based on free energy considerations, that an exact characterisation of the discrete set of internal variables was not necessary. Instead, only the changes occurring in the "pattern" of the internal structural rearrangements could be taken to describe the deformational response (strains) caused by the arrangements. This aspect of volume-averaging discrete internal state variables will be discussed later.

3. Lemaitre and Chaboche treatment of damage

Lemaitre and Chaboche extended the Kachanov-Rabotnov concept of damage in the context of creep of metals by generalising it to other damage modes, for example, fatigue damage, and to damage in nonmetallic materials such as concrete; see their book Lemaitre and Chaboche (1994) for a comprehensive treatment. Lemaitre (1984) described a damage variable by considering a "macroscale" volume element in a "damaged body," Fig. 4.4.1. The volume element was assumed to be large enough to contain many "defects" but small enough to be considered as a material point.

Following the concept of a Cauchy stress, Lemaitre (1984) placed a unit normal vector n on an arbitrary differential plane passing through the point. He then assumed that the "overall section area" S of this plane is effectively reduced due to the presence of the distributed defects, shown symbolically in Fig. 4.4.1 as S_D. Denoting the "effective resisting area" as S^*, a damage variable associated with the vector n was defined as

$$D_n = \frac{S - S^*}{S} = 1 - \frac{S^*}{S} \tag{4.4.3}$$

If the defects are oriented randomly with a uniform distribution, then the damage variable is independent of the unit normal $\boldsymbol{n}$, and $D_n = D$, an isotropic damage variable. It should be noted that the surface area of defects S_D indicated in Fig. 4.4.1 is not viewed as a measurable quantity and both S and $S^* < S$ are notional areas meant to suggest that $0 \leq D_n \leq 1$. In their work, Lemaitre and Chaboche did not address the actual surface area S_D of the assumed "defects" collectively representing the damage variable D_n. Consequently, the unquantified S_D, and its equivalent variable S^*, left the damage variable unquantified, other than its specified bounding values 0 and 1.

3.1 Quantification of damage

The Lemaitre-Chaboche treatment of damage consisted largely of regarding the variable D as a means of formulating material degradation rates for the purpose of determining the induced changes in deformational and failure responses, and subsequently, lifetimes, under given loading modes. Thus, D was not quantified in terms of the internal "defects" such as microcracks and microvoids, that is, their size, shape, and

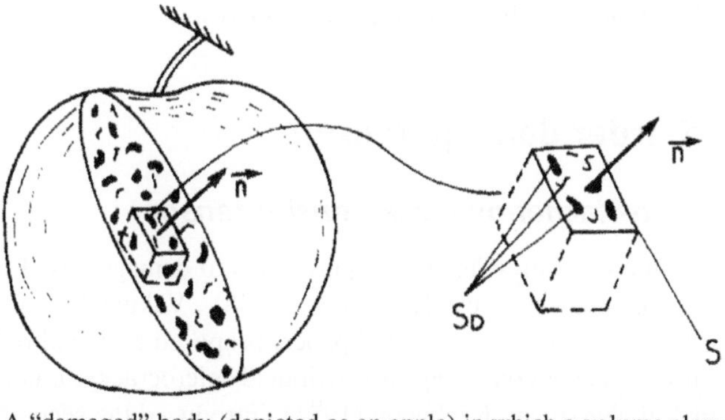

Figure 4.4.1 A "damaged" body (depicted as an apple) in which a volume element (shown to the right) exists that is large enough to contain many "defects" but is small enough to be considered as a material point.
Reproduced with permission from Lemaitre (1984), Nuclear Engineering and Design, Elsevier.

orientation characteristics were not used for that purpose. Instead, for each given phenomenon of material degradation such as fatigue or creep, or fatigue-creep interaction, the damage variable D and its rate of change were expressed in terms of the macroscale quantities such as plastic strain and creep strain; see Chaboche (1992) for details. In fact, an experimental characterisation of microcrack population was viewed as a "daunting task" (Krajcinovic, 2000) and using that path to quantifying D was not considered feasible as late as in the year 2000.

Lemaitre and Chaboche recognised that the microcracks in the material volume would induce changes in the initial material symmetry if the cracks were not randomly oriented. As a general case, Chaboche (1981) proposed that a fourth-order tensorial function of D based on transformation of the second-order initial (undamaged) stress tensor to its effective value when damage is present would be needed. Thus,

$$\sigma_{ij} = M_{ijkl}(D)\sigma^*_{kl} \tag{4.4.4}$$

where $\boldsymbol{\sigma}$ and $\boldsymbol{\sigma}^*$ are the pristine state and damage state stress tensors, respectively, and $\boldsymbol{M}$ is the fourth-order tensor that depends on the damage state. Note that D is still a scalar quantity described by Eq. (4.4.3) and remains unquantified.

The stress tensor in the damage state was viewed as an effective stress tensor following Kachanov's suggestion, Eq. (4.4.1). Specifically, the stress $\boldsymbol{\sigma}^*$ is assumed to produce the same strain in an undamaged volume as the stress $\boldsymbol{\sigma}$ does in the same volume with damage D. Depending on which material degradation phenomenon is addressed, various versions of Eq. (4.4.4) are possible (Chaboche, 1992).

A noteworthy characteristic of the approach to characterisation of damage in materials advocated by Lemaitre and Chaboche and others who followed that line of approach is that the physical aspects of damage (geometrical features of microcracks and their local effects) were not addressed explicitly. Instead, the effects of damage on the material's response (deformation and failure) were viewed as ways to characterise damage. At the time of those activities, this was an inevitable approach because the observational techniques and computational methodologies were not advanced enough to explicitly address the damage at the microscale.

4. Second-order damage tensors

4.1 Vakulenko-Kachanov crack density tensor

Independently of the Lemaitre-Chaboche approaches to damage based on the effective stress concept suggested by Kachanov (1958), Vakulenko and Kachanov (1971) (son of L.M. Kachanov of the 1958 paper) proposed a second-order tensorial characterisation of damage consisting of distributed microcracks. Later, Kachanov, the coauthor of Vakulenko and Kachanov (1971), introduced the second-order crack density tensor in the English literature and described its use in calculating the effective elastic properties of initially isotropic solids containing nonrandom distributions of microcracks (Kachanov, 1999). The crack density tensor is essentially a generalisation of the familiar volume-averaged crack density scalar for randomly distributed cracks. The crack orientations in a randomly oriented cracks of the same size are immaterial in their influence on the effective properties. However, for nonrandom (arbitrary) distributions that is not the case. The appropriately modified crack density measure must then account for the orientations of the crack planes. For the simple case of penny-shaped cracks of radius r in a volume V the crack density tensor then takes the following form (Kachanov, 1999).

$$\alpha_{ij} = \frac{r^3}{V} \sum n_i n_j \tag{4.4.5}$$

Quantification of this tensor presents challenges in experimentally determining the nonrandom distributions of microcracks. Thus, like for the scalar measure of damage D, α_{ij} is essentially a notional description of damage.

4.2 Murakami-Ohno damage tensor

Another second-order tensor to represent the nonrandom orientations of microcracks was proposed by Murakami and Ohno (1981) using the reduced area concept described by Lemaitre (1984), as depicted in Fig. 4.4.1. The description is explained simply by considering a bar containing microcracks and subjected to uniaxial tensile force F. The nominal stress produced by the force is σ, and it is related to the force by $F = \sigma. S$, where S is the cross-sectional area of the bar. If σ^* is the effective stress, then it can be related to the same force by $F = \sigma^*. S^*$, where S^* is the effective resisting area described above (Lemaitre, 1984) and is related to the damage variable D_n as in Eq. (4.4.1). Murakami and Ohno (1981) considered an elemental area dA of an arbitrarily oriented plane with a unit normal $\boldsymbol{n}$ lying in the damaged configuration (left in Fig. 4.4.2) at a generic point and assumed it to be related to an elemental area dA^* with

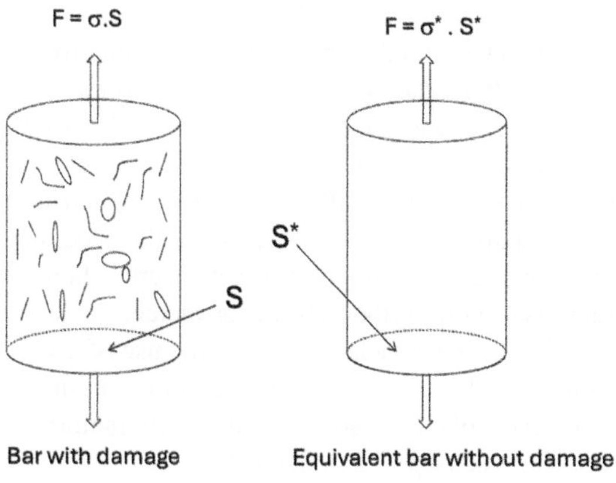

Figure 4.4.2 Equivalency of a uniaxially loaded bar with distributed microcracks (left) and a fictitious bar without damage (right), both producing the same axial strain. Reprinted with permission from Talreja and Singh (2010).

a unit normal $\boldsymbol{n}^*$ at the same point in the fictitious undamaged bar (to the right) by the relationship,

$$dA^* n_i^* = dA\beta_{ij} n_j \tag{4.4.6}$$

where β_{ij} is the second-order tensor representing the field of nonrandomly distributed microcracks. It is noted that this tensor like the Vakulenko-Kachanov tensor (Vakulenko and Kachanov, 1971) is notional, and its components are not possible to directly be measured from observations of microcracks.

5. Damage characterisation for composite laminates

The difficulties in experimental characterisation of microcracks and microvoids in metals, where these entities are often arbitrarily oriented, forced the researchers to find ways to indirectly evaluate damage. As will be described later, the changes in the material's deformational response induced by the damage were equated to the unquantified damage. This approach not only introduces nonuniqueness, as different damage entities can possibly induce the same deformational response, but it also limits the treatment of damage evolution by failing to express what drives the growth of damage locally.

In the first paper dealing with damage in composites, Talreja (1985a) noted the following points based on observations of microcracking in composite laminates reported in the literature, for example, for cross ply laminates of glass/epoxy by Parvizi and Bailey (1978) and for carbon/epoxy laminates by Jamison et al. (1984).

(i) Damage occurs in the form of different multiple-cracking modes. There is no isolated single crack that dominates the development of damage. (Free-edge delaminations are not relevant damage modes, as these are boundary layer effects and must therefore be considered separately.)

(ii) In composite laminates, damage develops along preferred orientations. For instance, matrix cracks in off-axis plies are usually channelled by the fibres in the plies, and interlaminar planar cracks grow along the fibres in the neighbouring plies.

The existence of different modes of damage noted in point (i) applies due to the highly heterogeneous interior structure of laminates consisting of fibre-reinforced layers in different configurations. In subsequent papers, Talreja (1985b, 1989, 1991) treated the damage modes at different length scales. The preferred orientation characteristic of microcracking in composite laminates allowed for the explicit and quantifiable characterisation of individual damage modes. This removed the necessity of defining damage in terms of its effects on the material's response measures, thereby avoiding the nonuniqueness of characterisation described above for metals.

The preferred orientation characteristic of microcracks suggested the use of a vector to represent a crack plane orientation, see Fig. 4.4.3. As depicted in the figure, a vector $\boldsymbol{p}^{(r)}$ is placed normal to the rth crack plane of area $A^{(r)}$ and a characteristic dimension $a^{(r)}$ in a family of n cracks. The vector $\boldsymbol{p}^{(r)}$ is defined as

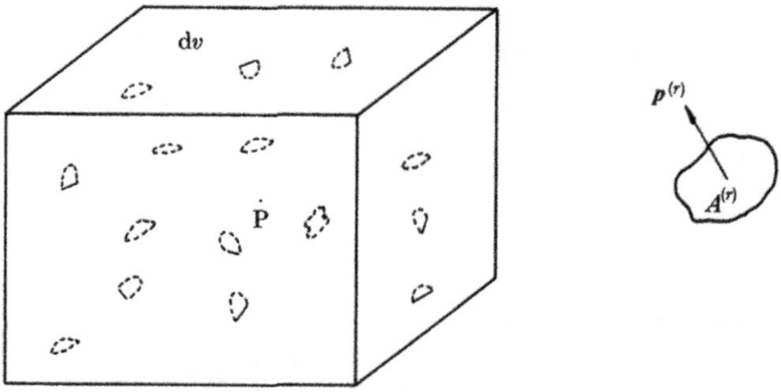

Figure 4.4.3 A volume element dv associated with a generic point P in a composite containing microcracks. The figure to the right shows a vector $\boldsymbol{p}^{(\mathrm{r})}$ pointing normal to the crack plane of surface area $A^{(\mathrm{r})}$.

$$\boldsymbol{p}^{(\mathrm{r})} = d^{(\mathrm{r})}\, \boldsymbol{n}^{(\mathrm{r})} \tag{4.4.7}$$

where $d^{(\mathrm{r})}$ can be suitably expressed as a function of $A^{(\mathrm{r})}$ and $a^{(\mathrm{r})}$.

Denoting by α a **damage mode** representing cracks of a given orientation, and assuming m such modes, $\alpha = 1, 2, \ldots, m$, a damage mode vector $\boldsymbol{V}^{(\alpha)}$ can be defined as

$$\boldsymbol{V}^{(\alpha)} = D^{(\alpha)} \cdot \boldsymbol{n}^{(\alpha)} \tag{4.4.8}$$

where $\boldsymbol{n}^{(\alpha)}$ is a unit vector normal to the crack planes of the α-orientation, and $D^{(\alpha)}$ is a suitably defined average over the volume dv of all $d^{(\alpha)}$ corresponding to $\boldsymbol{n}^{(\mathrm{r})} = \boldsymbol{n}^{(\alpha)}$.

It was realised that using a vector to represent a crack's orientation introduced inherent ambiguity because of the dual possibility of placing a vector normal to the crack plane, that is, pointing one way or the opposite way. Although Talreja (1985a) accounted for this ambiguity by imposing the condition that the material's response must remain independent of the sense of the vector, it was found prudent to resort to a second-order tensorial representation of cracks in which the ambiguity does not arise. This was done in Talreja (1990) where a comprehensive treatment of damage in composite materials was presented. A brief review of that treatment follows next.

Figure 4.4.4 depicts the continuum damage characterisation process for composite materials with a two-vector representation of damage. As indicated, Step 1 of the process consists of homogenisation of the so-called stationary microstructure, which for composite laminates involves smearing of fibres and matrix into a lamina and then homogenising the laminate of multidirectional layup into an anisotropic solid, while leaving in it all damage entities (e.g., matrix microcracks, fibre-matrix debonds, and interlaminar cracks) intact. In the homogenised laminate, for example, obtained by the classical laminate plate theory, Step 2 further homogenises the discrete damage entities into a field. A generic point P in the fully homogenised continuum is then assigned appropriate field functions in which damage enters as internal variables.

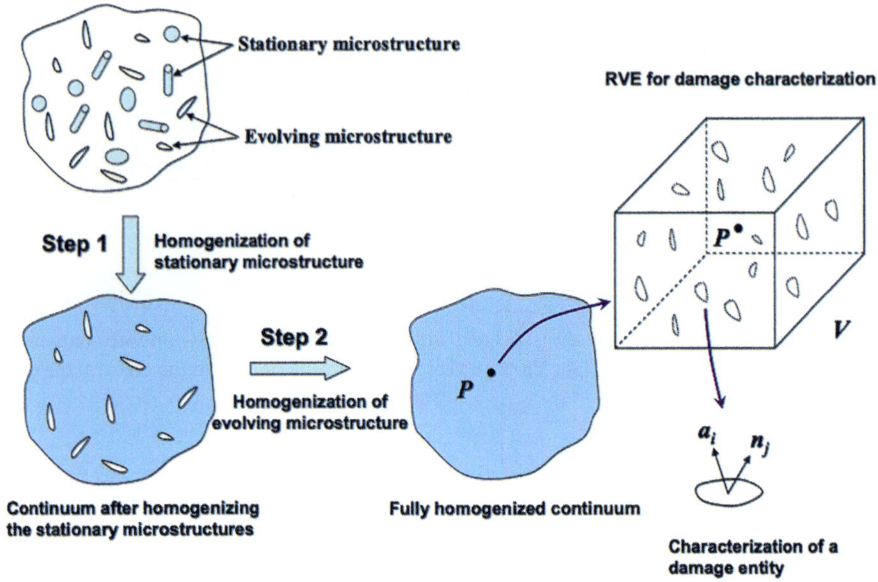

Figure 4.4.4 The continuum damage process with two-vector representation of damage. Reprinted with permission from Talreja and Singh (2010).

The internal variable characterisation of damage begins by considering a representative volume element (RVE) at the generic point P. The discrete damage entities that were smeared into a field become "visible" in the RVE. To each damage entity, two vectors are assigned: an outward normal $\boldsymbol{n}$ and a vector $\boldsymbol{a}$ representing an appropriately chosen "influence" of the damage entity on its surroundings.

A **damage entity tensor** is now defined as a dyadic product of the vectors $\boldsymbol{a}$ and $\boldsymbol{n}$ integrated over the surface S of the damage entity. Thus,

$$d_{ij} = \int_S a_i n_j \mathrm{d}S \tag{4.4.9}$$

The dyadic product avoids the ambiguity caused by the sense of $\boldsymbol{n}$ mentioned above as it always remains positive.

The physical significance of this characterisation is that it retains the oriented nature of the presence of internal surfaces. In composite laminates, the common internal surfaces are cracks (flat or curved) generated by matrix failure and fibre-matrix debonding. A unit normal vector at a point on the damage entity carries the information on orientation of the surface (with respect to the frame of reference), while the other vector represents an appropriate influence induced by activation of the considered point on the surface. This influence is generally also directed in some sense. For the case of mechanical response, the appropriate influence would be the displacement of the activated point on the damage entity surface. For a nonmechanical response, such as thermal or electrical conductivity, the relevant perturbation induced by an internal surface can also be cast as a vector-valued quantity.

The RVE associated with a generic point P carries a sufficiently large number of discrete damage entities to represent the collective effect on the homogenised constitutive response at the point. The number of damage entities needed for this representation, and the consequent RVE size, depend on the distribution of the entities. For instance, if the entities are sparsely distributed, then the RVE size would be large, while for a densely distributed case a small RVE would suffice. Furthermore, for uniformly distributed entities of the same geometry, a repeating unit cell containing a single entity can replace the RVE, while for the cases of nonuniform distribution of unequal entities, the RVE size will increase until a statistically homogeneous representation is attained. This implies that further increasing the RVE size will have no impact on the averages of the selected local fields. The minimum RVE size beyond which no appreciable change in the considered average is found may be taken as the appropriate RVE size. It is apparent that the RVE is not unique but is subject to the choice made of the variables for the formulation of the constitutive response of a continuum with damage. Consequently, there is no unique constitutive theory of a continuum with damage; however, the use of the concept of an internal state in each theory requires specifying RVE in a consistent manner and assuring that the conditions for its existence are present.

For damage in composite laminates observations show that it tends to occur as sets of parallel cracks within the plies, each oriented along fibres in a given ply. It is therefore convenient to separate each set of ply cracks according to its orientation, referred to a fixed frame of reference, and assign it a damage mode number. Denoting damage mode by $\alpha = 1, 2, \ldots, n$, a damage mode tensor can be defined as

$$D_{ij}^{(\alpha)} = \frac{1}{V} \sum_{k_\alpha} (d_{ij})_{k_\alpha} \tag{4.4.10}$$

where k_α is the number of damage entities in the αth mode, and V is the RVE volume. As noted above, if the ply cracks of a given orientation are uniformly spaced, then the RVE will reduce to the unit cell containing one crack. For nonuniform distribution of ply cracks, V must be large enough to provide a steady average of the damage mode tensor components.

As defined by Eq. (4.4.10), the damage mode tensor will in general be asymmetrical. Decomposing the influence vector $\boldsymbol{a}$ along directions normal and tangential to the damage entity surface S gives,

$$a_i = a n_i + b m_i \tag{4.4.11}$$

where $\boldsymbol{m}$ is a unit tangential vector on S such that $n_i m_i = 0$.

Using Eq. (4.4.11) in Eq. (4.4.10), the damage entity tensor can be written in two parts as

$$d_{ij} = d^1_{ij} + d^2_{ij} \tag{4.4.12}$$

where

$$d^1_{ij} = \int_S a n_i n_j \mathrm{d}S, d^2_{ij} = \int_S b m_i n_j \mathrm{d}S \tag{4.4.13}$$

Here, a and b are the magnitudes of the resolved components of $\boldsymbol{a}$ along the normal and tangential directions, respectively. The damage mode tensor for a given mode can now be written as

$$D^{(\alpha)}_{ij} = D^{1(\alpha)}_{ij} + D^{2(\alpha)}_{ij} \tag{4.4.14}$$

where

$$D^{1(\alpha)}_{ij} = \frac{1}{V} \sum_{k_\alpha} \left(d^1_{ij} \right)_{k_\alpha}, D^{2(\alpha)}_{ij} = \frac{1}{V} \sum_{k_\alpha} \left(d^2_{ij} \right)_{k_\alpha} \tag{4.4.15}$$

This separation of the damage mode tensor in two parts allows simplifying the analysis so that dealing with asymmetric tensors is avoided. For instance, for damage entities consisting of flat cracks, the two parts of the damage mode tensor represent the two crack surface separation modes, namely, the crack opening displacement (COD) and crack sliding displacement (CSD). If an assumption can be made that only the symmetrical crack surface separation, COD, is significant, then the second term in Eq. (4.4.15) can be neglected. This will render the damage mode tensor symmetrical, and it can then be written as

$$D^{(\alpha)}_{ij} = D^{1(\alpha)}_{ij} = \frac{1}{V} \sum_{k_\alpha} \left[\int_S a n_i n_j \mathrm{d}S \right]_{k_\alpha} \tag{4.4.16}$$

The consequence of this assumption was examined by Varna (2008) for one class of laminates, and it was found that not including CSD for ply cracks inclined to the laminate symmetry directions results in errors in estimating degradation of average elastic properties of laminates. However, these errors were found to be small in absolute values while being significant in percentages. In fact, for those ply crack orientations where CSD dominates, the cracks are difficult to initiate until high loads close to the failure load are applied.

6. Material response caused by damage

The field of damage mechanics emerged from the need to describe how discrete distributed entities generated within the volume of a material affect its response to external loads. Since the applied thermomechanical loads are also responsible for developing those entities, the evolution rates of those entities as functions of the applied loads also need to be described. The two tasks, namely, describing the response of a material containing a given damage and describing the evolution of the damage, taken together, constitute the scope of damage mechanics. A constitutive framework that can address both the stated tasks is thermodynamics with internal variables. Such a framework falls in the regime often called macro damage mechanics (MADM) or continuum damage mechanics (CDM). An alternative approach that considers the internal damage entities explicitly and analyzes their local effects is a version of micromechanics, described similarly as micro damage mechanics (MIDM). Both approaches will be addressed here, beginning with CDM.

6.1 Continuum damage mechanics (CDM)

At the foundation of CDM are the first and second laws of thermodynamics. In addition, use is made of the concept of an internal state, which is identified here as the evolving microstructure depicted in Fig. 4.4.4. As discussed before, this microstructure is homogenised into a damage field characterised by the set of damage mode tensors $D_{ij}^{(\alpha)}$. A constitutive framework of thermodynamics with internal state can be constructed with quantities that are placed in two categories: state variables and response functions. Table 4.4.1 lists those quantities for bodies undergoing small strains. As shown, the thermodynamic state of the body is given by the strain tensor $\varepsilon_{ij} = \frac{1}{2}\left(u_{i,j} + u_{j,i}\right)$, u_i being the displacement vector, the absolute temperature T, the temperature gradient $g_i = T_{,i}$, and a set of damage tensors $D_{ij}^{(\alpha)}$. The corresponding response functions are the Cauchy stress tensor σ_{ij}, the specific Helmholtz free energy ψ, the specific entropy η, the heat flux vector q_i, and a set of damage rate tensors $\dot{D}_{ij}^{(\alpha)}$, $\alpha = 1, 2, \ldots, n$.

Following Truesdell's equi-presence principle, which states that all state variables should be present in all response functions unless thermodynamics or other relevant considerations preclude their dependency, one can write,

Table 4.4.1 State variables and response functions in a thermodynamics-based constitutive framework.

Thermodynamic state variables	Thermodynamic response functions
1. Strain tensor $\varepsilon_{ij} = \frac{1}{2}\left(u_{i,j} + u_{j,i}\right)$	**1.** Cauchy stress tensor σ_{ij}
2. Temperature T	**2.** Specific Helmholtz free energy ψ
3. Temperature gradient $g_i = T_{,i}$	**3.** Specific entropy η
4. Damage tensors $D_{ij}^{(\alpha)}$	**4.** Heat flux vector q_i
	5. Damage rate tensors $\dot{D}_{ij}^{(\alpha)}$

$$\begin{aligned}
\sigma_{ij} &= \sigma_{ij}\left(\varepsilon_{kl}, T, g_k, D_{kl}^{(\alpha)}\right)\\
\psi &= \psi\left(\varepsilon_{kl}, T, g_k, D_{kl}^{(\alpha)}\right)\\
\eta &= \eta\left(\varepsilon_{kl}, T, g_k, D_{kl}^{(\alpha)}\right)\\
q &= q\left(\varepsilon_{kl}, T, g_k, D_{kl}^{(\alpha)}\right)\\
\dot{D}_{kl}^{(\alpha)} &= \dot{D}_{kl}^{(\alpha)}\left(\varepsilon_{kl}, T, g_k, D_{kl}^{(\beta)}\right)
\end{aligned} \tag{4.4.17}$$

The following balance laws should hold for the continuum body.

Balance of linear momentum:

$$\sigma_{ij,j} + \rho b_j = \rho \ddot{x}_j \tag{4.4.18}$$

where b_j are components of the body force per unit mass and ρ is the mass density.

The balance of angular momentum:

$$\sigma_{ij} = \sigma_{ji} \tag{4.4.19}$$

The balance of energy:

$$\rho \dot{u} - \sigma_{ij}\dot{\varepsilon}_{ij} + q_{i,i} = \rho r \tag{4.4.20}$$

where u is the specific internal energy per unit mass and r is the heat supply per unit mass.

The second law of thermodynamics in the form of the Clausius-Duhem inequality is,

$$\sigma_{ij}\dot{\varepsilon}_{ij} - \rho\dot{\psi} - \rho\dot{T}\eta - \frac{q_i g_i}{T} \geq 0 \tag{4.4.21}$$

where $\psi = u - T\eta$.

Time differentiation of ψ in Eq. (4.4.17) gives,

$$\dot{\psi} = \frac{\partial \psi}{\partial \varepsilon_{kl}}\dot{\varepsilon}_{kl} + \frac{\partial \psi}{\partial T}\dot{T} + \frac{\partial \psi}{\partial g_i}\dot{g}_i + \frac{\partial \psi}{\partial D_{kl}^{(\alpha)}}\dot{D}_{kl}^{(\alpha)} \tag{4.4.22}$$

Substitution of Eq. (4.4.22) into Eq. (4.4.21) gives,

$$\left(\sigma_{ij} - \rho\frac{\partial \psi}{\partial \varepsilon_{kl}}\right)\dot{\varepsilon}_{ij} - \rho\left(\eta + \frac{\partial \psi}{\partial T}\right)\dot{T} - \rho\frac{\partial \psi}{\partial g_i}\dot{g}_i - \rho\sum_{\alpha}\frac{\partial \psi}{\partial D_{kl}^{(\alpha)}}\dot{D}_{kl}^{(\alpha)} - \frac{q_i g_i}{T} \geq 0 \tag{4.4.23}$$

where the summation sign is shown on α to include all damage modes. Now requiring Eq. (4.4.23) to hold for the independently varying strain, temperature, and temperature gradient gives the following results.

$$\sigma_{ij}=\rho\frac{\partial\psi}{\partial\varepsilon_{kl}} \tag{4.4.24}$$

$$\eta=-\frac{\partial\psi}{\partial T} \tag{4.4.25}$$

$$\frac{\partial\psi}{\partial g_i}=0 \tag{4.4.26}$$

Eq. (4.4.26) states that the Helmholtz free energy function does not depend on the temperature gradient, and consequently, Eqs. (4.4.24) and (4.4.25) eliminate this dependency from stress and entropy. The last two equations in the response function set (4.4.17) remain unaffected.

Eqs. (4.4.26)–(4.4.27) lead now to the following functional dependencies.

$$\begin{aligned}\sigma_{ij} &= \sigma_{ij}\left(\varepsilon_{kl},T,D_{kl}^{(\alpha)}\right)\\ \psi &= \psi\left(\varepsilon_{kl},T,D_{kl}^{(\alpha)}\right)\\ \eta &= \eta\left(\varepsilon_{kl},T,D_{kl}^{(\alpha)}\right)\\ q &= q\left(\varepsilon_{kl},T,g_k,D_{kl}^{(\alpha)}\right)\\ \dot{D}_{kl}^{(\alpha)} &= \dot{D}_{kl}^{(\alpha)}\left(\varepsilon_{kl},T,g_k,D_{kl}^{(\beta)}\right)\end{aligned} \tag{4.4.27}$$

From Eq. (4.4.23), the following restriction (also known as the internal dissipation inequality) results.

$$\sum_{\alpha}R_{kl}^{(\alpha)}\dot{D}_{kl}^{(\alpha)}-\frac{q_ig_i}{T}\geq 0 \tag{4.4.28}$$

where $R_{kl}^{(\alpha)}$ are the thermodynamic forces conjugate to $D_{kl}^{(\alpha)}$ and are given by,

$$R_{kl}^{(\alpha)}=-\rho\frac{\partial\psi}{\partial D_{kl}^{(\alpha)}} \tag{4.4.29}$$

Each of these forces is analogous to the crack extension force (i.e., energy release rate) for a single crack. As an example, for a damage mode component, say D_{11} of the mode $\alpha=1$, the quantity $R_{11}^{(1)}$ can be interpreted as the "force" causing an infinitesimal change in the internal state represented by $D_{11}^{(1)}$. Eq. (4.4.28) expresses the condition these forces must satisfy as damage evolves under thermo-mechanical impulses.

The complete thermo-mechanical response for the composite body is governed by the set of functions (4.4.27) and the internal dissipation inequality (4.4.28). In view of the experimental data, which are mostly available for polymer matrix composites at room temperature, the thermo-mechanical framework will be developed further for the purely mechanical response. Thus, for isothermal conditions ($T = 0$, $g_i = 0$) the set of response functions is reduced to the following

$$\begin{aligned} \sigma_{ij} &= \sigma_{ij}\left(\varepsilon_{kl}, D_{kl}^{(\alpha)}\right) \\ \psi &= \psi\left(\varepsilon_{kl}, D_{kl}^{(\alpha)}\right) \\ \dot{D}_{kl}^{(\alpha)} &= \dot{D}_{kl}^{(\alpha)}\left(\varepsilon_{kl}, g_k, D_{kl}^{(\beta)}\right) \\ \text{and } & \sum_{\alpha} R_{kl}^{(\alpha)} \dot{D}_{kl}^{(\alpha)} \geq 0 \end{aligned} \tag{4.4.30}$$

Since σ_{ij} are derivable from ψ, according to Eq. (4.4.24), it suffices for a purely mechanical response to construct ψ and $\dot{D}_{kl}^{(\alpha)}$. Dealing with the scalar-valued function ψ as the sole response function for a given internal state of damage provides a favourable situation for further development of the theory. The form of the Helmholtz free energy function can be chosen in different ways. A powerful way is by use of the theory of invariants for polynomial functions (Spencer, 1971).

To derive the rate equations for σ_{ij}, we differentiate Eq. (4.4.24) with respect to time and use the functional dependency given in (4.4.30), which yields,

$$\dot{\sigma}_{ij} = \rho \frac{\partial^2 \psi}{\partial \varepsilon_{ij} \partial \varepsilon_{kl}} \dot{\varepsilon}_{kl} + \rho \sum_{\alpha} \frac{\partial^2 \psi}{\partial \varepsilon_{ij} \partial D_{mn}^{(\alpha)}} \dot{D}_{mn}^{(\alpha)} \tag{4.4.31}$$

Substitution of (4.4.29) gives,

$$\dot{\sigma}_{ij} = \rho \frac{\partial^2 \psi}{\partial \varepsilon_{ij} \partial \varepsilon_{kl}} \dot{\varepsilon}_{kl} - \sum_{\alpha} \frac{\partial R_{mn}^{(\alpha)}}{\partial \varepsilon_{ij}} \dot{D}_{mn}^{(\alpha)} \tag{4.4.32}$$

which may be rewritten as

$$\dot{\sigma}_{ij} = C_{ijkl} \dot{\varepsilon}_{kl} - \sum_{\alpha} K_{ijmn}^{(\alpha)} \dot{D}_{mn}^{(\alpha)} \tag{4.4.33}$$

where,

$$C_{ijkl} = \rho \frac{\partial^2 \psi}{\partial \varepsilon_{ij} \partial \varepsilon_{kl}} \tag{4.4.34}$$

and

$$K_{ijmn}^{(\alpha)} = \frac{\partial R_{mn}^{(\alpha)}}{\partial \varepsilon_{ij}} = -\rho \frac{\partial^2 \psi}{\partial \varepsilon_{ij} \partial D_{mn}^{(\alpha)}} \tag{4.4.35}$$

The matrix C_{ijkl} is the stiffness coefficients matrix of the composite at a given material state, whereas coefficients $K_{ijmn}^{(\alpha)}$ are functions determining the change in state caused by the internal dissipative mechanisms. The rate equations Eq. (4.4.33) require formulation of the scalar function ψ and the tensor components $\dot{D}_{kl}^{(\alpha)}$.

The constitutive response of a material with damage given by Eq. (4.4.33) has been elaborated for composite laminates by Talreja (1985b, 1986, 1989, 1990, 1991, 2010). An interested reader can find more details in these references.

6.2 The Ladeveze damage mechanics approach

The constitutive response of composite laminates with damage was addressed by Ladeveze and coworkers (Ladeveze, 1989; Ladeveze and Dantec, 1992; Ladeveze and Lubineau, 2001, 2002, 2003) using a model they described as a mesomodel. The model consists of decomposing a laminate into meso-level constituents, that is, plies and interfaces. These constituents represent homogenised layers that contain distributed damage entities. More specifically, the plies contain intralaminar cracks while the interface layers have distributed delaminations. It is assumed that the layers are degraded in their elastic deformational response by the presence of distributed cracks. To contrast this with the actual material response, it is noted that the *laminate* elastic response, when averaged over a representative volume, changes from the undamaged state due to the presence of distributed intralaminar cracks and internal delaminations (interlaminar cracks not at free edges). The response of a ply containing cracks depends on the crack surface displacements that in turn depend on the presence of adjacent plies. This effect, described as the "constraint effect," cannot be captured by considering a ply in isolation and homogenising it.

Regarding interlaminar cracks, modelled in the Ladeveze approach as cracks in the interface layers, it is observed that they do not form independently but are consequences of the intralaminar cracks diverting into the interfaces. Ladeveze and coworkers recognise this, but by placing the interlaminar cracks within the interface layers of finite thickness for the sake of homogenising the layers, they introduce the uncertainty of determining the layer thicknesses. To proceed further, the Ladaveze approach to determining the laminate elastic response with progressing damage is to compute changes in the elastic response of the meso-level constituents with finite element models using assumed scenarios of cracking patterns. While details of their computations can be found in the references cited above, it is noted that the internal state of damage conceived by Kachanov in the celebrated paper of 1958 differs from the way damage is treated in the mesomodel. The state of damage as physical entities (intralaminar and interlaminar cracks) is not characterised by RVE-averaged damage measures for *laminates*. Furthermore, the internal variables are assigned separately to the two meso-level constituents. These internal variables are not measures of the presence (e.g., intensity) of the damage entities but are *consequences* on the elastic response of the meso-level constituents. These consequences are separated as Young's modulus changes and shear modulus changes, resulting in two different damage variables.

6.3 The Matzenmiller damage mechanics approach

This approach appeared first in the paper Matzenmiller et al. (1995), where they consider a single UD ply to formulate its response caused by damage. It differs from the Ladeveze approach described above in that it aims to describe the failure of a homogenised ply from formation of a localised macrocrack. Thus, it combines the CDM approach with failure theories for UD composites. In formulating CDM it assumes that the initial orthotropic symmetry of a ply remains unchanged by damage from initiation until localisation in a macrocrack. Focussing on the failure of a ply, Matzenmiller et al. (1995) define individual failure modes, viz., fibre failure in tension, fibre buckling and kinking in compression, transverse matrix cracking, and transverse matrix crushing. To each failure mode, they assign scalar-valued damage variables. There is no incentive to connect the damage variables to the damage entities since the damage variables are only a means to achieve description of ply failure under loading. This treatment of damage is unconventional in the sense that it does not confine damage to distributed entities such as microcracks but also includes localised failure such as in compression-induced buckling as a damage mode.

Furthermore, in the Matzenmiller et al. (1995) paper there is no pathway described to go from a lamina damage to a laminate damage. In a later paper, Chatiri and Matzenmiller (2013) extended the lamina approach to laminates. The extension consists of extending the earlier two-dimensional formulation to three dimensions, thereby allowing consideration of multiple stacked plies of different orientations. The localised mode of damage was still included, but only in a ply. However, in one example of cross ply laminates used for illustration of the approach only distributed intralaminar cracks formed in transverse tension were considered; interlaminar cracks were not included. It was claimed that the single-ply approach could still be used as a conservative approach based on the observation that plies constrained within a laminate are usually capable of carrying higher loads than when loaded in isolation.

6.4 The Camanho and coworkers damage mechanics approach

This approach appeared first in the paper by Maimí et al. (2007a). There are striking similarities between this approach and that of Matzenmiller et al. (1995). The difference lies in the latter considering an isolated ply as a UD composite, while the former viewing the same as embedded in a laminate. The motivating failure mechanisms are still the same, renaming them as longitudinal and transverse failure in the fibre direction, as well as transverse failure in tension or compression. The transverse multiple cracking is correctly described in the context of a laminate. Also here, scalar-valued damage variables are assigned to each damage mode. The damage variables are arbitrarily assigned and denoted as d_1, d_2, and d_6, associated with longitudinal normal stress, transverse normal stress, and in-plane shear stress, respectively. As in the previous approaches described in Sections 6.2 and 6.3, these damage variables represent reductions in the longitudinal, transverse, and shear moduli, that is, they represent *consequences* of damage rather than damage itself. Thus, the damage variables are not in the sense of Kachanov (1958) as volume-averaged measures of the distributed damage entities. Obviously, this implies exclusion of localised damage such as buckling or formation of macrocracks. In a follow-on paper, Maimí et al. (2007b) implemented their model to quasi-isotropic laminates with central holes subjected to longitudinal tension. The focus was on predicting the failure load.

As a remark relevant to the approaches described above in Sections 6.2–6.4, it is noted that the CDM approach is meant only to treat *distributed* damage entities such that their volume-averaged measures can be used as internal variables. The criticality of material response that can be viewed as *failure* can then be appropriately chosen as a certain threshold of the volume-averaged response. Other approaches, such as fracture mechanics, are properly suited to deal with failure from localised regions in a composite.

6.5 Micro damage mechanics (MIDM)

Micromechanics of elastic heterogeneous solids is a mature field that can be viewed to have begun with the classical works of Voigt (1889) and Reuss (1929) whose names are associated with bounds to the effective elastic properties. A comprehensive treatment of elastic solids with voids and cracks can be found in Nemat-Nasser and Hori (2013), for instance. A direct application of micromechanics to cracked composite laminates was first reported by Hashin (1985) who used a variational approach to obtain lower bounds for the axial modulus and the in-plane shear modulus of a cross ply laminate $[0/90]_s$ containing multiple transverse cracks. Later, many contributions using the variational approach followed, for example, Varna and Berglund (1991), Nairn and Hu (1992); Joffe and Varna (1999); Li and Hafeez (2009); Vinogradov and Hashin (2010); Hajikazemi and Sadr (2014); Hajikazemi et al. (2014). McCartney (1992) proposed a generalised plane strain approach which was later compared with Hashin's approach by Hajikazemi and McCartney (2018). The variational approach continues to produce results, see recent work by Hajikazemi et al. (2022).

The variational approach has the advantage of being analytical, but it requires assumptions concerning crack geometry and local stress distribution. The results obtained are therefore necessarily approximate and their accuracy needs to be assessed against test data or computationally simulated results. For instance, Hashin (1985) compared his prediction of the axial modulus reduction with number density of transverse cracks reported in the literature for $[0/90_3]_s$ and $[0/90]_s$ laminates of glass/

epoxy and found inconsistent results. While the predictions matched perfectly with the first laminate, they came out higher than the data for the second laminate. The predictions are, in fact, lower bounds and their accuracy is affected by the assumptions made in solving the boundary value problem formulated to obtain local stress fields in a representative region of the cracked composite (usually a repeating unit cell). The main assumption made by Hashin (1985) is that the axial stress in the region between two neighbouring cracks is uniform along the ply thickness. The axial stress in fact deviates from uniformity depending on the transverse ply thickness and the constraint to crack opening induced by the axial plies. The inaccuracy induced by the assumption of uniform axial stress was examined by Varna and Berglund (1991).

Although Hashin (1985) showed a rigorous way to find bounds of reduced elastic moduli due to transverse cracking in cross ply laminates, and it was indeed a significant advancement compared with the approximate prior approaches based on the shear lag concept, in time it became clear that Hashin's approach was essentially limited to cross ply laminates. Much later, Vinogradov and Hashin (2010) extended the approach to inclined cracks in laminates and reported elastic moduli bounds for this case and compared their prediction of the axial Young's modulus of $[0/\pm45]_s$ with test data for a glass/epoxy laminate. They also reported their calculated results for other off-axis angles without comparing those with available test data in the literature, for instance, in Varna et al. (1999). In that paper, it was shown that off-axis cracks do not initiate in plies where the stress normal to the fibres is compressive and that the moduli reductions are affected by the shear modulus reduction in the uncracked inclined plies. It was also cautioned in Varna et al. (1999) that relating moduli reductions to crack density is not generally correct; one must first assure that cracks can be initiated.

An alternative approach to determining elastic moduli reductions in symmetric cracked laminates was developed by focussing on crack surface displacements. A full account of this approach has been given by Varna (2024). Beginning with a study of crack opening displacement (COD) of transverse cracks in cross ply laminates in Varna et al. (1993), a systematic approach described as GLOB-LOC model has been developed. This approach expresses the average thermoelastic properties of symmetric cracked laminates in terms of ply properties, laminate layup, and intralaminar crack densities. COD and crack sliding displacement (CSD) are calculated analytically, whenever possible, or by using finite element models and some curve-fitting parameters. Predictions have been successfully compared with data for cross ply laminates and for $[0/\pm45/90]_s$ laminates.

A general micromechanics approach to determining elastic moduli of cracked composite laminates by stress transfer models has been summarised in Hajikazemi and McCartney (2024). These models use generalised plane strain and variational approximations for determining local stress fields. The basic idea in these models is that the stresses in a cracked material can be represented by superposition of the stresses in the undamaged material and some unknown stress transfer functions due to the presence of cracks. A range of situations consisting of complex loading involving out-of-plane shear and bending has been treated.

6.5.1 Computational micromechanics

The increase in the power of computers and their widespread availability prompted development of approaches that compute local stresses with finite element models representing microstructural details explicitly, rather than smearing them out in a homogeneous continuum. An early example of this is Herráez et al. (2015) for transverse cracking in cross ply laminates. The initiation and progression of the local failure events in such approaches are treated by failure criteria for the constituents and their interfaces. Herráez et al. (2015) assumed a damage-plasticity model for the epoxy matrix failure and a cohesive crack model for the fibre-matrix interface failure. The material properties entering into the deformation and failure models need to be estimated indirectly from bulk tests or extracted from a comparison of simulation results with test data. The overall response of a composite undergoing damage is calculated by averaging the local stress fields over the volume of an RVE. This modelling approach is evolving and is driven by the need to perform multiscale modelling. Another chapter in this book dealing with multiscale modelling will provide a historical perspective of computational micromechanics.

6.6 Synergistic damage mechanics

The more-or-less simultaneous development of CDM, initiated by Talreja (1985a), and the first paper on MIDM by Hashin (1985) led in time to the realisation that both approaches have limitations. Talreja (1996) proposed to enrich the CDM approach by entering results obtained from MIDM into it. The most suitable MIDM results were the CODs calculated by Varna and associates and described comprehensively in Varna (2024). The enriched CDM approach was labelled synergistic damage mechanics (SDM) recognising the synergistic effect of combining CDM with MIDM. An assessment by Varna (2008) of the material parameters in SDM from the perspective of crack surface displacements provided clarification and a strong support for SDM. It was found that neglecting CSD in the SDM approach to retain symmetry of the stiffness change matrix was acceptable because of the resulting small error. Several papers have reported SDM results for multiple damage modes, for example, Varna et al. (2001, 2004), Singh and Talreja (2009, 2013) and Montesano and Singh (2015).

Figure 4.4.5 depicts the conceptual flow of SDM. The region in a composite structure that contains damage is identified as a substructure. Viewing the substructure as a homogenised continuum, CDM is applied as illustrated in Fig. 4.4.4. Thus, at a point

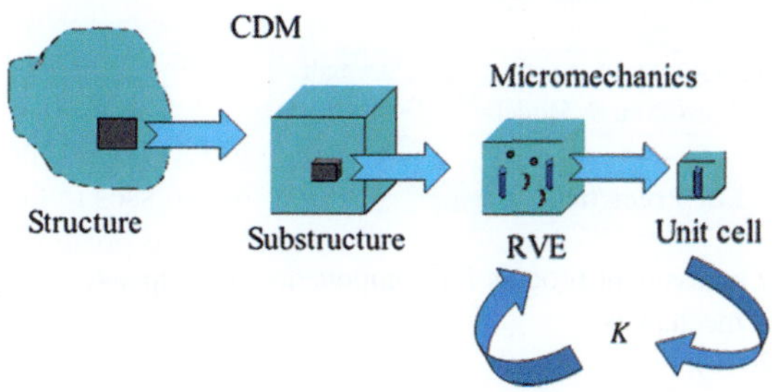

Figure 4.4.5 The flowchart of multiscale synergistic damage mechanics approach.

of interest in the substructure, an RVE is constructed. Micromechanics is then used to determine the constrained COD of the damage entity, given by the constraint parameter κ. This κ value is the micromechanics input to CDM via the damage mode tensor. For full details, see Singh and Talreja (2009, 2013) and Talreja (2010).

7. Damage mechanics for durability assessment

Damage mechanics has developed today as a field that provides tools for durability assessment of composite structures. A typical process of durability assessment is depicted in Fig. 4.4.6. A component, like the substructure shown in Fig. 4.4.5, is analysed by appropriate computational methods to obtain stress/strain/temperature histories at critical sites. These sites, also described as "hot spots," are often known from prior similar structures as sources of failure under service loading. The stress analysis conducted uses the specified service loading and the pristine state deformation models of the composite material to determine the desired local fields in the critical sites. Numerous studies of damage mechanisms under such fields for the selected composite material system have either been conducted or are performed in laboratory environments to reveal the mechanisms. Damage mechanics is (or should be) then applied as a mechanisms-based approach. This type of approach falls under physical modelling as against empirical phenomenological approaches.

As indicated in the figure, damage mechanics can consist of modelling at different length scales labelled as micro-, meso-, and macro-levels. MIDM discussed above operates at the micro level understood as the scale of constituents. The modelling here targets local stress fields, the averages of which provide the higher-level response. Models of this nature are typically described as computational

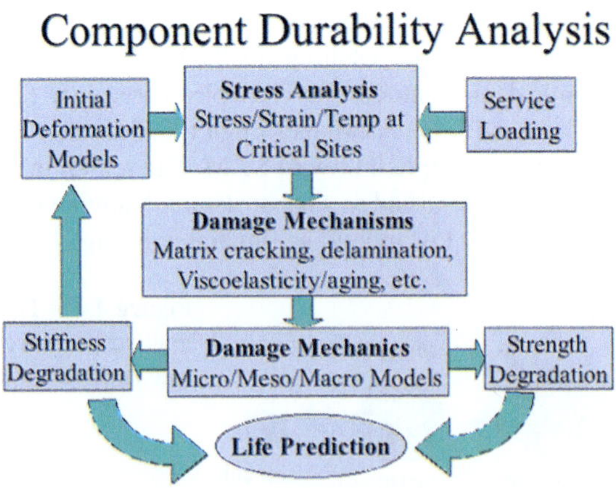

Figure 4.4.6 Flowchart of a durability assessment process for components in composite structures depicting the role of damage mechanics.
Reproduced from Talreja and Singh (2010).

micromechanics (see Section 6.5.1 above). The higher level can be meso or macro level depending on how the models are set up. For instance, if the response calculated is at the ply level of a laminate, then the level would be a meso (intermediate) level. Further analysis of a volume element consisting of multiple plies and the resulting stress fields can provide a macro (laminate) level response.

The starting scale of a model can be at the level of a ply that has been homogenised. The micro-level models in this case seek local fields within a ply. The variational models cited in Section 6.2 above are typical of this type of model. The response obtained by averaging over a volume element with multiple plies is then the macro-level response.

In CDM models, the baseline scale is the macro scale such as an RVE of a laminate with multiple plies. This can be the only scale involved if the damage is represented by certain internal state variables that are not specifically related to the damage entities. Such variables, either scalars or tensors, were described in Eq. 4.4.3 and 4.4.4 above. CDM approaches commonly keep to the macro-level only. An exception is the SDM approach developed by this author and described in Section 3 above. As stated there, the lower-level (micro or meso) input is taken via the constraint parameter which is evaluated by separate micromechanics models. Specifically, the constraint parameter is obtained from the calculated values of CODs.

Referring to Fig. 4.4.6, the output of damage mechanics models is the macro-level response that can be cast as deformational (stiffness degradation) response or strength degradation response. The current stiffness should be taken to update the initial deformation models used to calculate the pristine state stress fields. To arrive at the durability assessment, either the stiffness degradation path or the strength degradation path (or in some cases, both) are taken, and the design limits to these properties then provide the lifetime of a component under the specified service loading.

8. The timeline of damage mechanics

It is difficult to associate a timeline to the development of any field as it is in the nature of science that concepts seldom emerge without being influenced by previous ones. For this exposition, however, it may be useful to provide an historical timeline as an overview of the concepts and methods that have been consequential in evolving the field of damage mechanics as it relates to composite materials. The following is a tabulation for this purpose. The references have been cited above in the text.

1958: Kachanov's concept of material degradation viewed as loss of internal continuity in the volume.

1971: Rice's concept of internal structural rearrangements described as patterns of distributed discrete entities causing irreversible material response.

1971 onwards: Lemaitre and Chaboche concept of damage as multitude of microcracks that transform the actual stress at a point to an effective stress by reducing the area of a plane passing through the point.

1971: The Vakulenko-Kachanov second order microcrack density tensor as a measure of damage.
1981: The Murakami-Ohno second-order damage tensor derived by considering the Lemaitre-Chaboche concept of effective stress.
1985–90: The Talreja concept of distributed damage entities in a representative volume element characterised by damage entity vectors or tensors and the volume-averaged damage mode tensors.
1996: Talreja's development of synergistic damage mechanics that utilises output of micro-level analyses into internal state variables in continuum damage mechanics.

The current trends in damage mechanics consist of computational micromechanics as a methodology for multiscale modelling. A likely further development of damage mechanics is in the direction of machine learning.

9. Conclusion

The exposition here has been with the aim of providing a rendition of the historical development of the damage concept, its severity characterisation, and its relation to the field known as damage mechanics. The focus has been on the developments as they apply to composite materials. Considerations of viscoelasticity and plasticity have not been included to keep a sharper focus on damage in the form of multiple distributed cracks of different sizes and orientations observed in composite laminates. Modelling approaches in damage mechanics, conveniently described as micro damage mechanics (MIDM), macrodamage mechanics (MADM), and a special version labelled as synergistic damage mechanics (SDM), have been discussed. This has not been intended as a literature survey, but rather as a narrative from the perspective of the author who has been active in the field beginning with the first contribution in 1985. With a steady development over the past four decades, the field of damage mechanics has attained the character of a mature field but with still room for further developments.

References

Chaboche, J.L., 1981. Continuous damage mechanics—a tool to describe phenomena before crack initiation. Nucl. Eng. Des. 64 (2), 233–247.

Chaboche, J.L., 1992. Damage induced anisotropy: on the difficulties associated with the active/passive unilateral condition. Int. J. Damage Mech. 1 (2), 148–171.

Chatiri, M., Matzenmiller, A., 2013. A damage-mode based three dimensional constitutive model for fibre-reinforced composites. Comput. Mater. Continua (CMC) 35 (3), 255–283.

Hajikazemi, M., McCartney, L.N., 2018. Comparison of variational and generalized plane strain approaches for matrix cracking in general symmetric laminates. Int. J. Damage Mech. 27 (4), 507–540.

Hajikazemi, M., McCartney, L.N., 2024. Semi-analytical techniques for analysis of cracked composite laminates. In: Talreja, R., Varna, J. (Eds.), Modeling Damage, Fatigue and Failure of Composite Materials, second ed. Elsevier.

Hajikazemi, M., Sadr, M.H., 2014. Stiffness reduction of cracked general symmetric laminates using a variational approach. Int. J. Solid Struct. 51 (7–8), 1483–1493.

Hajikazemi, M., Sadr, M.H., Hosseini-Toudeshky, H., Mohammadi, B., 2014. Thermo-elastic constants of cracked symmetric laminates: a refined variational approach. Int. J. Mech. Sci. 89, 47–57.

Hajikazemi, M., Ahmadi, H., McCartney, L.N., Van Paepegem, W., 2022. A variational approach for accurate prediction of stress and displacement fields and thermo-elastic constants in general symmetric laminates containing ply cracking and delamination under general triaxial loading. Int. J. Solid Struct. 254, 111917.

Hashin, Z., 1985. Analysis of cracked laminates: a variational approach. Mech. Mater. 4 (2), 121–136.

Herráez, M., Mora, D., Naya, F., Lopes, C.S., González, C., Llorca, J., 2015. Transverse cracking of cross-ply laminates: a computational micromechanics perspective. Compos. Sci. Technol. 110, 196–204.

Jamison, R.D., Schulte, K., Reifsnider, K.L., Stinchcomb, W.W., 1984. Characterization and analysis of damage mechanisms in tension–tension fatigue of graphite/epoxy laminates. In: Effects of Defects in Composite Materials, ASTM STP 836. ASTM, Philadelphia, PA, pp. 21–55.

Joffe, R., Varna, J., 1999. Analytical modeling of stiffness reduction in symmetric and balanced laminates due to cracks in 90 layers. Compos. Sci. Technol. 59 (11), 1641–1652.

Kachanov, L.M., 1958. Rupture time under creep conditions. Izvestia Akademii Nauk SSSR, Otdelenie tekhnicheskich nauk, no. 8. pp. 26–31, (in Russian). English translation: International Journal of Fracture 97: xi–xviii, 1999.

Kachanov, M., 1999. Solids with cracks and non-spherical pores: proper parameters of defect density and effective elastic properties. Int. J. Fract. 97, 1–32.

Krajcinovic, D., 2000. Damage mechanics: accomplishments, trends and needs. Int. J. Solid Struct. 37 (1–2), 267–277.

Ladeveze, P., 1989. About a damage mechanics approach. In: Baptiste, D. (Ed.), Mechanics and Mechanisms of Damage in Composite and Multimaterials. MEP, pp. 119–142.

Ladeveze, P., Dantec, E.L., 1992. Damage modeling of the elementary ply for laminated composites. Compos. Sci. Technol. 43 (3), 257–267.

Ladeveze, P., Lubineau, G., 2001. On a damage mesomodel for laminates: micro–meso relationships, possibilities and limits. Compos. Sci. Technol. 61 (15), 2149–2158.

Ladeveze, P., Lubineau, G., 2002. An enhanced mesomodel for laminates based on micromechanics. Compos. Sci. Technol. 62, 533–541.

Ladevèze, P., Lubineau, G., 2003. On a damage mesomodel for laminates: micromechanics basis and improvement. Mech. Mater. 35 (8), 763–775.

Leckie, F.A., Hayhurst, D.R., 1977. Constitutive equations for creep rupture. Acta Metall. 25, 1059–1070.

Lemaitre, J., 1984. How to use damage mechanics. Nucl. Eng. Des. 80, 233–245.

Lemaitre, J., Chaboche, J.L., 1994. Mechanics of Solid Materials. Cambridge University Press.

Li, S., Hafeez, F., 2009. Variation-based cracked laminate analysis revisited and fundamentally extended. Int. J. Solid Struct. 46 (20), 3505–3515.

Maimí, P., Camanho, P.P., Mayugo, J.A., Dávila, C.G., 2007a. A continuum damage model for composite laminates: Part I–constitutive model. Mech. Mater. 39 (10), 897–908.

Maimí, P., Camanho, P.P., Mayugo, J.A., Dávila, C.G., 2007b. A continuum damage model for composite laminates: Part II–computational implementation and validation. Mech. Mater. 39 (10), 909–919.

Matzenmiller, A., Lubliner, J., Taylor, R.L., 1995. A constitutive model for anisotropic damage in fiber-composites. Mech. Mater. 20 (2), 125–152.
McCartney, L.N., 1992. Theory of stress transfer in a 0—90—0 cross-ply laminate containing a parallel array of transverse cracks. J. Mech. Phys. Solid. 40 (1), 27–68.
Montesano, J., Singh, C.V., 2015. A synergistic damage mechanics based multiscale model for composite laminates subjected to multiaxial strains. Mech. Mater. 83, 72–89.
Murakami, S., Ohno, N., 1981. A continuum theory of creep and creep damage. In: Ponter, A.R. S., Hayhurst, D.R. (Eds.), Creep in Structures, 3rd IUTAM Symposium. Springer-Verlag, Berlin, Germany, pp. 422–444.
Nairn, J.A., Hu, S., 1992. The formation and effect of outer-ply microcracks in cross-ply laminates: a variational approach. Eng. Fract. Mech. 41 (2), 203–221.
Nemat-Nasser, S., Hori, M., 2013. Micromechanics: Overall Properties of Heterogeneous Materials. Elsevier.
Parvizi, A., Bailey, J.E., 1978. Multiple transverse cracking in glass-fiber epoxy cross-ply laminates. J. Mater. Sci. 13 (10), 2131–2136.
Rabotnov, Y.N., 1969. Creep Problems in Structural Members. North-Holland, Amsterdam.
Reuss, A., 1929. Berechnung der fließgrenze von mischkristallen auf grund der plastizitätsbedingung für einkristalle. ZAMM-Journal of Applied Mathematics and Mechanics/ Zeitschrift für Angewandte Mathematik und Mechanik 9 (1), 49–58.
Rice, J.R., 1971. Inelastic constitutive relations for solids: an internal-variable theory and its application to metal plasticity. J. Mech. Phys. Solid. 19 (6), 433–455.
Singh, C.V., Talreja, R., 2009. A synergistic damage mechanics approach for composite laminates with matrix cracks in multiple orientations. Mech. Mater. 41 (8), 954–968.
Singh, C.V., Talreja, R., 2013. A synergistic damage mechanics approach to mechanical response of composite laminates with ply cracks. J. Compos. Mater. 47 (20–21), 2475–2501.
Spencer, A.J.M., 1971. Theory of invariants. In: Eringen, C. (Ed.), Continuum Physics. Academic Press, New York, pp. 239–353.
Talreja, R., 1985a. A continuum mechanics characterization of damage in composite materials. Proc. R. Soc. Lond., Ser. A, Math. Phys. Eng. Sci. 399 (1817), 195–216.
Talreja, R., 1985b. Transverse cracking and stiffness reduction in composite laminates. J. Comp. Mats. 19, 355–375.
Talreja, R., 1986. Stiffness properties of composite laminates with matrix cracking and interior delamination. Eng. Fract. Mech. 25 (5/6), 751–762.
Talreja, R., 1989. Damage development in composites: mechanisms and modelling. J. Strain Anal. 4, 215–222.
Talreja, R., 1990. Internal variable damage mechanics of composite materials. In: Boehler, J.P. (Ed.), Yielding, Damage, and Failure of Anisotropic Solids. Mechanical Engineering Publications, London, pp. 509–533. EGF5.
Talreja, R., 1991. Continuum modelling of damage in ceramic matrix composites. Mech. Mater. 12, 165–180.
Talreja, R., 1996. A synergistic damage mechanics approach to durability of composite material systems. In: Progress in Durability Analysis of Composite Systems, pp. 117–129.
Talreja, R., 2010. On multi-scale approaches to composites and heterogeneous solids with damage. Philos. Mag. 90 (31–32), 4333–4348.
Talreja, R., Singh, C.V., 2010. Damage, Fatigue and Failure of Composite Materials. Cambridge University Press.
Vakulenko, A.A., Kachanov, M.L., 1971. Continuum theory of medium with cracks. Izv. A. N. SSSR, MTT 59–166 (in Russian).

Varna, J., 2008. Physical interpretation of parameters in synergistic continuum damage mechanics model for laminate. Compos. Sci. Technol. 68 (13), 2592–2600.

Varna, J., 2024. Thermoelastic constants of damaged laminates: COD-and CSD-based methods. In: Modeling Damage Fatigue and Failure of Composite Materials. Woodhead Publishing, pp. 295–341.

Varna, J., Berglund, L., 1991. Multiple transverse cracking and stiffness reduction in cross-ply laminates. Compos. Technol. Res. 13 (2), 97–106.

Varna, J., Berglund, L.A., Talreja, R., Jakovics, A., 1993. A study of opening displacement of cracks in cross ply laminates. Int. J. Damage Mech. 2, 272–289.

Varna, J., Joffe, R., Akshantala, N.V., Talreja, R., 1999. Damage in composite laminates with off-axis plies. Compos. Sci. Technol. 59 (14), 2139–2147.

Varna, J., Joffe, R., Talreja, R., 2001. A synergistic damage-mechanics analysis of transverse cracking in [±θ/904] s laminates. Compos. Sci. Technol. 61 (5), 657–665.

Varna, J., Krasnikovs, A., Kumar, R.S., Talreja, R., 2004. A synergistic damage mechanics approach to viscoelastic response of cracked cross-ply laminates. Int. J. Damage Mech. 13 (4), 301–334.

Vinogradov, V., Hashin, Z., 2010. Variational analysis of cracked angle-ply laminates. Compos. Sci. Technol. 70 (4), 638–646.

Voigt, W., 1889. Über die Beziehung zwischen den beiden Elastizitätskonstanten isotroper Körper. Wied. Ann. 38 (12), 573–587.

Evolution of experimental methods for characterisation of the mechanical properties of composites

4.5

Jan Ivens[1] *and Mahoor Mehdikhani*[2]
[1]Department of Materials Engineering, KU Leuven, Sint-Katelijne-Waver, Belgium;
[2]Composite Materials Group, Department of Materials Engineering, Katholieke Universiteit Leuven (KU Leuven), Leuven, Belgium

1. Introduction

Mechanical testing is essential for determining the relevant properties of materials or components. However, the results can be influenced by various factors, including specimen preparation and test conditions. This is particularly true for fibre-reinforced composites. These highly anisotropic materials are sensitive to stress concentrations, leading to stresses in weaker, matrix-dominated directions and potential failure in the grips, misalignment causing unwanted stresses and failure modes, end effects affecting the presence of a uniform stress field (Saint-Venant's principle, Choi and Horgan, 1977), and edge effects causing damage and premature failure (Pipes and Pagano, 1970).

The development of composite testing is driven by advances in the scientific understanding of composite behaviour as well as in the relevant technologies. These advances provide new insights that lead to the development of new test methods and the improvement of existing ones. Consequently, the testing and standardisation community is continually working toward more accurate, reliable, and scientifically-sound test methods. As a result, the mechanical testing of composites has evolved over time to ensure more accurate and reproducible results. This evolution, based on increasing insight into the mechanics of composites and critical analysis of testing methods, is ongoing, with further research being conducted in all fields of composite testing. A timeline representing this historical evolution is illustrated in Fig. 4.5.1.

The key mechanical properties of any materials include modulus, strength, ductility, and fracture toughness. For polymer matrix composites, viscoelastic properties such as creep, stress relaxation, impact, and fatigue are also important, though secondary to the effects of anisotropy. The complexity introduced by anisotropy necessitates multiple experiments, including bending, compression, and shear testing, in addition to tensile testing as well as accurate in situ techniques for monitoring the test. The layered nature of laminated continuous fibre composites also requires

The History of Composites. https://doi.org/10.1016/B978-0-443-34119-9.00028-0

Figure 4.5.1 Historical evolution of composites mechanical test development.

additional testing of interlaminar shear properties. Moreover, the fibre–matrix interaction at the microscale, governed by their interface, plays a crucial role in the mechanical performance of composites, highlighting the need for characterising the interface.

2. Quasi-static properties

While measuring two elastic constants (E-modulus and Poisson's ratio) and one limit stress (yield strength) suffices for elastic design in isotropic materials, anisotropic composite materials require independent measurements of tension, compression, and shear modulus and strength in different directions. The significant differences between longitudinal, transverse, and shear strengths often lead to premature failure caused by secondary stresses in the test specimen.

2.1 Tensile testing

Despite its apparent simplicity, tensile testing of fibre-reinforced composites is complex due to their highly anisotropic nature. This complexity has led to numerous discussions and modifications regarding specimen shape, preparation, and testing procedures.

A tensile test specimen typically comprises three regions:

- Gripping region: Where the specimen is clamped.
- Central gauge region: Where a uniform stress state is present.
- Transition region: Between the gripping and central regions

In the gripping region, clamping forces needed for load introduction create stress concentrations. To mitigate these local stresses and prevent failure in the grips, an increase in the specimen's cross-section is often used. In the transition region, dimensions gradually change to the nominal cross-section of the central region. Initially, the dog-bone specimen shape, adapted from plastics and metals testing, was used but quickly found to be unsuitable for anisotropic materials. Other specimen shapes were used, as shown in Fig. 4.5.2, based on the presentations of the first American Society for Testing and Standardisation (ASTM) conference on Composites testing and design in 1969 (Godwin 2000). Shouldered and waisted specimens like the elongated bow-tie specimen have become obsolete because dimensional changes in the width direction usually lead to splitting failures in the transition and gripping region caused by the low in-plane shear strength of anisotropic composites.

This problem was resolved with straight-sided specimens with dimensional changes in the thickness direction using end tabs. This approach reduced splitting failures but did not completely eliminate grip-related failures, leading to concerns about underestimating tensile strength and high variability in strength data. Consequently, ASTM D3039 recommends describing the specimen failure mode using a three-letter code.

Continuous efforts aim to improve tab design, cutting methods and preparation accuracy, and the selection of materials for end tabs and adhesives. End-tab materials

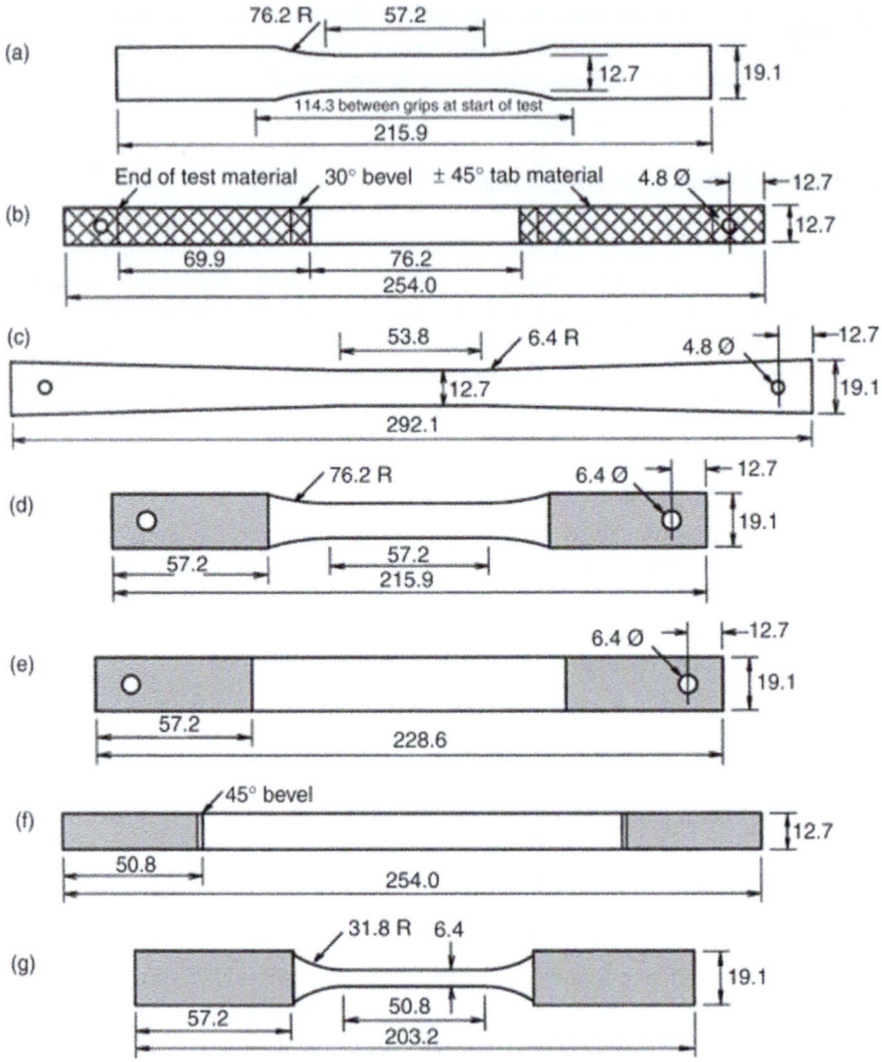

Figure 4.5.2 Different tensile test configurations as discussed during the ASTM conference STP 460.
Reproduced with permission from Godwin (2000), Woodhead Publishing Series in Composites Science and Engineering, Elsevier

need to balance compliance (to homogenise stresses) and stiffness (to transfer load), along with toughness to alleviate stress concentrations from rough grip surfaces. Initially, metal end tabs were common due to their high toughness and low cost, but were later abandoned because the steel end tabs had to be applied to individual test specimens. Aluminium end tabs require surface preparation to ensure a good bond strength. Glass fabric reinforced composites with a ±45 degree orientation offer a cost-effective alternative, as finite element analysis indicates that low-stiffness end tabs reduce stress concentrations in the composite specimen (Adams and Adams 2002).

While ASTM D3039 suggests both straight and tapered end tabs, ISO 527 recommends straight end tabs. Despite simulations indicating that tapered end tabs reduce stress concentrations (Cunningham et al., 1985; De Baere et al., 2011), experimental results do not support this, due to scatter (Kural and Flaggs, 1983; Hojo et al., 1994).

Extreme tapered end tabs have shown reduced scatter and increased tensile strength in some studies (Wisnom and Maheri, 1994) but are not practical for routine testing. Continuous glass end tabs eliminate stress concentrations and result in higher tensile strength (Czél et al., 2016), though at a higher cost and with the need for correction due to residual stresses. Arrow-shaped end tabs have shown promise in obtaining high tensile strengths comparable to the hybrid laminates of Czél et al. without the need for correction, as shown in Fig. 4.5.3 (Fazlali et al., 2023).

Despite the convergence of test standards over time, differences remain in aspects like end-tab design, specimen dimensions, and data analysis. Accurate displacement or strain measurements are critical due to the low strains in composite materials, especially for the determination of Poisson's ratio. Extensometers with high accuracy must be removed before specimen failure to avoid damage, and strain gauges, while precise, are expensive. Video extensometers and digital image correlation (DIC) techniques are now preferred for strain measurements.

Though composites are often considered linear elastic, they exhibit nonlinear behaviour due to both the matrix and the fibres. Thus, standards suggest specific strain intervals (0.05%–0.25% or 0.1%–0.3%) to determine the Young's modulus. The early practice of eliminating nonlinearity at the test start by extrapolating the E-modulus slope to 0 MPa stress is no longer recommended.

2.2 Compression testing

Compression testing holds a number of difficulties and risks of undesired failure modes because it faces contradictory requirements: because of Saint-Venant's principle, long and slender specimens are required to achieve a uniform stress level in the central gauge length of the sample. Long and slender specimens are prone to buckling, which has to be prevented by reducing the free length or by lateral supports. Specimens are typically straight and end-tabbed, like in tensile testing. Specimen alignment is very critical, since no self-alignment is possible. These contradictory requirements have over time resulted in different test methods, which can be divided according to the load introduction method: shear-loaded, end-loaded, or a combination of both.

Before 1970, there was little interest in compressive testing of composites. Nevertheless, the determination of the compressive properties by using a sandwich beam was already proposed in 1966. The method (first part of ASTM D3410, since 1993 ASTM D5467) uses a honeycomb core sandwich panel, loaded in 4-point bending, with a metal face sheet on the tensile side and the test coupon on the compressive side. Because the test method suppresses microbuckling failure, it tends to produce higher strength values. Moreover, specimen preparation is difficult.

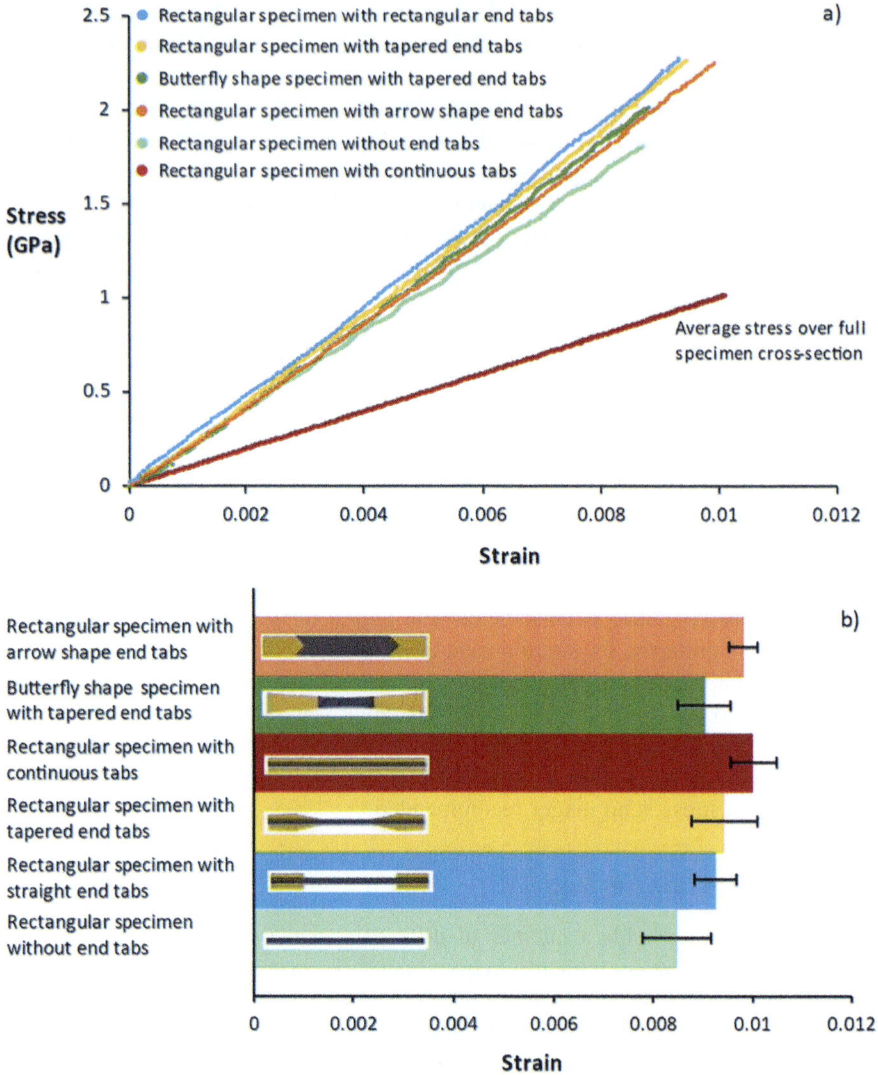

Figure 4.5.3 Measured stress–strain curves and average failure strain for six different end-tab configurations.
Reproduced with permission from Fazlali et al. (2023), Composites Part A: Applied Science and Manufacturing, Elsevier

Park proposed the Celanese compression test method (Park 1971): it consists of a long and slender specimen held in conical wedge grips placed in matching sleeves and then fit into a cylindrical shell for alignment (Fig. 4.5.4a). It was the first compression test standard for composite materials (ASTM D 3410, 1975). The setup required accurate manufacturing of the metallic components, and inevitably, frictional forces affected the test result, overestimating stiffness and strength. If the required dimensions

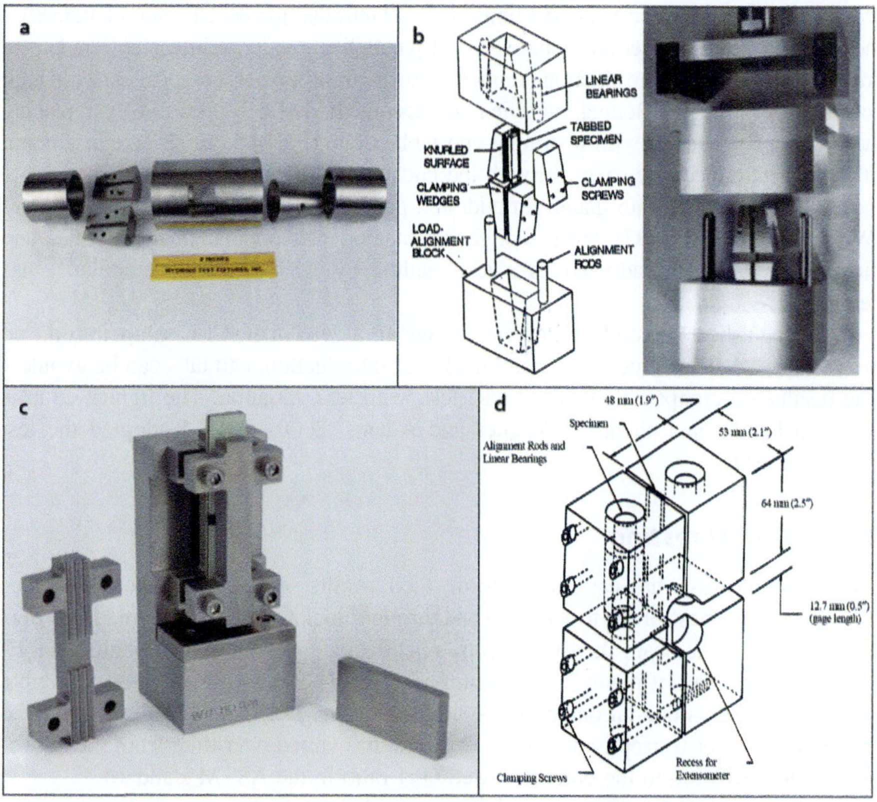

Figure 4.5.4 Different test setups for compression testing: (A) Celanese test setup (picture with permission from Wyoming Test Fixtures); (B) Illinois Institute of Technology Research Institute (IITRI) test setup (Wegner and Adams 2000); (C) The end-loading test method (Boeing standard BSS 7260, picture with permission from Wyoming Test Fixtures); (D) Combined loading compression (CLC) test setup (Wegner and Adams 2000).

are not met, the alignment is not perfect, resulting in instability. It was not well accepted, and research continued for new test methods or modified versions of the Celanese test rig (EN 2850).

In 1977, the IITRI proposed an alternative fixture for compression testing, the IITRI-fixture (Hofer and Rao 1977). It used wedge grips that fit in heavy metal blocks, aligned with pillar guides and linear ball bearings to minimise friction. Even though the specimens are long, the free length is limited; it is recommended to measure the strain on both sides of the specimen to detect buckling (Fig. 4.5.4b). The test was added to ASTM D3410 in 1987 and is still considered one of the best compression tests (Berg and Adams 1989). The main disadvantage is the heavy fixture (approximately 25 kg). A slightly modified version is described in ISO 14126.

While the latter test methods are shear-loaded methods, the ASTM D695 for compression properties of plastics—first introduced in 1942—uses end loading in combination with lateral supports to prevent buckling. The load introduction posed strict tolerances on the specimen edges in terms of flatness and parallelism, and specimens suffered from crushing at the load introduction, rendering them unsuited for composites with high compressive strength. Boeing modified the test method (BSS 7260 1982), replacing the dog-bone specimen with a straight-end tabbed one (Fig. 4.5.4c). A certain amount of load is introduced by shear through the end tabs. The distance between the tabs is too small to place a strain gauge. In addition, part of the load is carried by the fixture (Welsh and Adams 1997). Even though the method was popular in the eighties and nineties, it was never adopted as a test method by ASTM, and its popularity has decreased.

Combined shear and end loading was further employed in the CLC setup, introduced in 1995 (Fig. 4.5.4d). Due to the combined load introduction, end tabs can be avoided. The results are comparable to the IITRI test, with less variation. The lighter setup is easier to handle and cheaper (Wegner and Adams, 2000). ASTM adopted the test method in 2001.

2.3 Flexural testing

While bending tests are easy to perform, their results suffer from inhomogeneities through the thickness of the test sample and through-thickness shear deformations. Bernoulli–Euler elastic beam theory: deflection is affected by the in-plane longitudinal E-modulus and the out-of-plane shear modulus. For the shear deformation to be negligible in unidirectional carbon laminates, the span-to-thickness ratio has to be higher than the 16:1 ratio used for isotropic materials. This ratio has varied over time from the above-mentioned 16:1 ratio to the currently used 60:1 ratio in the ASTM standard.

The Bernoulli–Euler effect is exploited in sandwich panel testing to determine the skin E-modulus and the core out-of-plane shear modulus by using different span-to-thickness ratios. Under the assumption of a weak core and thin skins, the midpoint deflection δ of a symmetric sandwich panel loaded in 3-point bending can be written as:

$$\delta = \frac{PL^3}{48D} + \frac{PL}{4N} = \frac{PL^3}{24E_f t_f d^2}\left(1 + 6\left(\frac{E_f}{G_c}\right)\left(\frac{t_f t_c}{L^2}\right)\right) \tag{4.5.1}$$

where P is the central load, L is the span length, D is the bending rigidity, N is the shear rigidity, E_f is the skin's E-modulus, G_c is the core's out-of-plane shear modulus, and t_f and t_c are the thickness of the skin and the core, respectively. By measuring the deflection for a given load at different span lengths, D and N can be obtained.

2.4 Shear testing

2.4.1 In-plane shear

The goal of in-plane shear testing is to determine the shear modulus and shear strength. Over the years, many different test methods have been developed, some of which became standardised; many others disappeared due to the test complexity and/or the inability to produce consistent and correct values for the shear strength.

A simple way to determine the shear modulus is by performing a tensile test on a (±45 degrees) specimen, as proposed by Rosen (1972) and adopted as ASTM standard in 1976 (ASTM D3518 and EN14129). This test is accurate and generates less scatter than other shear test methods (Fig. 4.5.5a). However, the stress state is not pure shear, and edge effects occur, inducing failure, which results in an underestimation of the shear strength. The main advantage is that it is a cheap and simple test (Lee and Munro 1986).

The thin-walled tube torsion (ASTM D5448) was proposed by Whitney et al. (1971). With the wall thickness small compared to the tube diameter and the application of a pure torque, a pure and uniform shear stress is induced. The main drawbacks are the cumbersome test specimen production and preparation, and the special test apparatus needed. Hence, the tube torsion test did not become successful except for filament-wound composites.

In 1977, Chamis and Sinclair (1977) proposed to use a 10-degree off-axis tensile test. Their analysis showed that the high in-plane shear stress induces specimen failure. While it is a simple test to perform, the presence of shear coupling, out-of-plane bending, and twist accounts for lower shear strength results (Pindera and Herakovich 1986).

Two-rail shear and three-rail shear tests were used extensively in the seventies and eighties. They became an ASTM standard in 1983 (ASTM D4255). Even though the test specimens contain regions of uniform shear, both test methods suffer from stress concentrations and uneven loading at the load introduction points, resulting in an underestimation of the shear strength and high scatter in the results (Hussain and Adams 2004).

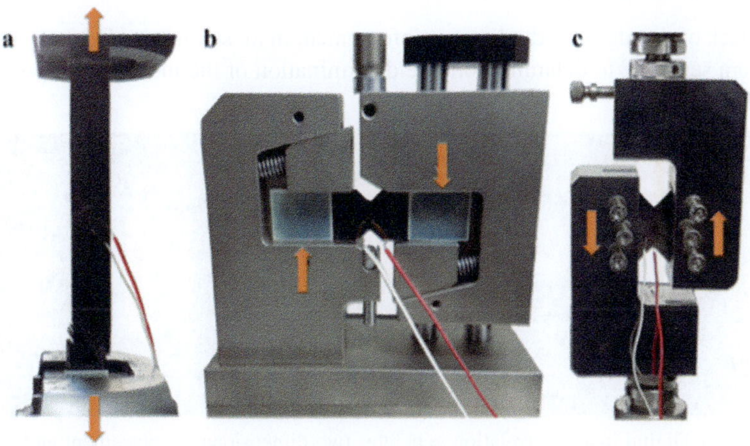

Figure 4.5.5 Three types of shear tests: (A) 45 degree tension, (B) Iosipescu shear test, and (C) V-notched rail shear test.
Reproduced with permission from Choi et al. (2019), Composite structures, Elsevier

A V-notched beam method was first introduced by Iosipescu in 1967 for metals testing. In 1983, the method was adapted for composite materials and gained popularity since its introduction (Fig. 4.5.5b). A state of pure shear stress is obtained at the midplane by applying an asymmetric 4-point bending load on the test sample (Walrath and Adams, 1983). By reducing the cross-section at the centreline by means of notches, failure is induced at that location, ASTM D5379, first approved in 1993. The shear strain is measured using two orthogonal strain gauges at 45 degree orientation placed at the centre line, preferably on both sides of the specimen. Even though the Iosipescu test yields good results (Swanson et al., 1985), some issues have to be considered. The edge loading of the specimen can induce local crushing in the case of high shear strength materials. Finite element analysis showed that the stress is not uniform across the specimen width, and correction factors were proposed (Broughton et al., 1990). The development of DIC methods (more details in section 5.2 and Chapter 4.6) has resulted in replacing the strain gauges with displacement field measurements, as shown in Fig. 4.5.6 (Choi et al., 2019).

The V-notched rail shear test can be considered a combination of the two-rail shear test and the Iosipescu shear test and was introduced in 2003 (Fig. 4.5.6c). It was adopted as an ASTM standard shortly after that (ASTM D7078), and it is considered the most appropriate shear test for composites, together with the Iosipescu shear test. It allows higher loads because of the face loading (compared to Iosipescu's edge loading). While the shear stress is uniform (Choi et al., 2019), observed high nonuniformity at low shear stress levels, which could affect the measurement of the shear modulus (Fig. 4.5.6). In 2009, Adams proposed a modification of the V-notched rail shear test, the combined loading shear test, in which edge and face loading are combined; it is larger than the V-notched rail shear test, has a larger and more uniform gripping region, and can hence achieve higher load levels before slipping or crushing occurs (Adams, 2009).

2.4.2 Out-of-plane shear

As the lack of through-the-thickness reinforcement in most laminated composites renders them sensitive to delamination, the determination of the interlaminar properties,

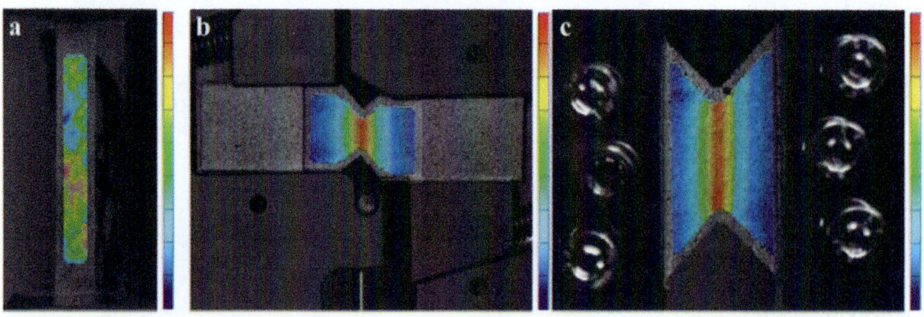

Figure 4.5.6 Digital image correlation generates two-dimensional displacement and strain fields, allowing the evaluation of the homogeneity of the stress field during (A) the Tensile test, (B) the Iosipescu shear test, and (C) the V-notched rail shear test.
Reproduced with permission from Choi et al. (2019), Composite structures, Elsevier

like the interlaminar shear strength, is important. The short beam shear test was introduced as a standard in 1965, using a small span-to-thickness ratio (typically 3.5:1), and interlaminar shear failure was induced during 3-point bending. It was very popular because of its simplicity. But the very high transverse forces at the load introduction point and the support points often cause unwanted damage at the load introduction point, resulting in undesired failure modes (Whitney and Browning, 1985). While the test is still used as a quick and dirty quality control test, its scientific merit is very limited, and preference is given to interlaminar fracture toughness tests.

2.5 Interlaminar fracture toughness

The importance of delaminations in damage development and structural integrity of composites became apparent in the work of Pipes and Pagano (1970). The severe shortcomings of the short beam shear test and the preference to use a fracture mechanics approach to experimentally determine the delamination resistance of laminated composites have triggered alternative interlaminar test methods. A new set of test methods was developed, based on the determination of the interlaminar fracture toughness, typically expressed by the critical strain energy release rate Gc. A delamination can grow under three loading conditions or a combination, where mode I and mode II are dominant. A discussion on fracture mechanics and strain energy release rates is outside the scope of this text; it can be found elsewhere. While in reality, delaminations will occur between layers with dissimilar ply orientations near geometric changes like ply drop-offs, free edges, and holes, or be induced by impact or other kinds of out-of-plane loads, the interlaminar fracture toughness is typically measured on unidirectional test samples with an artificial defect at the midplane as a starter delamination. Efforts to apply the method to multidirectional laminates inherently suffer from crack deviation and branching.

2.5.1 Mode I testing

For mode I testing, the double cantilever beam (DCB) test was introduced for composites in the early eighties by Bascom et al. (1980) and Wilkins et al. (1982); they adopted a test method used for adhesives, like that used by Brussat et al. (1977; Fig. 4.5.7). Straight-sided specimen geometries quickly became the standard, and research focused on the starter crack and the data reduction scheme. For the calculation of the interlaminar fracture toughness, load, displacement, and crack extension need to be measured. The DCB test was confronted with three issues: the detection of crack initiation, the deviation from the linear elastic beam theory behaviour, and fibre bridging during delamination growth. The latter is caused by nesting of the unidirectional plies and is a test artifact overestimating the propagation fracture toughness. The other two issues were addressed during the eighties and nineties through worldwide round-robin experiments. The linear elastic beam theory solution was corrected using a compliance calibration method or modified beam theory (Hashemi et al., 1989; Davies et al., 1992). Major attention was dedicated to the determination of the initiation fracture toughness. The influence of the starter crack thickness was

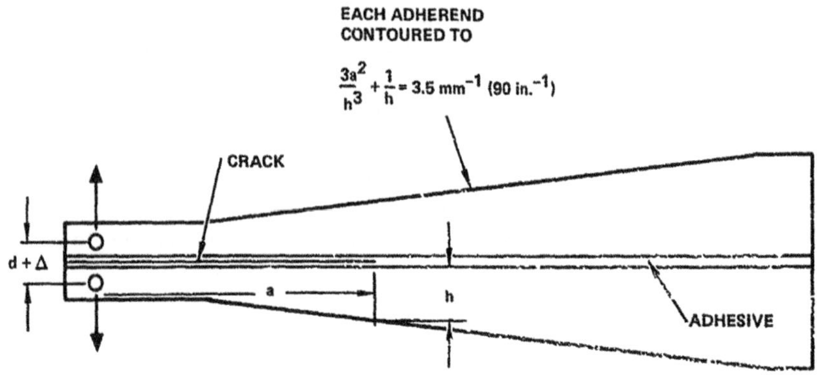

Figure 4.5.7 Tapered double cantilever beam test specimen for mode I testing of adhesives (Brussat et al., 1977).

investigated. Different crack initiation detection methods like visual inspection, onset of nonlinearity, or 5% offset of the maximum load were proposed (O'Brien and Martin 1993). The work resulted in the first test standard for DCB testing in 1994 (ASTM D5528), incorporating the different detection methods. Other standardisation bodies further investigated wedge precracking (Hojo et al., 1995), which was incorporated as an optional pretreatment in the ISO 15024 in 2001.

2.5.2 Mode II testing

Mode II fracture toughness testing, though more relevant as mode II is often the dominant mode in delamination growth, received less attention. The end-notched flexure test was first introduced by Russell and Street (1982). Since the test produces unstable crack growth except for large precrack lengths $a/L > 0.7$ (Carlsson et al., 1986), alternative test methods were intensively investigated, like the stabilised end notched flexure (SENF) test in Japan (Kageyama et al., 1991) and the end-loaded split (ELS) test in Europe (Davies et al., 1992, 1999). The latter produces more stable crack growth ($a/L > 0.55$). While it is less affected by friction than the ENF test, correction factors for friction can be applied. A modified version was standardised by ISO (ISO 15114) in 2014. Similar to mode I, corrections for the linear elastic beam theory are applied, and appropriate precracking is required. Mode II crack growth is much more difficult to detect.

3. Impact testing

The earliest work on impact testing of composites was conducted in the 1970s by Butcher, who impacted a unidirectional clamped test specimen in a universal testing machine using a pendulum (Butcher 1979). In the 1980s, Cantwell et al. (1986) used a drop weight impact tower for low and high velocity impact tests, combined with damage assessment using ultrasonic C-scanning, radiography and optical microscopy, and

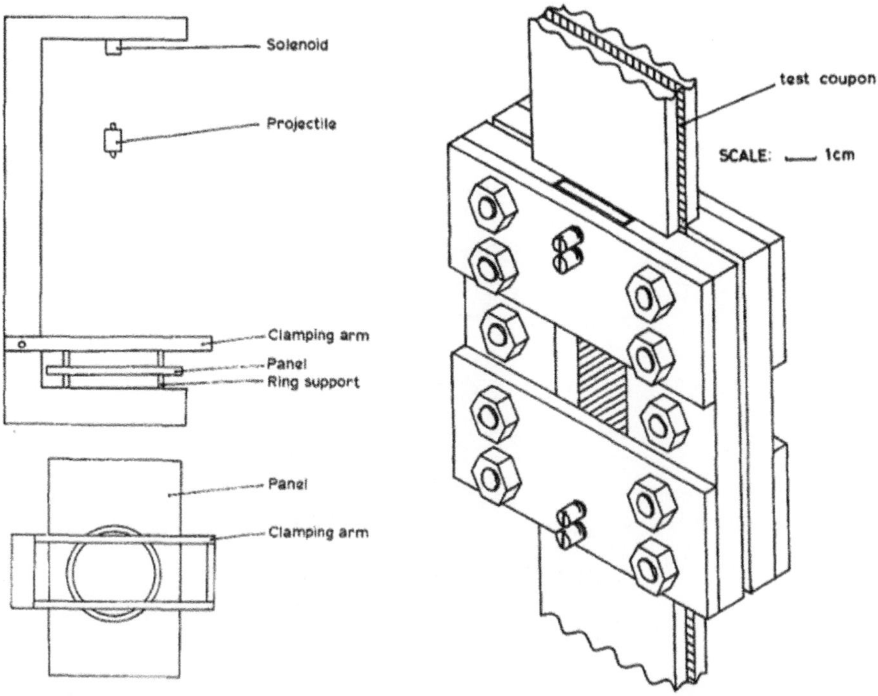

Figure 4.5.8 Drop weight impact setup and anti-macrobuckling test rig as used by. Reproduced with permission from Cantwell et al. (1986), Composites Science and Technology, Elsevier

postimpact property measurement in tension and compression (Fig. 4.5.8). For the latter, a special clamping device was designed to avoid macrobuckling, favouring impact damage-induced failure. Since then, the combination of drop weight impact testing, damage assessment, and subsequent compression after impact (CAI) testing has become common practice, even though other postimpact properties like fatigue have been studied as well (Stellbrink, 1983). During the eighties and nineties, the impact test methods differed in specimen size, boundary conditions, and impactor geometry. Nevertheless, both the drop weight impact and the CAI tests converged to standards, and in 2005, ASTM adopted a standard for drop weight impact (ASTM D7136) and for CAI (ASTM D7137). The most recent developments focus on the use of high-speed cameras and DIC techniques (Flores et al., 2017).

4. Characterisation of fibre–matrix interface in composites

In this section, a historical evolution of tests developed to study the fibre–matrix interface is presented. The timeline of this historical evolution is illustrated in Fig. 4.5.9. A

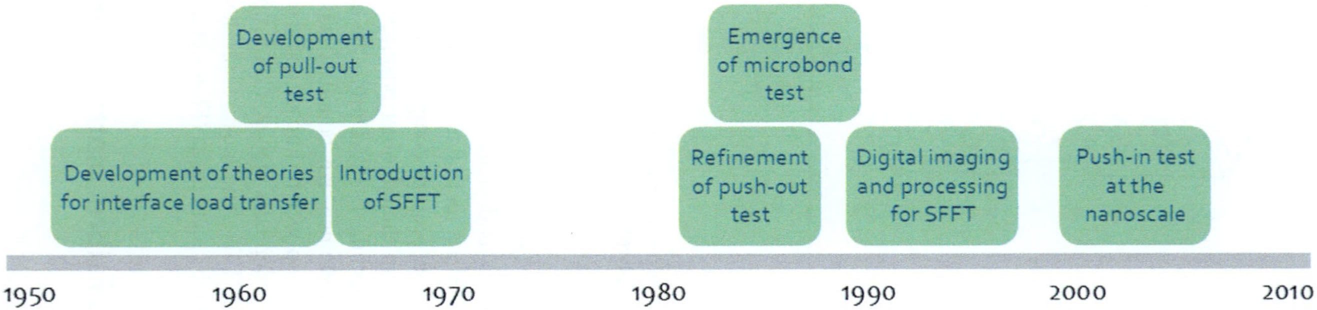

Figure 4.5.9 Historical timeline of developments of techniques for characterisation of fibre–matrix interface in composites—years are just indicative and hence not exact.

comprehensive and detailed review of their scientific and technical aspects can be found in AhmadvashAghbash et al. (2023). Moreover, the characterisation of other composite constituents, i.e., fibre and matrix, is not reviewed here.

4.1 The emergence of interface characterisation (1960s–1970s)

In the early days of composite materials research, scientists and engineers were captivated by the promise of combining strong fibres with versatile resin matrices to create materials with superior properties. As they explored this new frontier, they soon realised that the key to unlocking the full potential of composites lies in understanding the interface where the fibre meets the matrix. This realisation set the stage for the development of innovative testing methods to characterise this critical region, which some researchers have even regarded as a third constituent.

4.1.1 Single-fibre fragmentation test: the most popular

It was during this period that the single-fibre fragmentation test (SFFT) began to take shape. Researchers envisioned a method where a single fibre could be embedded in a (high failure strain) resin matrix and subjected to tensile loading. As the fibre fractured multiple times under stress, forming progressively smaller fragments, the pattern and frequency of breaks were used to infer the interfacial shear strength (IFSS) between the fibre and matrix. Initially, data reduction methods were simple, assuming constant shear stress at the fibre–matrix interface. Over time, advanced analyses that considered more realistic stress distributions were developed. Foundational contributions came from researchers like Cox (1952), Cottrell (1964), and Kelly and Tyson (1965), whose theoretical work provided the basis for SFFT analysis (Fig. 4.5.10). Moreover, later, more advanced tools were used to monitor the SFFT, including laser Raman spectroscopy and photoelasticity (see section 5.4).

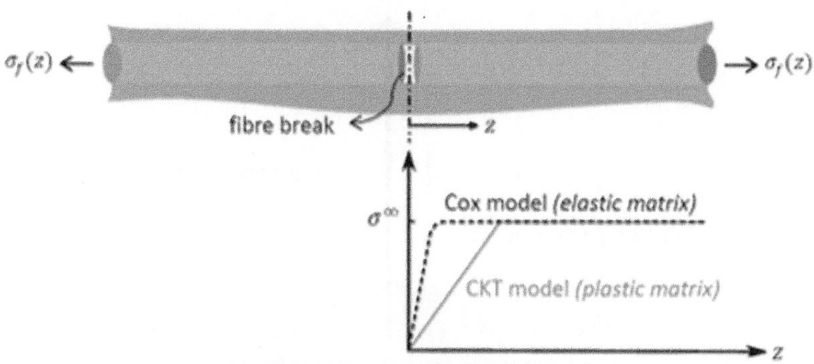

Figure 4.5.10 Stress recovery profiles upon single-fibre fragmentation predicted by Cox and Cottrell–Kelly–Tyson (CKT) models.
Reproduced with permission from AhmadvashAghbash et al. (2023), International Materials Reviews, SAGE Publications

4.2 Refining the techniques (1980s–1990s)

As composite materials found their way into more applications, the need for reliable and precise testing methods grew. This era saw significant advancements in the techniques used to characterise the fibre–matrix interface.

4.2.1 Pull-out test: The pioneer

The pull-out test, one of the earliest methods devised to measure fibre–matrix adhesion, saw significant refinement during this period. Initially developed by Shiriajeva and Andreevskaya (1962) and later modified by Favre and Perrin (1972), Piggott et al. (1985), and Hampe (1988), the test involved embedding a fibre in a matrix and then pulling it out while recording the force–displacement curve (Fig. 4.5.11). Theoretical models by researchers like Cox (1952) and Lawrence (1972) helped interpret the results in terms of IFSS. Improved gripping techniques and precision load frames reduced the variability and damage associated with the test, making it more reliable.

4.2.2 Microbond test: delicate, but accurate

In the early years of interfacial testing, the weaker reinforcing fibres were thicker and favoured by the pull-out test. However, more modern and commercially available thin fibres demanded the conception of new methods. In the 1980s, the microbond test emerged as a promising alternative to the fragmentation and pull-out tests. Researchers like Miller et al. (1987) developed a meticulous technique where a small droplet of resin was cured around a single fibre. The microbond test enabled interfacial characterisation of relatively thin fibres (5–15 μm diameter). This method allowed for direct measurement of the force required to debond the droplet, providing valuable insights into the IFSS. As precision micromanipulators and load cells became more accessible, the microbond test gained more popularity.

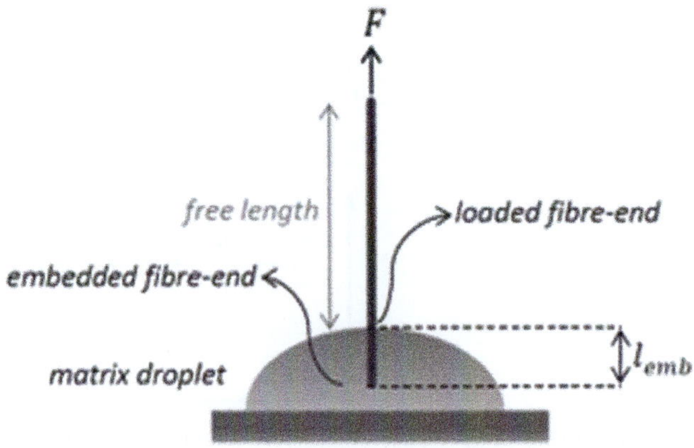

Figure 4.5.11 Schematics of a single-fibre pull-out test, illustrating the relevant parameters. Reproduced with permission from AhmadvashAghbash et al. (2023), International Materials Reviews, SAGE Publications

4.3 Increasing complexity (1990s–2000s)

As the understanding of composite materials deepened, the tests themselves and their data analysis became more sophisticated. The focus shifted from merely obtaining measurements to understanding the underlying mechanisms at play.

4.3.1 Push-out test: A new perspective

In a real composite, the interfacial stress state of the debonding fibre is affected by the presence of the neighbouring intact fibres. The deficient resemblance of the single-fibre tests to real composite stress states highlights the importance of bulk composite test methods for evaluating the interfacial parameter. The push-out test, developed by Marshall (1984) and advanced in the 1990s, offered a different approach by pushing a fibre out of a matrix in a thin slice rather than pulling it (Fig. 4.5.12). This method was particularly useful for brittle matrices, such as ceramics, where traditional pull-out tests could cause catastrophic failure. The introduction of microindentation equipment allowed for higher precision and control, making the push-out test a valuable tool in the characterisation arsenal. A similar test, called "push-in", was also introduced, which looked very much like nanoindentation (Mandell et al., 1980).

4.4 Integration of digital tools (1990s–Present)

With the advent of digital technology and advanced imaging techniques, the field of composite material testing entered a new era. These innovations brought extraordinary accuracy and insights into the fibre–matrix interface characterisation. The SFFT benefited greatly from advancements in digital image analysis and microscopy. High-resolution cameras and software tools allowed researchers to capture and analyse the fragmentation process in real-time, providing detailed information about stress distribution and damage modes. This period also saw the test being applied to a wider range of composite materials, from traditional glass and carbon fibres to advanced nanocomposites.

Both the microbond and pull-out tests continued to evolve with technological advancements. High-speed cameras and data acquisition systems provided real-time monitoring and analysis of the debonding process, offering deeper insights into interfacial failure mechanisms. Computational modelling became an integral part of the testing process, allowing researchers to simulate and predict the behaviour of fibre–matrix interfaces under various conditions. The push-out test, too, embraced the digital revolution. Microfabrication and nanotechnology advancements enabled tests on smaller scales (i.e., nanoindentation), allowing for the characterisation of nanocomposites and other advanced materials. The ability to measure interfacial properties at the nanoscale opened up new possibilities for designing and optimising composite materials for specific applications.

4.5 Interface characterisation history: conclusion

The evolution of the single-fibre fragmentation, pull-out, microbond, and push-out tests reflects the dynamic nature of composite science and engineering. From their humble beginnings in the 1960s to the sophisticated, high-tech methods of today,

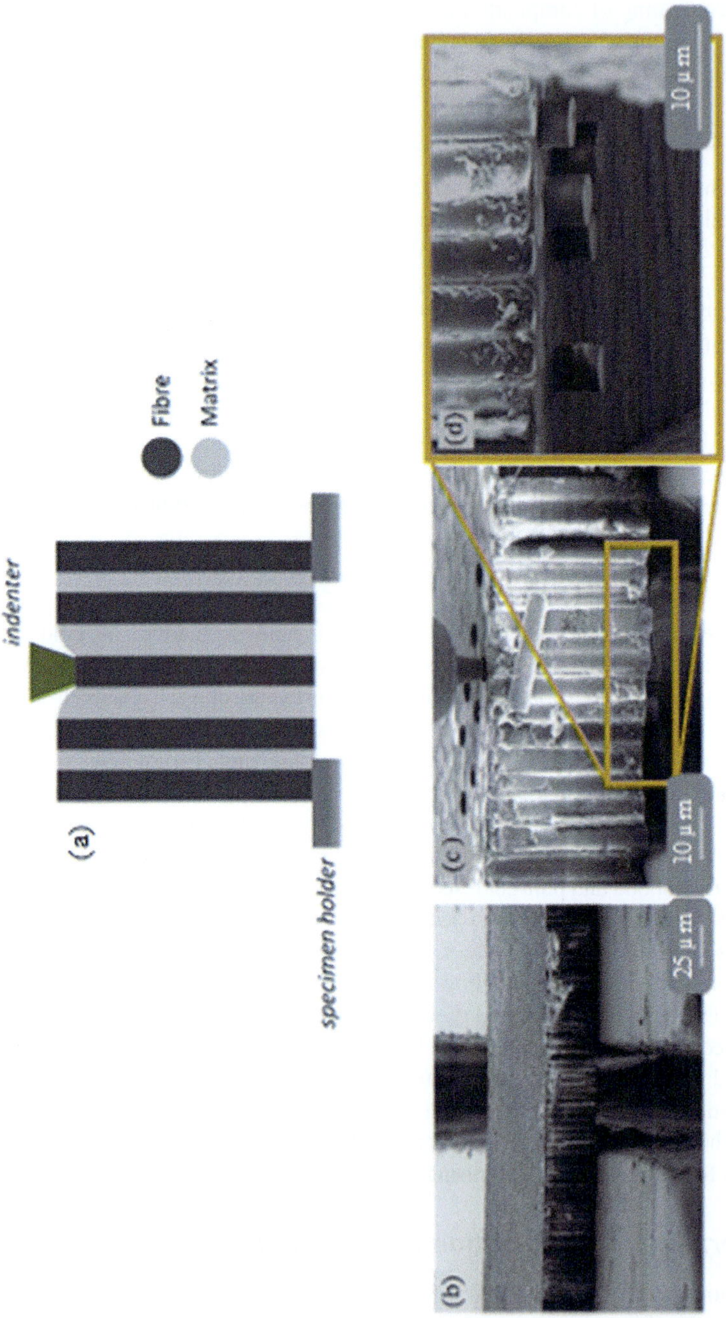

Figure 4.5.12 Fibre push-out test: (a) Schematics of the test; SEM images of the specimen (b) before the test and (c) and (d) after the test. Reproduced with permission from Ghaffari et al. (2021), Materials & Design, Elsevier

these tests have played a crucial role in advancing the understanding and application of composite materials. Each breakthrough, documented by influential researchers over the decades, has brought scientists and engineers closer to unlocking the full potential of these remarkable materials, driving innovation and shaping the future of technology. Still, these tests are criticised for characterising "model" rather than "real" composites. Moreover, as demonstrated in "A round-robin program on interfacial test methods" by Pitkethly et al. (1993), significant variability exists across different laboratories and testing methods. Hence, research in this field continues to advance, aiming to bring interface characterisation as close to reality as possible. Modern in situ damage monitoring methods like 4D X-ray computed tomography (XCT) are going to contribute significantly toward this aim.

5. In situ analysis techniques for composites

This part traces the historical evolution of the most significant in situ analysis techniques that have enabled us to explore the deformation and damage inside composites in real-time. These techniques have played a vital role in the experimental characterisation of composites. Therefore, they are reviewed briefly in this chapter, while a more extensive historical report of these techniques is provided in Chapter 4.6. Fig. 4.5.13 displays a historical timeline of the development of these techniques. Methods less commonly used for in situ investigation of composites in the laboratory are *not* reviewed here; they include fibre Bragg grating sensors, infrared thermography, in situ scanning electron microscopy, micro-Raman spectroscopy, ultrasonic testing, and electrical resistance measurement.

5.1 Acoustic emission (AE)

It all began in the 1950s with the "Kaiser Effect", discovered by Joseph Kaiser, which revealed that materials emit acoustic signals after a primary load level is exceeded (Kaiser, 1950). This discovery marked the birth of AE as a nondestructive testing method. While Kaiser is widely recognised for identifying AE in 1950, it wasn't until the mid-1960s, when Dunegan and colleagues (e.g., Dunegan et al., 1968) began exploring ultrasonic frequency ranges, that AE truly emerged as a practical tool in scientific and engineering applications.

Initially applied to metals, the potential of AE for composite materials was soon recognised in the 1970s (e.g., Mehan and Mullin, 1971; Pollock, 1973). The development of the first AE standards, such as ASTM E610 (1982), laid the groundwork for its use in material science. The 1980s saw the application of AE in composites gaining momentum, with studies by researchers like Hamstad (1986) and Sachse and Kim (1987) demonstrating its effectiveness in detecting delamination, fibre breakage, and matrix cracking.

By the 1990s, AE had become a standard tool in composite testing (e.g., Barré and Benzeggagh, 1994; De Groot et al., 1995). Improved algorithms for signal analysis and pattern recognition, as developed by Rao (1990), Bhat et al. (1994), Suzuki et al. (1996), and Wevers (1997), enabled more accurate damage assessment. Entering

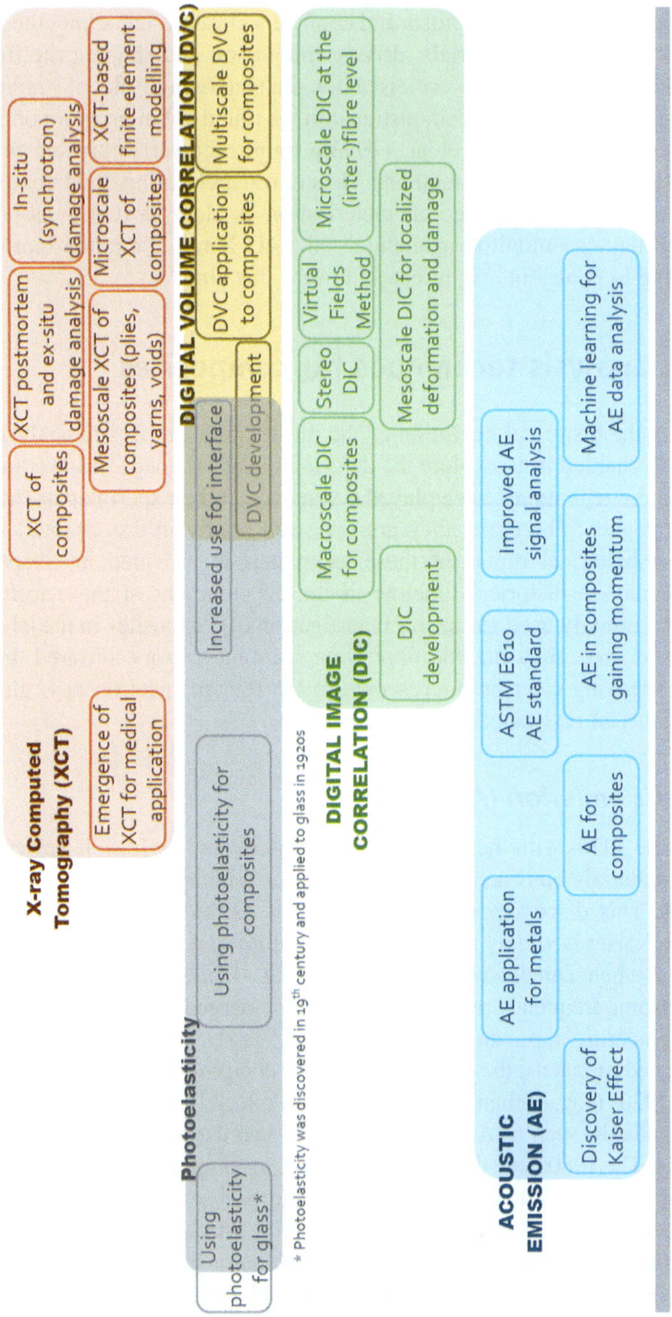

Figure 4.5.13 Historical timeline of developments of in situ analysis techniques for composites—years are just indicative and hence not exact.

the 21st century, AE systems became more sophisticated, leveraging advancements in digital signal processing and machine learning (Godin et al., 2004; Gutkin et al., 2011; Li et al., 2014; Loutas et al., 2017). Researchers in composite damage mechanics have utilised AE in two primary ways: (a) to detect the initiation or critical progression of damage, serving as a warning for imminent failure; and (b) as a tool for identifying the type of damage, particularly when integrated into multi-instrument damage monitoring methods (see next section). Real-time monitoring and automated damage assessment became possible, making AE an indispensable tool in the field of materials and structures.

5.2 Digital image correlation

DIC, developed in the 1980s at the University of South Carolina by researchers like M. A. Sutton, W. F. Ranson, T. C. Chu, and W. H. Peters III, is a noncontact, optical full-field method used to measure surface deformations. In fibre-reinforced composites, it has been used across various scales. Initially applied at the macroscale (Peters et al., 1989), DIC tracked global deformation and strain fields in large composite structures, supporting aerospace and automotive industries. The full-field strain measurements were applied to the analysis of the mechanical testing of coupons by M. Grediac, J. Mollimard, F. Pierron, and A. Vautrin since the end of the 1990s (see, e.g., Grédiac et al., 1994; Pierron et al., 1998). The virtual fields method was developed by Grédiac (2004) and Pierron et al. (2007) to identify mechanical parameters using full-field measurement. This research direction played its role in the development of precise DIC algorithms. This period also introduced stereo-DIC, facilitating out-of-plane strain measurement. In the 2000s, DIC advanced to mesoscale analysis (e.g., Lomov et al., 2008), allowing researchers to investigate localised deformation, cracks, and delamination, as well as the effects of fibre orientation on composite mechanics.

By the 2010s, improvements in high-resolution cameras and integration with scanning electron microscopy enabled DIC at the microscale (Canal et al., 2012; Mehdikhani et al., 2016a, 2016b)—see Fig. 4.5.14. This allowed for strain and damage analysis at the fibre and matrix level, revealing detailed interfibre deformation insights and validating micromechanical models. For more detailed and comprehensive damage analysis, DIC can be coupled with other tools like AE, leading to multi-instrument in situ damage monitoring (Oz et al., 2018).

The development of *DIC software* has advanced significantly since its inception in the late 1980s and early 1990s, when researchers like M. A. Sutton and H. W. Schreier (University of South Carolina) established the core algorithms for DIC. Commercialisation began in the mid-1990s, with *Correlated Solutions* (USA) creating tools for experimental mechanics. Later, *Gesellschaft für Optische Messtechnik (GOM)* (Germany) focused on optical metrology and DIC, and *LaVision* (Germany) and *Dantec Dynamics* (Denmark) developed high-precision strain measurement solutions for industrial and research applications. More recently, *MatchID* (Belgium) and *EikoSim* (France) entered the commercial DIC software market, establishing themselves as key players in the field of DIC development. These companies led advancements in DIC through the 2000s, leveraging better imaging technology and computational

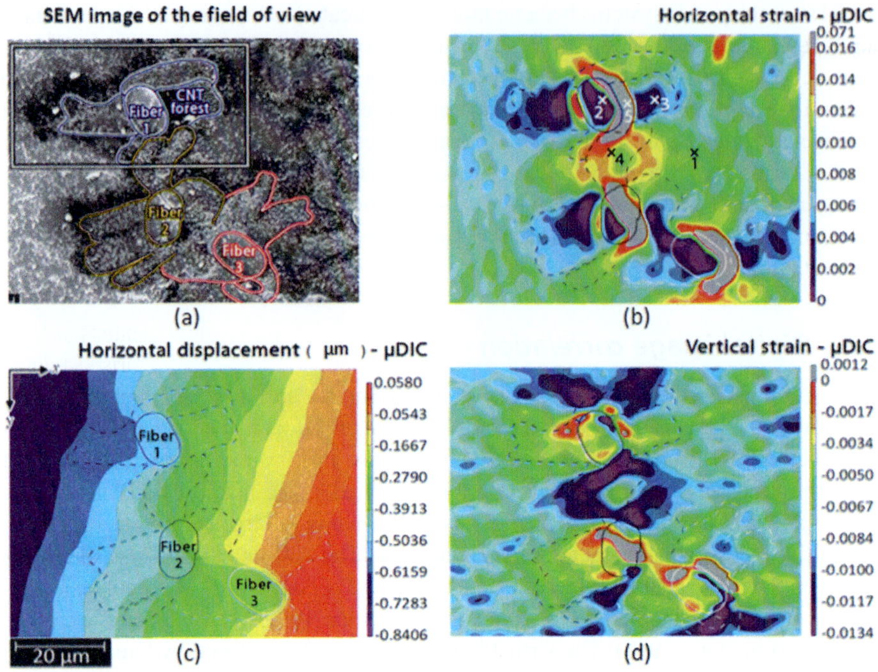

Figure 4.5.14 Microscale DIC on a nano-engineered composite under 3-point bending: (A) The SEM field of view at a working magnification of 1200×, (B) horizontal and (D) vertical DIC resulting strain maps, and (C) horizontal displacement map.
Adapted with permission from Mehdikhani et al. (2016b), Composites Science and Technology, Elsevier

power to enable out-of-plane strain measurements and integration with finite element models. Besides, *NCorr* is a common open-source MATLAB-based software, offering free access to 2D DIC for academic and research applications. By the 2010s, DIC software had expanded applications across aerospace, automotive, and composites testing, with innovations such as real-time analysis and AI-enhanced processing pushing the field into new frontiers.

Recent advancements have expanded DIC into 3D through digital volume correlation (DVC), which leverages 3D images from sources like X-ray tomography to analyse internal material deformation, offering a comprehensive view within the bulk of materials (see next section). With these developments, DIC has become an essential tool for multiscale analysis of composites, helping to link microstructural behaviour to overall material performance across various structural applications.

5.3 Digital volume correlation

DVC is a natural extension of DIC, adapted for three-dimensional analysis. Developed in the late 1990s, DVC uses volumetric images, often from in situ CT scans, to create 3D strain fields for inside the bulk. It offers a window into the complex internal deformation and damage mechanisms of composite materials, deepening our knowledge of their behaviour. With advancements in computational methods and high-resolution CT scanners in the 2010s, DVC has become a powerful tool in the study of composite materials. The first practical applications of DVC in composite science demonstrated its ability to provide detailed internal strain maps of composites under load (Brault et al., 2013). More recent studies like (Mehdikhani et al., 2021a) exploit DVC to detect various damage types in composite laminates at different length scales.

5.4 Photoelasticity

The roots of photoelasticity stretch back to the early 19th century, when scientists like David Brewster and Augustin-Jean Fresnel realised that they could use polarised light to study stress patterns in transparent materials. The early studies provided both qualitative and quantitative insights into stress distribution, laying the foundation for future developments. Between the 1930s and 1950s, the development of synthetic materials and better optical equipment made photoelasticity a standard tool for stress analysis in engineering. The technique has been adapted for use with composite materials since the 1960s. Researchers like Marloff and Daniel (1969) and Dally and Prabhakaran (1971) developed methodologies and composite models to employ photoelasticity to study stress concentrations and damage mechanisms in transparent (glass fibre–epoxy) composites.

In the 1990s and beyond, photoelasticity was used widely for fibre–matrix interface characterisation (Fig. 4.5.15; e.g., Herrera-Franco and Drzal, 1992; Kim and Nairn,

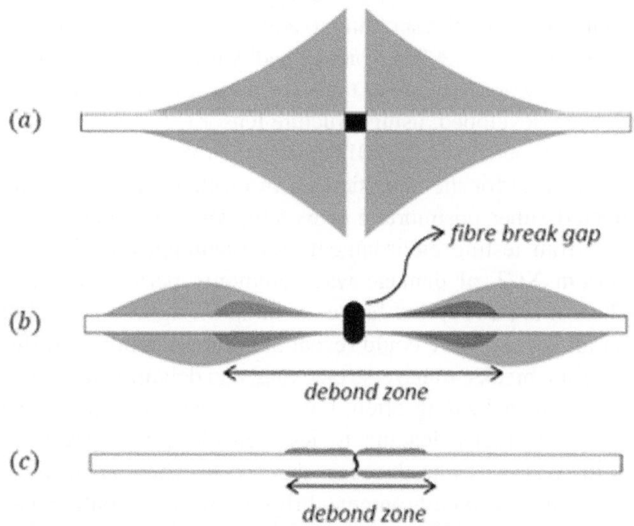

Figure 4.5.15 Schematics of photoelastic birefringence patterns illustrating shear stress from an E-glass fibre break within an epoxy matrix: (a) With no interfacial debonding, (b) with debonding present, and (c) in an unloaded state with diminished birefringence.
Reproduced with permission from AhmadvashAghbash et al. (2023), International Materials Reviews, SAGE Publications (redrawn from Kim and Nairn, 2002)

2002). It helped explore the mechanics of fibre–matrix adhesion and debonding in composite materials and characterise interfacial properties like shear strength and fracture toughness. Applying photoelasticity to experimental techniques such as fibre pull-out and fragmentation, researchers analyse the behaviour of the fibre–matrix interface under stress, examining factors like stress distribution, crack propagation, and distinct damage mechanisms, such as fibre breakage and matrix cracking.

5.5 X-ray computed tomography

In situ XCT has profoundly impacted composite materials research by enabling nondestructive, real-time 3D imaging of internal structures under varying stress conditions. The roots of this technology date back to the 1970s when Allan M. Cormack and Godfrey N. Hounsfield won the Nobel Prize in Physiology or Medicine in 1979 "for the development of computer-assisted tomography"—the first clinical CT scanner was invented in 1971 by Sir Godfrey Hounsfield. Although initially developed for medical imaging, XCT soon expanded to materials science in the 1980s as its potential for nondestructive internal inspection of materials became evident.

Initially, XCT applications in composites focused on the characterisation of their static internal structure (1990s). This started with studying features such as yarn geometry (Desplentere et al., 2005) and orientation (Pandita and Verpoest, 2003), voids (Schell et al., 2006), and fibre orientation (Krause et al., 2010), which was later used in building accurate finite element models (Naouar et al., 2014; Sencu et al., 2016; Liu et al., 2017). A key challenge of XCT characterisation of composites is the segmentation of microstructural features in the resulting 3D volumes. This has evolved from the "conventional" grayscale thresholding to advanced deep-learning methods for the detection of fibres (Guo et al., 2023), voids (Upadhyay et al., 2024), and cracks (Mehdikhani et al., 2023). Moreover, novel tools for the extraction of fibre orientation in XCT data have been developed, using structure tensor (Straumit et al., 2015), supervised segmentation (Emerson et al., 2017), and DIC (Mehdikhani et al., 2021b).

Later, XCT was used for the identification of mechanical damage in composites. This was conducted either postmortem or ex situ, where loading was interrupted to remove samples from testing environments for scanning, e.g., by Schilling et al. (2005). Postmortem XCT of damage was commonly performed for impact (Bull et al., 2013; Chen et al., 2019) and fatigue loading (Wright et al., 2008; Arif et al., 2014; Garcea et al., 2014). XCT could reveal 3D representation of different damage mechanisms like fibre breaks, transverse cracking, and delamination. The ex situ analysis of composites under loading often, however, creates errors like altered microstructures and closed cracks, leading to less reliable data (Garcea et al., 2018). Therefore, in situ XCT was introduced in the 2000s, when advancements in high-flux synchrotron sources and environmental rigs allowed for continuous 3D imaging of composites directly under load or heat, preserving their structural integrity and capturing critical internal changes in real-time. This enabled the study of the evolution of fast and small-scale damage mechanisms like fibre fracture (Scott et al., 2011; Swolfs et al., 2015)—Fig. 4.5.16.

Figure 4.5.16 Various damage mechanisms segmented in the XCT image of a carbon–epoxy laminate close to final failure.
Reproduced with permission from Scott et al. (2011), Composites Science and Technology, Elsevier

5.6 *In situ tools for composites: Conclusion*

In conclusion, the evolution of in situ analysis techniques has greatly enhanced our ability to study and understand the internal behaviour of composite materials under stress. Starting with AE in the 1950s, each technique—from DIC and photoelasticity to modern DVC and XCT—has progressively contributed to revealing damage mechanisms in composites with increasing precision. These advancements have made it possible to analyse composites at multiple scales, yielding even deeper insights and innovations, allowing for an increase in their application in high-performance industries like aerospace and automotive.

6. Conclusion

The historical journey of experimental techniques for characterising the mechanical properties of composite materials has been a fascinating one, marked by significant advancements and innovations. From the early days of simple tensile tests to the sophisticated in situ techniques of today, these methods have played a crucial role in unravelling the complex behaviour of composite materials. The development of test techniques has been driven by the evolving needs of various industries, particularly aerospace, automotive, and construction. As composite materials continue to push the boundaries of engineering and technology, the demand for accurate and reliable characterisation methods will only increase.

While significant progress has been made, there are still challenges to be addressed. One such challenge concerns the anisotropic nature of composites, which poses significant challenges, especially in tensile testing. Achieving a uniform stress state and accurately measuring strain remains a critical issue. The development of advanced testing techniques and standardised procedures is essential to ensure reliable and reproducible results. Moreover, new or updated standards are needed for interlaminar fracture toughness testing, as current methods face challenges such as crack tip sensitivity at initiation, the unrepresentativeness of delamination in unidirectional laminates compared to real-life composite parts, and the complexity of mixed-mode and mode III fracture toughness tests.

Another challenge is the accurate characterisation of the interfacial properties between fibres and matrix, which significantly influence the overall mechanical behaviour of composites. Developing advanced techniques to measure interfacial stresses and strains remains an active area of research. Moreover, real-time assessment of the damage state of composites still needs enhancement. AE and DIC have shown promise in this regard, but further advancements are needed to improve their sensitivity and reliability. The integration of XCT with DVC provides a more comprehensive understanding of the internal damage mechanisms in composites. Additionally, the development of advanced computational models that can accurately simulate the behaviour of composites will further enhance our ability to design and optimise these materials.

As we continue to explore the frontiers of composite materials, the development of innovative experimental techniques will remain essential. By building upon the rich history of these methods and embracing emerging technologies, we can unlock the full potential of composites and shape the future of lightweight structural materials.

References

Adams, D.F., 2009. Shear Test Methods: Iosipescu vs. V-Notched Rail, CompositesWorld. https://www.compositesworld.com/articles/shear-test-methods-iosipescu-vs-v-notched-rail.

Adams D.O., Adams D.F., 2002. Tabbing Guide for Composite Test Speicmens, Federal Aviation Administration Report DOT/FAA/AR-02/106. http://www.tc.faa.gov/its/worldpac/techrpt/ar02-106.pdf.

AhmadvashAghbash, S., Verpoest, I., Swolfs, Y., Mehdikhani, M., 2023. Methods and models for fibre–matrix interface characterisation in fibre-reinforced polymers: a review. Int. Mater. Rev. 68 (8), 1245–1319. https://doi.org/10.1080/09506608.2023.2265701.

Arif, M.F., Saintier, N., Meraghni, F., Fitoussi, J., Chemisky, Y., Robert, G., 2014. Multiscale fatigue damage characterization in short glass fibre reinforced polyamide-66. Compos. B Eng. 61, 55–65. https://doi.org/10.1016/j.compositesb.2014.01.019.

Barré, S., Benzeggagh, M.L., 1994. On the use of acoustic emission to investigate damage mechanisms in glass-fibre-reinforced polypropylene. Compos. Sci. Technol. 52 (3), 369–376. https://doi.org/10.1016/0266-3538(94)90171-6.

Bascom, W.D., Bitner, R.J., Moulton, R.J., Siebert, A.R., 1980. The interlaminar fracture of organic-matrix woven reinforced composites. Composites 11, 9–18. https://doi.org/10.1016/0010-4361(80)90016-6.

Berg, J.S., Adams, D.F., 1989. An evaluation of composite material compression test methods. J. Compos. Tcchnol. Rcs. 11, 41–46. https://doi.org/10.1520/CTR10150J.

Bhat, M.R., Majeed, M.A., Murthy, C.R.L., 1994. Characterization of fatigue damage in unidirectional GFRP composites through acoustic emission signal analysis. NDT & E International 27 (1), 27–32. https://doi.org/10.1016/0963-8695(94)90007-8.

Brault, R., Germaneau, A., Dupré, J.C., Doumalin, P., Mistou, S., Fazzini, M., 2013. In-situ analysis of laminated composite materials by X-ray micro-computed tomography and digital volume correlation. Exp. Mech. 53, 1143–1151. https://doi.org/10.1007/s11340-013-9730-9.

Broughton, W.R., Kumosa, M., Hull, D., 1990. Analysis of the Iosipescu shear test as applied to unidirectional carbon-fibre reinforced composites. Compos. Sci. Technol. 38, 299–325. https://doi.org/10.1016/0266-3538(90)90018-Z.

Brussat, T.R., Chiu, S.T., Mostovoy, S., 1977. Fracture Mechanics for Structural Adhestve Bonds, Technical Report AFML-77-163.

Bull, D.J., Helfen, L., Sinclair, I., Spearing, S.M., Baumbach, T.J.C.S., 2013. A comparison of multi-scale 3D X-ray tomographic inspection techniques for assessing carbon fibre composite impact damage. Compos. Sci. Technol. 75, 55–61. https://doi.org/10.1016/j.compscitech.2012.12.006.

Butcher, B.R., 1979. The impact resistance of unidirectional CFRP under tensile stress. Fibre Sci. Technol. 12, 295–326. https://doi.org/10.1016/0015-0568(79)90038-1.

Canal, L.P., González, C., Molina-Aldareguía, J.M., Segurado, J., Llorca, J., 2012. Application of digital image correlation at the microscale in fibre-reinforced composites. Compos. Appl. Sci. Manuf. 43 (10), 1630–1638. https://doi.org/10.1016/j.compositesa.2011.07.014.

Cantwell, W.J., Curtis, P.T., Morton, J., 1986. An assessment of the impact performance of CFRP reinforced with high-strain carbon fibres. Compos. Sci. Technol. 25, 133–148. https://doi.org/10.1016/0266-3538(86)90039-4

Carlsson, L.A., Gillespie Jr., J.W., Pipes, R.B., 1986. On the analysis and design of the end notched flexure (ENF) specimen for mode II testing. J. Compos. Mater. 20, 594–604. https://doi.org/10.1177/002199838602000606.

Chamis, C.C., Sinclair, J.H., 1977. 10° Off-Axis Test for Shear Properties in Fiber Composites, Nasa-TM-X-73550.

Chen, D., Luo, Q., Meng, M., Li, Q., Sun, G., 2019. Low velocity impact behavior of interlayer hybrid composite laminates with carbon/glass/basalt fibres. Compos. B Eng. 176, 107191. https://doi.org/10.1016/j.compositesb.2019.107191.

Choi, J.H., Jang, J., Shim, W., Cho, J.M., Yoon, S.J., Choi, D.H., Han, H.N., Yu, W.R., 2019. Determination of the in-plane shear modulus of unidirectional carbon fibre reinforced plastics using digital image correlation and finite-element analysis. Compos. Struct. 229, 111392. https://doi.org/10.1016/j.compstruct.2019.111392.

Choi, I., Horgan, C.O., 1977. Saint-Venant's principle and end effects in anisotropic elasticity. J. Appl. Mech. 44, 424–430. https://doi.org/10.1115/1.3424095.

Cottrell, A.H., 1964. Strong solids. Proc. Roy. Soc. Lond. Math. Phys. Sci. 282 (1388), 2–9. https://doi.org/10.1098/rspa.1964.0206.

Cox, H.L., 1952. The elasticity and strength of paper and other fibrous materials. Br. J. Appl. Phys. 3 (3), 72. https://doi.org/10.1088/0508-3443/3/3/302.

Cunningham, M.E., Schoultz, S.V., Toth, J.M., Jr., 1985, Effect of End-Tab Design on Tension Specimen Stress Concentrations. ASTM STP, vol. 864, pp. 253–262. https://doi.org/10.1520/STP32793S.

Czél, G., Jalalvand, M., Wisnom, M.R., 2016. Hybrid specimens eliminating stress concentrations in tensile and compressive testing of unidirectional composites. Composites Part A 91, 436–447. https://doi.org/10.1016/j.compositesa.2016.07.021.

Dally, J.W., Prabhakaran, R., 1971. Photo-orthotropic-elasticity: a new and improved method of fabricating transparent birefringent model material is presented. Also, stress-strain models are employed to predict the three fundamental photoelastic constants for unidirectional composites and are experimentally verified. Exp. Mech. 11, 346–356. https://doi.org/10.1007/BF02320536.

Davies, P., Kausch, H.H., Williams, J.G., Kinloch, A.J., Charalambides, M.N., Pavan, A., Moore, D.R., Prediger, R., Robinson, I., Burgoyne, N., Friedrich, K., Wittich, H., Rebelo, C.A., Marques, A.T., Ramsteiner, F., Melve, B., Fischer, M., Roux, N., Martin, D., Czarnocki, P., Neville, D., Verpoest, I., Goffaux, B., Lee, R., Walls, K., Trigwell, N., Partridge, I.K., Jaussaud, J., Andersen, S., Giraud, Y., Hale, G., McGrath, G., 1992. Round-Robin interlaminar fracture toughness testing of carbon-fibre-reinforced epoxy and PEEK composites. Compos. Sci. Technol. 43, 129–136. https://doi.org/10.1016/0266-3538(92)90003-L.

Davies, P., Sims, G.D., Blackman, B.R.K., Brunner, A.J., Kageyama, K., Hojo, M., Tanaka, K., Murri, G., Rousseau, C., Gieseke, B., Martin, R.H., 1999. Comparison of test configurations for determination of mode II interlaminar fracture toughness results from international collaborative test programme. Plast, Rubber Compos. 28, 432–437. https://doi.org/10.1179/146580199101540600.

De Baere, I., Van Paepegem, W., Quaresimin, M., Degrieck, J., 2011. On the tension–tension fatigue behaviour of a carbon reinforced thermoplastic part I: limitations of the ASTM D3039/D3479 standard. Polym. Test. 30, 625–632. https://doi.org/10.1016/j.polymertesting.2011.05.004.

De Groot, P.J., Wijnen, P.A., Janssen, R.B., 1995. Real-time frequency determination of acoustic emission for different fracture mechanisms in carbon/epoxy composites. Compos. Sci. Technol. 55 (4), 405–412. https://doi.org/10.1016/0266-3538(95)00121-2.

Desplentere, F., Lomov, S.V., Woerdeman, D.L., Verpoest, I., Wevers, M., Bogdanovich, A., 2005. Micro-CT characterization of variability in 3D textile architecture. Compos. Sci. Technol. 65 (13), 1920–1930. https://doi.org/10.1016/j.compscitech.2005.04.008.

Dunegan, H.L., Harris, D.O., Tatro, C.A., 1968. Fracture analysis by use of acoustic emission. Eng. Fract. Mech. 1 (1), 105–122. https://doi.org/10.1016/0013-7944(68)90018-0.

Emerson, M.J., Jespersen, K.M., Dahl, A.B., Conradsen, K., Mikkelsen, L.P., 2017. Individual fibre segmentation from 3D X-ray computed tomography for characterising the fibre orientation in unidirectional composite materials. Compos. Appl. Sci. Manuf. 97, 83–92. https://doi.org/10.1016/j.compositesa.2016.12.028.

Favre, J.P., Perrin, J., 1972. Carbon fibre adhesion to organic matrices. J. Mater. Sci. 7, 1113–1118. https://doi.org/10.1007/BF00550192.

Fazlali, B., Upadhyay, S., Ashodia, S.A., Mesquita, F., Lomov, S.L., Carvelli, V., Swolfs, Y., 2023. Specimen designs for accurate tensile testing of unidirectional composite laminates. Compos. Appl. Sci. Manuf. 175, 107799. https://doi.org/10.1016/j.compositesa.2023.107799.

Flores, M., Mollenhauer, D., Runatunga, V., Beberniss, T., Rapking, D., Pankow, M., 2017. High-speed 3D digital image correlation of low-velocity impacts on composite plates. Composites, Part B 131, 153–164. https://doi.org/10.1016/j.compositesb.2017.07.078.

Garcea, S.C., Mavrogordato, M.N., Scott, A.E., Sinclair, I., Spearing, S.M., 2014. Fatigue micromechanism characterisation in carbon fibre reinforced polymers using synchrotron radiation computed tomography. Compos. Sci. Technol. 99, 23–30. https://doi.org/10.1016/j.compscitech.2014.05.006.

Garcea, S.C., Wang, Y., Withers, P.J., 2018. X-ray computed tomography of polymer composites. Compos. Sci. Technol. 156, 305–319. https://doi.org/10.1016/j.compscitech.2017.10.023.

Ghaffari, S., Seon, G., Makeev, A., 2021. In-situ SEM based method for assessing fibre-matrix interface shear strength in CFRPs. Mater. Des. 197, 109242. https://doi.org/10.1016/j.matdes.2020.109242.

Godin, N., Huguet, S., Gaertner, R., Salmon, L., 2004. Clustering of acoustic emission signals collected during tensile tests on unidirectional glass/polyester composite using supervised and unsupervised classifiers. Ndt & E International 37 (4), 253–264. https://doi.org/10.1016/j.ndteint.2003.09.010.

Godwin, E.W., 2000. Tension. In: Hodgkinson, J.M. (Ed.), Mechanical Testing of Advanced Fibre Composites. Woodhead Publishing Ltd.

Grédiac, M., Pierron, F., Vautrin, A., 1994. The Iosipescu in-plane shear test applied to composites: a new approach based on displacement field processing. Compos. Sci. Technol. 51 (3), 409–417. https://doi.org/10.1016/0266-3538(94)90109-0.

Grédiac, M., 2004. The use of full-field measurement methods in composite material characterization: interest and limitations. Compos. Appl. Sci. Manuf. 35 (7–8), 751–761. https://doi.org/10.1016/j.compositesa.2004.01.019.

Guo, R., Stubbe, J., Zhang, Y., Schlepütz, C.M., Gomez, C.R., Mehdikhani, M., Villanueva-Perez, P., 2023. Deep-learning image enhancement and fibre segmentation from time-resolved computed tomography of fibre-reinforced composites. Compos. Sci. Technol. 244, 110278. https://doi.org/10.1016/j.compscitech.2023.110278.

Gutkin, R., Green, C.J., Vangrattanachai, S., Pinho, S.T., Robinson, P., Curtis, P.T., 2011. On acoustic emission for failure investigation in CFRP: pattern recognition and peak frequency analyses. Mech. Syst. Signal Process. 25 (4), 1393–1407. https://doi.org/10.1016/j.compositesb.2020.108039.

Hampe, A., 1988. Grundlegende untersuchungen an faserverstärkten polymeren. Amts-und Mitteilungsblatt der BAM 18 (1), 3–7.

Hamstad, M.A., 1986. A review: acoustic emission, a tool for composite-materials studies. Exp. Mech. 26, 7–13. https://doi.org/10.1007/BF02319949.

Hashemi, S., Kinloch, A.J., Williams, J.G., 1989. Corrections needed in double cantilever beam tests for assessing the interlaminar failure of fibre-composites. J. Mater. Sci. Lett. 8, 125–129. https://doi.org/10.1007/BF00730701.

Herrera-Franco, P.J., Drzal, L.T., 1992. Comparison of methods for the measurement of fibre/matrix adhesion in composites. Composites 23 (1), 2–27. https://doi.org/10.1016/0010-4361(92)90282-Y.

Hofer Jr., K.E., Rao, P.N., 1977. A new static compression fixture for advanced composite materials. J. Test. Eval. 5, 278–283. https://doi.org/10.1177/002199839502901307.

Hojo, M., Kageyama, K., Tanaka, K., 1995. Prestandardization study on mode I interlaminar fracture toughness test for CFRP in Japan. Composites 26, 243–255. https://doi.org/10.1016/0010-4361(95)93668-A.

Hojo, M., Sawada, Y., Miyairi, H., 1994. Influence of clamping method on tensile properties of unidirectional CFRP in 0° and 90° directions—round robin activity for international standardization in Japan. Composites 25, 786–796. https://doi.org/10.1016/0010-4361(94)90139-2.

Hussain, A.K., Adams, D.F., 2004. Analytical evaluation of the two-rail shear test method for composite materials. Compos. Sci. Technol. 64, 221–238. https://doi.org/10.1016/S0266-3538(03)00254-9.

Iosipescu, N., 1967. New accurate procedure for single shear testing of metals. J. Mater. 2, 537–566.

Kageyama K., Kikuchi M., Yanagisawa N., 1991. Stabilized End Notched Flexure Test: Characterization of Mode II Interlaminar Crack Growth. ASTM STP, vol. 1110, pp. 210–225. https://doi.org/10.1520/STP17720S.

Kaiser, J., 1950. Untersuchungen uber das auftreten gereausen bein Zugersuch (Doctoral dissertation, Technische Hutchshule).

Kelly, A., Tyson, A.W., 1965. Tensile properties of fibre-reinforced metals: copper/tungsten and copper/molybdenum. J. Mech. Phys. Solid. 13 (6), 329–350. https://doi.org/10.1016/0022-5096(65)90035-9.

Kim, B.W., Nairn, J.A., 2002. Observations of fibre fracture and interfacial debonding phenomena using the fragmentation test in single fibre composites. J. Compos. Mater. 36 (15), 1825–1858. https://doi.org/10.1177/0021998302036015243.

Krause, M., Hausherr, J.M., Burgeth, B., Herrmann, C., Krenkel, W., 2010. Determination of the fibre orientation in composites using the structure tensor and local X-ray transform. J. Mater. Sci. 45, 888–896. https://doi.org/10.1007/s10853-009-4016-4.

Kural, M.H., Flaggs, D.L., 1983. A finite element analysis of composite tension specimens. J. Compos. Technol. Res. 5, 11–17. https://doi.org/10.1520/CTR10785J.

Lawrence, P., 1972. Some theoretical considerations of fibre pull-out from an elastic matrix. J. Mater. Sci. 7, 1–6. https://doi.org/10.1007/BF00549541.

Lee, S., Munro, M., 1986. Evaluation of in-plane shear test methods for advanced composite materials by the decision analysis technique. Composites 17, 13–22. https://doi.org/10.1016/0010-4361(86)90729-9.

Li, L., Lomov, S.V., Yan, X., Carvelli, V., 2014. Cluster analysis of acoustic emission signals for 2D and 3D woven glass/epoxy composites. Compos. Struct. 116, 286–299. https://doi.org/10.1016/j.compstruct.2014.05.023.

Liu, Y., Straumit, I., Vasiukov, D., Lomov, S.V., Panier, S., 2017. Prediction of linear and non-linear behavior of 3D woven composite using mesoscopic voxel models reconstructed from X-ray micro-tomography. Compos. Struct. 179, 568–579. https://doi.org/10.1016/j.compstruct.2017.07.066.

Lomov, S.V., Ivanov, D.S., Verpoest, I., Zako, M., Kurashiki, T., Nakai, H., Vautrin, A., 2008. Full-field strain measurements for validation of meso-FE analysis of textile composites. Compos. Appl. Sci. Manuf. 39 (8), 1218–1231. https://doi.org/10.1016/j.compositesa.2007.09.011.

Loutas, T., Eleftheroglou, N., Zarouchas, D., 2017. A data-driven probabilistic framework towards the in-situ prognostics of fatigue life of composites based on acoustic emission data. Compos. Struct. 161, 522–529. https://doi.org/10.1016/j.compstruct.2016.10.109.

Mandell, J.F., Chen, J.H., McGarry, F.J., 1980. A microdebonding test for in situ assessment of fibre/matrix bond strength in composite materials. Int. J. Adhesion Adhes. 1 (1), 40–44. https://doi.org/10.1016/0143-7496(80)90033-0.

Marloff, R.H., Daniel, I.M., 1969. Three-dimensional photoelastic analysis of a fibre-reinforced composite model: standard stress-freezing technique is used to determine the stress distributions in the matrix of an unidirectionally fibre-reinforced composite model subjected to matrix shrinkage and normal transverse loading. Exp. Mech. 9, 156–162. https://doi.org/10.1007/BF02326564.

Marshall, D.B., 1984. An indentation method for measuring matrix-fibre frictional stresses in ceramic composites. J. Am. Ceram. Soc. 67 (12), C-259. https://doi.org/10.1111/j.1151-2916.1984.tb19690.x.

Mehan, R.L., Mullin, J.V., 1971. Analysis of composite failure mechanisms using acoustic emissions. J. Compos. Mater. 5 (2), 266–269. https://doi.org/10.1177/002199837100500213.

Mehdikhani, M., Aravand, M., Sabuncuoglu, B., Callens, M.G., Lomov, S.V., Gorbatikh, L., 2016a. Full-field strain measurements at the micro-scale in fibre-reinforced composites using digital image correlation. Compos. Struct. 140, 192–201. https://doi.org/10.1016/j.compstruct.2015.12.020.

Mehdikhani, M., Matveeva, A., Aravand, M.A., Wardle, B.L., Lomov, S.V., Gorbatikh, L., 2016b. Strain mapping at the micro-scale in hierarchical polymer composites with aligned carbon nanotube grafted fibres. Compos. Sci. Technol. 137, 24–34. https://doi.org/10.1016/j.compscitech.2016.10.021.

Mehdikhani, M., Breite, C., Swolfs, Y., Soete, J., Wevers, M., Lomov, S.V., Gorbatikh, L., 2021a. Digital volume correlation for meso/micro in-situ damage analysis in carbon fibre reinforced composites. Compos. Sci. Technol. 213, 108944. https://doi.org/10.1016/j.compscitech.2021.108944.

Mehdikhani, M., Breite, C., Swolfs, Y., Wevers, M., Lomov, S.V., Gorbatikh, L., 2021b. Combining digital image correlation with X-ray computed tomography for characterization of fibre orientation in unidirectional composites. Compos. Appl. Sci. Manuf. 142, 106234. https://doi.org/10.1016/j.compositesa.2020.106234.

Mehdikhani, M., Upadhyay, S., Soete, J., Swolfs, Y., Smith, A.G., Aravand, M.A., Lomov, S. V., February 2023. Damage identification in 4D images of a nano-engineered composite via a deep-learning tool. In: Twenty-third International Conference on Composite Materials (ICCM23): Proceedings.

Miller, B., Muri, P., Rebenfeld, L., 1987. A microbond method for determination of the shear strength of a fibre/resin interface. Compos. Sci. Technol. 28 (1), 17–32. https://doi.org/10.1016/0266-3538(87)90059-5.

Naouar, N., Vidal-Sallé, E., Schneider, J., Maire, E., Boisse, P., 2014. Meso-scale FE analyses of textile composite reinforcement deformation based on X-ray computed tomography. Compos. Struct. 116, 165–176. https://doi.org/10.1016/j.compstruct.2014.04.026.

O'Brien, T.K., Martin, R.H., 1993. Results of ASTM round robin testing for mode I interlaminar fracture toughness of composite materials. J. Compos. Technol. Res. 15, 269–281. https://doi.org/10.1520/CTR10379J.

Oz, F.E., Ersoy, N., Mehdikhani, M., Lomov, S.V., 2018. Multi-instrument in-situ damage monitoring in quasi-isotropic CFRP laminates under tension. Compos. Struct. 196, 163–180. https://doi.org/10.1016/j.compstruct.2018.05.006.

Pandita, S.D., Verpoest, I., 2003. Prediction of the tensile stiffness of weft knitted fabric composites based on X-ray tomography images. Compos. Sci. Technol. 63 (2), 311–325. https://doi.org/10.1016/S0266-3538(02)00253-1.

Park, I.K., 1971. Tensile and compressive test methods for high-modulus graphite fibre-reinforced composites. In: Proceedings of the International Conference on Carbon Fibres, Their Composites, and Applications, Paper No. 23, Plastics Institute, London.

Peters, W.H., Sutton, M.A., Ranson, W.F., Poplin, W.P., Walker, D.M., 1989. Whole-field experimental displacement analysis of composite cylinders. Exp. Mech. 29, 58–62. https://doi.org/10.1007/BF02327782.

Pierron, F., Alloba, E., Surrel, Y., Vautrin, A., 1998. Whole-field assessment of the effects of boundary conditions on the strain field in off-axis tensile testing of unidirectional composites. Compos. Sci. Technol. 58 (12), 1939–1947. https://doi.org/10.1016/S0266-3538(98)00027-X.

Pierron, F., Green, B., Wisnom, M.R., 2007. Full-field assessment of the damage process of laminated composite open-hole tensile specimens. Part I: methodology. Compos. Appl. Sci. Manuf. 38 (11), 2307–2320. https://doi.org/10.1016/j.compositesa.2007.01.010.

Piggott, M.R., Chua, P.S., Andison, D., 1985. The interface between glass and carbon fibres and thermosetting polymers. Polym. Compos. 6 (4), 242–248. https://doi.org/10.1002/pc.750060409.

Pindera, M.J., Herakovich, C.T., 1986. Shear characterization of unidirectional composites with the off-axis tension test. Exp. Mech. 26, 103–112. https://doi.org/10.1007/BF02319962.

Pipes, R.B., Pagano, N.J., 1970. Interlaminar stresses in composite laminates under uniform axial extension. J. Compos. Mater. 4, 538–548. https://doi.org/10.1177/002199837000400409.

Pitkethly, M.J., Favre, J.P., Gaur, U., Jakubowski, J., Mudrich, S.F., Caldwell, D.L., Verpoest, I., 1993. A round-robin programme on interfacial test methods. Compos. Sci. Technol. 48 (1–4), 205–214. https://doi.org/10.1016/0266-3538(93)90138-7.

Pollock, A.A., 1973. Acoustic emission-2: acoustic emission amplitudes. Non-destructive testing 6 (5), 264–269. https://doi.org/10.1016/0029-1021(73)90074-1.

Rao, A.K., 1990. Acoustic emission and signal analysis. Def. Sci. J. 40 (1), 55. https://doi.org/10.14429/dsj.40.4450.

Rosen, B., 1972. A simple procedure for experimental determination of the longitudinal shear modulus of unidirectional composites. J. Compos. Mater. 6, 552–554. https://doi.org/10.1177/002199837200600310.

Russell, A.J. and Street, K.N., 1982, Factors affecting the interlaminar fracture energy of graphite/epoxy laminates. In: Progress in Science and Engineering of Composites, Proceedings ICCM-IV, Tokyo, ISBN 4-931136-02-X, pp. 279–286.

Sachse, W., Kim, K.Y., 1987. Quantitative acoustic emission and failure mechanics of composite materials. Ultrasonics 25 (4), 195–203. https://doi.org/10.1016/0041-624X(87)90033-3.

Schell, J.S.U., Renggli, M., van Lenthe, G.H., Müller, R., Ermanni, P., 2006. Micro-computed tomography determination of glass fibre reinforced polymer meso-structure. Compos. Sci. Technol. 66 (13), 2016–2022. https://doi.org/10.1016/j.compscitech.2006.01.003.

Schilling, P.J., Karedla, B.R., Tatiparthi, A.K., Verges, M.A., Herrington, P.D., 2005. X-ray computed microtomography of internal damage in fibre reinforced polymer matrix composites. Compos. Sci. Technol. 65 (14), 2071–2078. https://doi.org/10.1016/j.compscitech.2005.05.014.

Scott, A.E., Mavrogordato, M., Wright, P., Sinclair, I., Spearing, S.M., 2011. In situ fibre fracture measurement in carbon–epoxy laminates using high resolution computed tomography. Compos. Sci. Technol. 71 (12), 1471–1477. https://doi.org/10.1016/j.compscitech.2011.06.004.

Sencu, R.M., Yang, Z., Wang, Y.C., Withers, P.J., Rau, C., Parson, A., Soutis, C., 2016. Generation of micro-scale finite element models from synchrotron X-ray CT images for multidirectional carbon fibre reinforced composites. Compos. Appl. Sci. Manuf. 91, 85–95. https://doi.org/10.1016/j.compositesa.2016.09.010.

Shiriajeva, G.V., Andreevskaya, G.D., 1962. No title. Soviet Plastics 4, 40.

Stellbrink, K., 1983. On the behaviour of impact damaged CFRP laminates. Fibre Sci. Technol. 18, 81–94. https://doi.org/10.1016/0015-0568(83)90006-4.

Straumit, I., Lomov, S.V., Wevers, M., 2015. Quantification of the internal structure and automatic generation of voxel models of textile composites from X-ray computed tomography data. Compos. Appl. Sci. Manuf. 69, 150–158. https://doi.org/10.1016/j.compositesa.2014.11.016.

Suzuki, H., Kinjo, T., Hayashi, Y., Takemoto, M., Ono, K., Hayashi, Y., 1996. Wavelet transform of acoustic emission signals. J. Acoust. Emiss. 14, 69–84.

Swanson, S.R., Messick, M., Toombes, G.R., 1985. Comparison of torsion tube and Iosipcscu in-plane shear test results for a carbon fibre-reinforced epoxy composite. Composites 16, 220–224. https://doi.org/10.1016/0010-4361(85)90605-6.

Swolfs, Y., Morton, H., Scott, A.E., Gorbatikh, L., Reed, P.A., Sinclair, I., Verpoest, I., 2015. Synchrotron radiation computed tomography for experimental validation of a tensile strength model for unidirectional fibre-reinforced composites. Compos. Appl. Sci. Manuf. 77, 106–113. https://doi.org/10.1016/j.compositesa.2015.06.018.

Upadhyay, S., Smith, A.G., Vandepitte, D., Lomov, S.V., Swolfs, Y., Mehdikhani, M., 2024. Deep-learning versus greyscale segmentation of voids in X-ray computed tomography images of filament-wound composites. Compos. Appl. Sci. Manuf. 177, 107937. https://doi.org/10.1016/j.compositesa.2023.107937.

Walrath, D.E., Adams, D.F., 1983. The Iosipescu shear test as applied to composite materials. Exp. Mech. 23 (1), 105–110. https://doi.org/10.1007/BF02328688.

Wegner, P.M., Adams D.F., 2000. Verification of the Combined Load Compression (CLC) Test Method, Technical Report DOT/FAA/AR-00/26. https://rosap.ntl.bts.gov/view/dot/42292.

Welsh, J.S., Adams, D.F., 1997. Current status of compression test methods for composite materials. SAMPE J. 33, 35–43.

Wevers, M., 1997. Listening to the sound of materials: acoustic emission for the analysis of material behaviour. Ndt & E International 30 (2), 99–106. https://doi.org/10.1016/S0963-8695(96)00051-5.

Whitney, J.M., Browning, C.E., 1985. On short-beam shear tests for composite materials. Exp. Mech. 25, 294–300. https://doi.org/10.1007/BF02325100.

Whitney, J.M., Pagano, N.J., Pipes, R.B., 1971. Design and Fabrication of Tubular Specimens for Composite Characterization. ASTM STP, vol. 497, pp. 52–67. https://doi.org/10.1520/STP27740S.

Wilkins, D.J., Eisenmann, J.R., Camin, R.A., Margolis, W.S., Benson, R.A., 1982. Characterizing Delamination Growth in Graphite-Epoxy. ASTM STP, vol. 775, pp. 168–183, https://doi.org/10.1520/STP775-EB.

Wisnom M.R., Maheri M.R., 1994, Tensile strength of unidirectional carbon fibre-epoxy from tapered specimens. In: Proc. Second European Conference on Composites Testing and Standardisation, Hamburg, vol. 1994, pp. 239–247.

Wright, P., Fu, X., Sinclair, I., Spearing, S.M., 2008. Ultra high resolution computed tomography of damage in notched carbon fibre—epoxy composites. J. Compos. Mater. 42 (19), 1993–2002. https://doi.org/10.1177/0021998308092211.

The history of nondestructive testing (NDT) on fibre-reinforced polymer composites 4.6

Martine Wevers[1], *Jeroen Soete*[1], *Helge Pfeiffer*[1], *Christ Glorieux*[2,1] and *Mahoor Mehdikhani*[3]

[1]Department of Materials Engineering, KU Leuven, Leuven, Belgium; [2]Department of Physics and Astronomy, KU Leuven, Leuven, Belgium; [3]Composite Materials Group, Department of Materials Engineering, Katholieke Universiteit Leuven (KU Leuven), Leuven, Belgium

1. Introduction

1.1 How composites fail: Not just a crack, but different damage types

Unlike metals, fibre-reinforced polymer composites are often multilayered, heterogeneous, anisotropic, or quasi-isotropic and can be tailored to a particular application. In metals, only cracks develop during service loading, often from stress concentrations, and one dominant crack will lead to failure. In composite materials, because of their complex structure, different kinds of damage can develop during service: fibre–matrix debonding, matrix cracks, delaminations, and fibre fractures, and the failure is often the result of the interaction between these damage mechanisms. Nondestructive testing (NDT) detects these damages without affecting the integrity of the part.

While NDT is essential for assessing the quality of metals, it becomes even more critical and challenging for composite materials. They require NDT not only to ensure the production quality with regard to manufacturing defects but also to evaluate their complex damage states during use. This assessment is vital to determine their structural integrity and serviceability and to feed reliable models for lifetime assessment or prediction. Thus, the development and implementation of reliable NDT methods play an important role in the evolution and utilisation of composite materials.

This contribution will not attempt to provide a comprehensive and detailed review of NDT of composites, since over the years many reviews have been written, for example, by Scott and Scala (1982) as an old reference, and by Gholizadeh (2016), Wang et al. (2020), and Zhong and Nsengiyumva (2022) as recent references. Instead, we bring the story of how different NDT techniques were introduced and evolved for composites.

The History of Composites. https://doi.org/10.1016/B978-0-443-34119-9.01029-9

1.2 Defect and damage types in composites

Defects and damage types in composites are generated during two different situations: the production process and the service loading of the composite, respectively. The following defects in high-end fibre-reinforced composites with continuous fibres are studied with NDT after production: voids, inclusions and contaminations, fibre misalignment and inaccurate stacking sequence, overlay or gaps between fibre tapes, resin-rich and resin-poor regions, insufficient curing or overcuring, and missing fibre rovings or knots in fibre rovings (Bar-Cohen, 1986). Damage types include fibre–matrix debonding, matrix cracks and matrix crazing, delaminations, and broken fibres, which can be introduced due to loading (static, impact, fatigue, etc.) or environmental effects (hygrothermal conditions, erosion, lightning, etc.) (De Meester and Wevers, 1988).

1.3 What we learnt from metals

As indicated by Wevers (1991), the NDT of composites has been built on the knowledge and experience gained with metals, but often the nature of the composite materials, being nonmagnetic, heterogeneous, multilayered, and anisotropic, impacted the choice of an NDT technique or imposed a combination of techniques. Usually, a multitude of possible defects and damages is to be investigated simultaneously. Moreover, the outcome of an NDT investigation must account for the fact that a feature, which constitutes a serious defect in one component, might be entirely acceptable in another composite component, which makes the setting of standards for quality control and inspection of composites not a simple matter.

The six most common NDT techniques for metals, being visual inspection, X-rays, ultrasound, eddy currents, dye penetrants, and magnetic particle inspection, can also be used for fibre-reinforced composites, except magnetic particle inspection. However, additional techniques had to be found to characterise the complex damage state of composites. The acoustic emission (AE) technique, for instance, which had been abandoned for proof loading of metals because it was found to be unreliable for certain metals, gained a new life when proof loading of a composite component was allowed.

The first step to take after the production of a composite material is to investigate the quality of the part with techniques that have no influence on the further performance of the material in service. For this purpose, visual inspection, X-rays, and ultrasound are most often used. The next step involves applying NDT at specific intervals during the service life of the composite material. This must be done without compromising its performance, enabling data-driven decisions about the material's continued use. Since the 1990s, NDT techniques have been introduced and developed for in situ monitoring of damage development in lab specimens mounted in a mechanical test rig or being exposed to certain (aggressive) environments, as well as for online monitoring of the damage development in large composite structures, called structural health monitoring (SHM). For the latter, the interest in 'smart composite structures' continues to grow. Finally, even during postmortem investigation, when attempting to identify the cause of failure, applying NDT is a critical first step before cutting or physically altering the material.

2. NDT techniques for composites: Look, probe, listen, and feel

2.1 Looking at defects and damage in composites

2.1.1 Visual inspection with or without penetrants

The first and oldest inspection technique for composites is visual inspection. After production or during service, signs of damage to the structure can be visually observed. Features like burns, discolouration due to chemical exposure, fibre waviness or misalignment, cracks open to the surface, and delaminations can be identified. In translucent (nonpigmented) composites like glass fibre–epoxy laminates, inspection by transmitted light can give a good indication of the presence of voids, dry spots, delaminations, and gross inclusions. The loss of transparency in glass fibre–epoxy composites is associated with the development of fibre–matrix debonding and matrix cracking, the phenomenon being known as stress whitening.

In nontranslucent composites—most composites, in fact—visual inspection techniques are limited to the observation of surface defects and damage. The detection of surface cracking (which does not reflect the true state of deterioration of a composite) and of deeper cracks that are open to the surface, if barely visible, may be enhanced by dye penetrant methods. Often, a fluorescent penetrant is used to increase the probability of detection. These methods can be used to indicate areas where a more detailed study needs to be carried out. However, one must be able to remove this penetrant upon further use of the composite or before repair, since there is a risk of the penetrant leading to more damage or jeopardising any bonding action during repair. In this perspective, dye penetrants are often considered unsuitable for composites.

2.1.2 Microscopy as an NDT tool for composites

Optical and electron microscopy have played pivotal roles in the development and analysis of fibre-reinforced composites, both as NDT tools and for understanding failure mechanisms. Initially, researchers used light microscopes to study fibre alignment, surface flaws, and resin distribution in newly developed composite materials. As composite technologies advanced, optical microscopy became an essential method for analysing polished cross-sections, revealing microstructural details such as fibre–matrix interfaces, voids, and crack patterns. The development of electron microscopy in the 1940s and 1950s was a significant step forward, offering unique resolutions to visualise the detailed microstructural features of composites. Scanning electron microscopy (SEM) was particularly transformative, allowing researchers to observe fracture surfaces, fibre pull-out mechanisms, and interfacial debonding at the micro/nanoscale (Shuh-Huei et al., 1993). By the 1970s, SEM was widely used to investigate damage mechanisms such as delaminations and matrix cracking, especially for the failure analysis of critical aerospace and structural components.

Historically, these techniques were primarily used for postfailure analysis, with SEM providing detailed insights into the damage mechanisms, difficult to observe otherwise. However, by the late 20th century, advancements in in situ optical and electron microscopy began to extend their application. Researchers started using these tools to study composites under mechanical loading, providing real-time insights into damage initiation and progression, including crack formation, fibre–matrix debonding, and strain distributions (Mehdikhani et al., 2016). This development points out the evolving role of microscopy in composite research, transitioning from static imaging to dynamic, high-resolution evaluation.

2.1.3 Optical stress-coating, dye paint, and printed/painted grid techniques

These three techniques are traditional NDT methods for visualising and analysing stress patterns on composite surfaces, used mostly in the 1970s and 1980s. An *optical stress-coating technique* can be used to locate regions of relatively greater stress at the surfaces of the composite. In this case, the coating fails before the composite, indicating regions of high stress just underneath due to flaws. However, several variables, including the nonuniform surface, the geometry, and the temperature sensitivity of the coating material, can contribute to the inability of the technique to detect specific flaw conditions.

Another simple way to reveal that surface damage might have been introduced, for instance by an impact, is to paint the accessible surface with a *microencapsulated dye*, which is released when crushed (Stone and Clark, 1987). This approach is very attractive and provides a simple visual indication of areas requiring further inspection. However, it may be difficult to ensure that the response is unambiguous.

For critical components and areas previously identified as being susceptible to damage, the *Moiré fringes method* may be suitable: photographic grids are printed/painted on the surface of the composite and irradiated by coherent light. The resulting interference fringes give clear indications of local stress concentrations and deformation, including those arising from subsurface damage. Moiré fringe patterns in early investigations also gave interesting information on the mechanical behaviour along the edges of a laminate, but to increase the resolution of damage at the edges of a laminate, the *replica technique* was later introduced (Oplinger et al., 1974; Highsmith and Reifsnider, 1982; Wevers, 1987; Highsmith, 1989).

2.1.4 Thermography, holography, and shearography

NDT techniques such as thermography, holography, and shearography have advanced significantly, providing innovative solutions for inspecting composite materials. These methods leverage unique physical principles to identify subsurface defects, delaminations, and other anomalies, contributing to the structural integrity assessment of composites.

Thermography's origins trace back to ancient times (400 B.C.), when it was used by physicians to detect temperature variations by observing patterns in the drying rates of mud applied to the human body.[1] In 1800, Sir William Herschel discovered infrared (IR) radiation, which laid the groundwork for modern thermographic applications. The use of thermography for industrial applications became possible with the introduction of portable systems in the 1970s. Advances in technology during the 20th century, including the development of focal plane array (FPA) detectors in the late 1980s, significantly enhanced image resolution and data interpretation. Moreover, the progress in computer technology and image processing helped a lot to simplify data collection and improved the interpretation of the images. By 1992, the American Society for Nondestructive Testing officially adopted IR thermography as a standard method.[2]

Thermography is a time-efficient technique for detecting damage or anomalies in composite structures, even while the specimen remains in a test frame. This avoids potential handling-induced damage. Active thermography methods, such as pulsed and lock-in thermography, create thermal contrasts to detect defects, while passive thermography is used for applications like water ingress detection in aircraft structures (Ciampa et al., 2018; Swiderski, 2015). Furthermore, Sunetchiieva (2024) demonstrated lock-in thermography's capability to visualise water ingress in impacted carbon fibre-reinforced plastic (CFRP) sandwich structures and assess moisture sensor functionality. Advanced autonomous systems like the C-CheckIR portable unit to detect delaminations, water, inclusions, etc., in composite materials according to aircraft testing procedures and NASA's robotic implementations for real-time flaw inspection on-site (Cramer, 2018) further illustrate thermography's potential for on-site inspections in aerospace applications.

While thermography can detect delaminations and cracks, its sensitivity and resolution are limited compared to ultrasonic methods. However, its noncontact nature and ability to monitor crack growth in real time remain advantageous for composite applications (Henneke et al., 1979; Scott and Scala, 1982).

Holography, developed in 1947 by Dennis Gabor, is an NDT technique that uses visible light waves coupled with photographic equipment to create a three-dimensional image. This method is particularly useful for rapid surface flaw assessment in composite structures. The process uses two laser beams, one called a reference beam and the other called an object beam, and to reveal anomalies, the component is stressed. Defects or stresses are revealed through fringe patterns in the hologram, making holography effective for detecting barely visible impact damage (Adams and Cawley, 1988; Hockley, 1972). Optical holography has been highlighted for its ability to detect delaminations in vibration-mode shapes of composite structures. The method requires a vibration exciter to obtain holograms of the entire structure, making it efficient for large-scale inspections. However, its reliance on vibration isolation can limit its applicability unless pulse systems are employed (Fagot et al., 1980).

[1] https://www.nde-ed.org/NDETechniques/Thermography/IR_History.xhtml.

[2] https://www.nde-ed.org.

Shearography emerged in the late 1960s as a method for full-field strain measurement (Yung, 1982) using laser-generated speckle patterns. This technique excels in detecting subsurface defects such as disbonds and delaminations, offering rapid inspection over large areas. When the part is properly stressed, even the size and shape of the subsurface anomalies are shown. Shearography's simplicity and adaptability for on-aircraft inspections, while meeting all laser safety concerns, make it a valuable tool for aerospace applications.[3] Shearography can be applied statically, where speckle images before and after stress application reveal defect-induced changes, or dynamically, through stroboscopic methods for full-field vibrometry, by synchronising laser pulses with the applied vibration. Features of sample surface vibrational patterns can reveal subsurface delaminations (Fig. 4.6.1) or cracks, due to the related changes of stiffness or the local defect resonant behaviour (Sarens et al., 2010). Moreover, De Angelis et al. (2012) developed a new numerical-experimental procedure to detect the size and depth of flat-bottom holes in CFRP laminates by digital shearography. Considering its simplicity and rapidity, laser shearography methodology was evaluated as a reliable NDT method. However, the method's reliance on optical interference introduces challenges in displacement measurement and phase unwrapping, which require precise control of testing conditions.

2.1.5 Digital image correlation

Digital image correlation (DIC), developed at the University of South Carolina in the 1980s by pioneers like Sutton et al. (1983), is a noncontact optical technique used to measure surface deformations of a deforming object. Its versatility and precision have made it a widely used method for analysing deformation and damage in fibre-reinforced composites across various length scales—macro, meso, and micro. The evolution of DIC applications in composites reflects advancements in camera technology, computational techniques and power, and composite material research.

DIC initially found its place in the *macroscale analysis (1980s–1990s)* of composite structures, focussing on global deformation and strain fields. Early applications, such as those documented by Peters et al. (1989), demonstrated DIC's potential for full-field surface displacement tracking, which proved particularly valuable in industries like aerospace and automotive, where fibre-reinforced composites were increasingly being used. During this period, DIC was primarily used for mechanical testing under different loading conditions, such as tension, compression, and bending. The main application of DIC in this era was the measurement of global deformation and strain distribution in composite structures or large specimens.

Full-field strain measurements have been utilised in the analysis of mechanical testing of coupons since the 1990s (Grédiac et al., 1994). With improvements in DIC systems, researchers began employing the technique for *mesoscale analysis*

[3] https://www.faa.gov/documentlibrary/media/advisory_circular/ac_43.13-1b_w-chg1.pdf.

[4] Nondestructive Evaluation Techniques: Acoustic Emission Testing: https://www.nde-ed.org/NDE-Techniques/AcousticEmission/AE_History.xhtml.

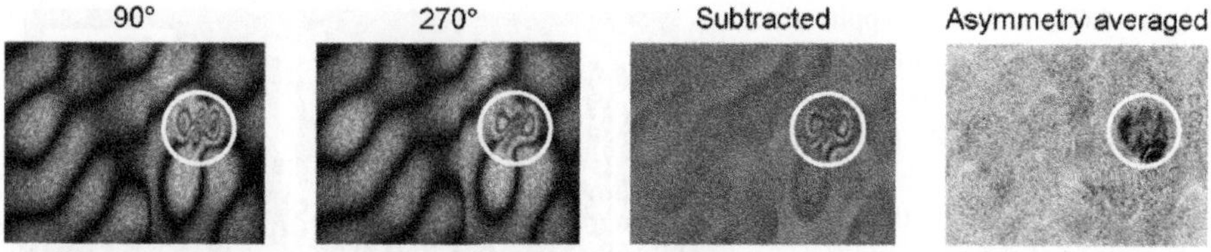

Figure 4.6.1 Example of shearographic images (field of view 150 × 200 mm^2) visualising the vibrational patterns of a piezoelectric transducer excited vibrating a 2-mm-thick, 300 × 300 mm^2 carbon fibre-reinforced composite plate at a frequency of 20 kHz. The plate consisted of 16 layers of unidirectional carbon fibres embedded in epoxy and containing a delamination between ply 2 and 3, at 0.25 mm depth below the surface.
From the Ph.D. thesis of Sarens (2009)

(2000s) to better understand the mechanics of fibre-reinforced composites and validate their numerical models (e.g., Lomov et al., 2008). This era saw a shift toward using DIC for analysing localised deformation and damage phenomena in composites, allowing a deeper study of the effect of meso-structure and fibre orientation, as well as the analysis of cracks and delamination. Moreover, it helped parameter identification and fatigue experiments in composites and even facilitated deformability analysis of textiles in providing (local) strain fields. Furthermore, stereo-DIC was introduced, enabling strain measurement in the out-of-plane direction.

Microscale DIC (2010s and beyond), which was applied to metals in the 2000s with advances in high-resolution cameras and microscopes, was introduced to fibre-reinforced composites as well. Pioneers like Canal et al. (2012) and Mehdikhani et al. (2016) began integrating DIC with techniques such as SEM to study strain localisation and damage evolution at the fibre and matrix level. A key challenge at this length scale is the creation of a high-quality random speckle pattern on the surface. The analysis of deformation at this scale revealed new insights into interfibre deformation and interface behaviour, offering experimental validation for micromechanical models that had not been previously verified. Meanwhile, DIC became increasingly valuable for *multi-scale analysis*, correlating deformation behaviours from the micro to macro levels (Mehdikhani et al., 2018). examined the interplay between microstructural features (e.g., voids) and global performance (e.g., transverse cracking), which is a key link to characterise in composites—see Fig. 4.6.2.

Recent *advancements in DIC* and the extension toward 3D, i.e., *digital volume correlation*, were introduced in the 2020s because both hardware and software have advanced significantly. Cameras now feature improved optics, higher resolutions (up to tens of megapixels), and faster imaging rates (up to ~100 Hz). Meanwhile, DIC algorithms have become more powerful and efficient, enabling the processing of large volumes of high-resolution images.

Obviously, DIC has advanced considerably since its inception in the 1980s, becoming a critical tool for analysing deformation and damage in fibre-reinforced composites. Initially focused on macroscale applications, DIC has since evolved to explore localised deformation at the meso- and microscale levels, offering deeper insights into composite mechanics and material behaviour. With improvements in

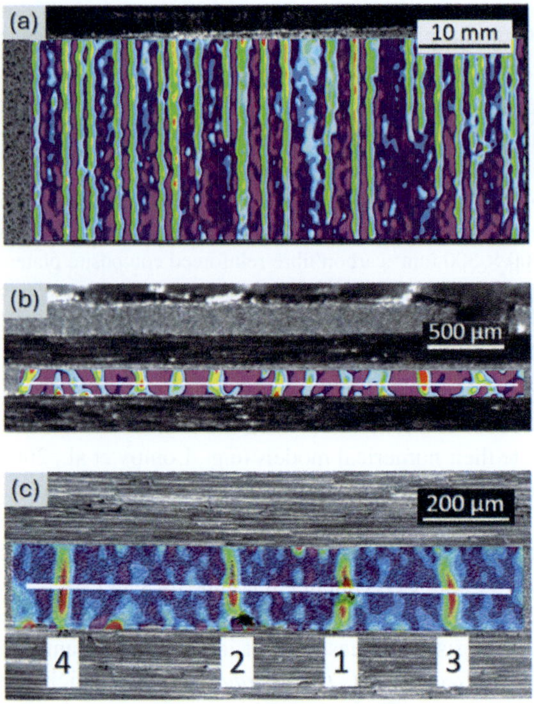

Figure 4.6.2 Multiscale application of DIC for detection of transverse cracks, which appear as localisation in strain maps on a (A) macroscale image of a tensile specimen's top surface, (B) mesoscale thickness view image of the side, and (C) microscale (SEM) image of the side displaying fibres and voids. Adapted with permission from Mehdikhani et al. (2018), Composites Part B: Engineering, Elsevier

camera technology and computational algorithms, DIC continues to play a pivotal role in understanding and validating the performance of composites across multiple length scales.

2.1.6 X-ray radiography, penetrant-enhanced radiography, and X-ray computed tomography

Since their discovery on the 28th of December in 1895 by Willem Conrad Röntgen, X-rays have been used in many fields, not only in health sciences but also in engineering. In X-ray radiography of composites, general fibre orientation, fibre misorientation, fibre kinks or wrinkles, and foreign objects or inclusions are detected. Cracks aligned parallel to the direction of the X-ray beam are also readily revealed as long as the difference in X-ray absorption between them and the composite material is high enough. The resolution that can be obtained is dependent on the X-ray system (Prakash, 1980; Scott and Scala, 1982; Teagle, 1983). When an adequate contrast is not obtained, liquid penetrants (organic or inorganic) are used to reveal delaminations and cracks, if penetration is possible in the damaged zones (Characterization, analysis and significance of defects in composite materials, 1983). The penetrant must be safe to use, induce no damage to the material, have suitable radio-opacity, propagate to the full extent of damage, and be removable. The *penetrant-enhanced X-ray radiography* can also be used in situ. For instance, from the X-ray images taken at different applied strain levels, the number of cracks open to the surface and the total crack length can be obtained and plotted versus the applied strain, giving more quantitative information on the damage state of the specimen (Wevers, 1987).

Since the interpretation of damage existing at different depths in the sample, causing overlap in the X-ray image, has been a challenge. The method of *stereo X-ray radiography* has been used to effectively resolve the radiographic details through the specimen thickness (Sendeckyj et al., 1982; Characterization, analysis and significance of defects in composite materials, 1983).

The contrast in conventional contact radiographic techniques was insufficient to permit resolution of individual fibres. Therefore, *projection radiography* has been proposed. The method has the advantage of considerably reducing the nonimage-forming scatter, which is normally recorded on a contact radiography, thus improving the resolution. In this way, with primary magnifications of up to 100 times and with subsequent further optical enlargement, it is even possible to resolve single carbon fibres in thin sections and very fine distributions of subsurface cracks in the damage zone near crack tips (Harris, 1982).

The transition from 2D radiography to *X-ray computed tomography* (XCT) represents a major technological leap in imaging technology. In the early 1970s, Godfrey Hounsfield and Allan Cormack independently developed the first XCT scanner, earning them the 1979 Nobel Prize in Medicine. X-ray tubes evolved over the years from mm focal spots to microfocus and nanofocus X-ray tubes. Simultaneously, the development of powerful detectors, reconstruction algorithms, and ever-improving computational capabilities has allowed the reconstruction of detailed cross-sectional images of fibre architectures in a nondestructive manner, providing unique insights into material structures and properties (Fig. 4.6.3).

XCT imaging is capable of resolving fibres, fibre bundles, the matrix, manufacturing defects, and in-service damage in composites (Garcea et al., 2018) and is commonly used to quantify the microstructure or to generate a digital twin of the material, which is used for the prediction of the mechanical properties of composites.

XCT has applications dealing with random fibre composites, laminated composites, and textile composites, as reviewed by Wevers et al., 2018. Sophisticated tools like Avizo (ThermoFisher), VGStudio MAX, VoxTex (KU Leuven), and FibreScout (Weissenbock et al., 2014) are applied for the quantitative analysis of fibrous composite materials, supporting advanced simulations like damage progression in composites or permeability calculations (see an example in Fig. 4.6.4). These developments mark a leap from simple imaging to comprehensive material characterisation, enhancing our ability to optimise composite designs for diverse applications.

Besides XCT, *Computed laminography* (CL) based on X-rays was developed to visualise a three-dimensional geometry of the internal structure of platelike structures, like composite laminates. Laminographic scanning is similar to CT, but it uses only a limited amount of tomographic information to reconstruct the volumetric information of the full component, thus removing the size limitations of conventional CT systems.

In situ loading and *time-resolved XCT* (4D-XCT) emerged from further technological advancements in XCT hardware, software algorithms, and computational power.

Figure 4.6.3 Micro-CT images of textiles and textile composites: (A) flax and (B) glass fibre woven fabrics, (C) glass and (D) steel fibre knitted fabrics, (E) glass and (F) carbon fibre 3D woven fabrics, and (G) alumina fibre woven laminate.
Reproduced with permission from Wevers et al. (2018), Industrial X-ray Computed Tomography, Springer

4D-XCT, i.e., repeated acquisition of 3D images in a stepwise or dynamic manner (Garcea et al., 2018), is crucial for understanding dynamic changes in materials during their manufacturing or load applications, providing invaluable insights into the microstructural changes over time (Scott et al., 2011; Gayot et al., 2024). This technique is used to perform real-time monitoring of void formation in fibre-reinforced composites (Chatziathanasiou et al., 2023). used an in situ double cantilever beam test to characterise the crack front in 3D and to track it as it evolved as a function of the applied load.

A notable recent breakthrough is the extension of DIC into 3D, known as digital volume correlation (DVC). Combining 4D-XCT and DVC has been instrumental in

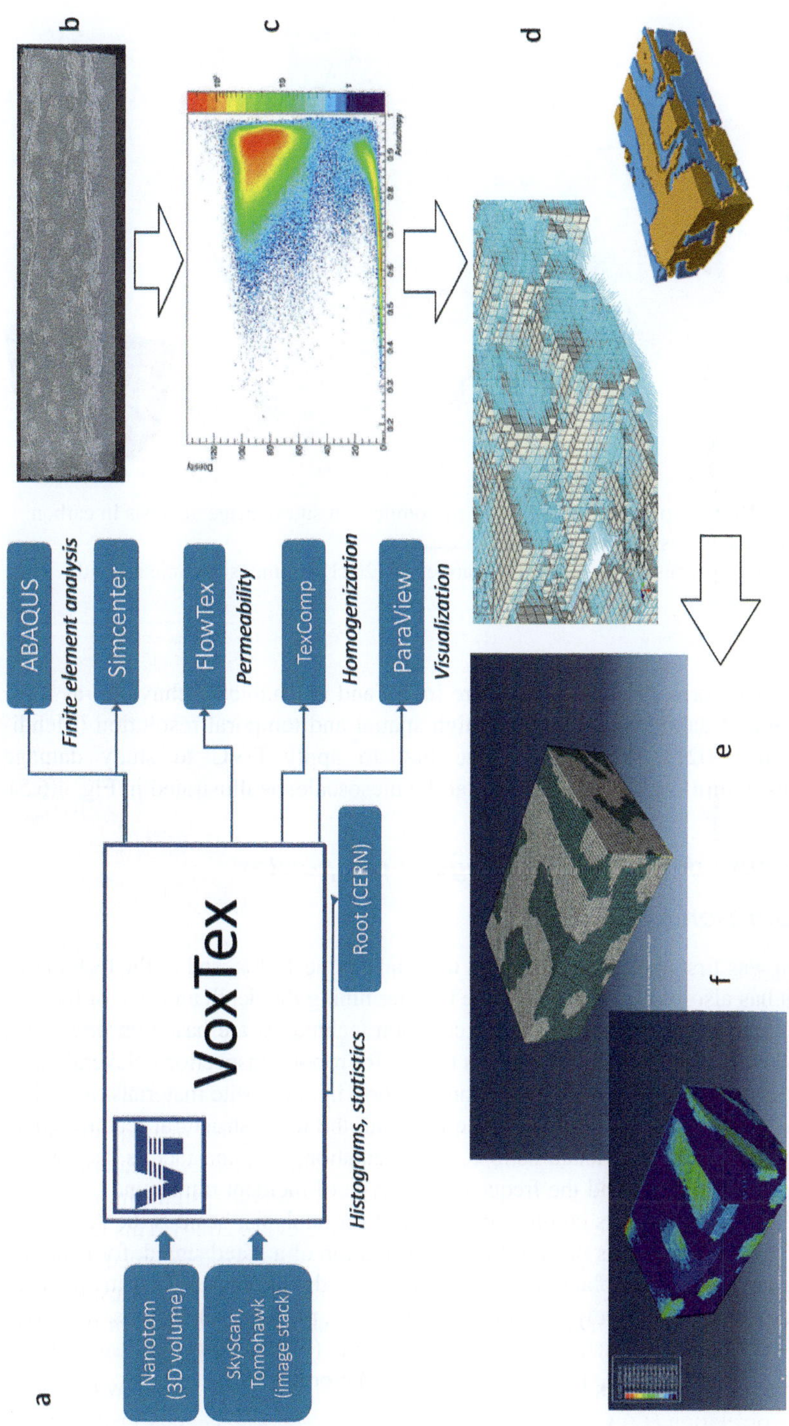

Figure 4.6.4 VoxTex software for image segmentation and fibre orientation analysis: (A) sources of the input and use of the output; data flow inside VoxTex: (B) micro-CT image, (C) segmentation, (D) the segmented image and voxel directions, (E) the finite element (FE) model, and (F) the solution.

Reproduced with permission from Wevers et al. (2018), Industrial X-ray Computed Tomography, Springer

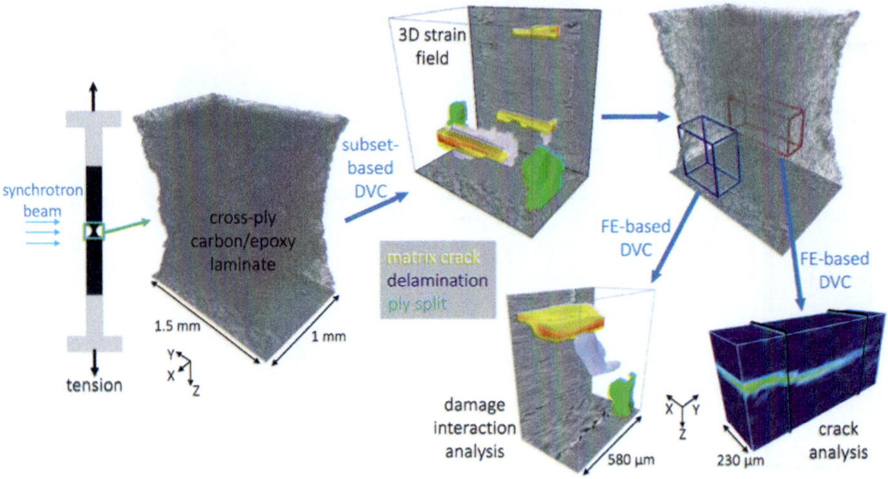

Figure 4.6.5 Digital volume correlation for meso/micro in situ damage analysis in carbon fibre-reinforced composites.
Reproduced with permission from Mehdikhani et al. (2021), Composites Science and Technology, Elsevier

composites for correlating microstructure in 3D and mechanical behaviour, as it reveals strain and damage patterns with high spatial and temporal resolution (Mehdikhani et al., 2021). were among the first to apply DVC to study damage interactions within composite laminates at the mesoscale, as illustrated in Fig. 4.6.5).

2.2 Probing defects and damage in composites

2.2.1 Ultrasonic inspection

Ultrasound was first applied to find flaws in metals in the 1930s, and as the technology evolved, it has also proven to be effective in determining the depth and size of flaws in composite materials, too. The two most common techniques are based on *immersion testing* and *waterjet through transmission* (easier for robotic inspection). Several material parameters influence the ultrasound propagation in composite materials, including the stiffness properties and density of the material, the microstructural features (fibre volume fraction, porosity, delaminations, ply orientation, ...), and factors such as the condition of the surfaces and the frequency and type of incident ultrasound.

The *ultrasonic C-scan* is commonly preferred for analysis, besides the A- and B-scan (Fig. 4.6.6). It presents the amplitude information of a gated signal, for instance, the reflection of the glass plate or of the backwall of the laminate. The ultrasonic C-scan technique can effectively measure delaminations, ply gaps, inclusions, resin-rich and resin-starved areas, surface scratches, and voids (Nevadunsky and Luca, 1975; Liber et al., 1979; Prakash, 1980; Harris, 1982; Mahoon, 1983).

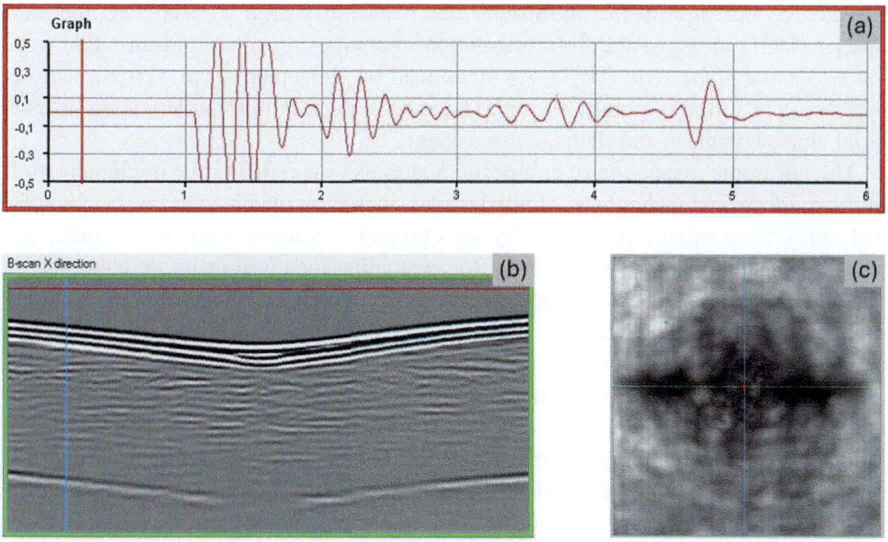

Figure 4.6.6 Ultrasound results viewing (A) A-scan in green point, (B) B-scan in X-direction (green line), and (C) C-scan of impact damage (area 65 × 65 mm) in a 5-mm thick glass fibre-reinforced epoxy plate (100 × 100 mm) scanned with Panametrics transducer V309 5 MHz and HFUS 2000 (Ingenieurbüro dr Hillger).
Images from MTM, KU Leuven

Ultrasonic velocity measurements, being simple to characterise accurately in all but very thin specimens, have been correlated to the fibre volume fraction, the matrix porosity, and also to the state of cure for some resin systems (Reynolds, 1985). In the *ultrasonic spectroscopy technique*, the frequency content of a pulse that has travelled through a material is analysed by spectrum analysis or fast Fourier transform methods (Cousins and Markham, 1977). succeeded in using this technique to determine the depth of delaminations in CFRPs with artificial delaminations.

The *acousto-ultrasonic technique* seeks to characterise a composite by investigating the way in which it modifies an ultrasonic pulse that travels through it. A sending transducer injects a repeating series of ultrasonic pulses into the composite material. The stress wave factor (SWF) can be correlated with the residual strength of damaged carbon fibre–epoxy composites (Williams and Lee, 1985). This SWF also depends on variations of fibre–resin bonding as well as fibre orientation (Vary and Lark, 1979). found that the SWF correlates strongly with ultimate tensile and shear strengths that accompany these variations.

The *Fokker Bond Tester* is also a well-established NDT tool based on ultrasound (Van Bochove and Somers, 1985). By appropriate calibration of the technique, it can give sensitive indications of the presence of voids, weak bonds, delaminations, and incorrect curing conditions. However, preparation of calibration curves for the specific composite material is needed.

Another technique using ultrasound has been introduced by Van Dreumel and Speijer (1981) to assess the fibre orientation in composites in a nondestructive way: the *ultrasonic polar scan* (UPS). As mentioned by Kersemans et al. (2016), various modifications to the original UPS setup have been successfully implemented for material characterisation and damage assessment.

The ultrasonic inspection of composite materials is also facing the implementation of *the guided wave technology*. Guided waves, including plate waves, surface waves, and subsurface waves, generated by an obliquely incident ultrasonic source, are considered to have excellent potential for evaluating localised gross defects such as delaminations, voids, etc., in composite laminates. Moreover, quantitative estimations of porosity levels, fibre volume fraction, and hygrothermal degradation have been successfully achieved (Rose et al., 1989).

Disbonds can be inspected with greater certainty by using not only the amplitude of the ultrasonic signal but also the polarity of the reflection from the disbond. The polarity of a reflected ultrasonic pulse can be derived from the material's impulse response, i.e., the sparse reflection series that results from the deconvolution of the wavelet from the ultrasonic A-scan (McRae, 1990). developed a real-time deconvolution technique accurate enough for imaging without extensive deconvolution procedures. Further advancements include the implementation of thickness compensation techniques presented by Bashyam (1990).

When used for the support of damage mechanics investigations in composites, high resolution and high reproducibility are required (Hillger, 1989). pioneered the high-frequency ultrasonic testing with dynamic range and high resolution. Using a 50 MHz polyvinylidene fluoride (PVDF) foil transducer, he detected 15-layer echoes in 2 mm thick defect-free laminate. Additionally, the signal-to-noise ratio of a 0.3 mm delamination echo amplitude exceeded 20 dB, enabling direct use in a B- or C-scan.

2.2.2 Laser ultrasonics

Laser-based generation of elastic waves was first proposed by White (1963). A comprehensive overview of laser ultrasonic testing (LUT) and its applications in composite materials is given by Zarei and Pilla (2024), highlighting the noncontact and broadband nature of LUT and the generation of bulk and guided ultrasonic waves. This technique leverages the photoacoustic effect for fast, accurate, in situ, and real-time NDT. By virtue of the flexibility of shaping light patterns, wave packets of arbitrary shapes and scales can be generated, allowing the analysis of the propagation and damping of optically detected mechanical waves in materials up to very high frequencies and with very high spatial resolution.

LUT-generated surface acoustic waves (SAWs) have the interesting feature of a wavelength-dependent penetration depth, offering the possibility for depth profiling of mechanical properties as presented and discussed by Glorieux et al. (2000). It is also interesting that SAWs are highly sensitive to surface and subsurface mechanical defects, such as delaminations and cracks. This has led to a plethora of NDT applications (Kromine et al., 2000), mainly for metal components (Gao et al., 2003) but also for composites (Park et al., 2014). The major challenges of LUT are the low signal-to-noise ratio, the challenging determination of elastic constants, and signal processing. However, for the latter, advancements in AI-based techniques have opened new doors to the applications of LUT, as indicated by Zarei and Pilla (2024).

2.3 Listening to growing damage in composites with acoustic emission

As early as 6500 B.C., potters were known to listen for audible sounds during the cooling of their ceramics, signifying structural failure. In metal working, the term "tin cry" (audible emissions produced by the mechanical twinning of pure tin during plastic deformation) was coined around 3700 B.C. by tin smelters in Asia Minor. The first documented observations of AE appear to have been made in the 8th century by the alchemist, Jabir ibn Hayyan. In a book, Hayyan wrote that *Jupiter (tin) gives off a 'harsh sound' when worked, while Mars (iron) 'sounds much' during forging.*[4]

AE refers to transient elastic waves generated by sudden strain energy release within a material due to the occurrence of microstructural changes. If enough energy is released, even audible sounds are produced. The AE technique extends the human auditory capabilities by allowing detection of higher frequency and lower intensity sounds. Measurements with Lead Zirconate Titanate (PZT) sensors today are typically carried out within a range of 50 kHz to 2 MHz. Higher frequencies are often absorbed by the material, or the AE signal lacks intensity, while lower frequencies are prone to background noise interference, e.g., vibrations from vehicles or equipment (Williams, 1980; Matthews, 1983). A distinction can be made between two types of AE: continuous emission and burst emission (Eisenblätter, 1988). The AE technique has the following peculiarities: (i) it is a passive method for in situ monitoring of the response of a material to an applied load, ensuring a 100% volumetric control; (ii) the technique is nondirectional, the emitting sources radiate the energy in every direction; (iii) AE is sensitive to defect growth and changes in the material rather than to the static presence of defects; (iv) AE is noninvasive (Wevers, 1997).

Fibre-reinforced polymer composites are perhaps the most widely studied material class via AE. AE even played an important role in the development of glass fibre-reinforced rocket motor cases by Aerojet for Polaris A3 in the 1960s (Green et al., 1964). Moreover, it is also one of the most promising methods for SHM. There exist many publications in the domain of AE of composites, as listed by Drouillard and Hamstad (1983), Hamstad (1986a, b), Ono and Gallego (2012), Wevers and Surgeon (2000), and many others.

A significant discovery in fibre-reinforced composites is the *Felicity effect*, or the breakdown of the *Kaiser effect*, the latter being known for metals. The Kaiser effect states that when a sample is reloaded in the same direction after unloading, AE often does not start until the previously achieved stress level is reached (Kaiser, 1953). For fibre-reinforced thermoset plastics, researchers (Liptai, 1972; Kim et al., 1972; Takehana and Kimpara, 1972; Fowler, 1977, 1979) noticed AE activities before reaching the maximum load, which is indicative of the degradation of the material. The absence of the Kaiser effect is often expressed in the Felicity ratio: the ratio between the applied load, at which the AE activity reappears during the next application of loading, and the previous maximum applied load. The history of the Felicity effect is well described by Fowler (2014). Numerous studies, e.g., Hamstad 1986a, Hamstad (1986b), Awerbuch et al. (1983), and Yuan (1984), studied the Felicity ratio for different types of composites (lay-up and fibre type) and loading cases, also indicating a lack of understanding of the Felicity effect and stress rupture.

The Committee on Acoustic Emission from Reinforced Plastics (CARP, 1987) conducted extensive research and development programs, culminating in the publication of "The Recommended Practice for Acoustic Emission Testing of Fibreglass Reinforced Plastic Tanks/Vessels." These efforts contributed to the establishment of ASME Boiler and Pressure Vessel Codes and ASTM standards for composite vessels. Procedures for an AE test typically specify the loads that must be applied (relative to the working load or design load) and the upper and lower limits on the loading rate.

The *AE activity analysis* has often been used to study the initiation and propagation of damage in composite structures when adequate precautions for noise rejection have been taken (Carlsson and Norrbom, 1983; Selden and Gustafson, 1985; Ziola and Gorman, 1989; Takeda et al., 1991; Surgeon et al., 1997; Brunner et al., 1995). These are examples of literature sources that indicate how an AE activity analysis can be used to study the behaviour of different composite types under a variety of loading regimes. An extension of the activity analysis that is often used in structural integrity testing is based on the Felicity effect, which has been documented above and led to a decreased number of catastrophic failures of FRP vessels, as shown in Fig. 4.6.7a.

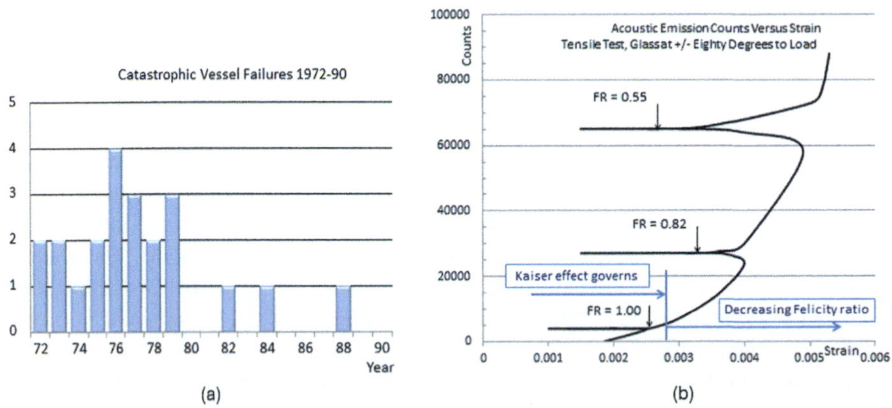

Figure 4.6.7 (A) Catastrophic failures of Monsanto Company FRP vessels for 1972–1990, indicating a drop when AE inspection was added as a quality control test (1979), (B) Acoustic emission counts versus applied strain in a tensile test of a specimen cut from the wall of a large cylindrical glass filament wound vessel and tested by McDonnell Aircraft.
Adapted from Fowler (2014), in the 31st Conference of the European Working Group on Acoustic Emission (EWGAE), under the terms of the Creative Commons CC BY-ND 3.0 license

Fig. 4.6.7b is a plot of cumulative emission counts versus applied strain. Loading is periodically interrupted with unload/reload cycles. The onset of emission during reload and the corresponding Felicity ratio are shown. Initially, the specimen exhibits the Kaiser effect. As the load increases and the material suffers damage, the Felicity effect becomes more marked.

A more detailed investigation is based on the *analysis of time and frequency domain parameters* that are extracted from the raw waveform. The parameter-based analysis has been mainly used in laboratory studies to distinguish or discriminate between various types of damage in composite materials. This type of analysis assumes that different damage phenomena (e.g., debonding, matrix cracking, delamination, fibre fracture, etc.) generate different AE waveforms having distinctly different features (Favre and Laizet, 1990; Tsamtsakis et al., 1996; Valentin et al., 1983; Berthelot and Rhazi, 1990; Wevers et al., 1991; de Groot et al., 1995). These are examples of literature sources where parameter-based analysis techniques are used. It must be noted that while literature results dealing with laboratory experiments have shown that, e.g., an amplitude-based analysis can result in a discrimination between damage phenomena, some of the literature results are apparently in contradiction. Moreover, it has proven very difficult to extrapolate the results to larger, practical, industrial structures. The main reason for this has to be found in the wave propagation effects. The main wave propagation effects are attenuation, reflection, mode conversion, and dispersion. In large structures, the AE parameters will mainly exhibit features of the wave propagation and not of the source phenomena.

One should also realise that reducing the information contained within an AE waveform to a limited number of parameters neglects a lot of information that could be useful for the analysis of the results. Generally, a parameter-based analysis is based on empirical correlations, which suffer from a lack of theoretical background and general validity. This suggests that analysis procedures taking into account the complete waveform and theoretical wave propagation aspects are preferable.

Since AE waves are mechanical in nature, it seems logical to also look for *waveform-based analysis procedures* resulting from the general theoretical concepts of wave propagation in solid materials. In the field of AE, the first attempts in this type of theoretical analysis used the three-dimensional equations of elasticity and tried to calculate, starting from the source phenomenon and the material properties, what the wave looked like once it had propagated to the sensor. Extensive use was made of Green's function (Hsu et al., 1977; Kline et al., 1981; Wadley et al., 1981). It was not until the beginning of the 1990s that AE researchers realised a practical compromise could be established between the theoretical concepts of wave mechanics and the need for a practically applicable analysis procedure. In a number of papers (Gorman, 1991; Ziola and Gorman, 1991; Prosser et al., 1992), it was suggested to use a simplified version of wave propagation theory, namely classical plate wave theory, to analyse AE signals.

The use of modal content and, more generally, the complete waveform of AE signals in data analysis has become a widely accepted practice, and the field of AE dealing with this type of analysis is now known as modal AE. As has been demonstrated in the literature by Surgeon and Wevers (1999a) and Surgeon (1999), modal

AE offers the AE user a good compromise between the oversimplified analysis procedures based on the classic AE parameters and the overcomplicated techniques based on the general equations of elasticity. The advent of faster personal computers with large computational power in the last decade has also made the gathering of complete AE waveforms more feasible. It is therefore expected that modal analysis techniques will become increasingly important in the domain of AE and that they will allow for a more consistent, intelligent, and quantitative analysis of AE results, which will in turn lead to more practical applications and a wider acceptance of the technique.

One of the main advantages of the AE technique is that it also allows for obtaining information about the spatial location of the damage. *Source location techniques* are generally performed by comparing the arrival time of an AE signal at a number of sensors that are distributed over the structure under examination. Traditional location procedures use one fixed velocity, but this can cause substantial errors, and the general conclusion states that accurate source location can only be performed if the modal content of the AE waves is taken into account (Ziola and Gorman, 1991; Surgeon and Wevers, 1999a). A modal location procedure can also use the velocities of the separate wave modes in a linear location based on the difference in arrival time between the extensional and the flexural mode at one sensor. The practical feasibility of this one-sensor linear location procedure was demonstrated using signals generated by transverse matrix cracking in CFRP laminates (Surgeon and Wevers, 1999b).

Automated Classification Procedures are used for efficient separation of the complete data set into a number of subsets, e.g., separating between noise signals and real damage signals, or to discriminate between signals generated by different damage phenomena. The traditional way of achieving this separation in the 1970s was through an operator looking at the data and examining whether signal classes having distinctly different characteristics are present. Over time, signal analysis procedures for AE data were also developed using artificial intelligence with classification algorithms and neural network technology. Some examples of the use of this type of analysis in the domain of composite materials can be found in Ohtsu and Ono (1987), Hoa and Smith (1989), Ono and Kawamoto (1990), and Anastassopoulos and Philippidis (1995). More recently, a review paper (Ciaburro and Iannace, 2022) on machine learning-based methods for AE testing indicated that the extraction of the characteristics is still based on the intervention of experts, with the consequent associated empirical risk. For instance, Almeida et al. (2023) proposed a method for the classification of AE signals based on 'supervised' machine learning (Fig. 4.6.8). Often, data are cross-checked with other complementary NDT techniques to obtain a reliable training dataset for the supervised classification model. The collected training dataset contained key details about the different breaking behaviours of fibres and matrices, as well as their interface. A correct identification of AE signals also gives essential information on the location, time, frequency, and intensity of each damage mechanism.

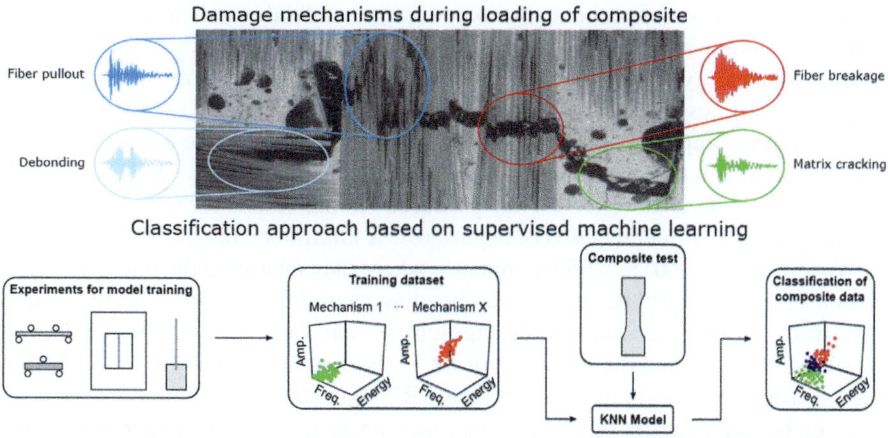

Figure 4.6.8 Damage mechanisms during loading of composites with their corresponding typical AE signal, followed by the chart representation of the supervised pattern recognition method for AE data analysis.
Reproduced from Almeida et al. (2023), Materials & Design, Elsevier, under the terms of the Creative Commons CC-BY license

2.4 Feel defects and damage

2.4.1 Resistance strain gauges

In the overview of Arena and Viscardi (2020), it is mentioned that since the 1940s, the resistance strain gauges have been powerful tools for experimental strain evaluation and a common choice for monitoring material deformations around defects and "damage" in composites. However, the physical environment of the strain gauge is also a crucial parameter that needs to be considered in the selection of the gauge, and it is the protective coating. Owing to their relatively high surface area and the need for protective coatings, electrical resistance strain gauges have faced a lot of difficulties in producing reliable results for laminated composite materials. With directed fibres, it is also essential to correctly align the strain gauge owing to the orthotropic material behaviour. Another main drawback is the large quantity of them required to monitor an entire structure, yielding a complex network with a lot of wires and cables. For this reason, the strain gauges' location is often optimised for the most critical areas and for lab testing.

2.4.2 Optical fibres for strain and damage measurements

Optical fibres have been introduced to overcome the weaknesses of conventional strain gauges. In many approaches, fibre optic sensors are used to achieve static or quasidynamic strain (Takeda et al., 2013), providing better resolution of up to two to three orders of magnitude compared with conventional strain gauges. When it comes to damage sensing, there exist different options. The first systems using optical fibres to detect damage in composite materials used embedded optical fibres that should break before the host material. Therefore, by checking the light intensity of an array of fibres embedded in the composite material, its integrity could be monitored (Glossop et al., 1990). embedded 125 μm multimode optical fibres in a Kevlar/epoxy composite specimen, and several lay-ups were tested with different optical fibre configurations. Other studies were also performed by Waite and Sage (1988) and Kitade et al. (1995). They produced similar results for tensile, bending, and fatigue tests with embedded optical fibres. A multimode optical fibre was used by Wevers et al. (2000) in CFRP laminates, where they measured the intensity modulation of the optical fibre due to microbending caused by AE stress waves generated from developing damage during tensile and fatigue tests (Surgeon and Wevers, 2001). studied the influence of the embedment of optical fibres on the fatigue damage progress in quasi-isotropic CFRP laminates. It was found that optical fibres embedded near the stress-carrying plies can accelerate the damage evolution. To prevent this, a multimode and later also a single-mode optical fibre, embedded in a SmartTape attached to the composite, was developed by those researchers within a European project, PDT-COIL NNE5-2001-887. They also developed advanced signal processing tools to identify the light transients caused by the strain energy being released when damage is created. More information can be found in (Vandenplas et al., 2004, 2005; Wevers et al., 2006).

A polarimetric optical fibre sensor was used by Murukeshan et al. (1999) to detect damage progression in composites with artificially produced damage (delamination and fibre cuts) during three-point bending tests. The stiffness reduction due to damage could be measured by precise strain measurements (Jensen et al., 1994). used a polarimetric sensor and an electronic speckle pattern interferometry (ESPI) instrument to measure the natural frequencies of a composite plate. An interfero-polarimetric sensor was used by Bonniau et al. (1991) to successfully detect impact-induced delamination. Two types of optical fibres were compared. The intensity of the light transmitted by the optical fibre was found to be related to the impact energy.

Most later studies use fibre Bragg grating (FBG) or interferometric sensors to obtain very precise strain and stress measurements to monitor the damage. However, Okabe et al. (2004) showed that the strain distribution along the length of the grating of an FBG sensor influences the form of the reflection spectrum. They found some indications of crack density and could detect the critical damage state, i.e., saturation of the transverse crack density.

3. Structural health monitoring of composites

The emergence of optical fibre communication technologies in the 1970s has enabled the development of *optical fibre sensors* for process condition monitoring and for smart material/structure applications. Meanwhile, thanks to the evolution of computer technology, powerful data analysis tools have appeared. With an optical fibre embedded in a SmartTape attached to composite material/structure and advanced

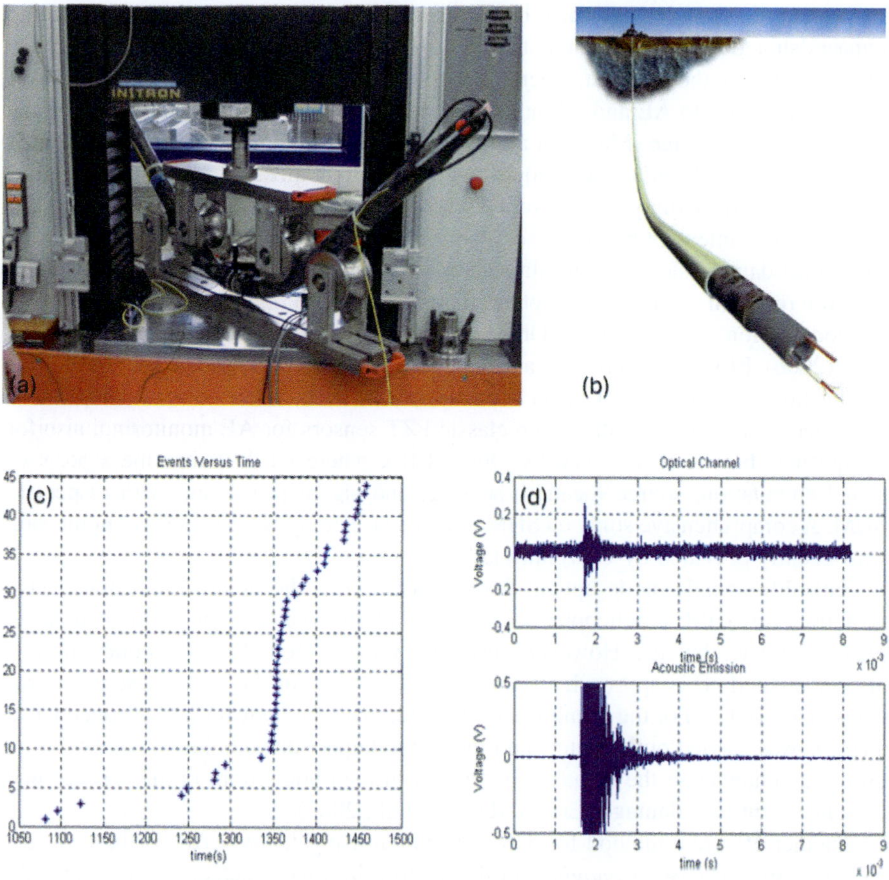

Figure 4.6.9 (A) Bending test on part of power data transmission coil shown in (B), which has a monitoring single-mode optical fibre embedded and AE sensors to validate, (C) the cumulative number of transient events in the optical fibre versus time during bending, and (D) one transient event in the optical fibre and of the AE sensor (MTM PDT-COIL report).

data processing of the optical fibre signals (Fig. 4.6.9), an NDT system can be integrated, similar to the neural system in a human body (Rippert et al., 2000; Wevers et al., 2007).

The development of these new sensors, for AE monitoring, for instance, with a higher sensitivity and robustness, could lead to more applications for real-time health monitoring. An example of such a sensor is the optical fibre sensor (Kageyama et al., 2005) with the following advantages over conventional piezoelectric sensors: small diameter and light weight, flexibility, high strength, heat resistance, immunity to electromagnetic interference, durability, and corrosion resistance. Furthermore, these sensors are believed to have a very high sensitivity in an extremely broad frequency range and can also be used for reinforced concrete structures.

Optical fibre sensors have also been reported as potential candidates for detecting impact damage in filament-wound tubes (Hofer, 1987; LeBlanc and Measures, 1992). They range from the simplest, where the optical fibre breaks when a crack propagates through the fibre, to AE and vibration sensors, which can pick up and determine the location of the damage (Martin et al., 1997). Simplicity, though, is often overlooked for more advanced sensing techniques, which require expensive and sensitive equipment (e.g., optical time domain reflectometers with high-resolution scanning). For these reasons, intensity-based sensors have been identified as a cost-effective means for impact damage detection in filament-wound tubes.

More research on the use of optical fibres for AE and related damage monitoring and localisation can be found in Ohsawa et al. (2003), Cho et al. (2005), and Matsuo et al. (2006). FBG sensors rely on a different principle and can also be used for source localisation (Tsuda, 2011; Takeda and Minakuchi, 2009).

Structural health evaluation with classic PZT sensors for AE monitoring, also for composites, has been reviewed by Ono (2018), where he describes the aspects of signal attenuation, source location, sensing, and signal processing with respect to SHM. A comprehensive study on SHM of aerospace composites using AE monitoring was recently provided by Giurgiutiu (2015).

Embedded single-mode optical fibres, operating in the polarisation mode, have been used for SHM of helicopter tail booms via Lamb waves to probe the composite plate for impact damage. However, during the impact, the light of the optical fibre is also modulated. This approach allows for an estimation of the impact energy (Pitropakis et al., 2012). For the Lamb waves that modulate the optical fibre, advanced data analysis was carried out to detect and locate the impact damage in the composite by observing changes in the signals measured with the optical fibre in the absence and presence of impact damage (Ercsey-Ravasz et al., 2007).

Another effective method for defect detection using *Lamb waves* is the full-field analysis through *C-scan imaging*. In this approach, a fixed actuator is permanently positioned on the composite part, and with single-sided access, the propagation of Lamb waves is measured via the C-scan (Hillger et al., 2012). Yet another promising approach for using Lamb waves in composite inspection involves the detection of *leaky Lamb waves* with an *acoustic camera*, as demonstrated by Pfeiffer et al. (2013). Although this is limited to the lower frequency range, a feasibility study has demonstrated that impact damage in sandwich composites can be identified by mapping the interaction between Lamb waves and the structural damage.

Moisture accumulation in composites compromises the material's structural integrity—particularly during freeze–thaw cycles of water trapped in sandwich structures. While *thermography* is the current gold standard for detecting water ingress, *permanent moisture sensors* provide interesting options for more tailored maintenance strategies, especially by providing early detection of moisture issues before they evolve into severe damage. To address this, *percolation sensors* based on polyvinyl alcohol and TiCN have been embedded in various configurations within the prepreg layers. These sensors monitor moisture uptake by monitoring the characteristic changes in electrical resistance (Sunetchiieva et al., 2019). Current research finally aims to functionalise standard epoxy resin within prepreg by carbon black to impart sensing capabilities (Fauche et al., 2024).

4. Conclusion

As it is for all materials, including composites, the need to manufacture, control, and use materials in a safe and reliable way has resulted in continuous development of test techniques/protocols and has turned NDT into an indispensable tool. This is also reflected in Fig. 4.6.10, where the number of publications related to the NDT of composites is normalised to the number of publications related to composites for the last half a century. This figure indicates an overall increasing trend of publications. Some ups and downs—a hesitation!—occur within 1970–1990, followed by a remarkable increase up to 2004. The curve now seems to resume its upward trend after a stable output over the past 20 years.

From the early days of visual inspection and hammer strikes (Fokker bond tester) to today's multitude of techniques that often complement each other and are blessed with highly sophisticated digital and automated systems, NDT has evolved in its use for composites. The history of NDT also reflects human ingenuity and the questing spirit of humans for safety and reliability in an increasingly complicated world.

We can now find the defect or damage with higher resolution and probability and are developing high-end SHM systems that cope with the demands of reliable and safe use of composites characterised by a complex damage development process. New technologies like artificial intelligence and the Internet of Things will only speed up this evolution and will further revolutionise the field by allowing highly precise and real-time inspections. The NDT inspections will be further enabled in predictive maintenance with the integration of machine learning and big data analytics to reduce downtime and enhance safety.

We are now at the beginning of even greater breakthroughs in the field, with the future of NDT promising unprecedented precision, efficiency, and innovation. This progress will ensure that our world's critical infrastructure, including composite materials, remains safe and reliable for generations to come.

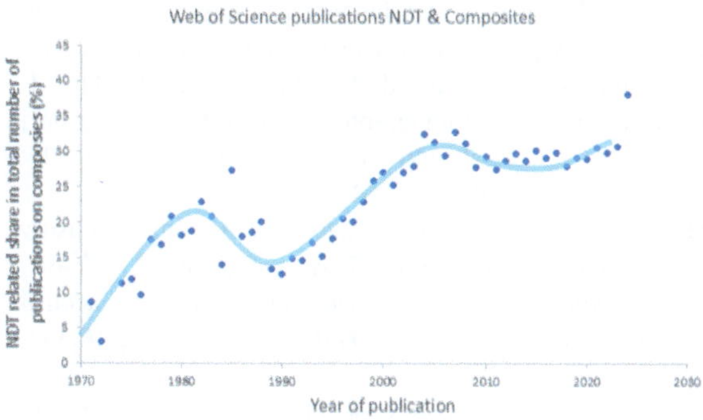

Figure 4.6.10 Evolution of the literature on NDT of composites, illustrated via the number of publications on NDT of composites normalised with total publications on fibre-reinforced composites.

References

Adams, R.D., Cawley, P., 1988. A review of defect types and nondestructive testing techniques for composites and bonded joints. NDT Int. 21 (4), 208–222.

Almeida, R.S.M., Magalhães, M.D., Karim, M.D.N., Tushtev, K., Rezwan, K., 2023. Identifying damage mechanisms of composites by acoustic emission and supervised machine learning. Mater. Des. 227. https://doi.org/10.1016/j.matdes.2023.111745.

Anastassopoulos, A.A., Philippidis, T.P., 1995. Clustering methodology for the evaluation of acoustic emission from composites. J. Acoust. Emiss. 13, 11–22.

Arena, M., Viscardi, M., 2020. Strain state detection in composite structures: review and new challenges. J. Compos. Sci. 4 (2), 60. https://doi.org/10.3390/jcs4020060.

Awerbuch, J., Gorman, M.R., Madhukar, M., 1983. Monitoring acoustic emission during quasistatic loading/unloading cycles of filament-wound graphite/epoxy laminate coupons. In: Proceedings of 1st to 6th International Symposium on AE from Reinforced Plastics/Composite Materials, SPI/ASNT, pp. 139–157.

Bar-Cohen, Y., 1986. NDE of fibre-reinforced composite materials: a review. Mater. Eval. 44, 446–454.

Bashyam, M., 1990. Thickness-compensation technique for ultrasonic evaluation of composite materials. Mater. Eval. 1360–1364.

Berthelot, J.M., Rhazi, J., 1990. Acoustic emission in carbon fibre composites. J. Compos. Sci. Technol. 37, 411–428.

Bonniau, P., Chazelas, J., Lecuellet, J., Gendre, F., Turpin, M.J., Le Pesant, J.-P., Brevignon, M., 1991. Damage detection in woven-composite materials using embedded fibre-optic sensors. Proc. SPIE 1588 (1), 52–63. https://doi.org/10.1117/12.50164.

Brunner, A.J., Nordstrom, R., Flueler, P., 1995. A study of acoustic emission–rate behavior in glass fibre-reinforced plastics. J. Acoust. Emiss. 13, 67–77.

Canal, L.P., González, C., Molina-Aldareguía, J.M., Segurado, J., Llorca, J., 2012. Application of digital image correlation at the microscale in fibre-reinforced composites. Compos. Appl. Sci. Manuf. 43 (10), 1630–1638. https://doi.org/10.1016/j.compositesa.2011.07.014.

Carlsson, L., Norrbom, B., 1983. Acoustic emission from graphite/epoxy composite laminates with special reference to delamination. J. Mater. Sci. 18, 2503–2509.

CARP, 1987. Recommended Practice for Acoustic Emission Testing of Fibreglass Reinforced Plastic Resin (RP) Tanks/Vessels, 1982. The Society of the Plastics Industry, Inc., New York, p. 13. Revised and republished, 1987.

Characterization, analysis and significance of defects in composite materials. https://www.sto.nato.int/document/characterization-analysis-and-significance-of-defects-in-composite-materials.

Chatziathanasiou, T., Soete, J., Vanhulst, J., Carrella-Payan, D., Gorbatikh, L., Mehdikhani, M., 2023. In-situ X-ray computed tomography of mode I delamination in carbon-epoxy composites: the effect of the interface ply orientation. Compos. B Eng. 260, 110761. https://doi.org/10.1016/j.compositesb.2023.110761.

Cho, H., Arai, R., Takemoto, M., 2005. Development of stabilized and high sensitive optical fibre acoustic emission system and its application. J. Acoust. Emiss. 23, 72–80.

Ciaburro, G., Iannace, G., 2022. Machine-learning-based methods for acoustic emission testing: a review. Appl. Sci. 12 (20), 10476. https://doi.org/10.3390/app122010476.

Ciampa, F., Mahmoodi, P., Pinto, F., Meo, M., 2018. Recent advances in active infrared thermography for non-destructive testing of aerospace components. MDPI: Sensors 18 (2), 609. https://doi.org/10.3390/s18020609.

Cousins, F.R., Markham, M.F., 1977. The use of ultrasonic spectroscopy in the location of delaminations in fibre-reinforced polymers. Composites 8, 145–152.

Cramer, K.E., 2018. Current and future needs and research for composite materials NDE. In: SPIE Smart Structures and Materials & Nondestructive Evaluation and Health Monitoring, Denver, US.

De Angelis, G., Meo, M., Almond, D.P., Pickering, S.G., Angioni, S.L., 2012. A new technique to detect defect size and depth in composite structures using digital shearography and unconstrained optimization. NDT & E International 45, 91–96. https://doi.org/10.1016/j.ndteint.2011.07.007.

de Groot, P.J., Wijnen, P.A.M., Janssen, R.B.F., 1995. Real-time frequency determination of acoustic emission for different fracture mechanisms in carbon/epoxy composites. Compos. Sci. Technol. 55, 405–412. https://doi.org/10.1016/0266-3538(95)00121-2.

De Meester, P.J.A., Wevers, M.G.T., 1988. Non-destructive testing of composites. In: Farley, J. M., Nichols, R.W. (Eds.), Non-destructive Testing, Proc. Of the 4th European Conference in London, U.K., Sept. 13-17, 1987, 1. Pergamon Press PLC, Oxford, U.K, pp. 95–117.

Drouillard, T.F., Hamstad, M.A., 1983. A comprehensive guide to the literature on acoustic emission from composites. In: 1st Int. Symposium on Acoustic Emission from Reinforced Plastics, SPI, Inc., New York, pp. 1–60.

Eisenblätter, J., 1988. Acoustic Emission Analysis: Introduction, Present Status and Future Development. DGM Informations-gesellschaft-Verlag, Oberursel.

Ercsey-Ravasz, M., Fransens, F., Pfeiffer, H., Monavon, G., Korosec, C., Hillger, W., Van Huffel, S., Wevers, M., 2007. Structural Health Monitoring of a helicopter tail boom using lamb waves - advanced data analysis of results obtained with integrated optical fibre sensing technology. In: EU Project Meeting on Aircraft Integrated Structural Health Assessment (AISHA), (Paper No. ID: 6963), (1-8). Presented at the Project Meeting on Aircraft Integrated Structural Health Assessment (AISHA), Leuven, Belgium, June 2007.

Fagot, H., Aibe, F., Smigielski, P., Arnaud, J.L., 1980. Holographic nondestructive testing of materials using pulsed lasers. Institut Franco-Allemand de Recherche de Saint-Louis, Report No. 223 80.

Fauche, R., Windey, R., Pfeiffer, H., Seveno, D., Wevers, M., 2024. Effects of water uptake on the electrical properties of carbon black-epoxy nanocomposites. ACS Appl. Nano Mater. 7, 3610–3619.

Favre, J.P., Laizet, J.C., 1990. Acoustic emission analysis of the accumulation of cracks in CFRP cross-ply laminates under tensile loading. J. Acoust. Emiss. 1990 (9), 97–102.

Fowler, T.J., 1977. Acoustic emission testing of fibre reinforced plastics. American Society of Civil Engineers (ASCE) Fall Convention and Exhibit, Preprint 3092, ASCE, New York, 1977.

Fowler, T.J., 1979. Acoustic emission testing of fibre reinforced plastics. In: Proceedings Paper 15021, J. Technical Council, ASCE, vol. 105 (TC2), pp. 281–289.

Fowler, T.J., 2014. The origin of CARP and the term "felicity effect". In: 31st Conference of the European Working Group on Acoustic Emission (EWGAE) – Commemorative Speech.

Gao, W.M., Glorieux, C., Thoen, J., 2003. Laser ultrasonic study of Lamb waves: determination of the thickness and velocities of a thin plate. Int. J. Eng. Sci. 41 (2), 219.

Garcea, S.C., Wang, Y., Withers, P.J., 2018. X-ray computed tomography of polymer composites. Compos. Sci. Technol. 156, 305–319. https://doi.org/10.1016/j.compscitech.2017.10.023.

Gayot, S.F., Soete, J., Vanhulst, J., Gérard, P., Pardoen, T., 2024. 4D-XCT monitoring of void formation in thick methacrylic composites produced by infusion. Acta Mater., 120449 https://doi.org/10.1016/j.actamat.2024.120449.

Gholizadeh, S., 2016. A review of non-destructive testing methods of composite materials. Procedia Struct. Integr. 1, 50–57. https://doi.org/10.1016/j.prostr.2016.02.008.

Giurgiutiu, V., 2015. Structural Health Monitoring of Aerospace Composites. Academic Press, New York, p. 470.

Glorieux, C., Gao, W., Kruger, S.E., Van de Rostyne, K., Lauriks, W., Thoen, J., 2000. Surface acoustic wave depth profiling of elastically inhomogeneous materials. J. Appl. Phys. 88 (7), 4394–4400. https://doi.org/10.1063/1.1290457.

Glossop, N.D.W., Dubois, S., Tsaw, W., Leblanc, M., Lymer, J., Measures, R.M., Tennyson, R. C., 1990. Optical fibre damage detection for an aircraft composite leading edge. Composites 21 (1), 71–80.

Gorman, M.R., 1991. Plate wave acoustic emission. J. Acoust. Soc. Am. 90, 358–364.

Grédiac, M., Pierron, F., Vautrin, A., 1994. The Iosipescu in-plane shear test applied to composites: a new approach based on displacement field processing. Compos. Sci. Technol. 51 (3), 409–417. https://doi.org/10.1016/0266-3538(94)90109-0.

Green, A.T., Lockman, C.S., Steele, R.K., 1964. Acoustic verification of structural integrity of Polaris chambers. Mod. Plast. 41 (July), 137–139, 178, 180.

Hamstad, M.A., 1986a. A discussion of the basic understanding of the Felicity effect in fibre composites. J. Acoust. Emiss. 5, 95–102.

Hamstad, M.A., 1986b. A review: acoustic emission, a tool for composite-materials studies. Exp. Mech. 26, 7–13. https://doi.org/10.1007/BF02319949.

Harris, B., 1982. Assessment of Structural Integrity of Composites by Non-Destructive Methods. AGARD. LS-124, paper published by the School of Materials Science, University of Bath, United Kingdom, p. 22.

Henneke, E.G., Reifsnider, K.L., Stinchcomb, W.W., 1979. Thermography — an NDI method for damage detection. JOM 31, 11–15. https://doi.org/10.1007/BF03354475.

Highsmith, A., Reifsnider, K.L., 1982. Measurement of nonuniform microstrain in composite laminates. Compos. Technol. Rev. 4 (1), 20–22.

Highsmith, A.L., 1989. Edge replication for laminated composites. In: Pendleton, R.L., Tuttle, M.E. (Eds.), Manual on Experimental Methods for Mechanical Testing of Composites. Springer, Dordrecht, pp. 165–169. https://doi.org/10.1007/978-94-009-1129-1_21.

Hillger, W., 1989. US-detection of CFRP-laminates with high resolution. In: Boogaard, J., van Dijk, G.M. (Eds.), Proceedings of the 12th World Conference on Non-destructive Testing. Elsevier Science Publishers B.V., Amsterdam, The Netherlands, pp. 1561–1566.

Hillger, W., Szewieczek, A., Schmidt, D., 2012. Damage detection in a helicopter composite tailboom by mode conversion of Lamb waves. In: Boller, C. (Ed.), Proceedings of the 6th European Workshop on Structural Health Monitoring Dresden. Germany.

Hoa, S.V., Smith, I.C., 1989, In: Proceedings of the 3rd International Symposium on Acoustic Emission from Composite Materials, AECM-3, Paris, American Society for Nondestructive Testing, Columbus, OH, pp. 240–249.

Hockley, B.S., 1972. Non-destructive testing by holography. Britt J NDI 14, 115–121.

Hofer, F., 1987. Fibre optic damage detection in composite structures. Composites 18 (4), 309–316. https://doi.org/10.1016/0010-4361(87)90294-1.

Hsu, N.N., Simmons, J.A., Hardy, S.C., 1977. An approach to acoustic emission signal. Analysis - theory and experiment. Mater. Eval. 35 (10), 100–106.

Jensen, A.E., Storoy, H., Irgens, F., Lokberg, O.J., Wang, G., 1994. Monitoring vibration of sandwich plates and delamination detection with fibre optic sensors. In: Gardiner, P.T., McEwen, R.S., Culshaw, B. (Eds.), Second European Conference on Smart Structures and Materials, Alaster McDonach, 2361. Proc. SPIE, pp. 140–143.

Kageyama, K., Murayama, H., Ohsawa, I., Kanai, M., Nagata, K., Machijima, Y., Matsumura, F., 2005. Acoustic emission monitoring of a reinforced concrete structure by applying new fibre-optic sensors. Smart Mater. Struct. 14, S52–S59.

Kaiser, J., 1953. Erkenntnisse und Folgerungen aus der Messung von Geräuschen bei Zugbeanspruchung von Metallischen Werkstoffen. Arch. Eisenhüttenwes. 24 (1–2), 43–45.

Kersemans, M., Martens, A., Degrieck, J., Van Den Abeele, K., Delrue, S., Pyl, L., Zastavnick, F., Sol, H., Van Paepegem, W., 2016. The ultrasonic polar scan for composite characterization and damage assessment: past, present and future. Appl. Sci. 6 (2), 58. https://doi.org/10.3390/app6020058.

Kim, H.C., Ripper Neto, A.P., Stephens, R.W.B., 1972. Some observations on acoustic emission during continuous tensile cycling of a carbon fibre/epoxy composite. Nat. Phys. Sci. 237, 78–80.

Kitade, S., Fukuda, T., Osaka, K., 1995. Fibre optic method for detection of impact induced damage in composite laminates. J. Soc. Mater. Sci. 44, 1196–1200.

Kline, R.A., Green, R.E., Palmer, C.H., 1981. Acoustic emission waveforms from cracking steel: experiment and theory. J. Appl. Phys. 52, 141–146.

Kromine, A.K., Fomitchov, P.A., Krishnaswamy, S., Achenbach, J.D., 2000. Laser ultrasonic detection of surface breaking discontinuities: scanning laser source technique. Mater. Eval. 58 (2), 173.

LeBlanc, M., Measures, R.M., 1992. Impact damage assessment in composite materials with embedded fibre-optic sensors. Compos. Eng. 2 (Issues 5–7), 573–577. https://doi.org/10.1016/0961-9526(92)90044-7, 579-596.

Liber, T., Daniel, I.M., Schramm, S.W., 1979. Ultrasonic techniques for inspecting flat and cylindrical composite specimens. In: Pipes, R.B. (Ed.), Nondestructive Evaluation and Flaw Criticality for Composite Materials, ASTM STP, 696. American Society for Testing and Materials, Philadelphia, pp. 5–25.

Liptai, R.G., 1972. Acoustic emission from composite materials. Compos. Mater. Test. Des. 497, 285–298.

Lomov, S.V., Ivanov, D.S., Verpoest, I., Zako, M., Kurashiki, T., Nakai, H., Vautrin, A., 2008. Full-field strain measurements for validation of meso-FE analysis of textile composites. Compos. Appl. Sci. Manuf. 39 (8), 1218–1231. https://doi.org/10.1016/j.compositesa.2007.09.011.

Mahoon, A., 1983. Non-destructive testing of aircraft composite structures. In: Proceedings of the International Conference on Testing, Evaluation and Quality Control of Composites. Butterworth & Co, Kent, pp. 93–101.

Martin, A.R., Fernando, G.F.†, Hale, K.F., 1997. Impact damage detection in filament wound tubes using embedded optical fibre sensors. Smart Mater. Struct. 6, 470–476. https://doi.org/10.1088/0964-1726/6/4/012.

Matsuo, T., Cho, H., Takemoto, M., 2006. Utilization of cascade optical fibre AE system for source location of Lamb waves through a cross-ply CFRP plate. J. Acoustic Emission. 24, 84–96.

Matthews, J.R., 1983. Acoustic Emission, vol. 2. Gordon and Breach, New York.

McRae, K.I., 1990. Deconvolution techniques for ultrasonic imaging of adhesive joints. Mater. Eval. 1380–1384.

Mehdikhani, M., Aravand, M., Sabuncuoglu, B., Callens, M.G., Lomov, S.V., Gorbatikh, L., 2016. Full-field strain measurements at the micro-scale in fibre-reinforced composites using digital image correlation. Compos. Struct. 140, 192–201. https://doi.org/10.1016/j.compstruct.2015.12.020.

Mehdikhani, M., Steensels, E., Standaert, A., Vallons, K.A., Gorbatikh, L., Lomov, S.V., 2018. Multi-scale digital image correlation for detection and quantification of matrix cracks in carbon fibre composite laminates in the absence and presence of voids controlled by the cure cycle. Compos. B Eng. 154, 138–147. https://doi.org/10.1016/j.compositesb.2018.07.006.

Mehdikhani, M., Breite, C., Swolfs, Y., Soete, J., Wevers, M., Lomov, S.V., Gorbatikh, L., 2021. Digital volume correlation for meso/micro in-situ damage analysis in carbon fibre reinforced composites. Compos. Sci. Technol. 213, 108944. https://doi.org/10.1016/j.compscitech.2021.108944.

Murukeshan, V.M., Chan, P.Y., Ong, L.S., Asundi, A., 1999. On-line health monitoring of smart composite structures using fibre polarimetric sensor. Smart Mater. Struct. 8, 544. https://doi.org/10.1088/0964-1726/8/5/303.

Nevadunsky, J.J., Luca, J.J., 1975. Early fatigue damage detection in composite materials. J. Compos. Mater. 9, 394–408.

Ohsawa, I., Kageyama, K., Tsuchida, Y., Kanai, M., 2003. Development of a novel optical fibre sensor for AE detection in composites. J. Acoust. Emiss. 21, 176–186.

Ohtsu, M., Ono, K., 1987. Pattern recognition analysis of acoustic emission from unidirectional carbon fibre-epoxy composites by using autoregressive modelling. J. Acoust. Emiss. 6, 61–71.

Okabe, Y., Tsuji, R., Takeda, N., 2004. Application of chirped fibre Bragg grating sensors for identification of crack locations in composites. Compos. Appl. Sci. Manuf. 35 (1), 59–65. https://doi.org/10.1016/j.compositesa.2003.09.004.

Ono, K., Kawamoto, K., 1990. Digital signal analysis of acoustic emission from carbon fibre/epoxy composites. J. Acoust. Emiss. 1990 (9), 109–116.

Ono, K., Gallego, A., 2012. Research and applications of AE on advanced composites. J. Acoust. Emiss. 30, 180–229.

Ono, K., 2018. Review on structural health evaluation with acoustic emission. Appl. Sci. 8 (6), 958. https://doi.org/10.3390/app8060958.

Oplinger, D.W., Parker, B.S., Chiang, F.P., 1974. Edge-effect studies in fibre-reinforced laminates. Exp. Mech. 14, 147–354.

Park, B., An, Y.K., Sohn, H., 2014. Visualization of hidden delamination and debonding in composites through noncontact laser ultrasonic scanning. Compos. Sci. Technol. 100, 10.

Peters, W.H., Sutton, M.A., Ranson, W.F., Poplin, W.P., Walker, D.M., 1989. Whole-field experimental displacement analysis of composite cylinders. Exp. Mech. 29, 58–62. https://doi.org/10.1007/BF02327782.

Pfeiffer, H., Hillger, W., Böck, M., Pitropakis, I., Szewieczek, A., Glorieux, C., Wevers, M., 2013. Identification of impact damage in sandwich composites by acoustic camera detection of leaky Lamb wave mode conversions. E-Journal of Nondestructive Testing. https://www.ndt.net/?id=14173.

Pitropakis, I., Pfeiffer, H., Wevers, M., 2012. The use of a polarimetric fibre sensor to detect impact damage in aircraft composites. In: Paipetis, et al. (eds.) Emerging Technologies in Non-Destructive Testing V. London, Taylor &Francis Group, pp. 205–209.

Prakash, R., 1980. Non-destructive testing of composites. Composites 11 (4), 217–224.

Prosser, W.H., Dorighi, J., Gorman, M.R., 1992. Extensional and flexural waves in a thin-. Walled graphite/epoxy tube. J. Compos. Mater. 26, 2016–2027.

Reynolds, W.N., 1985. Nondestructive Testing (NDT) of Fibre-Reinforced Composite Materials. SAMPE Quarterly 1–16.

Rippert, L., Wevers, M., Van Huffel, S., 2000. Optical and acoustic damage detection in laminated CFRP composite materials. Compos. Sci. Technol. 60, 2713–2724.

Rose, J.L., Balasubramaniam, K., Pilarski, A., Ditri, J., 1989. The utility of guided waves in the ultrasonic NDE of composite materials. In: Boogaard, J., van Dijk, G.M. (Eds.), Proceedings of the 12th World Conference on Non-destructive Testing. Elsevier Science Publishers B.V., Amsterdam, The Netherlands, pp. 1567–1572.

Sarens, B., 2009. Full-Field Optical Methods for Imaging Ultrasonic Waves and the Detection of Defects (Ph.D. Thesis). KU Leuven.

Sarens, B., Glorieux, C., Kalogiannakis, G., Van Hemelrijck, D., 2010. Investigation of contact acoustic nonlinearity in delaminations by shearographic imaging, laser doppler vibrometry scanning and finite difference modeling. IEEE Trans. Ultrason. Ferroelectrics Freq. Control 57, 1383–1395.

Scott, A.E., Mavrogordato, M., Wright, P., Sinclair, I., Spearing, S.M., 2011. In situ fibre fracture measurement in carbon–epoxy laminates using high resolution computed tomography. Compos. Sci. Technol. 71 (12), 1471–1477. https://doi.org/10.1016/j.compscitech.2011.06.004.

Scott, I.G., Scala, C.M., 1982a. A review of non-destructive testing of composite materials. NDT International 15 (2), 75–86. https://doi.org/10.1016/0308-9126(82)90001-3.

Selden, R.B., Gustafson, C.G., 1985. Transverse cracking and delamination in (0/±45/90)s CFRP detected through acoustic emission. J. Reinforc. Plast. Compos. 4, 365–382.

Sendeckyj, G.P., Maddux, G.E., Porter, E., 1982. Damage documentation in composites by stereo radiography. In: Reifsnider, K.L. (Ed.), Damage in Composite Materials, ASTM STP, 775. American Society for Testing and Materials, Philadelphia, pp. 16–26.

Shuh-Huei, L., Shah, S.P., Zongjin, L., Toshio, M., 1993. Micromechanical analysis of multiple fracture and evaluation of debonding behavior for fibre-reinforced composites. Int. J. Solid Struct. 1429–1459. https://doi.org/10.1016/0020-7683(93)90070-N.

Stone, D.E.W., Clark, B., 1987. Non-destructive evaluation of composite structures — an overview. In: Matthews, F.L., Buskell, N.R.C., Hodgkinson, J.M., Morton, J. (Eds.), Proc. of the ICCM6/ECCM2. Elsevier Applied Science, Oxford, pp. 1.28–1.59.

Sunetchiieva, S., Pfeiffer, H., Wevers, M., 2019. Structural health monitoring system for moisture ingress detection in sandwich composite structures. In: 9th EASN Conference, Athens, Greece.

Sunetchiieva, S., 2024. New Methods for Structural Health Monitoring in Aeronautics (Ph.D. Thesis). KU Leuven.

Surgeon, M., 1999, Continuous Damage Monitoring Techniques for Laminated CFRP Composite Materials (Ph.D. Thesis). KU, Leuven.

Surgeon, M., Wevers, M., 1999a. Modal analysis of acoustic emissions from CFRP laminates. NDT&E International 32, 311–322.

Surgeon, M., Wevers, M., 1999b. One sensor linear location of acoustic emission events using plate wave theories. Mater. Sci. Eng., A 265, 254–261.

Surgeon, M., Wevers, M., 2001. The influence of embedded optical fibres on the fatigue damage progress in quasi-isotropic CFRP laminates. J. Compos. Mater. 35, 931–940.

Surgeon, M., Vanswijgenhoven, E., Wevers, M., Van Der Biest, O., 1997. Acoustic emission during tensile testing of SiC-fibre-reinforced BMAS glass-ceramic composites. Compos. Appl. Sci. Manuf. 28 (5), 473–480. https://doi.org/10.1016/S1359-835X(96)00147-9.

Sutton, M.A., Wolters, W.J., Peters, W.H., Ranson, W.F., McNeill, S.R., 1983. Determination of displacements using an improved digital correlation method. Image Vis Comput. 1 (3), 133–139. https://doi.org/10.1016/0262-8856(83)90064-1.

Swiderski, W., 2015. IR thermography nondestructive testing methods of composite materials used in aerospace applications. In: Proceeding of the 2015 Asia International Conference on Quantitative InfraRed Thermography. https://doi.org/10.21611/qirt.2015.0016.

Takeda, N., Chen, O., Kishi, T., Tredway, W., Prewo, K., 1991. Acoustic emission characterization of the fracture mechanism of a high compliant, glass-matrix composite. Eng. Fract. Mech. 40, 791–799. https://doi.org/10.1016/0013-7944(91)90236-T.

Takeda, N., Minakuchi, S., 2009. Recent development of structural health monitoring technologies for aircraft composite structures in Japan. In: Proc. 17th Int. Conf. on Composite Materials (ICCM-17), Edinburgh, UK, 27–31 July 2009.

Takeda, N., Minakuchi, S., Nadabe, T., 2013. Distributed strain monitoring for damage evolution in CFRP bolted structures with embedded optical fibres. Key Eng. Mater. 558, 252–259.

Takehana, M., Kimpara, I., 1972. AE from FRP fracture processes. Proc. US-Japan Symposium on Acoustic Emission, Japanese Volume, Sec. 4, pp. 1–22.

Teagle, P.R., 1983. The quality control and non-destructive evaluation of composite aerospace components. Composites 14 (April), 115–128.

Tsamtsakis, D., Wevers, M., De Meester, P., 1996. Damage monitoring during fatigue loading of quasi-isotropic carbon epoxy laminates. In: Van Hemelrijck, D., Anastassopoulos, A. (Eds.), Non-Destructive Testing. Balkema Publishers, Rotterdam, pp. 127–135.

Tsuda, H., 2011. A Bragg wavelength-insensitive fibre Bragg grating ultrasound sensing system that uses a broadband light and no optical filter. Sensors 11, 6954–6966.

Valentin, D., Bonniau, P., Bunsell, A.R., 1983. Failure mechanisms discrimination in carbon fibre reinforced epoxy composites. Composites 14, 345–351.

Van Bochove, G.F.J., Somers, P.A.A.M., 1985. Research and development on automation of non-destructive testing methods. In: Bartelds, G., Schliekelmann, R.J. (Eds.), Progress in Advanced Materials and Processes: Durability, Reliability and Quality Control. Elsevier Science Publishers B.V., Amsterdam, pp. 29–39.

Van Dreumel, W.H.M., Speijer, J.L., 1981. Non-destructive composite laminate characterization by means of ultrasonic Polar-scan. Mater. Eval. 39 (10), 922–925.

Vandenplas, S., Papy, J.-M., Wevers, M., Van Huffel, S., 2004. Acoustic emission monitoring using a multimode optical fibre sensor. Insight - Non-Destructive Testing and Condition Monitoring 46 (4), 203–209. https://doi.org/10.1784/insi.46.4.203.55640.

Vandenplas, S., Papy, J.-M., Wevers, M., Van Huffel, S. and Inaudi, D. 2005. Acoustic emission monitoring using a polarimetric single mode optical fibre sensor. In: Proc. SPIE 5855, 17th International Conference on Optical Fibre Sensors, p. 1064.

Vary, A., Lark, R.F., 1979. Correlation of fibre composite tensile strength with the ultrasonic stress wave factor. J. Test. Eval. 7 (4), 185–191.

Wadley, H.N.G., Scruby, C.B., Shrimpton, G., 1981. Quantitative acoustic emission source characterisation during low temperature cleavage and intergranular fracture. Acta Metall. Mater. 29, 399–414.

Waite, S.R., Sage, G.N., 1988. The failure of optical fibres embedded in composite materials. Composites 19 (4).

Wang, B., Zhong, S., Lee, T.-L., Fancey, K.S., Mi, J., 2020. Non-destructive testing and evaluation of composite materials/structures: a state-of-the-art review. Adv. Mech. Eng. 12 (4). https://doi.org/10.1177/1687814020913761.

Weissenbock, J., Amirkhanov, A., Li, W.M., Reh, A., Groller, E., Kastner, J., Heinzl, C., 2014. Fibrescout: an interactive tool for exploring and analyzing fibre reinforced polymers. In: IEEE Pacific Visualization Symposium, pp. 153–160. https://doi.org/10.1109/PacificVis.2014.52.

Wevers, M., 1987. In situ NDT during laboratory testing of CFRP-composites. In: Verpoest, I., Wevers, M. (Eds.), Proceedings of the European Symposium on Damage Development and Failure Processes in Composite Materials, Leuven, Belgium, 4-6 May 1987, pp. 91–106.

Wevers, M., Verpoest, I., De Meester, P., Aernoudt, E., 1991. Identification of fatigue failure modes in carbon fibre reinforced composites with the energy discriminating acoustic emission method. In: Sachse, W., Roget, J., Yamaguchi, K. (Eds.), Acoustic Emission: Current Practice and Future Directions, ASTM STP 10779. American Society for Testing Materials, Philadelphia, PA, pp. 416–423.

Wevers, M., 1991. New developments on non-destructive inspection techniques for advanced composites. In: Kwakernaak, A., van Arkel, L. (Eds.), Advanced Materials: Cost Effectiveness, Quality Control, Health and Environment. SAMPE/Elsevier Science Publishers, B.V., pp. 243–254

Wevers, M., 1997. Listening to the sound of materials: acoustic emission for the analysis of material behaviour. NDT&E International 30 (2), 99–106.

Wevers, M., Surgeon, M., 2000. Acoustic emission and composites. In: Comprehensive Composite Materials, an Elsevier Reference Collection in Materials Science and Materials Engineering, Vol. 5, pp. 345–357.

Wevers, M., Rippert, L., Van Huffel, S., 2000. Optical fibres for in situ monitoring the damage development in composites and the relation with acoustic emission measurements. J. Acoustic Emission. 18, 41–50.

Wevers, M., Rippert, L., Papy, J.-M., Van Huffel, S., 2006. Processing of transient signals from damage in CFRP composite materials monitored with embedded intensity-modulated fibre optic sensors. NDT&E Int. 39, 229–235.

Wevers, M., Van Huffel, S., Vandenplas, S., Papy, J.-M., 2007. Acoustic emission monitoring using a polarimetric Single Mode optical fibre sensor. Int. J. Mater. Struct. Integr. 1 (1/2/3), 148–160.

Wevers, M., Nicolaï, B., Verboven, P., Swennen, R., Roels, S., Verstrynge, E., Lomov, S., Kerckhofs, G., Van Meerbeek, B., Mavridou, A.M., Bergmans, L., 2018. Applications of CT for non-destructive testing and materials characterization. In: Carmignato, S., Dewulf, W., Leach, R. (Eds.), Industrial X-Ray Computed Tomography. Springer, Cham. https://doi.org/10.1007/978-3-319-59573-3_8.

White, R.M., 1963. Generation of elastic waves by transient surface heating. J. Appl. Phys. 34 (12), 3559–3567.

Williams, R.V., 1980. Acoustic Emission. Adam Hilger Ltd, Bristol, UK.

Williams, J.H., Lee, S.S., 1985. Promising quantitative nondestructive evaluation techniques for composite materials. Mater. Eval. 28, 561–565.

Yuan, Z., 1984. Acoustic emission characterizations of internal damage for GFRP. In: Progress in Acoustic Emission II (Proc. IAES-7), JSNDI, Tokyo, pp. 458–463.

Yung, Y.Y., 1982. Shearography - a new optical method for strain-measurement and non-destructive testing. Opt. Eng. 21 (3), 391.

Zarei, A., Pilla, S., 2024. Laser ultrasonics for nondestructive testing of composite materials and structures: a review. Ultrasonics 136 (January), 107163. https://doi.org/10.1016/j.ultras.2023.107163.

Zhong, S., Nsengiyumva, W., 2022. Other NDT methods for fibre-reinforced composite structures. In: Nondestructive Testing and Evaluation of Fibre-Reinforced Composite Structures. Springer, Singapore. https://doi.org/10.1007/978-981-19-0848-4_8.

Ziola, S.M., Gorman, M.R., 1989. Acoustical detection of transverse cracking in a crossply composite. In: Proceedings of AECM-3, Paris, 1989, pp. 286–297.

Ziola, S.M., Gorman, M.R., 1991. Source location in thin plates using cross-correlation. J. Acoust. Soc. Am. 90, 2551–2556.

Further reading

Desplentere, F., Lomov, S.V., Woerdeman, D.L., Verpoest, I., Wevers, M., Bogdanovich, A., 2005. Micro-CT characterization of variability in 3D textile architecture. Compos. Sci. Technol. 65 (13), 1920–1930. https://doi.org/10.1016/j.compscitech.2005.04.008.

Gondrom, S., Schröpfer, S., 1999. Digital computed laminography and tomosynthesis - functional principles and industrial applications. In: International Symposium on Computerized Tomography for Industrial Applications and Image Processing in Radiology, Berlin.

Gorman, M.R., Prosser, W.H., 1991. AE source orientation by plate wave. Analysis. J. Acoust. Emiss. 9, 283–288.

Gorman, M.R., Ziola, S.M., 1991. Plate waves produced by transverse matrix cracking. Ultrasonics 29, 245–251.

Graff, K.F., 1975. Wave Motion in Elastic Solids. Dover Publications, New York.

Straumit, I., Lomov, S.V., Wevers, M., 2015. Quantification of the internal structure and automatic generation of voxel models of textile composites from X-ray computed tomography data. Compos. Appl. Sci. Manuf. 69, 150–158. https://doi.org/10.1016/j.compositesa.2014.11.016.

Wood, C.E., O'Brien, N., Denysov, A., Blumensath, T., 2019. Computed laminography of CFRP using an X-ray cone-beam and robotic sample manipulator systems. IEEE Trans. Nucl. Sci. 66 (3), 655–663.

History of composites organisations

4.7

Marino Quaresimin
Department of Management and Engineering, University of Padova, Vicenza, Italy

1. Introduction

This chapter aims to discuss and illustrate the associations within the scientific community of composite materials.

As a first point, it should be clarified that a comprehensive recognition of all the composite organisations worldwide would have been impossible due to the limited space and the difficulties in collecting the required information. It was then decided to present some representative examples, namely two continental associations, the American Society for Composites (ASC) and the European Society for Composite Materials (ESCM) as well as one international organisation, International Committee on Composite Materials (ICCM) because of its worldwide representativeness of the composites community. Apologies are therefore offered to all the societies, organisations, and associations, with a small or large number of members, that were not included in the chapter for the reasons described above.

The analysis was restricted, again for practical reasons, to academic and scientific groups, without including industrial company organisations. Continuous and trustworthy connections and exchanges between industry and academia are, however, of fundamental importance for the development of the composite scientific community and, fortunately, are continuously increasing at a very good pace! Therefore, again, as a representative example, a final section will be devoted to briefly present JEC group (JEC group Website, 2023), born basically as industrial association and grown over the years exactly as a bridge between industry and academia (JEC group Website, 2023). At the present time, JEC group organises probably the largest worldwide event for the composite community, the JEC World, held every year at the beginning of March in Paris.

2. The American Society for Composites (ASC)

The society was formed in 1985. As the funding for composite materials was increasing substantially (from many sources, including the Department of Defense and the National Science Foundation), those involved thought the time was right to create a national professional society focused on composite materials, primarily on research into the mechanics of composites. While the Society for the Advancement

The History of Composites. https://doi.org/10.1016/B978-0-443-34119-9.01030-5

of Materials and Process Engineers (SAMPE) as well as other societies (such as American Society for Mechanical Engineers, ASME, and American Institute of Aeronautics and Astronautics, AIAA) included composite materials in their scope of activities, composites were one of many topics covered by these large organisations. The SAMPE president at the time contacted the founding group, wanting to know why they were creating a new society. He thought SAMPE could fill the need and offered to have special sessions on composites and mechanics-related topics at SAMPE conferences. However, the founders considered SAMPE more applied than what was desired, and so ASC was founded and the first conference planned in 1986.

Founding members, among others, included Dr. James M. Whitney (Air Force Materials Laboratory, now a director of the Air Force Research Lab), Dr. Som Soni (AdTech Research LLC, a composites research organisation), and members of the Center for Composite Materials (CCM) at the University of Delaware. Mark Deshon from CCM created the ASC Logo, which includes the ASC moniker and founding year on a background depicting a textile reinforcement form used in composites. Moreover, present from the beginning were members of Technomic Publishing Company (later DEStech Publications, Inc.), who agreed to publish the first proceedings in bound form. Technomic was publishing the archival Journal of Composite Materials at the time and agreed to give the first proceedings away at no charge to conference attendees. Through the relationship with Technomic, ASC became the sponsoring professional society for the Journal of Composite Materials, a position it still holds today.

The founders organised the society as a "501(c) (3)" organisation chartered as nonprofit in the State of Ohio. A committee was formed for the purposes of creating a constitution and bylaws. The governing body (Executive Committee) would consist of a president, vice-president, treasurer, recording and membership secretaries, an editor (responsible for executing the society's publishing policies and maintaining the society's website), and two members-at-large. The term of all positions was (and remains) 2 years, with the vice-president automatically becoming president during the next 2-year cycle. The society members select the officers in biennial elections. The founding president, Dr. Whitney, served until 1988. He was succeeded by Dr. Keith Kedward from the University of California–Santa Barbara, who served for 1 year in 1989. The list of past presidents serves as a "who's who" in composite materials research and development (see Table 4.7.1).

The first conference was held in Dayton, Ohio, in September 1986, at what is now the Marriott Hotel at the University of Dayton. This and the first several conferences were administered by AdTech Research, although the locations and host institutions changed each year. The CCM at the University of Delaware was instrumental in helping to publicise and socialise the new society through its industry connections (CCM, which had a large consortium of industrial sponsors, had just celebrated the 10th anniversary of its founding in 1974 and realised the important need of a professional society dedicated to composite materials). The second conference, in 1987, was in fact hosted by the University of Delaware. Since then, host venues included institutions from coast-to-coast in the United States. The 2025 conference will return to Dayton

Table 4.7.1 Annual technical conferences, location and ASC presidents.

Annual technical conference	Year	Organiser/city	Chair	President ASC
1	1986	Dayton, Ohio	Charles E. Browning	James M. Whitney
2	1987	University of Delaware, Newark Delaware	Roy L. McCullough, Dale W. Wilson	James M. Whitney
3	1988	Seattle, Washington	James C. Seferis, J. T. Quinlivan	James M. Whitney
4	1989	Virginia Tech, Blacksburg, Virginia	Kenneth L. Reifsnider	Keith T. Kedward
5	1990	East Lansing, Michigan	Lawrence T. Drzal	Dick J. Wilkins
6	1991	Albany, New York	S. S. Sternstein	Dick J. Wilkins
7	1992	Penn State, University Park, Pennsylvania	H. Thomas Hahn	James H. Starnes
8	1993	Cleveland, Ohio	Golam M. Newaz	James H. Starnes
9	1994	University of Delaware, Newark, Delaware	Tsu-Wei Chou, Jack R. Vinson	Robert M. Jones
10	1995	Santa Monica, California	Ravi Deo	Robert M. Jones
11	1996	Atlanta, Georgia	W. Steven Johnson	Thomas Hahn
12	1997	Dearborn, Michigan	Ronald F. Gibson, Golam M. Newaz	Thomas Hahn
13	1998	Baltimore, Maryland	A. J. Vizzini	Jack Vinson
14	1999	Dayton, Ohio	J. M. Whitney	Jack Vinson
15	2000	Texas A&M University, College Station	Ozden O. Ochoa	Michael W. Hyer
16	2001	Virginia Tech, Blacksburg	Michael W. Hyer, Alfred C. Loos	Michael W. Hyer
17	2002	Purdue University, west Lafayette	C.T. Sun, Wade C. Jackson	Robert L. Sierakowski
18	2003	University of Florida, Gainesville	Bhavani V. Sankar, P. G. Ifju, Thomas S. Gates	Robert L. Sierakowski

Continued

Table 4.7.1 Annual technical conferences, location and ASC presidents.—cont'd

Annual technical conference	Year	Organiser/city	Chair	President ASC
19	2004	Georgia Tech, Atlanta	Erian Armanios, James Reeder	Ronald F. Gibson
20	2005	Drexel University, Philadelphia	Frank Ko	Ronald F. Gibson
21	2006	University of Michigan–Dearborn	P.K. Mallick	Ozden O. Ochoa
22	2007	University of Washington, Seattle	Paolo Feraboli, Patrick Stickler	Ozden O. Ochoa
23	2008	Memphis	Anthony J. Vizzini	Alfred C. Loos
24 [a]	2009	University of Delaware, Newark	J.W. Gillespie, S.V. Hoa	Alfred C. Loos
25 [b]	2010	University of Dayton	S. L. Donaldson, G.A. Schoeppner	Anthony J. Vizzini
26	2011	Concordia University	S.V. Hoa, M. Hyer, O. Ochoa	Anthony J. Vizzini
27	2012	University Texas Arlington	Wen chan	Gregory A. Schoeppner
28	2013	Penn State University	Charles E. Bakis	Gregory A. Schoeppner
29	2014	University of California, San Diego	Hyonny Kim	Charles E. Bakis
30	2015	Michigan State University	Xinran Xiao	Charles E. Bakis
31	2016	Williamsburg, VA	Barry Davidson	Thomas E. Lacy
32	2017	Purdue University	Wenbin Yu	Thomas E. Lacy
33 [c]	2018	University of Washington	Tony Waas	Hyonny Kim
34	2019	Georgia Tech	Kyriaki Kalaitzidou	Hyonny Kim

35	2020	Stevens Institute of technology/NYU Tandon school of engineering (Virtual)	Kishore Pochiraju	Kishore Pochiraju
36	2021	Texas A&M University (virtual)	Anastasia Muliana & John Whitcomb	Kishore Pochiraju
37	2022	University of Arizona	Olesya Zhupanska	Erian Armanios
38	2023	University of Massachusetts Lowell	Marianna Mariaru	Erian Armanios
39 [d]	2024	University of California, San Diego	Mehran Tehrani	Olesya Zhupanska
40	2025	University of Dayton	Thomas Whitney	Olesya Zhupanska

[a]in combination with the first US–Canada Conference.
[b]in combination with the 14th USA–Japan Conference on Composite Materials.
[c]in combination with the 18th USA–Japan Conference on Composite Materials.
[d]in combination with the 20th USA–Japan Conference on Composite Materials.

for the society's 40th anniversary. The 2026 conference will be in Madison, WI, and hosted by the University of Wisconsin (see again Table 4.7.1).

The society membership fluctuates between 400 and 500 members. Membership dues were initially set at $35/year and have remained unchanged. The annual conference attracts 300–450 attendees and features 150–250 technical presentations, keynote addresses, and special sessions. Conference proceedings are now peer-reviewed and published by Springer. While the annual technical conference is the primary activity of ASC, the society also provides annual awards for outstanding achievements in composites research and organises other activities that promote knowledge and use of composites. Scholarships are awarded to highly qualified PhD students to foster their development as composite researchers. Recently, several competitions have been introduced by the society, including a sustainable composites design competition, a 4-minute doctoral research impact competition, and a composites simulation competition.

The text of this section is based on the private notes taken at the end of 2024 by Prof. Thomas Whitney from the conversations with Dr. James M. Whitney, founding president of ASC and Prof. R. Byron Pipes.

3. The European Society for Composite Materials (ESCM)

3.1 The beginning and the first step toward the European Association for Composite Materials (EACM)

Fibre reinforced composites first became available in Europe in the 70s, which led to an incredible interest from the industrial and scientific communities. However, in the same period, major incidents occurred, such as the failure of the carbon-fibre Hyfil fan blades of the high bypass turbofan RB221 by Rolls Royce (which eventually led to the company's bankruptcy). The extreme engineering curiosity about this new class of high-performance materials and the need to guarantee higher safety in their applications triggered the first research projects and collaborative efforts to share knowledge between the European scientists.

On this wave of interest, the first steps were taken in the late 70s until the early 80s for the birth of the "Association Européenne pour les Matériaux Composites" (European Association for Composite Materials). One of the driving forces behind this process was Prof. Anthony Kelly FRS, Vice Chancellor at Surrey University, in England, convinced Europhile and a pioneering father figure for the composites community worldwide. He was the first to identify and suggest the urgent need to organise the rising research activity into a European composites society.

Antony Kelly and another Anthony, Prof. Bunsell, from Ecole des Mines de Paris, started the process by spreading a questionnaire around Europe asking the European scientists about their interest in the initiative and details of their activities in the field of composites. To their surprise, confirming the strong interest in this emerging field, they received many more answers than the number of questionnaires originally distributed! The results of the survey were later published as the "Directory of European Research" by Butterworths in 1985.

It is probably worth remembering that the French composites community was already organised at national level at that time, in the "Association des Matériaux Composites (AMAC)" which organised, on biannual basis, "Les Journées Nationales de Composites" (JNC) from 1978 until today. AMAC was also coorganizing the first International Conference on Composite Materials (ICCM-1) in Geneva and Boston in 1975 (the whole conference was moving from one city to the other at midterm!)—see Section 3. At the same time, the Italian community was also coming together through the "Centro Materiali Compositi" in Naples, which had been active since 1972. Several other national initiatives were also growing across Europe.

It this context of great international activity, the interest in creating a European Composites Group received significant attention by a secretariat in Bordeaux, which had strong political support from Jacques Valade. He was a scientist turned politician with an important position in the French government. Financial support was also provided by the Aquitaine Region. Then, the head of the secretariat, Albert Massiah, enthusiastically lent his support to create a secretariat for this new European group. After several meetings between Bordeaux and London, the "Association Européenne pour les Matériaux Composites (AEMC)", also called the "European Association for Composite Materials (EACM)", was eventually established and registered in Bordeaux in 1984. Fig. 4.7.1 shows the page of the "Journal Officiel de la Republique Francaise" of the fifth of June 1984, publishing the formal establishment.

The aim of the Association was "étudier les conditions du développement économique, de la production et de l'utilisation des matériaux composites, en Europe; promouvoir, coordonner, réaliser les actions nécessaires à ce développement, en Europe; mettre en place un service d'information scientifique, technique et économique dans le domaine des matériaux composites en Europe (banque de données), de favoriser les actions de recherche et de transfert de technologie dans ce domaine" or, in English, "study the conditions of economic development, production and use of composite materials in Europe; promote, coordinate and carry out the actions necessary for this development in Europe; set up a scientific, technical and economic information service in the field of composite materials in Europe (data bank), to promote research and technology transfer actions in this area".

Anthony Bunsell was elected as president, supported by a Standing Committee consisting of himself, Ignazio Crivelli-Visconti, Antony Kelly, Hans Lilholt, and Klaus Friedrich, who later stepped down to be replaced by Karl Schulte.

With the great enthusiasm of a newborn association and great contributions of the scientists from all over Europe, EACM started its activity and established a series of scientific events, called European Conference on Composite Materials (ECCM), still running very successfully to this day. The first meeting of the European Conference on Composite Materials (ECCM-1) was held in Bordeaux in 1985 with ECCM-2 held in London in 1987, conjoined with ICCM-VI. The following conferences until ECCM-7 were run under the umbrella of EACM, with a continuous increase in participation and interest from the academia and the industry (see Table 4.7.2 for the complete list).

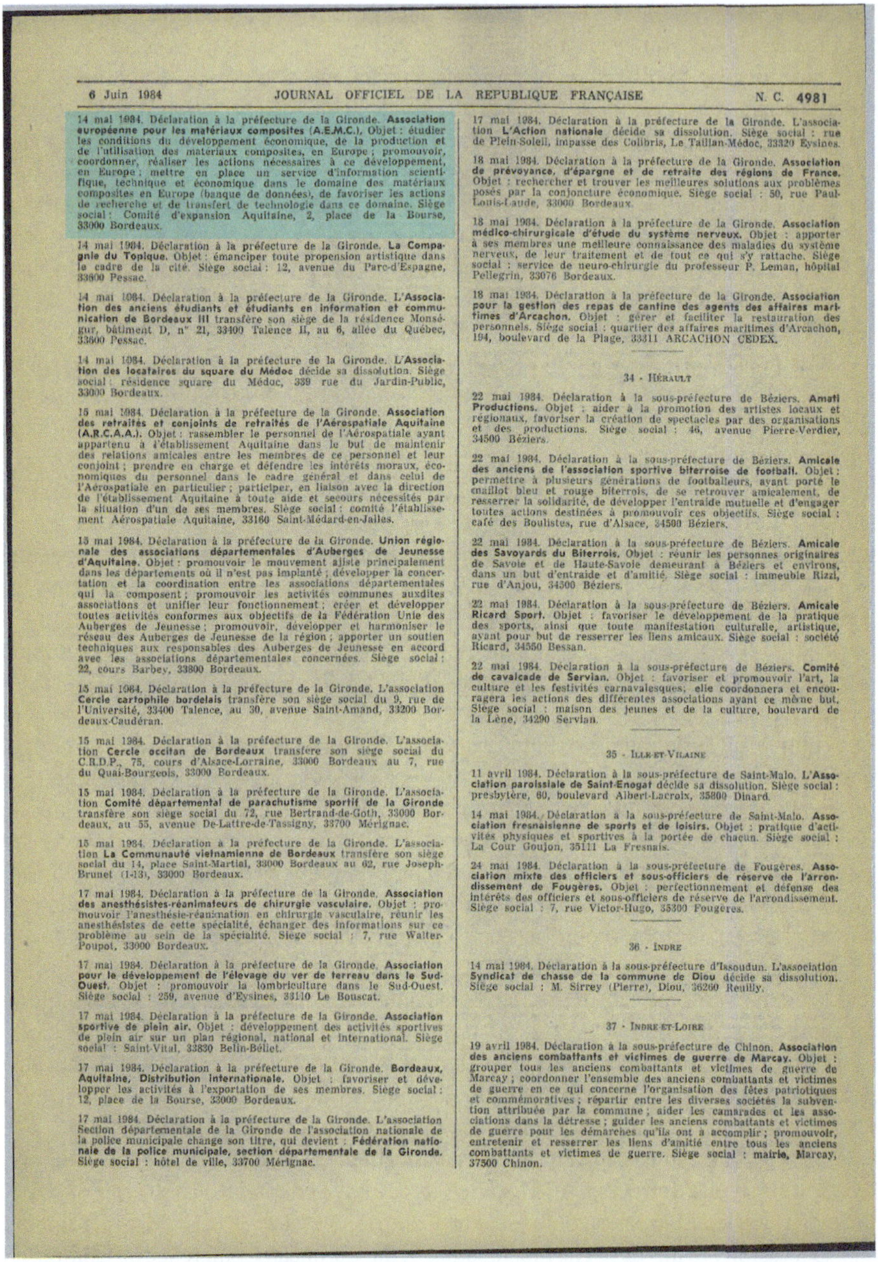

6 Juin 1984 JOURNAL OFFICIEL DE LA REPUBLIQUE FRANÇAISE N. C. 4981

14 mai 1984. Déclaration à la préfecture de la Gironde. **Association européenne pour les matériaux composites (A.E.M.C.).** Objet : étudier les conditions du développement économique, de la production et de l'utilisation des matériaux composites, en Europe ; promouvoir, coordonner, réaliser les actions nécessaires à ce développement, en Europe ; mettre en place un service d'information scientifique, technique et économique dans le domaine des matériaux composites en Europe (banque de données), de favoriser les actions de recherche et de transfert de technologie dans ce domaine. Siège social : Comité d'expansion Aquitaine, 2, place de la Bourse, 33000 Bordeaux.

14 mai 1984. Déclaration à la préfecture de la Gironde. **La Compagnie du Topique.** Objet : émanciper toute propension artistique dans le cadre de la cité. Siège social : 12, avenue du Parc-d'Espagne, 33600 Pessac.

14 mai 1984. Déclaration à la préfecture de la Gironde. **L'Association des anciens étudiants et étudiants en information et communication de Bordeaux III** transfère son siège de la résidence Monségur, bâtiment D, n° 21, 33400 Talence II, au 6, allée du Québec, 33600 Pessac.

14 mai 1984. Déclaration à la préfecture de la Gironde. **L'Association des locataires du square du Médoc** décide sa dissolution. Siège social : résidence square du Médoc, 339 rue du Jardin-Public, 33000 Bordeaux.

15 mai 1984. Déclaration à la préfecture de la Gironde. **Association des retraités et conjoints de retraités de l'Aérospatiale Aquitaine (A.R.C.A.A.).** Objet : rassembler le personnel de l'Aérospatiale ayant appartenu à l'établissement Aquitaine dans le but de maintenir des relations amicales entre les membres de ce personnel et leur conjoint ; prendre en charge et défendre les intérêts moraux, économiques du personnel dans le cadre général et dans celui de l'Aérospatiale en particulier ; participer, en liaison avec la direction de l'établissement Aquitaine à toute aide et secours nécessités par la situation d'un de ses membres. Siège social : comité l'établissement Aérospatiale Aquitaine, 33160 Saint-Médard-en-Jalles.

15 mai 1984. Déclaration à la préfecture de la Gironde. **Union régionale des associations départementales d'Auberges de Jeunesse d'Aquitaine.** Objet : promouvoir le mouvement ajiste principalement dans les départements où il n'est pas implanté ; développer la concertation et la coordination entre les associations départementales qui la composent ; promouvoir les activités communes auxdites associations et unifier leur fonctionnement ; créer et développer toutes activités conformes aux objectifs de la Fédération Unie des Auberges de Jeunesse ; promouvoir, développer et harmoniser le réseau des Auberges de Jeunesse de la région ; apporter un soutien techniques aux responsables des Auberges de Jeunesse en accord avec les associations départementales concernées. Siège social : 22, cours Barbey, 33800 Bordeaux.

15 mai 1984. Déclaration à la préfecture de la Gironde. L'association **Cercle cartophile bordelais** transfère son siège social du 9, rue de l'Université, 33400 Talence, au 30, avenue Saint-Amand, 33200 Bordeaux-Caudéran.

15 mai 1984. Déclaration à la préfecture de la Gironde. L'association **Cercle occitan de Bordeaux** transfère son siège social du C.R.D.P., 75, cours d'Alsace-Lorraine, 33000 Bordeaux au 7, rue du Quai-Bourgeois, 33000 Bordeaux.

15 mai 1984. Déclaration à la préfecture de la Gironde. L'association **Comité départemental de parachutisme sportif de la Gironde** transfère son siège social du 72, rue Bertrand-de-Goth, 33000 Bordeaux, au 55, avenue De-Lattre-de-Tassigny, 33700 Mérignac.

15 mai 1984. Déclaration à la préfecture de la Gironde. L'association **La Communauté vietnamienne de Bordeaux** transfère son siège social du 14, place Saint-Martial, 33000 Bordeaux au 62, rue Joseph-Brunet (1-13), 33000 Bordeaux.

17 mai 1984. Déclaration à la préfecture de la Gironde. **Association des anesthésistes-réanimateurs de chirurgie vasculaire.** Objet : promouvoir l'anesthésie-réanimation en chirurgie vasculaire, réunir les anesthésistes de cette spécialité, échanger des informations sur ce problème au sein de la spécialité. Siège social : 7, rue Walter-Poupot, 33000 Bordeaux.

17 mai 1984. Déclaration à la préfecture de la Gironde. **Association pour le développement de l'élevage du ver de terreau dans le Sud-Ouest.** Objet : promouvoir la lombriculture dans le Sud-Ouest. Siège social : 259, avenue d'Eysines, 33110 Le Bouscat.

17 mai 1984. Déclaration à la préfecture de la Gironde. **Association sportive de plein air.** Objet : développement des activités sportives de plein air sur un plan régional, national et international. Siège social : Saint-Vital, 33830 Belin-Béliet.

17 mai 1984. Déclaration à la préfecture de la Gironde. **Bordeaux, Aquitaine, Distribution internationale.** Objet : favoriser et développer les activités à l'exportation de ses membres. Siège social : 12, place de la Bourse, 33000 Bordeaux.

17 mai 1984. Déclaration à la préfecture de la Gironde. L'association Section départementale de la Gironde de l'association nationale de la police municipale change son titre, qui devient : **Fédération nationale de la police municipale, section départementale de la Gironde.** Siège social : hôtel de ville, 33700 Mérignac.

17 mai 1984. Déclaration à la préfecture de la Gironde. L'association **L'Action nationale** décide sa dissolution. Siège social : rue de Plein-Soleil, impasse des Colibris, Le Taillan-Médoc, 33320 Eysines.

18 mai 1984. Déclaration à la préfecture de la Gironde. **Association de prévoyance, d'épargne et de retraite des régions de France.** Objet : rechercher et trouver les meilleures solutions aux problèmes posés par la conjoncture économique. Siège social : 50, rue Paul-Louis-Laude, 33000 Bordeaux.

18 mai 1984. Déclaration à la préfecture de la Gironde. **Association médico-chirurgicale d'étude du système nerveux.** Objet : apporter à ses membres une meilleure connaissance des maladies du système nerveux, de leur traitement et de tout ce qui s'y rattache. Siège social : service de neuro-chirurgie du professeur P. Leman, hôpital Pellegrin, 33076 Bordeaux.

18 mai 1984. Déclaration à la préfecture de la Gironde. **Association pour la gestion des repas de cantine des agents des affaires maritimes d'Arcachon.** Objet : gérer et faciliter la restauration des personnels. Siège social : quartier des affaires maritimes d'Arcachon, 194, boulevard de la Plage, 33311 ARCACHON CEDEX.

34 - Hérault

22 mai 1984. Déclaration à la sous-préfecture de Béziers. **Amati Productions.** Objet : aider à la promotion des artistes locaux et régionaux, favoriser la création de spectacles par des organisations et des productions. Siège social : 46, avenue Pierre-Verdier, 34500 Béziers.

22 mai 1984. Déclaration à la sous-préfecture de Béziers. **Amicale des anciens de l'association sportive biterroise de football.** Objet : permettre à plusieurs générations de footballeurs, ayant porté le maillot bleu et rouge biterrois, de se retrouver amicalement, de resserrer la solidarité, de développer l'entraide mutuelle et d'engager toutes actions destinées à promouvoir ces objectifs. Siège social : café des Boulistes, rue d'Alsace, 34500 Béziers.

22 mai 1984. Déclaration à la sous-préfecture de Béziers. **Amicale des Savoyards du Biterrois.** Objet : réunir les personnes originaires de Savoie et de Haute-Savoie demeurant à Béziers et environs, dans un but d'entraide et d'amitié. Siège social : immeuble Rizzi, rue d'Anjou, 34500 Béziers.

22 mai 1984. Déclaration à la sous-préfecture de Béziers. **Amicale Ricard Sport.** Objet : favoriser le développement de la pratique des sports, ainsi que toute manifestation culturelle, artistique, ayant pour but de resserrer les liens amicaux. Siège social : société Ricard, 34550 Bessan.

22 mai 1984. Déclaration à la sous-préfecture de Béziers. **Comité de cavalcade de Servian.** Objet : favoriser et promouvoir l'art, la culture et les festivités carnavalesques; elle coordonnera et encouragera les actions des différentes associations ayant ce même but. Siège social : maison des jeunes et de la culture, boulevard de la Lène, 34290 Servian.

35 - Ille-et-Vilaine

11 avril 1984. Déclaration à la sous-préfecture de Saint-Malo. **L'Association paroissiale de Saint-Enogat** décide sa dissolution. Siège social : presbytère, 60, boulevard Albert-Lacroix, 35800 Dinard.

14 mai 1984. Déclaration à la sous-préfecture de Saint-Malo. **Association fresnaisienne de sports et de loisirs.** Objet : pratique d'activités physiques et sportives à la portée de chacun. Siège social : La Cour Goujon, 35111 La Fresnais.

24 mai 1984. Déclaration à la sous-préfecture de Fougères. **Association mixte des officiers et sous-officiers de réserve de l'arrondissement de Fougères.** Objet : perfectionnement et défense des intérêts des officiers et sous-officiers de réserve de l'arrondissement. Siège social : 7, rue Victor-Hugo, 35300 Fougères.

36 - Indre

14 mai 1984. Déclaration à la sous-préfecture d'Issoudun. L'association **Syndicat de chasse de la commune de Diou** décide sa dissolution. Siège social : M. Sirrey (Pierre), Diou, 36260 Reuilly.

37 - Indre-et-Loire

19 avril 1984. Déclaration à la sous-préfecture de Chinon. **Association des anciens combattants et victimes de guerre de Marcay.** Objet : grouper tous les anciens combattants et victimes de guerre de Marcay ; coordonner l'ensemble des anciens combattants et victimes de guerre en ce qui concerne l'organisation des fêtes patriotiques et commémoratives ; répartir entre les diverses sociétés la subvention attribuée par la commune ; aider les camarades et les associations dans la détresse ; guider les anciens combattants et victimes de guerre pour les démarches qu'ils ont à accomplir ; promouvoir, entretenir et resserrer les liens d'amitié entre tous les anciens combattants et victimes de guerre. Siège social : mairie, Marcay, 37500 Chinon.

Figure 4.7.1 Extract from the journal Officiel de la Republique Francaise of the 5th of June 1984 (supplied by Prof Philippe Olivier)

Table 4.7.2 ECCM conferences and EACM/ESCM Presidents.

Conference	Year	City	Country	President	Organisation
ECCM-1	1985	Bordeaux	France	Anthony R.Bunsell	EACM
ECCM-2	1987	London	United Kingdom	Anthony R.Bunsell	EACM
ECCM-3	1989	Bordeaux	France	Anthony R.Bunsell	EACM
ECCM-4	1990	Stuttgart	Germany	Anthony R.Bunsell	EACM
ECCM-5	1992	Bordeaux	France	Anthony R.Bunsell	EACM
ECCM-6	1993	Bordeaux	France	Anthony R.Bunsell	EACM
ECCM-7	1996	London	United Kingdom	Anthony R.Bunsell	EACM
	1996 1998		Provisional Exec. Committee	Ignaas Verpoest (provisional)	Transition EACM to ESCM
ECCM-8	1998	Naples	Italy	Ignaas Verpoest	ESCM
ECCM-9	2000	Brighton	United Kingdom	Ignazio Crivelli-Visconti	ESCM
ECCM-10	2002	Bruges	Belgium	Albert Cardon	ESCM
ECCM-11	2004	Rhodes	Greece	Costas Galiotis	ESCM
ECCM-12	2006	Biarritz	France	Jacques Lamon	ESCM
ECCM-13	2008	Stockholm	Sweden	Leif Asp	ESCM
ECCM-14	2010	Budapest	Hungary	Laszlo Kollar	ESCM
ECCM-15	2012	Venice	Italy	Marino Quaresimin	ESCM
ECCM-16	2014	Seville	Spain	Federico Paris	ESCM
ECCM-17	2016	Munich	Germany	Klaus Drechsler	ESCM
ECCM-18	2018	Athens	Greece	Vassilis Kostopoulos	ESCM
ECCM-19[a]	2020	Nantes	France	Vassilis Kostopoulos	ESCM
ECCM-20	2022	Lausanne	Switzerland	Anastasios Vassilopoulos	ESCM
ECCM-21	2024	Nantes	France	Christophe Binetruy	ESCM

[a](cancelled because of COVID)—special sessions and short training courses were held online.

Tony Bunsell remained president of the EACM until 1997, when he stepped down. Because of political changes in Bordeaux, as will be explained later, the EACM secretariat was relocated to London, at the Institute of Materials (IoM).

3.2 The birth of the European Society for Composite Materials (ESCM)

In 1996, Mike Bader and Ignaas Verpoest were invited to join the Standing Committee of EACM (originally composed by Anthony Bunsell, Ignazio Crivelli-Visconti, Anthony Kelly, Hans Lilholt, Albert Massiah and Karl Schulte), and it was immediately planned to have an in-depth discussion on the future of the European Association during the following ECCM 7 conference to be held the in May of the same year in London.

There were two main points to discuss. The first was related to the financial sustainability of the Association as the local financial support from the Région Aquitaine was dwindling. The second was the wish of many members to have ECCM as a peripatetic event, travelling in Europe to spread more effectively the composite culture.

To revive the local interest, it was proposed to organise an ECCM-8 in Bordeaux already in 1997, only 1 year after the previous ECCM-7 in London, held in 1996. According to most members of the Standing Committee, this would be too early and not in line with the wish from many to have ECCM travelling in Europe.

This was basically the start of reconsidering the link with the area of Bordeaux and exploring potential alternatives. In order to coordinate and lead these discussions, an "Inaugural Council" was elected, and a "Preliminary Executive Committee (PEC)" was established during ECCM-7 in London. Ignaas Verpoest was elected Provisional President, Ignazio Crivelli-Visconti Provisional Vice-President and Clive Phillips secretary. It was agreed that ECCM would be organised every other year in a different European country. The first conference of this "new" cycle would be ECCM-8 in Naples at the beginning of June 1998, and not in 1997 in Bordeaux.

The PEC convened several times in the period between ECCM-7 (May 1996) and ECCM-8 (June 1998), in Leuven, London and Paris.

In June 1997, a Provisional Council, composed of the previous EACM Council, completed by several well-known experts from the European composites community, gathered in London to discuss the transition from EACM to a new organisation and the general principles governing this new society. A Provisional President (Ignaas Verpoest) and a Provisional Executive Committee (Bader, Baptiste, Crivelli-Visconti, Massard, Peijs, Phillips and Schulte) were elected.

Following this, four PEC-meetings were held in Leuven and London, mainly discussing the future of the organisation, the establishment of a Constitution and Bylaws and the possible transfer of the EACM office from Bordeaux to the IoM in London. Already in October 1996, a first version of such a possible agreement between EACM and IoM had been drafted, but it took until April 1998, during a Provisional Executive Committee meeting in Paris, to have a final discussion with IoM, in preparation of ECCM-8 in Naples to be held in June of the same year. During the Naples conference, the Constitution and Regulations of the new organisation, called ESCM, were presented for approval to the Provisional General Assembly. This Provisional General Assembly comprised all Europeans registered at ECCM-8. The (Provisional) General Assembly also elected the first Council of ESCM. Ignaas Verpoest was elected as the first president of the new society, with Ignazio Crivelli-Visconti as vice-president.

Upon approval of the Constitution and Regulations, the ESCM was born.

As an interesting note, the ESCM Constitution was inspired by that of the ICCM, drafted almost simultaneously by Paul Lagace, drawing in turn from the Constitution of the International Olympic Committee. As one can read in the preamble of the ESCM Constitution, in recognition of the fundamental role played during the creation of EACM, the Society conferred the lifelong title of Honorary Members upon its original founders: Antony Kelly, Antony Bunsell, Hans Lilholt, Ignazio Crivelli-Visconti, Karl Schulte and Albert Massiah.

One of the first and important decisions of the new Council was to organise ECCM-9 in Brighton in 2000 and ECCM-10 in Brugge in 2002, and to further investigate the collaboration with IoM. For this, PEC meetings were held in London (September 1998) and Paris (July 1999), leading to the first "real" Council and Executive Committee meetings during ECCM in Brighton, where Ignazio-Crivelli-Visconti was elected as the new ESCM President.

Discussions on whether to continue the collaboration with IoM continued, and also the possibility to become a section of FEMS (Federation of European Materials Societies) was seriously investigated. Finally, at the end of May 2001, during a superb workshop on Structural Integrity of Composite Materials and Structures organised by Prof. Peter Beaumont in the beautiful island of Capri, it was decided not to continue the collaboration with IoM, but to have, in the future, ESCM's own office connected to the President of the Society, hence in principle, moving every 2 years. It was also established as a sort of "non-written rule", in the following years, that the organiser of an ECCM conference would also act as president in the 2 years following their ECCM.

In 2013, under the presidency of Prof. Marino Quaresimin, the ESCM Council started to provide also practical service to the members with the promotion of grants for the organisation of scientific events, workshops, summer schools and courses on composites, with particular attention to the younger members.

In 2016, during ECCM-17 in Seville under the presidency of Prof. Federico Paris, the regulations were updated to further improve ESCM operations. Additional modifications to the Constitution and rule updates were proposed more recently by the Council under the presidency of Prof. Anastasios (Tassos) Vassilopoulos in ECCM-20 in Lausanne (2022) and ratified during ECCM-21 in Nantes (2024).

Since the birth of the ESCM, at ECCM eight in Naples, ECCM conferences were in the range of 500–600 participants, until the ECCM-15 organised at the Lido of Venice, when the participation exploded with more than 1600 attendees, generating probably the largest and one of the most successful scientific events in the history of composite materials. ECCM-15 was indeed a landmark and a turning point for the ECCM conferences: since then, the average number of conference participants has remained stable at well over a 1000 with a consistent representation of the European and worldwide academic community and a continuously increasing participation from industrial representatives.

The text of this section was prepared on the basis of the interviews with and the material provided by Prof. Antony Bunsell and Prof. Ignazio Crivelli-Visconti and with the invaluable contribution of the personal archive of Prof. Ignaas Verpoest.

4. The International Committee on Composite Materials (ICCM)

The ICCM is an international, nongovernmental, not-for-profit, scientific and engineering organisation dedicated to composite materials. The mission of ICCM is rather broad and one can refer to the website for a detailed description. Very briefly, ICCM aims to provide a forum for discussion and interchange of information to promote the development and the understanding of composite materials science and technology within scientific and engineering community as well as to promote education, exchanges and international cooperation in the field.

The main mechanism for pursuing these aims is the organisation of the International Conferences on Composite Materials (ICCM), a conference series founded in 1971 by Pete Scala, a Professor of Metallurgical and Materials Sciences at Cornell University. The first edition, ICCM I, took place conjointly in Geneva and Boston, Massachusetts, in 1975, to bring together, with the support of AMAC in France, the young composite communities in Europe (EACM) and USA (ASC).

Those were indeed the good old days of serious conferencing! The conference commenced on the seventh of April, in Geneva and ran for 5 days, before travelling across the Atlantic Ocean to the USA, where it recommenced in Boston on the 14th of April, and ran for another 5 days. Since then, unfortunately, the ICCM conferences (and not only those) have always run for just 1 week and have been held in a single location. ICCM II was held in Toronto in 1977 (with 195 participants from 11 countries), while ICCM III was held in Paris, in 1980, in combination with the second JNC (Journées Nationales des Composites). This conference showed a significant change of scale, with about 650 participants from 26 countries, assuring also the continuity of the series. The sequence of the peripatetic conferences held around the world until today is presented in Table 4.7.3. From the hundreds of participants in the event in the early editions, the average number of participants is now always well above a 1000 attendees, a similar size to that seen for ESCM conferences, suggesting a significant development and a stable consolidation of the worldwide composite community.

Talking now about the people at the origin and behind the organisation, in the early, heady days of the late 1970s till the mid-1980s, the names of the other two "founders", alongside Pete Scala, have definitely to be mentioned: Profs. Anthony Kelly and Stephen Tsai. Tony and Steve supported an organic growth of ICCM in light of their scientific prominence and with the help of a period where composites were seen to have an enormous growth potential, thus facilitating the overall process.

Table 4.7.3 ICCM conferences and ICCM presidents.

Conference	Year	City	Country	President
ICCM-1	1975	Geneva and Boston	Switzerland and USA	Pete Scala
ICCM-2	1978	Toronto	Canada	B.R. Noton
ICCM-3	1980	Paris	France	Anthony Bunsell
ICCM-4	1982	Tokyo	Japan	T. Hayashi
ICCM-5	1985	San Diego	USA	W.C. Harrigan
ICCM-6	1987	London	UK	F. Matthews
ICCM-7	1989	Guangzhou	China	Stephen Tsai
ICCM-8	1991	Honolulu	USA	George Springer
ICCM-9	1993	Madrid	Spain	Paul Lagace
ICCM-10	1995	Whistler, British Columbia	Canada	Paul Lagace
ICCM-11	1997	City of Gold coast	Australia	Paul Lagace
ICCM-12	1999	Paris	France	Paul Lagace
ICCM-13	2001	Beijing	China	Thierry Massard
ICCM-14	2003	San Diego, California	USA	Murray Scott
ICCM-15	2005	Durban	S. Africa	Ignaas Verpoest
ICCM-16	2007	Kyoto	Japan	H. Thomas Hahn
ICCM-17	2009	Edinburgh	Scotland	Anoush Poursartip
ICCM-18	2011	Jeju island	Korea	Michael Wisnom
ICCM-19	2013	Montreal	Canada	Ozden Ochoa
ICCM-20	2015	Copenhagen	Denmark	Paul Smith
ICCM-21	2017	Xi'an	China	Nobuo Takeda
ICCM-22	2019	Melbourne	Australia	Lei Asp
ICCM-23	2023	Belfast	UK	Woo Il Lee/Pascal Hubert
ICCM-24	2025	Baltimore	USA	Pascal Hubert

The late 80s and early 90s represented instead a rather difficult period for the organisation, with political and financial issues. In 1993, Paul Lagace was elected Chair (President) and then reelected until 1999. Learning from other international organisations, including the Olympic Committee, he was the driving force behind a full rewriting of the Constitution and the Bylaws, continuously interacting with representatives from the main regions, like Anoush Poursartip (USA), Ignaas Verpoest (Europe) and Murray Scott (Australasia). In parallel, he led a deep reorganisation of the ICCM which continued after his presidency, where he remained the heart and the soul of the Committee, until his untimely death in 2021. Together with Paul Lagace, Anoush Poursartip and Murray Scott had also a substantial impact on the organisation and life of ICCM. Anoush Poursartip was the Chair of the ICCM-10 in Whistler, British Columbia, Canada and Murray Scott of the ICCM-11 in Gold Coast Australia. Paul, Ignaas, Anoush, and Scott, were indeed the key players in the rebirth of ICCM after those difficult years at the beginning of the 90s, joined thereafter by Paul Smith and Michael Wisnom.

The new ICCM Constitution was ratified in July 1999, under the presidency of Paul Lagace at the General Assembly of ICCM-12 in Paris. In the following years, under the presidency of Ignaas Verpoest (2003–05) and many colleagues, a further significant activity of fine-tuning the Constitution and Bylaws has continued on an ongoing basis to adapt the rules to the complex reality of such an international organisation.

Part of the mission of ICCM is the recognition of individuals who have made outstanding achievements in the science, technology, engineering and application of composite materials. Since 2005 at every ICCM, the outstanding contribution of one individual to the field is recognised through the Scala Award, which carries with it the designation of ICCM World Fellow and there is an associated Scala Lecture, which forms the opening plenary of the ICCM conference.

The designation of World Fellow is awarded by ICCM also in view of an international recognition and contribution to the work of ICCM or other high contributions to the field through education, contributions to other major conference series, development of standards/design codes, editorial works. World Fellows are designated as Life Members.

In 2001, in recognition of their lifetime achievements, Pete Scala, Anthony Kelly, Steve Tsai and Paul Lagace were nominated the first World Fellows and Life Members of ICCM. The complete list of the World Fellows of ICCM is presented in Table 4.7.4.

The text of this section was prepared based on the material and information provided by Prof. Anoush Poursartip, Prof. Murray Scott and Prof. Pascal Hubert and Prof. Steve Tsai.

Table 4.7.4 World Fellows of ICCM.

Recipient	Year	
Anthony Kelly	2001	Life member (in memoriam 2014)
Pete Scala	2001	Life member (in memoriam 2010)
Stephen Tsai	2001	Life member
Paul Lagace	2001	Life member (in memoriam 2021)
Anthony Bunsell	2003	Life member
Klaus Friedrich	2003	Scala award (In Memoriam, 2021)
Isao Kimpara	2005	Scala award
Anoush Poursartip	2005	Life member
Murray Scott	2005	Life member—Scala award 2017
Albert Cardon	2007	Life member (In Memoriam, 2007)
Thierry Massard	2007	Life member
Ignaas Verpoest	2007	Life member
Frank Jones	2007	Scala award (In Memoriam, 2022)
Tsu-Wei Chou	2009	Life member
H. Thomas Hahn	2009	Scala award

Table 4.7.4 World Fellows of ICCM.—cont'd

Recipient	Year	
Takashi Ishikawa	2011	Life member
Ken Street	2011	Scala award
R. Byron Pipes	2013	Life member
Karl Schulte	2013	Life member (In Memoriam, 2023)
C. T. Sun	2013	Life member
Ramesh Talreja	2013	Scala award
Allan Baker	2015	Life member
Shanyi du	2015	Life member
Ozden Ochoa	2015	Life member
Michael Wisnom	2015	Life member
Yiu-Wing Mai	2015	Scala award
Jinsong Leng	2017	Life member
Jan-Anders Månsson	2017	Life member
Paul Smith	2017	Life member
Mike Hinton	2019	Scala award
Suong Hoa	2019	Life member
Adrian Mouritz	2019	Life member (In Memoriam, 2023)
Nobuo Takeda	2019	Life member
Anthony Waas	2019	Life member
Nancy Sottos	2023	Scala award

5. JEC group

The JEC Group is a nonprofit organisation entirely dedicated to promoting composite materials and fostering their applications globally. The association JEC was created in 1963 in France and published the first information bulletin which continued until today and has become the *JEC Composites Magazine*, dedicated to composite materials, their technologies and application sectors.

In 1965, the first conference on Composites took place and since then the event evolved over the years to become what it is now, probably the largest worldwide event for the composite material community, the JEC World, held every year at the beginning of March in Paris. This year the event celebrates the 60th anniversary, and the numbers are indeed significant, with more than 45.000 participants from about 100 countries. The JEC World trade show is able to bring together in the same place the most important companies in the field of composites, from big industries to small innovative startups, leading academics and scientists, experts and R&D leaders.

Looking a bit more in details into the history of JEC, the first conference on Composites was organised in 1965 as two technical days held in the Salons de la Maison des Centraux in Paris. Then, in March 1969, a new event was organised: two Technical days of the "Centre de documentation du verre textile et des Plastiques

Renforcés" (JEC's previous name). In 1975, the Technical Days moved to the Centre International de Paris (Palais des Congrès, Porte Maillot), where the JEC tradeshow started in 1979, and JEC Europe took place in 1980. In March 1985, the first edition of "Journées Européennes des composites" (new name for the Technical Days and exhibition) took place again at the Centre International de Paris.

At the end of April 1992, JEC moved to CNIT in the shadow of the Grande Arche de La Défense (west of Paris), probably the best location ever, for sure my favourite! After about 10 years of growth in size, the tradeshow moved first to the area of Paris Expo at Porte de Versailles in 2002 and, eventually, in 2016 to the present location at the Parc des Expositions in Villepinte, near the Charles De Gaulle Airport.

The event was also successfully spreading around the world with tradeshows and forums starting in Singapore in 2008, America (Boston) in 2012, Bangkok in 2019, DACH region (Germany, Austria and Switzerland) in 2021, Italy in 2023 and 2025 in Poland.

Among the many initiatives developed by JEC to promote the composite culture and market, two in particular are worth mentioning: the Innovation Awards, launched in 1998 to reward cutting-edge creative projects capable of showing the full potential of composite materials, and the Startup Booster program, launched in 2017, which has become one of the industry's largest startup competitions, open to young companies building innovative projects in the field of composites and advanced materials.

6. Conclusions

The histories of key composite societies and associations, namely ASC, ESCM, ICCM, and JEC, have been presented. As a final remark, it is important to highlight the fundamental role these organisations have played in the development of the composite science and technology over the years. They are indeed the fundamental means to promote the exchange of information and diffusion of knowledge during the conferences, fostering interactions among researchers within the international communities and strengthening collaboration between industry and academia. Composite organisations are also the place where tradition and history can be protected, ensuring that the achievements of outstanding individuals are properly recognised and preserved as a future memory for the next generations of scientists.

References

American Society for Composites (ASC). Website. https://www.asc-composites.org.

European Society for Composite Materials (ESCM). Website. https://www.escm-eu.org/.

International Committee on Composite Materials (ICCM). Website. http://www.iccm-central.org.

JEC group Website. https://www.jeccomposites.com/about-jec-group/.

Index

'*Note:* Page numbers followed by "f" indicate figures, "t" indicate tables and "b" indicate boxes."

Printed and bound by CPI Group (UK) Ltd, Croydon, CR0 4YY

27/03/20[illegible]

020800[illegible]002